Hyperbolic Problems: Theory, Numerics, Applications

Springer
Berlin
Heidelberg
New York
Hong Kong
London
Milan
Paris
Tokyo

Thomas Y. Hou
Eitan Tadmor
Editors

Hyperbolic Problems: Theory, Numerics, Applications

Proceedings
of the Ninth International Conference
on Hyperbolic Problems
held in CalTech, Pasadena,
March 25-29, 2002

Springer

Editors

Thomas Y. Hou

Department of Applied and
Computational Mathematics 217-50
California Institute of Technology
Pasadena, CA 91125, USA
e-mail: hou@acm.caltech.edu

Eitan Tadmor

Center for Scientific Computation
and Mathematical Modeling (CSCAMM)
Department of Mathematics
and
Institute for Physical Science & Technology (IPST)
University of Maryland
College Park, MD 20742-3289, USA
e-mail: tadmor@cscamm.umd.edu

Cataloging-in-Publication Data applied for

A catalog record for this book is available from the Library of Congress.

Bibliographic information published by Die Deutsche Bibliothek
Die Deutsche Bibliothek lists this publication in the Deutsche Nationalbibliografie;
detailed bibliographic data is available in the Internet at <http://dnb.ddb.de>.

Cover figure courtesy of Björn Sjögreen

Mathematics Subject Classification (2000): 35Lxx, 35M10, 35Q30, 35Q60, 35R35, 65Mxx, 65Nxx, 65Txx, 65Yxx, 65Z05, 74B20, 74Jxx, 76L06, 76Rxx, 76Txx, 80A32, 80Mxx, 83F05

ISBN 3-540-44333-9 Springer-Verlag Berlin Heidelberg New York

Springer-Verlag Berlin Heidelberg New York
a member of BertelsmannSpringer Science+Business Media GmbH

http://www.springer.de

Printed in Germany

Typeset by the authors
Cover design: *design & production* GmbH, Heidelberg

Printed on acid-free paper 46/3142LK - 5 4 3 2 1 0

Preface

The International Conference on "Hyperbolic Problems: Theory, Numerics and Applications" was held in CalTech on March 25-30, 2002. The Hyp2002 conference was the ninth meeting in the bi-annual international series which became one of the highest quality and most successful conference series in Applied Mathematics.

This series originated in 1986 at Saint-Etienne, with an earlier focus on the theoretical aspects of hyperbolic conservation laws. As computers became more powerful in the late eighties, and as the interplay between new models, theory and modern numerical algorithms has gained considerable impact during the nineties, the scope of the Hyperbolic conference series was expanded to its present format.

This trend is demonstrated, for example, by many effective numerical methods developed originally in the context of Computational Fluid Dynamics, which in recent years have found new applications outside their traditional areas. Consequently, in addition to its traditional areas, the Hyp2002 has added new focal points. These included multiscale modeling and simulations, e.g., in deriving and simulating meso- and nano-scale material properties in micro devices, geophysical applications such as wave propagation in random media, and coarsening of multi-phase flows through porous media, and free boundary problems arising from materials science and multi-component fluid dynamics, including thin films, crystal growth, multi-fluid interfaces, solid/liquid interfaces.

A total of about 230 participants attended the Hyp2002 conference. It provided a unique forum to exchange and to stimulate new ideas from different avenues in academia, industry and government labs. We were very pleased to see that the Hyp2002 has brought together leading researchers from different disciplines to address theoretical, modeling and computational issues in solving hyperbolic PDEs and more generally, nonlinear evolution equations arising from different areas of applications. Future cooperations should prove the success of Hyp2002 in promoting new cross disciplinary interactions among researchers and practitioners.

A highlight of Hyp2002 was the conference banquet hosted at the historical Caltech faculty club, the Athenaeum, with keynote speeches by Professor Peter Lax and Professor Heinz-Otto Kreiss. Professors Lax and Kreiss are credited with many pioneering contributions in the general areas of theory and numerics of hyperbolic equations and it was indeed a unique occasion to share their excellent review of the early days as well as their projections into the future of the field.

More than one hundred contributions were assembled in this book, reflecting the high-level quality presentations made in Hyp2002, while covering a diverse range of topics in theory, numerics and applications assembled under the umbrella of 'hyperbolic problems'.

Over 30 young participants were provided with financial support to attend the Hyp2002 conference, and we would like to acknowledge in this context, the partial support for Hyp2002 provided by

- The National Science Foundation,
- The office of Naval Research,
- The Institute of Pure and Applied Mathematics at UCLA,
- The DFG Priority Research Program on Analysis and Numerics for Conservation Laws in Germany,
- The Center of Integrative Multiscale Modeling and Simulation,
- The Division of Engineering and Applied Science at Caltech
- The Caltech ASCI Center.

Finally, we would like to take this opportunity to thank Andrew Westhead for his dedicated work in assembling these proceedings.

Tom Hou
CalTech

Eitan Tadmor
University of Maryland College Park

February 2003

Contents

Plenary Talks

Invited Talks

Contributed Talks

Plenary Talks

Diffusion Limits of Kinetic Models

N. Ben Abdallah[1], P. Degond[1], F. Deluzet[1], V. Latocha[2], R. Talaalout[3] and M. H. Vignal[1]

[1] MIP, UMR 5640 (CNRS-UPS-INSA), Université Paul Sabatier, 118, route de Narbonne, 31062 TOULOUSE cedex, FRANCE *noufel@mip.ups-tlse.fr, degond@mip.ups-tlse.fr, deluzet@mip.ups-tlse.fr, mhvignal@mip.ups-tlse.fr*
[2] Department of Aeronautical Engineering, Kyoto University, Kyoto 606-8501, Japan *Vladimir.Latocha@gmm.insa-tlse.fr*
[3] Centre National d'Études Spatiales, 18 Avenue Edouard Belin, 31401, Toulouse cedex 4, France *Rachid.Talaalout@cnes.fr*

Summary. This paper reports on recent developments in diffusive limits of kinetic systems. Usually, collision operators in kinetic theory exhibit multiple relaxation scales and before a full relaxation towards a Maxwellian equilibrium has been achieved, the system passes through a series of states that can be described by partial equilibria. The paper describes various models describing the dynamics of these partial equilibria, namely the SHE (Spherical Harmonics Expansion) and the ET (Energy Transport) models. Various examples of applications of these models to plasma problems are presented.

Acknowledgements: Work partially supported by the Commissariat l'Energie Atomique and by Centre National d'Etudes Spatiales
Key words: Boltzmann equation, diffusive limits, Spherical Harmonics Expansion model, Energy Transport model, Drift-Diffusion model.

1 Introduction

Let us consider a particle system (e.g a gas, a plasma, or a semiconductor, etc). A fundamental question is how the elementary particle properties translate onto the large-scale behaviour of the system. This question is an old problem (which goes back to Maxwell, Boltzmann, Einstein, ...). However, recent developments have revived this question as it is central to the problem of model reduction in many scientific or industrial areas. Indeed, the use of microscopic models of kinetic type is too expensive. However, resorting to standard macroscopic models such as the drift-diffusion model leads to unacceptable inacurracies, as the account for the microscopic physics in such models is too coarse.

The aim of this paper is to present a strategy in order to derive a hierarchy of intermediate models which provide a better description of the physics at still an affordable numerical cost.

The outline of the paper is as follows. In section 2, we shall review the standard derivation of macroscopic diffusion models from microscopic kinetic models. Then, section 3 will present a first 'improved' model: the Spherical Harmonics Expansion (SHE) model. Section 4 will deal with a second such 'improved' model: the Energy-Transport model (ET) model, which is derived from the SHE model by a suitable averaging. Then, we shall conclude in section 5.

2 From micro to macro: standard theory and Drift-Diffusion limits

The microscopic description of particle systems uses the concept of distribution function (or phase-space density) $f(x, v, t)$, which describes the number of particles at position x, with velocity v, at time t. Each individual particle moves according to a succession of free flights in a given (or self-consistent) force field $F(x, t)$ and of collisions (among themselves and with surrounding obstacles). During their free flight, each individual particle follows Newton's equations

$$\frac{dx}{dt} = v \,, \quad m\frac{dv}{dt} = F(x,t) \,,$$

where m is the particle mass.

The time evolution of the distribution function f is such that the total time derivative of f while following free-flight trajectories is given by a collision operator $Q(f)$:

$$\frac{d}{dt} f(x(t), v(t), t) = Q(f)|_{(x(t),v(t),t)} \,.$$

The collision operator $Q(f)$ is local in (t, x) (as collisions are supposed instantaneous and the distance travelled by the particle during a collision is supposed small compared with the distance travelled during free-flights). Therefore, $Q(f)$ operates on the velocity variable v only. Applying the chain rule immediately leads to the Boltzmann equation:

$$\frac{\partial f}{\partial t} + v \cdot \nabla_x f + \frac{1}{m} F \cdot \nabla_v f = Q(f) \,. \tag{1}$$

A prototype model for the collision operator, which is suitable for plasmas or semiconductors is

$$Q(f)|_v = \int_{v'} \phi(v, v')[M(v)f(v') - M(v')f(v)]dv' \,, \tag{2}$$

where $\phi(v, v')$ is related to the collision scattering cross section. ϕ contains most of the information about the nature of the interaction which causes the collision event. The function

$$M(v) = \left(\frac{m}{2\pi k_B T}\right)^{3/2} e^{-\frac{mv^2}{2k_B T}} \,,$$

is the Maxwellian. It is the equilibrium distribution function towards which the system is going to relax under the influence of the collisions.

The description of a particle system through (1), (2) is fairly expensive from a numerical viewpoint because of the high dimensionality of the model (3 space coordinates, 3 velocity coordinates, plus time). In most industrial applications, it is necessary to reduce the dimensionality of the model and to use fluid equations.

The first idea for deriving fluid equations is to use moments. Example of such moments are the density n, the flux j and the energy W, given by:

$$\begin{pmatrix} n \\ j \\ W \end{pmatrix} = \int f \begin{pmatrix} 1 \\ v \\ m|v|^2/2 \end{pmatrix} dv \,.$$

Evolution equations for (n, j, W) can be obtained by integrating the Boltzmann equation with respect to v. However, in such a procedure, one faces a closure problem as the evolution system for n, j, W involves unknown moments, i.e. quantities obtained by integrating f against other functions of v than just $1, v$ or $|v|^2$. Therefore, since f is a priori not known, an arbitrary prescription is necesary to express these moments in terms of n, j and W. Several approahes have been proposed, e.g. [33], [41] and more recently, [37]. However, the so-obtained moment systems may be ill-posed. Also, the prescription may sometimes seem arbitrary as it is difficult to relate it to some properties of the system and to the associated physical scales.

To overcome the closure problem, a second idea is to consider the physical scales. Indeed, the Boltzmann equation is suitable for the microscopic scale, where the time scale is fixed by the mean time between collisions while the space scale is connected with the distance travelled by particles between two collisions (the so-called mean free path). The macroscopic description however deals with larger scales.

In our context, it is appropriate to introduce a diffusion rescaling. Let us introduce the small parameter η being the ratio of the mean free path to the typical macroscopic distance. The assumption that we are interested in the large scale behaviour of the system translates into $\eta \ll 1$. We now introduce the following change of variables:

$$x' = \eta x \,, \quad t' = \eta^2 t \,, \quad F = \eta F' \,. \tag{3}$$

The η^2 in the time rescaling is characteristic of diffusion rescaling, where diffusion occurs on time scales much larger than transport (or hydrodynamic) phenomena.

With this change of variables, the Boltzmann equation is now written:

$$\eta^2 \frac{\partial f^\eta}{\partial t} + \eta \left(v \cdot \nabla_x f^\eta + \frac{1}{m} F \cdot \nabla_v f^\eta \right) = Q(f^\eta) \tag{4}$$

The limit $\eta \to 0$ in (4) describes the large scale behaviour of the system. The following presentation will stay at a formal level. We refer the interested reader to [42], [32] for rigorous proofs.

The determination of the $\eta \to 0$ limit depends crucially on the properties of Q, as it appears in the leading order term in (4). We list below the most important of these properties:

(i) Conservation of particle number: $\int Q(f)dv = 0$
(ii) Null set $N(Q)$ (equilibria) :

$$Q(f) = 0 \Longleftrightarrow \exists n \in \mathbb{R} \text{ such that } f = nM(v)$$

(iii)$-Q$ is a self-adjoint non-negative operator:
(iv)$-Q$ is positive-definite and invertible on $N(Q)^{\perp}$

Property (iii) is related to entropy decay and states that the manifold of equilibria (i.e. the Maxwellians) is stable.

The formal derivation of the limit $\eta \to 0$ follows easily from these properties and leads to the so-called Drift-Diffusion (DD) model. More precisely, as $\eta \to 0$, $f^\eta \longrightarrow n(x,t)M(v)$ where $n(x,t)$ satisfies the system:

$$\frac{\partial n}{\partial t} + \nabla_x \cdot j = 0\,, \tag{5}$$

$$j = -A(\nabla_x n - nF)\,. \tag{6}$$

The first equation is called the continuity equation while the second one is the current equation.

The diffusion tensor A is related to Q and ϕ according to

$$A = -\int vQ^{-1}(vM)\,dv\,,$$

where Q^{-1} denotes the pseudo-inverse (acting from $N(Q)^{\perp}$ into itself), $Q^{-1}(vM)$ denotes the vector of components $Q^{-1}(v_i M)$ and $vQ^{-1}(vM)$ is a tensor product. As a consequence of (iv), A is a symmetric, positive-definite tensor.

The proof of this result follows the following steps: First, we note that if $f^\eta \to f$ smoothly, then, by (4), $Q(f) = 0$, i.e. by property (ii), $f = n(x,t)M(v)$ where the function $n(x,t)$ is still to be determined. Then, the continuity eq. (5) follows from integrating eq. (4), with respect to v, using the conservation of particle number (property (i)) and letting $\eta \to 0$. Finally, to derive the current eq (6), it is necessary to use a perturbation approach and to write

$$f^\eta = nM + \eta f^1 + o(\eta)\,, \tag{7}$$

from which, $j = \int f^1 v\,dv$ is obtained. f^1 may be viewed as the microscopic response of the system to gradients of macroscopic variables. To obtain f^1, one inserts the Hilbert expansion (7) into (4) and restricts to the leading order term. This leads to an equation for $Q(f^1)$ which is solved, thanks to property (iv). Details can be found in the above-cited references.

The Drift-Diffusion model is widely used by the engineers and has motivated considerable studies in both the semiconductor, plasma or gas discharge physics contexts (see e.g. [38], [44], [27]). However, it has proved unsuitable to describe strongly non equilibrium phenomena. One major reason for this is that it supposes instantaneous relaxation of the particle temperature to the background temperature T. This assumption fails for fast phenomena in which energy relaxation may be slower than the time evolution of the system.

The question which we are going to investigate in the next sections is the derivation of other, more elaborate macroscopic models with a broader range of physical applicability, through the same kind of methodology.

3 The Spherical Harmonics Expansion (SHE) model

Our approach relies on the observation that most collision operators exhibit a wide range of relaxation scales. Indeed, collisions may relax some moments faster than other ones: typically, momentum relaxation is usually faster than energy relaxation. This is particularly the case in plasmas, because of the very small electron mass compared with the ion mass, or in semiconductors, where the phonon energy is much smaller than the electron energy.

Therefore, our aim is to derive a model for the intermediate scale, i.e. after momentum relaxation has occurred but before energy relaxation has taken place. In such a case, we can expect that the distribution function f is isotropic with respect to v as a consequence of momentum relaxation, but is not yet Maxwellian.

Our method is to split the collision operator Q into an elastic operator $\mathcal{E}$ (the one which is going to produce momentum relaxation) and an inelastic operator $\mathcal{I}$ (which is responsible for energy relaxation), and then, to consider situations such that $\mathcal{E} \gg \mathcal{I}$. Then we shall proceed like in the first section, with the elastic operator $\mathcal{E}$ replacing Q.

The splitting is performed through a passage to spherical coordinates in velocity space $v = (\varepsilon, \omega)$, where $\varepsilon = m|v|^2/2$ is the energy and $\omega = v/|v|$ is the velocity angle). We then decompose the 'real' collision $v' = (\varepsilon', \omega') \rightarrow v = (\varepsilon, \omega)$ into 2 fictitious processes $(\varepsilon', \omega') \rightarrow (\varepsilon, \omega')$ and $(\varepsilon, \omega') \rightarrow (\varepsilon, \omega)$, introducing a fictitious state (ε, ω'). The first process changes the energy and gives rise to the inelastic operator while the second one only changes the velocity angle and gives rise to the elastic operator. We shall not detail the expressions of these operators and refer to [11] for details. We note $Q(f) = \mathcal{E}(f) + \mathcal{I}(f)$ the resulting decomposition of the collision operator into elastic and inelastic operators.

We now suppose that elastic scattering dominates over inelastic scattering: $\mathcal{E} \gg \mathcal{I}$. We are therefore entitled to rescale the operators according to

$$\mathcal{E} = O(1)\,, \quad \mathcal{I} = O(\eta^2)\,,$$

and we use again a diffusive scaling (3) for the variables x, t and F. Linking the scaling of the inelastic operator to the parameter η can be viewed as somewhat arbitrary. However, other scaling hypotheses lead to similar models [21].

The resulting rescaled Boltzmann equation is now written:

$$\eta^2 \frac{\partial f^\eta}{\partial t} + \eta \left(v \cdot \nabla_x f^\eta + \frac{1}{m} F \cdot \nabla_v f^\eta \right) = \mathcal{E}(f^\eta) + \eta^2 \mathcal{I}(f^\eta)\,. \tag{1}$$

Again, the investigation of the limit $\eta \to 0$ requires the determination of the properties of the leading term, namely the collision operator $\mathcal{E}$. First, we note that $\mathcal{E}$ is given by

$$\mathcal{E}(f)|_{(\varepsilon,\omega)} = \int_{\omega'} \psi(\varepsilon;\omega,\omega')[f(\varepsilon,\omega') - f(\varepsilon,\omega)]d\omega'\,,$$

where $\psi(\varepsilon;\omega,\omega')$ is related to ϕ by integration with respect to ε'. In particular, it is seen that $\mathcal{E}$ operates on the ω variable only. Now we have:

(i) Conservation of particle number on each energy surface:

$$\int \mathcal{E}(f)|_{(\varepsilon,\omega)} d\omega = 0\,, \quad \forall \varepsilon\,.$$

(ii) The null set $N(\mathcal{E})$ consists of constant functions with respect to ω (i.e. all functions of the energy only: $f(v) = G(\varepsilon)$).
(iii) $-\mathcal{E}$ is a self-adjoint non-negative operator and is positive-definite and invertible on $N(\mathcal{E})^\perp$.

Then, the limit $\eta \to 0$ follows in the same way as in the previous section. We have $f^\eta \longrightarrow G(x,\varepsilon,t)$ solving the SHE model:

$$N(\varepsilon)\frac{\partial G}{\partial t} + \tilde{\nabla} J = \tilde{\mathcal{I}}(G)\,, \tag{2}$$

$$J = -D(\varepsilon)\tilde{\nabla} G\,, \tag{3}$$

where $\tilde{\nabla}$ denotes the operator $\tilde{\nabla} = \nabla_x + F\frac{\partial}{\partial \varepsilon}$. Equation (2) is a continuity equation in the 'extended space' (x,ε) while eq. (3) is a current equation. The diffusion tensor $D(\varepsilon)$ is related with $\mathcal{E}$ through

$$D(\varepsilon) = -\sqrt{2\varepsilon} \int v \mathcal{E}^{-1}(v)\, d\omega\,,$$

while $N(\varepsilon) = \sqrt{2\varepsilon}$ is related with the geometry of the energy surfaces. Finally, $\tilde{\mathcal{I}}$ is the contribution of the inelastic collision operator $\mathcal{I}$ and is given by

$$\tilde{\mathcal{I}} = \sqrt{2\varepsilon} \int \mathcal{I}(G)\, d\omega\,.$$

The SHE model has received considerable attention lately. On can find for instance in [7], [8] a derivation of the model via stochastic processes. It is currently used by semiconductor engineers on a routine basis [30], [31]. The present derivation which does not make use of any expansion of the Boltzmann eq. in spherical harmonics can be extended to non spherically symmetric situations, which are common in semiconductor physics. It has first been derived in this way by [29] for simplified relaxation time collision operators. This derivation has later been extended to arbitrary collision operators in [2] (see also [10] for a review in the semiconductor physics context) and by [43] (where additionally, an approximation of the inelastic collision operator in terms of a Fokker-Planck operator is proposed). The proof of convergence of solutions from the Boltzmann to the SHE model is done in [11] in a slightly more general setting.

The SHE model is particularly attractive where relaxation towards an isotropic state results from collisions with the boundary. Indeed, the process in which particles are reflected by the boundary with conservation of energy and randomization of the velocity angle bears similarities with elastic collisions as described above. If particles undergo some confinement along the boundary which makes encounters with the boundary extremely frequent, then, the lateral dynamics (in the directions parallel to the boundary) can be described by a SHE type model. In this case, the passage from Boltzmann to SHE not only saves 2 velocity dimensions (the coordinates of the velocity angle), but also one space dimension (the coordinate in the direction normal to the boundary) and is particularly efficient.

Two situations have been investigated in which such an occurrence happens: first the motion between two parallel plates seperated by a very small gap. Second, the case of a confinement force which reflects the particles back to the boundary. Both cases correspond to situations of practical interest. The first case gives rise to a model for electron transport in Stationary Plasma Thrusters, a device used in satellite propulsion. The precise physical model is described in [18] and [36] as well as in [35]. As a matter of example of the efficiency of the SHE model, we show in figure 1 a comparison between the SHE model and the solution of the original Boltzmann equation by means of a standard Monte-Carlo simulation. The SHE model is about a hundred times faster.

The case where the confinement is due to a confining force has been investigated in [12]. The resulting SHE model has been used to derive an Energy-Transport model (see next section) for certain kinds of surface discharges that occur for instance on satellite solar pannels [25]. Indeed, in physically complex situations, the SHE model, with its additional energy variable, is still too costly compared with a standard fluid model. This is why it is attractive to investigate if the SHE model can be reduced to such a standard model. This is the aim of the next section.

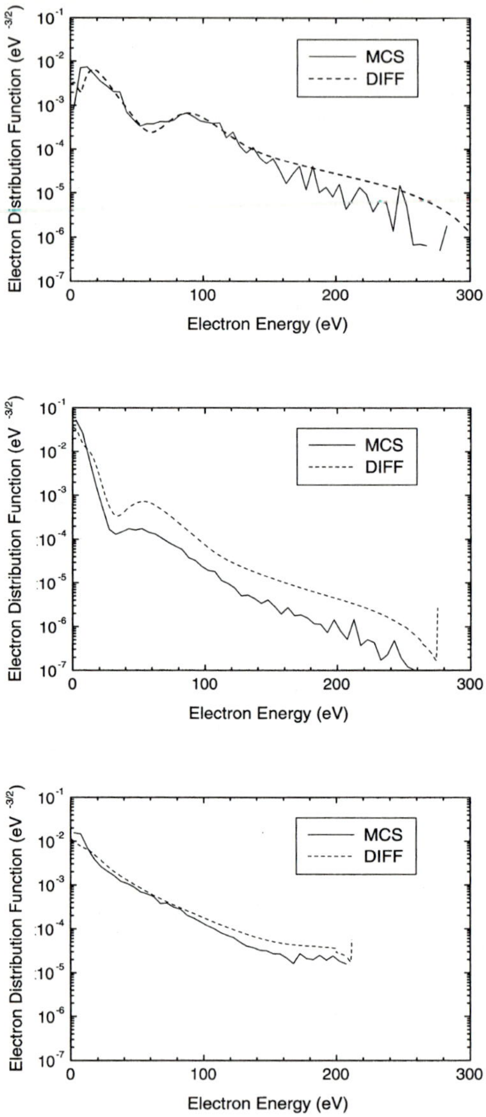

Fig. 1. Electron distribution function as a function of the energy at various locations of a plasma thruster. Solid line: Monte-Carlo simulation; dashed line: SHE model. From top to bottom: respectively at 0.1 cm, 2 cm and 3.4 cm from the anode. The total length of the device is 4 cm (see [18] and [36])

4 The Energy-Transport (ET) model

We are going to consider a relaxation limit of the SHE model in which the inelastic operator $\tilde{\mathcal{I}}$ tends to relax $G(x,\varepsilon,t)$ to a thermodynamical equilibrium. Such a relaxation mechanism may be due to e.g. binary collision, turbulence, etc. For that purpose, we assume that $\tilde{\mathcal{I}}$ satisfies the following hypotheses:

(i) The Null Set $N(\tilde{\mathcal{I}})$ consists of Maxwellian equilibria, with arbitrary temperature T:

$$G_{\mu,T}(\varepsilon) = e^{-((\varepsilon-\mu)/k_B T)}\,, \tag{1}$$

where μ is the chemical potential (related with the density).

(ii) $\tilde{\mathcal{I}}$ decreases the entropy (i.e. it relaxes G to the equilibrium $G_{\mu,T}$).

(iii) $\tilde{\mathcal{I}}$ preserves the local density and energy, i.e.

$$\int \tilde{\mathcal{I}}(G)d\varepsilon = 0\,, \qquad \int \tilde{\mathcal{I}}(G)\varepsilon d\varepsilon = 0\,.$$

Now, we rescale the SHE model as follows: let α denote the ratio of the inelastic mean free path to the macroscopic scale ($\alpha \ll 1$). We change to the macroscopic variables:

$$x' = \alpha x\,, \quad t' = \alpha^2 t\,, \quad F = \alpha F'\,, \quad J = \alpha J'\,,$$

and obtain the rescaled SHE model, which reads (after dropping the primes):

$$\alpha^2 \left(N(\varepsilon)\frac{\partial G^\alpha}{\partial t} + \tilde{\nabla} J^\alpha \right) = \tilde{\mathcal{I}}(G^\alpha)\,, \tag{2}$$

$$J^\alpha = -D(\varepsilon)\tilde{\nabla} G^\alpha\,, \quad \tilde{\nabla} = \nabla_x + F\frac{\partial}{\partial \varepsilon}\,. \tag{3}$$

Now, the formal limit $\alpha \to 0$ of (2), (3) leads to the Energy-Transport model, i.e. $G^\alpha \to G_{\mu,T}$ where the chemical potential μ and the temperature T satisfy the Energy-Transport system:

$$\frac{\partial}{\partial t}\begin{pmatrix} n(\mu,T) \\ W(\mu,T) \end{pmatrix} + \nabla \cdot \begin{pmatrix} j_n \\ j_W \end{pmatrix} + \begin{pmatrix} 0 \\ -F\cdot j_n \end{pmatrix} = 0\,, \tag{4}$$

$$\begin{pmatrix} j_n \\ j_W \end{pmatrix} = -\mathbb{D}\begin{pmatrix} \nabla(\mu/T) - F/T \\ \nabla(-1/T) \end{pmatrix}\,. \tag{5}$$

In (4) The density n and the energy W are local functions of μ and T, given by the integration of the Maxwellian (1) with respect to the energy ε, premultiplied by $N(\varepsilon)$ and $\varepsilon N(\varepsilon)$ respectively, while j_n and j_W are the particle and energy fluxes respectively.

The Energy Transport (ET) model is a nonlinear diffusion system. The diffusion matrix $\mathbb{D}$ is symmetric, positive-definite (and translates a well-known thermodynamic property, known as Onsager's reciprocity relation).

It is derived from the SHE-diffusivity D by integration with respect to ε [3], [2].

The ET model expresses the time evolution of conservative variables (the density n and the energy W) in terms of fluxes j_n and j_W which in turn are expressed in terms of gradients of the state (or entropic) variables μ and T. There exists a one-to-one correspondence between (n, W) and (μ, T) through the entropy (more precisely, between (n, W) and $(\mu/T, -1/T)$) (see e.g. [10] and [14]). This relation can be exploited to obtain existence results (see e.g. [15], [16]). The ET system can also be seen as a relaxation limit of the Navier-Stokes system [3].

The ET model is increasingly used for semiconductor or plasma discharge modeling. The present derivation stems from [3], [2]. A rigorous convergence result of the Boltzmann solution to the ET model has been proved in [5]. A formal derivation in the plasma physics context has been carried through in [22] and leads to the coupling of Euler equations for the ions with the ET model for electrons.

As an example of the use of the ET model, we present an application to certain kinds of surface discharges which occur on satellite solar panels. This model is derived from the SHE model obtained through collisions with the boundary and confinement along the boundary by a confining force, as presented in [12]. In this case, the occurence of the discharge and the confinement along the boundary originates from the charging of the satellite by the spatial environement like e.g. the solar wind, the impact by UV photons, etc. The model is adapted from a scenario originally proposed by M. Cho in [9]. The impact of the electrons on the boundary also generates secondary emission, thus increasing the number of electrons involved in the process and eventually leading to an avalanche phenomenon. Additionally, secondary emission leaves positive charges along the boundary which increase the confinement force and the collision frequency with the boundary, leading to a positive feedback which enhances the electron avalanche. Let us stress that, through the derivation of the diffusion constants, the ET model retains information about the hopping dynamics of the electrons along the boundary. The model and the results are described in more detail in [25]. Here, we only show in figure 2 the electron density as a function of the distance along the boundary for various times. The picture clearly displays the avalanche breakdown phenomenon.

Another example of use of the ET model in plasma modeling is for 2-fluid plasmas. According to the model derived in [22] (see also [28]), a convenient model for plasmas is obtained by coupling the ET model for electrons with the Euler equations for the ions. Compared with a more standard 2-fluid Euler model for both electrons and ions, the present model is obtained by letting the ratio of the electron to ion mass tend to zero. This amounts to assuming that the electron collisions against the ions are almost elastic, which they are up to terms of the order of this mass ratio. The advantage in doing such an

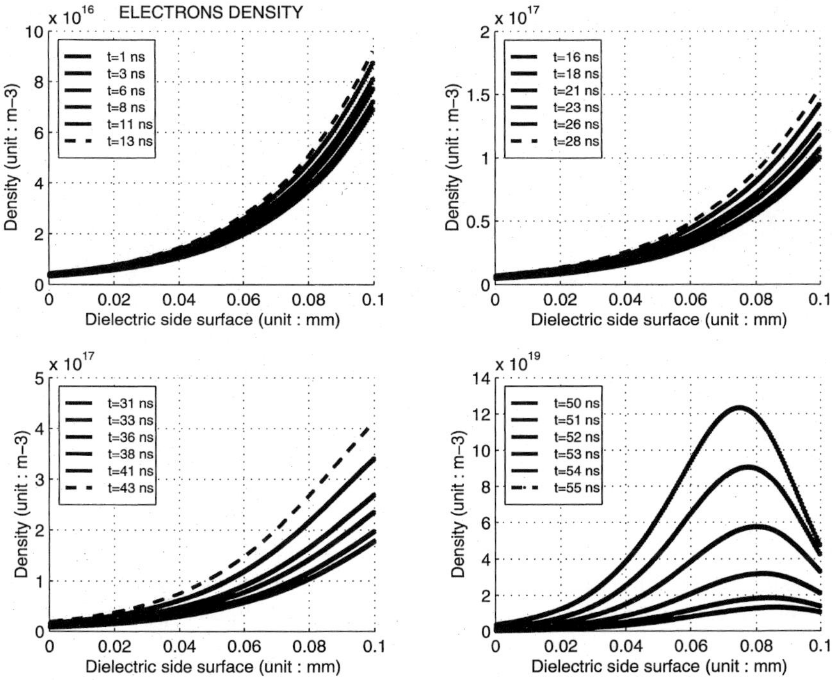

Fig. 2. Electron density along the dielectric surface of a solar cell during a discharge event. Snapshots at times ranging from 1 ns up to 55 ns are shown and displayed in 4 seperate pictures for clarity. Note that the density scale changes by 3 orders of magnitude between the begining and the end of the simulation, clearly showing that an avanlanche is occuring. Details are found in [25].

approximation is that no time-scale limitation in the computation is linked with the plasma frequency, which is fortunate since the plasma frequency is a very short scale phenomenon. We refer to [28] for details about this point.

Such a model is very useful for the modeling of devices such as the Plasma Opening Switch. In these devices, the plasma exhibits a strong departure from quasi-neutrality which makes the use of a 2-fluid model necessary. The particle dynamics is coupled with Maxwell equations in order to describe the interaction of the plasma with an electromagnetic wave. Simulations of such devices can be found in [39], [34] as well as references therein. However, in many instances, the ion to electron mass ratio has been artificially modified, in order to cope with the very different time scales associated with the two species of particles. By using an ET model for the electrons, we directly place ourselves in the time scale of the ion dynamics and avoid unphysical results that could result from an artificial change of the mass ratio.

Results of simulations of the POS by means of Particle-in-Cell type methods for the coupled electron-ET and ions-Euler system and with finite-difference discretization of the Maxwell equations are presented in figures 3 and 4. They clearly show the appearance of a non quasineutral region close to the cathode (sheath region). The test problem and the numerical results are presented in detail in [28].

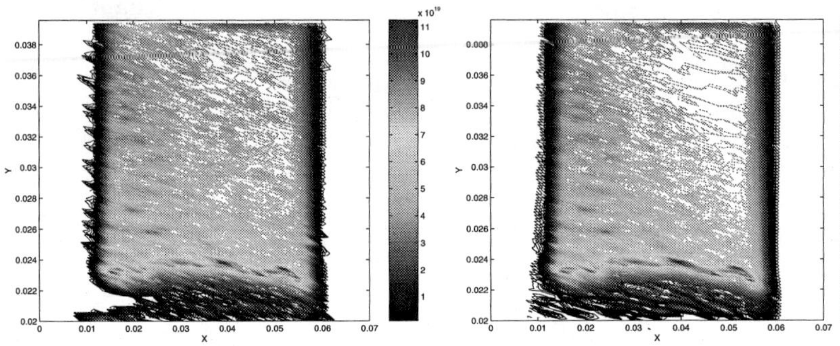

Fig. 3. Electron (left) and Ion (right) densities in the POS during the erosion of the plasma by the electromagnetic wave. Differences in density clearly show that quasineutrality breaks down close to the cathode (lower boundary). See also [28].

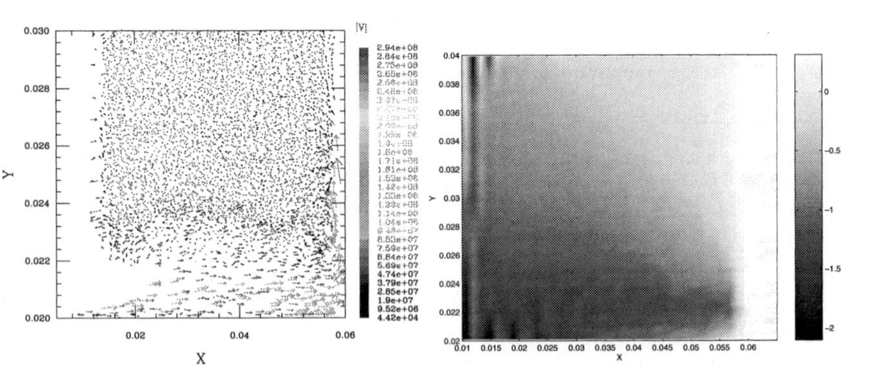

Fig. 4. Electron velocity (left) and magnetic field intensity (right) during the erosion phase of the POS.

5 Conclusion

In this paper, we have summarized the derivation of intermediate models between kinetic and fluid models from diffusion limits of kinetic models. The key idea is to use 'incomplete' relaxation by the physical dissipation mechanisms (such as e.g. collisions). More precisely, a first intermediate model, the SHE model, has been derived under the assumption of dominant elastic scaling. Then, the SHE model can be futher reduced to a second intermediate model, the Energy-Transport model, which appears as a diffusion system for the fluid density and internal energy. These models have been practically implemented in physically realistic situation and they have proven to be good compromises between physical accuracy and solution efficiency

Further developments have been or are currently being undertaken concerning boundary conditions and kinetic boundary layers [24], coupling with quantum models [13], scattering by interfaces [23], homogenization [4], [26], [1] (see also [40] and references therein).

Future research directions concern finding 'new' applications of these models, their careful numerical validation [6], the design of new 'artificial' collision operators in order to enrich the set of available macroscopic models.

Note that these systems have their counterpart in the context of hydrodynamic limits [19] and can be applied to the kinetic modeling of turbulence [20]. Also, these models can themselves be improved e.g. by considering SHE models with energy levels coupling in the diffusion operator [11] or by considering collisions with no principle of detailed balance [17] or [40].

References

1. Aoki, K., Degond, P., Sone, Y. (submitted): Homogenization of a flow in a periodic channel of small section
2. Ben Abdallah, N., Degond, P. (1996): On a hierarchy of macroscopic models for semiconductors. J. Math. Phys., **37**, 3306–3333
3. Ben Abdallah, N., Degond, P., Génieys, S. (1996): An energy-transport model for semiconductors derived from the Boltzmann equation. J. Stat. Phys., **84**, 205–231
4. Ben Abdallah, N., Degond, P., Mellet, A., Poupaud, F. (to appear): Electron transport in semiconductor superlattices. Quarterly Appl. Math.
5. Ben Abdallah, N., Desvillettes, L., Genieys, S. (2000): On the convergence of the Boltzmann equation for semiconductors towards an energy-transport model. J. Stat. Phys., **98**, 835–870
6. Bourgade, J. P., Mellet, A., Mieussens, L. (in preparation)
7. Bringuier, E. (1998): Kinetic theory of high-field transport in semiconductors. Phys. Rev. B, **57**, 2280-2285
8. Bringuier, E. (2000): Nonequilibrium statistical mechanics of drifting particles. Phys. Rev. E, **61**, 6351-6358
9. Cho, M., Hastings, D. E. (1990): Dielectric charging process and arcing rates of high-voltage solar arrays. J. Spacecrafts and Rockets, **28**, 698–706.

10. Degond, P. (2000): Mathematical modelling of microelectronics semiconductor devices. In AMS/IP Studies in Advanced Mathematics, Vol 15, AMS Society and International Press
11. Degond, P. (2001) An infinite system of diffusion equations arising in transport theory: the coupled spherical harmonics expansion model. Math. Methods and Models in the Applied Sciences, **11**, 903–932
12. Degond, P. (2002): Transport of trapped particles in a surface potential. In Studies in Mathematics and its applications, vol 31, D. Cioranescu and J. L. Lions (eds), Elsevier.
13. Degond, P., El Ayyadi, A. (2002): A coupled Schrödinger Drift-Diffusion model for quantum semiconductor device simulations. J. Comput. Phys., **181**, 1–38
14. Degond, P., Génieys, S., Jüngel, A. (1997): Symmetrization and entropy inequality for general diffusion equations. C. R. Acad. Sci. Paris, **325**, 963–968
15. Degond, P., Génieys, S., Jüngel, A. (1997): A system of parabolic equations in nonequilibrium thermodynamics including thermal and electrical effects. Journal de Mathématiques Pures et Appliquées, **76**, 991–1015
16. Degond, P., Génieys, S., Jüngel, A. (1998): A steady-state system in nonequilibrium thermodynamics including thermal and electrical effects. Math. Methods in the Applied Sciences, **21**, 1399–1413
17. Degond, P., Goudon, T., Poupaud, F. (2000) Diffusion limit for non homogeneous and non micro-reversible processes. Indiana University Math. J., **49**, 1175–1198
18. Degond, P., Latocha, V., Garrigues, L., Boeuf, J. P. (1998) Electron Transport in Stationary Plasma Thrusters. Transp. Theory and Stat. Phys., **27**, 203–221
19. Degond, P., Lemou, M. (2001): On the viscosity and thermal conduction of fluids with multivalued internal energy. European J. Mech. B/fluids, **20**, 303–327
20. Degond, P., Lemou, M. (2002): Turbulence models for incompressible fluids derived from kinetic theory. J. Math. Fluid Mech., 257–284
21. Degond, P., Levermore, C. D., Schmeiser, C. (to appear): A note on the energy-transport limit of the semiconductor Boltzmann equation. In the Proceedings of the Institute of Mathematics and Applications, Minneapolis
22. Degond, P., Lucquin-Desreux, B. (1996): Transport coefficients of plasmas and disparate mass binary gases. Transp. Theory and Stat. Phys., **25**, 595–633
23. Degond, P., Schmeiser, C. (1998): Macroscopic models for semiconductor heterostructures. J. Math. Phys **39**, 1–30
24. Degond, P., Schmeiser, C. (1999): Kinetic boundary layers and fluid-kinetic coupling in semiconductors. Transp. Theory and Stat. Phys. **28**, 31–55
25. Degond, P., Talaalout, R., Vignal, M. H. (2000): Electron transport and secondary emission in a surface of a solar cell. In Proceedings of the conference on 'multipactor, RF and DC corona and passive intermodulation in space RF hardware', ESTEC, Noordwijk, The Netherlands, Sept 4-6
26. Degond, P., Zhang, K. (to appear): Diffusion approximation of a scattering matrix model of a semiconductor superlattice. SIAM J. Appl. Math
27. Delcroix, J. L., Bers, A. (1994): Physique des plasmas, CNRS éditions, Paris
28. Deluzet, F. (2002): PhD dissertation, Institut National des Sciences Appliquées, Toulouse
29. Dmitruk, P., Saul, A., Reyna, L. (1992) High electric field approximation in semiconductor devices. Appl. Math. Letters **5**, 99–102

30. Gnudi, A., Ventura, D., Baccarani, G., Odeh, F. (1993): Two-dimensional MOSFET simulation by means of a multidimensional spherical harmonic expansion of the Boltzmann transport equation. Solid State Electron, **36**, 575–581
31. Goldsman, N., Henrickson, L., Frey, J. (1991): A physics based analytical numerical solution to the Boltzmann transport equation for use in device simulation. Solid State Electron., **34**, 389–396
32. Golse, F., Poupaud, F. (1992): Limite fluide des équations de Boltzmann des semiconducteurs pour une statistique de Fermi-Dirac. Asymptotic Analysis, **6**, 135–160
33. Grad, H. (1958): Principles of the kinetic theory of gases. In Handbuch der Physik, S. Flügge (ed.), vol XII, Springer, Berlin
34. Grossmann, G. M., Ottinger, P. F., Neri, J. M., Drobot, A. T. (1986): Numerical simulation of a low density plasma erosion switch. Phys. Fluids, **29**, 2724–2735
35. Latocha, V. (2001): PhD dissertation, Institut National des Sciences Appliquées, Toulouse
36. Latocha, V., Garrigues, L., Degond, P., Boeuf, J. P. (2002): Numerical Simulation of Electron Transport in the Channel Region of a Stationary Plasma Thruster. Plasma sources sci. Technol **11**, 104–114
37. Levermore, C. D. (1996): Moment closure hierarchies for kinetic theories. J. Stat. Phys., **83**, 1021–1065
38. Markowich, P. A., Ringhofer, C., Schmeiser, C. (1990): Semiconductor Equations, Springer, Wien
39. Mason, R. J., Wallace, J. M., Grossmann, J. M., Ottinger, P. F. (1987): Implicit collisional three-fluid simulation of the plasma opening switch. IEEE Trans. Plasma Science, **PS 15**, 715–724
40. Mellet, A. (2002): PhD dissertation, University Paul Sabatier, Toulouse
41. Müller, I., Ruggeri, T. (1993): Extended thermodynamics, Springer, Berlin
42. Poupaud, F. (1991): Diffusion approximation of the linear semiconductor equation: analysis of boundary layers. Asymptotic Analysis **4**, 293–317
43. Schmeiser, C., Zwirchmayr, A. (1998): Elastic and drift-diffusion limits of electron-phonon interaction in semiconductors. Math. Mod. Meth. Appl. Sci., **8**, 37–54
44. Selberherr, S. (1984): Analysis and simulation of semiconductor devices, Springer, Wien

Viscosity Solutions for Nonlinear Hyperbolic Systems

Alberto Bressan

S.I.S.S.A., Via Beirut 4, Trieste 34014, Italy. *bressan@sissa.it*

1 Introduction

A system of conservation laws in one space dimension takes the form

$$u_t + f(u)_x = 0. \tag{1.1}$$

We shall always assume that the system is *strictly hyperbolic*, so that each $n \times n$ Jacobian matrix $A(u) = Df(u)$ has real distinct eigenvalues

$$\lambda_1(u) < \lambda_2(u) < \cdots < \lambda_n(u)$$

and a basis of eigenvectors $r_1(u), \ldots, r_n(u)$ with unit length. Given an initial condition

$$u(0, x) = \bar{u}(x), \tag{1.2}$$

the existence of weak solutions to the Cauchy problem was proved in the celebrated paper of Glimm [14]. This global existence result is valid for initial data with small total variation and under the additional assumption

- **(H)** For each $i \in \{1, \ldots, n\}$, the i-th characteristic field is either *linearly degenerate*, so that
$$D\lambda_i(u) \cdot r_i(u) = 0 \qquad \forall u\,,$$
or else it is *genuinely nonlinear*, i. e.
$$D\lambda_i(u) \cdot r_i(u) > 0 \qquad \forall u\,.$$

The uniqueness of entropy weak solutions and their continuous dependence on the initial data was later established in a series of papers [3], [5], [7], [11], [9], [8], [10]. The main results can be summarized as follows (see [4] for a comprehensive account).

- The solutions obtained as limits of Glimm or front tracking approximations are unique and depend Lipschitz continuously on the initial data, in the $\mathbf{L}^1$ norm.
- Every small BV solution of the Cauchy problem (1.1)-(1.2) which satisfies the Lax entropy conditions [18] coincides with the unique limit of front tracking approximations.

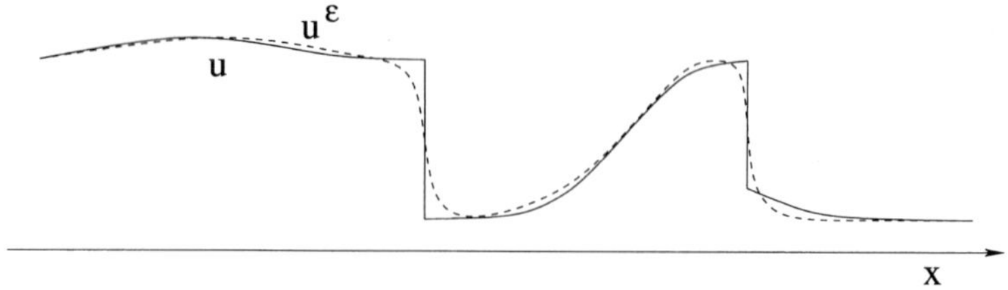

Fig. 1. Vanishing viscosity approximation.

It is natural to expect that the entropic solutions of the hyperbolic system (1.1) actually coincide with the limits of solutions to the parabolic system

$$u_t^\varepsilon + f(u^\varepsilon)_x = \varepsilon\, u_{xx}^\varepsilon \tag{1.3}$$

letting the viscosity coefficient $\varepsilon \to 0$. For smooth solutions, this convergence result is easy to show. However, one should keep in mind that a weak solution of the hyperbolic system (1.1) in general is only a function with bounded variation, possibly with a countable number of discontinuities. In this case, as the smooth functions u^ε approach the discontinuous solution u, near points of jump their gradients u_x^ε must tend to infinity (Fig. 1), while their second derivatives u_{xx}^ε become even more singular. Therefore, the convergence $u^\varepsilon \to u$ of the vanishing viscosity limit is a highly nontrivial matter, in the general situation. In earlier literature, results in this direction relied on three different approaches:

1 - **Comparison principles for parabolic equations.** For a scalar conservation law, the existence, uniqueness and global stability of vanishing viscosity solutions was first established by Oleinik [25] in one space dimension. The famous paper by Kruzhkov [16] covers the more general class of $\mathbf{L}^\infty$ solutions and is also valid in several space dimensions.

2 - **Singular perturbations.** This technique was developed by Goodman and Xin [15], and covers the case where the limit solution u is piecewise smooth, with a finite number of non-interacting, entropy admissible shocks. See also [31] and [26], for further results in this direction.

3 - **Compensated compactness.** With this approach, introduced by Tartar and DiPerna [13], one first considers a weakly convergent subsequence $u^\varepsilon \rightharpoonup u$. For a class of 2×2 systems, one can show that this weak limit u actually provides a distributional solution to the nonlinear system (1.1). The proof relies on a compensated compactness argument, based on the representation of the weak limit in terms of Young measures, which must reduce to a Dirac mass due to the presence of a large family of entropies.

In view of recent results on the well-posedness of the hyperbolic Cauchy problem within the space of BV functions, it is natural to seek uniform BV bounds also on the viscous approximations u^ε in (1.3). This is indeed the main goal accomplished in [2]. As soon as these BV bounds are established, the existence of a vanishing viscosity limit follows by a standard compactness argument. The uniqueness of the limit can then be deduced from the uniqueness theorem in [8]. By further analysis, one can also prove the continuous dependence on the initial data for the viscous approximations u^ε, in the $\mathbf{L}^1$ norm. These results are valid, more generally, for hyperbolic systems not necessarily in conservation form. In this case, however, the unique limits of vanishing viscosity approximations cannot be interpreted as weak solutions in a distributional sense. Their characterization thus requires more care. The main results in [2] can be stated as follows.

Theorem 1. *Consider the Cauchy problem for the hyperbolic system with viscosity*

$$u_t^\varepsilon + A(u^\varepsilon)u_x^\varepsilon = \varepsilon\, u_{xx}^\varepsilon \qquad u^\varepsilon(0,x) = \bar u(x)\,. \tag{1.4}$$

Assume that the matrices $A(u)$ are strictly hyperbolic, smoothly depending on u in a neighborhood of a compact set $K \subset \mathbb{R}^n$. Then there exist constants C, L, L' and $\delta > 0$ such that the following holds. If

$$Tot.Var.\{\bar u\} < \delta\,, \qquad \lim_{x\to-\infty} \bar u(x) \in K\,,$$

then for each $\varepsilon > 0$ the Cauchy problem $(1.4)_\varepsilon$ has a unique solution u^ε, defined for all $t \geq 0$. Adopting a semigroup notation, this will be written as $t \mapsto u^\varepsilon(t,\cdot) \doteq S_t^\varepsilon \bar u$. In addition, one has:

BV bounds :
$$Tot.Var.\{S_t^\varepsilon \bar u\} \leq C\; Tot.Var.\{\bar u\}\,. \tag{1.5}$$

$\mathbf{L}^1$ stability :
$$\left\|S_t^\varepsilon \bar u - S_t^\varepsilon \bar v\right\|_{\mathbf{L}^1} \leq L\left\|\bar u - \bar v\right\|_{\mathbf{L}^1}, \tag{1.6}$$

$$\left\|S_t^\varepsilon \bar u - S_s^\varepsilon \bar u\right\|_{\mathbf{L}^1} \leq L'\left(|t-s| + \left|\sqrt{\varepsilon t} - \sqrt{\varepsilon s}\,\right|\right). \tag{1.7}$$

Convergence: *As $\varepsilon \to 0+$, the solutions u^ε converge to the trajectories of a semigroup S such that*

$$\left\|S_t \bar u - S_s \bar v\right\|_{\mathbf{L}^1} \leq L\,\|\bar u - \bar v\|_{\mathbf{L}^1} + L'\,|t-s|\,. \tag{1.8}$$

These vanishing viscosity limits can be regarded as the unique **vanishing viscosity solutions** *of the hyperbolic Cauchy problem*

$$u_t + A(u)u_x = 0, \qquad u(0,x) = \bar u(x)\,. \tag{1.9}$$

In the conservative case $A(u) = Df(u)$, every vanishing viscosity solution is a weak solution of

$$u_t + f(u)_x = 0, \qquad u(0,x) = \bar u(x)\,, \tag{1.10}$$

satisfying the Liu admissibility conditions [19].

Remark 1. In nearly all previous works on BV solutions for systems of conservation laws, the basic estimates on the total variation were obtained by a careful study of the Riemann problem and elementary wave interactions [14]. The Riemann problem also takes the center stage in all earlier proofs of the stability of solutions [5], [7], [11]. In this connection, the hypothesis (H) introduced by Lax [18] is widely adopted in the literature. It guarantees that solutions of the Riemann problem have a simple structure, consisting of at most n elementary waves (shocks, centered rarefactions or contact discontinuities). If the assumption (H) is dropped, some results on global existence [20], [22] and continuous dependence [1] are still available, but their proofs become far more technical.
On the other hand, the new approach developed in [2] marks the first time where uniform BV estimates are obtained without any reference to Riemann problems. Global existence and stability of weak solutions are obtained for the whole class of strictly hyperbolic systems, regardless of the hypothesis (H).

Remark 2. For the viscous system of conservation laws

$$u_t + f(u)_x = u_{xx}\,, \tag{1.11}$$

previous results in [21], [28], [29], [31] have established the stability of special types of solutions, such as travelling viscous shocks or viscous rarefactions, w.r.t. suitably small perturbations. Taking $\varepsilon = 1$, the above theorem yields at once the uniform Lipschitz stability (w..r.t. the $\mathbf{L}^1$ distance) of ALL viscous solutions with sufficiently small total variation.

2 Outline of the Proof

We describe here the main ideas involved in the proof of Theorem 1, in several steps.

1. (Rescaling) The rescaling $t \mapsto t/\varepsilon$, $x \mapsto x/\varepsilon$ transforms the problem (1.4) into

$$u_t + A(u)u_x = u_{xx} \qquad\qquad u(0,x) = \bar{u}(\varepsilon x)\,. \tag{2.1}$$

Clearly, the total variation of a solution u is not affected by this rescaling. Notice however that the values of a solution u^ε of (1.4) on a time interval $[0,T]$ correspond to the value of a solution u of (2.1) on the much longer interval $[0,\ T/\varepsilon]$. To obtain BV estimates uniformly valid for all $\varepsilon > 0$, we thus need to prove similar estimates for the problem

$$u_t + A(u)u_x = u_{xx} \tag{2.2}$$

uniformly valid for all $t \in [0, \infty[$ and all initial data $u(0,\cdot)$ with suitably small total variation.

2. (Parabolic estimates) From now on we thus consider the parabolic system (2.2). On an initial time interval $[0, T]$, the solution can be represented as

$$u(t) = G(t) * u(0) - \int_0^t G(t-s) * A(u(s)) u_x(s)\, ds\,, \tag{2.3}$$

in terms of convolutions with the standard heat kernel

$$G(t,x) \doteq (4\pi t)^{-1/2} e^{-x^2/4t}.$$

Even if the initial data is discontinuous, the presence of viscosity has an immediate regularizing effect and the solution becomes smooth for all times $t > 0$. The smoothing effect of the parabolic equation is summarized in Fig. 2. Let δ_0 be the total variation of the initial data. Then, on an initial interval $t \in]0, T]$ with $T = \mathcal{O}(1) \cdot \delta_0^{-2}$ one can show that

- The total variation of $u(t, \cdot)$ remains bounded, i.e.

$$\big\| u_x(t,\cdot) \big\|_{\mathbf{L}^1} = \mathcal{O}(1) \cdot \delta_0\,. \tag{2.4}$$

- The norms of all higher order derivatives decay at algebraic rates. In particular

$$\begin{aligned} \big\| u_x(t,\cdot) \big\|_{\mathbf{L}^\infty} &\le \big\| u_{xx}(t,\cdot) \big\|_{\mathbf{L}^1} = \mathcal{O}(1) \cdot \delta_0 t^{-1/2}, \\ \big\| u_{xx}(t,\cdot) \big\|_{\mathbf{L}^\infty} &\le \big\| u_{xxx}(t,\cdot) \big\|_{\mathbf{L}^1} = \mathcal{O}(1) \cdot \delta_0 t^{-1}. \end{aligned} \tag{2.5}$$

All these bounds can be derived by looking at integral representations of the form (2.3).

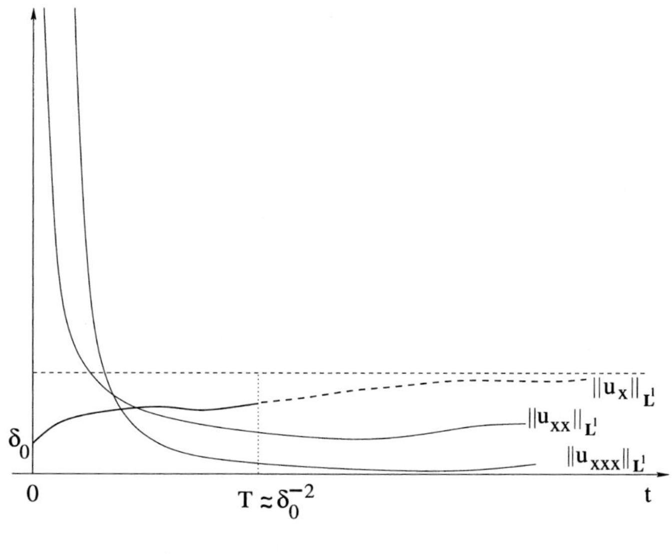

Fig. 2. Decay of higher derivatives.

For $t \in [T, \infty[$, these parabolic-type estimates yield only partial informations. What can be proved is the following: as long as the bound (2.4) on the total variation remains valid, also the norms of all higher derivatives remain small. Namely

$$\begin{aligned} \big\|u_x(t,\cdot)\big\|_{\mathbf{L}^\infty} &\leq \big\|u_{xx}(t,\cdot)\big\|_{\mathbf{L}^1} = \mathcal{O}(1)\cdot\delta_0^2\,, \\ \big\|u_{xx}(t,\cdot)\big\|_{\mathbf{L}^\infty} &\leq \big\|u_{xxx}(t,\cdot)\big\|_{\mathbf{L}^1} = \mathcal{O}(1)\cdot\delta_0^3\,. \end{aligned} \tag{2.6}$$

3. (Outline of the BV estimates) The main part of the proof is to establish the uniform bounds (2.4) on the total variation, for all times $t \in [0, \infty[$. By the previous step, we can assume that the bounds (2.6) on higher order derivatives already hold. Our basic approach is the following.

- At each point (t, x) we decompose the gradient along a suitable basis of unit vectors $\tilde{r}_i$, say
$$u_x = \sum v_i \tilde{r}_i\,. \tag{2.7}$$
- We then derive an equation describing the evolution of these gradient components
$$v_{i,t} + (\tilde{\lambda}_i v_i)_x - v_{i,xx} = \phi_i\,. \tag{2.8}$$
- Finally, we show that all source terms $\phi_i = \phi_i(t, x)$ are integrable, hence
$$\big\|v_i(\tau,\cdot)\big\|_{\mathbf{L}^1} \leq \big\|v_i(0,\cdot)\big\|_{\mathbf{L}^1} + \int_0^\infty \int_{I\!R} \big|\phi_i(t,x)\big|\,dxdt < \infty \tag{2.9}$$
for all $\tau > 0$. Of course, this will imply
$$\big\|u_x(\tau,\cdot)\big\|_{\mathbf{L}^1} \leq \sum \big\|v_i(\tau,\cdot)\big\|_{\mathbf{L}^1} < \infty\,.$$

In this connection, it seems natural to decompose the gradient u_x along the eigenvectors of the hyperbolic matrix $A(u)$. This approach however does NOT work. In the case where the solution u is a travelling viscous shock profile, we would obtain source terms which are not identically zero. Hence they are certainly not integrable over the domain $\{t \in [0, \infty[\,, x \in I\!R\}$.

An alternative approach, proposed by S. Bianchini, is to decompose u_x as a *sum of gradients of viscous travelling waves*. By a viscous travelling i-wave we mean a solution of (2.2) having the form

$$w(t, x) = U(x - \sigma t)\,, \tag{2.10}$$

where the speed σ is close to the i-th eigenvalue λ_i of the hyperbolic matrix A. Clearly, the function U must provide a solution to the second order O.D.E.

$$U'' = \big(A(U) - \sigma\big)U'. \tag{2.11}$$

The basic idea for the decomposition is as follows. At each point (t, x), given (u, u_x, u_{xx}), we seek travelling wave profiles $U_1, \ldots, U_n$ such that

$$U_i(x) = u(x) \qquad\qquad i = 1, \ldots, n, \tag{2.12}$$

$$\sum_i U_i'(x) = u_x(x), \qquad\qquad \sum_i U_i''(x) = u_{xx}(x). \tag{2.13}$$

We notice that the system of algebraic equations (2.12)-(2.13) is underdetermined, i.e. it admits infinitely many solutions. Indeed, for a given state u, one can assign arbitrarily a speed $\sigma \approx \lambda_i(u)$ and a vector $\mathbf{v} \in \mathbb{R}^n$. Then the O.D.E. (2.11) can be uniquely solved with initial data

$$U(x) = u(x), \qquad\qquad U'(x) = \mathbf{v}.$$

As a result, for each $i = 1, \ldots, n$, there exists a family of viscous travelling i-waves satisfying (2.12), depending on $n + 1$ scalar parameters. In all, we have $n(n + 1)$ parameters to fit the data $(u_x, u_{xx}) \in \mathbb{R}^{n+n}$ in (2.13). Too many! The first thing we need to do is to select a special subfamily of viscous travelling waves depending on the right number of parameters to fit the data. This will be achieved by the center manifold theorem.

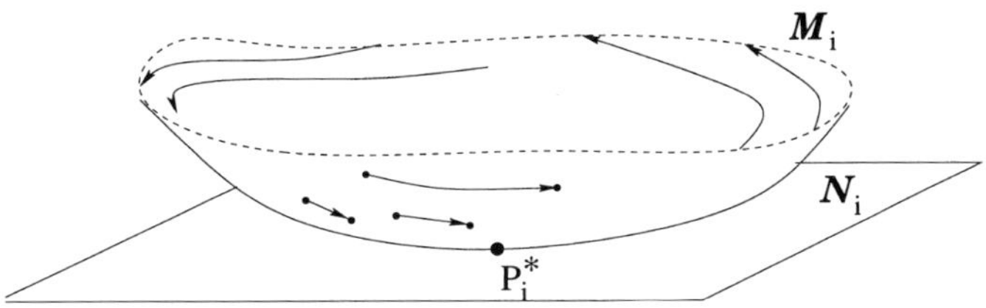

Fig. 3. A center manifold.

4. (A family of travelling wave profiles) The second order O.D.E. in (2.11) can be rewritten as a first order system:

$$\begin{cases} \dot{u} = v, \\ \dot{v} = \big(A(u) - \sigma\big)v, \\ \dot{\sigma} = 0. \end{cases} \tag{2.14}$$

Linearizing at a point $P_i^* = \big(u^*, 0, \lambda_i(u^*)\big)$ we obtain the linear system with constant coefficients

$$\begin{cases} \dot{u} = v, \\ \dot{v} = \big(A(u^*) - \lambda_i(u^*)\big)v, \\ \dot{\sigma} = 0. \end{cases} \tag{2.15}$$

This system has a null space $\mathcal{N}_i$ of dimension $n + 2$. By the center manifold theorem [30], there exists a manifold $\mathcal{M}_i$, tangent to $\mathcal{N}_i$ at the point P_i^*

(Fig. 3), locally invariant for the flow of the nonlinear system (2.14). Instead of looking at all viscous travelling profiles, we can now restrict ourselves to only those profiles corresponding to a trajectory on the center manifold $\mathcal{M}_i$. Since this manifold is $(n+2)$-dimensional, through each state $u \in \mathbb{R}^n$ there is a 2-parameter family of such profiles, say $U_i(u, v_i, \sigma_i)$. Here v_i denotes the scalar amplitude of the wave, and σ_i is the speed. Calling $\tilde{r}_i(u, v_i, \sigma_i)$ the unit vector parallel to U_i', this profile U_i is characterized by the equations

$$U_i'' = \big(A(U_i) - \sigma_i\big)U_i', \qquad U_i(0) = u\,, \qquad U_i'(0) = v_i\tilde{r}_i\,. \tag{2.16}$$

The manifold can then be defined as

$$\mathcal{M}_i = \big\{(u, v, \sigma_i) \in \mathbb{R}^{n+n+1}\,; \quad v = v_i\tilde{r}_i(u, v_i, \sigma_i)\big\}\,. \tag{2.17}$$

We can now try to solve the system (2.12)-(2.13) using not all travelling wave profiles in (2.11), but only those U_i which correspond to a trajectory on the center manifold $\mathcal{M}_i$. Notice that in this case we get the correct number of parameters: in all, the n profiles $U_1, \ldots, U_n$ through the point $u(x)$ depend on the $2n$ scalar parameters $v_1, \sigma_1, \ldots, v_n, \sigma_n$. Exactly the right number to fit the data $\big(u_x(x),\, u_{xx}(x)\big) \in \mathbb{R}^{n+n}$.

5. (Gradient decomposition) We now implement the program outlined at (2.7)–(2.9), decomposing the gradient u_x along the basis of unit vectors $\tilde{r}_i = \tilde{r}_i(u, v_i, \sigma_i)$ constructed in the previous step. To better understand the procedure, consider first the special case where u is precisely the profile of a viscous travelling wave of the j-th family. In this case, our decomposition should clearly contain one single component:

$$u_x = v_j\tilde{r}_j(u, v_j, \sigma_j)\,.$$

It is easy to guess what v_j and σ_j should be. Indeed, since by construction $|\tilde{r}_j| = 1$, the quantity

$$v_j = \pm|u_x|$$

is the signed strength of the wave. We also notice that, for a travelling wave, the vectors u_x and u_t are always parallel. Indeed, $u_t = -\sigma_j u_x$, where σ_j is the speed of the wave. We can thus write

$$u_t = u_{xx} - A(u)u_x = \omega_j\tilde{r}_j(u, v_j, \sigma_j)$$

for some scalar ω_j. The speed of the wave is now obtained as $\sigma_j = -\omega_j/v_j$.

Motivated by the previous analysis, as a first attempt we define our decomposition in terms of the equations

$$\begin{cases} u_x = \sum_i v_i\,\tilde{r}_i(u, v_i, \sigma_i), \\ u_t = \sum_i \omega_i\,\tilde{r}_i(u, v_i, \sigma_i), \end{cases} \qquad \sigma_i = -\frac{\omega_i}{v_i}\,. \tag{2.18}$$

The trouble with (2.18) is that the vectors $\tilde{r}_i$ are defined only for speeds σ_i close to the i-th characteristic speed $\lambda_i^* \doteq \lambda_i(u^*)$. However, when $u_x \approx 0$ one has $v_i \approx 0$ and the ratio ω_i/v_i may become arbitrarily large. To overcome this problem, we introduce a smooth cutoff function θ such that

$$\theta(s) = \begin{cases} s & \text{if} \quad |s| \le \delta_1 \\ 0 & \text{if} \quad |s| \ge 3\delta_1 \end{cases} \tag{2.19}$$

for some small constant $\delta_1 > 0$. We now rewrite (2.18) in terms of the new variable w_i, related to ω_i by $\omega_i \doteq w_i - \lambda_i^* v_i$. We require that σ_i coincides with $-\omega_i/v_i$ only when this ratio is sufficiently close to $\lambda_i^* \doteq \lambda_i(u^*)$. Our basic equations thus take the form

$$\begin{cases} u_x = \sum_i v_i \, \tilde{r}_i(u, v_i, \sigma_i), \\ u_t = \sum_i (w_i - \lambda_i^* v_i) \, \tilde{r}_i(u, v_i, \sigma_i), \end{cases} \tag{2.20}$$

where

$$u_t = u_{xx} - A(u) u_x \,, \qquad \sigma_i = \lambda_i^* - \theta\left(\frac{w_i}{v_i}\right). \tag{2.21}$$

The system (2.20)-(2.21) can be written in the more compact form

$$\begin{pmatrix} u_x \\ u_t \end{pmatrix} = \Lambda(u, v, w)\,. \tag{2.22}$$

Using the implicit function theorem one can prove that, for each given state $u \approx u^*$ and u_x, u_t sufficiently small, there exists unique $v = (v_1, \ldots, v_n)$ and $w = (w_1, \ldots, w_n)$ such that (2.20)-(2.21) hold. Furthermore, the map $(u, u_x, u_t) \mapsto (v, w)$ is $\mathcal{C}^1$ with Lipschitz continuous first derivative. This achieves the desired decomposition of the gradient u_x.

Remark 3. There above equations (2.20)-(2.21) are closely related to (2.13). Indeed, recalling (2.16), from the first equation in (2.20) it follows

$$u_x(x) = \sum_i U_i'(x)\,.$$

If $\sigma_i = \lambda_i^* - w_i/v_i$ for all $i = 1, \ldots, n$, i.e. if none of the cutoff functions is active, then

$$\begin{aligned} u_{xx}(x) = u_t + A(u) u_x \; &= \; \sum_i (w_i - \lambda_i^* v_i) \tilde{r}_i + A(u) \sum_i v_i \tilde{r}_i \\ = \sum_i \left(A(u) - \sigma_i\right) v_i \tilde{r}_i \; &= \; \sum_i U_i''(x)\,. \end{aligned} \tag{2.23}$$

In this case, both of the equalities in (2.13) hold. Notice however that the second equality in (2.13) may fail if $|w_i/v_i| > \delta_1$ for some i, so that $\theta(w_i/v_i) \neq w_i/v_i$.

Remark 4. It is important to appreciate the difference between the two decompositions

$$u_x = \sum_i v_i r_i(u)\,, \qquad\qquad u_x = \sum_i v_i \tilde r_i(u, v_i, \sigma_i),$$

which are used in the hyperbolic and in the parabolic case, respectively.

- In the hyperbolic case, the unit vectors $r_i(u)$ are the eigenvectors of the matrix $A(u)$, hence they depend only on the state u. On the other hand, the *viscous eigenvectors* $\tilde r_i(u, v_i, \sigma_i)$ which are introduced in (2.17) in terms of a center manifold, depend not only on u but also on two additional scalar parameters: the amplitude v_i and the speed σ_i of a viscous travelling profile through u. These can be determined as functions of u_x, u_{xx}, through the implicit equations (2.20)-(2.21).
- For each $i = 1, \ldots, n$, in addition to the eigenvalue $\lambda_i(u)$ of the matrix $A(u)$, we can define the *viscous eigenvalue* $\tilde\lambda_i(u, v_i, \sigma_i)$ by setting

$$\tilde\lambda_i \doteq \big\langle \tilde r_i\,,\ A(u)\tilde r_i\big\rangle\,. \tag{2.24}$$

Instead of the standard relation

$$\big(A(u) - \lambda_i\big) r_i = 0\,,$$

one can show that these viscous eigenvectors satisfy

$$\big(A(u) - \tilde\lambda_i\big)\tilde r_i = v_i\big(\tilde r_{i,u}\tilde r_i + \tilde r_{i,v}(\tilde\lambda_i - \sigma_i)\big)\,. \tag{2.25}$$

Here $\tilde r_{i,u} = \partial\tilde r_i/\partial u$ and $\tilde r_{i,v} = \partial \tilde r_i/\partial v_i$ denote partial derivatives. In the special case $v_i = 0$, an easy consequence of (2.25) is the identity

$$\tilde r_i(u, 0, \sigma_i) = r_i(u).$$

The presence of non-zero terms on the right hand side of (2.25) is of fundamental importance. Indeed, in the evolution equation for the component v_i, these terms achieve a crucial cancellation with other terms that would otherwise not be integrable.

6. (Evolution of scalar components) We now consider a smooth solution $u = u(t, x)$ of the parabolic system (2.2). By the previous step, we can define the scalar functions $v_i = v_i(t, x)$, $w_i = w_i(t, x)$, according to the decomposition (2.20)-(2.21). The equations governing the evolution of these $2n$ components can be written in the form

$$\begin{cases} v_{i,t} + (\tilde\lambda_i v_i)_x - v_{i,xx} = \phi_i\,, \\ w_{i,t} + (\tilde\lambda_i w_i)_x - w_{i,xx} = \psi_i\,. \end{cases} \tag{2.26}$$

We recall that $\tilde{\lambda}_i$ is the speed defined at (2.24). The source terms ϕ_i, ψ_i can be computed by inserting (2.22) in the evolution equation

$$\begin{pmatrix} u_x \\ u_t \end{pmatrix}_t + \left(\begin{bmatrix} A(u) & 0 \\ 0 & A(u) \end{bmatrix} \begin{pmatrix} u_x \\ u_t \end{pmatrix} \right)_x - \begin{pmatrix} u_x \\ u_t \end{pmatrix}_{xx}$$
$$= \begin{pmatrix} 0 \\ (DA(u)\cdot u_x)u_t - (DA(u)\cdot u_t)u_x \end{pmatrix}$$

and differentiating various times w.r.t. t, x, using the fundamental relation (2.25). For all details we refer to [2]. Rather than the exact form of the source terms, it is important to know their magnitude. This can be estimated as follows.

Lemma 1. *The source terms in (2.26) satisfy the bounds*

$$\phi_i,\ \psi_i = \mathcal{O}(1)\cdot \sum_j \big(|v_{j,x}| + |w_{j,x}|\big)\cdot |w_j - \theta_j v_j| \qquad \textbf{(cutoff error)}$$
$$+\mathcal{O}(1)\cdot \sum_j |v_{j,x}w_j - v_j w_{j,x}| \ +\ \mathcal{O}(1)\cdot \sum_j \left| v_j \left(\frac{w_j}{v_j}\right)_x \right|^2 \cdot \chi_{\{|w_j/v_j|<3\delta_1\}}$$
$$\textbf{(interaction of waves of the same family)}$$
$$+\mathcal{O}(1)\cdot \sum_{j\neq k} \big(|v_j v_k| + |v_{j,x}v_k| + |v_j w_k| + |v_{j,x}w_k| + |v_j w_{k,x}| + |w_j w_k|\big)$$
$$\textbf{(interaction of waves of different families)} \tag{2.27}$$

A rigorous proof of Lemma 1 requires lengthy calculations. Here we shall only provide an intuitive explanation of how the various terms arise. Consider first the special case where u is precisely one of the travelling wave profiles on the center manifold (Fig. 4), say $u(t,x) = U_j(x - \sigma_j t)$. In this case one has

$$u_x = v_j \tilde{r}_j\,, \qquad u_t = (w_j - \lambda_j^* v_j)\tilde{r}_j\,, \qquad v_i = w_i = 0 \quad \text{for } i \neq j\,,$$

and all source terms ϕ_i, ψ_i vanish identically.

Next, consider the case of a general solution $u = u(t,x)$ of (2.2). The sources on the right hand sides of (2.26) arise for three different reasons (Fig. 5).

- The ratio $|w_j/v_j|$ is large and hence the cutoff function $\theta_j \doteq \theta(w_j/v_j)$ is active. Typically, this will happen near a point x_0 where $u_x = 0$ but $u_t = u_{xx} \neq 0$. In this case the second identity in (2.13) fails because of a "wrong" choice of the speed: $\sigma_j \neq \lambda_j^* - (w_j/v_j)$.
- Waves of two different families $j \neq k$ are present at a given point x. These will produce quadratic source terms, due to transversal interactions.
- Since the decomposition (3.10) is defined pointwise, it may well happen that the travelling j-wave profile U_j at a point x is not the same as the

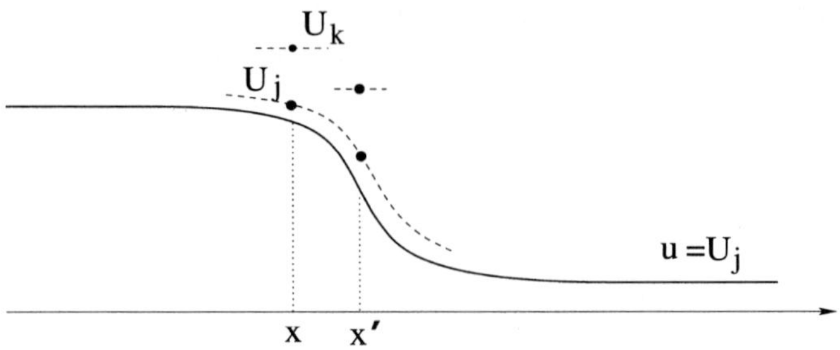

Fig. 4. Gradient decomposition for a viscous travelling wave.

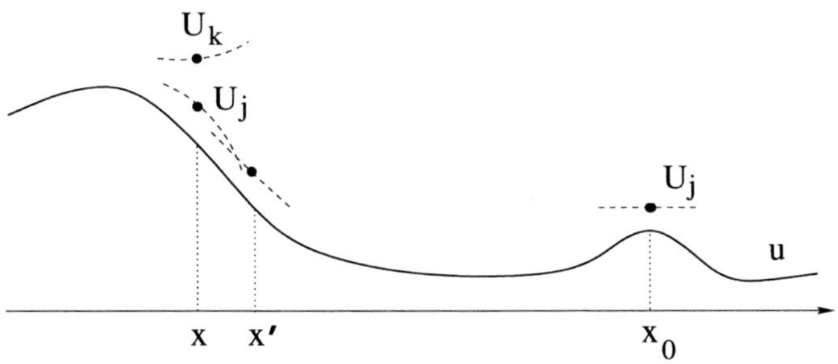

Fig. 5. Gradient decomposition for a general viscous profile.

profile U_j at a nearby point x'. Indeed, these two travelling waves may have slightly different speeds. It is the rate of change in this speed: $\sigma_{j,x} = -(w_j/v_j)_x$, that determines the infinitesimal interaction between nearby waves of the same family. A detailed analysis shows that the corresponding source terms can only be linear or quadratic w.r.t. $\sigma_{j,x}$, with the square of the strength of the wave always appearing as a factor. These terms can thus be estimated as $\mathcal{O}(1) \cdot v_j^2 \sigma_{j,x} + \mathcal{O}(1) \cdot v_j^2 \sigma_{j,x}^2$.

To complete the proof of the BV bounds, it remains to show that all terms on the right hand side of (2.27) are integrable over the domain $\{t > T\,,\ x \in I\!R\}$. For this purpose, one has to introduce four types of Lyapunov functionals, called *energy, transversal interaction, curve length* and *area functional*, monotonically decreasing along solutions. Each of the terms that we need to estimate will be dominated by the rate of decrease of one of these functionals. The next four steps describe these constructions.

7. (Tranversal interaction potential) Consider two independent, scalar diffusion equations with strictly different drifts (Fig. 6):

$$\begin{cases} z_t + \big[\lambda(t,x)z\big]_x - z_{xx} = 0\,, \\ z_t^* + \big[\lambda^*(t,x)z^*\big]_x - z_{xx}^* = 0\,, \end{cases}$$

assuming that

$$\inf_{t,x} \lambda^*(t,x) - \sup_{t,x} \lambda(t,x) \geq c > 0\,.$$

We can think z as the density of waves with a slow speed λ and z^* as the density of waves with a fast speed λ^*. We seek an a priori bound on the total amount of interaction

$$\int_0^\infty \int_{I\!R} \big|z(t,x)\big| \cdot \big|z^*(t,x)\big|\, dxdt\,. \tag{2.28}$$

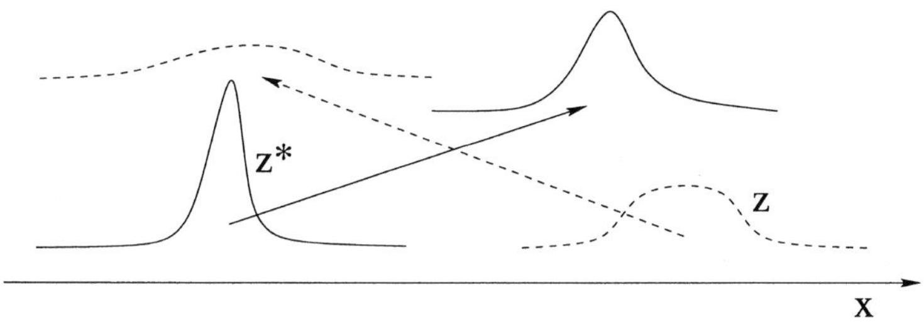

Fig. 6. Interaction of waves of different families.

This can be achieved by means of a *transversal interaction potential*, defined by

$$Q(z,z^*) \doteq \frac{1}{c} \iint_{I\!R^2} K(x_2 - x_1)\big|z(x_1)\big|\,\big|z^*(x_2)\big|\, dx_1 dx_2\,, \tag{2.29}$$

$$K(y) \doteq \begin{cases} e^{-cy/2} & \text{if} \quad y > 0\,, \\ 1 & \text{if} \quad y \leq 0\,. \end{cases} \tag{2.30}$$

Notice that, to the product of a slow wave located at x_1 and a fast wave at x_2, in (2.29) we assign weight $K(x_2 - x_1) = 1$ if the two waves are approaching (i.e. if $x_1 > x_2$), and an exponentially small weight if they are moving away from each other (i.e. if $x_1 < x_2$). It turns out that this functional is monotonically decreasing along every couple of solutions z, z^*. Moreover, the rate of decrease controls the integrand in (2.28). Hence

$$\int_0^\infty \int_{I\!R} \big|z(t,x)\big|\,\big|z^*(t,x)\big|\, dxdt \;\leq\; -\int_0^\infty \left[\frac{d}{dt} Q\big(z(t),\, z^*(t)\big)\right] dt$$

$$\leq Q\big(z(0),\, z^*(0)\big) \;\leq\; \frac{1}{c} \int_{I\!R} \big|z(0,x)\big|\, dx \cdot \int_{I\!R} \big|z^*(0,x)\big|\, dx\,.$$

By introducing Lyapunov functionals of this type, we can bound the integrals of all terms of the form

$$|v_j v_k| \quad |v_{j,x} v_k|\,, \quad |v_j w_k|\,, \quad |v_{j,x} w_k|\,, \quad |w_j w_k|$$

with $j \neq k$, due to the interaction of waves of different families.

Remark 5. In connection with the scalar conservation laws without diffusion

$$\begin{cases} z_t + \big[\lambda(t,x)z\big]_x = 0\,, \\ z_t^* + \big[\lambda^*(t,x)z^*\big]_x = 0\,, \end{cases}$$

the total amount of interaction (2.28) could be controlled in terms of the Glimm functional

$$\widehat{Q}(z,z^*) \doteq \frac{1}{c} \iint_{x_1 > x_2} \big|z(x_1)\big|\,\big|z^*(x_2)\big|\, dx_1 dx_2\,.$$

This has the same form as (2.29), except that now $K(y) = 0$ for $y > 0$. To appreciate the difference, assume that the two components z, z^* initially have disjoint compact supports, and are moving away from each other (Fig. 7). In the hyperbolic case, the supports will remain compact and disjoint for all times $t > 0$. Hence, no interaction ever takes place. On the other hand, in the presence of diffusion, at any positive time the tails of the distributions overlap, and a non-zero amount of interaction occurs. This is taken into account by our interaction potential (2.29)-(2.30), where the kernel K is exponentiallly small, but not zero, for $y > 0$.

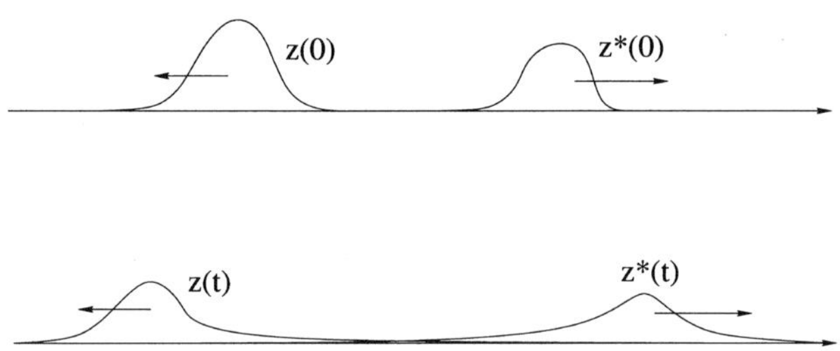

Fig. 7. Interaction of exponentially decreasing tails.

8. (Curve length functional) To bound interactions between waves of the same family, we seek Lyapunov functionals which are decreasing along every solution of the scalar viscous conservation law

$$u_t + f(u)_x = u_{xx}\,. \tag{2.31}$$

For this purpose, to a scalar function u we associate the curve (Fig. 8)

$$\gamma \doteq \begin{pmatrix} u \\ f(u) - u_x \end{pmatrix} = \begin{pmatrix} \text{conserved quantity} \\ \text{flux} \end{pmatrix} \tag{2.32}$$

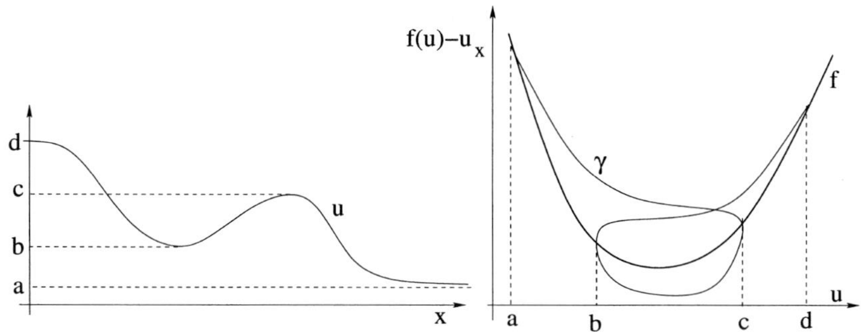

Fig. 8. The planar curve γ corresponding to a scalar function u.

In connection with a solution $u = u(t,x)$ of (2.31), the curve γ evolves according to

$$\gamma_t + f'(u)\gamma_x = \gamma_{xx}\,. \tag{2.33}$$

Notice that the vector $f'(u)\gamma_x$ is parallel to γ, hence the presence of this term in (2.33) only amounts to a reparametrization of the curve, and does not affect its shape. The curve thus evolves in the direction of curvature. An obvious Lyapunov functional is thus the *length* of the curve. In terms of the variables

$$\gamma_x = \begin{pmatrix} v \\ w \end{pmatrix} \doteq \begin{pmatrix} u_x \\ -u_t \end{pmatrix}, \tag{2.34}$$

this length is given by

$$L(\gamma) \doteq \int |\gamma_x|\,dx = \int \sqrt{v^2+w^2}\,dx\,. \tag{2.35}$$

We can estimate the rate of decrease of the length as

$$\begin{aligned} -\frac{d}{dt}L(\gamma(t)) &= \int_{I\!R} \frac{|v|\,\big[(w/v)_x\big]^2}{\big(1+(w/v)^2\big)^{3/2}}\,dx \\ &\geq \frac{1}{(1+\delta^2)^{3/2}} \int_{|w/v|\leq\delta} |v|\,\big[(w/v)_x\big]^2\,dx \end{aligned}$$

for any given constant $\delta > 0$. Therefore

$$\int_0^\infty \int_{|w/v|\leq\delta} |v|\,\big[(w/v)_x\big]^2\,dx\,dt = \mathcal{O}(1)\cdot \int_0^\infty \left|\frac{d}{dt}L(\gamma(t))\right| dt = \mathcal{O}(1)\cdot L(\gamma(0)).$$

This argument allows us to estimate all source terms in (2.27) of the form

$$\mathcal{O}(1)\cdot \left| v_j \left(\frac{w_j}{v_j}\right)_x \right|^2 \cdot \chi_{\{|w_j/v_j|<3\delta_1\}}\,.$$

9. (Area Functional) In connection with the same curve γ in (2.32), we now introduce a second functional, defined in terms of a wedge product.

$$Q(\gamma) \doteq \frac{1}{2} \iint_{x<x'} \big|\gamma_x(x) \wedge \gamma_x(x')\big|\, dx\, dx' . \tag{2.36}$$

For any curve that moves in the plane in the direction of curvature (Fig. 9, left), this functional is monotone decreasing and its decrease controls the area swept by the curve: $|dA| \le -dQ$. An intuitive proof of this fact can be obtained by looking at polygonal approximations.

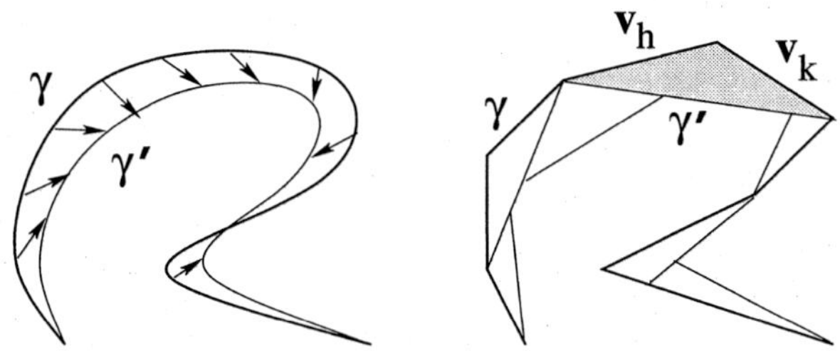

Fig. 9. Motion of a planar curve in the direction of curvature.

In the special case (Fig. 9, right) where γ is a polygonal with edges $\mathbf{v}_1, \ldots, \mathbf{v}_m$, the functional (2.36) reduces to

$$Q(\gamma) = \frac{1}{2} \sum_{i<j} \big|\mathbf{v}_i \wedge \mathbf{v}_j\big| .$$

Consider now a second polygonal γ', obtained from γ by replacing two consecutive edges $\mathbf{v}_h, \mathbf{v}_k$ by a single one. The area of the triangle between γ and γ' is then

$$\text{Area} \;=\; \frac{1}{2}\big|\mathbf{v}_h \wedge \mathbf{v}_k\big| \;\le\; Q(\gamma) - Q(\gamma').$$

Observing that every continuous curve moving in the direction of curvature can be approximated by a suitable sequence of polygonals, the relation $|dA| \le -dQ$ is obtained in the limit.

Using (2.33)-(2.34) we now compute

$$-\frac{dQ}{dt} \ge \left|\frac{dA}{dt}\right| = \int |\gamma_t \wedge \gamma_x|\, dx = \int |\gamma_{xx} \wedge \gamma_x|\, dx = \int |v_x w - v w_x|\, dx .$$

Therefore, integrating w.r.t. time one finds

$$\int_0^\infty \int |v_x w - v w_x|\, dx\, dt \;\le\; \int_0^\infty \left|\frac{d}{dt} Q\big(\gamma(t)\big)\right| dt \;\le\; Q\big(\gamma(0)\big) .$$

This argument allows us to estimate all source terms in (2.27) of the form

$$\mathcal{O}(1) \cdot \left| v_{j,x} w_j - v_j w_{j,x} \right| .$$

Remark 6. Defining the speed of a scalar viscous wave as $\sigma(x) \doteq -u_t(x)/u_x(x)$, the area functional $Q(\gamma)$ in (2.36) can be written as

$$\begin{aligned} Q(\gamma) &= \frac{1}{2} \iint_{x<y} \left| u_x(x) u_t(y) - u_t(x) u_x(y) \right| dx dy \\ &= \frac{1}{2} \iint_{x<y} \left| u_x(x)\, dx \right| \cdot \left| u_x(y)\, dy \right| \cdot \left| \sigma(x) - \sigma(y) \right| \\ &= \frac{1}{2} \iint_{x<y} [\text{wave at } x] \times [\text{wave at } y] \times [\text{difference in speeds}] . \end{aligned}$$

It now becomes clear that the area functional can be regarded as an interaction potential between waves of the same family. This provides a counterpart to the interaction functional introduced in [20] in connection with strictly hyperbolic systems.

Remark 7. In the case where u is precisely a viscous travelling wave, the curve γ reduces to a segment. Assume now that the flux f is genuinely nonlinear, say with $f'' \geq c > 0$. Consider a solution u which initially consists of two viscous travelling waves, far apart from each other (Fig. 10). To fix the ideas, let the first wave join a left state a with a middle state b, and the second wave join the middle state b with a right state c, with $a > b > c$. The strength of the two waves can be measured as $s = a - b$, $s' = b - c$. The corresponding curve γ is approximately given by two segments, joining the points $P \doteq \big(a, f(a)\big)$, $Q \doteq \big(b, f(b)\big)$, $P' \doteq \big(c, f(c)\big)$. After a long time $\tau >> 0$, the two shocks will interact, merging into one single viscous shock. The curve $\gamma(\tau)$ thus reduces to one single segment, joining P with P'. The area swept by the curve is approximately the area of the triangle PQP'. The assumption of genuine nonlinearity implies

$$\text{area swept} \quad = \quad \mathcal{O}(1) \cdot |ss'| \left(|s| + |s'| \right).$$

In this case, the decrease in the area functional is of cubic order w.r.t. the strengths s, s' of the interacting waves. This is indeed the correct order of magnitude needed to control the strength of new waves generated by the interaction in the genuinely nonlinear case. It is remarkable that this area functional gives the correct order of magnitude of waves generated by an interaction also for a general flux function f, not necessarily convex.

10. (Energy functional) Recalling the constant δ_1 in (2.19), we define a second cutoff function η such that

$$\eta(s) = \begin{cases} 0 & \text{if} \quad |s| \leq 3\delta_1/5 \\ 1 & \text{if} \quad |s| \geq 4\delta_1/5. \end{cases}$$

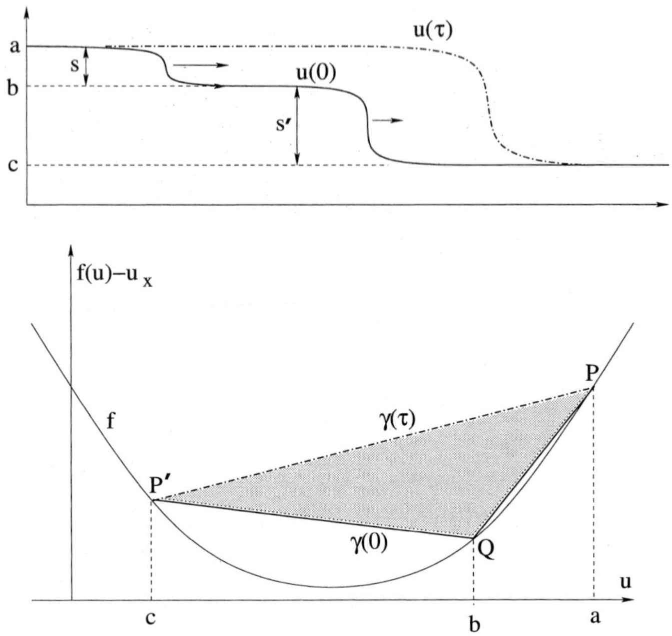

Fig. 10. Interaction of two viscous shocks.

In connection with a solution (v_i, w_i) of (2.26) we define $\eta_i \doteq \eta(w_i/v_i)$. The integral of the first terms on the right hand side of (2.27), due to the cutoff error, is then controlled by means of the *weighted energy functionals*

$$E_i(u) \doteq \int \eta_i \cdot v_{i,x}^2 \, dx \,, \qquad \widetilde{E}_i(u) \doteq \int \eta_i \cdot w_{i,x}^2 \, dx \,. \tag{2.37}$$

For all details we always refer to [2].

Remark 8. Intuitively, we expect that the cutoff function θ will be active when $|u_x| << |u_{xx}|$. In these regions, the diffusion term u_{xx} dominates over the convection term $A(u)u_x$ and the system (2.2) essentially reduces to the heat equation. It is thus natural to estimate the error due to cutoff in terms of an energy functional, similar to the one commonly used for the heat equation.

By means of the above functionals we obtain the $\mathbf{L}^1$ estimates on the source terms ϕ_i, ψ_i in (2.26), proving the uniform bounds on the total variation of a solution u of (2.2).

11. (Stability) To prove the uniform stability of all solutions of the parabolic system (2.2) having small total variation, we consider the linearized system describing the evolution of a first order variation. Inserting the formal expansion $u = u_0 + \epsilon z + O(\epsilon^2)$ in (2.2), we obtain

$$z_t + \big[DA(u) \cdot z\big] u_x + A(u) z_x = z_{xx} \,. \tag{2.38}$$

Our basic goal is to prove the bound

$$\big\|z(t)\big\|_{\mathbf{L}^1} \le L \, \big\|z(0)\big\|_{\mathbf{L}^1} \,, \tag{2.39}$$

for some constant L and all $t \ge 0$ and every solution z of (2.38). By a standard homotopy argument, from (2.39) one easily deduces the Lipschitz continuity of the solution of (2.2) on the initial data. Namely, for every couple of solutions $u, \tilde u$ with small total variation one has

$$\big\|u(t) - \tilde u(t)\big\|_{\mathbf{L}^1} \le L \, \big\|u(0) - \tilde u(0)\big\|_{\mathbf{L}^1} \,. \tag{2.40}$$

Toward a proof of (2.39) we follow the same strategy as in step 3. We decompose the vector z as a sum of scalar components: $z = \sum_i h_i \tilde r_i$, write an evolution equation for these components:

$$h_{i,t} + (\tilde\lambda_i h_i)_x - h_{i,xx} = \hat\phi_i \,,$$

and prove that the source terms are integrable on the domain $\{t > 0\,, x \in I\!R\}$.

Calling $\Upsilon \doteq z_x - A(u) z$, the decomposition takes the form

$$\begin{cases} z = \displaystyle\sum_i h_i \tilde r_i \big(u, v_i, \lambda_i^* - \theta(g_i/h_i)\big) \,, \\ \Upsilon = \displaystyle\sum_i (g_i - \lambda_i^* h_i) \tilde r_i \big(u, v_i, \lambda_i^* - \theta(g_i/h_i)\big) \,, \end{cases} \tag{2.41}$$

where θ is the cutoff function introduced at (2.19). These components satisfy the system of parabolic equations

$$\begin{cases} h_{i,t} + (\tilde\lambda_i h_i)_x - h_{i,xx} = \hat\phi_i \,, \\ g_{i,t} + (\tilde\lambda_i g_i)_x - g_{i,xx} = \hat\psi_i \,, \end{cases} \tag{2.42}$$

By similar arguments as in steps 6–10, we can provide uniform bounds on all source terms $\hat\phi_i, \hat\psi_i$, proving (2.39), and hence the stability of all solutions of (2.2) with small total variation.

Remark 9. The evolution equations (2.38) for a first order perturbation z are linear. It thus seems natural to use a decomposition $(z, z_x) \mapsto (h, g)$ where the components h_i, g_i are linear functions of z, z_x. This is the case if in (2.41) one chooses the same vectors $\tilde r_i = \tilde r_i(u, v_i, \sigma_i)$ as in (2.18), depending only on the reference solution u and not on the perturbation z. Unfortunately, this linear decomposition does not seem to work, since it leads to a system of the form (2.42) with possibly non-integrable source terms.

On the contrary, our decomposition (2.41) is positive homogeneous of degree one, but not linear. Indeed, the unit vectors $\tilde r_i$ depend on the reference solution u through the amplitude v_i, but also on the perturbation z implicitly through the speed $\sigma_i = \lambda_i^* - \theta(g_i/h_i)$. With this choice, we can prove that all source terms $\hat{\phi}_i, \hat{\psi}_i$ in (2.42) are integrable. In turn, this yields the basic estimate (2.39).

12. (The vanishing viscosity limit) For every initial data $u(0, \cdot) = \bar{u}$ with small total variation, the previous steps yield the existence of a unique global solution to the parabolic system (2.2), depending Lipschitz continuously on the initial data, in the $\mathbf{L}^1$ norm. Performing the rescaling $t \mapsto t/\varepsilon$, $x \mapsto x/\varepsilon$, we immediately obtain the same results for the Cauchy problem

$$u_t^\varepsilon + A(u^\varepsilon)u_x^\varepsilon = \varepsilon\, u_{xx}^\varepsilon\,, \qquad\qquad u^\varepsilon(0,x) = \bar{u}(x)\,.$$

Adopting a semigroup notation, this solution will be written as $u^\varepsilon(t,\cdot) = S_t^\varepsilon \bar{u}$. Thanks to the uniform bounds on the total variation, a compactness argument yields the existence of a strong limit in $\mathbf{L}^1_{\mathrm{loc}}$

$$u = \lim_{\varepsilon_m \to 0} u^{\varepsilon_m}\,, \tag{2.43}$$

at least for some subsequence $\varepsilon_m \to 0$. Since the u^ε depend continuously on the initial data, with a uniform Lipschitz constant, the same is true of $u(t,\cdot) = S_t \bar{u}$. In the conservative case where $A(u) = Df(u)$, it is not difficult to show that this limit u actually provides a weak solution to the Cauchy problem (1.10).

The only remaining issue is to show that the limit in (2.43) is unique, i.e. it does not depend on the subsequence $\{\varepsilon_m\}$. In the standard conservative case, this fact can already be deduced from the uniqueness criterion in [8]. In the general case, uniqueness is proved in two steps. First we show that, in the special case of a Riemann problem, the solution obtained as vanishing viscosity limit is unique and can be completely characterized. To conclude the proof, we then rely on the same general argument as in [3]: if two Lipschitz semigroups S, S' provide the same solutions to all Riemann problems, then they must coincide.

3 Concluding Remarks

Remark 10. In standard textbooks, the analysis of first order hyperbolic systems is based on the *method of characteristics.* A solution of the system

$$u_t + A(u)u_x = 0\,,$$

is decomposed along the eigenvectors of the matrix $A(u)$. The evolution of these scalar components is then described by a family of O.D.E's along the *characteristic curves,* i.e. the curves in the t-x plane which satisfy $dx/dt = \lambda_i\big(u(t,x)\big)$. In essence, the local decomposition (2.12)-(2.13) in terms of viscous travelling waves shows that this "hyperbolic" point of view retains its validity also in connection with the parabolic system (2.2). In this case, the projections are taken along the vectors $\tilde{r}_i$ while the characteristic curves are defined as $dx/dt = \sigma_i$, where σ_i is the speed of the i-th travelling wave. Notice that in the hyperbolic case the projections and the wave speeds depend only on the state u, through the eigenvectors $r_i(u)$ and the eigenvalues $\lambda_i(u)$ of the matrix $A(u)$. On the other hand, in the parabolic case the construction involves the derivatives u_x, u_{xx} as well.

Remark 11. It remains an open problem to establish the convergence of vanishing viscosity approximations of the form

$$u_t + A(u)u_x = \varepsilon\big(B(u)u_x\big)_x \tag{3.1}$$

for more general viscosity matrices B, as in [24]. In the present paper we are exclusively concerned with the case where B is the identity matrix. For systems which are not in conservative form, we expect that the limit of solutions of (3.1), as $\varepsilon \to 0$, will be heavily dependent on the choice of the matrix B. Studying the case where the viscosity matrix $B(u)$ is only semidefinite is particularly important: for applications to more realistic physical models and also toward the understanding of general relaxation approximations [23].

Remark 12. In the present paper we only consider initial data with small total variation. This is a convenient setting, adopted in much of the current literature, which guarantees the global existence of BV solutions of (1.1) and captures all basic features of the problem. A recent example of Jenssen [17] shows that, for initial data with large total variation, the solution can blow up in finite time. In this more general setting, one expects that the existence and uniqueness of weak solutions, together with the convergence of vanishing viscosity approximations, will hold locally in time as long as the total variation remains bounded. For the hyperbolic system (1.1), results on the local existence and stability of solutions with large BV data can be found in [27] and [6].

Remark 13. For initial data in $\mathbf{L}^\infty$, on the other hand, one cannot expect to have any general theorem on uniqueness and stability of vanishing viscosity solutions. A simple example of non-uniqueness was given in [12].

References

1. F. Ancona and A. Marson, Well posedness for general 2×2 systems of conservation laws, *Amer. Math. Soc. Memoir*, to appear.
2. S. Bianchini and A. Bressan, Vanishing viscosity solutions to nonlinear hyperbolic systems, Preprint S.I.S.S.A., Trieste 2001.
3. A. Bressan, The unique limit of the Glimm scheme, *Arch. Rational Mech. Anal.* **130** (1995), 205-230.
4. A. Bressan, *Hyperbolic Systems of Conservation Laws. The One Dimensional Cauchy Problem.* Oxford University Press, 2000.

5. A. Bressan and R. M. Colombo, The semigroup generated by 2×2 conservation laws, *Arch. Rational Mech. Anal.* **133** (1995), 1-75.
6. A. Bressan and R. M. Colombo, Unique solutions of 2×2 conservation laws with large data, *Indiana Univ. Math. J.* **44** (1995), 677-725.
7. A. Bressan, G. Crasta and B. Piccoli, Well posedness of the Cauchy problem for $n \times n$ conservation laws, *Amer. Math. Soc. Memoir* **694** (2000).
8. A. Bressan and P. Goatin, Oleinik type estimates and uniqueness for $n \times n$ conservation laws, *J. Diff. Equat.* **156** (1999), 26-49.
9. A. Bressan and P. LeFloch, Uniqueness of weak solutions to systems of conservation laws, *Arch. Rat. Mech. Anal.* **140** (1997), 301-317.
10. A. Bressan and M. Lewicka, A uniqueness condition for hyperbolic systems of conservation laws, *Discr. Cont. Dynam. Syst.* **6** (2000), 673-682.
11. A. Bressan, T. P. Liu and T. Yang, L^1 stability estimates for $n \times n$ conservation laws, *Arch. Rational Mech. Anal.* **149** (1999), 1-22.
12. A. Bressan and W. Shen, Uniqueness for discontinuous O.D.E. and conservation laws, *Nonlinear Analysis, T. M. A.* **34** (1998), 637-652.
13. R. DiPerna, Convergence of approximate solutions to conservation laws, *Arch. Rational Mech. Anal.* **82** (1983), 27-70.
14. J. Glimm, Solutions in the large for nonlinear hyperbolic systems of equations, *Comm. Pure Appl. Math.* **18** (1965), 697-715.
15. J. Goodman and Z. Xin, Viscous limits for piecewise smooth solutions to systems of conservation laws, *Arch. Rational Mech. Anal.* **121** (1992), 235-265.
16. S. Kruzhkov, First order quasilinear equations with several space variables, *Math. USSR Sbornik* **10** (1970), 217-243.
17. H. K. Jenssen, Blowup for systems of conservation laws, *SIAM J. Math. Anal.* **31** (2000), 894-908.
18. P. Lax, Hyperbolic systems of conservation laws II, *Comm. Pure Appl. Math.* **10** (1957), 537-566.
19. T. P. Liu, The entropy condition and the admissibility of shocks, *J. Math. Anal. Appl.* **53** (1976), 78-88.
20. T. P. Liu, Admissible solutions of hyperbolic conservation laws, *Amer. Math. Soc. Memoir* **240** (1981).
21. T. P. Liu, Nonlinear stability of shock waves, *Amer. Math. Soc. Memoir* **328** (1986).
22. T. P. Liu and T. Yang, Weak solutions of general systems of hyperbolic conservation laws, preprint.
23. R. Natalini, Recent results on hyperbolic relaxation problems, in *Analysis of Systems of Conservation Laws*, H. Freistuehler Ed., Chapman and Hall/CRC (1998), 128-198.
24. A. Majda and R. Pego, Stable viscosity matrices for systems of conservation laws, *J. Differential Equations* **56** (1985), 229-262.
25. O. Oleinik, Discontinuous solutions of nonlinear differential equations (1957), *Amer. Math. Soc. Translations* **26**, 95-172.
26. F. Rousset, Viscous approximation of strong shocks of systems of conservation laws, *SIAM J. Math.Anal.*, submitted.
27. S. Schochet, Sufficient conditions for local existence via Glimm's scheme for large BV data, *J. Differential Equations* **89** (1991), 317-354.
28. A. Szepessy and Z. Xin, Nonlinear stability and viscous shocks, *Arch. Rational Mech. Anal.* **122** (1993), 53-103.

29. A. Szepessy and K. Zumbrun, Stability of rarefaction waves in viscous media, *Arch. Rational Mech. Anal.* **133** (1996), 249-298.
30. A. Vanderbauwhede, Centre manifolds, normal forms and elementary bifurcations, *Dynamics Reported, Vol. 2* (1989), 89-169.
31. S. H. Yu, Zero-dissipation limit of solutions with shocks for systems of hyperbolic conservation laws, *Arch. Rational Mech. Anal.* **146** (1999), 275-370.

Phase Plane Behavior of Solitary Waves in Nonlinear Layered Media

Randall J. LeVeque[1] and Darryl H. Yong[2]

[1] Department of Applied Mathematics, University of Washington, Box 352420, Seattle, WA 98195-2420 *rjl@amath.washington.edu*

[2] Department of Applied and Computational Mathematics, California Institute of Technology, 1200 E California Blvd, MC 217-50 Pasadena, CA 91125 *dyong@caltech.edu*

1 Introduction

The one-dimensional elastic wave equations for compressional waves have the form

$$\begin{aligned} \epsilon_t(x,t) - u_x(x,t) &= 0 \\ (\rho(x)u(x,t))_t - \sigma(\epsilon(x,t),x)_x &= 0 \end{aligned} \tag{1}$$

where $\epsilon(x,t)$ is the strain and $u(x,t)$ the velocity. We consider a heterogeneous material with the density specified by $\rho(x)$ and a nonlinear constitutive relation for the stress given by a function $\sigma(\epsilon,x)$ that also varies explicitly with x. This is a hyperbolic system of conservation laws with a spatially-varying flux function, $q_t + f(q,x)_x = 0$.

In particular, we consider a periodic layered medium (a laminate) in which $\rho(x)$ and $\sigma(\epsilon,x)$ are given by

$$(\rho(x),\sigma(\epsilon,x)) = \begin{cases} (\rho_A,\sigma_A(\epsilon)) & \text{if } j\delta < x < (j+\alpha)\delta \text{ for some integer } j \\ (\rho_B,\sigma_B(\epsilon)) & \text{otherwise.} \end{cases} \tag{2}$$

The two layers have width $\alpha\delta$ and $(1-\alpha)\delta$ respectively. Here we consider the case $\alpha = 1/2$ and $\delta = 1$.

For sufficiently small strains, the response can be modeled by linear constitutive relations

$$\sigma_A(\epsilon) = K_A\epsilon, \qquad \sigma_B(\epsilon) = K_B\epsilon, \tag{3}$$

where the bulk moduli K_A and K_B of each material are constants. The equations are then equivalent to acoustics equations in the periodic medium, a case discussed for example in [2], [6]. Here we primarily consider nonlinear equations with

$$\sigma_A(\epsilon) = \exp(K_A\epsilon) - 1, \qquad \sigma_B(\epsilon) = \exp(K_B\epsilon) - 1. \tag{4}$$

A numerical method for solving more general systems of conservation laws with spatially-varying flux functions has been introduced in [1] and applied to the nonlinear systems given above in [4] and [5]. See also [3] for more discussion of the numerical methods used.

An interesting phenomenon has been observed from computations of this nonlinear system: the appearance of solitary waves that arise due to the combination of nonlinearity in the equations and the dispersion induced by reflections at the material interfaces. These solitary waves are investigated in our recent paper [5], where multi-scale homogenization is also applied to obtain a set of homogenized equations that do an excellent job of modeling the behavior seen in the layered medium. The layered medium is also related to the Toda lattice, a discrete nonlinear system that has exact soliton solutions.

Here we present some observations on the structure of these waves that arise from viewing the solitary waves in the phase plane. We observe that Riemann invariants of a related autonomous nonlinear system (with appropriately averaged values of ρ and K) appear to be significant to the study of wave propagation in this layered medium.

2 Waves in layered media

In Fig. 1 we show results of a numerical experiment that is discussed in more detail in [5]. Initially $q(x,0) = 0$ and a wave is generated at the left boundary by the boundary conditions

$$u(0,t) = \begin{cases} a(1+\cos(\pi(t-10)/10)) & \text{if } 0 \le t \le 20, \\ 0 & \text{if } t > 20. \end{cases} \tag{5}$$

The left edge is pulled outwards for $0 < t < 20$, generating a strain wave that propagates to the right. Once the wave has been generated, we switch to periodic boundary conditions so that the wave loops through the domain and the long-time behavior can be studied. In the plot at $t = 40$ we see that the smooth hump starts to steepen into a shock followed by a rarefaction, but that oscillations begin to appear behind the shock due to the dispersive nature of the layered medium. At later times these oscillations cause the hump to break up into a train of solitary waves.

Fig. 1 shows the stress, which is continuous at the layer interfaces. Fig. 2 shows a close up view of the first two solitary waves at $t = 600$ from Fig. 1, and also shows the corresponding strain ϵ. The strain is discontinuous at the interfaces and so we observe that each solitary wave has a width of about ten layers. These waves are called “stegotons” in [5] because of their ridged appearance. It is not clear whether these waves could possibly be solitons in the technical sense, but they do appear to interact in the same manner as classical solitons. Taller stegotons propagate faster than shorter ones, and when two collide they later separate cleanly into stegotons of the original amplitudes with only a shift in phase (see [5]).

These waves are not smooth when viewed as functions of x at a fixed time t. However, if we fix a location x_0 and record $q(x_0,t)$ as a function of time, then the solution is seen to be smoothly varying, as seen in Fig. 3. This approach is used in [5] to study the scaling properties of these waves.

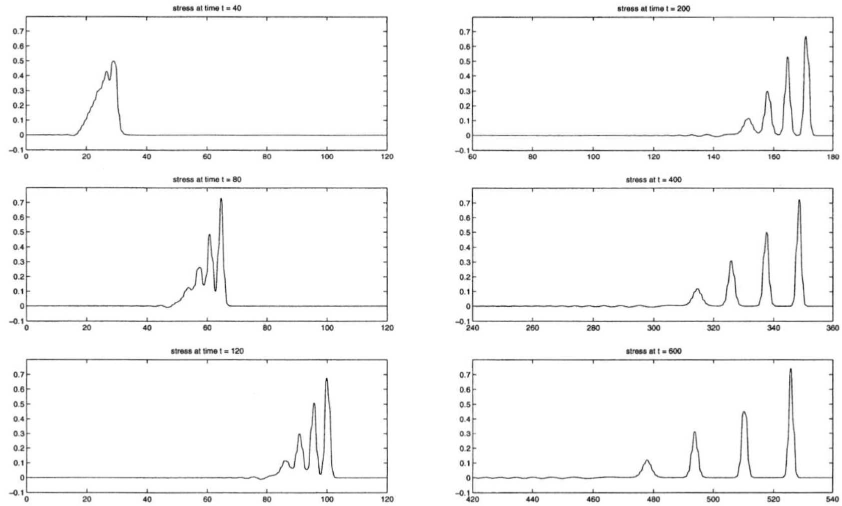

Fig. 1. Propagation in a layered nonlinear medium with impedance mis-match at the interfaces. The left column shows the stress at three times and the right column at three later times, illustrating breakup into a train of solitary waves.

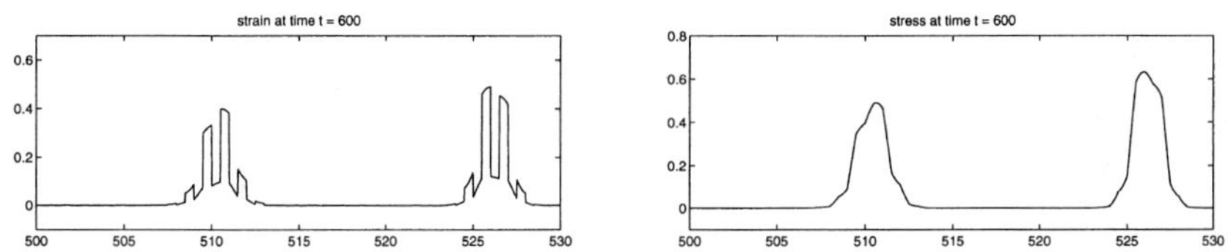

Fig. 2. Close-up view of the first two solitary waves from Fig. 1 at time $t = 600$, showing both the strain and the stress.

Here we begin to explore a different view of these waves that we hope may ultimately lead to new insights into their nature. We plot the waves from Fig. 3 in the σ-u plane, a variant of the phase plane for this system. What we observe depends on where the point x_0 lies within the layered structure of the medium. Fig. 4 shows the curves obtained for various locations of x_0. This computation was performed with 19 grid cells in each layer and the solution was recorded in grid cells 1 through 38 (throughout an A layer and a B layer) over the time period $660 \leq t \leq 700$ as shown in Fig. 3. (Fig. 3 shows the trace for $i = 10$, in the middle of the A layer.) The left column of

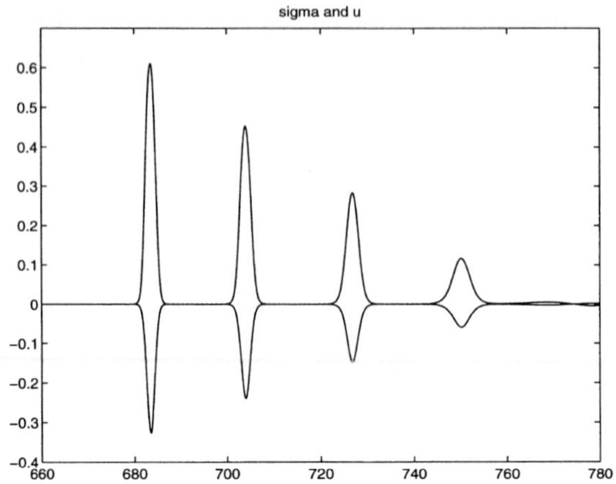

Fig. 3. Time trace of the stress $\sigma(x_0, t) \geq 0$ and velocity $u(x_0, t) \leq 0$ at a fixed point x_0 as four stegotons pass by.

Fig. 4 shows results at five locations in the A layer, while the right column shows 5 locations in the B layer. In each case four excursions from the origin are observed, corresponding to the four solitary waves.

Note that in the middle of each layer ($i = 10$ and $i = 19$), the solitary waves are symmetric in t and the same curve is traced out during the rising and falling portions of the the wave. Away from the center of a layer, the excursion traces out a loop, indicating that the wave would not looks so symmetric in t. In each case, however, the loops appear quite similar for the four stegotons, and scale with the height of the wave.

Note that if we solved the linear equations in a homogeneous medium (i.e., using $\sigma(\epsilon) = K\epsilon$ with K and ρ constant), then a right-going wave would be a simple wave with $u = -\sigma/Z$ everywhere, where $Z = \sqrt{\rho K}$ is the impedance (see [5]). Hence, plots corresponding to Fig. 4 in the homogeneous linear case would show that all values $(\sigma(x_0, t), u(x_0, t))$ lie on a line through the origin with slope $-1/Z$.

For a homogeneous nonlinear material (with $\sigma(\epsilon) = \exp(K\epsilon) - 1$ and K, ρ constant), the system (1) becomes

$$\begin{aligned} \epsilon_t - u_x &= 0 \\ \rho u_t - \exp(K\epsilon)_x &= 0. \end{aligned} \tag{6}$$

A right-going simple wave would have values (σ, u) lying on an integral curve of the eigenvector r^2 of the Jacobian matrix

$$f'(q) = \begin{bmatrix} 0 & 1/\rho \\ \sigma'(\epsilon) & 0 \end{bmatrix} = \begin{bmatrix} 0 & 1/\rho \\ Ke^{K\epsilon} & 0 \end{bmatrix}. \tag{7}$$

The eigenvectors of $f'(q)$ are

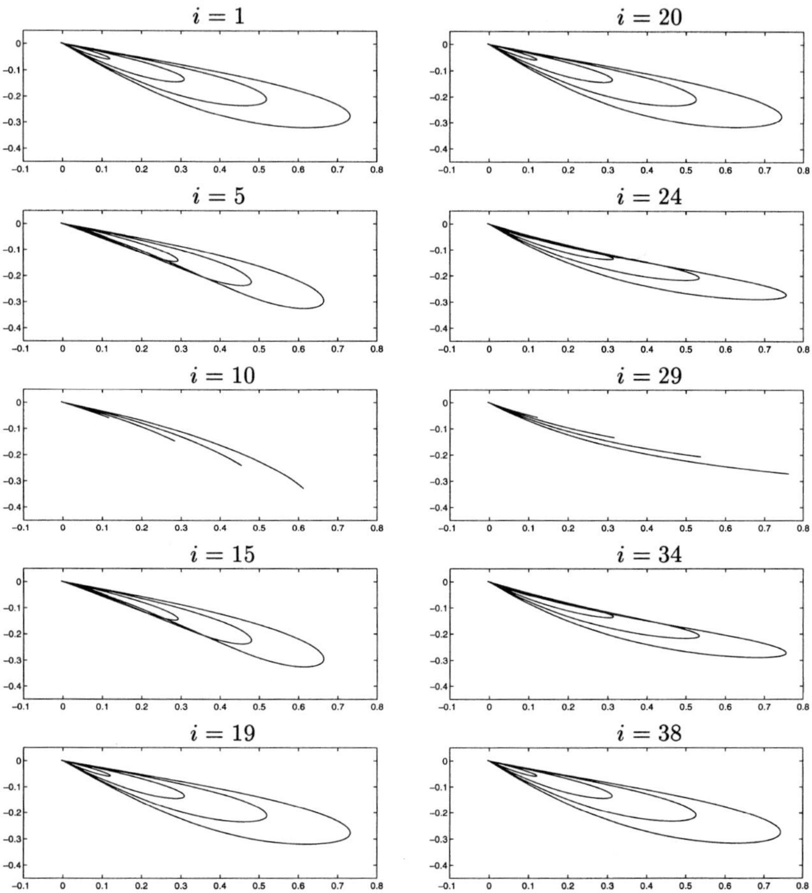

Fig. 4. Solutions of the form shown in Fig. 3 when plotted in the σ-u plane. The resulting curves are shown at 5 different locations within each of the two layers (A layer on the left, B layer on the right). The time trace shown in Fig. 3 is at $i = 10$, in the middle of an A-layer.

$$r^1 = \begin{bmatrix} 1 \\ Z \end{bmatrix}, \qquad r^2 = \begin{bmatrix} 1 \\ -Z \end{bmatrix},$$

where $Z = \sqrt{\rho\sigma'(\epsilon)} = \sqrt{\rho K e^{K\epsilon}}$ varies with $\epsilon(x,t)$ in the nonlinear case.

The Riemann invariants are

$$w^1 = \rho u - 2\sqrt{\rho/K} e^{K\epsilon/2} = \rho\left(u - \frac{2}{\sqrt{\rho K}}\sqrt{\sigma+1}\right) \tag{8}$$

and

$$w^2 = \rho u + 2\sqrt{\rho/K} e^{K\epsilon/2} = \rho\left(u + \frac{2}{\sqrt{\rho K}}\sqrt{\sigma+1}\right). \tag{9}$$

So a right-going simple wave has $w^2 \equiv$ constant. For a solitary wave that makes an excursion from $u = \epsilon = 0$, we must have $w^2 \equiv 2\sqrt{\rho/K}$ and so

$$u = \frac{2}{\sqrt{K\rho}}\left(1 - \sqrt{\sigma + 1}\right). \tag{10}$$

Note that $du/d\sigma|_{\sigma=0} = -1/\sqrt{K\rho} = -1/Z_0$, where Z_0 is the impedance of the corresponding linearized problem, so this is consistent with the linear case for small amplitude waves.

Homogenization theory (see [5] and [6]) suggests that we might compare the results from the layered medium with integral curves corresponding to a homogenized nonlinear medium with parameters

$$\bar{\rho} = \frac{1}{2}(\rho_A + \rho_B) = 5/2, \qquad \hat{K} = \left(\frac{1}{2K_A} + \frac{1}{2K_B}\right)^{-1} = 8/5, \tag{11}$$

and $\sqrt{\bar{\rho}\hat{K}} = 2$. There is in fact a close connection. Fig. 5 shows a comparison of two sets of curves from Fig. 4 (for $i = 10$ and $i = 19$, in the center of each layer) with the integral curve of r^2 obtained using the values $\bar{\rho}$ and $\hat{K}$ from (11), i.e., the curve $u = 1 - \sqrt{\sigma + 1}$.

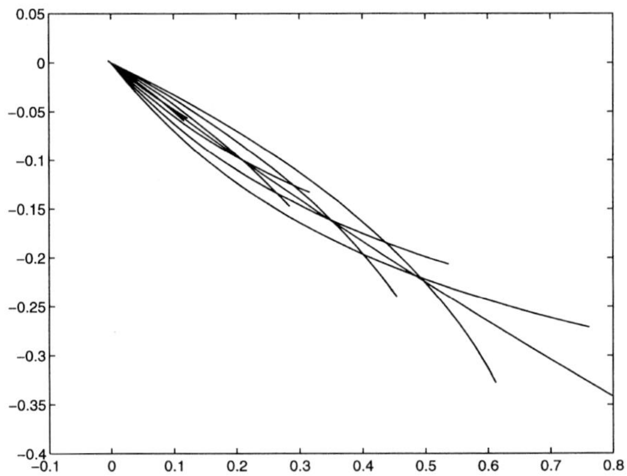

Fig. 5. The curves from Fig. 4 for $i = 10$ and $i = 29$, together with the integral curve $u = 1 - \sqrt{\sigma + 1}$.

3 Homogenization theory

Nonlinear homogenization theory is developed in [5]. For the case considered here, the following approximate equations are derived:

$$u_t = \frac{2\sigma_x}{5} + \frac{3\sigma_{xxx}}{500} + \left(\frac{3\sigma_{xxx}\sigma_x^2}{15625(\sigma+1)^2} - \frac{72u_{xx}^2\sigma_x}{15625(\sigma+1)} - \frac{12\sigma_{xxxx}\sigma_x}{15625(\sigma+1)} - \frac{96u_{xx}u_{xxx}}{15625} - \frac{12\sigma_{xx}\sigma_{xxx}}{15625(\sigma+1)} - \frac{357\sigma_{xxxxx}}{1000000}\right) + \cdots \tag{12a}$$

$$\sigma_t = \frac{8(\sigma+1)u_x}{5} + \left(\frac{3(\sigma+1)u_{xxx}}{125} + \frac{3u_{xx}\sigma_x}{50}\right) + \left(\frac{48u_x u_{xx}^2}{15625} - \frac{48\sigma_x\sigma_{xx}u_{xx}}{15625(\sigma+1)} - \frac{4761\sigma_{xxx}u_{xx}}{500000} - \frac{72u_{xxx}\sigma_x^2}{15625(\sigma+1)} - \frac{357(\sigma+1)u_{xxxxx}}{250000} - \frac{3543u_{xxxx}\sigma_x}{500000} - \frac{3891u_{xxx}\sigma_{xx}}{500000}\right) + \cdots . \tag{12b}$$

These equations have been solved using a pseudospectral method and shown to give results that agree well with solutions to the original layered media equations. Fig. 6 shows solutions to (12) with initial data

$$u(x,0) = 0, \qquad \sigma(x,0) = \text{sech}^2((x-100)/5) \quad \text{on} \quad 0 \le x \le 200,$$

with periodic boundary conditions. This hump splits up into left-going and right-going pulses that break up into solitary waves in exactly the same manner as observed in the layered medium. At time $t = 200$ these wave trains are approaching each other again due to the periodic boundary conditions.

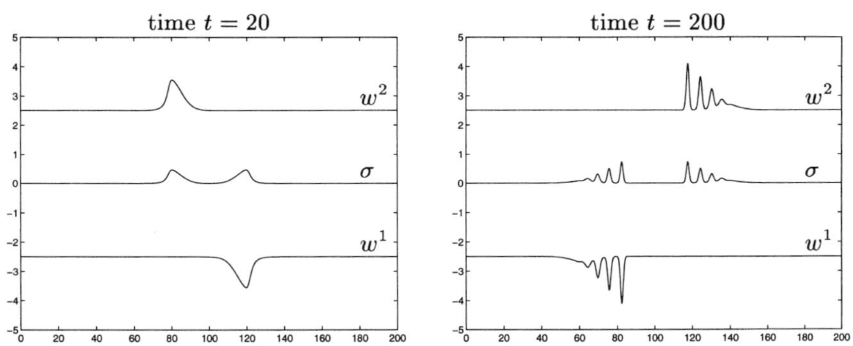

Fig. 6. Solution $\sigma(x,t)$ to the homogenized system of equations (12) at two different times. The Riemann invariants w^1 and w^2 corresponding to the first-order hyperbolic system (6) with parameters (11) are computed from this solution and also plotted.

In these plots, along with $\sigma(x,t)$ we have also plotted the Riemann invariants w^1 and w^2 evaluated from the u and σ values. Note that the expressions (8) and (9) are based on the simple first-order hyperbolic system for a homogeneous medium (using the parameters (11)), while u and σ result from solving the dispersive homogenized equations (12), a system that has very different behavior from the purely hyperbolic system. There is no reason a *priori* to expect that a left-going train of solitary waves appearing in the solution to (12) will exhibit a constant value of w^1, but in fact this appears to be true. We see that w^1 is essentially constant (to about 10^{-3}) through the left-going wave train while w^2 is essentially constant through the right-going wave train. In Fig. 7 we plot the solution values (σ, u) computed from the homogenized equations at $t = 200$ together with the integral curves of r^1 and r^2, the curves $u = \pm(1 - \sqrt{\sigma + 1})$. The solutions values lie very close to these curves.

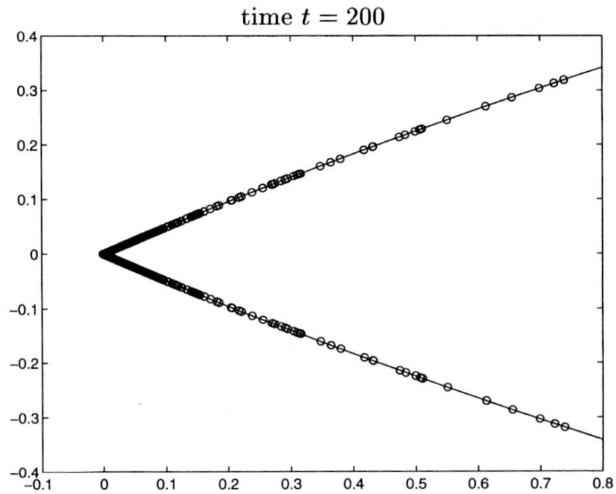

Fig. 7. Solution values from solving the homogenized equations (12), plotted in the σ-u plane at time $t = 200$. The solid lines are the integral curves $u = \pm(1 - \sqrt{\sigma + 1})$.

4 Conclusions

In the linear case, the constant coefficient system with averaged parameters (11) gives the correct homogenized system to leading order. A pulse propagating in the layered medium is well described by the constant coefficient first-order system for small time, though eventually dispersive effects are visible (see [2], [5], [6]). For the nonlinear problem, however, the first-order nonlinear hyperbolic system (6) with parameters (11) does not adequately describe the behavior seen in the layered medium, where the pulse rapidly breaks up into solitary waves. In spite of this, the Riemann invariants of the first order system with averaged parameters appear to be preserved in the wave trains computed using the proper homogenized equations. We are continuing to explore the implications of this observation.

Acknowledgments. This work was supported in part by DOE grant DE-FG03-96ER25292 and NSF grants DMS-9803442 and DMS-0106511.

References

1. Bale, D., LeVeque, R.J., Mitran, S., Rossmanith, J.R. (2002): A wave-propagation method for conservation laws and balance laws with spatially varying flux functions. SIAM J. Sci. Comput., to appear. (`ftp://amath.washington.edu/pub/rjl/papers/fwave.ps.gz`).
2. Fogarty, T., LeVeque, R.J. (1999): High-resolution finite volume methods for acoustics in periodic or random media. J. Acoust. Soc. Am. **106**, 17–28.
3. LeVeque, R.J. (2002): Finite Volume Methods for Hyperbolic Problems. Cambridge University Press.
4. LeVeque, R.J. (2001): Finite volume methods for nonlinear elasticity in heterogeneous media. Int. J. Numer. Meth. Fluids, to appear. (`ftp://amath.washington.edu/pub/rjl/papers/rjl:ox01.ps.gz`).
5. LeVeque, R.J., Yong, D.H. (2002): Solitary waves in layered nonlinear media. Submitted, 2002. (`ftp://amath.washington.edu/pub/rjl/papers/stegoton.ps.gz`).
6. Santosa, F., Symes, W. (1991): A dispersive effective medium for wave propagation in periodic composites. SIAM J. Appl. Math. **51**, 984–1005.

Energy Method for Equations in Gas Dynamics

Tai-Ping Liu[1] and Shih-Hsien Yu[2]

[1] Institute of Mathematics, Academia Sinica, Taipei & Department of Mathematics, Stanford University, USA *tpliu@math.sinica.edu.tw*, *liu@math.stanford.edu*

[2] Department of Mathematics, City University of Hong Kong *mashyu@cityu.edu.hk*

Summary. We briefly review the energy methods for hyperbolic and viscous conservation laws. These motivate the recent studies of the energy method for the Boltzmann equation. The new energy method is based on a macro-micro decomposition of the Boltzmann equation as well as the H-theorem. Previous energy method involves the spectral analysis and regularity of collision operator through sophisticated weighted norms.

1 Hyperbolic Conservation Laws

The energy method for the hyperbolic conservation laws

$$u_t + \nabla_x \cdot f(u) = 0$$

is based on the existence of entropy pairs

$$\eta(u)_t + \nabla_x \cdot q(u) = 0$$

here $\eta(u)$ is a scalar convex function and the entropy flux is compatible with the conservation laws,

$$\nabla q(u) = \nabla\eta(u)\nabla f(u).$$

The energy estimate is based on the observation that the system of conservation laws is a particular type of symmetric hyperbolic system in the dependent variables $w\nabla\eta(u)$. This is seen as follows: Consider the Legendre transformation $\xi(w)$ of the entropy function $\eta(u)$:

$$\eta(u) + \xi(w) = u \cdot w.$$

The entropy flux $q(u)$ is also transformed to

$$q(u) + p(w) = w \cdot f(u).$$

With these, the conservation laws become symmetric in the transformed variables w:

$$\nabla^2\xi(w)w_t + \nabla^2 p(w) \cdot \nabla_x w = 0,$$

[4],[3],[1]. The first order energy estimate is obtained by integrating the system times $\bar{u}$. This yields the entropy estimate. To close the estimates using the Sobolev embedding, the energy estimates are carried to higher order exceeding the order of $1+m/2$, m the space dimension,[9]. Because of the eventual appearance of shocks in a general solution, the energy method provides only the local existence of solutions.

2 Viscous Conservation Laws

Physical systems of viscous conservation laws are of the form

$$u_t + \nabla_x \cdot f(u) = \nabla_x \cdot (B(u)\nabla_x u)$$

where B is the viscosity matrix. Such a system is not uniformly parabolic, but hyperbolic-parabolic. This is reflected in the entropy estimate obtained, as for the hyperbolic conservation laws, by multiplying the system by $\nabla\eta(u)$ and integrate to obtain

$$\frac{d}{dt}\int \eta(u)dx = -\int (\nabla_x u)\cdot \nabla^2\eta(u)B(u)\cdot\nabla_x u dx.$$

The term on the right hand side represents the dissipation, which does not contain all the first differentiations of the independent variables. The simplest example is the so called p-system:

$$v_t - u_x = 0, u_t + p(v)_x = (\frac{u_x}{v})_x.$$

The entropy pair is

$$\eta = \frac{u^2}{2} - \int^v p(y)dy,\ q = up(v).$$

and the entropy estimate, obtained by multiplying the system by $\nabla\eta$ and integrating,

$$\int \eta dx|_0^t + \int_0^t\int \frac{{u_x}^2}{v}dxdt = 0.$$

In this case the dissipation term

$$\int_0^t\int \frac{{u_x}^2}{v}dxdt$$

contains ${u_x}^2$ but not ${v_x}^2$, which is obtained from integrating the second equation multiplying the first equation by u_x and the second equation by $-v_x$, making use of the constitutive requirement that $p'(v) < 0$. In other words, one makes uses of the coupling of the two equations and the hyperbolicity of the corresponding inviscid system.

Notice that the above does not work for the non-physical system such as

$$v_t - v_x = 0, u_t = u_{xx}.$$

Unlike the physical p-system above, there is no coupling between the two equations and the first equation is hyperbolic and not dissipative. For general physical systems, the coupling is expressed by the hypothesis that the eigenvectors of the flux $A(u) = f'(u)$ is not in the null space of the viscosity matrix $B(u)$. It is shown that there is then a skew-symmetric matrix K with the property that

$$(KA)' + B > 0$$

where $(KA)'$ is the symmetric part of KA. And the entropy estimate is complemented by the energy estimate obtained by integrating the system times Ku. There is then the global in time existence theory when the initial data is small in the Sobolev norm up to second order, [6].

3 Boltzmann Equation

The Boltzmann equation is of the form

$$f_t + \xi \cdot \nabla_x f = \frac{Q(f,f)}{\kappa}, \ (f, x, t, \xi) \in \mathbb{R} \times \mathbb{R}^3 \times \mathbb{R}^+ \times \mathbb{R}^3, \tag{1}$$

where the positive constant κ is the Knudsen number. For simplicity, consider the hard sphere model, for which the bilinear collision operator $Q(f,g)$ is of the following form:

$$Q(f,g)(\xi) \equiv$$

$$\frac{1}{2} \int_{\substack{\mathbb{R}^3 \times S^2 \\ (\xi-\xi_*)\cdot\Omega \geq 0}} \left(-f(\xi)g(\xi_*) - g(\xi)f(\xi_*) + f(\xi')g(\xi'_*) + g(\xi')f(\xi'_*)\right) |(\xi - \xi_*) \cdot \Omega| \; d\xi_* d\Omega, \tag{2}$$

where

$$\begin{cases} \xi' = \xi - [(\xi - \xi_*) \cdot \Omega] \, \Omega, \\ \xi'_* = \xi_* + [(\xi - \xi_*) \cdot \Omega] \, \Omega. \end{cases}$$

There are two steps needed to generalize the above energy method for conservation laws to Boltzmann equation. The first step is the derivation of the Navier-Stokes equations as part of the Boltzmann equation. The second step is the derivation of the entropy inequality as part of the H-theorem. These are done based on the macro-micro decomposition of a general function in ξ. We first describe the decomposition.

The macroscopic, fluid part of a given Boltzmann solution f is the local Maxwellian $\boldsymbol{M} = \boldsymbol{M}(x,t,\xi) = \boldsymbol{M}_{[\rho,u,\theta]}(\xi)$, and the microscopic, non-fluid part $\boldsymbol{G} = \boldsymbol{G}(x,t,\xi)$ of the solution:

$$f = \boldsymbol{M} + \boldsymbol{G}.$$

The local Maxwellian is constructed from the fluid variables, the five conserved quantities, the mass density $\rho(x,t)$, momentum $m(x,t) = \rho u(x,t)$ and energy $\mathscr{E} + |u|^2/2$ of the Boltzmann equation:

$$\begin{cases} \rho(x,t) \equiv \int_{\mathbb{R}^3} f(x,t,\xi)d\xi, \\ m^i(x,t) \equiv \int_{\mathbb{R}^3} \psi_i f(x,t,\xi)d\xi \text{ for } i = 1,2,3, \\ \rho(\mathscr{E} + \frac{1}{2}|u|^2)(x,t) \equiv \int_{\mathbb{R}^3} \psi_4 f(x,t,\xi)d\xi, \end{cases} \tag{3}$$

$$\boldsymbol{M} \equiv \boldsymbol{M}_{[\rho,u,\theta]}(\xi) \equiv \frac{\rho}{\sqrt{(2\pi R\theta)^3}} \exp\left(-\frac{|\xi - u|^2}{2R\theta}\right). \tag{4}$$

The five fluid variables are conserved quantities because of the following property of the collision invariants ψ_α, [Boltzmann]:

$$\int_{\mathbb{R}^3} \psi_\alpha Q(h,g)d\xi = 0 \text{ for any } \alpha = 0,1,2,3,4, \text{ and for any functions } h, g,$$

$$\begin{cases} \psi_0 \equiv 1, \\ \psi_i \equiv \xi^i \text{ for } i = 1,2,3, \\ \psi_4 \equiv \frac{1}{2}|\xi|^2. \end{cases} \tag{5}$$

With respect to the local Maxwellian, we define an inner product in $\xi \in \mathbb{R}^3$ as:

$$\langle h, g \rangle \equiv \int_{\mathbb{R}^3} \frac{1}{\boldsymbol{M}} h(\xi)g(\xi)d\xi,$$

for functions h, g of ξ. The collision invariants are orthogonalized as follows:

$$\begin{cases} \chi_0(\xi;\rho,u,\theta) \equiv \frac{1}{\sqrt{\rho}}\boldsymbol{M}, \\ \chi_i(\xi;\rho,u,\theta) \equiv \frac{\xi^i - u^i}{\sqrt{R\theta\rho}}\boldsymbol{M} \text{ for } i = 1,2,3, \\ \chi_4(\xi;\rho,u,\theta) \equiv \frac{1}{\sqrt{6\rho}}\left(\frac{|\xi-u|^2}{R\theta} - 3\right)\boldsymbol{M}, \\ \langle \chi_\alpha, \chi_\beta \rangle = \delta_{\alpha\beta}\ \alpha, \text{ for } \beta = 0,1,2,3,4. \end{cases} \tag{6}$$

The macroscopic projection $\boldsymbol{P}_0$ is the projection onto the the above fluid space and the microscopic projection $\boldsymbol{P}_1$ is the complementary projection:

$$\begin{cases} \boldsymbol{P}_0 h \equiv \sum_{\alpha=0}^4 \langle h, \chi_\alpha \rangle \chi_\alpha, \\ \boldsymbol{P}_1 h \equiv h - \boldsymbol{P}_0 h. \end{cases} \tag{7}$$

Thus the above decompositions are the linearization around the local Maxwellian states with the corresponding linear collision operator $L_{[\rho,u,\theta]}$

$$L = L_{[\rho,u,\theta]}g \equiv Q(\boldsymbol{M}_{[\rho,u,\theta]} + g, \boldsymbol{M}_{[\rho,u,\theta]} + g) - Q(g,g). \tag{8}$$

It is clear that for the solution $f(x,t,\xi)$ of the Boltzmann equation,

$$\boldsymbol{P}_0 f = \boldsymbol{M}, \ \boldsymbol{P}_1 f = \boldsymbol{G},$$

and the Boltzmann equation becomes:

$$(\boldsymbol{M}+\boldsymbol{G})_t + \xi \cdot \nabla_x(\boldsymbol{M}+\boldsymbol{G}) = \frac{1}{\kappa}(2Q(\boldsymbol{G},\boldsymbol{M}) + Q(\boldsymbol{G},\boldsymbol{G})). \tag{9}$$

We now decompose the Boltzmann equation. The macroscopic part are the conservation laws, obtained by integrating with respect to ξ of the Boltzmann equation times the collision invariants $\psi_\alpha(\xi)$:

$$\begin{cases} \rho_t + div\ m = 0, \\ m^i_t + (\sum_{j=1}^3 u^j m^i)_{x^j} + p_{x^i} + \int_{\mathbb{R}^3} \psi_i(\xi \cdot \nabla_x \boldsymbol{G}) d\xi = 0 \text{ for } i = 1,2,3, \\ [\rho(\frac{|u|^2}{2} + \mathscr{E})]_t + \sum_{j=1}^3 [u^j[\rho(\frac{|u|^2}{2} + \mathscr{E}) + p]]_{x^j} + \int_{\mathbb{R}^3} \psi_4(\xi \cdot \nabla_x \boldsymbol{G}) d\xi = 0. \end{cases} \tag{10}$$

The microscopic equation is obtained by applying the microscopic projection $\boldsymbol{P}_1$ to the Boltzmann equation:

$$\boldsymbol{G}_t + \boldsymbol{P}_1(\xi \cdot \nabla_x \boldsymbol{G} + \xi \cdot \nabla_x \boldsymbol{M}) = \frac{1}{\kappa} L\boldsymbol{G} + \frac{1}{\kappa} Q(\boldsymbol{G},\boldsymbol{G}). \tag{11}$$

The decomposition was first introduced in [7] and then refined in [8]. From (11) we have

$$\boldsymbol{G} = \kappa L^{-1}(\boldsymbol{P}_1 \xi \cdot \nabla_x \boldsymbol{M}) + L^{-1}(\kappa(\partial_t \boldsymbol{G} + \boldsymbol{P}_1 \xi \cdot \nabla_x \boldsymbol{G}) - Q(\boldsymbol{G},\boldsymbol{G}))\,; \tag{12}$$

and substitute this into (10) to result in

$$\begin{cases} \rho_t + div\ m = 0, \\ m^i_t + (\sum_{j=1}^3 u^j m^i)_{x^j} + p_{x^i} + \kappa \int_{\mathbb{R}^3} \psi_i(\xi \cdot \nabla_x L^{-1} \boldsymbol{P}_1 \xi \cdot \nabla_x \boldsymbol{M}) d\xi \\ \qquad + \int_{\mathbb{R}^3} \psi_i(\xi \cdot \nabla_x L^{-1}(\kappa[\boldsymbol{G}_t + \boldsymbol{P}_1 \xi \cdot \nabla_x \boldsymbol{G}] - Q(\boldsymbol{G},\boldsymbol{G}))\, d\xi = 0 \text{ for } i = 1,2,3, \\ [\rho(\frac{|u|^2}{2} + \mathscr{E})]_t + \sum_{j=1}^3 [u^j[\rho(\frac{|u|^2}{2} + \mathscr{E}) + p]]_{x^j} + \kappa \int_{\mathbb{R}^3} \psi_4(\xi \cdot \nabla_x L^{-1} \boldsymbol{P}_1 \xi \cdot \nabla_x \boldsymbol{M}) d\xi \\ \qquad + \int_{\mathbb{R}^3} \psi_4(\xi \cdot \nabla_x L^{-1}(\kappa[\boldsymbol{G}_t + \boldsymbol{P}_1 \xi \cdot \nabla_x \boldsymbol{G}] - Q(\boldsymbol{G},\boldsymbol{G}))\, d\xi = 0. \end{cases} \tag{13}$$

The fluid equations, the Euler and Navier-Stokes equations, are in fact part of the above equations: When the Knudsen number κ is set zero, the system (13) becomes the Euler equations as in the Hilbert expansion. When the microscopic part G is set to be zero in (13), it becomes the Navier-Stokes equations as in the Chapman-Enskog expansion.

In the above system,the terms

$$\begin{cases} -\kappa \int_{\mathbb{R}^3} \psi_i(\xi \cdot \nabla_x L^{-1} \boldsymbol{P}_1 \xi \cdot \nabla_x \boldsymbol{M}) d\xi = -\kappa \int_{\mathbb{R}^3} \psi_i(\xi \cdot \nabla_x L^{-1}_{[\rho,u,\theta]} \boldsymbol{P}_1 \xi \cdot \nabla_x \boldsymbol{M}_{[\rho,u,\theta]}) d\xi \\ = -\kappa \int_{\mathbb{R}^3} \psi_i(\xi \cdot \nabla_x L^{-1}_{[1,u,\theta]} \boldsymbol{P}_1 \xi \cdot \nabla_x M_{[1,u,\theta]}) d\xi, \ i = 1,2,3 \\ -\kappa \int_{\mathbb{R}^3} \psi_4(\xi \cdot \nabla_x L^{-1} \boldsymbol{P}_1 \xi \cdot \nabla_x \boldsymbol{M}) d\xi = -\kappa \int_{\mathbb{R}^3} \psi_4(\xi \cdot \nabla_x L^{-1}_{[\rho,u,\theta]} \boldsymbol{P}_1 \xi \cdot \nabla_x \boldsymbol{M}_{[\rho,u,\theta]}) d\xi \\ = -\kappa \int_{\mathbb{R}^3} \psi_4(\xi \cdot \nabla_x L^{-1}_{[1,u,\theta]} \boldsymbol{P}_1 \xi \cdot \nabla_x \boldsymbol{M}_{[1,u,\theta]}) d\xi \end{cases}$$

are the viscosity and heat conductivity terms for the Navier-Stokes equations; and they are independent of the density gradient $\nabla_x \rho$.

The H-theorem is obtained by multiplying the Boltzmann equation by $logf$ and integrate with respect to ξ:

$$\int_{\mathbb{R}^3} flogf d\xi + \nabla \cdot \int_{\mathbb{R}^3} \xi flogf d\xi = \kappa \int_{\mathbb{R}^3} Q(f,f) logf d\xi \le 0, \tag{14}$$

with the equality only when in thermo-equilibrium, $f = \boldsymbol{M}$. This implies that the Boltzmann equation is irreversible and dissipative. the simplest way to see this is that the linearized collision operator is negative definite:There exists $\sigma > 0$ such that for any function satisfying ($\boldsymbol{P}_0 h = 0$),

$$\langle h, Lh \rangle \le -\sigma \langle \nu h, h \rangle. \tag{15}$$

This was proved by Carleman [2] for hard sphere model, and by Grad [5] for cutoff hard potentials. This implies that the non-fluid part $\boldsymbol{G}$ of the Boltzmann solution,

$$\langle \boldsymbol{G}, L\boldsymbol{G} \rangle \le -\sigma \langle \nu \boldsymbol{G}, \boldsymbol{G} \rangle.$$

When the solution is space homogeneous and close to a Maxwellian, it follows from (15) that it will converge to the equilibrium exponentially in time. In other words, it gives a quantitative expression of the H-theorem. This is a microscopic version of the H-theorem and does not take into account of the space inhomogeneity of the solutions.

We next derive the macroscopic version of the H-theorem. This version is to study the dissipative property of the Boltzmann equation with regards to the space inhomogeneity and corresponds to the notion of entropy in gas dynamics. Set the macroscopic entropy S by:

$$-\frac{3}{2}\rho S \equiv \int_{\mathbb{R}^3} \boldsymbol{M} \log \boldsymbol{M} \ d\xi. \tag{16}$$

Instead of multiplying by $\log f$ as in the derivation of the H-theorem, multiply (9) by $\log \boldsymbol{M}$ and integrate in ξ to obtain, after some calculations,

$$
\begin{aligned}
-\frac{3}{2}(\rho S)_t - \frac{3}{2}\nabla_x \cdot (\rho S u) + \nabla_x \cdot \left(\int_{\mathbb{R}^3} \xi \boldsymbol{G} \log \boldsymbol{M} d\xi \right) \\
= \int_{\mathbb{R}^3} \frac{\boldsymbol{P}_1 \xi \cdot \nabla_x \boldsymbol{M}}{\boldsymbol{M}} L^{-1}_{[\rho,u,\theta]}(\kappa \boldsymbol{P}_1 \xi \cdot \nabla_x \boldsymbol{M}) d\xi \\
+ \int_{\mathbb{R}^3} \frac{\boldsymbol{P}_1 \xi \cdot \nabla_x \boldsymbol{M}}{\boldsymbol{M}} L^{-1}_{[\rho,u,\theta]} \left(\kappa \boldsymbol{G}_t + \kappa \boldsymbol{P}_1 \xi \cdot \nabla_x \boldsymbol{G} - Q(\boldsymbol{G}, \boldsymbol{G}) \right) d\xi. \quad (17)
\end{aligned}
$$

¿From (4) and (16)

$$
-\frac{3}{2}\rho S = \rho \log \rho - \frac{3}{2}\rho \log(2\pi R \theta) - \frac{1}{2}\rho. \quad (18)
$$

In the above identity, the term

$$
\kappa \int_{\mathbb{R}^3} \frac{\boldsymbol{P}_1 \xi \cdot \nabla_x \boldsymbol{M}}{\boldsymbol{M}} L^{-1}_{[\rho,u,\theta]}(\boldsymbol{P}_1 \xi \cdot \nabla_x \boldsymbol{M}) d\xi = \kappa \langle \boldsymbol{P}_1 \xi \cdot \nabla_x \boldsymbol{M}, L^{-1}_{[\rho,u,\theta]}(\boldsymbol{P}_1 \xi \cdot \nabla_x \boldsymbol{M}) \rangle
$$

represents the entropy dissipation. Since the non-fluid functions $\boldsymbol{P}_1 \xi \cdot \nabla_x \boldsymbol{M}$ belong to a finite dimensional space in the ξ variables, we have from the microscopic version of the H-theorem (15) that the term satisfies, for some positive constants σ_1 and σ_2,

$$
\begin{aligned}
\sigma_1 \int_{\mathbb{R}^3} \frac{|\boldsymbol{P}_1 \xi \cdot \nabla_x \boldsymbol{M}|^2}{\boldsymbol{M}} \le - \int_{\mathbb{R}^3} \frac{\boldsymbol{P}_1 \xi \cdot \nabla_x \boldsymbol{M}}{\boldsymbol{M}} L^{-1}_{[\rho,u,\theta]}(\boldsymbol{P}_1 \xi \cdot \nabla_x \boldsymbol{M}) d\xi \\
\le \sigma_2 \int_{\mathbb{R}^3} \frac{|\boldsymbol{P}_1 \xi \cdot \nabla_x \boldsymbol{M}|^2}{\boldsymbol{M}} d\xi. \quad (19)
\end{aligned}
$$

Simple calculation shows that

$$
\int_{\mathbb{R}^3} \frac{|\boldsymbol{P}_1 \xi \cdot \nabla_x \boldsymbol{M}|^2}{\boldsymbol{M}} d\xi = O(1) \sum_{j=1}^{3} [|\partial_{x^j} u|^2 + |\nabla_x \theta|^2], \quad (20)
$$

for some positive function $O(1)$. This is the same as the partial dissipation, without the term $\nabla_x \rho$, for the Navier-Stokes equations,

We may construct an entropy pair (η, q) around a given state $\bar{\boldsymbol{M}}$ from the above physical entropy function. The first order energy estimate is done by integrating the entropy equation (17). This yields the L_2 estimate for the fluid variables as well as the partial dissipations, same as for the Navier-Stokes equations. However, now the estimate is not closed within the fluid variables, because it is coupled with the non-fluid variable $\boldsymbol{G}$. Thus, besides similar higher order estimates for viscous conservation laws, one needs to make uses of the microscopic dissipation (15) to study the coupling of the macroscopic part $\boldsymbol{M}$ and the microscopic part $\boldsymbol{G}$. This new energy estimate is more basic than the more sophisticated ones based on the spectral analysis, e.g. [10], [11].

Acknowledgment: The research of the first author was supported by Institute of Mathematics, Academia Sinica and NSF Grant DMS-9803323. The research of the third author was supported by the Competitive Earmarked Research Grant of Hong Kong # 9040645.

References

1. Boillat, G. (1996): *Non Linear Fields and Waves* in CIME Course *Recent Mathematical Methods in Nonlinear Wave Propagation*, Montecatini 1994. Lecture Notes in Mathematics **1640**, pp. 1-47 T. RUGGERI Ed. Springer-Verlag.
2. Carleman, T. (1957) *Sur La Théorie de l'Équation Intégrodifférentielle de Boltzmann.* Acta Mathematica, **Vol 60**, pp 91-142
3. Friedrichs, K.O., Lax, P.D. (1971): *Systems of conservation equation with a convex extension.* Proc. Nat. Acad. Sci. USA, **68**, 1686.
4. Godunov, S. K. (1961): *An interesting class of quasilinear systems.* Sov. Math. **2**, 947.
5. Grad, H. (1963): *Asymptotic Theory of the Boltzmann Equation*, Rarefied Gas Dynamics, J. A. Laurmann, Ed. **Vol** 1, 26, pp 26-59 Academic Press, New York
6. Kawashima, S. (1987): *Large-time behaviour of solutions to hyperbolic-parabolic systems of conservation laws and applications.* Proc. Roy. Soc. Edinburgh Sect. A 106 no. 1-2, 169–194
7. Liu, T.-P., Yu, S.-H. *Boltzmann Equation: Micro-Macro decompositions and positivity of shock profiles*, to appear
8. Liu, T.-P., Yang, T., Yu, S.-H. *Energy estimate for Boltzmann equation*, to appear
9. Kato, T. (1975): *Quasi-linear equations of evolution, with applications to partial differential equations.* Spectral theory and differential equations (Proc. Sympos., Dundee, 1974; dedicated to Konrad Jrgens), pp. 25–70. Lecture Notes in Math., Vol. 448, Springer, Berlin.
10. Nishida, T. (1978): *Fluid dynamical limit of the nonlinear Boltzmann equation to the level of the compressible Euler equation.* Comm. Math. Phys. 61, **no.** 2, 119–148
11. Ukai, S. (1974): *On the existence of global solutions of mixed problem for nonlinear Boltzmann equation.* Proc. Japan Acad. 50, 179–184.

Convergence of the Upwind Interface Source Method for Hyperbolic Conservation Laws

Benoît Perthame and Chiara Simeoni

Département de Mathématiques et Applications - École Normale Supérieure
45, rue d'Ulm - 75230 Paris Cedex 05 - France
Benoit.Perthame@ens.fr, Chiara.Simeoni@ens.fr

Summary. This paper deals with typical questions arising in the analysis of numerical approximations for scalar conservation laws with a source term. We focus our attention on semi-discrete finite volume schemes, in the general case of a nonuniform spatial mesh. To define appropriate discretizations of the source term, we introduce the formalism peculiar to the Upwind Interface Source method and we establish conditions on the numerical functions so that the discrete solver preserves the steady state solutions. Then we formulate a rigorous definition of consistency, adapted to the class of well-balanced schemes, for which we are able to prove a Lax-Wendroff type convergence theorem. Some examples of numerical methods are discussed, in order to validate the arguments we propose.

1 Introduction

We consider a scalar conservation law with a source term, in one space dimension,

$$\frac{\partial u}{\partial t} + \frac{\partial A(u)}{\partial x} + B(x,u) = 0, \quad t \in \mathbb{R}_+,\ x \in \mathbb{R}, \tag{1}$$

with $u(t,x) \in \mathbb{R}$ and a real-valued flux function A, associated with a Cauchy problem by introducing the initial condition

$$u(0,x) = u^0(x) \in L^1(\mathbb{R}) \cap L^\infty(\mathbb{R}). \tag{2}$$

We set

$$a(u) = A'(u) \in \mathcal{C}^1(\mathbb{R}) \tag{3}$$

and we restrict our analysis to a particular form of source term,

$$B(x,u) = z'(x)b(u), \quad z' \in L^1(\mathbb{R}),\ b \in \mathcal{C}^1(\mathbb{R}). \tag{4}$$

This is suggested by the usual application of hyperbolic conservation laws as simple mathematical models in continuum mechanics: in the Saint-Venant system for shallow waters, for instance, $z(x)$ describes the bottom topography. The equation (1) is endowed with the family of entropy inequalities

$$\frac{\partial S(u)}{\partial t} + \frac{\partial \eta(u)}{\partial x} + S'(u)B(x,u) \le 0, \quad \eta'(u) = S'(u)a(u), \tag{5}$$

for any pair of a convex entropy function S and the corresponding entropy flux η (see [21] and [24]). Under stronger assumptions on the source term, Kružkov proved in [21] existence and uniqueness of the entropy solution for the initial value problem (1)-(2), in the functional space $L^\infty([0,T); L^1(\mathbb{R}))$, for all $T \in \mathbb{R}_+$. Many results concerning the convergence of numerical approximations for the entropy solution of hyperbolic conservation laws are inspired by this fundamental theory.
In the case of singular source terms (i.e. the function $z(x)$ is discontinuous), a remarkable uniqueness result has recently been proved by Vasseur [31].

The presence of source terms modifies the analytical properties of the equation (1), in comparison with the homogeneous case. More specifically, a fundamental change is the occurrence of other kinds of steady state solutions, resulting from the balance between source terms and internal forces, given by the formula

$$D(u) + z(x) = C^{st}, \quad D'(u) = \frac{a(u)}{b(u)}. \tag{6}$$

This fact also influences the numerical approach to the problem, as it was pointed out by several authors (refer to [15] and [27]), in order to investigate discrete approximations preserving the properties of the continuous system.

A well-known difficulty encountered in the numerical treatment of hyperbolic conservation laws with a source term relates to the approximation of such a source term, to assure that the scheme preserves the steady state solutions at discrete level.
Initially for scalar problems, Greenberg, LeRoux and others introduced the notion of well-balanced schemes (see [15],[16] for details). This definition has been further developed by Gosse and LeRoux [11], which used a reformulation of the source terms by means of non-conservative products to derive numerical fluxes at the interfaces of an unstructured mesh. A recent approach by LeVêque [27] is based on the Godunov scheme extended for an appropriately modified system. Botchorishvili, Perthame and Vasseur present in [2] a kinetic scheme, that maintains steady states and which is proved to converge when stiff source terms are considered. Using interfacial values, instead of the cell-averages, for the source term, Jin proposes in [17] a rather simple method for capturing steady state solutions with a high order accuracy. Previous schemes have also been modified for this target by Bermudez and Vasquez [1] and some different approaches are developed in [22],[23] and [18]. Quite recently, these kinds of numerical processing have been extended to hyperbolic systems of balance laws (like the Saint-Venant system for shallow waters), to obtain stable schemes which preserve the steady states (see [8], [13],[14] and [28], for instance). In particular, one of the main conclusions in [28] is that, while preserving steady states, well-balanced schemes can also enjoy stability under the usual CFL condition (independent of z').

The aim of this paper is to present a general consistency condition for discrete approximations of equation (1). In fact, to analyze theoretical properties of numerical solvers for a conservation law with a source term, the only classical condition on the flux function is not enough and specific definitions for the discrete source term are required.

The outline of the paper is the following. In Section 2, we illustrate the Upwind Interface Source method, which consists in upwinding source terms at the interfaces of the mesh cells, as usual for the fluxes according to the finite volume formalism. Then, in Section 3, we consider discretizations which preserve steady state solutions (well-balanced schemes) and we review several classical methods to build such schemes. The question of consistency is addressed in Section 4 and we show that well-balanced schemes are consistent in the sense we have established. By using these arguments, in Section 5, we finally prove an extension of the Lax-Wendroff theorem.

2 Upwind Interface Source method

The finite volume method is possible for treating numerically hyperbolic systems of conservation laws, it is robust and presents the advantage to be conservative (we refer to [6] for a survey of its properties). We look at the semi-discrete scheme, called method of lines, where only a space discretization of equation (1) is performed.

We consider a mesh of $\mathbb{R}$ made up of cells C_i, with center x_i, $i \in \mathbb{Z}$, and nonuniform length Δx_i. We denote by $x_{i+\frac{1}{2}}$ the cell interfaces, so that the control volume can be identified as $C_i = [x_{i-\frac{1}{2}}, x_{i+\frac{1}{2}})$ and $x_i = \frac{x_{i-\frac{1}{2}}+x_{i+\frac{1}{2}}}{2}$. Then, we construct a piecewise constant representation of the function $z(x)$ on the mesh, whose coefficients are $z_i = \frac{1}{\Delta x_i}\int_{C_i} z(x)\,dx$ for example.

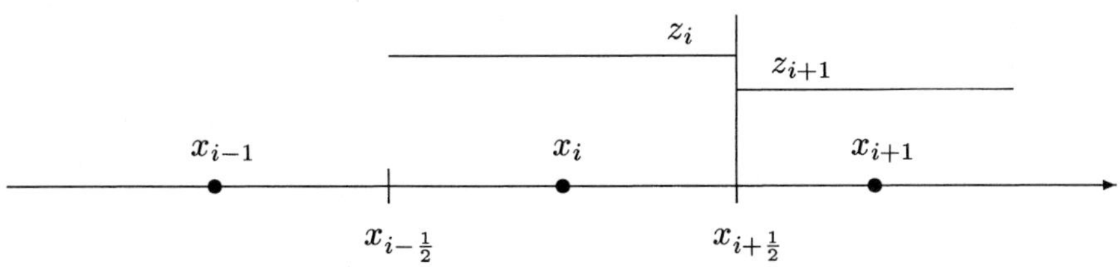

In this context, the discrete unknowns are expected to be approximations of the mean values of u on the mesh cells (the conservative quantities are cell centered),

$$u_i(t) \approx \frac{1}{\Delta x_i}\int_{C_i} u(t,x)\,dx, \quad t \in \mathbb{R}_+,\ i \in \mathbb{Z},$$

while the numerical fluxes are defined at the interfaces of the mesh.

To correctly treat the source term is more difficult than it seems and centered schemes give unsatisfactory results, as it is well reported in the literature: a direct discretization of the source term by cell-averages, for instance, can not preserve the steady state solutions. A better approach is based on the Upwind Interface Source method (upwinding of external terms was originally formulated by Roe [29]), where the source term is also upwinded at the interfaces.

The general finite volume scheme for equation (1) can thus be written in the following explicit form,

$$\Delta x_i \frac{du_i}{dt} + (A_{i+\frac{1}{2}} - A_{i-\frac{1}{2}}) + \mathcal{B}^+_{i-\frac{1}{2}} + \mathcal{B}^-_{i+\frac{1}{2}} = 0, \tag{7}$$

dropping the time dependence of the numerical functions for simplicity.
We proceed to explain the notation in the previous formula. We first introduce the discrete fluxes

$$A_{i+\frac{1}{2}} = \mathcal{A}(u_i, u_{i+1}), \quad \mathcal{A} \in \mathcal{C}^1, \tag{8}$$

where the numerical function $\mathcal{A}$ is chosen as a consistent approximation of the analytical flux,

$$\mathcal{A}(u, u) = A(u). \tag{9}$$

Because of the choice of particular source terms (4), the function $z(x)$ is defined up to a constant. Therefore, without loss of generality, we suppose the source term is discretized at the cell interfaces by means of functions

$$\mathcal{B}^{\pm}_{i+\frac{1}{2}} = \mathcal{B}^{\pm}(u_i, u_{i+1}, z_{i+1} - z_i), \qquad \mathcal{B}^{\pm} \in \mathcal{C}^2, \tag{10}$$

and, in view of (4), it is natural to impose

$$\mathcal{B}^+(u, v, 0) = \mathcal{B}^-(u, v, 0) = 0. \tag{11}$$

The last condition refers to the interpretation of the numerical solver (7), applied to the model (1)-(4). According to the finite volume formalism, we can identify

$$\mathcal{B}^+_{i-\frac{1}{2}} + \mathcal{B}^-_{i+\frac{1}{2}} \approx \int_{C_i} z'(x) b(u) dx; \tag{12}$$

formally, this leads to deduce

$$\mathcal{B}^-_{i+\frac{1}{2}} + \mathcal{B}^+_{i+\frac{1}{2}} \approx \int_{x_i}^{x_{i+\frac{1}{2}}} z'(x) b(u) dx + \int_{x_{i+\frac{1}{2}}}^{x_{i+1}} z'(x) b(u) dx, \tag{13}$$

that is a way to perform an interfacial approximation of zero order terms. Such a discretization is also upwinded, in the sense that $\mathcal{B}^-_{i+\frac{1}{2}}$ represents the contribution of the waves coming from the left of the interface $x_{i+\frac{1}{2}}$ and moving towards the cell C_i if they have a nonpositive velocity, while $\mathcal{B}^+_{i+\frac{1}{2}}$ represents the waves moving forwards from the right of $x_{i+\frac{1}{2}}$ and counted only if they have nonnegative velocity. Notice that, when the problem becomes homogeneous (for example, $z'(x) = 0$ in equation (4), motivated by the analogy with the Saint-Venant model), this scheme reduces to the usual finite volume approximation for a scalar conservation law.

We observe that all what is stated in this section and in the following is also valid for a fully explicit scheme (obtained, for instance, using a standard forward Euler method for the time discretization),

$$\frac{\Delta x_i}{\Delta t}(u_i^{n+1} - u_i^n) + (A_{i+\frac{1}{2}}^n - A_{i-\frac{1}{2}}^n) + \mathcal{B}_{i-\frac{1}{2}}^{n,+} + \mathcal{B}_{i+\frac{1}{2}}^{n,-} = 0,$$

where we introduce a time-step Δt and set $t_n = n\Delta t$, $n \in \mathbb{N}$. We then have to consider an additional restriction on the size of the ratio $\frac{\Delta t}{\Delta x_i}$, the usual CFL condition, to guarantee numerical stability.

3 Well-balanced schemes

We define general conditions on the discretizations $\mathcal{B}_{i+\frac{1}{2}}^{\pm}$ so that the numerical scheme (7) preserves the steady state solutions. Note that all the methods developed in the references mentioned above are compatible with the formalism introduced in Section 2 and can be put in form (7), as we will do later for some particular cases.

By integrating the stationary equation associated with (1)-(4), we obtain the algebraic relation (6) for smooth steady state solutions. A discrete version is given by

$$D(u_i) + z_i = C^{st}, \quad \forall\, i \in \mathbb{Z}. \tag{14}$$

We consider appropriate hypotheses on D, to ensure the existence of a unique Lipschitz continuous solution of that problem, namely that D is strictly monotonic. This assumption is restrictive and not always satisfied in realistic situations, but it is usually made (refer to [2], for instance).

For all initial data defined according to (14), a solver preserving steady states has to verify

$$(A_{i+\frac{1}{2}} - A_{i-\frac{1}{2}}) + \mathcal{B}_{i-\frac{1}{2}}^{+} + \mathcal{B}_{i+\frac{1}{2}}^{-} = 0.$$

This last statement can be formulated in terms of numerical functions, thanks to definition (8) and (10), so that it writes

$$\begin{aligned}\mathcal{A}(u_i, u_{i+1}) - \mathcal{A}(u_{i-1}, u_i) + \mathcal{B}^+(u_{i-1}, u_i, z_i - z_{i-1})&\\ + \mathcal{B}^-(u_i, u_{i+1}, z_{i+1} - z_i) &= 0,\end{aligned}$$

for all u_{i-1}, u_i, u_{i+1}, such that $D(u_j) + z_j = H$, $j = i-1, i, i+1$. In particular, we choose alternatively $u_{i-1} = u_i$ and $u_i = u_{i+1}$ (then we deduce, respectively, $z_{i-1} = z_i$ and $z_i = z_{i+1}$), to obtain equivalent conditions at the interfaces, also exploiting properties (9) and (11),

$$\begin{aligned}\mathcal{A}(u_i, u_{i+1}) - A(u_i) + \mathcal{B}^-(u_i, u_{i+1}, z_{i+1} - z_i) &= 0,\\ A(u_{i+1}) - \mathcal{A}(u_i, u_{i+1}) + \mathcal{B}^+(u_i, u_{i+1}, z_{i+1} - z_i) &= 0.\end{aligned}$$

We summarize the previous statements in the following proposition.

Lemma 1. *A numerical scheme in form (7)-(11) is well-balanced, i.e. it preserves the steady state solutions (14), if and only if the equalities*

$$\mathcal{A}(u,v) - A(u) + \mathcal{B}^-(u,v,z_+ - z_-) = 0, \tag{15}$$

$$A(v) - \mathcal{A}(u,v) + \mathcal{B}^+(u,v,z_+ - z_-) = 0, \tag{16}$$

hold true, for all u, v, z_-, z_+ such that

$$D(u) + z_- = D(v) + z_+. \tag{17}$$

We call well-balanced or Steady State Preserving schemes the numerical solvers for the problem (1)-(2) which satisfy those conditions.

We present some numerical schemes to which Lemma 1 applies. We check these approaches enable to preserve the discrete steady state solutions, according to the result stated above.

B.P.V. method In [2], the authors introduce their solver in a compact form, taking into account the source term directly in the definition of the numerical fluxes,

$$\Delta x_i \frac{du_i}{dt} + (A^-_{i+\frac{1}{2}} - A^+_{i-\frac{1}{2}}) = 0, \tag{18}$$

with

$$A^-_{i+\frac{1}{2}} = \mathcal{A}(u_i, u^-_{i+1}), \quad A^+_{i-\frac{1}{2}} = \mathcal{A}(u^+_{i-1}, u_i). \tag{19}$$

The numerical flux used in [2] is given by a standard Engquist-Osher function, but one readily finds out that similar methods can be formulated with any consistent flux function $\mathcal{A}$. The points u^-_{i+1} and u^+_{i-1} are defined by means of the relations

$$D(u^-_{i+1}) + z_i = D(u_{i+1}) + z_{i+1}, \tag{20}$$

$$D(u^+_{i-1}) + z_i = D(u_{i-1}) + z_{i-1}. \tag{21}$$

To reproduce this scheme in form (7), we identify

$$\mathcal{B}^+_{i-\frac{1}{2}} = \mathcal{A}(u_{i-1}, u_i) - \mathcal{A}(u^+_{i-1}, u_i), \quad \mathcal{B}^-_{i+\frac{1}{2}} = \mathcal{A}(u_i, u^-_{i+1}) - \mathcal{A}(u_i, u_{i+1}).$$

For a steady state, we immediately deduce from (14), (20) and (21) that

$$u^-_{i+1} = u_i, \quad u^+_{i-1} = u_i,$$

then resulting in the conditions (15) and (16).

This method extends to more general classes of function D, such as quadratic functions, and it also applies to hyperbolic systems of conservation laws endowed with a kinetic interpretation (see [28]).

We remark that, combined with specific approximate Riemann solvers, the algorithm (18)-(19) can be interpreted as the well-balanced scheme derived by Greenberg and LeRoux [15] or by Gosse and LeRoux [11], for which the condition (15)-(17) is verified.

The quasi-steady wave-propagation algorithm The basic idea of the method developed by LeVêque [27] is to introduce a new Riemann problem in the center of each mesh cell, with values u_i^- on the left half of the cell and u_i^+ on the right half, whose flux difference exactly cancels the effect of the source term. These artificial states are defined so that the cell-average is preserved,

$$u_i^- = u_i - \delta_i, \quad u_i^+ = u_i + \delta_i, \quad \frac{1}{2}(u_i^- + u_i^+) = u_i; \tag{22}$$

moreover, if δ_i is chosen according to the in-cell balance condition

$$A(u_i^+) - A(u_i^-) + z_i' b(u_i) \Delta x_i = 0, \tag{23}$$

then the jumps occurring at the cell interfaces will correspond to perturbations from the steady states. Note that (23) represents a discrete version of the stationary problem associated with equation (1). The explicit formula for the scheme thus obtained looks like the classical Godunov solver,

$$\Delta x_i \frac{du_i}{dt} + \left(\Delta^+ A(u_{i-1}^+, u_i^-) + \Delta^- A(u_i^+, u_{i+1}^-)\right) = 0,$$

with

$$\Delta^+ A(u_{i-1}^+, u_i^-) = A(u_i^-) - A(u_{i-\frac{1}{2}}^*), \tag{24}$$

$$\Delta^- A(u_i^+, u_{i+1}^-) = A(u_{i+\frac{1}{2}}^*) - A(u_i^+), \tag{25}$$

where $u_{i+\frac{1}{2}}^*$ now denotes the solution to the modified Riemann problem at the cell interfaces, between values u_i^+ and u_{i+1}^-. If the solution we are looking for is quasi-steady then we deduce from (22) and (23) that $u_i^+ \approx u_{i+1}^-$, as δ_i tends to 0, so that the steady states are asymptotically preserved.

As for previous examples, this method can extend to any consistent numerical flux function $\mathcal{A}$, by rewriting the flux differences (24)-(25) in the more general form

$$\Delta^+ A(u_{i-1}^+, u_i^-) = A(u_i^-) - \mathcal{A}(u_{i-1}^+, u_i^-),$$

$$\Delta^- A(u_i^+, u_{i+1}^-) = \mathcal{A}(u_i^+, u_{i+1}^-) - A(u_i^+).$$

Jin's formulas A simple scheme for handling hyperbolic systems of conservation laws with source terms is proposed in [17], which preserves the steady state solutions exactly at the cell interfaces. For methods based on generalized Riemann solvers, it takes form (7) and the source term is discretized by

$$\mathcal{B}_{i-\frac{1}{2}}^+ + \mathcal{B}_{i+\frac{1}{2}}^- = \frac{A_{i+\frac{1}{2}} - A_{i-\frac{1}{2}}}{D_{i+\frac{1}{2}} - D_{i-\frac{1}{2}}}(z_{i+\frac{1}{2}} - z_{i-\frac{1}{2}}), \tag{26}$$

using interface values $A_{i+\frac{1}{2}} = A(u_{i+\frac{1}{2}})$ and $D_{i+\frac{1}{2}} = D(u_{i+\frac{1}{2}})$, rather than the cell-averages. Consequently, if one gets a steady state at the interfaces,

$$D_{i+\frac{1}{2}} + z_{i+\frac{1}{2}} = C^{st}, \quad \forall i \in \mathbb{Z},$$

a direct computation leads to verify that it is preserved, as we have

$$(A_{i+\frac{1}{2}} - A_{i-\frac{1}{2}}) + \frac{A_{i+\frac{1}{2}} - A_{i-\frac{1}{2}}}{D_{i+\frac{1}{2}} - D_{i-\frac{1}{2}}}(z_{i+\frac{1}{2}} - z_{i-\frac{1}{2}}) = 0.$$

A more generic scheme, again proposed in [17], is defined by

$$\Delta x_i \frac{du_i}{dt} + (A_{i+\frac{1}{2}} - A_{i-\frac{1}{2}}) + \frac{b_{i-\frac{1}{2}} + b_{i+\frac{1}{2}}}{2}(z_{i+\frac{1}{2}} - z_{i-\frac{1}{2}}) = 0. \tag{27}$$

Although it is not possible to derive an explicit expression of D for a general flux function A, some applications considered by the author (shallow water equations, for instance) show this method yields formally second order approximations to the steady states at the interfaces, as suggested by an asymptotic expansion of (27).
The numerical discretizations formulated by Jin are called Steady State Capturing schemes, that is a weaker definition since only the interface values are preserved. According to the idea to process the source term by making use explicitly of relations on the steady states, a Steady State Preserving variation of (26), which agrees with the general formalism (7)-(10), is given by

$$\mathcal{B}^{+}_{i-\frac{1}{2}} = \frac{\mathcal{A}(u_{i-1}, u_i) - A(u_i)}{D(u_{i-1}) - D(u_i)}(z_i - z_{i-1}),$$

$$\mathcal{B}^{-}_{i+\frac{1}{2}} = \frac{\mathcal{A}(u_i, u_{i+1}) - A(u_i)}{D(u_{i+1}) - D(u_i)}(z_{i+1} - z_i).$$

Again we can readily check the condition (15)-(17) for this method.

4 Consistency

In order to investigate theoretical properties of the Upwind Interface Source method, a crucial question to discuss is that equation (7) verifies the consistency with the continuous equation (1)-(4).

We indicate a rigorous definition of consistency, which also results to be satisfied by well-balanced schemes.

Definition 1. *A numerical scheme in form (7)-(11) is said to be consistent with (1) if the following limit is verified, locally uniformly in u,*

$$\lim_{\lambda \to 0} \frac{\mathcal{B}^{+}(u, u, \lambda) + \mathcal{B}^{-}(u, u, \lambda)}{\lambda} = b(u). \tag{28}$$

We point out that the above definition of consistency for the source term, in finite volume sense, does not imply that the consistency error vanishes just as for the flux terms. Indeed, because of the choice of an arbitrary nonuniform spatial mesh, the space-step Δx_i could be very different from the length of an interfacial interval $\Delta x_{i+\frac{1}{2}} = |x_{i+1} - x_i| = \Delta x_i/2 + \Delta x_{i+1}/2$; therefore, the corresponding interfacial discretizations (12) and (13) may have very different values. The condition (28) we have established is closer to (13), which is the most appropriate interpretation of the discrete source term for the general method illustrated in this paper.

As it will be seen clearly in next section, consistency plays an important role to achieve convergence properties of a numerical solver, in particular to prove that the strong limit of discrete approximations (as the mesh is refined) is the suitable weak solution of the continuous problem.

The following result guarantees consistency for the numerical schemes described in Section 3.

Lemma 2. *We assume D is monotonic. Let a numerical solver for the system (1)-(4) satisfy the conditions (15)-(17) in Lemma 1, then the property (28) is verified. In other words, all Steady State Preserving schemes are consistent.*

Proof. We perform a Taylor expansion of the relation (17) and we deduce that, for some $\xi \in (u,v)$,

$$D'(\xi)(u-v) \;=\; z_+ - z_-. \tag{29}$$

After adding equality (15) to (16), thanks to the definition (3) and (6), this leads to

$$\begin{aligned} \mathcal{B}^+(u,v,z_+ - z_-) + \mathcal{B}^-(u,v,z_+ - z_-) &= A(u) - A(v) \\ &= \frac{a(\zeta)}{D'(\xi)}(z_+ - z_-), \end{aligned} \tag{30}$$

for some $\zeta \in (u,v)$. We also note that the regularity assumed for the numerical functions enables to perform the general approximations

$$\begin{aligned} \mathcal{B}^\pm(u,v,z_+ - z_-) &= \mathcal{B}^\pm(u,u,z_+ - z_-) \\ &\quad + \frac{\partial}{\partial v}\mathcal{B}^\pm(u,u,z_+ - z_-)(v-u) + O(|v-u|^2). \end{aligned}$$

It thus follows from (11) that

$$\frac{\partial}{\partial v}\mathcal{B}^+(u,u,0) = \frac{\partial}{\partial v}\mathcal{B}^-(u,u,0) = 0$$

and, in view of (29), we obtain that

$$\begin{aligned} &\lim_{z_+ - z_- \to 0} \frac{\mathcal{B}^+(u,u,z_+ - z_-) + \mathcal{B}^-(u,u,z_+ - z_-)}{z_+ - z_-} \\ &= \lim_{z_+ - z_- \to 0} \frac{\mathcal{B}^+(u,v,z_+ - z_-) + \mathcal{B}^-(u,v,z_+ - z_-)}{z_+ - z_-}. \end{aligned} \tag{31}$$

By combining relation (30) with (31), since

$$\frac{a(\zeta)}{D'(\xi)} \longrightarrow \frac{a(u)}{D'(u)} = b(u),$$

we finally conclude that the property (28) is satisfied.

5 A Lax-Wendroff type convergence theorem

We are now interested in the convergence of the numerical scheme (7), as the mesh size tends to zero, by analyzing the convergence properties of its solution $\{u_i(t)\}_{i\in\mathbb{Z}}$.
A discretization of the initial condition is given, for instance, by the sequence

$$u_i^0 = \frac{1}{|C_i|}\int_{C_i} u^0(x)dx, \quad i \in \mathbb{Z}.$$

As a measure of mesh refinement, we consider the parameter

$$h = \sup_{i\in\mathbb{Z}} \Delta x_i.$$

For our purpose, we introduce the piecewise constant function u_h, defined a.e. in $\mathbb{R}_+ \times \mathbb{R}$ by

$$u_h(t,x) = \sum_{i\in\mathbb{Z}} u_i(t)\mathbb{1}_{C_i}(x), \tag{32}$$

and we study the convergence towards a solution to the problem (1)-(2), as h tends to 0.

Theorem 1. *Assume $z \in W^{1,1}$ for the source term (4). Let u_h be obtained from a numerical scheme in form (7)-(11), which satisfies the consistency condition (28). Suppose there exists a constant C such that, uniformly in h,*

$$||u_h||_{L^\infty_{loc}(\mathbb{R}_+\times\mathbb{R})} \leq C \tag{33}$$

and that u_h converges to a function u in $L^1_{loc}(\mathbb{R}_+ \times \mathbb{R})$, as h tends to 0. Moreover, we assume either that, for all bounded intervals I of $\mathbb{R}$,

$$\sum_{i\in K} \Delta x_{i+\frac{1}{2}} |u_{i+1}(t) - u_i(t)| \overset{h\to 0}{\longrightarrow} 0, \quad \text{in } L^1_{loc}(\mathbb{R}_+), \tag{34}$$

where K denotes the set of indices such that $x_i \in I$; or a geometrical constraint on the spatial mesh, that is

$$\exists\ \alpha, \beta > 0 \ \text{ so that } \ \alpha\Delta x_{i+1} \leq \Delta x_i \leq \beta\Delta x_{i+1}, \ \ \forall i \in \mathbb{Z}. \tag{35}$$

Then u is a weak solution to the initial value problem (1)-(2), i.e.

$$\frac{\partial u}{\partial t} + \frac{\partial A(u)}{\partial x} + z'(x)b(u) = 0, \quad u(0,x) = u^0(x), \quad \text{in } \mathcal{D}'(\mathbb{R}_+ \times \mathbb{R}).$$

Proof. The proof is an adaptation of the classical Lax-Wendroff theorem [25] for homogeneous systems of conservation laws.
Let $\varphi \in \mathcal{C}_0^1(\mathbb{R}_+ \times \mathbb{R})$ be a test function and set

$$\varphi_i(t) = \varphi(t, x_i), \quad i \in \mathbb{Z}. \tag{36}$$

After multiplying equation (7) by φ_i, we sum over i and integrate in dt, to obtain

$$\int_{\mathbb{R}_+} \sum_{i\in\mathbb{Z}} \Delta x_i \frac{du_i}{dt} \varphi_i dt + \int_{\mathbb{R}_+} \sum_{i\in\mathbb{Z}} \left(\mathcal{A}_{i+\frac{1}{2}} - \mathcal{A}_{i-\frac{1}{2}}\right) \varphi_i dt$$
$$+ \int_{\mathbb{R}_+} \sum_{i\in\mathbb{Z}} \left(\mathcal{B}^+_{i-\frac{1}{2}} + \mathcal{B}^-_{i+\frac{1}{2}}\right) \varphi_i dt = 0.$$

An integration by parts in the first term and a summation by parts in the other ones give

$$\int_{\mathbb{R}_+} \sum_{i\in\mathbb{Z}} \Delta x_i u_i \frac{d\varphi_i}{dt} dt + \int_{\mathbb{R}_+} \sum_{i\in\mathbb{Z}} \mathcal{A}_{i+\frac{1}{2}} (\varphi_{i+1} - \varphi_i) dt$$
$$- \int_{\mathbb{R}_+} \sum_{i\in\mathbb{Z}} \left(\mathcal{B}^+_{i+\frac{1}{2}} \varphi_{i+1} + \mathcal{B}^-_{i+\frac{1}{2}} \varphi_i\right) dt + \sum_{i\in\mathbb{Z}} \Delta x_i u_i^0 \varphi_i(0) = 0. \tag{37}$$

We define a.e. in $\mathbb{R}_+ \times \mathbb{R}$ the piecewise constant functions $\mathcal{A}_h$ and $\mathcal{B}_h$, associated with the numerical flux and source term by

$$\mathcal{A}_h(t,x) = \mathcal{A}(u_i, u_{i+1}), \tag{38}$$

$$\mathcal{B}_h(t,x) = \frac{1}{\Delta x_{i+\frac{1}{2}}} \left(\mathcal{B}^+(u_i, u_{i+1}, z_{i+1} - z_i) + \mathcal{B}^-(u_i, u_{i+1}, z_{i+1} - z_i)\right), \tag{39}$$

for $t \in \mathbb{R}_+$ and $x \in [x_i, x_{i+1})$, recalling that $\Delta x_{i+\frac{1}{2}} = |x_{i+1} - x_i| = \frac{\Delta x_i}{2} + \frac{\Delta x_{i+1}}{2}$.
Next, according to (36), we introduce the piecewise constant approximation of the test function

$$\varphi_h(t,x) = \varphi_i(t), \quad t \in \mathbb{R}_+, \ x \in \mathcal{C}_i,$$

which converges to φ (together with $\frac{\partial \varphi_h}{\partial t}$ towards $\frac{\partial \varphi}{\partial t}$) uniformly in $\mathcal{C}_0(\mathbb{R}_+ \times \mathbb{R})$, as h tends to 0. We also consider a continuous piecewise linear function ψ_h such that

$$\psi_h(t, x_i) = \varphi_i(t), \quad i \in \mathbb{Z},$$
$$\frac{\partial \psi_h}{\partial x}(t,x) = \frac{\varphi_{i+1}(t) - \varphi_i(t)}{\Delta x_{i+\frac{1}{2}}}, \quad t \in \mathbb{R}_+, \ x \in [x_i, x_{i+1}),$$

so that ψ_h and $\frac{\partial \psi_h}{\partial x}$ converge respectively to φ and $\frac{\partial \varphi}{\partial x}$ in $\mathcal{C}_0^1(\mathbb{R}_+ \times \mathbb{R})$, as h tends to 0. As a direct consequence of these definitions, we have

$$\varphi_j = \frac{\varphi_i + \varphi_{i+1}}{2} + O(h), \quad j = i, i+1,$$

$$\int_{x_i}^{x_{i+1}} \psi_h(t,x)dx = \frac{\varphi_i + \varphi_{i+1}}{2} \, \Delta x_{i+\frac{1}{2}}.$$

Taking into account all the relations stated above, we can put the discrete sum (37) into the integral form

$$\begin{aligned} &\int_{\mathbb{R}_+}\int_{\mathbb{R}} u_h(t,x)\frac{\partial \varphi_h}{\partial t}(t,x)dxdt + \int_{\mathbb{R}_+}\int_{\mathbb{R}} A_h(t,x)\frac{\partial \psi_h}{\partial x}(t,x)dxdt \\ &- \int_{\mathbb{R}_+}\int_{\mathbb{R}} \mathcal{B}_h(t,x)\left(\psi_h(t,x) + O(h)\right)dxdt + \int_{\mathbb{R}} u_h^0(x)\varphi_h(0,x)dx = 0. \end{aligned} \tag{40}$$

As h tends to 0, passing to the limit in (40), we claim that it turns out

$$\begin{aligned} &\int_{\mathbb{R}_+}\int_{\mathbb{R}} \left[u(t,x)\frac{\partial \varphi}{\partial t}(t,x) + A(t,x)\frac{\partial \varphi}{\partial x}(t,x) - B(t,x)\varphi(t,x)\right] dxdt \\ &\qquad + \int_{\mathbb{R}} u^0(x)\varphi(0,x)dx = 0. \end{aligned} \tag{41}$$

The computation is obvious for the first and the last terms, by exploiting the convergence properties of approximations u_h and φ_h.
For the other integrals of (40), the process is less straightforward: as we remarked in Section 4, due to the presence of a variable space-step, the interfacial interval $[x_i, x_{i+1})$ could be really different from the mesh cell C_i (where the conservative unknowns are discretized); so, standard techniques do not work in this case and proving convergence requires the additional hypotheses on the structures that we have imposed.

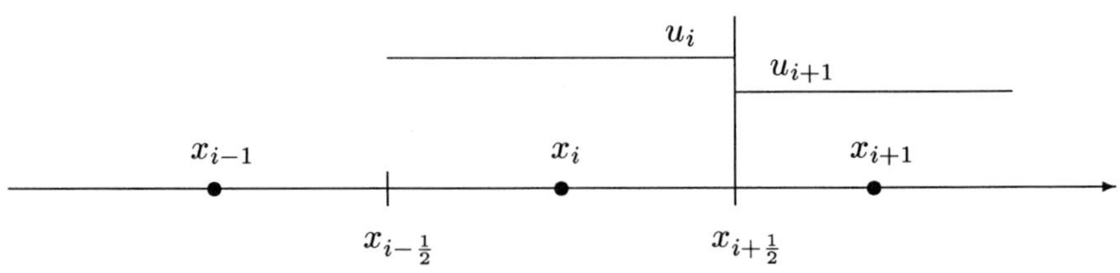

We need to characterize the functions $A(t,x)$ and $B(t,x)$ in equation (41) as the weak limits of $A_h(t,x)$ and $\mathcal{B}_h(t,x)$, for reproducing them in terms of the numerical unknowns.
We first observe that, thanks to (33) and the properties (8)-(9) of the numerical flux, A_h is locally bounded on $\mathbb{R}_+ \times \mathbb{R}$ (uniformly in h). Coming back to discrete notation (38), we decompose on the subintervals $[x_i, x_{i+\frac{1}{2}})$ and $[x_{i+\frac{1}{2}}, x_{i+1})$, rearranging terms as follows,

$$\begin{aligned} A_h(t,x) &= \sum_{i\in\mathbb{Z}} A(u_i) 1\!\!1_{[x_{i-\frac12},x_{i+\frac12})}(x) \\ &+ \sum_{i\in\mathbb{Z}} [\mathcal{A}(u_i,u_{i+1}) - A(u_i)] 1\!\!1_{[x_i,x_{i+\frac12})}(x) \\ &+ \sum_{i\in\mathbb{Z}} [\mathcal{A}(u_{i-1},u_i) - A(u_i)] 1\!\!1_{[x_{i-\frac12},x_i)}(x) \\ &= A(u_h(t,x)) + \mathcal{R}_h^+(t,x) + \mathcal{R}_h^-(t,x). \end{aligned} \tag{42}$$

We conclude that $A(u(t,x))$ is the expected value for the limit $A(t,x)$, as h tends to 0, provided that the two remainders in (42) vanish. We only treat with $\mathcal{R}_h^+$, the other one results in the same way. According to definition (32), since u_h converges to a function u in $L^1_{loc}(\mathbb{R}_+ \times \mathbb{R})$, we also derive

$$\sum_{i\in\mathbb{Z}} |u_i(t) - \bar{u}_i(t)| 1\!\!1_{C_i}(x) \underset{L^1_{loc}(\mathbb{R}_+\times\mathbb{R})}{\longrightarrow} 0, \tag{43}$$

where the sequence $\{\bar{u}_i(t)\}_{i\in\mathbb{Z}}$ is defined by the cell-averages of u on the discretization mesh. Let I be any bounded interval of $\mathbb{R}$ and we denote by $C_{\mathcal{A}}$ a Lipschitz constant for $\mathcal{A}$, then the assumptions on the numerical flux lead to estimate (with the notation set out above in the theorem)

$$\begin{aligned} \int_I \left|\mathcal{R}_h^+(t,x)\right| dx &\le \int_I \sum_{i\in\mathbb{Z}} |\mathcal{A}(u_i,u_{i+1}) - A(u_i)| 1\!\!1_{[x_i,x_{i+\frac12})}(x)\, dx \\ &\le C_{\mathcal{A}} \sum_{i\in K} \frac{\Delta x_i}{2} |u_{i+1} - u_i|. \end{aligned} \tag{44}$$

Under the hypothesis (34), this last term vanishes and the conclusion is done. Otherwise, we can further manipulate the previous bound by introducing appropriate quantities, that is

$$\begin{aligned} \sum_{i\in K} \Delta x_i |u_{i+1} - u_i| &\le \sum_{i\in K} \Delta x_i |u_{i+1} - \bar{u}_{i+1}| \\ &+ \sum_{i\in K} \Delta x_i |\bar{u}_{i+1} - \bar{u}_i| + \sum_{i\in K} \Delta x_i |\bar{u}_i - u_i|. \end{aligned} \tag{45}$$

In particular, for the alternative hypothesis of nondegeneracy (35) made on the mesh, we have

$$\sum_{i\in K} \Delta x_i |u_{i+1} - \bar{u}_{i+1}| \le \beta \sum_{i\in K} \Delta x_{i+1} |u_{i+1} - \bar{u}_{i+1}|,$$

so that property (43) ensures that first and last sum in the right-hand side of (45) vanish, as h tends to 0. It remains the second term to be studied, involving only the function u, which immediately converges to 0 if $u \in BV$ (or smooth enough); this result also holds for $u \in L^1$, by applying standard

regularization arguments (we define $u^\epsilon \in BV$, $u^\epsilon \to u$ in L^1, then we perform an estimation on the cell-averages $\bar{u}_i$ and $\bar{u}_i^\epsilon$ like in (45) and we finally conclude by combining convergence properties).
We now pass to the source term, to which a similar procedure applies. Taking into account the definition (39), setting $\mathcal{B} = \mathcal{B}^+ + \mathcal{B}^-$, we can write

$$\mathcal{B}_h(t,x) = \sum_{i\in\mathbb{Z}} \frac{\mathcal{B}(u_i, u_{i+1}, z_{i+1}-z_i)}{z_{i+1}-z_i} \cdot \frac{z_{i+1}-z_i}{\Delta x_{i+\frac{1}{2}}} \mathbb{1}_{[x_i, x_{i+1})}(x).$$

Notice that the hypothesis $z \in W^{1,1}$ guarantees that discrete differences converge to the derivative $z' \in L^1$; together with condition (28), this leads to justify the assertion that $\mathcal{B}_h(t,x)$ is $L^1 - weak$ bounded in $\mathbb{R}_+ \times \mathbb{R}$. We proceed as in (42), by performing the following decomposition

$$\begin{aligned} \mathcal{B}_h(t,x) &= \sum_{i\in\mathbb{Z}} \frac{\mathcal{B}(u_i, u_i, z_{i+1}-z_i)}{z_{i+1}-z_i} \cdot \frac{z_{i+1}-z_i}{\Delta x_{i+\frac{1}{2}}} \mathbb{1}_{[x_i, x_{i+\frac{1}{2}})}(x) + \mathcal{E}_h^+(t,x) \\ &+ \sum_{i\in\mathbb{Z}} \frac{\mathcal{B}(u_i, u_i, z_i - z_{i-1})}{z_i - z_{i-1}} \cdot \frac{z_i - z_{i-1}}{\Delta x_{i-\frac{1}{2}}} \mathbb{1}_{[x_{i-\frac{1}{2}}, x_i)}(x) + \mathcal{E}_h^-(t,x). \end{aligned} \tag{46}$$

The sum of the two principal terms of (46) converges to $z'(x)b(u)$, in view of the arguments just mentioned and the strong convergence of u_h towards u, by means of Lebesgue's theorem. For the remainders, we give details in the case of $\mathcal{E}_h^+(t,x)$, for instance. We then have

$$\begin{aligned} \mathcal{E}_h^+(t,x) = \sum_{i\in\mathbb{Z}} \frac{1}{\Delta x_{i+\frac{1}{2}}} &(\mathcal{B}(u_i, u_{i+1}, z_{i+1}-z_i) \\ &- \mathcal{B}(u_i, u_i, z_{i+1}-z_i)) \mathbb{1}_{[x_i, x_{i+\frac{1}{2}})}(x). \end{aligned} \tag{47}$$

The analogue of estimation (44) for (47) becomes

$$\begin{aligned} \int_I |\mathcal{E}_h^+(t,x)| dx &\le \sum_{i\in K} |\mathcal{B}(u_i, u_{i+1}, z_{i+1}-z_i) - \mathcal{B}(u_i, u_i, z_{i+1}-z_i)| \\ &= \sum_{i\in K} \left| \int_{u_i}^{u_{i+1}} \frac{\partial\mathcal{B}}{\partial v}(u_i, v, z_{i+1}-z_i) dv \right| \\ &= \sum_{i\in K} \left| \int_{u_i}^{u_{i+1}} \left(\frac{\partial\mathcal{B}}{\partial v}(u_i, v, z_{i+1}-z_i) - \frac{\partial\mathcal{B}}{\partial v}(u_i, v, 0) \right) dv \right| \\ &\le C_{\mathcal{B}} \sum_{i\in K} |u_{i+1}-u_i| |z_{i+1}-z_i|, \end{aligned}$$

where we exploited property (11) and the regularity assumed in (10). Then we can regularize the function z by introducing $z^\epsilon \in W^{1,\infty}$, $z^\epsilon \to z$ in $W^{1,1}$, for which we write

$$\begin{aligned}\sum_{i\in K}|u_{i+1}-u_i||z_{i+1}-z_i| &\le \sum_{i\in K}|u_{i+1}-u_i||z_{i+1}-z_{i+1}^{\epsilon}| \\ &\quad + \sum_{i\in K}|u_{i+1}-u_i||z_i^{\epsilon}-z_i| + \sum_{i\in K}|u_{i+1}-u_i||z_{i+1}^{\epsilon}-z_i^{\epsilon}| \\ &\le 4C\sum_{i\in K}|z_i^{\epsilon}-z_i| + C_\epsilon \sum_{i\in K}\Delta x_{i+\frac{1}{2}}|u_{i+1}-u_i|,\end{aligned} \tag{48}$$

with C defined as in (33) and C_ϵ only depending on the regularization. The same procedure we have considered before allow us to conclude that the upper bound (48) vanishes, as h tends to 0.
Thanks to the previous computations, we have proved that the limit function u satisfies the weak formulation of the Cauchy problem (1)-(2),

$$\int_{\mathbb{R}_+}\int_{\mathbb{R}}\left[u(t,x)\frac{\partial\varphi}{\partial t}(t,x) + A(u(t,x))\frac{\partial\varphi}{\partial x}(t,x) - z'(x)b(u(t,x))\varphi(t,x)\right]dx\,dt +$$

$$+\int_{\mathbb{R}} u^0(x)\varphi(0,x)dx = 0,$$

so that the proof of the theorem is completed.

Remark. *For the particular case of uniform spatial mesh, i.e. $\Delta x_i = h$, $\forall i \in \mathbb{Z}$, sets of weaker hypotheses can be considered and the proof is simplified, due to the fact that the consistency error vanishes for the source term (refer to [9] and [26]). The version presented above is compatible with those general statements (see also [10]) and actually extends the classical Lax-Wendroff theorem to scalar conservation laws with a source term.*

6 Conclusion

We have proposed a consistency condition for hyperbolic conservation laws with a source term $z'(x)b(u)$, discretized according to the Upwind Interface Source method. We have proved that numerical schemes which preserve the steady state solutions are consistent in that sense. Moreover, a strong limit of discrete solutions satisfies the continuous equation, as the mesh is refined. Theorem 1 thus constitutes a fundamental result in the theoretical analysis of the numerical method.

Nevertheless, the conditions established in the previous sections do not guarantee that discrete approximations given by (7)-(11) do converge and we do not know whether the limit weak solution is the unique physical solution of the Cauchy problem (1)-(4). For that we need to precise some criteria of stability for the approximate solution and we have to consider further assumptions on the discrete functions to derive suitable error estimates.

In order to ensure that a weak solution obtained as limit of (32) satisfies the family of entropy inequalities (5), it suffices to show that a discrete entropy inequality,

$$\Delta x_i \frac{d}{dt} S(u_i) + \left(\eta^S_{i+\frac{1}{2}} - \eta^S_{i-\frac{1}{2}}\right) + \mathcal{B}^{S,+}_{i-\frac{1}{2}} + \mathcal{B}^{S,-}_{i+\frac{1}{2}} \leq 0, \tag{49}$$

holds for the numerical scheme, with the usual formalism

$$\eta^S_{i+\frac{1}{2}} = \eta^S(u_i, u_{i+1}), \quad \mathcal{B}^{S,\pm}_{i+\frac{1}{2}} = \mathcal{B}^{S,\pm}(u_i, u_{i+1}, z_{i+1} - z_i), \tag{50}$$

where η^S and $\mathcal{B}^{S,\pm}$ are some numerical entropy flux function and source term, which must be consistent with $\eta(u)$ and $S'(u)B(x,u)$ in the same way that we required $\mathcal{A}$ and $\mathcal{B}^{\pm}$ to be consistent with $A(u)$ and $B(x,u)$ in (1). Therefore, mimicking the proof of the Lax-Wendroff theorem, we can prove that the weak form of the entropy inequality (5) is also verified.
Note that the possibility to write formulas (49)-(50) relies only on the characterization of the numerical flux (namely, at least the condition of E-scheme has to be assumed); for the source term, the definition is automatically made, since we have

$$\mathcal{B}^{S,+}_{i-\frac{1}{2}} = S'(u_i)\mathcal{B}^{+}_{i-\frac{1}{2}}, \qquad \mathcal{B}^{S,-}_{i+\frac{1}{2}} = S'(u_i)\mathcal{B}^{-}_{i+\frac{1}{2}}.$$

The question to determine the order of accuracy of the numerical scheme by means of error estimates is more delicate than for the homogeneous system, due to the presence of source terms. The approach formulated by Kruzkov is used henceforth in the literature (see [11],[12] and [19],[20] for instance), providing a method to convert any discrete entropy inequality into an error estimate. The general procedure consists in the following formulation for the approximate solution, in $\mathcal{D}'(\mathbb{R}_+ \times \mathbb{R})$,

$$\frac{\partial S(u_h)}{\partial t} + \frac{\partial}{\partial x}\eta^S(u_h) + S'(u_h)z'(x)b(u_h) \leq \frac{\partial}{\partial x}\mathrm{Err}_2(t,x) + \mathrm{Err}_1(t,x),$$

where we set

$$\frac{\partial}{\partial x}\mathrm{Err}_1(t,x) = \frac{\partial}{\partial x}\eta^S(u_h) - \sum_{i\in\mathbb{Z}} \frac{1}{\Delta x_i}\left(\eta^S_{i+\frac{1}{2}} - \eta^S_{i-\frac{1}{2}}\right)\mathbb{1}_{C_i}(x),$$

$$\mathrm{Err}_2(t,x) = S'(u_h)z'(x)b(u_h) - \sum_{i\in\mathbb{Z}} \frac{1}{\Delta x_i}\left(\mathcal{B}^{S,+}_{i-\frac{1}{2}} + \mathcal{B}^{S,-}_{i+\frac{1}{2}}\right)\mathbb{1}_{C_i}(x).$$

Then the results of [3] apply to this particular problem, to deduce stability bounds and conclude the convergence properties we have assumed in Theorem 1. We remark that regularity hypotheses like (34) and (35) are necessary to control the variations of the numerical solution in comparison with the space-step (refer to [30] for the homogeneous problem).

In order to avoid BV estimates, which are not available in case of insufficiently smooth source terms and for multidimensional systems on an unstructured mesh, arguments based on the measure-valued method and the so-called weak BV estimates are developed (see [7] and [4],[5], for instance) or the kinetic approach presented in [2].

References

1. Bermudez A., Vasquez M.E., Upwind methods for hyperbolic conservation laws with source terms, *Comput. & Fluids*, **23** (1994), no. 8, 1049-1071
2. Botchorishvili R., Perthame B., Vasseur A., Equilibrium Schemes for Scalar Conservation Laws with Stiff Sources, *Math. Comp.*, to appear
3. Bouchut F., Perthame B., Kružkov's estimates for scalar conservation laws revisited, *Trans. Amer. Math. Soc.*, **350** (1998), no. 7, 2847-2870
4. Cockburn B., Coquel F., LeFloch P., An error estimate for finite volume methods for multidimensional conservation laws, *Math. of Comp.*, **63** (1994), 77-103
5. Cockburn B., Coquel F., LeFloch P., Convergence of the finite volume method for multidimensional conservation laws, *SIAM J. Numer. Anal.*, **32** (1995), no. 3, 687-705
6. Eymard R., Gallouët T., Herbin R., *Finite Volume Methods*, Handbook of numerical analysis, vol. VIII, P.G.Ciarlet and J.L.Lions editors, Amsterdam, North-Holland, 2000
7. Eymard R., Gallouët T., Ghilani M., Herbin R., Error estimates for the approximate solutions of a nonlinear hyperbolic equation given by some finite volume schemes, *I.M.A. Journal of Numer. Anal.*, **18** (1998), 563-594
8. Gallouët T., Hérard J.M., Seguin N., Some approximate Godunov schemes to compute shallow-water equations with topography, *AIAA-2001* (2000)
9. Godlewski E., Raviart P.A., *Hyperbolic systems of conservation laws*, Mathématiques & Applications, n. 3/4, Paris, Ellipses, 1991
10. Godlewski E., Raviart P.A., *Numerical approximation of hyperbolic systems of conservation laws*, Applied Mathematical Sciences **118**, New York, Springer-Verlag, 1996
11. Gosse L., LeRoux A.Y., A well-balanced scheme designed for inhomogeneous scalar conservation laws, *C.R. Acad. Sci. Paris Sér.I Math.*, **323** (1996), no. 5, 543-546
12. Gosse L., A priori error estimate for a well-balanced scheme designed for inhomogeneous scalar conservation laws, *C.R. Acad. Sci. Paris Sér.I Math.*, **327** (1998), no. 5, 467-472
13. Gosse L., A well-balanced flux-vector splitting scheme designed for hyperbolic systems of conservation laws with source terms, *Comput. Math. Appl.*, **39** (2000), no. 9-10, 135-159
14. Gosse L., A well-balanced scheme using non-conservative products designed for hyperbolic systems of conservation laws with source terms, *Math. Models Methods Appl. Sci.*, **11** (2001), no. 2, 339-365
15. Greenberg J.M., LeRoux A.Y., A well balanced scheme for the numerical processing of source terms in hyperbolic equations, *SIAM J. Numer. Anal.*, **33** (1996), 1-16

16. Greenberg J.M., LeRoux A.Y., Baraille R., Noussair A., Analysis and approximation of conservation laws with source term, *SIAM J. Numer. Anal.*, **34** (1997), no. 5, 1980-2007
17. Jin S., A steady-state capturing method for hyperbolic systems with geometrical source terms, *M2AN Math. Model. Numer. Anal.*, **35** (2001), no. 4, 631-645
18. Karni S., personal communications
19. Katsaounis T., Makridakis C., Finite volume relaxation schemes for multidimensional conservation laws, *Math. Comp.*, **70** (2001), n. 234, 533-553
20. Katsoulakis M.A., Kossioris G., Makridakis C., Convergence and error estimates of relaxation schemes for multidimensional conservation laws, *Comm. Partial Differential Equations*, **24**(1999), n. 3-4, 395-424
21. Kružkov S.N., First order quasilinear equations in several independent space variables, *Math. USSR Sb.*, **10** (1970), 217-243
22. Kurganov A., Central-upwind schemes for balance laws. Application to the Broadwell model, *Proceedings of the Third International Symposium on Finite Volumes for Complex Applications* (2002), to appear
23. Kurganov A., Levy D., Central-Upwind Schemes for the Saint-Venant system, *M2AN Math. Model. Numer. Anal.* (2001), to appear
24. Lax P.D., Shock waves and entropy, in *Contributions to nonlinear functional analysis*, E.H.Zarantonello editor, New York, Academic Press, 1971, 603-634
25. Lax P.D., Wendroff B., Systems of conservations laws, *Comm. Pure Appl. Math.*, **13** (1960), 217-237
26. LeVêque R.J., *Numerical Methods for Conservation Laws*, Lectures in Mathematics, ETH Zurich, Birkhauser, 1992
27. LeVêque R.J., Balancing source terms and flux gradients in high-resolution Godunov methods: the quasi-steady wave-propagation algorithm, *J. Comput. Phys.*, **146** (1998), no. 1, 346-365
28. Perthame B., Simeoni C., A kinetic scheme for the Saint-Venant system with a source term, *Calcolo*, **38** (2001), no. 4, 201-231
29. Roe P.L., Upwind differencing schemes for hyperbolic conservation laws with source terms, in *Nonlinear Hyperbolic Problems*, C.Carasso, P.A.Raviart and D.Serre editors, Lecture Notes in Math., vol. 1270, Berlin, Springer-Verlag, 1987, pp. 41-51
30. Sanders R., On Convergence of Monotone Finite Difference Schemes with Variable Spatial Differencing, *Math. Comp.*, **40** (1983), 91-106
31. Vasseur A., Well-posedness of scalar conservation laws with singular sources, in preparation

An Overview on High Order Numerical Methods for Convection Dominated PDEs

Chi-Wang Shu

Division of Applied Mathematics, Brown University, Providence, RI 02912,
shu@cfm.brown.edu

Summary. In this presentation we give an overview of some recent progress in high order finite difference and finite volume weighted essentially non-oscillatory (WENO) and finite element discontinuous Galerkin (DG) methods for solving hyperbolic conservation laws and in general convection dominated PDEs, such as convection diffusion equations with small diffusion, and KdV type equations with small dispersion.

1 Introduction

In this presentation, we are concerned with the numerical solution of the so-called convection dominated PDEs, i.e. PDEs of the form

$$u_t + f(u)_x + g(u)_y = \alpha\, h \tag{1}$$

where we have written it in two spatial dimensions as an example, h may contain second and higher derivatives of u, but α is either 0 or very small. The difficulty of the numerical solution for such problems is due to the fact that the solution may contain discontinuities (when $\alpha = 0$) or sharp gradient regions. Traditional numerical methods are either only first order accurate or oscillatory in the presence of such discontinuities or sharp gradient regions.

We present an an overview of some recent progress in high order finite difference and finite volume weighted essentially non-oscillatory (WENO) and finite element discontinuous Galerkin (DG) methods for solving equation (1). We refer the readers to [23] for a more comprehensive review of these three methods when applied to computational fluid dynamics (CFD).

For all the methods discussed in this paper, time discretization is achieved by the TVD high order Runge-Kutta method first developed in [24]. A distinctive feature of this class of time discretizations is that they are convex combinations of first order forward Euler steps, hence they maintain strong stability properties in any semi-norm (total variation norm, maximum norm, entropy condition, etc.) of the forward Euler step. Thus one only needs to prove nonlinear stability for the first order forward Euler step, which is relatively easy in many situations (e.g. TVD schemes), and one automatically obtains the same strong stability property for the higher order time discretizations in this class. Typically, the third method in [24] is used, which, when used to solve an ODE

$$u_t = L(u(t), t),$$

has the following particularly simple form:

$$\begin{aligned} u^{(1)} &= u^n + \Delta t L(u^n, t^n) \\ u^{(2)} &= \frac{3}{4}u^n + \frac{1}{4}u^{(1)} + \frac{1}{4}\Delta t L(u^{(1)}, t^n + \Delta t) \\ u^{n+1} &= \frac{1}{3}u^n + \frac{2}{3}u^{(2)} + \frac{2}{3}\Delta t L(u^{(2)}, t^n + \frac{1}{2}\Delta t). \end{aligned}$$

In section 2 we will review the finite difference WENO schemes, and briefly discuss some recent work on the resolution study of using high order methods, and on multi-domain methods suitable for more general geometry and meshes. In section 3 we will review the finite volume WENO schemes, and briefly discuss a technique to treat negative linear weights in WENO schemes which arise in multi-dimensional finite volume cases, especially for those on unstructured meshes. In section 4 we will review the discontinuous Galerkin methods, and briefly discuss a class of stable and accurate local discontinuous Galerkin method for (1) with higher derivatives in h, and a post-processing technique which can effectively double the order of accuracy for the method on locally uniform meshes, with small extra computational cost.

2 WENO finite difference methods

The finite difference WENO methods are very simple to formulate and program and are relatively inexpensive in CPU costs, especially for multi-dimensional systems. The method can be formulated for any order of accuracy. The third order version is obtained in [13]. The fifth order version with a general framework for the smoothness indicator, which is the key ingredient of a WENO scheme, is given in [11]. Higher order versions are developed in [2]. The two lecture notes [20] and [21] contains many details of such schemes.

We will use the fifth order version for (1) in one space dimensional scalar case with $\alpha = 0$ to demonstrate the idea. We further assume $f'(u) \geq 0$, i.e. the wind direction is positive. The case of negative wind direction $f'(u) \leq 0$ can be handled by mirror symmetry with respect to the point $x_{j+1/2}$ where the numerical flux $\hat{f}_{j+1/2}$ is evaluated. More general cases can be handled by a flux splitting. The computational domain is discretized into a uniform mesh of N grid points:

$$x_j = j\Delta x; \qquad i = 1, 2, ..., N,$$

where Δx is the uniform mesh size. A smooth nonuniform mesh could also be used to concentrate grid points near certain regions to obtain better resolution, but arbitrary unstructured meshes cannot be used in this finite difference framework if the scheme is third order or higher in accuracy, and one must resort to the more complex and more expensive finite volume framework to be discussed in next section. A conservative numerical approximation $u_j(t)$ to the exact solution $u(x_j, t)$ of (1) in 1D with $\alpha = 0$ satisfies the following ordinary differential equation (ODE) system:

$$\frac{du_j(t)}{dt} + \frac{1}{\Delta x}\left(\hat{f}_{j+1/2} - \hat{f}_{j-1/2}\right) = 0 \tag{2}$$

where $\hat{f}_{j+1/2}$ is called the numerical flux, whose design is the key ingredient for a successful scheme. For the fifth order WENO scheme, the numerical flux $\hat{f}_{j+1/2}$ is defined as follows:

$$\hat{f}_{j+1/2} = \omega_1 \hat{f}^{(1)}_{j+1/2} + \omega_2 \hat{f}^{(2)}_{j+1/2} + \omega_3 \hat{f}^{(3)}_{j+1/2}$$

where $\hat{f}^{(m)}_{j+1/2}$ are the three third order fluxes on three different stencils given by

$$\begin{aligned}
\hat{f}^{(1)}_{j+1/2} &= \frac{1}{3} f_{j-2} - \frac{7}{6} f_{j-1} + \frac{11}{6} f_j, \\
\hat{f}^{(2)}_{j+1/2} &= -\frac{1}{6} f_{j-1} + \frac{5}{6} f_j + \frac{1}{3} f_{j+1}, \\
\hat{f}^{(3)}_{j+1/2} &= \frac{1}{3} f_j + \frac{5}{6} f_{j+1} - \frac{1}{6} f_{j+2},
\end{aligned}$$

and the nonlinear weights ω_m are given by

$$\omega_m = \frac{\tilde{\omega}_m}{\sum_{l=1}^{3} \tilde{\omega}_l}, \qquad \tilde{\omega}_l = \frac{\gamma_l}{(\varepsilon + \beta_l)^2},$$

with the linear weights γ_l given by

$$\gamma_1 = \frac{1}{10}, \qquad \gamma_2 = \frac{3}{5}, \qquad \gamma_3 = \frac{3}{10},$$

and the smoothness indicators β_l given by

$$\begin{aligned}
\beta_1 &= \frac{13}{12}\left(f_{j-2} - 2f_{j-1} + f_j\right)^2 + \frac{1}{4}\left(f_{j-2} - 4f_{j-1} + 3f_j\right)^2 \\
\beta_2 &= \frac{13}{12}\left(f_{j-1} - 2f_j + f_{j+1}\right)^2 + \frac{1}{4}\left(f_{j-1} - f_{j+1}\right)^2 \\
\beta_3 &= \frac{13}{12}\left(f_j - 2f_{j+1} + f_{j+2}\right)^2 + \frac{1}{4}\left(3f_j - 4f_{j+1} + f_{j+2}\right)^2 .
\end{aligned}$$

Finally, ε is a parameter to avoid the denominator to become 0 and is typically fixed at $\varepsilon = 10^{-6}$. The choice of ε does not affect accuracy: the numerical errors can go down to much lower than ε, reaching machine zeros (around 10^{-13} for double precision). Notice that we have used the short notation f_j to denote $f(u_j(t))$. We remark that the stencil for the scheme is biased to the left because of the positive wind direction.

For systems of hyperbolic conservation laws, the nonlinear part of the WENO procedure (i.e. the determination of the smoothness indicators β_k and hence the nonlinear weights w_i) should be carried out in local characteristic fields. Thus one would first find an average $u_{j+1/2}$ of u_j and u_{j+1} (e.g. the Roe average [15] which exists for many physical systems), and compute the left and right eigenvectors of the Jacobian $f'(u_{j+1/2})$ and put them into the rows of $R^{-1}_{j+1/2}$ and the columns of $R_{j+1/2}$, respectively, such that $R^{-1}_{j+1/2} f'(u_{j+1/2}) R_{j+1/2} = \Lambda_{j+1/2}$ where $\Lambda_{j+1/2}$ is a diagonal matrix containing the real eigenvalues of $f'(u_{j+1/2})$. One then transforms all the quantities needed for evaluating the numerical flux $\hat{f}_{j+1/2}$ to the local characteristic fields by left multiplying them with $R^{-1}_{j+1/2}$, and then computes the numerical fluxes by the scalar procedure in each characteristic field. Finally, the flux in the original physical space is obtained by left multiplying the numerical flux obtained in the local characteristic fields with $R_{j+1/2}$. This procedure is costly, but is often necessary to avoid spurious oscillations, especially when the order of the scheme is very high. See, e.g., [14].

With this elementary introduction to illustrate the basic idea of finite difference WENO schemes, we now turn our attention to a brief discussion of two recent developments of the finite difference WENO methods.

The first is a numerical study on the resolution of high order finite difference WENO schemes by Shi, Zhang and Shu [18]. The issue addressed in [18] is the numerical resolution and efficiency of high order WENO finite difference schemes for computing solutions containing both discontinuities and complex solution features. This study was performed through two representative numerical examples: the double Mach reflection problem and the Rayleigh-Taylor instability problem. The following conclusions are obtained in [18]: (i) The ninth order WENO method needs only about half the number of mesh points in each direction compared with the fifth order WENO method to achieve a comparable numerical resolution and hence to have the same small numerical viscosity, for such problems containing both discontinuities and complex solution structures. (ii) This translates to a factor of 66% to 80% reduction in CPU time in our implementation when a ninth order WENO method is used instead of a fifth order version, to achieve comparable resolution. Our implementation of both the fifth order and the ninth order WENO schemes are on parallel computers. (iii) The parallel efficiency of the WENO method can reach over 90% when the operation per processor is kept constant, i.e. when the number of mesh points increases together with the number of processors used.

The second is a multi-domain finite difference WENO method, developed by Sebastian and Shu [16]. As mentioned above, a high order WENO finite difference method can only be applied on a uniform rectangular or smooth curvilinear mesh, thus restricting the application of such methods for complex geometry or local adaptation. The multi-domain finite difference WENO method developed in [16] generalizes the flexibility of the method for more complex geometry and meshes. The method uses an interpolation procedure at the sub-domain interfaces. A simple Lagrange interpolation procedure is implemented and compared to a WENO interpolation procedure. Extensive numerical examples are shown to indicate the effectiveness of each procedure, including the measurement of conservation errors, orders of accuracy, essentially non-oscillatory properties at the domain interfaces, and robustness for problems containing strong shocks and complex geometry. Our numerical experiments have shown that the simple and efficient Lagrange interpolation suffices for the sub-domain interface treatment in the multi-domain WENO finite difference method, to retain essential conservation, full high order of accuracy, essentially non-oscillatory properties at the domain interfaces even for strong shocks, and robustness for problems containing strong shocks and complex geometry. The method developed in [16] can be used to solve problems in relatively complex geometry at a much smaller CPU cost than the finite volume version of the same method for the same accuracy.

3 WENO finite volume methods

If the geometry or adaptive mesh requirements are complicated, one cannot use the finite difference methods discussed in the previous section and would have to resort to finite volume methods. A finite volume scheme for a conservation law such as (1) (with $\alpha = 0$) approximates an integral version of it. Thus, the computational domain is partitioned into a collection of cells $\triangle_i$, which in 2D could be rectangles, triangles, etc., and the cell averages of the solution u

$$\bar{u}_i(t) \equiv \frac{1}{|\triangle_i|} \int_{\triangle_i} u(x, y, t)\,dxdy$$

are the numerically approximated quantities. If $\triangle_j$ is a control volume, the semi-discrete finite volume scheme of equation (1) is:

$$\frac{d}{dt}\bar{u}_j(t) + \frac{1}{|\triangle_j|} \int_{\partial\triangle_j} F \cdot n\,ds = 0 \tag{3}$$

where $F = (f, g)$, and n is the outward unit normal of the cell boundary $\partial\triangle_j$. The line integral in (3) is typically discretized by a Gaussian quadrature of sufficiently high order of accuracy,

$$\int_{\partial\triangle_j} F \cdot n\,ds \approx |\partial\triangle_j| \sum_{k=1}^{q} \omega_k F(u(G_k, t)) \cdot n$$

and $F(u(G_k, t)) \cdot n$ is replaced by a numerical flux (approximate or exact Riemann solvers). For example, one could use the simple Lax-Friedrichs flux, which is given by

$$F(u(G_k,t))\cdot n \approx \frac{1}{2}\left[\left(F(u^-(G_k,t)) + F(u^+(G_k,t))\right)\cdot n \right.$$
$$\left. -\bar{\alpha}\left(u^+(G_k,t) - u^-(G_k,t)\right)\right]$$

where $\bar{\alpha}$ is taken as an upper bound for the eigenvalues of the Jacobian in the n direction, and u^- and u^+ are the values of u inside the cell $\triangle_j$ and outside the cell $\triangle_j$ (inside the neighboring cell) at the Gaussian point G_k.

Clearly, the success of the finite volume scheme depends crucially on a good "reconstruction" procedure, which is the procedure to obtain high order and non-oscillatory approximations to the solution u at the Gaussian points along the cell boundary, $u^{\pm}(G_k,t)$, from the neighboring cell averages. Usually, this reconstruction problem is handled in the following way: given a stencil of $R = \frac{(r+1)(r+2)}{2}$ cells, find a polynomial of degree r, whose cell average in each cell within the stencil agrees with the given cell average of u in that cell. This gives a linear system of R equations and R unknowns (the coefficients of the polynomial when expanded in a certain basis), and, if it has a unique solution, the polynomial can be evaluated at the Gaussian point to get the approximation to $u^{\pm}(G_k,t)$. In practice, there are a lot of complications in this procedure, as not all stencils result in a solvable or well conditioned linear system. One would often resort to a least square procedure with more than the necessary number of cells in the stencil to solve this problem, see, e.g. [10]. If the cells are rectangles rather than triangles, then a tensor product polynomial and a tensor product stencil would be much easier to work with [17]. A WENO procedure similar to the one described in the previous section for the finite difference schemes is used here for the reconstruction.

With this very simple sketch of the idea of finite volume WENO schemes, we now turn our attention to a brief discussion of a recent development of the finite volume WENO methods, namely a technique to handle negative linear weights in a finite volume WENO scheme, developed in Shi, Hu and Shu [17].

A key idea in WENO schemes is a linear combination of lower order fluxes or reconstructions to obtain a higher order approximation. The combination coefficients, also called linear weights, are determined by local geometry of the mesh and order of accuracy and may become negative, such as in the central WENO schemes using staggered meshes, high order finite volume WENO schemes in two space dimensions, and finite difference WENO approximations for second derivatives. WENO procedures cannot be applied directly to obtain a stable scheme if negative linear weights are present. Previous strategy for handling this difficulty is by either regrouping of stencils (e.g. in [10]) or reducing the order of accuracy (e.g. in [12]) to get rid of the negative linear weights. In [17], a simple and effective technique for handling negative linear weights without a need to get rid of them is presented. The key idea is a decomposition of the weights into a positive weight group and a negative weight group, such as to make sure that every stencil has a significant representation in both the positive and the negative weight groups. Within each group, the WENO idea of redistributing the weights subject to a fixed sum according to the smoothness of the approximation is still followed as before. While in the unsplit case the nonlinear weights may become extremely large or small in magnitudes due to the lack of convexity, the split positive and negative nonlinear weights stay bounded as they separately sum to fixed constants of the linear weights. Moreover, it is expected that the distribution of the magnitudes of nonlinear weights in each of the positive and negative group follows the smoothness of the solution, as every stencil has its "fair share" in each of the positive and negative linear weight groups. Extensive numerical experiments have been performed in [17] to explore the effectiveness of this treatment. This treatment has negligible additional computational cost.

4 Discontinuous Galerkin Methods

Similar to a finite volume scheme, a discontinuous Galerkin (DG) method for a conservation law such as (1) with $\alpha = 0$ also approximates an integral version of it. The computational domain is again partitioned into a collection of cells $\triangle_i$, which in 2D could be rectangles, triangles, etc., and the numerical solution is a polynomial of degree r in each cell $\triangle_i$. The degree r could change with the cell, and there is no continuity requirement of the two polynomials along an interface of two cells. Thus, instead of only one degree of freedom per cell in a finite volume scheme, namely the cell average of the solution, there are $R = \frac{(r+1)(r+2)}{2}$ degrees of freedom per cell for a DG method using piecewise r-th degree polynomials in 2D. These R degrees of freedom are chosen as the coefficients of the polynomial when expanded in a local basis. One could use a locally orthogonal basis to simplify the computation, but this is not essential.

The DG method is obtained by multiplying (1) by a test function $v(x,y)$ (which is also a polynomial of degree r in the cell), integrating over the cell $\triangle_j$, and integrating by parts:

$$\frac{d}{dt}\int_{\triangle_j} u(x,y,t)v(x,y)dxdy - \int_{\triangle_j} F(u)\cdot\nabla v\, dxdy + \int_{\partial\triangle_j} F(u)\cdot n\, v\, ds = 0$$

where the notation and the treatment of the line integral are the same as in the finite volume scheme (3). The extra volume integral term $\int_{\triangle_j} F(u)\cdot\nabla v\, dxdy$ can be computed either by a numerical quadrature or by a quadrature free implementation [1] for special systems such as the Euler equations of compressible gas dynamics. Notice that if a locally orthogonal basis is chosen, the time derivative term $\frac{d}{dt}\int_{\triangle_j} u(x,y,t)v(x,y)dxdy$ would be explicit and there is no mass matrix to invert. However, even if the local basis is not orthogonal, one still only needs to invert a small $R \times R$ local mass matrix (by hand) and there is never a global mass matrix to invert as in a typical (continuous) finite element method.

When applied to problems with smooth solutions, the DG method as briefly described above can already be used as is. For problems containing discontinuous solutions, however, a nonlinear total variation bounded (TVB) limiter might be needed. For details, see [19, 6, 4, 7]. For a recent review of discontinuous Galerkin method, see [9].

With this very simple sketch of the idea of the discontinuous Galerkin method, we now turn our attention to a brief discussion of two recent developments of discontinuous Galerkin methods.

The first development is for solving problems containing higher derivative terms, namely (1) with h containing higher derivatives of u and α being small but nonzero. For such equations containing higher order spatial derivatives, discontinuous Galerkin methods cannot be directly applied. This is because the solution space, which consists of piecewise polynomials discontinuous at the element interfaces, is not regular enough to handle higher derivatives. This is a typical "non-conforming" case in finite elements. A naive and careless application of the discontinuous Galerkin method directly to the heat equation containing second derivatives could yield a method which behaves nicely in the computation but is inconsistent with the original equation and has O(1) errors to the exact solution [22, 9].

In [8] Cockburn and Shu, motivated by the successful numerical experiments of Bassi and Rebay [3] for the compressible Navier-Stokes equations, developed and analyzed the so-called local discontinuous Galerkin methods for solving convection diffusion equations, namely (1) with h depending on the second derivatives of u. A key ingredient of the local discontinuous Galerkin method is the introduction of an auxiliary variable q, approximating the first derivative of u, and a discontinuous Galerkin method is then applied to both u and q. However, unlike a mixed finite element method which also introduces such a q, here the auxiliary variable q can be locally eliminated, thus essentially the scheme is still for u only. The success of this method depends on a careful choice of the numerical fluxes. An incorrect choice of the numerical fluxes could lead to instability or loss of accuracy. The method is proven in [8] to be L^2 stable for a large class of nonlinear convection diffusion PDEs. Error estimates are also given in [8]. Later, Jan and Shu generalized the local discontinuous Galerkin methods to KdV type equations involving third derivatives [26], and to time dependent by-harmonic type equations involving fourth derivatives and PDEs involving fifth derivatives [27], with similar stability and accuracy properties and excellent numerical behavior.

The second development is an efficient local post-processing technique, which is applied only at the end of the calculation to the numerical solution at $t = T$, and involves only a local linear operation (convolution) using the information from a fixed (independent of the mesh size Δx) number of neighboring cells. This technique depends on two ingredients. The first is a negative norm estimate for the numerical solution, to show that the numerical solution converges in a higher rate for a negative norm than for the L^2 norm. This is typically true for Galerkin type schemes but needs to be proved for each individual case (different types of schemes, different PDEs). In [5] this is proven for the linear hyperbolic system cases. The second ingredient is the recovery of pointwise accurate information based on the negative norms, using a locally translation invariant meshes and a special spline based convolution kernels, see [25] for details. Numerical examples in [5] and [27] indicate that this post-processing technique is very effective in increasing (essentially doubling) the order of accuracy, for linear problems and even for some nonlinear problems such as the classical KdV equations with soliton solutions.

Acknowledgments: The research of the author is supported by ARO grant DAAD19-00-1-0405, NSF grants DMS-9804985 and ECS-9906606, NASA Langley grant NCC1-01035 and AFOSR grant F49620-02-1-0113.

References

1. H. Atkins and C.-W. Shu (1998): Quadrature-free implementation of the discontinuous Galerkin method for hyperbolic equations. AIAA Journal, **36**, 775–782.
2. D. Balsara and C.-W. Shu (2000): Monotonicity preserving weighted essentially non-oscillatory schemes with increasingly high order of accuracy. Journal of Computational Physics, **160**, 405–452.
3. F. Bassi and S. Rebay (1997): A high-order accurate discontinuous finite element method for the numerical solution of the compressible Navier-Stokes equations. Journal of Computational Physics, **131**, 267–279.
4. B. Cockburn, S. Hou, and C.-W. Shu (1990): TVB Runge-Kutta local projection discontinuous Galerkin finite element method for conservation laws IV: the multidimensional case. Mathematics of Computation, **54**, 545–581.
5. B. Cockburn, M. Luskin, C.-W. Shu and E. Süli (2001): Enhanced accuracy by post-processing for finite element methods for hyperbolic equations. Mathematics of Computation, to appear.
6. B. Cockburn and C.-W. Shu (1989): TVB Runge-Kutta local projection discontinuous Galerkin finite element method for scalar conservation laws II: general framework. Mathematics of Computation, **52**, 411–435.
7. B. Cockburn and C.-W. Shu (1998a): TVB Runge-Kutta local projection discontinuous Galerkin finite element method for scalar conservation laws V: multidimensional systems. Journal of Computational Physics, **141**, 199–224.
8. B. Cockburn and C.-W. Shu (1998b): The local discontinuous Galerkin method for time-dependent convection diffusion systems. SIAM Journal on Numerical Analysis, **35**, 2440–2463.
9. B. Cockburn and C.-W. Shu (2001): Runge-Kutta Discontinuous Galerkin methods for convection-dominated problems. Journal of Scientific Computing, **16**, 173–261.
10. C. Hu and C.-W. Shu (1999): Weighted essentially non-oscillatory schemes on triangular meshes. Journal of Computational Physics, **150**, 97–127.

11. G. Jiang and C.-W. Shu (1996): Efficient implementation of weighted ENO schemes. Journal of Computational Physics, **126**, 202–228.
12. D. Levy, G. Puppo and G. Russo (1999): Central WENO schemes for hyperbolic systems of conservation laws. Mathematical Modelling and Numerical Analysis, **33**, 547–571.
13. X.-D. Liu, S. Osher and T. Chan (1994): Weighted essentially non-oscillatory schemes. Journal of Computational Physics, **115**, 200–212.
14. J. Qiu and C.-W. Shu (2001): On the construction, comparison, and local characteristic decomposition for high order central WENO schemes. Journal of Computational Physics, submitted.
15. P. Roe (1978): Approximate Riemann solvers, parameter vectors and difference schemes. Journal of Computational Physics, **27**, 1–31.
16. K. Sebastian and C.-W. Shu (2002): Multi domain WENO finite difference method with interpolation at sub-domain interfaces. Journal of Scientific Computing, submitted.
17. J. Shi, C. Hu and C.-W. Shu (2002): A technique of treating negative weights in WENO schemes. Journal of Computational Physics, **175**, 108–127.
18. J. Shi, Y.-T. Zhang and C.-W. Shu (2002): Resolution of high order WENO schemes for complicated flow structures. Journal of Computational Physics, submitted.
19. C.-W. Shu (1987): TVB uniformly high-order schemes for conservation laws. Mathematics of Computation, **49**, 105–121.
20. C.-W. Shu (1998a): Essentially non-oscillatory and weighted essentially non-oscillatory schemes for hyperbolic conservation laws. In *Advanced Numerical Approximation of Nonlinear Hyperbolic Equations*, B. Cockburn, C. Johnson, C.-W. Shu and E. Tadmor (Editor: A. Quarteroni), Lecture Notes in Mathematics, volume 1697, Springer, 325–432.
21. C.-W. Shu (1998b): High order ENO and WENO schemes for computational fluid dynamics. In *High-Order Methods for Computational Physics*, T.J. Barth and H. Deconinck, editors, Lecture Notes in Computational Science and Engineering, volume 9, Springer, 439–582.
22. C.-W. Shu (2001a): Different formulations of the discontinuous Galerkin method for the viscous terms. In *Advances in Scientific Computing*, Z.-C. Shi, M. Mu, W. Xue and J. Zou, editors, Science Press, 144–155.
23. C.-W. Shu (2001b): High order finite difference and finite volume WENO schemes and discontinuous Galerkin methods for CFD. International Journal of Computational Fluid Dynamics, to appear.
24. C.-W. Shu and S. Osher (1988): Efficient implementation of essentially non-oscillatory shock capturing schemes. Journal of Computational Physics. **77**, 439–471.
25. L.B. Wahlbin (1995): Superconvergence in Galerkin finite element methods. Lecture Notes in Mathematics, vol. 1605, Springer Verlag.
26. J. Yan and C.-W. Shu (2001): A local discontinuous Galerkin method for KdV type equations. SIAM Journal on Numerical Analysis, to appear.
27. J. Yan and C.-W. Shu (2002): Local discontinuous Galerkin methods for partial differential equations with higher order derivatives. Journal of Scientific Computing, **17**, 27–47.

Analysis of High Order Difference Methods for Multiscale Complex Compressible Flows

Björn Sjögreen[1] and H. C. Yee[2]

[1] Royal Institute of Technology, Sweden *bjorns@nada.kth.se*
[2] NASA Ames Research Center *yee@nas.nasa.gov*

1 Introduction

Accurate numerical simulations of complex multiscale compressible viscous flows, especially high speed turbulence combustion and acoustics, demand high order schemes with adaptive numerical dissipation controls. Standard high resolution shock-capturing methods are too dissipative to capture the small scales and/or long-time wave propagations without extreme grid refinements and small time steps. An integrated approach for the control of numerical dissipation in high order schemes with incremental studies was initiated in [15, 16, 9] and summarized in [17]. Here we further refine the analysis on, and improve the understanding of the adaptive numerical dissipation control strategy.

Basically, the development of these schemes focuses on high order nondissipative schemes and takes advantage of the progress that has been made for the last 30 years in numerical methods for conservation laws, such as techniques for imposing boundary conditions, techniques for stability at shock waves, and techniques for stable and accurate long-time integration. We concentrate on high order centered spatial discretizations and a fourth-order Runge-Kutta temporal discretizations as the base scheme. Near the boundaries, the base scheme has stable boundary difference operators [12]. To further enhance stability, the split form of the inviscid flux derivatives [5] is frequently used for smooth flow problems. To enhance nonlinear stability, linear high order numerical dissipations are employed away from discontinuities, and nonlinear filters are employed after each time step in order to suppress spurious oscillations near discontinuities to minimize the smearing of turbulent fluctuations.

Although these schemes are built from many components, each of which is well-known, it is not entirely obvious how the different components be best connected. For example, the nonlinear filter could instead have been built into the spatial discretization, so that it would have been activated at each stage in the Runge-Kutta time stepping. We could think of a mechanism that activates the split form of the equations only at some parts of the domain. Another issue is how to define good sensors for determining in which parts of the computational domain a certain feature should be filtered by the appropriate numerical dissipation. For the present study we employ a wavelet technique introduced in [9] as sensors. Here, the method is briefly described with selected numerical experiments.

2 Base Schemes and Linear Numerical Dissipations

Consider a conservation law

$$u_t + f(u)_x = 0.$$

By a semi-discrete approximation

$$\frac{du_j(t)}{dt} + Df(u_j) = 0,$$

where D is a centered finite-difference operator, and $u_j(t)$ is the approximation of $u(x_j, t)$. A sixth-order spatial central operator

$$Du_j = (-u_{j-3} + 9u_{j-2} - 45u_{j-1} + 45u_{j+1} - 9u_{j+2} + u_{j+3})/(60\Delta x) \quad (1)$$

is used in our numerical examples. The resulting base scheme with a fourth-order Runge-Kutta in time is denoted by CEN66-RK4. Studies also were perform on the fourth and eighth-order centered schemes. The former is less accurate and the latter exhibits minor improvement and with a wider stencil and more complicated boundary operators over their sixth-order counterpart.

2.1 Boundary Operators

On the domain $j = 1, 2, \ldots, N$, the operator (1) traditionally applied at the points $j = 4, 5, \ldots, N-3$ with lower order centered and/or one-sided operators for $j = 1, 2, 3$ and $j = N-2, N-1, N$. In order to have a stable initial-boundary value problem, non-traditional boundary operators that have the summation by parts (SBP) property are used. This means that in a weighted discrete scalar product,

$$(u, v)_h = \sum_{i=1,j=1}^{N,N} \sigma_{i,j} u_i v_j \Delta x$$

with $\sigma_{i,j}$ a positive definite matrix and $h = \Delta x$ the fixed grid spacing, we have the property

$$(u, Dv)_h = -(Du, v)_h + u_N v_N - u_1 v_1. \quad (2)$$

This implies that whenever stability can be obtained for the PDE by use of integration by parts, the same stability estimate can be made for the difference approximation by use of (2). Property (2) involved boundary operators for grid points further away from the boundary than the traditional approach [12].

In the computations below we use boundary operators having the SBP property in a diagonal scalar product, and which are third-order accurate. Development of SBP operators was done by Kreiss in the 1970s. Later in [7], [12], the theory was revived and extended.

2.2 Entropy Splitting of the Inviscid Flux Derivatives

If the above conservation law can be symmetrized by a change of variables, $u = u(w)$, then

$$u(w)_t + f_w w_x = 0,$$

with $f_w = f_u u_w$ symmetric. If we furthermore assume that the flux function is homogeneous $f(\lambda w) = \lambda^\beta f(w)$, with $\beta \neq -1$, one can show that $\beta f = f_w w$. The thermally perfect gas dynamics have a family of variable transformations $u = u(w, \beta)$, defined by $w = E_u$, which give a homogeneous flux function. Here $E(u, \beta)$ is an entropy function of the conservation law.

Under the above assumption ($\beta < \frac{\gamma}{1-\gamma}$ or $\beta > 0$ for a perfect gas), the flux derivative can be split, resulting in the following form

$$u_t + \frac{\beta}{\beta+1} f_x + \frac{1}{\beta+1} f_w w_x = 0.$$

Taking the inner product with w, gives

$$-(\beta+1)(w, u_t) = \beta(w, f_x) + (w, f_w w_x) = \beta(w, f_x) + (f_w w, w_x).$$

Integration by parts in space and the homogeneity property $\beta f = f_w w$, give

$$(\beta+1)(w, u_t) = -[w^T f_w w]_a^b.$$

Homogeneity of the change of variables gives

$$\frac{d}{dt}(w, u_w w) = (\beta+1)(u_t, w) = -[w^T f_w w]_a^b. \tag{3}$$

Remark: The estimate (3) can be shown to be identical to

$$\frac{d}{dt}\int_a^b E(u)\, dx = -[F]_a^b$$

obtained by integrating the entropy equation, $E_t + F_x = 0$, in space. Here F is the so called the entropy flux function.

It follows that $w u_w w = (\beta+1)E(u)$, see [13]. In [5] the entropy splitting is extended to conservation laws with nonhomogeneous flux functions. Formulas for symmetrizing the compressible Euler equations are given in [3]. Formulas for entropy splitting of a perfect gas case are given in [2], and for a thermally perfect gas case are given in [16].

The above entropy splitting was used in a numerical method in [2], where the semi-discrete approximation was written as

$$\frac{du_j(t)}{dt} = -\frac{\beta}{\beta+1} Df(u_j) - \frac{1}{\beta+1} f_w(u_j) D w_j. \tag{4}$$

Here, D is a difference operator, having the summation by parts property. Due to the SBP property, the relation

$$\frac{d}{dt}(w, u_w w)_h = -w_N^T f_w(w_N) w_N + w_1^T f_w(w_1) w_1 \tag{5}$$

in the discrete scalar product follows in the same way as was shown for the PDE. In the special case of periodic boundary conditions, the boundary terms disappear, and we conclude that the integrated entropy $(w, u_w w)_h$ is constant.

2.3 High Order Linear Numerical Dissipation/Filter

Numerical dissipation is normally needed in conjunction with entropy splitting. To minimize the lost of accuracy, we employ linear high order numerical dissipation obtained by adding an operator of the form

$$d(-1)^{q-1} \Delta x^{2q-1} (D_+ D_-)^q u_j$$

to the base scheme of order $q+2$ (or as a linear filter after the completion of the full step of the time discretization). Here d is a tunable parameter, and Δx is the grid spacing. The divided difference operators are defined by $D_+ u_j = (u_{j+1} - u_j)/\Delta x$, and $D_- u_j = D_+ u_{j-1}$. The dissipation operator has an order of accuracy $2q-1$, and dissipation of order $2q$. It is well known [6], [8], how to make a boundary modification to obtain an operator which is semi-bounded, i.e., which has the property

$$(-1)^{q-1}((D_+ D_-)^q u_j, u_j)_h \leq 0. \tag{6}$$

If an extra stabilizing mechanism is desired for difference approximations on the split form, it is natural to define the dissipation on the entropy variables. The approximation (4) then becomes

$$\frac{du_j(t)}{dt} = -\frac{\beta}{\beta+1} Df(u_j) - \frac{1}{\beta+1} f_w(u_j) D w_j + d(-1)^{q-1} \Delta x^{2q-1} (D_+ D_-)^q w_j. \tag{7}$$

If the SBP property and the semi-boundedness for the dissipation both hold in the same scalar product, we take the scalar product with w_j, and obtain as previously

$$\frac{d}{dt}(w, u_w w)_h = -w_N^T f_w(w_N) w_N + w_1^T f_w(w_1) w_1 - d||(D_+ D_-)^{q/2} w||_h^2. \quad (8)$$

The dissipation thus helps to control the increase of the entropy. An example where this applies is for periodic problems. For general boundary conditions, the dissipation has to be modified somewhat in order to obtain (8), because the scalar product for the summation by parts property is not the same as the standard scalar product used for the dissipation estimate. For the sixth-order base scheme, eighth-order linear dissipation ($q = 4$) is used and is denoted by CEN66-D8 without entropy splitting, and CEN66-ENT-D8 with the splitting.

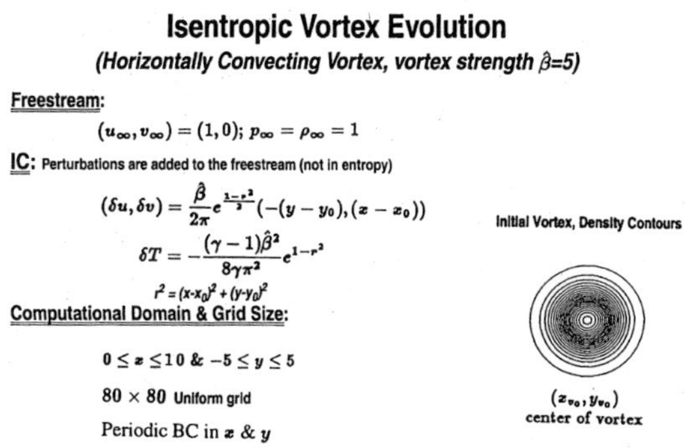

Fig. 1. Vortex convection problem description.

3 Long-Time Integration of Smooth Flows

Extensive numerical experiments on a 2-D compressible inviscid vortex convection with or without entropy splitting and/or nonlinear filter were conducted in [16, 9]. They showed that entropy splitting alone and/or nonlinear filter helps stabilize the scheme for a much longer convection time with almost perfect vortex preservation. Here we study the same problem for the behavior of the high order linear numerical dissipation and compare the solution with other well known methods. This is a pure convection problem and the exact solution should retain its shape. The problem is described in Fig. 1. The boundary conditions are periodic, and the solution is smooth, so that difficulties coming from boundary conditions and from non-smoothness of the solution are avoided. The computational domain is 10×10 with a uniform 80×80 grid. The time scale is such that one period corresponds to 10 time units. Density contours without entropy splitting and without any added dissipation (CEN66-RK4), are shown in Fig. 2. The solution after 5

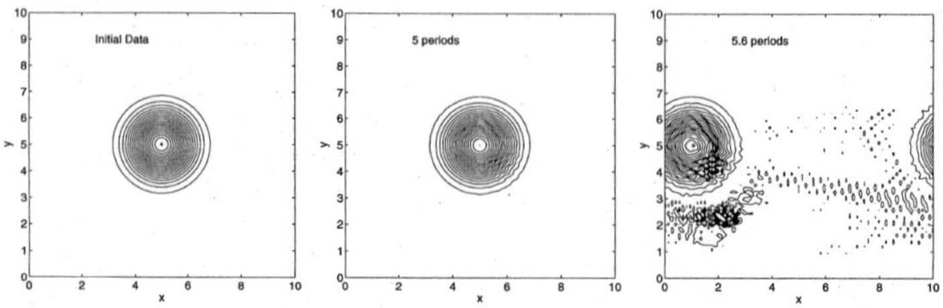

Fig. 2. Density contours after 0, 5, and 5.6 periods using CEN66-RK4.

periods has developed small, hardly visible, oscillations, and the computation breaks down before reaching 6 periods. The oscillations are due to nonlinear instabilities. Density contours using the entropy split form (4) (CEN66-ENT-

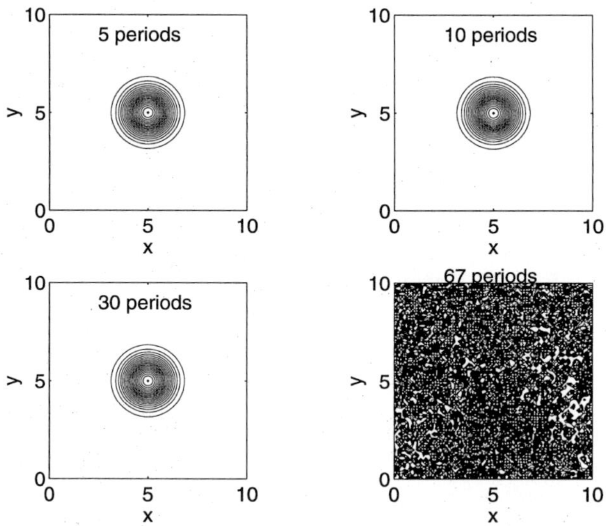

Fig. 3. Density contours using CEN66-ENT-RK4 with entropy split parameter, $\beta = 1$.

RK4) is shown in Fig. 3 with $\beta = 1$ (equal weight on the conservative and the non-conservative part). For this problem, $\beta = 1$ appears to be the best choice. Almost perfect vortex preservation is nearly 10 times longer than the unsplit case. However, the solution eventually breaks down (after 68 periods). For the entropy split method, it follows from (5) that the integrated entropy should be constant. Fig. 4 shows the entropy integral as function of time for the computation in Fig. 3. The integrated entropy does decrease, and the estimate holds, but is too weak. However, we have found that there are other values of β where the computed entropy integral increases. β does not appear in the semi-discrete estimate, but is apparently important for the time-discrete computations. It can be seen that entropy splitting retain the

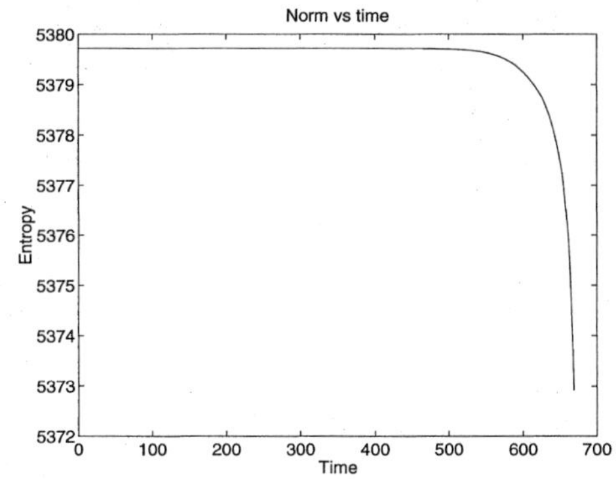

Fig. 4. Entropy integral vs. time of CEN66-ENT-RK4 with $\beta = 1$.

vortex shape from 5 periods to approximately 40 periods. However, in many applications, for example computation of rotorcraft flows, one needs to follow vortices for several hundred periods. We next investigate whether the use of high order linear numerical dissipation can be used to further increase the number of periods. Fig. 5 shows the density contours of the sixth-order base

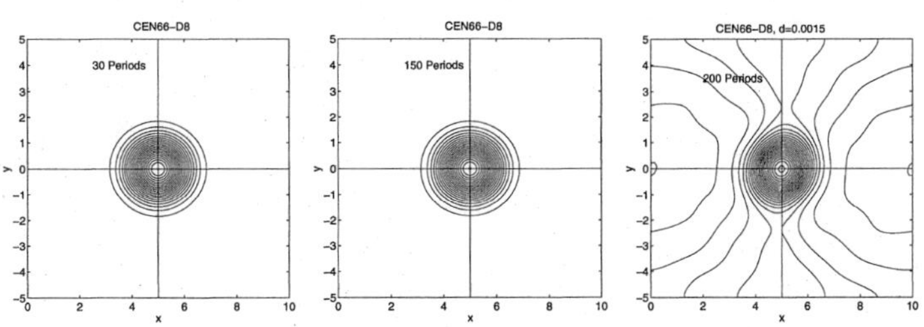

Fig. 5. Density contours by CEN66-D8-RK4 with $d = 0.0015$ after 30, 150 and 200 periods.

scheme with the eighth-order numerical dissipation added (CEN66-D8-RK4). The solution is very accurate up to approximately 150 periods, but the accuracy degenerates at later times. Although the accuracy becomes very poor, the solution did not break down. The computation could be run up to 300 periods, the maximum time we used. The time step used in this computation was 0.011. The coefficient d in front of the dissipation operator influences the behavior of the solution. The L^2 errors of the solution as function of time for different values of the dissipation parameter d are shown in Fig. 6. The error norm is computed as the sum of the error norms of the four components. From Fig. 6, we conclude that there is an initial phase where the lower the dissipation, the better the accuracy, and the error increases in a "regular" way. At later times the larger dissipations have two points where the error increases quickly. For example, for $d = 8 \times 10^{-3}$ the first error increase comes at 70 periods and the second at 190 periods. We also conclude that for long-time integration, using the smallest possible dissipation is not optimal. After 120 periods the dissipation 1×10^{-3} gives a more accurate solution than the dissipation 5×10^{-4}. The smaller dissipation is insufficient to suppress the nonlinear instabilities. After 200 periods, all dissipations end up in a state where all accuracy is lost. The norm of the difference between the solution and the constant state with no vortex is approximately 0.1. Results using

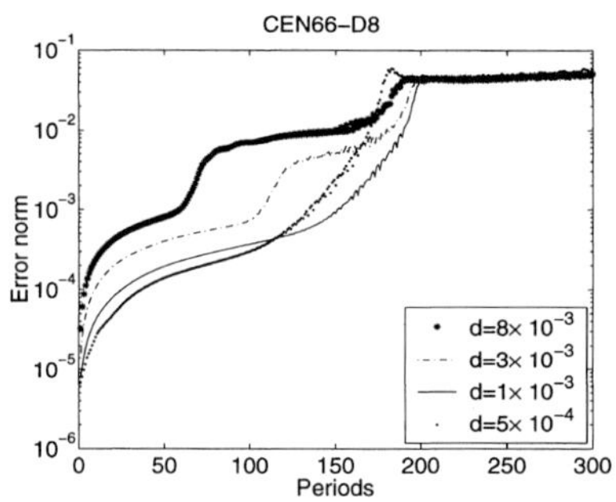

Fig. 6. Error vs. time for different dissipation coefficient d for CEN66-D8-RK4.

both entropy splitting and the linear numerical dissipation CEN66-ENT-D8 exhibit very minor improvement and no improvement at larger period of the vortex convection over the CEN66-D8 case. A possible explanation is that the very delicate cancellation properties, which keep the entropy split method stable, are destroyed by the numerical dissipation as time progresses. A fifth order WENO scheme (WENO5-RK4) [4], the so called DRP scheme [14], and a second-order accurate MUSCL TVD scheme (MUSCL-RK2) were computed for comparison. The WENO5 scheme is a general purpose scheme suited for flows with shock waves with no parameters to tune. The DRP scheme is designed for aeroacoustics. It is fourth-order accurate in space and second-order in time. Nonlinear instabilities are suppressed by a low order numerical dissipation term with a tunable parameter. We have tried to choose this dissipation strength as small as possible while keeping the scheme stable. Density contours are shown in Fig. 7 after 150 periods, and in Fig. 8

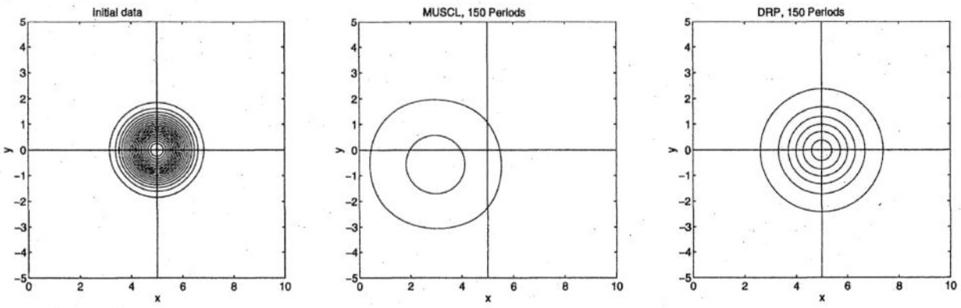

(a) IC (b) MUSCL (c) DRP

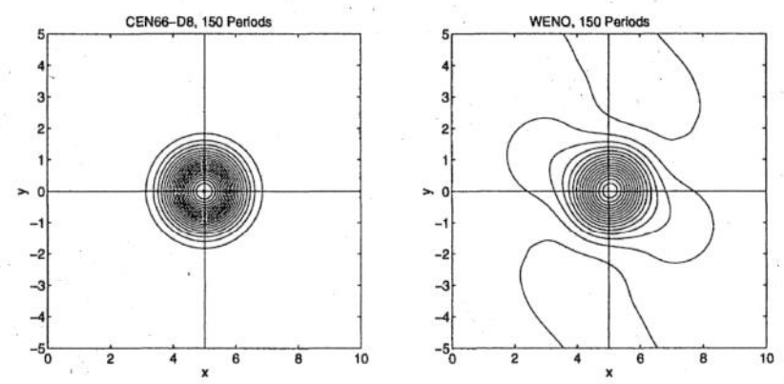

(d) CEN66-D8 (e) WENO5

Fig. 7. Comparison of different methods: Density contours at 150 Periods

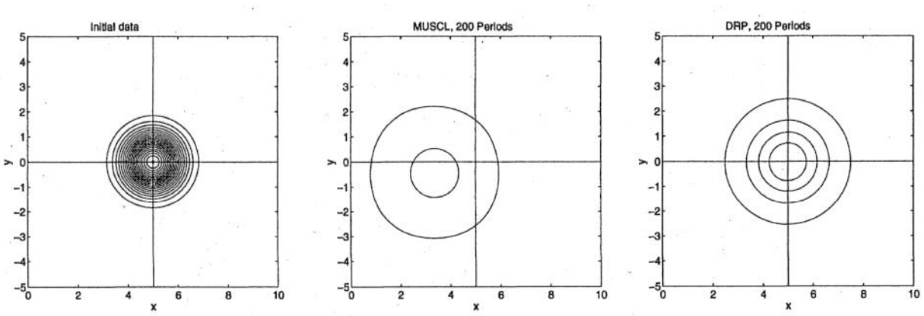

(a) IC (b) MUSCL (c) DRP

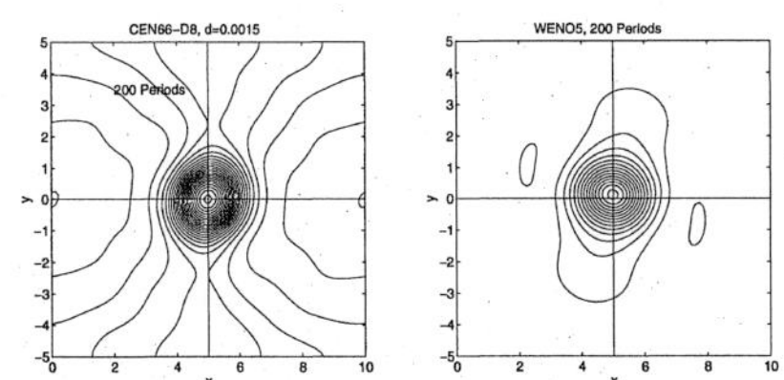

(d) CEN66-D8 (e) WENO5

Fig. 8. Comparison of different methods: Density contours at 200 Periods after 200 periods using the same contour levels. The MUSCL-RK2, and the DRP scheme are extremely diffusive. Both the WENO5 and the CEN66-D8 give better results. The error as function of time for the different methods is shown in Fig. 9. The error of MUSCL is not shown, instead the dashed curve shows the error of the entropy split sixth-order scheme, with $\beta = 1$, and no numerical dissipation (CEN66-ENT). After 200 periods, the result from

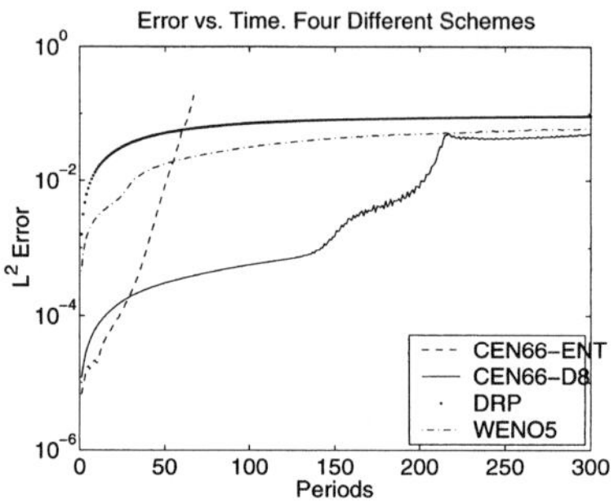

Fig. 9. Error vs. time for different methods.

the WENO5 scheme and the CEN66-D8 give similar results, but as shown in Fig. 9, the error norm of the WENO5 is larger than that of the centered scheme. The centered scheme loses accuracy due to dispersive errors, while for the other schemes, diffusive errors dominate. The performance of nonlinear filter scheme (including the use of the dissipative portion of WENO5 as a nonlinear filter) computations (same base scheme) are slightly more diffusive than CEN66-D8, and is reported in [9, 16, 18]. Studies on the blending of the linear dissipations with the nonlinear filters are reported in [17].

4 Nonlinear Filter/Numerical Dissipation for Discontinuities

The linear numerical dissipation used in the previous section is not strong enough to suppress the spurious oscillations due to the presence of shock and shear waves. A filter approach using nonlinear dissipation as a post processing filter, which is applied after each time step, is employed [15]. The filter consists of a conservative step

$$u_j^{n+1} = u_j^* - \lambda(h_{j+1/2}^* - h_{j-1/2}^*),$$

where u^* is the solution computed by the high order base scheme at time level $n+1$, and λ is the ratio of time step over Δx. For the numerical example the filter flux $h^*$is obtained from a second-order TVD scheme by subtracting the centered difference part and inserting a sensor, i.e.,

$$h^*_{j+1/2} = s_{j+1/2}(h^{TVD}_{j+1/2} - (f_{j+1} + f_j)/2).$$

The flux $h^{TVD}_{j+1/2}$ is the numerical flux of a second-order accurate TVD scheme. This TVD-flux typically contains flux limiters and Riemann solvers. The sensor $s_{j+1/2}$ is made such that the dissipation is switched on only near discontinuities. One example is to compute the local wavelet coefficients of the flow field by letting $s_{j+1/2}$ be close to unity where the wavelet coefficients show that the solution has poor regularity, and taking $s_{j+1/2}$ to be zero otherwise. See [9] for details. Computations using the dissipative portion of the WENO5 schemes or seventh-order high-resolution schemes were also considered. See [18] for an example using the WENO5 nonlinear dissipation.

5 2-D Shock/Shear/Boundary-Layer Interaction

Extensive grid refinement for shock-turbulence interactions, a complex multi-scale viscous combustion flows and a shock/shear/boundary-layer interaction are reported [17, 10, 11, 1]. Here we illustrate selected results for the complex shock/shear/boundary-layer interaction. An ideal gas is at rest in a 2-D box $0 \leq x, y \leq 1$. A planar shock of Mach 2.37 located at $x = 1/2$ at time zero separates two different states of the gas. The 2-D compressible Navier-Stokes equation with no slip BCs at the adiabatic walls is used. The solution will develop complex 2-D shock/shear/boundary-layer interactions, which depend on the Reynolds number. The dimensionless initial states given in [1] are

$$\rho_L = 120, \quad p_L = 120/\gamma, \qquad \rho_R = 1.2, \quad p_R = 1.2/\gamma,$$

where ρ_L, p_L are the density and pressure respectively to the left of $x = 1/2$, and ρ_R, p_R are the same quantities to the right of $x = 1/2$. All velocities are equal to zero, $\gamma = 1.4$, the Prandtl number is 0.73, and the Reynolds number is 1000. All the computations stop at the dimensionless time 1 after the shock wave has reflected from the right wall and has almost reached the middle of the domain, $x = 1/2$. Due to the rapidly developing symmetry flow structure, a uniform Cartesian grid with the lower half of the domain is used. A grid convergence study of the sixth-order method with wavelet as the sensor to filter the solution using the nonlinear numerical dissipation (WAV66-RK4) is shown in Fig. 11. The same computation using the MUSCL-RK2 is shown in Fig. 10. Except for the largest vortical structure, the 1000×500 grid solution of WAV66-RK4 captures the overall flow structure of a 4000×2000 grid, whereas the MUSCL scheme is still not convergent even at a finer grid of 4000×2000. Since the filter is only applied once after each time step, the computational cost of WAV66-RK4 is almost the same as the computational cost of MUSCL-RK2 with the added advantage that the physical viscosities were taken into consideration by WAV66-RK4.

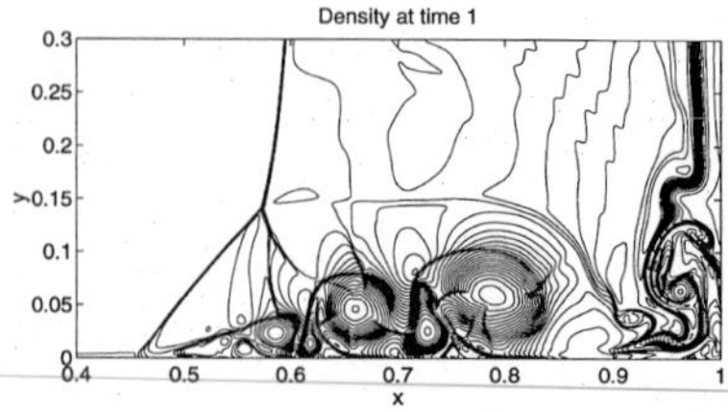

(a) 1000 × 500 grid

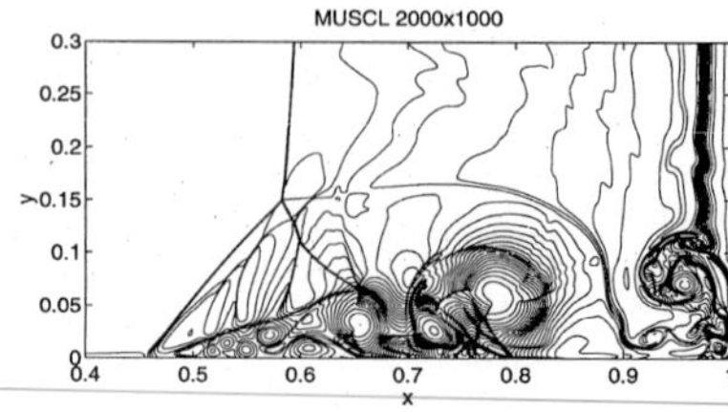

(b) 2000 × 1000 grid

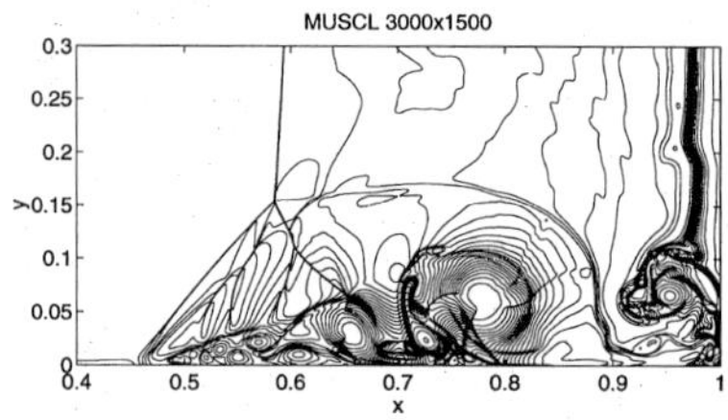

(c) 3000 × 1500 grid

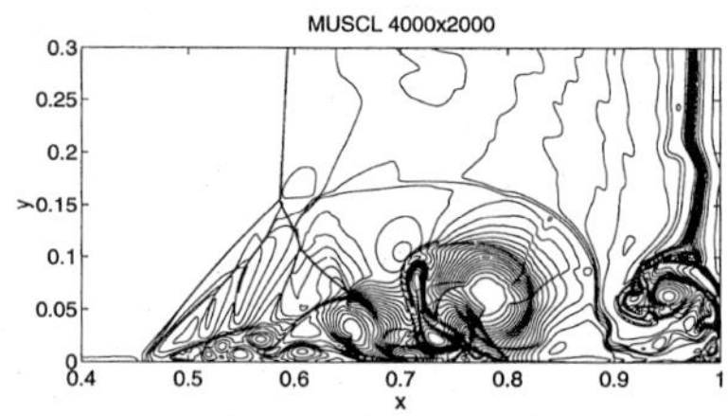

(d) 4000 × 2000 grid

Fig. 10. Grid refinement study: Density contours of MUSCL-RK2.

References

1. V. Daru and C. Tenaud, *Evaluation of TVD High Resolution Schemes for Unsteady Viscous Shocked Flows*, Computers Fluids, **30** 89–113 (2000).
2. M. Gerritsen and P. Olsson, *Designing an Efficient Solution Strategy for Fluid Flows I.* J. Comput. Phys., **129** 245–262 (1996).
3. A. Harten, *On the Symmetric Form of Systems of Conservation Laws with Entropy*, J. Comput. Phys.,**49** (1983), pp. 151–164.
4. G.-S. Jiang and C.-W. Shu, *Efficient Implementation of Weighted ENO schemes*, ICASE Report No. 95-73, (1995).
5. P. Olsson and J. Oliger, J., *Energy and Maximum Norm Estimates for Nonlinear Conservation Laws*, RIACS Technical Report 94.01 (1994).
6. P. Olsson, *High-order Difference Methods and Data Parallel Implementation*, Doctoral Thesis, Department of Scientific Computing, Uppsala University, 1992.
7. P. Olsson, P., *Summation by Parts, Projections, and Stability. I*, Math. Comp. **64** 1035–1065 (1995).
8. B. Sjögreen, *High Order Centered Difference Methods for the Compressible Navier-Stokes Equations*, J. Comput. Phys., **117** 67–78 (1995).
9. B. Sjögreen and H. C. Yee, *Multiresolution Wavelet Based Adaptive Numerical Dissipation Control for Shock-Turbulence Computation*, RIACS Technical Report TR01.01, NASA Ames research center (Oct 2000).
10. B. Sjögreen and H. C. Yee, *Grid Convergence of High Order Methods for Multiscale Complex Unsteady Viscous Compressible Flows*, RIACS Technical Report

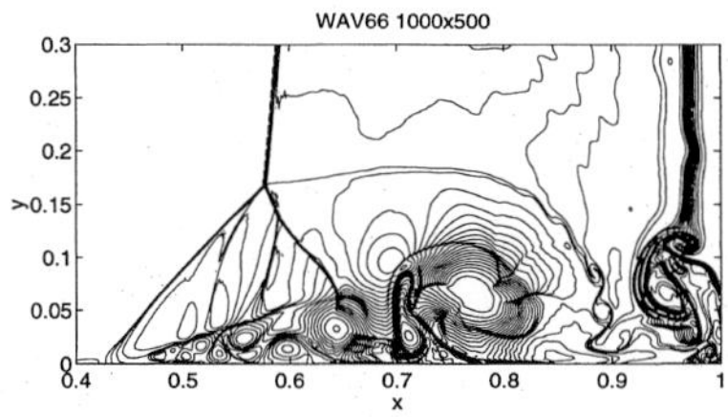

(a) 1000 × 500 grid

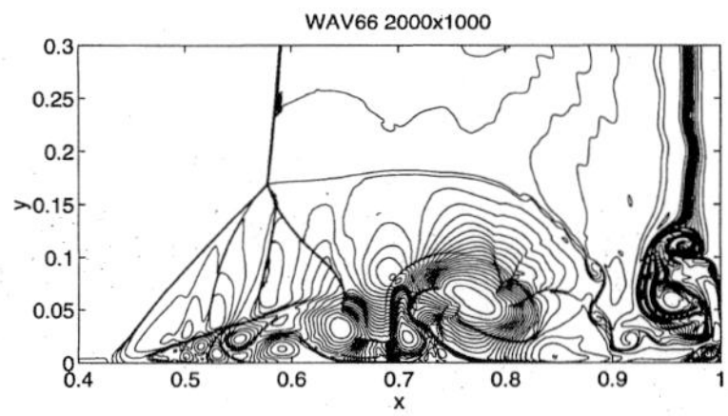

(b) 2000 × 1000 grid

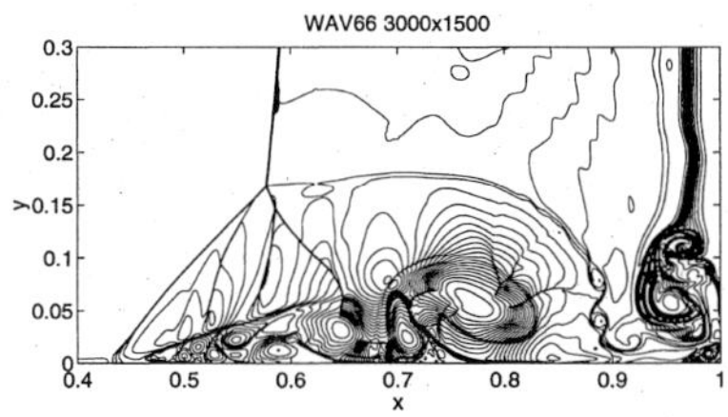

(c) 3000 × 1500 grid

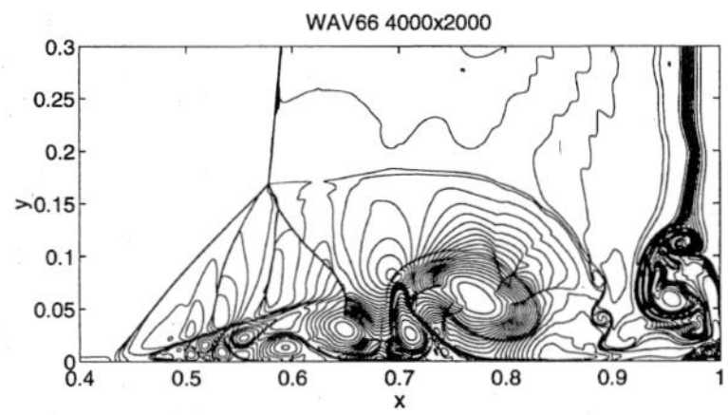

(d) 4000 × 2000 grid

Fig. 11. Grid refinement study: Density contours of WAV66-RK4.

TR01.06, April, 2001, NASA Ames research center; AIAA 2001-2599, Proceedings of the 15th AIAA CFD Conference, June 11-14, 2001, Anaheim, CA.

11. B. Sjögreen and H. C. Yee, *Low Dissipative High Order Numerical Simulations of Supersonic Reactive Flows*, RIACS Technical Report TR01-017, NASA Ames Research Center (May 2001); Proceedings of the ECCOMAS Computational Fluid Dynamics Conference 2001, Swansea, Wales, UK, September 4-7, 2001.
12. B. Strand, *Summation by Parts for Finite Difference Approximations for d/dx*, J. Comput. Phys. **110** 47–67 (1994).
13. E. Tadmor, *Skew-Selfadjoint Form for Systems of Conservation Laws*, J. Math. Anal. Appl., **103** 428–442 (1984).
14. C. K. W. Tam, *Computational Aeroacoustics: Issues and Methods*, AIAA journal, **33** (1995), pp. 1788–1796.
15. H.C. Yee, N.D. Sandham, N.D., and M.J. Djomehri, *Low Dissipative High Order Shock-Capturing Methods Using Characteristic-Based Filters*, J. Comput. Phys., **150** 199–238 (1999).
16. H.C. Yee, M. Vinokur, M., and M.J. Djomehri, *Entropy Splitting and Numerical Dissipation*, J. Comput. Phys., **162** 33-81 (2000).
17. H.C. Yee and B. Sjögreen, *Designing Adaptive Low Dissipative High Order Schemes for Long-Time Integrations*, **Turbulent Flow Computation**, (Eds. D. Drikakis & B. Geurts), Kluwer Academic Publisher (2002); also RIACS Technical Report TR01-28, Dec. 2001.
18. H.C. Yee and B. Sjögreen *Designing Adaptive Low Dissipative High Order Schemes*, Proceedings of the ICCFD2, Sydney, Australia, July 15-19, 2002.

Nonlinear Boundary Layers of the Boltzmann Equation

Seiji Ukai[1], Tong Tang[2] and Shih-Hsien Yu[2]

[1] Department of Applied Mathematics, Yokohama National University
79-5 Tokiwadai, Hodogaya, Yokohama 240-8501, Japan
ukai@mathlab.sci.ynu.ac.jp

[2] Department of Mathematics, City University of Hong Kong
83 Tat Chee Avenue, Kowloon Tong, Hong Kong
mashyu@cityu.edu.hk, *matyang@math.cityu.edu.hk*

1 Introduction and Main Result

The Dirichlet problem of the nonlinear Boltzmann equation in the half-space arises in the analysis of the kinetic boundary layer, the condensation-evaporation problem and other problems related to the kinetic behavior of gas near the wall, [5],[12]. The main concern is to find a solution which tends to an assigned *Maxwellian* at infinity.

An interesting feature of this problem is that not all Dirichlet data are admissible and the number of admissible conditions changes with the far Maxwellian. This has been shown for the linear case by many authors [3],[6],[7],[9], mainly in the context of the classical Milne's and Kramer's problems. Recently, a nonlinear addmissible condition was derived for the discrete velocity model in [13] and the stability of steady solutions was proven in [11]. The full nonlinear problem was solved on the exisitence of solutions in [8] for the case of the specular reflection boundary condition, whose proof, however, does not work for the Dirichlet boundary condition, and in [2] for this case, but with the ambiguity that the far Maxwellian cannot be fixed a priori. Here, we will establish the admissible conditions for each fixed far Maxwellian. Our proof provides also a new aspect of the linear problem.

It should be mentioned that K. Aoki, Y. Sone and their group ([1], [12], [?] and references therein) made an extensive numerical computation on the same nonlinear problem. Our result gives a partial explanation of their numerical results.

We are concerned with the steady state of a gas in the 3-dimensional half-space $D = \{(x, y, z) \in \mathbb{R}^3 \,|x > 0\}$, in which the mass density F of gas particles is assumed constant on each plane plarallel to the boundary ∂D although the particle motion is 3-dimensional, that is, F is assumed to be a function of position x (but not of y, z) and particle velocity $\xi = (\xi_1, \xi_2, \xi_3) \in \mathbb{R}^3$. Let ξ_1 stand for the velocity component along the x-axis. Then, F is governed by

$$\begin{cases} \xi_1 F_x = Q(F,F), & x > 0,\ \xi \in \mathbb{R}^3, \\ F|_{x=0} = F_0(\xi), & \xi_1 > 0,\ (\xi_2, \xi_3) \in \mathbb{R}^2, \\ F \to M_\infty(\xi) \quad (x \to \infty), & \xi \in \mathbb{R}^3. \end{cases} \tag{1}$$

Here, Q, the *collision operator*, is a bilinear integral operator in ξ. We do not use its explicit form here but need the following two classical properties, [4], [5].

(i) $Q(F) = 0$ if and only if

$$F = M[\rho, u, T](\xi) \equiv \frac{\rho}{(2\pi T)^{3/2}} \exp\Big(-\frac{|\xi - u|^2}{2T}\Big), \tag{2}$$

for any constants $\rho, T > 0$ and $u = (u_1, u_2, u_3) \in \mathbb{R}^3$. This is a *Maxwellian* and is the distribution function of a gas in the equlibrium state with the mass density ρ, flow velocity u and temperature T.

(ii) A function $\phi(\xi)$ is called a *collision invariant* of Q if

$$\langle \phi, Q(F, F) \rangle = 0 \qquad \text{for all } F,$$

$\langle , \rangle$ being the inner product of $L^2(\mathbb{R}^3_\xi)$. Q has five collision invariants

$$\phi_0 = 1, \quad \phi_i = \xi_i \ (i = 1, 2, 3), \quad \phi_4 = |\xi|^2, \tag{3}$$

which indicate the conservations of mass, momentum and energy in the course of the binary collision of particles.

The second equation in (1) is the Dirichlet boundary condition. The Dirichlet data $F_0(\xi)$ can be assigned only for incoming particles ($\xi_1 > 0$), but not for outgoing ones ($\xi_1 < 0$) because, then, the problem becomes ill-posed as will be seen from the a priori estimate stated in §3. This corresponds to the physical situation that one can control the incoming distribution but not the outgoing one, on the wall.

It is clear that the far field $M_\infty(\xi)$ in the third equation of (1) cannot be assigned arbitrarily but must be a zero of Q, and hence a Maxwellian. Thus, we must take

$$M_\infty = M[\rho_\infty, u_\infty, T_\infty], \tag{4}$$

where the constants $\rho_\infty > 0, u_\infty = (u_{\infty,1}, u_{\infty,2}, u_{\infty,3}) \in \mathbb{R}^3$, and $T_\infty > 0$ are the only quantities which we can control. By a shift of the variable ξ along the boundary ∂D, we can assume without loss of generality that $u_{\infty,2} = u_{\infty,3} = 0$. Therefore, the sound speed and Mach number of this equilibrium state are given by

$$c_\infty = \sqrt{\frac{5}{3} T_\infty}, \qquad \mathscr{M}_\infty = \frac{u_{\infty,1}}{c_\infty}, \tag{5}$$

respectively, [5]. Note that the flow at infinity is incoming (resp. outgoing) if $\mathscr{M}_\infty < 0$ (resp. > 0) and supersonic (resp. subsonic) if $|\mathscr{M}_\infty| > 1$ (resp. < 1).

We will see that the Mach number $\mathscr{M}_\infty$ provides significant changes on the solvability of (1). Indeed, since M_∞ is the "Dirichlet" data at $x = \infty$ and is imposed for all $\xi_1 \in \mathbb{R}$, it is over-determined (ill-posed) and as a consequence, (1) may not be solvable unconditionally. Actually, we will show that the number of solvability conditions changes with $\mathscr{M}_\infty$. To state this precisely, set

$$n^+ = \begin{cases} 0, & \mathscr{M}_\infty \in (-\infty, -1), \\ 1, & \mathscr{M}_\infty \in (-1, 0), \\ 4, & \mathscr{M}_\infty \in (0, 1), \\ 5, & \mathscr{M}_\infty \in (1, \infty), \end{cases} \tag{6}$$

and introduce the weight function

$$W_\beta(\xi) = (1 + |\xi|)^{-\beta} \Big(M[1, u_\infty, T_\infty](\xi) \Big)^{1/2}, \tag{7}$$

with $\beta \in \mathbb{R}$. Our main result is

Theorem 1. *Let M_∞ be given as in (4) with $\mathscr{M}_\infty \neq 0, \pm 1$, and let $\beta > 3/2$. Then, there exist positive numbers ϵ_0, σ, C_0, and a C^1 map*

$$\Psi : L^2(\mathbb{R}^3_+, \xi_1 d\xi) \longrightarrow \mathbb{R}^{n+}, \quad \Psi(0) = 0, \tag{8}$$

such that the following holds.
(i) *For any F_0 satisfying*

$$|F_0(\xi) - M_\infty(\xi)| \leq \epsilon_0 W_\beta(\xi), \quad \xi \in \mathbb{R}^3_+, \tag{9}$$

and

$$\Psi(F_0 - M_\infty) = 0, \tag{10}$$

the problem (1) *has a unique solution F in the class*

$$|F(x, \xi) - M_\infty(\xi)| + |\xi_1 F_x(x, \xi)| \leq C_0 e^{-\sigma x} W_\beta(\xi), \quad x > 0, \xi \in \mathbb{R}^3. \tag{11}$$

(ii) *The set of F_0 satisfying* (9) *and* (10) *forms a (local) C^1 manifold of codimension n^+.*

Remark 1. This theorem does not cover the cases $\mathscr{M}_\infty = 0, \pm 1$.

Remark 2. For each given M_∞, (9) is a smallness condition on the deviation of F_0 from M_∞ whereas (10) gives restrictions on F_0 however small it may be, if $n^+ \neq 0$. Thus, our theorem says that the problem (1) is solvable unconditionally for any F_0 sufficiently close to M_∞ if $\mathscr{M}_\infty < -1$ but otherwise not. A physical explanation of this is that if $\mathscr{M}_\infty < -1$, any phenomena near the boundary cannot affect the far field while a part of them can propagate to infinity and affect the far field if $\mathscr{M}_\infty > -1$.

Remark 3. In the numerical works made in [1], [12] and references therein, the Dirichlet data F_0 is fixed to be the standard Maxwellian $M[1, 0, 1](\xi)$ (of course for $\xi_1 > 0$), and values of 5 parameters $(\rho_\infty, \mathscr{M}_\infty, u_{2,\infty}, u_{3,\infty}, T_\infty)$ of M_∞ in (4) are sought numerically which admit smooth solutions connecting F_0 and M_∞. It is concluded that the set of such admissible values is, in the parameter space $\mathbb{R}^5$, a union of a 5-dimensional subdomain in the domain $\mathscr{M}_\infty < -1$, a 4-dimensional surface in $-1 < \mathscr{M}_\infty < 0$ and a 1-dimensional curve in $0 < \mathscr{M}_\infty < 1$ whereas no solutions are found for $\mathscr{M}_\infty > 1$. Our theorem agrees with this for the case $\mathscr{M}_\infty < 1$ in the sense that the above mentioned regions of admissible values have the codimension just equal to n^+ of (6) in $\mathbb{R}^5$. For the case $\mathscr{M}_\infty > 1$, $F_0 = M[1,0,1]$ may not be on the manifold defined by (10) and hence, no solutions.

Remark 4. The stability of the stationary solutions obtained in Theorem 1 is an important issue. We will show the exponentially asymptotic stability for the case $\mathscr{M}_\infty < -1$ in a forthcoming paper.

2 A Remark on the Linearized Problem

Our proof, which relies on the analysis of the corresponding linearized problem at M_∞, provides a new aspect of the linear case.

We shall look for the solution of (1) in the form

$$F(x,\xi) = M_\infty(\xi) + W_0(\xi) f(x,\xi), \tag{12}$$

where W_0 is the weight of (7) with $\beta = 0$. Then, the problem (1) reduces to

$$\begin{cases} \xi_1 f_x - Lf = \Gamma(f), & x > 0,\ \xi \in \mathbb{R}^3, \\ f|_{x=0} = a_0(\xi), & \xi \in \mathbb{R}^3_+, \\ f \to 0\ (x \to \infty), \xi \in \mathbb{R}^3, \end{cases} \tag{13}$$

where

$$Lf = W_0^{-1}\Big\{Q(M_\infty, W_0 f) + Q(W_0 f, M_\infty)\Big\}, \qquad \Gamma(f) = W_0^{-1} Q(W_0 f, W_0 f),$$

$$a_0 = W_0^{-1}\Big(F_0 - M_\infty\Big).$$

The operator L is linear while the remainder Γ is quadratic.

The linearized problem of (1) at M_∞ is just (13) with the term $\Gamma(f)$ dropped;

$$\begin{cases} \xi_1 f_x - Lf = 0, & x > 0,\ \xi \in \mathbb{R}^3, \\ f|_{x=0} = a_0(\xi), & \xi \in \mathbb{R}^3_+, \\ f \to 0\ (x \to \infty), \xi \in \mathbb{R}^3, \end{cases} \tag{14}$$

We can get the following linear version of Theorem 1.

Theorem 2. *Suppose $\mathscr{M}_\infty \neq 0, \pm 1$. Then, there exist n^+ functions r_i, $1 \leq i \leq n^+$, of $L^2(\mathbb{R}^3_+, \xi_1 d\xi)$ and positive constants σ, C_0 such that for any $a_0 \in R^\perp$ with*

$$R = \mathrm{span}\{r_1, r_2, \cdots, r_{n^+}\},$$

(14) has a unique L^2 solution of the form $f = e^{-\sigma x} g$ with g satisfying

$$||\xi_1|^{1/2}g^0|_- + ||(1+|\xi|)^{1/2}g|| + |||\xi_1|(1+|\xi|)^{-1/2}g_x|| \le C_0||\xi_1|^{1/2}a^0|_+, \quad (15)$$

where $g^0 = g|_{x=0}$ *while* $|\cdot|_\pm$ *and* $||\cdot||$ *denote the norms of* $L^2(\mathbb{R}^3_\pm)$, *and* $L^2((0,\infty)\times\mathbb{R}^3)$, *respectively.*

Remark 5. This theorem says that for the linear problem (14), the map Ψ of Theorem 1 is linear and the manifold of admissible a_0 is the hyperplane $R^\perp$. Actually, Ψ has the form

$$\Psi_{lin}(a) = (<\xi_1 r_i, a>_+)_{i=1,2,\cdots,n^+}, \quad (16)$$

where $<,>_+$ denote the inner product of $L^2(\mathbb{R}^3_+)$.

In order to compare our result with the ones known so far, [3],[6],[7],[9], recall the linear operator L and put $N = kerL$. It is classical that

$$N = \mathrm{span}\{W_0(\xi)\phi_i(\xi)\}_{i=0,1,\cdots,4} \quad (17)$$

where ϕ_i are as in (3) and thus N can be taken a 5 dimensional subspace of $L^2(\mathbb{R}^3)$. Let

$$P: L^2(\mathbb{R}^3) \to N, \qquad P^\perp: L^2(\mathbb{R}^3) \to N^\perp$$

be the orthogonal projections and define the linear operator

$$A = P\xi_1 P,$$

which is a 5-dimesional linear bounded self-adjoint operator and has the eigenvalues

$$\lambda_1 = u_{\infty,1} - c_\infty, \quad \lambda_i = u_{\infty,1}\ (i=2,3,4), \quad \lambda_5 = u_{\infty,1} + c_\infty, \quad (18)$$

on N. Define

$$I^+ = \{j|\lambda_j > 0\}, \qquad I^- = \{j|\lambda_j < 0\}.$$

Note that n^+ of (6) is just $\#I^+$. Let χ_j be the eigenfunction corresponding to the eigenvalue λ_j. In [7], the following is proved.

Theorem 3. ([7]) *For any* $a_0 \in L^2(\mathbb{R}^3_+, \xi_1 d\xi)$ *and for any constants* $c_j, j \in I^-$, *there exists a unique* L^2 *solution* f *satisfying the first two equations of* (14) *and instead of the last one, the auxiliary condition*

$$<\chi_j, f(x,\cdot)> = c_j, \qquad x > 0,\ j \in I^-. \quad (19)$$

Moreover, there exists an element $q_\infty \in N$ *such that*

$$f \to q_\infty \quad (x \to \infty) \quad in \ L^2(\mathbb{R}^3).$$

The proof in [7] does not tell us how to determine the limit q_∞. However, since (14) is linear and that $q_\infty \in N = kerL$, we see that $\tilde{f} = f - q_\infty$ solves all of three equations in (14) with a_0 replaced by $a_0 - q_\infty$. Now, Theorem 2 concludes

Theorem 4. $a_0 - q_\infty \in R^\perp$.

Remark 6. This and (19) determines uniquely q_∞.

3 Outline of the Proof

There are two ingredients in our proof. One is to add an artificial "damping" term and the other is to introduce the spatial weight function $e^{\sigma x}, \sigma > 0$.

To construct the damping term, decompose the operator A on N into the positive and negative parts A^+, A^-, and denote the corresponding eigenprojections by P^+, P^-. Note that if $\mathscr{M}_\infty \neq 0, \pm 1$, then A has not zero eigenvalues (see (18)), so that

$$A = A^+ + A^-, \qquad P = P^+ + P^-.$$

We modify (13) by adding to its right hand side the damping term defined by

$$-\gamma P^+ \xi_1 f, \quad \gamma > 0,$$

and then rewrite it by putting $f = e^{-\sigma x} g$, to deduce

$$\begin{cases} \xi_1 g_x - \sigma \xi_1 g - Lg = h - \gamma P^+ \xi_1 g, \; x > 0, \; \xi \in \mathbb{R}^3, \\ g|_{x=0} = a_0(\xi), \qquad \xi \in \mathbb{R}^3_+, \\ g \to 0 \; (x \to \infty), \; \xi \in \mathbb{R}^3, \end{cases} \tag{20}$$

with

$$h = e^{-\sigma x} \Gamma(g). \tag{21}$$

Note that for the case $\mathscr{M}_\infty < -1$, we have $n^+ = 0$ and $P^+ = 0$, and hence no extra damping term.

For a moment, we discard (21) and take (20) to be a linear problem with a inhomogeneous term h. Take the inner product of (20) and g in $L^2(\mathbb{R}_+ \times \mathbb{R}^3)$ and integrate by parts, to deduce

$$< |\xi_1| g^0, g^0 >_- + (Bg, g) - (Lg, g) = < \xi_1 a_0, a_0 >_+ + (h, g), \tag{22}$$

where $g^0 = g|_{x=0}$ while

$$B = -\sigma \xi_1 + \gamma P^+ \xi_1, \quad \sigma, \gamma > 0.$$

At the first glance, B has not a good sign, but it *does* on N if $\gamma > \sigma > 0$ as seen from

$$PBP = -\sigma A + \gamma A^+ = -\sigma A^- + (\gamma - \sigma) A^+,$$

whence we get

$$(Bg, g) \geq C_1 \sigma \Big(|Pg|^2 - ||\xi_1|^{1/2} P^\perp g|^2 \Big),$$

with some constant $C_1 > 0$ and the choice, say, $\gamma = 2\sigma$. On the other hand, it is classical that L is negative definite on $N^\perp$, and moreover for the case of the hard sphere gas, it holds that

$$-(Lg, g) \geq C_2 |(1 + |\xi_1|)^{1/2} P^\perp g|^2,$$

with some constant $C_2 > 0$. Thus, we have the a priori estimate

Proposition 1. *Let $\sigma > 0$ be sufficiently small. Then, any classical solution g to (20) satisfies*

$$||\xi_1|^{1/2}g^0|_- + ||(1+|\xi|)^{1/2}g|| + |||\xi_1|(1+|\xi|)^{-1/2}g_x|| \le C_0\Big(||\xi_1|^{1/2}a^0|_+ + ||h||\Big), \tag{23}$$

with some constant $C_0 > 0$ independent of a_0 and h.

Clearly, the estimate (15) follows from (23). Note that our argument is just the micro-macro or kinetic-fluid dynamic decomposition argument used in [10].

The estimate (23) is enough to construct our solutions. First, the same estimate can be derived for the adjoint problem to (20), which then enable us, together with the Hahn-Banach and Riesz representation theorms, to conclude the existence of a weak L^2 solution g to (20). Furthermore, taking suitable test functions, we can establish the "weak=strong" theorem, and thus this g is a unique strong solution and satisfies the estimate (23). Finally, starting from this estimate and using the bootstrap argument, we can get the L^∞ estimate of g which is just (23) with L^2-norms replaced by L^∞ norm weighted by the weight W_β.

Now, the contraction argument applies to the nonlinear problem (20) with h defined by (21), proving the existence of L^∞ nonlinear solutions for sufficently small a_0.

In the case $\mathcal{M}_\infty < -1$, this gives the solutions to (13) and hence to the original problem (1). It remains to discuss the case $\mathcal{M}_\infty > -1$. Clearly, if

$$P^+\xi_1 g = 0, \quad x > 0,\ \xi \in \mathbb{R}^3, \tag{24}$$

then g is also a solution of the original problem without the extra damping term. We can show that the condition (24) reduces to

$$P^+\xi_1 g^0 = 0, \quad \xi \in \mathbb{R}^3. \tag{25}$$

Clearly, g and hence g^0 are determined uniquely by the boundary data a_0. Write

$$\Psi(a_0) = P^+\xi_1 g^0. \tag{26}$$

Identifying the space P^+N with $\mathbb{R}^{n^+}$, we can show that (26) defines a C^1 map

$$\Psi : L^2(\mathbb{R}^3_+, \xi_1 d\xi) \to \mathbb{R}^{n^+}, \tag{27}$$

and $\Psi(0) = 0$. Moreover, we can prove

Proposition 2. *The Fréchet derivative of Ψ at $a_0 = 0$ is given by* (16).

This and the implicit function theorem, then, prove that the set of a_0's satisfying (25) forms a C^1 manifold of codimention n^+, whence Theorem 1 follows. The detail will be given elsewhere.

References

1. Aoki K., Nishino, K., Sone, Y., Sugimoto, H.(1991): Numerical analysis of steady flows of a gas condensing on or evaporating from its plane condensed phase on the basis of kinetic theory: Effect of gas motion along the condensed phase, Phys. Fluids A, **3**, 2260–2275
2. Arkeryd, L., Nouri, A. (2000): On the Milne problem and the hydrodynamic limit for a steady Boltzmann quation model, J. Stat. Phys., **99**, 993–1019
3. Bardos, C., Caflish, R. E., Nicolaenko, B. (1986): The Milne and Kramers problems for the Boltzmann equation of a hard sphere gas, Comm. Pure Appl. Math. **49**, 323–352
4. Carleman, T., (1932): Sur La Théorie de l'Équation Intégrodifférentielle de Boltzmann, Acta Mathematica, **60**, 91–142
5. Cercignani, C., Illner, R., Purvelenti, M. (1994): The Mathematical Theory of Dilute Gases, Springer-Verlag, Berline,
6. Cercignani, C. (1986): Half-space problem in the kinetic theory of gases, in: Kröner, E., Kirchgässner, K. (eds.) Trends in Applications of Pure Mathematics to Mechanics, Springer-Verlag, Berlin, 35–50
7. Coron, F., Golse, F., Sulem, C. (1988): A classification of well-posed kinetic layer problems, Commun. Pure Appl. Math., **41**, 409–435.
8. Golse, F., Perthame, B., Sulem, C. (1988): On a boundary layer problem for the nonlinear Boltzmann equation, Arch. Rational Mech. Anal.. **103** , 81–96
9. Golse, F., Poupaud, F.(1989): Stationary solutions of the linearized Boltzmann equation in a half-space, Math. Methods Appl. Sci., **11**, 483-502
10. Liu, T.-P., Yu, S.-H. (2002): Boltzmann Equation: Micro-Macro Decompositions and Positivity of Shock Profiles, to appear
11. Nikkuni, S., Kawashima, S. (2000): Stability of stationary solutions to the half-space problem for the discrete Boltzmann equation with multiple collisions, Kyushu J. Math., **54**, 233-255
12. Sone, Y. (2002): Kinetic Theory and Fluid Dynamics, Birkhäuser, Basel
13. Ukai, S. (1998): On the half-space problem for the discrete velocity model of the Boltzmann equation, in Kawashima, S., Yangisawa, T. (eds), Advances in Nonlinear Partial Differential Equations and Stochastic Series on Advances in Mathematics for Applied Sciences-Vol. **48**, World Scientific, Singapore-New York, 160–174.

Invited Talks

An Homogenized Hyperbolic Model of Multiclass Traffic Flow: a Few Examples

Patrizia Bagnerini[1] and Michel Rascle[2]

[1] Laboratoire de Mathématiques CNRS n. 6621, Université de Nice, Parc Valrose B.P. 71, 06108 Nice Cedex 2, France *pbagneri@math.unice.fr*

[2] Laboratoire de Mathématiques CNRS n. 6621, Université de Nice, Parc Valrose B.P. 71, 06108 Nice Cedex 2, France *rascle@math.unice.fr*

Summary. We introduce a new homogenized hyperbolic (multiclass) traffic flow model which allows to take into account the behaviors of different type of vehicles (cars, trucks, buses, etc.). We discretize the system with a Godunov Lagrangian scheme with oscillating initial data which describe the non homogeneity of the traffic, and we study the propagation of oscillations as time goes. We show the convergence of the scheme and the existence and uniqueness of the entropy solution to the homogenized system, which allows to completely identify the Young measure for all (x, t). Moreover, we show that this limit system is the *hydrodynamic limit* of the corresponding multi-class discrete system. Simulations are also presented.

1 Description of the model

In [1], Aw et Rascle developed a new macroscopic model of traffic flow which allows to avoid some inconsistencies of the models inspired by the gas dynamic system. In [2], a connection is established between the microscopic "Follow-the-Leader model", and (a semi-discretization of) the macroscopic model introduced in [1], which is its hydrodynamic limit.

In this Note, we introduce a new macroscopic (homogenized) hyperbolic model for a single lane multiclass flow, described by a hyperbolic non linear system of three conservation laws. The details are given in [3]. Here, some numerical simulations are presented.

The starting model, written in Lagrangian mass coordinates, is the following:

$$\begin{cases} \partial_t \tau - \partial_x v = 0, \\ \partial_t w = 0, \;\; \partial_t a = 0, \\ (\tau, w, a)(x, 0) = (\tau_0, w_0, a_0)(x), \end{cases} \tag{1}$$

where $\tau = 1/\rho$ is the specific volume, i.e the inverse of the (normalized) density of vehicles (i.e. of the fraction of space occupied by the cars), v the velocity, and $a \in [0, 1]$ is a dimensionless coefficient which allows to take into account the behaviors of different types of vehicles and/or of drivers. Here, up to a constant,

$$w = v - V(\tau, a), \tag{2}$$

where $V(\tau, a)$ is an "equilibrium" velocity for a given τ and for a given class a. We assume that for all a, $V(\cdot, a)$ is strictly increasing and strictly concave. Good prototypes of function V could be, e.g. (see resp. [2] and [7]), up to a constant

$$V(\tau, a) := \begin{cases} v_{ref}(a)\, c\, \ln(\tau), & (i) \\ v_{ref}(a)\, \exp(-(\tau_{ref}^2/(2\tau^2)) & (ii), \end{cases} \tag{3}$$

(note that the latter is only concave for $\tau \geq \tau_{ref}\sqrt{3}$, which we can enforce by using invariant regions), where c is a constant, $v_{ref}(a)$ is a given reference velocity for a class of vehicles a, e.g. 60 km/h for the slowest car-driver pairs and 130 km/h for the fastest ones, and τ_{ref} is a common parameter for all user classes, e.g. $\tau_{ref} := 1/\rho_{ref}$ with $\rho_{ref} = 50$ veh/km. For simplicity, *all the results are stated with* $V(\tau, a)$ given by

$$V(\tau, a) = aP(\tau), \;\; 0 < a_{min} \leq a \leq a_{max} \leq 1, \tag{4}$$

with $P'(\tau) < 0 \leq P''(\tau)$, but *they remain valid* as soon as $\frac{\partial V}{\partial \tau}(a, \tau) > 0 > \frac{\partial^2 V}{\partial \tau^2}(a, \tau)$. The last two equations of (1) express the fact that w and a are characteristic of each vehicle and then do not depend on time in Lagrangian coordinates. Moreover w represents the difference between the velocity of a vehicle at time t and the velocity which it would have for the density $1/\tau$ at the equilibrium.

The system is non-strictly hyperbolic, but the associated matrix is diagonalizable: the eigenvalue λ_1 is genuinely nonlinear (GNL) i.e. $\forall U$, $\nabla_U \lambda_1 \cdot r^1(U) = -a\, P''(\tau) < 0$, and $\lambda_2 = \lambda_3$ is linearly degenerate (LD), i.e. $\forall U$, $\nabla_U \lambda_k \cdot r^k(U) \equiv 0$, $k = 2, 3$.

The Riemann Problem associated to the system (1) consists of two waves: a rarefaction or a shock wave associated to λ_1, followed by a contact discontinuity associated to $\lambda_2 = \lambda_3$. The system (1) is a Temple system in the extended sense: the shock curves and rarefaction curves coincide. Moreover, v and (w, a) are strict Riemann invariants, respectively associated to λ_1 and $\lambda_2 = \lambda_3 = 0$.

The variables $(\tau, w, a)(x, t)$ and $v(x, t)$ remain in an invariant bounded region $\mathcal{R}$, away from the vacuum:

$$\mathcal{R} := \{(\tau, w, a); (v, w, a) \in [v_{min}, v_{max}] \times [w_{min}, w_{max}] \times [a_{min}, a_{max}]\}, \tag{5}$$

The qualitative properties of the solution are as expected: braking corresponds to a shock, accelerating to a rarefaction, no information travels faster than the velocity of cars, the velocity and the density remain nonnegative and bounded, etc.

1.1 The Godunov scheme

Now we discretize the system (1), written in general form as

$$\partial_t U + \partial_x G(U) = 0,$$

with the Godunov scheme. We introduce a grid in time and space, with step-size Δx and Δt and grid point $x_{j-1/2}$ and t_n. Let I_j be the *open* interval $(x_{j-1/2}, x_{j+1/2})$, and, $h := (\Delta x, \Delta t)$, with $\Delta t/\Delta x = C$. The solution U_j^{n+1} of the Godunov scheme is given by

$$U_j^{n+1} := U_j^n - \frac{\Delta t}{\Delta x}(G_{j+1/2}^n - G_{j-1/2}^n) = U_j^n + \frac{\Delta t}{\Delta x}(v_{j+1}^n - v_j^n), \tag{6}$$

with $\forall j$, $v_j^n := w_j^n - a_j^n P(\tau_j^n) = w_j^n + V(\tau_j^n, a_j^n)$.

1.2 Scaling

The macroscopic models are only valid if we consider a large number of vehicles on a long stretch of the road. Therefore, we introduce a scaling (zoom) such that the size of the domain and the number of vehicles is going to infinity, whereas the length of the vehicles tends to 0 (see also [2]). Given the Lagrangian coordinates x, t, we consider the new rescaled coordinates

$$x' := hx, \quad t' := ht,$$

where h is a small parameter, proportional to the inverse of the maximal possible number of cars per new unit length. In the new coordinates x', t', the normalized variables ρ, τ and the Riemann invariants v, w, a are unchanged, whereas the length of a vehicle becomes $\Delta x' = h\Delta x$, which tends to 0 as $h \to 0$.

Therefore, the step-size h of the discretization can be viewed as a scaling parameter $(x, t) \to (x', t') = (hx, ht)$. Assuming typically [1] that in each cell there is a unique vehicle, h letting to 0 corresponds to considering a large number of vehicles on a long stretch of the road, with a vanishing length of the vehicles.

For instance, assume that the initial units are meter and second and the new one, 1000 m and 60 seconds, with $\Delta x = 5m$. In the rescaled coordinates,

$$x' := \frac{1}{200}x, \quad t' := \frac{1}{60}t, \tag{7}$$

a reference velocity of 90 km/h, i.e. 25 m/s becomes 1.5 in the new units and the length of the car $\Delta x' = 1/200$, which is a "reasonable" step-size.

2 The Homogenized Problem

Practically, the distribution of the different types of vehicle-driver pairs on the road can be highly oscillatory: we are thus led to studying an *homogenized*

[1] This is not necessary for practical applications, where on the contrary one cell should contain several vehicles, otherwise there is no need for a fluid model!

system. We consider a *sequence* of initial data $(v_h^0, w_h^0, a_h^0, \tau_h^0)$, with $w_h^0 = v_h^0 - V(\tau_h^0, a_h^0)$, and $h \to 0$. We assume that the sequence (v_h^0) converges boundedly almost everywhere to some v_0^*, whereas (w_h^0, a_h^0) and therefore τ_h^0 only converge *weakly* to (w_0^*, a_0^*) and τ_0^*. In rescaled coordinates (x', t'), a good prototype of initial data would be

$$(w_h^0(x'), a_h^0(x')) := (W_0, A_0)(x', \frac{x'}{h}), \tag{8}$$

for some given functions (W_0, A_0). In the sequel we will drop the primes and write (x, t) instead of (x', t').

We then study the weak-star limit (v^*, w^*, a^*, τ^*) of the solution (v_h, w_h, a_h, τ_h) of system (1)-(2) as $h \to 0$. Using the notion of Young measure and the compensated compactness theory, we study the propagation of these initial oscillations: we proved in [3] that the Young measure $\nu_{x,t}$ associated to the variables (v, w, a) is a.e. a tensor product

$$\nu_{x,t} = \delta(v - v^*(x,t)) \otimes \mu_x(w, a),$$

where δ is the Dirac mass at the origin and $\mu_x(w, a)$ a probability measure defined a.e. in $\mathbb{R}_x$ by the initial data, and v^* is defined by the following Theorems:

Theorem 1. *Let $U_h = (v_h, w_h, a_h)$ be constructed by the Godunov scheme. We assume that U_h^0 is bounded in $L^\infty(\mathbb{R})$ and (v_h^0) bounded in $BV(\mathbb{R})$. We also assume that the whole sequence (w_h^0, a_h^0, τ_h^0) (resp. (v_h)) converges in L^∞ weak* (resp. a.e.) to a unique limit.*

(i) Under the CFL condition (with $\Delta t/\Delta x$ = constant) when $h \to 0$, there exists a subsequence, still denoted by (U_h), such that

$$(w_h, a_h)(x,t) \equiv (w_0^h, a_0^h)(x) \stackrel{*}{\rightharpoonup} (w_0^*, a_0^*), \quad \tau_h(\cdot,\cdot) \stackrel{*}{\rightharpoonup} \tau^*(\cdot,\cdot),$$
$$v_h(\cdot,\cdot) \to v^*(\cdot,\cdot) \text{ in } L^1_{loc}(\mathbb{R} \times \mathbb{R}_+) \text{ strong and in } L^\infty(\mathbb{R} \times \mathbb{R}_+) \text{ weak*}.$$

(i) At least for a subsequence, in fact (by Theorem 2) for the whole sequence, this sequence (U_h) constructed by the Godunov scheme converges to a weak solution of the system

$$\begin{cases} \partial_t \tau^* - \partial_x v^* = 0, \\ \partial_t w^* = 0, \;\; \partial_t a^* = 0, \end{cases} \tag{9}$$

with initial data

$$\begin{cases} \tau_0^* =< \mu_x(w,a), \mathcal{T}(v_0^*(x), w, a) >, \\ w_0^* =< \mu_x(w,a), w >, \;\; a_0^* =< \mu_x(w,a), a >, \end{cases} \tag{10}$$

and

$$\tau^*(x,t) =< \mu_x(w,a), \mathcal{T}(v^*(x,t),w,a) >:= \mathcal{T}^*(x, v^*(x,t)), \tag{11}$$

where $\mathcal{T}$ is defined by

$$\tau := \mathcal{T}(v,w,a) := (V(\cdot,a))^{-1}(v-w) = P^{-1}((v-w)/a). \tag{12}$$

(ii) The family of probability measures $\nu_{x,t}$ is a measure-valued (mv) solution in the following sense: for any entropy-flux pair (η, q) of the system (1),

$$\eta(v,w,a) := \tilde{\eta}(\tau,w,a)\big|_{\tau=\mathcal{T}(v,w,a)},$$

with $\tilde{\eta}$ convex with respect to τ *(or - this is equivalent - $\forall q$ concave with respect to v) and for any test-function $\varphi \geq 0$, we have*

$$\begin{aligned}&\int_0^\infty \int_{\mathbb{R}} (< \nu_{x,t}(v,w,a), \eta(v,w,a) > \partial_t \varphi + < \nu_{x,t}, q(v) > \partial_x \varphi)\; dxdt \\ &+ \int_{\mathbb{R}} < \nu_{x,0}(v,w,a), \eta(v,w,a) > \; \varphi(x,0)\; dx \geq 0.\end{aligned} \tag{13}$$

(iii) This inequality (with arbitrary entropy flux pairs $(\eta(w,a),0)$) implies (10). *In addition, $\forall \varphi \geq 0$, $\forall k \in \mathbb{R}$, we have*

$$\begin{aligned}&\int_0^\infty \int_{\mathbb{R}} (< \mu_x(w,a), |\mathcal{T}(v^*(x,t),w,a) - \mathcal{T}(k,w,a)| > \partial_t \varphi - |v^*(x,t) - k| \partial_x \varphi) dxdt \\ &+ \int_{\mathbb{R}} < \mu_x(w,a), |\mathcal{T}(v_0^*(x),w,a) - \mathcal{T}(k,w,a)| > \varphi(x,0) dx \geq 0.\end{aligned} \tag{14}$$

The existence of an entropy solution of (9) follows directly from the convergence of the Godunov method in Theorem 1. Now, since (w_h) and (a_h) only converge weakly as $h \to 0$, the Young measure $\nu_{x,t}$ is not a δ-function in (w,a). However, the system (9) can be rewritten as the two trivial equations $\partial_t w^* = \partial_t a^* = 0$, coupled with a scalar equation where the flux explicitly depends on x, through (11):

$$\partial_t \tau^* - \partial_x \mathcal{V}(x, \tau^*) = 0, \tag{15}$$

or in the more convenient form:

$$\partial_t \mathcal{T}^*(x, v^*) - \partial_x v^* = 0. \tag{16}$$

Therefore, the weak solution of system (9) satisfies (16), (11). Using the monotonicity in v^* of $\mathcal{T}^*(x,v^*)$ (the result would be wrong without this property), it turns out that v^* is also a weak solution "à la Kružkov" of (16): for any k in $\mathbb{R}$, (14) *is equivalent* to

$$\int (|\mathcal{T}^*(x,v^*) - \mathcal{T}^*(x,k)| \varphi_t - |v^* - k| \varphi_x) dxdt + \int_{\mathbb{R}} |\mathcal{T}^*(x,v_0^*) - \mathcal{T}^*(x,k)| \varphi(x,0) dx \geq 0, \tag{17}$$

where $\tau^*(\cdot,\cdot)$ is defined by (11). The low regularity in x of the flux - here through the Young measure (ν_x) - does not allow to use directly the uniqueness results available in the literature, see e.g. [4] or [5]. Since in (16) the function $\mathcal{T}^*(x, v^*)$ (explicitly depending on x) appears through its time-derivative, we have *exchanged* the roles of x and t so that inequality (17) is in conservative form. We have the following Theorem (see [3]):

Theorem 2. *We consider the scalar equation* (16), (11) *(or the equivalent equation* (15), (11)*), with initial data* (10) *and under the assumption that* V*, defined by* (2), *is strictly increasing and strictly concave. We also assume that the whole sequence* (w_h^0, a_h^0, τ_h^0) *converges in* L^∞ *weak* to a unique limit and* $v_0^* \in L^\infty(\mathbb{R}) \cap BV(\mathbb{R})$*. Then, there is a unique weak entropy solution "à la Kružkov"* v^* *of* (16).
Therefore the whole sequence (τ_h, w_h, a_h, v_h) *converges to the unique solution of the limit system described in Theorem 1.*

3 The microscopic model

Microscopic models of vehicular traffic are usually based on so-called "Follow-the Leader models", which usually consist, in Eulerian coordinates, of a system of second order ordinary differential equations. The basic idea is that the acceleration at time t depends on the relative speeds of the vehicle and its leading vehicle at time t and the distance between the cars. We consider the corresponding microscopic multi-class "Follow-the-Leader model":

$$\begin{cases} \dot{\tau}_j = \frac{v_{j+1} - v_j}{\Delta x}, \\ \dot{w}_j = \dot{v}_j - \frac{\partial V}{\partial \tau}(\tau_j, a_j) = 0, \ \dot{a}_j = 0, \\ v_j(0) = v_j^0, \ w_j(0) = w_j^0, \ a_j(0) = a_j^0. \end{cases} \tag{18}$$

where v_j is the speed of the vehicles at time t, Δx the length of the vehicle, $\tau_j = 1/\rho_j$ the local "specific volume around vehicle j" and ρ_j the local density, whereas a_j is a coefficient which depends on each type of vehicles. The system (18) is a Follow-the Leader type of model, in which the function V also depends on the coefficients a_j, characteristic of the different types of vehicles and their heterogeneous response to their leading vehicle, and, as in Section 1, w_j describes a difference between the actual velocity and the equilibrium one.

3.1 Hydrodynamic limit of the microscopic model

We consider the infinite system of ordinary differential equations (18), written in general form as

$$\begin{cases} \frac{dU(t)}{dt} = F(U(t)), \\ U(0) = U_0, \end{cases} \tag{19}$$

where $U := (U_1, U_2, \ldots, U_j, \ldots)$, $F(U) := (F_1(U), F_2(U), \ldots, F_j(U), \ldots)$, $U_j := (\tau_j, w_j, a_j)$ and

$$F_j(U) := \left(\frac{v_{j+1} - v_j}{\Delta x}, 0, 0\right) = \left(\frac{w_{j+1} - w_j}{\Delta x} - \frac{a_{j+1}P(\tau_{j+1}) - a_j P(\tau_j)}{\Delta x}, 0, 0\right).$$

We introduce its first order explicit Euler discretization in time:

$$\begin{cases} \tau_j^{n+1} = \tau_j^n + \frac{\Delta t}{\Delta x}\,(v_{j+1}^n - v_j^n), \\ w_j^{n+1} = w_j^n, \\ a_j^{n+1} = a_j^n, \end{cases} \tag{20}$$

which is exactly the Godunov scheme (6) *for system* (1). Now define

$$U_\Delta(x,t) := \sum_{j\in\mathbb{Z}} U_j(t)\chi_j(x),$$

where χ_j is the characteristic function of the interval I_j. In the following Theorem [3], we prove that, for any fixed Δx, the previous discretization is stable and consistent and therefore convergent as $\Delta t \to 0$, We also show that the microscopic multi-class model (18) is the semi-discretization of the Lagrangian system (1).

Theorem 3. *We consider the system* (20), *with initial data* $U_j^0 = (v_j^0, w_j^0, a_j^0)$ *in the invariant region* $\mathcal{R}$ *defined by* (5), *away from vacuum. We assume that the initial data are constant for* x *large enough, so that there is a "first" vehicle. Then,*

(i) The first order approximation (20) *is stable and consistent in the* l^∞ *space. Therefore, for any fixed* Δx, *the sequences* $U_j^n := (\tau_j^n, w_j^n, a_j^n)$ *and* v_j^n *converge as* $\Delta t \to 0$. *The limit functions* $U_\Delta(x,t)$ *and* $v_\Delta(x,t)$ *stay in the invariant region* $\mathcal{R}$ *and satisfies the uniform* BV *estimates inherited from the Godunov scheme. The microscopic model* (18) *is then the semi-discretization of the macroscopic system* (1), (2).

(iii) $U_j(t)$ *satisfies a semi-discrete entropy inequality, i.e. for any entropy* η convex with respect to τ_j, *associated to the entropy flux* q *and for any* j,

$$\frac{d\eta(U_j(t))}{dt} + (1/\Delta x)(q(U_{j+1}^n) - q(U_j^n)) \leq 0. \tag{21}$$

We rewrite (19) under the form

$$\begin{cases} \frac{dU_j(t)}{dt} + \frac{G(U_{j+1}(t)) - G(U_j(t))}{\Delta x} = 0, \\ U_j(0) = U_j^0, \end{cases} \tag{22}$$

with $G(U_j) := (-v_j, 0, 0) = (-(w_j - a_j P(\tau_j)), 0, 0)$. As in [2], we then show in [3] that the entropy solution of the system (9) is the limit as $\Delta x \to 0$ (i.e. when the number of vehicles goes to infinity), of the solution of the infinite-dimensional system of ordinary differential equations (22) (see [3]).

Theorem 4. *Under the same assumptions of the Theorem 3, at least a subsequence, still denoted by U_Δ, converges, as $l \to 0$, almost everywhere to an entropy weak solution to the macroscopic system* (9)-(11). *In fact, by Theorem 2, the whole sequence u_Δ converges to the unique solution.*

In conclusion, starting from the fully discrete system (6), we obtain the *same* limit (i.e. the macroscopic system (9)), either by letting $(\Delta x, \Delta t) \to 0$ with a fixed ratio and the CFL condition, or by first letting $\Delta t \to 0$ with a fixed Δx, and then letting $\Delta x \to 0$. In this sense, the homogenized model (9) is the *hydrodynamic limit* of the microscopic "Follow-the-Leader" system.

4 Numerical simulations

We consider the homogenized system in Lagrangian mass coordinates (9), (11) with initial data (10). As to the initial data, we choose a prototype situation in which there is only an arbitrary finite number N of classes a_i, associated to $w_i = v_i - V(\tau_i, a_i)$, and μ_x is given by

$$\mu_x(w,a) = \sum_{i=1}^{N} \alpha_i(x)\delta_{w_i,a_i}(w,a), \text{ with } \sum_{i=1}^{N} \alpha_i(x) \equiv 1, \tag{23}$$

where the nonnegative coefficients $\alpha_i(x)$ are the local proportion (possibly 0) of each class of car-driver pair and δ the Dirac measure. When we then compute an approximation of the *average* velocity v^* by the Godunov scheme, not only we know the average specific volume $\tau^*(x,t) = \sum_{i=1}^{N} \alpha_i(x)\mathcal{T}(v^*(x,t), w_i, a_i)$ introduced in equation (11), but also the specific volume τ_i (and then the density) of *every class* a_i for a given v^*.

We write $\tau, v, etc.$ instead of $\tau^*, v^*, etc.$. Let $\tau^0 = 1/\rho^0$ and v^0 be the initial specific volume and velocity. We introduce a grid in time and space, with step-size ΔX (the length of the vehicles) and Δt, and we compute τ_j^{n+1} by the Godunov scheme (6) in Lagrangian coordinates, i.e. by

$$\tau_j^{n+1} = \tau_j^n + \frac{\Delta t}{\Delta X}(v_{j+1}^n - v_j^n). \tag{24}$$

We take the equilibrium velocity $V(\tau, a)$ as in (3, i) and then

$$\mathcal{T}(v,w,a) := (V(\cdot,a))^{-1}(v-w) = \exp((v-w)/(cv_{ref}(a))).$$

By (23), the relation (11) between τ and v is then given by

$$\tau_j^{n+1} =< \mu_{x_j}(w,a), \mathcal{T}(v_j^{n+1}, w, a) >=< \mu_{x_j}(w,a), \exp((v_j^{n+1} - w)/(c\, v_{ref}(a))) >=$$

$$= \sum_{i=1}^{N} \alpha_i(x_j) \exp((v_j^{n+1} - w_i)/(c\, v_{ref}(a_i))),$$

$$\text{with } w_i(x_j) = v^0(x_j) - V(\tau^0(x_j), a_i) = v^0(x_j) - c\, v_{ref}(a_i) \ln(\tau^0(x_j)) \tag{25}$$

Knowing τ_j^{n+1} by (24), we compute v_j^{n+1} by solving the nonlinear equation (25) by a few iterations of the Newton method, and the location of the vehicles at time t_{n+1} by $x_{j-1/2}^{n+1} = x_{j-1/2}^{n} + \Delta t v_j^n$. Using the scaling (7), the new units are 1 km and 60 seconds and the length ΔX of the vehicles becomes $\Delta X' = \frac{1}{200}$. In the sequel we drop the primes.

We consider a domain $[a, b] = [-30 \text{ km}, 30 \text{ km}]$ and an initial number of vehicles equal to 2400. In Eulerian coordinates, the initial density and velocity are as in [7] (see Figure 1):

$$\rho_0(x_j) = \exp(-(x_j/5)^2/3.5)/1.5 + 0.025, v_0(x_j) = 1.5\exp(-0.5(\rho_0^2(x_j)/0.0625)),$$

and $c = 1/3$. The vehicles are initially distributed such that $x_{j+1} = x_j + \Delta X/\rho(x_j)$ and they are grouped in nine different classes a_i characterized by the free-flowing speeds $v_{ref}(a_i)$ of 1 (60 km/h), 1.125, ..., 1.25, 2 (120 km/h). At point x_j the proportion $\alpha_i(x_j)$ of each class a_i is chosen randomly.

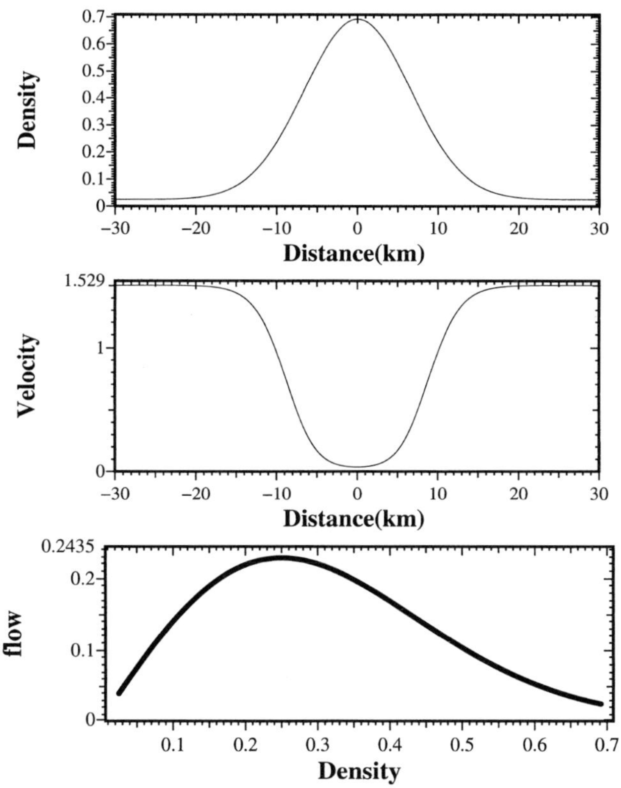

Fig. 1. Initial density, velocity and fundamental diagram

In Figure 2, we show the density, the velocity and the flow-density distribution after a time of two minutes. The density develops small oscillations due to the different behaviors of each vehicle-driver pairs, but the velocity remains with a uniformly bounded total variation, as seen in Theorem 1.

The flow-density distribution shows that at low densities, there is a linear relationship between the vehicle flow and the density and the slope corresponds to the average free velocity. With growing density, the velocity decreases monotonically, and it vanishes together with the flow at some jam density ρ_{jam}.

The flow-density data in the congested part are widely scattered within a two-dimensional area (see e.g. [6]): this complexity in traffic flow is usually interpreted as an effect of fluctuations, of jam formation, or of an instability of vehicle dynamics. If we take in (23) a unique class, these oscillations in the fundamental diagram do not appear, although in this case the initial flow density-distribution separates in two curves, which correspond to a (frequently observed) situation of two different regimes in the traffic flow. Already, studying a double Riemann problem provides a very nice cartoon of this remarkable fact, which we will investigate in a future work. Also, in Figure 3, we show an example of oscillating velocity at time 0 and after two minutes: as expected, the oscillations in v^* have disappeared.

References

1. A. Aw and M. Rascle. Resurrection of "second order" models of traffic flow. *SIAM J. Appl. Math.*, 60(3):916–938, 2000.
2. A. Aw, M. Rascle, Materne T., and Klar A. Derivation of continuum traffic flow models from microscopic follow-the-leader models. *SIAM J. Appl. Math.*, to appear, 2002.
3. P. Bagnerini and M. Rascle. A multi-class homogenized hyperbolic model of traffic flow. *Preprint Laboratoire Dieudonné 660*, 2002.
4. P. Baiti and H. K. Jenssen. Well-posedness for a class of 2×2 conservation laws with L^∞ data. *J. Differential Equations*, 140(1):161–185, 1997.
5. S. Bianchini. Stability of L^∞ solutions for hyperbolic systems with coinciding shocks and rarefactions. *SIAM J. Math. Anal.*, 33(4):959–981, 2001.
6. D. Helbing. Traffic and related self-driven many particle systems. *Reviews of Modern Physics*, 73, 2001.
7. G. C. K. Wong and S. C. Wong. A multi-class traffic flow model - an extension of LWR model with heterogeneous drivers. *to appear, Transportation Research Part A*, 36(9):827–841, 2002.

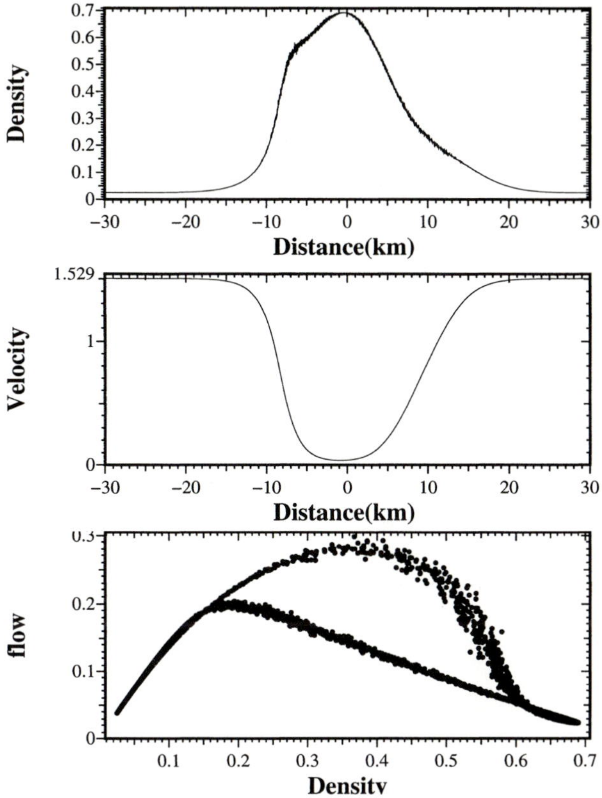

Fig. 2. Density, velocity and flow-density distribution after two minutes

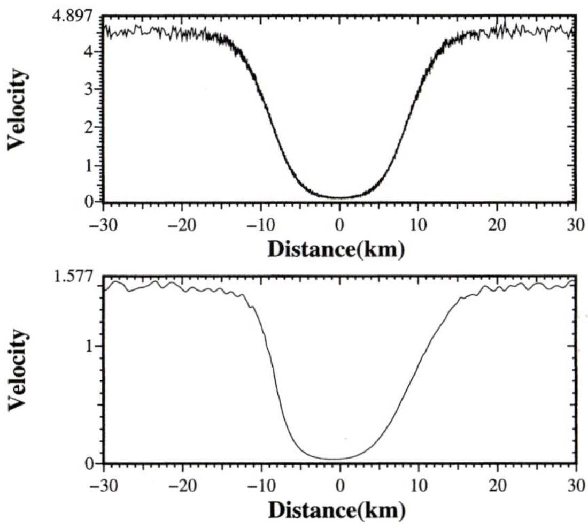

Fig. 3. Velocity at time 0 and after two minutes

The GRP Treatment of 2-D Complex Wave Structures

M. Ben-Artzi and J. Falcovitz

Institute of Mathematics, Hebrew University of Jerusalem, Israel
mbartzi@math.huji.ac.il, *ccjf@math.huji.ac.il*

1 Introduction

Solutions to initial value problems of hyperbolic conservations laws often involve discontinuities satisfying appropriate jump conditions. In the case of fluid dynamics, the 1-D equations admit two types of discontinuous waves – a shock discontinuity and a contact discontinuity. When fluid dynamics in two space dimensions is considered, even for the simplified "Riemann-type" problems (where the data are piecewise constant in sectors of the plane), a surprisingly rich variety of wave structure may evolve. Generally speaking, exact solutions of such fluid dynamical wave structures are not available and one must resort to approximations, such as those obtained by the high-resolution GRP scheme for hyperbolic conservation laws.

Studies of the family of 2-D scalar (nonlinear) conservation laws with Riemann-type initial data, for example the so-called Guckenheimer equation (see [3], and [1]), have shown that the resulting wave structures appear similar to those obtained in the fluid dynamical case. For some particular initial data [1], the wave interaction gives rise to a quite complex wave structure that includes an oblique shock, a sonic shock, a triple point, and a centered rarefaction wave (CRW). Here an exact solution is available, so that the GRP finite-difference approximation can be validated by verifying that it correctly produces all features of the solution.

In the present paper we therefore concentrate on two sample problems, one being the (scalar) Guckenheimer equation, the other a fluid dynamical interaction of a centered rarefaction wave with a converging nozzle. In both cases we present the GRP finite-difference solution. However, whereas in the scalar case it is compared to the exact solution, in the fluid dynamical case we compare it to the corresponding quasi-one-dimensional approximation – with interesting conclusions regarding this simplified approximation.

The GRP conservation law scheme is based on a solution (at each cell interface) of a "generalized Riemann problem", which is the initial value problem with piecewise linear data. Fluxes obtained this way lead to a second-order accurate, "high-resolution" integration of the conservation law. The analytic solution of the GRP at cell-boundaries is used not only for the evaluation of numerical fluxes but also, independently, for the evaluation of the new slopes. Thus, the slopes are updated by using the underlying differential equations, rather than resorting to a "reconstruction" procedure. This proves to be an important feature of the GRP-scheme, which distinguishes it from practically all other existing "high-resolution" schemes. Indeed, the GRP-scheme used here can be viewed as a predecessor to the emerging "discontinuous Galerkin" schemes, which emphasize the use of "higher-order" approximations in computational cells, based on an analytic treatment of the differential equations.

In two space dimensions, we use a Strang-type operator splitting, employing the (1-D) GRP as the one-dimensional integration operator. We refer to the recent monograph [1] for a detailed description of the GRP scheme. We point out the fact that while the split scheme is formally second-order, it is not at all clear that, in complex wave-patterns, it retains the same quality of resolution as achieved in 1-D calculations. This aspect of the approximation is also examined in this paper, using the exact (in the scalar case) solution.

The plan of this article is to start with a description of the Guckenheimer problem and its exact solution. We then present the GRP solution to this problem, and compare it to the exact solution. This is followed by a problem of fluid dynamical wave interaction, where two types of (GRP) finite-difference solutions are considered – the quasi-one-dimensional approximation, and the full 2-D solution.

2 The Guckenheimer Equation

We consider the initial value problem (IVP) for the equation,

$$u_t + f(u)_x + g(u)_y = 0, \tag{1}$$

$$u(x, y, 0) = \phi(x, y), \quad (x, y) \in \mathbb{R}^2, \tag{2}$$

where $u(x, y, t)$ is a real (scalar) function and $f(u)$, $g(u)$ are real smooth flux functions. In the "Guckenheimer equation" $f(u) = \frac{1}{2}u^2$ and $g(u) = \frac{1}{3}u^3$.

A "Riemann type" problem for (1) is the IVP where $\phi(x, y)$ is finitely valued and homogeneous of order zero,

$$\phi(x, y) = u_0(\theta), \quad \theta = \arg(x, y) \quad (= \arctan \frac{y}{x}), \tag{3}$$

Here we take the initial data

$$u_0(\theta) = \begin{cases} 0, & 0 < \theta < \frac{3\pi}{4}, \\ 1, & \frac{3\pi}{4} < \theta < \frac{3\pi}{2}, \\ -1, & \frac{3\pi}{2} < \pi < 2\pi. \end{cases} \tag{4}$$

The solution is self-similar, i.e., it is a function of the two variables $\xi = x/t$, $\eta = y/t$. Referring to the wave structure of the exact solution shown in Figure 1, we notice that outside of a large disk it consists of the following three shocks:

(a) A shock emanating (initially) from the line ($y = 0$, $x > 0$), moving at speed 1/3 in the positive y direction (note that $g(u) = u^3/3$ is concave on $[-1, 0]$).
(b) A standing shock along the line ($x = 0$, $y < 0$).
(c) A shock emanating (initially) from the line ($x + y = 0$, $x < 0$). The self-similar analysis shows that at $t > 0$ this line is given by $x + y = (5/6)t$.

The interaction of these three shocks in a disk around $(0, 0)$ gives rise to a very complex wave structure. At time $t = 1$ it can be described as follows (see Figure 1(b)). The shock **(b)** extends to a segment of the positive y-axis $0 \le y \le b$, $b = 0.2823057$. At the point $(0, b)$ it bifurcates into a CRW whose tail characteristic is a sonic shock, across which the solution $u(x, y, 1)$ jumps from -1 to the value $\tilde{v} = 0.6087418$. Then u increases across the rarefaction from $\tilde{v}$ to 1, and it is constant along each (straight) characteristic line. The rarefaction wave modifies shock **(c)** causing its front to curve. Note that the head characteristic of the CRW carries the value $u = 1$. It intersects the shock **(c)** at the point (x_0, y_0) given by

$$x_0 = \frac{\frac{5}{6} - b}{2 - b}, \qquad y_0 = \frac{\frac{5}{6} + \frac{b}{6}}{2 - b}. \tag{5}$$

The tail characteristic (sonic shock) intersects the shock **(a)** at the point $(\tilde{\xi}, \frac{1}{3})$, where $\tilde{\xi} = 0.3519610$. The result of the interaction between the CRW and the shock **(c)** leads, as noted above, to a "bending" of the latter, forming a shock branch $y = y(x)$ connecting (x_0, y_0) to $(\tilde{\xi}, \frac{1}{3})$. It can be determined by solving an ordinary differential equation (see [2]). Thus, we obtain a wave pattern that includes a shock wave bifurcating into a CRW and a sonic shock which serves as a tail characteristic of the CRW. It intersects with the other two shocks at the triple point $(\tilde{\xi}, \frac{1}{3})$. This wave pattern provides for a good test of finite-difference schemes.

Two numerical tests were performed, one using the Godunov scheme and the other with the GRP scheme. The computation domain was the square $[-1 \le x \le 1,\ -1 \le y \le 1]$ which was divided into 320×320 square cells. The time step was $\Delta t = 0.003125$ (i.e., $\mu_{CFL} = 0.5$ since $max|u| = 1$ and $f'(u) = u$, $g'(u) = u^2$), and the computation was performed to final time $t = 1$. The boundary conditions were specified by calculating the exact solution on the outer segments of boundary cells. This is possible as long as the domain boundary is intersected only by the three shocks **(a)**, **(b)**, **(c)**, which according to Figure 1(b) is still true at $t = 1$.

The results are shown in the sub-domain $[-0.05 \le x \le 0.60,\ 0 \le y \le 0.65]$ [see Figure 2(a) for the Godunov scheme, and Figure 2(b) for the GRP scheme]. Recall that inside the rarefaction fan u is constant along the (straight) characteristic lines, so that numerical U-level curves approximate the fan structure. The U-level sequence (6) given below is designed to show the shock fronts and the rarefaction fan. The five levels $L = 9, ..., 13$ correspond to the tail, head and three inner characteristic lines of the rarefaction fan [as shown in Figure 1(b)].

$$U_L = \begin{cases} -1 + 0.2\,L & \dots \quad L = 0, 1, ..., 8 \\ 0.60874,\ 0.68295,\ 0.76366,\ 0.86089,\ 1, & \dots \quad L = 9, ..., 13. \end{cases} \tag{6}$$

In order to enable interpolation at the lowest and highest U-levels, they were slightly shifted to -0.990 and 0.997, respectively. For comparison of the exact and numerical solutions, we represent the exact solution [Figure 1(b)] by discrete "marker points" situated on shock fronts, as shown in Figure 2. Additional marker points are located at points (x, y) inside the rarefaction fan, where the exact solution takes on the values U_L, $L = 9, ..., 13$ given by (6).

Our primary observation with respect to the numerical solution is that both finite-difference schemes (applied by Strang-type operator splitting) produce a correct approximation to this complex 2-D wave-interaction pattern [Figure 2]. The GRP solution agrees quite well with the exact one, whereas the Godunov solution shows a nearly equal agreement for the shock fronts, but a lesser agreement in the rarefaction fan. In this centered fan, the characteristic line that coincides with the sonic shock front corresponds to a constant value of $u = \tilde{v}$, and it is one of the U-level lines plotted ($L = 9$). In the GRP solution this line is seen very near the sonic shock front (Figure 2(b)), whereas in the Godunov case its stand-off distance is perceptibly higher [Figure 2(a)]. The captured sonic shock is represented by the cluster of level lines $L = 0, ..., 9$ (since the jump across this shock is from $u = U_0$ to $u = U_9$). At the other end of the rarefaction fan, the head characteristic line is plotted with $U_{13} = 0.997$ (close to the exact value of $U_{13} = 1$, for a clear U-level interpolation). In the Godunov solution this line extends well beyond the exact solution, whereas in the GRP solution it agrees well with the exact marker points. The rarefaction fan is the only region of the solution where $u(x, y, 1)$ varies smoothly with a non-zero gradient. Hence, these observations indicate that in such regions the (second-order accurate) GRP scheme produces considerably smaller errors than the (first-order accurate) Godunov scheme. In what concerns the bifurcation point $(0, b)$ and the triple point $(\tilde{\xi}, \frac{1}{3})$, resulting from the two-dimensional setting, we observe that they are well replicated by both schemes.

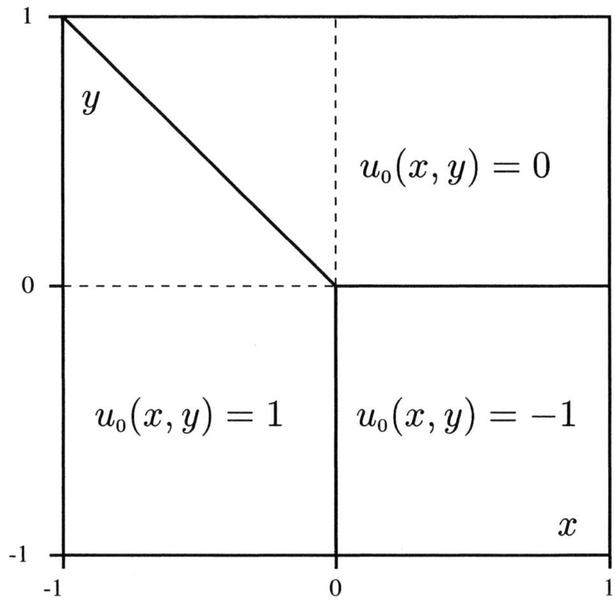

(a) Initial data for Guckenheimer Structure

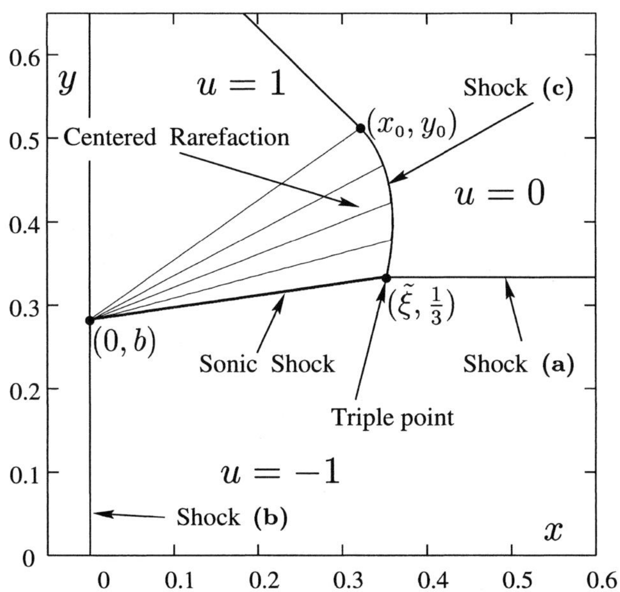

(b) Exact Guckenheimer Structure at $t=1$

Fig. 1. The Guckenheimer Structure for $u_t+(u^2/2)_x+(u^3/3)_y=0$.

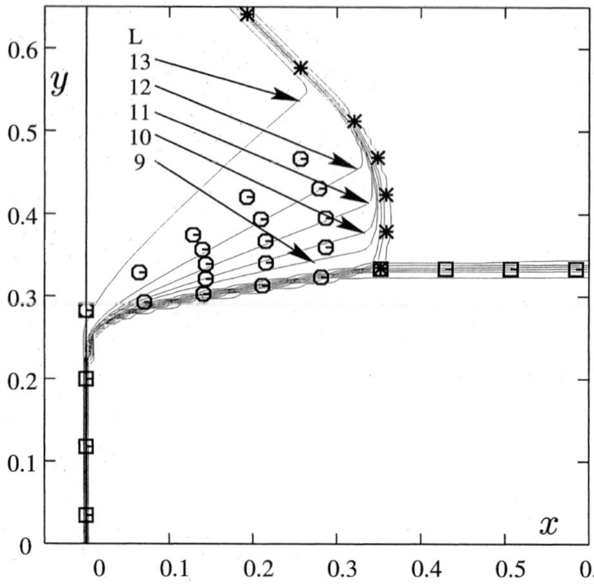

(a) First-order (Godunov $\Delta x = \Delta y = 0.00625$)

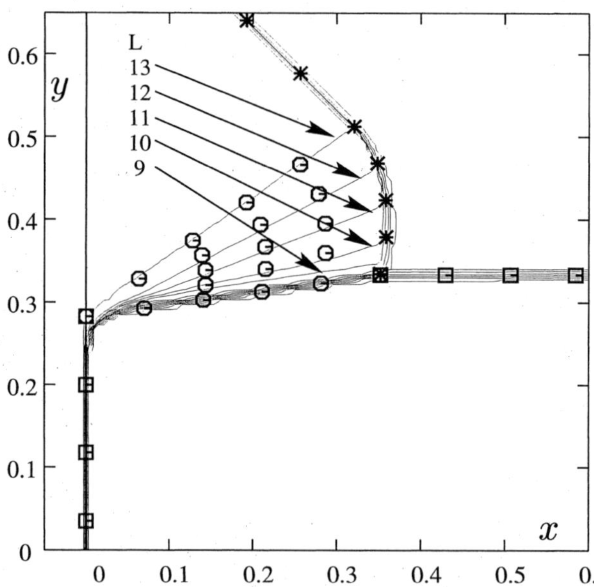

(b) Second-order (GRP $\Delta x = \Delta y = 0.00625$)

Legend
- ✱ curved or oblique shock
- ⊟ planar shock (x-facing or y-facing)
- ⊖ characteristic lines (or sonic shock)

Fig. 2. U-level curves for Guckenheimer equation at $t = 1$
$u_t + (u^2/2)_x + (u^3/3)_y = 0$. Initial data in Figure 1(a).

3 Interaction of a Centered Rarefaction Wave with a Converging Nozzle

Consider a centered rarefaction wave that propagates in a planar duct comprising two long segments of uniform cross-section area joined by a smooth converging nozzle. Such processes take place in numerous systems of industrial and scientific interest. The significance of this sample flow lies in the comparison we make between a full multidimensional solution and the corresponding quasi-1-D approximation. The reason is that the latter (often referred to as the "duct flow" approximation) is commonly employed as an engineering design tool. Hence, studying the bounds of its validity as a simplified approximation to the full multidimensional solution is highly significant to engineering design and analysis.

For the multidimensional computation of the wave interaction with a converging nozzle we employ the operator-split 2-D GRP scheme where the symmetric duct contour intersects an underlying two-dimensional Cartesian grid, using the "moving boundary" scheme (where the boundary is stationary). We refer to [1] for details of the MBT scheme and the nozzle contour, as well as to a more comprehensive analysis of the present wave interaction problem.

We consider the case of a 1:10 pressure ratio centered rarefaction wave (CRW) in a fluid assumed to be a perfect gas with $\gamma = 1.4$. The CRW is initially located in the wider part of the duct, and it propagates towards a (short) converging nozzle of $2:1$ cross-section area ratio. The initial data is that of a Riemann problem designed to produce a right-facing CRW. It consists of two uniform states $\mathbf{U}_L=[\rho_L, p_L, u_L]=[0.27030, 0.1, -1.4016]$, $x < 1.3$, and $\mathbf{U}_R=[\rho_R, p_R, u_R]=[1.4, 1, 0]$, $x > 1.3$. The location of the initial discontinuity ($x = 1.3$) is just ahead of the converging segment that occupies the interval $1.6 \le x \le 2.6$ (see Figure 3 at $T = 0$).

The 2-D computation is conducted in the (symmetric) upper half of the duct, embedding it in the rectangular domain $(x, y) \in [-1.6, 9.4] \times [0, 1]$, which is divided into a grid of 550×50 square cells ($\Delta x = \Delta y = 0.02$). A rigid-wall boundary condition (due to symmetry) is imposed at the nozzle contour and at the centerline ($y = 0$). On the left and right sides of the computational rectangle we impose "non-reflecting" boundary conditions, designed to allow waves to pass through these endplanes (almost) undisturbed. The computation was performed in the time interval $[0, 9]$, with time steps adjusted (at each integration cycle) to have a nearly constant CFL coefficient $\mu_{CFL} = 0.7$.

The quasi-1-D computation was conducted in the same x interval, with the same spatial grid, the same contour area function $A(x)$, and the same inflow/outflow boundary conditions as in the 2-D case. The computation was performed with time steps adjusted to have the same $\mu_{CFL} = 0.7$ and in the same time interval $[0, 9]$.

The results of the 2-D computation are shown as a time-sequence maps of isobars ($p = const.$) in Figure 3. Notice that in the initial state $\mathbf{U}_L$ the flow is already supersonic since $u_L = -1.947c_L$, so that $|u_L| > c_L$. Thus, a supersonic expansion flow gradually evolves in the *diverging* direction of the nozzle.

From the time-sequence of isobars plots (Figure 3), we observe that the flow over-expands in the nozzle, so that an upstream-facing oblique shock wave is formed (marked by $*$ in Figure 3). It serves to align the flow velocity vector with the (straight) duct contour, and to match the pressure to that prevailing further left in the wider duct segment. Another complex shock interaction forms at the centerline ($y = 0$), and evolves into a full Mach reflection.

In Figure 4 we compare the results of the 2-D and quasi-1-D computations at the final time $t = 9$, showing an isobars map (as in the last frame of Figure 3), followed by profiles of density, pressure, and flow Mach number, as functions of the centerline coordinate (Figures 3(b)–(d)).

The two solutions are in close agreement throughout the narrower duct segment ($x > 2.6$), but disagree elsewhere. At the entrance to the nozzle ($x = 2.6$), the flow speed is sonic ($M = 1$ in 4(d)), enabling the nozzle flow to approach a *steady* supersonic expansion flow, commencing at a virtual sonic plane ("nozzle entrance"), which serves to "match" the *unsteady* rarefaction wave on its right to the steady flow on its left. In other words, upon passing through the nozzle, the CRW is "split" into a "transmitted" part and a "reflected" part, the two being separated by a (nearly) steady flow through the diverging nozzle.

Now the source of disagreement between the quasi-1-D and the 2-D solutions is evident. The fluid expands as it (reversely) flows through the nozzle at supersonic speed, and a full 2-D description of this flow involves an oblique shock system at the nozzle exit (Figure 4(a)), which is poorly approximated by the cross-section-averaged quasi-1-D solution that relies on a normal shock for matching the over-expanded supersonic nozzle flow to the pressure ahead. Moreover, the flow passing through the Mach reflection shock structure is separated by the slip-line into two streams having distinctly different thermodynamic properties. Naturally, the 1-D "averaging" of these two streams involves significant deviations from the 2-D flow field.

It is concluded that although quasi-1-D calculations may generally be adequate as an engineering approximation, a verification by comparison to the appropriate multidimensional solution is required in order to make sure that the disagreement between the two remains within acceptable bounds. Our test case analysis thus brings out the significance of full multidimensional numerical solutions; the (simpler) quasi-1-D solutions may not always serve as adequate approximations.

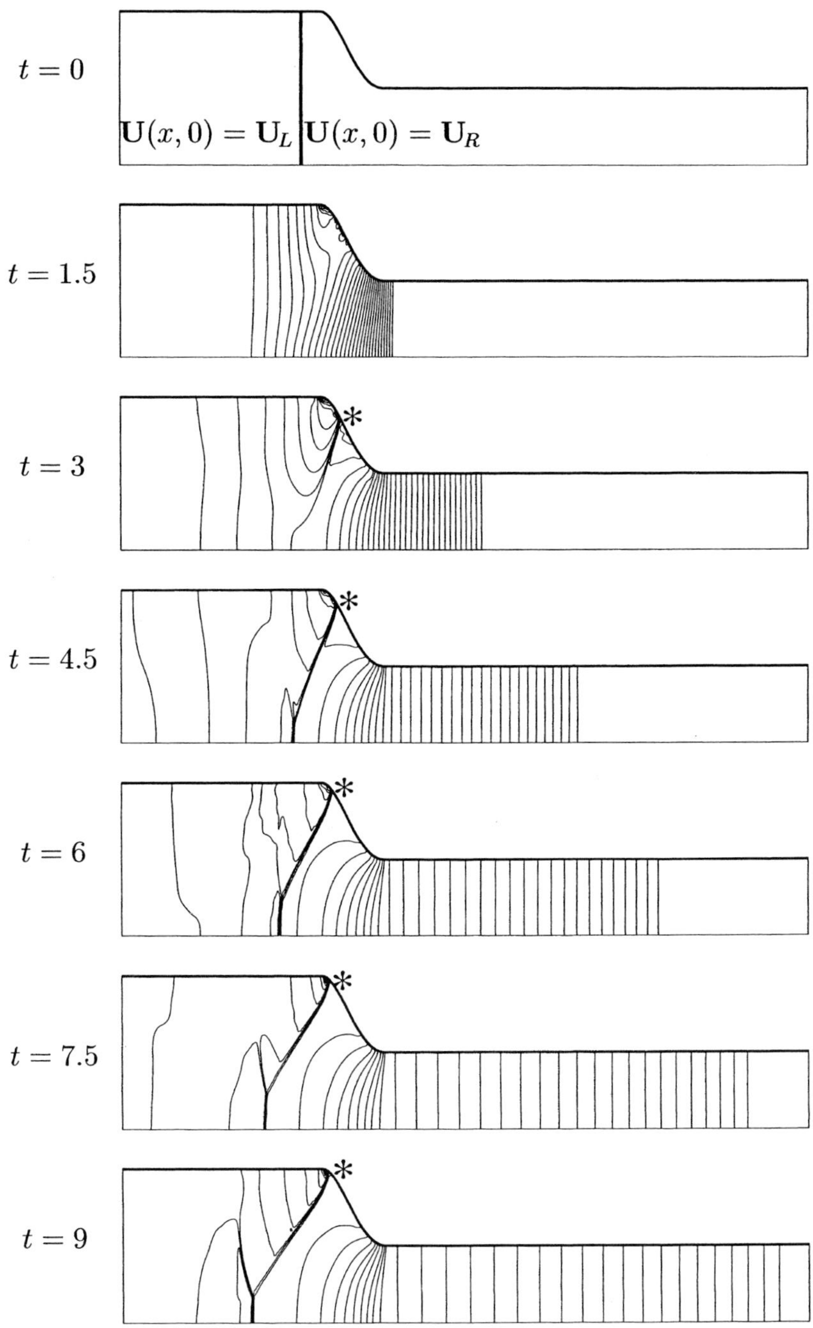

Fig. 3. Time sequence of CRW interaction with a converging segment. Isobars map. The * marks shock formation at duct wall. (Duct width here is twice the true size, for better visibility).

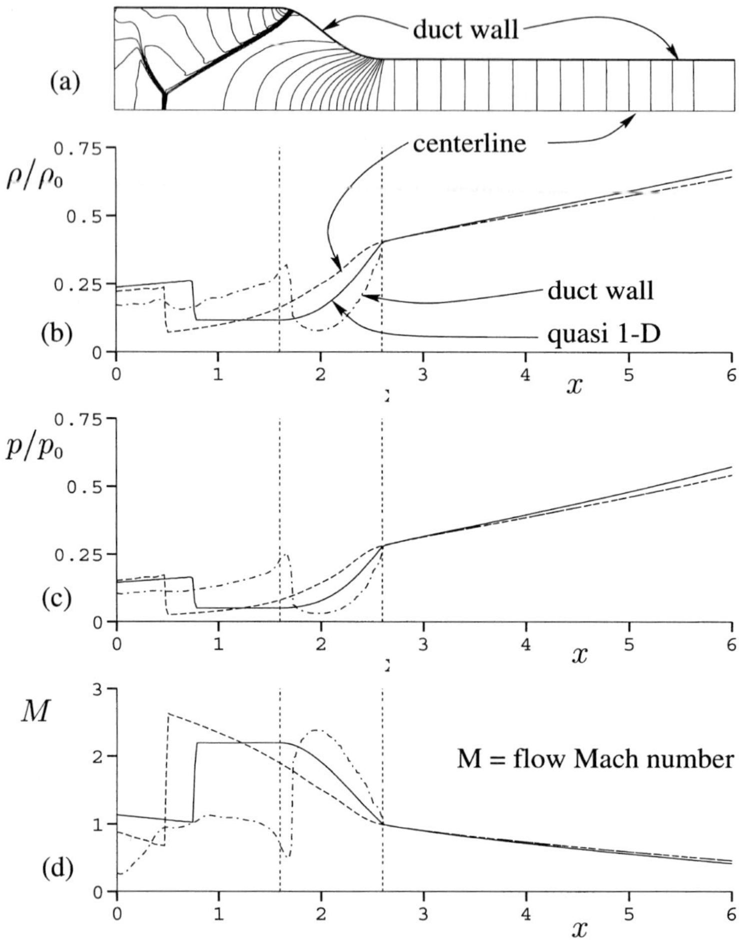

Fig. 4. CRW interaction with a converging segment; comparison of 2-D and quasi-1-D results at time $t = 9$.
(a) Isobars of 2-D calculation.
(b)–(d) Quasi 1-D solution. Distribution of density, pressure and flow Mach number (taken as positive).

References

1. M. Ben-Artzi and J. Falcovitz (2002): Generalized Riemann Problems in Computational Fluid Dynamics. Cambridge University Press, London, New York (in press).
2. M. Ben-Artzi, J. Falcovitz, and J. Li (2001): Wave interactions and numerical approximations for two-dimensional scalar conservation laws. Preprint, Institute of Mathematics, Hebrew University of Jerusalem.
3. J. Guckenheimer (1975): Shocks and rarefactions in two space dimensions. Arch. Rational Mech. Anal., **59**, 281–291.

BV Solutions of Semidiscrete Upwind Scheme

Stefano Bianchini

Istituto per le Applicazioni del Calcolo "M. Picone" - CNR, Viale del Policlinico 137, 00161 Roma (ITALY) *bianchin@iac.rm.cnr.it*

1 Introduction

Consider the strictly hyperbolic system of conservation laws

$$u_t + f(u)_x = 0, \tag{1}$$

where $u \in I\!R^N$, $f : I\!R^N \mapsto I\!R^N$ smooth. The system is said to be strictly hyperbolic provided that the Jacobian matrix $A(u) \doteq Df(u)$ has N distinct eigenvalues, $\lambda_1(u) < \ldots < \lambda_N(u)$.

In [14] weak solutions to (1) are constructed under the assumption that the initial data

$$u(0,x) = \bar{u}(x), \tag{2}$$

has small total variation and that for each $m \in \{1,\ldots,N\}$, the m-th characteristic field is either *linearly degenerate* or else it is *genuinely nonlinear.*

The idea in Glimm's proof is to obtain an a priori estimate on the total variation of the approximate solutions by introducing a *wave interaction potential.* In turn, the control of the total variation yields the compactness of the family of approximate solutions, and hence the existence of a strongly convergent subsequence in $L^1(I\!R, I\!R^N)$. Alternative constructions of approximate solutions, based on front tracking approximations, were subsequently developed in [8], [13].

The well posedness of the Cauchy problem was established in a series of papers [9], [11], [12]. For comprehensive account of the recent uniqueness and stability theory we refer to [10].

Recently in [4] it is proved that the solution u^ϵ of the parabolic system

$$u_t + f(u)_x - \epsilon u_{xx} = 0 \tag{3}$$

with initial data (2) converges to the unique viscosity solution of (1), as $\epsilon \to 0$. This solution depends Lipschitz continuously on the initial data $\bar{u}$ in the L^1 norm and can be characterized by defining a Riemann solver for (1). It turns out that this Riemann solver is the extension of Liu's Riemann solver for general strictly hyperbolic systems, see [3], [4], [16].

In this contribution we consider a semidiscrete approximation to (1), obtained by discretising only the space variable x:

$$\frac{\partial}{\partial t}u(t,x) + \frac{1}{\epsilon}\Big(f\big(u(t,x)\big) - f\big(u(t,x-\epsilon)\big)\Big) = 0. \tag{4}$$

For linear stability, we assume that $\lambda_1(u) > 0$.

The solution of (4) is defined on the lattice $\mathbb{R} \times \epsilon\mathbb{Z}$. In the following we shall denote with $u^j(t)$ the value of the solution in $j\epsilon$,

$$u^j(t) \doteq u(t, j\epsilon).$$

Our aim is to prove that if $\bar{u}$ has sufficiently small total variation, then the solution $u^j(t)$ to (4) has bounded BV norm uniformly in t and ϵ. As in the parabolic case, it is clear that the rescaling

$$t \mapsto \frac{t}{\epsilon}, \qquad x \mapsto \frac{x}{\epsilon}, \tag{5}$$

leaves the total variation of the initial data $\bar{u}$ and of the solution $u^j(t)$ unchanged, and (4) becomes

$$u^j_t + f(u^j) - f(u^{j-1}) = 0. \tag{6}$$

The main result presented in the talk is the following theorem:

Theorem 1. *Consider the semidiscrete upwind scheme (6) with initial data $u^j(0) = \bar{u}^j$. Assume that*

$$\text{Tot.Var.}(u^j) = \sum_j \left|u^j - u^{j-1}\right| \leq \delta_0/4$$

for some constant δ_0 sufficiently small. Then the solution to (6) exists for all $t \geq 0$ and has uniformly bounded total variation:

$$\text{Tot.Var.}\big(u^j(t)\big) \leq 4N\delta_0, \quad \forall t \in \mathbb{R}^+. \tag{7}$$

Moreover, the semigroup $S_t\bar{u}^j = u^j(t)$ generated by (6) is Lipschitz continuous w.r.t. the ℓ^1-norm: there exist constants L, L' such that if $u^j(t)$, $z^j(t)$ are two solutions, then

$$\sum_j \left|u^j(t) - z^j(s)\right| \leq L \sum_j \left|u^j(0) - z^j(0)\right| + L'\left|t - s\right|. \tag{8}$$

Note that the continuous dependence w.r.t t follows trivially from (6).

Using the above estimates, it is possible to prove that, as $\epsilon \to 0$, the solutions $u^{j,\epsilon}$ of (4) yield a weak solution to (1).

It is now possible to prove that the sequence u^j, solution of (4), converges to a weak solution to (1). In fact, define the function u^ϵ by

$$u^\epsilon(t,x) = u^{j,\epsilon}(t) \qquad (j-1)\epsilon < x \leq j\epsilon. \tag{9}$$

Since u^ϵ solves (4) with initial data

$$u^\epsilon(0,x) = \frac{1}{\epsilon}\int_{(j-1)\epsilon}^{j\epsilon} \bar{u}(x)dx \qquad (j-1)\epsilon < x \leq j\epsilon,$$

we can write in weak form

$$\iint_{\mathbb{R}^+\times\mathbb{R}} \left\{ u^\epsilon(t,x)\varphi_t(t,x) + f\left(u^\epsilon(t,x)\right)\frac{\varphi(t,x+\epsilon)-\varphi(t,x)}{\epsilon} \right\} dtdx \quad (10)$$

$$+\sum_j \int_{(j-1)\epsilon}^{j\epsilon} \bar{u}(x) \left(\int_{(j-1)\epsilon}^{j\epsilon} \varphi(0,y)dy \right) dx = 0,$$

where φ is a smooth compact supported function. From (7) and the fact that by construction

$$\lim_{j\to-\infty} u^j(t) = \lim_{j\to-\infty} \bar{u}^j,$$

it follows that, up to a subsequence, u^ϵ converges in L^1_{loc} to a BV function $u(t,x)$, so that passing to the limit in (10) one obtains

$$\iint_{\mathbb{R}^+\times\mathbb{R}} \left\{ u(t,x)\varphi_t(t,x) + f\left(u(t,x)\right)\varphi_x(t,x) \right\} dtdx + \int_{\mathbb{R}} \bar{u}(0,x)\varphi(0,x)dx = 0.$$

Since it is easy to check that

$$\left\|u^\epsilon\right\|_{L^1} = \epsilon\left\|u^{j,\epsilon}\right\|_{\ell^1} = \epsilon\left\|u^j\right\|_{\ell^1},$$

from (8), a simple argument shows that up to a subsequence the limit solutions of (1) satisfy

$$\left\|u(t)-z(s)\right\|_{L^1} \le L\left\|u(0)-z(0)\right\|_{L^1} + L'\left|t-s\right|.$$

In particular they form a Lipschitz continuous semigroup in L^1.

At this stage, we still must rule out the possibility that different subsequences may give rise to different semigroups. To characterize the limit, we follow the approach of [4], i.e. we first identify the Riemann solver compatible with the semidiscrete approximation (4).

Consider an initial data of the form

$$u(0,x) = \begin{cases} u^- & x < 0 \\ u^+ & x \ge 0 \end{cases} \quad (11)$$

with $u^+ - u^-$ sufficiently small. We prove that when $\epsilon \to 0$, the solution $u^\epsilon(t)$ of (4) tends to a particular self-similar function, which can be precisely described by means of the dynamics on the center manifold of the retarded functional equation

$$\begin{cases} -\sigma\phi'(\xi) + f\left(\phi(\xi)\right) - f\left(\phi(\xi-1)\right) = 0 \\ \sigma'(\xi) = 0 \end{cases} \quad (12)$$

In [3] it is shown how to construct a Riemann Solver compatible with the semidiscrete scheme (4). In particular it is shown that this Riemann Solver coincides with the Riemann Solver obtained by means of the vanishing viscosity approximations, and can be identified by saying that all jumps in the solution $u(t)$ satisfy Liu's stability condition. Due to the particular choice of the initial data (11), one can show using the same arguments of [4] that the whole sequence $u^\epsilon(t)$ converges to $u(t)$.

At this point the limiting solution $u(t)$ is identified by proving that it is a viscosity solution to

$$u_t + Df(u)u_x = 0. \tag{13}$$

The definition of viscosity solution is given in [9], and relies on local integral estimates: in the first estimate one compares $u(t)$ with the solution of a local Riemann problem, while in the second estimate one compare the solution $u(t)$ with the with the solution to the linear system

$$u_t + Df\big(u(\tau,\xi)\big)u_x = 0,$$

obtained by freezing the coefficients at some point $u(\tau, xi)$. Note that, while the weak solution to a linear system is uniquely defined, different Riemann Solvers define different viscosity solutions. The viscosity solution we are considering here is the one corresponding to the Riemann Solver described above.

As in [9], one can show that the semigroup trajectories corresponding to a given Riemann solver are precisely the viscosity solutions. Since we show that each limiting solution $u(t)$ of (13) obtained by the semidiscrete approximations $u^\epsilon(t)$ is a viscosity solution to (13), the uniqueness of the limit follows.

We prove then the following theorem:

Theorem 2. *As $\epsilon \to 0$, the solution $u^\epsilon(t)$ defined in (9) converges in L^1 to a unique limit $u(t)$. This limit coincides with the viscosity solution characterized by the Riemann solver defined in [3], [4], i.e. with the unique Riemann solver such that each shock satisfies Liu's stability condition. In particular, $u(t)$ coincides with the vanishing viscosity limit of (3).*

We now sketch the main steps for the proof of Theorem 1 and Theorem 2.

One of the fundamental points is the study of the center manifold of travelling profiles, i.e. the center manifold $\mathcal{C}_m$ of the Retarded Functional Differential Equation

$$\begin{cases} -\sigma\phi'(\xi) + f\big(\phi(\xi)\big) - f\big(\phi(\xi-1)\big) = 0 \\ \sigma'(\xi) = 0 \end{cases}$$

near the equilibrium point $(\bar{u}, 0, \lambda_m(\bar{u}))$. As it is shown in [1], this manifold is $n+2$-dimensional and contains all the long term dynamics near a fixed point $u \equiv u_0$, $\sigma \equiv \lambda_m(u_0)$, where $\lambda_m(u)$ is the m-th eigenvalue of $Df(u)$. In particular it contains all the small bounded travelling profiles with speed close to that eigenvalue. We can parameterize the manifold by $\phi(0)$, $-\sigma_m\phi_m'(0)$,

σ, i.e. the value of the profile, the m-th component of the time derivative at $\xi = 0$, and the speed σ:

$$(u, v_m, \sigma) \quad \mapsto \quad \phi(\xi; u, v_m, \sigma),$$

where $\xi \in [-1, 0]$. In particular we obtain the functions

$$-\sigma\phi'(0) = \omega_m(u, v_m, \sigma), \qquad -\sigma\phi'_m(-1) = \pi_{mm}(u, v_m, \sigma),$$

i.e. the derivative at $\xi = 0$ (not just to m-th component) and the m-th component at $\xi = -1$. One can show that the knowledge of the functions ω_m, π_{mm} is equivalent to the knowledge of the center manifold. In fact, since the points $(u, v_m = 0, \sigma)$ correspond to equilibria on the center manifold $\mathcal{C}_m$, on obtains

$$\begin{aligned} -\sigma\phi'(0) &= \omega_m(u, v_m, \sigma) = v_m \tilde{r}_m(u, v_m, \sigma) \\ -\sigma\phi'_m(-1) &= \pi_{mm}(u, v_m, \sigma) = v_m \tilde{p}_m(u, v_m, \sigma), \end{aligned}$$

for some smooth function $\tilde{r}_m$, $\tilde{p}_m$. ¿From (6) for a travelling profile ϕ we obtain

$$\phi(-1) = f^{-1}(v + f(u)) = f^{-1}\Big(v_m(x)\tilde{r}_m(u, v_m, \sigma_m) + f(u)\Big),$$

and the differentiation w.r.t. t yields

$$v_t = A(\phi(-1))\phi_t(-1) - A(u)v_m\tilde{r}_m = A(\phi(-1))v_m\tilde{p}_m\tilde{r}_m(\phi(-1)) - A(u)v_m\tilde{r}_m.$$

Projecting along the m-th component gives the ODE

$$v_{m,t} = v_m\Big(\tilde{\lambda}_m(\phi(-1), v_m\tilde{p}_m, \sigma)\tilde{p}_m - \tilde{\lambda}_m(u, v_m, \sigma)\Big) = v_m\phi(u, v_m, \sigma),$$

where

$$\tilde{\lambda}_m(, v_m, \sigma) = \big\langle l_m(\bar{u}), A(u)\tilde{r}_m(u, v_m, \sigma)\big\rangle.$$

This ODE gives the dynamics on the manifold $\mathcal{C}_m$.

Using a Taylor expansion, we prove some properties of the functions $\tilde{r}_m$, $\tilde{p}_m$ which are needed when we want to answer to the following question: how to identify a travelling profile looking at the solution near the j-th point. In particular the fact that

$$\frac{\partial \tilde{p}}{\partial v_m} \neq 0,$$

allows to identify the speed σ_m of a travelling profile of the m-th family as a function of the ratio v_m^{j-1}/v_m^j.

Using also the property

$$\frac{\partial \tilde{r}_m}{\partial \sigma} = \mathcal{O}(1)v_m,$$

which is a consequence of the fact that if the system is linearized (i.e. $v_m \to 0$), then it is diagonalizable using the eigenvectors $r_m(u)$, it is possible to show that in each j it is possible to decompose u_t^j as a sum of N time derivatives of travelling profiles ϕ_m, i.e.

$$u_t^j = \sum_m \phi_{m,t} = \sum_m v_m \tilde{r}_m.$$

The important property of the above decomposition is that if u^j is a single travelling profile, then we identify it exactly.

This situation is similar to the parabolic or hyperbolic case, where by looking at the first and second derivatives or the jump we can specify the travelling profile or shock which best fits the data.

The main difference from the parabolic or hyperbolic case is that, to identify a wave, we must look at the solution in 3 consecutive points, i.e. u^j, u^{j-1}, u^{j-2}. This implies that we cannot decompose the solution u^j to (6) locally, but we must decompose it as a whole. A consequence of this fact is that, when we write the equations satisfied by the scalar components of our decomposition, we will get a source term which depends on all the previous waves. This means that the source in j depends on all $k \leq j$. Remember in fact that the kernel of the linearized equation moves only forward, i.e. the solution in $k > j$ has no influence in the point j.

When u^j is not a travelling profile, we obtain errors in the scalar equation satisfied by the components v_m^j,

$$v_m^j + \tilde{\lambda}_m(u^j, v_m^j, \sigma_m^j) v_m^j - \tilde{\lambda}_m^{j-1}(u^{j-1}, v_m^{j-1}, \sigma_m^{j-1}) v_m^{j-1} = \phi_m^j(t)$$

We can collect the source terms appearing in $\phi_m^j(t)$ in 3 categories:

1. terms corresponding to the interaction among waves of different families. Following [5], we will refer to these terms as *transversal terms.*
2. terms corresponding to the interaction of waves of the same family. We will call them *non transversal terms.*
3. terms due to the fact that the center manifold is defined for speed close to $\lambda_m(u_0)$, so that in some cases we cannot give the right speed. This means that the numerical diffusion of (6) is greater than the drift. These terms will be called the *energy terms.*

By study the source terms arising in the above decomposition, it is possible to prove that they are integrable in $\mathbb{R}^+ \times \mathbb{Z}$ and that their integral is of the order of the total variation of u^j squared. To achieve this result, we introduce 4 functionals. The first functional is related to the transversal terms, and was studied in [2]: this is the semidiscrete analog of the Glimm functional for wave of different families. The other three are the semidiscrete analog of the functionals introduced in [6], [7], [4] for studying interaction among waves of the same family. Using these functionals we will prove that the source terms of total variation is of the order of the total variation squared, hence the total variation remains uniformly bounded if it is sufficiently small at $t = 0$:

$$\text{Tot.Var.}\big(u^j(t)\big) \leq 16N\text{Tot.Var.}(\bar{u}^j).$$

The construction of the functionals for the same family surprisingly depends on the dispersion relation of (6). In fact, due the dispersion, the weight that we assign to each wave in the interaction functional is a nonlinear function of the speed, and not the speed itself. This nonlinear function is obtained directly from the dispersion relation.

This can be easily seen by considering a linear scalar semidiscrete equation

$$u_t^j + u^j - u^{j-1}, \qquad u^j \in I\!R.$$

In this case it can be shown that the Glimm functional is given by

$$Q(t) = \frac{1}{2}\sum_{j<k}\left|u_t^j u_t^{k-1} - u_t^k u_t^{j-1}\right| = \frac{1}{2}\sum_{j<k}|v^j||v^k|\left|\frac{v^{j-1}}{v^j} - \frac{v^{k-1}}{v^k}\right|.$$

Note that the weight we assign to each wave is v^{j-1}/v^j, which is related to the real speed σ by

$$\sigma = \frac{1 - v^{j-1}/v^j}{\log\big(v^{j-1}/v^j\big)}.$$

We also observe that, differently from the vanishing viscosity case, we have to consider the non-transversal functionals and the energy together, i.e. the time derivative of each functional contains the derivative of the others, multiplied by a small constant. This implies that we cannot show the decrease of a single functional.

Using similar techniques, one can prove that a perturbation ζ^j of (6), satisfying

$$\zeta_t^j + A(u^j)\zeta^j - A(u^{j-1})\zeta^{j-1} = 0,$$

can be studied by means of a decomposition in travelling profiles:

$$\zeta = \sum_m \zeta_m \tilde{r}_m\big(u^j, v_m^j, \sigma_m^j\big).$$

In particular on obtains that the ℓ^1 norm of a perturbation ζ is uniformly bounded:

$$\big\|\zeta(t)\big\|_{\ell^1} \leq L\big\|\zeta(0)\big\|_{\ell^1},$$

for some constant L depending only on the total variation of u^j. A simple homotopy argument then shows that the solutions to the semidiscrete upwind scheme (6) form a Lipschitz continuous semigroup: if $u(t)$, $z(t)$ are two solutions of (6), then

$$\big\|u(t) - z(t)\big\|_{\ell^1} \leq L\big\|u(0) - z(0)\big\|_{\ell^1}.$$

This concludes the proof of Theorem 1.

Finally one can show that the limit of u^ϵ defined in (9) coincides with the vanishing viscosity solution. We prove that in the limit $\epsilon \to 0$ the solution u has finite speed of propagation. Next, using the results of [3], [4], we can identify the Riemann Solver corresponding to the limiting solution $u(t)$. Finally we prove that $u(t)$ can be identified as a viscosity solution to (13).

This will yield a proof of Theorem 2.

References

1. S. Benzoni-Gavage. Semidiscrete shock profiles for hyperbolic systems of conservation laws. *Physica D*, 115:109–123, 1998.
2. S. Bianchini. A note on singular limits to hyperbolic systems. preprint SISSA, 2000.
3. S. Bianchini. On the Riemann problem for non-conservative hyperbolic systems. preprint IMS nr. 85, 2002.
4. S. Bianchini and A. Bressan. Vanishing viscosity solutions of nonlinear hyperbolic systems. preprint SISSA.
5. S. Bianchini and A. Bressan. BV solution for a class of viscous hyperbolic systems. *Indiana Univ. J.*, 49:1673–1713, 2000.
6. S. Bianchini and A. Bressan. A case study in vanishing viscosity. *Disc. Cont. Dyn. Sys.*, 7(3):449–476, 2001.
7. S. Bianchini and A. Bressan. A center manifold technique for tracing viscous travelling waves. *Comm. Pure Appl. Analysis*, 1(2):161–190, 2002.
8. A. Bressan. Global solutions to systems of conservation laws by wave-front tracking. *J. Math. Anal. Appl.*, 170:414–432, 1992.
9. A. Bressan. The unique limit of the Glimm scheme. *Arch. Rational Mech. Anal.*, 130:205–230, 1995.
10. A. Bressan. *Hyperbolic systems of conservation laws.* Oxford Univ. Press, 2000.
11. A. Bressan, G. Crasta, and B. Piccoli. *Well posedness of the Cauchy problem for $n \times n$ systems of conservation laws.* Memoir A.M.S., 2000. to appear.
12. A. Bressan, T.P. Liu, and T. Yang. l^1 stability estimates for $n \times n$ conservation laws. *Arch. Rat. Mech. Anal.*, 149:1–22, 1999.
13. R.J. DiPerna. Global existence of approximate solutions to conservation laws. *J. Diff. Eq.*, 20:187–212, 1976.
14. J. Glimm. Solutions in the large for nonlinear hyperbolic systems of equations. *Comm. Pure Appl. Math.*, 18:697–715, 1965.
15. J.K. Hale and S.M. Verduyn Lunel. *Introduction to functional differential equation.* Springer-Verlag, New York, 1993.
16. T.-P. Liu. *Admissible solutions of hyperbolic conservation laws*, volume 240. Memoir A.M.S., 1981.

A Supplement to Entropy Condition

Shuxing Chen

Center of Mathematics of Zhejiang University, *sxchen@mail.hz.zj.cn*, and
Institute of Mathematics, Fudan University, *sxchen@fudan.ac.cn*

1 Introduction

The purpose of this article is to study the uniqueness and stability of steady flow with a shock front. When a uniform supersonic flow past a plane wedge with the vertex angle less than a critical value, there will be a plain shock front attached at the edge of the wedge. The location of the shock can be determined by the intersection of the ray with a given angle and the shock polar determined by the parameters of the coming flow. However, in most cases there will be two such intersections: one corresponds to the weaker oblique shock, the other corresponds to the stronger oblique shock. Both two satisfy the Rankine-Hugoniot condition and the entropy condition. Hence a natural question arises, which one is actually occurs? It is generally understood that the weaker one is stable and the stronger one is not, so that only the weak one could occur. However, there is not a convincing proof in mathematics so far. Since in the shock theory many problems are based on the study of this fundamental problem, for instance, the study of supersonic flow past a given body, the reflection or the interaction of shock waves etc., then, it is important and desirable to find a criterion as a supplement of the entropy condition to single out a physically reasonable shock front.

In this paper we will give a way to distinguish these two shocks. The main spirit of the method is to study the asymptotic behavior of the shock in the perturbed problem obtained by smoothing the boundary. That is, we first replace the broken-line boundary by a smooth curved boundary, then the flow also becomes a smooth Prandtl-Meyer flow near the boundary. Due to the bending of the boundary the flow will produce a shock front, the strength of which will gradually increases. It can be proved that the shock can extend to infinity and approaches the weaker oblique shock in the original problem. Meanwhile, the stronger shock disappears forever. Such a phenomenon leads us to the conclusion that the oblique weaker shock is stable under the perturbation of the boundary. In the last part of the paper we will emphasize this point and put forward our criterion to choice a unique reasonable shock front.

2 The problem and its smoothing version

Let us use the system of isentropic and irrotational flow equation to describe the flow. In two dimensional case the system is

$$\begin{cases} \dfrac{\partial(\rho u)}{\partial x} + \dfrac{\partial(\rho u)}{\partial y} = 0, \\ \dfrac{\partial v}{\partial x} - \dfrac{\partial u}{\partial y} = 0 \end{cases} \tag{2.1}$$

where (u, v) is the components of the velocity, ρ is the density, and the pressure p is determined by $p = A\rho^\gamma$ $(1 \le \gamma \le 3)$. The parameters of the flow satisfy the Bernoulli relation $\frac{1}{2}(u^2 + v^2) + \dfrac{\gamma\rho^\gamma}{\gamma - 1} = const.$ If in the flow field the curve $y = \phi(x)$ is the equation of a shock front, then $\sigma = \phi'(x)$ must satisfies the Rankine-Hugoniot condition

$$\sigma = \frac{[\rho v]}{[\rho u]}, \qquad \sigma = -\frac{[u]}{[v]} \tag{2.2}$$

The explanation of the entropy condition on the shock front is given in [3,4], it can be simply described as the increasing of the pressure, or the decreasing of the normal component of the velocity across the shock.

Assume that the velocity of the upstream supersonic flow is $(q_0, 0)$, and the density is ρ_0. The flow is moving along a rigid wall $y = f(x)$ with $f(x) = 0 (x \le 0), f(x) = bx (x > 0)$ with $b > 0$ less than a critical value. The problem is called (P) in this paper. It is well known that in this case there will be a plane shock $y = \sigma x$ through the origin. Denote $q = \sqrt{u^2 + v^2}$, $\theta = \arctan \frac{v}{u}$, then we have

$$\cos\theta = \frac{\rho q^2 + \rho_0 q_0^2}{(\rho + \rho_0) q_0 q}. \tag{2.3}$$

The graph of (2.3) on the plane (u, v) is called shock polar, which is denote by Γ. When b is small (less than a critical value), the ray $v = bu$ intersects with Γ at two points. According to the distance from the point $A(q_0, 0)$, the nearer one $B(u_w, v_w)$ corresponds to the weaker shock, while the farther one $B'(u_s, v_s)$ corresponds to the stronger shock.

As mentioned in the previous section, we will perturb the boundary $y = f(x)$ by its smoothing $y = f_m(x)$, where

$$\begin{cases} f_m(x) = 0 (x \le x_+), \ f_m(x) = bx (x \ge x_+), \\ f_m(x) > 0, f_m'(x) > 0, f_m''(x) > 0 \quad \text{in} \quad x_- < x < x_+. \end{cases} \tag{2.4}$$

Since $f_m(x)$ is smooth, then near the wall $y = f_m(x)$ there is a Prandtl-Meyer flow, the second characteristics of the flow are straight (see [1,2]). The bending of the wall causes the second straight characteristics to be compressed.

Define

$$\lambda_\pm = \frac{uv \pm a\sqrt{u^2 + v^2 - a^2}}{u^2 - a^2)}, \tag{2.5}$$

where $-, +$ corresponds to the first and second characteristic direction respectively. Along the first (second) characteristics the Riemann invariants $r = \theta + \int \frac{\sqrt{q^2 - a^2}}{aq} dq$ ($s = \theta - \int \frac{\sqrt{q^2 - a^2}}{aq} dq$) keeps constant. In the region of Prandtl-Meyer flow near the smoothing wall the flow can be described by $r = const$. Choose $x_0 > x_+$ and near x_+, we consider a boundary value problem of (2.1) in $\{x \geq x_0, y \geq bx\}$ with the initial condition

$$u = u_0(y), v = v_0(y) \qquad \text{on} \quad x = x_0 \tag{2.6}$$

and the boundary condition

$$v = bu, \tag{2.7}$$

which is equivalent to $\frac{r+s}{2} = \arctan b$. We notice that $f_m(x)$ is nonlinear only on $x_- < x < x_+$, correspondingly, $\lambda_+(y) = \lambda_+(u_0(y), v_0(y))$ are nonlinear on (y_α, y_β), and is constant outside.

To simplify our analysis, we also require $\lambda_+(y)$ satisfies ND condition as follows:

$$\lambda'_+(y_*) < 0,\ \lambda''_+(y_*) = 0,\ \lambda'''_+(y_*) > 0, (y - y_*)\lambda''_+(y) > 0, y \in (y_\alpha, y_\beta) \tag{2.8}$$

(2.8) can be reduced to a requirement for $f_m(x)$. It is not difficult to find a function $f_m(x)$ satisfying the above requirement and (2.4).

The ND condition indicates that the y coordinate on the envelope attain its minimum only at one point, which is nothing but the cusp $P_{**}(x_{**}, y_{**})$ of the envelope.

In [4] we proved the local existence of the solution of (2.1) with shock front near the cusp, as well as the estimates of the solution. The conclusion is:

Theorem 1. *Under above assumptions the problem (2.1),(2.6),(2.7) admits an weak entropy solution with a shock $y = \phi(x)$ near (x_{**}, y_{**}). Meanwhile, the following estimates hold:*

$$\begin{cases} \phi(x) = y_{**} + \alpha(x - x_{**}) + O((x - x_{**})^2), \\ r(x,y) = O((x - x_{**})^3 + (y - y_{**} - \alpha(x - x_{**}))^2)^{1/2} \\ s(x,y) = s_0(y_*) + O((x - x_{**})^3 + (y - y_{**} - \alpha(x - x_{**}))^2)^{1/6}, \end{cases} \tag{2.9}$$

where α is the slope of the second characteristics of (2.1) at (x_{**}, y_{**}).

3 Global existence

To carry on our analysis we are going to prove that the local solution obtained in Section 2 , as well as the corresponding shock front $y = \phi(x)$, can be extended to infinity. To do it we first study the asymptotic behavior of the envelope.

Lemma 1. *Denote two branches of the envelope of the family of second characteristics of (1.2) by $\Gamma_\alpha : y = \eta_\alpha(x), \Gamma_\beta : y = \eta_\beta(x) \quad (\eta_\beta < \eta_\alpha)$ then Γ_α takes $L_\alpha : y = y_\alpha + (x - x_0)\lambda_+(y_\alpha)$ as its asymptote. Same conclusion holds for Γ_β with y_β replaced by y_β.*

Proof In view of the ND condition, we have $\lambda_+''(y_\alpha) = 0$ and $\lambda_+''(y) = c_2(y)(y-y_\alpha)^k$ for some $k \geq 1$ with $c_2(y_\alpha) \neq 0$. Therefore

$$\lambda_+'(y) = c_1(y)(y-y_\alpha)^{k+1}, \lambda_+(y) = \lambda_+(y_\alpha) + c_0(y)(y-y_\alpha)^{k+2},$$

where $c_1(y_\alpha), c_0(y_\alpha)$ do not vanish. On the other hand the equation of the envelope Γ_α is

$$\begin{cases} -1 = \lambda_+'(z)(x-x_0), \\ y = z - \frac{\lambda_+(z)}{\lambda_+'(z)} \end{cases} \tag{3.1}$$

with z as parameter. Obviously, $z \to y_\alpha$ as $x \to \infty$. Hence we have $x - x_0 = -\dfrac{1}{c_1(z)(z-y_\alpha)^{k+1}}$ and $z - y_\alpha = O((x-x_0)^{-\frac{1}{k+1}})$. This leads us to

$$\eta_\alpha = z - \frac{\lambda_+(y_\alpha) + c_0(z)(z-y_\alpha)^{k+2}}{c_1(z)(z-y_\alpha)^{k+1}} = y_\alpha + \lambda_+(y_\alpha)(x-x_0) + O((x-x_0)^{-\frac{1}{k+1}}) \tag{3.2}$$

It means that L_α is the asymptote of Γ_α.

To extend the solution $s(x,y), r(x,y)$ and $\phi(x)$ we use the same approximation procedure as we applied in the proof of local existence, and prove that all approximate solutions can be extended to infinite, which are uniformly bounded and locally compact in the whole domain.

Indeed, the characteristic form of (2.1) is

$$\begin{cases} \frac{\partial s}{\partial x} + \lambda_+ \frac{\partial s}{\partial y} = 0, \\ \frac{\partial r}{\partial x} + \lambda_- \frac{\partial r}{\partial y} = 0. \end{cases} \tag{3.3}$$

Denote $G(\tau)$ the inverse function of $\tau = F(q)$ $(= \int \frac{\sqrt{q^2-a^2}}{aq} dq)$, we define

$$r^{(0)}(x,y) = r_0(y)(=0)$$

$s^{(0)}(x,y)$ is the weak solution of the first equation of (3.1) with $r \equiv 0$, and $s(x_0,y) = s_0(y)$. For general ν, we define

$$\sigma^{(\nu+1} = -\frac{\left[G(\frac{r^{(\nu+1)}-s^{(\nu+1)}}{2})\cos(\frac{r^{(\nu+1)}+s^{(\nu+1)}}{2})\right]}{\left[G(\frac{r^{(\nu+1)}-s^{(\nu+1)}}{2})\sin(\frac{r^{(\nu+1)}+s^{(\nu+1)}}{2})\right]} \tag{3.4}$$

$$\phi^{(\nu+1)} = \begin{cases} y_{**} + \int_{x_{**}}^{x} \sigma^{(\nu+1)} dx, & x > x_{**}, \\ y_* + \lambda_+(y_*)(x-x_0), & x_{**} \geq x \geq x_0 \end{cases} \tag{3.5}$$

$$\begin{cases} \partial_x s^{(\nu+1)} + \lambda_+^{(\nu)} \partial_y s^{(\nu+1)} = 0, & x > x_0, bx < y < \phi^{(\nu)}(x), \\ s^{(\nu+1)}(x_0,y) = s_0(y), & x = x_0, \\ s^{(\nu+1)} = 2\arctan b - r^{(\nu)}, y = bx, x > x_0 \end{cases} \tag{3.6}$$

$$\begin{cases} \partial_x r^{(\nu+1)} + \lambda_-^{(\nu)} \partial_y r^{(\nu+1)} = 0, & x > x_0, bx < y < \phi^{(\nu)}(x), \\ r^{(\nu+1)}(x_0, y) = 0, & x = x_0, \\ r^{(\nu+1)} = F(s^{(\nu+1)}(x), s_-(x))[s^{(\nu+1)}(x)]^3 & y = \phi^{(\nu)}, x > x_0 \end{cases} \tag{3.7}$$

where the boundary condition on $y = bx$ in (3.6) is derived from (2.7), and the condition on the approximate shock in (3.7) is derived from (2.2). The latter means that the jump of the Riemann invariant r is the third order of the jump of the Riemann invariant s. The deduction of the last equation of (3.7) can be found in [6].

Lemma 2. *The approximate solution $r^{(\nu)}, s^{(\nu)}$ obtained are uniformly bounded.*

Proof The proof can be carried out on (u, v) plane by induction. On (u, v) plane $r = const$ and $s = const$ are two families of epicycloids. When $(r(x, y), s(x, y))$ is the solution of (3.3), the epicycloid in the first family (resp. second family) is perpendicular to the characteristics of second class (resp. first class) on the (x, y) plane (see [1]).

Denote by Γ the shock polar with the point $A(q_0, 0)$ as its intersection, by L the ray $v = bu$, by ℓ_0 the first epicycloid passing through A. Denote by B, E the intersection $L \cap \Gamma, L \cap \ell_0$, by Δ the curved triangle formed by Γ, L and ℓ_0. Let $\langle r, s\rangle$ stands for the point (u, v) determined from $r = \theta + F(q), s = \theta - F(q)$. We are going to prove

$$s_A \le s^{(\nu)} \le s_B, \quad r_B \le r^{(\nu)} \le r_A \tag{3.8}$$

by induction.

Indeed, $\langle r^{(0)}, s^{(0)}\rangle \in \ell_0(AE)$. Now Assume that (3.7) is valid for index ν. The equation in (3.5) means that $s^{(\nu+1)}$ is constant along any λ_+ characteristics. Since the initial data implies $s_A \le s^{(\nu)} \le s_E$, and the boundary condition implies $\langle r^{(\nu)}, s^{(\nu+1)}\rangle \in L(BE)$, then $s^{(\nu+1)}$ on $x = x_0$ and on $L(BE)$ is in between s_A and s_B, so is $s^{(\nu+1)}$ in the whole domain.

Furthermore, the equation (3.6) means $r^{(\nu+1)}$ is constant along any λ_- characteristics. The initial data means $r^{(\nu+1)} = r_A = 0$ initially, while the boundary condition implies $\langle r^{(\nu+1)}, s^{(\nu+1)}\rangle$ is located at some point on $\tilde{\Gamma}$, which is a shock polar starting from $\langle r_A, s_-\rangle \in \ell_0$. Due to the restriction on $s^{(\nu+1)}$ we know $r^{(\nu+1)}$ must greater than r_B, while $r^{(\nu+1)} \le r_A$ is an obvious fact. Hence we have

$$r_B \le r^{(\nu+1)} \le r_A$$

The uniform boundedness is thus established.

From Lemma 3.2 we know that if b is not large, then the difference of the slope of the shock (including the approximate shock) and the slope of the characteristics (including the approximate characteristics) is away from zero, so is the difference of the slope of the wall and the slope of the characteristics. Therefore, we can establish the following fact in geometry. For any point (x, y) in the region right to $x = x_{**}$ and between the approximate shock $y = \phi^{(\nu+1)}(x)$ and the wall $y = f_m(x)$ we draw a leftward broken approximate characteristics, which turns from first (resp. second) class to second (resp. first) class at the reflected point by shock (resp. wall except by the first reflection). Then the broken characteristics must finally intersect $x = x_{**}$ after finite times of reflection. Denote the times of reflection by N, we have

$$C_1 \ln(x - x_{**}) \leq N \leq C_2 \ln(x - x_{**}). \tag{3.9}$$

Another fact is we can find a constant $D > 0$ for any $x_1 > x_{**}$, the leftward broken characteristics starting from some point on $x = x_1 + D$ will reach $x = x_1$ through at most one reflection.

Theorem 2. *(2.1) admits a global piecewise smooth solution in the domain $-\infty < x < \infty, y > f_m(x)$, which satisfies the R-H condition and the entropy condition on shock and the boundary condition (2.5) on $y = f_m(x)$.*

We are going to show that the domain of existence can be extended to infinity. As we stated in [4] that the sequences $\{r^{(\nu)}(x,y), s^{(\nu)}(x,y)\}$ (as well as $\{\phi^{(\nu)}\}$ are convergent near the point P_{**}. Next, we indicated that for the constant $D > 0$ mentioned above (related the condition of one reflection), so that for any $\delta > 0$, the sequences $\{r^{(\nu)}(x,y), s^{(\nu)}(x,y)\}$ are convergent in $x_{**} + \delta < x < x_{**} + \delta + D$.

Introduce the transformation

$$T^{(\nu)} : \tilde{x} = x, \tilde{y} = \frac{y - \phi^{(\nu)}(x)}{\phi^{(\nu)}(x) - f(x)}, \tag{3.10}$$

depending on the free boundary. Obviously, $T^{(\nu)}$ transforms $y = f(x)$ into $\tilde{y} = -1$, and transforms the approximate shock $y = \phi^{(\nu)}$ into $\tilde{y} = 0$. Correspondingly, $r^{(\nu)}(x,y), s^{(\nu)}(x,y)$ are transformed into $\tilde{r}^{(\nu)}(\tilde{x},\tilde{y}), \tilde{s}^{(\nu)}(\tilde{x},\tilde{y})$ satisfying the equation similar to (3.5),(3.6) with the initial data on $x = x_0$ replaced by the value of $r^{(\nu)}, s^{(\nu)}$ on $x = x_{**} + \delta$.

Denote $\psi^{(\nu)} = \tilde{s}^{(\nu+1)} - \tilde{s}^{(\nu)}$,$\chi^{(\nu)} = \tilde{r}^{(\nu+1)} - \tilde{r}^{(\nu)}$. By subtracting we easily obtain the equations satisfied by $\tilde{\psi}^{(\nu)}$ and $\tilde{\chi}^{(\nu)}$:

$$\begin{cases} \partial_x \tilde{\psi}^{(\nu)} + \lambda_+^{(\nu)} \partial_y \tilde{\psi}^{(\nu)} + (\lambda_+^{(\nu)} - \lambda_+^{(\nu-1)}) \tilde{s}_y^{(\nu)} = 0, \\ \tilde{\psi}^{(\nu)}(x_{**} + \delta, y) = g^{(\nu)}(y), \\ \tilde{\psi}^{(\nu)}(\tilde{x}, -1) = 2 \arctan b - \chi^{(\nu-1)}(\tilde{x}, -1), \end{cases} \tag{3.11}$$

$$\begin{cases} \partial_x \tilde{\chi}^{(\nu)} + \lambda_-^{(\nu)} \partial_y \tilde{\chi}^{(\nu)} + (\lambda_-^{(\nu)} - \lambda_-^{(\nu-1)}) \tilde{r}_y^{(\nu)} = 0, \\ \tilde{\chi}^{(\nu)}(x_{**} + \delta, y) = h^{(\nu)}(y), \\ \tilde{\chi}^{(\nu)}(\tilde{x}, 0) = \tilde{\psi}^{(\nu)}(\tilde{x}, 0) G, \end{cases} \tag{3.12}$$

where G is a bounded quantity.

Define $a^{(\nu)}(\tilde{x}) = \sup_{\tilde{y}} \max\{|\psi^{(\nu)}|, |\chi^{(\nu)}|\}$ and $\epsilon_\nu = \sup_{\tilde{y}} \max\{|g^{\nu)}|, |h^{(\nu)}|\}$. Assume that $\tilde{s}^{(\nu)}, \tilde{r}^{(\nu)}$ are convergent on $x = x_{**} + \delta$, then we have $\Sigma \epsilon_\nu < \infty$. Integrating $\psi^{(\nu)}$ and $\chi^{(\nu)}$ along characteristics we gave

$$a^{(\nu)}(\tilde{x}) \leq \int_{x_{**}}^{\tilde{x}} K_1(a^{(\nu-1)}(\xi) + a^{(\nu-2)}(\xi))d\xi + K_2\epsilon_\nu. \tag{3.13}$$

Hence by using the iteration method we can establish

$$\sum_\nu a^{(\nu)}(\tilde{x}) \leq C(\sum_\nu \epsilon_\nu + 1)(\sum_\nu \frac{K^\nu(\tilde{x} - x_{**})^\nu}{\nu!}) < \infty \tag{3.14}$$

for suitable constant K, C. Therefore, $\tilde{s}^{(\nu)}, \tilde{r}^{(\nu)}$ are convergent in $\{x_{**} + \delta \leq x \leq x_{**} + \delta + D, 0 \leq \tilde{y} \leq 1$. Similarly, we can obtain $\nabla_{\tilde{x},\tilde{y}}\tilde{s}^{(\nu)}, \nabla_{\tilde{x},\tilde{y}}\tilde{r}^{(\nu)}$. Hence the limit of $\tilde{s}^{(\nu)}, \tilde{r}^{(\nu)}$ are C^1 functions. Return to the original coordinates, we obtain the convergence of $\tilde{s}^{(\nu)}$ and $\tilde{r}^\nu$. The limit solves the problem (2.5).(2.6) in the domain $x_{**} + \delta, x_{**} + \delta + D$. Since D is not depending on δ, we are led to the global existence.

Remark 1. We indicate here that in the above iteration process the influence of in initial data gradually propagates to infinity by reflection of characteristics. Hence the convergence of $s^{(\nu)}, r^{(\nu)}$ in the whole domain $\Omega : \{x_{**} < x < \infty, f(x) < y < \phi(x)\}$ is not uniform.

4 Asymptotic behavior

Consider the asymptotic behavior of the solution given in Theorem 3.1, we have

Theorem 3. *When $x \to +\infty$, the solution given in Theorem 3.1 approaches the solution of Problem (P) with rate x^{-1}, and the slope of the shock starting from (x_{**}, y_{**}) tends to the slope of the weaker shock given in Theorem 3.1.*

To prove this conclusion we start with analyzing the property of broken characteristics. For any point $M(x, y)$ in the domain$\{x > x_{**}, f(x) < y < \phi(x)\}$, we draw a broken characteristics. As we state in the last section, after finite times of reflection on shock or wall, the broken characteristics will reach $x = x_{**}$ at some point P. Denote the reflection point by $Q_1, Q_2, \cdots, Q_N$, the broken curve can be written as $MQ_1 \cdots Q_N P$.

Using the transformation $u = u(x, y), v = v(x, y)$, the image of the broken characteristics $MQ_1 \cdots Q_N P$ on (u, v) plane is $\tilde{M}\tilde{Q}_1 \cdots \tilde{Q}_N\tilde{P}$, whose each segment is epicycloid (in first family or second family alternatively). Each reflection of broken characteristics on (x, y) plane corresponds to a reflection of broken epicycloid on (u, v) plane. If we choose two families of epicycloid as coordinate curves, then r and s are constant on these curves respectively. Meanwhile, the shock polar Γ, the ray $v = bu$ are transversal to each coordinate curves. Therefore, when the point $\tilde{P}$ moves to $\tilde{M}$ along the broken epicycloid, it will approach B gradually. Moreover, there are positive constants C_3, C_4, α, β, such that

$$C_3 e^{-\alpha N} < |\tilde{M}B| < C_4 e^{-\beta N} \tag{4.1}$$

Combining with (3.9) , we have

$$|r(x,y) - r_0|, |s(x,y) - s_0| \leq Cx^{-1} \tag{4.2}$$

for some constant C. Correspondingly, the estimate $|\sigma(x) - \sigma_0| \leq Cx^{-1}$ can also be established.

Since the point B stands for (u_w, v_w) corresponding to the state behind the weaker oblique shock front, then we are led to the conclusion of Theorem 4.1.

By taking zoom, i.e. by carrying on a self-similar transformation $x_1 = \alpha x, y_1 = \alpha y$ with $\alpha > 1$, we can obtain the solution of the same supersonic flow above a wall $y = \frac{1}{\alpha} f_m(\alpha x)(= f_{m\alpha}(x)$. The solution in this case is $u_\alpha(x,y) = u(\alpha x, \alpha y)$,$v_\alpha(x,y) = v(\alpha x, \alpha y)$, and the shock front is $y = \phi_\alpha(x)(= \frac{1}{\alpha}\phi(\alpha x))$ starting from $(\frac{1}{\alpha}x_{**}, \frac{1}{\alpha}y_{**})$. Let $\alpha \to \infty$, the function $f_{m\alpha}(x)$ tends to $f(x)$. Theorem 4.1 indicates that (u_α, v_α) tends to (u_w, v_w), while the shock front $y = \phi_\alpha(x)$ approaches the weaker oblique shock starting from the corner of the unperturbed wall.

5 Vanishing perturbation principle

From the above analysis we see that in the problem of a supersonic flow past a turned wall, the weaker shock is stable with respect to the perturbation of the surface of the wall, and by taking zoom we can recover the weaker shock solution. However, such a perturbation causes the stronger shock disappear. It means that the stronger shock is at least unstable with respect to such perturbation. The analysis enlighten us to introduce an auxiliary condition to single out a unique physically reasonable weak solution.

For the problem describing a flow past a turned wall, the turning angle is the basic parameter causes the appearance of the shock. Write the strength $S_w(\theta)$ or $S_s(\theta)$ of the shock as a function of the angle θ, we have $S_w(0) = 0$, and $S_s(0) \neq 0$. The last inequality is nothing but the reason why the stronger oblique shock is unstable. Therefore, for any shock configuration we suggest to introduce a principle: **The strength of the shock must be continuous with respect to any vanishing parameters, whose non-zero value causes the appearance of the shock**. According to the above discussion this principle can be regarded as a supplement of the entropy condition to derive uniqueness in the study of the weak solution of hyperbolic conservation laws. Correspondingly, if we take this principle as a supplement of the entropy condition, then the uniqueness of the problem of a steady supersonic flow past a wedge or a cone can be derived with the help of this principle (see [9.10]).

References

1. Courant, R., Friedrichs, K.O. (1948): Supersonic Flow and Shock Waves. Interscience Publishers Inc., New York
2. Whitham, G.B. (1974): Linear and Nonlinear Waves. John Wiley & Sons, New York
3. Keyfitz, B.L., Warneck, G.G. (1991): The existence of viscous profile and admissibility for transonic shocks. Comm. PDEs, **16**, 1197-1221
4. Chen, S.X., Dong, L.M. (1998): Formation of shock in potential flow. Advances in Nonlinear PDEs and Related Areas. World Scientific, 45-66.
5. Lebaud, M.P.(1994) Description de la formation d'un choc dans le p-systems. Jour. Math. Pures. Appl. **73**, 523-565
6. Chen, S.X., Dong, L.M. (2001): Formation and construction of shocks for p-system. Science in China **44**, 1139-1147
7. Chen, S.X. (2001): How does a shock in supersonic flow grows out of nothing. Journal of Mathematical Physics **42**, 1154-1162.
8. Chen, S.X. (2002): Stability of weak oblique shock front. Science in China **45**
9. Chen, S.X. (1992): Existence of local solution to supersonic flow around a three dimensional wing. Advances in Appl. Math., **13**, 273-304.
10. Chen, S.X. (2001): Existence of stationary supersonic flow past a pointed body. Archive Rat. Mech. Anal. **156**, 141-181.

Robust Treatment of Interfaces for Fluid Flows and Computer Graphics

Doug Enright[1] and Ron Fedkiw[2]

[1] Mathematics Department, University of California, Los Angeles, *enright@math.ucla.edu*
[2] Computer Science Department, Stanford University, *fedkiw@cs.stanford.edu*

1 Introduction

Researchers have used numerical techniques to solve partial differential equations describing physical phenomena for many years. One challenging area, the numerical treatment of interfaces, motivated the creation of a topologically robust interface capturing algorithm, the level set method of Osher and Sethian [29]. The level set method has been used to track interfaces in a wide variety of applications. Utilizing geometrical information about the interface, which is naturally obtained from the level set function, an accurate treatment of material discontinuities across the interface can be obtained via the Ghost Fluid Method [15]. Discontinuities are implicitly enforced with the ghost fluid method, avoiding any numerical smoothing of discontinuous quantities across the interface. The ghost fluid method and related techniques have been used to model discontinuities in compressible and incompressible flows [15, 24, 22, 3], flames and detonations [27, 16], solid fluid coupling [14] and Stefan problems [19, 18, 4]. A newly proposed, fully conservative ghost fluid method has been used to track contact discontinuities, inert shocks and detonation waves [25]. Accurate modeling of the motion of a contact discontinuity itself for incompressible flows has been a challenge for level set methods. Recently a new method, the "particle level set method" [10], has been proposed to accurately track contact discontinuities for incompressible flows. The particle level set method conserves mass to an accuracy comparable to explicit front tracking and volume of fluid methods. Due to the robustness and ease of programming of these interface methods in three spatial dimensions combined with the ever increasing speed and memory of desktop computers, physics-based animation algorithms to model fire and water [26, 17, 11] have taken advantage of these methods in order to produce realistic looking behavior on the coarse computational grids commonly used in a production animation environment. In this article we give a brief overview of the level set method, the use of the ghost fluid method for modeling discontinuities across the interface, the "particle level set method" for tracking contact discontinuities, and illustrate the use of these methods in the context of computer animation. Additional details concerning these methods can be found in the recently published book, "Level Set Methods and Dynamic Implicit Surfaces" [28].

2 Level Set Method

In order to robustly deal with topological changes to a dynamically evolving interface, a simple and versatile method to treat this important problem is obtained by embedding the interface of an open region Ω as the level set of a smooth (at least Lipschitz continuous) higher dimensional function $\phi(\mathbf{x},t)$. The level set function ϕ has the properties:

$$\begin{aligned}\phi(\mathbf{x},t) &< 0 \text{ for } \mathbf{x} \in \Omega \\ \phi(\mathbf{x},t) &> 0 \text{ for } \mathbf{x} \notin \bar{\Omega} \\ \phi(\mathbf{x},t) &= 0 \text{ for } \mathbf{x} \in \partial\Omega = \Gamma(t).\end{aligned}$$

The description of the interface in this manner allows for the natural merging or separation of the interface without any additional involvement by the user.

The interface Γ is evolved in time by a velocity field $\mathbf{V}(\mathbf{x},t)$ according to the simple advection equation

$$\frac{\partial \phi}{\partial t} + \mathbf{V} \cdot \nabla \phi = 0. \tag{1}$$

High order accurate WENO methods [21] methods can be used to discretize the spatial derivatives in equation 1 combined with an explicit TVD Runge-Kutta method utilizing convex combinations of simple forward Euler updates [32, 22] in order to integrate equation 1 forward in time. Local level set methods [1, 30] can substantially reduce the spatial complexity of equation 1 by reducing the calculation to a banded region about the interface.

Geometrical quantities can be easily calculated from the level set function. Unit normals are given by

$$\mathbf{N} = \frac{\nabla \phi}{|\nabla \phi|} \tag{2}$$

and the curvature by

$$\kappa = \nabla \cdot \left(\frac{\nabla \phi}{|\nabla \phi|} \right). \tag{3}$$

Two commonly performed operations using level set functions include the reintialization of ϕ to be the signed distance to the interface Γ and the extrapolation of quantities across the interface from one side of the domain to the other. Reinitialization of ϕ can be achieved by solving to steady state (as fictitious time $\tau \to \infty$) the equation

$$\phi_\tau + S(\phi_0)(|\nabla \phi| - 1) = 0, \tag{4}$$

where $S(\phi_0) = \phi_0/\sqrt{\phi_0^2 + (\Delta x)^2}$ [36]. Extrapolation of a variable I across the interface is obtained by again solving to steady state

$$I_\tau \pm \mathbf{N} \cdot \nabla I = 0 \tag{5}$$

[15, 5]. By making clever use of the way the information in equations 4 and 5 propagates, fast heap-based methods [37, 31, 2] may be used to solve these equations in $O(N \log N)$ time, where N is the number of grid points.

Theoretical justification of the level set method for geometrically-based motion came through the theory of viscosity solutions for scalar time-dependent partial differential equations [6, 12]. The notion of having a vanishing viscosity solution guarantees the existence of a unique solution which is consistent with equation 1 [9]. While this behavior is certainly comforting to a computational user, the vanishing viscosity solution approaches the true solution in the limit as $\Delta x \to 0$, a case never truly obtained in practice. The implications of calculating a vanishing viscosity solution to equation 1 on coarse computational grids, or in under-resolved regions of the flow (such as a sharp corner in a geometry driven flow) is discussed further in section 4.

3 Ghost Fluid Method

Spurious oscillations in material fields resulting from discontinuities due to shocks or contact discontinuities have been a source of difficulty in the numerical solution of hyperbolic conservation laws. In order to obtain correct shock speeds and strengths, the Lax-Wendroff theorem [23] states that a numerical method used should be fully conservative. An explicit way to deal with this requirement is by solving multidimensional Riemann problems at the location of the interface. This approach has been used by Glimm et al. [20] in conjunction with an explicit representation of the interface. However, complicated interfacial geometry along with changes in topology and a lack of a proper entropy condition built into the interface representation, place an onerous burden on any explicit method to capture these details in a robust manner. The Ghost Fluid Method on the other hand implicitly captures the Rankine-Hugoniot jump conditions at the interface in a manner similar to the implicit capturing of the location of an interface by the level set method. The result is an accurate, easy-to-implement, and topologically robust numerical algorithm.

Conservation of mass, momentum, and energy fluxes ($F_\rho, \mathbf{F}_{\rho \mathbf{V}}$, and F_E) across the interface results in the Rankine-Hugoniot jump conditions for the relevant physical variables. For an interface moving at speed D in the normal direction to the interface, $F_\rho, \mathbf{F}_{\rho \mathbf{V}}$, and F_E describing an inviscid compressible fluid are given by

$$F_\rho = \rho(V_N - D) \tag{6}$$

$$\mathbf{F}_{\rho \mathbf{V}} = \rho(\mathbf{V}^T - D\mathbf{N}^T)(V_N - D) + p\mathbf{N}^T \tag{7}$$

$$F_E = \left(\rho e + \frac{\rho|\mathbf{V} - D\mathbf{N}|^2}{2} + p\right)(V_N - D), \tag{8}$$

where ρ is the density, $\mathbf{V}$ is the fluid velocity, $\mathbf{N}$ is the normal to the interface, $V_N = \mathbf{V} \cdot \mathbf{N}$, e is the internal energy per unit mass, and p is the pressure.

Equations 6, 7, and 8 allow for chemical reactions at the interface, resulting in an interface velocity different from the underlying fluid velocity. For contact discontinuities with $\mathbf{V}_N = D$, two of these fluxes are zero across the interface, and the interface separates two gases (or materials) with possibly different equations of state.

The key to the ghost fluid method is that by defining a set of ghost cells on each side of the interface, the ghost cells can implicitly capture the physically correct boundary conditions as defined by equations 6, 7, and 8 in such a manner as to avoid any finite differencing across a discontinuity. The method also avoids the common approach of numerically smoothing a discontinuity with the aim of preventing the creation of nonphysical oscillations. The location of the ghost cells is the same as the real grid cell locations. Since both fluids are defined in a neighborhood about the front, one can solve for each fluid *independently* using standard schemes, regardless of the geometry of the front. After updating each fluid, the choice as to which of the two values, the "ghost fluid" or "real fluid", to take near the interface is determined by the updated level set function describing the new location of the interface.

At each time step ghost cells are populated in a node-by-node fashion in order to preserve the continuity of the mass, momentum, and energy fluxes across the interface. This is achieved by solving the system of equations: $F_\rho^G = F_\rho^R, \mathbf{F}_{\rho\mathbf{V}}^G = \mathbf{F}_{\rho\mathbf{V}}^R$, and $F_E^G = F_E^R$ at each grid point with "R" representing the known real fluid values and "G" for the ghost values. The solution of this system of equations is simple compared to applying the Rankine-Hugoniot conditions at the interface. In the case of a contact discontinuity, we identify the continuous and discontinuous quantities and set them appropriately in order to ensure that the jump conditions are satisfied as follows. In this case, $[V] = 0$ and $[p] = 0$, where $[\cdot]$ is the jump in value across the interface, while the entropy, S, is discontinuous across the interface front. Since the pressure and velocity are continuous across the interface, we can take $V^G = V^R$ and $p^G = p^R$, however we need to extrapolate the value of the entropy across the interface to avoid differencing across the discontinuity and to ensure that the ghost fluid is indeed representing the real fluid it is standing in for. This is comparable to the ghost fluid taking on the correct equation of state. In multiple spatial dimensions, extrapolation of the discontinuous variables can be implemented according to equation 5.

Besides capturing discontinuous boundary conditions at an irregular interface for compressible flows, discontinuities in physical variables, e.g. pressure and density, can exist in incompressible flows. These discontinuities at the interface need to be accounted for when solving the resulting Poisson equation for the pressure. As shown in [24, 22, 27], a ghost fluid method approach for capturing these discontinuities across the interface is possible as well. The resulting numerical method for a variable coefficient Poisson equation in the presence of interfaces where the coefficients and the solution itself may be discontinuous is robust and easy to implement in multiple spatial dimensions. In addition, the resulting coefficient matrix of the associated linear system is symmetric, allowing for the use of fast "black-box" solvers such as a preconditioned conjugate gradient method.

4 Particle Level Set Method

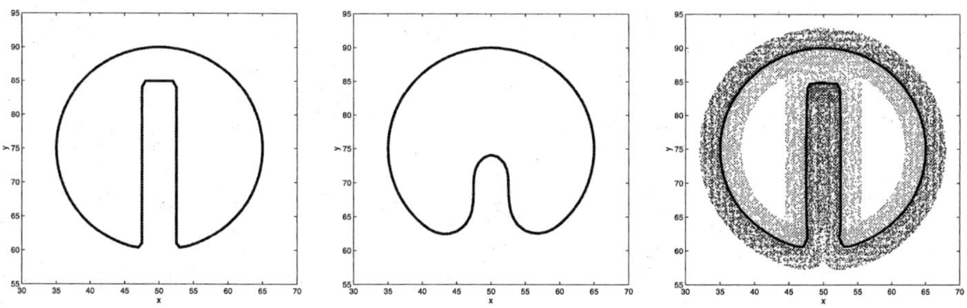

(a) Initial (b) Level Set Only (c) Particle Level Set

Fig. 1. Rigid Body Rotation of a Notched Disk

The robustness of the level set method is due to the regularization by curvature property that a numerical implementation the level set method intrinsically possesses. This regularization property allows for effortless changes in interface topology, e.g. the pinching off or merging together of the interface, a key aspect of the level set method. An upwind discretization of equation 1 results in a numerical truncation error of the form $\epsilon\Delta\phi$, with $\epsilon \approx O((\Delta x)^r)$, r being the order of accuracy of the discretization used. So instead of equation 1 being solved exactly, we actually are obtaining a solution to

$$\phi_t + \mathbf{V} \cdot \nabla\phi = \epsilon\Delta\phi. \tag{9}$$

The ϕ obtained from the above equation is actually the vanishing viscosity solution to equation 1. The viscosity term on the right hand side of equation 9 is proportional to the curvature of the interface and goes to zero as $\Delta x \to 0$. The effect of this unmodeled, but always present viscosity term can be seen in figure 1. Here a notched disk (with a notch width of 5 grid cells), undergoes a rigid body rotation. After one revolution, a level set only representation of the interface is seen in figure 1(b). The thin notch region along with the high curvature convex corners at the bottom of the disk have experienced large amounts of numerical diffusion. One solution to this problem is to increase the grid resolution, thereby decreasing the effect of numerical viscosity. However, this solution can dramatically increase the computational time needed, especially when the level set method is used to track a contact discontinuity, e.g. an air-water interface, in a computational fluid dynamics simulation.

An alternative solution, is to use an error correction mechanism along portions of the interface which are suseptible to large amounts of diffusion. The error correcting mechanism we propose to use are diffusionless particles placed near the $\phi = 0$ isocontour. The results of this new "particle level set method" can be seen in figure 1(c) where the thin notch along with the sharp corners have maintained their original shape with little to no diffusion. The particles move according to $d\mathbf{x}_p/dt = \mathbf{V}(\mathbf{x}_p)$, and each particle possesses a radius r_p and a sign s_p. Since the level set is tracking a contact discontinuity, particles which correspond to the $\phi > 0$ region should always remain in the $\phi > 0$ region and vice versa, however excessive amounts of numerical diffusion can cause positive particles, i.e. particles with $s_p = +1$, to end up in a $\phi < 0$ region according to the level set function. These particles are said to have "escaped" from their respective side of the interface and indicate that a first order error in the location of the interface has occurred. This first order accurate error in ϕ can be corrected for by the particles since the radius of each particle defines a local level set function, ϕ_p, which we can compare against ϕ at the corners of the grid cell containing the particle. We take the value closest to zero as the new more accurate value of ϕ. After iterating through all the escaped particles and determining corrected ϕ values, the particles then resample their distance to the interface and adjust their radii accordingly. This error reduction technique can also be used to correct errors made when ϕ is reinitialized to be a signed distance function as well. In this case the particle velocity is assumed to be zero since the interface should not move. A complete description of this error reduction technique can be found in [10].

The robustness of the level set method is maintained by the particle level set method since the marker particles do not *explicitly* delineate the location of the interface. Rather, they locally capture the location of the interface through the ϕ_p function, and the level set function itself is used to automatically treat connectivity (merging and pinching of fronts). The ease-of-implementation of the level set method is maintained since the particles are disconnected and communicate with the level set function only during the error reduction stage described above. Since particles are placed within a band about the $\phi = 0$ isocontour, the interface is resolved on multiple scales by the particles. This multi-resolution approach is quite successful in preserving the volume of the level set when the interface undergoes large amounts of stretching induced by an incompressible flow field as seen in figure 2. A level set only approach as seen in figure 2(a) can not maintain regions of high curvature and the thin (approximately one grid cell thick) pancake region formed during the deformation process. On the other hand, the particle level set method can resolve these regions on the 100^3 grid used. Also, while tearing of the interface is seen in the thin pancake region with the particle level set method, particles

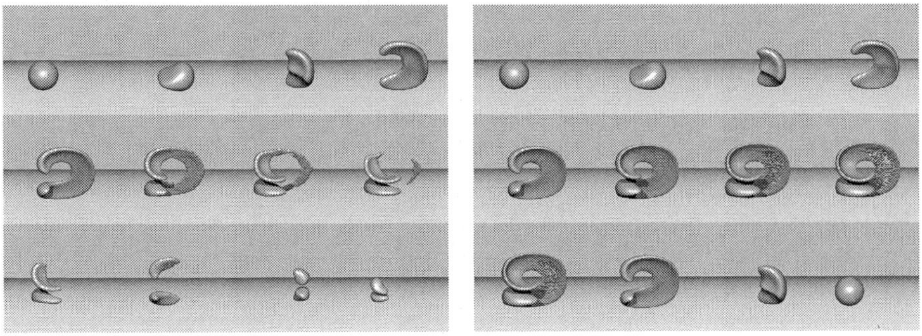

(a) Level Set Only (b) Particle Level Set

Fig. 2. 3D Deformation Test

which remain escaped are not deleted and can contribute to the rebuilding of the interface as seen in the last row of frames in figure 2(b). The sphere which loses over 80% of its volume by the end of the deformation process when represented with a level set only method, loses only 2% of its volume with the particle level set method. The additional cost of placing particles near the interface is offset by the ability to use much coarser volumetric grids when calculating the pressure during flow calculations without sacrificing a faithful representation of the interface.

5 Computer Graphics

The modeling of natural phenomena such as water and fire for computer graphics applications remains a major challenge. The complexity of the motion exhibited by these phenomena defies the ability of animators to realistically animate by hand. The ever increasing use of computer animation in feature films to create photorealistic effects in their own right and to supplement practical elements previously filmed have motivated researchers in computer graphics (CG) to examine the extensive computational fluid dynamics (CFD) literature for algorithms which can be adapted for use in an animation environment. An important criteria for the use of such algorithms is the ability to robustly model fully three dimensional effects on the coarse computational grids commonly used in a CG environment. Recent research [13, 17, 11, 26] has shown promise that when appropriate CFD algorithms are coupled with the level set related methods, the long sought after goal of the CG community of photorealistic fire and water behavior can be attained.

For the purposes of CG, the motion of water and fire (low speed deflagrations) can be modeled using the inviscid, incompressible Euler equations,

$$\nabla \cdot \mathbf{V} = 0 \tag{10}$$

$$\mathbf{V}_t + (\mathbf{V} \cdot \nabla)\mathbf{V} + \frac{\nabla p}{\rho} = \mathbf{f}, \tag{11}$$

where $\mathbf{f}$ can be a variety of body forces including gravity and buoyancy as appropriate. Additional transport equations for the reaction coordinate, temperature, and the density of soot resulting from the chemical reaction at the flame front need to be modeled in order to obtain the necessary information for the visualization of fire. These equations are discussed in detail in [26]. Due to the 1000 to 1 density ratio between water and air, the dynamics of the air on the water can be safely neglected, requiring only the solution of equations 10 and 11 on the water side of the interface. Fire on the other hand requires the solution of the Euler equations on both sides of the interface and more importantly the accurate capturing of the jump conditions resulting from equations 6 and 7.

A projection method [7] is used to update equation 11, where an intermediate velocity field $\mathbf{V}^*$ is first obtained by neglecting the pressure term,

$$\frac{\mathbf{V}^* - \mathbf{V}^n}{\Delta t} + (\mathbf{V} \cdot \nabla)\mathbf{V} = \mathbf{0}. \tag{12}$$

Unconditional stability of a numerical scheme is important for its use in an animation environment, leading the CG community to adopt a semi-Lagrangian method [8, 34, 33] to discretize the convective term in equation 11. Use of a semi-Lagrangian method may introduce large amounts of numerical dissipation, especially on the coarse computational grids used. The method of choice to reduce this dissipation is the "vorticity confinement" method [35] (discussed below). To enforce mass conservation, the pressure is determined by the Poisson equation,

$$\nabla \cdot \left(\frac{1}{\rho}\nabla p\right) = \frac{\nabla \cdot \mathbf{V}^*}{\Delta t}. \tag{13}$$

The gradient of the pressure is then used to advance the velocity field to the $n+1$ time level according to

$$\frac{\mathbf{V}^{n+1} - \mathbf{V}^*}{\Delta t} + \frac{\nabla p}{\rho} = \mathbf{0}. \tag{14}$$

The movement of the contact discontinuity describing the air-water interface in the pouring of a glass of water shown in figure 3(a) can be described by equation 1, where $\mathbf{V}$ is the underlying liquid velocity at the interface. Being able to obtain realistic looking merging and pinching off of the water interface while at the same time limiting the amount of numerical diffusion resulting from the $55 \times 55 \times 120$ grid used for this simulation is a necessary requirement of any interface method used. The particle level set method is able to represent the complex liquid surface shown and maintain the "liveliness"

(a) CG Water

(b) CG Fire

Fig. 3. Physics Based Animation of Water and Fire

of the motion of the interface by limiting the amount of interface dissipation present. In addition, an extrapolation of the liquid velocity field into the unmodeled air using equation 5 can be used to provide a velocity field for the "air" particles associated with particle level set method and plausible velocity boundary conditions which satisfy equation 10, all of which result in a smoothly moving and visually pleasing liquid surface.

An important part of the visual appearance of fire is the expansion of the gas as it undergoes a transformation from unburnt fuel into hot gaseous products. The Rankine-Hugoniot jump conditions at the interface naturally capture this outward expansion of the gas that is next to impossible to achieve through the use of low level hacks and random numbers usually resorted to by the computer graphics animation community. The jump conditions at the interface are

$$[\mathbf{V}] = -\rho_{fuel} S \left[\frac{1}{\rho}\right] \mathbf{N} \tag{15}$$

$$[p] = -\rho_{fuel}^2 S^2 \left[\frac{1}{\rho}\right], \tag{16}$$

where $[\rho] = \rho_{prod} - \rho_{fuel}$ and S is the flame speed. The flame front is not a contact discontinuity, rather the interface moves over the unreacted gas with a speed $S = S_o + \sigma\kappa$, where κ is the curvature of the interface. The overall speed of the interface in a reference frame at rest with respect to the moving fuel is $V_N + S$, where V_N is the normal velocity of the underlying fuel. Appropriate ghost velocities determined by equation 15 are used in the update step given in equation 12. The jump in pressure is incorporated in a boundary condition capturing manner into the solution of equation 13. To combat numerical dissipation and the resulting loss of small scale rolling features characteristic of fire and smoke on the coarse computational grid used, a numerically consistent "vorticity confinement" [35, 13] body force term is used. This term introduces additional vorticity in regions of the flow which posses large gradients in vorticity and are thus sensitive to excessive amounts of artificial damping. As illustrated by the campfire in figure 3(b), a robust three dimensional interface method is required to attain the correct visual look since fire can physically wrap around objects like the top unlit log at the base of the campfire and it is a participating medium, acting as an unsteady volumetric light source. This aspect of fire can be detected by an observer due to the reflection and scattering of the emitted light off other objects in the scene such as the rocks surrounding the campfire. A complete description of the physics, numerical calculation, and visual appearance of fire can be found in [27] and [26].

6 Acknowledgments

Research supported in part by an ONR YIP and PECASE award (ONR N00014-01-1-0620), a Packard Foundation Fellowship, ONR N00014-03-1-0071, NSF DMS-0106694 and NSF ITR-0121288. In addition, D.E. was supported in part by an NSF postdoctoral fellowship (NSF DMS-0202459).

References

1. Adalsteinsson, D. and Sethian, J., *A Fast Level Set Method for Propagating Interfaces*, J. Comp. Phys. 118, 269–277 (1995).
2. Adalsteinsson, D. and Sethian, J., *The Fast Construction of Extension Velocities in Level Set Methods*, J. Comp. Phys. 148, 2–22 (1999).
3. Caiden, R., Fedkiw, R. and Anderson, C., *A Numerical Method For Two-Phase Flow Consisting of Separate Compressible and Incompressible Regions*, J. Comp. Phys. 166, 1–27 (2001).
4. Chen, S., Merriman, B., Kang, M., Caflisch, R., Ratsch, C., Cheng, L.-T., Gyure, M., Fedkiw, R., Anderson, C. and Osher, S., *Level Set Method for Thin Film Epitaxial Growth*, J. Comp. Phys. 167, 475–500 (2001).
5. Chen, S., Merriman, B., Osher, S. and Smereka, P., *A Simple Level Set Method for Solving Stefan Problems*, J. Comp. Phys. 135, 8–29 (1997).
6. Chen, Y. G., Giga, Y. and Goto, S., *Uniqueness and existence of viscosity solutions of generalized mean curvature flow equations*, J. Differential Geom. 33, 749–786 (1991).
7. Chorin, A. J., *Numerical Solution of the Navier-Stokes Equations*, Math. Comp. 22, 745–762 (1968).

8. Courant, R., Issacson, E. and Rees, M., *On the Solution of Nonlinear Hyperbolic Differential Equations by Finite Differences*, Comm. Pure and Applied Math 5, 243–255 (1952).
9. Crandall, M. G., Ishii, H. and Lions, P.-L., *User's guide to viscosity solutions of second order partial differential equations*, Bull. Amer. Math. Soc. 27, 1–67 (1992).
10. Enright, D., Fedkiw, R., Ferziger, J. and Mitchell, I., *A Hybrid Particle Level Set Method for Improved Interface Capturing*, J. Comp. Phys. (2002), in press.
11. Enright, D., Marschner, S. and Fedkiw, R., *Animation and Rendering of Complex Water Surfaces*, ACM Trans. on Graphics (SIGGRAPH 2002 Proceedings) 21, 736–744 (2002).
12. Evans, Y. C. and Spruck, J., *Motion of level sets by mean curvature. I.*, J. Differential Geom. 33, 635–681 (1991).
13. Fedkiw, R., Stam, J. and Jensen, H. W., *Visual Simulation of Smoke*, in Fiume, E., ed., *Proceedings of SIGGRAPH 2001*, Computer Graphics Proceedings, Annual Conference Series, pp. 15–22, ACM, ACM Press / ACM SIGGRAPH, 2001.
14. Fedkiw, R. P., *Coupling an Eulerian Fluid Calculation to a Lagrangian Solid Calculation with the Ghost Fluid Method*, J. Comp. Phys. 175, 200–224 (2002).
15. Fedkiw, R. P., Aslam, T., Merriman, B. and Osher, S., *A Non-oscillatory Eulerian Approach to Interfaces in Multimaterial Flows (The Ghost Fluid Method)*, J. Comp. Phys. 152, 457–492 (1999).
16. Fedkiw, R. P., Aslam, T. and Xu, S., *The Ghost Fluid Method for Deflagration and Detonation Discontinuities*, J. Comp. Phys. 154, 393–427 (1999).
17. Foster, N. and Fedkiw, R., *Practical Animation of Liquids*, in Fiume, E., ed., *Proceedings of SIGGRAPH 2001*, Computer Graphics Proceedings, Annual Conference Series, pp. 23–30, ACM, ACM Press / ACM SIGGRAPH, 2001.
18. Gibou, F., Fedkiw, R., Caflisch, R. and Osher, S., *A Level Set Approach for the Numerical Simulation of Dendritic Growth*, J. Sci. Comput. (2002), in press.
19. Gibou, F., Fedkiw, R. P., Cheng, L.-T. and Kang, M., *A Second-Order-Accurate Symmetric Discretization of the Poisson Equation on Irregular Domains*, J. Comp. Phys. 176, 205–227 (2002).
20. Glimm, J., Grove, J. W., Li, X. and Zhao, N., *Simple Front Tracking*, Contemp. Math. 238, 133–149 (1999).
21. Jiang, G.-S. and Peng, D., *Weighted ENO Schemes for Hamilton-Jacobi Equations*, SIAM J. Sci. Comput. 21, 2126–2143 (2000).
22. Kang, M., Fedkiw, R. and Liu, X.-D., *A Boundary Condition Capturing Method for Multiphase Incompressible Flow*, J. Sci. Comput. 15, 323–360 (2000).
23. Lax, P. and Wendroff, B., *Systems of Conservation Laws*, Comm. Pure Appl. Math. 13, 217–237 (1960).
24. Liu, X.-D., Fedkiw, R. and Kang, M., *A Boundary Condition Capturing Method for Poisson's Equation on Irregular Domains*, J. Comp. Phys. 160, 151–178 (2000).
25. Nguyen, D., Gibou, F. and Fedkiw, R., *A Fully Conservative Ghost Fluid Method and Stiff Detonation Waves*, in *12th International Detonation Symposium*, ONR, 2002.
26. Nguyen, D. Q., Fedkiw, R. and Jensen, H. W., *Physically Based Modeling and Animation of Fire*, ACM Trans. on Graphics (SIGGRAPH 2002 Proceedings) 21, 721–728 (2002).

27. Nguyen, D. Q., Fedkiw, R. P. and Kang, M., *A Boundary Condition Capturing Method for Incompressible Flame Discontinuities*, J. Comp. Phys. 172, 71–98 (2001).
28. Osher, S. and Fedkiw, R., *Level Set Methods and Dynamic Implicit Surfaces*, Springer-Verlag, New York, 2002.
29. Osher, S. and Sethian, J., *Fronts Propagating with Curvature Dependent Speed: Algorithms Based On Hamiliton-Jacobi Formulations*, J. Comp. Phys. 79, 12–49 (1988).
30. Peng, D., Merriman, B., Osher, S., Zhao, H.-K. and Kang, M., *A PDE-Based Fast Local Level Set Method*, J. Comp. Phys. 155, 410–438 (1999).
31. Sethian, J., *A Fast Marching Level Set Method for Monotonically Advancing Fronts*, Proc. Natl. Acad. Sci. 93, 1591–1595 (1996).
32. Shu, C. and Osher, S., *Efficient Implementation of Essentially Non-Oscillatory Shock Capturing Schemes*, J. Comp. Phys. 77, 439–471 (1988).
33. Stam, J., *Stable Fluids*, in *Proceedings of SIGGRAPH 99*, Computer Graphics Proceedings, Annual Conference Series, pp. 121–128, ACM, ACM SIGGRAPH / Addison Wesley Longman, 1999.
34. Staniforth, A. and Côté, J., *Semi-Lagrangian Integration Schemes for Atmospheric Models - A Review*, Monthly Weather Review 119, 2206–2223 (1991).
35. Steinhoff, J. and Underhill, D., *Modification of the Euler Equations for "vorticity confinement": Application to the computation of interacting vortex rings*, Phys. Fluids 6, 2738–2744 (1994).
36. Sussman, M., Smereka, P. and Osher, S., *A Level Set Approach for Computing Solutions to Incompressible Two-Phase Flow*, J. Comp. Phys. 114, 146–159 (1994).
37. Tsitsiklis, J., *Efficient Algorithms for Globally Optimal Trajectories*, IEEE Trans. on Automatic Control 40, 1528–1538 (1995).

Asymptotic Convergence to Diffusive Wave of Bipolar Hydrodynamical Model for Semiconductors

Ingenuin Gasser[1], Ling Hsiao[2], and Hailiang Li[3]

[1] Fachbereich Mathematik, Universität Hamburg, Germany
[2] AMSS, CAS, Beijing 100080, P. R. China, *hsiaol@math.ac.cn*
[3] Institute of Mathematics, University of Vienna, Austria, *hailiang.li@univie.ac.at*

1 Introduction and main results

We consider the large time behavior of solutions of the one-dimensional isentropic bipolar hydrodynamical model for semiconductors:

$$n_t + J_x = 0, \tag{1}$$

$$J_t + \left(\frac{J^2}{n} + p(n)\right)_x = nE - \frac{J}{\mu_n \tau}, \tag{2}$$

$$m_t + I_x = 0, \tag{3}$$

$$I_t + \left(\frac{I^2}{m} + q(m)\right)_x = -mE - \frac{I}{\mu_m \tau}, \tag{4}$$

$$\lambda^2 E_x = n - m, \tag{5}$$

where $n > 0$, $m > 0$, J, I and E denote the densities, current densities, and electric field respectively. $p = p(n)$ and $q = q(m)$ are the pressure-density functions which satisfy

$$\left(\rho^2 p'(\rho)\right)' > 0, \quad \left(\rho^2 q'(\rho)\right)' > 0, \quad \rho > 0. \tag{6}$$

And $\tau_n > 0$, $\tau_m > 0$ are the momentum relaxation times, λ is the re-scaled Debye number. The device domain is chosen to be the whole real line. The equations (1)–(5) are used in the modelling of semiconductors device to describe the to model hot electron effects and can be derived by applying the moment method to the bipolar semiconductor Boltzmann equations for electron and hole.

In real micro-electron devices the Debye number and relaxation time are small compared with other physical relevant scalings [10]. Combining the following Debye number and relaxation limit [3]

$$t \to \tau t, \ \lambda = \tau^{1+\alpha}, \ -1 < \alpha,$$

$$n^\tau = n(\frac{t}{\tau}, x), \ J^\tau = \frac{1}{\tau} J(\frac{t}{\tau}, x), \ m^\tau = m(\frac{t}{\tau}, x), \ I^\tau = \frac{1}{\tau} I(\frac{t}{\tau}, x),$$

we obtain the porous media equation for the limiting density w of both n^τ and m^τ

$$w_t = P_{xx}, \quad P(w) = \frac{\mu_n \mu_m}{\mu_n + \mu_m} (p(w) + q(w)), \tag{7}$$

which admits a unique self-similar solution [2], the nonlinear diffusive wave, $w(x,t) = W(\xi)$, ($\xi = \frac{x}{\sqrt{1+t}}$) of (7) with boundary conditions:

$$W(\pm\infty) = \rho_\pm. \tag{8}$$

In this paper, we prove that the solutions of bipolar hydrodynamical equations converge to the nonlinear diffusive wave $W(\xi)$ time-asymptotically for fixed coefficients. We first consider the initial value problems for (1)–(5) with initial data

$$\begin{aligned} &(n, J, m, I)(x, 0) = (n_0, J_0, m_0, I_0)(x), \ x \in \mathbb{R}, \\ &(n_0, J_0, m_0, I_0)(\pm\infty) = (\rho_\pm, J_\pm, \rho_\pm, I_\pm), \end{aligned} \tag{9}$$

and the electric field at $x = -\infty$, $E_-(t)$, is set to be zero for simplicity. We assume that the pressure-density function satisfy

$$p(\rho) = q(\rho) = \frac{1}{\gamma}\rho^\gamma, \quad \gamma \geq 1, \tag{10}$$

and set μ_n, μ_m, τ and λ to be one for simplicity. And hence, $P(\rho)$ is replaced by $p(\rho)$ in (7). Moreover, we assume that the initial densities n_0 and m_0 are perturbations of the nonlinear diffusive wave. Such perturbations cause a shift of the nonlinear diffusive wave in the following sense:

$$\int_{-\infty}^{\infty} [(n_0(x) - W(x + x_0, t = 0)]dx = \mu_n \tau (J_+ - J_-),$$
$$\int_{-\infty}^{\infty} [(m_0(x) - W(x + y_0, t = 0)]dx = \mu_m \tau (I_+ - I_-),$$

with x_0, $y_0 \in \mathbb{R}$ two constants. For convenience we only consider the case that $I_\pm = J_\pm = 0$.

Set

$$\phi_0 = \int_{-\infty}^{x} (n_0(y) - W(y + x_0, 0))dy, \quad \phi_1 = J_0 + p(W(x + x_0, 0))_x, \tag{11}$$

$$\psi_0 = \int_{-\infty}^{x} (m_0(y) - W(y + y_0, 0))dy, \quad \psi_1 = I_0 + p(W(x + y_0, 0))_x. \tag{12}$$

Then, we have:

Theorem 1. *Assume that* (6) *and* (10) *hold. Suppose that* $(\phi_0, \phi_1, \psi_0, \psi_1) \in H^3 \times H^2 \times H^3 \times H^2$ *with* $x_0 = y_0$, *and* $\delta_0 = |\rho_+ - \rho_-| \ll 1$. *Then, there is a* $\eta_0 > 0$ *such that if* $\|(\phi_0, \psi_0)\|_3 + \|(\phi_1, \psi_1)\|_2 \leq \eta_0$, *the global classical solutions* (n, J, m, I, E) *of IVP* (1)-(5) *and* (9) *exist and satisfy*

$$\sum_{k=0}^{2}(1+t)^{k+1}||\partial_x^k(n-W,m-W)(.,t)||^2$$
$$+\sum_{k=0}^{2}(1+t)^{k+2}||\partial_x^k(J+p(W)_x,I+p(W)_x)(.,t)||^2 \le C(\Phi_0+\delta_0), \quad (13)$$
$$||(n-m,E)(.,t)||_2^2 \le C(\Phi_0+\delta_0)e^{-\alpha t}, \quad (14)$$

with $\alpha > 0$ *a constant and* $\Phi_0 = ||(\phi_0,\psi_0)||_3^2 + ||(\phi_1,\psi_1)||_2^2$.

Remark 1. One can obtain the global existence of classical solutions for general pressure density functions instead, for example when initial densities are around a positive constant (i.e., $W \equiv$ constant). Here, however we focus on the case (10) so as to simplify the discussion on the large time behavior of solutions.

The proof of above theorem consists of two step. The first step is to establish the uniform a-priori estimates so as to prove the global existence of classical solutions by applying energy estimates to the reformulated damped wave equations for the perturbation ϕ and ψ respectively and using the ODE of electric field. The second step is concerned with the proof of large time behavior. The exponential decay of electric field is obtained by applying energy method to the damped Klein-Gordon equation for electric field. Then, the two damped wave equations can be de-coupled and the algebraic decay rate of ϕ and ψ is obtained by applying Nishihara's methods [12, 13]. Furthermore, more careful resolution should be done in order to control the convection terms, and more regularity of the initial perturbations (ϕ,ψ) are required in order to obtain the desired time-decay rates of solutions.

There are other results on the existence of solutions for (bipolar) hydrodynamical model for semiconductors, the reader can refer to [1, 3, 4, 7, 8, 9, 11, 14] and references therein.

The rest part of this paper is devoted to the proof of Theorem 1. In subsection 2.1, we state some known results on the diffusive wave solution W of (7) and (8). After the reformulation of the original problem in subsection 2.2, we establish the uniform a-priori estimates on local solution in subsection 2.3, and finally in subsection 2.4, we prove the algebraic convergence of two densities to the nonlinear diffusive wave and the exponential decay of their difference and the electric field to zero.

Notation. Throughout this paper C always denotes generic positive constant. $L^2(\mathbb{R})$ is the space of square integrable real valued function defined on $\mathbb{R}$ with the norm $||\cdot||$, and $H^k(\mathbb{R})$ (H^k without any ambiguity) denotes the usual Sobolev space with the norm $||\cdot||_k$, especially, $||\cdot||_0 = ||\cdot||$.

2 Proof of main result

2.1 Nonlinear diffusive waves

Assume that the pressure-density functions satisfy (10) and μ_n, μ_m, τ and λ are set to be one. Then the nonlinear parabolic equation (7) reads:

$$w_t = p(w)_{xx}, \qquad p'(w) > 0,$$

which possesses a unique self-similar solution $w(x,t)$ [2]

$$w(x,t) \triangleq W(\zeta), \quad \zeta = \frac{x}{\sqrt{t+1}},$$

satisfying

$$\begin{cases} W''(\zeta) + \dfrac{p''(W(\zeta))W'(\zeta) - \frac{1}{2}\zeta}{p'(W(\zeta))} W'(\zeta) = 0, \\ W(\pm\infty) = \rho_\pm, \quad (\rho_+, \rho_- > 0). \end{cases}$$

This solution is increasing if $\rho_- < \rho_+$ and decreasing if $\rho_- > \rho_+$, and satisfies

$$\sum_{k=1}^{6} |\frac{d^k}{d\zeta^k}\Phi(\zeta)| + |W(\zeta) - v_+|_{\zeta>0} + |W(\zeta) - v_-|_{\zeta<0} \le C\delta_0 e^{-c\zeta^2},$$
$$|W_t(x,t)| \le C\delta_0(1+t)^{-1}, \ |W_x(x,t)| \le C\delta_0(1+t)^{-\frac{1}{2}},$$

where and throughout $\delta_0 = |\rho_+ - \rho_-|$. Moreover, we can obtain the L^2-estimates and L^p-estimates of the derivatives of W as ([5, 6, 12, 13]):

Lemma 1. *Let W be the self-similar solution of (7) and (8). Then it holds*

$$||\partial_t^l \partial_x^k W(.,t)||_{L_p} \le C\delta_0(1+t)^{-l-\frac{k}{2}+\frac{1}{2p}},$$

for $1 \le l+k \le 6$ and $p \in [2,\infty]$.

2.2 Reformulation of original problems

For any $T > 0$, assume that (n, J, m, I, E) is a classical solution of the IVP (1)–(5) and (9). Set

$$\begin{cases} \phi(x,t) = \displaystyle\int_{-\infty}^{x} [n(x,t) - W(x+x_0,t)]dx, \\ \eta(x,t) = J(x,t) + p(W(x,t))_x, \end{cases} \tag{19}$$

$$\begin{cases} \psi(x,t) = \displaystyle\int_{-\infty}^{x} [m(x,t) - W(x+y_0,t)]dx, \\ \zeta(x,t) = I(x,t) + p(W(x,t))_x. \end{cases} \tag{20}$$

Then by (1)–(5) and (19)–(20), the corresponding IVP for $(\phi, \eta, \psi, \xi, E)$ are obtained as

$$\phi_t + \eta = 0, \tag{21}$$

$$\eta_t + \left[\frac{(p(W)_x + \eta)^2}{W + \phi_x} + p(W + \phi_x) - p(W) \right]_x = (W + \phi_x)E - \eta - p(W)_{xt}, \tag{22}$$

$$\psi_t + \zeta = 0, \tag{23}$$

$$\zeta_t + \left[\frac{(p(W)_x + \zeta)^2}{W + \psi_x} + p(W + \psi_x) - p(W) \right]_x = -(W + \psi_x)E - \zeta - p(W)_{xt}, \tag{24}$$

$$E = \phi - \psi, \tag{25}$$

$$(\phi, \eta, \psi, \zeta)(x, 0) = (\phi_0, \phi_1, \psi_0, \psi_1)(x), \ x \in R. \tag{26}$$

By (21)–(22) and (23)–(24), we obtain the damped "wave equation" for ϕ and ψ respectively as

$$\phi_{tt} - (p(W + \phi_x) - p(W))_x + \phi_t + (W + \phi_x)E = -p(W)_{xt} + (f_1)_x, \tag{27}$$

$$\psi_{tt} - (p(W + \psi_x) - p(W))_x + \psi_t - (W + \psi_x)E = -p(W)_{xt} + (f_2)_x, \tag{28}$$

where

$$f_1 = \frac{(p(W)_x - \phi_t)^2}{W + \phi_x}, \ f_2 = \frac{(p(W)_x - \psi_t)^2}{W + \psi_x}. \tag{29}$$

The corresponding initial data are

$$(\phi, \phi_t)(x, 0) = (\phi_0, -\phi_1)(x), \ (\psi, \psi_t)(x, 0) = (\psi_0, -\psi_1)(x). \tag{30}$$

By (27), (28) and (25), we have the IVP for the damped "Klein-Gordon" equation for the electric field E as

$$E_{tt} + E_t + 2WE = f_3(x, t), \tag{31}$$

$$E(x, 0) = E_0(x) := (\phi_0 - \psi_0)(x), \quad E_t(x, 0) = E_1 := -(J_0 - I_0)(x), \tag{32}$$

where

$$f_3 = (p(W + \phi_x) - p(W + \psi_x))_x - (\phi_x + \psi_x)E + \left(\frac{(p(W)_x - \phi_t)^2}{W + \phi_x} - \frac{(p(W)_x - \psi_t)^2}{W + \psi_x} \right)_x =: (f_{31})_x + f_{32} + (f_{33})_x. \tag{33}$$

For $T > 0$, denote the basic space for the IVP (21)–(26) as

$$X(T) = \{(\phi, \phi_t, \psi, \psi_t, E) \in H^3 \times H^2 \times H^3 \times H^2 \times H^2, \ 0 \le t \le T\},$$

and assume that the following assumption holds

$$N(T) = \max_{0 \le t \le T} \{\|(\phi, \psi)(., t)\|_3 + \|(\phi_t, \psi_t)(., t)\|_2\} \ll 1. \tag{34}$$

2.3 The a-priori estimates

To prove global existence of classical solutions of (21)–(26), it is sufficient to establish the uniform a-priori estimates for IVP (27)–(30):

Lemma 2. *For $T>0$ let $(\phi,\psi,E)\in X(T)$. Then, it holds*

$$\begin{aligned}&\|(\phi,\psi)(.,t)\|_3^2+\|(\phi_t,\psi_t,E)(.,t)\|_2^2+\int_0^T\|(\phi_t,\phi_x,\psi_t,\psi_x,E)(.,s)\|_2^2ds\\&\le C(\|(\phi_0,\psi_0)\|_3^2+\|(\phi_1,\psi_1)\|_2^2+\delta_0),\quad 0<t<T,\end{aligned}\tag{35}$$

provided that $N(T)+\delta_0$ is small enough.

Proof. Multiplying (27) with ϕ and integrating over $\mathbb{R}$, we have, after integration by parts and recombination of the integer terms, that

$$\begin{aligned}&\frac{d}{dt}\left(\int_{-\infty}^{\infty}[\phi_t\phi+\frac{1}{2}\phi_t^2]dx\right)-(1+\kappa)\int_{-\infty}^{\infty}\phi_t^2dx+(p_*-2\kappa)\int_{-\infty}^{\infty}\phi_x^2dx\\&+\int_{-\infty}^{\infty}(W+\phi_x)E\phi dx\le C\int_{-\infty}^{\infty}(W_x^4+W_t^2)dx,\end{aligned}\tag{36}$$

provided that $C(N(T)+\delta_0)\le\kappa$. Here and after $p_*=\min_{y\in[\frac{1}{2}\rho_-,\frac{3}{2}\rho_+]}p'(y)$, $\kappa\le\frac{1}{8}\min\{p_*,1\}$, and $p^*=\max_{y\in[\frac{1}{2}\rho_-,\frac{3}{2}\rho_+]}p'(y)$.
Multiplying (27) with ϕ_t and integrating it by parts over $\mathbb{R}$, we have

$$\begin{aligned}&\frac{1}{2}\frac{d}{dt}\left(\int_{-\infty}^{\infty}\left[2\sigma(\phi_x,W)-\frac{(p(W)_x-\phi_t)^2}{(W+\phi_x)^2}\phi_x^2+\phi_t^2\right]dx\right)\\&+\int_{-\infty}^{\infty}\left[(1-\kappa)\phi_t^2-\frac{\partial\sigma(\phi_x,W)}{\partial W}W_t\right]dx+\int_{-\infty}^{\infty}(W+\phi_x)E\phi_tdx\\&\le C(N(T)+\delta_0)\int_{-\infty}^{\infty}[\phi_x^2+\phi_t^2]dx+C\delta_0,\end{aligned}\tag{37}$$

where $\sigma(\xi,W)=\int_0^\xi[p(W+\tau)-p(W)]d\tau$ satisfying $p_*\,\xi^2\le\sigma(\xi,W)\le p^*\,\xi^2$ (cf. [6]).
Using $[(36)+2(1+\kappa)\times(37)]$, we have

$$\begin{aligned}&\frac{d}{dt}\left(\int_{-\infty}^{\infty}\left[(1+\kappa)\phi_t^2+\frac{1}{2}\phi^2+\phi\phi_t\right]dx\right)\\&+\frac{d}{dt}\left(\int_{-\infty}^{\infty}\left[2(1+\kappa)\sigma(\phi_x,W)-\frac{2(1+\kappa)(p(W)_x-\phi_t)^2}{(W+\phi_x)^2}\phi_x^2\right]dx\right)\\&+\int_{-\infty}^{\infty}\left[(1+\kappa)(1-2\kappa)\phi_t^2+(p_*-2\kappa)\phi_x^2-2(1+\kappa)\frac{\partial\sigma(\phi_x,W)}{\partial W}W_t\right]dx\\&+\int_{-\infty}^{\infty}(W+\phi_x)E(\phi+2(1+\kappa)\phi_t)dx\end{aligned}$$

$$\leq C(N(T)+\delta_0)\int_{-\infty}^{\infty}[\phi_x^2+\phi_t^2]dx + C\delta_0. \tag{38}$$

Multiplying (28) by $[\psi + 2(1+\kappa)\psi_t]$, applying the similar procedures as above, noticing the negative symbol of the electric term of (28), we obtain

$$\begin{aligned}
&\frac{d}{dt}\left(\int_{-\infty}^{\infty}\left[(1+\kappa)\psi_t^2+\frac{1}{2}\psi^2+\psi\psi_t\right]dx\right)\\
&+\frac{d}{dt}\left(\int_{-\infty}^{\infty}\left[2(1+\kappa)\sigma(\psi_x,W)-\frac{2(1+\kappa)(p(W)_x-\psi_t)^2}{(W+\psi_x)^2}\psi_x^2\right]dx\right)\\
&+\int_{-\infty}^{\infty}\left[(1+\kappa)(1-2\kappa)\psi_t^2+(p_*-2\kappa)\psi_x^2-2(1+\kappa)\frac{\partial\sigma(\psi_x,W)}{\partial W}W_t\right]dx\\
&-\int_{-\infty}^{\infty}(W+\psi_x)E(\psi+2(1+\kappa)\psi_t)dx\\
\leq\, & C(N(T)+\delta_0)\int_{-\infty}^{\infty}[\psi_x^2+\psi_t^2]dx + C\delta_0,
\end{aligned} \tag{39}$$

provided that $N(T)+\delta_0 \ll 1$.

Taking [(38) + (39)], integrating it over $[0,T]$, using the following facts

$$\begin{aligned}
&\|E_0\|_2^2+\|E_1\|_1^2 \leq C(\|(\phi,\psi)\|_3^2+\|(\phi_1,\psi_1)\|_2^2)\leq C\Phi_0^2,\\
&\left|\frac{\partial\sigma(\xi,W)}{\partial W}W_t\right| \leq \kappa\xi^2 + C(\kappa)W_t^2, \quad \xi=\phi_x,\ \psi_x,\\
&\textstyle\int_{-\infty}^{\infty}\left[(W+\phi_x)E(\phi+2(1+\kappa)\phi_t)-(W+\psi_x)E(\psi+2(1+\kappa)\psi_t)\right]dx\\
&\geq \frac{d}{dt}\left(\textstyle\int_{-\infty}^{\infty}(1+\kappa)WE^2dx\right)+\textstyle\int_{-\infty}^{\infty}(W-(1+\kappa)W_t)E^2dx\\
&-CN(T)\textstyle\int_{-\infty}^{\infty}[E^2+\phi_x^2+\phi_t^2+\psi_x^2+\psi_t^2]dx,
\end{aligned}$$

we obtain

$$\begin{aligned}
&\|(\phi_t,\phi_x,\phi,\psi_t,\psi_x,\psi,E)(.,t)\|^2+\int_0^T\|(\phi_t,\phi_x,\psi_t,\psi_x,E)(.,s)\|^2ds\\
&\leq C(\|(\phi_0,\psi_0)\|_3^2+\|(\phi_1,\psi_1)\|_2^2+\delta_0),
\end{aligned}$$

provided that $(N(T)+\delta_0)\ll 1$.

Differentiating (27) and (28) with respect to x and using (25), in analogy, we can obtain higher order estimates (we omit the details). Combining these estimates, we obtain (35) finally. One can refer to [3] for details.

Now, we turn to prove the global existence of classical solutions of IVP (27)–(30). By standard argument, we can prove the local existence of classical solutions. By Lemma 2 and a standard continuity argument, we can extend the local classical solutions of IVP (27)–(30) globally in time.

2.4 The time decay rate of solutions

By Lemma 2, we know that the classical solutions (ϕ, ψ, E) satisfy

$$\|(\phi, \phi_x, \phi_{xx}, \phi_t, \phi_{xt}, \phi_{tt}, \psi, \psi_x, \psi_{xx}, \psi_t, \psi_{xt}, \psi_{tt}, E, E_x)(t)\|_{L^\infty} \leq C(\Phi_0 + \delta_0).$$

We first have the exponential decay of electric field:

Lemma 3. *Let (ϕ, ψ, E) be the classical solutions of IVP (27)–(32) with initial data satisfying $\|(\phi_0, \psi_0)\|_3 + \|(\phi_1, \psi_1)\|_2 \ll 1$. Then it holds for electric field E that*

$$\|(E, E_t, E_x, E_{xt}, E_{xx}, E_{tt})\|^2 \leq C(\Phi_0 + \delta_0) \exp\{-\beta_0 t\} \tag{41}$$

for $t > 0$.

Proof. Multiply (31) by $[E + 2E_t]$ and integrate it by parts over $\mathbb{R}$, we obtain

$$\frac{d}{dt}\left(\int_{-\infty}^{\infty} [E_t^2 + \frac{1}{2}E^2 + EE_t + 2WE^2]dx\right) + \int_{-\infty}^{\infty} [E_t^2 + 2(W - W_t)E^2]dx$$
$$= \int_{-\infty}^{\infty} f_3(E + 2E_t)dx = \int_{-\infty}^{\infty} [(f_{31})_x + f_{32} + (f_{33})_x](E + 2E_t)dx. \tag{42}$$

The right hand side terms in (42) can be estimated as follows.

$$\begin{aligned}&\int_{-\infty}^{\infty} (f_{31})_x (E + 2E_t)dx \\ &\leq -\frac{d}{dt}\left(\int_{-\infty}^{\infty} p'(W + \phi_x)E_x^2 dx\right) - \int_{-\infty}^{\infty} p'(W + \phi_x)E_x^2 dx \\ &\quad + C(N(T) + \delta_0)\|(E_x, E_t)(t)\|^2,\end{aligned} \tag{43}$$

$$\left|\int_{-\infty}^{\infty} f_{32}(E + 2E_t)dx\right| \leq C(\Phi_0 + \delta_0)\int_{-\infty}^{\infty} [E^2 + E_t^2]dx, \tag{44}$$

and

$$\begin{aligned}\int_{-\infty}^{\infty} (f_{33})_x (E + 2E_t)dx \leq &\frac{d}{dt}\left(\int_{-\infty}^{\infty} \frac{(p(W)_x - \phi_t)^2}{2(W + \phi_x)^2} E_x^2 dx\right) \\ &+ C(\Phi_0 + \delta_0)\|(E_x, E_t)(t)\|^2.\end{aligned} \tag{45}$$

Substituting (43), (44), (45) into (42) and using Gronwall's inequality, we obtain

$$\|(E, E_x, E_t)\|^2 \leq C(\Phi_0 + \delta_0) \exp\{-\beta_1 t\}. \tag{46}$$

Differentiating (31) with respect to x, multiplying the resulted equation by $[E_x + 2E_{xt}]$ and integrating it by parts over $\mathbb{R}$. In analogy, we obtain, after a computation, that

$$\|(E_x, E_{xx}, E_{xt})\|^2 \leq C(\|E_0\|_2^2 + \|E_1\|_1^2 + \delta_0) \exp\{-\beta_2 t\}. \tag{47}$$

By (46), (47), and the estimate $|E_{tt}| \leq C|E_{xt} + E_{xx} + E_t + E_x + E|$, we can prove (41) for $\beta_0 = \min\{\beta_1,\ \beta_2\}$. One can refer to [3] for details.

Then, we prove the algebraical decay rate of (ϕ, ψ) for large time by applying Nishihara's methods [13, 12] (we omit the details here):

Lemma 4. *Let (ϕ, ψ, E) be the classical solutions of IVP (27)–(32) with initial data satisfying $\|(\phi_0, \psi_0)\|_3 + \|(\phi_1, \psi_1)\|_2 \ll 1$. If it holds for (ϕ, ψ) that*

$$\sum_{k=0}^{3}(1+t)^k\|\partial_x^k(\phi,\psi)(.,t)\|^2 + \sum_{k=0}^{2}(1+t)^{k+2}\|\partial_x^k(\phi_t,\psi_t)(.,t)\|^2 \le C,$$

then we have

$$\begin{aligned}
&\sum_{k=0}^{3}(1+t)^k\|\partial_x^k(\phi,\psi)(.,t)\|^2 \\
&\qquad + \sum_{k=1}^{3}\int_0^t (1+s)^{k-1}\|\partial_x^k(\phi,\psi)(.,s)\|^2 ds \le C\Phi_1, \\
&\sum_{k=0}^{2}(1+t)^{k+2}\|\partial_x^k(\phi_t,\psi_t)(.,t)\|^2 \\
&\qquad + \sum_{k=0}^{2}\int_0^t (1+s)^{k+1}\|\partial_x^k(\phi_t,\psi_t)(.,s)\|^2 ds \le C\Phi_1, \\
&(1+t)^5\|(\phi_{ttt},\psi_{ttt},\phi_{xtt},\psi_{xtt})(.,t)\|^2 \\
&\qquad + \int_0^t (1+s)^5\|(\phi_{ttt},\psi_{ttt})(.,s)\|^2 ds \le C\Phi_1,
\end{aligned}$$

provided that $\Phi_1 =: \Phi_0 + \delta_0 \ll 1$.

Proof of Theorem 1. By Lemma 2, the solutions of IVP (1)–(5) and (9) exist globally in time and are uniformly bounded. By Lemma 3–4, we can prove (13)–(14) for the classical solutions. The L^p decay rate can be obtained by applying the method of approximate Green's function [13] and a more complicated resolution on the convection term. We omit the details.

Acknowledgments: I. Gasser acknowledges the support from the EU-funded TMR-network 'Asymptotic Methods in Kinetic Theory' (Contract # ERB FMRX CT970157). L. Hsiao's research is supported by the Special Funds of State Major Basic Research Projects (Grant No.1999075107), the Innovation funds of AMSS, CAS of China, and the Austrian–Chinese Scientific–Technical Collaboration Agreement. H.-L. Li acknowledges the support from the Wittgenstein Award 2000 of P. A. Markowich, funded by the Austrian FWF.

References

1. Chen, G., Wang, D., (1996): Convergence of shock schemes for the compressible Euler-Poisson equations. Comm. Math. Phys., **179**, 333-364.
2. Duyn, C.T., Peletier, L.A. (1977): A class of similarity solutions of the nonlinear diffusion equation. Nonlinear Analysis, TMA, **1**, 223–233.
3. I. Gasser, L. Hsiao, and H.-L. Li, Large time behavior of solutions of the bipolar hydrodynamical model for semiconductors. preprint.
4. Hattori, H., Zhu, C. (2000): Stability of steady state solutions for an isentropic hydrodynamic model of semiconductors of two species, J. Differential Equations, **166**, 1–32.
5. Hsiao, L., Liu, T.-P. (1992): Convergence to nonlinear diffusive waves for solutions of a system of hyperbolic conservation laws with damping, Comm. Math. Phys., **143**, 599–605.
6. Hsiao, L., Liu, T.-P. (1993): Nonlinear diffusive phenomena of nonlinear hyperbolic systems, Chin. Ann. Math., **14B**, 465-480.
7. Hsiao, L., Zhang, K. (2000): The global weak solution and relaxation limits of the initial boundary value problem to the bipolar hydrodynamic model for semiconductors. Math. Models Methods Appl. Sci., **10**, 1333–1361.
8. Li, H.-L., Markowich, P. (2001): A review of hydrodynamical models for semiconductors: asymptotic behavior. Bol. Soc. Bras. Mat., **32**, 321–342.
9. Marcati, P., Natalini, R. (1995): Weak solutions to a hydrodynamic model for semiconductors and relaxation to the drift-diffusion equation. Arch. Rational Mech. Anal., **129**, 129–145.
10. Markowich, P.A., Ringhofer, C., Schmeiser, C. (1990): Semiconductor Equations. Springer, Wien.
11. Natalini, R. (1996): The bipolar hydrodynamic model for semiconductors and the drift-diffusion equations. J. Math. Anal. Appl., **198**, 262–281.
12. Nishihara, K., Wang, W., Yang, T. (2000): L_p-convergence rate to nonlinear diffusion waves for p-system with damping. J. Differential Equations, **161**, 191–218.
13. Nishihara, K. (1996): Convergence rates to nonlinear diffusion waves for solutions of system of hyperbolic conservation laws with damping. J. Differential Equations, **131**, 171–188.
14. Zhang, K. (2001): On the initial-boundary value problem for the bipolar hydrodynamic model for semiconductors. J. Differential Equations, **171**, 257–293.

Simplification, Conservation and Adaptivity in the Front Tracking Method

E. George[1], J. Glimm[1,2], J. W. Grove[3], X. L. Li[1], Y. J. Liu[1], Z. L. Xu[1], and N. Zhao[1]

[1] SUNY at Stony Brook, Stony Brook, NY 11794 *linli@ams.sunysb.edu*
[2] Brookhaven National Laboratory, Upton, NY 11973 *glimm@ams.sunysb.edu*
[3] Los Alamos National Laboratory, Los Alamos, NM 87545 *jgrove@lanl.gov*

1 Introduction

The front tracking method was first introduced by Richtmyer for the numerical solution of hyperbolic equations. The solution of hyperbolic equations contains discontinuities such as contact discontinuities and shocks. The former exist even in linear equations when the initial condition is discontinuous. The latter are associated with the nonlinearity of the hyperbolic system. Finite difference and finite volume methods give a satisfactory numerical approximation and convergence rate in the region where the solution is smooth. However they fail to deliver physically correct solutions at discontinuities, especially when the equation of state across the discontinuity is sharply different. By separating the smooth regions at the discontinuity through an interface, the front tracking method overcomes this numerical difficulty by applying finite difference solvers to each smooth subdomain while treating the discontinuity with special care and propagating the front using the exact solution of the Riemann problem.

In the last few years, progress has been made to extend the front tracking method to three dimensional space and to address the issue of conservation in the coupling between the front solution and the interior solver. There has also been an effort to combine tracking and adaptive mesh refinement in the computation of the hyperbolic equations. In this paper, we will review the work done in this field. We will describe the addition of the grid-based tracking and hybrid methods, the use of dynamic flux and control volume for conservative tracking and the combination of the adaptive computational code Overture and the front tracking code FronTier.

A major thrust of the front tracking program has been its application to scientific problems. In this paper, we report on one of the most successful applications of the front tracking method to fluid physics. The clear separation of different fluid components in the computation of gravity driven Rayleigh-Taylor instability has maintained correct buoyancy driven acceleration throughout the numerical simulation, thus giving a dimensionless acceleration consistent with that observed in experiments.

Sections 2, 3, and 4 will discuss recent developments of the front tracking method, its simplification and extension to three dimensional space, implementation of the conservative tracking scheme and the merging of FronTier with the Overture code, while Section 5 will discuss a successful application of the front tracking method. A short conclusion is given in Section 6.

2 Geometrical Simplification

The classical front tracking method in which the front geometry is independent of the rectangular grid of the computational domain has encountered an increased difficulty in its extension to the three dimensional space. The main reason for this fact is: (1) the determination and correction of the interface topology requires the information of the full interface and (2) the interface could evolve into a configuration with high complexity, even within one time step and with small probability. This has limited the robustness of the numerical algorithm and increased the complexity of the code in order to handle cases which may rarely occur (but when they happen, they terminate the computation). Such complexity had already occurred in two dimensional front tracking, but became a significant obstacle when the front tracking is extended to three dimensions.

To resolve this problem, a new method, called the grid-based tracking method [5], was introduced to augment the classical or the grid-free tracking method. This method makes a closer link between the interface and the rectangular grid of the computational domain. The interface is grouped into segments which each fit into a mesh block in the rectangular grid. With an artificially imposed constraint that the interface is allowed to intercept each mesh block edge only once, the topology of the interface within each mesh block is greatly simplified. In two dimensions, there are only three isomorphically distinct cases and in three dimensions, only 14 isomorphically distinct cases. The construction of the block interface is the same as in the marching cube method by Lorensen and Cline [8], but the reconstruction is dynamic. Since the level set method [9] also borrows the marching cube based graphical tools to show the interface, the simplicity as well as the resolution level of the interface is the same for both methods. Figures 1 and 2 show the comparison of the interface in the classical (or grid-free) tracking method and the grid-based tracking method.

While gaining robustness, the grid-based tracking method loses resolution. The quality of its elements is also decreased due to the constraint imposed by the rectangular grid blocks. In two dimension, these interface elements are bonds and in three dimension, they are triangles. To combine the merits of the two tracking method, we also propose a hybrid method. The hybrid algorithm can vary according to the application needs. One method is the failure backing-up algorithm in which the grid based method is used only when the grid-free method fails to resolve the topological changes. In that case, the

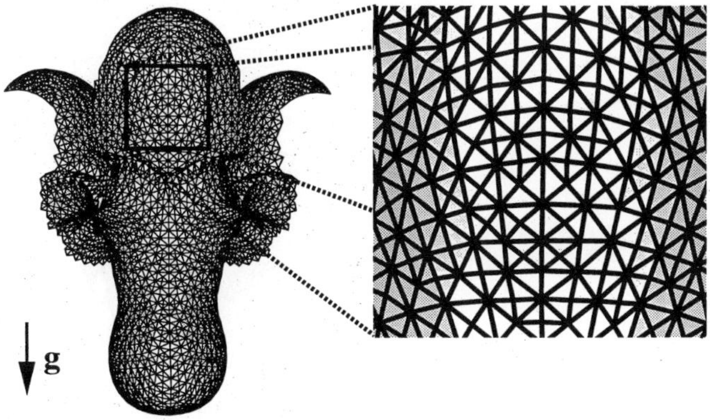

Fig. 1. The the grid-free interface in the simulation of Rayleigh-Taylor instability

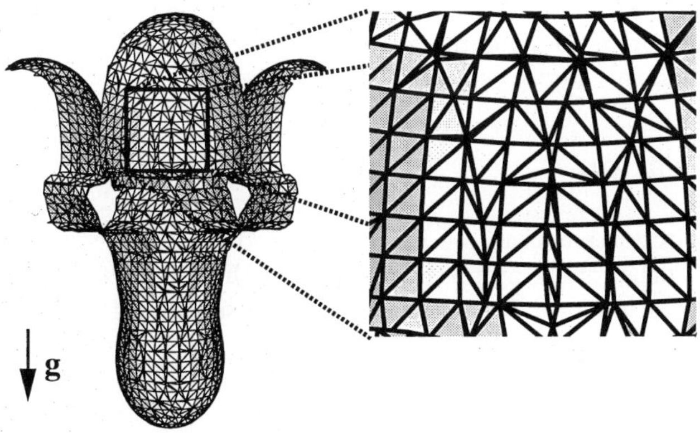

Fig. 2. The reconstruction of the grid-based interface in the simulation of Rayleigh-Taylor instability

whole interface will be reconstructed through the grid-based method. Another one is the locally grid-based method. In such a method when a topological bifurcation is detected, a rectangular box of varying size is constructed which totally confines the region of the interface tangling and only interface pieces inside the box will undergo a grid-based reconstruction.

The grid-based reconstruction of the interface mesh consists of four major steps: (1) the propagation of the interface, (2) intersection of the interface with grid edges, (3) removal of the unphysical or topologically incorrect intersections, and (4) the block-wise reconstruction of the interface. The details of this algorithm are presented in [5]. Figure 3 shows the evolution of a 3-D Rayleigh-Taylor unstable interface in a front tracking simulation.

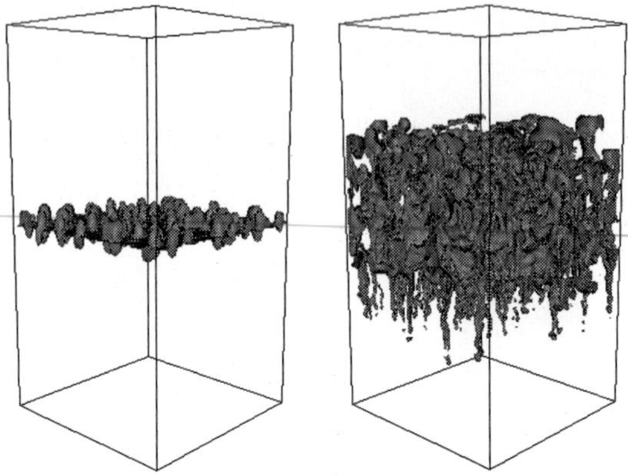

Fig. 3. Evolution of the fluid interface in the simulation of Rayleigh-Taylor instability with randomly perturbed initial interface

3 Conservation

The second important issue we have addressed in the front tracking method as well as in other tracking methods (such as the level set tracking method), is the conservation property of the numerical scheme. To separate different material components and prevent the cross-interface interpolation (such interpolation is the source of numerical diffusion and dispersion), the front tracking method used extrapolated states from the interface to create the numerical stencil to update the near-interface grid points [7]. This algorithm was later adopted to use in the level set tracking and was called the ghost-cell method [3]. For example, to use a five-point MUSCL or a Godunov-type scheme to update the states at two grid points j and $j+1$ between which there is an interface, we calculate the flux at the right cell edge of point j through $F_{j+1/2}(u^n_{j-1}, u^n_j, u^{\bar{n}}_{j+1}, u^{\bar{n}}_{j+2})$, and the flux at the left cell edge of point $j+1$ through $F_{j+1/2}(u^{\bar{n}}_{j-1}, u^{\bar{n}}_j, u^n_{j+1}, u^n_{j+2})$, where $u^{\bar{n}}_{j+1}$, $u^{\bar{n}}_{j+2}$ are ghost states extrapolated from the left side of the interface, and $u^{\bar{n}}_{j-1}$,$u^{\bar{n}}_j$ are the ghost states extrapolated from the right side of the interface. Since the two fluxes calculated at the right cell edge of j and the left cell edge of $j+1$ cannot cancel with each other, the numerical scheme is not conservative. The lack of conservation occurs in the same way in the level set tracking, the difference between the two methods lying in the algorithm of extrapolation.

Our new numerical scheme [6] is based on the concept of dynamic flux defined as:

$$f^d_J(u) = f(u_J) - su_J, \qquad J = L,\ R$$

where $f(u)$ is the stationary flux and s is the front speed. When the dynamic flux is applied at a dynamical cell edge (the interface), with complete sep-

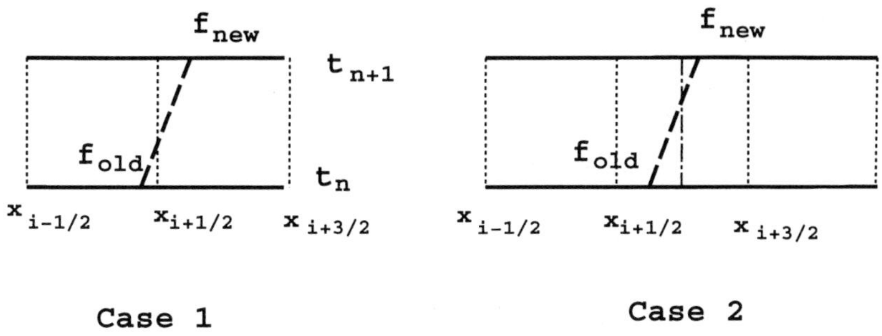

Fig. 4. The 1-D time space cells. The left is without merging and the right is with merging with neighboring cells.

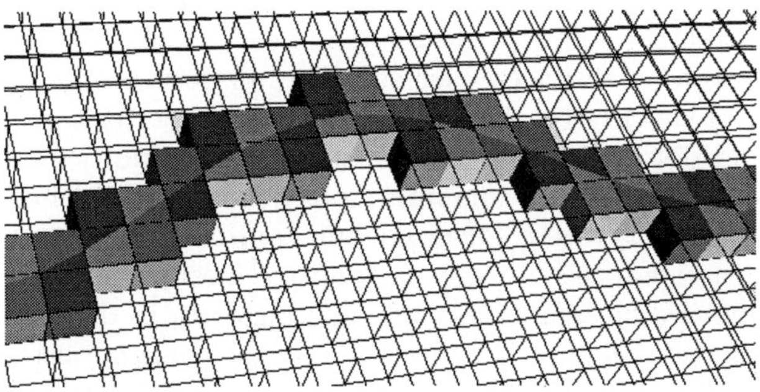

Fig. 5. The 2-D time space cells for the computation of conservative front tracking

aration of fluid states on each side of the interface, we can still have flux cancellation so long as the state variables u_L and u_R are the solution of the Riemann problem and thus satisfy the Rankine-Hugoniot condition. The implementation of the new scheme is based on the construction of the space-time control volume. Figures 4 and 5 show the control volumes for one and two space dimensions.

The conservation law for a control volume with a dynamic boundary can be written as

$$(u_+ - u_-)n_t + \sum_{j=1}^{d}(F_j(u_+) - F(u_-))n_{xj}) = 0$$

where n_t and n_{xj} are the time and space components of the unit vector at the boundary of the space-time control volume. This condition is satisfied at any space-time control volume boundary, should it be discontinuous or not. In particular, we use the space-time interface as one side of the control volume so that cross-interface interpolation can be avoided. The accuracy of the scheme is then dependent on the algorithm to evaluate u_+ and u_- on each side of the control volume.

4 Adaptivity

Front tracking is most effective in separating the fluid components, especially those with different equation of states. Tracking of shock waves is more easily replaced by other methods such as capturing with the use of limiters. The interaction of different waves can be easily solved in one dimension. In two dimensions, although the shock polar solution allows tracking of both contacts and shocks, there are cases in which the 2D required Riemann problem does not have a simple solution. In three dimensions, the problem of wave interactions in gas dynamics is yet to be studied. Another reason for not tracking shocks is that such discontinuities are compressive. With converging characteristics, numerical errors tend to be annihilated at the discontinuity and diffusion is self-adjusted. Capturing methods with the application of limiters provide a satisfactory solution to such nonlinear discontinuities.

Unlike the contact surface which separates materials with different equation of states, a shock wave does not have to be prescribed initially. Even if the initial condition does not contain a shock discontinuity, one can be generated in the interior of the computational domain after a finite time. The resolution required by these localized but captured waves is typically distributed very nonuniformly in space. Adaptive mesh refinement [1] provides a very effective and efficient computational platform to deal with this problem. A combination of the front tracking and adaptive mesh refinement will compliment each other and become a very powerful tool to study problems in fluid physics with both prescribed material boundary and interior wave dynamics.

The adaptive mesh refinement code we use is adopted from the Lawrence Livermore National Laboratory. The algorithm used in this code follows Berger. The concentration of dynamical changes are detected by the gradient of the physical quantities such as the density and pressure. Grid patches at different levels of refinement are created according to the user prescribed input parameters. Since the material interface is always discontinuous and has the largest gradient, it will always be included in patches of the finest level. Therefore we embed the tracked interface only in the most refined level of the adaptive grid. Figure 6 shows the adaptive patches with a tracked front in the finest level of the grid.

5 A Scientific Application

One of the successful applications of the front tracking method is its application to the gravity driven fluid interface instability, the Rayleigh-Taylor instability. Over the last decade, scientists have been trying to reproduce through computation the experiments done by Read and Youngs on the RT instability with randomly perturbed fluid interface. Read and Youngs first

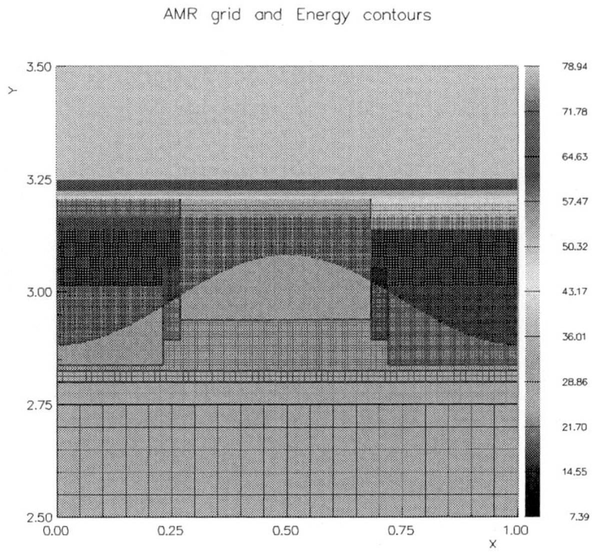

Fig. 6. The adaptively refined patches and the tracked front in a combined Overture-FronTier initialization.

observed that there exists a dimensionless acceleration rate of the mixing layer

$$\alpha = \frac{h}{Agt^2}$$

where h, A, g and t are the height of the mixing layer, the Atwood number of the two fluids, gravity and time respectively. The measurement by Read [10] on the acceleration rate α of the mixing layer on the bubble side gives its values at about $0.06 - 0.077$. However, most of the numerical simulation with untracked fluid interface gives the acceleration rate in the range of $0.015 - 0.04$. The sharp contrast between experiment and computation stimulated intense debate among scientists as to what could be wrong with the experiment or the simulation.

To answer this question, we have performed a comparative study of the RT instability using both an untracked numerical code and the front tracking method. Our front tracking simulation showed an [4] acceleration rate lying within the range of experimental variation. The average value of the α in the front tracking simulation is about 0.07 while the untracked code (using the TVD scheme) gives much smaller acceleration rate of about 0.04, which agrees with most of the simulations by other computational scientists using untracked code. The front tracking value for α has been reanalyzed using lower than leading order asymptotics, yielding a corrected $\alpha = 0.0625$ [2]. A careful study of the fluid density profile revealed a sharp contrast between the tracked and untracked simulations. The former maintains a clear boundary of the two fluids with a sharp density discontinuity across the fluid interface, while the latter has significant smearing-out of the fluid from one side to the other.

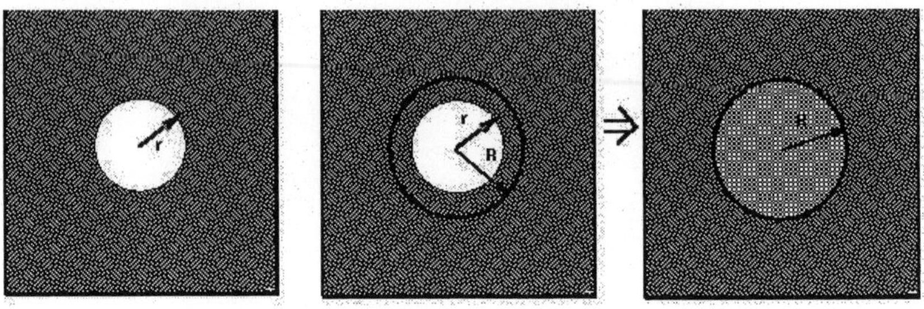

(a) (b) (c)

Fig. 7. The buoyancy acceleration for undiffused and diffused bubbles, see text for explanation.

An analysis of the net buoyancy (acceleration) acting on the bubbles (and spikes) revealed that the diffusive layer between the two fluids can significantly retard the relative motion between the two fluids. The reduced mixing rate due to unphysical numerical diffusion can be understood from Figure 7. The left frame represents an immiscible bubble of radius r. The central and the right frame assume that this bubble is smeared out numerically to a radius R while the total mass inside the sphere of radius R is conserved. The buoyancy forces

$$f_1 = f_2 = \frac{4\pi r^3}{3}(\rho_H - \rho_L)g \tag{1}$$

for the bubbles in frames (a) and (c) are the same. However, due to the difference between the mass in the nondiffused bubble (a) and the diffused bubble (c), the two acceleration rates

$$a_1 = \frac{\rho_H - \rho_L}{\rho_L}g \quad > \quad a_2 = \frac{\rho_H - \rho_L}{\rho_L + \left(\frac{R^3}{r^3} - 1\right)\rho_H}g \tag{2}$$

are different. In the diffused fluids, the buoyancy force is distributed to a larger amount of mass, thus reducing the acceleration of the bubble.

We then computed an effective Atwood number $A(t)$ as a function of time for the untracked simulations. This is determined from the highest and lowest densities in a horizontal slice, with the resulting time and space dependent Atwood number averaged over heights in the upper third of the mixing zone at a fixed time to get an Atwood number dependent on time alone. The time dependence of $A(t)$ in the *FronTier* simulation is caused purely by (small) compressibility effects. For the mass diffusive TVD simulation, the initial density contrast, $A(t = 0) = 0.5$, is almost completely washed out; the earliest time displayed shows $A(t = 2) \approx 0.15$. As new pure (heavy and light) fluid is injected into the mixing region, the effective Atwood number increases, but it is still reduced to about $A \approx 0.3$ on a time averaged basis, or nearly a 50% reduction relative to its initial value.

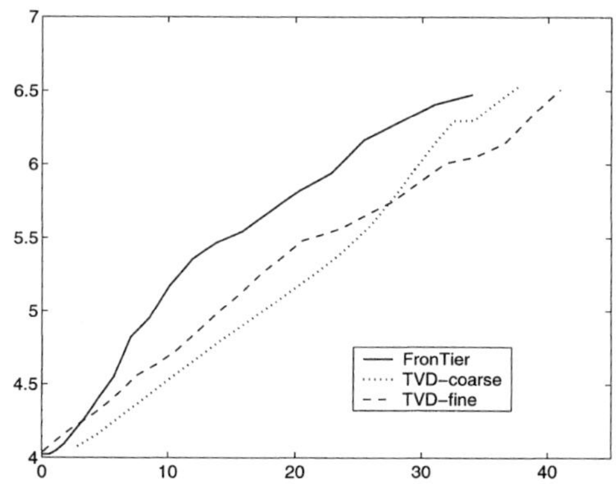

Fig. 8. Bubble height vs. $A_{eff}gt^2$ in tracked and untracked numerical simulation

To compensate for the time dependent Atwood number $A(t)$, we define an effective alpha, $\alpha_{\text{eff}} \approx h/2 \int \int A(s)gdsdt$. Specifically, α or α_{eff} is defined here as the slope of the straight line joining the beginning and end of the $h(t)$ curve in Figure 8. This definition, although somewhat arbitrary, is conventional, and thus allows comparison to the results of others. We observe an improved comparison between *FronTier* and TVD and between TVD and experiment. On this basis, we can state that the diffusive buoyancy renormalization of α is capable of resolving existing discrepancies among simulations, among diffusive and non-diffusive simulations and with theory.

6 Conclusion

We conclude that front tracking method can overcome its complexity in handling the interface geometry with the help of the rectangular grid. Using the grid edges as the basis for the interface, we can reduce most of the geometrical calculation to a single coordinate and limit the topological configuration of the interface to small number of isomorphically distinct cases. While combining with the classical grid-free tracking, we can also have the flexibility of enhancing the quality of the interface meshing and maintaining higher resolution of the interface for certain physical problems.

Conservation of the front tracking method can be achieved through construction of space-time control volumes and application of the dynamic flux at its moving boundary. In combining the front tracking method and the adaptive mesh refinement, we will create a powerful tool for the solution of hyperbolic conservation laws with contact surfaces separating different materials and efficiently resolve the internal wave interactions.

Front tracking method is shown to be able to deliver solutions with sharp contrast to the untracked numerical methods in comparison with experiments. The prediction of the correct acceleration rate in the simulation of RT instability with randomly perturbed interface provides a good example of its success.

7 Acknowledgements

The authors of this work were supported in part by the grants from DOE DEFC0201ER25461, DEFG0398DP00206, DEFG0290ER25084, the grant by the Army Research Office DAAD190110642, and the contract from Los Alamos National Laboratory 26730001014L.

References

1. M. Berger and P. Colella, *Local adaptive mesh refinement for shock hydrodynamics*, J. Comp. Phys., 82 (1989), pp. 64–84.
2. B. Cheng, J. Glimm, and D. H. Sharp, *Dynamical evolution of the Rayleigh-Taylor and Richtmyer-Meshkov mixing fronts*, Submitted to: Phys. Rev. E, (2001). SUNYSB-AMS-01-07 State Univ. of New York at Stony Brook.
3. R. P. Fedkiw, T. Aslam, B. Merriman, and S. Osher, *A non-oscillatory Eulerian approach to interfaces in multimaterial flows (the ghost fluid method)*, J. Comp. Phys., 152 (1999), pp. 457–492.
4. E. George, J. Glimm, X. L. Li, A. Marchese, and Z. L. Xu, *A comparison of experimental, theoretical, and numerical simulation Rayleigh-Taylor mixing rates*, Proc. National Academy of Sci., 99 (2002), pp. 2587–2592.
5. J. Glimm, J. W. Grove, X.-L. Li, and D. C. Tan, *Robust computational algorithms for dynamic interface tracking in three dimensions*, SIAM J. Sci. Comp., 21 (2000), pp. 2240–2256.
6. J. Glimm, X.-L. Li, and Y.-J. Liu, *Conservative front tracking in one space dimension*, Contemporary Mathematics, (2002). Proceedings of the Joint Summer Research Conference on Fluid Flow and Transport in Porous Media: Mathematical and Numerical Treatment. In Press, Report SUNYSB-AMS-01-16.
7. J. Glimm, D. Marchesin, and O. McBryan, *A numerical method for two phase flow with an unstable interface*, J. Comp. Phys., 39 (1981), pp. 179–200.
8. W. E. Lorensen and H. E. Cline, *Marching cubes: A high resolution 3D surface construction algorithm*, Computer Graphics, 21 (1987), pp. 163–169.
9. S. Osher and J. Sethian, *Fronts propagating with curvature-dependent speed: Algorithms based on Hamilton-Jacobi equations*, Jour. Comp. Phys, 79 (1988), pp. 12–49.
10. K. I. Read, *Experimental investigation of turbulent mixing by Rayleigh-Taylor instability*, Physica D, 12 (1984), pp. 45–58.

Discrete Adjoint Approximations with Shocks

M.B. Giles[1]

Oxford University Computing Laboratory, Oxford, U.K. *giles@comlab.ox.ac.uk*

1 Introduction

In recent years there has been considerable research into the use of adjoint flow equations for design optimisation (e.g. [Jam95]) and error analysis (e.g. [PG00, BR01]). In almost every case, the adjoint equations have been formulated under the assumption that the original nonlinear flow solution is smooth. Since most applications have been for incompressible or subsonic flow, this has been valid, however there is now increasing use of such techniques in transonic design applications for which there are shocks. It is therefore of interest to investigate the formulation and discretisation of adjoint equations when in the presence of shocks.

The reason that shocks present a problem is that the adjoint equations are defined to be adjoint to the equations obtained by linearising the original nonlinear flow equations. Therefore, this raises the whole issue of linearised perturbations to the shock. The validity of linearised shock capturing for harmonically oscillating shocks in flutter analysis was investigated by Lindquist and Giles [LG94] who showed that the shock capturing produces the correct prediction of integral quantities such as unsteady lift and moment provided the shock is smeared over a number of grid points. As a result, linearised shock capturing is now the standard method of turbomachinery aeroelastic analysis [HCL94], benefitting from the computational advantages of the linearised approach, without the many drawbacks of shock fitting.

There has been very little prior research into adjoint equations for flows with shocks. Giles and Pierce [GP01] have shown that the analytic derivation of the adjoint equations for the steady quasi-one-dimensional Euler equations requires the specification of an internal adjoint boundary condition at the shock. However, the numerical evidence [GP98] is that the correct adjoint solution is obtained using either the "fully discrete" approach (in which one linearises the discrete equations and uses the transpose) or the "continuous" approach (in which one discretises the analytic adjoint equations). It is not clear though that this will remain true in two dimensions, for which there is a similar adjoint boundary condition along a shock.

In this paper, we consider unsteady one-dimensional hyperbolic equations with a convex scalar flux, and in particular obtain numerical results for Burgers equation. Tadmor [Tad91] developed a Lip' topology for the formulation of adjoint equations for this problem, with application to linear post-processing functionals. Building on this and the work of Bouchut and James [BJ98], Ulbrich has very recently introduced the concept of shift-differentiability [Ulb02a, Ulb02b] to handle nonlinear functionals of the type considered in this paper. This supplies the analytic adjoint solution against which the numerical solutions in this paper will be compared. An alternative derivation of this analytic solution is presented in an expanded version of this paper [Gil02].

2 Analytic adjoint solution

Let $u(x,t)$ be the solution of the scalar equation

$$\frac{\partial u}{\partial t} + \frac{\partial f(u)}{\partial x} = 0, \qquad 0<x<1, \;\; 0<t<T$$

subject to initial conditions $u(x,0) = u_0(x)$. Numerical results will be presented later for the Burgers equation for which $f(u) \equiv \frac{1}{2}u^2$, but here we consider a general convex function $f(u)$. If the solution $u(x,t)$ is differentiable, then $u(x,t)$ is constant along straight characteristics defined by $\mathrm{d}x/\mathrm{d}t = \mathrm{d}f/\mathrm{d}u$. We will assume that $\mathrm{d}f/\mathrm{d}u > 0$ at $x=0$ and $\mathrm{d}f/\mathrm{d}u < 0$ at $x=1$, and therefore the value of $u(x,t)$ is specified on the two side boundaries.

When one is interested in the output functional

$$J(u) = \int_0^1 G(u(x,T))\, \mathrm{d}x,$$

the corresponding adjoint solution satisfies the equation

$$\frac{\partial v}{\partial t} + \frac{\mathrm{d}f}{\mathrm{d}u}\frac{\partial v}{\partial x} = 0, \tag{1}$$

subject to the final condition $v = \mathrm{d}G/\mathrm{d}u$ on $t=T$. The linearised functional corresponding to a linearised source term $s(x,t)$ can then be expressed as

$$\tilde{J} = \iint_{\Omega} v s \,\mathrm{d}x\,\mathrm{d}t, \tag{2}$$

which can be evaluated without computing the linearised solution to the original equation. This is the basis of the use of adjoint solutions in both design optimisation and error analysis for specific output functionals.

When there is a shock, as illustrated in Figure 1, the adjoint equation remains valid on either side of the shock, but along the shock the adjoint solution has value $([G]/[u])_{t=T}$, the ratio of the jumps in $G(u)$ and u across the shock at the terminal time $t=T$. This results in the adjoint solution having a uniform value along all characteristics leading backwards from the shock, as well as a constant value along each individual characteristic coming backwards in time from $t=T$.

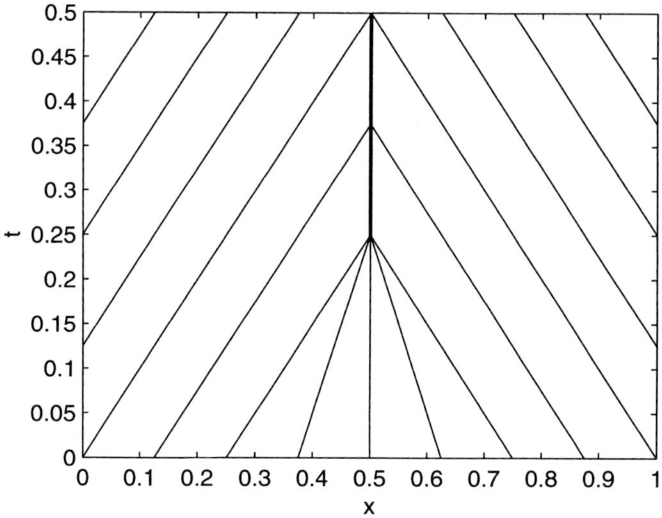

Fig. 1. Characteristics with a shock forming along $x=0.5$.

3 Numerical discretisation

We consider a class of explicit finite volume discretisations of the form

$$\frac{1}{\Delta t} M \left(U^{n+1} - U^n\right) + \Delta F^n = 0.$$

Here U^n is the vector of solution values $U_j^n, 0 \leq j \leq J$ at the nth timestep. M is a diagonal mass matrix whose entries are

$$M_{jj} = \begin{cases} \frac{1}{2}(x_1 - x_0) & j = 0, \\ \frac{1}{2}(x_{j+1} - x_{j-1}) & 0 < j < J \\ \frac{1}{2}(x_J - x_{J-1}) & j = J, \end{cases}$$

Given a numerical flux $F_{j+1/2}$ which is a function of both U_j and U_{j+1}, the flux difference ΔF^n is defined as

$$\Delta F_j^n = \begin{cases} F_{1/2}^n - f(u(0, t^n)) & j = 0, \\ F_{j+1/2}^n - F_{j-1/2}^n & 0 < j < J \\ f(u(1, t^n)) - F_{J-1/2}^n & j = J. \end{cases}$$

Note that this uses a weak implementation of the Dirichlet boundary conditions, as opposed to explicitly setting the values of U_0^n and U_J^n. This weak treatment is preferable because it leads to a cleaner formulation of the adjoint discretisation.

Having computed the numerical solution, the discrete form of the nonlinear output functional is evaluated as

$$J = \sum_j M_{jj} G(U_j^N).$$

The linearised equations with the inclusion of the source term can be written as

$$\frac{1}{\Delta t} M \left(\tilde{U}^{n+1} - \tilde{U}^n \right) + A^n \, \tilde{U}^n = M \; S^n. \tag{3}$$

In addition, the linearised output functional is

$$\tilde{J} = g^T M \tilde{U}^N \equiv \sum_j \left(\frac{\mathrm{d}G}{\mathrm{d}u} \right)_j^N M_{jj} \tilde{U}_j^N.$$

In formulating the discrete adjoint equations, we follow what is often termed the "fully discrete" approach in which the goal is to define the adjoint equations in such a way as to obtain exactly the same value for the discrete linearised functional. This is in contrast to the "continuous adjoint" approach which directly discretises the adjoint differential equation, independently of the discretisation of the original nonlinear equation.

Considering to begin with the case in which $S^n = 0$ for $n > 0$, the linear discrete equations, (3), may be solved to obtain

$$\tilde{J} = \Delta t \; g^T M \left(I - \Delta t \, M^{-1} A^{N-1} \right) \ldots \left(I - \Delta t \, M^{-1} A^2 \right) \left(I - \Delta t \, M^{-1} A^1 \right) S^0.$$

This may be re-arranged as

$$\begin{aligned} \tilde{J} &= \Delta t \; g^T \left(I - \Delta t \, A^{N-1} M^{-1} \right) \ldots \left(I - \Delta t \, A^2 M^{-1} \right) \left(I - \Delta t \, A^1 M^{-1} \right) M S^0 \\ &= \Delta t \; (V^1)^T M S^0, \end{aligned}$$

where V^1 is obtained by solving the discrete adjoint equation

$$\frac{1}{\Delta t} \; M \left(V^{n+1} - V^n \right) + (A^n)^T \; V^{n+1} = 0, \tag{4}$$

subject to the final condition $V^N = g$.

Extending to the general case in which S^n is non-zero at all time levels, the definition of the adjoint variables is unchanged and the resulting expression for the functional is

$$\tilde{J} = \Delta t \sum_{n=0}^{N-1} (V^{n+1})^T M S^n.$$

Note that this is a discrete equivalent of equation (2).

We will continue to use the same adjoint discretisation when the flow solution contains a shock. The question to be investigated is whether this will automatically capture the correct adjoint solution in the limit of increasing grid resolution. As a prelude, we note that only $g \equiv \mathrm{d}G/\mathrm{d}u$ enters into the adjoint calculation as initial data, not $[G]$ and so it is not clear that the adjoint calculation has the information necessary to correctly predict the adjoint solution in the neighbourhood of the shock.

4 Numerical tests

The numerical tests are all performed with the Burgers equation, $f(u) \equiv \frac{1}{2}u^2$, with initial conditions

$$u(x,0) = \begin{cases} 1, & x < 0.25 \\ 2-4x, & 0.25 \leq x \leq 0.75 \\ -1, & x > 0.75 \end{cases}$$

and boundary conditions $u(0,t)=1,\ u(1,t)=-1$. As shown in Figure 1, a stationary shock forms at $x=0.5$ at time $t=0.25$.

To assess the degree to which the solutions are grid converged, numerical results are obtained on two uniform grids with $\Delta x = 0.0025, 0.005$. The corresponding timesteps are $\Delta t = 0.4\Delta x$ giving a maximum CFL number of 0.4.

The output functional uses $G(u) = u^5 - u$. This gives $g(x) = 4$ on either side of the shock. Furthermore, the jump $[G]$ across the shock is equal to zero, so the analytic solution has $v=0$ for all backward travelling characteristics emanating from the shock. Hence the complete adjoint solution is

$$v(x,t) = \begin{cases} 4, & x < t \\ 0, & t < x < 1-t \\ 4, & x > 1-t \end{cases}$$

4.1 Riemann flux

The first results use a first order Riemann numerical flux function,

$$F(u_1,u_2) = \frac{1}{2}\max\left((\max(0,u_1))^2, (\min(0,u_2))^2 \right).$$

The upper two plots in Figure 2 show the nonlinear and adjoint solution at times $t = 0.1, 0.4$. There is very little difference between the solutions for the two grids. However it is very clear that the adjoint solution is completely wrong in the region emanating from the shock, where the computed value is approximately equal to -1.

The cause for this incorrect value can be seen in the lower two plots which show the nonlinear and adjoint solutions at the final time $t = 0.5$, plotted versus node number relative to the central node at $x = 0.5$. It is seen that on both grids the nonlinear solution has a single shock point at which $u = 0$. For this point the corresponding adjoint value is $g = \mathrm{d}G/\mathrm{d}u(0) = -1$, and a detailed examination of the matrix A reveals that this value is propagated backward in time along the length of the shock, and along any characteristic which propagates out of the shock.

An even more dramatic example of incorrect behaviour would be obtained by using an odd number of cells instead of an even number, so that the shock

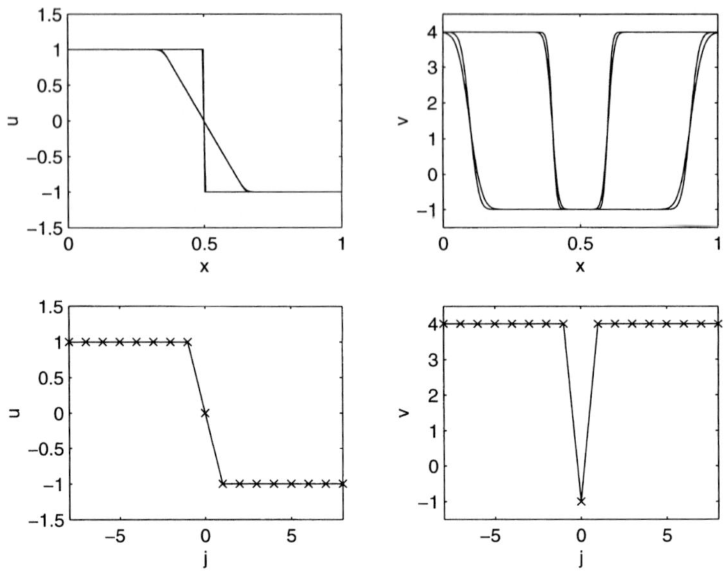

Fig. 2. Nonlinear and adjoint solutions obtained with the Riemann flux function, for $t=0.1, 0.4$ in the upper plots, and $t=0.5$ in the lower plots.

centre lies half-way between two nodes. In that case, the final solution would have no interior shock point, and so all of the elements of g would have the value 4, leading to the entire discrete adjoint solution having value 4.

4.2 Lax-Friedrichs flux

The rest of the results all use a simple Lax-Friedrichs flux, combining a central average flux with additional first order smoothing,

$$F(u_1, u_2) = \frac{1}{2}\left(f(u_2)+f(u_1)\right) - \mu\,(u_2-u_1)\,.$$

Figure 3 shows results for $\mu=0.25$. The values computed on the two grids are almost identical. In the vicinity of the shock, the nonlinear solution is very close to a self-similar steady-state solution which depends solely on μ and the grid ratio $\Delta t/\Delta x$, and with this level of smoothing there is again only one grid point in the middle of the shock. The adjoint solution appears grid converged, but to a value which is incorrect.

Figure 4 shows results for $\mu=1.0$. There are now many grid points across the shock, and therefore fairly good resolution of the differing values of $g = \mathrm{d}G/\mathrm{d}u$ for u ranging from 1 on the left of the shock to -1 on the right of the shock. The numerical adjoint solution now has a value very close to the analytic value of zero in the central part of the domain.

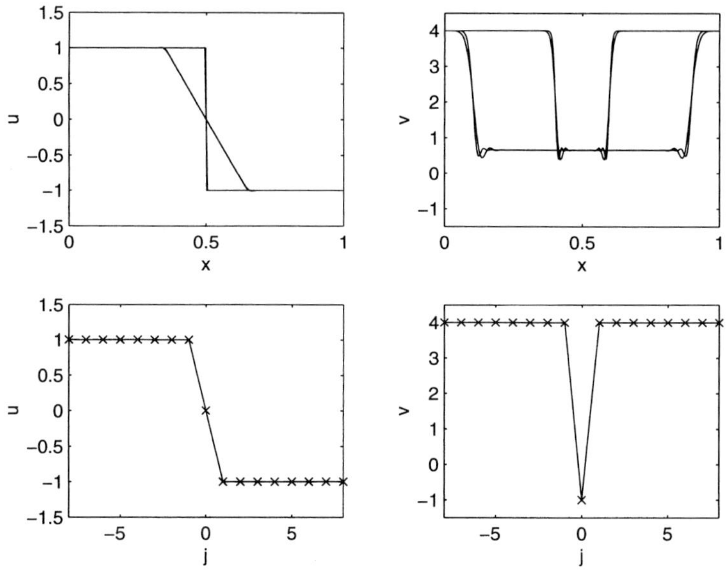

Fig. 3. Nonlinear and adjoint solutions obtained with the Lax-Friedrichs flux with smoothing $\mu = 0.25$, for $t = 0.1, 0.4$ in the upper plots, and $t = 0.5$ in the lower plots.

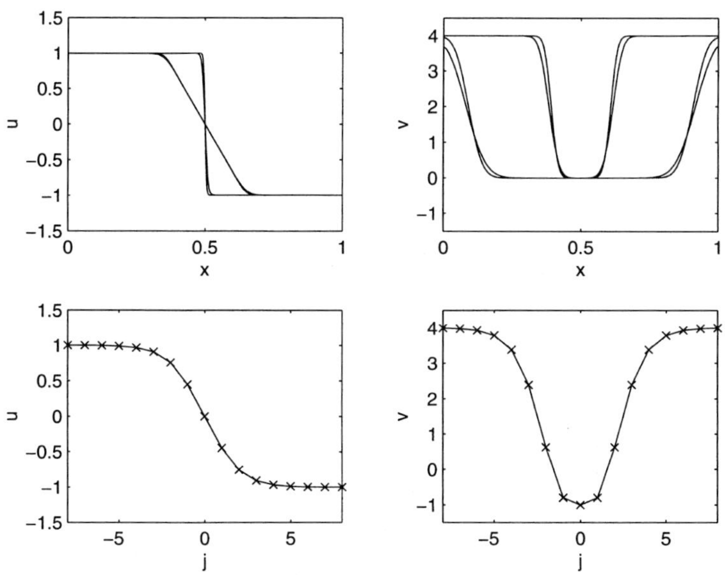

Fig. 4. Nonlinear and adjoint solutions obtained with the Lax-Friedrichs flux with smoothing $\mu = 1.0$, for $t = 0.1, 0.4$ in the upper plots, and $t = 0.5$ in the lower plots.

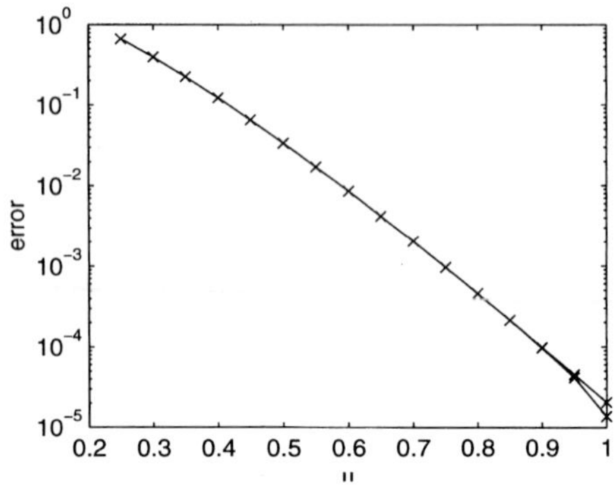

Fig. 5. Error in the computed value for $v(0.5, 0)$ as a function of the numerical smoothing coefficient μ.

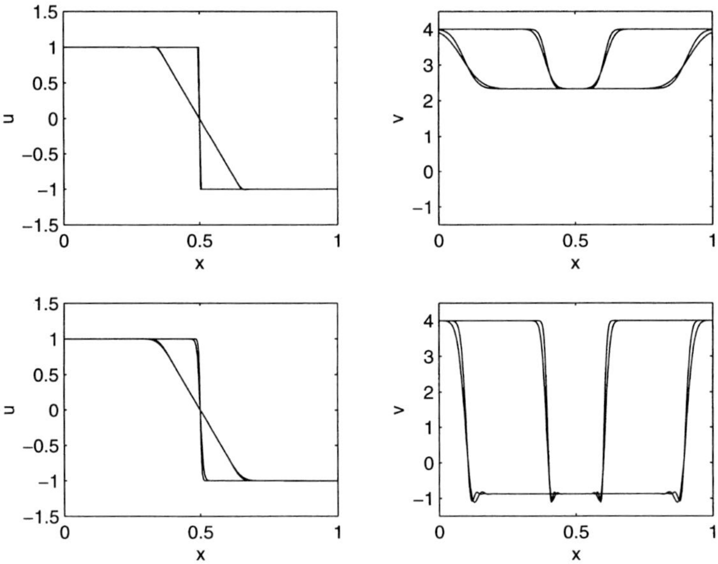

Fig. 6. Solutions at $t = 0.1, 0.4$ obtained with the Lax-Friedrichs flux with different levels of smoothing in the nonlinear and adjoint calculations. Upper results: $\mu(\text{nonlinear}) = 0.25$, $\mu(\text{adjoint}) = 1.0$. Lower results: $\mu(\text{nonlinear}) = 1.0$,, $\mu(\text{adjoint}) = 0.25$.

Figure 5 plots the error in the computed value for $v(0.5, 0)$ versus the value of the smoothing coefficient μ. It appears from these results that the error decreases exponentially with the value of μ and hence the number of grid points across the shock.

Figure 6 presents results obtained by using different values for μ in the nonlinear and adjoint calculations. The upper results use $\mu = 0.25$ for the nonlinear calculations, and $\mu = 1.0$ for the adjoint calculation. The higher

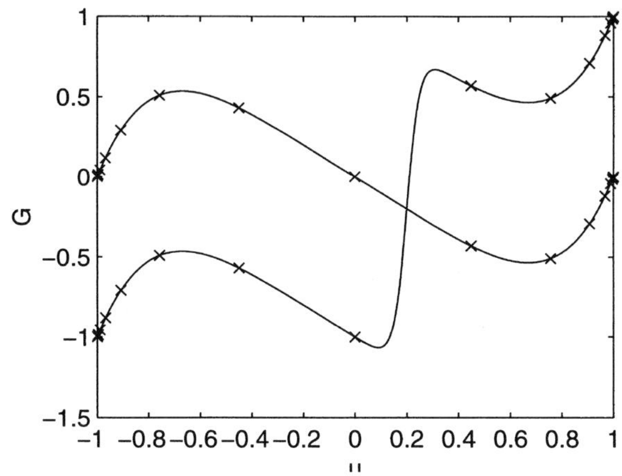

Fig. 7. Two objective functions $G(u)$.

value for μ in the adjoint calculation leads to rapid diffusion bringing into the shock region the larger values for the adjoint solution $v(x,t)$ on either side of the shock, leading to incorrect values in the shock region. The lower results use $\mu=1.0$ for the nonlinear calculations, and $\mu=0.25$ for the adjoint calculation. The lower value for μ in the adjoint calculation leads to very little diffusion, and so the adjoint solution value $g = -1$ at the centre of the smeared shock is convected backwards in time leading to $v(x,t) \approx -1$ throughout the shock region. These results show the importance of using the same value of μ in both calculations, so that the adjoint discretisation corresponds correctly to the linearisation of the nonlinear discretisation.

5 Discussion

One clear conclusion from these numerical results is that there must be consistency between the nonlinear and adjoint calculations regarding the level of numerical smoothing. Also, for convergence it is necessary that as the grid resolution improves, the numerical smoothing varies in a way which increases the number of points across the shock, while at the same time the overall width of the shock decreases.

To understand why this latter point is fundamental, and not just a feature of the particular numerical experiments conducted, we need to consider the information supplied to the adjoint code. The analytic solution has a value along the shock which depends on the jump $[G(u)]$ across the shock at the final time $t=T$. However, the end conditions for the numerical adjoint solution are given by the values of $\mathrm{d}G/\mathrm{d}u$ for the final values of u obtained from the nonlinear calculation. These means that the numerical solution must implicitly evaluate $[G(u)]$ by some process which effectively integrates $\mathrm{d}G/\mathrm{d}u$

across the smeared shock. For this to be done accurately requires adequate resolution of the variation in $\mathrm{d}G/\mathrm{d}u$.

This point is illustrated in Figure 7. The smoother of the two curves is $G(u)=u^5-u$, the objective function in the numerical experiments. The symbols correspond to the values of u at the final time $t=T$ in Figure 4. The second curve is $G(u) = u^5 - u + \tanh 20(u-0.2)$. This function has almost identical gradient values at the indicated sampling points, and therefore produces a numerical adjoint solution which is visually indistinguishable from Figure 4. However, the analytic solution has a different jump in $G(u)$ across the shock, and so the analytic solution is quite different. This shows that for any numerical discretisation with a fixed number of points across the shock, it is easy to construct an objective function for which the numerical adjoint solution will not converge.

References

[BJ98] F. Bouchut and F. James. One-dimensional transport equations with discontinuous coefficients. *Nonlinear Analysis*, 32:891–933, 1998.

[BR01] R. Becker and R. Rannacher. An optimal control approach to error control and mesh adaptation. In A. Iserles, editor, *Acta Numerica 2001*. Cambridge University Press, 2001.

[Gil02] M.B. Giles. Adjoint equations and discrete approximations in the presence of shocks. Technical Report NA02/10, Oxford University Computing Laboratory, 2002.

[GP98] M.B. Giles and N.A. Pierce. On the properties of solutions of the adjoint Euler equations. In M. Baines, editor, *Numerical Methods for Fluid Dynamics VI*. ICFD, Jun 1998.

[GP01] M.B. Giles and N.A. Pierce. Analytic adjoint solutions for the quasi-one-dimensional Euler equations. *J. Fluid Mech.*, 426:327–345, 2001.

[HCL94] K.C. Hall, W.S. Clark, and C.B. Lorence. A linearized Euler analysis of unsteady transonic flows in turbomachinery. *J. Turbomachinery*, 116:477–488, 1994.

[Jam95] A. Jameson. Optimum aerodynamic design using control theory. In M. Hafez and K. Oshima, editors, *Computational Fluid Dynamics Review 1995*, pages 495–528. John Wiley & Sons, 1995.

[LG94] D.R. Lindquist and M.B. Giles. Validity of linearized unsteady Euler equations with shock capturing. *AIAA J.*, 32(1):46, 1994.

[PG00] N.A. Pierce and M.B. Giles. Adjoint recovery of superconvergent functionals from PDE approximations. *SIAM Rev.*, 42(2):247–264, 2000.

[Tad91] E. Tadmor. Local error estimates for discontinuous solutions of nonlinear hyperbolic equations. *SIAM J. Numer. Anal.*, 28:891–906, 1991.

[Ulb02a] S. Ulbrich. Adjoint-based derivative computations for the optimal control of discontinuous solutions of hyperbolic conservation laws. *Systems & Control Letters*, to appear, 2002.

[Ulb02b] S. Ulbrich. A sensitivity and adjoint calculus for discontinuous solutions of hyperbolic conservation laws with source terms. *SIAM J. Control and Optim.*, to appear, 2002.

Far-field Evaluation via Nonreflecting Boundary Conditions

M. J. Grote[1] and C. Kirsch[2]

[1] Department of Mathematics, University of Basel, Rheinsprung 21, CH-4051 Basel, Switzerland. Email: *grote@math.unibas.ch*
[2] Department of Mathematics, University of Basel, Rheinsprung 21, CH-4051 Basel, Switzerland. Email: *kirsch@math.unibas.ch*

Summary. A new far-field evaluation formula is presented which enables the efficient evaluation of the far-field solution for wave propagation problems, if the exact nonreflecting boundary condition in [5, 6] is used for the numerical computation in the near-field. In particular, the evaluation formula permits to take advantage of the fast decay with distance of selected modes and thus to store only the minimal amount of information necessary from the past. The accuracy of this formula is illustrated via numerical experiments.

1 Introduction

The simulation of waves in unbounded media arises in many applications from acoustics, electromagnetics, or elasticity. Typically, the local phenomenon of interest contains complicated geometric features, inhomogeneity, and possibly nonlinear effects. Modern numerical methods can handle complicated geometry, inhomogeneous media, and nonlinearities. However, they require an artificial boundary $\mathcal{B}$, which truncates the unbounded exterior domain and restricts the region of interest to a finite computational domain Ω. It then becomes necessary to impose a boundary condition at $\mathcal{B}$, which ensures that the solution in Ω coincides with the restriction to Ω of the solution in the unbounded region. Usually various approximate boundary conditions are used, such as the Bayliss-Turkel [2] or Engquist-Majda [1] boundary conditions, which produce some spurious reflection. To eliminate spurious reflection from the artificial boundary, Grote and Keller devised *exact nonreflecting boundary conditions* for the wave equation [5, 6], Maxwell's equations [7], and the elastic wave equation [9, 10]. These boundary conditions are *local in time* and involve only first derivatives of the solution. Therefore, they are easy to use with standard finite difference or finite element methods.

During or after the computation values of solutions outside the computational domain Ω are often needed. We will show that the auxiliary functions used by the nonreflecting boundary condition in [5, 6] enable us to efficiently compute the far-field solution anywhere outside Ω.

The derivation of the far-field evaluation formula is presented in sections 2, 3 and 4. In section 5 we shall illustrate its accuracy via numerical experiments.

2 Nonreflecting boundary condition

We consider a complicated scattering region in unbounded three-dimensional space. The scatterer may contain obstacles, inhomogeneities and nonlinearity. Outside the ball of radius R, however, we assume that the scattered field satisfies the homogeneous time dependent wave equation in the unbounded region $\{(r,\phi,\theta) \mid r \geq R,\ \phi \in [0,2\pi),\ \theta \in (0,\pi)\}$ with constant speed of propagation $c > 0$:

$$u_{tt} - c^2 \Delta u = 0, \quad t > 0,\ r \geq R \tag{1}$$

$$u(r,\phi,\theta,0) = 0,\ u_t(r,\phi,\theta,0) = 0, \quad r \geq R. \tag{2}$$

We now restrict the computation to the finite region of interest

$$\Omega = \{(r,\phi,\theta) \mid r \leq R,\ \phi \in [0,2\pi),\ \theta \in (0,\pi)\}$$

by introducing an artificial boundary at $r = R$, which effectively truncates the surrounding unbounded region. At the artificial boundary a boundary condition is needed, which ensures that the solution inside Ω, with that boundary condition imposed at $r = R$, coincides with the restriction of the solution in the original unbounded region. Such an exact boundary condition was derived by Grote and Keller [5, 6]. It is given by

$$\left(\frac{1}{c}\frac{\partial}{\partial t} + \frac{\partial}{\partial r}\right)[ru](r,\phi,\theta,t) = -\frac{1}{R}\sum_{n=1}^{\infty}\sum_{m=-n}^{n} \mathbf{d}_n \cdot \psi_{nm}(t) Y_{nm}(\phi,\theta) \tag{3}$$

for $r = R$ where the functions ψ_{nm} are solutions of the systems of ordinary differential equations

$$\begin{aligned} \frac{d}{dt}\psi_{nm}(t) &= \frac{1}{R}\mathsf{A}_n \psi_{nm}(t) + (u(r,\phi,\theta,t)|_{r=R}, Y_{nm}(\phi,\theta))\mathbf{e}_n, \quad t > 0 \\ \psi_{nm}(0) &= \mathbf{0}. \end{aligned} \tag{4}$$

Here the $\psi_{nm}(t) = \{\psi_{nm}^k(t)\}_{k=1}^n$ are n-vector-valued functions, $\mathbf{d}_n$ and $\mathbf{e}_n$ are constant n-vectors and A_n are constant $(n \times n)$-matrices. The functions $Y_{nm}(\phi,\theta)$ denote the spherical harmonics while $(\cdot,\cdot)$ denotes the standard L^2-inner product on the sphere of radius R. The boundary condition in (3) is called nonreflecting because it ensures that no spurious reflection occurs as the propagating waves leave the computational domain Ω.

3 Alternative representations for the Fourier coefficients

The derivation of the nonreflecting boundary condition (3) is based on the expansion of u in spherical harmonics outside Ω:

$$u(r,\phi,\theta,t)=\sum_{n=0}^{\infty}\sum_{m=-n}^{n}u_{nm}(r,t)Y_{nm}(\phi,\theta),\quad r\geq R,\ t>0. \tag{5}$$

We note that each Fourier coefficient u_{nm}, $0\leq|m|\leq n$, satisfies the equation

$$\left(\frac{1}{c^2}\frac{\partial^2}{\partial t^2}-\frac{\partial^2}{\partial r^2}-\frac{2}{r}\frac{\partial}{\partial r}+\frac{\lambda_n}{r^2}\right)u_{nm}(r,t)=0,\quad r\geq R,\ t>0,\ \lambda_n=n(n+1). \tag{6}$$

We now solve (6) explicitly via the Laplace transform. This yields a representation of each Fourier coefficient u_{nm} as a linear combination of one-dimensional outgoing waves f_{nm}^k:

$$u_{nm}(r,t)=\sum_{k=0}^{n}\frac{f_{nm}^k(t-\frac{r}{c})}{r^{k+1}},\quad 0\leq|m|\leq n,\ r\geq R,\ t>0, \tag{7}$$

with $f_{nm}^k(t)=0$, $t\leq 0$. A similar expression was derived by Hariharan and Hagstrom [8]. Moreover, if we introduce (7) into (5) and rearrange terms, we obtain the representation used by Bayliss and Turkel [2]. We now use (7) in (6), which yields the following recursion for the functions f_{nm}^k:

$$\left(f_{nm}^{k+1}\right)'(t)=\frac{(n+k+1)(n-k)}{2(k+1)}f_{nm}^k(t),\quad 0\leq k\leq n-1. \tag{8}$$

To eliminate f_{nm}^0 we evaluate u_{nm} at $r=R$:

$$\frac{f_{nm}^0(t-\frac{R}{c})}{R}=u_{nm}(R,t)-\sum_{k=1}^{n}\frac{f_{nm}^k(t-\frac{R}{c})}{R^{k+1}},\quad 0\leq|m|\leq n,\ t>0. \tag{9}$$

We now use (8) in (9) to find that ψ_{nm}^k is related to f_{nm}^k via

$$\psi_{nm}^k(t)=\frac{2R}{\lambda_n}\frac{f_{nm}^k(t-\frac{R}{c})}{R^{k+1}},\quad 1\leq k\leq n. \tag{10}$$

Finally, we combine (7) with (9) and (10) to obtain a representation for the Fourier coefficient u_{nm}, which uses only past values of ψ_{nm}^k and $u_{nm}(R,\cdot)$ on the artificial boundary:

$$\begin{aligned}u_{nm}(r,t)&=\frac{R}{r}[u_{nm}(R,\cdot)]+\frac{\lambda_n}{2r}\sum_{k=1}^{n}\left(\left(\frac{R}{r}\right)^k-1\right)[\psi_{nm}^k],\quad t>\frac{r-R}{c}\\ u_{nm}(r,t)&=0,\quad 0<t\leq\frac{r-R}{c},\end{aligned} \tag{11}$$

Here $[\cdot]$ denotes the retarded value operator

$$[f]:=f(t-\frac{r-R}{c}). \tag{12}$$

We remark that (11) holds everywhere outside Ω and for all time.

4 Far-field evaluation

Multiplication of (11) by Y_{nm} and summation over n and m yields an exact expression for the three-dimensional wave field u anywhere outside the computational domain and at all times. Moreover, it uses only past values of the Fourier coefficients $u_{nm}(R,\cdot)$ on the boundary and of the functions ψ^k_{nm}. All values in this expression but $u_{00}(R,\cdot)$ are known because they were computed at earlier times to impose the boundary condition (3) at $r = R$.

In practice, we need to truncate the sum over n at a certain $N > 0$ (typically the same N which is used for the implementation of the nonreflecting boundary condition (3)). To further reduce the amount of storage required for the far-field evaluation, we also truncate the second sum over k at a certain K, $1 \leq K \leq N$:

$$
\begin{aligned}
u_{N,K}(r,\phi,\theta,t) = {} & \frac{R}{r}\sum_{n=0}^{N}\sum_{m=-n}^{n}[u_{nm}(R,\cdot)]Y_{nm}(\phi,\theta) + \\
& -\frac{1}{r}\sum_{n=1}^{N}\frac{\lambda_n}{2}\sum_{m=-n}^{n}\left(\sum_{k=1}^{n}[\psi^k_{nm}]\right)Y_{nm}(\phi,\theta) + \\
& +\frac{1}{r}\sum_{n=1}^{N}\frac{\lambda_n}{2}\sum_{m=-n}^{n}\left(\sum_{k=1}^{\min\{n,K\}}\left(\frac{R}{r}\right)^k[\psi^k_{nm}]\right)Y_{nm}(\phi,\theta),
\end{aligned}
\tag{13}
$$

for $t > (r-R)/c$ (and $u_{N,K}(r,\phi,\theta,t) = 0$ for $0 \leq t \leq (r-R)/c$). Indeed, the L^2-error for the approximation $u_{N,K}$ can be estimated from above by

$$
E(r,t) := \|u(r,\phi,\theta,t) - u_{N,K}(r,\phi,\theta,t)\|_{L^2(S^2)} \leq E_N(r,t) + E_{N,K}(r,t), \tag{14}
$$

where the error components E_N and $E_{N,K}$ have the following asymptotic behavior:

$$
E_N(r,t) = O(r^{-1}), \quad r \to \infty \tag{15}
$$

$$
E_{N,K}(r,t) = O(r^{-(K+2)}), \quad r \to \infty,\ K < N, \quad E_{N,N}(r,t) = 0. \tag{16}
$$

Since $E_{N,K}$ rapidly decreases with increasing r, we can choose K much smaller than N. This greatly reduces storage requirements and operation counts.

For the evaluation of $u_{N,K}$ at any specific point in space over time, we need to store for each n and m a fixed amount of past values of $u_{nm}(R,\cdot)$, of ψ^k_{nm}, $k = 1,\ldots,\min\{n,K\}$, and of the sums $\sum_{k=1}^{n}\psi^k_{nm}$ for $n = K+1,\ldots,N$. Therefore the storage needed is

$$
S_{N,K} = (K+2)N^2 + 2(K+2)N - \frac{1}{3}K^3 - \frac{3}{2}K^2 - \frac{7}{6}K + 1 \tag{17}
$$

values per time step. The number of operations per time step is

$$O_{N,K} = 3KN^2 - K^3 + \text{ lower order terms.} \tag{18}$$

We now compare the above approach with the standard Kirchhoff formula. In fact Ting and Miksis derived a nonreflecting boundary condition based on this formula [3]; it was implemented by Cohen and Givoli [4]. If the artificial boundary is a sphere with radius $R > 0$ centered at the origin, Kirchhoff's formula reduces to
$u(r, \phi, \theta, t) =$

$$-\frac{1}{4\pi}\int_0^{2\pi}\int_0^{\pi}\left([u]\frac{\partial}{\partial r}\left(\frac{1}{d}\right) - \frac{1}{d}\left[\frac{\partial u}{\partial r}\right] - \frac{1}{cd}\frac{\partial d}{\partial r}\left[\frac{\partial u}{\partial t}\right]\right) R^2 \sin\theta_S \, d\theta_S d\phi_S, \tag{19}$$

where d corresponds to the distance between the location (r, ϕ, θ) and the point (R, ϕ_S, θ_S) on the sphere,

$$d = \sqrt{r^2 - R^2 - 2rR\left(\sin\theta\sin\theta_S\cos(\phi - \phi_S) + \cos\theta\cos\theta_S\right)}, \tag{20}$$

$$[f] = f(R, \phi_S, \theta_S, t - \frac{d}{c}). \tag{21}$$

If we discretize the artificial boundary by $2M^2$ equispaced points on the sphere both the storage and the number of operations per time step for the numerical evaluation of the integral in (19) asymptotically behave like $O(M^2)$. In contrast the representation formula in (13) needs no additional integral to be evaluated.

5 Numerical examples

In this section we present two numerical examples to illustrate the accuracy of the far-field solution obtained by using the evaluation formula (13). For both examples the artificial boundary is a sphere with radius $R = 1$ centered at the origin. The inner boundary is located at $r = 1/2$. Because the solutions are independent of ϕ our computational domain reduces to

$$\Omega = \{(r, \theta) \mid r \in [1/2, 1], \theta \in [0, \pi]\} \subset \mathbf{R}^2.$$

The computational domain Ω is discretized by an equispaced grid with 40×240 points while the initial-boundary-value problem is solved using second-order finite differences in both space and time together with the nonreflecting boundary condition (3). Once the numerical solution has been computed up to some time $T > 0$, we evaluate the far-field solution at two points with $r = 2$ and $r = 8$ along the ray $\theta = 3$ by using (13) and compare the values to a reference solution.

First we consider a spherical wave originating from the point $(r, \theta) = (d, 0)$, $0 \leq d < 1/2$:

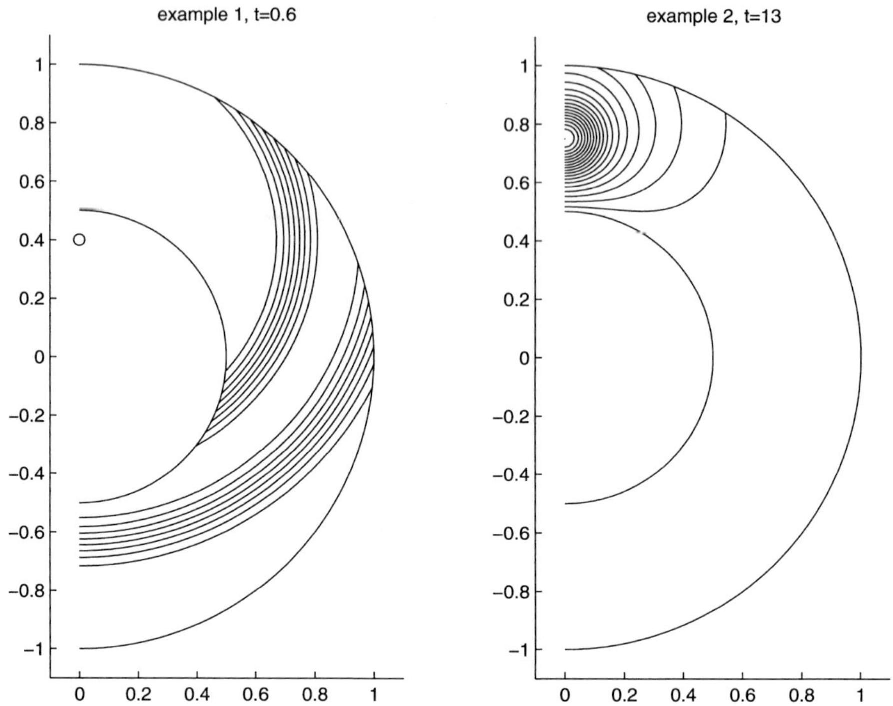

Fig. 1. Contour lines of the near-field solutions for both examples at fixed times. Left: A spherical wave originating from the point marked by a small circle. Right: An oscillating source near a spherical obstacle.

$$
\begin{aligned}
u_{tt} - \Delta u &= 0, \quad r > 1/2,\ t > 0 \\
u(r,\theta,0) &= \frac{g(\overline{r})}{\overline{r}}, \quad r \geq 1/2 \\
u_t(r,\theta,0) &= -\frac{g'(\overline{r})}{\overline{r}}, \quad r \geq 1/2 \\
u(r,\theta,t) &= \frac{g(\overline{r}-t)}{\overline{r}}, \quad r = 1/2,\ t > 0 \\
\overline{r} &= \sqrt{r^2 + d^2 - 2rd\cos\theta},
\end{aligned}
\tag{22}
$$

where $g \in C^2(\mathbf{R})$ is a piecewise polynomial function which vanishes outside $[0, R-d]$. Here the reference solution is simply given by the exact solution to this problem, namely

$$
u(r,\theta,t) = \frac{g(\overline{r}-t)}{\overline{r}}, \quad \overline{r} = \sqrt{r^2 + d^2 - 2rd\cos\theta} > 0. \tag{23}
$$

We choose $d = 0.4$ and $T = 12$. The contour lines of the near-field solution at time $t = 0.6$ are shown in Fig. 1 (left). The far-field solution is evaluated by using (13) with $N = K = 35$. It is shown in Fig. 2 (upper) while the error with respect to the reference solution is shown in Fig. 2 (middle). We observe that the error decreases, as does the solution, with increasing r. The relative error is about 0.5%. In Fig. 2 (lower), we show the maximal absolute error of the far-field solution evaluated by (13) compared to the reference solution for varying K. The value of K is critical for the amount of storage needed. We observe that the value K can be chosen much lower than the maximal value at $N = 35$, while maintaining accuracy.

Next we consider an oscillating source near a spherical obstacle of radius 1/2:

$$\begin{aligned} u_{tt} - \Delta u &= f, \quad r > 1/2,\ t > 0 \\ u(r,\theta,0) &= 0, \quad r > 1/2 \\ u_t(r,\theta,0) &= 0, \quad r > 1/2 \\ u(r,\theta,t) &= 0, \quad r = 1/2,\ t > 0. \end{aligned} \tag{24}$$

Here the source term f is given by

$$f(r,\theta,t) = \begin{cases} \alpha \sin^2(\omega t) \sin^2((1 - |\mathbf{x} - \mathbf{x}_f|/r_f)\pi/2), & |\mathbf{x} - \mathbf{x}_f| < r_f \\ 0, & \text{otherwise} \end{cases} . \tag{25}$$

We choose the parameter values $\mathbf{x}_f = (0.75, 0)$, $r_f = 0.15$, $\alpha = 9 \cdot 10^4$ and $\omega = \pi/2$. The solution is computed up to time $T = 16$. Here the reference solution is computed on a domain which is large enough, so that no reflected waves re-enter the computational domain during the entire simulation. The contour lines of the near-field solution at the time $t = 13$ are shown in Fig. 1 (right). The error in the far-field solution evaluated by using (13) with $N = K = 35$ is shown vs. time in Fig. 3 (upper). The relative error is about 0.05%. The far-field solution itself is shown for two different values of K, namely for $K = 35$ in Fig. 3 (middle) and for $K = 5$ in Fig. 3 (lower). For $K = 5$, the relative error is still about 1% and we do not see any difference in the solution. Hence most auxiliary functions ψ_{nm}^k can be neglected in the far-field evaluation.

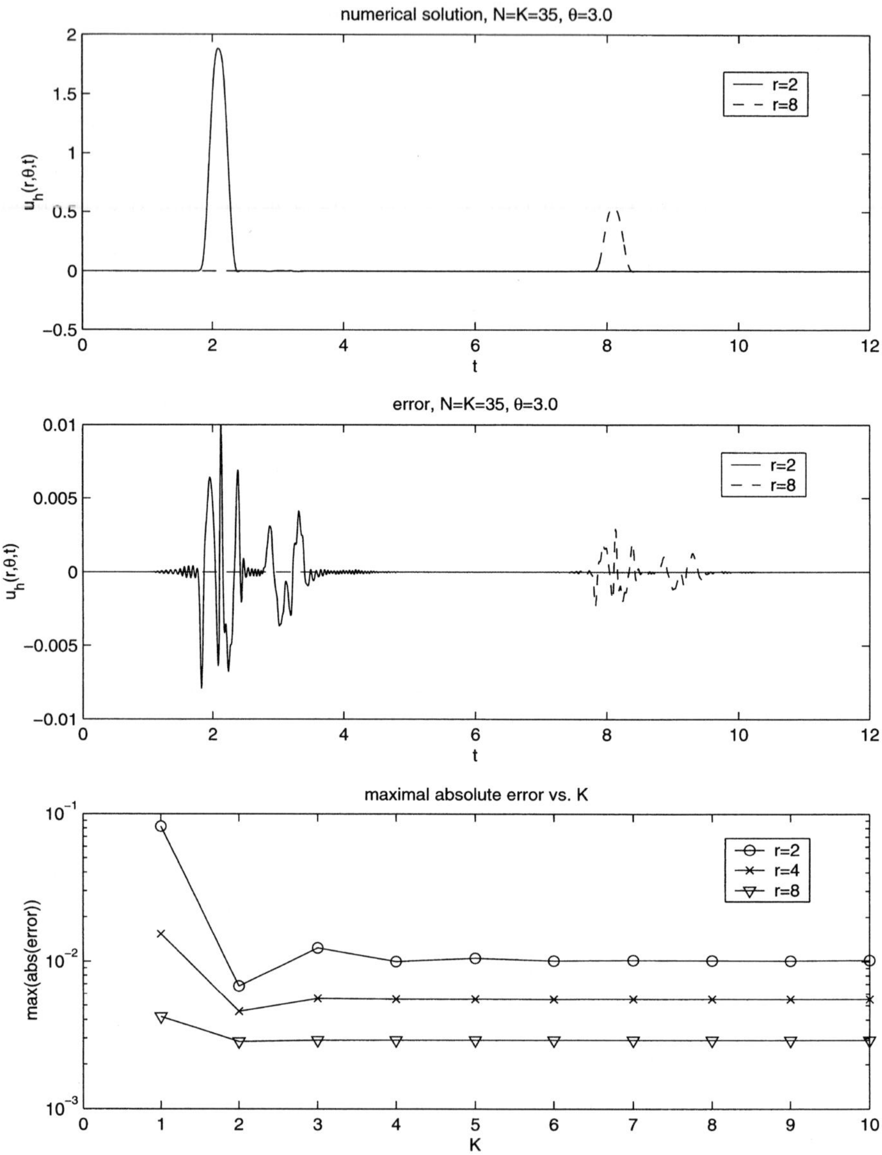

Fig. 2. The far-field solution for the spherical wave (example 1) evaluated by (13) (upper), the error compared to the reference solution (middle), and the maximal absolute error vs. K (lower). We observe that the far-field formula (13) gives an accurate result and that we can neglect most of the auxiliary functions ψ_{nm}^{k} with little loss of accuracy.

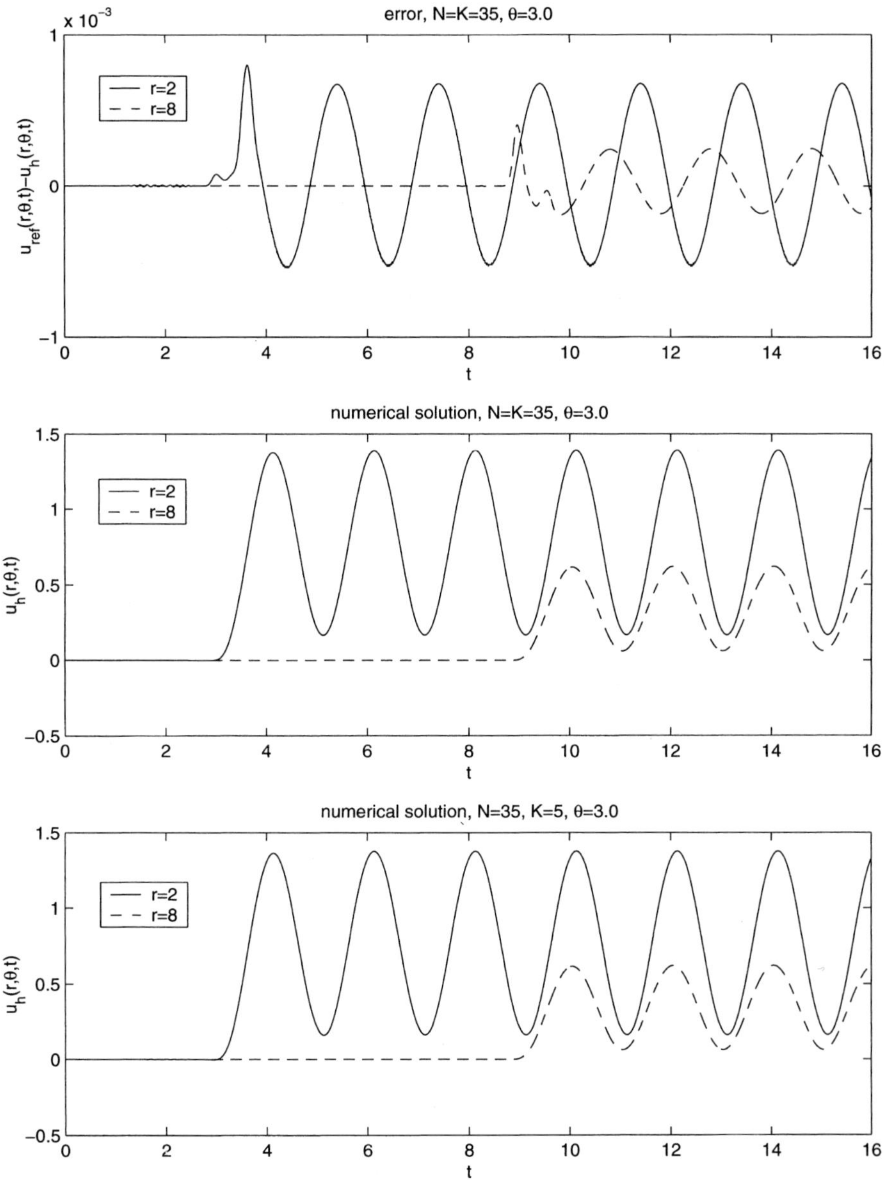

Fig. 3. The error of the far-field solution for the oscillating source (example 2) evaluated by (13) compared to the reference solution vs. time (upper) and the far-field solution itself, for $K = 35$ (middle) and $K = 5$ (lower). We observe that the truncation at $K = 5$ has but little effect upon the accuracy for $r \geq 2$.

6 Conclusion

We have presented an exact formula for the far-field evaluation of solutions to the wave equation, which can be used if the numerical solution in the bounded domain is computed by using the nonreflecting boundary condition in [5, 6]. The formula is more efficient and requires less storage than Kirchhoff-type evaluation formulas. Indeed, it only uses auxiliary quantities computed by the numerical scheme for the near-field solution. In particular, the evaluation formula permits to take advantage of the fast decay with distance of selected modes and thus to store only the minimal amount of information necessary from the past. The accuracy of the evaluation formula has been illustrated via numerical experiments. For multiple scattering problems the use of a single artificial boundary surrounding all scatterers involved becomes prohibitively expensive in memory requirement. Instead, it is necessary to enclose each scatterer within a single separate computational domain. Clearly waves that leave a certain domain, Ω_1, will impinge upon a different domain, Ω_2, at later times; hence they are no longer purely outgoing waves. To transfer the time-retarded information from Ω_1 to Ω_2 an analytical representation of the solution in the unbounded medium becomes necessary. This analytical representation is inherent to the exact nonreflecting boundary conditions described above and thus naturally leads to *exact transmission boundary conditions* for multiple scattering problems. The authors are currently investigating these issues and will report on them elsewhere in the near future.

References

1. Engquist, B., Majda, A. (1977): Absorbing Boundary Conditions for the Numerical Simulation of Waves. Math. Comp. **31** (139), pp. 629–651
2. Bayliss, A., Turkel, E. (1980): Radiation Boundary Conditions for Wave-Like Equations. Comm. Pure and Appl. Math. **23**, pp. 707–725
3. Ting, L., Miksis, M. J. (1986): Exact boundary conditions for scattering problems. J. Acoust. Soc. Am. **80** (6), pp. 1825–1827
4. Givoli, D., Cohen, D. (1995): Nonreflecting Boundary Conditions Based on Kirchhoff-Type Formulae. J. Comp. Phys. **117**, pp. 102–113
5. Grote, M. J., Keller, J. B. (1995): Exact Nonreflecting Boundary Conditions for the Time Dependent Wave Equation. SIAM J. Appl. Math. **55** (2), pp. 280–297
6. Grote, M. J., Keller, J. B. (1996): Nonreflecting Boundary Conditions for Time-Dependent Scattering. J. Comp. Phys. **127**, pp. 52–65
7. Grote, M. J., Keller, J. B. (1998): Nonreflecting Boundary Conditions for Maxwell's Equations. J. Comp. Phys. **139**, pp. 327–342
8. Hagstrom, T., Hariharan, S. I. (1998): A formulation of asymptotic and exact boundary conditions using local operators. Appl. Num. Math. **27**, pp. 403–416
9. Grote, M. J., Keller, J. B. (2000): Exact Nonreflecting Boundary Conditions for Elastic Waves. SIAM J. Appl. Math. **60** (3), pp. 803–819
10. Grote, M. J. (2000): Nonreflecting Boundary Conditions for Elastodynamic Scattering. J. Comp. Phys. **161**, pp. 331–353

L^1 Stability for the One-dimensional Broadwell Model of a Discrete Velocity Gas

Seung-Yeal Ha[1] and Athanasios E. Tzavaras[2]

[1] Department of Mathematics, University of Wisconsin-Madison, *ha@math.wisc.edu*,
[2] Department of Mathematics, University of Wisconsin-Madison, *tzavaras@math.wisc.edu*

Summary. We prove L^1 stability for the one-dimensional Broadwell model for a discrete velocity gas. For initial data in $L^1(R) \cap L^\infty_+(R)$ with small mass, we show that bounded mild solutions are L^1-stable. For this, we employ a nonlinear functional $\mathcal{H}(t)$ that is equivalent to the L^1 distance between two mild solutions and non-increasing in time t.

1 Introduction

The celebrated Boltzmann equation describes the evolution of the density function $f(x, v, t)$ for a dilute gas:

$$\partial_t f(x, v, t) + v \cdot \partial_x f(x, v, t) = Q(f, f)(x, v, t). \tag{1}$$

The transport operator $\partial_t f(x, v, t) + v \cdot \partial_x f(x, v, t)$ describes the free motion of non-interacting particles while the integral operator $Q(f, f)$ accounts for binary collisions between particles. The discretization of the velocity space in the Boltzmann equation (1) allows to replace the integro-differential equation by a system of semilinear hyperbolic equations. The paradigm for one-dimensional discrete Boltzmann equations is the one-dimensional Broadwell model:

$$\begin{aligned} \partial_t f_1 - \partial_x f_1 &= f_2^2 - f_1 f_3, \\ \partial_t f_2 &= -\frac{1}{2}(f_2^2 - f_1 f_3), \\ \partial_t f_3 + \partial_x f_3 &= f_2^2 - f_1 f_3. \end{aligned} \tag{2}$$

Here, f_1, f_2 and f_3 are densities of particles moving with velocities $-1, 0$ and 1 respectively. This system was proposed by Broadwell [5] as one of the simplest models for the description of a dilute gas, with molecules moving with finite speeds. Let $Q_i(f, f)$ denote the collision terms appearing in the equation for the f_i,

$$Q_1(f, f) = -2Q_2(f, f) = Q_3(f, f) \equiv f_2^2 - f_1 f_3.$$

The definition of solution in the mild sense can be stated as follows.

Definition 1. *The function* $f = (f_1, f_2, f_3) \in C([0,T]; (L^1(R))^3 \cap (L^\infty(R))^3)$ *is a mild solution for (2) with initial data* $f^0 \in (L^1(R))^3 \cap (L^\infty(R))^3$ *if and only if, for* $t \in [0,T]$*, a.e* $x \in R$ *and* $i = 1, 2, 3$*,* $f(x,t)$ *satisfies the integral equation*

$$f_i(x,t) = f_i^0(x - v_i t)) + \int_0^t Q_i(f,f)(x - v_i(t-s), s) ds,$$

where $v_1 = -1, \quad v_2 = 0$, and $v_3 = 1$. Global existence of mild solutions for (2) is proved by Nishida and Mimura [12] for small L^1-initial data, and by Tartar [15, 16], Bony [3] for one-dimensional discrete Boltzmann equations with large L^1 data. Large time behavior of mild solutions was studied in Beale [1, 2]. We refer for further references to the survey article by Platkowski and Illner [13].

For discrete velocity Boltzmann equations the kinetic function f is expected to be positive, and the mass is invariant in time. The L^1 topology is thus a natural framework for the analysis. Our stability analysis will monitor the L^1 norm, it is based on appropriate nonlinear functionals and is motivated by the Liu-Yang's functional [11] utilized in stability for hyperbolic systems of conservation laws. In the sequel, we denote the weighted L^1 distance by $||\cdot||$. For a mild solution $f = (f_i)_{i=1}^3$, $||f||$ is defined by

$$||f(\cdot,t)|| \equiv ||f_1(\cdot,t)||_{L^1(R)} + 4||f_2(\cdot,t)||_{L^1(R)} + ||f_3(\cdot,t)||_{L^1(R)}.$$

One obtains by a direct calculation

$$\frac{d}{dt}||f(\cdot,t) - \bar{f}(\cdot,t)|| \leq \bar{c}\mu ||f(\cdot,t) - \bar{f}(\cdot,t)||,$$

where $\mu = \sup_{t\geq 0,i} ||f_i(\cdot,t)||_{L^1(R)} + \sup_{t\geq 0,i} ||\bar{f}_i(\cdot,t)||_{L^1(R)}$ and $\bar{c}$ is a positive constant. As long as the solutions are uniformly bounded in x and t (which is the case for general classes of one-dimensional discrete velocity Boltzmann models [1, 2, 3]), we have local in time L^1 stability:

$$||f(\cdot,t) - \bar{f}(\cdot,t)|| \leq e^{\bar{c}\mu t} ||f^0(\cdot) - \bar{f}^0(\cdot)||, \tag{3}$$

The objective of this article is to obtain uniform L^1 stability for (2) by introducing a nonlinear functional which effectively accounts for dispersive effects of the evolution. The nonlinear functional $\mathcal{H}(t) \equiv \mathcal{H}[f(\cdot,t), \bar{f}(\cdot,t)]$ is defined as a linear combination of two subfunctionals $\mathcal{L}(t)$ and $\mathcal{Q}_d(t)$. The functional $\mathcal{L}(t)$ measures the L^1 distance between two mild solutions. Complementing this, the functional $\mathcal{Q}_d(t)$ measures the nonlinear coupling between particles with different velocities. In Section 3, we show that the functional $\mathcal{H}(t)$ satisfies the following two key properties:

1. $\frac{1}{C_0}||f(\cdot,t) - \bar{f}(\cdot,t)|| \leq \mathcal{H}(t) \leq C_0 ||f(\cdot,t) - \bar{f}(\cdot,t)||_{L^1(R)}$,
2. $\mathcal{H}(t)$ is non-increasing in time t, i.e., $\mathcal{H}(t) \leq \mathcal{H}(0), \quad t \geq 0$,

where f and $\bar{f}$ are mild solutions of (2) emanating from data f_0 and $\bar{f}_0$ respectively, and C_0 is a positive constant independent of time. ¿From these properties L^1 stability follows:

$$||f(\cdot,t) - \bar{f}(\cdot,t)|| \le C_0\mathcal{H}(t) \le C_0\mathcal{H}(0) \le C_0^2||f^0(\cdot) - \bar{f}^0(\cdot)||.$$

The time-decay estimate of the functional $\mathcal{H}(t)$ is based on the following conditions:

- Strict hyperbolicity of the system (2).
- Smallness of initial data in L^1.
- Conservation of mass $\sum_{i=1}^3 Q_i(f,f)(x,t) = 0$, for given $(x,t) \in R \times R^+$.

Remark 1. As in [3, 4], we do not use the H-theorem in our L^1-analysis. Since our initial data are a small perturbation of a vacuum state and the system is strictly hyperbolic, the particles will eventually decouple so that time-asymptotically the solutions tend to non-interacting states.

For a class of discrete velocity Boltzmann equations with transversal interaction terms (which do not include the Broadwell model), L^1 stability was obtained by Tartar [16].

The main theorem is as follows.

Theorem 1. *Let f and $\bar{f} \in C(R_+;(L^1(R))^3 \cap (L^\infty_+(R))^3)$ be two mild solutions of (2) corresponding to initial data f^0 and $\bar{f}^0 \in (L^1(R))^3 \cap (L^\infty_+(R))^3$ such that $||f^0|| + ||\bar{f}^0|| < 2$. Then, we have*

$$||f(\cdot,t) - \bar{f}(\cdot,t)|| \le C||f^0(\cdot) - \bar{f}^0(\cdot)||,$$

where C is a positive constant independent of time t.

This paper is organized as follows. In Section 2, we briefly review the basics of discrete Boltzmann equations and global existence results for mild solutions to the system (2). In Section 3, we explicitly construct the nonlinear functional for the one-dimensional Broadwell model, we study its time-evolution eventually proving L^1 stability for the one-dimensional Broadwell model. For more general one-dimensional models, of the type pursued in [2], L^1 stability analysis will be addressed in the forthcoming paper [6].

2 Preliminaries

In this section, we review the basics of one-dimensional discrete velocity Boltzmann equations (see [2, 3, 7] and [10, Appendix]),

$$\partial_t f_i + v_i \partial_x f_i = \sum_{j,k,l} (A_{ij}^{kl} f_k f_l - A_{kl}^{ij} f_i f_j), \tag{4}$$

where the collision coefficients A_{ij}^{kl} are positive and satisfy the symmetry and micro-reversibility conditions:

$$A_{ij}^{kl} = A_{ij}^{lk} = A_{ji}^{kl}, \quad A_{ij}^{kl} = A_{kl}^{ij}. \tag{5}$$

Such models are derived by considering a three-dimensional discrete velocity model and then restricting to motions that are one dimensional (*e.g.* plane waves). The three-dimensional model is equipped with microscopic conservation of mass, momentum and energy between the pre-collisional and post-collisional velocities. When restricting to one-dimensional motions, the projections of the pre-collisional velocities (in the propagation direction) v_i, v_j and the projections of the post-collisional velocities v_k, v_l satisfy microscopic conservation laws of mass and momentum,

$$v_i + v_j = v_k + v_l, \tag{6}$$

From the viewpoint that the one-dimensional discrete model (4) is a special case of the three-dimensional discrete model, the one-dimensional discrete model does not have to satisfy microscopic conservation of energy. By the same reason, the system does not need to be strictly hyperbolic even though the original three-dimensional system may well be strictly hyperbolic. By assuming the same initial data for particles moving with the same (projections of) velocities and suitable congruence conditions on the transitional probabilities A_{ij}^{kl}, Beale [2] transforms a subclass of the discrete velocity models (4) to

$$\partial_t f_i + v_i \partial_x f_i = \sum_{j,k} B_i^{jk} f_j f_k\,, \tag{7}$$

subject to certain structural assumptions on the interaction coefficients B_i^{jk}. In particular, this class encompasses the model (2). In Section 3, we will discuss stability for (2) while the more general model (7) will be discussed in [6].

Returning to (4), we briefly review the properties of the collision operator:

$$Q_i(f,f) \equiv \sum_{j,k,l} A_{ij}^{kl}(f_k f_l - f_i f_j).$$

Let ϕ be any measurable function. Then,

$$\partial_t \left(\sum_i \phi(v_i) f_i \right) + \partial_x \left(\sum_i v_i \phi(v_i) f_i \right) = \sum_{i,j,k,l} \phi(v_i) A_{ij}^{kl} (f_k f_l - f_i f_j).$$

As in the theory of the Boltzmann equation, the properties (5) imply that the right hand side can be rearranged so that

$$\text{R.H.S.} = \frac{1}{4} \sum_{i,j,k,l} A_{ij}^{kl}(\phi(v_i) + \phi(v_j) - \phi(v_k) - \phi(v_l))(f_k f_l - f_i f_j). \tag{8}$$

Any choice of collisional invariants will then produce an associated macroscopic conservation law. For a model satisfying microscopic mass and momentum conservation, we have macroscopic conservation of mass, momentum:

$$\frac{d}{dt}\sum_{i=1}^{n}\int_R f_i(x,t)dx = 0, \qquad \frac{d}{dt}\sum_{i=1}^{n}\int_R v_i f_i(x,t)dx = 0.$$

The entropy $H(t)$ is defined by

$$H(t) = \sum_{i=1}^{n}\int_R f_i(x,t)\log(f_i(x,t))dx,$$

and satisfies an analog of the Boltzmann H-theorem,

$$\begin{aligned}
\partial_t\left(\sum_i f_i\log f_i\right) + \partial_x\left(\sum_i v_i f_i\log f_i\right) &= \sum_{i,j,k,l} A_{ij}^{kl}\log f_i(f_k f_l - f_i f_j) \\
= \frac{1}{4}\sum_{i,j,k,l} A_{ij}^{kl}(\log f_i + \log f_j - \log f_k - \log f_l)&(f_k f_l - f_i f_j) \\
= -\frac{1}{4}\sum_{i,j,k,l} A_{ij}^{kl}\log\left(\frac{f_k f_l}{f_i f_j}\right)&(f_k f_l - f_i f_j) \le 0,
\end{aligned} \tag{9}$$

where we have used $A_{ij}^{kl} \ge 0$, (8) and $(\log x - \log y)(x-y) \ge 0$.

A paradigm, for discrete velocity models is the Broadwell model. This corresponds to a specific interaction mechanism [5], and it reads

$$\begin{aligned}
\partial_t f_1^+ + \partial_x f_1^+ &= \frac{1}{2}(f_2^+ f_2^- + f_3^+ f_3^-) - f_1^+ f_1^-, \\
\partial_t f_1^- - \partial_x f_1^- &= \frac{1}{2}(f_2^+ f_2^- + f_3^+ f_3^-) - f_1^+ f_1^-, \\
\partial_t f_2^+ + \partial_y f_2^+ &= \frac{1}{2}(f_1^+ f_1^- + f_3^+ f_3^-) - f_2^+ f_2^-, \\
\partial_t f_2^- - \partial_y f_2^- &= \frac{1}{2}(f_1^+ f_1^- + f_3^+ f_3^-) - f_2^+ f_2^-, \\
\partial_t f_3^+ + \partial_z f_3^+ &= \frac{1}{2}(f_1^+ f_1^- + f_2^+ f_2^-) - f_3^+ f_3^-, \\
\partial_t f_3^- - \partial_z f_3^- &= \frac{1}{2}(f_1^+ f_1^- + f_2^+ f_2^-) - f_3^+ f_3^-,
\end{aligned}$$

Here $f_1^\pm$, $f_2^\pm$ and $f_3^\pm$ are densities of particles moving with velocities ± 1 in the direction of x, y, and z axis respectively. If we consider one-dimensional motions of particles, independent of the y and z coordinates, and motivated by symmetry considerations, set

$$f_1^+ = f_1, \quad f_1^- = f_3, \quad f_2^+ = f_2^- = f_3^+ = f_3^- = f_2.$$

the above system reduces to the one-dimensional Broadwell model (2). For (2) we have the following global existence theorem (see [2, 3] and [12] and [15] for earlier variants).

Theorem 2. *For f^0 in $(L^1(R))^3 \cap (L^\infty_+(R))^3$ there exist a unique globally defined nonnegative solution of (2) in the sense of Definition 1.*

3 L^1 stability

In this section, we construct an explicit nonlinear functional $\mathcal{H}(t)$ which is equivalent to the L^1 distance and non-increasing in time along solutions.

Let f and $\bar{f}$ be two solutions of the Broadwell model which for the time are taken to be C^1. We use the notation $f(x)$, $f(y)$ for the evaluation of f at the points (x,t), (y,t) respectively; the t dependence is mostly suppressed. ¿From (2), we derive the conservations for the partial masses,

$$\partial_t(2f_2 + f_3)(x) + \partial_x f_3(x) = 0. \tag{10}$$

$$\partial_t(f_1 + 2f_2)(y) - \partial_y f_1(y) = 0, \tag{11}$$

In the sequel, we omit the t-dependence and use the simple notation

$$\delta_i(x,t) = \mathrm{sgn}(f_i(x,t) - \bar{f}_i(x,t))$$

for the sign of the difference. We also define a potential of interaction functional $Q(t)$,

$$Q(t) \equiv \int_{R^2} \mathbf{1}_{x<y}[2f_2(x) + f_3(x)][f_1(y) + 2f_2(y)]dydx,$$

whose definition is motivated in the following lemma. This lemma appears in Tartar [17] where it is attributed to Varadhan.

Proposition 1. *Along solutions f of (2), we have*

$$\frac{dQ(t)}{dt} = -2\int_R (f_1 f_3 + f_1 f_2 + f_2 f_3)(x,t)\,dx.$$

Proof. We multiply (10) by $(f_1 + 2f_2)(y)$ and (11) by $(2f_2 + f_3)(x)$. Adding and multiplying the resulting identity by $\mathbf{1}_{x<y}$, we arrive at the identity

$$\begin{aligned}\partial_t&\Big[(f_1 + 2f_2)(y)\,(2f_2 + f_3)(x)\,\mathbf{1}_{x<y}\Big]\\ &+\mathrm{div}_{(x,y)}\Big[\Big(f_3(x)(f_1 + 2f_2)(y), -(2f_2 + f_3)(x)f_1(y)\Big)\mathbf{1}_{x<y}\Big]\\ &\quad+\delta(x-y)\Big(f_3(x)(f_1 + 2f_2)(y) + f_1(y)(2f_2 + f_3)(x)\Big) = 0\end{aligned}$$

Note that the last term is positive and provides some decay by dispersion. Integrating the above equation over R^2, we obtain

$$\frac{dQ(t)}{dt} = -2\int_R [f_1(x)f_3(x) + f_1(x)f_2(x) + f_2(x)f_3(x)]dx.$$

This completes the proof. □

Next, we define certain nonlinear functionals and study their time-variations.

$$\mathcal{L}(t) \equiv \int_R \Big(|f_1 - \bar{f}_1| + 4|f_2 - \bar{f}_2| + |f_3 - \bar{f}_3|\Big)(x)dx,$$

$$\mathcal{Q}_d(t) \equiv \int_{R^2} \mathbf{1}_{x<y}\Big((2f_2 + 2\bar{f}_2 + f_3 + \bar{f}_3)(x)(|f_1 - \bar{f}_1| + 2|f_2 - \bar{f}_2|)(y)\Big)dxdy$$
$$+ \int_{R^2} \mathbf{1}_{x<y}\Big((2|f_2 - \bar{f}_2| + |f_3 - \bar{f}_3|)(x)(f_1 + \bar{f}_1 + 2f_2 + 2\bar{f}_2)(y)\Big)dxdy,$$

$$\mathcal{H}(t) \equiv \mathcal{L}(t) + K\mathcal{Q}_d(t),$$

where K is positive constant to be determined later. We also define the instantaneous interaction productions as follows.

$$\Lambda(f,\bar{f})(t) = \Lambda_s(f,\bar{f})(t) + \Lambda_d(f,\bar{f})(t), \tag{12}$$

$$\Lambda_s(f,\bar{f}) \equiv \int_R \left(2 - \frac{\delta_1(x)}{\delta_2(x)} - \frac{\delta_3(x)}{\delta_2(x)}\right)|f_2(x) - \bar{f}_2(x)|(f_2(x) + \bar{f}_2(x))dx \tag{13}$$

$$\Lambda_d(f,\bar{f})(t) \equiv \int_R \sum_{m=1}^{3} |f_m(x) - \bar{f}_m(x)| \sum_{n\neq m} (f_n(x) + \bar{f}_n(x))dx. \tag{14}$$

Note that the above functionals are all positive and that for positive and bounded solutions $\mathcal{H}(t)$ is equivalent to the L^1 norm. Furthermore, $\mathcal{L}(t)$ denotes the weighted L^1 distance (and the total mass at time t) while $\mathcal{Q}_d(t)$ represents the potential of interactions between particles.

Next, we study the time-evolution of these functionals.

Lemma 1. *Let f and $\bar{f}$ be two solutions of (2) corresponding to initial data f^0 and $\bar{f}^0$ with $||f^0 + \bar{f}^0|| < 2$. Then K can be selected so that the functionals satisfy the Lyapunov type estimates:*

$$\frac{d\mathcal{L}(t)}{dt} \le -\Lambda_s(f,\bar{f})(t) + \Lambda_d(f,\bar{f})(t),$$
$$\frac{d\mathcal{Q}_d(t)}{dt} \le (-2 + ||f + \bar{f}||)\Lambda_d(f,\bar{f})(t),$$
$$\frac{d\mathcal{H}(t)}{dt} \le -C_1\Lambda(f,\bar{f})(t),$$

where C_1 is a positive constant independent of time t.

Proof. First, we derive the equations for the differences $|f_i(x,t) - \bar{f}_i(x,t)|$, $1 \le i \le 3$, in the form

$$\partial_t|f_1 - \bar{f}_1| - \partial_x|f_1 - \bar{f}_1| = \frac{\delta_1}{\delta_2}|f_2 - \bar{f}_2|(f_2 + \bar{f}_2) - |f_1 - \bar{f}_1|\frac{f_3 + \bar{f}_3}{2}$$

$$
\begin{aligned}
&\quad -\frac{\delta_1}{\delta_3}|f_3-\bar f_3|\frac{f_1+\bar f_1}{2},\\
\partial_t|f_2-\bar f_2| &= -\frac{1}{2}|f_2-\bar f_2|(f_2+\bar f_2)+\frac{\delta_2}{2\delta_1}|f_1-\bar f_1|\frac{f_3+\bar f_3}{2}\\
&\quad +\frac{\delta_2}{2\delta_3}|f_3-\bar f_3|\frac{f_1+\bar f_1}{2}\\
\partial_t|f_3-\bar f_3|+\partial_x|f_3-\bar f_3| &= \frac{\delta_3}{\delta_2}|f_2-\bar f_2|(f_2+\bar f_2)-\frac{\delta_3}{\delta_1}|f_1-\bar f_1|\frac{f_3+\bar f_3}{2}\\
&\quad -|f_3-\bar f_3|\frac{f_1+\bar f_1}{2}.
\end{aligned}
\tag{15}
$$

Step 1: We consider the functionals separately. ¿From (15) we have

$$
\begin{aligned}
&\partial_t\Big(|f_1-\bar f_1|+4|f_2-\bar f_2|+|f_3-\bar f_3|\Big)+\partial_x\left(|f_3-\bar f_3|-|f_1-\bar f_1|\right)\\
&\quad +(2-\frac{\delta_1}{\delta_2}-\frac{\delta_3}{\delta_2})|f_2-\bar f_2|(f_2+\bar f_2)\\
&= (2\frac{\delta_2}{\delta_1}-\frac{\delta_3}{\delta_1}-1)|f_1-\bar f_1|\frac{f_3+\bar f_3}{2}+(2\frac{\delta_2}{\delta_3}-\frac{\delta_1}{\delta_3}-1)|f_3-\bar f_3|\frac{f_1+\bar f_1}{2}
\end{aligned}
\tag{16}
$$

By the definitions of $\mathcal{L}(t)$, Λ_s and Λ_d, we have

$$
\begin{aligned}
\frac{d\mathcal{L}(t)}{dt} &\le -\Lambda_s(f,\bar f)(t)+\int_R |f_1-\bar f_1|(f_3+\bar f_3)+|f_3-\bar f_3|(f_1+\bar f_1)\,dx\\
&\le -\Lambda_s(f,\bar f)(t)+\Lambda_d(f,\bar f)(t)\,.
\end{aligned}
\tag{17}
$$

Step 2: By a direct yet cumbersome calculation, we obtain from (10), (11) and (15) the identity

$$
\begin{aligned}
\frac{dQ_d(t)}{dt} &= -2\int_R \sum_{m=1}^{3}|f_m-\bar f_m|(x)\left(\sum_{n\neq m}(f_n+\bar f_n)(x)\right)\\
&\quad +\int_{R^2}\mathbf{1}_{x<y}\left(\frac{\delta_1}{\delta_2}(y)-1\right)(f_2+\bar f_2)(y)|f_2-\bar f_2|(y)|(f_3+\bar f_3)(x)\,dxdy\\
&\quad +\int_{R^2}2\mathbf{1}_{x<y}\left(\frac{\delta_1}{\delta_2}(y)-1\right)(f_2+\bar f_2)(y)|f_2-\bar f_2|(y)(f_2+\bar f_2)(x)\,dxdy\\
&\quad +\int_{R^2}\mathbf{1}_{x<y}\left(\frac{\delta_3}{\delta_2}(x)-1\right)(f_2+\bar f_2)(x)|f_2-\bar f_2|(x)|(f_1+\bar f_1)(y)\,dxdy\\
&\quad +\int_{R^2}2\mathbf{1}_{x<y}\left(\frac{\delta_3}{\delta_2}(x)-1\right)(f_2+\bar f_2)(x)|f_2-\bar f_2|(x)|(f_2+\bar f_2)(y)\,dxdy\\
&\quad +\int_{R^2}\mathbf{1}_{x<y}\left(\frac{\delta_2}{\delta_1}(y)-1\right)\frac{f_3+\bar f_3}{2}(y)|f_1-\bar f_1|(y)(f_3+\bar f_3)(x)\,dxdy\\
&\quad +\int_{R^2}\mathbf{1}_{x<y}\left(\frac{\delta_2}{\delta_1}(y)-1\right)(f_3+\bar f_3)(y)|f_1-\bar f_1|(y)(f_2+\bar f_2)(x)\,dxdy
\end{aligned}
$$

$$
\begin{aligned}
&+\int_{R^2} \mathbf{1}_{x<y}\left(\frac{\delta_2}{\delta_1}(x)-\frac{\delta_3}{\delta_1}(x)\right)\frac{f_3+\bar{f}_3}{2}(x)|f_1-\bar{f}_1|(x)(f_1+\bar{f}_1)(y)\,dxdy\\
&+\int_{R^2} \mathbf{1}_{x<y}\left(\frac{\delta_2}{\delta_1}(x)-\frac{\delta_3}{\delta_1}(x)\right)(f_3+\bar{f}_3)(x)|f_1-\bar{f}_1|(x)(f_2+\bar{f}_2)(y)\,dxdy\\
&+\int_{R^2} \mathbf{1}_{x<y}\left(\frac{\delta_2}{\delta_3}(x)-1\right)\frac{f_1+\bar{f}_1}{2}(x)|f_3-\bar{f}_3|(x)(f_1+\bar{f}_1)(y)\,dxdy\\
&+\int_{R^2} \mathbf{1}_{x<y}\left(\frac{\delta_2}{\delta_3}(x)-1\right)(f_1+\bar{f}_1)(x)|f_3-\bar{f}_3|(x)(f_2+\bar{f}_2)(y)\,dxdy\\
&+\int_{R^2} \mathbf{1}_{x<y}\left(\frac{\delta_2}{\delta_3}(y)-\frac{\delta_1}{\delta_3}(y)\right)\frac{f_1+\bar{f}_1}{2}(y)|f_3-\bar{f}_3|(y)(f_3+\bar{f}_3)(x)\,dxdy\\
&+\int_{R^2} \mathbf{1}_{x<y}\left(\frac{\delta_2}{\delta_3}(y)-\frac{\delta_1}{\delta_3}(y)\right)(f_1+\bar{f}_1)(y)|f_3-\bar{f}_3|(y)(f_2+\bar{f}_2)(x)\,dxdy\\
&\le -2\int_R \sum_{m=1}^{3}|f_m-\bar{f}_m|(x)\left(\sum_{n\neq m}(f_n+\bar{f}_n)(x)\right)\\
&+\left(||f_1+\bar{f}_1||_{L^1(R)}+2||f_2+\bar{f}_2||_{L^1(R)}\right)\int_R (f_3+\bar{f}_3)(x)|f_1-\bar{f}_1|(x)dx\\
&+\left(2||f_2+\bar{f}_2||_{L^1(R)}+||f_3+\bar{f}_3||_{L^1(R)}\right)\int_R (f_1+\bar{f}_1)(x)|f_3-\bar{f}_3|(x)dx\\
&\le (-2+||f+\bar{f}||)\Lambda_d(f,\bar{f})(t).
\end{aligned}
$$

Step 3: By the definition of $\mathcal{H}(t)$, we have

$$
\begin{aligned}
\frac{d\mathcal{H}(t)}{dt} &= \frac{d\mathcal{L}(t)}{dt}+K\frac{d\mathcal{Q}_d(t)}{dt}\\
&\le -\Lambda_s(f,\bar{f})(t)+[1+K(-2+||f+\bar{f}||)]\Lambda_d(f,\bar{f})(t).
\end{aligned}
$$

Since $||f+\bar{f}||<2$, we can choose K sufficiently large so that

$$1+K(-2+||f+\bar{f}||)<0.$$

We then have

$$\frac{d\mathcal{H}(t)}{dt}\le -C_1\Lambda(f,\bar{f})(t),$$

where C_1 is a positive constant independent of time t. This completes the proof. □

From Lemma 1, we have the following L^1 stability estimate.

Proof of Theorem 1 Let $f_0^{(k)}$ and $\bar{f}_0^{(k)}$ be smooth approximations of the given initial data f_0 and $\bar{f}_0$ such that

$$f_0^{(k)}\to f_0,\quad \bar{f}_0^{(k)}\to \bar{f}_0 \quad \text{in } L^1(R) \quad \text{as } k\to\infty$$

Then we can construct smooth solutions $f^{(k)}(x,t)$ and $\bar{f}^{(k)}(x,t)$ corresponding to the data $f_0^{(k)}$ and $\bar{f}_0^{(k)}$ respectively. It follows from (3) that

$$f^{(k)}(x,t) \to f(x,t), \quad \bar{f}^{(k)}(x,t) \to \bar{f}(x,t) \quad \text{in } L^1(R) \ \text{ as } k \to \infty.$$

Define the nonlinear functional $\mathcal{H}(t)$ for f and $\bar{f}$ as follows.

$$\mathcal{H}(t) = \mathcal{H}[f(\cdot,t), \bar{f}(\cdot,t)] \equiv \lim_{k\to\infty} \mathcal{H}[f^{(k)}(\cdot,t), \bar{f}^{(k)}(\cdot,t)].$$

Then by the two key-properties of $\mathcal{H}(t)$ for $f^{(k)}$ and $\bar{f}^{(k)}$, we have

$$||f^{(k)}(\cdot,t) - \bar{f}^{(k)}(\cdot,t)||_{L^1(R)} \le C||f_0^{(k)}(\cdot) - \bar{f}_0^{(k)}(\cdot)||_{L^1(R)},$$

where C is a positive constant independent of time t and k. Letting $k \to \infty$, we have

$$||f(\cdot,t) - \bar{f}(\cdot,t)||_{L^1(R)} \le C||f_0(\cdot) - \bar{f}_0(\cdot)||_{L^1(R)}.$$

This completes the proof. □

References

1. Beale, J.T. (1985): *Large-time behavior of the Broadwell model of a discrete velocity gas.* Comm. Math. Phys., 102, 217-235.
2. Beale, J. T. (1986): *Large-time behavior of discrete velocity Boltzmann equations.* Comm. Math. Phys., 106, no. 4, 659-678.
3. Bony, J.-M. (1987): *Solutions globales bornées pour les modèles discrete de l'équation de Boltzmann en dimension 1 d'espace.* Actes Journees E.D.P.St. Jean de Monts, n° XVI.
4. Bony, J.-M. (1991): *Existence globale à données de Cauchy petites pour modèles discrets de l'équation de Boltzmann.* (French) Comm. Partial Differential Equations, 16, no. 4-5, 533-545.
5. Broadwell, J.E. (1964): *Shock structure in a simple discrete velocity gas.* Phy. Fluids, 7, 1243-1247.
6. Ha, S.-Y., Tzavaras, A.E. (2002): *Lyapunov functionals and L^1 stability for discrete velocity Boltzmann equations.* In preparation.
7. Gatignol, R. (1975): *Théorie Cinétique des gaz à Répartition Discrète de vitesses.* (French) Lectures Notes in Physics, vol. 36, Springer-Verlag.
8. Illner, R. (1982): *Global existence results for discrete velocity models fo the Boltzmann equation.* J.Meca. Th. Appl., 1, 611-622.
9. Kawashima, S. (1981): *Global solution of the initial value problem for a discrete velocity model of the Boltzmann equation.* Proc. Japan Acad., 57, 19-24
10. Kawashima, S. (1987): *Large-time behavior of solutions of the discrete Boltzmann equation.* Commun. Math. Phys., 109, 563-589.
11. Liu, T.-P., Yang, T. (1999): *Well posedness theory for hyperbolic conservation laws.* Comm. Pure Appl. Math. 52, no. 12, 1553-1586.
12. Nishida, T., Mimura, M. (1974): *On the Broadwell model of the Boltzmann equation for a simple discrete velocity gas.* Proc. Japan. Acad., 50, 812-817.
13. Platkowski, T., Illner, R. (1988): *Discrete velocity models of the Boltzmann equation: A survey on the mathematical aspects of the theory.* SIAM Review., vol. 30, no. 2.

14. Slemrod, M., (1990): *Large time behavior of the Broadwell model of a discrete velocity gas with specularly reflective boundary conditions.* Arch. Rational Mech. Anal. 111, 323-342.
15. Tartar, L. (1976): *Existence globale pour un système hyperbolique semi-linéaire de la théorie cinétique des gaz.* (French) Séminarie Goulaouic-Schwartz, École Polytech., Palaiseau, no 1.
16. Tartar, L. (1980): *Some existence theorems for semilinear hyperbolic systems in one space variable.* MRC technical summary report, University of Wisconsin.
17. Tartar, L. (1987): *Oscillations and asymptotic behaviour for two semilinear hyperbolic systems.* In Dynamics of infinite-dimensional systems, NATO Adv. Sci. Inst. Ser. F, 37, Springer, Berlin, 341–356.
18. Tzavaras, A. E. (2000): *On the mathematical theory of fluid dynamic limits to conservation laws.* CMS technical summary report 00 – 06.

Recent Progress on Quantum Hydrodynamic Models for Semiconductors

Ansgar Jüngel[1], Hailiang Li[2], Peter A. Markowich[2], and Shu Wang[2]

[1] Fachbereich Mathematik und Statistik, Universität Konstanz, 78457 Konstanz, Germany, *ansgar.juengel@uni-konstanz.de*

[2] Institut für Mathematik, Universität Wien, 1090 Wien, Austria, *hailiang.li@univie.ac.at, peter.markowich@univie.ac.at, shu.wang@univie.ac.at*

1 Introduction

1.1 General Remarks on the QHD Model and Its Derivation

For ultra-small semiconductor devices in which quantum effects are present, the corresponding mathematical models have to incorporate the quantum mechanical phenomena. Roughly speaking, these models can be divided into two classes: microscopic and macroscopic quantum models.

Microscopic models include *quantum kinetic equations*, like the Wigner equation and the Wigner-(Poisson-)Fokker-Plank model, and Schrödinger-Poisson systems [4, 30]. It has been known since the early days of quantum mechanics [27, 34] that these microscopic models can be written in hydrodynamic form using macroscopic quantities like the electron density or current density. The hydrodynamic form of quantum kinetic equations is termed *quantum hydrodynamic models.*

The advantage of quantum hydrodynamic models is that in semiconductor applications, the numerical simulation of microscopic models is quite expensive [32] while the numerical approximation of macroscopic models seems to be easier than for the microscopic ones.

Quantum hydrodynamic models (QHD models for short) can be derived from the quantum Wigner-Boltzmann equation using a velocity-moment method [10] or from the Schrödinger equation [9, 26]. It governs the evolution of the electron density ρ, the electron current density J and the electrostatic potential V. Assuming that the typical time scale is given by the momentum relaxation time constant τ (see [13]), we obtain the following unipolar scaled QHD equations:

$$\partial_t \rho + \mathrm{div} J = 0, \tag{1}$$

$$\partial_t J + \mathrm{div}\left(\frac{J \otimes J}{\rho}\right) + \nabla p(\rho) - \rho \nabla V - \frac{\varepsilon^2}{2} \rho \nabla \left(\frac{\Delta\sqrt{\rho}}{\sqrt{\rho}}\right) = -\frac{J}{\tau}, \tag{2}$$

$$\lambda^2 \Delta V = \rho - C(x), \tag{3}$$

for $x \in \mathbb{R}^d$ and $t > 0$. Here, $\frac{\varepsilon^2}{2}\rho\nabla\left(\Delta\sqrt{\rho}/\sqrt{\rho}\right)$ is the quantum Bohm correction and $p(\rho)$ is the pressure term. The scaled parameters are the Planck constant

ε (the ratio of the de Broglie length and the device length) and the Debye length λ. The doping profile $C(x)$ models fixed background charges. Also small mean-free-path asymptotics have been used to derive QHD models in the case of high electric fields. The obtained models have linear quantum corrections [28].

1.2 Connection of the QHD Model with the Schrödinger-Poisson and the Wigner Equation

We start with a single state nonlinear Schrödinger-Poisson system in $\mathbb{R}^d$:

$$i\epsilon\partial_t\Psi = -\frac{\epsilon^2}{2}\Delta\Psi + \left(V + Th(|\Psi|^2)\right)\Psi, \qquad x \in \mathbb{R}^d,\ t > 0, \tag{4}$$

$$\lambda^2\Delta V = |\Psi|^2 - C(x), \qquad x \in \mathbb{R}^d,\ t > 0. \tag{5}$$

The variables are the wave function $\Psi = \Psi(t,x)$, and the electrostatic potential $V = V(t,x)$. The function $h(s)$ can be interpreted as a self-interacting potential, and $T > 0$ is a constant (see below for the physical meaning of these quantities).

Now introduce the Wigner transform [2, 26, 29]

$$w(t,x,v) = \frac{1}{(2\pi)^d}\int_{\mathbb{R}^d}\overline{\Psi}\Big(t,x+\frac{\eta}{2}\Big)\Psi\Big(t,x-\frac{\eta}{2}\Big)e^{i\eta\cdot v}d\eta.$$

A simple calculation shows that w satisfies the so-called Wigner equation

$$\partial_t w + v\cdot\nabla w - \Theta[V + h(\rho)]w = 0, \qquad x,v \in \mathbb{R}^d,\ t > 0, \tag{6}$$

where $\rho = \int w dv$ and $\Theta[V + h(\rho)]$ is the pseudo-differential operator

$$\begin{aligned}&(\Theta[V + h(\rho)]w)(t,x,v)\\ &= \frac{i}{(2\pi)^d}\int_{\mathbb{R}^d}\int_{\mathbb{R}^d}\frac{i}{\varepsilon}\Big\{(V + h(\rho))\Big(t,x+\varepsilon\frac{\eta}{2}\Big) - (V + h(\rho))\Big(t,x-\varepsilon\frac{\eta}{2}\Big)\Big\}\\ &\quad\times w_j(t,x,v')e^{i(v-v')\cdot\eta}d\eta dv'.\end{aligned}$$

Multiplying the Wigner equation (6) by 1 and v and integrating over $\mathbb{R}^d$, we obtain the continuity equation (1) and the momentum equation (2), respectively, where the pressure $p(s)$ is related to $h(s)$ by $p'(s) = Tsh'(s)$, T can be interpreted as a temperature (constant), and we also assume $\tau = \infty$ in the momentum equation (2). For details, we refer to [8, 9, 26].

Clearly, a *physical* derivation of the QHD model has to start with the *linear* Schrödinger equation (4) without the self-interacting potential $h(\rho)$. Performing the same calculations as above, we arrive to the QHD equations (1)-(2) without the pressure term $p(\rho)$. A non-zero pressure can be obtained, for instance, by a mixed-state approach; see [8].

1.3 Relaxation-time Limit and the QDD Model

Rescaling the current density via $J \to \tau J$ and the time via $t \to t/\tau$ and performing the formal limit $\tau \to 0$, we obtain the scaled *quantum drift-diffusion equations* (see, for instance, [15]):

$$\partial_t \rho + \operatorname{div} J = 0, \quad J = \rho \nabla \left(\frac{\varepsilon^2}{2} \frac{\Delta\sqrt{\rho}}{\sqrt{\rho}} - Th(\rho) + V \right), \tag{7}$$

$$\lambda^2 \Delta V = \rho - C(x), \qquad x \in \Omega,\ t > 0. \tag{8}$$

The QDD model is quite well understood both analytically and numerically. The existence of solutions to the stationary and the time-dependent equations has been proved, and the semi-classical limits performed too. For more details, we refer to [1, 21, 22] and references therein.

This paper is organized as follows. In Section 2 we present some results on the existence and large-time behavior of solutions to the QHD models, and Section 3 is concerned with the corresponding semi-classical limits.

2 Existence and large-time behavior of solutions

2.1 The Stationary QHD Equations

The third-order term $\rho\nabla(\Delta\sqrt{\rho}/\sqrt{\rho})$ in (2) is the main difficulty in the analytical treatment of the QHD equations. Two strategies have been considered in the mathematical literature: First a potential-flow formulation has been assumed and the stationary momentum equation (2) is integrated once, yielding a *second-order* elliptic equation. Second, the stationary equation (2) is differentiated with respect to the spatial variable which gives a *fourth-order* elliptic equation. We review these strategies now in more detail.

Assume that the current density can be written as $J = \rho\nabla S$ with the quantum quasi-Fermi (or velocity) potential S. This means that we assume a potential flow with irrotational velocity $J/\rho = \nabla S$.

After division by ρ and integration of (2) we obtain the stationary equations:

$$\tfrac{\varepsilon^2}{2}\Delta\sqrt{\rho} = \sqrt{\rho}\left(\tfrac{1}{2}|\nabla S|^2 + Th(\rho) - V + \tfrac{1}{\tau}S\right), \tag{9}$$

$$\operatorname{div}(\rho\nabla S) = 0, \quad \lambda^2\Delta V = \rho - C(x) \quad \text{in } \Omega, \tag{10}$$

subject to Dirichlet boundary conditions:

$$\rho = \rho_D,\ S = S_D,\ V = V_D \quad \text{on } \partial\Omega. \tag{11}$$

For a different choice of boundary conditions, see below. In the above formulation, the function h, defined by $T\rho h'(\rho) = p'(\rho)$ and $h(1) = 0$, is called *enthalpy*.

We assume that $\Omega \subset \mathbb{R}^d$ ($d \geq 1$) is a bounded domain with $\partial\Omega \in C^{1,1}$. Using a Stampacchia truncation method and elliptic estimates, the following result has been proved in [14] (also see [16]).

Theorem 2.1. *Let $h(\rho) = \log\rho$ (to simplify) and assume that $\sqrt{\rho_D} \in W^{2,p}(\Omega)$ with $p > d/2$ and $\inf_\Omega \rho_D > 0$, $S_D \in C^{1,1}(\overline{\Omega})$, $V_D \in H^1(\Omega) \cap L^\infty(\Omega)$, and $C \in L^2(\Omega)$. Then there exists $\eta > 0$ such that if*

$$T^{-1}\|S_D\|^2_{C^{1,1}(\overline{\Omega})} \leq \eta, \tag{12}$$

there exists a solution (ρ, S, V) of (9)-(11) satisfying

$$\sqrt{\rho} \in W^{2,p}(\Omega), \;\; S \in C^1(\overline{\Omega}), \;\; V \in H^1(\Omega) \cap L^\infty(\Omega), \;\; \text{and} \; \inf_\Omega \rho > 0.$$

The assumption (12) can be interpreted as a subsonic-type condition. Indeed, (12) and elliptic estimates imply that

$$|J/\rho| = |\nabla S| \leq \|S\|_{C^1(\overline{\Omega})} \leq c\|S_D\|_{C^1(\overline{\Omega})} \leq c\sqrt{\eta} \cdot \sqrt{T},$$

and thus, for sufficiently small $\eta > 0$, $|J/\rho| < \sqrt{T}$.

It is well known that uniqueness of solutions to semiconductor equations cannot hold in general as there are devices which are based on multiple steady states [33]. Generally, uniqueness of solutions is to be expected close to the *thermal equilibrium state* (defined by $J = 0$) or, equivalently, for sufficiently small applied voltage. The uniqueness of solutions of (9)-(11) is shown in [14] under the condition (12) and for sufficiently large $\varepsilon > 0$. Notice that S_D is related to the applied voltage such that the first condition is natural; the second condition is needed for technical reasons.

An existence and *non-existence* result based on the same strategy but with different boundary conditions has been presented in [5]. In this paper, the one-dimensional model in $\Omega = (0,1)$ has been analyzed using the set of boundary conditions

$$\rho(0) = \rho_0 > 0, \;\; \rho(1) = \rho_1 > 0, \;\; V(0) = V_0, \;\; V_x(0) = -E_0, \;\; S(0) = K, \tag{13}$$

where $K < 0$ is a specific (given) constant depending on the data [5]. Notice that in one space dimension, $J = \text{const}$. The following result has been proved in [5].

Theorem 2.2. *Let $\Omega = (0,1)$ and $C \in L^2(\Omega)$. Then there exist $J_0, J_1 > 0$ such that*

(i) if $h(\rho) = \log\rho$ (to simplify) and $|J| \leq J_0$ then there exists a classical solution (ρ, V) of the stationary equations (1)-(2), (13) satisfying $\min_\Omega \rho > 0$, *and*

(ii) if $h(\rho) = \rho^\gamma - 1$ with $\gamma > 2$ and $J \geq J_1$ then the stationary QHD equations cannot have a weak solution for any $K \in \mathbb{R}$.

It is unknown if the QHD equations (9)-(10) subject to the boundary conditions (11) possess a (weak) solution for sufficiently large S_D (in an appropriate norm). For a simplified QHD model, it is shown in [14] by a simple example that the electron density vanishes for sufficiently large S_D and consequently, no weak solution can be expected in this case.

The second strategy to analyze the one-dimensional QHD model is to differentiate (2) with respect to x and to use the Poisson equation (3). This leads to an equation for the variable $u = \log \rho$:

$$\frac{\varepsilon^2}{2}\left(u_{xx} + \frac{u_x^2}{2}\right)_{xx} + J^2(e^{-2u}u_x)_x - Tu_{xx} + \lambda^{-2}(e^{-u} - C(x)) = \frac{J}{\tau}(e^{-u})_x, \quad (14)$$

for $x \in \Omega = (0,1)$ where we have taken $h(\rho) = \log \rho$. The boundary conditions are

$$u(0) = u_0, \ u(1) = u_1, \ u_x(0) = u_x(1) = 0. \quad (15)$$

In [12] the following result is shown.

Theorem 2.3. *Let $C \in L^2(\Omega)$. Then there exists $J_0 > 0$ such that if $|J| \leq J_0$ there exists a unique solution $u \in H^4(\Omega)$ of (14)-(15). Moreover, (ρ, V) with $\rho = e^u$ (and V satisfying (3)) is a solution of the stationary problem (1)-(3) such that* $\min_\Omega \rho > 0$.

The positivity of the particle density cannot be proven using the maximum principle since the equation (14) is of *fourth* order. Instead, Sobolev estimates are derived for the solution u (using the free energy of the problem) such that $u \in L^\infty(\Omega)$. This implies the positivity of $n = e^u$.

The assumption $h(\rho) = \log \rho$ in the above theorem can be substantially weakened. In fact, the pressure p does not need to be a monotone or convex function; for details we refer to [17].

2.2 The Time-Dependent QHD Equations

There are only few results in the mathematical literature concerning the transient QHD equations. Here we review two results. Local-in-time solutions have been studied in [19]. More precisely, existence of strong solutions of the potential-flow formulation

$$\partial_t \rho + \operatorname{div}(\rho \nabla S) = 0, \quad \lambda^2 \Delta V = \rho - C(x), \quad (16)$$

$$\partial_t S + \tfrac{1}{2}|\nabla S|^2 + Th(\rho) - V - \tfrac{\varepsilon^2}{2}\tfrac{\Delta\sqrt{\rho}}{\sqrt{\rho}} = -\tfrac{S}{\tau} \quad \text{in } \Omega \times (0,\infty) \quad (17)$$

subject to the initial-boundary conditions

$$\rho = \rho_D, \ S = S_D, \ V = V_D \ \text{ on } \partial\Omega \times (0,\infty), \quad (18)$$

$$\rho(\cdot,0) = \rho_I, \ S(\cdot,0) = S_I \ \text{ in } \Omega, \quad (19)$$

has been proved. Here, $\Omega \subset \mathbb{R}^d$ ($d \leq 3$) is supposed to be a bounded domain with $\partial\Omega \in C^2$.

Theorem 2.4. *Assume that* $p \in C^{2,1}(0,\infty)$, $C \in L^2(\Omega)$, ρ_D, S_D, V_D, ρ_I, $S_I \in H^2(\Omega)$ *with* $\rho_I = \rho_D$, $S_I = S_D$ *on* $\partial\Omega$ *and* $\min_\Omega \rho_I > 0$. *Then there exist* $T^* > 0$ *and a unique solution* (ρ, S, V) *of (16)-(19) satisfying*

$$\rho, S, V \in C^0([0,T^*]; H^2(\Omega)) \quad \text{and} \quad \min_\Omega \rho(\cdot,t) > 0 \text{ for } 0 \le t \le T^*.$$

The proof of this result is based on a reformulation of (16)-(17) as the nonlinear Schrödinger equation

$$i\varepsilon\partial_t\psi = -\frac{\varepsilon^2}{2}\Delta\psi - V\psi + h(|\psi|^2)\psi + \frac{1}{\tau}S\psi$$

and the use of semigroup theory.

To our knowledge, the existence of *global-in-time* solutions has been shown only in the case of small perturbations from the steady state in one space dimension [18]. The initial and boundary conditions are as follows:

$$\rho(0,t) = \rho_0,\ \rho(1,t) = \rho_1,\ \rho_x(0,t) = \rho_x(1,t) = 0, \tag{20}$$

$$V(0,t) = V_0,\ V(1,t) = V_1 \quad \text{for } t > 0, \tag{21}$$

$$\rho(x,0) = \rho_I(x),\ J(x,0) = J_I(x) \quad \text{for } x \in \Omega. \tag{22}$$

Theorem 2.5. *Assume that* $\Omega = (0,1)$, $p \in C^4(0,\infty)$ *is non-decreasing (to simplify),* $C \in H^2(\Omega)$, $\sqrt{\rho_I} \in H^6(\Omega)$ *with* $\min_\Omega \rho_I > 0$, *and* $J_I \in H^5(\Omega)$. *Then there exist a classical solution* (ρ_0, J_0, V_0) *of the stationary QHD equations and a number* $\eta > 0$ *such that if*

$$\|J_I/\rho_I\|_{C^1(\overline{\Omega})} + \|C_x\|_{H^1(\Omega)} + \|\rho_I - \rho_0\|_{H^6(\Omega)} + \|J_I - J_0\|_{H^5(\Omega)} \le \eta,$$

there exists a classical solution (ρ, J, V) *of (1)-(3), (20)-(22) for all time and, moreover,*

$$\|\rho(\cdot,t) - \rho_0\|_{H^6(\Omega)} + \|J(\cdot,t) - J_0\|_{H^5(\Omega)} + \|V(\cdot,t) - V_0\|_{H^4(\Omega)} \le ce^{-\alpha t},$$

where $c, \alpha > 0$ *are some constants.*

Clearly, the initial and boundary data have to satisfy appropriate compatibility conditions ensuring the needed regularity of the solutions. The idea of the proof is to write (1)-(2) as a fourth-order wave equation and to obtain a-priori estimates which allow to extend the local-in-time solution. The existence of local-in-time solutions is based on a construction of a sequence of approximate solutions converging to the local solution of the transient problem.

3 Semi-classical limits

In this section we review two results on the semi-classical limit in the quantum models leading to classical fluid-dynamical equations, in particular the *combined* semi-classical and quasi-neutral limit in the Schrödinger-Poisson system and the semi-classical limit in the QHD equations.

3.1 From the Schrödinger-Poisson to the Compressible Euler Equation

The combined semi-classical quasi-neutral limit in the bipolar Schrödinger-Poisson system gives the compressible Euler equations. This limit has been made rigorous in [23]. More precisely, consider the Schrödinger-Poisson system

$$i\varepsilon\partial_t\psi_j = -\frac{\varepsilon^2}{2}\Delta\psi_j + (q_jV + h(|\psi_j|^2))\psi_j, \quad \psi_j(\cdot,0) = \psi_{I,j}, \tag{23}$$

$$\lambda^2\Delta V = |\psi_1|^2 - |\psi_2|^2 \quad \text{in } \mathbb{R}^d \times (0,\infty),\ j = 1,2, \tag{24}$$

where $d \geq 1$, $q_1 = -1$ and $q_2 = +1$. Let $(\psi_1^\delta, \psi_2^\delta, V^\delta)$ be a solution of (23)-(24) with $\delta := (\varepsilon, \lambda)$. In the limit $\delta \to 0$, it is possible to prove that the particle densities ρ_j^δ and the current densities J_j^ε, defined by

$$\rho_j^\delta = |\psi_j^\delta|^2, \quad J_j^\delta = \varepsilon \mathrm{Im}(\overline{\psi_j^\delta}\nabla\psi_j^\delta),$$

converge to ρ and ρu, respectively, where (ρ, u) is a (smooth) solution of

$$\partial_t\rho + \mathrm{div}(\rho u) = 0, \quad \partial_t u + u\cdot\nabla u + \nabla h(\rho) = 0 \quad \text{in } \mathbb{R}^d \times (0,T^*), \tag{25}$$

$$\rho(\cdot,0) = \rho_I, \quad u(\cdot,0) = u_0 \quad \text{in } \mathbb{R}^d, \tag{26}$$

and $(0,T^*)$ is the maximal time interval in which there exists a smooth solution (ρ,u) of (25)-(26).

Theorem 3.6. *Let $h(\rho) = \rho^{\gamma-1} - 1$ with $\gamma \geq 2$, $\rho_0 := |\psi_{I,1}^\delta|^2 = |\psi_{I,2}^\delta|^2 \in L^\gamma(\mathbb{R}^d)$, $u_0 \in H^s(\mathbb{R}^d)$, where $s > d/2 + 2$. Furthermore, assume that $\|\psi_j^\delta\|_{H^s(\mathbb{R}^d)}$ is sufficiently small and that*

$$\|u_0\psi_{I,j}^\delta + i\varepsilon\nabla\psi_{I,j}^\delta\|_{L^2(\mathbb{R}^d)} \to 0 \tag{27}$$

as $\delta \to 0$ $(j = 1,2)$. Then, as $\delta = (\varepsilon,\lambda) \to 0$,

$$\rho_j^\delta \to \rho \quad \text{weakly in } L^\infty(0,T^*;L^\gamma(\mathbb{R}^d)),$$
$$J_j^\delta \to \rho u \quad \text{in the sense of } L^\infty(0,\infty;\mathcal{M}(\mathbb{R}^d)),$$

where $\mathcal{M}(\mathbb{R}^d)$ is the d-vector-valued space of Radon measures and (ρ,u) is the unique $H^s(\mathbb{R}^d)$ solution of (25)-(26).

If ψ_j^δ can be decomposed as $\psi_j^\delta = \sqrt{\rho_j^\delta}\exp(iS_j^\delta/\varepsilon)$ the assumption (27) can be written in the form

$$\varepsilon\|\nabla\sqrt{\rho_j^\delta}\|_{L^2(\Omega)}^2 + \|\sqrt{\rho_j^\delta}(u_0 - \nabla S_j^\delta)\|_{L^2(\mathbb{R}^d)}^2 = \|u_0\psi_{I,j} + i\varepsilon\nabla\psi_{I,j}^\delta\|_{L^2(\mathbb{R}^d)}^2 \to 0,$$

as $\delta \to 0$. In particular, if $\rho_0 = \rho_{I,j}^\delta$ and $u_0 = \nabla S_j^\delta$ $(j = 1,2)$, then (27) is trivially satisfied since

$$\|u_0\psi_{I,j}^\delta + i\varepsilon\nabla\psi_{I,j}^\delta\|_{L^2(\mathbb{R}^d)}^2 = \varepsilon\|\nabla\sqrt{\rho_0}\|_{L^2(\Omega)}^2 \to 0 \quad \text{as } \delta \to 0.$$

For the proof of Theorem 3.6, the Schrödinger-Poisson system (23)-(24) is formulated as a quantum hydrodynamic model in which the convection term is replaced by an expression which contains the Wigner function. Using estimates from a modulated energy functional, it is possible to prove that $\rho_j^\delta \to \rho$ weakly* in $L^\infty(0,T^*;L^\gamma(\mathbb{R}^d))$. Moreover, the Wigner function converges weakly to some Radon measure $w(x,v,t) = \rho(x,t)\delta(v - u(x,t))$ as shown in [9, 26]. Here, δ denotes the Dirac δ-distribution. From this result, the proof can be concluded (see [23] for details).

The combined semi-classical quasi-neutral limit from the Schrödinger–Poisson equation (for one carrier type) to the *incompressible* Euler equations has been shown in [31] for the *linear* case and for *"well-prepared"* initial data and in [25] for the *nonlinear* case and for *general* initial data.

There are only a few results on the *quasi-neutral* limit $\lambda \to 0$. We refer to [3, 11] on the limit $\lambda \to 0$ in Vlasov-Poisson systems and to [7, 20] in drift-diffusion equations.

Another way to perform the semi-classical limit rigorously is to make a WKB ansatz and to reformulate the (defocusing) Schrödinger-Poisson system as a quantum hydrodynamic model. This model can be seen as a dispersive perturbation of the Euler-Poisson problem written in the form of a quasi-linear symmetric hyperbolic system to which the Lax-Friedrich-Kato theory can be applied. In [24] it is proved that the solution of the (unipolar) Schrödinger-Poisson system converges to a (local-in-time smooth) solution of the compressible Euler-Poisson system as $\varepsilon \to 0$.

3.2 From the Quantum Hydrodynamic to the Subsonic Hydrodynamic Equations

Let $(\rho_\varepsilon, V_\varepsilon) \in H^4(\Omega) \times H^2(\Omega)$ be the solution of the one-dimensional stationary QHD model (1)-(3) with $\Omega = (0,1)$ and boundary conditions (15) where $\rho = e^u$. Then it is shown in [12] that $(\rho_\varepsilon, V_\varepsilon)$ converges to the solution (n, V) of the hydrodynamic equations

$$\left(\frac{J^2}{\rho} + Tp(\rho)\right)_x - \rho V_x = -\frac{J}{\tau}, \quad \lambda^2 V_{xx} = \rho - C(x), \tag{28}$$

$$\rho(0) = \rho_0 > 0,\ \rho(1) = \rho_1 > 0,\ V(0) = V_0, \tag{29}$$

if the (constant) current density is small enough and if $p(\rho) = \rho$.

Theorem 3.7. *Let $C \in L^2(\Omega)$ and $p(\rho) = \rho$. Then, for sufficiently small $J > 0$, as $\varepsilon \to 0$,*

$$\begin{aligned} \rho_\varepsilon \to \rho \quad & \text{weakly in } H^1(\Omega) \text{ and strongly in } C^0(\overline{\Omega}), \\ V_\varepsilon \to V \quad & \text{weakly in } L^2(\Omega), \end{aligned}$$

and $(\rho, V) \in H^1(\Omega) \times H^1(\Omega)$ *is the unique solution of (28)-(29) satisfying* $\min_\Omega \rho > 0$.

The idea of the proof is to use $\log \rho_\varepsilon - \log \tilde{\rho}$ with $\tilde{\rho}(0) = \rho_0$ and $\tilde{\rho}(1) = \rho_1$ as a test function in the stationary equation (2). The pressure term provides an $H^1(\Omega)$ estimate for $\log \rho_\varepsilon$ independent of ε from which we can conclude the result. A similar result using different boundary conditions has been presented in [6].

Acknowledgments: A.J. acknowledges partial support from the Deutsche Forschungsgemeinschaft, grants JU 359/3-1 (Gerhard-Hess Award) and JU 359/5-1 (Priority Program "Multiscale Problems"), and by the AFF Project of the University of Konstanz. H.L. and S.W. are supported by the Wittgenstein Award 2000 of Peter A. Markowich, funded by the Austrian FWF. A.J and P.M. acknowledge partial support from the TMR Project "Asymptotic Methods in Kinetic Theory", grant ERB-FMBX-CT97-0157. S.W acknowledges partial support from the National Youth Natural Science Fundation (grant 10001034) of China and the FWF-Projekt 14876-MAT "Fokker-Planck- und Mittlere-Feld-Gleichungen" (Austria).

References

1. Ben Abdallah, N., Unterreiter, A. (1998): On the stationary quantum drift-diffusion model. Z. Angew. Math. Phys. **49**, 251-275.
2. Brezzi, F., Markowich, P. (1991): The three dimensional Wigner-Poisson problem: existence, uniqueness and approximation. Math. Methods Appl. Sci. **14**, 35-62.
3. Brenier, Y. (2000): Convergence of the Vlasov-Poisson system to the incompressible Euler equations. Commun. Part. Diff. Eqs. **25**, 737-754.
4. Castella, F., Erdös, L., Frommlet, F., Markowich, P. (2000): Fokker-Planck equations as scaling limits of reversible quantum systems. J. Stat. Phys. **100**, 543-601. 1999.
5. Gamba, I., Jüngel, A. (2001): Positive solutions to singular second and third order differential equations for quantum fluids. Arch. Rat. Mech. Anal. **156**, 183-203.
6. Gamba, I., Jüngel, A. (2002): Asymptotic limits in quantum trajectory models. Commun. Part. Diff. Eqs. **27**, 669-691.
7. Gasser, I., Levermore, C. D., Markowich, P., Schmeiser, C. (2001): The initial time layer problem and the quasi-neutral limit in the semiconductor drift-diffusion model. Europ. J. Appl. Math. **12**, 97-116.
8. Gasser, I., Lin C. K., Markowich, P. (2000): A review of dispersive limits of Nonlinear Schrödinger-type equations. Taiwanese J. Math. **4**, 501-529.
9. Gasser, I., Markowich, P. (1997): Quantum hydrodynamics, Wigner transforms and the classical limit. Asympt. Anal. **14**, 97-116.
10. Gardner, C. (1994): The quantum hydrodynamic model for semiconductor devices. SIAM J. Appl. Math. **54**, 409-427.
11. Grenier, E. (1996): Oscillations in Vlasov-Poisson systems. Commun. Part. Diff. Eqs. **21**, 363-394.

12. Gyi, M.T., Jüngel, A. (2000): A quantum regularization of the one-dimensional hydrodynamic model for semiconductors. Adv. Diff. Eqs. **5**, 773-800.
13. Jüngel, A. (1997): A note on current-voltage characteristics from the quantum hydrodynamic equations for semiconductors. Appl. Math. Lett. **10**, 29-34.
14. Jüngel, A. (1998): A steady-state potential flow Euler-Poisson system for charged quantum fluids. Commun. Math. Phys. **194**, 463-479.
15. Jüngel, A. (2001). Nonlinear problems in quantum semiconductor modeling. Nonlin. Anal. **47**, 5873-5884.
16. Jüngel, A. (2001). Quasi-hydrodynamic Semiconductor Equations. Birkhäuser, Basel.
17. Jüngel, A., Li, H.-L. (2001): Quantum Euler-Poisson system: existence of stationary states. Submitted for publication.
18. Jüngel, A., Li, H.-L. (2002): Quantum Euler-Poisson system: global existence and exponential decay. Submitted for publication.
19. Jüngel, A., Mariani, M. C., Rial, D. (2001): Local existence of solutions to the transient quantum hydrodynamic equations. Math. Models Meth. Appl. Sci. **12**, 485-495.
20. Jüngel, A., Peng, Y.-J. (2001): A hierarchy of hydrodynamic models for plasmas: quasineutral limits in the drift-diffusion equations. Asympt. Anal. **28**, 49-73.
21. Jüngel, A., Pinnau, R. (2000): Global non-negative solutions of a nonlinear fourth order parabolic equation for quantum systems. SIAM J. Math. Anal. **32**, 760-777.
22. Jüngel, A., Pinnau, P. (2001): A positivity preserving numerical scheme for a nonlinear fourth-order parabolic system. SIAM J. Num. Anal. **39**, 385-406.
23. Jüngel, A., Wang, S. (2002): Convergence of nonlinear Schrödinger-Poisson systems to the compressible Euler equations. Submitted for publication.
24. Li., H.-L., Lin, C.-K. (2001): Semiclassical limit and well-posedness of nonlinear Schrödinger-Poisson. Submitted for publication.
25. Li, H.-L., Lin, C.-K., Masmoudi, N. (2002): From Schrödinger-Poisson system to the incompressible Euler for general initial data. Submitted for publication.
26. Lions, P. L., Paul, T. (1993): Sur les mésures de Wigner. Revista Mat. Iberoamericana **9**, 553-618.
27. Madelung, E.(1927): Quantentheorie in hydrodynamischer Form. Z. Physik **40**, 322.
28. Markowich, P. A., Ringhofer, C. (1994): Quantum hydrodynamics for semiconductors in the high-field case. Appl. Math. Lett. **7**, 37-41.
29. Markowich, P., Mauser, N. (1993): The classical limit of a self-consistent quantum-Vlasov equation in 3D. Math. Models Meth. Appl. Sci. **3**, 109-124.
30. Markowich, P. A., Ringhofer, C., Schmeiser, C. (1990): Semiconductor equations. Springer, Vienna.
31. Puel, M. (2000): Etudes variationnelles et asymptotiques des problèmes de la mécanique des fluides et des plasmas. PhD thesis, Université Paris 6.
32. Ringhofer, C. (1997): Computational methods for semiclassical and quantum transport in semiconductor devices. Acta Numerica, **6**, 485-521.
33. Ward, M., Reyna, L., Odeh, F. (1991): Multiple steady-state solutions in a multijunction semiconductor device. SIAM J. Appl. Math. **51**, 90-123.
34. Wigner, E. (1932): On the quantum correction for thermodynamic equilibrum. Phys. Rev. **40**, 749-759.

Critical Thresholds and Conditional Stability for Euler Equations and Related Models

Hailiang Liu[1] and Eitan Tadmor[2]

[1] UCLA, Department of Mathematics, Los Angeles, CA 90095-1555, USA. Current address: Iowa State University, Department of Mathematics, Ames, IA 50011. Email: hliu@iastate.edu

[2] UCLA, Department of Mathematics, Los Angeles, CA 90095-1555, USA. Current address: University of Maryland, Department of Mathematics, Center for Scientific Computation and Mathematical Modeling (CSCAMM) and Institute for Physical Science & Technology (IPST), College Park, MD 20742-4015. Email: tadmor@math.umd.edu

1 Introduction

We are concerned with time regularity of the velocity field for a class of so-called restricted flows, where the velocity field, u, is governed by the Newtonian law,

$$\partial_t u + u \cdot \nabla u = F, \quad x \in \mathbb{R}^n, \tag{1}$$

which shows up in different contexts dictated by the different modeling of F's, where F stands for a general forcing acting on the flow. The examples range from Euler/Navier-Stokes equations to Euler-Poisson equations. There has been an enormous amount of papers related to the study of global behavior of solutions to (1) with possible associated laws (say, conservation of mass, energy etc). For the question of global behavior of strong solutions the choice of the initial data and/or damping forces is decisive. The classical methods of analysis include energy method (for small initial perturbations due to the nonlinearity) and the singularity prediction (the finite life span is often due to a *global* condition of large enough initial (generalized) energy, staying outside a critical threshold ball).

Indeed when dealing with the questions of time regularity for the above Euler-related equations, one encounters several limitations with the classical stability analysis. Among other issues, we mention that

(i) the stability analysis does not tell us how large perturbations are allowed before losing stability, say with the incompressible Navier-Stokes equations;

(ii) the steady solution may be only conditionally stable due to the weak dissipation in the system, say in certain Euler-Poisson models, [ELT].

In order to address these difficulties we propose a new notion of critical threshold (CT) in [ELT], which serves to describe the conditional stability for underlying physical problem. It is shown that the CT phenomena does reflect the delicate balance among various forcing mechanisms. Little or no attention has been paid to this remarkable phenomena, and our goal is to bridge the gap left in previous studies on the solution behavior of Euler-related equations in the small and in the large.

We first illustrate the CT phenomena by analyzing a 2×2 ODE system, which governs the time dynamics of 1-D Euler-Poisson equations along the particle path; The result on the CT phenomena for a 1-D convolution model [LT1] is also recorded.

We then discuss this remarkable CT phenomena associated with the Euler-Poisson equations [ELT], where the answer to questions of global smoothness vs. finite time breakdown depends on whether the initial configuration crosses an intrinsic, $O(1)$ critical threshold.

We investigate various one-dimensional problems with or without forcing mechanisms as well as multi-dimensional isotropic models with geometric symmetry. The critical thresholds for these essentially 1-D problems are shown to depend on the relative size of the initial velocity slope and the initial density.

We then extend our discussion of the CT phenomena for multi-dimensional systems of the form (1). Here we utilize a novel description for the spectral dynamics of the (possibly complex) eigenvalues, $\lambda = \lambda(\nabla u)$ which are shown to be governed by the Ricatti-like equation $\lambda_t + u \cdot \nabla\lambda + \lambda^2 = \langle l, \nabla Fr \rangle$. Restricting attention to restricted Euler-Poisson equations driven by localized forcing we identify the set of their $[n/2]$ global invariants, which in turn yield (i) sufficient conditions for finite time breakdown, and (ii) characterization of a large class of 2-dimensional initial configurations leading to global smooth solutions.

Moreover, the critical thresholds for 2D REPs are shown to depend on the relative size of three quantities at initial time: density, divergence and the spectral gap $\lambda_2 - \lambda_1$.

The results on CT phenomena reviewed here are obtained in a series of papers [ELT, LT1, LT2, LT3, LT4], which provides an account on recent developments of conditional stability theory for a class of Eulerian flow models. The corresponding methods in obtaining these results will be briefly illustrated.

2 Critical Threshold Phenomena

It is known that a fundamental question for equation (1) is whether the smooth solution develops singularity in finite time, and a key issue in this line of research is the control of the velocity gradient tensor $\nabla_x u$. As pointed earlier the classical stability analysis often fails to detect the delicate balance between the convection and the underlying forcing. Our approach to address these difficulties is the use of a new notion of critical threshold (CT) in order to describe the conditional stability for the underlying problem.

In order to highlight the idea of CT notion we start with a 2×2 ODE system

$$u' = -uv, \tag{1}$$

$$v' = ku - v^2, \quad k = Const. > 0. \tag{2}$$

Observe that for the global solution behavior the linear part of the system is of no help since both eigenvalues of the coefficient matrix of linearized system are zero. Stability theory does NOT work even locally around the equilibrium point $(0,0)$.

A manipulation like $[-v \times (1) + u \times (2)]/u^2$ leads to

$$\left(\frac{v}{u}\right)' = \frac{uv' - vu'}{u^2} = k, \quad \frac{v}{u} = kt + \frac{v(0)}{u(0)}.$$

Inserting back into (2) we obtain a decoupled system

$$v' + v^2 = \frac{kv}{kt + v(0)/u(0)},$$

from which we find that

$$v(t) = \frac{v(0) + ku(0)t}{1 + v(0)t + ku(0)t^2/2}.$$

Clearly this solution is globally bounded if and only if the initial data lie in the region

$$\{(u,v), \quad v > -\sqrt{2ku}, \quad u \geq 0\}.$$

The boundary of this region in the phase space is called the critical threshold (CT). The global behavior of the solution depends only on whether the initial data cross such critical threshold.

In [LT1] such CT phenomena is justified for a 1-D scalar PDE model:

$$u_t + uu_x = Q * u - u, \tag{3}$$

where $Q \geq 0$ is a regular symmetric kernel, monotonically decreasing on $\mathbb{R}^+$, subject to initial data

$$u(0,x) = u_0(x), \quad u_0 \in C_b^1(\mathbb{R}). \tag{4}$$

Theorem 2.1 *[LT1] Consider the Cauchy problem* (3)–(4) *with initial data* $u_0 \in C_b^1(\mathbb{R})$. *Let the kernel* Q *be as stated above; then we have the following:*

– *If* $V(u_0) < \frac{1}{4Q(0)}$ *and*

$$\inf_{x \in \mathbb{R}} \partial_x u_0(x) > -\frac{1}{2}\left[1 + \sqrt{1 - 4Q(0)V(u_0)}\right], \quad V(u_0) := \max u_0 - \min u_0,$$

then the smooth solution exists for all time.

– *If*

$$\inf_x \partial_x u_0(x) < -\frac{1}{2}\left[1+\sqrt{1+4Q(0)V(u_0)}\right],$$

then the solution u *must break down at finite time* T.

The above upper and lower-threshold clearly confirm the existence of the critical threshold. ¿From these results we see that if the magnitude of the initial profile is small, both thresholds given in Theorem 2.1 are close to $\inf_{x\in \mathbb{R}} \partial_x u_0(x) = -1$, which is exactly the critical threshold for the damped Burgers' equation:

$$u_t + uu_x = -u.$$

Indeed, along the particle path $x(\alpha, t)$ defined by

$$\frac{d}{dt}x(\alpha,t) = u(t, x(\alpha,t)), \quad x(\alpha,0)=\alpha, \quad \alpha \in \mathbb{R},$$

the gradient of the solution to the damped Burgers' equation above can be written explicitly as

$$u_x(t,x) = [e^t(1+(\partial_x u_0(\alpha))^{-1}) - 1]^{-1},$$

which is bounded from below for all time if and only if $\inf_{x\in \mathbb{R}} \partial_x u_0(x) \geq -1$. This remarkable critical threshold phenomenon explains why (3) admits narrower shock layers than those in the viscous Burgers' equation, see e.g. [KN, Ro, ST].

3 Isotropic Euler-Poisson Equations

In this section we present the results obtained in [ELT] on a new phenomena associated with the Euler-Poisson equations —the so called critical threshold phenomena. To this end we focus our attention on the n-dimensional isotropic model,

$$\begin{aligned} &r^\nu \rho_t + (\rho u r^\nu)_r = 0, \quad r>0, \\ &\rho(u_t + uu_r) = k\rho\phi_r + \text{viscosity} + \text{relaxation}, \\ &(r^\nu \phi_r)_r = \rho r^\nu + \text{background}, \quad \nu = n-1. \end{aligned} \tag{1}$$

It is well known that finite time breakdown is a generic phenomena for nonlinear hyperbolic convection equations, which is realized by the formation of shock discontinuities. In the context of Euler-Poisson equation, however, there is a delicate balance between the forcing mechanism (governed by Poisson equation), and the nonlinear focusing (governed by Newton's second law), which supports a critical threshold phenomena.

3.1 The basic model with zero background

We consider the 1D Euler-Poisson equation of the form

$$\rho_t + (\rho u)_x = 0, \tag{2}$$
$$u_t + uu_x = -k\phi_x = kE, \tag{3}$$
$$E_x = -\phi_{xx} = \rho. \tag{4}$$

Note that the density ρ and the velocity slope u_x satisfy the ODE (1)-(2) along the particle path $x = x(\alpha, t)$ governed by $\frac{dx}{dt} = u(x,t)$. The critical threshold observed for (1)-(2) enables us to conclude the following

Theorem 3.1 *[ELT] The system of Euler-Poisson equations (2)-(4) admits a global smooth solution if and only if*

$$u_0'(\alpha) > -\sqrt{2k\rho_0(\alpha)}, \quad \forall \alpha \in \mathrm{IR}. \tag{5}$$

This result reveals a phenomenon very much like the conditional breakdown of waves on the beach; only waves above certain initial critical threshold experience finite-time breakdown, but otherwise they propagate smoothly.

3.2 The basic model with constant background

We now consider the system

$$\rho_t + (\rho u)_x = 0, \tag{6}$$
$$u_t + uu_x = -k\phi_x = kE, \tag{7}$$
$$E_x = -\phi_{xx} = \rho - c, \tag{8}$$

with constant "background" state $c > 0$. Here we require that $\int_{-\infty}^{\infty}(\rho(\xi) - c)\, d\xi = 0$. The presence of a constant background c in the Poisson equation changes the "physical situation." However, the CT phenomena for this system are still observed and justified in [ELT].

Theorem 3.2 *Consider the system of Euler-Poisson equations (6)-(8) with constant background charge c. Then*

(i) for the repulsive force $k > 0$, it admits a global smooth solution if and only if

$$|u_0'(\alpha)| < \sqrt{k(2\rho_0(\alpha) - c)}, \quad \forall \alpha \in \mathrm{IR}. \tag{9}$$

In this case, the density oscillates around the nonzero background charge c, and the velocity gradient does not decay in time.

(ii) for attractive force, $k < 0$. Then, it admits a global smooth solution if and only if

$$u_0'(\alpha) \geq -\left(1 - \frac{\rho_0(\alpha)}{c}\right)\sqrt{-ck}, \quad \forall \alpha \in \mathrm{IR}. \tag{10}$$

In this case, the density approaches the zero exponentially in time, and the velocity gradient remains bounded uniformly in time.

The above results indicate that the presence of the background can balance the attractive forcing ($k < 0$). The fact that (9) when $c \downarrow 0$ does not yield (5) implies that the limit of $c \downarrow 0$ is a sort of singular limit.

3.3 Euler-Poisson system with relaxation

We consider a further modification of our problem (6), (8), where (7) is now augmented by a relaxation term

$$u_t + uu_x = -k\phi_x - \frac{u}{\epsilon} = kE - \frac{u}{\epsilon}, \quad \epsilon > 0. \tag{11}$$

We still require that $\int_{-\infty}^{\infty} (\rho(\xi) - c)\, d\xi = 0$.

Theorem 3.3 *Consider the system of Euler-Poisson equations (6), (11), (8) with a constant background charge $c > 0$, a repulsive force, $k > 0$ and a strong relaxation term, $\epsilon < \frac{1}{2\sqrt{ck}}$. If at all points $\alpha \in \mathbb{R}$*

$$u_0'(\alpha) > \min\left\{0, -(1 - \frac{\rho_0(\alpha)}{c})\left(\sqrt{\frac{1}{4\epsilon^2} - ck} + \frac{1}{2\epsilon}\right)\right\}, \tag{12}$$

then the solution of (6), (11), (8) remains smooth for all time. In this case, $u(x,t) \to 0$ and $\rho(x,t) \to c$ exponentially fast as $t \to \infty$.

If the relaxation is weak then the solution becomes oscillatory. However, a critical threshold can still be detected. We summarize by stating that

Theorem 3.4 *Consider the system of Euler-Poisson equations (6), (11), (8) with a constant background charge $c > 0$, a repulsive force $k > 0$ and weak relaxation $\epsilon > \frac{1}{2\sqrt{ck}}$. Then there exists a critical time $t^* > 0$ such that if at all points $\alpha \in \mathbb{R}$,*

$$\left|u_0'(\alpha) + \frac{c - \rho_0(\alpha)}{2\epsilon c}\right| < \sqrt{k - \frac{1}{4c\epsilon^2}}\sqrt{\frac{\rho_0^2(\alpha)}{c}(e^{t^*/\epsilon} - 1) + 2\rho_0(\alpha) - c}, \tag{13}$$

the solution of (6), (11), (8) remains smooth for all time. In this case, $u(x,t) \to 0$ and $\rho(x,t) \to c$ exponentially as $t \to \infty$.

Remark 3.5 *Note that as $\epsilon \to \infty$ we recover the local condition (10) for the case without relaxation.*

In addition to the above mentioned example we also justify the CT phenomena for the 1-D Euler-Poisson equation with viscosity mechanism, we refer to [ELT] for further details.

3.4 Critical Thresholds–Multi-D model with Geometric Symmetry

Let us consider the Euler-Poisson equations governing the $\nu+1$ dimensional isotropic ion expansion in the electrostatic fluid approximation for cold ions

$$r^\nu \rho_t + (\rho u r^\nu)_r = 0, \quad r > 0, \tag{14}$$
$$u_t + u u_r = -k\phi_r = kE, \tag{15}$$
$$(r^\nu \phi_r)_r = -\rho r^\nu, \tag{16}$$

subject to the initial data

$$(\rho, u)(r, 0) = (\rho_0, u_0)(r), \quad \rho_0(r) \geq 0. \tag{17}$$

We recall some results obtained in [ELT] for the multi- D case, $\nu \geq 1$, where we confirm the remarkable persistence of critical threshold phenomena in the multidimensional problem with geometric symmetry.

Theorem 3.6 (Global existence of smooth solutions for the cylindrical case $\nu = 1$)
A global smooth solution of Euler- Poisson equations (14)-(16) with $\nu = 1$ exists provided the initial data (u_0, ρ_0) with $E_0 = \alpha^{-1}\int_0^\alpha \rho_0(\xi)\xi d\xi$ satisfy

$$u_0' > -\frac{k}{u_0}[\alpha\rho_0 h(\alpha) - E_0], \quad \forall \alpha \in \mathsf{IR}^+, \tag{18}$$

where $h(\alpha)$ is determined by

$$k\alpha^2\rho_0 u_0 \int_0^{h(\alpha)} \frac{h(\alpha) - \eta}{[u_0^2 + 2kE_0\alpha\eta]^{3/2}} e^\eta d\eta \equiv 1, \quad \forall \alpha \in \mathsf{IR}^+. \tag{19}$$

Condition (18) could be viewed as an upper threshold in the sense of providing a sufficient condition leading to global smooth solutions, though the permissible class of the initial data for global smooth solutions is clearly larger. However, the existence of the critical threshold can be ensured by combining this upper threshold with the following lower threshold for the finite time breakdown.

Theorem 3.7 (Breakdown of smooth solutions for cylindrical case $\nu = 1$)
The smooth solution to the Euler- Poisson equations (14)-(16) with $\nu = 1$ breaks down in finite time if the condition, $u_0'(\alpha) > -\sqrt{2k\rho_0(\alpha)}$, fails, i.e., if

$$\exists \alpha \in \mathsf{IR}^+ \quad s.t. \quad u_0'(\alpha) \leq -\sqrt{2k\rho_0(\alpha)}. \tag{20}$$

For the 3-dimensional case we obtain the lower threshold for finite time breakdown.

Theorem 3.8 (Breakdown of smooth solutions for spherical case $\nu = 2$)
The solution of Euler-Poisson equations (14)-(16) for $\nu = 2$ blows up in finite time if the condition, $u_0' \geq -\frac{k}{u_0}[\alpha\rho_0 - E_0]$, fails, i.e.,

$$\exists \alpha \in \mathrm{IR}^+ \quad s.t. \quad u_0' < -\frac{k}{u_0}[\alpha\rho_0 - E_0]. \tag{21}$$

Again an immediate critical threshold is confirmed by combining the above lower threshold and the upper threshold for the existence of global smooth solution below.

Theorem 3.9 (Global existence of smooth solutions for general $\nu \geq 2$ cases)
A global smooth solution of the Euler-Poisson equations (14)-(16) with $\nu \geq 2$ exists provided the initial data (u_0, ρ_0) with $E_0 = \alpha^{-\nu}\int_0^\alpha \rho_0(\xi)\xi^\nu d\xi$ is prescribed such that for all $\alpha \in \mathrm{IR}^+$

$$u_0' > -\frac{k}{u_0}\left[\frac{\rho_0\alpha^\nu h_\nu}{\nu - 1} - E_0\right]. \tag{22}$$

Here, h_ν is determined by

$$\frac{ku_0\rho_0\alpha}{(\nu-1)^2}\int_0^{h_\nu} \frac{h_\nu(\alpha) - \eta}{[u_0^2 + \frac{2kE_0\alpha^\nu}{\nu-1}\eta]^{3/2}}(\alpha^{1-\nu} - \eta)^{\frac{\nu}{1-\nu}} d\eta \equiv 1. \tag{23}$$

The above results for various models indicate that the critical thresholds for these essentially 1-D problems are shown to depend on the relative size of the initial velocity slope and the initial density. For truly multi-D models the velocity gradient is a $n \times n$ square matrix, the key is to seek an appropriate characterization and control of its entries, which is the main topic of next section.

4 Spectral Dynamics—Multi-D Models

As noted earlier a key issue of studying the time regularity is the control of the velocity gradient $\nabla_x u$, and a classical approach in this context, is to consider linear combinations of the entries of ∇u, controlling physically relevant quantities like vorticity, divergence, etc. [BKM, CLM, M].

The novelty of the analysis taken in [LT2] is the use of the eigenvalues of the velocity gradient field. The eigenvalues, $\lambda = \lambda(\nabla u)$, exhibit of course a strong nonlinear dependence on the entries of ∇u, and are shown to play a crucial role in governing the behavior of the flow. Indeed, the dynamics of these eigenvalues $\lambda(\nabla u)$, is shown, in [LT2], to be governed by Ricatti-like equation

$$\partial_t \lambda + u \cdot \nabla \lambda + \lambda^2 = < l, \nabla F r >, \tag{1}$$

with $l(r)$ being the left (right) eigenvectors of ∇u. Equipped with this description for the spectral dynamics of ∇u, we are able to study the time dynamics for several physical models with different forcing.

In [LT2] we focus our investigation on four prototype models associated with different forcing F, ranging from simple linear damping and a viscous dusty medium models to the main thrust of the paper — the restricted models of Euler/Navier-Stokes equations and Euler-Poisson equations.

In particular, we address the question of the time regularity for these models, that is, whether they admit a finite time breakdown, a global smooth solution, or an intermediate scenario of critical threshold phenomena where global regularity depends on initial configurations.

Using the spectral dynamics as our essential tool in these investigations, we obtain a simple form of a critical threshold for the linear damping model and we identify the 2D vanishing viscosity limit for the viscous irrotational dusty medium model. Moreover, for the n-dimensional restricted Euler equations we obtain $[n/2] + 1$ global invariants, interesting for their own sake, which enable us to precisely characterize the local topology at breakdown time, extending previous studies in the $n = 3$-dimensional case, see [Vi]. Finally, as a forth model we introduce the n-dimensional restricted Euler-Poisson (REP)system, identifying a set of $[n/2]$ global invariants, which in turn yield (i) sufficient conditions for finite time breakdown, and (ii) characterization of a large class of 2-dimensional initial configurations leading to global smooth solutions. Consequently, the 2D restricted Euler-Poisson equations are shown to admit a critical threshold.

Here we highlight the main idea by presenting the result for so-called Restricted Euler-Poisson dynamics introduced in [LT2], consult [LT2, LT3] for results about other related models.

The Euler-Poisson equations

$$\rho_t + \nabla \cdot (\rho u) = 0, \quad x \in \mathrm{IR}^n, \quad t \in \mathrm{IR}^+, \tag{2}$$

$$(\rho u)_t + \nabla \cdot (\rho u \otimes u) = k \rho \nabla \phi, \tag{3}$$

$$\Delta \phi = \rho, \quad x \in \mathrm{IR}^n, \tag{4}$$

are the usual statements of the conservation of mass, Newton's second law, and the Poisson equation defining, say, the electric field in terms of the charge.

Passing to Lagrangian coordinates, that is, using the change of variables $\alpha \mapsto x(\alpha, t)$ with $x(\alpha, t)$ solving

$$\frac{dx}{dt} = u(x, t), \quad x(\alpha, 0) = \alpha,$$

then Euler-Poisson equations have the equivalent formulation in terms of velocity gradient field $M := \nabla_x u$ and the density ρ

$$\frac{d}{dt}M + M^2 = kR[\rho], \quad \frac{d}{dt} := \partial_t + u \cdot \nabla, \tag{5}$$

$$\frac{d}{dt}\rho + \rho tr M = 0. \tag{6}$$

Here $R[w]$ denotes the $n \times n$ matrix whose entries are given by $(R[w])_{ij} := R_i R_j(w)$ where R_j denote the Risez transforms, $R_j = -(-\Delta)^{-1/2}\partial_j$, i.e.,

$$\widehat{[R_j w]}(\xi) = -i\frac{\xi_j}{|\xi|}\hat{w}(\xi) \quad \text{for} \quad 1 \leq j \leq n.$$

It is the *global* term in the above equations, $R[\rho]$, which makes the problem rather intricate to solve, both analytically and numerically. A localized alternative to (5)-(6) was introduced in [LT2] as a restricted Euler-Poisson dynamics. Specifically, we consider a gradient flow governed by

$$\partial_t M + u \cdot \nabla M + M^2 = \frac{k}{n}\rho I_{n \times n}, \tag{7}$$

$$\partial_t \rho + u \cdot \nabla \rho + \rho tr M = 0. \tag{8}$$

If we let $\lambda_i(x,t)$ denote the eigenvalues of velocity gradient tensor ∇u, then by the spectral dynamics relation (1), the eigenvalues and the density ρ are coupled through

$$\partial_t \rho + u \cdot \nabla \rho + \rho \sum_{j=1}^{n} \lambda_j = 0, \tag{9}$$

$$\partial_t \lambda_i + u \cdot \nabla \lambda_i + \lambda_i^2 = \frac{k\rho}{n}, \quad i = 1, \cdots, n. \tag{10}$$

This is a closed system governing the restricted Euler-Poisson (REP) equations, which serves as a simple approximation for the evolution of the full Euler-Poisson system (2)-(4).

Theorem 4.1 *[LT2]* **(Global material invariants for $n \geq 2$)** *Consider the restricted Euler-Poisson dynamics (9)-(10) with real initial data $(\rho_0, \lambda_1(0), \cdots, \lambda_n(0))$. Then, there exist $\left[\frac{n}{2}\right]$ global invariants*

$$\frac{\Pi_{(i,j)\in\mathcal{I}}(\lambda_i - \lambda_j)}{\rho^N} = \text{Const.}, \quad N = \begin{cases} 1, & n \text{ even}, \\ 2, & n \text{ odd}. \end{cases} \tag{11}$$

The general solution may break down at finite time in the orthant $\{+, -, \cdots, -\}$.

This result indicates that the spectral gap $\lambda_i - \lambda_j$ play important roles in the evolution of the flow. Also in [LT2] we identify a class of 2D initial configurations for global smoothness.

Theorem 4.2 (Global existence)
The solutions of the 2-D restricted Euler-Poisson equations (9),(10) ($n = 2$) remains smooth for all time $t > 0$ if both $\lambda_i(0)$, $i = 1, 2$ are complex, i.e., $Im(\lambda_i(\alpha, 0)) \neq 0$.

As a consequence, it follows that the 2D restricted Euler-Poisson equations admit a critical threshold which distinguishes between initial configurations leading to finite time breakdown and global smooth solutions. A detailed study of this 2-dimensional critical threshold phenomena in this context is provided in [LT3] for the 2-D REP with zero and nonzero background. This complements the study of critical threshold phenomena for isotropic configurations in the general (global) Euler-Poisson equations presented in [ELT].

We recall the result for zero background case, and refer to [LT3] for nonzero background case.

To state the main result, we introduce two quantities with which we characterize the behavior of the velocity gradient tensor $M = \nabla U$, namely, the divergence

$$d := trM = \lambda_1 + \lambda_2$$

and the nonlinear quantity

$$\Gamma := (trM)^2 - 4detM = (\lambda_2 - \lambda_1)^2,$$

which serves as an index for the spectral gap.

Theorem 4.3 *[LT3]* **(2D REP with zero background).**
Consider the 2D repulsive REP system (9),(10)($n = 2$) with $k > 0$ and with zero background $c = 0$. Then the solution of 2D REP remains smooth for all time if and only if the initial data (ρ_0, M_0) lies in one of the following two regions, $(\rho_0, d_0, \Gamma_0) \in S_1 \cap S_2$:

(i) Either

$$(\rho_0, d_0, \Gamma_0) \in S_1, \quad S_1 := \left\{ (\rho, d, \Gamma) | \Gamma \leq 0 \text{ and } \begin{cases} d \geq 0 & \text{if} \quad \rho = 0 \\ d \quad \text{arbitrary} & \text{if} \quad \rho > 0 \end{cases} \right\};$$

(ii) or

$$(\rho_0, d_0, \Gamma_0) \in S_2, \quad S_2 := \{(\rho, d, \Gamma) \quad | \quad \rho > 0, \quad \Gamma > 0, \quad \text{and} \quad d \geq g(\rho, \Gamma)\}$$

where

$$g(\rho, \Gamma) := \text{sgn}(\Gamma - 2k\rho)\sqrt{\Gamma - 2k\rho + 2k\rho ln\left(\frac{2k\rho}{\Gamma}\right)}.$$

Remarks:

1. The above result shows that the global smooth solution is always ensured if the velocity gradient tensor has complex eigenvalues, as stated in Theorem 4.2, which applies, for example, for a class of initial configurations

with sufficiently large vorticity $|u_{0y} - v_{0x}| >> 1$. With other initial configurations, however, the finite time breakdown of solutions may – and actually does occur, unless the initial divergence is above a critical threshold, expressed in terms of the initial density and initial spectral gap. Hence global regularity depends on whether the initial configuration crosses an intrinsic, $O(1)$ critical threshold.

2. The critical threshold in the 1D Euler-Poisson equations depends on the relative size of the initial velocity slope and the initial density, consult [ELT]. In contrast the critical threshold presented here depends on three initial quantities: density ρ_0, divergence $\nabla \cdot U_0$ and initial spectral gap $\Gamma_0 = (u_{0x} - v_{0y})^2 + 4u_{0y}v_{0x}$.

3. Theorem 4.3 tells us that the size of initial sub-critical range which gives rise to regular solution is decreasing as the initial ratio Γ_0/ρ_0 is increasing. In particular when this ratio is larger than $2k$, then the initial divergence must stay above a positive critical threshold to avoid the finite time breakdown. □

5 And there's more ...

In the previous sections we show how the persistence of the global features of the solutions for various models hinges on a delicate balance between the nonlinear convection and the underlying forcing. This remarkable CT phenomena is expected in various contexts. Also the CT as a new notion precisely characterizes the conditional stability of the corresponding physical problems.

As another application we recall that in Geophysical flows the rotational forcing plays an important role. Indeed in [LT4] we show that the rotational forcing also supports the CT phenomena, therefore serves as a candidate forcing to prevent the finite time breakdown.

Once a critical threshold is detected for a given system, the scaling limit in the different regime can be further justified. We recall here a result obtained in [LT5], where the link of the Euler-Poisson equation and the quantum description of the physical process–Schödinger-Poisson equation is discussed. The semiclassical limit is justified in the regime where the corresponding Euler-Poisson equation has global smooth solution.

Consider the nonlinear Schrödinger-Poisson (NLSP) equation,

$$i\epsilon\psi_t^\epsilon = -\frac{\epsilon^2}{2}\Delta_x\psi^\epsilon - k\Big(\Delta_x^{-1}(|\psi^\epsilon|^2 - c)\Big)\psi^\epsilon, \quad x \in \mathbb{R}^n, \quad t \in \mathbb{R}^+, \tag{1}$$

subject to the initial condition

$$\psi_0^\epsilon(x) = \sqrt{\rho_0(x)}e^{iS_0(x)/\epsilon}, \quad x \in \mathbb{R}^n. \tag{2}$$

Here, $\psi^\epsilon(\cdot, t)$ is a complex-valued wavefunction depending on the scaled Planck constant $\epsilon \sim \hbar$, with k and $c > 0$ being scaled physical constants.

The NLSP equation (1) has been studied in different contexts, and in particular, as the fundamental equation in semiconductors applications. We refer the reader to the recent review [GLM] and references therein.

Indeed, introducing S^ϵ as the phase of the wave function

$$\psi^\epsilon = \sqrt{\rho^\epsilon(x,t)}e^{iS^\epsilon(x,t)/\epsilon}$$

and separating real and imaginary parts in the NLSP equation (1), one obtains the following irrotational flow equations for the density-velocity pair, $(\rho^\epsilon, u^\epsilon := \nabla_x S^\epsilon)$, consult e.g., [LL]

$$\rho_t^\epsilon + \nabla \cdot (\rho^\epsilon u^\epsilon) = 0, \tag{3}$$

$$u_t^\epsilon + u^\epsilon \cdot \nabla u^\epsilon = k\nabla \Delta_x^{-1}(\rho^\epsilon - c) + \frac{\epsilon^2}{2}\nabla\left[\frac{\Delta\sqrt{\rho^\epsilon}}{\sqrt{\rho^\epsilon}}\right]. \tag{4}$$

Ignoring $O(\epsilon^2)$ term on the right of (4) for small ϵ one then obtains the Euler-Poisson equations with nonzero background, whose 1D version is exactly (6)-(8) we met before.

The above argument is, of course, only formal. The challenging issue here is to justify such a dispersive limit. The asymptotics of observable quantities as $\epsilon \to 0$ is known as 'semi-classical', expressing the passage from quantum to Newton mechanics, where the time and distance scales become large enough relative to Planck's constant. The 'semi-classical' limit problem associated with the NLSP equation (1) can be summarized as follows. We consider the family of solutions for the NLSP Cauchy problem (1), $\{\psi^\epsilon, \epsilon > 0\}$, subject to the WKB-type initial data (2), where ρ_0 and S_0 are given smooth functions. The purpose is to determine the behavior of the wavefunction ψ^ϵ as $\epsilon \to 0$, through the dynamics governing semi-classical limits of its observables, the density, $\rho = \lim |\psi^\epsilon|^2$, and the velocity $u = \lim \epsilon\nabla_x \arg\psi^\epsilon$.

Theorem 5.1 *[LT5]* **(Global convergence)**
Consider the 1-dimensional NLSP equation (1) ($n = 1$) subject to WKB initial data (2), $\psi_0^\epsilon(x) = \sqrt{\rho_0(x)}exp\{iS_0(x)/\epsilon\}$, with $(\rho_0(x), S_0(x))$ bounded in $(H^2(\mathrm{IR}), H^4(\mathrm{IR}))$. Assume that the initial data satisfy the sub-critical condition

$$|\partial_x^2 S_0(\alpha)| < \sqrt{k(2\rho_0(\alpha) - c)}, \quad \alpha \in \mathrm{IR}.$$

Then, for ϵ small enough, (1) admits a global solution of the form

$$\psi^\epsilon = \sqrt{\rho^\epsilon(x,t)}exp\{iS^\epsilon(x,t)/\epsilon\}$$

for all $t < Const.ln\frac{1}{\epsilon}$, with the following semi-classical limit, $(\rho^\epsilon, u^\epsilon := S_x^\epsilon) \longrightarrow (\rho, u)$ where (ρ, u) is the solution of the Euler-Poisson equations (6)-(8). Moreover, the following error estimate holds

$$||\rho^\epsilon(\cdot,t) - \rho(\cdot,t)||_{H_2} + ||u^\epsilon(\cdot,t) - u(\cdot,t)||_{H_1} \le Const_T \cdot \epsilon, \qquad t \le T \sim \ln\frac{1}{\epsilon}.$$

ACKNOWLEDGMENTS

Research was supported in part by ONR Grant No. N00014-91-J-1076 (ET) and by NSF grant #DMS01-07917 (ET, HL).

References

[BKM] J.T. BEALE, T. KATO AND A. MAJDA, *Remarks on the breakdown of smooth solutions for the 3D Euler equations*, Commun. Math. Phys., **94** (1984), 61.

[CLM] P. CONSTANTIN, P.D. LAX AND A. MAJDA, *A simple one-dimensional model for the three-dimensional vorticity equation*, Comm. Pure Appl. Math., **38** (1985), 715-724.

[ELT] S. ENGELBERG, H. LIU AND E. TADMOR, *Critical Thresholds in Euler-Poisson Equations* , Indiana Univ. Math. J., **50**, (2001), 109–157.

[GLM] I. GASSER, C.-K LIN AND P.A. MARKOWICH, *A review of dispersive limits of (non)linear Schrödinger-type equations*, preprint, 2000.

[KN] S. KAWASHIMA AND S. NISHIBATA, *Shock waves for a model system of the radiating gas*, SIAM J. math. Anal. **30** (1999), 95–117.

[LL] L.D. Landau and E.M. Lifschitz, *Course of Theoretical Physics. Quantum Mechanics : Nonrelativistic Theory*, vol. III, Oxford, Butterworth-Heinemann, 1976.

[LT1] H. LIU AND E. TADMOR, *Critical thresholds in a convolution model for nonlinear conservation laws*, SIAM J. Math. Anal. **33** (2001), 930–945.

[LT2] H. LIU AND E. TADMOR, *Spectral dynamics of velocity gradient field in restricted flows*, to appear in Commun. Math. Phys.

[LT3] H. LIU AND E. TADMOR, *Critical Thresholds in 2D Restricted Euler-Poisson Equations*, submitted, 2002.

[LT4] H. LIU AND E. TADMOR, in preparation.

[LT5] H. LIU AND E. TADMOR, *Semiclassical Limit of the Nonlinear Schrödinger-Poisson Equation with Subcritical Initial Data*, submitted, 2002.

[M] A. MAJDA, *Vorticity and the mathematical theory of incompressible fluid flow*, Comm. Pure Appl. Math., **39** (1984), 187.

[Ro] P. ROSENAU, *Extending hydrodynamics via the regularization of the Chapman-Enskog expansion,* Phys. Rev. A, **40** (1989), 7193-6.

[ST] S. SCHOCHET AND E. TADMOR, *Regularized Chapman-Enskog expansion for scalar conservation laws*, Arch. Rational Mech. Anal. **119** (1992), 95-107.

[Vi] P. VIEILLEFOSSE, *Local interaction between vorticity and shear in a perfect incompressible flow*, J. Phys. (Paris) **43**, (1982), 837.

High Order Asymptotically Strong-Stability-Preserving Methods for Hyperbolic Systems with Stiff Relaxation

Lorenzo Pareschi[1] and Giovanni Russo[2]

[1] Department of Mathematics, University of Ferrara, Via Machiavelli 35, 44100 Ferrara, Italy. *pareschi@dm.unife.it*

[2] Department of Mathematics and Computer Science, University of Catania, Via A.Doria 6, 95125 Catania, Italy. *russo@dmi.unict.it*

Summary. In this note we report some recent result on the time discretization of hyperbolic systems of conservation laws with stiff relaxation terms. The formalism of Implicit-Explicit (IMEX) Runge-Kutta methods is essential in the derivation and analysis of the schemes. Here we restrict to diagonally implicit schemes and consider the development of schemes up to order 3 that are asymptotic-preserving (AP) and strong-stability-preserving (SSP) for the limiting system of conservation laws.

1 Introduction

Several physical phenomena of great importance for applications are described by hyperbolic systems with relaxation terms, for example we mention discrete kinetic theory of rarefied gases, hydrodynamical models for semiconductors, viscoelasticity, linear and nonlinear waves, multiphase and phase transitions, radiations hydrodynamics, etc.

In one space dimension these systems have the form

$$\partial_t U + \partial_x F(U) = \frac{1}{\varepsilon} R(U), \quad x \in \mathbb{R}, \tag{1}$$

where $U = U(x,t) \in \mathbb{R}^N$, $F : \mathbb{R}^N \to \mathbb{R}^N$, the Jacobian matrix $F'(U)$ has real eigenvalues and admits a basis of eigenvectors $\forall U \in \mathbb{R}^N$ and $\varepsilon > 0$ is the relaxation parameter.

The development of efficient numerical schemes for such systems is challenging, since in many applications the relaxation time varies from values of order one to very small values if compared to the time scale determined by the characteristic speed of the system. In this second case the hyperbolic system with relaxation is said to be stiff and typically its solutions are well approximated by solutions of a suitable reduced set of conservation laws called equilibrium system [4]

The construction of schemes that work for all ranges of the relaxation time, using coarse grids that do not resolve the small relaxation time, has been studied mainly in the context of upwind methods using a method of line approach combined with suitable operator splitting techniques [3, 8] and more recently in the context of central schemes [9, 10].

In this note we will present some recent results on the development of high order, underresolved Runge-Kutta time discretization for such systems. In particular, using the formalism of implicit-explicit (IMEX) Runge-Kutta schemes [1, 2, 11] we derive schemes up to order 3 that are strong-stability-preserving (SSP) for the limiting system of conservation laws (such methods were originally referred to as total variation diminishing (TVD) methods, see [5, 6, 14]).

To this aim, we derive general conditions that guarantee the asymptotic preserving property, i.e. the consistency of the scheme with the equilibrium system and the asymptotic accuracy, i.e. the order of accuracy is maintained in the stiff limit.

2 IMEX Runge-Kutta schemes

An IMEX Runge-Kutta scheme consists of applying an implicit discretization to the source terms and an explicit one for the flux in the form

$$U^{(i)} = U^n - h\sum_{j=1}^{i-1} \tilde{a}_{ij}\partial_x F(U^{(j)}) + h\sum_{j=1}^{\nu} a_{ij}\frac{1}{\varepsilon}R(U^{(j)}), \tag{2}$$

$$U^{n+1} = U^n - h\sum_{i=1}^{\nu} \tilde{w}_i\partial_x F(U^{(i)}) + h\sum_{i=1}^{\nu} w_i\frac{1}{\varepsilon}R(U^{(i)}). \tag{3}$$

The matrices $\tilde{A} = (\tilde{a}_{ij})$, $\tilde{a}_{ij} = 0$ for $j \geq i$ and $A = (a_{ij})$ are $\nu \times \nu$ matrices such that the resulting scheme is explicit in F, and implicit in R. An IMEX Runge-Kutta scheme is characterized by these two matrices and the coefficient vectors $\tilde{w} = (\tilde{w}_1, \ldots, \tilde{w}_\nu)^T$, $w = (w_1, \ldots, w_\nu)^T$.

Since the simplicity and efficiency of solving the algebraic equations corresponding to the implicit part of the discretization at each step is of paramount importance, it is natural to consider diagonally implicit Runge-Kutta (DIRK) schemes [7] for the source terms ($a_{ij} = 0$, for $j > i$).

They can be represented by a double *tableau* in the usual Butcher notation,

$$\begin{array}{c|c} \tilde{c} & \tilde{A} \\ \hline & \tilde{w}^T \end{array} \qquad \begin{array}{c|c} c & A \\ \hline & w^T \end{array}$$

where the coefficients $\tilde{c}$ and c used for the treatment of non autonomous systems, are given by the usual relation

$$\tilde{c}_i = \sum_{j=1}^{i-1} \tilde{a}_{ij}, \quad c_i = \sum_{j=1}^{i} a_{ij}, \tag{4}$$

The use of a DIRK scheme for R is a sufficient condition to guarantee that F is always evaluated explicitly.

In the case of system (1), in order to obtain a numerical scheme a suitable space discretization of equations (2)-(3) is required. This discretization can be performed at the level of the original system (1) in which case one has a system of ODEs that is then solved in time by the IMEX scheme (method of lines). Alternatively one can apply a suitable space discretization directly to the time discrete equations (2)-(3).

In view of high order accuracy and considering that the source term does not contain derivatives and requires an implicit treatment, it appears natural to use finite-difference type space discretizations, rather than finite-volume ones. See, for example, [13] for high order finite volume and finite difference high order space discretizations, frequently used in the construction of shock capturing schemes for conservation laws.

2.1 Relation with previous schemes

As already observed, here we will restrict to diagonally implicit Runge-Kutta (DIRK) schemes. One step of such a scheme can be performed in splitting form as follows.

For $i = 1, \ldots, \nu$

1. Evaluate

$$U_*^{(i)} = U_0 - h \sum_{j=1}^{i-1} \tilde{a}_{ij} \partial_x F(U^{(j)}). \tag{5}$$

2. Solve for $U^{(i)}$

$$U^{(i)} = U_*^{(i)} + h \sum_{j=1}^{i-1} a_{ij} \frac{1}{\varepsilon} R(U^{(j)}) + h\, a_{ii} \frac{1}{\varepsilon} R(U^{(i)}). \tag{6}$$

Finally evaluate

$$U_1 = U_0 - h \sum_{i=1}^{\nu} \tilde{w}_i \partial_x F(U^{(i)}) + h \sum_{i=1}^{\nu} w_i \frac{1}{\varepsilon} R(U^{(i)}). \tag{7}$$

Note that, in many cases the implicit step (6) based on a DIRK solver for R can be solved explicitly for an hyperbolic system of the form (1). This is relevant for the simplicity and efficiency of the resulting schemes.

Similarly, every Runge-Kutta splitting scheme can be written as an IMEX Runge-Kutta scheme based on a DIRK solver for R.

We shall use the notation Name(s, σ, p), where the triplet (s, σ, p) characterizes the number s of stages of the implicit scheme, the number σ of stages of the explicit scheme and the order p of the scheme.

For example we report in Tables 1-6 the tableau corresponding to the simple splitting scheme SP(1,1,1) based on forward and backward Euler, Jin splitting Jin(2,2,2) [8] defined only in the stiff regime $h \gg \varepsilon$. We will also report the IMEX Midpoint(1,2,2) scheme in Table 3 which has the nice feature of being second order accurate in the nonstiff regime using only one evaluation of the stiff source term, and the IMEX Crank-Nicolson scheme CN(2,2,2) (see Table 4). Finally in Table 5-6 we give the Butcher tableau of the "uniformly" second order accurate Caflisch, Jin, Russo scheme CJR(3,2,2) [3] and Liotta, Romano, Russo scheme LRR(3,2,2) [9].

$$
\begin{array}{c|c}
0 & 0 \\
\hline
 & 1
\end{array}
\qquad
\begin{array}{c|c}
1 & 1 \\
\hline
 & 1
\end{array}
$$

Table 1. Tableau for the explicit (left) implicit (right) SP(1,1,1) scheme

$$
\begin{array}{c|cc}
0 & 0 & 0 \\
1 & 1 & 0 \\
\hline
 & 1/2 & 1/2
\end{array}
\qquad
\begin{array}{c|cc}
-1 & -1 & 0 \\
2 & 1 & 1 \\
\hline
 & 1/2 & 1/2
\end{array}
$$

Table 2. Tableau for the explicit (left) implicit (right) Jin(2,2,2) scheme

$$
\begin{array}{c|cc}
0 & 0 & 0 \\
1/2 & 1/2 & 0 \\
\hline
 & 0 & 1
\end{array}
\qquad
\begin{array}{c|cc}
0 & 0 & 0 \\
1/2 & 0 & 1/2 \\
\hline
 & 0 & 1
\end{array}
$$

Table 3. Tableau for the explicit (left) implicit (right) Midpoint(1,2,2) scheme

$$
\begin{array}{c|cc}
0 & 0 & 0 \\
1 & 1 & 0 \\
\hline
 & 1/2 & 1/2
\end{array}
\qquad
\begin{array}{c|cc}
0 & 0 & 0 \\
1 & 1/2 & 1/2 \\
\hline
 & 1/2 & 1/2
\end{array}
$$

Table 4. Tableau for the explicit (left) implicit (right) CN(2,2,2) scheme

$$
\begin{array}{c|cccc}
0 & 0 & 0 & 0 & 0\\
1/2\mu & 1/2\mu & 0 & 0 & 0\\
1/2\mu & 1/2\mu & 0 & 0 & 0\\
1 & 1-\mu & \mu & 0 & 0\\
\hline
1 & 1-\mu & \mu & 0 & 0
\end{array}
\qquad
\begin{array}{c|cccc}
0 & 0 & 0 & 0 & 0\\
0 & 0 & 0 & 0 & 0\\
1/2\mu & \gamma & 0 & \beta & 0\\
1 & \gamma\eta & 0 & \beta\eta & \mu\\
\hline
1 & \gamma\eta & 0 & \beta\eta & \mu
\end{array}
$$

$$
\beta = \frac{2\mu-1}{2(\mu-1)}, \gamma = -\frac{2\mu^2-2\mu+1}{2\mu(\mu-1)}, \eta = -2\mu(\mu-1), \quad 0<\mu<1/2.
$$

Table 5. Tableau for the explicit (left) implicit (right) CJR(3,2,2) scheme

$$
\begin{array}{c|cccc}
0 & 0 & 0 & 0 & 0\\
1/2 & 1/2 & 0 & 0 & 0\\
1/3 & 1/3 & 0 & 0 & 0\\
1 & 0 & 1 & 0 & 0\\
\hline
 & 0 & 1 & 0 & 0
\end{array}
\qquad
\begin{array}{c|cccc}
0 & 0 & 0 & 0 & 0\\
1/2 & 0 & 1/2 & 0 & 0\\
1/3 & 0 & 0 & 1/3 & 0\\
1 & 0 & 0 & 3/4 & 1/4\\
\hline
1 & 0 & 0 & 3/4 & 1/4
\end{array}
$$

Table 6. Tableau for the explicit (left) implicit (right) LRR(3,2,2) scheme

2.2 Order conditions

The general technique to derive order conditions for Runge-Kutta schemes is based on the Taylor expansion of the exact and numerical solution. In particular, conditions for schemes of order p are obtained by imposing that the solution of system (1) at time $t = t_0 + h$, with a given initial condition at time t_0, agrees with the numerical solution obtained by one step of a Runge-Kutta scheme with the same initial condition, up to order h^{p+1}.

Here we report the order conditions for IMEX Runge-Kutta schemes up to order $p = 3$, which is already considered high order for PDEs problems.

We apply scheme (2)-(3) to system (1), with $\varepsilon = 1$. We assume that the coefficients $\tilde{c}_i$, c_i, $\tilde{a}_{ij}$, a_{ij} satisfy conditions (4. Then the order conditions are the following

First order

$$
\sum_{i=1}^{\nu} \tilde{w}_i = 1, \quad \sum_{i=1}^{\nu} w_i = 1. \tag{8}
$$

Second order

$$
\sum_i \tilde{w}_i \tilde{c}_i = 1/2, \quad \sum_i w_i c_i = 1/2, \tag{9}
$$

$$
\sum_i \tilde{w}_i c_i = 1/2, \quad \sum_i w_i \tilde{c}_i = 1/2, \tag{10}
$$

Third order

$$\begin{aligned} &\sum_{ij} \tilde{w}_i \tilde{a}_{ij} \tilde{c}_j = 1/6, \quad \sum_{i} \tilde{w}_i \tilde{c}_i \tilde{c}_i = 1/3, \\ &\sum_{ij} w_i a_{ij} c_j = 1/6, \quad \sum_{i} w_i c_i c_i = 1/3, \end{aligned} \tag{11}$$

$$\begin{aligned} &\textstyle\sum_{ij} \tilde{w}_i \tilde{a}_{ij} c_j = 1/6, \quad \sum_{ij} \tilde{w}_i a_{ij} \tilde{c}_j = 1/6, \quad \sum_{ij} \tilde{w}_i a_{ij} c_j = 1/6, \\ &\textstyle\sum_{ij} w_i \tilde{a}_{ij} c_j = 1/6, \quad \sum_{ij} w_i a_{ij} \tilde{c}_j = 1/6, \quad \sum_{ij} w_i \tilde{a}_{ij} \tilde{c}_j = 1/6, \\ &\textstyle\sum_{i} \tilde{w}_i c_i c_i = 1/3, \quad \sum_{i} \tilde{w}_i \tilde{c}_i c_i = 1/3, \\ &\textstyle\sum_{i} w_i \tilde{c}_i \tilde{c}_i = 1/3, \quad \sum_{i} w_i \tilde{c}_i c_i = 1/3. \end{aligned} \tag{12}$$

Conditions (8), (9), (11) are the standard order conditions for the two *tableau*, each of them taken separately. Conditions (10) and (12) are new conditions that arise because of the coupling of the two schemes.

The order conditions will simplify a lot if $\tilde{c} = c$. For this reason only such schemes are considered in [1]. In particular, we observe that, if the two tableau differ only for the value of the matrices A, i.e. if $\tilde{c}_i = c_i$ and $\tilde{w}_i = w_i$, then the standard order conditions for the two schemes are enough to ensure that the combined scheme is third order. Note, however, that this is true only for schemes up to third order.

3 Asymptotic behavior

In this section we give sufficient conditions for asymptotic preserving and asymptotic accuracy properties of IMEX schemes. This properties are strongly related to $L-$stability of the implicit part of the scheme.

3.1 Zero relaxation limit

Let us consider here one-dimensional hyperbolic systems with relaxation of the form (1). The operator $R : \mathbb{R}^N \to \mathbb{R}^N$ is said a relaxation operator, and consequently (1) defines a relaxation system, if there exists a constant $n \times N$ matrix Q with $\text{rank}(Q) = n < N$ such that

$$QR(U) = 0 \quad \forall\, U \in \mathbb{R}^N. \tag{13}$$

This gives n independent conserved quantities $u = QU$. Moreover we assume that equation $R(U) = 0$ can be uniquely solved in terms of u, i.e.

$$U = \mathrm{e}(u) \text{ such that } R(\mathrm{e}(u)) = 0. \tag{14}$$

The image of e represents the manifold of local equilibria of the relaxation operator R.

Using (13) in (1) we obtain a system of n conservation laws which is satisfied by every solution of (1)

$$\partial_t(QU) + \partial_x(QF(U)) = 0. \tag{15}$$

For vanishingly small values of the relaxation parameter ε from (1) we get $R(U) = 0$ which by (14) implies $U = \mathrm{e}(u)$. In this case system (1) is well approximated by the equilibrium system [4].

$$\partial_t u + \partial_x G(u) = 0, \tag{16}$$

where $G(u) = QF(\mathrm{e}(u))$.

3.2 Asymptotic preserving and asymptotically accurate schemes

We start with the following

Definition 1. *We say that an IMEX scheme for system (1) in the form (2)-(3) is* asymptotic preserving *(AP) if in the limit $\epsilon \to 0$ the scheme becomes a consistent discretization of the limit equation (16).*

Note that this definition does not imply that the scheme preserves the order of accuracy in t in the stiff limit $\epsilon \to 0$. In the latter case the scheme is said *asymptotically accurate.*

For example let us consider the results of a direct consistency analysis of the schemes of the previous section. Scheme SP(1,1,1), shown in Table 1, is clearly AP. Scheme Jin(2,2,2) is AP, but it is not uniformly valid in ε, since the absolute-stability region of the implicit part of the scheme has a hole in the middle of the complex half plane $\Re z < 0$. Schemes Midpoint(1,2,2) and CN(2,2,2) are not AP even if both implicit parts of the schemes are A-stable [3]. On the contrary, schemes CJR(3,2,2) and LRR(3,2,2) are AP and uniformly valid in ε, but only scheme LRR(3,2,2) preserves the second order accuracy in the stiff limit.

In order to give sufficient conditions for the AP and asymptotically accurate property, we make use of the following simple

Lemma 1. *If all diagonal element of the triangular coefficient matrix A that characterize the DIRK scheme are non zero, then*

$$\lim_{\epsilon \to 0} R(U^{(i)}) = 0.$$

Proof. In the limit $\epsilon \to 0$ from (6) we have

$$\sum_{j=1}^{i} a_{ij} R(U^{(j)}) = 0, \quad i = 1, \dots, \nu.$$

Since the matrix A is non-singular, this implies $R(U^{(i)}) = 0$, $i = 1, \dots, \nu$.

In order to apply the previous Lemma, the vectors of c and $\tilde{c}$ cannot be equal. In fact $\tilde{c}_1 = 0$ whereas $c_1 \neq 0$. The condition $c = \tilde{c}$ is commonly used by several authors [1] since it simplifies the analysis of the schemes. Note that if $c_1 = 0$ but $a_{ii} \neq 0$ for $i > 1$ then we still have $\lim_{\epsilon \to 0} R(U^{(i)}) = 0$ for $i > 1$ but $\lim_{\epsilon \to 0} R(U^{(1)}) \neq 0$ in general. The corresponding scheme may be inaccurate if the initial condition is not "well prepared" ($R(U_0) \neq 0$). In this case the scheme is not able to treat the so called "initial layer" problem and degradation of accuracy in the stiff limit is expected [3, 11].

Next we can state

Theorem 1. *If* $\det A \neq 0$, *then in the limit* $\epsilon \to 0$, *the IMEX scheme (2)-(3) applied to system (1) becomes the explicit RK scheme characterized by* $(\tilde{A}, \tilde{w}, \tilde{c})$ *applied to the limit equation (16).*

Proof. As in the case of the continuous system, multiplying equations (2) and (3) by the matrix Q, and making use of the relation $QR(U) = 0 \, \forall \, U$, we obtain

$$u^{(i)} = u_0 - h \sum_{j=1}^{i-1} \tilde{a}_{ij} \partial_x QF(U^{(j)}), \tag{17}$$

and

$$u_1 = u_0 - h \sum_{i=1}^{\nu} \tilde{w}_i \partial_x QF(U^{(i)}), \tag{18}$$

where

$$u^{(i)} = QU^{(i)}, \; i = 1, \dots \nu, \quad u_0 = QU_0, \quad u_1 = QU_1.$$

¿From the previous lemma, in the stiff limit one has

$$R(U^{(i)}) = 0, \; i = 1, \dots \nu,$$

and, from property (14) of the relaxation operator R, this implies

$$U^{(i)} = \mathrm{e}(u^{(i)}), \; i = 1, \dots \nu.$$

Substituting this expression in (17)-(18) one obtains

$$u^{(i)} = u_0 - h \sum_{j=1}^{i-1} \tilde{a}_{ij} \partial_x G(u^{(j)}), \tag{19}$$

and

$$u_1 = u_0 - h \sum_{i=1}^{\nu} \tilde{w}_i \partial_x G(u^{(i)}), \tag{20}$$

where $G(u) = \partial_x QF(\mathrm{e}(u))$.

Clearly one may claim that if the implicit part of the IMEX scheme is A-stable or L-stable the previous theorem is satisfied. Note however that this is true only if the tableau of the implicit integrator does not contain any column of zeros that makes it reducible to a simpler A-stable or L-stable form.

Finally we observe that this result does not guarantee the accuracy of the solution for the $N-n$ non conserved quantities. In fact, since the very last step in the scheme it is not a projection towards the local equilibrium, a final layer effect occurs. It is easy to show that [12]

Corollary 1. *If* $\det A \neq 0$ *and* $w_j = a_{\nu j}$, $j = 1, \dots \nu$ *then in the limit* $\epsilon \to 0$, *the IMEX scheme (2)-(3) is* asymptotically accurate, *that is it provides the order of accuracy of the explicit RK scheme characterized by* $(\tilde{A}, \tilde{w}, \tilde{c})$ *for both conserved and non conserved variables*

We recall that the additional condition $w_j = a_{\nu j}$, $j = 1, \dots \nu$ makes an $A-$stable method $L-$stable. Usually these methods are referred to as *stiffly accurate* [7].

3.3 Asymptotically SSP schemes

Theorem 1 guarantees that in the stiff limit the numerical scheme becomes the explicit RK scheme applied to the equilibrium system, and therefore the order of accuracy of the limiting scheme is greater or equal to the order of accuracy of the original IMEX scheme. In particular this implies that if the explicit part of the IMEX scheme is SSP then, in the stiff limit, we will obtain an SSP method for the limiting conservation law. This property is essential to avoid spurious oscillations in the limit scheme for equation (16). We recall that if U^n represents a vector of solution values (for example obtained from a method of line approach in solving (16)) we recall the following [6]

Definition 2. *A sequence* $\{U^n\}$ *is said to be* strongly stable *in a given norm* $||\cdot||$ *provided that* $||U^{n+1}|| \leq ||U^n||$ *for all* $n \geq 0$.

The most commonly used norms are the TV-norm and the infinity norm.

In this section we give the Butcher tableau for second and third order IMEX schemes that satisfy the conditions of theorem 1. In all these schemes the implicit tableau corresponds to an $L-$stable scheme, that is $w^T A^{-1} e = 1$, e being a vector whose components are all equal to 1 [7], whereas the explicit tableau is SSPk, where k denotes the order of the SSP scheme. We refer to [12] for some application of these schemes to hyperbolic systems with relaxation. Several examples of asymptotically SSP schemes are reported in tables (7)-(11).

$$
\begin{array}{c|cc}
0 & 0 & 0 \\
1 & 1 & 0 \\
\hline
 & 1/2 & 1/2
\end{array}
\qquad
\begin{array}{c|cc}
\gamma & \gamma & 0 \\
1-\gamma & 1-2\gamma & \gamma \\
\hline
 & 1/2 & 1/2
\end{array}
\qquad
\gamma = 1 - \frac{1}{\sqrt{2}}
$$

Table 7. Tableau for the explicit (left) implicit (right) IMEX-SSP2(2,2,2) L-stable scheme

$$
\begin{array}{c|ccc}
0 & 0 & 0 & 0 \\
0 & 0 & 0 & 0 \\
1 & 0 & 1 & 0 \\
\hline
 & 0 & 1/2 & 1/2
\end{array}
\qquad
\begin{array}{c|ccc}
1/2 & 1/2 & 0 & 0 \\
0 & -1/2 & 1/2 & 0 \\
1 & 0 & 1/2 & 1/2 \\
\hline
 & 0 & 1/2 & 1/2
\end{array}
$$

Table 8. Tableau for the explicit (left) implicit (right) IMEX-SSP2(3,2,2) stiffly accurate scheme

$$
\begin{array}{c|ccc}
0 & 0 & 0 & 0 \\
1/2 & 1/2 & 0 & 0 \\
1 & 1/2 & 1/2 & 0 \\
\hline
 & 1/3 & 1/3 & 1/3
\end{array}
\qquad
\begin{array}{c|ccc}
1/4 & 1/4 & 0 & 0 \\
1/4 & 0 & 1/4 & 0 \\
1 & 1/3 & 1/3 & 1/3 \\
\hline
 & 1/3 & 1/3 & 1/3
\end{array}
$$

Table 9. Tableau for the explicit (left) implicit (right) IMEX-SSP2(3,3,2) stiffly accurate scheme

$$
\begin{array}{c|ccc}
0 & 0 & 0 & 0 \\
1 & 1 & 0 & 0 \\
1/2 & 1/4 & 1/4 & 0 \\
\hline
 & 1/6 & 1/6 & 2/3
\end{array}
\qquad
\begin{array}{c|ccc}
\gamma & \gamma & 0 & 0 \\
1-\gamma & 1-2\gamma & \gamma & 0 \\
1/2 & 1/2-\gamma & 0 & \gamma \\
\hline
 & 1/6 & 1/6 & 2/3
\end{array}
\qquad
\gamma = 1 - \frac{1}{\sqrt{2}}
$$

Table 10. Tableau for the explicit (left) implicit (right) IMEX-SSP3(3,3,2) L-stable scheme

$$
\begin{array}{c|cccc}
0 & 0 & 0 & 0 & 0 \\
0 & 0 & 0 & 0 & 0 \\
1 & 0 & 1 & 0 & 0 \\
1/2 & 0 & 1/4 & 1/4 & 0 \\
\hline
 & 0 & 1/6 & 1/6 & 2/3
\end{array}
\qquad
\begin{array}{c|cccc}
\alpha & \alpha & 0 & 0 & 0 \\
0 & -\alpha & \alpha & 0 & 0 \\
1 & 0 & 1-\alpha & \alpha & 0 \\
1/2 & \beta & \eta & 1/2-\beta-\eta-\alpha & \alpha \\
\hline
 & 0 & 1/6 & 1/6 & 2/3
\end{array}
$$

$$\alpha = 0.24169426078821, \quad \beta = 0.06042356519705 \quad \eta = 0.12915286960590$$

Table 11. Tableau for the explicit (left) implicit (right) IMEX-SSP3(4,3,3) L-stable scheme

References

1. U. Asher, R. J. Spiteri, S. J. Ruuth, *Implicit-explicit Runge-Kutta methods for time dependent Partial Differential Equations*, Appl. Numer. Math. 25, (1997), pp. 151-167.
2. U. Asher, S. J. Ruuth, B. Wetton, *Implicit-explicit methods for time dependent PDE's*, SIAM J. Numer. Anal., 32, (1995) pp. 797-823.
3. R. E. Caflisch, S. Jin, G. Russo, *Uniformly accurate schemes for hyperbolic systems with relaxation*, SIAM J. Numer. Anal., 34, (1997) pp. 246-281.
4. G. Q. Chen, D. Levermore, T. P. Liu, *Hyperbolic conservations laws with stiff relaxation terms and entropy*, Comm. Pure Appl. Math., 47, (1994), pp. 787-830.
5. S. Gottlieb, C.-W. Shu, *Total Variation Diminishing Runge-Kutta schemes*, Math. Comp. 67 (1998) pp. 73-85.
6. S. Gottlieb, C.-W. Shu, E. Tadmor, *Strong-stability-preserving high order time discretization methods*, SIAM Review, 43 (2001) pp. 89-112.
7. E. Hairer, G. Wanner, *Solving ordinary differential equations, Vol.2 Stiff and differential-algebraic problems*, Springer-Verlag, New York (1987).
8. S. Jin, *Runge-Kutta methods for hyperbolic systems with stiff relaxation terms* J. Comput. Phys., 122, (1995) pp. 51-67.
9. S. F. Liotta, V. Romano, G. Russo, *Central schemes for balance laws of relaxation type*, SIAM J. Numer. Anal., 38, (2000), pp. 1337-1356.
10. L. Pareschi, *Central differencing based numerical schemes for hyperbolic conservation laws with stiff relaxation terms*, SIAM J. Num. Anal., 39, (2001), pp. 1395-1417.
11. L. Pareschi, G. Russo *Implicit-Explicit Runge-Kutta schemes for stiff systems of differential equations*, Recent Trends in Numerical Analysis, Edited by L.Brugnano and D.Trigiante, Vol. 3, pp.269-289 (2000).
12. L. Pareschi, G. Russo, *Implicit-explicit Runge-Kutta schemes and applications to hyperbolic systems with relaxation*, preprint (2002).
13. C.-W. Shu, *Essentially Non Oscillatory and Weighted Essentially NOn OScillatory Schemes for Hyperbolic Conservation Laws*, in Advanced numerical approximation of nonlinear hyperbolic equations, Lecture Notes in Mathematics, 1697, (2000).
14. R. J. Spiteri, S. J. Ruuth, *A new class of optimal strong-stability-preserving time discretization methods*, SIAM. J. Num. Anal. (to appear).

Contributed Talks

Some Results on the Boundary Control of Systems of Conservation Laws

Fabio Ancona[1], Alberto Bressan[2] and Giuseppe Maria Coclite[3]

[1] Dipartimento di Matematica and C.I.R.A.M., P.zza Porta S. Donato, n. 5, 40123 - Bologna, Italy, e-mail: *ancona@ciram3.ing.unibo.it*
[2] SISSA-ISAS, via Beirut 2-4, 34014 - Trieste, Italy, e-mail: *bressan@sissa.it*
[3] SISSA-ISAS, via Beirut 2-4, 34014 - Trieste, Italy, e-mail: *coclite@sissa.it*

1 Introduction

Consider an $n \times n$ system of conservation laws on a bounded interval

$$u_t + f(u)_x = 0 \qquad\qquad t \geq 0, \quad x \in]a, b[\,, \tag{1}$$

with the initial condition

$$u(0, x) = \varphi(x), \qquad a \leq x \leq b, \tag{2}$$

and a weak form of the Dirichlet boundary conditions

$$u_i(t, a) = \alpha_i(t), \qquad u_i(t, b) = \beta_i(t), \qquad t > 0 \tag{3}$$

(see [13, 14, 19] and reference therein for several weak formulations of (3)).

We want to study the effect of boundary conditions on the solution of (1) from the point of view of control theory. Namely, following the same approach adopted in [5] for scalar conservation laws, we take the initial data φ fixed, and, regarding the measurable maps α_i, β_i in (3) as *control functions*, we want to investigate the property of the *attainable set* for (1)-(2), i.e. of the set

$$\mathcal{A}(T) \doteq \Big\{ u(T, \cdot)\ ; \quad u \ \text{ is a sol. to } \ (1) - (3) \Big\}, \tag{4}$$

which consists of all profiles that can be attained at a fixed time $T > 0$, by entropy weak solutions of (1)-(3). For the definitions and the basic properties of weak solutions we refer to [8]. See also [17, 18, 1, 2] for results concerning the existence and $\mathbf{L}^1$ stability of entropy weak solutions of the mixed problem taking values in the space BV of functions with bounded variation, and [10, 4] for the case of $\mathbf{L}^\infty$ solutions (with possibly unbounded variation) of Temple class systems.

Throughout, we shall assume:

(H_1) the map $f : \Omega \longrightarrow \mathbf{R}^n$ is smooth and $\Omega \subset \mathbf{R}^n$ is open;
(H_2) the system (1) is strictly hyperbolic, i.e. the Jacobian matrix $Df(u)$ has n real and distinct eigenvalues $\lambda_1(u) < \ldots < \lambda_n(u)$, $u \in \Omega$ (with a corresponding basis of eigenvectors $\{r_1(u), \ldots, r_n(u)\}$);

(H_3) each characteristic field r_i is linearly degenerate or genuinely nonlinear in the sense of Lax [15];

(H_4) there exist $p \in \{1, ..., n\}$ and $c_0 > 0$ such that

$$\lambda_1(u) < ... < \lambda_p(u) \leq -c_0 < 0 < c_0 \leq \lambda_{p+1}(u) < ... < \lambda_n(u), \qquad (5)$$

for each $u \in \Omega$.

By (H_4), for a solution defined on the strip $t \geq 0$, $x \in]a, b[$, there will be $n - p$ characteristics entering at the boundary point $x = a$, and p characteristics entering at $x = b$. The initial-boundary value problem is thus well posed if we prescribe $n - p$ scalar conditions at $x = a$, and p scalar conditions at $x = b$ (see [16]).

Definition 1. *Given $\varphi \in L^1([a, b])$, $v \in \Omega$, and $T > 0$, we say that the problem (1)-(2) is exact controllable at time T to the state v if and only if there exist measurable maps α_i, β_i such that the solution of (1)-(3) satisfies*

$$u(T, \cdot) \equiv v, \qquad a.\ e.\ in\ [a, b].$$

Definition 2. *Given $\varphi \in L^1([a, b])$, $v \in \Omega$, we say that the problem (1)-(2) is asymptotic stabilizable near the state v if and only if there exist measurable maps α_i, β_i such that the solution of (1)-(3) satisfies*

$$u(t, \cdot) \longrightarrow v, \qquad in\ L^1([a, b]) \qquad as\ t \longrightarrow +\infty.$$

In this note we present some recent results obtained by the authors [3, 9] concerning both the problem of exact controllability and of asymptotic stabilization near a constant state. We will first consider the case of Temple systems [20], for which it was obtained a characterization of the corresponding attainable set (4) in terms of suitable Oleinik-type estimates, which is a natural extension of the results in [5, 6] concerning scalar conservation laws. For general nonlinear systems, one cannot expect such an exact controllability result. Indeed, even if all wave-fronts in the initial data exit from the interval $[a, b]$ within finite time, they can generate new waves by interacting among themselves. In turn (figure 4), further interactions can produce a sequence of wave-fronts remaining within the interval $[a, b]$ for all times $t > 0$. Therefore, the effect of the initial data on the solution $u(T, \cdot)$ may never be completely erased, no matter how large we choose the terminal time T. Hence, we will present a result concerning the asymptotic stabilization of a general system of conservation laws near a constant state. Finally, we discuss a counterexample to the exact controllability concerning a class of 2×2 systems for which, in general, a constant state u^* cannot be attained, in a finite time T.

An outline of these results established in [3, 9] is given in the following sections.

2 The Attainable Set for Temple Class Systems

Our first result is concerned with the problem of exact controllability of Temple class systems [19, 20], which are systems that satisfy the following additional assumption.

(H_5) There exists a complete set of Riemann coordinates $w = (w_1, \dots, w_n)$ such that each level set $\{u\,;\;\; w_i(u) = \text{constant}\}$ is an hyperplane.

As a consequence, all integral curves of the eigenvectors are straight lines and coincide with the Hugoniot curves. We shall also assume

(H_6) as w ranges within the product set $\Gamma \doteq [w_1^-, w_1^+] \times \cdots \times [w_n^-, w_n^+]$, the corresponding state u remains inside the domain Ω and each characteristic field is genuinely nonlinear.

In the case of systems of this type, the well-posedness theory for the mixed problem was established in [4] within domains of $\mathbf{L}^\infty$ functions (with possibly unbounded variation). Here, the boundary condition is formulated in terms of the strong $\mathbf{L}^1$ trace of the solution u at the boundary and, in the same spirit of [13], is based on the definition of a time-dependent set of *admissible boundary data*, that is related to the notion of Riemann problem. Moreover, for sake of uniqueness, it was introduced in [3, 4] a definition of *entropy admissible weak solution* to the mixed problem that includes an entropy admissibility condition of Oleinik type.

Notice that, for such systems, wave interactions can only change the speed of wave fronts, without modifying their amplitudes. Therefore, the only restriction to boundary controllability is the decay due to genuine nonlinearity. We thus consider a set of maps, defined in terms of the above Riemann coordinates, that satisfy certain Oleinik-type conditions on the decay of positive waves. Call K^ρ, $\rho > 0$, the set of the maps $\psi \in \mathbf{L}^\infty([a,b], \Gamma)$ such that

$$\begin{aligned} \frac{w_i(\psi(y)) - w_i(\psi(x))}{y-x} &\le \frac{\rho}{x-a} \qquad \text{if } i \in \{p+1, ..., n\}, \\ \frac{w_i(\psi(y)) - w_i(\psi(x))}{y-x} &\le \frac{\rho}{b-y} \qquad \text{if } i \in \{1, ..., p\}, \end{aligned} \tag{6}$$

for a.e. $a < x < y < b$. The inequalities in (6) reflect the fact that positive waves entering through the boundary at a or at b decay in time. Therefore, their density is inversely proportional to their distance from their entrance point on the boundary.

We can now state our first main result (see [3]).

Theorem 1. *Let (1) be a system of Temple class, and assume that* (H_1)-(H_6) *are verified. Then, letting $\mathcal{A}(T)$ be the attainable set defined in (4) (in which the solution is understood as an "entropy admissible weak solution"), the following hold:*

(i) for every fixed $\overline{\tau} > 0$*, there exists* $\rho = \rho(\overline{\tau}) > 0$ *such that*

$$\mathcal{A}(\tau) \subseteq K^{\rho}, \qquad \tau \geq \overline{\tau}; \tag{7}$$

(ii) there exist $T > 0$ *and* $\rho' < \rho(T)$*, such that*

$$K^{\rho'} \subseteq \mathcal{A}(\tau), \qquad \tau > T; \tag{8}$$

(iii) $\mathcal{A}(T)$ *is a compact subset of* $L^1([a,b])$ *for each* $T > 0$.

The first property (i) is an immediate consequence of the definition of entropy admissible solution that satisfies suitable Oleinik-type estimates.

The proof of (ii) is established in two steps.
1) Backward Construction of Front Tracking Solutions. We take

$$\tau > T \doteq 4\frac{b-a}{\lambda^{\min}}, \qquad \lambda^{\min} \doteq \min_i |\lambda_i|, \tag{9}$$

and, for a given function $\psi \in K^{\rho'}$, we construct a sequence of approximate solutions u^{ν} on the strip $[0,\tau] \times [a,b]$ such that

$$\begin{aligned} u_{\nu}(\tau,\cdot) &= \psi_{\nu} \qquad & \psi_{\nu} &\stackrel{\mathbf{L}^1}{\longrightarrow} \psi, \\ u_{\nu}(0,\cdot) &= \overline{u}_{\nu} \qquad & \overline{u}_{\nu} &\stackrel{\mathbf{L}^1}{\longrightarrow} \varphi, \end{aligned} \tag{10}$$

with the following procedure. We partition the strip $[0,\tau] \times [a,b]$ in three regions. On the rectangle $[(3/4)T, \tau] \times [a,b]$, starting from $t = \tau$, we construct backward in time the front tracking solution u_{ν} relying on the fact that the Oleinik estimates of the definition (6) of $K^{\rho'}$ guarantee that two rarefaction fronts of the same family never cross in Ω (see figure 1). The total number of wave-fronts in $u^{\nu}(t,\cdot)$ decreases as $t \downarrow (3/4)T$ whenever a (backward) front crosses the boundary points $x = a$, $x = b$. Therefore, since the maximum time taken by fronts of u^{ν} to cross the interval $[a,b]$ is $(b-a)/\lambda^{\min}$, the definition (9) of T guarantees that all the (backward) fronts of u^{ν} will hit the boundaries $x = a$, $x = b$ within some time $\tau' \in](3/4)T, \tau[$. Hence, there will be some constant state $\omega \in \Omega$ such that $u^{\nu}((3/4)T,\cdot) \equiv \omega$. We next define u_{ν} on the rectangle $[0, T/4] \times [a,b]$ as the restriction to $[0,T/4] \times [a,b]$ of the front tracking solution to the Cauchy problem for (1), with initial data

$$\overline{u}(x) = \begin{cases} \overline{u}^{\nu}(a+) & \text{if} \quad x < a, \\ \overline{u}^{\nu}(x) & \text{if} \quad a \leq x \leq b, \\ \overline{u}^{\nu}(b-) & \text{if} \quad x > b. \end{cases}$$

Since u^{ν} contains only fronts originated at the points of the segment $\{(0,x)\,;\, x \in [a,b]\}$, because of (9) these wave-fronts cross the whole interval $[a,b]$ and exit from the boundaries $x = a$, $x = b$ before time $T/4$. Hence, there will be

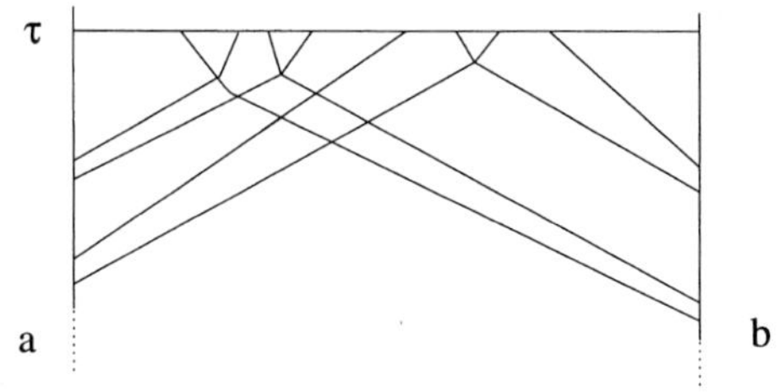

Fig. 1.

some state $\omega' \in \Omega$ such that $u^\nu(T/4, \cdot) \equiv \omega'$. Finally, we define $u^\nu(t, \cdot)$ for $t \in [T/4, (3/4)T]$ so that $u^\nu(T/4, \cdot) \equiv \omega'$, $u^\nu((3/4)T, \cdot) \equiv \omega$.

2) Convergence of Front Tracking Solutions. Since the sequence of approximate solutions constructed by our algorithm satisfy the Oleinik-type estimates on the decay of positive waves, they have a uniformly bounded variation on each interval of the type $[a + \varepsilon, b - \varepsilon]$, $\varepsilon > 0$. Thus, applying Helly's Theorem, and by a diagonal procedure, we can extract a subsequence that converges in $\mathbf{L}^1$ to a weak solution u of (1). We then extend a regularity property established by A.Bressan and P.G. LeFloch [11] for solutions with small total variation of general genuinely nonlinear systems, to the case of solutions with arbitrarily large variation of genuinely nonlinear Temple class systems.

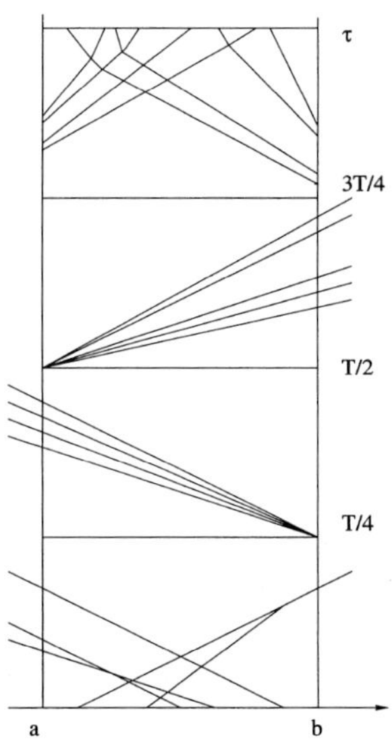

Fig. 2.

This property guarantees that, for Temple systems, solutions of the mixed problem (1)-(3) are continuous outsided a countable number of Lipschitz curves. As an immediate consequence we deduce that the solution u admits a strong $\mathbf{L}^1$ trace at the boundaries $x = a, x = b$ and satisfies the corresponding boundary conditions.

Concerning (iii), the compactness of $\mathcal{A}(T)$ is achieved with the same type of arguments used to establish the convergence of the approximate solutions in the proof of (ii).

3 Asymptotic Stabilization

We now consider a general $n \times n$ system and show that, starting with an initial data with small oscillations, the system can be asymptotically steered to any constant state (see [9, Theorem 1]).

Theorem 2. *Let K be a compact, connected subset of the open domain $\Omega \subset \mathbf{R}^n$ and assume that (1) satisfies $(H_1) - (H_4)$. Then there exist constants $C_0, \delta, \kappa > 0$ such that the following holds. For every constant state $u^* \in K$ and every initial data $\varphi : [a,b] \mapsto K$ with Tot.Var.$\{\varphi\} < \delta$, there exists an entropy weak solution $u = u(t,x)$ of (1)-(2) such that, for all $t > 0$,*

$$\textit{Tot.Var.}\{u(t,\cdot)\} \le C_0\, e^{-2^{\kappa t}}, \qquad \left\|u(t,\cdot) - u^*\right\|_{L^\infty} \le C_0\, e^{-2^{\kappa t}}. \tag{11}$$

The idea of the proof is as follows. Call $\lambda_* > 0$ a lower bound for the absolute value of all wave speeds and set $\tau \doteq (b-a)/\lambda_*$. In this way, all waves present in the solution at a given time t will exit through one the boundaries within time $t+\tau$. On the first interval $[0,\tau]$ we let all waves exit, arranging the boundary values at $x = a$ and at $x = b$ so that no reflected waves ever enter the domain $[a,b]$.

Therefore, the only waves present in the solution at time τ are those generated by interactions, in the interior of the interval $]a,b[$. They can be estimated as

$$\text{Tot.Var.}\{u(\tau,\cdot)\} \le C_0 \text{Tot.Var.}\{u(0,\cdot)\}^2.$$

The above estimate shows that the solution $u(\tau,\cdot)$ remains very close to some constant state, say $u^\dagger$. If $u^\dagger \neq u^*$, we suitably change the boundary conditions, producing new incoming waves at (τ, b), $(2\tau, a)$, achieving the bounds

$$\text{Tot.Var.}\{u(3\tau,\cdot)\} \le C_0 \text{Tot.Var.}\{u(0,\cdot)\}^2,$$

$$\left\|u(3\tau,\cdot) - u^*\right\|_{L^\infty} \le C_0 \left\|u(0,\cdot) - u^*\right\|_{L^\infty}^2.$$

Repeating inductively the same strategy in the time intervals $[3\tau, 6\tau]$, $[6\tau, 9\tau]$... we obtain the result.

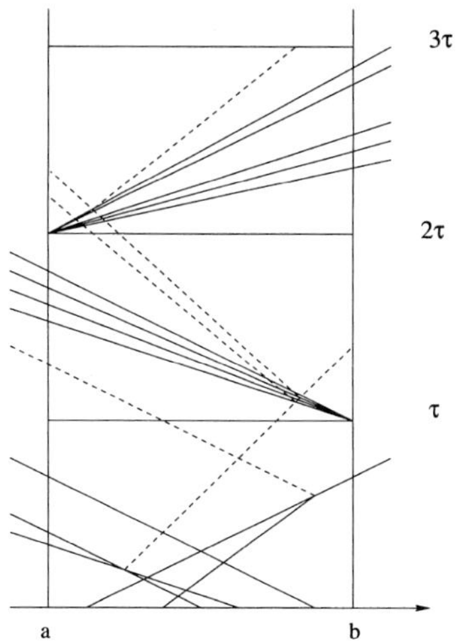

Fig. 3.

4 A Counterexample to Exact Controllability

An interesting question is whether one can reach exactly a constant state u^* within a finite time T. By Theorem 1, this is certainly the case for Temple class systems. In the final part of this paper, we show that this exact controllability cannot be attained in finite time, in general.

Our counterexample is concerned with a class of 2×2 strictly hyperbolic, genuinely nonlinear systems with the property that the interaction of two shocks of the same family generates a shock in the other family (see [9, Theorem 2]). This is the case for the system (see [12]):

$$\begin{cases} \rho_t + (u\rho)_x = 0, \\ u_t + \left(\dfrac{u^2}{2} + \dfrac{K^2}{\gamma - 1}\rho^{\gamma-1}\right)_x = 0. \end{cases} \tag{12}$$

with $1 < \gamma < 3$. Here $\rho > 0$ and u denote the density and the velocity of a gas, respectively.

Theorem 3. *Consider a 2×2 system of conservation laws satisfying (H_1) and the following*

(H_7) *there exist $0 < \lambda_* < \lambda^*$ such that*

$$-\lambda^* < \lambda_1(u) < -\lambda_* < 0 < \lambda_* < \lambda_2(u) < \lambda^*,$$

$$D\lambda_1 \cdot r_1 > 0, \qquad D\lambda_2 \cdot r_2 > 0,$$

$$r_1 \wedge r_2 < 0, \qquad r_1 \wedge (Dr_1 \cdot r_1) < 0, \qquad r_2 \wedge (Dr_2 \cdot r_2) < 0,$$

where $r_1(u), r_2(u)$ are the right eigenvectors of $Df(u)$.

Let $\varphi \in BV([a,b];\mathbf{R}^2)$ with small total variation and a dense set of shocks. Every entropic solution of (1)-(2) has a dense set of shocks in $u(t,\cdot)$, for each $t \geq 0$. In particular, $u(t,\cdot)$ cannot be a constant.

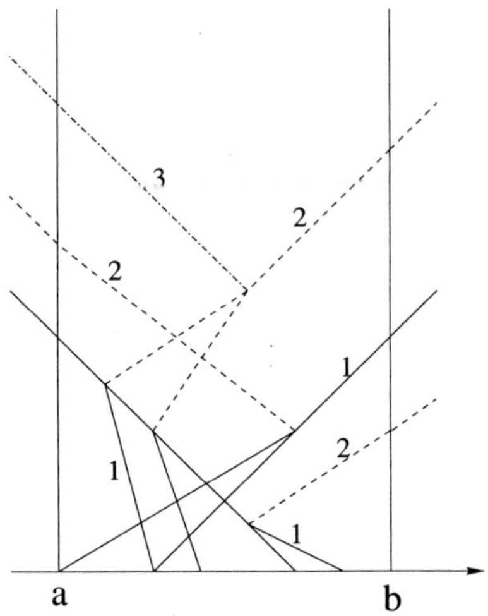

Fig. 4.

Toward a proof, we note that the geometric properties of the system imply that

a) the interaction of two shocks of the same family produces an outgoing shock of the other family,
b) the interaction of a shock with a rarefaction wave of the same family produces a rarefaction wave in the other family.

As in the scalar case we have (see [9, Section 3]).

Lemma 1. (Oleinik type estimate) *Let $u = u(t,x)$ be a solution of (1) with $n = 2$ and satisfying (H_1) and (H_7). There exist $k, \delta > 0$ such that, if*

$$Tot.\,Var.(u(t,\cdot)) < \delta,$$

then

$$\omega_i(t,y) - \omega_i(t,x) \leq \frac{k}{t} \cdot (y-x), \qquad x < y,\ t > 0,\ i = 1,2,$$

where ω_1, ω_2 are the Riemann coordinates associated to (1).

By the previous properties, a shock can never be completely canceled by interactions with rarefaction waves of the same family. Hence, it can only disappear by exiting from one of the boundaries $x = a$ or $x = b$. However, if the set of shocks at time $t = 0$ is everywhere dense, these shocks will interact among each other on a dense set of points in the domain $[a, b] \times [0, \infty[$, and give rise to a dense set of new shocks. One can arrange so that the total strength of these shocks quickly approaches zero, according to Theorem 3.1, but cannot become exactly zero within finite time. For all details we refer to [9, Section 3].

Remark 1. The previous analysis breaks down in the case of the $p-$system, because in this case the interaction of two shocks of the same family produces a centered rarefaction wave of the other family. In particular, an Oleinik type estimate cannot holds.

References

1. Amadori, D. (1997): Initial-boundary value problems for nonlinear systems of conservation laws. Nonlin. Diff. Equat. Appl., **4**, 1–42.
2. Amadori, D., Colombo, R.M. (1997): Continuous dependence for 2×2 conservation laws with boundary. J. Differential Equations, **138**, 229–266.
3. Ancona, F., Coclite, G.M. (2002): On the Attainable set for Temple Class Systems with Boundary Controls. Ref. S.I.S.S.A. 10/2002/M.
4. Ancona, F., Goatin, P. (2001): Uniqueness and Stability of $\mathbf{L}^\infty$ Solutions for Temple Class Systems with Boundary and Properties of the Attainable Sets. SIAM J. Math. Anal., to appear.
5. Ancona, F., Marson, A. (1998): On the attainable set for scalar non-linear conservation laws with boundary control. SIAM J. on Control and Optimization, **36**, 290–312.
6. Ancona, F., Marson, A. (1999): Scalar non-linear conservation laws with integrable boundary data. Nonlinear Anal., **35**, pp. 687–710.
7. Baiti, P., Bressan, A. (1997): The semigroup generated by a Temple class system with large data. Differ. Integ. Equat., **10**, 401–418.
8. Bressan, A. (2000): Hyperbolic Systems of Conservation Laws. The one dimensional Cauchy problem. Oxford Univ. Press.
9. Bressan, A., Coclite, G.M. (2001): On the Boundary Control of System of Conservation Laws. SIAM J. Control Optim., to appear.
10. Bressan, A., Goatin, P. (2000): Stability of $\mathbf{L}^\infty$ solutions of Temple class systems. Differ. Integ. Equat., **13**, 1503–1528.
11. Bressan, A., LeFloch, P.G. (1999): Structural stability and regularity of entropy solutions to hyperbolic systems of conservation laws. Indiana Univ. Math. J., **48**, 43–84.
12. DiPerna, R. (1973): Global solutions to a class of nonlinear hyperbolic systems of equations. Comm. Pure Appl. Math., **26** (1973), 1–28.
13. Dubois, F., LeFloch, P.G. (1988): Boundary conditions for non-linear hyperbolic systems of conservation laws. J. Differential Equations, **71**, 93–122.
14. Joseph, K.T., LeFloch, P.G. (1999): Boundary layers in weak solutions of hyperbolic conservation laws. Arch. Rational Mech. Anal., **147** , 47–88.

15. Lax, P.D. (1957): Hyperbolic systems of conservation laws II. Comm. Pure Appl. Math., **10** , 537–566.
16. Li, T., Yu, W. (1985): Boundary Value Problems for Quasilinear Hyperbolic Systems. Duke University Mathematics Series V.
17. Liu, T.P. (1977): Initial-boundary value problem for gas dynamics. Arch. Rational Mech. Anal., **64**, 137–168.
18. Sablé-Tougeron, M. (1993): Méthode de Glimm et probléme mixte. Ann. Inst. Henri Poincaré, **10**, 423–443.
19. Serre, D. (2000): Systems of Conservation Laws I, II. Cambridge University Press.
20. Temple, B. (1983): Systems of conservation laws with invariant submanifolds. Trans. Amer. Math. Soc., **280**, 781–795.

Dynamics of Collapsing Bubbles Near Walls

Sigrid Andreae[1], Josef Ballmann[1], Siegfried Müller[2], and Alexander Voß[2]

[1] Lehr- und Forschungsgebiet für Mechanik der RWTH Aachen, Germany
sigrid.andreae@lufmech.rwth-aachen.de, ballmann@lufmech.rwth-aachen.de

[2] Institut für Geometrie und Praktische Mathematik, RWTH Aachen, Germany
mueller@igpm.rwth-aachen.de, voss@igpm.rwth-aachen.de

Summary. The influence of an adjacent wall on a collapsing bubble is investigated numerically. For this purpose, the two phase fluid is modeled by a single equation of state using the van der Waals equation of state and applying the Maxwell construction in the mixture region. Thereby the possibility of phase transition is included. The numerical simulations are performed using an adaptive finite volume scheme where the grid refinement strategy is based on multiresolution techniques.

1 Introduction

The formation and collapse of vapor bubbles in a liquid is called cavitation. Lord Rayleigh discovered that pressure waves emitted during the process of cavitation [8] may damage solids, e.g., marine screw propellers. However, up to now little is known about the mechanisms causing the damage of the solid.

Cavitation is induced by a pressure drop in the liquid below vapor pressure. Such a pressure decrease may occur due to local acceleration of the liquid flow caused by geometrical constraints, e.g., if the liquid flows through a narrow orifice or around an obstacle. In case, the pressure drops below vapor pressure, the liquid bursts and creates a free surface filled with gas and vapor – the bubble. Due to changes in the flow field, the pressure in the liquid may increase again causing the bubble to collapse. The collapse is accompanied by strong shock and rarefaction waves running into the bubble and the surrounding liquid. The shock wave focuses in the center of the bubble. This leads to extreme physical states in the interior. In addition, the shrinking of the bubble leads to a compression of the vapor. Both effects evoke an increase of pressure which bulges the bubble. Hereby, a dynamic oscillation process is initiated which finally leads to the collapse of the bubble. If the collapse takes place next to a solid, the pressure distribution becomes asymmetric and a liquid jet develops [7] which is either directed towards or away from the solid boundary. The direction of the jet depends on the elasticity of the solid and on a ratio determined by the initial distance between the boundary and the center of the bubble divided by the maximum extension of the bubble. This has been shown experimentally by Brujan et al. in [1], [2].

In order to investigate experimentally the dynamics of a bubble collapse, the bubble is produced by a laser pulse. Thereby the fluid is heated in the focus of the laser and forms a small, hot gas bubble at very high temperature. This experimental setup provides an exact positioning of the bubble. The processes taking place in the interior of the collapsing and oscillating bubble and the prediction of onset and extent of the cavitation damaging are still subject of theoretical and experimental research. However, small time and space scales as well as the complicated dynamics make an experimental approach difficult. Therefore numerical investigations are needed to reveal information about the wave dynamics in the fluid as well as the damaging of the solid. Of particular interest are pressure contours and velocity vectors in the liquid phase as requested in [2].

The primary objective of the present work is to provide an accurate prediction of all occuring wave phenomena. This concerns waves running into the liquid and the center of the bubble as well as the interaction of these waves with a neighboring solid. Normally, a collapsing gas bubble in a liquid is modeled as a two phase flow problem which means for each fluid a different equation of state is used. This can be done by using two meshes, whereby each mesh has to be updated in every step. Other techniques are tracking the boundary using a level-set or a volume-of-fluid method. Consequently, the phase boundary is modeled as a mathematically sharp boundary without a transition zone which is not physical. Actually, there is a region where vapor as well as liquid may be present, as will be discussed in Sec. 2.2. To overcome that problem a homogeneous description of both phases with one equation of state is used. The equation of state has to provide a physically correct representation of the transition zone as well as liquid and gas phase, see Sec. 2.1. Since cavitation may occur in injection nozzles, fuels should be included as possible fluids. Some of them belong to the group of retrograde and even BZT (Bethe-Zel'dovich-Thompson) fluids which make certain demands on the equation of state and possibly generate additional nonclassical wave effects. For details on BZT fluids and nonclassical waves see Thompson [9] and Cramer [3], respectively, as well as the literature cited there. BZT fluids will be discussed in Sec. 2.3.

The small time scales of the unsteady problem require the numerical scheme to be highly efficient regarding computational time and memory requirements. This is realized by a local grid refinement strategy. Furthermore, all physical relevant phenomena have to be reliably detected and adequately resolved. Moreover, the scheme has to be robust and must not exhibit numerical oscillations, e.g., pressure oscillations at the phase boundary. The details of the numerical scheme are presented in Sec. 3.

In Sec. 4 numerical results for a bubble collapse near a rigid wall are presented and the arising dynamic wave patterns are discussed.

2 Physical Model

For our investigations the fluid is modeled by the van der Waals equation of state. This gives a qualitatively good representation of the fluid in various important regions, namely the regimes of high compression during the collapse of the bubble, the area with negative nonlinearity near the critical point on account of BZT fluids, and the mixture region to allow for phase transition.

2.1 Van der Waals Equation of State

A vital assumption of the ideal gas model is that the particles are non-interacting masspoints, i.e., an appreciable force acts on them only during a collision. Furthermore, the volume of the particles is negligible compared to the total volume occupied by the gas. This is no longer valid if the density is so high that the distance between two particles is of the order of their interaction diameter. Such extreme densities can be reached during the collapse of the bubble. In this case, the range of validity of the ideal gas is exceeded which is investigated in [4].

The van der Waals equation of state is an extension of the ideal gas model with two material parameters a and b to take into account the attraction of particles and the reduction of free volume. The thermal equation of state is determined by

$$p(v,T) = \frac{RT}{v-b} - \frac{a}{v^2} \tag{1}$$

with pressure p, temperature T, specific volume v, specific gas constant R, internal pressure a/v^2 and covolume b covolume. The internal pressure is subtracted from the pressure to take into account that the attraction of particles diminishes the pressure. The covolume b reduces the volume v to the available free volume $v - b$. The caloric equation of state reads

$$e(v,T) = e_0 + c_v(v,T)\,T - \frac{a}{v} \tag{2}$$

where c_v denotes the heat capacity at constant volume and e the internal energy.

2.2 Mixture Region

Since we want to apply the van der Waals equation of state to gases and liquids as well, there is a need for describing of what happens in the region of phase transition. In the p-v diagram in Fig. 1, three isotherms are plotted corresponding to (i) a temperature above the critical temperature (T_c), (ii) the critical temperature (the critical isotherm) and (iii) a temperature below

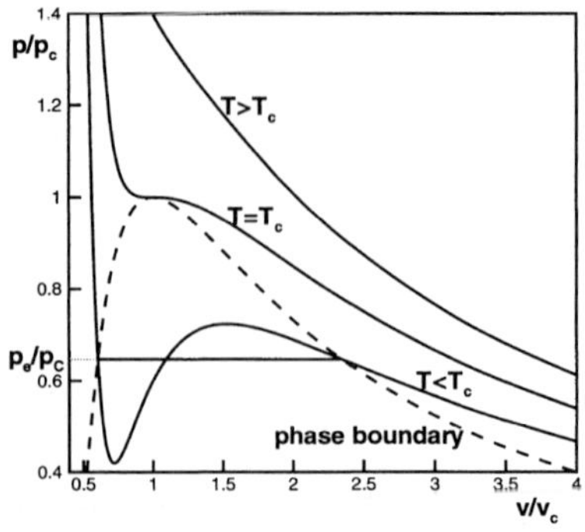

Fig. 1. Isotherms in p-v-diagram

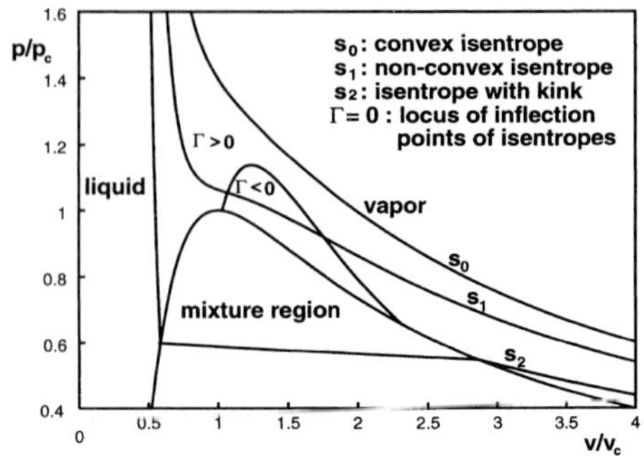

Fig. 2. Isentropes in p-v-diagram

T_c. In the mixture region, liquid and gas are present in a continuous changing fraction where the fraction of liquid rises when the volume decreases. In the mixture area, the van der Waals equation shows an unphysical inclination in the p-v-diagram. Considering an isotherm crossing the mixture region, we observe two local extrema called spinodal points. They correspond to the endpoints of supersaturated vapor (maximum) and overexpanded liquid (minimum), respectively. Between these two points the derivative $\frac{\partial p}{\partial v}|_T$ is positive and hence physically excluded. In order to provide physically meaningful data, we calculate the equilibrium pressure p_{e}, see Fig. 1, as a function of volume and temperature using the Maxwell construction. For details on the implementation see [10], [11].

2.3 BZT Fluids

BZT fluids are high molecular hydrocarbons or -fluorids. They are a subgroup of retrograde fluids which exhibit a characteristic overhang of the saturation curve in the T-s-diagram. Fig. 3 and 4 show the difference between a generic and a retrograde fluid, respectively. The T-s-diagram for the generic fluid shows that the fluid has to be cooled for an equilibrium condensation [9]. However, the overhang in the T-s-diagram for the retrograde fluid implies that the retrograde fluid has to be heated for an equilibrium condensation. For the BZT fluid the curvature of the saturation curve is so intense that it leads to a concave bending of the isentropes near the critical point in the p-v-plane, see Fig. 2. This is in contrast to the behavior of the most common fluids which isentropes are convex. A measure for the bending is the fundamental derivative of gas dynamics

$$\Gamma = \frac{v^3}{2c^2}\frac{\partial^2 p}{\partial v^2} \tag{3}$$

which becomes negative in the respective region. Here, c denotes the speed of sound. The gasdynamic peculiarity of the $\Gamma<0-$region implies the possibility

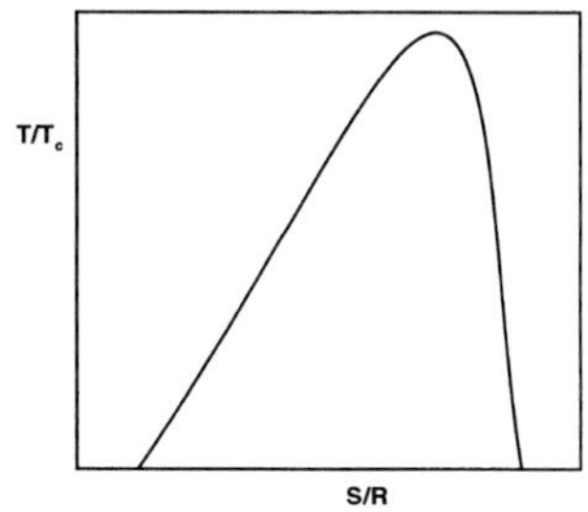

Fig. 3. T-s-diagram for C_2H_4

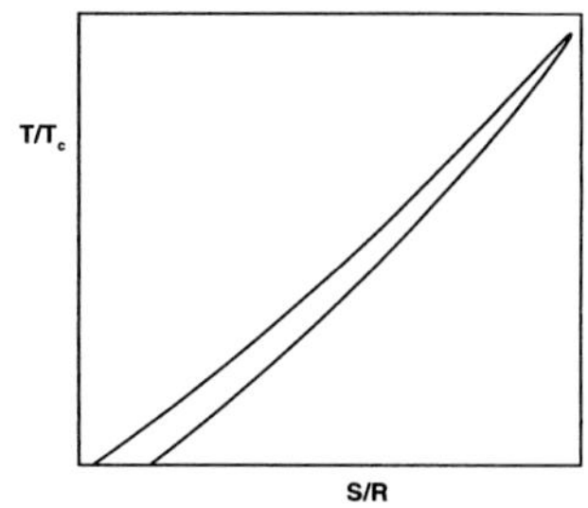

Fig. 4. T-s-diagram for PP_{10}

of *compression* waves and *expansion* shocks [3]. In the later case, the pressure jump has to be positive to fulfill the entropy condition. The chosen van der Waals equation of state provides a good representation of the effects in the region of negative Γ. For more details on BZT fluids see Thompson [9].

3 Governing Equations and Method of Solution

The fluid flow is modeled by the time-dependent 2D Euler equations for compressible fluids.

$$\frac{\partial}{\partial t}\int_V \mathbf{U}\ \mathbf{dV} + \oint_{\partial\mathbf{V}} (\mathbf{F^c} - \mathbf{F^v}) \cdot \mathbf{n}\ \mathbf{dS} = \mathbf{0} \quad \text{with} \tag{4}$$

$$\mathbf{U} = \begin{pmatrix} \rho \\ \rho\mathbf{u} \\ \rho E \end{pmatrix}, \qquad \mathbf{F^c} = \begin{pmatrix} \rho\mathbf{u} \\ \rho\mathbf{u} \circ \mathbf{u} + \mathbf{p}\mathbf{1} \\ \mathbf{u}(\rho\mathbf{E} + \mathbf{p}) \end{pmatrix}, \qquad \mathbf{F^v} = \begin{pmatrix} \mathbf{0} \\ \boldsymbol{\tau} \\ \mathbf{u} \cdot \boldsymbol{\tau} - \mathbf{q} \end{pmatrix}.$$

$\mathbf{U}$ is the array of the mean conserved quantities: density, momentum, and specific total energy. $\boldsymbol{\tau}$ is the viscous strain tensor and $\mathbf{q}$ the heat flux vector. The quantity V denotes a time-independent control volume with the boundary ∂V and the outer normal $\mathbf{n}$. The flux array is divided into its convective ($\mathbf{F^c}$) and diffusive ($\mathbf{F^v}$) part.

The transport equations are solved using an explicit finite volume method. For the convective terms a Godunov-type upwind scheme is applied. To improve the spatial accuracy we apply a higher order ENO reconstruction. The efficiency of the finite volume method is significantly improved by an adaptive grid refinement strategy which is based on multiresolution techniques that have been recently developed and investigated in [6].

Since BZT fluids exhibit wave phenomena different from common fluids, these have to be properly addressed by the numerical algorithms. For this purpose, the approximate Riemann solver due to Roe has to be modified with regard to a general equation of state, i.e., $p = p(v, e)$. The derivation is analogously to the approach of Liu and Vinokur, see [5]. As shown in [11] the numerical method is able to resolve expansion shocks as well as wave splitting phenomena.

4 Numerical Results

The focus of the current work is to verify the ability of the scheme to handle steep density gradients and to resolve the dynamic wave patterns accurately. Therefore, as a first step a generic fluid with initial conditions away from the phase transition is considered for the two-dimensional case. However, the initial conditions exhibit data differences which may be comparable to a phase boundary. Where a hot bubble at low pressure is enclosed by cold air at high pressure, see Fig. 5. The fluid is modeled by the van der Waals equation. Here

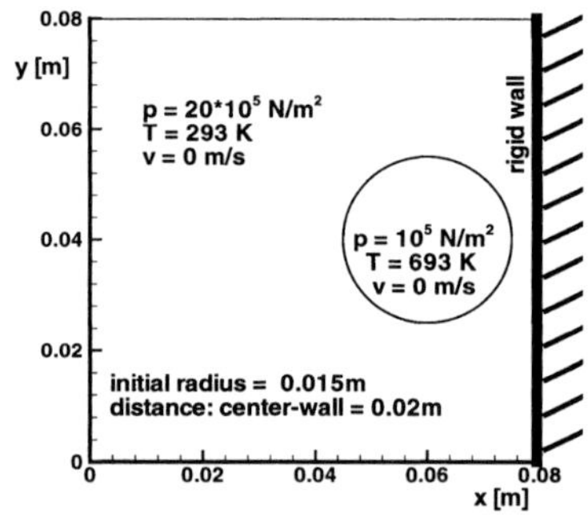

Fig. 5. Initial conditions and computational domain of the planar problem

we still assume constant heat capacity c_v, $\gamma = 1.4$ and $R = 287 J/mol/K$. The initial conditions correspond to a Riemann problem where three types of waves occur: an inward running compression shock, an outward running rarefaction wave and a contact discontinuity. At time $t = 0.01640ms$ the fastest wave - the inward running shock - has reached the center of the bubble and is reflected, see Fig. 7. The pressure in the focus exceeds $40 \times 10^6 N/m^2$. The outward running rarefaction is visible in the pressure as well as in the density. Whereas the contact discontinuity is only seen in the density (see Fig. 6 at 0.068m). It just reaches the solid wall. After the reflection in the center, the shock wave runs outward and passes the contact discontinuity. Hereby, the shock is partly transmitted and partly reflected. The reflected part runs again into the center, is reflected and passes the contact discontinuity again. The rarefaction wave is reflected at the solid wall as rarefaction. The corresponding doubling of the amplitude produces a region of low pressure which increases the shock pressure ratio and, therfore, accelerates the shock wave towards the wall, see Fig. 8 and 9, where it is reflected, see Fig. 11. The reflection area extends along the wall developing a Mach stem, see Fig. 12, at time $t = 0.14925ms$. In the density contours the slip line, a contact discontinuity originating at the triple point of the Mach stem, is clearly visible. The discontinuity separates two regions of different values of entropy that are caused in the material passing through shocks of different strength. The Mach stem The pressure distribution on the solid (right) and the density contours (left) are presented to display the loading jump on the solid evoked by the Mach stem (visible at $y = 0.074m$ and $y = 0.006m$). To identify the remainder of the bubble, Fig. 13 shows the pressure and density distribution on the symmetry line at $y = 0.04$ for the same time. The contact discontinuity is characterized by a jump in density and nearly constant pressure between $x = 0.062m$ and $x = 0.067m$.

As mentioned earlier, the initial conditions determine whether the bubble is repelled from or attracted towards the wall. In addition, the surrounding fluid is accelerated such that a jet of cold gas occurs in wall direction or opposite. In this configuration cold gas flows away from the wall, repells the hot bubble from the solid and finally splits it. This is observable by the velocity in x-direction and the integral curves of the velocity direction field which is marked by arrow-lines in Fig. 14 at time $t = 0.70417ms$. In the experiments [1] for rigid wall a jet directed versus the wall was observed. The opposite in this computation is caused by the discrepancy of the initial conditions concerning the velocity field on both sides of the bubble surface.

5 Conclusion and Future Work

The numerical method is able to handle the small time scales occuring during the collapse of a bubble due to the grid refinement strategy. Besides, it is quite stable even at steep density gradients. The interaction of the shock wave and the rarefaction wave as well as the reflection of both waves on the rigid wall are well resolved. In particular, the development of the Mach stem is traced accurately. The contact discontinuity exhibits instabilities which are due to the Cartesian grid.

In future work, the position of the bubble will be varied in order to compute the prediction of Brujan et al. [1] that the direction of the jet depends on the distance between boundary and bubble. Furthermore, initial states near the critical point will be considered to investigate the influence of retrograde behavior and effects caused by phase transition.

6 Acknowledgement

This work is financially supported by the DFG Priority Research Program "Analysis and Numerics for Conservation Laws".

References

1. Brujan, E.-A., Nahen, K., Schmidt, P., Vogel, A. (2001): Dynamics of laser-induced cavitation bubbles near an elastic boundary. J. Fluid Mech., **433**, 251–281
2. Brujan, E.-A., Nahen, K., Schmidt, P., Vogel, A. (2001): Dynamics of laser-induced cavitation bubbles near elastic boundaries: influence of the elastic modulus. J. Fluid Mech., **433**, 283–314
3. Cramer, M. S. (1991): Nonclassical dynamics of classical waves. In: Kluwick, A. (ed) Nonlinear Waves in Real Fluids. Springer Wien New York
4. Hanke, M., Ballmann, J. (1998): Strong changes of state in collapsing bubbles. ZAMM, **78**, 453–453
5. Liu Y. , Vinokur, M. (1989): Nonequilibrium flow computations. I. An analysis of numerical formulations of conservation laws. J. Comp. Phys., **83**, 373–397
6. Müller, S., (2002): Adaptive Multiscale Schemes for Conservation Laws. Accepted for publication in Lecture Notes on Computational Sciences and Engineering. Springer
7. Philipp, A., Lauterborn, W. (1998): Cavitation erosion by single laser-produced bubbles. J. Fluid Mech., **361**, 75–116
8. Lord Rayleigh (1917): On the pressure developed in a liquid during the collapse of a spherical cavity. Phil. Mag., **34**, 94–98
9. Thompson, P. A. (1991): Liquid-vapor adiabatic phase changes and related phenomena. In: Kluwick, A. (ed) Nonlinear Waves in Real Fluids. Springer Wien New York
10. Voß, A. (2002): A Numerical Library for Fluids and Convex and Nonconvex EOS Including Phase Transition. IGPM-report, RWTH Aachen, in preparation
11. Voß, A. : Flows in Fluids with Nonconvex Equation of State. PhD Thesis, RWTH Aachen, in preparation

7 Figures

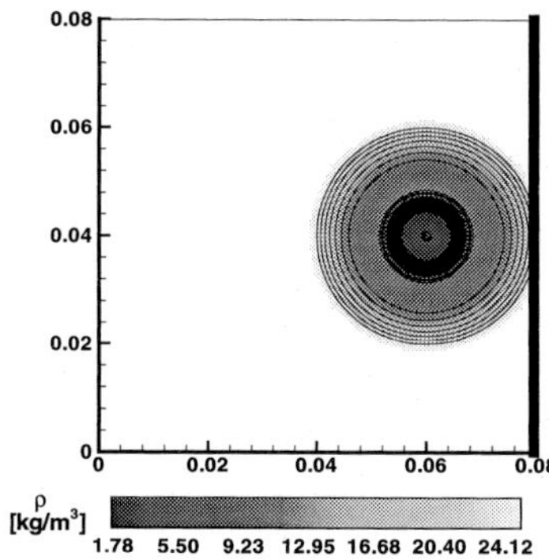

Fig. 6. $t = 0.01640ms$

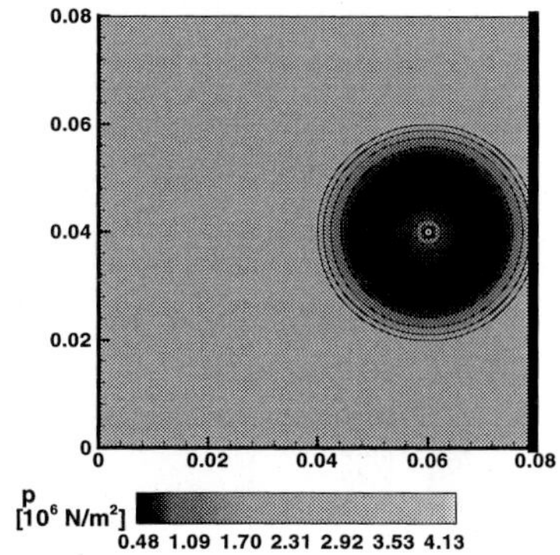

Fig. 7. $t = 0.01640ms$

Fig. 8. $t = 0.04851ms$

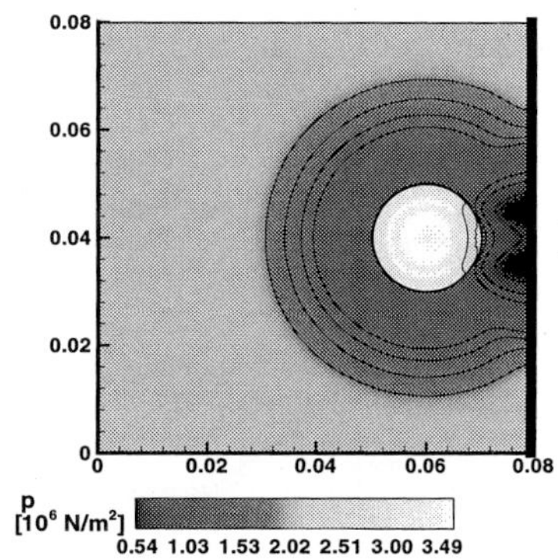

Fig. 9. $t = 0.04851ms$

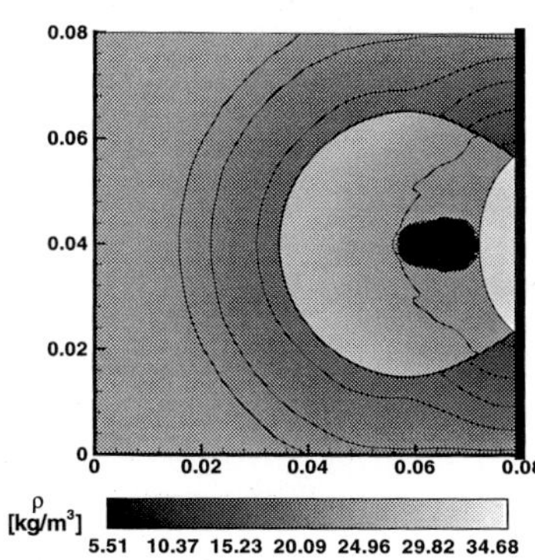

Fig. 10. $t = 0.10150ms$

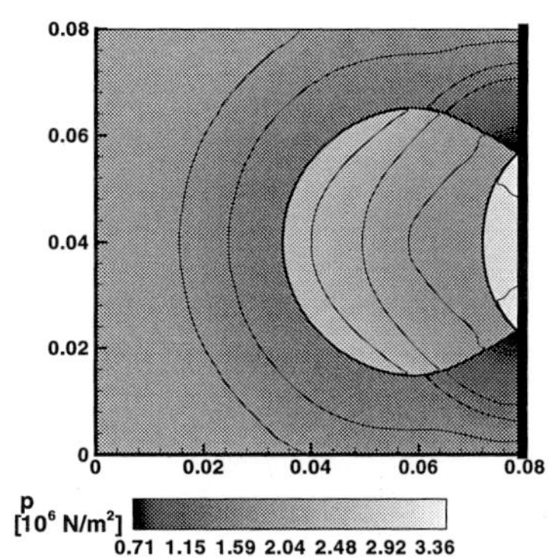

Fig. 11. $t = 0.10150ms$

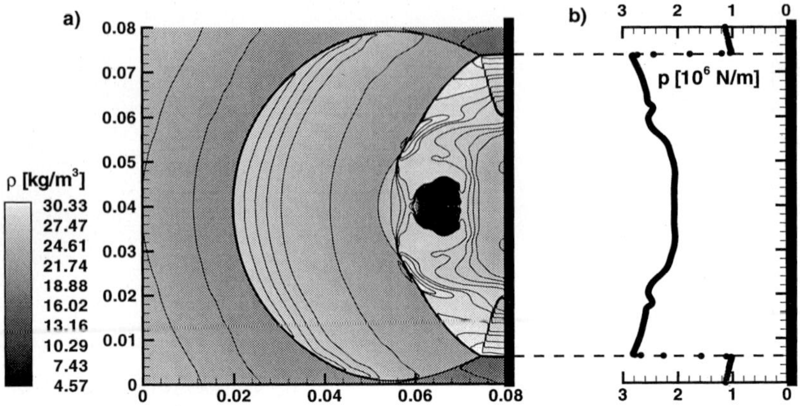

Fig. 12. $t = 0.14925ms$: a) density contours, b) pressure distribution at wall

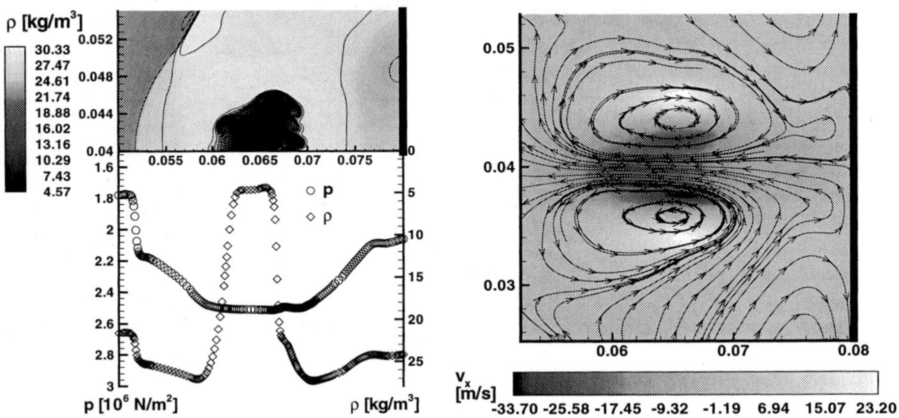

The Riemann Problem for a Two-phase Model

Nikolai Andrianov[1], Richard Saurel[2] and Gerald Warnecke[1]

[1] IAN, Otto-von-Guericke Universität Magdeburg, PSF 4120, D-39016 Magdeburg, Germany. *nikolai.andrianov|gerald.warnecke@mathematik.uni-magdeburg.de*

[2] Institut Universitaire des Systèmes Thermiques Industriels, 5 rue Enrico Fermi, 13453 Marseille Cedex 13, France *richard@iusti.univ-mrs.fr*

1 The Two-phase Model

The subject the present work is the study of the two-phase model, proposed by Baer and Nunziato [2], and modified by Saurel and Abgrall [8]. In what follows, the first model will be referred to as the BN-model, and the second one to as the SA-model. The BN-model describes the deflagration-to-detonation transition (DDT) in high-energy explosives. The modification of Saurel and Abgrall [8] concerns the interface and the relaxation terms. Due to these modifications, the model is able to describe a wide range of applications, e.g. the calculation of the shock waves in solid mixtures, cavitaion in fluids, and interface resolution.

The main assumptions of the model are the following. The model under consideration treats each phase as a continuum, and the phases coexist at each spatial location. Also, we assume that each phase is compressible. In this work, we restrict ourselves to inviscid, one-dimensional flow. Then, the governing equations are

$$
\begin{aligned}
\frac{\partial \alpha_a}{\partial t} + U\frac{\partial \alpha_a}{\partial x} &= \mathcal{F} + \frac{\mathcal{C}}{\rho_a} \\
\frac{\partial \alpha_a \rho_a}{\partial t} + \frac{\partial \alpha_a \rho_a u_a}{\partial x} &= +\mathcal{C} \\
\frac{\partial \alpha_a \rho_a u_a}{\partial t} + \frac{\partial \alpha_a \rho_a u_a^2 + \alpha_a p_a}{\partial x} &= +P\frac{\partial \alpha_a}{\partial x} + \mathcal{M} \\
\frac{\partial \alpha_a \rho_a E_a}{\partial t} + \frac{\partial \alpha_a u_a(\rho_a E_a + p_a)}{\partial x} &= +PU\frac{\partial \alpha_a}{\partial x} + \mathcal{E} \\
\frac{\partial \alpha_b \rho_b}{\partial t} + \frac{\partial \alpha_b \rho_b u_b}{\partial x} &= -\mathcal{C} \\
\frac{\partial \alpha_b \rho_b u_b}{\partial t} + \frac{\partial \alpha_b \rho_b u_b^2 + \alpha_b p_b}{\partial x} &= -P\frac{\partial \alpha_a}{\partial x} - \mathcal{M} \\
\frac{\partial \alpha_b \rho_b E_b}{\partial t} + \frac{\partial \alpha_b u_b(\rho_b E_b + p_b)}{\partial x} &= -PU\frac{\partial \alpha_a}{\partial x} - \mathcal{E}
\end{aligned}
\tag{1}
$$

Here, the two phases are denoted with subscripts a and b, respectively. The notations are classical: ρ_k is the density, u_k the velocity, p_k the pressure, E_k the total specific energy, and α_k the volume fraction of the phase $k = a, b$. The terms $\mathcal{F}$, $\mathcal{C}$, $\mathcal{M}$, and $\mathcal{E}$ are the source terms. They describe several processes, which lead the phases towards the equilibrium. In this work, we will not account for the source terms, and consider only the differential part of (1). See [2], [3], [6], [8] for different expressions for these terms.

Note that for the case $\alpha_a = \text{const}$, the system decouples to the two times Euler equations for the phases a and b. For the mixture, one can get the balance equations by summing the corresponding single-phase equations.

The parameters U and P have different expressions for the BN- and the SA-models. For the BN-model,

$$U = u_a, \qquad P = p_b. \tag{2}$$

For the SA-model,

$$U = \frac{a_a\rho_a u_a + \alpha_b\rho_b u_b}{\alpha_a\rho_a + \alpha_b\rho_b}, \qquad P = \alpha_a p_a + \alpha_b p_b. \tag{3}$$

To close the system, we use the saturation constraint on the two phases

$$\alpha_a + \alpha_b = 1.$$

We assume, that each phase is governed by the *Stiffened gas equation of state (EOS)*,

$$p_k = (\gamma_k - 1)\rho_k e_k - \gamma_k\pi_k, \quad k = a, b,$$

where γ_k, π_k are the thermodynamic constants, specific for each phase.

The coupling of the phases is achieved by the terms at the right hand side of the system (1). The differential terms at the right hand side of (1), and the transport equation for α_a are called the *non-conservative terms*, because due to their presence the system (1) can not be written in divergence form. Therefore, the usual notions of the weak solution, Rankine-Hugoniot conditions are not applicable for (1).

The characteristic analysis of the system (1) shows that it is hyperbolic, although not strictly hyperbolic. The characteristic speeds, which are the eigenvalues of the Jacobian matrix of (1), are

$$\begin{aligned} &\lambda_c = U \\ &\lambda_a^- = u_a - c_a \quad \lambda_a^0 = u_a \quad \lambda_a^+ = u_a + c_a \\ &\lambda_b^- = u_b - c_b \quad \lambda_b^0 = u_b \quad \lambda_b^+ = u_b + c_b \end{aligned} \tag{4}$$

We see that the characteristic speeds, associated with the different phases, can coincide with each other. Also, the corresponding eigenvectors can be linearly degenerate.

Now we investigate the particular characteristic fields. Naturally, the type of the first characteristic field (compaction field) depends on which value do we take for U. For the BN-model, the characteristic analysis was performed by Embid and Baer [4]. They have shown that the compaction field is linearly degenerate, and the Riemann invariants are

$$\begin{aligned}
\psi_1 &= u_a \\
\psi_2 &= \eta_b \\
\psi_3 &= \alpha_b \rho_b (u_a - u_b) \\
\psi_4 &= \alpha_a p_a + \alpha_b p_b + \alpha_b \rho_b (u_a - u_b)^2 \\
\psi_5 &= \frac{(u_a - u_b)^2}{2} + h_b,
\end{aligned}$$

where h_b is the specific enthalpy of the phase b.

For the SA-model, due to the complicated expression for U, we failed to determine the type of this field. The other characteristic fileds are analogous to those for the Euler equations. The fields $u \pm c$ are genuinely nonlinear, and the u^0 fields are linearly degenerate. We can easily find the corresponding Riemann invariants, they appear to be essentially the same as for the Euler equations. One can also show that the parameters for the phase a do not change across the waves of the phase b and vice versa. The volume fraction changes only across the compaction characteristic field.

2 Exact Solution to the Riemann Problem for the BN-model

In what follows, the phase a will be referred to as *solid*, and the phase b to as *gas*. The reason for it is that the BN-model [2] describes the interaction between the solid granular material and the burnt gases.

Consider the Riemann problem for the BN-model, e.g. the system of the governing equations (1), (2) and the initial data

$$\mathbf{u}(x,0) = \begin{cases} \mathbf{u}_L, \ x \leq 0 \\ \mathbf{u}_R, \ x > 0. \end{cases}$$

Since there are six distinct eigenvalues, cf. (2), (4), there are at most seven intermediate states in the solution of the Riemann problem. The usual "direct" solution of the Riemann problem is performed as follows: given the end states $\mathbf{u}_L, \mathbf{u}_R$, one has to find the intermediate states. However, it is problematic to find it for the two-phase model. Indeed, the Rankine–Hugoniot conditions are in general not determined. Then, the structure of the Riemann problem is quite complicated: for each phase, one has four possible configurations (rarefaction/shock, shock/rarefaction, rarefaction/rarefaction, shock/shock), and the waves can overlap with each other.

Our main interest in the exact solution to the Riemann problem is to get the benchmark for testing of numerical schemes. Therefore, instead of solving the "direct" Riemann problem, we can perform what we call the "inverse"

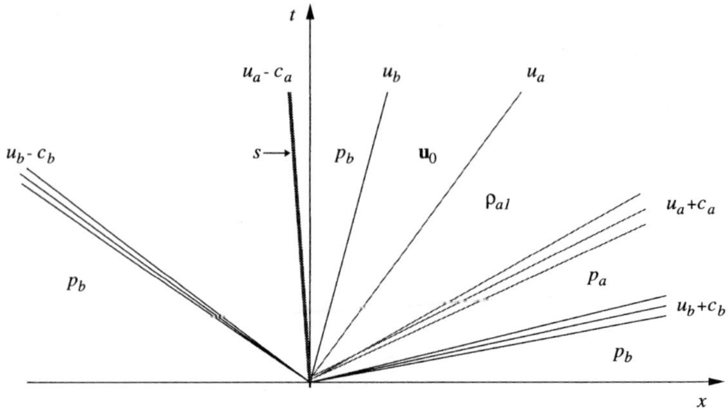

Fig. 1. "Inverse" solution of the Riemann problem

Riemann problem: Given the intermediate state $\mathbf{u}_0$ and the configuration of the Riemann problem, one finds the end states $\mathbf{u}_L, \mathbf{u}_R$, which can be connected to $\mathbf{u}_0$ by the admissible waves.

Let us concretize the procedure of the "inverse" solution, see Fig. 1. We choose the intermediate state $\mathbf{u}_0$ at the left to the solid contact u_a, and fix the volume fractions to the left and to the right of u_a. Given also the configuration, which means that we prescribe, which waves should be at the left of u_a, and which to the right. Depending on the wave, we determine the position of the wave by specifying the following quantities

- Shock: given shock speed, family
- Rarefaction: given pressure behind, family
- Contact: given density behind

We make an important assumption that the solid contact u_a does not coincide with any other wave. Therefore, we can use the usual Rankine-Hugoniot conditions across the shocks. Then, using the configuration data, for each wave we can find the state behind it.

It appears that across the solid contact u_a the solution is not trivial. The solid particle speed invariants are

$$\begin{aligned}
u_{a0} &= u_{a1} = u_a \\
\eta_{b0} &= \eta_{b1} = \eta_b \\
\phi_{b0}\rho_{b0}(u_a - u_{b0}) &= \phi_{b1}\rho_{b1}(u_a - u_{b1}) = M \qquad (5)\\
\phi_a p_{a0} + \phi_{b0} p_{b0} + \phi_{b0}\rho_{b0}(u_a - u_{b0})^2 &= \phi_a p_{a1} + \phi_{b1} p_{b1} + \phi_{b1}\rho_{b1}(u_a - u_{b1})^2 = P \\
\frac{(u_a - u_{b0})^2}{2} + h_{b0} &= \frac{(u_a - u_{b1})^2}{2} + h_{b1} = E.
\end{aligned}$$

Here

$$\eta_b = \frac{p_b + \pi_b}{\rho_b^{\gamma_b}}, \qquad h_b = \frac{\gamma_b(p_b + \pi_b)}{\rho_b(\gamma_b - 1)}$$

are respectively the entropy and the enthalpy of the gas phase, subscripts 0 and 1 denote the states to the left and the right of u_a. Note that the solid density ρ_a does not appear in (5), it acts as a free parameter.

Note that the system (5) is reminiscent of the mass, momentum, and energy balance across a discontinuity, propagating with the speed u_a, for the Euler equations for the gas phase. Consider the case

$$\phi_{b0} = \phi_{b1}.$$

Remember that this corresponds to the decoupled Euler equations for the gas and solid phases, see the system (1). So, the gas flow does not see the solid phase, and obviously the gas parameters do not change across u_a. Thus, there always exists the trivial solution

$$\begin{aligned} \rho_{b0} &= \rho_{b1} \\ u_{b0} &= u_{b1} \\ p_{b0} &= p_{b1}. \end{aligned} \tag{6}$$

In what follows, we will refer to this case as to "Euler", and the case

$$\phi_{b0} \neq \phi_{b1}$$

to as "multiphase".

Combining the equations of (5), we get the following nonlinear equation for the gas density to the right of u_a,

$$F(\rho_{b1}) = \frac{1}{2}\left(\frac{M}{\phi_{b1}\rho_{b1}}\right)^2 + \frac{\gamma_b \eta_b \rho_{b1}^{\gamma_b - 1}}{\gamma_b - 1} - E = 0. \tag{7}$$

It is easy to check that at the point ρ_*, the function $F(\rho)$ reaches its minimum,

$$\rho_*^{\gamma_b + 1} = \frac{M^2}{\phi_{b1}^2 \gamma_b \eta_b}.$$

Depending on the sign of $F(\rho_{b1})$, the equation (7) can have no, one, or two roots. We will investigate each case separately.

2.1 No Roots

Consider the solid particle speed invariants (5). We will fix the state 0, and look for the root of the equation (7) for the variable right gas volume fraction ϕ_{b1}.

We can rewrite the second equation of (5) as

$$\rho_{b0}(u_a - u_{b0}) = \frac{\phi_{b1}}{\phi_{b0}}\rho_{b1}(u_a - u_{b1}) = \frac{M}{\phi_{b0}} = M_E,$$

where the subscript E stands for Euler. Indeed, M_E is equal to the mass flow for the Euler case $\phi_{b0} = \phi_{b1}$. With a *constant* ϕ_{b0}, we have the constant M_E at the right hand side.

Let us rewrite the equation (7) as follows,

$$F(\rho_{b1}) = \frac{1}{2}M_E^2\left(\frac{\phi_{b0}}{\phi_{b1}}\right)^2\frac{1}{\rho_{b1}^2} + \frac{\gamma_b\eta_b\rho_{b1}^{\gamma_b-1}}{\gamma_b \quad 1} - E = 0. \tag{8}$$

For the Euler case

$$\frac{\phi_{b0}}{\phi_{b1}} = 1$$

and (8) always has at least one solution (6). For the variable ϕ_{b1} (multiphase case),

$$\frac{\phi_{b0}}{\phi_{b1}} \in (\phi_{b0}, \infty), \qquad \phi_{b1} \in (0, 1).$$

It is obvious that if for some positive A, A big enough,

$$\frac{\phi_{b0}}{\phi_{b1}} > A,$$

then (8) has no roots. Thus for all ϕ_{b1}, $\phi_{b1} < \phi_{b0}/A$, the equation (8) has no roots. Decreasing ϕ_{b1} (or increasing ϕ_{a1}, $\phi_{a1} = 1 - \phi_{b1}$), one necessarily gets no solution for some value of ϕ_{b1}.

2.2 One or Two Roots

Consider again the equations

$$F(\rho_{b1}) = \frac{1}{2}\left(\frac{M}{\phi_{b1}\rho_{b1}}\right)^2 + \frac{\gamma_b\eta_b\rho_{b1}^{\gamma_b-1}}{\gamma_b - 1} - E = 0 \tag{9}$$

$$F'(\rho) = -\frac{M^2}{\phi_{b1}^2\rho^3} + \gamma_b\eta_b\rho^{\gamma_b-2}. \tag{10}$$

Suppose first that (9) has two roots. Denote them $\tilde{\rho}_{b1}$, $\tilde{\tilde{\rho}}_{b1}$,

$$\tilde{\rho}_{b1} < \tilde{\tilde{\rho}}_{b1}.$$

Then

$$F'(\tilde{\rho}_{b1}) < 0, \qquad F'(\tilde{\tilde{\rho}}_{b1}) > 0.$$

Using

$$\frac{M^2}{\phi_{b1}^2} = (u_a - u_{b1})^2\rho_{b1}^2,$$

see the third equation of (5), we can rewrite (10) as

$$-(u_a - u_{b1})^2 + c_{b1}^2 = F'(\rho_{b1})\rho_{b1}, \tag{11}$$

where $c_{b1}^2 = \gamma_b(p_{b1} + \pi_b)/\rho_{b1}$ is the gas sound speed. We see that the volume fraction ϕ_b does not appear in this equation. This means that it should hold true for all values of ϕ_{b0}, ϕ_{b1} on the both sides of u_a. In particular, for the Euler case $\phi_{b0} = \phi_{b1}$, the admissible solution is (6),

$$\begin{aligned} \rho_{b0} &= \rho_{b1} \\ u_{b0} &= u_{b1} \\ p_{b0} &= p_{b1}. \end{aligned}$$

Thus, from the two roots $\tilde{\rho}_{b1}$, $\tilde{\tilde{\rho}}_{b1}$ of (9) we choose the one for which

$$\text{sign}(-(u_a - u_{b0})^2 + c_{b0}^2) = \text{sign}(-(u_a - u_{b1})^2 + c_{b1}^2).$$

By (11), it is equivalent to the condition

$$\text{sign}(F'(\rho_{b0})) = \text{sign}(F'(\rho_{b1})). \tag{12}$$

For the multiphase case $\phi_{b0} \neq \phi_{b1}$, we also choose the root of (9), which satisfies the condition (12). It corresponds to the admissible root for the Euler case, if ϕ_{b1} would be equal to ϕ_{b0}.

The case of the single root of (9) corresponds to the case, when the gas flow velocity relative to u_a is sonic to the right of u_a.

3 Numerical Method

For the solution of the hyperbolic part of the model, we employ the approach due to Gallouet and Masella [5]. In what follows, it will be referred to as *VFRoe*, which stands for *Volumes Finis Roe* in French. Consider the Riemann problem for the system of m hyperbolic conservation laws

$$\frac{\partial \mathbf{u}}{\partial t} + \frac{\partial \mathbf{f}(\mathbf{u})}{\partial x} = 0 \tag{13}$$

with the initial data

$$\mathbf{u}(x, 0) = \begin{cases} \mathbf{u}_L, \ x < 0 \\ \mathbf{u}_R, \ x > 0. \end{cases}$$

Then, the solution consists of at most m waves, which separate $m+1$ constant states. We linearize the system (13) at some state $\bar{\mathbf{u}}$,

$$\frac{\partial \mathbf{u}}{\partial t} + \mathbf{A}(\bar{\mathbf{u}}(\mathbf{u}_L, \mathbf{u}_R))\frac{\partial \mathbf{u}}{\partial x} = 0.$$

For $\bar{\mathbf{u}}$, we choose the arithmetic average of $\mathbf{u}_L$ and $\mathbf{u}_R$,

$$\bar{\mathbf{u}} = \frac{\mathbf{u}_L + \mathbf{u}_R}{2}$$

Then, we can find the intermediate state in the solution of the Riemann problem as it is done in Roe's method,

$$\mathbf{u}^* = \mathbf{u}_L + \sum_{\lambda_i < 0} \alpha_i \mathbf{r}_i$$

where λ_i are the eigenvalues of $\mathbf{A}$, $\mathbf{r}_i$ the corresponding eigenvectors, and α_i are the coefficients of the eigenvector decomposition of the difference $\mathbf{u}_R - \mathbf{u}_L$.

We define the numerical flux function as

$$\mathbf{F}(\mathbf{u}_L, \mathbf{u}_R) = \mathbf{f}(\mathbf{u}^*)$$

and the Godunov-type numerical scheme as

$$\mathbf{u}_i^{n+1} = \mathbf{u}_i^n - \frac{\Delta t}{\Delta x}\left[\mathbf{f}(\mathbf{u}^*_{i+1/2}) - \mathbf{f}(\mathbf{u}^*_{i-1/2})\right].$$

The properties of the scheme are the following. Obviously, the numerical flux function is consistent. The scheme ios written in conservative form so by the Lax-Wendroff theorem the numerical solution, if convergent, converges to some weak solution of (13). The numerical scheme is suitable for flows with complex physical flux functions. Finally, the entropy fix is needed when finding the intermediate state $\mathbf{u}^*$.

We apply this numerical method to the solution of the two-phase model as follows. The source terms are integrated with the Strang splitting method,

$$\mathbf{u}_i^{n+1} = L_a^{\Delta t/2} L_h^{\Delta t} L_a^{\Delta t/2} \mathbf{u}_i^n,$$

where L_h is the operator of the solution of the hyperbolic part of the system (1), and L_a is the operator of the integration of the source terms. See Lallemand and Saurel [7] for details on it.

The solution of the hyperbolic part is performed with the Godunov-type scheme of the form

$$\mathbf{u}_i^{n+1} = \mathbf{u}_i^n - \frac{\Delta t}{\Delta x}\left[\mathbf{F}(\mathbf{u}_i^n, \mathbf{u}_{i+1}^n) - \mathbf{F}(\mathbf{u}_{i-1}^n, \mathbf{u}_i^n)\right] + \Delta t \cdot \mathbf{H}_i \cdot \Delta.$$

Here $\mathbf{H}_i \cdot \Delta$ is some discretization of the non-conservative term at the right hand side of (1). To decide, which discretization should we take, we use the following principle: A flow, uniform in pressure and velocity, remains uniform in time. Writing the scheme componentwise, and substituting everywhere the constant velocity and pressure, we get the following discretizations for the non-conservative terms

$$\Delta = \frac{1}{\Delta x}\left[(\alpha_a)^*_{i+1/2} - (\alpha_a)^*_{i-1/2}\right]$$

$$(\alpha_a)_i^{n+1} = (\alpha_a)_i^n - \Delta t \cdot (U)_i \cdot \Delta$$

Depending on which model we are interested in, U will be determined either by (2) or (3. The scheme, built in such a way, preserves the moving contact disvontinuity by construction. It can be extended to the second order by the MUSCL approach.

4 Numerical Example

Here, we present only the comparison of the numerical solution with the exact solution to the Riemann problem for the BN-model. The numerical results for the SA-model can be found in [1].

For simplicity, we assume that both phases are governed just by the ideal gas EOS, i.e. $\gamma_k = 1.4$, $\pi_k = 0$, $k = a, b$. For example, we can choose the following data for the inverse Riemann problem,

$$\mathbf{u}_0 = \begin{bmatrix} \rho_{a0} \\ u_{a0} \\ p_{a0} \\ \rho_{b0} \\ u_{b0} \\ p_{b0} \end{bmatrix} = \begin{bmatrix} 2.0 \\ 0.2 \\ 5.0 \\ 1.0 \\ 0.1 \\ 1.0 \end{bmatrix} \qquad \begin{matrix} (\alpha_a)_L = 0.5 \\ (\alpha_a)_R = 0.6 \end{matrix}$$

and prescribe the configuration

Left: Gas contact, density behind: 0.5
Gas 1-rarefaction, pressure behind: 1.5
Solid 1-shock, speed: -1.0
Right: Gas 3-rarefaction, pressure behind: 1.5
Solid 3-shock, speed: 1.5

Then, the Riemann problem has the structure, depicted in Fig. 2. Now, we can solve the "direct" Riemann problem and compare the results with the exact solution. Some results are shown in Fig. 3. We see that the method computes the flow correctly.

References

1. Andrianov, N., Saurel, R., and Warnecke, G.: A simple method for compressible multiphase mixtures and interfaces, submitted to Int. J. Numer. Meth. Fluids. Also available as INRIA preprint at `http://www.inria.fr/rrrt/rr-4247.html`.
2. Baer, M.R. and Nunziato, J.W. (1986): A two-phase mixture theory for the deflagration-to-detonation transition (DDT) in reactive granular materials, Int. J. of Multiphase Flows, **12**, 861-889.
3. Bdzil, J.B., Menikoff, R., Son, S.F., Kapila, A.K., and Stewart, D.S. (1999): Two-phase modeling of DDT in granular materials: a critical examination of modeling issues, Phys. Fluids **11** (2), 378-402.

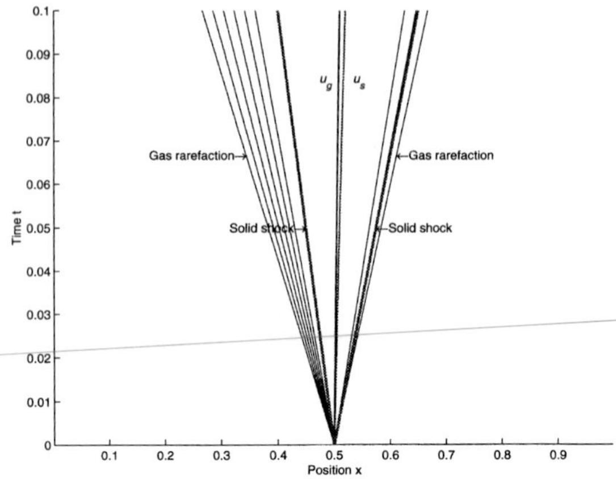

Fig. 2. Configuration of the Riemann problem

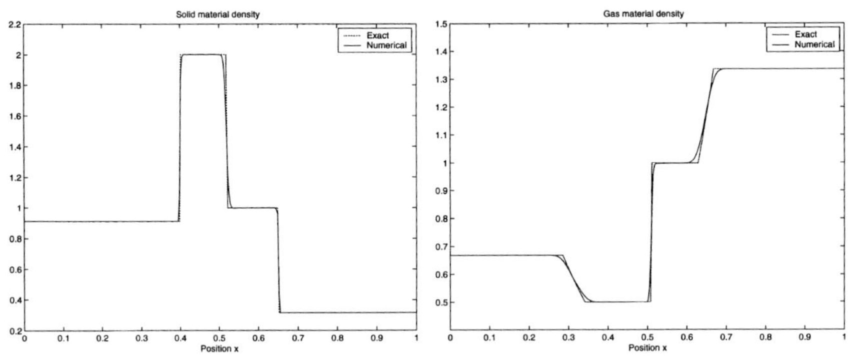

Fig. 3. Densities. Second order method, 200 mesh cells, CFL=0.9

4. Embid, P. and Baer, M. (1992): Mathematical analysis of a two-phase continuum mixture theory, Continuum Mech. Thermodyn. **4**, 279-312.
5. Gallouet, T. and Masella, J.-M. (1996): Un schéma de Godunov approché, C. R. Acad. Sci. Paris Sér. I **323**, 77-84.
6. Kapila, A.K., Son, S.F., Bdzil, J.B., Menikoff, R., Stewart, D.S. (1997): Two-phase modeling of DDT: structure of the velocity-relaxation zone, Phys. Fluids **9** (12), 3885-3897.
7. Lallemand, M.-H. and Saurel, R.: Pressure relaxation procedures for multiphase compressible flows, INRIA Rapport de recherche n^0 4038, available online at `http://www.inria.fr/rrrt/rr-4038.html`.
8. Saurel, R. and Abgrall R. (1999): A multiphase Godunov method for compressible multifluid and multiphase flows, J. Comput. Phys. **150**, 425-467.

Evaluation of a Well-posed Perfectly Matched Layer for Computational Acoustics

Daniel Appelö[1] and Gunilla Kreiss[1]

[1] NADA, KTH, Sweden *appelo@nada.kth.se*
[2] NADA, KTH, Sweden *gunillak@nada.kth.se*

1 Introduction

Accurate numerical simulations of wave phenomena are important in many fields, for example in electro magnetics, seismics and acoustics. We are interested in aeroacoustic problems with a non-vanishing mean flow. The governing equations are the Euler equations. In a typical aeroacoustic situation the solution consists of a slowly varying mean flow with an added small amplitude acoustic wave.

The simulations must in general be confined to truncated domains, much smaller than the physical space where the phenomena takes place. At the artificial boundary, where the truncation takes place, boundary conditions must be set. Ideally the solution of the problem in the truncated domain is identical to the solution of the original problem. However, often undesirable effects, such as reflections, limit the accuracy of the solution of the truncated problem. Boundary conditions that do not cause reflections are called absorbing or non-reflecting boundary conditions

Some of the earliest absorbing or non-reflecting boundary conditions were derived for the wave equation by Engquist and Majda [6] and Bayliss and Turkel [4]. For an overview of the subsequent development of absorbing boundary conditions for wave propagation problems, see for instance Hagstrom [8].

For Mawell's equations an important advance is the derivation of perfectly matched layers (PML) by Berenger [5] in 1994. The idea of the PML is to surround the computational domain with an highly absorbing layer in which the outgoing waves is attenuated. The governing equations of the absorbing layer must be matched with the equations of the computational domain, i.e. there should be no reflection at the interface of incoming waves, independent of their angle of incidence and frequency. Berenger introduced a splitting of the equations to obtain the additional degrees of freedom needed to match the equations at the interface. The resulting PML was orders of magnitude better than earlier Absorbing Boundary conditions (ABC's). One serious disadvantage of Berengers PML, discovered by Abarbanel and Gottlieb [1], was that the splitted set of PDE's was only weakly well-posed. Other derivations of a PML for Maxwel's equations which are well-posed have been presented [2], and are as efficient as Berengers PML. Today PML is commonly used in commercial CEM software.

In 1996 Hu [10] derived a PML for the Euler equations, linearized at constant flow, by using the splitting technique introduced by Berenger. An analysis of this PML showed that it is only weakly well-posed[9]. In 1998 Abarbanel, Gottlieb and Hesthaven [3] derived a well-posed PML. However, the PML technique has so far not been as succesfull for the linearized Euler equations as for the Maxwells equations. This is true even in the case of vanishing mean flow. One difference, which may be important is the presence of vorticity waves.

In this paper we present results from numerical experiments where we solve the linearized Euler equations and use the PML derived in [3] as ABC.

In one experiment we compare the performance of the PML for the two different types of waves, vorticity waves and sound waves, that are supported by the equations. We have also varied the angles of incidence with which the waves approach the boundary. To analyze the result we use the wavesplitting technique of Wilson [7], described in Jensen, Efraimsson [11], to decompose the solution into the different waves. The computations show that the PML is less efficient for vorticity waves than for sound waves.

In an other experiment we vary the mach number spatially. Computations show that varying Mach number does not effect the performance of the PML significantly.

2 The PML and test problems

The PML in [3] is derived for the Euler equations linearized around an isentropic uniform (aligned to the x-axis) flow on a rectangular domain represented in Cartesian coordinates. The equations are nondimensionalized with the Mach number M. In all cases with constant M we have used $M = 0.5$. The PML is derived in a transformed coordinate system. The transformation $(x, y, t) \to (\xi, \eta, \tau)$ is defined by

$$\xi = x, \qquad \eta = \sqrt{1 - M^2}y = \gamma y, \qquad \tau = Mx + \gamma^2 t. \tag{1}$$

Eigenvalues belonging to plane wave solutions of the transformed problem satisfy a simple dispersion relation which simplifies the derivation of the PML. The derivation is done separately for the ξ- and η-direction, resulting in two sets of well-posed perfectly matched layers, which are added together, overlapping each other in the corners. The PML is transformed back to the physical coordinate system (x, y, t) yielding,

$$
\begin{aligned}
\frac{\partial v}{\partial t} + M\frac{\partial v}{\partial x} + \frac{\partial \rho}{\partial y} =& - \sigma_x v + 2\sigma_x Q_x + \sigma_x^2 R_x + M\sigma_x^{'} P_x \\
& + 2\sigma_y Q_y + \sigma_y^2 R_y + \sigma_y^{'} P_y - \varepsilon_x v + 2\mu_y Q_y, \\
\frac{\partial u}{\partial t} + M\frac{\partial u}{\partial x} + \frac{\partial \rho}{\partial x} =& - \sigma_x u - \sigma_x M\rho - \mu_y u, \\
\frac{\partial \rho}{\partial t} + M\frac{\partial \rho}{\partial x} + \frac{\partial u}{\partial x} + \frac{\partial v}{\partial y} =& - \sigma_x \rho - \sigma_x M u - \mu_y \rho, \\
\frac{\partial Q_x}{\partial t} =& - \gamma^2 \frac{\partial \rho}{\partial y}, \\
\frac{\partial Q_y}{\partial t} =& \gamma^2 \Big[\frac{\partial \rho}{\partial y} - 2\sigma_y Q_y - \sigma_y^2 R_y - \sigma_y^{'} P_y + \sigma_x v \\
& - 2\sigma_x Q_x - \sigma_x^2 R_x - M\sigma_x^{'} P_x + \mu_x v - 2\mu_y Q_y\Big], \\
\frac{\partial P_x}{\partial t} =& \gamma^2 \left[v - \sigma_x P_x - \varepsilon_x P_x\right], \\
\frac{\partial P_y}{\partial t} =& \gamma^2 \left[\rho - \sigma_y P_y\right], \\
\frac{\partial R_x}{\partial t} =& \gamma^2 \left[Q_x - \mu_x R_x\right], \\
\frac{\partial R_y}{\partial t} =& \gamma^2 \left[Q_y - \mu_y R_y\right].
\end{aligned} \tag{2}
$$

The parameters $\sigma_x = \sigma_x(x)$ and $\sigma_y = \sigma_y(y)$ are polynomials starting at zero at the PML-interface and monotonically increasing through the PML. By setting the parameters σ_x and σ_y to zero the linearized Euler equations are recovered for the velocity components u and v and the density ρ.

The parameters ε_x, μ_x and μ_y are added to the PML to prevent algebraic growth. They are motivated in [3] by an analysis of a system of ODE's obtained by neglecting spatial variations. In chapter 3 we show that the addition of ε_x, μ_x and μ_y is insufficient to prevent growing solutions. We also suggest an ad hoc modification of the PML that prevent growing solutions.

2.1 Testproblem 1

The first test problem is adopted from [3]. The computational domain is bounded by $|x| \leq L_x, |y| \leq L_y$, we choose $L_x = L_y = 50$. Outside the computational domain a PML of thickness δ_x, δ_y respectively is added. The PML is terminated with characteristic boundary conditions. In the inner domain, the linearized Euler equations are solved, and in the outer domain the PDE's (2). In the inner domain ρ, u and v are subjected to a continuous forcing

$$\rho^f(x,y,t) = e^{-(\ln 2)\frac{(x-x_a)^2+(y-y_a)^2}{\delta_a^2}} sin\left(\frac{\pi t}{10}\right),$$
$$u^f(x,y,t) = 0.05(y-y_b)e^{-(\ln 2)\frac{(x-x_b)^2+(y-y_b)^2}{\delta_b^2}} sin\left(\frac{\pi t}{10}\right), \tag{3}$$
$$v^f(x,y,t) = -0.05(x-x_b)e^{-(\ln 2)\frac{(x-x_b)^2+(y-y_b)^2}{\delta_b^2}} sin\left(\frac{\pi t}{10}\right),$$

where (x_a, y_a) is the center of the sound source, which has the width δ_a, (x_b, y_b) is center of the vorticity source with width δ_b. The centers and widths are $(x_a, y_a) = (-25, 0)$, $(x_b, y_b) = (25, 0)$, $\delta_a = 3, \delta_b = 4$ respectively. The error is meassured as the l_2^2 error along the line $x = 48$. For testproblem 1 the PML thickness δ_x and δ_y are varied. The testproblem is extended to include spatially varying mach number. This is discussed in section 4.

2.2 Wavesplitting

Let $\mathbf{f} = [\rho, u, v]^T$. By Fourier transform the function $\mathbf{f}$ can be represented as $\hat{\mathbf{f}}(x, k_y, \omega) = \mathcal{F}\{\mathbf{f}(x, y, t)\}$ and then the linearized Euler equations may be written as

$$\mathbf{A}\frac{\partial \hat{\mathbf{f}}}{\partial x} + \mathbf{B}(x, i_k y, i\omega)\hat{\mathbf{f}} = 0. \tag{4}$$

The eigenvectors of the matrix $\mathbf{A}^{-1}\mathbf{B}$ form the columns of the matrix $\mathbf{L}$ used to define $\mathbf{q} = \mathbf{L}^{-1}\hat{\mathbf{f}}$. After multiplying (4) with $\mathbf{A}^{-1}$ and $\mathbf{L}^{-1}$ we obtain the decoupled system of ordinary differential equations

$$\frac{\partial \mathbf{q}}{\partial x} + \mathbf{\Lambda}\mathbf{q} = 0,$$

where $\mathbf{\Lambda} = \text{diag}([\lambda_{ds}, \lambda_{us}, \lambda_{vor}])$ contains the eigenvalues of $\mathbf{A}^{-1}\mathbf{B}$. The absolute value of the components of $\mathbf{q}$ are the amplitudes of the upstream and downstream sound wave and the vorticity wave.

2.3 Testproblem 2

The second testproblem is adopted from [11]. The test problem is designed to test the efficiency of an absorbing boundary condition for down-stream sound wave and vorticity wave, separately.

The computational domain is $0 < x < L_x$, $|y| < L_y$. At $x = 0$ we prescribe the solution and at $x = L_x$ we use PML terminated with characteristic boundary conditions. In the y-direction we use periodic boundary conditions.

By using the wavesplitting technique, described above, we derive boundary conditions for the inflow-boundary that corresponds to either a pure downstream plane sound wave solution or a pure vorticity plane-wave solution with amplitude 1.

Fixing the height of the computational domain to $|y| \leq L_y = 50$ and the wave number $k_y = 2\pi/100$ the solution is periodic in the y-direction. By specifying different frequencies ω we obtain plane wave solutions which propagate in different angles relative the y-axis. The width of the computational domain, L_x, is chosen such that one period of the solution is contained within. Further we use $\delta_x = L_x$.

2.4 Discretization

To simulate the testproblems we have used standard 2:nd order central differences in space and 4:th order Runge-Kutta to advance in time. In [3] a 4:th order scheme is used in space but for our evaluation this difference is not significant. For testproblem 2, since the exact solution was known, the discretization step was chosen to be 25 points per wavelength both in space and time. In testproblem 1 the discretization step was choosen empirically so that the computed solution was well resolved in space and time.

3 Analysis of corners

During simulations of the first testproblem using the discretization described above we experienced growth originating from the corners. To understand the growth we consider (2) with σ_x and σ_y constant and $\varepsilon_x = \mu_x = \mu_y = 0$ to obtain a constant coefficient problem.

Using the notation $w = [v, u, \rho, Q_x, Q_y, P_x, P_y, R_x, R_y]^T$ we write (2) as

$$\frac{\partial w}{\partial t} = A\frac{\partial w}{\partial x} + B\frac{\partial w}{\partial y} + Cw. \tag{5}$$

By Fourier transform in the x and y directions we obtain

$$\frac{\partial \hat{w}}{\partial t} = (i\xi A + i\eta B + C)\hat{w} \equiv \mathbf{P}(i\xi, i\eta, t)\hat{w}.$$

Here $\mathbf{P}$ is a 9×9 matrix. If the eigenvalues of $\mathbf{P}(i\xi, i\eta, t)$ have real part greater than zero $||w||$ will grow. To investigate this we numerically calculate the eigenvalues λ_i of $\mathbf{P}$ for fixed ξ and η.

Let $\lambda_{max} = \max[Re\{\lambda_i\}]$. In figure 1(a) λ_{max} is plotted for a set of frequencies. Clearly there exist $\lambda_{max} > 0$, explaining the growth experienced.

To investigate weather the analysis of the constant coefficient system (5) is relevant for the first testproblem we calculated λ_{max} for the spatial discretisation of (2) including non constant σ_x and σ_y and choosing the parameters μ_x, μ_y and ϵ_x according to [3]. Also for this case $\lambda_{max} > 0$. This result was obtained independent of the discretization step size. The growth rate of the norm of the solution could be connected to λ_{max}.

Investigation of the eigenfunctions, corresponding to growing modes, showed that they were nonzero only at nodes where σ_x and σ_y were simultaneously nonzero, i.e. in the corners. The simplest way to get rid of the growth

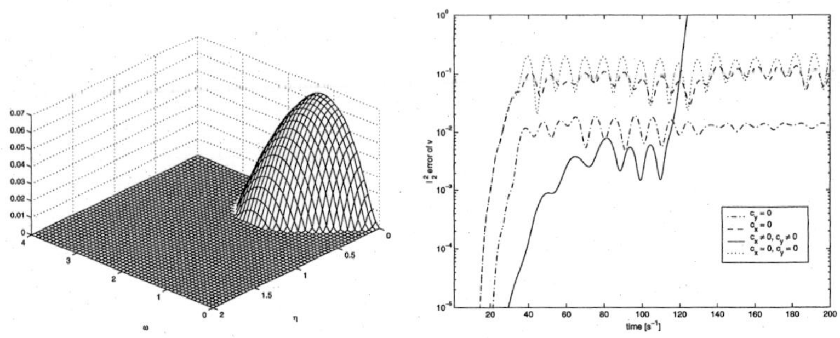

(a) Plot of $\mathbf{Re}\{\lambda_{max}\}$ for some frequencies. The bump corresponds to growing modes.

(b) Exponential growth for overlapping σ_x and σ_y, no growth with the suggested fix. The decrease in efficiency is about one half order of magnitude. Here $c_i = \sigma_i$.

Fig. 1.

was to prevent σ_x and σ_y from being simultaneously nonzero. This can be accomplished by putting either σ_x or σ_y or both zero in the corners where they overlap. In figure 1(b) performance for the three different choices and the original setup is showed.

The best choice seems to be keeping the PML at the in and outflow boundaries intact and putting $\sigma_y = 0$ inside that PML (the dash-dotted line). Note that the error level of the modified PML is only one half order of magnitude worse than the original. The modified approach has been integrated for long time and no growth have been seen.

4 Numerical results

4.1 Variable machnumber

To make an acoustic ABC practically useful it should be applicable to the linearized Euler equations with variable mach number and hopefully to the

Euler equations. We have considered the simpler of these two extensions by making experiments on a case where we let the machnumber vary spatially. A spatial variation of the machnumber may be used in simulations of acoustic waves in nozzles or subsonic jets.

We have made computations for a spatially varying machnumber,

$$M = M(z) = e^{\frac{-z^2}{10^2}}, \quad \text{where z is either x or y.} \tag{6}$$

The performance of the PML for these two cases seems to be nearly as good as for the constant mach number (compare figures (2(a)-2(c))). We conclude that the PML may be used for this kind of computations.

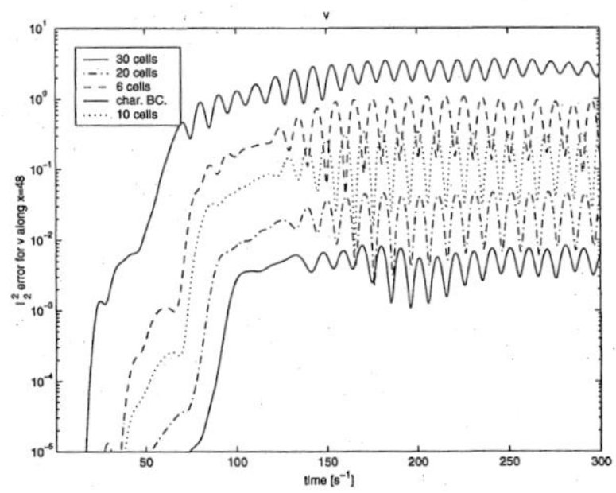

(a) Error along x=48 for v for constant Machnumber

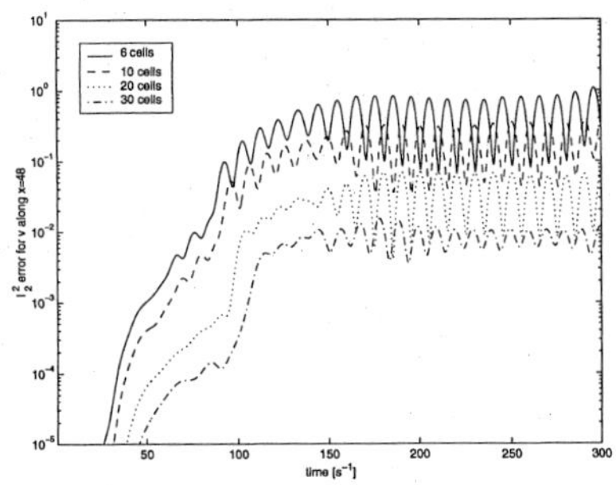

(b) Error along x=48 for v for a simulated subsonic jet

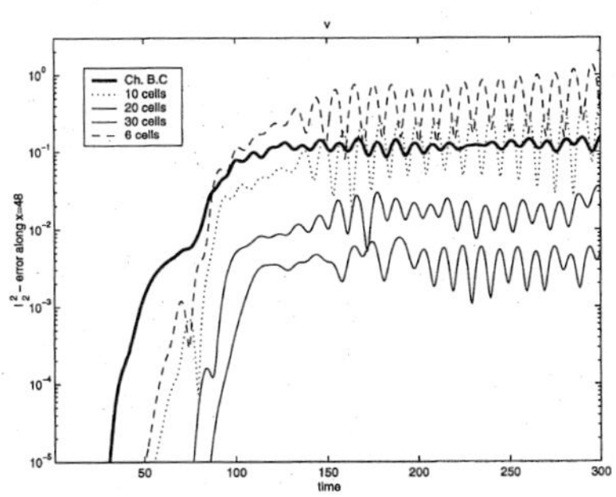

(c) Error along x=48 for v for a simulated slim nozzle

Fig. 2. The difference in performance for spatially varying Machnumber compared to constant is small. It should be noted that the form of the curves for the error coincides with [3] but the level is higher.

4.2 Efficiency for different waves

We used zero as initial condition for testproblem 2. To avoid transient noise the solution was first integrated in time for 100 periods. After that the solution was integrated and saved for another period.

Fourier transforming and wavesplitting. To compare different numerical solutions we use the Fourier transform in time and in the y-direction. We then use the wavesplitting technique (described above) to extract separate information on each of the waves of the solution. For each node in the x-direction the amplitude of the other two waves (which ideally should be zero) gives an estimate of the relative error in the computed solution.

In figures 4(a) - 5(b) some examples are shown. We like to stress that in all tests the upstream soundwave amplitude should be taken as the significant estimate of the performance of the absorbing boundary condition since it is the only one which is reflected from the PML. The other wave is due to the boundary treatment at the inflow boundary.

The results in figures 3(a) and 3(b) show that the PML performs best for waves which fall in normal to the outflow boundary. For soundwaves the performance deteriorates as the angle of incidence decreases. Surprisingly for vorticity waves the worst performance is found at an angle of 60 degrees. However the efficiency of the PML for different angles is always better than characteristic boundary conditions.

References

1. S. Abarbanel and D. Gottlieb. A mathematical analysis of the pml method. *J. Comput. Phys.*, 134:357, 1997.
2. S. Abarbanel and D. Gottlieb. On the construction and analysis of absorbing layers in cem. *Appl. Numer. Math.*, 27:331, 1998.
3. S. Abarbanel, D. Gottlieb, and J.S Hesthaven. Well-posed perfectly matched layers for advective acoustics. *Journal of Computational Physics*, 154:266–283, 1999.
4. A. Bayliss and E. Turkel. Radiation boundary conditions for wave-like equations. *Comm. Pure Appl. Math.*, 33:707, 1980.
5. J. P. Brernger. A perfectly matched layer for the absorption of electromagnetic waves. *Journal of Computational Physics*, 114:185, 1994.
6. B. Engquist and A. Majda. Absorbing boundary conditions for the numerical simulation of waves. *Math. Comp.*, 31:629, 1977.
7. Wilson. A. G. Method for deriving tone noise information from cfd calculations on the aeroengine fan stage. draft paper, RollsRoyce plc, United Kingdom, 2000.
8. Thomas Hagstrom. Radiation boundary conditions for the numerical simulation of waves. *Acta Numerica*, pages 47–106, 1999.
9. J.S. Hesthaven. On the analysis and construction of perfectly matched layers for the linearized euler equations. *Journal of computational Physics*, 142:129–147, 1998.

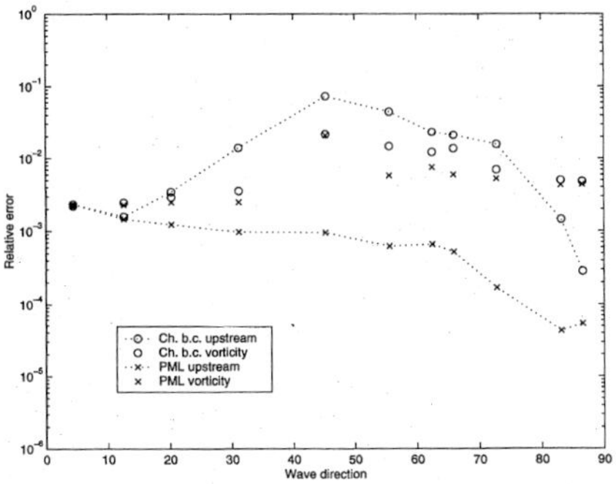

(a) *Downstream soundwave.* The curves display the mean relative error for the vorticity and upstream soundwave. Note that the PML is better than ch. b.c. for all angles except for almost glancing incidence where the ch. b.c. seems to perform as well.

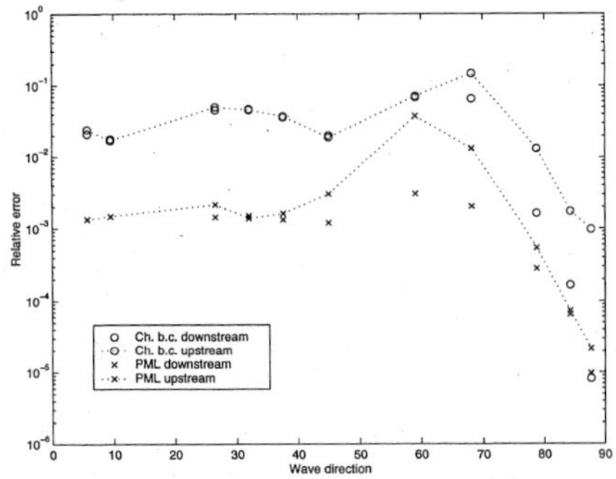

(b) *Vorticity wave.* Mean relative error for the vorticity and upstream soundwave. For this type of inflow b.c. the PML is better at all angles.

Fig. 3. The comparison between vorticity and soundwaves show that the PML work better for the soundwaves since the vorticity waves has a dip in performance at 60 degree. The error displayed on the y-axis is the mean error over all x-nodes in the computational domain.

10. F. Q. Hu. On absorbing boundary conditions for linearized euler equations by a perfectly matched layer. *Journal of Computational Physics*, 129:201–209, 1996.
11. Kristoffer Jensen and Gunilla Efraimsson. Aeroengine fan-noise resolution requirement in computational fluid dynamics. Technical report, Swedish Defence research agency, FOI, 2001.

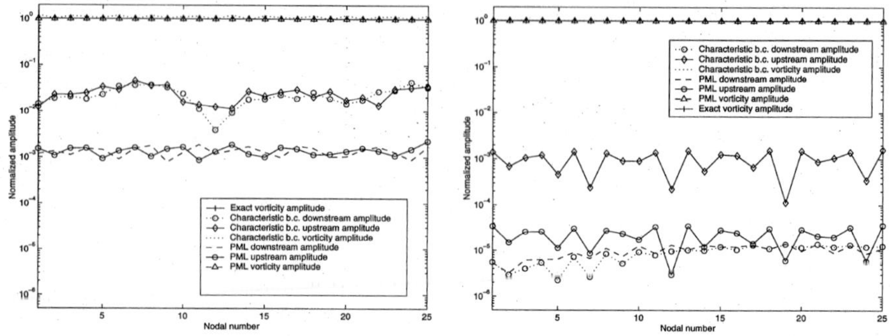

(a) Vorticity wave incoming at angle 5.7 degree (almost glancing incidence).

(b) Vorticity wave incoming at angle 87.7 degree (almost normal incidence).

Fig. 4. Comparison between PML and characteristic boundary condition for a Vorticity wave along the computational domain. Note that both boundary conditions work best at normal incidence.

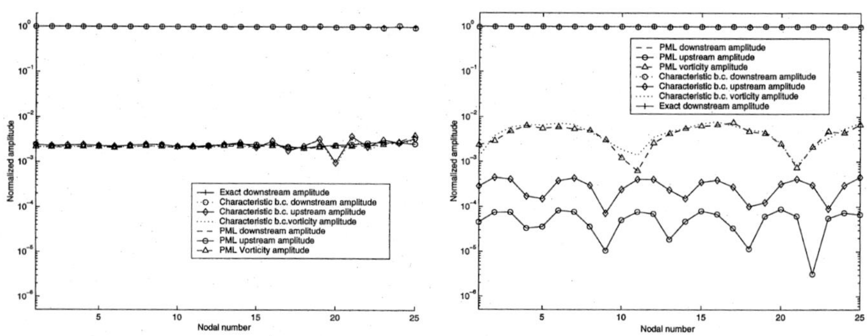

(a) Downstream soundwave incoming at angle 4.4 degree (almost glancing incidence).

(b) Downstream soundwave incoming at angle 86.6 degree (almost normal incidence)

Fig. 5. Comparison between PML and characteristic boundary condition for a downstream soundwave. Note that both boundary conditions work best at normal incidence.

New Space Staggered and Time Interleaved 2nd Order Finite Volume Methods

P. Arminjon[1,2] and A. St-Cyr[1,3,4]

[1] Centre de Recherches Mathématiques, Université de Montréal,C.P.6128, Succ. Centre-ville, Montréal, Québec, Canada, H3C 3J7.
[2] *arminjon@crm.umontreal.ca*
[3] *stcyr@dms.umontreal.ca*
[4] *Currently at*: CFD Lab, Department of Mechanical Engineering, McGill University, 688 Sherbrooke St. West, Montreal, Québec, Canada H3A 2S6 *amik@cfdlab.mcgill.ca*

Summary. A new modified version of the Nessyahu-Tadmor (NT) 1-dimensional finite volume central scheme is presented , as well as corresponding new versions for 2D-structured and 3D-unstructured grids inspired from the NT scheme. The modification avoids the intermediate predictor time step between t^n and t^{n+1}.The CPU gain is not very important in the 1D case, but becomes significant in the 2 and 3D cases. 3D comparative simulations for the shock tube problem and for a supersonic inviscid flow through a channel with a 4% circular bump are presented.

1 Introduction

We consider central methods for scalar hyperbolic conservation laws or corresponding systems:

$$u_t + f(u)_x = 0, \qquad u(x, t = 0) = u_0(x) \tag{1}$$

as well as their 2 and 3-dimensional analogue, to be described below. The non-oscillatory central difference scheme of Nessyahu and Tadmor may be interpreted as a Godunov-type scheme for one-dimensional hyperbolic conservation laws in which the resolution of the Riemann problems at the cell interfaces is by-passed thanks to the use of the staggered Lax-Friedrichs scheme. Piecewise linear MUSCL-type cell interpolants and slope limiters lead to an oscillation-free second-order resolution.

In earlier papers [5, 7, 6] a two- dimensional finite volume method was presented, generalizing the one-dimensional Lax-Friedrichs (LF) and Nessyahu - Tadmor (NT) [12] difference schemes for hyperbolic conservation laws to unstructured triangular grids, while in [3, 4] a corresponding extension in the case of 2-dimensional Cartesian grids was constructed. In [9], Jiang and Tadmor have presented a slightly different extension for Cartesian grids.

In [2, 4] new extensions of the LxF and NT schemes to central finite volume methods for 3-dimensional hyperbolic systems on **unstructured tetrahedral grids** and on **Cartesian grids**, respectively, were presented.

In this paper, we investigate another way to improve our 2 and 3-dimensional central finite volume methods inspired from the LF and NT schemes. These methods require the implementation of an intermediate half-time predictor step to compute the fluxes with the midpoint rule. Since the execution of this predictor step becomes more and more time consuming in 2 and 3D, we present a variant of our schemes which avoids the predictor step. The new method is inspired by multistep methods for ODE's, and computes the fluxes with the help of the values of the dependent variables at the last two time steps. Since the NT scheme and our finite volume extensions use two staggered grids at alternate time steps, the solution at time t^{n+1} is constructed using the fluxes at t^n and t^{n-2}, both obtained from the same grid.
Although the new method does not lead to significant computer time reduction in the one dimensional case, it really becomes advantageous in two and even more so in three spatial dimensions, where it can lead to reductions up to 40% in total computer times.

2 The 1D NT-type scheme without predictor step

2.1 The Lax-Friedrichs and Nessyahu-Tadmor difference schemes

We consider the initial value problem (1) and the staggered LF scheme

$$u_{i+1/2}^{n+1} = \frac{1}{2}(u_{i+1}^n + u_i^n) - \lambda(f(u_{i+1}^n) - f(u_i^n)) \qquad (\lambda \equiv \frac{\Delta t}{h} \equiv \frac{\Delta t}{\Delta x}) \tag{2}$$

Starting from initial values

$$u_i^0 = \frac{1}{h}\int_{x_{i-1/2}}^{x_{i+1/2}} u_0(x)dx \tag{3}$$

the first (or further odd) time step of (2), with $n = 0$ (or $n = 2n', n' = 1, 2, ...$) leads to staggered values $\{u_{i+1/2}^1\}$ (or $\{u_i^{2n'+1}\}$) which can be considered as piecewise constant on the cells $C_{i+1/2}$ of the staggered, dual grid:

$$C_{i+1/2} = \{x | x_i = ih < x < x_{i+1}\} \tag{4}$$

To obtain second-order accuracy, Nessyahu and Tadmor [12] introduced van Leer's MUSCL-type [16] piecewise linear reconstruction of the piecewise constant solution obtained at the previous step:

$$u(x, t^n) = L_i(x, t^n) = u_i^n + (x - x_i)\frac{u_i'}{\Delta x} \quad x \in C_i \tag{5}$$

where

$$u_i' = (u_i^n)' \cong h\frac{\partial}{\partial x}u(x, t^n)|_{x=x_i} + O(\Delta x^2) \tag{6}$$

Integrating (1) on $[x_i, x_{i+1}] \times [t^n, t^{n+1}]$ and using Green's formula gives

$$\int_{x_i}^{x_{i+1}} u(x, t^{n+1})dx = \int_{x_i}^{x_{i+1}} u(x, t^n)dx - \{\int_{t^n}^{t^{n+1}} f(u(x_{i+1}, t))dt - \int_{t^n}^{t^{n+1}} f(u(x_i, t))dt\} \quad (7)$$

Applying the midpoint rule to the flux integrals then leads to the NT formula:

$$u_{i+1/2}^{n+1} = \frac{1}{2}(u_{i+1}^n + u_i^n) + \frac{1}{8}((u_i^n)' - (u_{i+1}^n)') - \lambda(f(u_{i+1}^{n+1/2}) - f(u_i^{n+1/2})) \quad (8)$$

where $u_i^{n+1/2}$ is an approximate value defined by a **predictor step**[12].

2.2 One-dimensional NT-type scheme without predictor step

Instead of using the (NT)-predictor to obtain second order time accuracy, the flux function will be reconstructed with the help of previous values. We start from (7) and denote by $\mathcal{P}(x, t)$ the first degree Lagrange polynomial interpolating the function $f(u(x, t))$ at the nodes t^{n-2}, t^n (considering x as a parameter). Integrating $\mathcal{P}(x, t)$ instead of $f(u(x, t))$ in (7) leads to our modified form of the NT scheme

$$u_{i+1/2}^{n+1} = \frac{1}{2}(u_{i+1}^n + u_i^n) + \frac{1}{8}((u_i^n)' - (u_{i+1}^n)') - \frac{\Delta t^{n+1}}{h}\{\mathcal{F}(u_{i+1}^n, u_{i+1}^{n-2}) - \mathcal{F}(u_i^n, u_i^{n-2})\} \quad (9)$$

where

$$A_1 = 1 + \frac{\Delta t^{n+1}}{2(\Delta t^{n-1} + \Delta t^n)}, \qquad A_2 = 1 - A_1 \quad (10)$$

and $\Delta t^n = t^n - t^{n-1}$, and where

$$\mathcal{F}(u_i^n, u_i^{n-2}) \equiv A_1 f(u_i^n) + A_2 f(u_i^{n-2}) \quad (11)$$

is a modified numerical flux inspired by multistep methods for ODE's.

3 2D and 3D NT-type FV methods without predictor

Starting from the following model equation ($\mathbf{F} : \mathbb{R} \mapsto \mathbb{R}^{(2,3)}$):

$$u_t + \nabla \cdot \mathbf{F}(u) = 0, \ \ u(\mathbf{x}, 0) = u_0(\mathbf{x}) \quad (12)$$

we rewrite (12) in a method of lines fashion:

$$u_t = -\nabla \cdot \mathbf{F}(u) \equiv G(u) = \frac{\partial u}{\partial t} \tag{13}$$

In the original finite volume formulation for tetrahedral grids, (12) is integrated with respect to space on the dual cells L_{ij} and from t^n to t^{n+1} (Fig. 1). Instead of $G(u)$ we integrate its $\mathcal{P}_1$-Lagrange polynomial:

$$\mathcal{P}(\mathbf{x},t) = (\frac{(t-t^{n-2})G(u(\mathbf{x},t^n))}{\Delta t^n + \Delta t^{n-1}} + \frac{(t^n - t)G(u(\mathbf{x},t^{n-2}))}{\Delta t^n + \Delta t^{n-1}}) \tag{14}$$

First, integrating by parts with respect to time leads to

$$\int_{L_{ij}} u(x,t^{n+1})dV = \int_{L_{ij}} u(x,t^n)dV + \int_{L_{ij}}\int_{t^n}^{t^{n+1}} \mathcal{P}(\mathbf{x},t)dtdV + O(\Delta t)^2 \tag{15}$$

Proceeding as in section 2 for the cell values $u_{L_{ij}}(t^n) = u^n_{L_{ij}}$, then gives

$$\begin{aligned} u_{L_{ij}}(t^{n+1})vol(L_{ij}) = &\int_{L_{ij}\cap C_i} u(x,t^n)dV + \int_{L_{ij}\cap C_j} u(x,t^n)dV \\ &+ \int_{L_{ij}} \{A_1 G(u(\mathbf{x},t^n)) + A_2 G(u(\mathbf{x},t^{n-2}))\}dV \end{aligned} \tag{16}$$

with A_1, A_2 given by (10). Applying the divergence theorem leads to

$$\begin{aligned} u_{L_{ij}}(t^{n+1})vol(L_{ij}) = &\int_{L_{ij}\cap C_i} u(x,t^n)dV + \int_{L_{ij}\cap C_j} u(x,t^n)dV \\ -(&\int_{\partial L_{ij}\cap C_i} + \int_{\partial L_{ij}\cap C_j})\{A_1\mathbf{F}(u(\mathbf{x},t^n)) + A_2\mathbf{F}(u(\mathbf{x},t^{n-2}))\}\cdot \hat{n}dA \end{aligned} \tag{17}$$

which represents the **NT-type finite volume method without predictor** in its version valid for **3-dimensional unstructured tetrahedral grids**. Fig. 1 shows the part of the cell C_i (centered at node i) which is contained in the tetrahedron $ijkl$, and the dual cell L_{ij} constructed about the edge ij.

4 Stability

The study of stability for our new scheme (9) is more difficult than for the NT scheme. We have used a heuristic CFL condition obtained by multiplying by $\frac{1}{A_1}(\equiv \frac{4}{5}$ in the case of constant $\Delta t)$ the maximum CFL number used in the case of the corresponding schemes without occurrence of the values at time t^{n-2}. For instance, since the NT scheme is TVD for

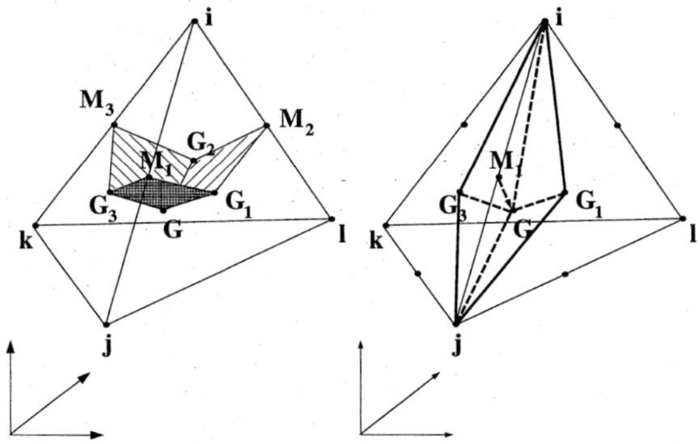

Fig. 1. Barycentric cell C_i and diamond cell L_{ij}

$$\lambda \max_i |f'(u_i)| \leq \beta, \quad \text{with} \quad \beta \leq \frac{1}{2}, \tag{18}$$

we have used a CFL number of the order of $\frac{4}{5} \cdot \frac{1}{2} = \frac{2}{5}$. Considering now the unstructured tetrahedral 3D case, recalling the original scheme presented in [2], for the Euler equations, and applying the condition for a monotone scheme $\frac{\partial H}{\partial u_j^n} \geq 0$, we had been using the following time step for the scheme with predictor

$$\Delta t_i = \frac{Vol(C_i)}{\lambda^i_{max} \int_{\partial C_i} d\sigma}$$

where

$$\lambda^i_{max} = \max(\lambda_i, \max_{j \in \mathcal{N}(i)} \lambda_j) \quad \text{and} \quad \lambda_i = ||\overrightarrow{V}_i|| + c_i$$

and $\overrightarrow{V}_i$, c_i refer to the values in the cell C_i of the velocity vector and sound speed, respectively. We chose the minimum Δt_i for all cell indices i : $(1 \leq i \leq nv)$ $\Delta t = \underset{1 \leq i \leq nv}{min} \{\Delta t_i\}$. We then used the multiplier $\frac{4}{5}$ to obtain our time step for the new scheme.

For a more detailed discussion of stability, consult Shu [13] and a forthcoming paper.

5 Numerical results

In all structured tests the *minmod-2* limiter (consult [12] p. 418) was used, while on unstructured 3D grids for the shock tube problem the new Venkatakrishnan limiter was applied. For the bump problem, van Leer's limiter was used because of its low computational cost. For details on those limiters, one can consult [16, 17, 4, 9]. It is worth mentioning that no artificial compression was applied on the NT scheme in any dimensions. Doing so would greatly improve the quality of the solution, particularly for the contact discontinuities, and could be applied to all of the newly proposed methods.

5.1 Cartesian 2D

Solving again Sod's shock tube problem [14] extended to 2 spatial dimensions and taking a transverse cut in the 2D plane gives the densities shown on figures (2) (to be compared with the results of the 1D scheme). The results are very similar for both schemes and generally good, except for the contact discontinuity which is somewhat smeared.

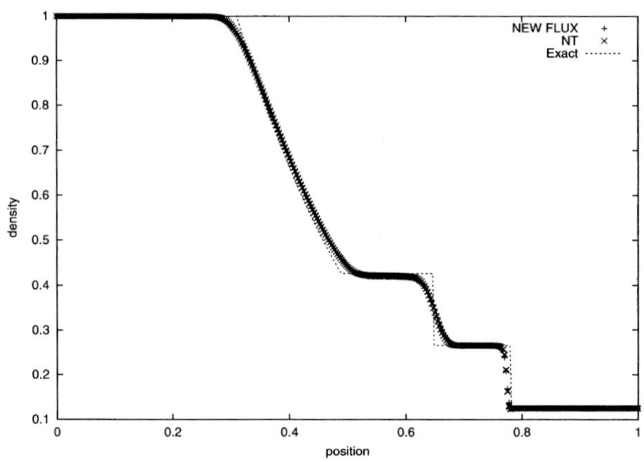

Fig. 2. Sod's shock tube 2D N=400 at $t = 0.16$

5.2 Unstructured 3D

Shock tube For comparison purposes, the shock tube problem is again solved here by applying the unstructured tetrahedral 3D schemes. Resulting densities obtained respectively by the use of the Nessyahu-Tadmor-like scheme (with predictor) and with the new proposed (faster) scheme (without predictor) are given in Fig.3. As in the 2-D case, the results for both schemes are practically indistinguishable. Here in 3D, the cost of performing the half time step is significant since 3 matrix-vector multiplications are needed at each node. The order of convergence of the scheme has been observed to be only 1.22 for nonlinear systems in [2], which has been confirmed by the recent theoretical results of [11] for a truly nonlinear scalar equation.

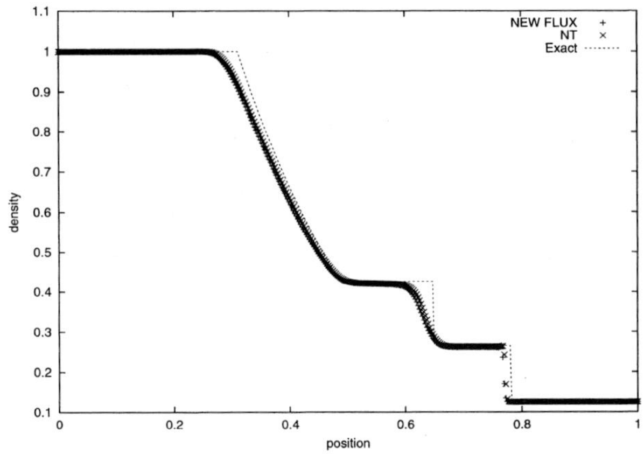

Fig. 3. Sod's shock tube 3D N=400 at $t = 0.16$

Channel with a 4% circular arc bump with $M_\infty = 1.65$ *Original 3D-unstructured Scheme*

An inviscid supersonic flow is considered over a circular 4% arc bump; this test case was first proposed as a benchmark problem by Eidelman et al. [8].The left boundary of the computational domain is considered as the inlet of the flow with a prescribed supersonic condition, namely : $(\rho, u, v, w, p) = (1.4, 1.65, 0, 0, 1)$ and the right boundary is considered as a supersonic outlet. For the sake of comparison 3 different meshes were used with 18079, 64099 and 138561 tetrahedra, respectively. For simplicity, we only show the results obtained with the final grid, which has a characteristic length of 1/30, 23279 nodes and 154535 edges (fig. 4). The results are comparable with those of the structured cases presented in [10] since the numbers of unknowns are about the same. An interesting property of the new scheme is the convergence rate to the stationary solution (Fig.6); the single precision machine epsilon is reached without requiring any special limiter like the one proposed by Venkatakrishnan, hence preventing any over/undershoot as observed in [1, 10].

3D-unstructured Interleaved Scheme

The new scheme does not need an predictor step as required by the original unstructured scheme proposed in [2]. Since the scheme is explicit and both edge based and vertex based data structures are used, the gain is evaluated on a tetrahedron basis (second/(cell × iteration)). The original scheme requires 2.5×10^{-5} second while the interleaved scheme requires only half of that time i.e. 1.25×10^{-5} second/(cell × iteration). This acceleration is directly reflected in the calculations since it is possible to use twice as many cells and

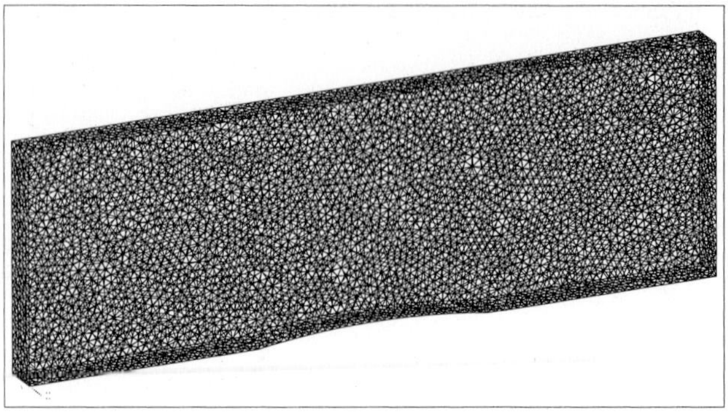

Fig. 4. Final mesh with characteristic length 1/30 (138561 Tetrahedra)

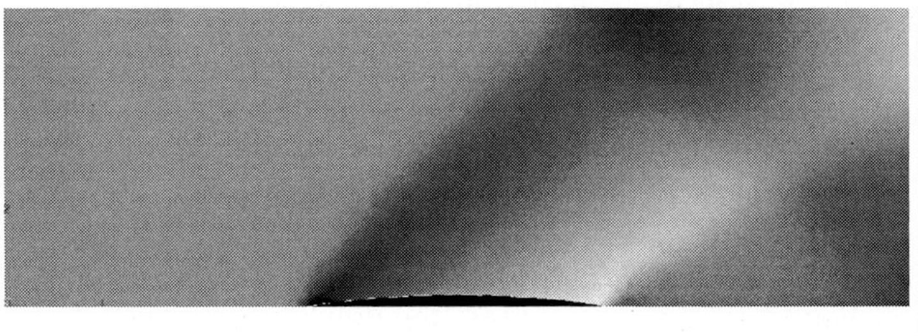

obtain a substantially finer solution in about the same time as that needed for the first method to solve the problem on a coarser mesh. The results are practically indistinguishable from those obtained with the scheme with predictor, so that Fig.5 is valid for both schemes. A plot of the isomach lines is given in Fig.6. As regards convergence to the steady state, the machine tolerance is reached in less than 2000 iterations for the original scheme, and 2500 iterations for the new scheme. Despite the fact that the CFL condition used with the interleaved scheme is more restrictive, thus requiring more time steps to reach the stationary state, the computing times are reduced by approximately 37% (Table 1).

Mesh 3 (138561 Elements)	iterations	time(sec)	Normalized
Original Scheme	1692	5868	1.00
Interleaved Scheme	2123	3681	0.63

Table 1. Comparison of the two schemes for a fixed residual

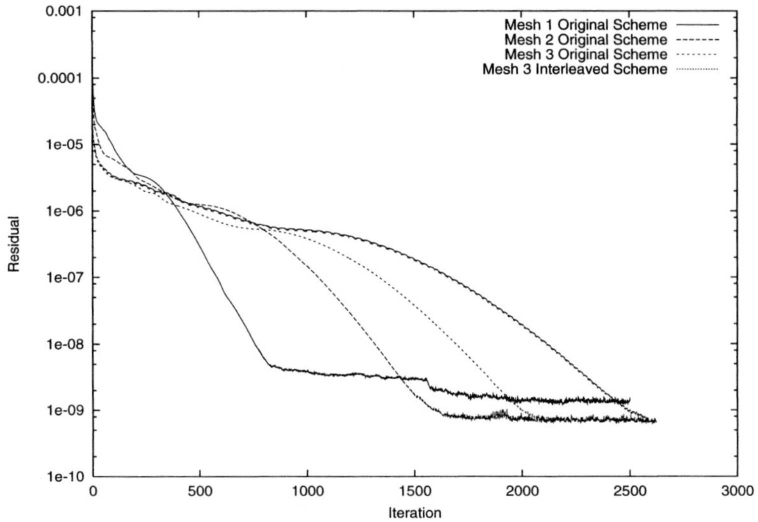

Fig. 6. Convergence history (residuals)

6 Conclusion

We have presented a new structured and unstructured Riemann solver-free centered finite-volume method for solving systems of conservation laws in 1,2 and 3D. The new scheme is significantly faster for 2 and 3D systems. The resolution is comparable in all dimensions with that of the NT scheme and its 2 and 3D extensions. This makes the new scheme computationally more efficient while being comparable, for accuracy, with the NT scheme and its 2 and 3D finite volume extensions for Cartesian or unstructured grids. The authors would like to thank J. F. Remacle at Rensselaer Polytechnic Institute for the freely available mesh generator GMSH.

References

1. M. Aftosmis, D. Gaitonde and T. S. Tavares (1995), *Behavior of Linear Reconstruction Techniques on Unstructured Meshes*, AIAA Journal, Vol. 33, No. 11, November 1995.
2. P. Arminjon, A. Madrane and A. St-Cyr, (2000), *Numerical simulation of 3-D flows with a non-oscillatory central scheme on staggered unstructured tetrahedral grids*, in Proc. 8th Int. Conf. on Hyperbolic Problems, Magdeburg (Germany) Feb. 28 - Mar. 3, 2000, H. Freistuehler and G. Warnecke, editors published by Birkhauser, Int Series of Num. Mathematics Vol. 140 (2001), pp. 59-68.

3. P. Arminjon, D. Stanescu and M. C. Viallon, (1995), *A two-dimensional finite volume extension of the Lax-Friedrichs and Nessyahu-Tadmor schemes for compressible flows*, Proc.of the 6th. Int. Symp. on Comp. Fluid Dynamics, Lake Tahoe (Nevada) September 4-8, 1995, M.Hafez and K. Oshima, editors, Vol. IV, pp. 7-14.
4. P. Arminjon, A. St-Cyr, A. Madrane (2002), *New 2 and 3-dimensional non-oscillatory central finite volume methods for staggered Cartesian grids*, Applied Numerical Mathematics, febuary 2002, Vol 40/3, pp 367-390.
5. P. Arminjon and M. C. Viallon, (1995), *Généralisation du schéma de Nessayahu-Tadmor pour une équation hyperbolique à deux dimensions d'espace*, C.R. Acad. Sci. Paris, t.320, serie I (1995), pp. 85 88.
6. P. Arminjon and M. C. Viallon, (1999), *Convergence of a finite volume extension of the Nessyahu-Tadmor scheme on unstructured grids for a two-dimensional linear hyperbolic equation*, SIAM J. Num. Anal., Vol. 36, No.3, pp. 738-771.
7. P. Arminjon, M.C. Viallon and A. Madrane, (1994), *A Finite Volume Extension of the Lax-Friedrichs and Nessyahu-Tadmor Schemes for Conservation Laws on Unstructured Grids*, revised version with numerical applications, Int. J. of Comp. Fluid Dynamics (1997), Vol. 9, No. 1, 1-22.
8. S. Eidelman, P. Collela and R. P. Shreeve (1984), *Application of the Godunov Method and its Second-Order Extension to Cascade Flow modeling*, AIAA Journal, Vol. 22, No. 11, 1984, pp. 1609-1615.
9. G. Jiang and E. Tadmor, (1998), *Non-oscillatory Central Schemes for Multidimensional Hyperbolic Conservation Laws*, SIAM J. on Scientific Computing, 19, pp.1892-1917.
10. M. H. Kobayashi and J. C. Pereira (1996), Characteristic-Based Pressure Correction at All Speeds, AIAA Journal, Vol. 34, No. 2, February 1996.
11. M. Küther (2001), *Error estimates for the staggered Lax-Friedrichs scheme on unstructured grids*, SIAM J. Numer. Anal. 39 (4) (2001) 1269-1301
12. H. Nessyahu and E. Tadmor, (1990), *Non-oscillatory central differencing for hyperbolic conservation laws*, J. Comp. Phys., 87, No. 2, pp. 408–463.
13. C.-W. Shu, (1988), *Total-variation diminishing time discretizations*, SIAM J. Sci. Stat. Comp., pp. 1073-1084.
14. G. A. Sod, (1978), *A survey of several finite difference methods for systems of nonlinear hyperbolic conservation laws*, J. Comp. Physisc, vol. 27, pp. 1-31.
15. G.D. van Albada, B. van Leer, W.W. Roberts, (1982), *A comparative study of computational methods in cosmic gas dynamics*, Astron. Astrophys. 108, pp 76-84.
16. B. van Leer, (1979), *Towards the ultimate conservative difference scheme V. A Second-Order Sequel to Godunov's Method*, J.Comp.Physics, 32, pp. 101-136.
17. V. Venkatakrishnan, (1995), *Convergence to Steady State Solutions of the Euler Equations on Unstructured Grids with Limiters*, J. Comp. Phys. 118, pp. 120-130.

High–resolution Riemann–solver–free Methods for Conservation Laws

Robert Artebrant and Hans Joachim Schroll

Centre for Mathematical Sciences, Lund University, Sweden
robert.artebrant@na.lu.se and *achim.schroll@na.lu.se*

Summary. Relaxed methods for hyperbolic conservation laws are reviewed. Similarities and differences to central schemes are described. A new relaxed upwind method for the shallow water system is introduced.

1 Introduction

During the 90th, based on the work by Nessyahu and Tadmor [8], so–called central schemes for conservation laws were developed. The beauty of these methods is their higher order accuracy despite the conceptional simplicity. Unlike traditional Godunov type methods, central schemes avoid the nonlinear Riemann problem and therefore became popular as easy–to–use general purpose methods. The first generation central schemes computed approximations to cell averages on staggered grids [8]. This grid staggering and the CFL condition make sure that the flux across cell interfaces is a continuous function in space and no Riemann problem needs to be solved. However, this approach has two consequences: First, the CFL number is reduced by a factor 1/2 compared to standard non–staggered methods. This is because of waves emanating from the discontinuous initial data have only half a step size to travel until they reach the nearest cell boundary. Consequently, a staggered grid central method requires twice as many time steps as non–staggered methods. Second, staggered grid methods suffer from a relatively large numerical diffusion. The numerical viscosity coefficient for the second order Nessyahu–Tadmor scheme is of order $\mathcal{O}(\cdot\S^{\triangle}/\cdot\sqcup)$, which leads to extensive diffusion in cells where the local Courant number is small [5]. However, grid staggering is a very effective approach for the development of schemes for multidimensional conservation laws, see [2] or the references listed on the web–site [16]. The second generation central schemes by Kurganov and Tadmor [5] avoid the staggered grids by introducing a second reconstruction step computing cell averages in the original unstaggered cells. In [5] they show that these new central schemes do in fact have a uniformly bounded numerical viscosity coefficient, independent of the time step.

High order resolution is implemented into the methods applying some kind of non–oscillatory reconstruction to the cell averages. Following the black–box concept of central schemes, it is straight forward to apply reconstructions componentwise to the conserved variables avoiding costly characteristic decompositions. However this simplification is questionable. Shu reports in [13] Sect.2.3 that the robustness of ENO schemes is greatly improved by applying characteristic decompositions. In the context with high order central schemes Pareschi, Puppo and Russo encountered difficulties with componentwise reconstructions. Therefore they apply an characteristic projection step in their central Runge–Kutta method [9].

An entirely different approach leading to similar high–resolution and Riemann–solver–free methods is based on the relaxation approximation by Jin and Xin [4]. In the following we describe the construction of relaxed schemes as an approximation to the original conservation law (without any source term). Furthermore, we discuss similarities and differences of central and relaxed schemes. In Section 3 we explain a third order scheme, based on piecewise hyperbolic reconstructions. Relaxed upwind methods, related to flux decompositions, are constructed in Section 4.

2 Riemann–solver–free methods via relaxation

Numerical schemes for conservation laws are derived from the integral form of the PDE. The central issue is to define an accurate and efficient numerical flux approximation. This is where Jin and Xin's relaxation approximation [4] becomes interesting from the numerical point of view.

$$\begin{aligned} u_t^\epsilon + \ v_x^\epsilon \ &= 0 \ , \\ v_t^\epsilon + a u_x^\epsilon &= \frac{1}{\epsilon}(f(u^\epsilon) - v^\epsilon) \ , \quad 0 < \epsilon \ll 1 \ . \end{aligned} \tag{1}$$

In the limit $\epsilon \to 0$, the v–component converges to the nonlinear flux $f(u)$ and the u–component is governed by the nonlinear conservation law

$$u_t + f(u)_x = 0 \ .$$

In other words: *Relaxation is flux approximation!* As the convective transport in the relaxation approximation (1) is described by a linear advection operator, discretizations based on (1) obviously avoid the nonlinear Riemann problem and thus lead to Riemann–solver–free schemes. First and second order relaxation schemes were originally introduced in [4]. A third order method was presented in [6] and higher order schemes were recently developed in [11, 12]. The later schemes are based on the following algorithm for the flux computation.

Algorithm (*relaxed, high resolution method*):

1. At time t_n, given cell averages u_i let $v_i = f(u_i)$.

2. In every cell $[x_{i-1/2}, x_{i+1/2})$, compute non–oscillatory, high–order reconstructions of $v \pm \sqrt{a}u$.
3. Evaluate the reconstructions at the cell boundaries

$$(v \pm \sqrt{a}u)_{i+1/2} = \lim_{h \to 0+} (v \pm \sqrt{a}u)(x_{i+1/2} \mp h) \ .$$

4. Interpret the average $v_{i+1/2} = \frac{1}{2}[(v + \sqrt{a}u)_{i+1/2} + (v - \sqrt{a}u)_{i+1/2}]$ as numerical flux and integrate $\partial_t u_i + \frac{1}{\Delta x}(v_{i+1/2} - v_{i-1/2}) = 0$ for one time step $\frac{\Delta t}{\Delta x}\sqrt{a} \leq 1$.

Some **remarks** are appropriate.

1. By a Chapman–Enskog expansion, the parameter a is related to the numerical viscosity of the scheme, see [12] for details. For stability reasons, a must be big enough: All eigenvalues of the Jacobian $Df(u)$ must lie between $\pm\sqrt{a}$. For accuracy reasons however, it is desirable to have a as small as possible, so that the optimal choice is

$$\sqrt{a} = \max\{|\lambda|, \lambda \in \sigma(Df(u)), \text{ for all relevant u } \} \ .$$

2. Note that $v \pm \sqrt{a}u$ are the characteristic variables of the advection operator in (1). Similar as for central methods, every time step of the relaxed method requires two reconstructions. However, it is natural to reconstruct $v \pm \sqrt{a}u$ instead of the conserved variables u. The way how reconstructions are applied is in fact the main difference between central and relaxed schemes.

3. It is well known that in the first order version, when the reconstructions are piecewise constant (and CFL = 1), the relaxed scheme reduces to the Lax–Friedrichs scheme, the prototype of staggered, central methods.

4. Relaxed schemes are stable for Courant numbers up to one $\frac{\Delta t}{\Delta x}\sqrt{a} \leq 1$. Note that $\pm\sqrt{a}$ are the characteristic speeds of (1).

5. Because of the positive and negative characteristic speeds of (1), all schemes based on the Jin–Xin approximation have cental stencils and can not simulate upwinding.

6. The numerical diffusion of relaxed method can be further reduced, by allowing for a local relaxation parameter $a = a(x)$. In this case $a(x_i) = a_i$ can be addapted to those eigenvalues involved in the local reconstruction (nearest neigbours). Such methods are called *locally relaxed* schemes, see [12]. Again, the first order version is known as Rusanov's "local Lax–Friedrichs" scheme [10]. In this sense locally relaxed schemes are related to unstaggered central schemes — the difference lies again in how the reconstructions are applied.

3 The locally relaxed, piecewise hyperbolic method

The algorithm in Section 2 describes a class of methods. To obtain any particular scheme it remains to select a reconstruction procedure. It turns out that in the relaxation context Marquina's *local hyperbolic reconstruction* (LHR) [7] is a stable and efficient procedure. Applying LHR in the relaxed, high–resolution algorithm with variable relaxation parameter $a = a(x)$, defines the *locally relaxed, piecewise hyperbolic method* (LRPHM). The time integration is performed by the third order nonlinear SSP–Runge–Kutta method from [3], Section 4.1. The LRPHM is tested and analyzed in [12]. In Section 5, higher order convergence is demonstrated in a numerical experiment. As LHR will be applied for relaxed upwinding, we summarize the method as follows:

Marquina proposed hyperbolas $r_i(x) = a_i + \frac{b_i}{(x-x_i)+c_i}$ to reconstruct the unknown functions from discrete data w_i within a computational cell. The parameters a_i, b_i and c_i are determined from three conditions:

i) Conservation of the average $w_i = \frac{1}{\Delta x}\int_{x_{i-1/2}}^{x_{i+1/2}} r_i(\xi)\, \mathrm{d}\xi$.
ii) Preservation of both lateral gradients $\pm\frac{1}{\Delta x}(w_{i\pm1} - w_i) = r_i'(x_{i\pm1/2})$.

Whenever the scalar data w_{i-1}, w_i and w_{i+1} is monotone, there is a unique hyperbola satisfying i) and ii). This hyperbola reconstructs the data with third order accuracy. As hyperbolas are monotone, the construction does not work for non–monotone data. In such transition cells, Marquina replaces the lateral derivative with larger absolute value by the other one multiplied by Δx^2. By this manipulation, the order of accuracy drops to two in these cells.

4 Relaxed upwinding via flux decompositions

One way to built upwinding into relaxation schemes is to decompose the flux into right– and left going parts, such that each part can be approximated seperately by relaxation. The *flux decomposition method* (FDM) is a particular realization of the class of discrete kinetic approximations [1], where the conserved variables are decomposed into N parts governed by a weakly coupled system of advection equations. Consider the D–dimensional system of conservation laws

$$\partial_t u + \sum_{d=1}^{D} \partial_{x_d} f_d(u) = 0 \ , \quad u(x,0) = u_0(x) \ . \tag{2}$$

The flux decomposition method uses $N = 2D + 1$ equations and a decomposition of the flux functions $f_d(u) = f_d^+(u) - f_d^-(u)$ with the Jacobians of the partial fluxes $f^\pm$ having nonnegative eigenvalues only. The FDM approximation of (2) reads $u^\epsilon = \sum_{n=1}^{2D+1} u_n^\epsilon$, where each of the components is governed by

$$\begin{cases} \partial_t u_n^\epsilon + \lambda_n \partial_{x_n} u_n^\epsilon & = \frac{1}{\epsilon}\left[\frac{1}{\lambda_n} f_n^+(u^\epsilon) - u_n^\epsilon\right] , \qquad n = 1, \dots, D \\ \partial_t u_{D+1}^\epsilon & = \frac{1}{\epsilon}\left[u^\epsilon - \sum_{d=1}^{D} \frac{1}{\lambda_d}(f_d^+(u^\epsilon) + f_d^-(u^\epsilon)) - u_{D+1}^\epsilon\right] \\ \partial_t u_n^\epsilon - \lambda_{n-(D+1)} \partial_{x_{n-D+1}} u_n^\epsilon & = \frac{1}{\epsilon}\left[\frac{1}{\lambda_{n-(D+1)}} f_{n-(D+1)}^-(u^\epsilon) - u_n^\epsilon\right] , \\ & \qquad n = D+2, \dots, 2D+1 . \end{cases} \tag{3}$$

The PDE's in (3) can be summed up to obtain

$$\partial_t u^\epsilon + \sum_{n=1}^{D} \lambda_n \partial_{x_n} u_n^\epsilon - \sum_{n=D+2}^{2D+1} \lambda_{n-(D+1)} \partial_{x_{n-(D+1)}} u_n^\epsilon = 0 . \tag{4}$$

In the relaxation limit $\epsilon \to 0$, we formally get the convergence

$$u^\epsilon \to u , \quad \lambda_d u_d^\epsilon \to f_d^+(u) , \quad \lambda_d u_{d+D+1}^\epsilon \to f_d^-(u) , \quad d = 1, \dots, D . \tag{5}$$

Using this convergence the relaxed FDM approximation becomes

$$\partial_t u + \sum_{d=1}^{D} \partial_{x_d} f_d^+(u) - \sum_{d=1}^{D} \partial_{x_d} f_d^-(u) = 0 . \tag{6}$$

The λ_d's do not appear explicitly but are needed to ensure that the CFL–condition is satisfied. Natalini's idea was to reconstruct the kinetic variables u_n^ϵ. By the convergence (5) we see that in the relaxation limit the partial fluxes are just the kinetic variables multiplied by a constant factor. Therefore we equivalently reconstruct the partial fluxes $f_d^\pm$. Using LHR and the third order SSP–RK method in the algorthm of Section 2 is refered to as *flux decomposition, piecewise hyperbolic method* (FDPHM).

If we consider the Euler system and use the van Leer decomposition of the fluxes with explicit first order time integration, (6) is up to a minus sign nothing else than van Leer's classical upwind scheme, [15].

4.1 A flux decomposition for the shallow water equations

In order to apply the FDM approximation to the shallow water system we have to find a decomposition of the flux function. Consider the shallow water flux function

$$\hat{f}(h, hu) = \begin{bmatrix} hu \\ hu^2 + h^2/2 \end{bmatrix} . \tag{7}$$

We use the regular transformation

$$t : \ (h, hu) \mapsto (h, \frac{hu}{h^{3/2}}) , \tag{8}$$

to write (7) as a function of the positive variable h and the Froude number $s = \frac{u}{\sqrt{h}}$. We obtain

$$f(h,s) = \begin{bmatrix} sh^{3/2} \\ s^2h^2 + h^2/2 \end{bmatrix} . \tag{9}$$

Lemma 1. *The Jacobian of (9) has only nonnegative (nonpositive) eigenvalues if $s \geq 1$ ($s \leq -1$).*

Proof. The Jacobian of (7) has eigenvalues $\lambda^{\pm} = u \pm \sqrt{h}$ so both eigenvalues are nonnegative (nonpositive) if $u \geq \sqrt{h}$ ($u \leq -\sqrt{h}$). The transformation (8) is regular so the zeros of the eigenvalues do not change. Compute an eigenvalue of the Jacobian of (9) at any point ($h > 0, s > 1$) (($h > 0, s < -1$)). □

For $s \in I = [-1, 1]$ we have the following ansatz for the decomposed fluxes

$$f^{\pm}(h,s;\alpha) = \begin{bmatrix} \pm\frac{1}{2(1+\alpha)}h^{3/2}(\pm s+1)(\pm\alpha s+1) \\ \frac{1}{4}h^2(\pm s+1)^2(\pm\frac{1}{2}s+1) \end{bmatrix} , \tag{10}$$

with $f(h,s) = f^+(h,s;\alpha) + f^-(h,s;\alpha)$ for all $h > 0$ and $s \in I$. Furthermore the ansatz ensures that $f^{\pm}$ are continuous with $f^+(h,s;\alpha) = f(h,s)$ for $s \geq 1$ and $f^-(h,s;\alpha) = f(h,s)$ for $s \leq -1$.

Lemma 2. *The Jacobian of f^+ (f^-) has nonnegative (nonpositive) eigenvalues only provided that*

$$-1/3 \leq \alpha \leq \alpha^* = \frac{4}{57}\sqrt{6} + \frac{13}{19} \approx 0.856 .$$

Proof. Consider f^+, the case f^- follows by symmetry. First we have to determine a range of α for which we have real and separated eigenvalues. The Jacobian of f^+ is

$$Df^+ = \begin{bmatrix} \frac{3}{4(1+\alpha)}\sqrt{h}(s+1)(\alpha s+1) & \frac{1}{2(1+\alpha)}h^{3/2}(2\alpha s+\alpha+1) \\ \frac{1}{4}h(s+1)^2(s+2) & \frac{1}{8}h^2(3s^2+8s+5) \end{bmatrix} =: \begin{bmatrix} a & b \\ c & d \end{bmatrix} .$$

The eigenvalues are real if $(\operatorname{tr} Df^+)^2 - 4 \det Df^+ \geq 0$, that is if $(a+d)^2 - 4(ad-bc) = (a-d)^2 + 4bc \geq 0$. Since $c \geq 0 \; \forall s \in I$ it is sufficient (but not necessary) that $b \geq 0$ to have real eigenvalues. We see that for $\alpha < -1, b \geq 0$ for some $s \in I$. For $\alpha > -1$ it is enough to consider the factor of b

$$g(s,\alpha) = 2\alpha s + \alpha + 1 .$$

We have $g(-1,\alpha) = 1-\alpha \geq 0$ if $\alpha \leq 1$ and $g(1,\alpha) = 3\alpha+1 \geq 0$ if $\alpha \geq -1/3$. For $\alpha \in J = [-1/3, 1]$ and $s \in I = [-1, 1]$, write $\Omega = I \times J$. On the boundary $\partial\Omega$ g is nonnegative, and furthermore there is a unique stationary (saddle) point $(s^0, \alpha^0) = (-1/2, 0)$ in Ω with $g(s^0, \alpha^0) = 1 > 0$. Now since g is continuous it is nonnegative on Ω and hence Df^+ has real eigenvalues on Ω.

Now we look for $\alpha \in J$ such that both eigenvalues are nonnegative. The idea is to find a set of α for which $\det Df^+ \geq 0$, then the eigenvalues have the same sign and it is enough to compute one eigenvalue at a specific point and see that it is positive. The determinant of Df^+ can be written as

$$\det Df^+ = p(h,s,\alpha) \cdot q(s,\alpha) \ ,$$

with

$$p(h,s,\alpha) = \frac{1}{32(1+\alpha)} h^{5/2}(s+1)^2 \geq 0 \ \forall h > 0 \ , \quad s \in I \ , \quad \alpha \in J$$

$$q(s,\alpha) = \alpha s^2 + 5(1-\alpha)s - 8\alpha + 7 \ .$$

Since $\partial_\alpha q = s^2 - 5s - 8 < 0$, $\forall s \in I$, we want to find the smallest α^* such that $q(s,\alpha^*) = 0$, because then we have $q(s,\alpha) > 0$, $\forall\, \alpha < \alpha^*$. Now

$$q(s,\alpha) = 0 \quad \Rightarrow \quad \alpha(s) = -\frac{5s+7}{s^2 - 5s - 8} \ .$$

Minimizing α for $s \in I$ gives $\alpha^* = \frac{4}{57}\sqrt{6} + \frac{13}{19} < 1$, so $\alpha^* \in J$. We have $g(s,\alpha) \geq 0$ and hence $\det Df^+ \geq 0$, $\forall s \in I$, $\alpha \in [-1/3, \alpha^*]$. □

In practice we use $\alpha = \alpha^*$ in the ansatz for the decomposed fluxes, because it gives the smallest possible jump in the eigenvalues for Froude numbers $s = 1$ and $s = -1$. In the 2D case the fluxes are written as

$$F(h,s,t) = \begin{bmatrix} sh^{3/2} \\ s^2h^2 + h^2/2 \\ sth^2 \end{bmatrix} \ , \quad G(h,s,t) = \begin{bmatrix} th^{3/2} \\ sth^2 \\ t^2h^2 + h^2/2 \end{bmatrix} \ ,$$

where $s = u/\sqrt{h}$ and $t = v/\sqrt{h}$. The decomposed fluxes are

$$F^\pm(h,s,t) = \begin{bmatrix} f_1^\pm(h,s) \\ f_2^\pm(h,s) \\ f_1^\pm(h,s)\sqrt{h}t \end{bmatrix} \ , \quad G^\pm(h,s,t) = \begin{bmatrix} f_1^\pm(h,t) \\ f_1^\pm(h,t)\sqrt{h}s \\ f_2^\pm(h,t) \end{bmatrix} \ .$$

5 Numerical experiments

Our first experiment is to check the order of convergence of LRPHM. We apply the method to the nonlinear wave equation

$$u_{tt} = c(u)u_{xx} \ , \quad c(u) = (1 + u_x^2)/2 \ ,$$

with smooth initial data. The error is estimated in L_1–norm, compared to a reference solution computed by the same method on a very fine grid. Figure 1 shows the smooth solution at the final time and a convergence rate

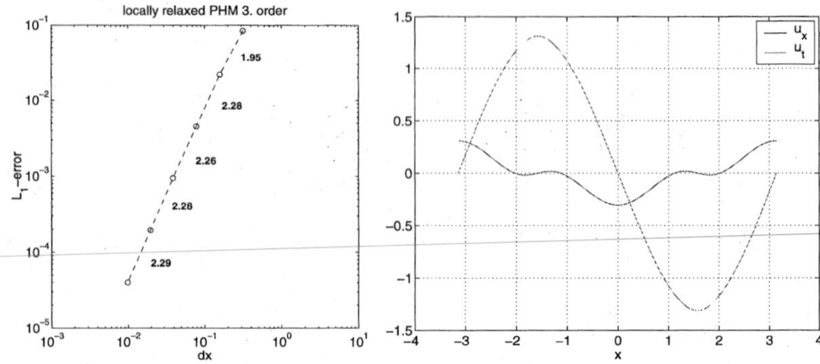

Fig. 1. Convergence test and the solution at $t = 1$ for LRPHM with CFL=1.

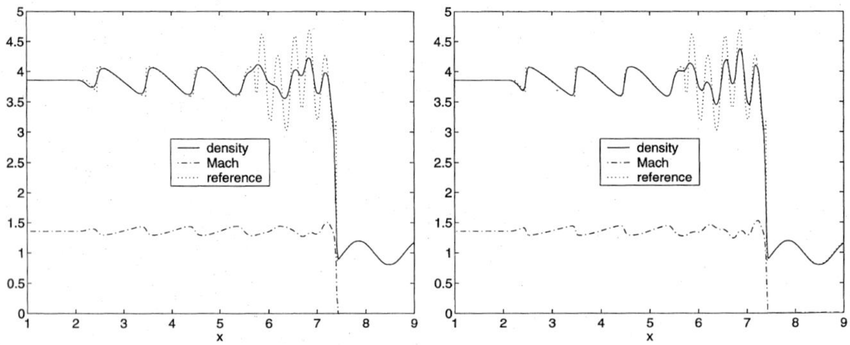

Fig. 2. Euler 1D. Shu-Osher acoustic-shock interaction. From left to right, the LRPHM and the FDPHM computations at $t = 1.8$.

of approximately 2.3 which makes sense since LHR reduces to second order accuracy at local extrema.

To illustrate the effect of upwinding, we compute the Shu–Osher acoustic–shock interaction [14] by LRPHM and FDPHM. The solutions were computed on 400 grid points using CFL=1. The maximal characteristic speed for van Leer's flux decomposition is estimated by

$$\lambda_n = \lambda = \max\left\{ (|u| + c)\, \frac{\gamma + 3}{2\gamma + |M|(3 - \gamma)} \right\} \;, \quad n = 1, \ldots, 2D + 1 \;, \tag{11}$$

where c is the speed of sound. The results displayed in Figure 2 show clearly the higher resolution of the upwind FDPHM approximation.

As a two dimensional test, we show a shock–bubble interaction. For initial data and a description of the problem, see the web–site [17]. Our computation is performed with FDPHM on a 640×400 grid with CFL=0.9. The result is displayed as a emulated Schlieren plot in Figure 3.

Finally, we present tests with our new splitting for the shallow water system. In analogy to (11), the maximal propagation speed is estimated by

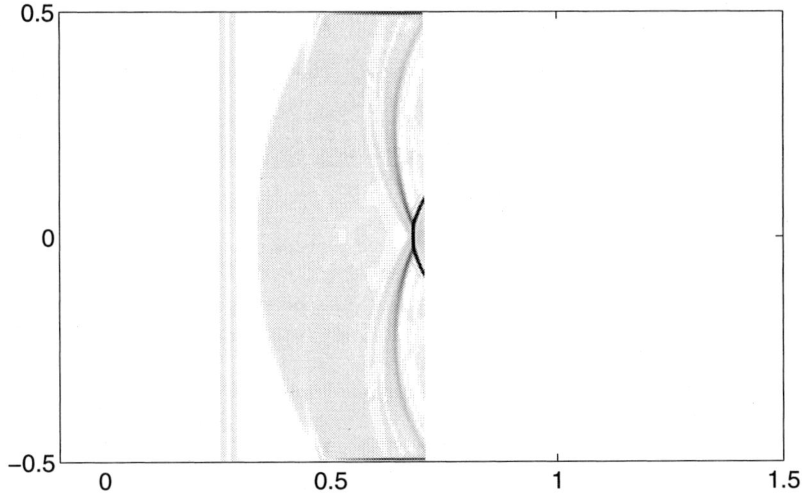

Fig. 3. Euler 2D. Shock-bubble interaction at $t = 0.4$. Computed by FDPHM

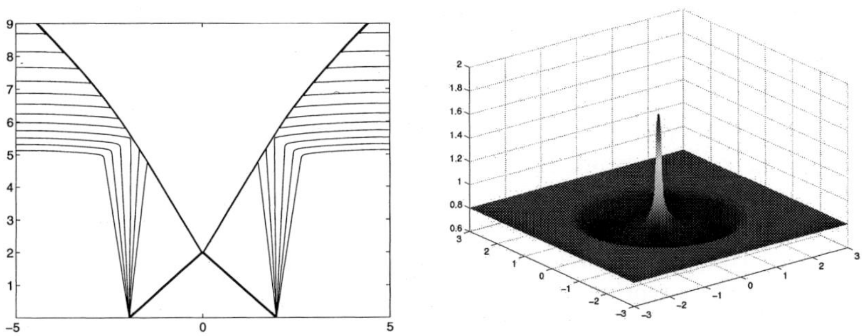

Fig. 4. Shallow water. To the left 1D double RP and to the right 2D implosion computed by FDPHM.

$$\lambda_n = \lambda = \max\left\{\left(|u| + \sqrt{h}\right)\frac{2}{1+|s|}\right\} \quad , \quad n = 1, \ldots, 2D+1 \ .$$

The examples consist of a 1D double Riemann problem and a 2D implosion test. The double RP was computed on 400 grid points with CFL=0.9. In Figure 4 we see the two waves and their interaction clearly resolved. The initial data for the 2D test was a centered circle with radius 1 and water height 0.4, surrounded by water of height 0.8. The water was initially at rest and the solution was computed at a 600×600 grid with CFL=0.8. Figure 4 shows the sharp peak at $t = 1.0$.

References

1. Aregba–Driollet D. and Natalini R. (2000): *Discrete Kinetic Schemes for Multidimensional Conservation Laws*, SIAM J. Num. Anal. **37**, pp. 1973-2004.
2. Arminjon P., Stanescu D. and Viallon M.-C. (1995): *A two–dimensional finite volue extension of the Lax–Friedrichs and Nessyahu–Tadmor schemes for compressible flow.* In: Hafez M. and Oshima K. (eds.) *Proc. 6th. Int. Symp. on CFD*, Vol. IV, Lake Tahoe, pp. 7–14.
3. Gottlieb S., Shu C.W. and Tadmor E. (2001): *Strong stability–preserving high order time discretization methods*, SIAM Review **43** pp. 89–112.
4. Jin S. and Xin Z.P. (1995): *The relaxation schemes for systems of conservation laws in arbitrary space dimensions*, Comm. Pure Appl. Math. **48** pp. 235–277.
5. Kurganov A. and Tadmor E. (2000): *New high–resolution central schemes for nonlinear conservation laws and convection–diffusion equations*, J. Comput. Phys. **160** pp. 241–282.
6. Li X.-G., Yu X.-J. and Chen G.-N. (2002): *The third–order relaxation schemes for hyperbolic conservation laws*, J. Comput. Appl. Math. **138**, no. 1, 93–108.
7. Marquina A. (1994): *Local piecewise hyperbolic reconstruction of numerical fluxes for nonlinear scalar conservation laws*, SIAM J. Sci. Comput. **15**, no. 4, pp. 892–915.
8. Nessyahu H. and Tadmor E. (1990): *Non–oscillatory central differencing for hyperbolic conservation laws*, J. Comput. Phys. **87** pp. 408–463.
9. Pareschi L., Puppo G. and Russo G. (2002): *Central Runge–Kutta schemes for conservation laws*, presented at HYP 2002.
10. Rusanov V.V. (1961): *Calculation of interaction of non–steady shock waves with obsracles*, J. Comput. Math. Phys. USSR, **1**, pp. 267–279.
11. Schroll H.J. (2002): *High resolution relaxed upwind schemes in gas dynamics*, J. Sci. Comp. **17(1-4)**, pp. 599-607.
12. Schroll H.J. (2002): *Relaxed High Resolution Schemes for Hyperbolic Conservation Laws*, preprint, Lund university.
13. Shu C.W. (1998): *Essentially non–oscillatory and weighted essentially non–oscillatory schemes for hyperbolic conservation laws*, In: Quateroni A. (ed.) *Advanced Numerical Approximation of Nonlinear Hyperbolic Equations*, Lecture Notes in Mathematics **1697**, Springer, Berlin, pp. 325–432.
14. Shu C.W. and Osher S. (1989): *Efficient implementation of essentially non–oscillatory shock–capturing schemes, II* J. Comput. Phys. **83**, pp. 32–78.
15. van Leer B. (1982): *Flux–vector splitting for the Euler equations*, In: Krause E., (ed.), Lecture Notes in Physics **170**, Springer, Berlin, pp. 507–512.
16. `www.math.ucla.edu/~tadmor/centralstation/`
17. `www.math.ntnu.no/~andreas/fronttrack/gas/sb/`

Riemann Problem for Conservation Laws with an Umbilic Point

Fumioki Asakura[1] and Mitsuru Yamazaki[2]

[1] Faculty of Engineering, Osaka Electro-Communication Univ. *asakura@isc.osakac.ac.jp*
[2] Institute of Mathematics, Univ. of Tsukuba *yamazaki@math.tsukuba.ac.jp*

Summary. We study the Riemann problems for 2×2 conservation laws with a hyperbolic singularity. The flux are a pair of quadratic functions where the characteristic speeds are equals and the Jacobian matrix is diagonal at the hyperbolic singularity i.e. umbilic point. Discontinuous solutions will be considered. They are characterized by 2 points on the Hugoniot curves which consist of 1-Hugoniot curve, 2-Hugoniot curve and a detached curve. The parts of compressible and overcompressible waves on the wave curves will be determined.

1 Introduction

We study a 2×2 system of conservation laws in 1 space dimension: $U = {}^t(u, v) \in \Omega \subset \mathbf{R}^2$,

$$U_t + F(U)_x = 0, \quad (x, t) \in \mathbf{R} \times \mathbf{R}_+ \tag{1}$$

where $F : \Omega \to \mathbf{R}^2$ is a smooth map.

We say that this system of equations is *hyperbolic*, when the Jacobian matrix $F'(U)$ has *real* eigenvalues $\lambda_1(U), \lambda_2(U)$ for any $U \in \Omega$. If, in particular, these eigenvalues are *distinct*: $\lambda_1(U) < \lambda_2(U)$, the system is called *strictly hyperbolic* at U. A state $U^* \in \Omega$ is called an *umbilic* point, if $\lambda_1(U) = \lambda_2(U)$ and $F'(U)$ is diagonal at $U = U^*$:

$$F'(U^*) = \lambda^* \mathrm{Id}_2; \; \lambda^* = \lambda_1(U^*) = \lambda_2(U^*). \tag{2}$$

The umbilic point makes a hyperbolic singularity.

In a strictly hyperbolic region, we have a pair of *characteristic fields* $R_1(U), R_2(U)$ which are right eigenvectors corresponding to $\lambda_1(U), \lambda_2(U)$. We choose left eigenvectors $L^1(U), L^2(U)$ such that

$$L^1(U)R_1(U) = L^2(U)R_2(U) = 1$$

$$L^2(U)R_1(U) = L^1(U)R_2(U) = 0.$$

Suppose that $U = U^*$ is an *isolated* umbilic point. ¿From Taylor expansion of $F(U)$ near $U = U^*$, it follows that

$$F(U) = F(U^*) + \lambda^*(U - U^*) + Q(U - U^*) + O(1)|U - U^*|^3 \tag{3}$$

where $Q : \mathbf{R}^2 \to \mathbf{R}^2$ is a homogeneous quadratic mapping. Change of variables: $x \to x - \lambda^* t$, $U \to U - U^*$ reduces the system (8) to

$$U_t + Q(U)_x = 0, \quad (x,t) \in \mathbf{R} \times \mathbf{R}_+ \tag{4}$$

modulo higher order terms. By changing unknown functions $V = S^{-1}U$ with a $S \in GL_2(\mathbf{R})$ the system (4) is transformed into a new system $V_t + P(V)_x = 0$ where $P(V) = S^{-1}Q(SV)$.

Definition 1. *Two quadratic mappings $Q_1(U)$ and $Q_2(U)$ are said to be* equivalent $Q_1(U) \sim Q_2(U) \Leftrightarrow \exists S \in GL_2(\mathbf{R})$ *such that*

$$Q_2(U) = S^{-1}Q_1(SU) \quad \text{for all} \quad U \in \mathbf{R}^2. \tag{5}$$

A general quadratic mapping $Q(U)$ has 6 coefficients and $GL_2(\mathbf{R})$ is a 4 dimensional group. Thus by the above equivalence transformations, we can eliminate four parameters.
These procedures induce *normal forms*(Schaeffer-Shearer [18]).

Let $Q(U)$ be a hyperbolic quadratic mapping with an isolated umbilic point $U = 0$.
$\Rightarrow \exists$ *2 parameters $a, b \in \mathbf{R}$ with $a \neq 1 + b^2$*
$Q(U) \sim \frac{1}{2}\nabla C$ where $\nabla = {}^t(\partial_u, \partial_v)$ and

$$C(U) = \tfrac{1}{3}au^3 + bu^2v + uv^2 \tag{6}$$

that is

$$Q(U) \sim \frac{1}{2}\nabla C(U) = \frac{1}{2}\begin{pmatrix} au^2 + 2buv + v^2 \\ bu^2 + 2uv \end{pmatrix}. \tag{7}$$

Moreover, if $(a,b) \neq (a',b') \Rightarrow \frac{1}{2}\nabla C \not\sim \frac{1}{2}\nabla C'$.

In the following argument, we shall confine ourselves to the system:

$$U_t + F(U)_x = 0, \quad (x,t) \in \mathbf{R} \times \mathbf{R}_+ \tag{8}$$

with

$$F(U) = Q(U) = \frac{1}{2}\begin{pmatrix} au^2 + 2buv + v^2 \\ bu^2 + 2uv \end{pmatrix}. \tag{9}$$

Geometric properties of the mapping $Q(U)$, for example the integral curves of characteristic vector fields, change as (a,b) varies in the ab-plane.

Schaeffer-Shearer's classification ([18]) :
Case I is $a < \frac{3}{4}b^2$; Case II is $\frac{3}{4}b^2 < a < 1 + b^2$; for $a > 1 + b^2$, the boundary between Case III and Case IV is $4\{4b^2 - 3(a-2)\}^3 = \{16b^3 + 9(1-2a)b\}^2$. We notice that these 2×2 system of hyperbolic conservation laws with an isolated umbilic point is a generalization of a three phase Buckley-Leverett

model for oil reservoir flow. Schaeffer–Shearer–Marchesin–Paes-Leme show that the quadratic approximation of the flux functions is either Case I or Case II.

The Riemann problem for (8) is the Cauchy problem with initial data of the form

$$U(x,0) = \begin{cases} U_L \text{ for } x < 0, \\ U_R \text{ for } x > 0 \end{cases} \tag{10}$$

where U_L, U_R are constant states in Ω. A jump discontinuity

$$U(x,t) = \begin{cases} U_L \text{ for } x < st, \\ U_R \text{ for } x > st \end{cases} \tag{11}$$

is a piecewise constant weak solution to the Riemann problem $\Leftrightarrow$ the *Rankine-Hugoniot condition* :

$$s(U_R - U_L) = F(U_R) - F(U_L). \tag{12}$$

The above discontinuity is a j-*compressive* shock wave ($j = 1, 2$) $\Leftrightarrow$ the *Lax entropy conditions*

$$\lambda_j(U_R) < s < \lambda_j(U_L) \tag{13}$$
$$\lambda_{j-1}(U_L) < s < \lambda_{j+1}(U_R). \tag{14}$$

Here the convention $\lambda_0 = -\infty$ and $\lambda_3 = \infty$. These shock waves are sometimes called j-*admissible* shock waves. In Case II, we shall also face with the *overcompressive* shock wave :

$$\lambda_1(U_R) < s < \lambda_1(U_L) \tag{15}$$
$$\lambda_2(U_R) < s < \lambda_2(U_L). \tag{16}$$

The *Hugoniot loci* over U_0 are the set of (U, s) satisfying

$$H_{U_0}(U,s) = s(U - U_0) - \{F(U) - F(U_0)\} = 0. \tag{17}$$

Their projections on to the U-plane are called the *Hugoniot curves* through U_0. If U_0 is not an umbilic point, Lax ([12]) shows: there exist 2 branches Hugoniot loci over U_0.

In this talk, we shall confine ourselves to Case I and II which involve complicated hyperbolic structure compared with other cases. Our aim is to determine rigorously a compressive section of the Hugoniot curves.

Although we have an extensive bibliography: Gomes ([5]), Isaacson-Marchesin-Plohr-Temple ([6]), Isaacson-Marchesin-Plohr ([7]), Isaacson-Marchesin-Palmeira-Plohr ([8]), Isaacson-Temple ([9, 10]), Schaeffer-Shearer ([18, 19]), Shearer ([20]), Schaeffer-Shearer-Marchesin-Paes-Leme ([21]), Schecter-Shearer ([23]) etc., the purely mathematical analysis of Hugoniot curves has not been done much and the determinations of the compressive and overcompressive parts of Hugoniot curves have been carried out mainly through numerical computations so far. On the other hand, Gomes ([5]) reports that there exist, on a detached branch of Hugoniot curves, compressive shock waves that do not have viscous profiles.

Our study is based on the fact that the Hugoniot curves are rational curves on the plane, which is already pointed out by Schaeffer-Shearer [19]. Our main tools are Wendroff's lemma, first proved Wendroff [24]. In section 3, we obtain parametrizations of these curves by rational functions. We also review Wendroff's lemma and its consequences. In section 4, we determine compressive and overcompressive parts of the Hugoniot curve. The wave curves through the hyperbolic singularity can be discussed and their compressive and overcompressive parts can be obtained but we omit our results in this talk owing to limited space.

2 Characteristic Fields

The characteristic equation of $F'(U)$ is:

$$\lambda^2 - \{(a+1)u + bv\}\lambda - \{v^2 + buv + (b^2 - a)u^2\} = 0. \tag{18}$$

On integral curves (u, v) to the right eigenvectors:

$$\begin{aligned}&(bu+v)\left(\frac{dv}{du}\right)^2 + \{(a-1)u + bv\}\left(\frac{dv}{du}\right)\\ &-(bu+v) = 0.\end{aligned} \tag{19}$$

Darboux ([4]) solved the equation by using the *Legendre transformation*:

$$p = \frac{dv}{du}, \quad q = u\frac{dv}{du} - v = pu - v. \tag{20}$$

The above equation (19) is equivalent to

$$\frac{dq}{dp} = \frac{q(p^2 + bp - 1)}{p^3 + 2bp^2 + (a-2)p - b}. \tag{21}$$

Lemma 1. *The points at which the gradients of integral curves $p = \frac{dv}{du}$ are equal constitute a line through the origin.*

Denote by $\Phi(p)$ the denominator of the expressions (21):

$$\Phi(p) = p^3 + 2bp^2 + (a-2)p - b. \tag{22}$$

Lemma 2. *Assume $a < 1 + b^2, b \geq 0$. Then $\Phi(p) = 0$ has 3 distinct roots $\mu_1, \mu_2, \mu_3 \in \mathbf{R}$.*

Definition 2. *The following three lines are called* medians .

$$M_1 : \ v = \mu_1 u, \quad M_2 : \ v = \mu_2 u, \quad M_3 : \ v = \mu_3 u. \tag{23}$$

Darboux showed

$$q = A(p-\mu_1)^{\alpha_1}(p-\mu_2)^{\alpha_2}(p-\mu_3)^{\alpha_3}$$

where A is a constant and

$$\alpha_l = -\frac{(1+\mu_l\mu_{l+1})(1+\mu_{l-1}\mu_l)}{(\mu_{l+1}-\mu_l)(\mu_{l-1}-\mu_l)} \quad (1 \le l \le 3)$$

with a convention $\mu_0 = \mu_3, \mu_4 = \mu_1$. Moreover, by setting

$$\kappa = A(p-\mu_1)^{\alpha_1-1}(p-\mu_2)^{\alpha_2-1}(p-\mu_3)^{\alpha_3-1},$$

$$\begin{aligned} u &= \kappa\{p^2 - (\mu_1\mu_2\mu_3 + \mu_1 + \mu_2 + \mu_3)p - 1\}, \\ v &= -\kappa\{\mu_1\mu_2\mu_3(p^2-1) \\ &\quad + (\mu_1\mu_2 + \mu_2\mu_3 + \mu_3\mu_1 + 1)p\}. \end{aligned}$$

For every state $U_0 \notin \cup_{k=1}^3 M_k$, there exists a unique j-integral ($j = 1, 2$) curve through U_0. This integral curve has three connected components and $p = \frac{dv}{du}$ ($p \neq \mu_l, 1 \le l \le 3$) is a regular parameter. Each median: M_k ($1 \le k \le 3$) is an asymptote for two components as $p \to \pm\mu_l$. For $U_0 \in M_k$ ($1 \le k \le 3$), the median itself is an integral curve; the one for other characteristic direction has two connected components.

The j-characteristic ($j = 1, 2$) direction is *genuinely nonlinear* at U if and only if

$$(\nabla\lambda_j \cdot R_j)(U) \neq 0. \tag{24}$$

The set of U satisfying $(\nabla\lambda_j \cdot R_j)(U) = 0$ is called the *j-inflection locus I_j*.

Proposition 1. *In Case I, there exist 3 inflection loci, while in Case II, there exists 1 inflection loci.*

3 Hugoniot Loci

The Hugoniot loci@$U = {}^t(u, v)$ through $U_0 = {}^t(u_0, v_0)$ is described by:

$$\begin{aligned} &\{a(u^2 - u_0^2) + 2b(uv - u_0v_0) + (v^2 - v_0^2)\}(v - v_0) \\ &\quad = \{b(u^2 - u_0^2) + 2(uv - u_0v_0)\}(u - u_0). \end{aligned} \tag{25}$$

Using parameter ξ: $v - v_0 = \xi(u - u_0)$,

Proposition 2. *we have the rational parametrization:*

$$u = \frac{1}{\Phi(\xi)}\{(\xi^3 - a\xi + b)u_0 + 2(1 - b\xi - \xi^2)v_0\},$$
$$v = \frac{1}{\Phi(\xi)}\{2\{b\xi - (a-1)\xi^2 - b\xi^3\}u_0$$
$$+(-b + a\xi - \xi^3)v_0\},$$
$$s = \frac{1}{\Phi(\xi)}\{(\xi^2 + b\xi + b^2 - a)(u_0\xi - v_0)\}.$$

The Hugoniot curve $\mathcal{H}(U_0)$ has 3 connected components with the shock speed $s = s(U_0, U)$:

1-Hugoniot curve, 2-Hugoniot curve and detached Hugoniot curve.

Lemma 3 (Wendroff [24]).
Let $U = U(\xi) \in \mathcal{H}(U_0)$. *Then* $\dot{s} = 0 \Leftrightarrow s = \lambda_j(U)$ $(j = 1$ *or* $2)$

Lemma 4. *Let* $U = U(\xi) \in \mathcal{H}(U_0)$. *If* $\dot{s} = 0$ *at* $\xi = \xi_1$. $\ddot{s} = 0 \Leftrightarrow \dot{\lambda}_j(U) = 0$. *Moreover, if* $\dot{s} = 0$ *and* $\ddot{s} = 0$ *at* $\xi = \xi_1$. *Then* $\dddot{s} = 0 \Leftrightarrow \ddot{\lambda}_j(U) = 0$.

Proposition 3. *The state* U_0 *is a bifurcation point of* $\mathcal{H}(U_0)$ *(the primary bifurcation point). There exists a secondary bifurcation point of* $\mathcal{H}(U_0)$ *if and only if* $U_0 \in \cup_{k=1}^3 M_k$.

4 Compressive Parts of the Hugoniot Curve

For a given U_0, the point U_1 is said to be a *j-limit point* if $s(U_0, U_1) = \lambda_j(U_1)$ and *j-overlap point*, if $s(U_0, U_1) = \lambda_j(U_0)$. We can show that there do not exist 1-limit point nor 2-overlap point.

The *j-double contact locus* D_j is the set of states U_0 such that $\mathcal{H}(U_0)$ has a point U_1 *limit-overlap point* such that U_1 is a j-overlap point with $s(U_0, U_1) = \lambda_j(U_0)$ and also a limit point with $s(U_0, U_1) = \lambda_k(U_1)$ $(k = 1$ or $2)$.

We also need the *Hysteresis locus* H that is the set of states U_0 such that there exists U_1 on $\mathcal{H}(U_0)$: $\dot{s}(U_0, U) = \ddot{s}(U_0, U) = 0$ at $U = U_1$.

Let us now consider Case I.

Proposition 4. *In Case I, the hysteresis loci consist of three lines.*

Assume that:

$$U_0 \notin \bigcup_{j=1}^{3}(M_j \cup I_j \cup H_j). \tag{26}$$

Assuming for simplicity

$$\frac{v_0}{u_0} > \mu_3 \text{ and } u_0 > 0, \tag{27}$$

we find that

1. 1-Hugoniot curve for $\mu_1 < \xi < \mu_2$,
2. 2-Hugoniot curve for $\mu_2 < \xi < \mu_3$,
3. Detached Hugoniot curve for $\xi < \mu_1$ and $\xi > \mu_3$.

Let us first consider the 1-Hugoniot curve. Let $U_0 = U(\xi_0)$, $\xi_0 \in (\mu_1, \mu_2)$. Since U_0 is not a inflection point, we have a classical configuration of Lax ([12]) in a neighborhood of U_0. Let the part for $\xi > \xi_0$ be compressive.

Theorem 1. *Under the above assumptions, the 1-Hugoniot curve is compressive for all $\xi : \xi_0 < \xi < \mu_2$. For $\xi < \xi_0$, this curve is eventually 1-compressive as $\xi \to \mu_1 + 0$.*

Remark 1. The above discussion also covers the case $U_0 \notin \bigcup_{j=1}^3 (M_j \cup H_j)$ but $U_0 \in \cup_{j=1}^3 I_j$. About $\xi = \xi_0$ we have an alternative: 1-Hugoniot curve is compressive for both sides or not for either side. We can show the curve is always compressive or eventually compressive for both sides.

Theorem 2. *Under the above assumptions, the 2-compressive part of the 2-Hugoniot curve is contained in a bounded region.*

Theorem 3. *Under the above assumptions, in the detached Hugoniot curve, the part: $s < 0$ is eventually 1-compressive as $\xi \to \mu_3 + 0$.*

Remark 2. We can easily check above all compressive shock waves are admissible in the sense that they satisfy the Liu-Oleĭnik condition:

$$s(\xi) \leq s(\xi') \quad \text{for any } \xi' \text{ between } \xi_0 \text{ and } \xi. \tag{28}$$

Moreover, there exist viscous profiles: traveling wave solutions to the parabolic system

$$U_t + F(U)_x = \varepsilon U_{xx} \quad (\varepsilon > 0)$$

connecting these states, even if they are not close (see Asakura [1]).

Outline of Proof: Due to the assumptions, the 1-Hugoniot curve is compressive for $\xi > \xi_0$ in a neighborhood of ξ_0. Owing to the geometry of the set $\{U; \lambda_j(U) = const.\}$, we can prove that there is neither 1-limit point nor 2-overlap point. Using this fact, we can show that the 1-Hugoniot curve is always compressive for $\xi \in (\xi_0, \mu_2)$. On the other hand, the 1-Hugoniot curve is not compressive for $\xi < \xi_0$ in a neighborhood of ξ_0. Owing to the Wendroff lemma, we can show that the graphs of $s(\xi)$ and $\lambda_1(U(\xi))$ cross transversely. Combining this result and the fact that $s(\xi) \to -\infty$ as $\xi \to \mu_1 + 0$, we can prove that the 1-Hugoniot curve is ultimately 1-compressive as $\xi \to \mu_1 + 0$.

Let us now consider Case II under the same assumption as in Case I. Recall that

$$s(\xi) = \frac{(\xi - \vartheta_1)(\xi - \vartheta_2)(u_0\xi - v_0)}{(\xi - \mu_1)(\xi - \mu_2)(\xi - \mu_3)}. \tag{29}$$

Proposition 5. *In Case II and* $b > 0$,

1. *the pieces of the double contact loci with* $u \geq 0$ *i.e.* $\{U; v = \vartheta_j u, u \geq 0, j = 1, 2\}$ *is exactly the set of* $\{U; \lambda_1(U) = 0\}$.
2. *the pieces of the double contact loci with* $u \leq 0$ *i.e.* $\{U; v = \vartheta_j u, u \leq 0, j = 1, 2\}$ *is exactly the set of* $\{U; \lambda_2(U) = 0\}$.

Theorem 4. *Under the above assumptions, the 1-Hugoniot curve for* $\xi < \xi_0$ *is eventually 1-compressive and its overcompressive part is contained in a bounded region.*

Theorem 5. *Under the above assumptions, the compressive part of the 2-Hugoniot curve is contained in a bounded region and the 2-Hugoniot curve is eventually overcompressive as* $\xi \to \mu_2 + 0$.

As pointed out in Remark 2, all compressible shock waves obtained here in Case II also satisfy Liu-Oleĭnik condition (28) and have corresponding viscous profiles.

References

1. Asakura, F. (preprint): Admissibility of strong shock waves in 2 × 2 systems of conservation laws
2. Čanić, S., Plohr, B.J. (1995): Shock Wave Admissibility for Quadratic Conservation Laws, J. Differential Equations, **118**, 293–335
3. Chen, G.Q., Kan, P.T. (1995): Hyperbolic Conservation Laws with Umbilic Degeneracy, I, Arch. Rational Mech. Anal., **130**, 231–276
4. Darboux, G. (1896): Sur la forme des lignes de courbure dans le voisinage d'un omblic. In: Leçon sur la Theéorie Générale des Surfaces, quatrième partie, Note VII. Gauthier-Villars Paris
5. Gomes, M.E.S. (1989): Riemann problems requiring a viscous profile entropy condition, Adv. Appl. Math. **10**, 285–323
6. Isaacson, E. , Marchesin, D., Plohr, B. Temple, B. (1988): The Riemann Problem near a hyperbolic singularity: The classification of solutions of quadratic Riemann problems I, SIAM J, Appl. Math. **48**, 1009–1032
7. Isaacson, E. , Marchesin, D., Plohr, B. (1990): Transitional waves for conservation laws, SIAM J, Math. Anal. **21**, 837–866
8. Isaacson, E. , Marchesin, D., Palmeira, C. , Plohr, B. (1992): A global formalism for nonlinear waves in conservation laws, Comm. Math. Phys. **146**, 505–552
9. Isaacson, E. , Temple, B. (1988): The Riemann Problem near the hyperbolic singularity II, SIAM J, Appl. Math. **48**, 1287–1301
10. Isaacson, E. , Temple, B. (1988): The Riemann Problem near the hyperbolic singularity III, SIAM J, Appl. Math. **48**, 1302–1318
11. Keyfitz, B. , Kranzer, H. (1978): Existence and uniqueness of entropy solutions to the Riemann problem for hyperbolic systems of two conservation laws, J. Differential Equations, **27**, 444–476

12. Lax, P.D. (1957): Hyperbolic systems of conservation laws II, Comm. Pure Appl. Math., **10**, 537–566
13. Lax, P.D. (1971): Shock waves and entropy. In: E. Zarantonello (ed) Contributions to nonlinear Functional Analysis. Academic Press New York
14. Liu, T.P. (1974): The Riemann problem for general 2×2 conservation laws, Trans. Amer. Math. Soc. **199**, 89–112.
15. Liu, T.P. (1975): Existence and uniqueness theorems for Riemann problems, Trans. Amer. Math. Soc. **212**, 375–382
16. Liu, T.P. (1975): The Riemann problem for general systems of conservation laws, J. Differential Equations, **18**, 218–234
17. Oleĭnik, O. (1957): Discontinuous solutions of nonlinear differential equations, Usp. Mat. Nauk, 12, , 3–73
18. Schaeffer, D., Shearer, M. (1987): The classification of 2×2 systems of nonstrictly hyperbolic conservation laws, with applications to oil recovery, Comm. Pure Appl. Math., **40**, 141–178
19. Schaeffer, D., Shearer, M. (1987): The Riemann problem for nonstrictly hyperbolic 2×2 systems of conservation laws, Trans. Amer. Math. Soc., **304**, 267–306
20. Shearer, M. (1989): The Riemann problem for 2×2 systems of hyperbolic conservation laws with case I quadratic nonlinearities, J. Diff. Equations, **80**, 343–363
21. Shearer, M., Schaeffer, D., Marchesin, D., Paes–Leme, P.L. (1987): Solution of the Riemann problem for a prototype 2×2 system of nonstrictly hyperbolic conservation laws, Arch. Rational Mech. Anal., **97**, 299–320
22. Smoller, J. (1994): Shock Waves and Reaction-Diffusion Equations (2nd ed.). Springer Verlag New York
23. Schecter, S., Shearer, M. (1991): Undercompressible shocks for nonstrictly hyperbolic conservation laws, J. Dyn. Diff. Equations **3**, 199–271
24. Wendroff, B. (1972): The Riemann problem for materials with nonconvex equations of state II: General flow, J. Math. Anal. Appl., **38**, 640–658

Admissibility of Shock Waves and Uniqueness of the Riemann Problem

Fumioki Asakura

Faculty of Engineering, Osaka Electro-Communication Univ.
asakura@isc.osakac.ac.jp

Summary. We study the admissibility of shock waves and the uniqueness of the Riemann problem for a general 2×2 hyperbolic system of conservation laws in one space dimension: $U_t + F(U)_x = 0$ with initial data having large amplitude. We assume that that the characteristic fields are *strictly separated*; that is: there exist two disjoint (open) conical neighborhoods such that the first characteristic field is confined to one of these neighborhoods and the second characteristic field to the other. We show that, together with some technical assumptions, the viscous profile exists for states U_-, U_+ not necessarily close and there exists at most one solution to the Riemann problem. These results will generalize admissibility and uniqueness theorems of Liu and Smoller, by giving descriptions free from particular choice of rectangular coordinates.

1 Introduction

We study a 2×2 system of conservation laws in one space dimension:

$$U_t + F(U)_x = 0, \quad (x,t) \in \boldsymbol{R} \times \boldsymbol{R}_+, \tag{1}$$

where $U = {}^t(u,v)$ denotes the state vector and $F(U) = {}^t(f,g)$ flux functions defined in a simply connected region $\Omega \subset \boldsymbol{R}^2$. This system is called *strictly hyperbolic*, if the Jacobian matrix $F'(U)$ has *real and distinct* eigenvalues: $\lambda_1(U) < \lambda_2(U)$ for $U \in \Omega$. If the domain Ω is simply connected, the corresponding (right) eigenvectors $R_1(U), R_2(U)$ are globally defined vector fields that are called the *characteristic fields*. The left eigenvectors $L^1(U), L^2(U)$ are so chosen that

$$L^1 R_1 = L^2 R_2 = 1, \quad L^2 R_1 = L^1 R_2 = 0.$$

The simplest self-similar discontinuity defined by

$$U(x,t) = \begin{cases} U_- \text{ for } x < st, \\ U_+ \text{ for } x > st. \end{cases} \tag{2}$$

makes a piecewise constant weak solution to the system of equations (1) if and only if these quantities satisfy the *Rankine-Hugoniot condition*:

$$s(U_+ - U_-) = F(U_+) - F(U_-). \tag{3}$$

The Riemann problem for (1) is the Cauchy problem with initial data of the form

$$U(x,0) = \begin{cases} U_L \text{ for } x < 0, \\ U_R \text{ for } x > 0 \end{cases} \tag{4}$$

where U_L, U_R are constant states in Ω.

The aim of this note is to propose geometric conditions for the flux functions $F(U)$ so that the viscous profile exists for states U_-, U_+ not necessarily close and there exists at most one solution to the Riemann problem. Liu [6], [7], [8] and Smoller [10] have studied these problems under the condition

$$f_v g_u > 0. \tag{5}$$

Our results will provide a description of this condition free from particular choice of rectangular coordinates.

2 Strictly Separated Characteristic Fields

Let α and β be two half-lines issued from the origin O such that their directions are linearly independent. We denote by $\Pi_{\alpha,\beta}$ the union of two open (smaller) sectors with the center O: one is made by α and β, the other is by half-lines with opposite directions $-\alpha, -\beta$. We denote by $\Pi_{\alpha,\beta}(U_0)$ the translation of $\Pi_{\alpha,\beta}$ by $\overrightarrow{OU_0}$ where O denotes the state corresponding to the origin; $\Pi_{\alpha,\beta} = \Pi_{\alpha,\beta}(O)$ by this notation. We say that the characteristic fields are *strictly separated*, if there exist two half-lines α, β from O such that the directions of α and β are linearly independent and one characteristic field satisfies $R_1(U) \in \Pi_{\alpha,\beta}$ and the other $R_2(U) \in \Pi_{-\beta,\alpha}$ at every state U. In Liu [6], [7] and Smoller [10], $\Pi_{\alpha,\beta}$ is the union of the first and the third quadrants. We first suppose that:

Assumption I The domain Ω is simply connected and the system of equations is hyperbolic in Ω. Moreover, the characteristic fields are strictly separated.

Let $\boldsymbol{a} = \begin{pmatrix} a_1 \\ a_2 \end{pmatrix}$ be a non-zero vector and $\boldsymbol{a}^\perp = \begin{pmatrix} -a_2 \\ a_1 \end{pmatrix}$ that is the vector orthogonal to $\boldsymbol{a}$ such that $|\boldsymbol{a}| = |\boldsymbol{a}^\perp|$ and $\det(\boldsymbol{a}, \boldsymbol{a}^\perp) > 0$. Let $\boldsymbol{e}_\alpha, \boldsymbol{e}_\beta$, respectively, be the unit vectors in the direction of α, β, respectively.

Lemma 1. *The characteristic fields are strictly separated, if and only if the following quantities are not null and have different signs for every U.*

$${}^t\boldsymbol{e}_\alpha^\perp F'(U)\boldsymbol{e}_\alpha, \quad {}^t\boldsymbol{e}_\beta^\perp F'(U)\boldsymbol{e}_\beta. \tag{6}$$

Proof. Let us denote by $e_\theta = \begin{pmatrix} \cos\theta \\ \sin\theta \end{pmatrix}$ the general unit vector. We will identify the symbols of half-lines α, β, respectively, with their argument $(-\beta \to \beta - \pi)$ in accordance with the notation e_θ. Note that the vector e_θ is a characteristic field if and only if $\Phi_U(\theta) = {}^t e_\theta^\perp F'(U) e_\theta = 0$. Hence the assumption implies that, for every U, there exist θ_1, θ_2 satisfying $\Phi_U(\theta_1) = \Phi_U(\theta_2) = 0, \alpha < \theta_1 < \beta$ and $-\beta < \theta_2 < \alpha$. Since the equation $\Phi_U(\theta) = 0$ is essentially a quadratic equation and the directions of α and β do not coincide with either characteristic field, we find that the quantities (6) are not null and have different signs. Conversely, if the above quantities have different signs, then there exist characteristic fields $e_{\theta_1}, e_{\theta_2}$ such that $\alpha < \theta_1 < \beta$ and $-\beta < \theta_2 < \alpha$. Thus the lemma follows.

In the above computation, we have

$$\Phi_U(\theta) = g_u \cos^2\theta - (f_u - g_v)\cos\theta\sin\theta - f_v \sin^2\theta.$$

If the Smoller-Liu condition (5) holds, then it follows that

$$\Phi_U(0)\Phi_U(\tfrac{\pi}{2}) = \Phi_U(-\pi)\Phi_U(\tfrac{\pi}{2}) = -f_v g_u < 0.$$

Hence we conclude that one characteristic field is confined in the first quadrant and another in the second.

Now, we can show that Assumption I is equivalent to a global property of the vector field:

$$H_{U_-}(U, s) = F(U) - F(U_-) - s(U - U_-). \tag{7}$$

Theorem 1. *Suppose that the characteristic fields are strictly separated and $R_1(U) \in \Pi_{\alpha,\beta}$, $R_2(U) \in \Pi_{-\beta,\alpha}$ at every state U. Then, for every U_0, the vector field $H_{U_0}(U, s)$ points inward (or outward) with respect to $\Pi_{\alpha,\beta}(U_0)$ on the boundary except for U_0. Conversely, suppose that the vector field $H_{U_-}(U, s)$ has the above property and the directions of α and β do not coincide with either characteristic fields. Then the characteristic fields are strictly separated.*

Proof. Differentiating quantities

$${}^t e_\alpha^\perp \{F(U_0 + \lambda e_\alpha) - F(U_0)\}, \quad {}^t e_\beta^\perp \{F(U_0 + \lambda e_\beta) - F(U_0)\}$$

with respect to λ, we obtain the quantities (6) with $U = U_0 + \lambda e_\alpha, U_0 + \lambda e_\beta$, respectively. By Lemma 1 lemma together with the mean value theorem, we find that the quantities

$$\lambda\, {}^t e_\alpha^\perp \{F(U_0 + \lambda e_\alpha) - F(U_0)\}, \quad \lambda\, {}^t e_\beta^\perp \{F(U_0 + \lambda e_\beta) - F(U_0)\} \tag{8}$$

have different signs for every U_0 and $\lambda \neq 0$, which implies the first part of the theorem.

Suppose conversely that above quantities have different signs; for simplicity assume that

$$\lambda^t e_\alpha^\perp \{F(U_0 + \lambda e_\alpha) - F(U_0)\} > 0, \quad \lambda^t e_\beta^\perp \{F(U_0 + \lambda e_\beta) - F(U_0)\} < 0.$$

Dividing the above quantities by λ^2 and letting $\lambda \to 0$, we have

$$^t e_\alpha^\perp F'(U) e_\alpha \geq 0, \quad ^t e_\beta^\perp F'(U) e_\beta \leq 0.$$

Since the directions of α and β are not equal to any characteristic fields, strict inequalities hold in the above expressions and hence we have strictly separated characteristic field. Thus the theorem is proved.

It is easy to see that states U_0, U satisfy the Rankine-Hugoniot condition if and only if

$$^t(U - U_0)^\perp \{F(U) - F(U_0)\} = 0. \tag{9}$$

This expression is due to Schaeffer-Shearer [9]. Theorem 1 says that the above left-hand quantity has different signs on α-side and β-side of $\Pi_{\alpha,\beta}$. Hence, for given R, there exist at least two U satisfying (9) and $|U - U_0| = R$. Obviously

$$\begin{aligned} L^1(U)(U - U_0) \neq 0 \text{ for } U \in \Pi_{\alpha,\beta}(U_0), \\ L^2(U)(U - U_0) \neq 0 \text{ for } U \in \Pi_{-\beta,\alpha}(U_0). \end{aligned} \tag{10}$$

The *Hugoniot loci* over U_0 are the set of (U, s) satisfying $H_{U_0}(U, s) = 0$. Their projections on to the U-plane are called the *Hugoniot curves* through U_0. We say that (U_*, s_*) is a *bifurcation point*, if

$$\operatorname{rank} H'_{U_0}(U_*, s_*) < 2.$$

Multiplying $H'_{U_0}(U)$ on the left with $L = \begin{pmatrix} L^1 \\ L^2 \end{pmatrix}$, we have

$$LH'_{U_0}(U, s) = \begin{pmatrix} (s - \lambda_1)L^1 & L^1(U - U_0) \\ (s - \lambda_2)L^2 & L^2(U - U_0) \end{pmatrix}.$$

We find by the above observation that there is no bifurcation point of H_{U_0} on the Hugoniot loci other than $U_* = U_0, s_* = \lambda_j(U_0),\ j = 1, 2$. Hence the Hugoniot loci consist of regular curves and the Hugoniot curves are regular except for $U = U_0$.

Proposition 1. *Suppose that Assumption I holds. Then, for every U_0, there exist two global Hugoniot curves $\mathcal{S}_1(U_0), \mathcal{S}_2(U_0)$ such that $U \in \Pi_{\alpha,\beta}(U_0)$ for $U \in \mathcal{S}_1(U_0)$ and $U \in \Pi_{-\alpha,\beta}(U_0)$ for $U \in \mathcal{S}_2(U_0)$.*

For each $j = 1, 2$, we choose the parameter μ of $\mathcal{S}_j(U_0)$ so that $L^j \dot{U} > 0$. Let us denote

$$\mathcal{S}_j^+(U_0) = \{U(\mu) \in \mathcal{S}_j;\ \mu > 0\}, \quad \mathcal{S}_j^-(U_0) = \{U(\mu) \in \mathcal{S}_j;\ \mu < 0\}.$$

Since $U - U_0 = \mu R_j(U_0) + O(\mu^2)$ for small μ and by (10), we have

$$\begin{aligned} L^j(U - U_0) > 0 \text{ if and only if } U \in \mathcal{S}_j^+(U_0), \\ L^j(U - U_0) < 0 \text{ if and only if } U \in \mathcal{S}_j^-(U_0). \end{aligned} \tag{11}$$

3 Admissibility of Shock Waves

Study of the vector field:

$$H_{U_-}(U,s) = F(U) - F(U_-) - s(U - U_-)$$

is the central subject of this section. Obviously U_- and U_+ are critical points of this vector field. We say that the states U_- and U_+ are connected by a *viscous profile* : a solution of the form $U(x,t) = \widehat{U}(\frac{x-st}{\epsilon})$ to a parabolic system

$$U_t + F(U)_x = \epsilon U_{xx} \quad (\epsilon > 0), \quad (x,t) \in \mathbf{R} \times \mathbf{R}_+ \tag{12}$$

satisfying

$$\lim_{\xi \to \pm\infty} \widehat{U}(\xi) = U_\pm. \tag{13}$$

The Rankine-Hugoniot condition (3) is necessarily satisfied and $\widehat{U}$ is a solution to the differential equation:

$$U_\xi = H_{U_-}(U,s). \tag{14}$$

The discontinuity (2) is said to be an *admissible* shock wave, if the states U_- and U_+ are connected by a viscous profile.

Now we impose technical assumptions:

Assumption II: The Hugoniot curves through U_0 consist of two curves $\mathcal{S}_1(U_0)$ and $\mathcal{S}_2(U_0)$ such that for any $U \in \mathcal{S}_1(U_0)$, the corresponding shock speed satisfies $s_1 < \lambda_2(U)$ and for $U \in \mathcal{S}_2(U_0)$, $s_2 > \lambda_1(U)$.

Assumption III: Let $\Omega_j^0 = \{U;\ \nabla\lambda_j R_j = 0\}$. Every integral curve intersects Ω_j^0 at discrete states. Moreover, if an integral curve and a Hugoniot curve intersect Ω_j^0 at $U = U_*$ having the same tangent, then the sign of $\nabla\lambda_j R_j$ and $\dot{\lambda}_j = \nabla\lambda_j \dot{U}$ are the same in a neighborhood of $U = U_*$.

By Assumption III, we have the following proposition whose proof is found in Asakura [1].

Proposition 2. *Along each Hugoniot curves, the point satisfying $\dot{s} = 0$ is isolated.*

We shall see easily that the next lemma is also true under our Assumption I, II and III.

Lemma 2 ([8] Lemma 1.2). *Assume that $U_1, U_2 \in \mathcal{S}_j^+(U_0)$ (or $\mathcal{S}_j^-(U_0)$, respectively) for some j and U_1 is located between U_0 and U_2. If $s(U_0,U_1) = s(U_0,U_2)$, then $U_1 \in \mathcal{S}_j^-(U_2)$ ($\mathcal{S}_j^+(U_2)$, respectively) and $s(U_1,U_2) = s(U_0,U_1) = s(U_0,U_2)$.*

Using this Lemma, we have

Proposition 3 ([7] Theorem 2). *Assume that $U_1 \in \mathcal{S}_1(U_0)$ and $U_2 \in \mathcal{S}_2(U_0)$. Then It follows that $s(U_0, U_1) < s(U_0, U_2)$.*

Suppose that the states U_-, U_+ satisfy the Rankine-Hugoniot condition (3) and the vector $\overrightarrow{U_-U_+}$ is limited to $\Pi_{\alpha,\beta}$. A parallelogram $\Delta(U_-, U_+) = \Pi_{\alpha,\beta}(U_-) \cap \Pi_{\alpha,\beta}(U_+)$ is called the *connection diamond* with respect to U_-, U_+ and the Hugoniot curve between U_- and U_+ is denoted simply by $U(\mu)$. Following lemma is important.

Lemma 3. *Suppose that Assumption I, II and III hold. If the strict Liu-Oleinik condition:*

$$s < s(\mu) \quad \textit{between } U_- \textit{ and } U_+ \tag{15}$$

is satisfied, then there is no critical point of the vector field $H_{U_-}(U, s)$ in the interior of $\Delta(U_-, U_+)$.

Proof. The Liu-Oleinik condition says that there is no critical point on the Hugoniot curve connecting U_- and U_+. Let U_* be an interior critical point. Assume, for example, that $U_+, U_* \in \mathcal{S}_1^+(U_-)$. Then U_+ is located between U_- and U_* and it follows from Lemma 2 that $U_+ \in \mathcal{S}_1^-(U_*)$. However, we find by (11) that $L^1(U_+ - U_*) < 0$, which contradicts the assumption $U_* \in \Delta(U_-, U_+)$.

Suppose that sates U_-, U_+ constitute a compressive shock wave with speed s. These states $U_\pm$ are hyperbolic critical points of $H_{U_-}(U, s)$ and it is easy to see:

states	U_-	U_+
1-shock wave	unstable node	saddle
2-shock wave	saddle	stable node

Hence it follows from Lemma 3 and the above observation

Theorem 2. *Suppose that Assumption I, II and III hold, and states U_-, U_+ constitute a compressive shock wave with speed s. If the strict Liu-Oleinik condition (15) is satisfied, then this shock wave is admissible having a viscous profile connecting these states.*

Proof. It follows from Theorem 1 that the vector field $H_{U_-}(U, s)$ points inward (or outward) with respect to $\Delta(U_-, U_+)$ on the boundary except for $U_\pm$. Changing the parameter $\xi \to -\xi$, if necessary, we can assume that $H_{U_-}(U, s)$ points inward on the boundary. Let us consider a compressive shock wave connecting U_- and U_+. There is an exceptional (unstable) locus issuing from the saddle U_- towards $\Delta(U_-, U_+)$. Since there is no critical point in $\Delta(U_-, U_+)$ by Lemma 3, the locus must be continued to U_+, which prove the theorem.

If $s = \lambda_j(U_-)$ or $s = \lambda_j(U_-)$, we have to study higher order terms, because the critical point is not hyperbolic. We can prove

Theorem 3. *Suppose that Assumption I, II and III hold, and j-characteristic field is genuinely nonlinear at U_-, U_+. If $s = \lambda_j(U_-)$ or $s = \lambda_j(U_-)$ and the strict Liu-Oleinik condition (15) is satisfied between these states, then this shock wave is admissible.*

4 Uniqueness of the Riemann Problem

We finally discuss the uniqueness of the solution to the Riemann problem. First we claim that the solution connecting U_L and U_R only by a series of 1-waves (or 2-waves) is unique. This solution contains a series of the states $U_L = U^0$ (or U_0), $U_1, U^1, \dots, U_n, U^n, U_{n+1} = U_R$ (or U^{n+1}), where U_i, U^i $(1 \leq i \leq n)$ are connected by rarefaction waves and U^i, U_{i+1} $(0 \leq i \leq n-1)$ are by contact discontinuities; U_L and U_1 (U^n and U_R, respectively) are connected by shock waves or rarefaction waves. Next theorem is originally due to Liu [6], [7] and also true under our assumptions.

Theorem 4. *Suppose that Assumption I, II and III hold. Then for any $U_0 \in \Omega$, there exist the C^1-curves $\mathcal{W}_j^F(U_0)$, $j = 1, 2$ through U_0 such that U_0 can be connected to every $U \in \mathcal{W}_j^F(U_0)$ to the right by a series of j-rarefaction waves, j-shocks and j-contact discontinuities. Moreover, the Liu-Oleinik condition is satisfied across any discontinuity. Conversely, if $U \in \Omega$ is connected to U_0 to the left by admissible j-waves, then $U \in \mathcal{W}_j^F(U_0)$ and the solution has a unique form.*

Using above discussion and assumptions, we shall prove the next global uniqueness theorem. Here Assumption I and Proposition 1 are decisive.

Theorem 5. *Suppose that Assumption I, II and III hold. Then for any Riemann data $U_L, U_R \in \Omega$, the solution composed of a series of 1-waves connecting U_L and U_M, and 2-waves connecting U_M and U_R is a unique solution.*

Proof. Assume that there exist more than one solutions for some given U_L, U_R. Then there exists two middle states U_{M_1} and U_{M_2} through which U_L and U_R are connected. The idea of our proof is similar to Keyfitz-Kranzer [4] Theorem 3.2 and the proof consists of several steps.

1. *Show that there exist states U_R^* and U^* such that the backward wave curve $\mathcal{W}_2^B(U_R^*)$ is tangent to $\mathcal{W}_1^F(U_L)$ at U^*.*

Let U_{M_3} be a state on $\mathcal{W}_1^F(U_L)$ located between U_{M_1} and U_{M_2}. Since U_R is located on both wave curves $\mathcal{W}_2^F(U_{M_1})$ and $\mathcal{W}_2^F(U_{M_2})$, the 2-wave curve issuing from U_{M_3} intersects $\mathcal{W}_2^F(U_{M_1})$ or $\mathcal{W}_2^F(U_{M_2})$ at a state U_{R_1}. Hence we can find another right state U_{R_1} for which there exist at least two solutions such that the distance between U_{M_1} and U_{M_3} (or U_{M_3} and U_{M_2}) is less than that between U_{M_1} and U_{M_2}. Repeating this argument, we can choose sequences of states $\{U_{M_n}\}$ converging to a state U^* on $\mathcal{W}_1^F(U_L)$. Similarly, by choosing a subsequence if necessary, right states $\{U_{R_n}\}$ converge to a state U_R^*. We may assume $U_R^* \neq U^*$. If $U_R^* = U^*$, then U_{M_n} and U_{R_n} are contained in a small neighborhood of U^*, which contradicts the local uniqueness theorem of Lax [5]. Rearranging the indices if necessary, we may assume that U_{R_n} is located on the both wave curves $\mathcal{W}_2^F(U_{M_{2n}})$ and $\mathcal{W}_2^F(U_{M_{2n+1}})$. In other words, the backward wave curve $\mathcal{W}_2^B(U_{R_n})$ contains both $U_{M_{2n}}$ and $U_{M_{2n+1}}$. Hence it follows from the limit process that the backward wave curve $\mathcal{W}_2^B(U_R^*)$ is tangent to $\mathcal{W}_1^F(U_L)$ at U^*.

2. *Show that there exist two states U_1^0, U_2^0 such that $U^* \in \mathcal{S}_1(U_1^0) \cap \mathcal{S}_2(U_2^0)$.*

The state U^* is located on both $\mathcal{W}_1^F(U_L)$ and $\mathcal{W}_2^B(U_R^*)$. These curves are composed of rarefaction, Hugoniot and mixed curves. Since at a state on the mixed curve a Hugoniot curve is tangent there, we may assume that each state on the wave curves is contained in a rarefaction or Hugoniot curves. It is impossible that U^* is contained in $\mathcal{R}_1 \cap \mathcal{R}_2$, because the characteristic fields R_1 and R_1 are linearly independent. Assume that $U^* \in \mathcal{S}_1 \cap \mathcal{R}_2$ or $U^* \in \mathcal{R}_1 \cap \mathcal{S}_2$. Then we have $L^1\dot{U} = 0$ or $L^2\dot{U} = 0$ at U^* respectively, which contradicts the assumption. Hence the claim follows.

We denote by U_1 the state on $\mathcal{S}_1(U_1^0)$ and U_2 on $\mathcal{S}_2(U_2^0)$. We choose the parameters satisfying $L^1\dot{U}_1 > 0$ and $L^2\dot{U}_2 < 0$, and assume for simplicity $L^1(U_1 - U_1^0) > 0$ and $L^2(U_2 - U_2^0) > 0$. The case: $L^1(U_1 - U_1^0) < 0$ and $L^2(U_2 - U_2^0) < 0$ can be treated in the same manner and the other two cases easily. We set $\dot{U}_2 = \eta_1 R_1 + \eta_2 R_2$ and $s_1 = s_1(U_1^0, U_1), s_2 = s_2(U_2^0, U_2)$.

3. *Show that $\eta_1 < 0, \eta_2 < 0$ and $\dot{s}_1 < 0, \dot{s}_2 < 0$.*

Clearly $\eta_2 < 0$ by $L^2\dot{U}_2 < 0$. Since $\dot{U}_1 \propto -\dot{U}_2$, we have $\dot{U}_1 = -\zeta\dot{U}_2$ for some $\zeta > 0$. Hence $L^1\dot{U}_1 = -\zeta\eta_1 > 0$ and the first part of the claim follows. By the Liu-Oleinik condition, we have $\dot{s}_1 \leq 0, \dot{s}_2 \leq 0$. If $\dot{s}_1 = 0$, $\mathcal{R}_1$ is tangent to $\mathcal{S}_1$, which is impossible by the argument in 2. Similarly, we have $\dot{s}_2 < 0$.

4. *Finish the proof.*

By differentiating the Rankine-Hugoniot condition, we have

$$\begin{aligned} \dot{s}_2(U_2 - U_2^0) &= \{F'(U_2) - s_2 I\}\dot{U}_2 \\ &= (\lambda_1 - s_2)\eta_1 R_1 + (\lambda_2 - s_2)\eta_2 R_2. \end{aligned}$$

Hence at $U_2 = U^*$

$$U^* - U_2^0 = \frac{\lambda_1 - s_2}{\dot{s}_2}\eta_1 R_1 + \frac{\lambda_2 - s_2}{\dot{s}_2}\eta_2 R_2. \tag{16}$$

Since $L^2(U_2 - U_2^0) > 0$, we have together with Assumption I-3

$$\frac{\lambda_1 - s_2}{\dot{s}_2}\eta_1 < 0, \quad \frac{\lambda_2 - s_2}{\dot{s}_2}\eta_2 > 0. \tag{17}$$

By the same argument, we have

$$U^* - U_1^0 = -\frac{\lambda_1 - s_1}{\dot{s}_1}\zeta\eta_1 R_1 - \frac{\lambda_2 - s_1}{\dot{s}_1}\zeta\eta_2 R_2, \tag{18}$$

$$-\frac{\lambda_1 - s_1}{\dot{s}_1}\zeta\eta_1 > 0, \quad -\frac{\lambda_2 - s_1}{\dot{s}_1}\zeta\eta_2 < 0. \tag{19}$$

These expressions show that U_2^0 is located in the sector with center U^* made by $R_1(U^*)$ and $-R_2(U^*)$ and U_1^0 in the sector by $-R_1(U^*)$ and $R_2(U^*)$.

On the other hand, we find by the Rankine-Hugoniot condition

$$\begin{aligned}
\dot{s}_1(U^* - U_1^0) &= \{F'(U^*) - s_1 I\}\dot{U}_1 \\
\dot{s}_2(U^* - U_2^0) &= \{F'(U^*) - s_2 I\}\dot{U}_2 \\
&= -\frac{1}{\zeta}\{F'(U^*) - s_2 I\}\dot{U}_1.
\end{aligned}$$

Hence by eliminating $F'(U^*)$, we have

$$U_2^0 - U^* = -\frac{\dot{s}_1}{\zeta\dot{s}_2}(U_1^0 - U^*) - \frac{s_2 - s_1}{\zeta\dot{s}_2}\dot{U}_1. \tag{20}$$

Notice that $s_2 > s_1$ by Proposition 3. The above expression says that U_2^0 is located in the sector with center U^* made by $-\overrightarrow{U^*U_1^0} = U^* - U_1^0$ and $\dot{U}_1$. Let us consider the states $U_1 \in \mathcal{S}_1(U_1^0)$ and $U_2 \in \mathcal{S}_2(U_2^0)$. We may assume that in a neighborhood of U^*, $\overrightarrow{U_1^0 U_1}$ and $\overrightarrow{U_2^0 U_2}$ rotate in the same way, clockwise for example, as the parameter μ increases. For $U_2 \in \mathcal{S}_2(U_2^0)$ located behind U^* with respect to our parameter, we can find a state $\widetilde{U}^*$ such that the directed line segment $\overrightarrow{U_1^0\widetilde{U}^*}$ is obtained by a clockwise rotation of $\overrightarrow{U_1^0 U^*}$ about U_1^0 and parallel to $\overrightarrow{U_2^0 U_2}$. In case there exists a state $U_1 \in \mathcal{S}_1(U_1^0)$ such that $\overrightarrow{U_1^0 U_1} \propto \overrightarrow{U_1^0\widetilde{U}^*}$ or an asymptote of $\mathcal{S}_1(U_1^0)$ parallel to $\overrightarrow{U_2 U_2^0}$, both are contrary to Assumption I. Otherwise, we observe that the boundary of $\Pi_{\alpha,\beta}$ must be located between $\overrightarrow{U_1^0 U^*}$ and $\overrightarrow{U_1^0\widetilde{U}^*}$, which contradicts the assumption that $R_1(U)$ is also situated in $\Pi_{\alpha,\beta}$. Thus the theorem is proved.

References

1. Asakura, F. (2001): Geometric uniqueness and existence theorems for the Riemann problem in a strictly hyperbolic region, preprint.
2. Conley, C., Smoller. J. (1970): Viscosity matrices for two-dimensional nonlinear hyperbolic systems, Comm. Pure Appl. Math., **23**, 876–884.
3. Gomes, M.E. (1989): Riemann problems requiring a viscous profile entropy condition, Adv. Appl. Math. **10**, 285–323.

4. Keyfitz, B., Kranzer, H. (1978): Existence and uniqueness of entropy solutions to the Riemann problem for hyperbolic systems of two conservation laws, J. Differential Equations, **27**, 444–476.
5. Lax, P.D. (1957): Hyperbolic systems of conservation laws II, Comm. Pure Appl. Math., **10**, 537–566.
6. Liu, T.-P. (1974): The Riemann problem for general 2×2 conservation laws, Trans. Amer. Math. Soc. **199**, 89–112.
7. Liu, T.-P. (1975): Existence and uniqueness theorems for Riemann problems, Trans. Amer. Math. Soc. **212**, 375–382.
8. Liu, T.-P. (1976): The entropy condition and the admissibility of shocks, J. Math. Anal. Appl., **53**, 78-88.
9. Schaeffer, D.G., Shearer, M. (1987): The Riemann problem for nonstrictly hyperbolic 2×2 systems of conservation laws, Trans. Amer. Math. Soc. **304**, 267–306.
10. Smoller, J. (1969): On the solution of the Riemann problem with general step data for an extended class of hyperbolic systems, Mich. Math. J., **16**, 201-210.
11. Smoller, J. (1994): Shock Waves and Reaction-Diffusion Equations (2nd ed.), Springer Verlag, New York.

Eulerian Approximate Ray Tracing and Applications to Grid Generation

Patrizia Bagnerini

Laboratoire de Mathématiques CNRS n. 6621, Université de Nice, Parc Valrose B.P. 71, 06108 Nice Cedex 2, France *pbagneri@math.unice.fr*

Summary. We introduce a scheme to compute the viscosity solution of the Riemannian Eikonal equation, on a regular grid or triangular mesh and which uses the order given by the Sweeping algorithm, to update the points. We also compute the bicharacteristic curves of the viscosity solution in a domain Ω in Eulerian way: instead of solving a system of ODE for every source point belongs to $\partial\Omega$, we label each ray by a parameter θ and thus we compute a function $\theta = \theta(s,t)$, $\theta : \Omega \subset \mathbb{R}^2 \longrightarrow \mathbb{R}$, whose level sets are the rays. We then present some numerical results.

1 Viscosity solution of the anisotropic Eikonal equation

Computing the solution of Hamilton-Jacobi equations has many applications in optimal control theory, computer science, geometrical optics, robotic motion planning, etc. An important set of numerical algorithms involves the fast solution of steady time independent Hamilton-Jacobi equations.

In particular, concerning the Eikonal equation $|\nabla\phi(x)| = c(x)$, with given $c(x)$, Tsitsiklis [16] combined the heap sort with a variant of the classical Dijkstra algorithm. Then, Sethian [14] and independently Helmsen-Puckett-Colella-Dorr [7] published the "Fast Marching method", which computes the viscosity solution of the Eikonal equation, with a complexity of $O(n\ logn)$, with n the number of grid points.

The Fast Marching is based on the fact that the characteristics of the Eikonal equation, i.e. the optimal trajectories for the isotropic min-time control problem coincide with the gradient lines of $\phi(x)$, and therefore a causality property exists: the value of ϕ at the grid point x only depends on the smaller "adjacent" values. The solution is then constructed by updating the points in the order of increasing values of ϕ, using a monotone upwind scheme à la Rouy-Tourin [12].

This causality property expresses that the values function ϕ is strictly increasing along the characteristics, which, in the isotropic case, coincide with the gradient of ϕ.

In this Note, we are interested in the class of time-independent anisotropic Hamilton-Jacobi equations:

$$H(\phi_x, \phi_y) = \sqrt{a\phi_x^2 + 2b\phi_x\phi_y + c\phi_y^2} = 1, \tag{1}$$

where $\phi : \mathbb{R}^2 \to \mathbb{R}$ is continuous and a, b, c are scalar functions defined on $\mathbb{R}^2$, satisfying $ab > c^2$.

Given a domain Ω and a metric $G \equiv G(x) = \left(\begin{smallmatrix} a & b \\ b & c \end{smallmatrix}\right)$, we consider the (anisotropic) Riemannian Eikonal equation (1) on Ω, with boundary conditions $\phi = 0$ on $\partial\Omega$. The viscosity solution of (1) at point x represents a Riemannian distance to the boundary of Ω in the Riemannian manifold given by the metric G. This anisotropic equation occurs in several fields in which there exists some preferred directions, as crystal growth, geophysics, tomography, etc.

The Fast Marching Method (FFM) is not directly applicable to equation (1), because a causality property does not hold: since the characteristics and the gradient directions are different, a simplex xx_ix_j may contain the characteristic for the point x, even if the gradient $\nabla\phi(x)$ is pointing from another simplex. The values of ϕ are increasing along the characteristics, but in such a case it can happen that $\phi(x) < \phi(x_i)$ or $\phi(x) < \phi(x_j)$, so that it is not possible to compute the grid points in ascending order. Sethian and Vladimirsky [13] therefore introduced a variant of the FFM for anisotropic problems, using a "weaker causality property": they update the value of ϕ at point x using a much larger set of points, depending on the degree of anisotropy of equation. Another possible algorithm to define an order to compute the value of ϕ, is

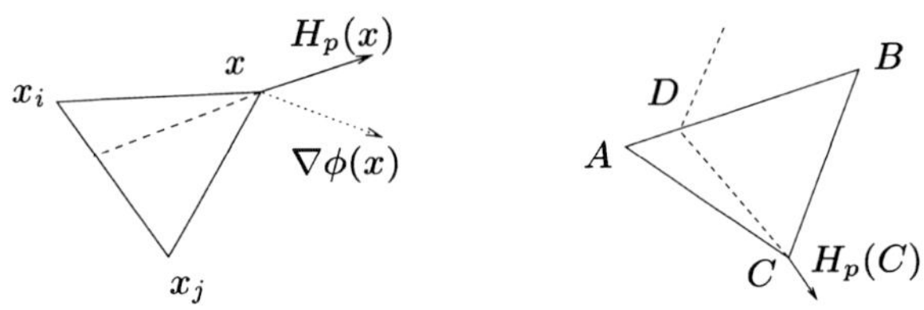

Fig. 1. Direction of the characteristic in a mesh element

the " Fast Sweeping method", motivated by [4], see also [18], [15]. The cost of each sweeping is $O(n)$, where n is the total number of grid points, and the number of iterations is independent of the grid size. Zhao [17] proved the convergence of the Sweeping algorithm on a rectangular grid for the "Euclidean" Eikonal equation: he showed that a number of 2^N sweeps is enough to compute, with first order accuracy, the distance function to an arbitrary set of discrete points in N dimension.

The main idea of the Sweeping method consists in using a Gauss-Seidel algorithm with alternating order: dividing the characteristics into a finite number of groups, according to their directions, each sweep of iterations takes into account a group of characteristics. The value at point x is then the minimum between the values computed from this finite number of directions.

Given a regular grid in two dimensions, denoting by $x_{i,j}$ the grid point, $\phi(x_{i,j})$ is computed from the value of its neighbors $\phi_{i\pm1,j}, \phi_{i,j\pm1}$. The sweeping iteration is given repeating the four alternating ordering:

$$\begin{array}{ll} 1)\ i=1:I,\ \ j=1:J & 2)\ i=I:1,\ \ j=1:J \\ 3)\ i=I:1,\ \ j=J:1 & 4)\ i=1:I,\ \ j=J:1, \end{array} \tag{2}$$

where I and J are the number of grid points respectively in x and in y.

In this paper, we introduce a scheme to compute the viscosity solution of the Riemannian Eikonal equation (1) on a regular grid or a triangular mesh, using the order given by the Sweeping algorithm to update the points.

2 Description of the method

Fomel [5] proposed a scheme for the "Euclidean" Eikonal equation, derived from Fermat's principle, which uses the Fast Marching method on a triangular mesh. Following [5], we consider a triangular mesh on Ω and for every mesh point C, we define $S(C)$ to be a set of all simplexes in the mesh, which have C as one of their vertices. We want to compute the viscosity solution ϕ of the equation (1), with boundary conditions $\phi = 0$ on $\partial\Omega$.

Fermat's principle states that the ray between two points C and D corresponds to a local extremum of the travel time with respect to raypath perturbations. We assume that the direction of this path does not change until it reaches the edge of a simplex belonging to $S(C)$.

Let A, B, C be the vertices of that simplex and D the point where the characteristic intersects the segment with end points A and B (see Figure 1). The value of ϕ at point D can be linearly approximated using the value of ϕ at point A and B, denoted by ϕ_A and ϕ_B:

$$\phi_D = \phi_A + (1-\xi)\phi_B,$$

for $\xi \in [0,1]$. We approximate the value of ϕ at point C by

$$\phi_C = \phi_D + d_G(D,C) = \phi_D + \sqrt{(D-C)^T G(D-C)} = \tag{3}$$

$$= \phi_A + (1-\xi)\phi_B + \sqrt{(A+\xi(B-A)-C)^T G(A+\xi(B-A)-C)}, \tag{4}$$

where $d_G(D,C)$ is the Riemannian distance from D to C for a metric G. We want to find a local extremum of ϕ_C with respect to the parameter ξ. Expressing that the ξ derivative vanishes, we obtain

$$\frac{\partial \phi_C}{\partial \xi} = (\phi_B - \phi_A) + \frac{(B-A)^T G(D-C)}{\sqrt{(D-C)^T G(D-C)}} = 0,$$

and then

$$\begin{aligned} &(\phi_B - \phi_A) + \sqrt{(A+\xi(B-A)-C)^T G(A+\xi(B-A)-C)} + \\ &+ (B-A)^T G(A+\xi(B-A)-C) = 0. \end{aligned} \tag{5}$$

Let (x_A, y_A), (x_B, y_B), (x_C, y_C) be the coordinates of A, B, C and $v_1 := (x_A - x_C) + \xi(x_B - x_A)$, $v_2 := (y_A - y_C) + \xi(y_B - y_A)$. The equation (5) can be rewritten as

$$\begin{aligned}(\phi_B - \phi_A) + \sqrt{av_1^2 + 2bv_1v_2 + cv_2^2} + av_1(x_B - x_A) + bv_2(x_B - x_A) + \\ + bv_1(y_B - y_A) + cv_2(y_B - y_A) = 0.\end{aligned} \tag{6}$$

We compute an explicit solution $\bar{\xi}$ of (6), considering the corresponding quadratic equation and selecting the appropriate branch of the square root. If the computed $\bar{\xi}$ belongs to the interval $[0, 1]$, ϕ_C is obtained substituting $\bar{\xi}$ in (3).

2.1 Algorithm on a regular grid

In the case where the mesh is a regular grid, we compute a value at point $x_{i,j}$ of the grid, using the four triangles belonging to $S(x_{i,j})$, formed with the neighboring points $x_{i\pm1,j}$ and $x_{i,j\pm1}$. First initialize ϕ at all grid points to be ∞, i.e. a large value, which should be larger than the maximum possible values. Compute the exact Riemannian distance (i.e. the exact solution of (1)) at those grid points in a neighborhood of the boundary of Ω. The values of these points will not be modified.

For the other grid points $x_{i,j}$, visited in the alternating order given by (2), consider the four triangles formed by $x_{i,j}$ and the neighboring points $x_{i\pm1,j}$ and $x_{i,j\pm1}$. For each triangle, compute a value (eventually ∞) of ϕ at $x_{i,j}$ using the above scheme (equations (3), (6)) and then update $\phi_{i,j}$ to be the smaller between the previous value of $\phi_{i,j}$ and the four possible computed values. In fact, the algorithm corresponds to approximate the cost to go out from each triangle belonging to $S(x_{i,j})$ and choose the minimum cost. We remark that we use the Gauss-Seidel iterations, i.e. for updating $\phi_{i,j}$, we use the current values at the neighboring points, and not the values computed at the previous sweep iteration.

The complexity of the algorithm is $O(n)$, where n is the total number of grid points. For the Riemannian Eikonal equation, the characteristics are curves instead of straight lines. Then, more than 2^N sweeps (with N the dimension) may be needed to approximate the solution, but the number of iterations is still independent of the grid size, [17].

Comparing the algorithm introduced here for instance with the Gonzales-Rofman scheme [6], iterations for computing the minimum are not necessary, the value of ϕ at the point $x_{i,j}$ is obtained by solving directly the quadratic equation (6).

In Figure 2 left, we represent the level sets of the viscosity solution of (1), with a metric G which allows to obtain curves concentrated near the curve of equation $x^3 - y^2 + 2 - 3\lambda x = 0$, for a given parameter λ.

2.2 Algorithm on a triangular mesh

Extending the sweeping algorithm on a triangular mesh requires to divide all characteristics into a finite number of groups according to their directions. In a general mesh, some preferred directions do not exist. We then propose to put the original mesh $\mathcal{T}$ in a background mesh. This mesh is used in mesh generation or adaptation, it must cover the entire domain and is often provided by a simple regular grid or an octree structure.

When we generate the mesh of the domain, we also construct a background mesh formed by a regular grid $\mathcal{G}$: each point of the original mesh $\mathcal{T}$ belongs to a cell of such a grid. Then, we sweep the whole domain covering the grid points $i, j \in \mathcal{G}$, using the sweeping algorithm (2): for every cell $C_{i,j}$, we compute the value of ϕ at all mesh points $m_i \in \mathcal{T}$ which are contained in the cell $C_{i,j}$. The characteristics are therefore divided in four groups according to their directions.

We compute the value of ϕ at point $m_i \in \mathcal{T}$ in the following way: we consider all simplexes belong to $S(m_i)$ and compute all possible value (eventually ∞) of $\phi(m_i)$ by using (3), (6). $\phi(m_i)$ is updated to be the minimum between the previous value and all possible computed values. In Figure 2 right we show an example of the viscosity solution computed on a mesh. The domain is the space around the Naca012 airfoil contained in a circle of rays 1.5 and the metric is given by an adaptive process by solving a compressible Euler equation at Mach number 0.96.

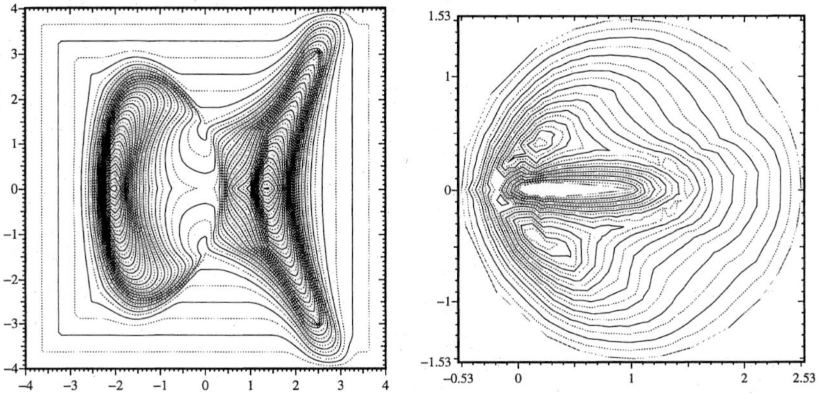

Fig. 2. Viscosity solution computed on a triangular grid and on a mesh

3 Eulerian ray tracing

Let us come back to the Hamilton-Jacobi equation

$$H(x, \nabla\phi) = 0 \text{ in } \Omega. \tag{7}$$

The bicharacteristics are the integral curves of the associated Hamiltonian system:

$$\begin{cases} \dot{x}(t) = H_p(x,p) = \frac{\partial H}{\partial p}(x,p), \\ \dot{p}(t) = -H_x(x,p). \end{cases} \tag{8}$$

$H(x,p)$ is assumed to be continuous up to its second partial derivatives and convex in the variable p. The rays are generally computed in a Lagrangian way, solving the system of ordinary differential equations (8) in the phase-space $\mathbb{R}_x \times \mathbb{R}_p$.

The aim here is to approximate the rays corresponding to the viscosity solution, i.e. the rays which correspond to the first arrival time, in a domain $\Omega \subset \mathbb{R}^2$. In order to generate meshes, but also for the other applications (rays in geometrical optics, etc ..), it would be useful to compute the rays in an Eulerian way. So, following e.g. [2], we label each ray by a parameter, say θ and thus we compute a function $\theta = \theta(s,t)$, $\theta : \Omega \subset \mathbb{R}^2 \longrightarrow \mathbb{R}$, whose level sets are the rays. At every point $x \in \Omega$ the direction of the characteristics is given by $H_p(x, \nabla\phi)$. Since $\nabla\theta(x)$ is orthogonal to the level set of θ at point x, then θ is solution to the problem:

$$\begin{cases} F(x, \nabla\theta) = H_p(x, \nabla\phi) \cdot \nabla\theta = 0 & \text{in } \Omega, \\ \theta(x) \text{ given } \in [0, 2\pi) & \text{on } \partial\Omega. \end{cases} \tag{9}$$

This approach allows in principle to obtain directly all the characteristic curves as the level sets of the function θ, instead of solving a system of ODE for every initial point on $\partial\Omega$.

Now, (9) is a stationary linear transport equation with non-smooth coefficients. The characteristics of the equation (1) are compressive i.e. they satisfy a one-sided Lipschitz condition, since the viscosity solution (the distance function) ϕ is quasi-concave, and they are pointing inside the domain Ω. So, conceptually, the corresponding initial boundary-value problem (IBVP) is well posed, see e.g. [11], [3], [10], ..., although *a priori* the product of distributions $(H_p(x, \nabla\phi) \cdot \nabla\theta)$ is not well defined at points where the characteristics merge. In any case, the characteristics enter this set of possible bad points, which is in many cases a null set. Therefore, without rigorous justification, it is possible to approximate a.e. in Ω the solution θ of the stationary problem (9).

In contrast, computing the rays starting from a point inside the domain by solving backward the EDO (8) would be an ill-posed problem. Consequently, we can compute the rays in an Eulerian way, by visualizing the level curves of θ. For each point x inside Ω we know directly the exit point, i.e. the point where one of the optimal trajectories reaches $\partial\Omega$: in fact, except for the smooth case, the backward characteristic is not unique. The exit points are given by the values of θ at point x.

3.1 Algorithm to compute the rays

Let us assume there is no change of topology in the level sets of ϕ. Given a regular grid, we approximate the solution of (9) as the limit for $t \to \infty$ of the corresponding initial boundary value problem, (e.g.) with initial data $\theta_0(x) \equiv 0$ in Ω.

Let (u, v) be the components of the vector $H_p(x, \nabla\phi)$ and $h = (\Delta x, \Delta y)$ be the discretization step of the grid. We approximate the equation of the (IBVP) with a Lax-Friedrichs type of scheme, whose numerical viscosity formally adds some nonlinear diffusion term to the right-hand side of (9).

The algorithm is the following: for $n = 0$, we choose $\theta^0 = (\theta^0_{ij})$ such that $\theta^0_{ij} \in [0, 2\pi)$ on $\partial\Omega$ and $\theta^0 \equiv 0$ in Ω. At the step $n+1$, we compute θ^{n+1} by the scheme

$$\theta^{n+1}_{ij} = -\Delta t F\left(x_{ij}, \frac{\theta^n_{i+1,j} - \theta^n_{i-1,j}}{2\Delta x}, \frac{\theta^n_{i,j+1} - \theta^n_{i,j-1}}{2\Delta y}\right) + \frac{1}{2\Delta x}\alpha(x_{ij})(\theta^n_{i+1,j}+ \\ + \theta^n_{i-1,j}) + \frac{1}{2\Delta y}\beta(x_{ij})(\theta^n_{i,j+1} + \theta^n_{i,j-1})), \tag{10}$$

with

$$\alpha(x_{ij}) = \sup_{u,v}\left|\frac{\partial F}{\partial u}(x_{ij}, u, v)\right| = u,\ \beta(x_{ij}) = \sup_{u,v}\left|\frac{\partial F}{\partial v}(x_{ij}, u, v)\right| = v,$$

the "optimized" coefficients as in [8]. We stop the iterations when the difference between $\theta^n = (\theta^n_{i,j})$ and $\theta^{n+1} = (\theta^{n+1}_{i,j})$ is smaller than a given value.

Remark 1. In the Euclidean case, one can use an obvious upwind scheme to solve directly the stationary equation (9), knowing ϕ, or even computing ϕ and θ at the same time, e.g. by the FMM. As we said in Section 1, *this is not possible in general in the Riemannian case*, see e.g. Fig. 1. This is why we have to use the above algorithm here.

3.2 Discontinuity of the boundary data

An additional numerical difficulty is that the problem (9) has (artificially) discontinuous boundary data, e.g. at $\theta = 0$ and $\theta = 2\pi$. More precisely, the boundary conditions can be considered as periodic with (e.g.) period 2π, and this discontinuity propagates inside Ω.

When we compute points near this curve of discontinuity Γ, and later when we visualize the characteristics, we have to *interpolate* θ between two points M_1 and M_2 on each side of Γ. If we do not know Γ, this interpolation will artificially produce arbitrary values between 0 and 2π. It is not clear how to overcome this difficulty, even by computing simultaneously an other determination of $\tilde{\theta}$.

Therefore we have chosen to first *track* this curve Γ, by using a simple algorithm: assume that the initial point $M_0 \in \partial\Omega$, for which $\theta(M_0) = 0$, is a grid point $M_0 = (i_0, j_0)$. The next point in the chain connecting M_0 to another point of Γ, is selected to be the neighbor point (k, l) on the grid in the direction of $H_p(i_0, j_0)$ and so forth. We obtain an ordered set of grid points which approximates Γ. Let us consider Γ oriented from p_0 to the last point of Γ.

The points near Γ have to be processed in a different way from the other grid points and therefore they are divided in three classes in the following way, see Figure 3, left. For every point $p \in \Gamma$, we consider the cells C_p which contains p. The four grid points $\in C_p$ are divided in three categories: *fixed*, *left* and *right*. The points $\in \Gamma$ are tagged as *fixed*, the points on the right of Γ, with respect its orientation, as *right* and the points on the left, as *left*.

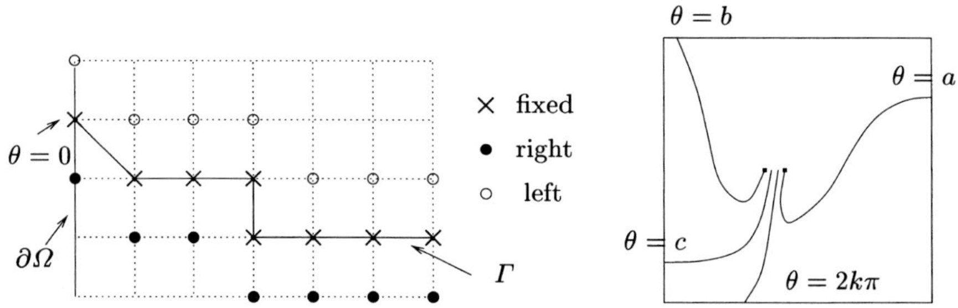

Fig. 3. Points near Γ and rays in the case of strong shocks

The algorithm to compute θ is the one described above: the only difference concerns the points *left* and *right*. θ^{n+1} is computed by (10) using the neighbor points $(i \pm 1, j)$, $(i, j \pm 1)$. If one of this points is tagged *fixed*, the value of θ^n at this point is 0 if (i, j) is a *right* point, and 2π if is a *left* point. The value of θ at *fixed* points takes therefore two values 0 and 2π, according to the circumstances and is not modified in the computation.

We obtain an approximation $\theta := \theta^n$ of the solution of (9) for a certain n. The rays are then computed as the level sets of θ, by interpolating in the two triangles which form each cell, chosen so that the segments of Γ are an edge of the triangles. The value of θ at a *fixed* point takes the values 0 or 2π, according to whether the triangle is resp. on the left or on the right of Γ.

It can happen (very rarely) that the same grid point is on the left for a part of Γ and on the right for another part. In this case, we can choose as Γ a characteristic starting from another point of $\partial\Omega$. The choice of the point of discontinuity on $\partial\Omega$ is, in fact, completely arbitrary.

This method is still Lagrangian, but *only* for one characteristic. In addition, in the case of real strong shocks for θ, if there is no change of topology of the level curves of ϕ, we can thus determine on which part of $\partial\Omega$ the intermediate rays come from, see Figure 3, right.

The values of θ are not perfectly computed near the curve Γ, since Γ is only approximated by grid points. Another method would be to approximate Γ by solving the ordinary differential equation (8) with a fixed step. Γ is then approximated by a finite number of segments. The function θ is therefore computed on a triangular mesh for which the segments which form Γ are a part of the boundary. We will test this method in a future report.

4 Application to mesh generation

In [9, 1], a method was introduced to generate anisotropic meshes which are automatically refined in some regions, where a given function $\tilde{z}$, that we want to approximate, has large derivatives (typically $\tilde{z}$ is the numerical solution of some unspecified partial differential equation). Given a previously computed approximation z of this function, the main idea was to generate an adapted mesh by solving the Eikonal equation on the graph of z, equipped with a suitable Riemannian metric.

Consider an initial closed curve, typically the boundary of the domain Ω, and move this curve with a constant Riemannian speed on the graph of z: when we project the curve on Ω in the regions where z is rapidly varying, its horizontal projection moves very slowly, so that the resulting mesh is automatically refined.

In [9, 1], the curve was represented as the zero-level set of a function $\Phi : (x,t) \in R^2 \times R^+ \mapsto R$: the successive positions of the curve define the basic structure of the mesh. Φ is the unique viscosity solution to the evolution Riemannian Eikonal equation:

$$\partial_t \Phi + |G^{-1/2} \nabla \Phi| = 0, \tag{11}$$

where $|G^{-1/2}\nabla\Phi|$ denotes the Euclidean norm of $G^{-1/2}\nabla\Phi$ (compare with (1)), whereas here, the same curves are computed as the level curves $\{\phi = t\}$, where ϕ is the viscosity solution of the stationary equation (1).

Moreover, in order to construct the mesh, in [1] the rays were computed in a Lagrangian way: given m points M_i belonging to the boundary of the domain to be meshed, compute for every edge $M_i^n M_{i+1}^n$ the mid-point $M_{i+1/2}^n$ and move this point in the direction of the characteristic. To construct the triangles, connect the new points $M_{i+1/2}^{n+1}$ with M_i^n and M_{i+1}^n, check the quality of the triangles, and repeat the same process. The aim is to obtain triangles as equilateral as possible with respect to the Riemannian metric G, and the edges of the same given Riemannian length.

Here, in contrast, the rays are directly obtained as the level sets of the function θ, which, together with the level sets of the viscosity solution of (1), form the basic structure of the mesh.

With this Eulerian generation of rays the method is more robust. In particular, the numerical viscosity of the Lax-Friedrichs scheme prevents the rays form numerically crossing each other, and even smooths out the shocks, see Figure 4. However, when the level curves of ϕ exhibit some changes of topologies, the value of θ is not well computed in the corresponding regions. It is then necessary to mesh these curves separately.

In Figure 4, we choose the Riemannian metric $G(x) = I + \nabla z(x) \otimes \nabla z(x)$, where z is given by $z(x, y) = c_1 \text{ arctan } (c_2 f(x, y))$, with suitable constants c_1 and c_2, so that the mesh is refined near $f(x, y) = 0$. The function f is given by $f(x, y) = y - a * x^3 - 1.5 * a * x^2 + (a/2 - 1) * x + 1$ with $a = -4$. In the first figure, we represent the level sets of the viscosity solution of (1) and of the function θ, and in the second one, a triangular mesh obtained using these two families of level curves.

References

1. P. Bagnerini, P. Hoch, and M. Rascle. The eikonal equation on a manifold. Applications to grid generation or refinement. In *Hyperbolic problems: theory, numerics, applications, Vol. I, II (Magdeburg, 2000)*, pages 109–118. 2001.
2. J.-D. Benamou. Big ray tracing: Multivalued travel time field computation using viscosity solutions of the eikonal equation. *J. Comput. Phys.*, 128(2):463–474, 1996.
3. F. Bouchut and F. James. One-dimensional transport equations with discontinuous coefficients. *Nonlinear Anal.*, 32(7):891–933, 1998.
4. M. Boué and P. Dupuis. Markov chain approximations for deterministic control problems with affine dynamics and quadratic cost in the control. *SIAM J. Numer. Anal.*, 36(3):667–695 (electronic), 1999.
5. S. Fomel. A variational formulation of the fast marching eikonal solver. *Stanford Exploration Project, Report*, 2000.
6. R. Gonzalez and E. Rofman. On deterministic control problems: An approximation procedure for the optimal cost. 1. The stationary problem. *SIAM J. Control Optimization*, 23:242–266, 1985.
7. J. Helmsen, E. Puckett, P. Colella, and M. Dorr. Two new methods for simulating photolithography development in 3d. *Proc. SPIE*, 2726:253–261, 1996.
8. Ph. Hoch. Approximation de problèmes hyperboliques non linéaires, équations de hamilton-jacobi et applications. *PhD Thesis*, 2000.
9. Ph. Hoch and M. Rascle. Hamilton–Jacobi equations on a manifold and applications to grid generation or refinement. *SIAM J. Sci. Comput.*, 23(6):2056–2074, 2002.
10. G. Petrova and B. Popov. Linear transport equations with μ-monotone coefficients. *J. Math. Anal. Appl.*, 260(2):307–324, 2001.
11. F. Poupaud and M. Rascle. Measure solutions to the linear multi-dimensional transport equation with non-smooth coefficients. *Comm. Partial Differential Equations*, 22(1-2):337–358, 1997.
12. E. Rouy and A. Tourin. A viscosity solutions approach to shape-from-shading. *SIAM J. Numer. Anal.*, 29(3):867–884, 1992.
13. J. A. Sethian and A. Vladimirsky. Ordered upwind methods for static Hamilton-Jacobi equations: theory and algorithms. *Berkeley PAM-792*, 2001.
14. J.A. Sethian. Fast marching level set methods for three dimensional photolithography development. *Proc. SPIE*, 2726:262–272, 1996.

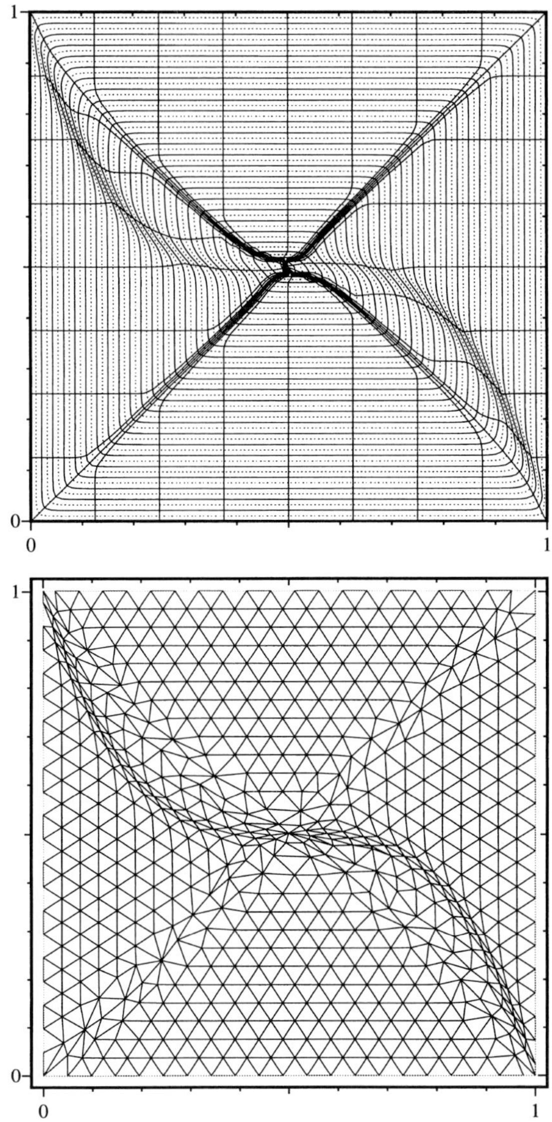

Fig. 4. Level curves of ϕ and θ and generation of a triangular mesh

15. Y.R. Tsai, L.-T. Cheng, S. Osher, and H.K. Zhao. Rapid and accurate computation of the distance function using grids. *UCLA CAM Report 01-27*, 2001.
16. J. N. Tsitsiklis. Efficient algorithms for globally optimal trajectories. *IEEE Trans. Automat. Control*, 40(9):1528–1538, 1995.
17. H.K. Zhao. Fast sweeping method for eikonal equation I: distance function. *preprint*, 2002.

18. H.K. Zhao, S. Osher, B. Merriman, and M. Kang. Implicit and non-parametric shape reconstruction from unorganized points using variational level set method. *Computer Vision and Image Understanding*, 80(3):295–319, 2000.

Non-convex Flux Functions and Compound Shock Waves in Sediment Beds

Gert Bartholomeeusen[1], Hans De Sterck[2], and Gilliane Sills[3]

[1] University of Oxford, Department of Engineering Science, Parks Road, Oxford OX1 3PJ, UK *gert.bartholomeeusen@eng.ox.ac.uk*
[2] University of Colorado at Boulder, Department of Applied Mathematics, Colorado CO-80309-0526 *desterck@colorado.edu*
[3] University of Oxford, Department of Engineering Science, Parks Road, Oxford OX1 3PJ, UK *gilliane.sills@eng.ox.ac.uk*

Summary. Sediment layers deposited under water undergo a deformation that for low soil concentrations can be described by a scalar nonlinear hyperbolic conservation law. The associated flux function is non-convex and compound shock waves arise, which are shocks followed or preceded by a rarefaction with the shock speed equal to the wave speed at the point of attachment. The paper describes an experimental study of compound shock waves in sediment beds and the numerical modelling of the sedimentation process using an experimentally obtained flux function for kaolinite soil. The work also fits in a broader investigation of the transition parameter between the physical processes of soil sedimentation and consolidation, as the data obtained allows identification of this parameter.

1 Introduction

In soil sedimentation processes soil particles settle in a fluid as a group, driven by a flux function which only depends on the particle concentration. The mathematical formulation is simply given by the continuity equation which is classified as a hyperbolic partial differential equation:

$$\frac{\partial n}{\partial t} + \frac{\partial f(n)}{\partial x} = 0; \tag{1}$$

where n is the porosity, defined as the volume of voids divided by the total volume. Given an initial uniform porosity in a settling experiment, different types of shock waves will propagate through the sediment depending on the precise form of the flux function, $f(n)$. For soil suspensions very little is known about the form of the flux function or about the convexity or non-convexity of the function. However, detailed experimental proof of discontinuities propagating in sediments dates from the early 1980's [3], and has later been confirmed by others [1]. This paper presents soil sedimentation experiments, in which shocks are tracked by carefully monitoring surface settlement and X-ray density or porosity profiles, to obtain a suitable flux function for the kaolinite soil used. The flux function is validated by using it in a numerical solution of equation 1, which is compared with experimental results.

The group settling of particles dependent only on the local concentration, often also called hindered settling, defines the sedimentation process, and the experiments performed in this study are focussed on the range of porosities where this physical mechanism is dominant. When suspensions become more dilute, drag forces around the larger particles will move the smaller ones up, leading to the process of differential settling. Next to an upper bound in porosity for the sedimentation process, there is a lower bound for denser suspensions, where the particle interactions provide a supporting framework which influences the settling process. This process is known to geotechnical engineers as large strain consolidation [7]. Neither the upper or lower bound processes are considered in this paper.

Kynch [8] was the first to formulate the theory of sedimentation, which was later mainly applied in chemical engineering problems, such as fluidisation. Often artificial particles were used, eg glass beads and rubber balls, but some work was also performed on soil sediments. Michaels & Bolger [10] showed that the theory of sedimentation applies to non-equisized particles and flocculated suspensions. In geotechnical engineering problems the volume reduction of a suspension bed by sedimentation is typically in the order of 50%, and therefore a good understanding is of great importance for design and management purposes. In spite of this in practice soil sedimentation is most often treated as an instantaneous collapse of the suspension bed.

2 Wave Structures

The soil sedimentation process starts from a uniformly mixed soil-water suspension, and subsequent temporal evolution is characterised by an increasing clear water layer appearing at the surface and a more compact soil bed building-up from the base. The experiments presented in this paper are carried out in settling columns with an internal diameter of 101 mm. The soil used is industrially exploited kaolinite, which has a variety of applications, eg in the paper industry. It is chosen here for its very fine grain size: 90 percent of the particles, d_{90}, have a diameter smaller than 8 micron and 10 percent, d_{10}, is smaller than 0.7 micron. Therefore, it is most likely that the region of porosities at which differential settling occurs will be fairly narrow and the sedimentation region can also be found not too far from water ($n = 1$).

Figure 1 shows the results of a settling experiment in which the evolution of the porosity profiles is monitored by measurement of X-rays passing through a settling column containing a kaolinite suspension. The method provides an accuracy of measurement of porosity of ± 0.0013 and a spatial resolution of ± 1 mm [3, 12]. At the 29 min profile a thin, more compact layer has formed at the base, which is divided from the initial porosity by a discontinuity. At the top another discontinuity separates the initial porosity from that of water. With the passage of time the two discontinuities move towards each other as shock waves, as seen in the 98 and 177 min profiles. Three distinct parts can be identified: a lower porosity bed, the initial porosity layer and a water layer. Eventually the two shock fronts meet each other, and further compaction occurs. This process of sedimentation in a settling column experiment, described by equation 1, can be modelled as a double Riemann problem with two wave fronts moving towards each other.

Shocks can occur in various ways depending on the flux function $f(n)$ and the position of the initial porosity on it. If porosity discontinuity, such as the water-suspension interface or the suspension-bed interface, is plotted in (x, t) space, shock waves can be classified in terms of the characteristic lines that intersect the shock. A regular shock wave occurs when the characteristic lines (with slope $f'(n)$) enter the discontinuity from both sides. Then the speed of the shock is determined by the Rankine Hugoniot Jump Condition:

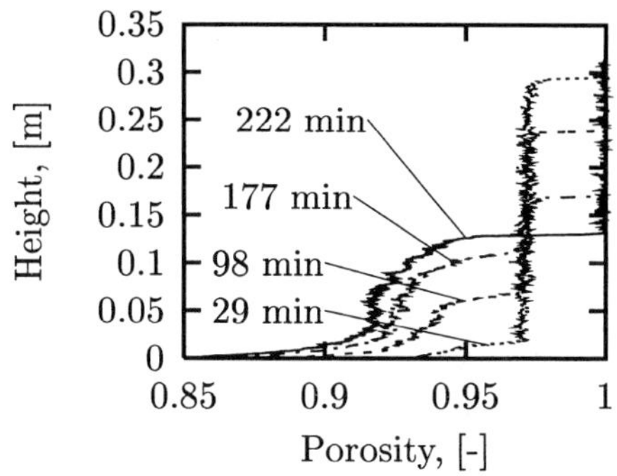

Fig. 1. Porosity profiles monitored by means of X-rays. (h_i=0.315m and n_i=0.972).

$$s = \frac{f(n_l) - f(n_r)}{n_l - n_r}. \tag{2}$$

Equivalently if, with n_l and n_r the porosities on the left and right of the shock, respectively, the entropy condition [11]

$$\frac{f(n) - f(n_l)}{n - n_l} \geqslant s \geqslant \frac{f(n) - f(n_r)}{n - n_r}; \tag{3}$$

is satisfied for all n between n_l and n_r, a regular shock occurs. When condition 3 is not fulfilled, the relevant solution is a rarefaction wave or a compound shock wave. When the local wave speed diverges away from the shock on the left and right side, a rarefaction wave is obtained. For a compound shock wave, characteristic lines go into the shock on one side, while on the other side the characteristic line or local wave speed is parallel to the shock.

In order to illustrate these three different types of shock waves, simple flux functions are assumed. For the regular shock wave, a convex flux function $f(n) = n(1-n)/2$ is taken. Expressions 2 and 3 applied to the sedimentation problem will, for any initial porosity, result in two regular shock waves, down- and upwards, as the characteristic lines go into the shock, see figure 2(a). Upon increasing the order by one, the now non-convex flux function, $f(n) = n^2(1-n)/2$, contains an inflection point. The downward moving shock remains regular. However, from the base a rarefaction (figure 2(b)) or a compound shock wave (figure 2(c)) travels upwards. If the initial porosity, n_i, is smaller than or equal to the porosity at the inflection point, n_I, a rarefaction will occur. A compound shock wave propagates when $n_i > n_I$. Note that all the porosities from the characteristic line of the top rarefaction down to $n = 0$ are included in the rarefaction. Different forms of compound shock waves can occur if the number of inflection points increases. As long as there is no inflection point between $n = 1$ and the porosity that corresponds to the maximum flux, a regular shock wave is always obtained for the sediment-water interface.

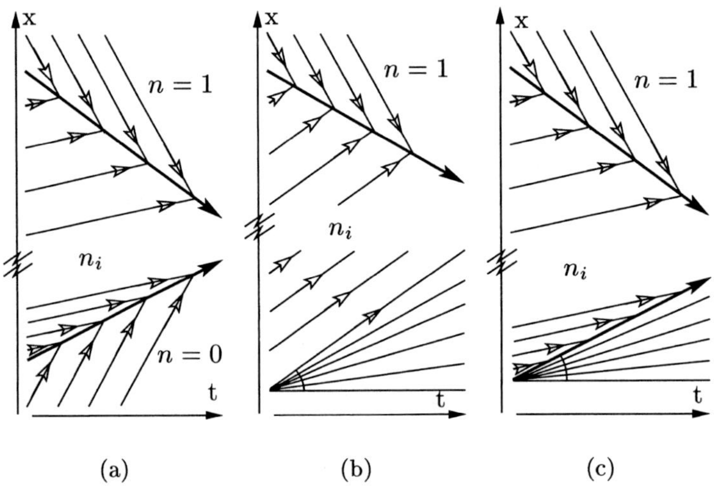

Fig. 2. Illustration of possible wave structures in sedimentation. (a) Regular shock wave down- and upwards. (b) Regular shock wave down and rarefaction wave up. (c) Regular shock wave down and compound shock wave up.

It is well-known that compound shocks arise in systems described by hyperbolic conservation laws with non-convex flux functions, i.e, for which wave speeds do not vary in a monotonic manner with relevant changes of the state variables. Many physical systems that can be described by hyperbolic conservation laws have flux functions with quadratic nonlinearities that are convex. For example, the scalar Burgers equation and the system of Euler equations for gas dynamics are convex such that compound shocks do not occur. More complex flux functions may be obtained when real gases are modelled, which may allow for the occurrence of compound shocks. An early example of a non-convex scalar flux function is given by the Buckley-Leveret equation which models two-phase flows that pertain to oil recovery problems [9]. This equation exhibits a cubic flux function. The system of magnetohy-

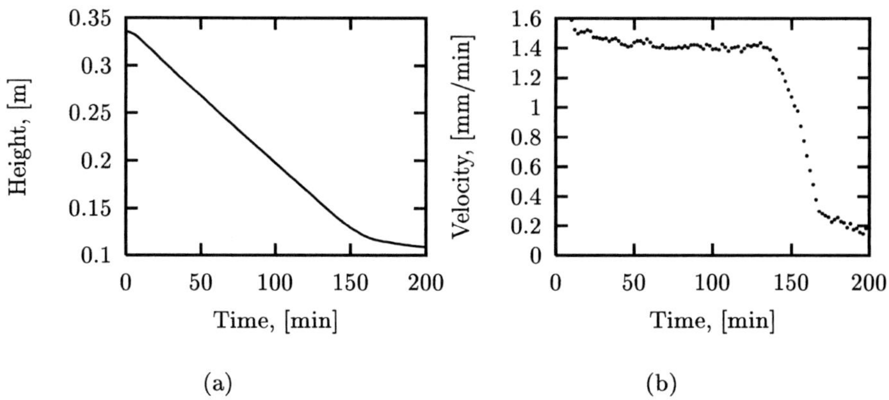

Fig. 3. Surface settlement (a) and surface velocity (b) - Kaolin suspension h_i =0.335 m or n_i = 0.979.

drodynamics equations has two non-convex modes [4] and exhibits compound shocks, which have been found in numerical simulations in one, two and three spatial dimensions [4, 5, 6].

3 Experimental Observations

The flux function of kaolinite suspensions can be determined from experimental observations by applying the Rankine-Hugoniot Jump condition, equation 2. For initial porosities between that corresponding to the maximum flux and unity (water), digital imaging has been used to locate the sediment-water interface with an accuracy better than 0.6 mm. The sediment surface settlement was monitored in thirteen such experiments, covering an initial porosity range from 0.94 to 0.98. Figure 3(a) shows an example of the surface settlement, and it can be clearly seen that the sediment first settles at a constant rate and then reduces to a much slower settling rate. The speed of the surface sediment is depicted in figure 3(b), which makes this behaviour even more apparent. Around 140 min the constant speed reduces from 1.4 $\frac{\text{mm}}{\text{min}}$ to 0.2 $\frac{\text{mm}}{\text{min}}$. At the time the surface velocity drops, the associated regular surface shock wave meets either the first characteristic line of a rarefaction or the shock of a compound shock wave, see figures 2(b) and 2(c). During the time of constant settling velocity, the average is taken as the shock speed, s, and with the porosities above and below the surface shock known, expression 2 can be solved for the flux of the initial porosity as the flux of water ($n = 1$) is zero. For the experiments for which the compound shock wave (propagating up from the base of the column) has been tracked by means of X-ray density

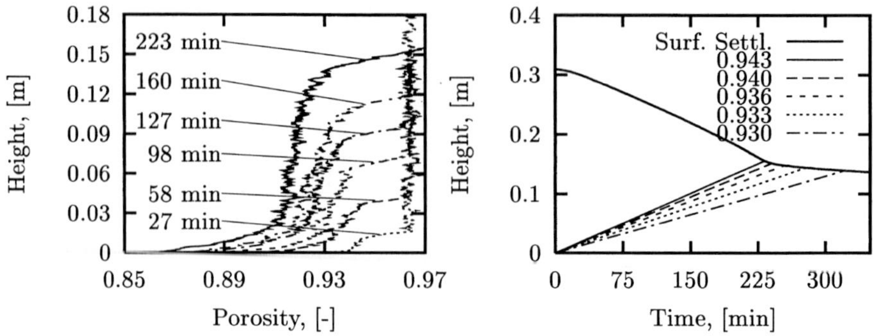

(a) (b)

Fig. 4. X-ray shock track experiment - Kaolin suspension h_i =0.309 m and n_i = 0.964. (a) Porosity profiles, (b) Surface settlement and characteristic lines.

measurements, lines of constant porosity or characteristic lines can be constructed. The Rankine-Hugoniot Jump condition applies locally when such characteristic lines enter the surface shock, and flux can be calculated from equation 2. Figure 4(a) focuses on the upward travelling porosity shock of a settling column experiment. The behaviour is very similar to the porosity profile presented in figure 1 with compound shocks clearly visible. Characteristic lines, which are obtained by linear regression of points of constant porosity, are shown in figure 4(b) combined with the surface settlement.

The flux data obtained by the methods described above is depicted in figure 5. The thirteen surface settlement experiments have given thirteen flux points, while three shock track experiments using the X-ray system together resulted in eighteen points. Within the porosity range of ~0.92 to ~0.95 the shock track experiments define the left flank of the flux function consistently. The surface track data and the shock track data overlap in the region of n=0.94 to 0.96. Between this maximum and water (n=1) a steady decrease in the flux is observed. The sedimentation experiments performed show that the flux function of kaolin is highly peaked. In order to perform numerical predictions of sedimentation experiments of kaolin suspensions a fitted function needs to be found. The basic condition for the function is that the flux needs to be zero for pure solids and water, respectively n=0 and n=1. Preferably the function should contain only one inflection point to avoid very complex wave structures [9]. By splitting the flux function up in three parts, simple power law expressions can be used:

$$n = 0 \rightarrow n = n_{\mathrm{I}} : \frac{f_1(n)}{f_1(n_{\mathrm{I}})} = \left(\frac{n}{n_{\mathrm{I}}}\right)^b ; \qquad (4)$$

$$n = n_{\rm I} \to n = n_{\rm M} : \frac{f_2(n_{\rm M}) - f_2(n)}{f_2(n_{\rm M}) - f_2(n_{\rm I})} = \left(\frac{n_{\rm M} - n}{n_{\rm M} - n_{\rm I}}\right)^{b'} ; \tag{5}$$

$$n = n_{\rm M} \to n = 1 : \frac{f_3(n_{\rm M}) - f_3(n)}{f_3(n_{\rm M}) - 0)} = \left(\frac{n_{\rm M} - n}{1 - n_{\rm M}}\right)^{b''} ; \tag{6}$$

where b, b' and b'' are determined by calibration and fulfill the condition for continuous differentiability:

$$\left(\frac{df_1}{dn}\right)_{n=n_I} = \left(\frac{df_2}{dn}\right)_{n=n_I} ; \tag{7}$$

$$\left(\frac{df_2}{dn}\right)_{n=n_M} = \left(\frac{df_3}{dn}\right)_{n=n_M} = 0. \tag{8}$$

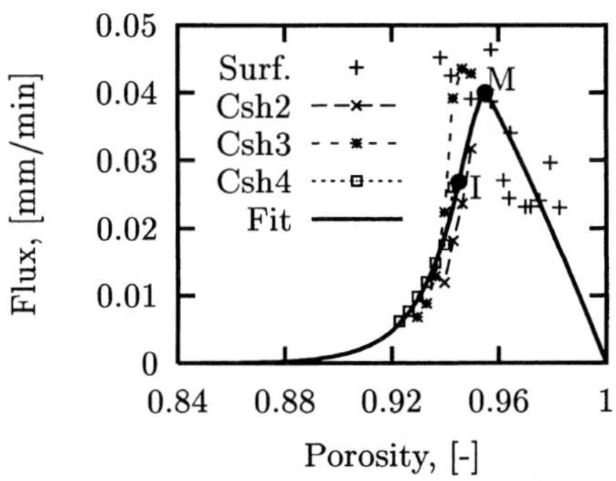

Fig. 5. Experimental Flux Data.

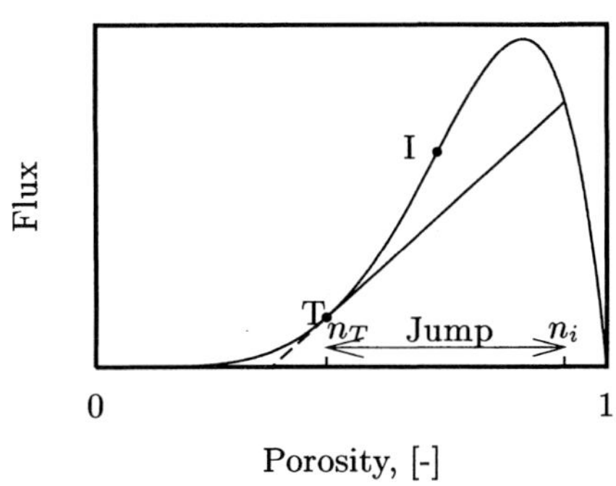

Fig. 6. Conceptual illustration of use of flux function and tangent lines.

The inflection point I and the maximum M are partly chosen by fitting functions $f_1(n)$ and $f_3(n)$ to the data and partly on an arbitrary basis. Having determined the inflection point I, compound shock waves will only occur for initial porosities larger than $n_{\rm I}$. The magnitude of the porosity jump can easily be determined by drawing a tangent line from the initial point $(n_{\rm i}, f(n_{\rm i}))$ to the flux function and the corresponding difference in porosity equals the step magnitude (Jump) as illustrated in figure 6. The speed of the shock attached to the rarefaction is determined as the slope of the tangent line from n_i to the inside of the flux curve [9].

4 Numerical Simulation Results

The double Riemann problem described by equation 1 is numerically approximated by a conservative second order accurate slope limited scheme [9]. The conservation property of the numerical scheme results in correctly calculated wave speeds.

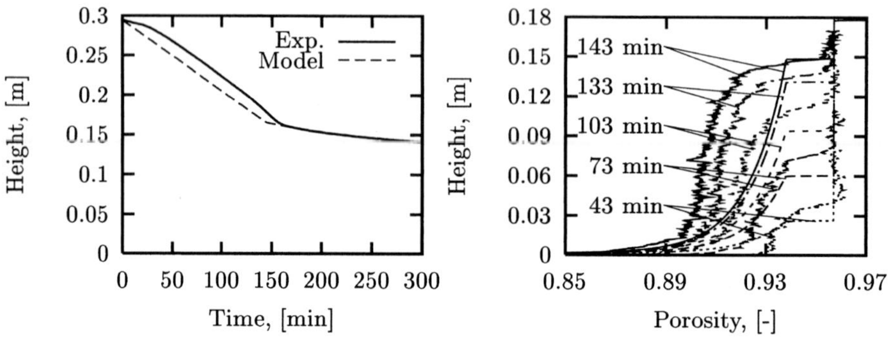

(a) (b)

Fig. 7. Comparison - Kaolin suspension h_i =0.295 m and n_i = 0.957. (a) Surface settlement, (b) Porosity profiles with compound shocks composed of a shock with an attached rarefaction.

Figure 7(a) shows the comparison between simulation and a kaolinite settling column experiment with h_i=0.295 m and n_i=0.957 using the experimentally obtained flux function. The model has clearly predicted the same constant settling rate before the slow down, but the experiment experiences a time-lag due to the turbulent energy of the mixing process, which cannot be predicted by the model. At the point the first characteristic meets the surface shock the settling slows down and the time-lag disappears and model and experiment are in accordance. The best way to evaluate the validity and quality of the sedimentation model is by comparing it to the upward travelling porosity shock. In order to make a realistic comparison of the porosity profiles, the time-lag of the experiment is shifted back to correspond with the model prediction. Figure 7(b) compares the experimental porosity profiles to the numerically simulated. At 43 and 73 minutes realistic approximations of the porosity shock moving upwards and the porosity profiles underneath the shock are obtained. At the base the model has a porosity that equals zero, though in the experimental profiles this is far greater. This feature becomes more apparent for the later profiles where the porosity decreases gradually. The predicted magnitude of the shock is good, while the tracking in time is relatively good and almost exact in the profiles of 133 and 143 minutes. The shapes of the experimental porosity profiles from the base up to the shock after 73 minutes are concavely shaped from the base upwards, but turn gradually over to a convex shape towards the shock. The model predicted profiles all have a concave shape from the base to the shock, with a significant amount of material contained at the base. Consolidation is a dissipative process and is thus not modelled by the dissipation-free hyperbolic equation 1. It is in this dissimilarity in predicted and experimental porosity profiles, that the neglected diffusive particle interaction and consolidation process in equation 1 and in the theory of sedimentation becomes most apparent. The theory of soil consolidation [7] is built-upon the deformation by particle interaction only, and concave-convex porosity profiles can be predicted [2], though this model does not handle the experimentally observed compound shock waves.

5 Conclusions

It has been shown that by using the Rankine-Hugoniot Jump Condition, equation 2, an experimental flux function can be derived from the experiments performed. Three similar power law expressions, which together differentiate continuously and define one inflection point and a maximum, give a simple but accurate representation of the experimental data. The region of porosities where compound shock waves occur, is situated between n_{I} and 1.

Numerical predictions of the surface settlement provide a good simulation of the constant settling rate and subsequent reduction in speed, despite the initially slower start of the experiment. The upward-shock tracking performance of the model is satisfactory and the size of the jump is adequate. Initially the soil bed porosity under the shock is also predicted well, but as the bed grows a differently shaped form is obtained due to the influence of inter-particle forces.

Overall it can be concluded that compound shock waves typify sedimentation and exist for kaolin suspensions only in a narrow region of initial porosities. It is clear from figure 7(b) that the dissipative effect of consolidation manifests itself in the whole rarefaction part of the compound shock. The question at which porosity consolidation first occurs for a given material has been investigated in the literature but has not been answered satisfactorily. The experiments and simulations reported in this paper suggest strongly that consolidation occurs in the whole rarefaction part of the compound shock. We thus conclude that, for a given material and a given experiment, the highest porosity for which consolidation occurs is the porosity at the characteristic line that connects the rarefaction to the compound shock (the tangent point in figure 6). This highest porosity does thus not only depend on the material, but also on the porosity above the compound shock, which is the uniform initial porosity in our sedimentation experiments. When varying this initial porosity in the experiments, the absolutely highest porosity for which consolidation can occur is the porosity corresponding to the inflection point n_{I} in figure 6. This identifies clearly for which porosity the transition between sedimentation and consolidation occurs.

Acknowledgements

Financial support from Dredging International and the Engineering and Physical Sciences Research Council is gratefully acknowledged. Prof Guy Houlsby of the Oxford University Engineering Department is thanked for many useful suggestions.

References

[1] ALVES, M.-C. 1992. *Compartamento de sedimentação e adensamento de uma lama vermelha.* Ph.D. thesis, Pontificia Universidade Católica do Rio de Janeiro.

[2] BARTHOLOMEEUSEN, G., SILLS, G.C., ZNIDARČIĆ, D., VAN KESTEREN, W., MERCKELBACH, L.M., PYKE, ROBERT, III, W.D. CARRIER, LIN, H., PENUMADU, D., MASALA, S., & CHAN, D. accepted for publication. Sidere: Prediction of large strain consolidation. *Géotechnique.*

[3] BEEN, K., & SILLS, G.C. 1981. Self-weight consolidation of soft soils. *Géotechnique*, **31**, 519–535.

[4] BRIO, M., & WU, C.C. 1988. An upwind scheme for the equations of ideal magnetohydrodynamics. *Journal of computational physics*, **75**, 400–422.

[5] DE STERCK, H., LOW, B.C., & POEDTS, S. 1999. Characteristic analysis of a complex two-dimensional magnetohydrodynamic bow shock flow with steady compound shocks. *Physics of plasmas*, **6**(3), 954–969.

[6] DE STERCK, H., LOW, B.C., & POEDTS, S. 2001. Overcompressive shocks and compound shocks in 2D and 3D magnetohydrodynamic flows. *In: Proceedings of the Eighth International Conference on Hyperbolic Problems.* International Series of Numerical Mathematics.

[7] GIBSON, R.E., SCHIFFMAN, R.L., & CARGILL, K.W. 1981. The theory of one-dimensional consolidation of saturated clays. 2. Finite non-linear consolidation of thick homogeneous layers. *Canadian geotechnical journal*, **18**, 280–293.

[8] KYNCH, G.J. 1952. A theory of sedimentation. *Transactions faraday society*, **48**, 166–176.

[9] LEVEQUE, R.J. 1992. *Numerical methods for conservation laws.* Birkhäuser Verlag, Basel.

[10] MICHAELS, A.S., & BOLGER, J.C. 1962. Settling rates and sediment volumes of flocculated kaolin suspensions. *Industrial and engineering chemistry fundamentals*, **1**(1), 24–33.

[11] OLEINIK, O.A. 1957. Discontinuous solutions of non-linear differential equations. *American Mathematical Society Translations*, **2**(26), 95–172.

[12] SILLS, G.C. 1998. Development of structure in sedimenting soils. *Philosophical transactions of the Royal Society of London A*, **356**, 2515–2534.

Derivation of Schrödinger Poisson as the Non-relativistic Limit of Klein-Gordon Maxwell

Philippe Bechouche[1], Norbert J. Mauser[2] and Sigmund Selberg[2]

[1] Wolfgang Pauli Inst. c/o Inst. f. Math., Univ. Wien, Strudlhofg. 4, A–1090 Wien and Dep. de Matemática aplicada, Facultad de Ciencias, Av. Fuentenueva, E–18071 Granada, Spain. E-mail: philippe.bechouche@univie.ac.at

[2] Wolfgang Pauli Inst. c/o Inst. f. Math., Univ. Wien, Strudlhofg. 4, A–1090 Wien. E-mail: mauser@courant.nyu.edu, sigmund.selberg@univie.ac.at

Summary. We deal with the derivation of the Schrödinger-Poisson system as a non-relativistic limit (i.e. $c \to \infty$ where c is the speed of light) of the Klein-Gordon-Maxwell or the Dirac-Maxwell system on $\mathbb{R}^{1+3}$. We deal with convergence in the energy space $C([0,T];H^1)$. We use and motivate a splitting of the scalar Klein-Gordon field into a sum of two fields, corresponding, in the physical interpretation, to electrons and positrons. The crucial technique is to use the Klainerman-Machedon machinery for the appropriately scaled system which has null form structure for the case of Coulomb gauge.

1 Introduction

This paper deals with rigorous asymptotic analysis of the "hierarchy of relativistic models" describing a "particle" coupled to its "self-consistent field". We work in a Hilbert space framework of relativisic quantum mechanics without "second quantization", where no particle-antiparticle creation/annihilation is considered and the fields are treated in a classical way.
Then the fundamental model one can consider for Fermions is the Dirac Maxwell system, where the "particle" is described by a "four spinor" subject to the Dirac equation. In the Klein-Gordon Maxwell system a scalar equation for the "particle" does not take into account "spin".
On the level of non-relativistic quantum mechanics we have the "scalar" Schrödinger Poisson system where no spin and no magnetic field are taken into account. On the "semi-non-relativistic" [10] level we have the selfconsistent Pauli equation for a "two spinor" , including spin and magnetic field in first order in 1/c. In [8] the "Pauli-Poiswell" system was motivated where both the electric and the magnetic potential are given by Poisson equations. Another "intermediate model" is the (magnetic) Schrödinger Maxwell system, where the non-relativistic limit has been only partially performed for the "particle" equation (cf [5] for non-unique global weak solutions).

We shall state rigorous results on the transition from Klein-Gordon Maxwell to Schrödinger Poisson for the time dependent case [2] and link them to the case of the transition from Dirac Maxwell to Schrödinger Poisson. We have been made aware of recent, completely independent work of Masmoudi and Nakanishi [9], who have obtained similar results to ours.

There are some earlier results concerning the non-relativistic limit of the linear Klein-Gordon equation with a given electromagnetic potential, see [4] and [12], but there are no previous results for the coupled Klein-Gordon-Maxwell system. Moreover, these papers only treat the static case, i.e. time-independent potential. The non-relativistic limit for the Dirac equation with time-dependent external potential was treated in [1].

The following diagram sketches the relation of the different nonlinear models in the asymptotic limits :

Dirac-Maxwell	$\longleftrightarrow$	**Klein-Gordon-Maxwell**
$\downarrow$		$\downarrow$
Pauli-"Poisswell"		$\downarrow$
$\downarrow$		$\downarrow$
$c \to \infty$		$c \to \infty$
$\downarrow$		$\downarrow$
Schrödinger-Poisson		**Schrödinger - Poisson**

Here we deal with the limit KGM to SP on the right hand side; the left hand side will be proved in upcoming paper [3].

1.1 The Klein-Gordon-Maxwell system

On the spacetime $\mathbb{R}^{1+3}$ we use relativistic coordinates $x^0 = ct \in \mathbb{R}, x = (x^1, x^2, x^3) \in \mathbb{R}^3$, where c is the speed of light. ∂_μ stands for $\frac{\partial}{\partial x^\mu}$. Thus, $\partial_0 = \frac{1}{c}\partial_t$, where $\partial_t = \frac{\partial}{\partial t}$. We also write $\nabla = (\partial_1, \partial_2, \partial_3)$. Indices are raised and lowered relative to the Minkowski metric with signature $\{-1, 1, 1, 1\}$, and the Einstein summation convention is in effect: Greek indices are summed over $0, 1, 2, 3$, roman indices over $1, 2, 3$. Thus, $\Delta = \partial_i \partial^i$ is the spatial Laplacian and $\Box = \partial_\mu \partial^\mu = -\frac{1}{c^2}\partial_t^2 + \Delta$ is the wave operator.

The Klein-Gordon-Maxwell (abbr. *KGM*) system in Coulomb gauge reads

$$D_\mu D^\mu \phi = c^2 \phi, \quad \partial^\nu F_{\mu\nu} = \frac{J_\mu}{c}, \quad J_\mu = \operatorname{Im}\left(\overline{\phi} D_\mu \phi\right), \quad \partial^i A_i = 0, \tag{1}$$

where $F_{\mu\nu} = \partial_\mu A_\nu - \partial_\nu A_\mu$ and $D_\mu = \partial_\mu - \frac{i}{c} A_\mu$. The unknowns are a scalar field $\phi(t, x) \in \mathbb{C}$ and a real potential $(A_\mu(t, x))_{\mu=0,1,2,3}$. The latter splits into its temporal part A_0 and its spatial part $\mathbf{A} = (A_1, A_2, A_3)$. The last equation in (1), the Coulomb condition, then reads $\operatorname{div} \mathbf{A} = 0$.

The first equation in (1) is the Klein-Gordon (abbr. *KG*) equation with unit rest mass. It is obtained from the free *KG* equation $\Box\phi = c^2\phi$ via the substitution $\partial_\mu \to D_\mu$. The second equation in (1) is Maxwell's equation. To write it in classical notation, define the electric and magnetic field vectors by, respectively, $\mathbf{E} = -\nabla A^0 - \partial_t \mathbf{A}/c$ and $\mathbf{B} = \nabla \times \mathbf{A}$. Then $F_{i0} = E_i$ and $F_{ij} = \epsilon_{ijk} B^k$, so $\partial^\nu F_{\mu\nu} = J_\mu / c$ becomes

$$\operatorname{div} \mathbf{E} = 4\pi\rho, \quad \nabla \times \mathbf{B} - \partial_t \mathbf{E}/c = \mathbf{J}/c,$$

where $\rho = J^0/c$ and $\mathbf{J} = (J^1, J^2, J^3)$. The equations

$$\operatorname{div} \mathbf{B} = 0, \quad \nabla \times \mathbf{E} + \partial_t \mathbf{B}/c = 0,$$

which follow from the definitions of $\mathbf{E}$ and $\mathbf{B}$, complete the Maxwell system in standard form. The third equation in (1) defines the four-current density associated to the *KG* equation; multiplying the *KG* equation by $\overline{\phi}$ and taking imaginary parts yields the conservation law $\partial^\mu J_\mu = 0$, i.e. $\partial_t \rho + \operatorname{div} \mathbf{J} = 0$.

The *KGM* system also has a conserved positive definite energy

$$\mathcal{E}(t) := \int_{\mathbb{R}^3} \sum\nolimits_{\mu=0}^{3} |D_\mu \phi|^2 + c^2 |\phi|^2 + \mathbf{E}^2 + \mathbf{B}^2 \, dx = \text{const.} \tag{2}$$

Klainerman and Machedon [6] proved that (1) is globally well-posed[1] for finite energy initial data

$$(\phi, \partial_t \phi)\big|_{t=0} \in H^1 \times L^2, \quad (\mathbf{A}, \partial_t \mathbf{A})\big|_{t=0} \in \mathcal{P}\dot{H}^1 \times \mathcal{P}L^2, \tag{3}$$

where $\mathcal{P}$ denotes the projection onto the divergence free vector fields on $\mathbb{R}^3$. Using linear Strichartz estimates for the homogeneous wave equation $\Box u = 0$, it is possible to prove local well-posedness of the *KGM* system, at fixed c, if one assumes slightly more L^2 regularity of the data than in (3). To get the sharp result proved in [6], however, requires certain bilinear generalizations of Strichartz' L^4 estimate (see [6, Section 2]) to handle the most dangerous bilinear terms. The crucial fact is that, due to the Coulomb gauge condition, these terms have a null form structure, without which the estimates would in fact fail.

On the other hand, the *KG* equation can be introduced as "the square" of the Dirac equation, avoiding the notorious problem of the negative spectrum of the latter. As we shall see, the underlying problem of the "electrons vs. positrons" is still present in the *KG* equation and the decomposition of the "wave function" requires essentially the same projectors (cf. [1] and Remark 2 below). This reflects the fact that the *KG* equation is fully relativistic.

The Dirac-Maxwell system in Coulomb gauge reads ,

$$i\gamma^\mu D_\mu \psi = c\psi, \quad \partial^\nu F_{\mu\nu} = \frac{J_\mu}{c}, \quad J_\mu = \langle \gamma^0 \gamma^\mu \psi, \psi \rangle, \quad \partial^i A_i = 0\,. \tag{4}$$

[1] In view of the conservation of energy, this reduces to local well-posedness.

Here the unknowns are the four-spinor $\psi(t,x) \in \mathbb{C}^4$, regarded as a column vector, and the potential $A_\mu(t,x)$. γ^μ are the Dirac matrices, and $\langle \cdot, \cdot \rangle$ is the standard inner product on $\mathbb{C}^4$.

Squaring the first equation in (4), the Dirac equation, yields

$$D_\mu D^\mu \psi = c^2 \psi - \frac{1}{c} B_j S^j \psi + \frac{i}{c} E_j \gamma^0 \gamma^j \psi, \tag{5}$$

where $S^m := i\gamma^k\gamma^l$ for cyclic permutations (k,l,m) of $(1,2,3)$. Observe also that the current can be expressed, using the Dirac equation, as $J^\mu = i \langle \gamma^0\gamma^\mu\gamma^\nu D_\nu\psi, \psi \rangle$. The *"squared Dirac-Maxwell"* system is therefore quite similar in structure to the *KGM* system (but not identical !), and one would expect, for example, that (4) is locally well-posed for data as in (3). This is in fact the case, as was recently proved by the last author [11]. A key fact is that the additional terms in (5) (compared to *KG*) can also be expressed using null forms, in an indirect way, once again due to the Coulomb gauge condition.

The limit from Dirac Maxwell (4) to Schrödinger Poisson can also be performed using the null form estimate machinery [3].

1.2 Main result

The question we are interested in here is: *What happens to the solutions of the KGM system* (1) *as $c \to \infty$?* Let us state our main result, and then in Sect. 1.3 we provide some motivation, based on the analysis of a simple model case, namely the free *KG* equation.

The main technical tools in the analysis of the limit $c \to \infty$ are versions of the null form estimates of Klainerman and Machedon in which the wave operator $\Box$ may be replaced by $i\partial_t \pm (M - c^2)$, where

$$M = M(c) = \sqrt{c^4 - c^2 \Delta}. \tag{6}$$

These new estimates are briefly discussed in Sect. 3.

Some notation: The O, o notation always refers to the limit $c \to \infty$. If X is a Banach space of functions on $\mathbb{R}^3_x$, we denote by $L^p_t X$ the space with norm $\|u\|_{L^p_t X} = (\int_{-\infty}^{\infty} \|u(t,\cdot)\|_X^p \, dt)^{1/p}$, with the usual modification if $p = \infty$. The localization of this norm to a time slab $S_T := [0,T] \times \mathbb{R}^3$ is denoted $\|u\|_{L^p_t X(S_T)}$.

Theorem 1. *(Non-relativistic limit of KGM.) Let $(\phi, A_0, \mathbf{A})$ be the global solution of (1), obtained by Klainerman-Machedon [6], corresponding to data (3) such that (see (6) for the definition of M)*

(a) $\alpha = \lim_{c\to\infty} \phi|_{t=0}$ *and* $\beta = \lim_{c\to\infty} M^{-1}\partial_t\phi|_{t=0}$ *exist in* H^1,
(b) $\|\mathbf{A}|_{t=0}\|_{\dot{H}^1} + \frac{1}{c}\|\partial_t \mathbf{A}|_{t=0}\|_{L^2} = O(1)$.

Thus, $(\phi, A_0, \mathbf{A})$ *depends implicitly on the parameter* c. *Write*

$$\phi = \phi^+ + \phi^- = e^{-itc^2}\psi^+ + e^{+itc^2}\psi^- \tag{7}$$

where

$$\phi^\pm = \frac{1}{2}\left\{\phi \pm iM^{-1}\partial_t\phi\right\} \quad \text{and} \quad \psi^\pm = e^{\pm itc^2}\phi^\pm. \tag{8}$$

Let (v^+, v^-, u) *be the solution of the Schrödinger-Poisson system*

$$\left(i\partial_t \pm \frac{\Delta}{2}\right)v^\pm = -uv^\pm, \quad \Delta u = \left|v^+\right|^2 - \left|v^-\right|^2, \tag{9}$$

with data $v^\pm\big|_{t=0} = \frac{1}{2}(\alpha \pm i\beta)$. *Then for every* $0 < T < \infty$,

$$(\psi^+, \psi^-, A_0) \longrightarrow (v^+, v^-, u) \quad \text{in} \quad L_t^\infty(H^1 \times H^1 \times \dot{H}^1)(S_T)$$

as $c \to \infty$. *Moreover, the KG current density* $\mathbf{J} = (J^1, J^2, J^3)$ *given by* (1) *converges weakly to the current density of the limit system:*

$$\mathbf{J} \longrightarrow \mathrm{Im}\left(\overline{v^+}\nabla v^+\right) + \mathrm{Im}\left(\overline{v^-}\nabla v^-\right) \quad \text{in} \quad \mathcal{D}'(\mathbb{R}^{1+3}) \tag{10}$$

as $c \to \infty$.

To motivate our result, we consider a simple but instructive example, namely the free Klein-Gordon equation :

1.3 Model case: Free *KG*

The free Klein-Gordon equation with unit mass,

$$\Box\phi = c^2\phi, \tag{11}$$

can be derived as a relativistic analogue of the free Schrödinger equation. Indeed, recall the quantum mechanical principle whereby classical quantities are replaced by operators: $E \to i\frac{\partial}{\partial t}$ and $\mathbf{p} \to \frac{1}{i}\nabla$. Thus, from the nonrelativistic energy of a free particle with unit rest mass,

$$E = \mathbf{p}^2/2,$$

one obtains the free Schrödinger equation $i\partial_t\psi = -\frac{\Delta}{2}\psi$. Proceeding instead from the relativistic energy-momentum relation

$$E = \sqrt{c^2\mathbf{p}^2 + c^4} \tag{12}$$

yields $i\partial_t\phi = M\phi$, where M is defined in (6). Squaring this gives $-\partial_t^2\phi = M^2\phi$, which we can write in the compact and obviously Lorentz invariant form (11).

Conversely, Klein-Gordon reverts to Schrödinger in the non-relativistic limit $c \to \infty$. Reversing the above steps, we see that we first have to formally take square roots of (11), written in the form $-\partial_t^2 \phi = M^2 \phi$. Then there will naturally be two separate fields $\phi^\pm$, corresponding to positive and negative square roots, and solving

$$i\partial_t \phi^\pm = \pm M \phi^\pm. \tag{13}$$

For ϕ^+, this corresponds to the relation (12), but before we can let $c \to \infty$ there, we clearly have to subtract the dominant term c^2, the rest energy. Thus, we note that if E is given by (12), then

$$E - c^2 = \mathbf{p}^2/(1 + E^2/c^2) \to \mathbf{p}^2/2 \tag{14}$$

as $c \to \infty$. In the case of negative energy, i.e., for ϕ^-, we have to *add* the rest energy, of course. Since the kinetic energy E corresponds to the Fourier variable of $-t$, while $\mathbf{p}$ corresponds to that of x, this procedure of subtracting or adding the rest energy corresponds in physical space to multiplication by the oscillating factors e^{+itc^2} or e^{-itc^2}, respectively.

The above heuristics suggest that to analyze the limit $c \to \infty$, a solution ϕ of (11) must be split according to $\phi = \phi^+ + \phi^-$, where $\phi^\pm$ solve (13), hence (11). But this forces the initial constraints

$$\phi_0 = \phi_0^+ + \phi_0^-, \quad i\phi_1 = M\phi_0^+ - M\phi_0^-,$$

where $\phi_0 = \phi|_{t=0}$, $\phi_1 = \partial_t \phi|_{t=0}$ and $\phi_0^\pm = \phi^\pm|_{t=0}$. Solving for $\phi_0^\pm$ gives

$$\phi_0^\pm = \frac{1}{2}\left\{\phi_0 \pm iM^{-1}\phi_1\right\}. \tag{15}$$

Next, we either subtract or add the rest energy. That is, we define $\psi^\pm = e^{\pm itc^2}\phi^\pm$. Then by (13) and (15),

$$i\partial_t \psi^\pm = \pm(M - c^2)\psi^\pm, \quad \psi^\pm\big|_{t=0} = \frac{1}{2}\left\{\phi_0 \pm iM^{-1}\phi_1\right\}.$$

Observe that the limits $\lim_{c\to\infty} \psi^\pm|_{t=0}$ exist in some space if and only if the limits in Theorem 1(a) exist. Since the symbol of $M - c^2$ is [cf. (14)]

$$\sqrt{c^2 |\xi|^2 + c^4} - c^2 = \frac{|\xi|^2}{1 + \sqrt{1 + |\xi|^2/c^2}} \longrightarrow \frac{|\xi|^2}{2} \quad \text{as} \quad c \to \infty, \tag{16}$$

we would therefore expect $\psi^\pm$ to converge to the solutions of

$$i\partial_t v^\pm = \mp\frac{\Delta}{2}v^\pm, \quad v^\pm\big|_{t=0} = \frac{1}{2}(\alpha \pm i\beta). \tag{17}$$

This is of course easy to verify directly here, since $\psi^\pm = e^{\mp it(M-c^2)}\phi_0^\pm$.

Remark 1. In this example we defined $\phi^{\pm}$ as the solutions of (13) with data (15), but this is equivalent to using (8) at each time t. Indeed, taking a time derivative in (8) and using the fact that ϕ solves (11), one obtains (13).

Remark 2. The splitting (8) of the Klein-Gordon field corresponds exactly to the splitting of the 4-spinor of the Dirac equation used in [1]. To see this, write (11) as a first order system

$$\partial_t \begin{pmatrix} \phi^{(0)} \\ \phi^{(1)} \end{pmatrix} = \begin{pmatrix} 0 & I \\ -M^2 & 0 \end{pmatrix} \begin{pmatrix} \phi^{(0)} \\ \phi^{(1)} \end{pmatrix},$$

where $\phi^{(0)} = \phi$ and $\phi^{(1)} = \partial_t \phi$. The 2×2 matrix on the right hand side has two eigenvalues, $\lambda^{\pm} = \mp iM$. The eigenspace projections corresponding to $\lambda^{\pm}$ are, respectively,

$$\Pi_{\pm} = \frac{1}{2} \begin{pmatrix} I & \pm iM^{-1} \\ \mp iM & I \end{pmatrix},$$

exactly as in [1, Eq. (1.21)].

2 A priori bounds for *KGM*

The main difficulty in proving Theorem 1 is to obtain *a priori* bounds as $c \to \infty$. The bounds obtained from the conservation of energy [cf. (2)] are not good enough. For example, energy conservation gives $\|\phi\|_{L_t^\infty H^1} = O(c)$, but this can be improved to $O(1)$ (on finite time intervals) using spacetime estimates of Strichartz type.

As we have seen, the splitting $\phi = \phi^+ + \phi^-$ used in Theorem 1 is motivated by the free *KG* equation. The substitution $\partial_\mu \to D_\mu$ transforms this to

$$\phi^{\pm} = \frac{1}{2} \left\{ \phi \pm M^{-1}(i\partial_t \phi + A_0 \phi) \right\}. \tag{18}$$

Since it turns out (see Theorem 2 below) that $M^{-1}(A_0\phi)$ is $O(c^{-1})$ in H^1 under the hypotheses of Theorem 1, it is clear that as far as Theorem 1 is concerned, it is immaterial which of the definitions of $\phi^{\pm}$ we use. For technical reasons we prefer to work with (18), however.

Theorem 1 essentially reduces to the following:

Theorem 2. ***(A priori estimates for KGM.)*** *Let $(A_0, \mathbf{A}, \phi)$ be the global solution, obtained in [6], of the system (1) with data (3) which are $O(1)$ as $c \to \infty$. Thus, $(A_0, \mathbf{A}, \phi)$ depends implicitly on c. Define $\phi^{\pm}$ as in (18) and set $\psi^{\pm} = e^{\pm itc^2}\phi^{\pm}$. Then*

$$L^{\pm}\psi^{\pm} = -A_0\psi^{\pm} \pm e^{\pm itc^2} R, \tag{19a}$$

$$\Delta A_0 = \left|\psi^+\right|^2 - \left|\psi^-\right|^2 + \tfrac{1}{c^2} R', \tag{19b}$$

$$\Box \mathbf{A} = \frac{1}{c}\mathcal{P}\left(\operatorname{Im}\left(\phi \overline{\nabla \phi}\right)\right) + \frac{1}{c^2}\mathcal{P}(|\phi|^2 \mathbf{A}), \tag{19c}$$

where $L^\pm = L^\pm(c) = i\partial_t \mp (M - c^2)$, *and as* $c \to \infty$ *we have the following bounds, for every* $0 < T < \infty$*:*

(a) $\|\phi^\pm\|_{L_t^\infty H^1(S_T)} = \|\psi^\pm\|_{L_t^\infty H^1(S_T)} = O(1)$,
(b) $\|\mathbf{A}\|_{L_t^\infty \dot{H}^1(S_T)} + \frac{1}{c}\|\partial_t \mathbf{A}\|_{L_t^\infty L_x^2(S_T)} = O(1)$,
(c) $\|L^\pm\psi^\pm\|_{L_t^1 H^1(S_T)} + c\|\Box\mathbf{A}\|_{L_t^1 L_x^2(S_T)} = O(1)$,
(d) $\|R\|_{L_t^1 H^1(S_T)} = O(c^{-1/2})$,
(e) $\|R'\|_{L_t^1 L_x^r(S_T)} = O(c^{-1/2})$ *for* $1 \le r \le (3/2)^+$,
(f) $\|R'\|_{L_t^\infty L_x^r(S_T)} = O(c^{-1})$ *for* $1 \le r \le 3/2$,
(g) $\left\|M^{-1}(A_0\phi)\right\|_{L_t^\infty H^1(S_T)} = O(c^{-1})$.

3 Spacetime estimates

The main technical tools needed to prove Theorem 2 are modifications of the null form estimates of Klainerman and Machedon, in which the wave operator $\Box$ may be replaced by $i\partial_t \pm (M - c^2)$. We first discuss the linear Strichartz type estimates for this operator, and then in Sect. 3.2 we state the null form estimates.

3.1 Linear estimates

The key observation is that the propagators $U^\pm(t) = e^{\mp it(M-c^2)}$ associated to $L^\pm$ behave like the Schrödinger propagators $V^\pm(t) = e^{\pm it\Delta/2}$ at low frequencies ($\lesssim c$) and like the wave equation propagators $e^{\mp itc\sqrt{-\Delta}}$ at high frequencies ($\gg c$). Indeed, $U^\pm(t)$ is a multiplier with Fourier symbol $e^{\mp ith_c(\xi)}$, where

$$h_c(\xi) = \frac{|\xi|^2}{1 + \sqrt{1 + |\xi|^2/c^2}} \sim \begin{cases} |\xi|^2/2 & \text{for } |\xi| \lesssim c, \\ c\,|\xi| & \text{for } |\xi| \gg c. \end{cases} \tag{20}$$

We say that a pair (q,r) of Lebesgue exponents is *sharp wave admissible* (for $\mathbb{R}^{1+3}$) if $\frac{1}{q} + \frac{1}{r} = \frac{1}{2}$ and $(q,r) \neq (2,\infty)$, and we say (q,r) is *Schrödinger admissible* (for $\mathbb{R}^{1+3}$) if $q, r \ge 2$ and $\frac{2}{q} + \frac{3}{r} = \frac{3}{2}$. Then we have the following estimates.

Proposition 1. *For every sharp wave admissible pair* (q,r),

$$\left\|U^\pm(t)f\right\|_{L_t^q L_x^r(S_T)} \lesssim \|f\|_{\dot{H}^{\frac{1}{q}}} + c^{-\frac{1}{q}}\|f\|_{\dot{H}^{\frac{2}{q}}} \lesssim \|f\|_{H^{\frac{2}{q}}}\,. \tag{21}$$

Using a smooth cut-off in Fourier space, we split functions $f(x)$ into its low ($\lesssim c$) and high ($\gg c$) frequency parts. Thus, $f = f_l + f_h$, where f_l (resp. f_h) has Fourier transform supported in $|\xi| \lesssim c$ (resp. $|\xi| \gg c$).

Proposition 2. *Let (q,r) and $(\tilde{q},\tilde{r})$ be Schrödinger admissible pairs. Then for the low frequency part of the solution of $L^{\pm}u = F$ with data $u|_{t=0} = f$,*

$$\|u_l\|_{L_t^q L_x^r(S_T)} + \|u_l\|_{L_t^\infty L_x^2(S_T)} \lesssim \|f_l\|_{L^2} + \|F_l\|_{L_t^{\tilde{q}'} L_x^{\tilde{r}'}(S_T)},$$

where $\frac{1}{\tilde{q}} + \frac{1}{\tilde{q}'} = 1$ and $\frac{1}{\tilde{r}} + \frac{1}{\tilde{r}'} = 1$.

Concerning the proof of these estimates, we make the following remarks. First, proceeding as in the standard proof of the Strichartz estimates for the homogeneous wave equation, we reduce Proposition 1 to the decay estimate

$$|K_{\mu,c}(t,x)| \lesssim \begin{cases} \mu\,|t|^{-1} & \text{for } \mu \lesssim c, \\ \mu^2(c\,|t|)^{-1} & \text{for } \mu \gg c, \end{cases}$$

for the convolution kernel $K_{\mu,c}(t,x) = \int e^{ix\cdot\xi} e^{ith_c(\xi)} \beta\left(\frac{\xi}{\mu}\right) d\xi$, where β is a Littlewood-Paley cut-off function. By scaling we reduce to the case $c = 1$, and use stationary phase estimates.

Secondly, Proposition 2 reduces to the decay estimate $|K_c(t,x)| \lesssim |t|^{-3/2}$ for the convolution kernel $K_c(t,x) = \int e^{ix\cdot\xi} e^{ith_c(\xi)} \chi(\xi/c)\, d\xi$. By scaling we reduce to $c = 1$, and we apply a standard decay estimate for Fourier transforms of surface carried measures.

3.2 Bilinear null form estimates

In [7], Klainerman and Machedon proved that the estimate

$$\|u\nabla v\|_{L^2(\mathbb{R}^{1+3})} \lesssim \|f\|_{H^1}\, \|g\|_{H^1} \tag{22}$$

fails for solutions of $\Box u = \Box v = 0$ on $\mathbb{R}^{1+3}$ with initial data $(f,0)$ and $(g,0)$. If the bilinear form $u\nabla v$ in (22) is replaced by one of the null forms $Q_{ij}(D^{-1}u, v)$ or $D^{-1}Q_{ij}(u,v)$, the estimate is true, however, as proved in [7]. Here $D^\alpha = (-\Delta)^{\alpha/2}$ and $Q_{ij}(u,v) = \partial_i u\, \partial_j v - \partial_j u\, \partial_i v$. This fact was used in [6] to control the bilinear terms with derivatives in the *KGM* system, which turn out to have this structure when the Coulomb gauge condition is used. In fact, straightforward calculations yield the identities

$$\mathcal{P}\,\langle\, u, \partial_i v\,\rangle = D^{-1} R^j Q_{ij}(u,v), \qquad \mathcal{P}\mathbf{A}\cdot\nabla\phi = \frac{1}{2} Q_{ij}(\phi, D^{-1}a^{ij}),$$

where $R_i = D^{-1}\partial_i$ and $a_{ij} = R_i A_j - R_j A_i$. These types of terms appear in the *KGM* and Dirac-Maxwell systems.

Here we prove versions of the Klainerman-Machedon null form estimates where one or both of u, v solve $L^{\pm}[\cdot] = 0$ instead of $\Box[\cdot] = 0$. For example, we have the following.

Proposition 3. *Suppose* $\Box u = 0$ *and* $L^{\pm} v = 0$ *with initial data* $u\big|_{t=0} = f$, $\partial_t u\big|_{t=0} = 0$ *and* $v\big|_{t=0} = g$. *Then*

$$\left\|Q(D^{-1}u, v)\right\|_{L^2(R^{1+3})} \lesssim c^{-1/2} \left\|f\right\|_{\dot{H}^1} \left\|g\right\|_{\dot{H}^1}.$$

Changing variables $t \to ct$, and applying Plancherel's theorem and the Cauchy-Schwarz inequality, this is reduced to proving that $\|I_{\pm}\|_{L^\infty(\mathbb{R}^{1+3})} = O(1)$ as $c \to \infty$, where

$$I^{\pm}(\tau,\xi) := \int_{\mathbb{R}^3} \frac{|\eta \times (\xi-\eta)|^2}{|\eta|^4 |\xi-\eta|^2} \delta\left(\tau - |\eta| \pm c\, h(\tfrac{\xi-\eta}{c})\right) d\eta \quad (\tau \in \mathbb{R},\ \xi \in \mathbb{R}^3)$$

with $h(\xi) := \alpha(|\xi|)$ and $\alpha(r) := \frac{r^2}{1+\sqrt{1+r^2}}$. To this end, we apply the following general result, with $k(r) = c\alpha(r/c)$.

Lemma 1. *Suppose* $k(r)$ *is positive and differentiable for* $r > 0$, *and that* $|k'(r)| \leq 1$. *Define*

$$I^{\pm}(\tau,\xi) = \int \frac{\sin^2\theta}{|\eta|^2} \delta(\tau - |\eta| \pm k(|\xi-\eta|))\, d\eta,$$

where θ *is the angle between* η *and* $\xi - \eta$. *Then* $\sup_{\tau,\xi} I^{\pm}(\tau,\xi) \leq 8\pi$.

To see that this applies with $k(r) = c\alpha(r/c)$, we need only observe that $k'(r) = \alpha'(r/c)$, and $0 < \alpha' < 1$. We remark that the above lemma also applies with $k(r) = r$, which corresponds to the Klainerman-Machedon estimates (then u and v both solve the homogeneous wave equation).

References

1. Bechouche, P., Mauser, N.J., Poupaud, F. (1998) : (Semi)-nonrelativistic limits of the Dirac equation with external time-dependent electromagnetic field. Comm. Math. Phys. **197**, no. 2, 405–425
2. Bechouche, P., Mauser, N.J., Selberg, S. (2002) : Nonrelativistic limit of Klein-Gordon Maxwell to Schrödinger-Poisson, submitted to American J. of Math.
3. Bechouche, P., Mauser, N.J., Selberg, S. (2002) : Nonrelativistic limit of Dirac Maxwell to Schrödinger-Poisson, manuscript
4. Cirincione, R., Chernoff, P.R. (1981) : Dirac and Klein-Gordon equations: convergence of solutions in the nonrelativistic limit. Comm. Math. Phys. **79**, no. 1, 33–46
5. Y. Guo, K, Nakamitsu and W. Strauss (1995) : Global Finite-Energy Solutions of the Maxwell-Schrödinger System, Comm. Math. Phys. **170**, 181–196
6. Klainerman, S., Machedon, M. (1994): On the Klein-Gordon-Maxwell equation with finite energy. Duke Math. J. **74**, 19–44
7. Klainerman, S., Machedon, M. (1993): Space-time estimates for null forms and the local existence theorem. Comm. Pure Appl. Math., **46**, 1221–1268

8. Masmoudi, N., Mauser, N.J. (2001): The selfconsistent Pauli equation. Mathematische Monatshefte **132**, 19-24
9. Masmoudi, N., Nakanishi, K. (2001): private communication
10. N.J. Mauser (2000) : Semi-relativistic approximations of the Dirac equation : first and second order corrections, Trans. Theor. Stat. Phys, **29**, 122-137
11. Selberg, S. (2002): On low regularity solutions of the Dirac-Maxwell system. Preprint
12. Veselić, K. (1983): On the nonrelativistic limit of the bound states of the Klein-Gordon equation. J. Math. Anal. Appl. **96**, no. 1, 63–84

Multigrid for Atmospheric Data Assimilation: Analysis

Achi Brandt[1,2] and Rima Gandlin[1,3]

[1] Weizmann Institute of Science, Rehovot 76100, Israel
[2] Email: *achi@wisdom.weizmann.ac.il*
[3] Email: *rima@wisdom.weizmann.ac.il*

Summary. Multigrid methods are known to provide the most efficient solvers to many well-posed boundary-value PDE problems. In the case of ill-determined problems they can supply several additional advantages. Unlike evolution problems with well-posed initial conditions which can be solved by direct marching in time, when only scattered data are known, each datum affects both earlier and later solution values, so simple marching cannot be used, and fast solvers would again require multigrid methods. Multigrid solver can provide natural regularization to the ill-posed problem, since the main ill-posedness is the long term and long range influence of fine-scale oscillations, while the multiscale large-scale interactions are mediated by coarse grids that omit those oscillations. As a model problem we treat a hyperbolic PDE: the wave equation with only approximately known coefficients. The results of a detailed Fourier analysis, comparing full-flow control with initial-value control are presented.

1 Introduction

Current assimilation methods require much more computer time and space than the direct solving of atmospheric flow equations. The main reason for this is that any measurement at any place and time should in principle affect the solution at any other place and time, which creates a huge dense matrix of influence.

Current methods are not only very slow but they are also based on highly questionable compromises such as assimilating only the data from one time interval at a time without fully correlating with other intervals and limiting control to only the initial value of the flow at some arbitrary chosen initial time instead of controlling the numerical equations at all times.

The fast multiscale algorithm could avoid such compromises and assimilate the data at a cost comparable to that of solving the direct problem because of three main reasons: (1) large-scale averages may be assimilated well enough on the coarser levels of the multiscale solver, which is not expensive; (2) deviations from any large scale average must be assimilated on some finer scale, but their correlation on that scale is local; (3) the measurements are usually less accurate than the numerical flow itself, hence their assimilation need not be done at the finest level.

In most current data assimilation approaches, the control parameters, i.e. those changed to obtain fitness of the solution to the data, are only the initial values of the solution. This makes it impossible to benefit from the details (the oscillating components) of the measurements far in time from the initial time because those details at those times are ill-determined by the initial solution, due to the fact that their phase error becomes practically random, in which case the amplitude of the entire solution which would best fit the data can be shown to tend to zero. Therefore, instead of controlling just initial values, i.e. solving *initial control problem (ICP)*, we suggest to control the *entire numerical solution*. We call this approach the *residual control problem (RCP)*. It can be effectively handled only by a *multiscale treatment*.

We present here the results of a detailed Fourier analysis, comparing full-flow control with initial-value control in a model case of 1D + time wave equation. Those results demonstrate the advantages of the RCP approach.

2 Model Problem

As a simple demonstration model we have studied a case of 1D + time wave equation:

$$u_{tt} = c\triangle u \quad in\ \Omega \times [0,T], \quad T \in R^+, \quad \Omega \subset R \tag{1a}$$

$$u|_{\partial\Omega} = p(t), \tag{1b}$$

$$u|_{t=0} = u_0(x), \tag{1c}$$

$$u_t|_{t=0} = g(x), \tag{1d}$$

except that part of the model, e.g. some of the functions $c(x,t)$, $u_0(x)$, $g(x)$, $p(t)$, are not known or known only partially, and instead various observations on the solution are given.

For example, let be given finitely many data $d_l \in R$, such that

$$P_l u = d_l. \tag{2}$$

The projection P_l represents a local averaging over some neigborhood, i.e.:

$$P_l u = \int p_l(x,t)u(x,t)dxdt, \qquad \int p_l(x,t)dxdt = 1,$$

when each p_l has some (small) compact support.

ICP here is to find an initial function $u_0(x)$ so that the resulting solution $u(x,t)$ of the wave equation $(1a,b,d)$ would satisfy (2) as well as possible, in the sense of minimizing the functional

$$F(u_0) = \sum_l \beta_l \Big[P_l u - d_l\Big]^2, \tag{3}$$

where $(\beta_l)^{-1}$ is an estimate of the expected square error in the measurements.

The ICP is *ill-posed* in the sense that there is no unique solution depending continuously on the data. The classical way to overcome this ill-posedness is to apply a regularization technique. The detailed description of an algorithm for solving the ICP stated above using *Tikhonov regularization* can be found in [8]. For the analysis of this formulation in terms of Fourier components see Sect. 3.2 below.

The RCP, on the other hand, involves minimizing the functional

$$F(\mathbf{u}) = \int_{\Omega\times[0,T]} \gamma R^2 dx dt + \int_{\partial\Omega} \gamma_1 R_1^2 dx + \int_0^T \gamma_2 R_2^2 dt + \\ + \int_0^T \gamma_3 R_3^2 dt + \sum_l \beta_l \left[P_l u - d_l\right]^2,$$

where γ, γ_1, γ_2, γ_3 are weight functions of x, t, u; $R = u_{tt} - \Delta u$ are the residuals of the interior equations; R_1 are residuals of the boundary conditions; R_2 and R_3 are residuals of the initial conditions. Fourier analysis for this method is reported in Sect. 3.3 below.

Before solving the problem numerically one should choose a suitable discretization scheme. Let h be a meshsize in x-direction and δt be a time step with $K+1$ gridpoints in the time direction, $K\delta t = T$. The discretization of the Laplacian at time $k\delta t$ is as follows:

$$\Delta^h u_j^k = \frac{u_{j+1}^k - 2u_j^k + u_{j-1}^k}{h^2}.$$

Consider the following second order discretizaion scheme for $(1a)$:

$$q\Delta^h u_j^{k+1} + (1-2q)\Delta^h u_j^k + q\Delta^h u_j^{k-1} = \frac{u_j^{k+1} - 2u_j^k + u_j^{k-1}}{\delta t^2}, \tag{4}$$

where $0 \le q \le 1$ is a parameter, which has to be determined. By applying the Fourier analysis, we found that this discretization scheme is *unconditionally stable* if and only if $q \ge \frac{1}{4}$. Let us take the smallest $q = \frac{1}{4}$ which still guarantees unconditional stability, since the smaller q, the more accurate the discretization.

3 Fourier analysis

3.1 General

Assume for simplicity that

a). Ω is an infinite domain. Hence $p(t)$ in $(1b)$ is replaced just by the condition that at each time t $u(x,t)$ is a <u>bounded</u> function of x;

b). $g(x)$ in $(1d)$ is known to vanish.

Let $u(x,t)$ be a Fourier component of frequency ω in x-direction and amplitude $A(t)$:

$$u(x,t) = A(t)e^{i\omega x}.$$

This function is discretized as follows:

$$u_j^k = A_k e^{i\Theta j}, \quad \Theta = \omega h, \quad j \in Z, \quad A_k = A(t_k), \quad k = 0,\dots,K, \quad K\delta t = T.$$

Let us apply the discretization scheme (4) with a choosen q:

$$A_{k+1}\left(\frac{1}{\delta t^2}+\frac{1}{h^2}\sin^2\frac{\Theta}{2}\right)-2A_k\left(\frac{1}{\delta t^2}-\frac{1}{h^2}\sin^2\frac{\Theta}{2}\right)+A_{k-1}\left(\frac{1}{\delta t^2}+\frac{1}{h^2}\sin^2\frac{\Theta}{2}\right) = 0.$$

3.2 ICP

$$A_k = C_1\left(\frac{1-im\sin\frac{\Theta}{2}}{1+im\sin\frac{\Theta}{2}}\right)^k + C_2\left(\frac{1+im\sin\frac{\Theta}{2}}{1-im\sin\frac{\Theta}{2}}\right)^k,$$

where $m = \frac{\delta t}{h}$.

From (1b) follows that $u|_{t=0} = A_0 e^{i\omega x}$. From (1$d$) with $g(x) \equiv 0$ follows that $u_t|_{t=0} = 0 \Rightarrow \frac{A_1 - A_0}{\delta t} = 0 \Rightarrow A_1 = A_0$. Thus

$$A_k = \frac{A_0}{2}\left[\left(1-im\sin\frac{\Theta}{2}\right)\left(\frac{1+im\sin\frac{\Theta}{2}}{1-im\sin\frac{\Theta}{2}}\right)^k + \left(1+im\sin\frac{\Theta}{2}\right)\left(\frac{1-im\sin\frac{\Theta}{2}}{1+im\sin\frac{\Theta}{2}}\right)^k\right]$$

Assume that the data $\{d_l\}$ are also given in terms of Fourier components, i.e. given a measurement $\overline{A}_k$ of the amplitude A_k for all $k \in I'$, I' is an index set. The functional (3) which we want to minimize now as a function of A_0 can be rewritten in the form:

$$F(A_0) = \sum_{k\in I'} \alpha_k (A_k - \overline{A_k})^2,$$

where α_k's are weight functions. α_k is proportional to an inverse of the estimated (upper bound) value of $(A_k - \overline{A_k})^2$, which can result from discretization and algebraic errors, measurements errors or modelization errors.
After solving the equation $F'(A_0) = 0$,

$$A_0 = 2\frac{\sum_{k\in I'} \alpha_k \overline{A_k}\left[\frac{(1+im\sin\frac{\Theta}{2})^k}{(1-im\sin\frac{\Theta}{2})^{k-1}} + \frac{(1-im\sin\frac{\Theta}{2})^k}{(1+im\sin\frac{\Theta}{2})^{k-1}}\right]}{\sum_{k\in I'} \alpha_k \left[\frac{(1+im\sin\frac{\Theta}{2})^k}{(1-im\sin\frac{\Theta}{2})^{k-1}} + \frac{(1-im\sin\frac{\Theta}{2})^k}{(1+im\sin\frac{\Theta}{2})^{k-1}}\right]^2}.$$

3.3 RCP

Instead of finding initial function by solving exactly $(1a-d)$ and minimizing (3), consider all A_0, A_1, …, A_K as unknowns which have to be found by minimizing a new functional:

$$F(A_0, A_1, \dots, A_K) = \sum_{k=1}^{K} \gamma_k R_k^2 + \overline{\gamma}\Big(\frac{A_1 - A_0}{\delta t}\Big)^2 + \sum_{k=1}^{K} \beta_k (A_k - \overline{A_k})^2,$$

where β_k's, $\overline{\gamma}$ and γ_k's are weight functions. γ_k and $\overline{\gamma}$ are of the order of the discretization error in the corresponding equations. For $k \notin I'$, $\beta_k = 0$. For $k \in I'$, β_k is proportional to an inverse of the estimated (upper bound) value of $(A_k - \overline{A_k})^2$, which results from the measurements errors. R_k is the residual of the *k-th* equation. Shortly speaking we want to reduce the residuals of (1) as much as possible while the solution fits the data best.

In order to minimize this functional the linear system of $K+1$ equations $\frac{\delta F}{\delta A_k} = 0$, $k = 0, \dots, K$ with $K+1$ unknowns has to be solved: $\mathcal{L}\mathbf{A} = \mathbf{D}$, where $\mathbf{A} = (A_0, A_1, \dots, A_K)$, $\mathbf{D}$ is a vector of data with components $\beta_k \overline{A}_k$ if $k \in I'$, 0 otherwise, and $\mathcal{L}$ is a $(K+1) \times (K+1)$ matrix.

Solve this linear system and find all A_0, A_1, …, A_K at once.

4 Numerical results: RCP vs. ICP

The exact solution of $(1a-d)$ is $u(x,t) = e^{i\omega x} \cos \omega t$. The goal is to recover the solution $\cos \omega t$ from the given data $\overline{A_k}$ for $k \in I'$ by two methods described above, the icp and rcp.

There are three main sources of errors: a). discretization and algebraic errors; b). data noise; c). modelization errors (the model doesn't describe exactly the real nature on which measurements are made). Let us consider the influence of each one on the result. The parameters on which the results depend, are:

a). frequency $\omega = \{0.1, 1, 10\}$;
b). meshsize $h = \{0.1, 0.01\}$;
c). time step $\delta t = \{0.1, 0.01\}$;
d). number of measurements $L = \{4, 12, 50, 200\}$;
e). measurements placement;
f). weight functions α_k, β_k, γ_k and $\overline{\gamma}$;
i). noise level in data;
j). modeling assumption.

First consider the data without the noise, i.e. $\overline{A_k} = \cos k\omega\delta t$. If this is a high frequency component ($\omega = 10$) and data are gathered *only* far from the initial time, then the result A_0 of the icp algorithm tends to 0 as the number of measurements L increases. This happens due to the discretization and algebraic errors (phase errors). The solution of the rcp algorithm behaves

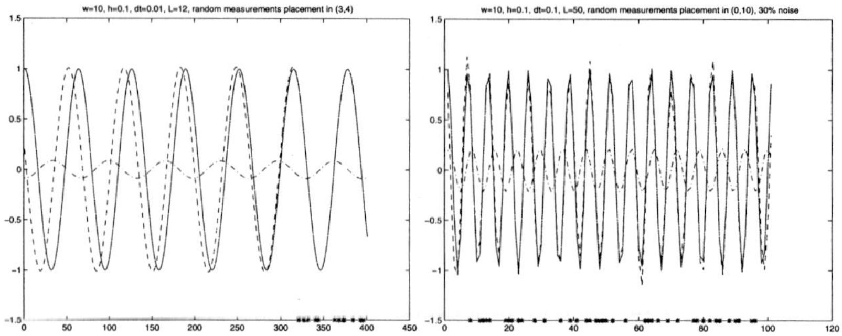

(a) (b)

Fig. 1.

(a) Data without noise. Measurements at random locations in interval (3,4). $\omega = 10$, $h = 0.1$, $\delta t = 0.01$, $L = 12$.
(b) Noised data. Noise is random number r in $(-0.5, 0.5) \times 0.3$. $\omega = 10$, $h = 0.1$, $\delta t = 0.1$, $L = 50$.

solid line: exact solution, dash dot: icp, dashed: rcp, * - measurement placement.

just in the opposite way: the result is increasingly accurate as L increases (see Fig. 1a).

Let us calculate the phase error. Consider the equation $u_{tt} = u_{xx}$. The velocity of the analytic wave propagation in this case is $c = 1$. The discretization scheme is (4) with $q = \frac{1}{4}$. For the discrete wave $\exp^{i\gamma k\delta t}\exp^{i\omega jh}$: $\tan\frac{\gamma\delta t}{2} = \frac{\delta t}{h}\sin\frac{\omega h}{2}$. As a result, the discrete velocity is $\tilde{c} = \frac{\gamma}{\omega} \sim 1 - \frac{\omega^2 h^2}{24} - \frac{\omega^2\delta t^2}{12}$. The error in the distance traveled by a given phase over time t is $t(c - \tilde{c}) = \frac{t\omega^2}{24}(h^2 + 2\delta t^2)$. Then the *phase error* is $\sim \frac{t\omega^3}{48\pi}(h^2 + 2\delta t^2)$.

Assume that the noise level in measurements is very small, $noise = 0.001$, we took $\beta_k = \frac{1}{0.001^2} = 1000000$. The local discretization error of the interior equation has the upper bound $err_d = \frac{h^2}{12} + \frac{2\delta t^2}{h^2} + \frac{\delta t^2}{12}$. Therefore, $\gamma_k = \frac{1}{err_d^2}$. The local discretization error of the initial condition has the upper bound $\frac{\delta t}{2}$. Therefore, $\overline{\gamma} = \frac{4}{\delta t^2}$.

Now consider another case. The measurements are placed at randomly chosen points and the data are *noised*: $\overline{A_k} = \cos k\omega\delta t + noise$, where $noise$ is 10% or 30% of a random number r between -0.5 and 0.5. Here also there are cases where the solution A_0 of icp tends to 0, while the solution of rcp gives a good approximation (see Fig. 1b).

The expected variance of a square of the random number r between -0.5 and 0.5 is $\int_{-0.5}^{0.5} r^2 dr = \frac{1}{12}$. Therefore, $\beta_k = \frac{1}{0.1^2 * \frac{1}{12}} = 1200$. Other weight functions were chosen as above.

Assume there is an error in modeling and the real equation we want to solve is $u_{tt} = c\Delta u$ with $c = 1.1$ or 1.2, while we use $c = 1$ in our numerical

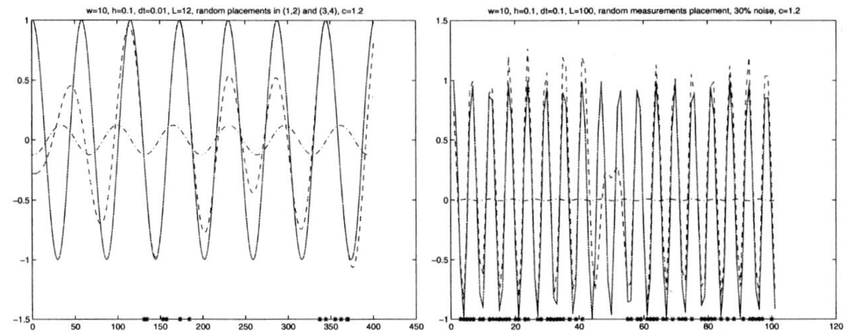

(a) (b)

Fig. 2.

(a) Modeling error. Data are solutions of $u_{tt} = c\Delta u$ at random locations. $c = 1.2$. $\omega = 10$, $h = 0.1$, $\delta t = 0.01$, $L = 50$.
(b) Three kinds of errors. Data at random locations, random r in $(-0.5, 0.5)$. $\omega = 10$, $h = 0.1$, $\delta t = 0.1$, $L = 100$, $noise = 0.3r$, $c = 1.2$.
solid line: exact solution, dash dot line: icp, dashed: rcp, $*$ - measurement placement.

solver. Then again there are cases where the solution A_0 of icp tends to 0, while the solution of rcp gives a good approximation (see Fig. 2a).

When all three kinds of errors exist, the result of the icp algorithm is practically zero-function, while rcp algorithm is still successful in approximation the real solution at least in the neighborhood of the measurements (see Fig. 2b).

5 Summary

In this paper we presented just one aspect of how multiscale computational methods can contribute to data assimilation problems (and similarly to other inverse problems).

Also multiscale formulations provide ill-posed problems with additional tools, including: one-shot solution of non-linearities; local refinement techniques; space+time parallel processing; unlimited correlation ranges and efficient representation of direct and inverse covariance matrices; multiscale organization of the observational data and in particular coarse-scale representation of oscillatory data; etc. (see Sect. 4.3 of [3]).

Multigrid solvers can provide natural regularization to many ill-posed problems, since the main ill-posedness in those problems is the long term and long range influence of fine-scale oscillations, while the multiscale large-scale interactions are mediated by coarse grids that omit those oscillations.

The general data assimilation RCP is an optimization problem which leads to a system of hyperbolic equations with corresponding boundary, initial and terminal conditions. Usually time-dependent hyperbolic problems are solved by forward marching in time, but the measurements introduce backward time couplings as well. Hence, a simple marching in time could not be applied here.

Traditional multigrid solvers are not effective for this problem either, because some "characteristic" high-frequency components are non-local (their size is determined by conditions many meshsizes away) exactly on all those grids which are fine enough to approximate such components. On the other hand, on coarser grids such components cannot be approximated, because the grid does not resolve their oscillations or resolves them with large phase errors.

Our next goal is to develop a more sophisticated efficient multigrid method for solving this kind of problems which embeds forward and backward time marching in a partly distributive relaxation process.

References

1. Brandt, A. (1981): Multi-grid Solvers for Non-Elliptic and Singular-Perturbation Steady-State Problems, Weizmann Institute of Science.
2. Brandt, A. (1984): Multigrid Techniques: 1984 Guide, with Applications to Fluid Dynamics, 191 pages, ISBN-3-88457-081-1, GMD-Studie Nr. 85, Available from GMD-AIW, Postfach 1316, D-53731, St. Augustin 1, Germany.
3. Brandt, A. (2002): Multiscale Scientific Computation: Review 2001. In: Barth, T.J., Chan, T., Haimes, R. (eds) Lecture notes in comp. science and eng. Springer. **20**, 3–95.
4. Brandt, A. (1994): Rigorous Quantitative Analysis of Multigrid: I. Constant Coefficients Two Level Cycle with L_2 norm. SIAM J. Numer. Anal., **31**, 1695–1730.
5. Brandt, A., Livshits, I. (1997): Way-ray multigrid method for standing wave equations. Electr. Trans. Numer. Anal., **6**, 162–181.
6. Brandt, A., Diskin, B. (1999): Multigrid solvers for the non-aligned sonic flow: constant coefficient case, Report WI/GC-8, 1997, Comp. and Fluids, **28**, 511–549.
7. Calderon, A. (1980): On an inverse boundary value problem, Seminar on Numerical Analysis and its Applications. In: Meyer, W., Raupp, M. (eds) Brazilian Mathematical Society, Rio de Janeiro, 1–7.
8. Czajkowski, M. (1996): Iterative solution of an initial control problem for the wave equation, Universität Düsseldorf, Germany.
9. Gandlin, R. (1999): Solvers for some Inverse Problems. Proposal for Ph.D. Thesis, Weizmann Institute of Science.
10. Gandlin, R. (2000): Multigrid solvers for some Inverse Problems. Intermediate report, Weizmann Institute of Science.
11. Morozov, V.A., (1984): Methods for Solving Incorrectly Posed Problems, Springer-Verlag.
12. Sylvester, J., Uhlmann, G. (1987): A global uniqueness theorem for an inverse boundary value problem, Annals of Math., **125**, 153–169.

A Theory of Implicit Methods for Scalar Conservation Laws

Michael Breuß

University of Hamburg, Germany

Summary. We present a theory of implicit methods for 1-D scalar conservation laws. The theory is centered around an implicit monotonicity criterion. We give rigorous theoretical results concerning the range of applicability of implicit schemes.

After a brief introduction, we describe the components of the implicit theory. We quickly investigate three implicit schemes and sketch the convergence proof for one of it. After a few numerical tests we finish this paper with some conclusive remarks and acknowledgements.

1 Introduction

Within the literature about numerical methods for conservation laws, attributes of numerical schemes like monotonicity are generally discussed for explicit schemes, see e.g. [3] and the references therein. Especially, it is usually suggested without a theoretical basis that implicit schemes have the same properties as their explicit counterparts who are identified by the use of the same numerical flux function.

This paper deals with a concept of implicit methods. Thereby, we do not restrict ourselves to Lipschitz continuous fluxes in order to make possible a numerical analysis which reflects the potential of implicit methods. As a consequence we do not rely on the TVD convergence theory. Instead we sketch a proof of convergence for which the key ingredient is an implicit monotonicity criterion.

The main assertions of this implicit theory look at first glance quite similar to their explicit counterparts since we use structures imposed by a monotonicity property; for explicit monotone schemes see especially Crandall and Majda [2] or Sanders [6]. However, there are strong differences within the techniques within the proofs – especially in the proof concerning monotonicity – and also in important details following from differences within the assumptions of the assertions.

We want to point out some developments documented within this work:

(a) We state an implicit monotonicity condition and prove that it is meaningful. In the case of implicit mappings, the usefulness of a condition has to be proven rigorously since the solution itself is part of the arguments and may a priori vary indefinitely. Note that our condition is also free of derivatives and relies simply on the comparison of data sets generated by an implicit method using induction, thus no analytic structures of the involved functions are of importance.

(b) The convergence proof does not rely on the compactness of a suitable function space as in the TVD case or as in [2], nor does it involve deducing sophisticated estimates in some function space. The concept of monotone convergence is used as the main tool to show convergence of the whole approximative sequence as the grid is refined, thus there is also no ambiguity in the limit.

(c) Since all the proofs are free of derivatives and because the convergence does not rely on a TVD idea, as a technical consequence we only need the continuity of the flux. Thus, we naturally include certain non-standard phenomena (which cannot be captured by explicit schemes) within the range of applicability of our methods, see the section on numerical tests.

(d) The application of the notions developed here shows, that the attributes of an implicit scheme may depend on the regularity of the flux as e.g. can be observed in the case of the implicit Lax-Friedrichs scheme investigated within this work. This is in contrast to the idea that the numerical characteristics of an implicit scheme include all possible transport situations.

This paper is organized as follows. At first, the cocept of implicit methods is developed. Three numerical methods are discussed in some detail. The proof of convergence to the entropy solution is illustrated. Finally, we discuss some numerical results followed by conclusive remarks.

2 Implicit Methods

In the following, u_m^l denotes the value of the numerical solution at $m\Delta x$ and $l\Delta t$. The presence of an uniform grid with a constant ratio $\lambda = \Delta t/\Delta x$ is assumed although this choice is not essential. For simplicity, implicit 3-point-methods of the form

$$u_0^{n+1} = H(u_{-1}^{n+1}, u_0^{n+1}, u_1^{n+1}; u_0^n) \tag{1}$$

are discussed. However, the whole presented theory is also valid for more general $(2k+1)$-point-methods ($k \geq 2$), meaning that the stencil can especially include the values $u_{-k}^{n+1}, \ldots, u_k^{n+1}$.

Within the first part of this section, notions for implicit methods are developed while the second part is concerned with examples. Convergence to the entropy solution is briefly discussed for one of the examined schemes.

2.1 Implicit notions

Definition 1 (Conservation Form, Numerical Flux Function). *A numerical scheme is in conservation form if a (continuous) numerical flux function* $g : \mathbf{R}^2 \to \mathbf{R}$ *exists, so that the method reads*

$$H(u_{-1}^{n+1}, u_0^{n+1}, u_1^{n+1}; u_0^n) = u_0^n - \lambda\big[g(u_0^{n+1}, u_1^{n+1}) - g(u_{-1}^{n+1}, u_0^{n+1})\big]. \tag{2}$$

The transfer of the notion of consistency to the implicit case is also trivial.

Definition 2 (Consistency). *A numerical method described by g is consistent if*

$$g(v,\ldots,v) \quad = \quad f(v) \tag{3}$$

holds for all $v \in \mathbf{R}$.

The key to nonlinear stability is the notion of monotonicity.

Definition 3 (Monotonicity). *Let two data sequences v^n and w^n be given. Let the investigated consistent and conservative numerical method produce new sequences of data v^{n+1} and w^{n+1} out of the given data v^n and w^n, respectively. Then the numerical method is monotone iff the implication*

$$v^n \geq w^n \quad \Rightarrow \quad v^{n+1} \geq w^{n+1} \tag{4}$$

holds in the sense of the comparison of components.

Note that the definition captures the essence of the notion of monotonicity and is free of derivatives of H.
Let the abbreviations $u_{j-1}^{n+1} =: a$, $u_j^{n+1} =: b$, $u_{j+1}^{n+1} =: c$ and $u_j^n =: d$ hold throughout the explanations concerning monotonicity. Now the main result is described.

Theorem 1 (Monotonicity of implicit 3-point-methods). *A consistent and conservative implicit 3-point-method is monotone iff the following conditions hold for all $\Delta a, \Delta c, \Delta d \geq 0$:*

$$H(a+\Delta a, b, c; d) \geq H(a,b,c;d), \tag{5}$$
$$H(a, b, c+\Delta c; d) \geq H(a,b,c;d), \tag{6}$$
$$H(a, b, c; d+\Delta d) \geq H(a,b,c;d). \tag{7}$$

Proof. Let two sequences v^n, w^n be given with $v^n \mapsto v^{n+1}$ and $w^n \mapsto w^{n+1}$ by application of the considered consistent and conservative numerical method. Furthermore, let $[v^n = w^n] \Rightarrow [v^{n+1} = w^{n+1}]$ hold.
"'$\Rightarrow$"': The proof follows by contraposition, using the attributes of H.
"'$\Leftarrow$"': To prove is the validity of (4). Let $\mathbf{I}$ be the set of indices k with

$$v^n \ni v_k^n \quad > \quad w_k^n \in w^n.$$

There are only a few possibilities for the composition of the set $\mathbf{I}$: It may consist of the empty set or a finite or infinite subset of $\mathbf{Z}$. The proof of the assertion follows by diversion of cases and induction by content of $\mathbf{I}$; for details see [1].

The L^∞-stability is needed as a technical component in the convergence proof. The proof is rather trivial.

Proposition 1 (L_∞-Stability). *Let an implicit 3-point-method be given which is also conservative and monotone. Then the numerical solution satisfies*

$$||u^n||_{L_\infty} \quad \leq \quad ||u_0||_{L_\infty}. \tag{8}$$

Proof. Let a sequence $u^n \in L_\infty$ be given. With $a_k := \inf_{j \in \mathbf{Z}}(u_j^k)$ and $b_k := \sup_{j \in \mathbf{Z}}(u_j^k)$ we define sequences a^n, a^{n+1} and b^n, b^{n+1} by $a_j^k := a_k$ and $b_j^k := b_k$ for all $j \in \mathbf{Z}$. Since the method is conservative,

$$a_{n+1} \; = \; u_j^n - \lambda \{g(a_{n+1}, a_{n+1}) - g(a_{n+1}, a_{n+1})\} \; \geq \; a_n$$

and $$b_{n+1} \; = \; u_j^n - \lambda \{g(b_{n+1}, b_{n+1}) - g(b_{n+1}, b_{n+1})\} \; \leq \; b_n$$

holds. Using a^k and b^k as minorant and maiorant of the data u^k, $k = n, n+1$, respectively, the L_∞-stability follows by the assumed monotonicity.

Concerning the important relation between the method and the discrete entropy inequality, we follow the lines of classical theory as presented e.g. in [3]. Nevertheless, the corresponding proofs have to be re-established for implicit methods, resulting in differences within the details [1].

Definition 4 (Entropy consistency). *An implicit numerical scheme is consistent with the entropy condition of Kruzhkov if there exists a continuous numerical entropy flux G which satisfies for all $l \in \mathbf{R}$ the following assertions:*

1. *Consistency with the entropy flux of Kruzhkov*

$$G(v, v; l) = F(v; l) \;\forall v \text{ with } F(v; l) = sgn(v - l)\,[f(v) - f(l)] \tag{9}$$

where sgn denotes the signum function.

2. *Validity of a discrete entropy inequality*

$$\frac{U(u_j^{n+1}; l) - U(u_j^n; l)}{\Delta t} \leq -\frac{G(u_j^{n+1}, u_{j+1}^{n+1}; l) - G(u_{j-1}^{n+1}, u_j^{n+1}; l)}{\Delta x} \tag{10}$$

where $U(v; l) = |v - l|$ is chosen due to Kruzhkov.

$$\text{Using} \quad a \vee b := \max(a, b) \quad \text{and} \quad a \wedge b := \min(a, b), \tag{11}$$

An important connection between the numerical entropy flux G and the numerical flux function g is now established, see also [2].

Lemma 1. *Let a consistent and conservative numerical scheme be given with a numerical flux function g. Then the numerical entropy flux defined by*

$$G(v, w; l) \; := \; g(v \vee l, w \vee l) - g(v \wedge l, w \wedge l) \tag{12}$$

is consistent with the entropy flux of Kruzhkov.

The proof relies on the consistency and conservation form of the underlying numerical scheme. The proceeding within the proof of the following assertion is a variation of a procedure employed in [2].

Theorem 2. *Let an implicit 3-point-scheme be given which is also consistent, conservative and monotone. Then the scheme is also consistent with the entropy condition of Kruzhkov.*

Proof. Since the method is assumed to be consistent and conservative, there exists a numerical flux function g so that one can construct a numerical entropy flux by Lemma 1 consistent with the entropy flux of Kruzhkov.
It is left to show the validity of a discrete entropy inequality (10). We proceed similarly as in [3, 2]. Therefore, let $l \in \mathbf{R}$ be arbitrary but fixed. It follows

$$\begin{aligned} &-\lambda\{G(u_j^{n+1}, u_{j+1}^{n+1}; l) - G(u_{j-1}^{n+1}, u_j^{n+1}; l)\} = \\ &\{H\left(u_{j-1}^{n+1} \vee l, u_j^{n+1} \vee l, u_{j+1}^{n+1} \vee l; u_j^n \vee l\right) - u_j^n \vee l\} \\ &-\{H\left(u_{j-1}^{n+1} \wedge l, u_j^{n+1} \wedge l, u_{j+1}^{n+1} \wedge l; u_j^n \wedge l\right) - u_j^n \wedge l\}. \end{aligned} \tag{13}$$

Using (11) one can derive

$$\begin{aligned} &-\lambda\{G(u_j^{n+1}, u_{j+1}^{n+1}; l) - G(u_{j-1}^{n+1}, u_j^{n+1}; l)\} \\ &= H\left(u_{j-1}^{n+1} \vee l, u_j^{n+1} \vee l, u_{j+1}^{n+1} \vee l; u_j^n \vee l\right) \\ &\quad -H\left(u_{j-1}^{n+1} \wedge l, u_j^{n+1} \wedge l, u_{j+1}^{n+1} \wedge l; u_j^n \wedge l\right) - \left|u_j^n - l\right|. \end{aligned} \tag{14}$$

One can show the validity of

$$H\left(u_{j-1}^{n+1} \vee l, u_j^{n+1} \vee l, u_{j+1}^{n+1} \vee l; u_j^n \vee l\right) \geq \left(u_j^{n+1} \vee l\right) \tag{15}$$

and
$$H(u_{j-1}^{n+1} \wedge l, u_j^{n+1} \wedge l, u_{j+1}^{n+1} \wedge l; u_j^n \wedge l) \leq u_j^{n+1} \wedge l \tag{16}$$

by using the monotonicity of the given scheme and by a suitable diversion by cases, see [1] for details concerning this. By (14), (15) and (16) we obtain the discrete entropy inequality.

2.2 Examples of Implicit Methods

Implicit Upwind
The implicit formulation of the Upwind scheme reads

$$u_j^{n+1} = u_j^n - \lambda\left[f(u_j^{n+1}) - f(u_{j-1}^{n+1})\right]. \tag{17}$$

To condition (5):

$$H(a + \Delta a, b, c; d) - H(a, b, c; d) = \lambda\left[f(a + \Delta a) - f(a)\right]. \tag{18}$$

Thus the method can only be monotone if f grows monotonously.
To condition (6):

$$H(a,b,c+\Delta c;d) - H(a,b,c;d) = 0 \quad (\geq 0)\,. \tag{19}$$

To condition (7):

$$H(a,b,c;d+\Delta d) - H(a,b,c;d) = \Delta d \quad (\geq 0)\,. \tag{20}$$

Because of (18), (19) and (20) the implicit Upwind scheme is monotone if f grows monotonously. Note that f does not need to be Lipschitz continuous.

Now we sketch the proof of convergence. Thereby, at first the convergence to a weak solution of the conservation law is shown, followed by the verification that this weak solution is the entropy solution. For brevity of the notation, we omit the arguments (x,t) when appropriate.
As the most important technical detail, a special discretization of the initial condition u_0 is employed. Without loss of generality, we assume that $u_0 \geq 0$ *a.e.* holds. We use nested grids, i.e. refined grids always inherit cell borders. The reason is that the crucial attribute of the corresponding sequences of numerical solutions is the monotonicity with respect to monotoneously growing sequences of numerical initial conditions.
On cell j the initial condition is discretized by

$$u_j^0 \;:=\; \inf_{x\in(\Delta x(j-1),j\Delta x]} u_0(x). \tag{21}$$

Multiplication of the Upwind scheme (17) with $\Delta x\Delta t$ as well as with the discrete test element ϕ_j^{n+1}, summation over $j \in \mathbf{Z}$ and $n \geq 0$ and finally summation by parts yields

$$\Delta x\Delta t \sum_{j\in\mathbf{Z}}\sum_{n\geq 0}\left[u_j^n\frac{\phi_j^{n+1}-\phi_j^n}{\Delta t} + f(u_j^{n+1})\frac{\phi_{j+1}^{n+1}-\phi_j^{n+1}}{\Delta x}\right] = -\Delta x\sum_{j\in\mathbf{Z}} u_j^0\phi_j^0. \tag{22}$$

Using the well-known *Theorem of Monotone Convergence* due to Beppo Levi together with the L^∞-stability of the scheme following from Proposition 1, one can prove the strong convergence of (22) to the form which implies that u is a weak solution of the original problem, namely

$$\int_0^\infty \int_{-\infty}^\infty [u\phi_t + f(u)\phi_x]\,dxdt = -\int_{-\infty}^\infty u_0\phi_0 dx.$$

Similarly, we derive a weak form of the discrete entropy condition (10) connected with the implicit Upwind scheme using Lemma 1 together with the assertion of Theorem 2.
The weak form of the discrete entropy condition composed via the implicit Upwind scheme reads

$$\Delta x\Delta t \sum_{j\in\mathbf{Z}}\sum_{n\geq 0}\left[|u_j^n - l|\frac{\phi_j^{n+1}-\phi_j^n}{\Delta t}\right.$$

$$+sgn\left(u_j^{n+1}-l\right)\left[f(u_j^{n+1})-f(l)\right]\frac{\phi_{j+1}^{n+1}-\phi_j^{n+1}}{\Delta x}\Bigg]$$

$$= -\Delta x \sum_{j\in\mathbf{Z}} |u_j^0 - l|\phi_j^0. \tag{23}$$

Using then the established strong convergence of the function sequence generated by the numerical method for $\Delta t, \Delta x \to 0$, one can prove convergence of (23) towards the weak form of the entropy condition due to Kruzhkov

$$\int_0^\infty \int_{-\infty}^\infty \left[|u-l|\phi_t + sgn\,(u-l)\left[f(u)-f(l)\right]\phi_x\right]dxdt = -\int_{-\infty}^\infty |u_0 - l|\phi_0 dx.$$

For more details, see [1].

Implicit Lax-Friedrichs
We investigate the implicit Lax-Friedrichs scheme

$$u_j^{n+1} = u_j^n + \frac{1}{2}\left[u_{j-1}^{n+1} - 2u_j^{n+1} + u_{j+1}^{n+1}\right] - \frac{\lambda}{2}\left[f(u_{j+1}^{n+1}) - f(u_{j-1}^{n+1})\right]. \tag{24}$$

Note that condition (7) is automatically satisfied.
To condition (5):

$$H(a+\Delta a, b, c; d) - H(a, b, c; d) = \frac{1}{2}\Delta a + \frac{\lambda}{2}\left[f(a+\Delta a) - f(a)\right]. \tag{25}$$

To condition (6):

$$H(a, b, c+\Delta c; d) - H(a, b, c; d) = \frac{1}{2}\Delta c - \frac{\lambda}{2}\left[f(c+\Delta c) - f(c)\right]. \tag{26}$$

The expressions (25) and (26) can be combined to

$$|f(x+\Delta x) - f(x)| \leq \frac{1}{\lambda}\Delta x.$$

Therefore, the implicit Lax-Friedrichs scheme is monotone only for Lipschitz-continuous flux functions with a Lipschitz constant $L \leq (1/\lambda)$. This is a surprising result because it is normally suggested that the numerical characteristics include the whole domain in the case of implicit methods. Note that by this obvious geometrical argument, the derived condition is not a classical CFL condition.

The convergence proof can be done in a similar way as already indicated.

An implicit Godunov-type method
The scheme is defined via the numerical flux function

$$g_G(v,w) = \begin{cases} \min_{v\leq u\leq w} f(u) : v \leq w, \\ \max_{w\leq u\leq v} f(u) : v > w. \end{cases} \tag{27}$$

Analogously to the proceeding before, the monotonicity property is checked. Since the relative values of the test variables have to be compared within the scheme, diversions by cases have to be employed. The validity of (7) is again trivial.

To condition (5):

Generally and before any diversions of cases,

$$H(a+\Delta a,b,c;d)-H(a,b,c;d)=\lambda\big[g_G(a+\Delta a,b)-g_G(a,b)\big] \tag{28}$$

holds. Since in (28) only the values b, a and $a+\Delta a$ are of importance, it is necessary to investigate three cases.

1. Case: $b \le a \le a+\Delta a$

$$\lambda\big[g_G(a+\Delta a,b)-g_G(a,b)\big]=\lambda\left[\max_{b\le u\le a+\Delta a} f(u)-\max_{b\le u\le a} f(u)\right]\ge 0. \tag{29}$$

2. Case: $a \le b \le a+\Delta a$

$$\lambda\big[g_G(a+\Delta a,b)-g_G(a,b)\big]=\lambda\left[\max_{b\le u\le a+\Delta a} f(u)-\min_{a\le u\le b} f(u)\right]\ge 0. \tag{30}$$

3. Case: $a \le a+\Delta a \le b$

$$\lambda\big[g_G(a+\Delta a,b)-g_G(a,b)\big]=\lambda\left[\min_{a+\Delta a\le u\le b} f(u)-\min_{a\le u\le b} f(u)\right]\ge 0. \tag{31}$$

With (29), (30) and (31) the validity of the condition (5) is proven without any additional condition on the flux function. This can also be verified within the validation of condition (6), so that the investigated Godunov-type scheme is monotone for general continuous flux functions. The convergence proof is again similar as in the case of the implicit Upwind method; see [1] for details.

3 Numerical Tests

In order to show the applicability of the implicit theory, we investigate a non-standard conservation law developed by Kruzhkov and Panov [5]. It features the flux function $f(u)=|u|^\alpha/\alpha$ with $\alpha\in(0,1)$. We take the initial condition

$$u_0(x) = \begin{cases} 0: & x<-1,\\ 1: & -1\le x\le 0,\\ 0: & x>0.\end{cases}$$

The exact solution is given in [5] and reads

$$u(x,t) = \begin{cases} 0 & : \quad t > \alpha(x+1), \\ 1 & : x < t \le \alpha(x+1), \\ \left(\frac{t}{x}\right)^{1/(1-\alpha)} & : \quad t \le x. \end{cases}$$

It especially features a rarefaction wave which goes to infinity after an arbitrarily small time step, thus no explicit scheme can be used. For the numerical solution we use a grid of 1000 points with $\Delta x = 0.005$, a time step size of $\Delta t = 0.00075$ and $\alpha = 0.5$ within the described flux function. The arising nonlinear systems of equations can be solved by very simple iterative methods, namely by bisection in the case of the implicit Upwind scheme and by a simple relaxation procedure in the other cases [1].
As expected, the implicit Lax-Friedrichs scheme does not give a sensible solution because the considered problem does not feature a Lipschitz continuous flux function. Since the numerical solution of the other two schemes are quite identical we only display the solution obtained by the implicit Upwind scheme, see Figure 1.

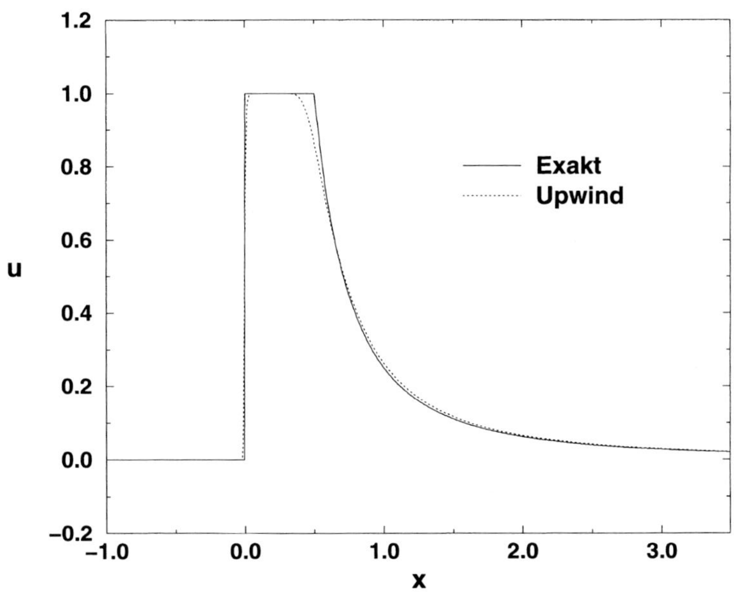

Fig. 1. Numerical solution of the described example due to Kruzhkov, obtained by the implicit Upwind method.

4 Conclusive Remarks

We have described the fundament of a theory of implicit methods for conservation laws, extending the rigorously verified range of applicability of numerical methods.

5 Acknowledgements

Since this contribution is about the content of his doctoral thesis, the author would like to express his gratitude to the supervisor of that work, Professor Rainer Ansorge from the University of Hamburg.

References

1. Breuß, M. (2001): Numerical Methods for Conservation Laws in Non-Standard-Situations. Doctoral Thesis, University of Hamburg, Germany, in german
2. Crandall, M.G., Majda, A. (1980): Monotone difference approximations for scalar conservation laws. Math. Comp., **34**, 1–21
3. Godlewski, H., Raviart, P.A. (1991): Hyperbolic Systems of Conservation Laws. Ellipses
4. A. Harten - On a class of high resolution total-variation-stable finite-difference schemes. *SIAM J. Num. Anal.* **21**, *pp. 1-23, (1984)*
5. Kruzhkov, S.N., Panov, E.Y. (1991): Conservative quasilinear first-order laws with an infinite domain of dependence on the initial data. Soviet Math. Doklady, **42**, 316–321
6. Sanders, R. (1983): On Convergence of Monotone Finite Difference Schemes with Variable Spatial Differencing. Math. Comp. **40**, 91–106
7. Yee, H.C. (1994): A class of high-resolution explicit and implicit shock-capturing methods. In: High Resolution (Upwind and TVD) Methods for the Compressible Flow Equations. VKI

High-Order Schemes for Multi-Dimensional Hamilton-Jacobi Equations

Steve Bryson[1] and Doron Levy[2]

[1] Program in Scientific Computing/Computational Mathematics, Stanford University and the NASA Advanced Supercomputing Division, NASA Ames Research Center, Moffett Field, CA 94035-1000; *bryson@nas.nasa.gov*

[2] Department of Mathematics, Stanford University, Stanford, CA 94305-2125; *dlevy@math.stanford.edu*

1 Introduction

We are interested in high-order numerical approximations for the solution of multi-dimensional Hamilton-Jacobi (HJ) equations of the form

$$\phi_t + H(\nabla\phi) = 0, \qquad \mathbf{x} = (x_1, \ldots x_d) \in \mathbb{R}^d.$$

Here, H is the Hamiltonian, which we assume depends on $\nabla\phi$ and possibly on x and t. The main difficulty in approximating the solutions of HJ equations is the discontinuous derivatives that develop in the solution in finite time even when the initial data is smooth. From an analytical point of view, solutions of HJ equations past the discontinuity are treated via the machinery of viscosity solutions (see [3, 4, 11] and the references therein). As far as the numerics is concerned, the first significant result were the converging first-order approximations by Souganidis [18]. High-order upwind methods, that are based on the ENO reconstruction of Harten *et al.* [6], were introduced by Osher, Sethian and Shu [16, 17]. Jiang and Peng combined the Weighted ENO (WENO) interpolant [8, 14] with a monotone flux obtaining the upwind-WENO schemes for HJ equations in [7].

Recently, Lin and Tadmor developed first- and second-order central schemes for HJ equations in [12, 13]. These schemes were based on the Nessyahu-Tadmor scheme for approximating solutions of hyperbolic conservation laws [15]. Semi-discrete versions of these schemes were introduced by Kurganov *et al.* in [9, 10]. In [1] we presented an efficient version of the fully-discrete central schemes for multi-dimensional problems.

In this work we introduce fifth-order accurate schemes for approximating solutions of multi-dimensional HJ equations. These are the *first* central schemes for such equations of order greater than two. We start in §2 with the derivation of our one-dimensional fifth-order scheme, based on a WENO reconstruction. In §3 we generalize this scheme to two dimensions. Since the solution at the next time step is computed at grid points that are different from those on which the data is given, we re-project the evolved solution back onto the original grid points. We conclude in §4 with several numerical examples in two space dimensions that confirm the expected order of accuracy the scheme. We also compare our results with the scheme of Jiang and Peng in [7].

A more detailed discussion of the methods developed in this paper, including a third-order method, other variants of the multi-dimensional reconstructions, and examples in one, two, and three dimensions, can be found in [2].

Acknowledgment: We would like to thank Volker Elling for helpful discussions throughout the early stages of this work. The work of D. Levy was supported in part by the National Science Foundation under Career Grant No. DMS-0133511.

2 One-Dimensional Schemes

2.1 One-Dimensional Central Schemes

Consider the one-dimensional Hamilton-Jacobi equation of the form

$$\phi_t(x,t) + H(\phi_x) = 0, \qquad x \in \mathbb{R}. \tag{1}$$

We are interested in approximating solutions of (1) subject to the initial data $\phi(x,t{=}0) = \phi_0(x)$. For simplicity we assume a uniform grid grid in space and time with mesh spacings, Δx and Δt, respectively. Denote the grid points by $x_i = i\Delta x$, $t^n = n\Delta t$, and the fixed mesh ratio by $\lambda = \Delta t/\Delta x$. Let φ_i^n denote the approximate value of $\phi(x_i, t^n)$.

Assume that the approximate solution at time t^n, φ_i^n is given. A Godunov-type scheme for approximating the solution of (1) starts with a continuous piecewise-polynomial $\tilde{\varphi}(x,t^n)$ that is reconstructed from the data, φ_i^n,

$$\tilde{\varphi}(x,t^n) = \sum_i P_{i+\frac{1}{2}}(x,t^n)\chi_{i+\frac{1}{2}}(x). \tag{2}$$

Here, $\chi_{i+1/2}(x)$ is the characteristic function of the interval $[x_i, x_{i+1}]$, and $P_{i+1/2}(x,t^n)$ is a polynomial of a suitable degree that satisfies the interpolation requirements

$$P_{i+\frac{1}{2}}(x_{i+\beta}, t^n) = \varphi_{i+\beta}^n, \quad \beta = 0, 1.$$

The reconstruction (2) is then evolved from time t^n to time t^{n+1} according to (1), and sampled at the half-integer grid-points, $\{x_{i+1/2}\}$, where the reconstruction is smooth (as long as the CFL condition $\lambda |H'(\varphi_x)| \leq 1/2$ is satisfied)

$$\varphi_{i+\frac{1}{2}}^{n+1} = \varphi_{i+\frac{1}{2}}^n - \int_{t^n}^{t^{n+1}} H\left(\tilde{\varphi}_x\left(x_{i+\frac{1}{2}}, \tau\right)\right) d\tau. \tag{3}$$

The point-value $\varphi_{i+1/2}^n$ is obtained by sampling (2), at $x_{i+1/2}$, i.e., $\varphi_{i+1/2}^n = \tilde{\varphi}(x_{i+1/2}, t^n)$. Since the evolution step (3) is done at points where the solution is smooth, we can approximate the time integral at the RHS of (3) using a sufficiently accurate quadrature rule. For example, for a third- and fourth-order method, this integral can be replaced by a Simpson's quadrature,

$$\int_{t^n}^{t^{n+1}} H\left(\tilde{\varphi}_x\left(x_{i+\frac{1}{2}},\tau\right)\right) d\tau \approx \frac{\Delta t}{6}\left[H\left(\varphi'^{\,n}_{i+\frac{1}{2}}\right) + 4H\left(\varphi'^{\,n+\frac{1}{2}}_{i+\frac{1}{2}}\right) + H\left(\varphi'^{\,n+1}_{i+\frac{1}{2}}\right)\right]. \quad (4)$$

The derivative at time t^n, $\varphi'^{\,n}_{i+1/2}$ is obtained by sampling the derivative of the reconstruction (2), i.e., $\varphi'^{\,n}_{i+1/2} = \tilde{\varphi}'(x_{i+1/2}, t^n)$. The intermediate values of the derivative in time, $\varphi'^{\,n+1/2}_{i+1/2}$, and $\varphi'^{\,n+1}_{i+1/2}$, which are required in the quadrature (4), can be predicted using a Taylor expansion or with a Runge-Kutta (RK) method such as the strong stability preserving (SSP) Runge-Kutta scheme of [5]. Alternatively, (1) can be treated as a semi-discrete equation by replacing the spatial derivatives with their numerical approximations and integrating in time via an RK method.

The only remaining ingredient to specify is the reconstruction (2). Below, we present a sixth-order reconstruction that results in a fifth-order scheme.

Remarks.

1. In order to return to the original grid, we project $\varphi^{n+1}_{i+1/2}$ back onto the integer grid points $\{x_i\}$ to end up with φ^{n+1}_i. This projection is accomplished with the same reconstruction used to approximate $\varphi^n_{i+1/2}$ from φ^n_i.
2. In order to maximize the size of the time-step, the evolution points should be taken as far as possible from the singularities in the reconstructed piecewise-polynomial. In one dimension the appropriate evolution point is located at $x_{i+1/2}$. In d-dimensions with a uniform grid with spacing Δx, the optimal evolution points are located at $x_{i+\alpha} = x_i + \alpha\Delta x$ in each direction, where $\alpha = 1/\left(d + \sqrt{d}\right)$ (see [1]). The multi-dimensional scheme we present in §3 is based on one-dimensional reconstructions. Hence, in order to prepare the grounds for the multi-dimensional setup, we write the one-dimensional reconstruction in this section assuming that the evolution points are $x_{i\pm\alpha}$. The reader should keep in mind that in one dimension, $\alpha = 1/2$.

2.2 A Fifth-Order Scheme

In order to obtain a fifth-order scheme, we need a sixth-order approximation of the point-values of φ and of the derivative φ'.

We start with the reconstruction of $\varphi_{i+\alpha}$ from φ_i. We write sixth-order interpolants as a convex combination of three cubic interpolants, each of which requires the evaluation of φ on a four-point stencil. We use the polynomials $\varphi_-(x)$ defined on the left-biased stencil $\{x_{i-2}, x_{i-1}, x_i, x_{i+1}\}$, $\varphi_0(x)$

defined on the centered stencil $\{x_{i-1}, x_i, x_{i+1}, x_{i+2}\}$ and $\varphi_+(x)$ defined on the right-biased stencil $\{x_i, x_{i+1}, x_{i+2}, x_{i+3}\}$. For smooth φ

$$\begin{aligned}
\varphi_{-,i+\alpha} &= a_1\varphi_{i-2} + a_2\varphi_{i-1} + a_3\varphi_i + a_4\varphi_{i+1} = \varphi\left(x_{i+\alpha}\right) + O\left(\Delta x^4\right), \qquad (5)\\
\varphi_{0,i+\alpha} &= a_5\varphi_{i-1} + a_6\varphi_i + a_7\varphi_{i+1} + a_8\varphi_{i+2} = \varphi\left(x_{i+\alpha}\right) + O\left(\Delta x^4\right),\\
\varphi_{+,i+\alpha} &= a_9\varphi_i + a_{10}\varphi_{i+1} + a_{11}\varphi_{i+2} + a_{12}\varphi_{i+3} = \varphi\left(x_{i+\alpha}\right) + O\left(\Delta x^4\right),
\end{aligned}$$

where the constants are given by

$$\begin{aligned}
a_1 &= \frac{1}{6}\alpha - \frac{1}{6}\alpha^3, & a_2 &= -\alpha + \frac{1}{2}\alpha^2 + \frac{1}{2}\alpha^3,\\
a_3 &= 1 + \frac{1}{2}\alpha - \alpha^2 - \frac{1}{2}\alpha^3, & a_4 &= \frac{1}{3}\alpha + \frac{1}{2}\alpha^2 + \frac{1}{6}\alpha^3,\\
a_5 &= -\frac{1}{3}\alpha + \frac{1}{2}\alpha^2 - \frac{1}{6}\alpha^3, & a_6 &= 1 - \frac{1}{2}\alpha - \alpha^2 + \frac{1}{2}\alpha^3,\\
a_7 &= \alpha + \frac{1}{2}\alpha^2 - \frac{1}{2}\alpha^3, & a_8 &= -\frac{1}{6}\alpha + \frac{1}{6}\alpha^3 = -a_1,\\
a_9 &= 1 - \frac{11}{6}\alpha + \alpha^2 - \frac{1}{6}\alpha^3, & a_{10} &= 3\alpha - \frac{5}{2}\alpha^2 + \frac{1}{2}\alpha^3,\\
a_{11} &= -\frac{3}{2}\alpha + 2\alpha^2 - \frac{1}{2}\alpha^3, & a_{12} &= \frac{1}{3}\alpha - \frac{1}{2}\alpha^2 + \frac{1}{6}\alpha^3.
\end{aligned}$$

At $x_{i-\alpha}$ we have

$$\begin{aligned}
\varphi_{-,i-\alpha} &= a_{12}\varphi_{i-3} + a_{11}\varphi_{i-2} + a_{10}\varphi_{i-1} + a_9\varphi_i = \varphi\left(x_{i-\alpha}\right) + O\left(\Delta x^4\right), \qquad (6)\\
\varphi_{0,i-\alpha} &= a_8\varphi_{i-2} + a_7\varphi_{i-1} + a_6\varphi_i + a_5\varphi_{i+1} = \varphi\left(x_{i-\alpha}\right) + O\left(\Delta x^4\right),\\
\varphi_{+,i-\alpha} &= a_4\varphi_{i-1} + a_3\varphi_i + a_2\varphi_{i+1} + a_1\varphi_{i+2} = \varphi\left(x_{i-\alpha}\right) + O\left(\Delta x^4\right).
\end{aligned}$$

A straightforward computation yields

$$c^-_{i\pm\alpha}\varphi_{-,i\pm\alpha} + c^0_{i\pm\alpha}\varphi_{0,i\pm\alpha} + c^+_{i\pm\alpha}\varphi_{+,i\pm\alpha} = \varphi\left(x_{i\pm\alpha}\right) + O\left(\Delta x^6\right),$$

where

$$\begin{aligned}
c^-_{i+\alpha} &= c^+_{i-\alpha} = \frac{1}{20}\alpha^2 - \frac{1}{4}\alpha + \frac{3}{10}, \qquad (7)\\
c^0_{i\pm\alpha} &= -\frac{1}{10}\alpha^2 + \frac{1}{10}\alpha + \frac{3}{5},\\
c^+_{i+\alpha} &= c^-_{i-\alpha} = \frac{1}{20}\alpha^2 + \frac{3}{20}\alpha + \frac{1}{10}.
\end{aligned}$$

A sixth-order WENO reconstruction of $\varphi_{i\pm\alpha}$ is therefore given by

$$\varphi_{i\pm\alpha} = w^-_{i\pm\alpha}\varphi_{-,i\pm\alpha} + w^0_{i\pm\alpha}\varphi_{0,i\pm\alpha} + w^+_{i\pm\alpha}\varphi_{+,i\pm\alpha}, \qquad (8)$$

where the weights w^k are given by

$$w^k_{i\pm\alpha} = \frac{\alpha^k_{i\pm\alpha}}{\sum_l \alpha^l_{i\pm\alpha}}, \qquad \alpha^k_{i\pm\alpha} = \frac{c^k_{i\pm\alpha}}{\left(\epsilon + S^k_{i\pm\alpha}\right)^p}, \qquad (9)$$

with $k, l \in \{+, 0, -\}$, and the constants c^k are given by (7). The oscillatory indicators are given by $S^-_{i\pm\alpha} = S_i[-2,0]$, $S^0_{i\pm\alpha} = S_i[-1,1]$ and $S^+_{i\pm\alpha} = S_i[0,2]$ where

$$S_i[r,s] = \Delta x \sum_{j=r}^{s} \left(\frac{1}{\Delta x}\Delta^+ \varphi_{i+j}\right)^2 + \Delta x \sum_{j=r+1}^{s} \left(\frac{1}{\Delta x^2}\Delta^+\Delta^- \varphi_{i+j}\right)^2 . \quad (10)$$

A sixth-order approximation of $\varphi'_{i\pm\alpha}$ from $\varphi_{i\pm\alpha}$ is written as a convex combination of four cubic interpolants. This reconstruction is similar to the third-order case, and is based on a non-central WENO reconstruction. We skip the details and summarize the result:

$$\varphi'_{i\pm\alpha} = w^1_{i\pm\alpha}\varphi'_{1,i\pm\alpha} + w^2_{i\pm\alpha}\varphi'_{2,i\pm\alpha} + w^3_{i\pm\alpha}\varphi'_{3,i\pm\alpha} + w^4_{i\pm\alpha}\varphi'_{4,i\pm\alpha}, \quad (11)$$

where

$$\begin{aligned}
\varphi'_{1,i\pm\alpha} &= \frac{1}{6\Delta x}(-2\varphi_{i-3\pm\alpha} + 9\varphi_{i-2\pm\alpha} - 18\varphi_{i-1\pm\alpha} + 11\varphi_{i\pm\alpha}),\\
\varphi'_{2,i\pm\alpha} &= \frac{1}{6\Delta x}(\varphi_{i-2\pm\alpha} - 6\varphi_{i-1\pm\alpha} + 3\varphi_{i\pm\alpha} + 2\varphi_{i+1\pm\alpha}),\\
\varphi'_{3,i\pm\alpha} &= \frac{1}{6\Delta x}(-2\varphi_{i-1\pm\alpha} - 3\varphi_{i\pm\alpha} + 6\varphi_{i+1\pm\alpha} - \varphi_{i+2\pm\alpha}),\\
\varphi'_{4,i\pm\alpha} &= \frac{1}{6\Delta x}(-11\varphi_{i\pm\alpha} + 18\varphi_{i+1\pm\alpha} - 9\varphi_{i+2\pm\alpha} + 2\varphi_{i+3\pm\alpha}).
\end{aligned}$$

Here the weights w^k are given by (9) with $c_1 = c_4 = 1/20, c_2 = c_3 = 9/20$, $S^1_{i\pm\alpha} = S_{i\pm\alpha}[-3,-1]$, $S^2_{i\pm\alpha} = S_{i\pm\alpha}[-2,0]$, $S^3_{i\pm\alpha} = S_{i\pm\alpha}[-1,1]$ and $S^4_{i\pm\alpha} = S_{i\pm\alpha}[0,2]$.

Notations.

1. We label the reconstruction of the point-values (8) as
$$\varphi_{i\pm\alpha} = \text{reconstruct\_}\varphi\text{\_1D\_5}\,(i, \pm\alpha, \varphi), \quad (12)$$
where φ is the one-dimensional array $(\varphi_1, \ldots, \varphi_N)$.
2. We label the reconstruction of $\varphi'_{i\pm\alpha}$ (11) as
$$\varphi'_{i\pm\alpha} = \text{reconstruct\_}\varphi'\text{\_1D\_5}\,(i, \pm\alpha, \varphi_{\pm\alpha}), \quad (13)$$
where $\varphi_{\pm\alpha}$ is the one-dimensional array $(\varphi_{1\pm\alpha}, \ldots, \varphi_{N\pm\alpha})$.

3 Multi-Dimensional Schemes

We develop a two-dimensional fifth-order method which is easily generalized to higher dimensions. Consider the two-dimensional HJ equation of the form

$$\phi_t + H(\nabla\phi) = 0, \qquad \mathbf{x} = (x_1, x_2) \in \mathbb{R}^2, \quad (14)$$

with initial data $\phi(\mathbf{x}, t = 0) = \phi_0(\mathbf{x})$. Denote $x_{i,j} := (x_1 + i\Delta x_1, x_2 + j\Delta x_2)$. Similarly to the one-dimensional setup, $\varphi_{i,j}$ will denote the approximation of

ϕ at $x_{i,j}$. We define the two sets of grid points, $I_+ = \{x_{i,j}, x_{i+1,j}, x_{i,j+1}\}$, and $I_- = \{x_{i,j}, x_{i-1,j}, x_{i,j-1}\}$, and denote by T_+, T_- the triangles with vertices I_+ and I_- respectively. For simplicity we assume a uniform grid $\Delta x_1 = \Delta x_2 = \Delta x$.

At $x_{i\pm\alpha,j\pm\alpha}$ our two-dimensional scheme will use the two reconstructions $\tilde{\varphi}(x_{i\pm\alpha,j\pm\alpha}, t^n)$ defined on $T_\pm$.

The point-values $\varphi^n_{i\pm\alpha,j\pm\alpha}$ are obtained by sampling $\tilde{\varphi}(x_{i\pm\alpha,j\pm\alpha}, t^n)$ at $x_{i\pm\alpha,j\pm\alpha}$, i.e., $\varphi^n_{i\pm\alpha,j\pm\alpha} = \tilde{\varphi}(x_{i\pm\alpha,j\pm\alpha}, t^n)$ and similarly for the derivatives. The evolved solution now reads

$$\varphi^{n+1}_{i\pm\alpha,j\pm\alpha} = \varphi^n_{i\pm\alpha,j\pm\alpha} - \int_{t^n}^{t^{n+1}} H\left(\nabla\tilde{\varphi}\left(x_{i\pm\alpha,j\pm\alpha}, \tau\right)\right) d\tau. \tag{15}$$

In two dimensions the choice $\alpha = 1/(2+\sqrt{2})$ guarantees that the solution remains smooth at the evolution point as long as the CFL condition $\frac{\Delta t}{\Delta x}\left|H'\left(\nabla\varphi\right)\right| < \alpha$ is satisfied. The integral in (15) is approximated as in the one-dimensional case.

Our two-dimensional reconstruction of the values of $\varphi_{i\pm\alpha,j\pm\alpha}$ is obtained by carrying out a sequence of one-dimensional reconstructions from §2.2. This dimension-by-dimension approach for the reconstruction step is similar in spirit to that of [7], but here, in order to generate a Godunov-type scheme (unlike [7]), we are forced to use evolution points that are not positioned in the same locations as the data $x_{i,j}$. An appropriately chosen sequence of one-dimensional reconstructions addresses this problem.

We use the subscript '$*$' to denote the full range of an array, such that $\varphi_{*,j}$ and $\varphi_{i,*}$ denote the one-dimensional arrays $\varphi_{*,j} = (\varphi_{1,j}, \ldots, \varphi_{N,j})$ and $\varphi_{i,*} = (\varphi_{i,1}, \ldots, \varphi_{i,N})$. With the notation for the one-dimensional fifth-order reconstruction, (12), we can express the dimension-by-dimension reconstruction at $x_{i\pm\alpha,j\pm\alpha}$ as

1. For each i,j: $\varphi_{i\pm\alpha,j} = \text{reconstruct\_}\varphi\text{\_1D\_5}\left(i, \pm\alpha, \varphi_{*,j}\right)$
2. For each i,j: $\varphi_{i\pm\alpha,j\pm\alpha} = \text{reconstruct\_}\varphi\text{\_1D\_5}\left(j, \pm\alpha, \varphi_{i+\alpha,*}\right)$.

Here, we first interpolate along the first coordinate axis and reconstruct φ at, *e.g.*, $x_{i+\alpha,j}$. The data at $x_{i+\alpha,j}$ is then interpolated along the second coordinate axis to the location $x_{i+\alpha,j+\alpha}$ to give $\varphi_{i+\alpha,j+\alpha}$ (see Figure 1). Obviously, the order in which the steps are performed is not important. We denote this reconstruction as

$$\varphi_{i\pm\alpha,j\pm\alpha} = \text{reconstruct\_}\varphi\text{\_2D\_5}\left(i, j, \pm\alpha, \varphi\right) \tag{16}$$

For the derivatives we have

1. For each i,j: $\varphi'_{i\pm\alpha,j} = \text{reconstruct\_}\varphi'\text{\_1D\_5}\left(i, \pm\alpha, \varphi_{*,j}\right)$
2. For each i,j: $\varphi'_{i\pm\alpha,j\pm\alpha} = \text{reconstruct\_}\varphi'\text{\_1D\_5}\left(j, \pm\alpha, \varphi_{i\pm\alpha,*}\right)$

which we denote as

$$\varphi'_{i\pm\alpha,j\pm\alpha} = \text{reconstruct\_}\varphi'\text{\_2D\_5}\left(i, j, \pm\alpha, \varphi\right). \tag{17}$$

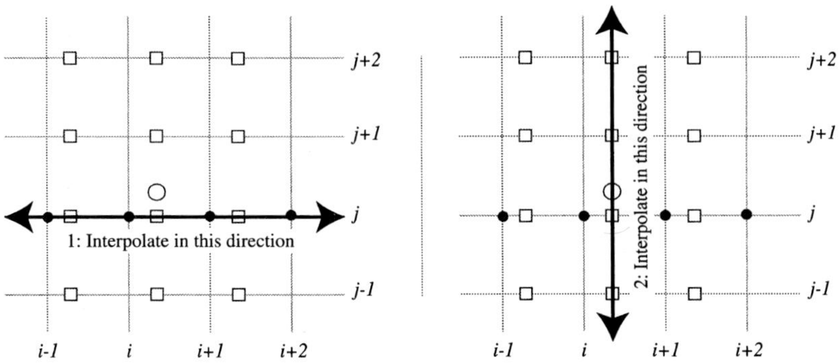

Fig. 1. The dimension-by-dimension reconstruction process in two dimensions. *Left:* the first step where the intermediate interpolants $\varphi_{i+\alpha,j}$ at $x_{i+\alpha,j}$ (open squares) are computed using the data $\varphi_{i,j}$ (black dots). *Right:* the second step, where $\varphi_{i+\alpha,j}$ is interpolated in the j direction, giving $\varphi_{i+\alpha,j+\alpha}$ at $x_{i+\alpha,j+\alpha}$ (open circle).

After evolving the solution to the next time step at the evolution points $x_{i\pm\alpha,j\pm\alpha}$ we would like to re-project $\varphi^{n+1}_{i+\alpha,j+\alpha}$ back onto the integer grid points $x_{i,j}$ to end up with $\varphi^{n+1}_{i,j}$.

The evolution points at $x_{i\pm\alpha,j\pm\alpha}$ have the same geometrical relationship to $x_{i,j}$ as $x_{i,j}$ has to $x_{i\mp\alpha,j\mp\alpha}$. Hence, in order to reconstruct $\varphi^{n+1}_{i,j}$ from $\varphi_{i\pm\alpha,j\pm\alpha}$, we can directly utilize the reconstruction (17), taking $\varphi_{i\pm\alpha,j\pm\alpha}$ as the input data, and reversing the sign of the parameter from $\pm\alpha$ to $\mp\alpha$:

1. For each i,j: $\varphi^{+}_{i,j} = \text{reconstruct\_}\varphi\text{\_2D\_5}\,(i,-\alpha,\varphi_{i+\alpha,j+\alpha})$
2. For each i,j: $\varphi^{-}_{i,j} = \text{reconstruct\_}\varphi\text{\_2D\_5}\,(i,\alpha,\varphi_{i-\alpha,j-\alpha})$
3. For each i,j: $\varphi^{n+1}_{i,j} = \frac{1}{2}\left(\varphi^{+}_{i,j}+\varphi^{-}_{i,j}\right)$.

4 Numerical Simulations

In this section we present simulations that test the schemes we developed in this paper. In §4.1 we examine stability issues in two dimensions, including comparisons with [7]. Some of these examples are standard test cases that can be found, e.g., in [10, 13, 17]. Additional examples using the methods described in this paper can be found in [2]. We do not follow the practice in [7] of masking singular regions from our error measurements.

A convex Hamiltonian In two dimensions we solve the problem

$$\phi_t + \frac{1}{2}\left(\phi_x + \phi_y + 1\right)^2 = 0. \tag{18}$$

The solutions of (18) develop a singularity at time $t = \pi^{-2}$. The results of the fifth-order calculations for the initial data $\phi(x,y,0) = -\cos(\pi(x+y)/2) = -\cos(\pi\xi)$ are shown in Figure 2. The convergence rates for the two-dimensional fifth-order scheme before and after the singularity are shown in Table 1.

Before singularity $T = 0.8/\pi^2$				
N	*relative* L^1*-error*	L^1*-order*	*relative* L^∞*-error*	L^∞*-order*
50	1.19×10^{-4}	–	7.78×10^{-7}	–
100	6.80×10^{-6}	4.13	1.64×10^{-8}	5.56
200	1.73×10^{-7}	5.30	1.12×10^{-10}	7.20
After singularity $T = 1.5/\pi^2$				
N	*relative* L^1*-error*	L^1*-order*	*relative* L^∞*-error*	L^∞*-order*
50	1.32×10^{-3}	–	2.07×10^{-5}	–
100	3.89×10^{-4}	1.76	3.60×10^{-6}	2.52
200	4.86×10^{-5}	3.00	1.69×10^{-7}	4.41

Table 1. Relative L^1- and L^∞-errors for the two-dimensional convex HJ problem (18) before and after singularity formation, computed via the fifth-order method.

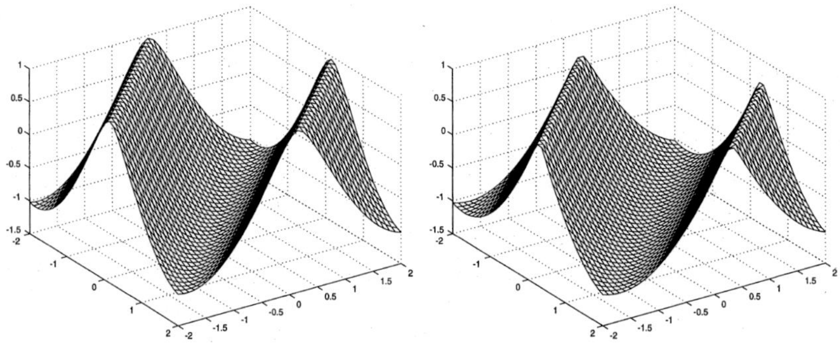

Fig. 2. Two-dimensional convex Hamiltonian, (18). *Left:* the solution before the singularity formation, $T = 0.8/\pi^2$. *Right:* the solution after the singularity formation, $T = 1.5/\pi^2$. $N = 40 \times 40$. The solution is computed with the fifth-order method.

A non-convex Hamiltonian The two-dimensional non-convex problem is

$$\phi_t - \cos(\phi_x + \phi_y + 1) = 0. \tag{19}$$

Here we assume initial data that is given by $\phi(x, y, 0) = -\cos(\pi(x+y)/2)$, and periodic boundary conditions. In this case (19) has a smooth solution for $t \lesssim 1.049/\pi^2$, after which a singularity forms. A second singularity forms at $t \approx 1.29/\pi^2$. The convergence results for the two-dimensional fifth-order scheme before and after the singularity formation are given in Table 2.

4.1 A Stability Study

In this section we check the stability properties of the two-dimensional fifth-order method of §3 by computing the L^1 errors for various examples while varying the CFL number. In Figure 3 we compare the results obtained with

Before singularity $T = 0.8/\pi^2$				
N	*relative* L^1*-error*	L^1*-order*	*relative* L^∞*-error*	L^∞*-order*
50	1.11×10^{-4}	–	1.26×10^{-6}	–
100	6.91×10^{-6}	4.00	2.42×10^{-8}	5.70
200	3.85×10^{-7}	4.17	6.27×10^{-10}	5.27
After singularity $T = 1.5/\pi^2$				
N	*relative* L^1*-error*	L^1*-order*	*relative* L^∞*-error*	L^∞*-order*
50	1.47×10^{-3}	–	8.58×10^{-6}	–
100	1.93×10^{-4}	2.93	9.27×10^{-7}	3.21
200	8.87×10^{-5}	1.12	3.09×10^{-7}	1.58

Table 2. Relative L^1- and L^∞-errors for the two-dimensional non-convex HJ problem (19) before and after the singularity formation, computed with the fifth-order method.

our fifth-order scheme with the fifth-order method of [7], for which we used a local Lax-Friedrichs flux. Similar results were found for other examples. The numerical tests indicate that larger CFL numbers can be used with our method.

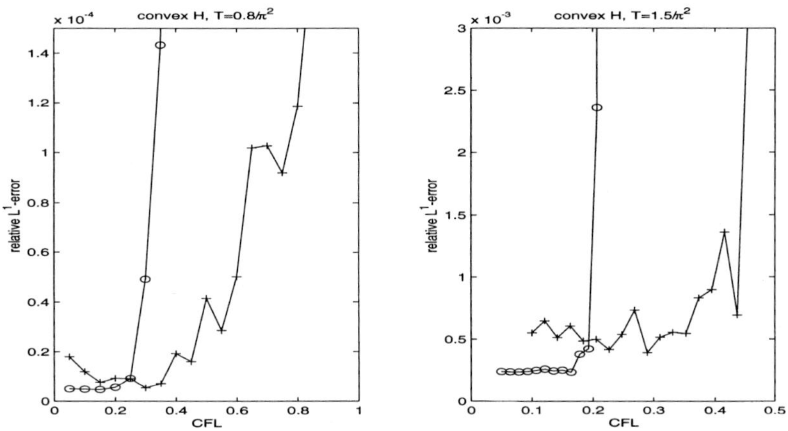

Fig. 3. Stability of the two-dimensional fifth-order method. $N = 100 \times 100$. "+": our fifth-order method. "o": the fifth-order method of [7] with a local Lax-Friedrichs flux. Convex Hamiltonian (18), before the singularity (left) and after the singularity (right).

References

1. Bryson S., Levy D., *Central schemes for multi-dimensional Hamilton-Jacobi equations*, NAS Technical Report NAS–01–014, 2001, SIAM J. Sci. Comp., to appear.
2. Bryson S., Levy D., *High-order central WENO schemes for multi-dimensional Hamilton-Jacobi equations*, NAS Technical Report NAS–02–004, submitted.
3. Crandall M.G., Evans L.C., Lions P.-L., *Some properties of viscosity solutions of Hamilton-Jacobi equations*, Trans. Amer. Math. Soc., **282**, (1984), pp.487–502.
4. Crandall M.G., Lions P.-L., *Viscosity solutions of Hamilton-Jacobi equations*, Trans. Amer. Math. Soc., **277** (1983), pp.1–42.
5. Gottlieb S., Shu C.-W., Tadmor E., *Strong stability-preserving high order time discretization methods*, SIAM Review, **43**, (2001), pp.89–112.
6. Harten A., Engquist B., Osher S., Chakravarthy S., *Uniformly high order accurate essentially non-oscillatory schemes III*, J. Comp. Phys., **71** (1987), pp.231–303.
7. Jiang G.-S., Peng D., *Weighted ENO schemes for Hamilton-Jacobi equations*, SIAM J. Sci. Comp., **21**, (2000), pp.2126–2143.
8. Jiang G.-S., Shu C.-W., *Efficient implementation of weighted ENO schemes*, JCP, **126**, (1996), pp.202–228.
9. Kurganov A., Noelle S., Petrova G., *Semi-discrete central-upwind schemes for hyperbolic conservation laws and Hamilton-Jacobi equations*, SIAM J. Sci. Comp., **23**, (2001), pp.707–740.
10. Kurganov A., Tadmor E., *New high-resolution semi-discrete central schemes for Hamilton-Jacobi equations*, JCP, **160**, (2000), pp.720–724.
11. Lions P.L., *Generalized solutions of Hamilton-Jacobi equations*, Pitman, London, 1982.
12. Lin C.-T., Tadmor E., *L^1-stability and error estimates for approximate Hamilton-Jacobi solutions*, Numer. Math., **87**, (2001), pp.701–735.
13. Lin C.-T., Tadmor E., *High-resolution non-oscillatory central schemes for approximate Hamilton-Jacobi equations*, SIAM J. Sci. Comp., **21**, no. 6, (2000), pp.2163–2186.
14. Liu X.-D., Osher S., Chan T., *Weighted essentially non-oscillatory schemes*, JCP, **115**, (1994), pp.200–212.
15. Nessyahu H., Tadmor E., *Non-oscillatory central differencing for hyperbolic conservation laws*, JCP, **87**, no. 2 (1990), pp.408–463.
16. Osher S., Sethian J., *Fronts propagating with curvature dependent speed: algorithms based on Hamilton-Jacobi formulations*, JCP, **79**, (1988), pp.12–49.
17. Osher S., Shu C.-W., *High-order essentially nonoscillatory schemes for Hamilton-Jacobi equations*, SIAM J. Numer. Anal., **28**, (1991), pp.907–922.
18. Souganidis P.E., *Approximation schemes for viscosity solutions of Hamilton-Jacobi equations*, J. Diff. Equations., **59**, (1985), pp.1–43.

On a Model for Continuous Sedimentation in Vessels with Discontinuous Cross-sectional Area

R. Bürger[1], K. H. Karlsen[2], N. H. Risebro[3], and J. D. Towers[4]

[1] Institute of Applied Analysis and Numerical Simulation, University of Stuttgart, Pfaffenwaldring 57, D–70569 Stuttgart, Germany.
E-mail: buerger@mathematik.uni-stuttgart.de

[2] Department of Mathematics, University of Bergen,
Johs. Brunsgt. 12, N–5008 Bergen, Norway. E-Mail: kennethk@mi.uib.no

[3] Department of Mathematics, University of Oslo,
P.O. Box 1053, Blindern, N–0316 Oslo, Norway. E-Mail: nilshr@math.uio.no

[4] MiraCosta College, 3333 Manchester Avenue,
Cardiff-by-the-Sea, CA 92007-1516, USA. E-Mail: jtowers@cts.com

Summary. We study a clarifier-thickener unit considering that its cross-sectional area is not constant in both the clarification and the thickening zones. A mathematical model of sedimentation in such a vessel can be formulated as an initial value problem for a scalar conservation law with a nonconvex spatially varying flux and several types of discontinuities with respect to the space variable. Weak solutions of this non-standard conservation law are approximated by an extension of the well-known Engquist-Osher scheme. The key new ingredient is the discretization of the discontinuous flux parameters on a grid staggered against that of the conserved (sought) variable. The mathematical model and the numerical scheme are illustrated by a number of simulations.

1 Introduction

The kinematic theory of one-dimensional sedimentation of ideal suspensions originated in Kynch's paper [26]. Its main assumption states that in a suspension of small equal-sized solid particles dispersed in a viscous fluid, considered as a superposition of two continuous phases, the solid-fluid relative or slip velocity v_{r} is a function of the local solids concentration u only. Under this assumption and defining the so-called Kynch batch flux density function $h(u) := u(1-u)v_{\mathrm{r}}(u)$, we can state the governing equation for batch settling of a suspension in a cylindrical closed vessel as the scalar conservation law $u_t + h(u)_x = 0$. The function $h(u)$ depends on the material specific properties of the mixture under study. A very frequently used function is [30]

$$h(u) = v_\infty u(1-u)^n, \quad v_\infty > 0,\ n \geq 1, \tag{1}$$

where $v_\infty > 0$ is the settling velocity of a single particle in an unbounded pure fluid. It should be pointed out that the kinematic sedimentation model forms an example of the general theory of kinematic waves by Lighthill and Whitham [28, 34], and is very similar to the kinematic theory of traffic flow on highways. It soon turned out that the kinematic wave theory bears severe shortcomings in the practical application to traffic flow, which essentially are due to the fact that cars equipped with anisotropic 'driver psychology' can not simply be described as a flowing continuum. This led to the (apparently, still controversial [16]) postulation of additional 'momentum' equations [2, 21]. On the other hand, numerous studies (see [10, 11] for overviews) and extensive use in industry have established the kinematic sedimentation theory as a useful description for the settling of a variety of materials such as mineral tailings, wastewater and blood. Due to its simplicity, the traffic flow model is used in many textbooks as an instructive applicative example for a nonlinear scalar conservation law. However, the application of the same equation (but with a non-convex flux) to sedimentation of suspensions has gained more practical importance.

In its 'raw' form, the equation $u_t + h(u)_x = 0$ applies to batch settling in a closed column. Practitioners very soon used the kinematic sedimentation theory for design calculations of continuously operated settling tanks, so-called thickeners, by adding a transport term $q(t)u$ to the flux density function $h(u)$, where $q(t)$ is a controllable mixture flow velocity [22, 32]. This results in the equation

$$u_t + \big(q(t)u + h(u)\big)_x = 0. \tag{2}$$

Several researchers recognized that information on whether a cylindrical continuous thickener is able to treat a given suspension with a known function $h(u)$ under stationary flow conditions can essentially be read off from plots of the continuous flux function $qu + h(u)$ (with $q = q(t)$ kept constant) [23, 24, 31, 35]. However, to use (2) for the simulation of continuous sedimentation, one needs to explicitly model the feed and discharge mechanisms and to provide (in the widest sense) boundary conditions. At the latest at this point engineering intuition had to give way to serious mathematical analysis.

Petty [29] was the first to propose Dirichlet boundary concentrations for Eq. (2), but recognized that the resulting model was not well-posed since overflow and underflow concentration waves may break through the feed and discharge boundary levels, such that in dependence of the solution u, prescribed boundary data are ignored. The well-posedness of a scalar conservation law with boundary conditions was recovered by the concept of (set-valued) entropy boundary conditions developed by Bardos et al. [3], Dubois and Le Floch [19] and Le Roux [27]. In a series of papers, Bustos and her co-workers utilized this new concept to provide a rigorous mathematical framework to continuous sedimentation, including existence and uniqueness results [14] and constructions of elementary solutions [12] and a control model [13].

In spite of its amenability to mathematical analysis, the Petty-Bustos model [12, 13, 14, 29] suffers from some drawbacks. Among them is the lack

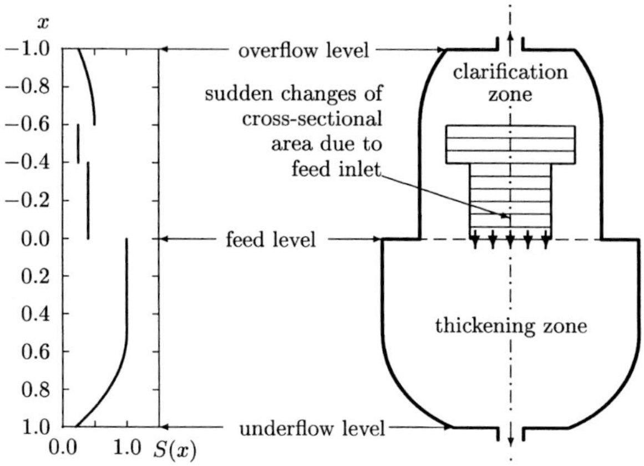

Fig. 1. Vertical cross-section of an axisymmetric clarifier-thickener unit characterized by its cross-sectional area function $S(x)$.

of a global conservation principle due to the use of Dirichlet boundary conditions, and the complete neglect of a clarification zone. It is preferable to replace the Dirichlet boundary conditions by transitions between the transport flux $q(x,t)u$ and the composite flux $q(x,t)u + h(u)$, which leads to a pure initial value problem. Moreover, the feed suspension should enter *between* the overflow outlet at the top and the discharge outlet at the bottom. This gives rise to an upwards-directed volume average velocity $q_\mathrm{l} \leq 0$ above and a downwards-directed velocity $q_\mathrm{r} \geq 0$ below the feed level. The feed source itself is modeled by a singular source term.

Thorough constructions and classifications of solutions of clarifier-thickener models were presented in a series of papers by Diehl, see [17, 18] and the references cited therein. We have adopted a more computational approach and are interested in devising numerical methods for clarifier-thickener units for proving existence of weak solutions and to provide a simulation tool. Thereby we can also include a quasi-one-dimensional formulation of a vertical vessel with a variable cross-sectional area $S(x)$, where $-1 \leq x \leq 1$ is the dimensionless depth variable. For such a configuration it is at least complicated [1] to construct exact solutions by the method of characteristics. Within our finite differencing framework, it is not even necessary to assume that $S(x)$ is a continuous function of depth x, i.e. we allow $S(x)$ to have jumps. Such jumps occur, for example, at the feed level due to the internal feed mechanism, see Figure 1.

The remainder of the paper is organized as follows. In Section 2 the governing equation for continuous sedimentation in a vessel with varying cross section is derived in some detail. In the formulation of the final initial value problem we add the feed source to the model for constant cross-sectional area formulated in [6, 7, 8]. Moreover, weak solutions are defined and the main convergence result is stated. In Section 3 the numerical method is outlined and its basic properties are stated. In Section 4 we present (and discuss) numerical examples in the application to clarifier-thickener models.

2 Mathematical model

We assume that $x = -1$ is the overflow level, $x = 0$ corresponds to the feed level, and $x = 1$ is the discharge level, and assume that the unknown volumetric solids concentration u depends on depth only, i.e., $u = u(x,t)$. Then the conservation of mass equations for the solids and the fluid are

$$S(x)u_t + \big(S(x)uv_{\mathrm{s}}\big)_x = 0, \tag{3}$$

$$-S(x)u_t + \big(S(x)(1-u)v_{\mathrm{f}}\big)_x = 0, \tag{4}$$

respectively, where t is time and v_{s} and v_{f} are the solids and the fluid phase velocities. The mixture flux, that is the volume average flow velocity weighted with $S(x)$, is given by $Q(x,t) := S(x)(uv_{\mathrm{s}} + (1-u)v_{\mathrm{f}})$.

The sum of (3) and (4) produces the continuity equation of the mixture, $Q_x = 0$ for $-1 < x < 0$ and $0 \leq x \leq 1$ and $t > 0$, which implies that $Q(\cdot,t)$ is constant as a function of x. Since Q only suffers a jump across the feed source level $x = 0$, we get for $t > 0$

$$Q(x,t) = \begin{cases} Q(-1,t) =: Q_{\mathrm{L}}(t) & \text{for } -1 \leq x < 0, \\ Q(1,t) =: Q_{\mathrm{R}}(t) & \text{for } 0 < x \leq 1, \end{cases} \tag{5}$$

where $Q_{\mathrm{L}}(t) \leq 0$ and $Q_{\mathrm{R}}(t) \geq 0$ are the signed volumetric suspension overflow and discharge rates, respectively, prescribed by the control of the suspension volume flows at $x = -1$ and $x = 1$. Note that we here identify a downwards flow with a flow to the right. Equation (5) is equivalent to one of the mass balance equations. We let (5) replace (4) and rewrite (3) in terms of the flow rate $Q(x,t)$ and the solid-fluid relative velocity or slip velocity $v_{\mathrm{r}} := v_{\mathrm{s}} - v_{\mathrm{f}}$, for which a constitutive equation will be formulated. Observing that

$$uv_{\mathrm{s}} = \big(uv_{\mathrm{s}} + (1-u)v_{\mathrm{f}}\big)u + u(1-u)(v_{\mathrm{s}} - v_{\mathrm{f}}) = \frac{Q(x,t)u}{S(x)} + u(1-u)v_{\mathrm{r}}, \tag{6}$$

and assuming for a moment that no solids enter the vessel at $x = 0$, we obtain from (3) $S(x)u_t + (Q(x,t)u + S(x)u(1-u)v_{\mathrm{r}})_x = 0$. Expressing v_{r} in terms of $h(u)$, we arrive at $S(x)u_t + (Q(x,t)u + S(x)h(u))_x = 0$. The function h is assumed to be piecewise differentiable with $h(u) = 0$ for $u \leq 0$ or $u \geq u_{\max}$, where $u_{\max}$ is the maximum solids concentration, $h(u) > 0$ for $0 < u < u_{\max}$, $h'(0) > 0$ and $h'(u_{\max}) \leq 0$. For simplicity we assume that $u_{\max} = 1$.

For the analysis we assume that $h \in C^2[0,1]$, and that h is genuinely nonlinear in the sense that $|h''| > 0$ except for finitely many inflection points. Several authors have studied one-dimensional sedimentation in non-cylindrical vessels [1, 5, 15, 17, 33] under the assumption that S is a smooth function of x. We here assume that S is only piecewise continuous, which means that the diameter jumps. In this case the available mathematical and numerical theory breaks down. We assume that $0 < S_{\min} \le S(x) \le S_{\max}$ for $x \in \mathbf{R}$ and $|S|_{BV} < \infty$, and that the initial data u_0 satisfies $u_0 \in L^1(\mathbf{R}) \cap BV(\mathbf{R})$, $0 \le u_0(x) \le 1$ for $x \in \mathbf{R}$. To simplify the problem further, we assume $Q_{\mathrm{L}}(t) = Q_{\mathrm{L}} = \text{Const.}$ and $Q_{\mathrm{R}}(t) = Q_{\mathrm{R}} = \text{Const.}$ As in [6, 7], we assume that at $x = 1$, the composite flux $Q_{\mathrm{R}}u + S(x)h(u)$ is changed to the discharge transport flux $Q_{\mathrm{R}}u$, and that at $x = 0$, a feed source is located. To model this feed source, we recall that a clarifier-thickener unit with $S \equiv S_0 = const.$ can be modeled by the equation $u_t + \hat{g}(u,x)_x = 0$ with the composite flux

$$\hat{g}(u,x) = \begin{cases} q_{\mathrm{L}}u & \text{for } x < -1, \\ q_{\mathrm{L}}u + h(u) & \text{for } -1 < x < 0, \\ q_{\mathrm{R}}u + h(u) + (q_{\mathrm{L}} - q_{\mathrm{R}})u_{\mathrm{F}} & \text{for } 0 < x < 1, \\ q_{\mathrm{R}}u + (q_{\mathrm{L}} - q_{\mathrm{R}})u_{\mathrm{F}} & \text{for } x > 1, \end{cases}$$

where $q_{\mathrm{L}} = Q_{\mathrm{L}}/S_0 \le 0$ and $q_{\mathrm{R}} = Q_{\mathrm{R}}/S_0 \ge 0$ are the prescribed volume overflow and discharge rates, respectively, divided by the cross-sectional area, and $u_{\mathrm{F}} \in [0, u_{\max}]$ is the concentration of the suspension fed into the unit through the singular source at $x = 0$ at the volumetric rate $Q_{\mathrm{F}} = Q_{\mathrm{R}} - Q_{\mathrm{L}}$. Subtracting the constant term $q_{\mathrm{L}}u_{\mathrm{F}}$ from the flux $\hat{g}(u,x)$, we can finally state our problem as the initial-value problem

$$S(x)u_t + g(u,x)_x = 0, \quad x \in \mathbf{R},\ t > 0, \tag{8a}$$

$$g(u,x) = \begin{cases} Q_{\mathrm{L}}(u - u_{\mathrm{F}}) & \text{for } x < -1, \\ Q_{\mathrm{L}}(u - u_{\mathrm{F}}) + S(x)h(u) & \text{for } -1 < x < 0, \\ Q_{\mathrm{R}}(u - u_{\mathrm{F}}) + S(x)h(u) & \text{for } 0 < x < 1, \\ Q_{\mathrm{R}}(u - u_{\mathrm{F}}) & \text{for } x > 1 \end{cases} \tag{8b}$$

$$u(x,0) = u_0(x), \quad x \in \mathbf{R}; \quad u_0(x) \in [0, u_{\max}] \quad \text{for all} \quad x. \tag{8c}$$

Independently of the smoothness of $g(u,\cdot)$ and of u_0, solutions of (8b)–(8c) are in general discontinuous, and so weak solutions must be sought.

Definition 1. *A bounded measurable function $u(x,t)$ is a* weak solution *of* (8) *if for all test functions $\phi \in \mathcal{D}(\Pi_T)$ such that $\phi|_{t=T} = 0$,*

$$\iint_{\Pi_T} \Big(S(x)u\phi_t + g(u,x)\phi_x \Big)\, dx\, dt + \int_{\mathbf{R}} u_0(x)\phi(x,0)S(x)\, dx = 0. \tag{9}$$

Our latest result, which is proved in [9], is stated in the following theorem:

Theorem 1. *There exists a weak solution to* (8), *and this weak solution can be constructed as a limit of a sequence of approximate solutions constructed by a finite difference scheme.*

3 Numerical method

To facilitate the notation and analysis, we define the composite flux function $f(\gamma(x), u) := g(u, x)$, where $\gamma(x) := (\gamma_1(x), \gamma_2(x))$ is the vector of the flux parameters $\gamma_1(x) := S(x)\chi_{[-1,1]}(x)$ and $\gamma_2(x) := Q_{\mathrm{L}}$ for $x < 0$ and $\gamma_2(x) := Q_{\mathrm{R}}$ for $x > 0$. In this notation, we have $f(\gamma, u) = \gamma_2(u - u_{\mathrm{F}}) + \gamma_1 h(u) + Q_{\mathrm{R}} u_{\mathrm{F}}$. Thus the initial value problem for the conservation law (8b) reads

$$S(x)u_t + f\big(\gamma(x), u\big)_x = 0, \quad u(x, 0) = u_0(x). \tag{10}$$

We now define a numerical method for (10). We choose a discretization Δx, Δt, set $x_j := j\Delta x$, $t^n := n\Delta t$,and then discretize the coefficient vector γ, the initial data, and the cross sectional area by setting

$$\gamma_{j+\frac{1}{2}} = \frac{1}{\Delta x}\int\limits_{x_j}^{x_{j+1}} \gamma(x)\,dx,\ U_j^0 = \frac{1}{\Delta x}\int\limits_{x_{j-\frac{1}{2}}}^{x_{j+\frac{1}{2}}} u_0(x)\,dx,\ S_j = \frac{1}{\Delta x}\int\limits_{x_{j-\frac{1}{2}}}^{x_{j+\frac{1}{2}}} S(x)\,dx.$$

Let Δ_- and Δ_+ denote backward and forward spatial difference operators. For $n > 0$ we define $U_j^n \approx u(x_j, t^n)$ by the explicit marching formula

$$U_j^{n+1} = U_j^n - \lambda_j \Delta_- f^{\mathrm{EO}}(\gamma_{j+\frac{1}{2}}, U_{j+1}^n, U_j^n), \tag{11}$$

where $\lambda_j = \Delta t/(S_j \Delta x)$, and f^{EO} is defined by

$$f^{\mathrm{EO}}(\gamma, v, u) = \frac{1}{2}\left(f(\gamma, u) + f(\gamma, v) - \int_u^v |f_u(\gamma, w)|\,dw\right), \tag{12}$$

which is a slight generalization of the standard Engquist-Osher flux [20].

An important feature of our scheme is that the discretization of the parameter vector γ is *staggered* with respect to that of the conserved variable u. This greatly simplifies both the scheme and its analysis. Were the two discretizations aligned, more complicated 2×2 Riemann problems would have to be solved. Mesh staggering allows us to use a simple scalar Riemann solver.

Our choice of the Engquist–Osher Riemann solver is motivated by two considerations. First, it results in an upwind scheme, allowing for relatively sharp resolution of discontinuities. Second, the Engquist–Osher flux has a close functional relationship with a certain singular mapping to be introduced below, which has enabled us to prove convergence (along a subsequence) of the approximate solutions generated by the scheme. The following lemma, proved in [9], summarizes some important properties of our difference scheme.

Lemma 1. *Let* $f_u^+(w) = \max(f_u(w), 0)$, $f_u^-(w) = \min(f_u(w), 0)$. *If the ratio* $\Delta x/\Delta t$ *is chosen so that the following CFL condition is satisfied:*

$$\lambda_j \left| f_u^+\left(\gamma_{j+\frac{1}{2}}, U_j^n\right) - f_u^-\left(\gamma_{j-\frac{1}{2}}, U_j^n\right)\right| \le 1, \quad \forall j \in \mathbf{Z}, \tag{13}$$

then the computed solutions remain in the interval $[0, 1]$, *the CFL condition* (13) *holds for each succeeding time step, and the scheme* (11) *is monotone. In addition, there exists a constant* C, *independent of* Δ, *such that*

$$\sum_{j \in \mathbf{Z}} |U_j^{n+1} - U_j^n| S_j \Delta x \leq \sum_{j \in \mathbf{Z}} |U_j^1 - U_j^0| S_j \Delta x \leq C \Delta t.$$

Let χ^n denote the characteristic function of $[t_n, t_{n+1})$ and χ_j the characteristic function of $[x_{j-\frac{1}{2}}, x_{j+\frac{1}{2}})$. We then define

$$u^\Delta(x,t) := \sum_{n \geq 0} \sum_{j \in \mathbf{Z}} U_j^n \chi_j(x) \chi^n(t), \quad \text{and} \quad \gamma^\Delta(x) := \sum_{j \in \mathbf{Z}} \gamma_{j+\frac{1}{2}} \chi_{j+\frac{1}{2}}(x).$$

As mentioned above, our main result is convergence of a subsequence of the approximations u^Δ to a weak solution u of the conservation law. The key ingredient in the proof is the use of a so-called singular mapping, defined by $\Psi(\gamma, u) := \int_0^u |f_u(\gamma, w)| dw$. For these problems, a bound on the total variation of the conserved quantity u is very difficult, if not impossible, which makes it necessary to use the singular mapping approach. We transform the numerical approximations according to $z^\Delta = \Psi(\gamma^\Delta, u^\Delta)$, and prove compactness for the the transformed sequence z^Δ. Since Ψ is continuous and strictly increasing in its second argument, it is possible to recover the limit of the conserved variable u by inverting Ψ. The scheme also satisfies a set of cell entropy inequalities. For limit solutions that are piecewise smooth, it follows from the results in [25] that certain geometric entropy conditions are satisfied, and that such solutions form an L^1 contraction semigroup.

4 Numerical examples

The physical problem and the finite difference scheme from Section 3 are now illustrated by some numerical examples. As in [4, 5], we assume that the function h is given by (1) with $v_\infty = 1$ and $n = 5$. Here, we let a clarifier-thickener unit be defined by the dimensionless function (see Figure 1)

$$S(x) = \begin{cases} 0.5 - 1.5625(x+0.6)^2 & \text{for } x \in (-1, -0.6], \\ 0.25 & \text{for } x \in (-0.6, -0.4], \\ 0.5 & \text{for } x \in (-0.4, 0], \\ 1 & \text{for } x \in (0, 0.5], \\ 1 - 3.2(x-0.5)^2 & \text{for } x \in (0.5, 1). \end{cases}$$

In the numerical examples (see Figure 2) we use $\Delta x = 1/100$ and $\lambda = 0.05$. Note that the computational grid is finer than the visual grid used to display the results. The four cases correspond to different parameters Q_{L}, Q_{R} and u_{F}, which are chosen as constants (with respect to t) and produce different modes of behavior of the clarifier-thickener unit. See [18] for detailed analytical predictions of the clarifier-thickener response to operating conditions.

The first case (Figure 2 upper left) corresponds to $Q_{\mathrm{L}} = -0.1$, $Q_{\mathrm{R}} = 0.1$ and $u_{\mathrm{F}} = 0.4$. The simulation shows that the solids initially settle exclusively into the thickening zone ($x > 0$) and form a rising sediment, which breaks through the feed level ($x = 0$) at about $t = 20$. The particles entering the clarification zone ($x < 0$) start to produce an overflow of clarified suspension at about $t = 29$. The shape of the concentration surface in that zone reflects the vessel geometry, and that there the solution becomes stationary. The system converges to a steady state of simultaneous clarification and thickening.

The second case (Figure 2 upper right) is produced by $Q_{\mathrm{L}} = -0.1$, $Q_{\mathrm{R}} = 0.4$ and $u_{\mathrm{F}} = 0.4$. This choice again leads to a steady state, but the solids immediately start at $t = 0$ to enter the clarification zone.

The third case (Figure 2 lower left) corresponds to $Q_{\mathrm{L}} = -0.2$, $Q_{\mathrm{R}} = 0.0$ and again $u_{\mathrm{F}} = 0.4$, i.e. the vessel is kept closed at its bottom. Thus, the thickening zone is slowly filled up. Moreover, as in the second example, particles start immediately to enter the clarification zone. An additional solids flux into that zone is produced when the rising sediment reaches the feed level. This leads to an increase of the concentrations in the clarification zone, which in contrast to the first and second case does not remain stationary.

In the first three examples, the solution at large times reflects the vessel geometry. This is not valid in the fourth case (Figure 2 lower right), where we apply a strong feed flux by letting $Q_{\mathrm{L}} = -0.2$, $Q_{\mathrm{R}} = 0.2$ and $u_{\mathrm{F}} = 0.8$. In this 'hydraulic forcing' situation, the feed slurry immediately breaks into the clarification zone, which it leaves undiluted. The thickening zone first fills up successively, with a dilution of the feed suspension and sediment forming, but then also replenishes with the feed slurry at its feed concentration.

References

1. Anestis, G. (1981): Eine eindimensionale Theorie der Sedimentation in Absetzbehältern veränderlichen Querschnitts und in Zentrifugen. Doctoral Thesis, Technical University of Vienna, Austria.
2. Aw, A. (2000): Resurrection of "second order" models of traffic flow. SIAM J. Appl. Math. **60**, 916–938.
3. Bardos, C., Le Roux, A.Y. and Nedelec, J.C. (1979): First order quasilinear equations with boundary conditions. Comm. Partial Diff. Eqns. **4**, 1017–1034.
4. Bürger, R. and Concha, F. (2001): Settling velocities of particulate systems: 12. Batch centrifugation of flocculated suspensions. Int. J. Mineral Process. **63**, 115–145
5. Bürger, R., Damasceno, J.J.R. and Karlsen, K.H. (2003): A mathematical model for batch and continuous thickening in vessels with varying cross section. Int. J. Mineral Process., to appear.
6. Bürger, R., Karlsen, K.H., Klingenberg, C. and Risebro, N.H. (2003): A front tracking approach to a model of continuous sedimentation in ideal clarifier-thickener units. Nonlin. Anal. Real World Appl., to appear.

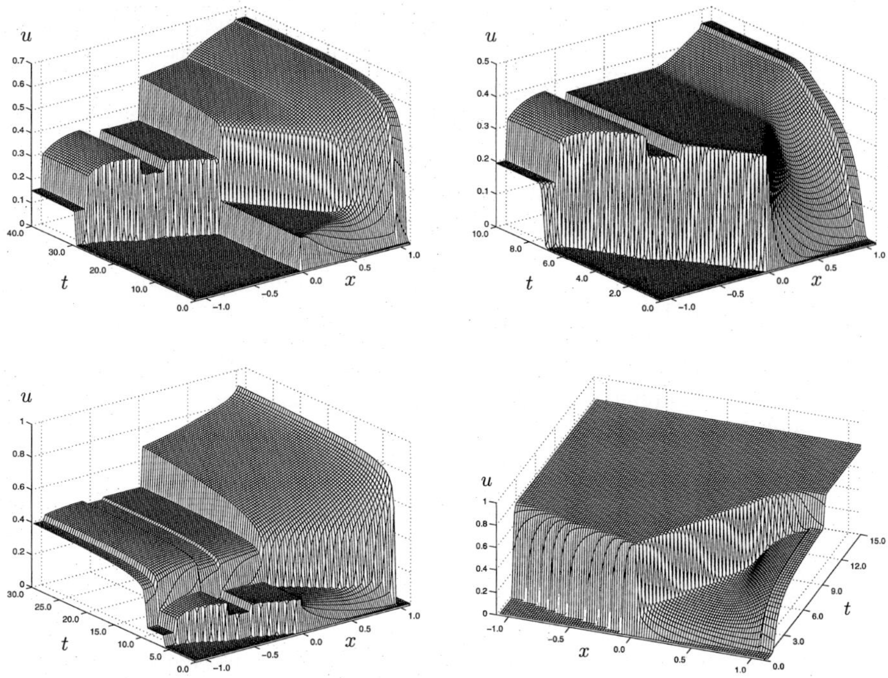

Fig. 2. Simulation of the operation of the continuous clarifier–thickener.

7. Bürger, R., Karlsen, K.H., Risebro, N.H. and Towers, J.D. (2003): Numerical methods for the simulation of continuous sedimentation in ideal clarifier-thickener units. Int. J. Mineral Process., to appear.
8. Bürger, R., Karlsen, K.H., Risebro, N.H. and Towers, J.D. (2002): Convergence of a difference scheme for continuous sedimentation in ideal clarifier-thickener units. In preparation.
9. Bürger, R., Karlsen, K.H., Risebro, N.H. and Towers, J.D. (2002): Monotone difference approximations for the simulation of clarifier-thickener units. Submitted to Comp. Visual. Sci.
10. Bürger, R. and Wendland, W.L. (2001): Sedimentation and suspension flows: Historical perspective and some recent developments. J. Eng. Math. **41**, 101–116.
11. Bustos, M.C., Concha, F., Bürger, R. and Tory, E.M. (1999): Sedimentation and Thickening. Kluwer Academic Publishers, Dordrecht, The Netherlands.
12. Bustos, M.C., Concha, F. and Wendland, W.L. (1990): Global weak solutions to the problem of continuous sedimentation of an ideal suspension. Math. Meth. Appl. Sci. **13**, 1–22.
13. Bustos, M.C., Paiva, F. and Wendland, W.L. (1990): Control of continuous sedimentation of ideal suspensions as an initial and boundary value problem. Math. Meth. Appl. Sci. **12**, 533–548.
14. Bustos, M.C., Paiva, F. and Wendland, W.L. (1996): Entropy boundary conditions in the theory of sedimentation of ideal suspensions. Math. Meth. Appl. Sci. **19**, 679–697.

15. Chancelier, J.P., Cohen de Lara, M., Joannis, C., Pacard, F. (1997): New insights in dynamic modeling of a secondary settler—I. Flux theory and steady-states analysis. Wat. Res. **31**, 1847–1856.
16. Daganzo, C.F. (1995): Requiem for second-order fluid approximations of traffic flow, Transp. Res. B **29B**, 277–286.
17. Diehl, S. (1997): Dynamic and steady-state behavior of continuous sedimentation. SIAM J. Appl. Math. **57**, 991–1018.
18. Diehl, S. (2001): Operating charts for continuous sedimentation I: Control of steady states. J. Eng. Math. **41**, 117–144.
19. Dubois, F. and Le Floch, P. (1988): Boundary conditions for non-linear hyperbolic systems of conservation laws. J. Diff. Eqns. **71**, 93–122.
20. Engquist, B. and Osher, S. (1980): Stable and entropy-satisfying approximations for transonic flow calculations. Math. Comp. **34**, 45–75.
21. Greenberg, J.M. (2001): Extensions and amplifications of a traffic model of Aw and Rascle. SIAM J. Appl. Math. **62**, 729–745.
22. Hassett, N.J. (1958): Design and operation of continuous thickeners. Ind. Chemist **34**, 116–120, 169–172, 489–494.
23. Hassett, N.J. (1964/65): Mechanism of thickening and thickener design. Inst. Min. Met. Trans. **74**, 627–656.
24. Hassett, N.J. (1968): Thickening in theory and practice. Min. Sci. Eng. **1**, 24–40.
25. Karlsen, K.H., Risebro, N.H., Towers, J.D. (2002): L^1 stability for piecewise smooth entropy solutions of nonlinear degenerate parabolic convection-diffusion equations with discontinuous coefficients. In preparation.
26. Kynch, G.J. (1952): A theory of sedimentation. Trans. Farad. Soc. **48**, 166–176.
27. Le Roux, A.Y. (1977): Etude du problème mixte pour une équation quasi-linéaire du premier ordre. C. R. Acad. Sci. Paris Sér. A **285**, 351–354.
28. Lighthill, M.J. and Whitham, G.B. (1955): On kinematic waves: I. Flood movement in long rivers, II. A theory of traffic flow on long crowded roads. Proc. Royal Soc. A **229**, 281–316 and 317–345.
29. Petty, C.A. (1975): Continuous sedimentation of a suspension with a nonconvex flux law. Chem. Eng. Sci. **30**, 1451–1458.
30. Richardson, J.F. and Zaki, W.N. (1954): Sedimentation and fluidization: Part I. Trans. Instn. Chem. Engrs. (London) **32**, 35–53.
31. Shannon, P.T. and Tory, E.M. (1966): The analysis of continuous thickening. SME Trans. **235**, 375–382.
32. Talmage, W.P. and Fitch, E.B. (1955): Determining thickener unit areas. Ind. Eng. Chem. **47**, 38–41.
33. White, D.A. Verdone, N. (2000): Numerical modelling of sedimentation processes. Chem. Eng. Sci. **55**, 2213–2222
34. Whitham, G.B. (1974): Linear and Nonlinear Waves. Wiley, New York.
35. Yoshioka, N., Hotta, Y., Tanaka, S., Naito, S. and Tsugami, S. (1957): Continuous thickening of homogeneous flocculated slurries. Chem. Engrg. Japan **21**, 66–75.

Numerical Approximation of the Navier-Stokes Equations with Several Independent Specific Entropies

C. Chalons[1,2] and F. Coquel[3]

[1] ONERA, BP 72, 29 avenue de la division Leclerc, 92322 Châtillon Cedex, France. *Christophe.Chalons@onera.fr*

[2] CMAP & CNRS, U.M.R. 7641, École Polytechnique, 91128 Palaiseau Cedex, France. *chalons@cmapx.polytechnique.fr*

[3] Laboratoire JLL & CNRS, Boite courrier 187, Université Pierre et Marie CURIE, 75252 Paris Cedex 05, France. *coquel@ann.jussieu.fr*

Summary. The present work is devoted to the numerical approximation of the solutions of convective-diffusive systems which can be understood as natural extensions of the classical Navier-Stokes (NS) equations. Solutions of these systems are governed by N ($N \geq 2$) independent pressure laws or equivalently N independent specific entropies. Several models from the physics actually enter the present framework. These extended forms are seen to share many properties with the usual NS equations (i.e. when $N = 1$) but with the very difference that, generally speaking, they cannot be recast in full conservation form. This property is known to make the definition of the endpoints of the Travelling Wave (TW) solutions sensitive with respect to the shape of the diffusive tensor (see [7]). Motivated by [1] devoted to $N = 2$, here we exhibit $(N - 1)$ Generalized Rankine-Hugoniot (GRH) conditions that reflect this sensitivity. We underline the reason why classical algorithms can only fail in satisfying these GRH relations and we show how to enforce them for validity at the discrete level when extending [1] to $N \geq 2$ general pressure laws. Several numerical evidences are proposed.

1 The mathematical model and some basic properties

In this work, we are interested in the numerical approximation of the solutions of the following convective-diffusive system (in one space dimension for simplicity) :

$$\begin{cases} \partial_t \rho + \partial_x \rho u = 0, \qquad x \in I\!R, \quad t > 0, \\ \partial_t \rho u + \partial_x (\rho u^2 + \sum_{i=1}^N p_i) = \partial_x ((\sum_{i=1}^N \mu_i) \partial_x u), \\ \partial_t \rho \epsilon_i + \partial_x (\rho \epsilon_i u) + p_i \partial_x u = \mu_i (\partial_x u)^2, \quad i = 1, ..., N, \end{cases} \tag{1}$$

where with classical notations, $\rho, \rho u$ and $\{\rho \epsilon_i\}_{i=1,...,N}$ respectively denote the density, the momentum and N independent internal energies. Before addressing the required closure relations, let us stress that (1) is nothing but a natural extension of the usual NS equations : namely when considering a single pressure law. Here, N ($N \geq 2$) independent pressure laws $p_i(\rho, \rho\epsilon_i)$ are

involved via N independent governing equations for each of the $\rho\epsilon_i$. In (1) and for $i = 1, \dots, N$, μ_i ($\mu_i \geq 0$) denotes a smooth mapping standing for the viscosity law associated with the internal energy $\rho\epsilon_i$. It will be always assumed that $\mu := \sum_{i=1}^{N} \mu_i > 0$. Depending on the choice of the closure equations for defining each of the μ_i, several models from the physics enter the present framework. Let us quote for instance compressible turbulence models like the $k - \epsilon$ model (see [1]), and more generally multi-scale models (see [2]) for which a laminar pressure and some turbulent pressures are involved, or models coming from the plasma physics (see [3], [4]).
Now considering the required closure equations for defining each of the N pressure laws, we shall assume that according to the second principle of the thermodynamics :

$$-T_i(\tau, \epsilon_i)ds_i = d\epsilon_i + p_i(\tau, \epsilon_i)d\tau, \quad \tau = 1/\rho, \tag{2}$$

defines a strictly convex smooth mapping $(\tau, \epsilon_i) \to s_i(\tau, \epsilon_i) > 0$ for any given $i \in \{1, .., N\}$ where the temperature T_i is classicaly assumed to stay positive. As a well known consequence [6], the well defined mapping $(\tau, s_i) \to \varepsilon_i(\tau, s_i)$ is then strictly convex, as well as the mapping $(\rho, \rho\varepsilon_i) \to \{\rho s_i\}(\rho, \rho\varepsilon_i) := \rho s_i(1/\rho, \rho\varepsilon_i/\rho)$ and $(\rho, \rho s_i) \to \{\rho\varepsilon_i\}(\rho, \rho s_i) := \rho\varepsilon_i(1/\rho, \rho s_i/\rho)$ with some little abuse in the notations. Each energy law is assumed to obey in addition the well-known Weyl's assumptions [6] and in particular the following asymptotic conditions :

$$\lim_{\rho s_i \to 0+} \rho\varepsilon_i(\rho, \rho s_i) = +\infty \text{ and } \lim_{\rho s_i \to +\infty} \rho\varepsilon_i(\rho, \rho s_i) = 0. \tag{3}$$

Notice that all the above assumptions are quite classical within the frame of the usual NS equations (e.g. with $N = 1$). Owing to these assumptions, our first statement highlights the relationships with the usual setting, when defining the following natural phase space for (1):

$$\Omega = \{\mathbf{v} := (\rho, \rho u, \{\rho\epsilon_i\}_{1 \leq i \leq N}) \in I\!R^{N+2} / \rho > 0,\ \rho u \in I\!R,\ \rho\epsilon_i > 0,\ 1 \leq i \leq N\}.$$

Lemma 1. *The first order extracted system from (1) is hyperbolic over Ω, with the following increasingly ordered eigenvalues :*

$$\lambda_1(\mathbf{v}) = u - c \leq \lambda_{j=2,\dots,N+1}(\mathbf{v}) = u \leq \lambda_{N+2}(\mathbf{v}) = u + c,\ c^2(\mathbf{v}) = \sum_{i=1}^{N} c_i^2(\mathbf{v}),$$

where each partial sound speed is such that $c_i^2(\mathbf{v}) := (\partial_\rho p_i)_{s_i} > 0$. The $1-$ and $(N+2)-$ fields are genuinely nonlinear, all the other fields being linearly degenerate.

As already put forward, the system (1) naturally writes in non conservation form. This rises the question of the existence of an admissible change of variables that recasts (1) in full conservation form. With this respect, the next claim gives all the additional non trivial equations for the smooth solutions of (1) (see [2] or [1] when $N = 2$):

Theorem 1. *Smooth solutions of (1) obey the following conservation law :*

$$\partial_t \rho E + \partial_x(\rho E + \sum_{i=1}^{N} p_i)u = \partial_x((\sum_{i=1}^{N} \mu_i) u \partial_x u), \tag{4}$$

where the total energy reads $\rho E = (\rho u)^2/2\rho + \sum_{i=1}^{N} \rho \epsilon_i$. *In addition, such solutions satisfy :*

$$\partial_t \rho s_i + \partial_x \rho s_i u = -\frac{\mu_i}{T_i}(\partial_x u)^2, \quad i = 1, ..., N. \tag{5}$$

Up to classical nonlinear transformations in the $\{s_i\}_{1\leq i\leq N}$, *these are the only additional non trivial equations (without special assumptions on the* $\{\mu_i\}_{1\leq i\leq N}$*).*

As a consequence, the system (1) cannot be recast generally speaking in full conservation form. However, let us rephrase the above result when investigating suitable change of variables for (1). Obviously, the three variables ρ, ρu and ρE are natural candidates since they are governed by conservation laws. Next and for completeness, let us consider without restriction the set $\{\rho s_i\}_{1\leq i\leq N-1}$ to define $\mathbf{u} := (\rho, \rho u, \rho E, \{\rho s_i\}_{1\leq i\leq N-1})$ as an admissible change of variables. Then, smooth solutions of (1) equivalently satisfy :

$$\begin{cases} \partial_t \rho + \partial_x \rho u = 0, \qquad x \in I\!R, \quad t > 0, \\ \partial_t \rho u + \partial_x(\rho u^2 + \sum_{i=1}^{N} p_i) = \partial_x((\sum_{i=1}^{N} \mu_i)\partial_x u), \\ \partial_t \rho E + \partial_x(\rho E + \sum_{i=1}^{N} p_i)u = \partial_x((\sum_{i=1}^{N} \mu_i) u \partial_x u), \\ \partial_t \rho s_i + \partial_x \rho s_i u = -\frac{\mu_i}{T_i}(\partial_x u)^2, \quad i = 1, ..., N-1. \end{cases} \tag{6}$$

For conciseness in the forthcoming developments, we right away introduce the following condensed notation for (6):

$$\partial_t \mathbf{u} + \partial_x \mathbf{f}(\mathbf{u}) = \mathcal{R}(\mathbf{u}, \partial_x(\mathcal{B}(\mathbf{u})\partial_x \mathbf{u})). \tag{7}$$

The definition of the tensor $\mathcal{B}$ follows when writing $(\partial_x u)^2 = (\partial_x(u\partial_x u) - u\partial_{xx}u)$. Now, let us emphasize that whitout further condition, smooth solutions of (6) automatically obey **with equality** the additional PDE :

$$\partial_t \rho s_N(\mathbf{u}) + \partial_x \rho s_N(\mathbf{u}) u = -\frac{\mu_N}{T_N}(\partial_x u)^2. \tag{8}$$

Notice that as a consequence of the convexity properties we have put forward at the beginning of this section, the mapping $\mathbf{u} \to \rho s_N(\mathbf{u})$ is strictly convex. Hence, the pair $(\rho s_N, \rho s_N u)$ is nothing else a Lax entropy pair but with a **prescribed** rate of entropy dissipation. To assess this issue, let us observe after [1] that despite the identities (5) might be understood as N fully independent equations, these do actually evolve **proportionally** according to :

Corollary 1. *Smooth solutions of (6) obey for $i = 1, \ldots, N-1$:*

$$\mu_N T_i \{\partial_t \rho s_i + \partial_x \rho s_i u\} = \mu_i T_N \{\partial_t \rho s_N(\mathbf{u}) + \partial_x \rho s_N(\mathbf{u}) u\}. \tag{9}$$

(9) expresses nothing but a trivial cancellation property in $(\partial_x u)^2$. However, this observation turns out to be crucial for our numerical purposes. Indeed when applied to TW solutions of (6), the identities (9) can be understood as $(N-1)$ Generalized Rankine-Hugoniot (GRH) conditions (see [2]). To exemplify this claim, let us temporarily adopt the following closure for the viscosity laws : $\mu_i(\mathbf{u}) = \mu_i^0 T_i$ where for each $i = 1, \ldots, N$, μ_i^0 denotes a given positive real number. Let us recall that such a closure can be inferred from the classical transport theory for dilute gases. Then, (9) are easily seen to boil down to :

$$\{\partial_t \rho(\frac{\mu_N}{\mu} s_i - \frac{\mu_i}{\mu} s_N(\mathbf{u})) + \partial_x \rho(\frac{\mu_N}{\mu} s_i - \frac{\mu_i}{\mu} s_N(\mathbf{u}))u\} = 0, \quad i = 1, ..., N-1. \tag{10}$$

Therefore, (10) clearly provide us with $(N-1)$ jump conditions that, with the three classical Rankine-Hugoniot conditions respectively for ρ, ρu and ρE, yield a complete set of jump relations for defining the endpoints of a TW solution of (6). Let us underline that such endpoints clearly depend on the ratios of the viscosities.

More generally, the proportionality in the evolution in time of all the entropy balance equations (as put forward in (9)) expresses a well-known property of general convective-diffusive systems in non conservation form. Namely, the endpoints of their TW solutions entirely depend on the shape of the diffusive tensor under consideration (see [5] and [7]). As underlined in the next section, this key property will make the numerical approximation of the TW solutions of (9) a non classical issue (see also [7]). In the present work, we show how to enforce numerically for validity the GRH relations (9).

2 Numerical approximation

In this section, we present a suitable numerical method for approximating the solutions of (1). This method can be understood as a prediction-correction strategy. In the prediction step, we just propose to solve the equivalent form (6) with appropriate initial data. Such an approach sounds quite natural but however will be seen to fail in preserving the validity of the GRH conditions at the discrete level. Highlighting the roots of such a failure will naturally motivate the correction step which precisely aims at restoring the validity of (9) at each time step.

Let be given a constant time step Δt and a constant space step Δx. Introducing $x_{j+1/2} = (j+1/2)\Delta x$ for $j \in Z$ and $t^n = n\Delta t$ for $n \in N$, the approximate solution of (6), say $\mathbf{u}_\lambda(x,t)$, with $\mathbf{u}_0$ as initial data is classicaly sought as a piecewise constant function on each slab $C_j^n = [x_{j-1/2}; x_{j+1/2}) \times [t^n; t^{n+1})$ for $(n,j) \in N \times Z$, and we set

$$\mathbf{u}_\lambda(x,t) = \mathbf{u}_j^n \quad \text{for all} \quad (x,t) \in C_j^n, \quad j \in Z, \quad n \in N.$$

Here λ stands for the ratio $\lambda = \Delta t/\Delta x$, and for completeness we set $\mathbf{u}_j^0 = \frac{1}{\Delta x}\int_{x_{j-1/2}}^{x_{j+1/2}} \mathbf{u}_0(x)dx$. Assuming that the approximate solution $\mathbf{u}_\lambda(x,t^n)$ is known, we propose to advance it to the next time level t^{n+1} according to a classical splitting strategy between the convective and diffusive parts of (6). We now describe these two steps into details.

First step ($t^n \to t^{n+1=}$) This step consists in solving the first order underlying system in (6), that is to say:

$$\partial_t \mathbf{u} + \partial_x \mathbf{f}(\mathbf{u}) = 0, \tag{11}$$

with $\mathbf{u}_\lambda(x,t^n)$ as initial data, and supplemented with the following Lax entropy inequality :

$$\partial_t \rho s_N(\mathbf{u}) + \partial_x \rho s_N(\mathbf{u})u \leq 0. \tag{12}$$

Let us underline that (12) is actually needed for uniqueness of the weak solutions of (11). To define $\mathbf{u}_\mathbf{j}^{\mathbf{n+1=}}$, we propose to use the celebrated Godunov method under the usual CFL condition

$$\lambda \max_{\mathbf{u}}(|\lambda_i(\mathbf{u})|, i = 1, ..., N+2) \leq \frac{1}{2}, \tag{13}$$

where the maximum is taken over all the $\mathbf{u}$ under consideration. Generically and for any given $\mathbf{u}_L$ and $\mathbf{u}_R$ in the phase space, we then denote by $\mathbf{u}(., \mathbf{u}_L, \mathbf{u}_R)$ the self similar solution of the Riemann problem (11)-(12) with $\mathbf{u}(x,0) = \mathbf{u}_L$ if $x < 0$ and $\mathbf{u}(x,0) = \mathbf{u}_R$ if $x > 0$ as initial data. Equipped with these notations, we state :

Theorem 2. *Under the CFL condition (13), the Godunov method applied to (11)-(12) with* $\mathbf{u}_\lambda(x,t^n)$ *as initial data leads to*

$$\mathbf{u}_j^{n+1=} = \mathbf{u}_j^n - \lambda \Delta \mathbf{f}_{j+1/2}^n, \quad \Delta \mathbf{f}_{j+1/2}^n = \mathbf{f}_{j+1/2}^n - \mathbf{f}_{j-1/2}^n, \quad j \in Z, \tag{14}$$

where we have classicaly set $\mathbf{f}_{j+1/2}^n = \mathbf{f}(\mathbf{u}(0^+; \mathbf{u}_j^n, \mathbf{u}_{j+1}^n))$ *for all* $j \in Z$.
In addition, the following cell entropy inequality holds true:

$$(\rho s_N)_j^{n+1=} := \rho s_N(\mathbf{u}_j^{n+1=}) \leq (\rho s_N)_j^n - \lambda \Delta \{\rho s_N u\}_{j+1/2}^n, \tag{15}$$

with $\{\rho s_N u\}_{j+1/2}^n = \{\rho s_N u\}(\mathbf{u}(0^+; \mathbf{u}_j^n, \mathbf{u}_{j+1}^n))$*, as well as the Tadmor's maximum principles on the specific entropies*

$$(s_i)_j^{n+1=} \leq \max((s_i)_{j-1}^n, (s_i)_j^n, (s_i)_{j+1}^n), \quad i = 1, ..., N. \tag{16}$$

Let us first underline that the analysis proposed by Tadmor in [8] readily applies to give the validity of the maximum principles (16) for the N specific entropies. Let us next recall that the inequality (15) comes from the entropy

selection principle (12) together with the Jensen inequality associated with the averaging procedure in the Godunov method. As a consequence, such an inequality is **strict** generally speaking.
To conclude this first step, observe that one could have used a suitable approximate Riemann solver instead of the exact one for solving the Cauchy problem (11)-(12). Indeed, the forthcoming analysis entirely stays valid provided that natural stability properties (like consistency, average conservation, average entropy dissipation, ...) are assumed to hold true for the approximate Riemann solver under consideration. We refer the reader to [2] for such considerations. Let us now address the second step.

Second step ($t^{n+1=} \to t^{n+1-}$) This step takes into account the diffusive part of (6). It amounts to solve

$$\partial_t \mathbf{u} = \mathcal{R}(\mathbf{u}, \partial_x(\mathcal{B}(\mathbf{u})\partial_x \mathbf{u})), \tag{17}$$

with $\mathbf{u}_\lambda(x, t^{n+1=})$ as initial data. Notice that the smooth solutions of this system automatically obey the following additional equation:

$$\partial_t \rho s_N(\mathbf{u}) = -\frac{\mu_N}{T_N}(\partial_x u)^2. \tag{18}$$

Several numerical techniques are at hand in the litterature to provide us with an updated solution, say $\mathbf{u}_\lambda(x, t^{n+1-})$, which satisfies (17) as well as (18) in the sense of the finite differences. Here and after [1], we propose the following implicit discretization, to avoid a too restrictive CFL-like condition:

$$\begin{cases} \rho_j^{n+1-} = \rho_j^{n+1=}, \\ (\rho u)_j^{n+1-} = (\rho u)_j^{n+1=} + \Delta t \overline{\partial_x(\mu \partial_x u)_j}^{n+1-}, \\ (\rho E)_j^{n+1-} = (\rho E)_j^{n+1=} + \Delta t \overline{\partial_x(\mu u \partial_x u)_j}^{n+1-}, \\ (\rho s_i)_j^{n+1-} = (\rho s_i)_j^{n+1=} - \Delta t \overline{\frac{\mu_i}{T_i}(\partial_x u)^2}_j^{n+1-}, \quad i = 1, ..., N-1. \end{cases} \tag{19}$$

Here the finite differences formulae read :

$$\begin{cases} \overline{\partial_x(\mu \partial_x u)_j}^{n+1-} \\ = \frac{1}{\Delta x^2}(\mu_{j+1/2}^n (M^n u_{j+1} - M^n u_j) - \mu_{j-1/2}^n (M^n u_j - M^n u_{j-1})), \\ \overline{\partial_x(\mu u \partial_x u)_j}^{n+1-} \\ = \frac{1}{2\Delta x^2}(\mu_{j+1/2}^n ((M^n u_{j+1})^2 - (M^n u_j)^2) - \mu_{j-1/2}^n ((M^n u_j)^2 - (M^n u_{j-1})^2)), \\ \overline{\frac{\mu_i}{T_i}(\partial_x u)^2}_j^{n+1-} \\ = \frac{1}{T_{i_j}^{n+1-}}((\mu_i)_{j+1/2}^n (M^n u_{j+1} - M^n u_j)^2 + (\mu_i)_{j-1/2}^n (M^n u_j - M^n u_{j-1})^2), \end{cases}$$

where $M^n X = (X^{n+1-} + X^{n+1=})/2$ and for all i from 1 to N :

$$\overline{T_{i_j}}^{n+1-} = \begin{cases} -\frac{\varepsilon_i(\tau_j^{n+1-}, (s_i)_j^{n+1-}) - \varepsilon_i(\tau_j^{n+1-}, (s_i)_j^{n+1=})}{(s_i)_j^{n+1-} - (s_i)_j^{n+1=}} & \text{if } (s_i)_j^{n+1-} \neq (s_i)_j^{n+1=}, \\ T_i(\tau_j^{n+1-}, (s_i)_j^{n+1-}) & \text{otherwise.} \end{cases}$$

Let us emphasize that since the density is kept unchanged, solving (19) just amounts to solve a linear problem in $M^n u$, with a positive definite symmetric matrix. The unknowns $(\rho E)_j^{n+1-}$ and $\{(\rho s_i)_j^{n+1-}\}_{i=1,...,N-1}$ then follow with explicit formulae. We do not discuss this step into details and refer the reader to [1] for more informations as well as for the proof of the following statement.

Lemma 2. *The updating formulae (19) lead to*

$$(\rho s_N)_j^{n+1-} := \{\rho s_N\}(\mathbf{u}_j^{n+1-}) = (\rho s_N)_j^{n+1=} - \Delta t \overline{\frac{\mu_N}{T_N}(\partial_x u)^2}_j^{n+1-}. \tag{20}$$

Moreover, the following Tadmor's maximum principles hold true:

$$(s_i)_j^{n+1-} \leq (s_i)_j^{n+1=} \leq \max((s_i)_{j-1}^n, (s_i)_j^n, (s_i)_{j+1}^n), \quad i = 1, ..., N. \tag{21}$$

Thus, at the end of this second step, (17) and (18) can be given a consistent definition in the sense of finite differences. To summarize the prediction step (*i.e.* the first step and the second step), we have with clear notations

$$\begin{aligned}
\rho_j^{n+1-} &= \rho_j^n - \lambda \Delta \{\rho u\}_{j+1/2}^n \\
(\rho u)_j^{n+1-} &= (\rho u)_j^n - \lambda \Delta \{\rho u^2 + \textstyle\sum_{i=1}^N p_i\}_{j+1/2}^n + \Delta t \overline{\partial_x(\mu \partial_x u)}_j^{n+1-} \\
(\rho E)_j^{n+1-} &= (\rho E)_j^n - \lambda \Delta \{(\rho u^2 + \textstyle\sum_{i=1}^N p_i) u\}_{j+1/2}^n + \Delta t \overline{\partial_x(\mu u \partial_x u)}_j^{n+1-} \\
(\rho s_i)_j^{n+1-} &= (\rho s_i)_j^n - \lambda \Delta \{\rho s_i u\}_{j+1/2}^n - \Delta t \overline{\tfrac{\mu_i}{T_i}(\partial_x u)^2}_j^{n+1-}, \quad i = 1, ..., N-1,
\end{aligned}$$

$$(\rho s_N)_j^{n+1-} \leq (\rho s_N)_j^n - \lambda \Delta \{\rho s_N u\}_{j+1/2}^n - \Delta t \overline{\frac{\mu_N}{T_N}(\partial_x u)^2}_j^{n+1-}, \tag{22}$$

where as a consequence of (15), the inequality is **strict** generally speaking. The classical approach would consist in choosing $\mathbf{u}_j^{n+1} = \mathbf{u}_j^{n+1-}$. But, because of the strict inequality in (22) the cancellation property, responsible for (9), is lost and the GRH conditions can no longer be satisfied. Obviously, such a failure persists when using a suitable approximate Riemann solver for (11)–(12). In the next section, numerical evidences will show that the corresponding approximate solutions grossly disagree with the exact ones. This discussion then clearly motivates the need for a correction step to define the updated solution $\mathbf{u}^h(x, t^{n+1})$ so that the required GRH conditions hold. We propose a correction step which extends the procedure introduced in [1] to arbitrary $N \geq 2$ general pressure laws and we prove its relevance.

Correction step ($t^{n+1-} \to t^{n+1}$) We keep unchanged the updated conservative variables, i.e. :

$$\rho_j^{n+1} = \rho_j^{n+1-}, \quad (\rho u)_j^{n+1} = (\rho u)_j^{n+1-}, \quad (\rho E)_j^{n+1} = (\rho E)_j^{n+1-},$$

and we propose to define the N entropies $\{(\rho s_i)_j^{n+1}\}_{i=1,...,N-1}$ as the solutions of the following nonlinear algebraic problem :

$$\overline{\mu_i}\overline{T_{N}}_j^{n+1-}(\{\rho s_N\}(\mathbf{u}_j^{n+1}) - (\rho s_N)_j^n + \lambda\Delta\{\rho s_N u\}_{j+1/2}^n) \\ = \\ \overline{\mu_N}\overline{T_{i}}_j^{n+1-}((\rho s_i)_j^{n+1} - (\rho s_i)_j^n + \lambda\Delta\{\rho s_i u\}_{j+1/2}^n), \\ i = 1, ..., N-1. \tag{23}$$

Here $\overline{\mu_i}$ is given a natural consistent definition for all $i = 1, ..., N$. Notice that (23) is nothing but a discrete version of the GRH conditions (9). The numerical fluxes and the definitions of $\overline{T_{i}}_j^{n+1-}$ entering (23) are directly inherited from the prediction step. The following result assesses the relevance of the correction step.

Theorem 3. *The algebraic problem (23) admits an unique solution* $\{(\rho s_i)_j^{n+1}\}_{i=1,\dots,N-1}$ *which obeys in addition the following entropy inequalities*

$$(\rho s_i)_j^{n+1} \leq (\rho s_i)_j^n - \lambda\Delta\{\rho s_i u\}_{j+1/2}^n, \quad i = 1, ..., N-1, \\ (\rho s_N)_j^{n+1} := \{\rho s_N\}(\mathbf{u}_j^{n+1}) \leq (\rho s_N)_j^n - \lambda\Delta\{\rho s_N u\}_{j+1/2}^n, \tag{24}$$

as well as the Tadmor's maximum principles on the specific entropies

$$(s_i)_j^{n+1} \leq \max((s_i)_{j-1}^n, (s_i)_j^n, (s_i)_{j+1}^n), \quad i = 1, ..., N, \quad j \in Z. \tag{25}$$

Proof

Let us introduce the following convenient notations

$$(\rho s_N)_j^\star = (\rho s_N)_j^n - \lambda\Delta\{\rho s_N u\}_{j+1/2}^n (\neq \{\rho s_N\}(\mathbf{u}_j^{n+1=})),$$

$$\Gamma_{m,n} = \overline{\mu_m}\overline{T_{n}}_j^{n+1-}/\overline{\mu_n}\overline{T_{m}}_j^{n+1-} \geq 0, \quad m = 1, ..., N, \quad ,n = 1, ..., N.$$

Let us observe that these quantities are explicitly known at this stage. The first step of the proof consists in recasting (23) as follows:

$$\begin{aligned}(\rho s_i)_j^{n+1} &= \rho s_i((\rho s_{N-1})_j^{n+1}) \\ &= (\rho s_i)_j^{n+1,=} + \Gamma_{i,N-1}((\rho s_{N-1})_j^{n+1} - (\rho s_{N-1})_j^{n+1,=}),\end{aligned} \tag{26}$$

with $i = 1, \dots, N-2$ and :

$$\{\rho s_N\}(\rho_j^{n+1}, (\rho u)_j^{n+1}, (\rho E)_j^{n+1}, \{\rho s_i((\rho s_{N-1})_j^{n+1})\}_{i=1,\dots,N-2}, (\rho s_{N-1})_j^{n+1}) \\ = \\ (\rho s_N)_j^\star + \Gamma_{N,N-1}((\rho s_{N-1})_j^{n+1} - (\rho s_{N-1})_j^{n+1,=}). \tag{27}$$

These identities express the property that the unknowns $(\rho s_i)_j^{n+1}$ for $i = 1, ..., N-2$ can be parametrized by $(\rho s_{N-1})_j^{n+1}$ so that the algebraic problem (23) can be solved with $(\rho s_{N-1})_j^{n+1}$ as the only scalar unknown. Thus, existence and uniqueness of the $\{(\rho s_i)_j^{n+1}\}_{i=1,...,N-1}$ as the solutions of (26)–(27) is equivalent to prove the existence of a non positive real number $\mathcal{M}_j \leq 0$ (namely $\mathcal{M}_j = (\rho s_{N-1})_j^{n+1} - (\rho s_{N-1})_j^{n+1,=}$) such that

$$\begin{aligned}
(\rho s_i)_j^{n+1} &= (\rho s_i)_j^{n+1,=} + \Gamma_{i,N-1}\mathcal{M}_j, \quad i = 1, ..., N-1, \\
f(\mathcal{M}_j) &= \{\rho s_N\}(\rho_j^{n+1}, (\rho u)_j^{n+1}, (\rho E)_j^{n+1}, \{(\rho s_i)_j^{n+1,=} \\
&\quad + \Gamma_{i,N-1}\mathcal{M}_j\}_{i=1,...,N-1}) \quad - \Gamma_{N,N-1}\mathcal{M}_j - (\rho s_N)_j^\star \quad (28) \\
&= 0. \quad (29)
\end{aligned}$$

Now, since $\partial_{\rho s_i}\{\rho s_N\} = -T_i/T_N < 0$, the mapping $\mathbf{u} \to \rho s_N(\mathbf{u})$ is strictly decreasing with respect to each of the entropies $\{\rho s_i\}_{i=1,...,N-1}$, so that the mapping $\mathcal{M} \to f(\mathcal{M})$ is also strictly decreasing thanks to the positivity of $\Gamma_{m,n}$. Arguing of the asymptotic Weyl's assumptions (3), let us prove that $\{\rho s_N\}(\mathcal{M}_j) \to +\infty$ (*i.e.* $f(\mathcal{M}_j) \to +\infty$) for a first sufficiently large negative value of $\mathcal{M}_j$, say $\overline{\mathcal{M}_j}$. In that aim, notice that $\rho s_i(\mathcal{M}_j) := (\rho s_i)_j^{n+1=} + \Gamma_{i,N-1}\mathcal{M}_j$ approaches 0^+ for negative enough values of $\mathcal{M}_j$ for each $i \in \{1, ..., N-1\}$. By the asymptotic condition $\lim_{\rho s_i \to 0^+} \rho\varepsilon_i(\rho, \rho s_i) = +\infty$, there exists necessarily a first value of $\mathcal{M}_j$, namely $\overline{\mathcal{M}_j}$, solution of

$$(\rho E)_j^{n+1} - \frac{((\rho u)_j^{n+1})^2}{2\rho_j^{n+1}} = \sum_{i=1}^{N-1} \rho\varepsilon_i(\rho_j^{n+1}, \rho s_i(\mathcal{M}_j)),$$

that is to say that $\rho\varepsilon_N(\mathcal{M}_j) = 0$. Note that

$$\rho\varepsilon_N(0) = \rho\varepsilon_N(\rho_j^{n+1=}, (\rho s_N)_j^{n+1=}) > 0$$

But the asymptotic condition $\lim_{\rho\varepsilon_N \to 0^+} \rho s_N(\rho, \rho\varepsilon_N) = +\infty$ just says that $\lim_{\mathcal{M}_j \to \overline{\mathcal{M}_j}^+} \rho s_N(\mathcal{M}_j) = +\infty$. Note that we have proved in passing that necessarily $\rho s_i(\mathcal{M}_j) > 0$ for all $\overline{\mathcal{M}_j} < \mathcal{M}_j \leq 0$ (since otherwise $f(\mathcal{M}_j) = +\infty$). Let us next observe the estimate :

$$\begin{aligned}
f(0) &= \{\rho s_N\}(\rho_j^{n+1}, (\rho u)_j^{n+1}, (\rho E)_j^{n+1}, \{(\rho s_i)_j^{n+1=}\}_{i=1,...,N-1}) - (\rho s_N)_j^\star \\
&= \{\rho s_N\}(\rho_j^{n+1-}, (\rho u)_j^{n+1-}, (\rho E)_j^{n+1-}, \{(\rho s_i)_j^{n+1=}\}_{i=1,...,N-1}) - (\rho s_N)_j^\star \leq 0,
\end{aligned}$$

as an immediate consequence of (22): indeed, notice that thanks to (21) and the decreasing property of $\{\rho s_N\}$ with respect to ρs_i, the following inequality holds :

$$\{\rho s_N\}(\rho_j^{n+1-}, (\rho u)_j^{n+1-}, (\rho E)_j^{n+1-}, \{(\rho s_i)_j^{n+1=}\}_{i=1,...,N-1}) \leq \{\rho s_N\}(\mathbf{u}_j^{n+1-}).$$

Since $f(0) \leq 0$ with $\lim_{\mathcal{M}_j \to \overline{\mathcal{M}_j}^+} f(\mathcal{M}_j) = +\infty$, we necessarily get from the strict monotonicity of the continuous function f the existence of an unique $\mathcal{M}_j$, solution of (28), with the property that $\overline{\mathcal{M}_j} < \mathcal{M}_j \leq 0$, so that necessarily :

$$(\rho s_i)_j^{n+1} = (\rho s_i)_j^{n+1=} + \Gamma_{i,N-1}\mathcal{M}_j > 0, \quad i = 1, ..., N-1.$$

Turning now establishing the maximum principles (25), the entropy inequalities we have just obtained rewrite $(\rho s_i)_j^{n+1} \leq (\rho s_i)_j^{n+1=}$ that is to say $(s_i)_j^{n+1} \leq (s_i)_j^{n+1=}$ for all $i = 1, ..., N-1$ since $\rho_j^{n+1} = \rho_j^{n+1=}$. Invoking now (16), we obtain the required conclusion for $i = 1, ..., N-1$. Concerning now ρs_N, we have just proved $(\rho s_N)_j^{n+1} \leq (\rho s_N)_j^\star$ which is known after Tadmor [8] to be sufficient to obtain the required maximum principle in the Godunov method setting. This concludes the proof.

3 Numerical experiments

In this section, we give some numerical evidences to illustrate the relevance of the prediction-correction scheme we have proposed. To that purpose, we deal with N internal energies of polytropic ideal gases (thermally and calorically perfect) associated with N constant adiabatic exponents $\gamma_i > 1$ for all i=1,...,N. Hence, the total energy reads $\rho E = (\rho u)^2/2\rho + \sum_{i=1}^N p_i/\gamma_i - 1$. For simplicity we consider constant viscosity laws with a Reynolds number equal to 10^5.

We consider as initial data two constant states, called left state and right state in the following, separated by a discontinuity at $x = 0$ and we approximate the solution on a uniform grid with $\Delta x = 1/300$.

Experiment 1 We set $N = 3$, $(\gamma_1, \gamma_2, \gamma_3) = (1.4, 1.6, 1.8)$ and $(\mu_2/\mu_1, \mu_3/\mu_1) = (1., 1.)$, while the left and right states of the initial data read:
$(\rho, u, p_1, p_2, p_3)_{left} = (4., 1., 1.2, 1.4, 1.6)$,
$(\rho, u, p_1, p_2, p_3)_{right} = (2.5568, -1.4305, 0.5162, 0.5103, 0.4999)$.

Experiment 2 We set $N = 3$, $(\gamma_1, \gamma_2, \gamma_3) = (1.4, 1.4, 1.4)$ and $(\mu_2/\mu_1, \mu_3/\mu_1) = (1., 1.)$ while the left and right states of the initial data read:
$(\rho, u, p_1, p_2, p_3)_{left} = (3., 1., 1., 1.2, 1.4)$,
$(\rho, u, p_1, p_2, p_3)_{right} = (2.6529, -1.1153, 0.8160, 0.9844, 1.1528)$.

In the Figures 1 and 2, we compare some of the corresponding pressure profiles for the exact solutions (see [2] for the derivation of exact solutions) together with the numerical solutions generated by the scheme both with and without the correction step. As expected, we observe that the classical approach (without correction) fails in capturing the correct solution while the correction step provides us with a numerical solution in good agreement with the exact one. We refer the reader to [2] for additional numerical experiments.

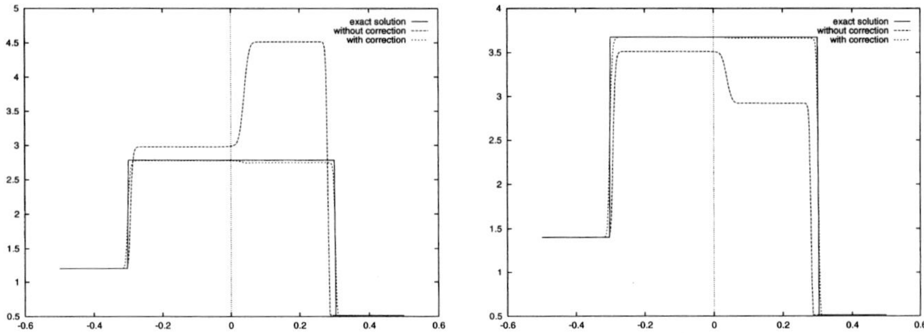

Fig. 1. Experiment 1 - Pressures 1 and 2

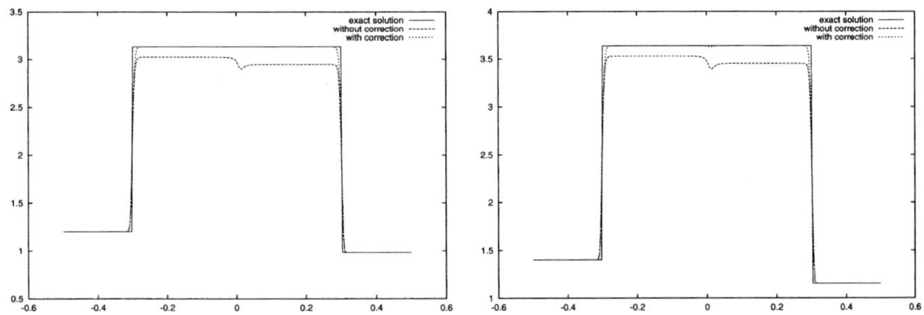

Fig. 2. Experiment 2 - Pressures 2 and 3

To conclude this paper, let us emphasize that several distinct methods have been actually developed and analysed (see [2]). Their design principle is always focused at preserving the validity of the GRH conditions we have put forward in (9). In addition, they all satisfy the same stability estimates but they come with a different cost of numerical evaluation. Here, we have restricted ourselves to the most natural one (along the lines of the first section) but at the expense of the largest numerical cost (we have to solve the **nonlinear** problem (26)–(27)). In [2], fully linear correction steps are proposed.

References

1. Berthon, C., Coquel, F. (2001) : Nonlinear projection scheme to solve systems in non conservation form. In : Innovative methods for numerical solutions of PDE, Eds Hafez–Chattot, World Scientific.
2. Chalons, C. (2002), Thesis, École Polytechnique, France
3. Coquel, F., Marmignon, C. (1999) : A Roe-type Linearization for the Euler Equations for Weakly Ionized Gases. In : Finite Volumes for Complex Applications II, Hermes
4. Cordier, S., al. (1995). In : Asymptotic Analysis, **10**
5. Dal Maso, G., Lefloch, P.G., Murat, F. (1995) : Definition and weak stability of a non conservative product. J. Math. Pures et Appli., **74**

6. Godlewski, E., Raviart, P.A. (1996) : Numerical Approximation of Hyperbolic Systems of Conservation Laws. Springer
7. LeFloch, P.G. (1998) : An introduction to Nonclassical Shocks. In : Lecture Notes in Computational Science and Engineering, **5**
8. Tadmor, E. (1986) : A minimum entropy principle in the gas dynamics equations. In : Applied Numerical Mathematics **2**, pp 211–219.

Existence and Stability of Multidimensional Transonic Shocks for the Euler Equations for Steady Potential Fluids in Unbounded Domains

Gui-Qiang Chen[1] and Mikhail Feldman[2]

[1] Department of Mathematics, Northwestern University, Evanston, IL 60208-2730, USA *gqchen@math.northwestern.edu*

[2] Department of Mathematics, University of Wisconsin, Madison, WI 53706-1388, USA *feldman@math.wisc.edu*

1 Introduction

We are concerned with the existence and stability of multidimensional transonic shocks in inviscid compressible fluid dynamics. In this paper, we focus on inviscid steady potential fluid flows, which are governed by the Euler equations consisting of the conservation law of mass and the Bernoulli law for velocity. Then the Euler equations for the velocity potential $\varphi : \Omega \subset \mathbf{R}^n \to \mathbf{R}$ can be formulated into the following second order nonlinear equation of mixed elliptic-hyperbolic type:

$$\operatorname{div}\left(\rho(|D\varphi|^2)D\varphi\right) = 0, \tag{1}$$

where the density function $\rho(q^2)$ has the form:

$$\rho(q^2) = \left(1 - \frac{\gamma - 1}{2} q^2\right)^{\frac{1}{\gamma-1}} \tag{2}$$

with the adiabatic exponent $\gamma > 1$.

The second order nonlinear equation (1) is elliptic at $D\varphi$ with $|D\varphi| = q$ if

$$\rho(q^2) + 2q^2\rho'(q^2) > 0, \tag{3}$$

and is hyperbolic if

$$\rho(q^2) + 2q^2\rho'(q^2) < 0. \tag{4}$$

The elliptic regions of the equation (1) correspond to the subsonic flow and the hyperbolic regions to the supersonic flow. Shocks are discontinuities in the velocity $D\varphi$. We study weak entropy solutions of (1) with transonic (hyperbolic-elliptic) shocks.

Shiffman [30], Bers [4], and Finn-Gilbarg [17] studied subsonic (elliptic) solutions of (1) outside an obstacle; also see Dong [16]. Alt-Caffarelli-Friedman [3] studied free boundary problems for subsonic flows.

Morawetz in [26] showed that the flows of (1) past an obstacle may contain transonic shocks in general. One exception is that, when the obstacle forms a wedge with small angle, the uniform outflow with supersonic speed may produce a non-transonic shock (hyperbolic-hyperbolic shock); see [13, 22, 23, 29, 31]. Transonic shocks also arise in many other situations of physical importance. For example, when a plane shock hits a wedge, a reflected flow may form a transonic shock; see Glimm-Majda [19] and Morawetz [27]. Steady transonic shocks are also very useful for solving some unsteady important problems; see Chen-Glimm [11] and Chen-Wang [12].

Čanic-Keyfitz-Lieberman [8] studied perturbations of steady transonic shocks between two uniform flows for the two-dimensional transonic small-disturbance (TSD) equation, which governs the behavior of the first nontrivial term in the geometric optics expansion to (1) near a certain physical point. The TSD model can be written as a second order nonlinear equation of mixed type with coefficients depending only upon the unknown function itself, but independent of its gradient as in (1). This feature is essential for the approach in [8] (also see [9]).

Majda [24] studied the existence and stability of multidimensional shock fronts, locally in time, for the Euler equations for compressible fluids (also see Métivier [25]).

In this paper, we discuss several recent results on the existence and stability of multidimensional transonic shocks for (1) in unbounded domains and present a nonlinear method for establishing these results. For the detailed proofs, see Chen-Feldman [10].

2 Transonic Shocks

In this section, we first introduce multidimensional transonic shocks for (1).

A function $\varphi \in W^{1,\infty}(\Omega)$ is a weak solution of (1) in an unbounded domain Ω if

(i) $|D\varphi(x)| \leq 1/\sqrt{\theta}$ a.e.
(ii) For any $\zeta \in C_0^\infty(\Omega)$,

$$\int_\Omega \rho(|D\varphi|^2) D\varphi \cdot D\zeta \, dx = 0. \tag{5}$$

We are interested in weak solutions with shocks. Let Ω^+ and $\Omega^- = \Omega \backslash \overline{\Omega^+}$ be nonempty open subsets of Ω, and $S = \partial\Omega^+ \cap \Omega$. Let $\varphi \in W^{1,\infty}(\Omega)$ be a weak solution of (1) and be in $C^2(\Omega^\pm) \cap C^1(\overline{\Omega^\pm})$ so that $D\varphi$ experiences a jump across S that is an $(n-1)$-dimensional smooth surface. Then the condition that φ is a weak solution of (1) implies the continuity of φ across S:

$$\varphi^+ = \varphi^- \qquad \text{on } S \tag{6}$$

and the following Rankine-Hugoniot condition on S:

$$\Big[\rho(|D\varphi|^2)D\varphi\cdot\nu\Big]_S = 0, \tag{7}$$

where ν is the unit normal to S in the direction of Ω^+, and the bracket denotes the difference between the values of the function along S on the $\Omega^\pm$ sides, respectively. We can also write (7) as

$$\rho(|D\varphi^+|^2)D\varphi^+\cdot\nu = \rho(|D\varphi^-|^2)D\varphi^-\cdot\nu \qquad \text{on } \ S. \tag{8}$$

Let $K > 0$. Then the function

$$\Phi_K(p) := \left(K - \theta p^2\right)^{\frac{1}{2\theta}} p, \tag{9}$$

defined for $p \in \left[0, \sqrt{K/\theta}\right]$, satisfies

$$\lim_{p\to 0+}\Phi_K(p) = \lim_{p\to\sqrt{K/\theta}-}\Phi_K(p) = 0, \tag{10}$$

$$\Phi_K(p) > 0 \quad \text{for } p \in \left(0, \sqrt{K/\theta}\right), \tag{11}$$

$$0 < \Phi_K'(p) < K^{\frac{1}{2\theta}} \text{ on } (0, p^K_{sonic}), \text{ and } \Phi_K'(p) < 0 \text{ on } \left(p^K_{sonic}, \sqrt{K/\theta}\right) \tag{12}$$

$$\Phi_K''(p) < 0 \ \text{ on } \ (0, p^K_{sonic}], \tag{13}$$

where

$$p^K_{sonic} = \sqrt{K/(\theta+1)}. \tag{14}$$

Suppose that a weak solution φ in Ω and nonempty open $\Omega^\pm \subset \Omega$ satisfy

$$\varphi \in C^2(\Omega^\pm)\cap C^1(\overline{\Omega^\pm}), \qquad |D\varphi| < p^1_{sonic} = \frac{1}{\sqrt{\theta+1}} \ \text{ in } \Omega^+, \tag{15}$$

$$|D\varphi| > p^1_{sonic} \ \text{ in } \Omega^- = \Omega\setminus\overline{\Omega^+}, \qquad D\varphi^\pm\cdot\nu > 0 \qquad \text{along } S = \partial\Omega^+\cap\Omega,$$

besides (6) and (7). Then φ is a *transonic shock solution* with *transonic shock* S dividing Ω into the *subsonic region* Ω^+ and the *supersonic region* Ω^-. The solution $\varphi(x)$ satisfies the physical entropy condition (see Courant-Friedrichs [14]; also see Lax [21]):

$$\rho(|D\varphi^-|^2) < \rho(|D\varphi^+|^2) \qquad \text{along } S, \tag{16}$$

that is, the density ρ increases across a shock in the flow direction. Note that the equation (1) is elliptic in the subsonic region and hyperbolic in the supersonic region.

In order to deal with multidimensional transonic shocks in an unbounded domain Ω, we define the following weighted Hölder semi-norms and norms in a domain $\mathcal{D} \subset \mathbf{R}^n$: Let $x \to \delta_x$ be a given non-negative function defined on $\mathcal{D}$, which will be specified in each case we consider below. Let $\delta_{x,y} := \min(\delta_x, \delta_y)$ for $x, y \in \mathcal{D}$. For $k \in \mathbf{R}$, $\alpha \in (0,1)$, and $m \in \mathbf{N}$ (the set of nonnegative integers), we define

$$
\begin{aligned}
&[u]^{(k)}_{m;0;\mathcal{D}} = \sum_{|\beta|=m} \sup_{x\in\mathcal{D}} \left(\delta_x^{m+k} |D^\beta u(x)|\right), \\
&[u]^{(k)}_{m;\alpha;\mathcal{D}} = \sum_{|\beta|=m} \sup_{x,y\in\mathcal{D}, x\neq y} \left(\delta_{x,y}^{m+\alpha+k} \frac{|D^\beta u(x) - D^\beta u(y)|}{|x-y|^\alpha}\right), \qquad (17) \\
&\|u\|^{(k)}_{m;0;\mathcal{D}} = \sum_{j=0}^{m} [u]^{(k)}_{j;0;\mathcal{D}}, \qquad \|u\|^{(k)}_{m;\alpha;\mathcal{D}} = \|u\|^{(k)}_{m;0;\mathcal{D}} + [u]^{(k)}_{m;\alpha;\mathcal{D}},
\end{aligned}
$$

where $D^\beta = \partial_{x_1}^{\beta_1} \cdots \partial_{x_n}^{\beta_n}$, $\beta = (\beta_1, \ldots, \beta_n)$ is a multi-index with $\beta_j \geq 0, \beta_j \in \mathbf{N}$, and $|\beta| = \beta_1 + \cdots + \beta_n$.

We denote by $\|u\|_{m;\alpha;\Omega}$ the (non-weighted) Hölder norms in a domain Ω, i.e., the norms defined as above with $\delta_x = \delta_{x,y} = 1$.

3 Multidimensional Transonic Shocks near Flat Shocks

Let (x', x_n) be the coordinates in $\mathbf{R}^n$, where $x_n \in \mathbf{R}$ and $x' = (x_1, \ldots, x_{n-1}) \in \mathbf{R}^{n-1}$.

Let $q_0^- \in \left(p^1_{sonic}, 1/\sqrt{\theta}\right)$ and $\varphi_0^-(x) := q_0^- x_n$. Then φ_0^- is a supersonic solution in Ω. According to (10)–(12), there exists a unique $q_0^+ \in (0, p^1_{sonic})$ such that

$$
\left(1 - \theta (q_0^+)^2\right)^{\frac{1}{2\theta}} q_0^+ = \left(1 - \theta (q_0^-)^2\right)^{\frac{1}{2\theta}} q_0^- . \qquad (18)
$$

Thus, the function

$$
\varphi_0(x) = \begin{cases} q_0^- x_n, & x \in \Omega_0^- := \Omega \cap \{x \,:\, x_n < 0\}, \\ q_0^+ x_n, & x \in \Omega_0^+ := \Omega \cap \{x \,:\, x_n > 0\} \end{cases} \qquad (19)
$$

is a plane transonic shock solution in Ω, Ω_0^+ and Ω_0^- are its subsonic and supersonic regions, respectively, and $S = \{x_n = 0\}$ is a transonic shock. Defining $\varphi_0^\pm(x) := q_0^\pm x_n$ in Ω, we have

$$
\varphi_0(x) = \min(\varphi_0^+(x), \varphi_0^-(x)) \qquad \text{for } x \in \Omega. \qquad (20)
$$

3.1 Transonic Shocks near Flat Shocks in Infinite Cylinders

In this section, we focus on an infinite cylinder $(0,a)^{n-1} \times (-\infty, \infty)$. We use the norms in (17) with the weight function

$$\delta_x = 1 + |x_n|.$$

Since it is not necessary to require that the supersonic perturbation φ^- be defined in the whole infinite cylinder, we introduce a finite subcylinder $\Omega_1 := (0,a)^{n-1} \times (-1,1)$ and focus on the cylinder domain $\Omega := (0,a)^{n-1} \times (-1,\infty)$ without loss of generality. Then our multidimensional transonic shock problem can be formulated into the following form:

Problem A. Given a supersonic weak solution φ^- of (1) in Ω_1, which is a $C^{1,\alpha}$ perturbation of φ_0^-, for some $\alpha \in (0,1)$:

$$\|\varphi^- - \varphi_0^-\|_{1,\alpha,\Omega_1} \le \sigma, \tag{21}$$

with $\sigma > 0$ small, and

$$\varphi_\nu^- = 0 \qquad \text{on} \quad \partial(0,a)^{n-1} \times [-1,1], \tag{22}$$

find a transonic shock solution φ in Ω such that $\Omega^- \subset \Omega_1$,

$$\varphi = \varphi^- \qquad \text{in } \ \Omega^-,$$

where Ω^- is the supersonic region (15) of φ, and

$$\varphi = \varphi^-, \ \varphi_{x_n} = \varphi^-_{x_n} \qquad \text{on } (0,a)^{n-1} \times \{-1\}, \tag{23}$$

$$\varphi_\nu = 0 \qquad \text{on } \partial(0,a)^{n-1} \times [-1,\infty), \tag{24}$$

$$\varphi(\cdot, x_n) - \omega x_n \to 0 \qquad \text{uniformly on } (0,a)^{n-1} \text{ as } x_n \to \infty, \tag{25}$$

for some constant $\omega \in (0, p^1_{sonic})$ which is not prescribed.

Then we have

Theorem 1. *Let $q_0^+ \in (0, p^1_{sonic})$ and $q_0^- \in \left(p^1_{sonic}, 1/\sqrt{\theta}\right)$ satisfy (18), and let φ_0 be the plane transonic shock solution (19). Then there exist $\sigma_0 > 0$, C_0 and C depending only on n, a, α, γ, and q_0^+ such that, for every $\sigma \in (0,\sigma_0)$ and any supersonic solution φ^- of (1) satisfying the conditions stated in Problem A, there exists a solution $\varphi(x)$ of Problem A satisfying*

$$\|\varphi - q_0^+ x_n\|^{(-1)}_{1;\alpha;\Omega^+(\varphi)} \le C_0 \sigma. \tag{26}$$

Moreover, $\varphi(x)$ satisfies the following properties:

(i) The constant ω in (25) must be q^+:

$$\omega = q^+, \tag{27}$$

where q^+ is the unique solution in the interval $(0, p^1_{sonic})$ of the equation

$$\rho((q^+)^2)q^+ = Q^+$$

with

$$Q^+ = \frac{1}{a^{n-1}} \int_{(0,a)^{n-1}} \rho(|D\varphi^-(x', -1)|^2)\varphi^-_{x_n}(x', -1)dx'. \tag{28}$$

The q^+ *satisfies*

$$|q^+ - q_0^+| \leq C\sigma. \tag{29}$$

(ii) *The subsonic region* $\Omega^+(\varphi) := \{x \in \Omega \ : \ |D\varphi(x)| < p^1_{sonic}\}$ *is of the form:*

$$\Omega^+(\varphi) = \{x_n > f(x')\} \cap \Omega \quad with \quad f \in C^{1,\alpha}([0,a]^{n-1}), \ Df|_{\partial(0,a)^{n-1}} = 0, \tag{30}$$

where f *satisfies*

$$||f||_{1,\alpha,[0,a]^{n-1}} \leq C\sigma. \tag{31}$$

(iii) *For every* $k = 1, 2, \ldots.$

$$||\varphi - q^+ x_n||^{(k)}_{1;\alpha;\Omega^+(\varphi)} \leq C_k \sigma, \tag{32}$$

where C_k *depends only on* k*,* n*,* a*,* γ*, and* q_0^+*. That is, when* $x_n \to \infty$*,* $\varphi(x)$ *uniformly converges to the linear function* $q^+ x_n$ *with respect to* $x' \in (0,a)^{n-1}$ *at a rate faster than any algebraic order.*

If $\varphi^- \in C^{2,\alpha}(\overline{\Omega_1})$ *in addition to the previous assumptions, then* $\varphi \in C^{2,\alpha}(\overline{\Omega^+(\varphi)})$ *and* $f \in C^{2,\alpha}([0,a]^{n-1})$*, and*

$$||D^2 f||_{0,\alpha,[0,a]^{n-1}} \leq C(n, a, \alpha, \gamma, q_0^+, \sigma)||D^2\varphi^-||_{0,\alpha,\Omega_1}, \tag{33}$$

$$||D^2\varphi||^{(k)}_{0;\alpha;\Omega^+(\varphi)} \leq C(n, a, \alpha, \gamma, q_0^+, \sigma, k)||D^2\varphi^-||_{0,\alpha,\Omega_1}, \tag{34}$$

for $k = 1, 2, 3, \ldots.$

Furthermore, if

$$||\varphi^- - \varphi_0^-||_{2,\alpha,\Omega_1} \leq \sigma, \tag{35}$$

then the solution $\varphi(x)$ *of Problem A satisfying* (26) *is unique.*

Remark 1. Theorem 2.1 indicates that, given any supersonic solution φ^- in the upstream region, the necessary and sufficient condition for the existence of a transonic shock solution in an infinite cylinder Ω is that the uniform velocity state $\omega\nu_0, \nu_0 = (0, \ldots, 0, 1)$, at infinity in the downstream direction must be $q^+\nu_0$ determined uniquely by φ^-. In other words, given a supersonic solution φ^- in the upstream region, there is no transonic shock solution that contains a transonic shock dividing Ω into the subsonic region and the supersonic region, provided that one expects a different velocity state at infinity from $q^+\nu_0$ in the downstream direction. This means that, for this problem, one can apriori prescribe the uniformity condition of the flow, but can not prescribe a velocity state, at infinity in the downstream direction in general, which otherwise the problem is overdetermined.

Furthermore, as a consequence of the uniqueness, non-degeneracy, and regularity of solutions of the free boundary problem, we have the following stability theorem.

Theorem 2. *There exist a nonnegative function $\Psi \in C([0,\infty))$ satisfying $\Psi(0) = 0$ and $\sigma_0 > 0$ depending only upon n, a, α, γ, and q_0^+ such that, if $\sigma < \sigma_0$, φ^- satisfies* (35), *and $\hat{\varphi}^-$ satisfies*

$$\|\varphi^- - \hat{\varphi}^-\|_{2,\alpha,\Omega_1} \le \kappa, \tag{36}$$

with $\kappa < \sigma$, then the unique solutions φ and $\hat{\varphi}$ of Problem A for φ^- and $\hat{\varphi}^-$, respectively, satisfy

$$\|f_\varphi - f_{\hat{\varphi}}\|_{2,\alpha,(0,a)^{n-1}} \le \Psi(\kappa), \tag{37}$$

where $f_\varphi(x')$ and $f_{\hat{\varphi}}(x')$ are the free boundary functions in (30) *with $\varphi(x)$ and $\hat{\varphi}(x)$, respectively.*

3.2 Transonic Shocks near Flat Shocks in $\mathbf{R}^n$

We continue to study perturbations of the uniform transonic shock solution (19). In this section we focus on the whole space $\mathbf{R}^n$ with $n \ge 3$. As in previous section, since it is enough to specify the supersonic perturbation φ^- only in a neighborhood of the nonperturbed shock surface $\{x_n = 0\}$, we introduce domains $\Omega := \mathbf{R}^{n-1} \times (-1, \infty)$ and $\Omega_2 := \mathbf{R}^{n-1} \times (-1, 1)$. Note that we expect the subsonic region Ω^+ to be close to the half-space $\Omega_0^+ = \{x_n > 0\}$.

We use the norms in (17) with the weight function

$$\delta_x = 1 + |x|,$$

and consider the following problem:

Problem B. Given a supersonic solution φ^- of (1) in $\mathbf{R}^n$ such that, for some $\alpha > 0$,

$$\|\varphi^- - \varphi_0^-\|_{1,\alpha,\Omega_2}^{(n-1)} \le \sigma, \tag{38}$$

with $\sigma > 0$ small, find a transonic shock solution φ in Ω such that $\Omega^- \subset \Omega_2$ and $\varphi = \varphi^-$ in Ω^-, where Ω^- is the supersonic region (15) of φ.

Remark 2. Note that, since $n \ge 3$, our assumptions imply that the perturbation is not only small in $C^{1,\alpha}$, but also "localized", that is, the perturbation has an appropriate polynomial decay at infinity.

Then we have the following theorem.

Theorem 3. *Let $q^+ \in (0, p^1_{sonic})$ and $q^- \in \left(p^1_{sonic}, 1/\sqrt{\theta}\right)$ satisfy (18), and let φ_0 be the transonic shock solution (19). Then there exist positive constants σ_0, C_1, and C_2 depending only on n, γ, and q^+ such that, for every $\sigma \le \sigma_0$ and any supersonic solution $\varphi^-(x)$ of (1) satisfying the conditions stated in Problem B, there exists a solution $\varphi(x)$ of Problem B such that*

$$\|\varphi - \varphi_0^+\|^{(n-2)}_{1,\alpha,\Omega^+} \le C_1\sigma. \tag{39}$$

Moreover, $\Omega^+ = \{x_n > f(x')\}$, and $f : \mathbf{R}^{n-1} \to \mathbf{R}$ satisfies

$$\|f\|^{(n-2)}_{1,\alpha,\mathbf{R}^{n-1}} \le C_2\sigma, \tag{40}$$

that is, the shock surface $S = \{(x', x_n) \in \mathbf{R}^n : x_n = f(x'), x' \in \mathbf{R}^{n-1}\}$ is $C^{1,\alpha}$ and converges at infinity, with an appropriate polynomial rate, to the hyperplane $S_0 = \{x_n = 0\}$.

If $\varphi^- \in C^{2,\alpha}(\Omega_2)$ in addition to the previous assumptions, then $\varphi \in C^{2,\alpha}(\overline{\Omega^+(\varphi)})$ and $f \in C^{2,\alpha}(\mathbf{R}^{n-1})$. In particular, if $\|D^2\varphi^-\|^{(n)}_{0,\alpha,\mathbf{R}^n} < \infty$, then

$$\|D^2 f\|^{(n)}_{0,\alpha,\mathbf{R}^{n-1}} \le C(n, \alpha, \gamma, q_0^+, \sigma)\|D^2\varphi^-\|^{(n)}_{0,\alpha,\Omega_2}, \tag{41}$$

$$\|D^2\varphi\|^{(n)}_{0;\alpha;\Omega^+(\varphi)} \le C(n, \alpha, \gamma, q_0^+, \sigma)\|D^2\varphi^-\|^{(n)}_{0,\alpha,\Omega_2}. \tag{42}$$

Furthermore, if

$$\|\varphi^- - \varphi_0^-\|^{(n-1)}_{2,\alpha,\Omega_2} \le \sigma, \tag{43}$$

then the solution $\varphi(x)$ of Problem B satisfying (39) is unique.

We also establish a similar stability theorem to Theorem 2 for the multidimensional transonic shocks near flat shocks in $\mathbf{R}^n$.

4 Multidimensional Transonic Shocks near Spherical Transonic Shocks

We first construct spherical transonic shocks in $\mathbf{R}^n, n \ge 3$. Choosing any

$$0 < R_1 < R_0 < \infty,$$

we consider the domain $\Omega = \{x \in \mathbf{R}^n : R_1 < |x| < \infty\}$ and show that there exists a weak solution $\varphi_0 \in W^{1,\infty}(\Omega)$ of (1) in the sense of (5) with $\varphi_0(x) = w(|x|)$ for some $w : \mathbf{R} \to \mathbf{R}$ such that $\varphi_0 \in C^\infty(\overline{\Omega_0^\pm})$, and

$$\Omega_0^- = \{x \in \mathbf{R}^n : R_1 < |x| < R_0\} \quad \text{and} \quad \Omega_0^+ = \{x \in \mathbf{R}^n : R_0 < |x| < \infty\}$$

are respectively supersonic and subsonic regions of φ_0, that is,

$$|D\varphi_0| > p^1_{sonic} \quad \text{in } \Omega_0^-, \qquad |D\varphi_0| < p^1_{sonic} \quad \text{in } \Omega_0^+, \tag{44}$$

and $|D\varphi_0|$ has a jump across $S_0 := \{|x| = R_0\}$.

It is easy to see that a function $\varphi_0(x) = w(|x|)$ satisfies (5) in Ω if and only if, for any $\zeta \in C_0^\infty((R_1, \infty))$,

$$\int_{R_1}^{\infty} \rho(|w'(r)|^2) w'(r) \zeta'(r) r^{n-1} dr = 0.$$

It follows that $w(|x|)$ is a smooth solution of (1) if

$$\left(\rho(|w'(r)|^2) w'(r) r^{n-1}\right)' = 0.$$

Thus $\rho(|w'(r)|^2) w'(r) = \frac{c}{r^{n-1}}$, which can be written as

$$\Phi_1(w'(r)) = \frac{c}{r^{n-1}}, \tag{45}$$

where $\Phi_1(\cdot)$ is defined by (9). From (10)–(13), it follows that there exist smooth functions

$$\Phi_+^{-1} : (0, \Phi_1(p^1_{sonic})) \to (0, p^1_{sonic}),$$
$$\Phi_-^{-1} : (0, \Phi_1(p^1_{sonic})) \to (p^1_{sonic}, \sqrt{1/\theta}),$$

which are the inverse functions of $\Phi_1(\cdot)$ in the sense that $\Phi_1(\Phi_\pm^{-1}(\tau)) = \tau$, for any $\tau \in (0, \Phi_1(p^1_{sonic}))$, such that $(\Phi_+^{-1})'(\tau) > 0$ and $(\Phi_-^{-1})'(\tau) < 0$ for $\tau \in (0, \Phi_1(p^1_{sonic}))$. Thus, in order to satisfy (44) and (45), we have to choose $c > 0$ such that

$$0 < \frac{c}{R_0^{n-1}} < \frac{c}{R_1^{n-1}} < \Phi_1(p^1_{sonic}),$$

and set

$$w'(r) = \begin{cases} \Phi_-^{-1}\left(\frac{c}{r^{n-1}}\right) & \text{for } R_1 < r < R_0, \\ \Phi_+^{-1}\left(\frac{c}{r^{n-1}}\right) & \text{for } R_0 < r < \infty. \end{cases} \tag{46}$$

Thus, we obtain a weak solution $\varphi_0(x) = w(|x|)$ of (1) in Ω, satisfying (44), by setting

$$w(r) = \begin{cases} -\int_r^{R_0} \Phi_-^{-1}\left(\frac{c}{\tau^{n-1}}\right) d\tau < 0 & \text{for } R_1 < r < R_0, \\ \int_{R_0}^{r} \Phi_+^{-1}\left(\frac{c}{\tau^{n-1}}\right) d\tau > 0 & \text{for } R_0 < r < \infty. \end{cases}$$

Note that, since $\Phi'(0+) > 0$, the function $w(r)$ is bounded on (R_1, ∞) when $n \geq 3$.

We can express the function φ_0 as

$$\varphi_0(x) = \min(\varphi_0^+(x), \varphi_0^-(x)), \tag{47}$$

where $\varphi_0^\pm \in C^\infty(\overline{\Omega})$ are defined as

$$\varphi_0^\pm(x) = w_\pm(|x|), \qquad w_\pm(r) = \int_{R_0}^{r} \Phi_\pm^{-1}\left(\frac{c}{\tau^{n-1}}\right) d\tau \quad \text{for } r > R_1. \tag{48}$$

The solution $\varphi_0(x)$ contains the spherical transonic shock and satisfies the entropy condition (16) across the transonic shock from the hyperbolic phase to the elliptic phase, which is the flow direction.

We now state our results on the existence and stability of multidimensional transonic shocks that are close to the solution $\varphi_0(x)$ for $n \geq 3$. We use the Hölder norms in (17) with the weight function:

$$\delta_x = 1 + |x|.$$

Since it does not require to define the perturbation on the whole Ω, we introduce a bounded domain

$$\Omega_3 := \{x \in \mathbf{R}^n \ : \ R_1 < |x| < R_0 + 1\}.$$

Problem C. Given a supersonic solution φ^- of (1) in Ω_3, which is a $C^{1,\alpha}$ perturbation of φ_0^-, for some $\alpha > 0$:

$$\|\varphi^- - \varphi_0^-\|_{1,\alpha,\Omega_3} \leq \sigma, \tag{49}$$

with $\sigma > 0$ small, find a transonic shock solution φ in Ω such that $\Omega^- \subset \Omega_3$ and $\varphi = \varphi^-$ in Ω^-, where Ω^- is the supersonic region (15) of φ, and

$$\varphi = \varphi^-, \quad \varphi_\nu = \varphi_\nu^- \qquad \text{on} \qquad \{|x| = R_1\}. \tag{50}$$

Theorem 4. *There exist positive constants σ_0, C_1, and C_2, depending only on n, γ, c, and Ω, such that, for every $\sigma \leq \sigma_0$ and any supersonic solution φ^- of (1) satisfying the conditions stated in Problem C, there exists a solution φ of Problem C satisfying*

$$\|\varphi - \varphi_0^+\|_{1,\alpha,\Omega^+}^{(n-2)} \leq C_1\sigma. \tag{51}$$

Moreover, the shock surface $S = \{x \in \mathbf{R}^n \ : \ x = f(\frac{x}{|x|})\} \cap \Omega$ is C^α, that is, $\Omega^+ = \{|x| > f(\frac{x}{|x|})\} \cap \Omega$, with $f : S^{n-1} \to \mathbf{R}$ and

$$\|f - R_0\|_{0,\alpha,S^{n-1}} \leq C_2\sigma.$$

If $\varphi^- \in C^{2,\alpha}(\Omega_3)$ in addition to the previous assumptions, then $\varphi \in C^{2,\alpha}(\overline{\Omega^+(\varphi)})$ and $f \in C^{2,\alpha}(S^{n-1})$ satisfy

$$\|D^2 f\|_{0,\alpha,S^{n-1}} \leq C(n,\alpha,\gamma,q_0^+,\sigma)\|D^2\varphi^-\|_{0,\alpha,\Omega_3}, \tag{52}$$

$$\|D^2\varphi - D^2\varphi_0^+\|_{0;\alpha;\Omega^+(\varphi)}^{(n)} \leq C(n,\alpha,\gamma,q_0^+,\sigma)\|D^2\varphi^-\|_{0,\alpha,\Omega_3}. \tag{53}$$

Furthermore, if

$$\|\varphi^- - \varphi_0^-\|_{2,\alpha,\Omega_3} \leq \sigma, \tag{54}$$

then the solution $\varphi(x)$ of Problem B satisfying (51) is unique.

We also establish a similar stability theorem to Theorem 2 for the multidimensional transonic shocks near spherical shocks in $\mathbf{R}^n$.

5 Methods of the Proofs: Free Boundary Problems

The transonic shock problems can be formulated into a one-phase free boundary problem for a nonlinear elliptic equation: Given $\varphi^- \in C^{1,\alpha}(\overline{\Omega})$, find a function φ that is continuous in Ω and satisfies

$$\varphi \leq \varphi^- \qquad \text{in } \bar{\Omega}, \tag{55}$$

the equation (1) and the ellipticity condition (3) in the non-coincidence set

$$\Omega^+ = \{\varphi < \varphi^-\},$$

the free boundary condition (8) on the boundary $S = \partial\Omega^+ \cap \Omega$, as well as the prescribed conditions on the fixed boundary $\partial\Omega$ and at infinity. These conditions are different in different cases (Problems A, B, C), say, conditions (23)–(25) for Problem A.

The free boundary is the location of the shock, and the free boundary condition (8) is the Rankine-Hugoniot condition (7). Note that the condition (55) is motivated by the similar properties (20) and (47) of non-perturbed shocks; and, locally on the shock, (55) is equivalent to the entropy condition (16). The condition (55) transforms the transonic shock problem, in which the subsonic region Ω^+ is determined by the gradient condition $|D\varphi(x)| < p^1_{sonic}$, into a free boundary problem, in which Ω^+ is the non-coincidence set.

In order to solve this free boundary problem, we first modify the equation (1) to make it uniformly elliptic and, correspondingly, modify the free boundary condition (8). Then we solve this modified free boundary problem. Since φ^- is a small $C^{1,\alpha}$ perturbation of φ_0^-, we show that the solution φ of the free boundary problem is a small $C^{1,\alpha}$ perturbation of the given subsonic shock solution φ_0^+ in Ω^+. In particular, the gradient estimate implies that φ in fact satisfies the original free boundary problem, hence the transonic shock problem, Problem A (Problem B and Problem C, respectively).

The modified free boundary problem does not directly fit into the variational framework of Alt-Caffarelli [1] and Alt-Caffarelli-Friedman [2] and the regularization framework of Berestycki-Caffarelli-Nirenberg [5]. Also, the nonlinearity of (56) makes it difficult to apply the Harnack inequality approach of Caffarelli [7]. In particular, a boundary comparison principle for positive solutions of nonlinear elliptic equations in Lipschitz domains is unavailable yet for the equations that are not homogeneous with respect to D^2u, Du, and u, which however is our case.

The approach we develop is an iteration scheme based on the nondegeneracy of the free boundary condition: the jump of the normal derivative of a solution across the free boundary has a strictly positive lower bound. Our iteration process is as follows: Suppose the domain Ω_k^+ is given so that $S_k := \partial\Omega_k^+ \setminus \partial\Omega$ is $C^{1,\alpha}$. We solve the oblique derivative problem in Ω_k^+ obtained by rewriting the (modified) equation (1) and free boundary condition (8) in terms of the function $u := \varphi - \varphi_0^+$. Then the problem has the following form:

$$\begin{aligned} \operatorname{div} A(x, Du) &= F(x) \quad \text{in } \Omega_k^+ := \{u > 0\}, \\ A(x, Du) \cdot \nu &= G(x, \nu) \quad \text{on } S := \partial\Omega_k^+ \setminus \partial\Omega, \end{aligned} \tag{56}$$

plus the fixed boundary conditions on $\partial\Omega_k^+ \cap \partial\Omega$ and the conditions at infinity. The equation is quasilinear, uniformly elliptic, $A(x, 0) \equiv 0$, while $G(x, \nu)$ has a certain structure. Let $u_k \in C^{1,\alpha}(\overline{\Omega_k^+})$ be the solution of (56). We estimate that $\|u_k\|_{C^{1,\alpha}(\Omega_k^+)}$ is small if the perturbation is small, where we actually use appropriate weighted Hölder norms in the unbounded domains. Then we extend the function $\varphi_k := \varphi_0^+ + u_k$ from Ω_k^+ to Ω so that the $C^{1,\alpha}$ norm of $\varphi_k - \varphi_0^+$ in Ω is controlled by $\|u_k\|_{C^{1,\alpha}}$. We define $\Omega_{k+1}^+ := \{x \in \Omega \mid \varphi^k(x) < \varphi^-(x)\}$ for the next step. Note that, since $\|\varphi_k - \varphi_0^+\|_{C^{1,\alpha}(\Omega)}$ and $\|\varphi^- - \varphi_0^-\|_{C^{1,\alpha}(\Omega)}$ are small, we have $|D\varphi^-| - |D\varphi_k| \geq \delta > 0$ in Ω, and this nondegeneracy implies that $S_{k+1} := \partial\Omega_{k+1}^+ \setminus \partial\Omega$ is $C^{1,\alpha}$ and its norm is estimated in terms of the data of the problem.

The fixed point Ω^+ of this process determines a solution of the free boundary problem since the corresponding solution φ satisfies $\Omega^+ = \{\varphi < \varphi^-\}$ and the Rankine-Hugoniot condition holds on $S := \partial\Omega^+ \cap \Omega$.

On the other hand, the elliptic estimates alone are not sufficient to get the existence of a fixed point, because the right-hand side of the boundary condition in the problem (56) depends on the unit normal ν of the free boundary. We use the following feature of the flat or spherical shocks:

$$\rho(|D\varphi_0^+|^2)D\varphi_0^+ = \rho(|D\varphi_0^-|^2)D\varphi_0^- \quad \text{in } \Omega \tag{57}$$

to obtain better estimates for the iteration and to prove the existence of a fixed point. Note that this is a vector identity, and the Rankine-Hugoniot condition (8) is the normal part of (57) on the non-perturbed free boundary S_0.

The uniqueness and stability of solutions for the transonic shock problems are obtained by using the regularity and nondegeneracy of solutions.

For more details, see Chen-Feldman [10].

Acknowledgments

Gui-Qiang Chen's research was supported in part by the National Science Foundation under Grants DMS-0204225, DMS-0204455, and INT-9987378. Mikhail Feldman's research was supported in part by the National Science Foundation under Grants DMS-0200644 and DMS-0074037.

References

1. H. W. Alt and L. A. Caffarelli, Existence and regularity for a minimum problem with free boundary, *J. Reine Angew. Math.* **325** (1981), 105–144.
2. H. W. Alt, L. A. Caffarelli, and A. Friedman, A free-boundary problem for quasilinear elliptic equations, *Ann. Scuola Norm. Sup. Pisa Cl. Sci. (4)*, **11** (1984), 1–44.
3. H. W. Alt, L. A. Caffarelli, and A. Friedman, Compressible flows of jets and cavities, *J. Diff. Eqs.* **56** (1985), 82–141.
4. L. Bers, Existence and uniqueness of subsonic flows past a given profile, *Comm. Pure Appl. Math.* **7** (1954), 441-504.
5. H. Berestycki, L. Caffarelli, and L. Nirenberg, Uniform estimates for regularization of free boundary problems, In: *Analysis and Partial Differential Equations*, pp. 567–619, Lecture Notes in Pure and Appl. Math. 122, Dekker, New York, 1990.
6. H. Berestycki, L. Caffarelli, and L. Nirenberg, Further qualitative properties for elliptic equations in unbounded domains, Dedicated to Ennio De Giorgi. *Ann. Scuola Norm. Sup. Pisa Cl. Sci.* (4) **25** (1997), 69–94 (1998); Monotonicity for elliptic equations in unbounded Lipschitz domains, *Comm. Pure Appl. Math.* **50** (1997), 1089–1111; Inequalities for second-order elliptic equations with applications to unbounded domains I, *Duke Math. J.* **81** (1996), 467–494.
7. L. A. Caffarelli, A Harnack inequality approach to the regularity of free boundaries, I. Lipschitz free boundaries are $C^{1,\alpha}$, *Rev. Mat. Iberoamericana*, **3** (1987), 139–162; II. Flat free boundaries are Lipschitz, *Comm. Pure Appl. Math.* **42** (1989), 55–78; III. Existence theory, compactness, and dependence on X, *Ann. Scuola Norm. Sup. Pisa Cl. Sci. (4)*, **15** (1989), 583–602.
8. S. Canić, B. L. Keyfitz, and G. Lieberman, A proof of existence of perturbed steady transonic shocks via a free boundary problem, *Comm. Pure Appl. Math.* **53** (2000), 0484-0511.
9. S. Canić, B. L. Keyfitz, and E. H. Kim, Free boundary problems for the unsteady transonic small disturbance equation: transonic regular reflection, *Methods Appl. Anal.* **7** (2000), 313–335; Free boundary problems for a quasilinear degenerate elliptic equation: regular reflection of weak shocks, *Comm. Pure Appl. Math.* **55** (2002), 71–92.
10. G.-Q. Chen and M. Feldman, Multidimensional transonic shocks and free boundary problems for nonlinear equations of mixed type, J. Amer. Math. Soc. 2003 (to appear); Steady Transonic shocks and free boundary problems in infinite cylinders for the Euler equations, Preprint, Northwestern University, May 2002; Transonic shocks and one-phase free boundary problems for nonlinear elliptic equations in unbounded domain, Preprint, Northwestern University, June 2002.
11. G.-Q. Chen and J. Glimm, Global solutions to the compressible Euler equations with geometrical structure, *Commun. Math. Phys.* **180** (1996), 153–193.
12. G.-Q. Chen and D. Wang, Convergence of shock capturing schemes for the compressible Euler-Poisson equations, *Commun. Math. Phys.* **179** (1996), 333–364.
13. S.-X. Chen, Existence of stationary supersonic flows past a point body, *Arch. Rational Mech. Anal.* **156** (2001), 141-181; Asymptotic behavior of supersonic flow past a convex combined wedge, *Chinese Ann. Math.* **19B** (1998), 255-264.

14. R. Courant and K. O. Friedrichs, *Supersonic Flow and Shock Waves*, Springer-Verlag: New York, 1948.
15. C. M. Dafermos, *Hyperbolic Conservation Laws in Continuum Physics*, Springer-Verlag: Berlin, 2000.
16. G. Dong, *Nonlinear Partial Differential Equations of Second Order*, Transl. Math. Monographs, **95**, AMS: Providence, RI, 1991.
17. R. Finn and D. Gilbarg, Three-dimensional subsonic flows, and asymptotic estimates for elliptic partial differential equations, *Acta Math.* **98** (1957), 265-296.
18. D. Gilbarg and N. Trudinger, *Elliptic Partial Differential Equations of Second Order,* 2nd Ed., Springer-Verlag: Berlin, 1983.
19. J. Glimm and A. Majda, *Multidimensional Hyperbolic Problems and Computations*, Springer-Verlag: New York, 1991.
20. C.-H. Gu, A method for solving the supersonic flow past a curved wedge (in Chinese), *Fudan J.* **7** (1962), 11-14.
21. P. D. Lax, *Hyperbolic Systems of Conservation Laws and the Mathematical Theory of Shock Waves*, CBMS-RCSM, SIAM: Philiadelphia, 1973.
22. T.-T. Li, On a free boundary problem, *Chinese Ann. Math.* **1** (1980), 351-358.
23. W.-C. Lien and T.-P. Liu, Nonlinear stability of a self-similar 3-dimensional gas flow, *Commun. Math. Phys.* **204** (1999), 525–549.
24. A. Majda, The stability of multidimensional shock fronts, *Mem. Amer. Math. Soc.* **275**, AMS: Providence, 1983; The existence of multidimensional shock fronts, *Mem. Amer. Math. Soc.* **281**, AMS: Providence, 1983.
25. G. Métivier, Stability of multi-dimensional weak shocks, *Comm. Partial Diff. Eqs.* **15** (1990), 983-1028.
26. C. S. Morawetz, On the non-existence of continuous transonic flows past profiles I–III, *Comm. Pure Appl. Math.* **9** (1956), 45–68; **10** (1957), 107–131; **11** (1958), 129–144.
27. C. S. Morawetz, Potential theory for regular and Mach reflection of a shock at a wedge. *Comm. Pure Appl. Math.*, **47** (1994), 593–624.
28. O. Oleinik, *Some Asymptotic Problems in the Theory of Partial Differential Equations*, Lezioni Lincee. [Lincei Lectures] Cambridge University Press: Cambridge, 1996.
29. D. G. Schaeffer, Supersonic flow past a nearly straight wedge, *Duke Math. J.* **43** (1976), 637-670.
30. M. Shiffman, On the existence of subsonic flows of a compressible fluid, *J. Rational Mech. Anal.* **1** (1952), 605–652.
31. Y. Zhang, Global existence of steady supersonic potential flow past a curved wedge with a piecewise smooth boundary, *SIAM J. Math. Anal.* **31** (1999), 166-183.

Analysis on a Model for the Dynamic Combustion of a Compressible, Reacting Fluid

Gui-Qiang Chen[1], David Hoff[2] and Konstantina Trivisa[3]

[1] Northwestern University, Evanston, IL 60208, USA
gqchen@math.northwestern.edu
[2] Indiana University, Bloomington, IN 47405, USA
hoff@indiana.edu
[3] University of Maryland, College Park, MD 20742, USA
trivisa@math.umd.edu

Summary. The global existence and large time behavior of solutions are established for a model which describes the dynamic combustion of a compressible, exothermically reacting fluid. Necessary and sufficient conditions for complete combustion are presented in certain cases. The adiabatic "constant" and specific heat depend on the mass fraction of the reactant and therefore vary in time and space. This model is formulated by the Navier-Stokes equations expressing the conservation of mass, the balance of momentum and energy, and two-species chemical kinetics. The results are obtained for large and discontinuous initial data (cf. Chen, Hoff and Trivisa [7]).

1 Introduction

The global existence and large time behavior of solutions are established for a model, which describes the dynamic combustion of a compressible, exothermically reacting fluid. Necessary and sufficient conditions are presented, which guarantee the complete combustion of the fluid in certain cases. The adiabatic constants and specific heats of the reactant (unburnt gas) and the product (burnt gas) differ, in general, and the initial data may be *large* and discontinuous. The model is formulated by the one-dimensional Navier-Stokes equations expressing the conservation of mass, the balance of momentum and energy, and two-species chemical kinetics as follows.

$$\begin{cases} v_t - u_x = 0, \\ u_t + p(v,\theta,Z)_x = \left(\frac{\varepsilon u_x}{v}\right)_x, \\ (e + \frac{1}{2}u^2 + qZ)_t + (up(v,\theta,Z))_x = \left(\dfrac{\lambda\theta_x + \varepsilon u u_x}{v}\right)_x, \\ Z_t + K\phi(\theta)Z = 0. \end{cases} \tag{1}$$

Here $v = 1/\rho, u, \theta$, and Z represent specific volume, velocity, temperature, and reactant mass fraction, respectively; e and p are the specific internal energy and pressure, and ϕ is the reaction-rate function; ε, λ, K, and q are positive constants describing viscosity, thermal conduction, reaction rate, and the difference in heats between reactant and product. The above equations are written in Lagrangian coordinates, so that $x =$ constant corresponds to the particle path.

The existence of global-in-time solutions with quite general initial data is presented in Theorem 1, yielding a variety of regularity results as well as estimates which in certain cases are independent of time. Results concerning the propagation of singularities in solutions for which v and Z are initially piecewise H^1 are given in Theorem 2. These results show that, while the parabolicity of the momentum and energy equations has a regularizing effect on u and θ, singularities in v, p, Z, e, u_x, and θ_x persist for all time, convecting along particle trajectories, and evolving according to a certain dynamics derivable from the corresponding Rankine-Hugoniot conditions. The large-time behavior of solutions is described in Theorem 3 under the assumption that the regularity estimates of Theorem 1.1 are independent of time, showing that the solution converges as $t \to \infty$ to an equilibrium determined by the initial mass and energy. Finally, in Theorem 4, we give both necessary conditions and sufficient conditions that complete combustion occurs, namely, $Z \equiv 0$ in the equilibrium state, in terms of the total initial energy. Here K is a rate constant, and the reaction function ϕ is assumed to be an increasing, Lipschitz function on $[0, \infty)$ satisfying

$$C^{-1}\phi(\theta) \le \theta^2 \phi'(\theta) \le C\phi(\theta) \tag{2}$$

for a constant C, and

$$\begin{cases} \phi(\theta) = 0, & 0 \le \theta \le \theta_i, \\ \phi(\theta) > 0, & \theta > \theta_i, \end{cases} \tag{3}$$

where $\theta_i > 0$ is the ignition temperature. These conditions are satisfied, for example, by the Arrhenius function $\phi(\theta) = e^{-A/\theta}, \theta \gg \theta_i$. Thus, when $Z = 1$, a given fluid particle consists entirely of reactant. When its temperature rises above the ignition temperature, combustion will occur, resulting in the conversion of some or all of its mass to the product species, and decreasing Z accordingly. When $Z = 0$, complete combustion has occurred, and the particle consists entirely of the product species.

We may regard the reactant and product as separate fluids with respective densities $\rho_1 = Z\rho$ and $\rho_2 = (1 - Z)\rho$, having the same velocities and temperatures at each point, and characterized by polytropic equations of state

$$\begin{cases} P_j = c_j(\gamma_j - 1)\theta, & \\ e_j = c_j\theta, & j = 1, 2. \end{cases} \tag{4}$$

Here P_j and e_j are pressures and specific internal energies, $\gamma_j > 1$ are adiabatic constants, and c_j are (constant) specific heats. Now define the macroscopic pressure and energy by

$$\begin{cases} P = P_1 + P_2 \quad \text{(Dalton's law)}, \\ \rho E = \rho_1 E_1 + \rho_2 E_2, \\ e = E - u^2/2, \end{cases} \tag{5}$$

and system parameters

$$\varepsilon = \varepsilon_1 + \varepsilon_2, \quad \lambda = \lambda_1 + \lambda_2, \quad q = q_2 - q_1, \tag{6}$$

where $E_j = e_j + u^2/2$, and ε_j, λ_j, and q_j are positive constants.
We define a variable specific heat $c = c(Z)$ and a variable adiabatic constant $\gamma = \gamma(Z)$ by

$$\begin{cases} c(Z) = c_1 Z + c_2(1 - Z), \\ \gamma(Z) = \frac{Zc_1\gamma_1 + (1-Z)c_2\gamma_2}{Zc_1 + (1-Z)c_2}, \end{cases} \tag{7}$$

so that, again by (4) and (5),

$$P(\rho, \theta, Z) = a(Z)\rho\theta, \tag{8}$$

where $a(Z) = c(Z)(\gamma(Z) - 1)$. Hence the states P and e for the composite fluid are exactly those of a polytropic fluid, as in (4), but with adiabatic "constant" γ and specific heat c depending on Z. In Lagrangian coordinates, taking

$$p(v, \theta, Z) = P(v^{-1}, \theta, Z) = a(Z)\theta/v, \tag{9}$$

we then obtain the system (1).

Here we consider an initial boundary value problem where a fluid of finite total mass is confined to a bounded region, so that, without loss of generality, $0 < x < 1$. We impose boundary conditions

$$u(x, t) = 0, \quad \theta_x(x, t) = 0, \quad x = 0, 1. \tag{10}$$

The solutions considered in the present article are weak solutions in the usual sense: writing the system (1) in the form

$$w_t^j + f^j(w)_x = (b^{jk}(w)w_x^k)_x + g^j(w), \tag{11}$$

we say that w is a weak solution provided that w is suitably integrable, and

$$\int_0^1 w^j(x, \cdot)\psi(x, \cdot)dx|_{t_1}^{t_2} = \int_{t_1}^{t_2} \int_0^1 (w^j\psi_t + (f^j - b^{jk}w_x^k)\psi_x + g^j\psi)dxdt \tag{12}$$

for all $t_2 \geq t_1 \geq 0$ and all $\psi \in C^1([0, 1] \times [0, \infty))$, with the exception that, when $w^j = u$, (12) is required to hold only for ψ satisfying the boundary conditions $\psi(0, t) = \psi(1, t) = 0$.

We consider initial data

$$(v, u, \theta, Z)|_{t=0} = (v_0, u_0, Z_0, \theta_0) \tag{13}$$

satisfying

$$\begin{cases} C_0^{-1} \le v_0(x) \le C_0, \ \theta_0(x) \ge C_0^{-1}, \ 0 \le Z_0(x) \le 1, \\ ||\theta_0||_{L^2} + ||u_0||_{L^4} \le C_0 \end{cases} \tag{14}$$

for a constant $C_0 > 0$, which may be arbitrarily large. Finally, we define the following functionals for a given weak solution (v, u, θ, Z) of (1) and (10):

$$\mathcal{D}(t) = \sup_{0\le s\le t} \left(||u(\cdot,s)^2||^2 + ||\theta(\cdot,s)||^2\right) \tag{15}$$

$$+ \int_0^1 \int_0^t \left(\theta^{-1}u_x^2 + u^2u_x^2 + \theta_x^2 + \phi(\theta)Z^2\right) dxdt, \tag{16}$$

$$\mathcal{E}(t) = \sup_{0\le s\le t} (\sigma(s)||u_x(\cdot,s)||^2 + \sigma^2(s)||\theta_x(\cdot,s)||^2)$$

$$+ \int_0^1 (||(u_x,\theta_x)(\cdot,s)||^2 + \sigma(s)||u_t(\cdot,s)||^2 + \sigma^2(s)||\theta_t(\cdot,s)||^2)ds,$$

$$\mathcal{F}(t) = \sup_{0\le s\le t} (\sigma^2(s)||u_t(\cdot,s)||^2 + \sigma^3(s)||\theta_t(\cdot,s)||^2)$$

$$+ \int_0^t (\sigma^2(s)||u_{xt}(\cdot,s)||^2 + \sigma^3(s)||\theta_{xt}(\cdot,s)||^2)ds, \tag{17}$$

where $\sigma(t) = \min\{1, t\}$, and $||\cdot||$ denotes the usual norm in $L^2([0,1])$.

The following theorem then gives our main results concerning the existence and regularity of solutions of the system (1).

Theorem 1. *Assume that the functions* $p(v,\theta,Z), e(\theta,Z)$, *and* $\phi(\theta)$ *are as described above in* (2)–(3) *and* (7)–(8), *and let the initial data* $(v_0, u_0, \theta_0, Z_0)$ *be given, as in* (14). *Then the system* (1), (10), *and* (13) *has a global weak solution* (v, u, θ, Z) *for which*

$$\begin{cases} v \in C([0,\infty); L^p([0,1])), 1 \le p < \infty, \\ u \in C([0,\infty); L^p([0,1])), 1 \le p < 4, \\ \theta \in C((0,\infty); L^2([0,1])), \theta(\cdot,t) \rightharpoonup \theta_0 \text{ weakly in } L^2 \text{ as } t \to 0, \\ Z \in C([0,\infty); L^p([0,1])), 1 \le p \le \infty. \end{cases} \tag{18}$$

Also, given $T > 0$, *there is a constant* M *depending on* T, C_0, *and on the system parameters, such that, for* $t \in (0,T]$,

$$\begin{cases} M^{-1} \le v(x,t) \le M, \quad a.e.\ x \in [0,1], \\ M^{-1} \le \theta(x,t) \le M \max\{1, t^{-1}\}, \quad x \in [0,1], \\ 0 \le Z(x,t) \le 1, \quad a.e.\ x \in [0,1], \end{cases} \tag{19}$$

and

$$\mathcal{D}(T) + \mathcal{E}(T) + \mathcal{F}(T) \le M. \tag{20}$$

In particular, u *and* θ *are Hölder continuous on compact sets in* $[0,1]\times(0,\infty)$, *and*

$$u(\cdot,t),\ \theta(\cdot,t),\ (\frac{\varepsilon u_x}{v}-p)(\cdot,t)\in H^1([0,1]) \tag{21}$$

for $t>0$. Finally, in the case that $c_1=c_2$ and $\gamma_1=\gamma_2$, the bounds in (19) *and* (20) *hold with a constant M which is independent of T.*

We prove Theorem 1 by deriving certain a priori bounds for suitable approximate solutions, then applying these bounds to obtain the solutions of the theorem as strong limits of the approximate solutions [7].

The analysis relies on pointwise bounds for the temperature and the specific volume. These estimates are achieved by a modification of the analysis in Chen, Hoff, and Trivisa [5] for nonreacting flows, which was motivated by the original ideas of Kazhikov and Shelukhin [21]. These pointwise estimates show that vacuum never appears for solutions to the present model (1). Having established these pointwise estimates, the higher-order regularity estimates (15) and (20) can be derived by some rather long and technical energy estimates.

The other interesting feature of Theorem 1 is the proof of strong compactness of approximate solutions in the absence of uniform regularity for the variables Z and v. Compactness for approximate Z's can be obtained from the corresponding compactness of approximate θ's using the evolution equation of Z. For the specific volume v, a more delicate treatment is required, and strong compactness is shown to result using the structure of the approximate solutions.

Results on the propagation of singularities in solutions of (1) are given in Theorem 2 below, and are similar to corresponding results for nonreacting flows (see [15] example). Still, the variability of the mass fraction give rise to interesting new features. For the description of these results, we first assume that (v,u,θ,Z) is a piecewise smooth solution, having one-sided limits along a curve $(y(t),t)$, and we derive at the heuristic level the various conditions that guarantee that (v,u,θ,Z) is a weak solution across $(y(t),t)$. Now, in light of the regularity estimates in Theorem 1 for u and θ,

$$[u(y,t)]=[\theta(y,t)]=0, \tag{22}$$

where $[u(y,t)] = u(y+,t) - u(y-,t)$, etc. Next, applying the Rankine-Hugoniot conditions to the mass equation in (1), we find that $\dot{y}(t)[v(y,t)] = [u(y,t)]=0$. Combining this with the observation that singularities in Z must remain at fixed points of the $x-$space, we see that $y(t)=y$ is constant. We then conclude that singularities remain stationary in the Lagrangian space, and therefore convect along particle trajectories in the physical space. Applying the Rankine-Hugoniot condition to the other equations in (1), we then find that, at (y,t),

$$\left[\frac{\varepsilon u_x}{v}-p\right]=0,\quad \left[\frac{\theta_x}{v}\right]=0,\quad [Z]_t+K\phi(\theta)[Z]=0. \tag{23}$$

We solve these equations formally as follows. First,

$$\frac{d}{dt}[\log v(y,t)] = \left[\frac{v_t}{v}\right] = \left[\frac{u_x}{v}\right] = \varepsilon^{-1}[p] = \alpha[\log v] + \beta[Z], \tag{24}$$

where

$$\alpha(y,t) = \varepsilon^{-1}\theta(y,t)\ \frac{c_+(\gamma_+ - 1) + c_-(\gamma_- - 1)}{2}\ \frac{[v^{-1}(y,t)]}{[\log v(y,t)]}, \tag{25}$$

and

$$\beta(y,t) = \varepsilon^{-1}\theta(y,t)\frac{v_+ + v_-}{2v_+v_-}\left(c_1(\gamma_1 - 1) + c_2(\gamma_2 - 1)\right) \tag{26}$$

(here $c_+ = c(Z(y+,t))$, etc.). Solving the ode's in (23) and (24), we then obtain

$$[\log v(y,t)] = e^{-\int_0^t \alpha(y,s)ds}[\log v_0(y)] \tag{27}$$

$$+ \int_0^t e^{-\int_s^t \alpha(y,\tau)d\tau}\beta(y,s)[Z(y,s)]ds, \tag{28}$$

and

$$[Z(y,t)] = e^{-\int_0^t K\phi(\theta(y,s))ds}[Z_0(y)]. \tag{29}$$

Explicit expressions for $[u_x]$ and $[\theta_x]$ can then be obtained from (23). Note that α is strictly negative, so that the strengths of jump discontinuities in v and Z, hence in u_x and θ_x, decrease in time. Unlike the case of nonreacting flows, however, these strengths may not decrease to zero as $t \to \infty$, because $\theta(y,t)$ may be asymptotically less than θ_i.

These observations are made precise in the following theorem.

Theorem 2. *Assume that the hypotheses of Theorem 1 are in force, and suppose in addition that v_0 and Z_0 are piecewise in H^1, having simple jump discontinuities at points $y_1 < y_2 < \cdots < y_n$ in $(0,1)$ (thus $v_0, Z_0 \in H^1((y_{j-1}, y_j))$ for each j). Then, for the solution (v,u,θ,Z) of Theorem 1, the quantities $v(\cdot,t), p(\cdot,t), Z(\cdot,t), u_x(\cdot,t)$, and $\theta_x(\cdot,t)$ are piecewise in H^1 for $t > 0$ and therefore have one-sided limits at each y_j for $t > 0$, and the jump conditions (22) and (23) and the representations (27) and (29) hold in a strict pointwise sense at each $y = y_j$ for $t > 0$.*

Next we turn to the study of the large-time behavior of solutions. This will require the assumption that the estimates of Theorem 1 hold independently of time, and as indicated in Theorem 1, this time-independence is known so far only for the special case that $c_1 = c_2$ and $\gamma_1 = \gamma_2$. We emphasize, however, that the large-time results given below in Theorems 3 and 4 do not depend directly upon the latter assumptions in any way, but instead are consequences of the weak forms of the equations (1) and time-independent estimates (19)–(20). In the following theorem, we apply these estimates to show that the solution tends to a time-asymptotic steady-state solution, and we characterize this steady-state in terms of the asymptotic mass-fraction and the initial mass and energy, given by

$$m_0 = \int_0^1 v_0(x)dx, \tag{30}$$

$$E_0 = \int_0^1 \left(c(Z_0(x))\theta_0(x) + \tfrac{1}{2}u_0(x)^2 + qZ_0(x)\right)dx. \tag{31}$$

Theorem 3. *Let (v, u, θ, z) be a weak solution of (1) as in Theorem 1, satisfying (19) and (20) with M independent of T, and let m_0 and E_0 be as above in (30)–(31). Then there is a function Z_∞ such that, as $t \to \infty$,*

$$Z(x,t) \to Z_\infty(x), \quad a.e.\ x; \tag{32}$$

and, if $v_\infty(x)$ and θ_∞ are defined by

$$\theta_\infty = \frac{E_0 - q\displaystyle\int_0^1 Z_\infty(x)dx}{\displaystyle\int_0^1 c(Z_\infty(x))dx}, \tag{33}$$

and

$$v_\infty(x) = \frac{\gamma(Z_\infty(x)) - 1}{\displaystyle\int_0^1 (\gamma(Z_\infty(x)) - 1)dx}\, m_0, \tag{34}$$

then, as $t \to \infty$,

$$||u(\cdot,t), \theta(\cdot,t) - \theta_\infty||_{H^1([0,1])} \to 0, \tag{35}$$

$$||v(\cdot,t) - v_\infty||_{L^p([0,1])} \to 0, \quad 1 \leq p < \infty. \tag{36}$$

We emphasize that when the asymptotic temperature θ_∞ is greater than the ignition temperature θ_i, the steady-state $(v_\infty, 0, \theta_\infty, Z_\infty)$ is constant in space, and is characterized by complete combustion, $Z_\infty \equiv 0$. This is indeed the case when the total energy E_0 in the system is sufficiently large.

Theorem 4. *Let the hypotheses and notations of Theorem 3 be in force and assume that the initial mass-fraction Z_0 is not identically zero. Then*

- *If $Z_\infty \equiv 0$, then $E_0 \geq c_2\theta_i$.*
- *Conversely, if*

$$E_0 > \max\{c_2\theta_i, \left(c_2 + (c_1 - c_2)\int_0^1 Z_0 dx\right)\theta_i + q\int_0^1 Z_0 dx\}, \tag{37}$$

then $Z_\infty \equiv 0$ and $v_\infty \equiv m_0$.

The gap between the necessary condition in (4) and the sufficient condition in (4) can be closed for the special class of solutions for which $v = v_0$ is constant, $u = 0$, and θ, Z, and e depend only on t. In this case, the fluid has neither kinetic energy nor elastic potential energy, and all effects are thermal:

$$\begin{cases} \frac{de}{dt} = Kq\phi(\theta)Z, \\ \frac{dZ}{dt} = -K\phi(\theta)Z, \\ \theta = \frac{e}{(c_1-c_2)Z+c_2}. \end{cases} \tag{38}$$

The solution (e, Z) thus remains on to the line segment

$$e + qZ = E_0 = \text{constant},$$

so that

$$e(t) = E_0 - qZ(t) \quad \text{and} \quad \theta(e(t), Z(t)) = \frac{E_0 - qZ(t)}{(c_1 - c_2)Z(t) + c_2}.$$

The trajectory $\{(e(t), Z(t)) : 0 \le t < \infty\}$ is therefore disjoint from the set of points $\{(e, Z) : \theta(e, Z) \le \theta_i\}$ if and only if

$$E_0 - qZ > (c_2 + (c_1 - c_2)Z)\theta_i$$

for $Z \in (Z_\infty, Z_0]$. It then follows easily that $Z_\infty = 0$ if and only if

$$E_0 \ge c_2\theta_i \quad \text{and} \quad E_0 > (c_2 + (c_1 - c_2)Z_0)\theta_i + qZ_0.$$

The results of this paper (the detailed analysis is presented in [7]) apply to nonreacting flows as well: if $Z_0 \equiv 0$, then $Z(\cdot, t) \equiv 0$ for all $t, c(Z) \equiv c_2$, and $\gamma(Z) \equiv \gamma_2$. The bounds in (19) and (20) are then independent of time, and the time-asymptotic state given in Theorem 3 is simply $(m_0, 0, E_0/c_2, 0)$. In particular, singularities disappear in the time-asymptotic limit (27). The same "asymptotic compactness" has been established for the case of barotropic flows, even in the presence of an external force, yielding a global attractor theory (Hoff and Ziane [18]). For the more general case of reacting flows considered here [7], the situation is quite a bit more complicated: when $\theta_\infty < \theta_i$ the steady-state will not be constant in space, and, more importantly, singularities in solutions will not disappear in the time-asymptotic limit (see (27) and (29)).

The existence and asymptotic behavior of global smooth solutions for reacting, compressible Navier-Stokes flows, but with constant adiabatic exponent and additional species diffusion, was first established by Chen [4]. The case of *nonreacting* flows in one space dimension has been treated by many authors. See for example Chen, Hoff, and Trivisa [5] on the well-posedness and asymptotic behavior of global solutions with large discontinuous initial data, Matsumura and Yanagi [24] for the isothermal case, Amosov and Zlotnick [1] for the existence of weak solutions for certain initial-boundary data, and Hoff [14]–[16] for the global well-posedness and large-time behavior for small discontinuous initial data, including the case of several space variables.

We also remark that these results carry over to the Dirichlet problem for (1), as well as to the corresponding system written in Eulerian coordinates. Analysis on various initial boundary value problems for a variant of this model in the presence of species diffusion is presented in [25].

Acknowledgments

Gui-Qiang Chen's research was supported in part by the National Science Foundation under Grants DMS-9971793, INT-9987378, and INT-9726215. David Hoff's research was supported in part by the National Science Foundation under Grant DMS-9986658. Trivisa was supported in part by the National Science Foundation under Grants DMS-0072496 and INT-9987378 and under the Alfred P. Sloan Foundation Research Fellowship.

References

1. A. A. Amosov and A. A. Zlotnick, A semidiscrete method for solving equations of the one-dimensional motion of a non-homogeneous viscous heat-conducting gas with nonsmooth data, Izv. Vyssh. Uchebn. Zaved. Mat. 1997, 3–19 (Russian); transl. in Russian Math. (Iz. VUZ) 41 (1997), 1–17.
2. A. Bourlioux, A. J. Majda, and V. Roytburd, Theoretical and numerical structure for unstable one-dimensional detonations, *SIAM J. Appl. Math.* **51** (1991), 303–343.
3. T. Chang and L. Hsiao, The Riemann problem and interaction of waves in gas dynamics, Pitman Monographs and Surveys in Pure and Applied Mathematics, 41, Longman Scientific & Technical, Harlow; copublished in the United States with John Wiley & Sons, Inc.: New York, 1989.
4. G.-Q. Chen, Global solutions to the compressible Navier-Stokes equations for a reacting mixture, *SIAM J. Math. Anal.* **23** (1992), 609–634.
5. G.-Q. Chen, D. Hoff, and K. Trivisa, Global solutions of the compressible Navier-Stokes equations with large discontinuous initial data, *Comm. Partial Diff. Eqs.* **25** (2000), 2233–2257.
6. G.-Q. Chen, D. Hoff, and K. Trivisa, On the Navier-Stokes equations for exothermically reacting compressible fluids, *Acta Math. Appl. Sinica*, **18** (2002), 15-36.
7. G. Q. Chen, D. Hoff and K. Trivisa. Global Solutions to a Model for Exothermically Reacting, Compressible Flows with Large Discontinuous Initial Data. Submitted to *Arch. Ration. Mech. Anal.*
8. G.-Q. Chen and D. Wagner, Global entropy solutions to exothermically reacting, compressible Euler equations, *J. Diff. Eqs.* (to appear).

9. P. Collela, A. Majda and V. Roytburd, Theoretical and numerical structure for reacting shock waves, *SIAM J. Sci. Stat. Comput.* **7** (1986), 1059–1080.
10. R. Courant and K. O. Friedrichs, Supersonic Flow and Shock Waves, *Interscience: New York*, 1948.
11. C. M. Dafermos, Global smooth solutions to the initial–boundary value problem for the equations of one–dimensional nonlinear thermoviscoelasticity, *SIAM J. Math. Anal.* **13** (1982), 397–408.
12. J. Glimm, The continuous structure of discontinuities, *Lecture Notes in Physics* **344** (1989), 177–186.
13. E. Godlewski and P. Raviart, Numerical Approximation of Hyperbolic Systems of Conservation Laws, *Appl. Math. Sc.* **118**, Springer-Verlag: New York, 1996.
14. D. Hoff, Global well–posedness of the Cauchy problem for nonisentropic gas dynamics with discontinuous initial data, *J. Diff. Eqs.* **95** (1992), 33-74.
15. D. Hoff, Discontinuous solutions of the Navier–Stokes equations for compressible flow, *Arch. Rational Mech. Anal.* **114** (1991), 15–46.
16. D. Hoff. Discontinuous solutions of the Navier-Stokes equations for multidimensional heat-conducting fluids, Arch. Rational Mech. Anal. **139** (1997), 303–354.
17. D. Hoff and D. Serre. The failure of continuous dependence on the initial data for the Navier Stokes equations of compressible flow, *SIAM J. Appl. Math.* **51** (1991), 887-898.
18. D. Hoff and M. Ziane, The global attraction and finite determining nodes for the Navier-Stokes equations of one dimensional compressible flow with discontinuous initial data, Indiana Univ. Math. J. **49(3)** (2000), 843-889.
19. Y. Kanel, On a model system of equations of one–dimensional gas motion, *J. Diff. Eq.* **4** (1968), 374–380.
20. A. V. Kazhikhov. On the theory of initial–boundary–value problems for the equations of one–dimensional nonstationary motion of a viscous heat-conductive gas, *Din. Sploshnoi Sredy* **50** (1981), 37–62 (Russian).
21. A. V. Kazhikhov and V. V. Shelukhin, Unique global solution with respect to time of initial-boundary-value problems for one–dimensional equations of a viscous gas, *J. Appl. Math. Mech.* **41** (1977), 273–282.
22. A. Majda. High Mach number combustion, *Lecture Notes in Appl. Math.* **24** (1986), 109–184.
23. G. S. S. Ludford. Low Mach number combustion, *Lecture Notes in Appl. Math.* **24** (1986), 3–74.
24. A. Matsumura and S. Yanagi, Uniform boundedness of the solutions for a one-dimensional isentropic model system of compressible viscous gas, *Comm. Math. Phys.* **175** (1996), 259–274.
25. K. Trivisa. Global Existence and Asymptotic Analysis on a Model for the Dynamic Combustion of a Compressible, Reacting Fluid. In preparation.
26. F. A. Williams, *Combustion Theory*, Addison-Wesley, Reading, MA, 1965.

Discontinuous Solutions of Hamilton-Jacobi Equations: Existence, Uniqueness, and Regularity

Gui-Qiang Chen[1] and Bo Su[2]

[1] Department of Mathematics, Northwestern University, 2033 Sheridan Road, Evanston, IL 60208-2730, USA
[2] Department of Mathematics, University of Wisconsin at Madison, 480 Lincoln Drive, Madison, WI 53706-1380, USA

1 Introduction

The theory of continuous viscosity solutions for Hamilton-Jacobi equations and fully nonlinear second-order elliptic and parabolic equations has been established (see [8]) since Crandall-Lions introduced the viscosity solutions in [6]. In this Note, we are concerned with global discontinuous solutions of the Cauchy problem for Hamilton-Jacobi equations:

$$u_t + H(t, x, u, Du) = 0, \qquad x \in \boldsymbol{R}^d, \ \ t > 0, \quad u(0, x) = \varphi(x). \tag{1}$$

The discontinuous solutions of Hamilton-Jacobi equations arise in many important situations. The study of geometrically based motions demands deep understanding of discontinuous solutions of Hamilton-Jacobi equations. (e.g. [2]). Many examples in the control theory and the differential game theory do not have continuous solutions. Another motivation is that Hamiltonians arising in the differential game theory and other areas are discontinuous with respect to some or all of t, x, Du (e.g. [10, 13]). The conventional theories of viscosity solutions do not apply.

In Ishii [15], the classical semicontinuous viscosity solutions were introduced and the Perron method was applied to show the existence of possible classical semicontinuous solutions. The classical semicontinuous solutions well fit the applications. The definition of classical semicontinuous solutions is a two-side notion with two differential inequalities. It was unclear whether the Ishii solution is unique even in the important special case that the Hamiltonian is convex. Also see [1, 3, 4, 11, 20] for several different notions of discontinuous solutions for (1) and related existence results on the discontinuous solutions.

The properties of the semicontinuous solutions , especially the uniqueness and regularity, are in need for further understanding. The proof of the existence, uniqueness, and regularity are shown in [4] and [5]. In this Note, we present a survey of the latest development on the uniqueness and regularity of the discontinuous solutions of (1) with globally Lipschitz convex Hamiltonians for discontinuous function $\varphi(x)$ that is continuous outside a set Γ of measure zero and satisfies the following condition.

$$\varphi(x) \geq \varphi_{**}(x) \equiv \liminf_{y \to x, y \in \boldsymbol{R}^d \setminus \Gamma} \varphi(y). \tag{2}$$

More precisely, we prove the classical semicontinuous solution of (1) is unique when $\varphi(x)$ is almost everywhere continuous (a.e.-continuous, for short) and satisfies (2) with the aid of a comparison principle. We show that the discontinuous solutions preserve the a.e.-continuity, which implies that the space of a.e.-continuous functions is well-posed for discontinuous solutions. We also conclude that classical semicontinuous solutions [15], bi-lateral solutions [3], L-solutions [11], and L^∞-solutions [4] are the same a.e.. Then we establish the Lipschitz regularity of the discontinuous solutions after finite time for (1) with globally Lipschitz continuous, *locally* strictly convex Hamiltonians $H = H(Du)$. That is, when the initial data are a.e.-continuous and satisfy (1), the almost everywhere continuous solutions become Lipschitz continuous after finite time. This regularization effect is a nonlinear feature and is new even for conventional continuous viscosity solutions. Then we show the instantaneous BV regularity for coercive Hamiltonians such as $H(Du) = |Du|$ with linear growth at infinity. We also clarify the connections among the discontinuous solutions from the different notions in the semicontinuous setting and show the L^1_{loc}-accessibility of initial data of these solutions for a general Hamiltonian. The comparison principle between continuous solutions and discontinuous solutions with a.e.-continuous initial data are extended to general Hamiltonians in the L^∞ setting.

2 Uniqueness and A.E.-Continuity

Consider the following Cauchy problem:

$$u_t + H(Du) = 0, \qquad x \in \boldsymbol{R}^d,\ t > 0, \quad u|_{t=0} = \varphi(x), \tag{3}$$

where $H(p)$ is a convex function with uniform Lipschitz constant. At first, we introduce the following sense of a.e.-continuity.

Definition 2.1 *A measurable function $f : \boldsymbol{R}^N \to \boldsymbol{R}$ is said to be a.e.-continuous if there is a set Γ satisfying $m^N(\Gamma) = 0$ such that $f(X)$ is continuous at every point $X \in \boldsymbol{R}^N \setminus \Gamma$. That is, $\lim_{Y \to X} f(Y) = f(X)$, for any $X \in \boldsymbol{R}^N \setminus \Gamma$. Denote*

$$f^*(X) := \limsup_{Y \to X} f(Y), \qquad f_*(X) := \liminf_{Y \to X} f(Y);$$

that is, $f^*(X)$ is the upper envelope of $f(X)$ and $f_*(X)$ the lower envelope of $f(X)$.

Remark 2.1. From the definition of a.e.-continuity, it follows that the space of a.e.-continuous functions with the L^∞-norm is a closed subspace in L^∞ and contains the well-known space of piecewise continuous functions with the L^∞-norm.

We recall the notion of the classical semicontinuous super(sub)-solutions of Ishii [15] in the context of the Cauchy problem (3).

Definition 2.2 *A function $u(t,x)$ is called a sub-solution of* (3) *if*

$$\varphi_t(\bar{t},\bar{x}) + H(D\varphi(\bar{t},\bar{x})) \leq 0$$

holds for all $(\bar{t},\bar{x}) \in [0,\infty) \times \boldsymbol{R}^d$, $\varphi \in C^1([0,\infty) \times \boldsymbol{R}^d)$ satisfying

$$\max_{[0,\infty)\times \boldsymbol{R}^d} (u^* - \varphi) = (u^* - \varphi)(\bar{t},\bar{x});$$

A function $u(t,x)$ is called a super-solution of (3) *if*

$$\varphi_t(\bar{t},\bar{x}) + H(D\varphi(\bar{t},\bar{x})) \geq 0$$

holds for all $(\bar{t},\bar{x}) \in [0,\infty) \times \boldsymbol{R}^d$, $\varphi \in C^1([0,\infty) \times \boldsymbol{R}^d)$ satisfying

$$\min_{[0,\infty)\times \boldsymbol{R}^d} (u_* - \varphi) = (u_* - \varphi)(\bar{t},\bar{x}).$$

The proof of the following comparison principle can be found in [6] and [14], in which the initial data are locally bounded measurable with exponential growth assumption. The general case is shown in [5].

Theorem 2.1 *Let $\varphi^\pm(x)$ satisfy $(\varphi^+)_*(x) \geq (\varphi^-)^*(x)$. Let $u^\pm(t,x)$ be Ishii's semicontinuous super(sub)-solutions of* (1) *with initial data $\varphi^\pm(x)$ Then $(u^+)_*(t,x) \geq (u^-)^*(t,x)$.*

Before we prove the uniqueness of classical semicontinuous solutions, we need the following interior cone property for a convex body.

Observe that any d-dimensional convex body Ω in $\boldsymbol{R}^d$ has the following interior cone property: For any $x \in \bar{\Omega}$, there exists a cone C_x with extreme point at x such that

$$C_x \backslash \{x\} \subset int(\Omega),$$

where

$$C_x = B(x, r_1(x)) \cap \{x + \lambda(y-x) \,|\, y \in B(z, r_2(x)), \lambda > 0, x \notin B(z, r_2(x))\}$$

for some $r_j(x) > 0, j = 1,2$. Then we use this fact to show the following uniqueness theorem in the measure sense.

Theorem 2.2 *Suppose that $H : \boldsymbol{R}^d \to \boldsymbol{R}$ is convex and $\varphi(x)$ is a.e.-continuous with $\varphi(x) \geq \varphi_{**}(x), x \in \boldsymbol{R}^d$. Then the classical semicontinuous* *solution of* (2) *is unique a.e. and a.e.-continuous with* $u(t,x) \geq u_{**}(t,x)$; *and the solution is determined by the Lax formula:*

$$u(t,x) = \inf_{y \in \boldsymbol{R}^d} \{\varphi(y) + tL(\frac{x-y}{t})\}, \tag{4}$$

where $L(q) := \sup_{p \in \boldsymbol{R}^d}\{\langle q,p\rangle - H(p)\}$ *is the Legendre transform of* $H(p)$.

Proof. We can construct a monotone decreasing (or increasing) sequence of continuous functions $\{\varphi_k^{\pm}\}_{k=1}^{\infty}$ satisfying

$$\mp(\varphi_k^{\mp}(x) - \varphi(x)) \geq 0, \qquad \lim_{k\to\infty} m(\{|\varphi_k^{\mp}(x) - \varphi(x)| > \frac{1}{k}\} \cap B(0,r)) = 0, \tag{5}$$

for $r > 0$ and, for $x \in \Gamma$ (the set of discontinuity points of $\varphi(x)$),

$$\lim_{r\to 0}(\varphi^{\mp})_*(x,r) = \varphi_{**}(x), \quad \lim_{r\to 0}(\varphi^{+})^*(x,r) = \varphi^*(x), \quad \lim_{r\to 0}(\varphi^{-})^*(x,r) \leq \varphi^*(x), \tag{6}$$

where

$$(\varphi^{\mp})_*(x,r) = \lim_{k\to\infty} \inf_{y\in B(x,r)} \varphi_k^{\mp}(y)$$

and

$$(\varphi^{\mp})^*(x,r) = \lim_{k\to\infty} \sup_{y\in B(x,r)} \varphi_k^{\mp}(y).$$

Let $\{u_k^{\pm}\}_{k=1}^{\infty}$ be the sequences of continuous viscosity solutions to (3) with initial data $\{\varphi_k^{\pm}\}_{k=1}^{\infty}$. By Theorem 2.1, for every k,

$$u_k^-(t,x) \leq u_*(t,x) \leq u(t,x) \leq u^*(t,x) \leq u_k^+(t,x), \tag{7}$$

where $u(t,x)$ is any classical semicontinuous solution with initial data $\varphi(x)$.

It is clear that $\{u_k^{\pm}\}_{k=1}^{\infty}$ are monotone decreasing and increasing sequences of continuous functions, respectively. Denote $\overline{u} = \lim_{k\to\infty} u_k^+(t,x)$ and $\underline{u} = \lim_{k\to\infty} u_k^-(t,x)$. Then these functions $\bar{u}$ and $\underline{u}$ are measurable in $\boldsymbol{R}^+ \times \boldsymbol{R}^d$ and in $\{t\} \times \boldsymbol{R}^d$ for every $t > 0$, and $\underline{u}(t,x) \leq \overline{u}(t,x)$. From (7), we have

$$\underline{u}(t,x) \leq u_*(t,x) \leq u(t,x) \leq u^*(t,x) \leq \overline{u}(t,x). \tag{8}$$

On the other hand, we can prove $\underline{u}(t,x) = \bar{u}(t,x)$ at almost everywhere Lebesgue point of $\underline{u}(t,x)$ and $\bar{u}(t,x)$; its proof is achieved by a contradiction to (6) by using the Lax formula for $\underline{u}_k(t,x)$ and $\bar{u}_k(t,x)$ for large k with the aid of the interior cone properties of $Dom(L)$ if we assume $m(\{\bar{u}(t,x) > \underline{u}(t,x)\}) > 0$. Then theorem follows by (8). For the details, see [5].

3 Regularity

In this section, we first prove the Lipschitz continuity after finite time for solutions of the Cauchy problem (3) in which $H(p)$ is locally strictly convex with positive growth at infinity. More precisely, when discontinuous initial data are almost everywhere continuous and satisfy (2), the classical semicontinuous solutions of (3) become Lipschitz continuous after finite time. Then we show the instantaneous BV regularity for coercive Hamiltonians with linear growth. We say a Hamiltonian H is locally strictly convex if $D^2H(p) > 0$ for any $p \in \boldsymbol{R}^d$ (the strict convexity may fail as $|p| \to \infty$). A typical locally strictly convex Hamiltonian is $H(p) = (|p|^2 + 1)^{\frac{1}{2}}$.

Theorem 3.1 *Suppose that globally Lipschitz continuous Hamiltonian $H(p)$ is C^2, locally strictly convex, and $\lim_{|p|\to+\infty} H(p) = +\infty$. Then any classical semicontinuous solution of (2) becomes Lipschitz continuous for $t \geq T_0 > 0$ with T_0 depending on $\sup_{x\in\boldsymbol{R}^d} |\varphi(x)|$ and $H(p)$, provided that the discontinuous initial data are uniformly bounded, a.e.-continuous, and satisfy* $(*)$.

Proof: We first observe that the Legendre transform $L(q) = \sup_{p\in\boldsymbol{R}^d}\{\langle q, p\rangle - H(p)\}$ of H is a convex function; $\Omega := Dom(L)$ is a bounded convex body in $\boldsymbol{R}^d$ containing the origin as its interior point; $\Omega_\alpha = \{q \in \boldsymbol{R}^d \mid L(q) \leq \alpha\}$ is a bounded closed convex subset in Ω; and L attains its minimum in $int(\Omega)$.

Set $K = \sup_{x\in\boldsymbol{R}^d} |\varphi(x)|$, $M = \inf_{q\in\partial\Omega} L(q)$, and $m = inf_{q\in\Omega}L(q)$. Since L attains its minimum in $int(\Omega)$, $M - m = a > 0$. It is easy to see that $\Omega_{m+\frac{ia}{4}}$ is a bounded convex body in $\boldsymbol{R}^d$ for $i = 1, 2, 3$. Set $d = dist(\Omega_{m+\frac{a}{2}}, \Omega\backslash\Omega_{m+\frac{3a}{4}}) > 0$. For any $q \in \Omega_{m+\frac{3a}{4}}$, $L(q)$ satisfies $|L(q_1) - L(q_2)| \leq C_1|q_1 - q_2|$, where C_1 is a constant depending only on C. For $t > \frac{8K}{a}$, we have

$$u(t,x) = \inf_{z\in\boldsymbol{R}^d}\{\varphi(z) + tL(\frac{x-z}{t})\} = \inf_{z\in x - t\Omega_{m+\frac{a}{2}}}\{\varphi(z) + tL(\frac{x-z}{t})\}.$$

For any two points x_1 and x_2 with $|x_1 - x_2| \leq \frac{4Kb}{a}$ and positive $\epsilon < 1$, there is a point $z_\epsilon \in x_1 - t\Omega_{m+\frac{a}{2}}$ satisfying $\varphi(z_\epsilon) + tL(\frac{x_1-z_\epsilon}{t}) - \epsilon|x_1 - x_2| \leq u(t, x_1)$, for $t > \frac{8K}{a}$. Hence,

$$\begin{aligned} u(t,x_2) - u(t,x_1) &\leq t(L(\frac{x_2 - z_\epsilon}{t}) - L(\frac{x_1 - z_\epsilon}{t})) + \epsilon|x_1 - x_2| \\ &\leq C_1|x_1 - x_2| + \epsilon|x_1 - x_2|, \quad \text{for } t > \frac{8K}{a}. \end{aligned}$$

Switching the roles of x_1 and x_2 and letting $\epsilon \to 0$, we finally obtain $|u(t,x_1) - u(t,x_2)| \leq C_1|x_1 - x_2|$, which implies the Lipschitz continuity of $u(t,x)$ with respect to x for $t > \frac{8K}{a}$. Using the equation, the Lipschitz continuity of $u(t,x)$ with respect to t immediately follows.

Remark 3.1. In general, one can not expect the instantaneous regularity in Theorem 3.1 since the speed of propagation is finite when the solution experiences a jump. Consider a one-dimensional example of (3): $H(p) = (|p|^2+1)^{\frac{1}{2}}$ and $\varphi(x)=0$, if $x<0$; $\varphi(x)=1$, if $x \geq 0$. Easy calculation shows that $L(q) = -(1-q^2)^{\frac{1}{2}}$, $u(t,x) = \inf_{z\in \boldsymbol{R}^1}\{\varphi(z) - (t^2-(x-z)^2)^{\frac{1}{2}}\}$. Then

$$u(t,x) = \begin{cases} -t, & \text{if } x \leq 0, \\ -(t^2-x^2)^{\frac{1}{2}}, & \text{if } 0 < x \leq t \text{ when } t \leq 1, \text{ or } 0 < x \leq (2t-1)^{\frac{1}{2}} \text{ when } t > 1, \\ 1-t, & \text{if } x > t, \text{ when } t \leq 1, \text{ or } x > (2t-1)^{\frac{1}{2}} \text{ when } t > 1. \end{cases}$$

which is discontinuous on the line $x=t$ for $0 \leq t \leq 1$, and, at $t=1$, $u_x(1,x) \to \infty$ as x goes to 1 from the left side. However, $u(t,x)$ is Lipschitz continuous after $t=1$.

Remark 3.2. When the Hamiltonian $H(Du)$ is strictly convex, the viscosity solutions instantaneously become Lipschitz continuous after initial time as shown by Kruzhkov [16]. The regularity result was generalized to superlinear convex Hamiltonians by Lions [17]. In the context of one-dimensional scalar conservation laws with strictly convex flux functions, Liu-Pierre [18] showed that the measure solutions in the Kruzhkov sense instantaneously become L^∞ functions with initial data involving δ-measures. Demengel-Serre [9] showed that, when the flux functions are strictly convex with linear growth at infinity, there are measure solutions with initial data of positive measures which become L^1 functions after finite time. The example in [5] shows that, even when $u_0 \in L^q$ for some $q>1$, one can not expect that the corresponding solution instantaneously become an L^∞ function for $t>0$.

In particular, the proof of Theorem 3.1 implies the following theorem.

Theorem 3.2 *Let the C^2 Hamiltonian $H(p)$ be locally strictly convex, and* $\lim_{|p|\to+\infty}$
$H(p) = +\infty$. *Then any continuous solution of* (2) *with uniformly bounded and a.e.-continuous initial data $\varphi(x)$ becomes Lipschitz continuous for $t \geq T_0 > 0$ with T_0 depending on* $\sup_{x\in\boldsymbol{R}^d}|\varphi(x)|$ *and $H(p)$.*

Now we show the instantaneous BV regularity for the coercive Hamiltonians such as $H(Du) = |Du|$ with linear growth. More precisely, we have the following theorem.

Theorem 3.3 *Suppose that global Lipschitz continuous $H = H(Du)$ is convex with linear growth.*

$$C_2|p| - C_3 \leq H(p) \leq C_1|p| + C_4, \tag{9}$$

where $0 < C_2 < C_1$. Let $u(t,x)$ be the unique semiclassical viscosity solution of (3)–(4) *with locally bounded initial data $\varphi(x)$ satisfying $\varphi(x) \geq \varphi_{**}(x)$. Then $u(t,x) \in BV_{loc}([0,T]\times\mathbb{R}^d)$.*

Proof: As shown in the proof of Theorem 2.2, there always exists a sequence of continuous viscosity solutions $\lim_{k\to\infty} u_k = u$ in the almost everywhere sense with continuous initial data $\lim_{k\to\infty} \varphi_k = \varphi$, a.e. It suffices to show the local BV regularity of continuous solution u as long as the local BV bound of u depends only on the local L^∞ bound of u. Without loss of generality, we assume that initial data φ is Lipschitz continuous with compact support in $B(0, R)$.

Consider the following approximate viscous equation:

$$u_t^\epsilon - \epsilon \Delta u^\epsilon + H(Du^\epsilon) = 0, \qquad \text{in } [0,T]\times \mathbb{R}^d, \tag{10}$$

$$u^\epsilon(0,x) = \varphi(x), \qquad \text{in } \mathbb{R}^d. \tag{11}$$

The existence of a smooth solution of (10)–(11) can be easily established by the Leray-Schauder fixed point theorem. We now derive some basic estimates of the solution. It follows from the maximum principle that

$$\|u^\epsilon\|_{L^\infty([0,T]\times\mathbb{R}^d)} \le \|\varphi\|_{L^\infty(\mathbb{R}^d)} + T|H(0)| \equiv C_5. \tag{12}$$

Then we have the spatial decay estimate of $u^\epsilon(t,x)$:

$$|u^\epsilon(t,x)| \le w_\epsilon(t,x) \equiv C_5 \exp(\frac{1}{\sqrt{\epsilon}}(C_\epsilon t + R - |x|)), \qquad \text{for } (t,x)\in \mathcal{C}_\epsilon, \tag{13}$$

where $C_\epsilon = L + 2\epsilon$ and $\mathcal{C}_\epsilon = \{(t,x) |\, |x| \ge C_\epsilon t + R\}$, L is the Lipschitz constant of H.

This fact can be proved as follows: Note that, on $\partial\mathcal{C}_\epsilon$,

$$|u(0,x)| = 0 \le w_\epsilon(0,x), \qquad \text{in } \mathbb{R}^d\backslash B(0,R),$$

$$|u(t,x)| \le C_5 \le w_\epsilon(t,x), \qquad \text{on } |x| = R + C_\epsilon t.$$

In addition, we have

$$|u_t^\epsilon - \epsilon\Delta u^\epsilon| = |H(Du^\epsilon) - H(0)| \le L|Du^\epsilon|, \qquad \text{in } \mathcal{C}_\epsilon,$$

$$\begin{aligned} w_t^\epsilon - \epsilon\Delta w^\epsilon - L|Dw^\epsilon| &= \frac{1}{\sqrt{\epsilon}} w^\epsilon\{C_\epsilon - L - \sqrt{\epsilon} - (n-1)\epsilon|x|^{-1}\} \\ &\ge \frac{1}{\sqrt{\epsilon}} w^\epsilon\{C_\epsilon - L - 2\sqrt{\epsilon}\} = 0. \end{aligned}$$

Thus the inequality (13) follows from the maximum principle. Similarly, we differentiate (10), repeat the above procedure, and conclude that $\lim_{|x|\to\infty} |Du^\epsilon| = 0$.

Integrating the equation (3) over $[0,T]\times\mathbb{R}^d$, we have

$$\int_{\mathbb{R}^d} u^\epsilon(T,x)\,dx + \int_0^T \int_{\mathbb{R}^d} H(t,x,Du^\epsilon)\,dt\,dx = \int_{\mathbb{R}^d} \varphi(x)\,dx.$$

It follows from (9), (12), and (13) that

$$\int_0^T \int_{\mathbb{R}^d} |Du^\epsilon| \, dt \, dx \le C(R, T, ||\varphi||_{L^\infty}). \tag{14}$$

Let $\epsilon \to 0$, $u^\epsilon \to u$ in $Lip([0,T] \times \mathbb{R}^d)$ where $u(t,x)$ is the unique continuous viscosity solution of (3)–(4). It follows from (14) that

$$\int_0^T \int_{\mathbb{R}^d} |Du(t,x)| \, dt \, dx \le C(R, T, ||\varphi||_{L^\infty}). \tag{15}$$

Since (3) holds for $u(t,x)$ at almost everywhere $(t,x) \in [0,T] \times \mathbb{R}^d$. Thus, it follows from (15) and (9) that

$$\int_0^T \int_{\mathbb{R}^d} |u_t(t,x)| \, dt \, dx = \int_0^T \int_{\mathbb{R}^d} |H(Du(t,x))| \, dt \, dx \le C(R, T, ||\varphi||_{L^\infty}). \tag{16}$$

Therefore

$$||u||_{BV([0,T]\times\mathbb{R}^d)} \le C(R, T, ||\varphi||_{L^\infty}). \tag{17}$$

The theorem is a straightforward consequence (17) and the dependence cone for the global Lipschitz Hamiltonians, the Theorem V.3 in [6]. That is,

$$||u||_{BV(\mathcal{C}\cap[0,T]\times\mathbb{R}^d)} \le C(R, T, ||\varphi||_{L^\infty}), \tag{18}$$

where $\mathcal{C} = \{(t,x) | \, |x| \ge R - Lt\}$.

Remark 3.3. Note that solutions of (1)–(2) with the linear Hamiltonians for $d = 1$ such as $H(u_x) = u_x$ do not have such BV regularity. The BV regularity is available for the coercive nonlinear Hamiltonians with at least linear growth since superlinear Hamiltonians enjoy instantaneous Lipschitz regularity. So the discontinuous solution of (3)-(4) with locally strictly convex Hamiltonian enjoys BV regularity before it is smoothed out.

For more general Lipschitz coercive Hamiltonian, we have the following theorem

Theorem 3.4 *Suppose that a global Lipschitz Hamiltonian $H = H(t,x,Du)$ with Lipschitz constant L satisfies the following growth condition:*

$$C_2|p| - f(t,x) \le H(t,x,p) \le C_1|p| + g(t,x), \tag{19}$$

where $f(t,x)$ and $g(t,x)$ are locally bounded, and $0 < C_2 \le C_1$. Let $u(t,x)$ be the unique continuous viscosity solution of (1)–(2) with locally continuous initial data $\varphi(x)$. Then $u(t,x) \in BV_{loc}([0,T] \times \mathbb{R}^d)$.

4 Comparison Principle, Initial Data Accessibility, and Connections

As usual, we consider the Hamiltonians satisfying:

(A1). $H(t,x,z,p)$ is continuous in (t,x,z,p) and increasing in z;

(A2). $|H(t,x,z,p_1) - H(t,x,z,p_2)| \le C_0(1+|x|)|p_1 - p_2|$, and $|H(t,x,z,0)| \le C_0(1+|x|+|z|)$, for all $t \in (0,T]$;

(A3). $|H(t,x_1,z,p) - H(t,x_2,z,p)| \le \lambda(L_0)(1+|p|)|x_1 - x_2|$, for $|x_1|, |x_2| \le L_0$;

(A4). $|H(t,x,z_1,p) - H(t,x,z_2,p)| \le C_0(1+|x|+|p|)|z_1 - z_2|$.

Define the essential infimum and supremum of an $L^\infty_{loc}(\boldsymbol{R}^d)$ function $v(x)$ at every point $x \in \boldsymbol{R}^d$:

$$I(v)(x) := \sup_{A\in S_x} ess \inf_{y\in A} v(y), \quad S(v)(x) := \inf_{A\in S_x} ess \sup_{y\in A} v(y),$$

where $S_x = \{A \subset \boldsymbol{R}^d \text{ measurable} \mid \lim_{r\to 0} \frac{m^d(A\cap B(x,r))}{m^d(B(x,r))} = 1\}$. It is clear that $I(v)(x)$ and $S(v)(x)$ are well-defined at every point $x \in \boldsymbol{R}^d$ and $I(v)(x) = S(v)(x)$ a.e.

The definitions of the winning and losing profit functions introduced in [4] are stated as follows.

Definition 4.1 *Fix $\tau \in [0,T]$ and $p(t,x) \in C([0,T]\times \boldsymbol{R}^d; \boldsymbol{R}^d)$. Given a measurable function $v(x)$ and a position (or value) function $f(t,x)$, define the winning and losing profit functions:*

$$\begin{aligned} &\Lambda^v_-(t,x,(\tau,f,p)) \\ &\quad = \inf\{I(v)(x(\tau)) - z(\tau) \,|\, (x(\cdot),z(\cdot)) \in Sol(t,f(t,x),p)\}, \end{aligned} \tag{20}$$

$$\begin{aligned} &\Lambda^v_+(t,x,(\tau,f,p)) \\ &\quad = \sup\{S(v)(x(\tau)) - z(\tau) \,|\, (x(\cdot),z(\cdot)) \in Sol(t,f(t,x),p)\}, \end{aligned} \tag{21}$$

where $Sol(t,f(t,x),p)$ denotes the set of solutions: $(x(\cdot),z(\cdot)) : [\tau,t] \to \boldsymbol{R}^d \times \boldsymbol{R}$, for $t \ge \tau$, of the characteristic inclusions: $(\dot{x}(\cdot),\dot{z}(\cdot)) \in E(t,x,z,p)$ satisfying the conditions: $x(t) = x$, $z(t) = f(t,x)$, where

$$E(t,x,z,p) := \{(h,g) \in \boldsymbol{R}^d \times \boldsymbol{R} \mid |h| \le C_0(1+|x|), g = \langle h,p\rangle - H(t,x,z,p)\}.$$

We now recall the definitions of L^∞ super-solutions and L^∞ sub-solutions for the Cauchy problem (1) in terms of profit functions in [4].

Definition 4.2 *A locally bounded measurable function $w(t,x) : [0,T]\times\boldsymbol{R}^d \to \boldsymbol{R}$ is called a super-solution (sub-solution) if $w(t,x)$ satisfies the following.*

(i) For any $p(t,x) \in C([0,T]\times \boldsymbol{R}^d; \boldsymbol{R}^d)$,

$$\Lambda^\varphi_-(t,x,(0,w,p)) \le 0 \quad (\Lambda^\varphi_+(t,x,(0,w,p)) \ge 0), \tag{22}$$

for almost all $(t,x) \in [0,T] \times \boldsymbol{R}^d$. *Additionally, for every* $t \in [0,T]$, (22) *holds for almost all* $x \in \boldsymbol{R}^d$.

(*ii*) *The semigroup property: For every* $\tau \in [0,T]$,

$$\Lambda_-^{w(\tau,x)}(t,x,(\tau,w,p)) \leq 0 \quad (\Lambda_+^{w(\tau,x)}(t,x,(\tau,w,p)) \geq 0), \tag{23}$$

for almost all $(t,x) \in [\tau,T] \times \boldsymbol{R}^d$. *Additionally, for every* $t \in [\tau,T]$, (23) *holds for almost all* $x \in \boldsymbol{R}^d$.

We say that $u(t,x)$ *is a solution of the Cauchy problem* (1) *if* $u(t,x)$ *is a sup-solution and also a subsolution.*

The global existence of L^∞ solutions was established for general Lipschitz Hamiltonians for arbitrary large L^∞ initial data in [4] under the conditions $(A1)$–$(A4)$. It was also shown that L^∞ solutions consist with viscosity solutions introduced by Crandall-Lions [6] and minimax solutions introduced by Subbotin [20]. In the semicontinuous setting, it is straightforward to verify that the definition of L^∞ solutions is equivalent to Subbotin's definition of semicontinuous super-solutions and sub-solutions. The notion of L^∞ solutions becomes the notion of semicontinuous solutions when the essential infimum and superemum in (20)-(21) are replaced by the standard infimum and superemum. The existence of discontinuous solutions can be shown by adopting the proof in [4] to the semicontinuous setting. In [20], it is shown that Subbotin's definition is equivalent to Ishii's definition for semicontinuous super-solutions and sub-solutions. Therefore, in the semicontinuous setting, Ishii's semicontinuous solutions are the same as L^∞ solutions. In [11], it is shown that the definition of bi-lateral solutions is equivalent to that of L-solutions when the Hamiltonian is convex, and L-solutions are exactly the maximal sup-semicontinuous sub-solutions in the sense of Ishii [15].

Theorem 4.1 *Assume that* $\varphi^\pm(x)$ *are a.e.-continuous and* $\varphi(x)$ *is continuous with* $\pm(\varphi^\pm(x) - \varphi(x)) \geq 0$. *Let* $u^\pm(t,x)$ *be* L^∞ *super(sub)-solutions of* (1) *with initial data* $\varphi^\pm(x)$, *and* $v(t,x)$ *the continuous solution with initial data* $\varphi(x)$. *Then* $\pm(u^\pm(t,x) - v(t,x)) \geq 0$ *a.e.*

For Ishii's semicontinuous solutions, if the assumption of $\varphi^\pm(x)$ in Theorem 4.1 is replaced by $(\varphi^+)_*(x) \geq (\varphi^-)^*(x)$, then $(u^+)_*(t,x) \geq (u^-)^*(t,x)$ holds everywhere. The L^∞ solutions of (1) access initial data in L^1_{loc}, provided the closure of the set of discontinuity points is measure zero.

Theorem 4.2 *Suppose that* $u(t,x)$ *is an* L^∞ *solution of* (1) *with a.e.-continuous initial data* $\varphi(x)$ *and* $m^d(\bar{\Gamma}) = 0$. *Then* $u(t,\cdot) \to \varphi(\cdot)$, *when* $t \to 0$, *in the* L^1_{loc} *topology.*

Acknowledgements. The authors thank Jim Glimm, Tai-Ping Liu, Takis Souganidis for stimulating conversations. Gui-Qiang Chen's research was supported in part by the National Science Foundation through grants DMS-9971793, INT-9987378, and INT-9726215.

References

1. Barles, G. and Perthame, B., Discontinuous solutions of deterministic optimal stopping problem, Math. Model. Num. Anal. **2** (1987), 557-579.
2. Barles, G., Soner, H. M., and Souganidis, P. E., *Front propagation and phase field theory*, SIAM J. Control and Optimization, **31** (1993), 439-469.
3. Barron, E. N. and Jensen, R., Semicontinuous viscosity solutions of Hamilton-Jacobi equations with convex Hamiltonians, Commun. Partial Diff. Eqs. **15** (1990), 1713-1742.
4. Chen, G.-Q. and Su, B., Discontinuous solutions in L^∞ for Hamilton-Jacobi equations, Chinese Annals Math. **2** (2000), 165-186.
5. Chen, G.-Q. and Su, B., On Discontinuous solutions of Hamilton-Jacobi equations, Preprint, June 2001.
6. Crandall, M. and Lions, P. L., Viscosity solutions of Hamilton-Jacobi equations, Trans. Amer. Math. Soc. **277** (1983), 1-42.
7. Crandall, M., Evans, L. C., and Lions, P. L., *Some properties of viscosity solutions of Hamilton-Jacobi equations*, Trans. Amer. Math. Soc. **282** (1984), 487-502.
8. Crandall, M., Ishii, H., and Lions, P. L., A user's guide to viscosity solutions of second-order partial differential equations, Bulletin Amer. Math. Soc. **27** (1992), 1-67.
9. Demengel, F. and Serre, D., Nonvanishing singular parts of measure valued solutions for scalar hyperbolic equations, Comm. Partial Diff. Eqs. **16** (1991), 221-254.
10. Friedman, A., Differential Games, John Wiley, New York, 1971.
11. Giga, Y. and Sato, M. H., A level set approach to semicontinuous solutions for Cauchy problems, Preprint, 2001.
12. Glimm, J., Kranzer, H. C., Tan, D., and Tangerman, F. M., Wave fronts for Hamilton-Jacobi equations: the general theory for Riemann solutions in $\mathbf{R}^n$, Comm. Math. Phys. **187** (1997), 647–677.
13. Issacs, R. P., Differential Games, Krieger, New York, 1967.
14. Ishii, H., Uniqueness of unbounded viscosity solution of Hamilton-Jacobi equations, Indiana Univ. Math. J. **33** (1984), 721-748.
15. Ishii, H., Perron's method for Hamilton-Jacobi equations, Duke Math. J. **55** (1987), 368-384.
16. Kruzhkov, S. N., Generalized solutions of nonlinear equations of the first order with several independent variables, II, (Russian) Mat. Sb. (N.S.) **114** (1967), 108-134.
17. Lions, P. L., Generalized Solutions of Hamilton-Jacobi Equations, Research Notes in Math. **69**, Pitman, Boston-London-Melbourne, 1982.
18. Liu, T.-P. and Pierre, M., Source solutions and asymptotic behavior in conservation laws, J. Diff. Eqs. **51** (1984), 419-441.
19. Schneider, R. Convex bodies: the Brunn-Minkowski theory. Encyclopedia of Mathematics and its Applications, 44. Cambridge University Press, Cambridge, 1993.
20. Subbotin, A. I., Generalized Solutions of First Order PDEs, Birkhauser, Boston, 1995.
21. Subbotin, A. I., *Generalized solutions of partial differential equations of the first order. The invariance of graphs relative to differential inclusions*, J. Math. Sci. **78** (1996), 594-611.

Conservation Laws and O.D.E.s. A Traffic Problem

Rinaldo M. Colombo[1] and Andrea Marson[2]

[1] Dipartimento di Matematica, Via Valotti 9, 25133 Brescia, Italy *rinaldo@ing.unibs.it*

[2] Dipartimento di Matematica Pura e Applicata, Via G. Belzoni 7, 35131 Padova, Italy *marson@math.unipd.it*

Summary. Motivated by a problem of traffic flow, we consider an ordinary differential equation whose vector field depends upon the solution of a conservation law. The Cauchy problem related to the O.D.E. does not fit in the results currently available in the literature. Here we address its well posedness.

1 Introduction

The classical Lighthill-Whitham [9] and Richards [11] traffic flow model reads

$$\begin{cases} \partial_t \rho + \partial_x \left[\rho \cdot v(\rho)\right] = 0 \\ \rho(0,x) = \bar{\rho}(x)\,. \end{cases} \tag{1}$$

Here ρ is the car density and it is assumed to take values in a compact interval $[0, R]$, $v(\rho)$ is the speed of cars at traffic density ρ and $\bar{\rho}$ is a suitable initial datum. We will denote by $f(\rho) = \rho \cdot v(\rho)$ the traffic flow. For a comparison of this model with other more recent ones see [7], while an extension of (1) that comprises phase transitions is presented in [4].

Assume that a driver travels along a road modeled by (1), being influenced by the traffic along the road but without influencing it significantly. The driver's position $p = p(t)$ then solves the Cauchy problem

$$\begin{cases} \dot{p} = w\left(\rho(t,p)\right) \\ p(0) = \bar{p} \end{cases} \tag{2}$$

for a suitable initial position $\bar{p}$ and a speed function w characteristic of the driver. A step that necessarily precedes the statement of any problem related to the optimization of the driver's behavior is the well posedness of (1)-(2), which is the object of the present note. Similar questions, but related to the optimization of the main traffic flow rather than of the individual path, are considered for instance in [5].

Furthermore, under assumptions reasonable from the point of view of the traffic flow models, (1)–(2) does not fit in the well posedness results usually considered in the literature. The main theorems on ordinary differential equations with discontinuous right hand side require either assumptions of

Caratheodory type [8, §1] or the bounded directional variation of the vector field [1]. As noted in [2], problems of type (1)-(2) fall in neither of the two categories above. In fact, differently from the usual results in the theory of ordinary differential equations, here the solutions turn out to be a Hölder (not Lipschitz) function of the initial data.

In [2], aiming at a result on the well posedness of a class of 2×2 systems of hyperbolic conservation laws, the authors prove an existence and uniqueness result for ordinary differential equations depending on the solution of a scalar conservation law. However, problem (1)-(2) does not fit in the result therein, for the assumption (A2) in [2] is unreasonable in the present framework. Here, this assumption would imply that the speed of the driver varies in an interval *disjoint* from that of the main traffic speed.

Moreover, in [2] the authors consider Caratheodory solutions to (2). Recall that a Caratheodory solution is defined by the integral equation

$$p(t) = \bar{p} + \int_0^t w\left(\rho(\tau, p(\tau))\right) d\tau .$$

¿From the traffic point of view, such a solution ceases to exists as soon as the driver is *trapped* along a discontinuity: for instance, the driver reaches the tail of a queue but can not go through it. The choice of Carathodory solutions is thus unsuitable to the present setting since drivers may well be trapped in a queue and this feature is described by (1)–(2). Consider the case

$$\bar{\rho}(x) = \begin{cases} \rho^\ell & \text{if } x < 1 \\ \rho^r & \text{if } x > 1 \end{cases} \qquad \rho^\ell < \rho^r , \qquad \text{and} \qquad \bar{p} = 0 . \tag{3}$$

If the flux function f is strictly concave, the solution to (1) with initial data (3) is a single shock wave that represents the tail of a queue. Whenever we choose w so that

$$w(\rho^\ell) > f'(\rho^\ell) \quad \text{and} \quad w(\rho^r) < f'(\rho^r) ,$$

the driver reaches the queue but then can not pass through it (see Figure 1). Hence a Caratheodory solution does not exist for all times. This motivates our choice of looking for a Filippov [8] solution to (2) defined on an interval $[0, T]$, $T > 0$. More precisely, we define $p(\cdot)$ to be a solution to (2) in $[0, T]$ if it is an absolutely continuous function such that

$$\begin{cases} \dot{p}(t) \in \overline{\mathrm{co}}\left\{w(\rho) \ : \ \rho \in \mathcal{I}\left([\rho(t, p(t)-), \, \rho(t, p(t)+)]\right)\right\} , \\ p(0) = \bar{p} . \end{cases}$$

Here $\overline{\mathrm{co}}$ stands for the closed convex hull; $\rho(t, x-)$, $\rho(t, x+)$ are respectively the left and right limits of $\rho(t, \cdot)$ at x and $\mathcal{I}\left([a, \, b]\right)$ is the smallest interval containing a and b. As it is well known, the choice of the wider class of Filippov solutions easily guarantees the existence. In the present setting, in spite of this choice, uniqueness and continuous dependence (also on the initial datum $\bar{\rho}$ of (1)) hold, as shown in the following section.

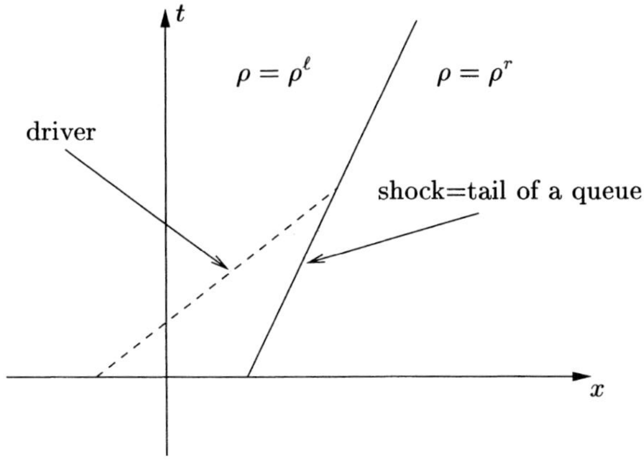

Fig. 1. A Caratheodory solution may fail to exist for all times

2 Main Results

The speed function $\rho \mapsto v(\rho)$ is assumed to satisfy the following usual assumption:

(V) $v \colon [0, R] \mapsto [0, +\infty[$ is smooth, decreasing, $v(R) = 0$ and $\dfrac{d^2(\rho\, v(\rho))}{d\rho} < 0.$

A few remarks are in order. v is decreasing since at higher density, drivers are slower. At the maximal density R, cars are bumper to bumper and no one can proceed. Finally, the hypothesis on the strict concavity of the flux is of technical nature and is relaxed in the related result [10].

Theorem 1. *Let v be as in* **(V)** *and $w \colon [0, R] \mapsto [0, +\infty[$ be smooth and decreasing. If there exists a constant $\alpha \in]0, 1[$ such that*

$$\frac{w(0) - w(\rho)}{w(0) - f'(\rho)} \in [0, 1 - \alpha], \qquad \forall\, \rho \in [0, R], \tag{4}$$

then

1. *for all initial data $\bar{\rho} \in \mathbf{L}^\infty$, problem (1)–(2) admits a unique Filippov solution on $[0, T]$;*
2. *there exists a constant $C > 0$ such that if $p_1(\cdot)$ and $p_2(\cdot)$ solve (2) with initial data respectively $\bar{p}_1$, $\bar{p}_2$, then*

$$\left|p_1(t) - p_2(t)\right| \leq C \cdot \left|\bar{p}_1 - \bar{p}_2\right|^\alpha, \tag{5}$$

   for all $t \in [0, T]$;
3. *if $\left\{\bar{\rho}_n\right\}_{n \in \mathbb{N}}$ is a sequence of $\mathbf{L}^\infty$ initial data to (1) converging in $\mathbf{L}^1(\mathbb{R})$ to $\bar{\rho}_\infty \in \mathbf{L}^\infty(\mathbb{R})$, then the solutions p_n to (2) corresponding to $\bar{\rho}_n$ converge in $\mathbf{C}^0$ to the solution p_∞ corresponding to $\bar{\rho}_\infty$.*

Note that the assumption (4) is satisfied, for instance, in each of the following situations:

1. $w = v$, together with **(V)**, meaning that the driver behaves as the main traffic;
2. $w(\rho) \geq v(\rho)$, $\forall\ \rho \in]0, R]$ and $w(0) > v(0)$, under **(V)**, meaning that the driver travels faster than the main traffic. This condition guarantees that Filippov solutions are also Caratheodory solutions since it implies that any discontinuity is crossed without the driver being trapped along it. In fact, by the Rankine-Hugoniot conditions, the speed Λ of a shock between the states ρ^ℓ and ρ^r is

$$\Lambda = v(\rho^r) + \frac{\rho^\ell}{\rho^r - \rho^\ell} \cdot \left(v(\rho^r) - v(\rho^\ell)\right) \leq v(\rho^r) \leq v(\rho^\ell)$$

and is lower than both $w(\rho^r)$ and $w(\rho^\ell)$.

If $w = w(\rho)$ attains values both lower and greater than $f'(\rho)$, then the continuous dependence of the solutions to (2) on the initial data may fail. Indeed, let $\bar{\rho}$ in (1) be defined by

$$\bar{\rho}(x) \doteq \begin{cases} \rho^\ell & \text{if } x < 0\,, \\ \rho^r & \text{if } x > 0\,, \end{cases} \tag{6}$$

with $\rho^\ell > \rho^r$, so that, whenever $f(\rho) = \rho \cdot v(\rho)$ is strictly concave, the solution of (1) with initial datum (6) is a centered rarefaction wave departing from $x = 0$. Assign w such that

$$w(\rho^\ell) < f'(\rho^\ell)\,, \qquad w(\rho^r) > f'(\rho^r)\,.$$

Then, problem (1)–(2) is solved as follows:

$$\begin{array}{ll} p(t) = \bar{p} + w(\rho^\ell)t & \text{if } \bar{p} < 0 \\ p(t) = \bar{p} + w(\rho^r)t & \text{if } p > 0 \end{array}$$

showing that continuous dependence may not hold (see Figure 2).

Under assumption (4), whenever $w(0) = f'(0)$ we can get at most Hölder continuous dependence for the Cauchy problem (2) even if the initial datum ρ_0 (and hence the solution ρ) of (1) has bounded total variation [1, 2]. This can be deduced by the following example. For the problem (1)–(2) consider the data

$$w(\rho) = v(\rho) = R - \rho\,, \qquad f(\rho) = \rho(R - \rho)\,, \qquad \rho_0(x) = \begin{cases} R & \text{if } \ x < 0 \\ 0 & \text{if } \ x > 0, \end{cases}$$

where $R > 0$. Then the corresponding solution to (1) is a rarefaction wave centered in $x = 0$

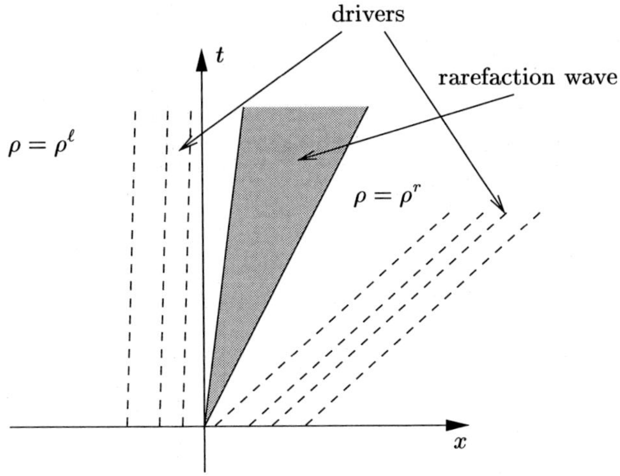

Fig. 2. How continuous dependence can be lost.

$$\rho(t,x) = \frac{Rt - x}{2t} \qquad \text{if} \qquad -Rt \le x < Rt\,,$$

while the solution to (2) with initial datum $p_0 < 0$ is given by

$$p(t) = R\big(t - 2\sqrt{t}\cdot\sqrt{|p_0|}\big)\,, \qquad t \ge |p_0|$$

Hence, the solution $p = p(t)$ depends Hölder and not Lipschitz continuously on p_0.

Remark that (4) is weaker than the assumption (A2) that in [2] plays a role analogous to the one played by (4) here.

Suitable technical modifications lead to an analogous result with the Cauchy problem for (1) substituted by the following initial-boundary value problem.

Theorem 2. *Let v as in **(V)** and $T > 0$. Fix an initial data $\bar\rho \in \mathbf{L}^\infty([0,L])$ and a boundary data $\tilde\rho \in \mathbf{L}^\infty([0,T])$ given. Let ρ: $[0,+\infty[\times[0,L] \mapsto \mathbb{R}$ solve the initial - boundary value problem*

$$\begin{cases} \partial_t\rho + \partial_x\left(\rho\cdot v(\rho)\right) = 0 \\ \rho(0,x) = \bar\rho(x) \\ \rho(\bar t,0) = \tilde\rho(t) \end{cases} \tag{7}$$

Assume the driver moves according to

$$\begin{cases} \dot p = w\left(\rho(t,p)\right) \\ p(t_0)) = 0 \end{cases} \tag{8}$$

with w as in (4). Then, (8) admits a unique solution p. If $p_1(\cdot)$ and $p_2(\cdot)$ solve (2), starting respectively at times $\bar t_1$, $\bar t_2$, then we have the Hölder dependence

$$\left|p_1(t) - p_2(t)\right| \leq C \cdot \left|\bar{t}_1 - \bar{t}_2\right|^\alpha , \tag{9}$$

for some $C > 0$, $t \in [0,T]$ and α as in (4).

The setting of Theorem 2 seems more suitable to traffic flow problem, where a standard situation is the assignment of the car inflow at a point for all t , rather than the distribution of cars over all the road at a given time.

3 Sketch of the Proofs

Since $\rho(t,\cdot) \in \mathbf{L}^\infty$, we can assume it to be right continuous. Following [2, 6], the core of the proof of Theorem 1 and 2 is the construction of a suitable approximation ρ_ε of the solution ρ to (1) such that

(P1) ρ_ε is piecewise constant with jumps along Lipschitz continuous time–like polygonal lines in (t,x)–plane;

(P2) if $x = \psi(t)$ is a polygonal line along which ρ_ε suffers a discontinuity, then there holds

$$\dot{\psi}(t) = \begin{cases} \dfrac{f\left[\rho(t,\phi(t))\right] - f\left[\rho(t,\phi(t)-)\right]}{\rho(t,\phi(t)) - \rho(t,\phi(t)-)} + \mathcal{O}(\varepsilon) & \text{if } \rho(t,\phi(t)) \neq \rho(t,\phi(t)-)\,, \\ f'\left[\rho(t,\phi(t))\right] + \mathcal{O}(\varepsilon) & \text{if } \rho(t,\phi(t)) = \rho(t,\phi(t)-)\,. \end{cases}$$

for some $x = \varphi(t)$ such that $\|\psi(t) - \varphi(t)\|_\infty = \mathcal{O}(\varepsilon)$.

(P3) there holds

$$\rho_\varepsilon(t,x) \in B\big(\rho(B((t,x),\varepsilon)),\varepsilon\big)\,.$$

As a second step, we consider the ordinary differential equation

$$\dot{p}^\epsilon(t) = w\big(\rho^\epsilon(t,p(t))\big)\,, \tag{10}$$

and get an estimate of the Hölder continuous dependence of the Filippov solutions to (10) on the initial data with constants independent on ϵ. To this aim, consider a reference trajectory $p^\epsilon(\cdot)$ of (10) with initial datum $\bar{p}$. Following [2], we introduce the tangent vector

$$\xi(t) = \lim_{h \to 0} \frac{p_h^\epsilon(t) - p^\epsilon(t)}{h}\,,$$

where $p_h^\epsilon(\cdot)$ is the solution to (10) with initial datum $\bar{p} + h$. Carrying out a careful analysis of the evolution of the tangent vector $\xi(\cdot)$ across discontinuities of ρ^ϵ, we get that there exist $C > 0$ independent on ϵ such that

$$\frac{\xi(t)}{\xi(s)} \leq C \left(\frac{t}{s}\right)^{1-\alpha}\,, \qquad 0 < s < t\,, \tag{11}$$

holds. From (11) we obtain that, if $p_1^\epsilon(\cdot)$ and $p_2^\epsilon(\cdot)$ are solutions to (10) with initial data respectively $\bar{p}_1$ and $\bar{p}_2$, then there holds

$$|p_1^\epsilon(t) - p_2^\epsilon(t)| \leq \begin{cases} (1+\sup w - \inf w) C t^{1-\alpha} |\bar{p}_1 - \bar{p}_2|^\alpha & \text{if } |\bar{p}_1 - \bar{p}_2| \leq t\,, \\ (1+\sup w - \inf w)|\bar{p}_1 - \bar{p}_2| & \text{otherwise,} \end{cases}$$

with all the constants above independent on ϵ. Letting $\epsilon \to 0$ proves the existence of a solution. To get uniqueness and Hölder continuous dependence it is essential to verify that all the solutions to (2) are obtained as limit of solutions to (10).

Regarding the third claim of Theorem 1, observe that if ρ_∞ is a $\mathbf{L}^\infty$ solution to (1) with initial datum $\bar{\rho} = \bar{\rho}_\infty$, then for all positive ε there exists $\delta > 0$ such that for any fix $(\bar{t}, \bar{x})$ in the (t,x)–plane there exists a set $\mathcal{A} \subseteq B_\delta(\bar{t}, \bar{x})$ such that $meas(\mathcal{A}) \geq \delta^3$ and

$$|\rho_\infty(\bar{t}, \bar{x}) - \rho_\infty(t,x)| < \varepsilon\,, \quad \forall (t,x) \in \mathcal{A}\,. \tag{12}$$

Let ρ_n be the solution to (1) with initial datum $\bar{\rho} = \bar{\rho}_n$. Combining Egoroff's Theorem with (12), we get that, up to a subsequence, there holds

$$\rho_n(t,x) \in B\big(\rho_\infty\big(B((t,x),\varepsilon)\big),\varepsilon\big)$$

for large n. Then the sequence of solutions $\{p_n(\cdot)\}_{n\in\mathbb{N}}$ to (2) with $\rho = \rho_n$ converges uniformly (up to a subsequence) to the solution p_∞ corresponding to ρ_∞. Due to the uniqueness of p_∞, we deduce that the entire sequence $\{p_n(\cdot)\}_{n\in\mathbb{N}}$ converges to p_∞ in the $\mathbf{C}^0$ norm.

All the details are collected in [6].

References

1. A. Bressan (1988): Unique solutions for a class of discontinuous differential equations. Proc. Amer. Math. Soc., **104**, 772–778
2. A. Bressan and W. Shen (1998): Uniqueness for discontinuous O.D.E. and conservation laws. Nonlinear Anal., **34**, 637–652
3. A. Bressan and W. Shen (1998): On Discontinuous Differential Equations, in Differential Inclusions and Optimal Control, J.Andres, L.Gorniewicz and P.Nistri editors, J. Schauder Center Lecture Notes in Nonlinear Analysis, 73–87
4. R.M. Colombo Hyperbolic Phase Transitions in Traffic Flow. SIAM J. Appl. Math., to appear
5. R.M. Colombo and A. Groli (2002): On the Optimization of a Conservation Law. Calc. Var. Partial Differential Equations, to appear
6. R.M. Colombo and A. Marson (2002): A Hölder Continuous ODE Related to Traffic Flow. Proc. Roy. Soc. Edinburgh Sect. A, to appear
7. C.F. Daganzo (1995): Requiem for high-order fluid approximations of traffic flow. Trans. Res., **29B**, 4, 277–287
8. A.F. Filippov (1973): Differential equations with discontinuous right hand sides. Kluwer Academic Publishers Group, Dordrecht
9. M. Lighthill and G. Whitham (1955): On kinematic waves. II. A theory of traffic flow on long crowded roads. Proc. Roy. Soc. London. Ser. A., **229**, 317–345
10. A. Marson (2002): Non Laxian conservation laws and O.D.Es., in preparation
11. P. Richards (1956): Shock waves on the highway. Operations Res., **4**, 42–51

Phase Transitions and Chapman-Jouguet Combustions

Rinaldo M. Colombo[1] and Andrea Corli[2]

[1] Department of Mathematics, University of Brescia, Via Valotti 9, 25133 Brescia, Italy *rinaldo@ing.unibs.it*
[2] Department of Mathematics, University of Ferrara, Via Machiavelli 35, 44100 Ferrara, Italy *crl@unife.it*

1 Introduction

We consider a system of conservation laws of the form

$$\partial_t u + \partial_x f(u) = 0 \ , \tag{1}$$

where $t \in [0, +\infty[$, $x \in \mathbf{R}$, $u \in \Omega \subseteq \mathbf{R}^n$ and the function $f\colon \Omega \mapsto \mathbf{R}^n$ is smooth. Ω consists of two connected components, called here *phases*:

$$\Omega = \Omega_0 \cup \Omega_1, \quad \Omega_0 \cap \Omega_1 = \emptyset, \quad \Omega_0 \neq \emptyset, \ \Omega_1 \neq \emptyset \,. \tag{2}$$

The system (1) is strictly hyperbolic in Ω and each eigenvalue is either genuinely nonlinear or linearly degenerate.

System (1) under assumption (2) can be used to model non-stationary two-phase flows. For instance, phase transitions in a real fluid, Ω_1 being the vapor phase and Ω_0 the liquid phase; phase transitions in elastodynamics, with Ω_1 and Ω_0 being the hard and soft phases; chemical reactions as combustions, where Ω_0 is the unburnt gas and Ω_1 is the burnt gas.

More precisely, phase transitions are solutions to (1) assuming values in both phases. Since the phases are disconnected, these solutions are discontinuous, and must then be intended as weak solutions. In the $t - x$ plane the discontinuity line $x = \Lambda(t)$ separating states in different phases is called a phase boundary.

In this paper we overview some recent and new results related to the long time stability of a piecewise constant solution $\underline{u}$ with respect to $\mathbf{BV}$-small perturbations, where $\underline{u}$ attains the value $\underline{u}^{\pm}$ according to $\pm(x - \underline{\Lambda} t) > 0$, $\underline{u}^{\pm}$ being constant states in different phases. This problem needs entirely different treatments and techniques depending on the number of characteristics diverging from the unperturbed phase boundary $x = \underline{\Lambda} t$, as it is briefly described here below.

(i) If the number of characteristics diverging from the phase boundary is $n - 1$, then small perturbations of the sonic initial data $(\underline{u}^l, \underline{u}^r)$ may lead either to a well-posed or to a 1-underdetermined Riemann problem for (1). In the case of reacting gases an example is provided by Chapman-Jouguet detonations.

(ii) If n characteristics diverge from the phase boundary then a small perturbation of the initial data may lead either to a 1-underdetermined problem or to a 2-underdetermined Riemann problem. An example of this situation is given by Chapman-Jouguet deflagrations.

Case (i) was studied in [7]. Under a suitable stability condition, there exists a unique solution to perturbed Riemann problems. In spite of the fact that eigenvalues are required not to change type, the solutions may contain compound waves (rarefactions attached to a sonic phase boundary), as in non convex system of conservation laws. It is of interest to note that the under determinacy is solved introducing a constraint in the *structure* of the solution, without asking for any other admissibility condition. This case is discussed in Section 3.

In (ii), recently tackled in [8], the Riemann problem is more severely underdetermined than above and a further admissibility condition needs to be imposed. Differently form the usual habits, we do not impose any condition motivated by physical *a priori* condition. On the contrary, we introduce a fairly general condition and impose only those assumptions that lead, *a posteriori*, to a well-posed problem. More precisely, we require the presence of a smooth function $\Psi: (\Omega_1 \times \Omega_0) \cup (\Omega_0 \times \Omega_1) \mapsto \mathbf{R}$ and say that the phase transition between u^l and u^r is *Ψ-admissible* if $\Psi(u^l, u^r) = 0$ holds, besides the usual Rankine-Hugoniot conditions. This criterion generalizes both the *kinetic* criterion of [1] and the *visco-capillarity* criterion of [13]. The Riemann solver so introduced provides unique solutions where compound waves may well appear. To obtain the well posedness of the Riemann problem we need both a compatibility condition and a stability assumption relating Ψ and f. This case is discussed in Section 4.

In both (i) and (ii), we exploit the wave-front tracking technique to prove the global existence of solutions to the Cauchy problem for (1) with data having suitable small **BV** norm. A sketch of the proofs is given in Section 5, while in Section 6 we apply the general results to the case of reacting gas dynamics.

2 Notations and Definitions

We consider the system of conservation laws (1). The function $f: \Omega \mapsto \mathbf{R}^n$ is of class $\mathbf{C}^3$ and satisfies (2). Moreover, we require:

(1)(i) $Df(u)$ is strictly hyperbolic, i.e. it has n real distinct eigenvalues, and its eigenvalues are either genuinely nonlinear or linearly degenerate;
(ii) for some k the k-characteristic family is genuinely nonlinear in Ω_1.

We denote by $\lambda_i(u)$ and $r_i(u)$, respectively, the i-th eigenvalue and the i-th right eigenvector of $Df(u)$, for $i = 1, \dots, n$ and $u \in \Omega$. Indexes are chosen so that $\lambda_{i-1}(u) < \lambda_i(u)$ for all u in Ω. Assumption **(1)**(ii) is essential in the following, since it allows the presence of rarefactions of the k family. If the k family is linearly degenerate we refer to [5, 9, 11].

Let $u: [0, +\infty[\times \mathbf{R} \mapsto \Omega$ be a weak solution to (1) such that $u(t, \cdot) \in \mathbf{BV}$ for all t. We say that a Lipschitz-continuous curve $x = \Lambda(t)$ is a *phase boundary* if for a.e. t the traces $u(t, \Lambda(t)\pm)$ are in different phases. A phase boundary $x = \Lambda(t)$ is *sonic on the left* at time t with respect to the k-th characteristic field if $\Lambda(t) = \lambda_k(u(t, \Lambda(t)-))$.

Fix now two constant states $\underline{u}^l \in \Omega_1$ and $\underline{u}^r \in \Omega_0$ and assume that the Riemann problem

$$\partial_t u + \partial_x [f(u)] = 0 \qquad u(0,x) = \begin{cases} \underline{u}^l & \text{if } x < 0 \\ \underline{u}^r & \text{if } x > 0 \end{cases} \tag{3}$$

has the solution

$$\underline{u}(t,x) = \begin{cases} \underline{u}^l & \text{if } x < \underline{\Lambda} t \\ \underline{u}^r & \text{if } x > \underline{\Lambda} t \, . \end{cases} \tag{4}$$

In the following, according to the number of characteristics diverging from the phase boundary, we shall introduce two different Riemann solvers for the *"perturbed"* Riemann problem

$$\partial_t u + \partial_x [f(u)] = 0 \qquad u(0,x) = \begin{cases} u^l & \text{if } x < 0 \\ u^r & \text{if } x > 0 \end{cases} \tag{5}$$

where $u^l \in \Omega_1$, $u^r \in \Omega_0$ are two constant states close to $\underline{u}^l$, $\underline{u}^r$ respectively. Once the Riemann solvers are defined, we consider the Cauchy problem

$$\partial_t u + \partial_x [f(u)] = 0 \qquad u(0,x) = \begin{cases} u^l + u_o(x) & \text{if } x < 0 \\ u^r + u_o(x) & \text{if } x > 0 \end{cases} \tag{6}$$

where the $u_o: \mathbf{R} \mapsto \Omega$ satisfies $\|u_o\|_{\mathbf{L}^\infty} + \mathrm{TV}(u_o) < \delta$ for δ sufficiently small.

3 The case of $n-1$ diverging characteristics

In this section we assume that

(2) the unperturbed phase transition (4) satisfies

$$\lambda_k(\underline{u}^r) < \lambda_k(\underline{u}^l) = \underline{\Lambda} < \lambda_{k+1}(\underline{u}^r) \, .$$

The phase boundary $x = \underline{\Lambda} t$ is thus sonic on the left with respect to the k characteristic field and the number of diverging characteristics is $n-1$. Perturbing slightly this phase boundary we see that the three cases shown in Figure 1 need to be considered.

While in the subsonic case the pattern of the characteristics is exactly as in the case of a k-shock wave, in the supersonic case the number of outgoing characteristics is n. Therefore the Riemann problem for states close to a

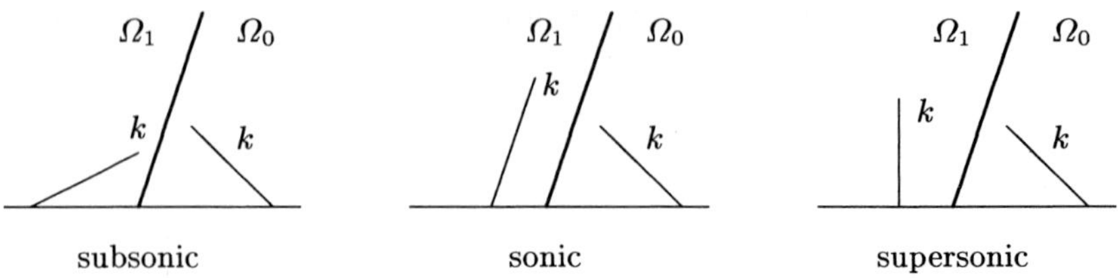

Fig. 1. Perturbation of a k-sonic phase boundary satisfying **(2)**. Here subsonic, sonic, supersonic refer to the phase boundary (thick line) with respect the the k-characteristics (thin lines) on the left.

supersonic phase boundary is 1-underdetermined, if we admit the maximal number of outgoing waves. To avoid this under determinacy we introduce the following Riemann solver.

Definition 1. *Assume that* **(1)**, **(2)** *hold and consider two states* $u^l \in \Omega_1$, $u^r \in \Omega_0$ *close to* $\underline{u}^l$, $\underline{u}^r$ *respectively. An* admissible solution *to the Riemann problem* (5) *is a function consisting (from the left to the right) of* $k-1$ *Lax waves in* Ω_1*, a phase boundary (with possibly a* k*-rarefaction in* Ω_1 *attached to its left), and* $(n-k)$ *Lax waves in* Ω_0.

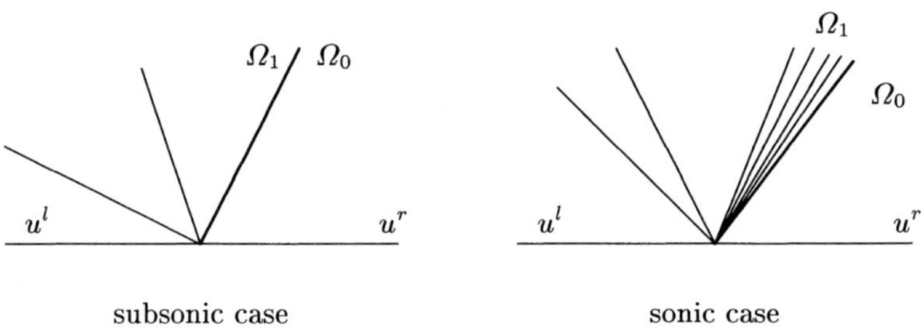

Fig. 2. Solution of the Riemann problem for $n = 3$, $k = 3$, under assumption **(2)**.

The phase boundary that is selected according to the definition above is then either subsonic or sonic on the left with respect to the k-th characteristic field, see Figure 2. In the former case, no k-wave is present in the solution. In the latter case, a k-rarefaction and the phase boundary constitute a compound wave, similarly to what happens in non-genuinely nonlinear systems of conservation laws, see [12].

To uniquely solve the Riemann problem we require

(3) the solution $\underline{u}$ satisfies

$$\det\left[r_1(\underline{u}^l), \ldots, r_{k-1}(\underline{u}^l), \underline{u}^r - \underline{u}^l, r_{k+1}(\underline{u}^r), \ldots, r_n(\underline{u}^r)\right] \neq 0 \, .$$

This condition coincides with Majda's stability condition for a large k-shock wave. An application of the implicit function theorem leads to:

Proposition 1. *Let assumptions* **(1)** *and* **(2)** *hold. If the Riemann problem* (3) *admits the solution* (4) *satisfying* **(3)**, *then for all* u^l, u^r *in suitable neighborhoods of* $\underline{u}^l$ *and* $\underline{u}^r$ *the Riemann problem* (5) *admits a unique admissible solution in the sense of Definition* 1.

Consider now the Cauchy problem (6).

Theorem 1. *Assume that* **(1)**, **(2)** *and* **(3)** *hold. If* u_o *satisfies*

$$\|u_o\|_{\mathbf{L}^\infty} + \mathrm{TV}(u_o) \leq \delta \tag{7}$$

for some sufficiently small δ, *then the Cauchy problem* (6) *has a global solution* $u \in \mathbf{BV}_{\mathrm{loc}}\left(]0,+\infty[\, \times \mathbf{R}\right)$. *For all times, the solution* u *has a Lipschitz-continuous phase boundary* $x = \Lambda(t)$ *which is either left-sonic or left-subsonic. In the region* $x < \Lambda(t)$ *(resp.* $x > \Lambda(t)$*)* u *is a weak entropic solution to* (1) *and* $u(t,x) \in \Omega_1$ *(resp.* $u(t,x) \in \Omega_0$*). Moreover*

$$\mathrm{TV}\left(u(t,\cdot);]-\infty,\Lambda(t)[\right) + \mathrm{TV}\left(\dot{\Lambda}\right) + \mathrm{TV}\left(u(t,\cdot);]\Lambda(t),+\infty[\right) = \mathcal{O}(1)\cdot\delta\,. \tag{8}$$

Above, $\mathrm{TV}(u;I)$ denotes the total variation of the function u over the (space) interval I, while $\mathrm{TV}(\dot{\Lambda})$ is the total variation of Λ over the whole (time) interval $[0,+\infty[$.

We refer to [7] for the full proof, while a sketch is given in Section 5.

4 The case of n diverging characteristics

In this section we keep assumption **(1)** but instead of **(2)** we assume:

(2') the unperturbed phase transition (4) satisfies

$$\lambda_{k-1}(\underline{u}^r) < \lambda_k(\underline{u}^l) = \underline{\Lambda} < \lambda_k(\underline{u}^r)\,.$$

Again, the phase boundary $x = \underline{\Lambda}t$ is sonic on the left with respect to the k characteristic field; the number of characteristics diverging from it is however n. As in the previous case we see that small perturbations of this phase boundary lead to the three cases shown in Figure 3.

The Riemann problem is underdetermined (if we admit the maximal number of outgoing waves) both for states close to a subsonic phase boundary and for those close to a supersonic one. More precisely, the Riemann problem is 1-underdetermined in the former case and 2-underdetermined in the latter. We solve this problem both by means of an exterior function, reducing the under determinacy degree by 1, and prescribing a structure to the solutions analogous to the one in previous case.

We consider then a smooth function $\Psi\colon (\Omega_1 \times \Omega_0) \cup (\Omega_0 \times \Omega_1) \mapsto \mathbf{R}$ and say that the phase transition between u^l and u^r is Ψ*-admissible* if

$$\Psi(u^l, u^r) = 0. \tag{9}$$

The Riemann solver is defined as follows.

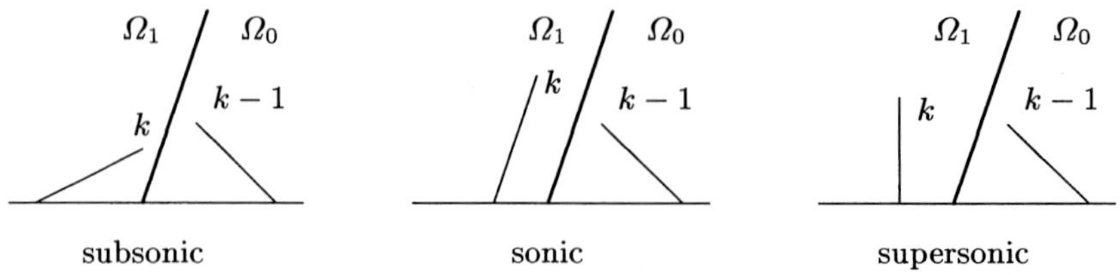

Fig. 3. Perturbation of a k-sonic phase boundary satisfying **(3')**. Subsonic, sonic, supersonic refer to the phase boundary and to the k-characteristic on the left.

Definition 2. *Assume* **(1)**, **(2')** *and consider two states* $u^l \in \Omega_1$, $u^r \in \Omega_0$ *close to* $\underline{u}^l$, $\underline{u}^r$ *respectively. A* Ψ-admissible solution *to the Riemann problem* (5) *is a function consisting (from left to right) of* $k-1$ *Lax waves in* Ω_1, *a* Ψ*-admissible phase boundary (with possibly a* k*-rarefaction in* Ω_1 *attached to its left), and* $n-k+1$ *Lax waves in* Ω_0.

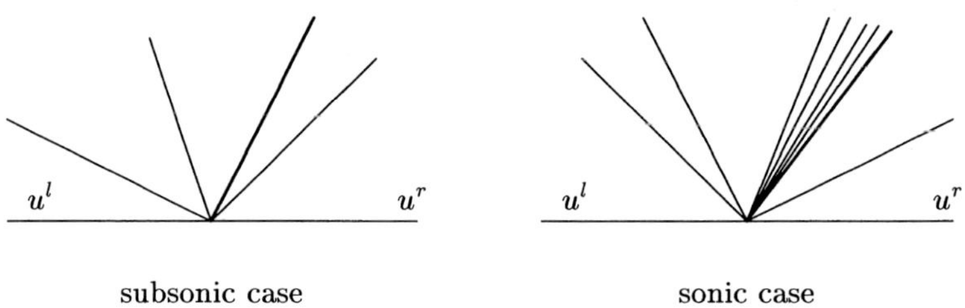

Fig. 4. Solution of the Riemann problem for $n = 3$, $k = 3$, under **(2')**. The phase boundary is Ψ-admissible.

Once again we consider in the solution only phase boundaries that are either subsonic or sonic on the left with respect to the k-th characteristic field, see Figure 4. Supersonic phase boundaries are ruled out. Compound waves may well arise in the solution.

For the well posedness of the Riemann problem (5) we need both a compatibility condition and a stability assumption relating Ψ and f:

(3')(i) if u^l and u^r are close to $\underline{u}^l$, $\underline{u}^r$ and are connected by a k-left-sonic phase boundary, then $D_1\Psi(u^l, u^r)r_k(u^l) = 0$.

(ii) the quantity $\Delta := D_2\Psi(\underline{u}^l, \underline{u}^r)\cdot\left(Df(\underline{u}^r) - \underline{\Lambda}\,\mathrm{Id}\right)^{-1}[\underline{u}]$ does not vanish;

(iii) the matrix

$$\left(r_1(\underline{u}^l) - \alpha_1[\underline{u}], \quad \ldots, \quad r_{k-1}(\underline{u}^l) - \alpha_{k-1}[\underline{u}], \quad r_k(\underline{u}^r), \quad \ldots, \quad r_n(\underline{u}^r)\right)$$

is nonsingular, where the numbers α_i, $i = 1, \ldots, k-1$, are defined by

$$\alpha_i = \frac{1}{\Delta}\left(D_2\Psi(\underline{u}^l,\underline{u}^r)\left(Df(\underline{u}^r) - \underline{\Lambda}\,\mathrm{Id}\right)^{-1} + \frac{D_1\Psi(\underline{u}^l,\underline{u}^r)}{\lambda_i(\underline{u}^l) - \underline{\Lambda}}\right) r_i(\underline{u}^l)\,.$$

Condition **(3')**(i) is the compatibility condition. Roughly speaking, it means that the function Ψ must be compatible with the sonic structure of the problem. We refer to Section 5 for a discussion.

Conditions **(3')**(ii) and (iii) are the stability condition. This assumption is somewhat intricate because of the presence of the exterior function Ψ, which *"breaks"* the geometric properties of the convective part. The first stability condition, namely $\Delta \neq 0$, assures the existence of a Ψ-admissible Hugoniot curve of phase boundaries: any state u^l sufficiently close to $\underline{u}^l$ can be connected to some other state u^r (close to $\underline{u}^r$) by a Ψ-admissible phase boundary. Remark that the invertibility of the matrix $Df(\underline{u}^r) - \lambda_k(\underline{u}^l)\,\mathrm{Id}$ holds because the unperturbed phase boundary is not characteristic on the right-hand side. The second stability condition is exploited in the implicit function theorem to prove the following result.

Proposition 2. *Let* **(1)**, **(2')** *hold. If the Riemann problem* (3) *admits the solution* (4) *satisfying* **(3')**, *then for all* u^l, u^r *in suitable neighborhoods of* $\underline{u}^l$ *and* $\underline{u}^r$, *the Riemann problem* (5) *admits a unique* Ψ*-admissible solution in the sense of Definition* 2.

The result on the Cauchy problem (6) now follows.

Theorem 2. *Assume that* **(1)** *and* **(2')** *hold. Assume moreover that the Riemann problem* (3) *admits the weak solution* (4) *consisting of a* Ψ*-admissible phase boundary and satisfying* **(3')**. *If* u_o *satisfies* (7) *for a sufficiently small* δ, *then there exists a global solution* $u\colon [0,+\infty[\ \mapsto \mathbf{BV}(\mathbf{R})$ *to the Cauchy problem* (6).

For all times, the solution u *has a Lipschitz-continuous* Ψ*-admissible phase boundary* $x = \Lambda(t)$ *which is either sonic or subsonic on the left for the* k*-characteristic family. In the region* $x < \Lambda(t)$ *(resp.* $x > \Lambda(t)$*)* u *is a weak entropic solution to* (1) *and* $u(t,x) \in \Omega_1$ *(resp.* $u(t,x) \in \Omega_0$*). The traces* $u(t,\Lambda(t)\pm)$ *satisfy* (9), *i.e.*

$$\Psi\left(u(t,\Lambda(t)-), u(t,\Lambda(t)+)\right) = 0$$

for a.e. t. *Moreover* (8) *still holds.*

5 Sketch of the proofs and comments

The proofs of both Theorems 1 and 2 exploit a suitable wave-front tracking algorithm [4] and are thus obtained through the construction of piecewise constant approximate solutions with uniformly bounded total variation. Several issues deserve however some comments that we collect in this section.

First, a consequence of the fact that the phase boundary is sonic on the left-hand side is that the Hugoniot function $H(u^l, \Lambda)$, providing the right state that can be connected to the left state u^l by a phase boundary of speed Λ, has a kernel since $H\big(u^l, \lambda_k(u^l)\big) r_k(u^l) = 0$. In the case considered in Section 4, condition **(3')**(i) can then be reformulated as being the analog of this condition for the function Ψ.

As already mentioned we consider the phase boundary and the rarefaction possibly attached to it as a single compound wave. By means of the Hugoniot function H, this compound wave is described by a generalized Lax function. However, its smoothness is in general merely $\mathbf{C}^{1,1}$ and not $\mathbf{C}^2$ as it happens for non convex Lax curves.

In the algorithm it is convenient to assign to each rarefaction wavelet (the small waves in the fans approximating continuous rarefactions) the speed of the state to its left. As a consequence, approximate sonic phase boundaries are assigned null size and propagate with the exact speed. A delicate item is the control of almost tangential interactions of small k-waves on the left with the phase boundary, where second order estimates are needed. As a consequence, the Glimm functionals controlling the total variation of the approximate solutions must be modified and suitable weights introduced. At last, a tricky point consists in checking that the Rankine-Hugoniot conditions (and the condition (9)) hold in the limit.

We close this section with some general remarks. The introduction of an exterior function Ψ in order to solve underdetermined Riemann problems was proposed independently in [1] and [15], as a further physical information due to the dissipation of entropy. This idea was then generalized in [6]. In the next section we make use of an exterior function which is by no means related to the entropy. This is the first time that the approach [1, 15] is exploited without reference to entropy.

The case of two or more phase boundaries, either subsonic or sonic, can be dealt with the techniques employed in [6].

The reader acquainted with the semigroup techniques developed in [4] should have noticed that continuous dependence on the data is missing in the previous results. This seems a very difficult issue; in fact, as we emphasized above, the presence of compound waves makes our problems very akin to non convex ones, where, to the knowledge of the authors, only the 2×2 case is solved, see [2]. The recent result [3] on the vanishing viscosity limit is here not applicable, due to the lack of connectedness of Ω.

6 An application to reacting gas dynamics

In Eulerian coordinates the gas dynamics equations for reacting gases are

$$\begin{cases} \rho_t + (\rho v)_x = 0 \\ (\rho v)_t + (\rho v^2 + p)_x = 0 \\ E_t + (Ev + pv)_x = 0 \end{cases} \tag{10}$$

where ρ is the density, v the velocity, $m = \rho v$ the momentum, p the pressure, $E = \rho\epsilon + \rho v^2/2$ the total energy, ϵ the total internal energy, see [10] for more details on this model. We assume that

$$\epsilon = \begin{cases} \epsilon_1 = e & \text{for burnt gas} \\ \epsilon_0 = e + q & \text{for unburnt gas} \end{cases}$$

where $e = \frac{1}{\gamma-1}\frac{p}{\rho}$ is the internal non chemical energy. The reaction is assumed exothermic, so that $q > 0$. The temperature is $T = (\gamma - 1)e$.

At time $t = 0$, burnt gas covers the half-line $x < 0$, while unburnt gas fills $x > 0$. For $t > 0$ there is a discontinuity curve in the $t - x$ plane separating the unburnt gas from the burnt gas. This curve, the reaction front, is what we called above a phase boundary. We use the subscript 1 for quantities related to burnt gas, and 0 for those of the unburnt gas.

Fix on the left a burnt state $\underline{u}_1 = (\underline{\tau}_1, \underline{v}_1, \underline{\epsilon}_1)$ and an unburnt state $\underline{u}_0 = (\underline{\tau}_0, \underline{v}_0, \underline{\epsilon}_0)$ on the right. The jump conditions are assumed to hold for $\underline{u}_1$ and $\underline{u}_0$. We assume that the reaction front is right-moving, so that $k = 3$. Then, assumption **(1)** holds. Let Ω_1 be a neighborhood of $\underline{u}_1$ and Ω_0 a neighborhood of $\underline{u}_0$, with $\Omega_1 \cap \Omega_0 = \emptyset$.

We deal first with the case of Section 3. In this case we choose $\underline{u}_1$ to be the Chapman-Jouguet detonation point related to $\underline{u}_0$; then **(2)** is satisfied with $n = k = 3$. In this example subsonic phase boundaries are simply strong detonations, whereas supersonic ones are weak detonations, which, for this model, are ruled out by physical considerations. This motivates the choice of our Riemann solver. A simple computation shows that also **(3)** holds, and then by Theorem 1 applies. Therefore we obtain the global existence of solutions to (10) for initial data close to a Chapman-Jouguet detonation. We refer to [7] for more details and other examples in phase transitions.

We consider now the case of Section 4. We take $\underline{u}_1$ to be the Chapman-Jouguet deflagration point related to $\underline{u}_0$; then **(2')** is satisfied ($n = k = 3$). In this case subsonic phase boundaries are strong deflagrations, while supersonic ones are weak deflagrations, which are ruled out by physical considerations. The choice of the exterior function Ψ can be made in different ways, see [8]; we consider here the approach proposed in [14]. In that paper it is assumed that the reaction front for deflagrations moves at the speed Λ given by

$$\Lambda = v_0 + KT_0^Q \tag{11}$$

for $K > 0$ and $Q \geq 1/2$. For simplicity we choose a reference frame where $v_0 = 0$ and consider only the case of *laminar* flames, i.e., $Q = 1/2$. Condition (11) can written under the form (9) by defining the function Ψ as

$$\Psi(u_1, u_0) = v_0 + KT_0^{1/2} - \Lambda = \frac{m_0}{\rho_0} + M\sqrt{\gamma(\gamma-1)}\sqrt{\frac{E_0}{\rho_0} - \frac{1}{2}\frac{m_0^2}{\rho_0^2} - q} - \Lambda \tag{12}$$

where $M = K/\sqrt{\gamma}$ is the Mach number with respect to the unburnt state. The speed Λ is a function of the states u_0, u_1 by means of the jump conditions.

We assume that the phase boundary of speed $\underline{\Lambda} = \underline{v}_1 + \underline{c}_1$ separating $\underline{u}_0$ from $\underline{u}_1$ satisfies $\Psi(\underline{u}_1, \underline{u}_0) = 0$, so that $\underline{v}_1 + \underline{c}_1 = M\underline{c}_0$.

Proposition 3. *Under the previous assumptions on* (10), *condition* **(3')***(i) holds for every* γ, *while* **(3')***(ii) holds for* $\gamma \in]1, 3]$.

We could not verify whether **(3')**(iii) holds or not. However we found a simple form for it. We refer to [8] for more details.

References

1. R. Abeyaratne and J. K. Knowles. Kinetic relations and the propagation of phase boundaries in solids. *Arch. Rational Mech. Anal.*, 114(2):119–154, 1991.
2. F. Ancona and A. Marson. A wavefront tracking algorithm for $N \times N$ nongenuinely nonlinear conservation laws. *J. Differential Equations*, 177(2):454–493, 2001.
3. S. Bianchini and A. Bressan. Vanishing viscosity solutions of nonlinear hyperbolic systems. Preprint.
4. A. Bressan. *Hyperbolic systems of conservation laws*. Oxford University Press, Oxford, 2000. The one-dimensional Cauchy problem.
5. A. Bressan and R. M. Colombo. Unique solutions of 2×2 conservation laws with large data. *Indiana Univ. Math. J.*, 44(3):677–725, 1995.
6. R. M. Colombo and A. Corli. Continuous dependence in conservation laws with phase transitions. *SIAM J. Math. Anal.*, 31(1):34–62 (electronic), 1999.
7. R. M. Colombo and A. Corli. Sonic hyperbolic phase transitions and Chapman-Jouguet detonations. *J. Differential Equations*, to appear, 2001.
8. R. M. Colombo and A. Corli. Sonic and kinetic phase transitions with applications to Chapman-Jouguet deflagrations. *Submitted*, 2002.
9. A. Corli and M. Sablé-Tougeron. Stability of contact discontinuities under perturbations of bounded variation. *Rend. Sem. Mat. Univ. Padova*, 97:35–60, 1997.
10. E. Godlewski and P.-A. Raviart. *Numerical approximation of hyperbolic systems of conservation laws*. Springer-Verlag, New York, 1996.
11. P. LeFloch. Propagating phase boundaries: formulation of the problem and existence via the Glimm method. *Arch. Rational Mech. Anal.*, 123(2):153–197, 1993.
12. T. P. Liu. The Riemann problem for general systems of conservation laws. *J. Differential Equations*, 18:218–234, 1975.
13. M. Slemrod. Admissibility criteria for propagating phase boundaries in a van der Waals fluid. *Arch. Rational Mech. Anal.*, 81(4):301–315, 1983.
14. Z. H. Teng, A. J. Chorin, and T. P. Liu. Riemann problems for reacting gas, with applications to transition. *SIAM J. Appl. Math.*, 42(5):964–981, 1982.
15. L. M. Truskinovskii. Dynamics of nonequilibrium phase boundaries in a heat conducting non-linearly lastic medium. *Prikl. Mat. Mekh.*, 51(6):1009–1019, 1987.

Weak Stability of Multidimensional Shocks

Jean-François Coulombel

ENS Lyon, 46 allée d'Italie, 69364 LYON CEDEX 07, France
jfcoulom@umpa.ens-lyon.fr

In [8], A. Majda studied the linear stability of multidimensional shock waves for general systems of conservation laws. His analysis relied on two main assumptions: the shock wave was assumed to satisfy the so-called uniform stability condition (we shall recall it in the sequel of this paper) and the system was assumed to satisfy a block structure condition. The linear study performed in [8] enabled Majda to prove the (local) existence of multidimensional shock waves in this context, see [7, 9, 14] for an overview. Up to our knowledge, the study of multidimensional shock waves has known little progress since these oustanding breakthroughs. Let us mention however two significant results. To carry out the study of the variable coefficients linear systems in [8], Majda used an H^s version of pseudodifferential calculus (essentially based on Moser's inequalities). Another approach for this type of problems is the paradifferential calculus of Bony and Meyer [2, 12]. Using these techniques, Mokrane precised the regularity assumptions under which Majda's theorems still hold, see [13] and [11]. More recently, Métivier has shown in [10] that the block structure condition defined by Majda in [8] was met by all symmetric hyperbolic systems with constant multiplicity. Whether the block structure condition is met by the magnetohydrodynamics equations (or by other nonconstant multiplicity systems) is still an open (but interesting) question.

In this paper, we show how weaker assumptions than Majda's uniform stability condition still enable to derive an a priori estimate on the linearized equations. Our approach is very close to Majda's analysis but takes the neutrally unstable modes into account through some modifications in the usual Kreiss' symmetrizers construction (see [6] and [3] for this construction). Examples of shock waves for which our analysis applies arise in isentropic gas dynamics and in phase transitions for Van der Waals type fluids, see [1, 4]. Shock waves for a multidimensional scalar conservation law provide us with another example of weakly stable shock waves, but this case can be treated completely with simpler arguments, see [9, 14].

We begin this paper with a rather general description of the problem and then recall Majda's main assumptions. We shall then detail our "weak stability condition" and state our main result. The proof will be briefly sketched.

1 The linear stability of shock waves: a quick overview

We consider a general system of N conservation laws ($N \geq 2$) in time-space $\mathbb{R} \times \mathbb{R}^d$ ($d \geq 2$):

$$\sum_{j=0}^{d} \partial_j \, f_j(u) = 0 \,, \tag{1}$$

where x_0 is the time variable, also denoted by t in the sequel, $(x_1, \dots, x_d)$ is the space variable. The fluxes $f_0, \dots, f_d$ are C^∞ functions defined on an open set U of $\mathbb{R}^N$ with values in $\mathbb{R}^N$. The jacobian matrix of f_j at point u will be denoted by $A_j(u)$. We make the following hyperbolicity assumption:

Assumption 1 *If we define a matrix $A(u,\xi)$ by*

$$\forall \, u \in U \,, \quad \forall \, \xi \in \mathbb{R}^d \,, \quad A(u,\xi) := \sum_{j=1}^{d} \xi_j \, A_0(u)^{-1} \, A_j(u) \,, \tag{2}$$

then $A(u,\xi)$ is diagonalizable over $\mathbb{R}$ with C^∞ eigenvalues $\lambda_1, \dots, \lambda_q$ (defined on $U \times \mathbb{R}^d \setminus \{0\}$) of fixed multiplicity $m_1, \dots, m_q$. Without loss of generality, we may assume that

$$\forall \, u \in U \,, \quad \forall \, \xi \in \mathbb{R}^d \setminus \{0\} \,, \quad \lambda_1(u,\xi) < \cdots < \lambda_q(u,\xi) \,.$$

We are interested in the existence of shock waves solutions to (1): we let $\Gamma := \{x_d - \varphi(x_0, \dots, x_{d-1}) = 0\}$ denote a smooth hypersurface in $\mathbb{R} \times \mathbb{R}^d$ and u be a smooth function on either side of Γ. Such a function is a weak solution of (1) if and only if:

$$\begin{cases} L(u_\sharp^\pm, \varphi) u_\sharp^\pm = 0 & \pm x_d > 0 \,, \\ B(u_\sharp^+, u_\sharp^-, \varphi) = 0 & x_d = 0 \,, \end{cases} \tag{3}$$

where

$$u_\sharp(x_0, \dots, x_d) := u(x_0, \dots, x_{d-1}, x_d + \varphi(x_0, \dots, x_{d-1})) \,,$$

functions $u_\sharp^+$, resp. $u_\sharp^-$, denote the restrictions of $u_\sharp$ to the half-space $\{x_d > 0\}$, resp. $\{x_d < 0\}$, and the operators L and B are defined by

$$L(v,\psi) \, w := \sum_{j=0}^{d-1} A_j(v) \, \partial_j \, w + \widetilde{A_d}(v, \nabla\psi) \, \partial_d \, w \tag{4a}$$

$$\text{with} \quad \widetilde{A_d}(v, \nabla\psi) := A_d(v) - \sum_{j=0}^{d-1} \partial_j \psi \, A_j(v) \,, \tag{4b}$$

$$B(w^+, w^-, \psi) := \sum_{j=0}^{d-1} \partial_j \psi \, [f_j(w)] \; - [f_d(w)] \,. \tag{4c}$$

For convenience, we drop the $\sharp$ subscript. The existence of multidimensional shock waves reduces to the following question: given an initial data u^0 that is smooth on either side of a hypersurface $\{x_d - \varphi^0(x_1, \dots, x_{d-1}) = 0\}$, does there exist a smooth solution (u, φ) of (3) with initial value (u^0, φ^0), at least locally in time? This question has received a positive answer in [7] when (u^0, φ^0) is a suitable perturbation of a uniformly stable planar shock wave. However, as was detailed in some previous works [1, 8], the uniform stability condition may break down and Majda's linear and nonlinear results can not be applied anymore. The aim of this paper is to investigate such a case where the uniform stability condition breaks down.

A first step in proving a nonlinear existence result is to investigate the linear stability of a planar shock wave: we let

$$u := \begin{cases} u_r & \text{if } x_d > 0, \\ u_l & \text{if } x_d < 0, \end{cases}$$

be a planar stationary shock wave solution to (1) (changing coordinates if necessary, all planar shock waves have this form). The states u_r and u_l satisfy the Rankine-Hugoniot conditions

$$f_d(u_r) - f_d(u_l) = 0\,.$$

We assume that u is a Lax-shock wave, that is satisfies the following assumption:

Assumption 2 *There exists an integer $p \in \{1, \dots, q\}$ such that*

$$\lambda_{p-1}(u_l, \xi) < 0 < \lambda_p(u_l, \xi) \quad \textit{and} \quad \lambda_p(u_r, \xi) < 0 < \lambda_{p+1}(u_r, \xi)$$

where $\xi := (0, \dots, 0, 1) \in \mathbb{R}^d$ is the vector of propagation of the shock wave u. Moreover, λ_p is a simple eigenvalue, that is $m_p = 1$.

Assumption 2 ensures that there are exactly $N+1$ characteristics (counted with multiplicity) that enter the shock front $\{x_d = 0\}$. Linearizing $L(u^\pm, \varphi)u^\pm$ and $B(u^+, u^-, \varphi)$ about the particular solution $u^+ \equiv u_r$, $u^- \equiv u_l$, $\varphi \equiv 0$, yields the following boundary value problem:

$$\sum_{j=0}^{d} A_j(u_r)\,\partial_j v^+ = f^+ \quad x_d > 0\,,$$

$$\sum_{j=0}^{d} A_j(u_l)\,\partial_j v^- = f^- \quad x_d < 0\,,$$

$$\sum_{j=0}^{d-1} \partial_j \psi\,[f_j(u)] - A_d(u_r)\,v^+ + A_d(u_l)\,v^- = g \quad x_d = 0\,,$$

where f^+, f^- and g are higher order terms in the expansion in $v^\pm, \psi$. Changing the space variable x_d into $-x_d$ in the evolution equation for v^-, we obtain a linear boundary value problem for $v := (v^+, v^-)$ and ψ in the half-space $\{x_d > 0\}$, that we write in the following way:

$$\begin{aligned} \mathcal{L}v &:= \sum_{j=0}^{d} \mathcal{A}_j\, \partial_j v = f \quad x_d > 0\,, \\ \mathcal{B}(v,\psi) &:= \sum_{j=0}^{d-1} \partial_j \psi\, b_j + M\, v = g \quad x_d = 0\,. \end{aligned} \tag{5}$$

We remark that (5) is a nonstandard boundary value problem (as in Majda's work [8]) because a partial differential equation on ψ is coupled with the boundary conditions for v and Kreiss' theory [6] can not be directly applied. We also remark that assumption 2 ensures that the boundary $\{x_d = 0\}$ is noncharacteristic for the operator $\mathcal{L}$, that is $\mathcal{A}_d$ is an invertible matrix.

2 The normal modes analysis

Recall that energy estimates for boundary value problems such as (5) are obtained via a Laplace transform in the time variable and a Fourier transform in the tangential space variables. The Laplace tansform can be seen as a Fourier transform of the new functions

$$\widetilde{v} := \exp(-\gamma x_0) v \quad \text{and} \quad \widetilde{\psi} := \exp(-\gamma x_0)\psi\,,$$

where γ is a nonnegative real number. Functions $(\widetilde{v}, \widetilde{\psi})$ satisfy the linear system:

$$\begin{cases} \mathcal{L}^\gamma \widetilde{v} := \gamma \mathcal{A}_0 \widetilde{v} + \mathcal{L}\widetilde{v} = F & x_d > 0\,, \\ \mathcal{B}^\gamma(\widetilde{v}, \widetilde{\psi}) := \gamma \widetilde{\psi} b_0 + \mathcal{B}(\widetilde{v}, \widetilde{\psi}) = G & x_d = 0\,. \end{cases} \tag{6}$$

We drop the tilda to facilitate the reading and perform a Fourier transform and obtain the following dynamical system:

$$\left(\gamma \mathcal{A}_0 + i \sum_{j=0}^{d-1} \eta_j \mathcal{A}_j \right) \widehat{v} + \mathcal{A}_d \frac{d\widehat{v}}{dx_d} = \widehat{F} \quad x_d > 0\,,$$

$$\left(\gamma b_0 + i \sum_{j=0}^{d-1} \eta_j b_j \right) \widehat{\psi} + M\widehat{v} = \widehat{G} \quad x_d = 0\,.$$

In short, this last boundary problem reads:

$$\begin{aligned} &\frac{d\widehat{v}}{dx_d} - \mathcal{A}(\eta,\gamma)\widehat{v} = \mathcal{A}_d^{-1}\widehat{F} \quad x_d > 0\,, \\ &b(\eta,\gamma)\, \widehat{\psi} + M\widehat{v}(0) = \widehat{G} \quad x_d = 0\,, \end{aligned} \tag{7}$$

where $\mathcal{A}$ and b are homogeneous of degree 1 in (η, γ). A first result, due to Hersh [5], is that the homogeneous dynamical system

$$\frac{dV}{dx_d} - \mathcal{A}(\eta, \gamma)V = 0$$

is of hyperbolic type as long as $\gamma > 0$. We thus define the stable subspace $\mathcal{E}^-(\eta, \gamma)$ as the set of the traces at $x_d = 0$ of the solutions of this system of ODEs that tend to zero at $+\infty$. Using assumption 2 and a continuity/connectedness argument, one can show that this stable subspace has constant dimension $N - 1$. Moreover, $\mathcal{E}^-(\eta, \gamma)$ admits a continuous extension to all points $(\underline{\eta}, 0) \neq (0, 0)$, that we still denote by $\mathcal{E}^-(\underline{\eta}, 0)$. In view of the homogeneity properties of the above symbols $\mathcal{A}$ and b, we define the hemisphere

$$\Sigma_+ := \{(\eta, \gamma) \in \mathbb{R}^{d+1} \text{ s.t. } \gamma \geq 0 \text{ and } |\eta|^2 + \gamma^2 = 1\} .$$

The following definition was introduced by Majda in [8]:

Definition 1 *The shock wave u will be said to satisfy the uniform stability condition if there exists a constant $c > 0$ such that for all $(\eta, \gamma) \in \Sigma_+$, one has*

$$\forall (\chi, Z) \in \mathbb{C} \times \mathcal{E}^-(\eta, \gamma) , \quad |\chi\, b(\eta, \gamma) + M\, Z| \geq c\,(|\chi| + |Z|) . \tag{8}$$

Using the compactness of Σ_+, one checks that a shock wave u is uniformly stable if and only if for all $(\eta, \gamma) \in \Sigma_+$

$$\{(\chi, Z) \in \mathbb{C} \times \mathcal{E}^-(\eta, \gamma) \text{ s.t. } \chi\, b(\eta, \gamma) + M\, Z = 0\} = \{0\} .$$

It is shown in [8] that under the uniform stability condition, operators $\mathcal{L}^\gamma$ and $\mathcal{B}^\gamma$ satisfy the so-called maximal L^2 energy estimate for all $\gamma \geq 1$:

$$\gamma |\!|\!| v |\!|\!|_{0,\gamma}^2 + ||v_{|x_d=0}||_{0,\gamma}^2 + ||\psi||_{1,\gamma}^2 \leq C \left(\frac{1}{\gamma} |\!|\!| \mathcal{L}^\gamma v |\!|\!|_{0,\gamma}^2 + ||\mathcal{B}^\gamma(v, \psi)||_{0,\gamma}^2 \right) ,$$

where the norms are weighted Sobolev ones. As in [8, 11], the Sobolev space $H^s(\mathbb{R}^d)$, $s \in \mathbb{R}$, is equipped with the norm

$$||u||_{s,\gamma}^2 := \frac{1}{(2\pi)^d} \int_{\mathbb{R}^d} (\gamma^2 + |\xi|^2)^s\, |\widehat{u}(\xi)|^2\, d\xi ,$$

and the Sobolev space $H^k(\{x_d > 0\})$, $k \in \mathbb{N}$, is equipped with the norm

$$|\!|\!| u |\!|\!|_{k,\gamma}^2 := \sum_{j=0}^{k} \int_0^{+\infty} ||\partial_d^j u(\cdot, x_d)||_{k-j,\gamma}^2\, dx_d .$$

We want to allow some weak "instability" phenomena to occur but also to keep some information of the same type as (8). With this idea in mind, we make the following assumption on the shock wave u:

Assumption 3 *There exists a positive constant c such that*

$$\forall\,(\eta,\gamma)\in\Sigma_+\,,\quad |b(\eta,\gamma)|\geq c\,. \tag{9}$$

The lower bound in (9) trivially follows from (8), when satisfied. It is crucial in our analysis that (8) fails. It is shown in [4] that (9) holds for shock waves in isentropic gas dynamics that violate the uniform stability condition. However, we emphasize that (9) does not hold for scalar conservation laws in several space variables, see [8, 9] (but as mentionned in the introduction, the scalar case is simpler than the system case).

Because the symbol b does not vanish when $\gamma\geq 0$ and $(\eta,\gamma)\neq 0$, we can define the orthognal projector $\Pi(\eta,\gamma)$ by

$$\forall\,h\in\mathbb{C}^N\,,\quad \Pi(u,\eta,\gamma)\,h:=h-\frac{\langle h,b(\eta,\gamma)\rangle}{|b(\eta,\gamma)|^2}\,b(\eta,\gamma)\,.$$

Multiplying (7) by Π and b yields the new boundary conditions

$$\Pi(\eta,\gamma)M\,\widehat{v}(0)=\Pi(\eta,\gamma)\,\widehat{G}\,,$$
$$|b(\eta,\gamma)|^2\,\widehat{\psi}=\langle\widehat{G}-M\,\widehat{v}(0),b(\eta,\gamma)\rangle\,.$$

Because of the homogeneity properties of b and of (9), we get from the last boundary condition

$$(\gamma^2+|\eta|^2)\,|\widehat{\psi}|^2\leq C\,(|\widehat{G}|^2+|\widehat{v}(0)|^2)\,.$$

Integrating with respect to η and using Plancherel's theorem, we get

$$||\psi||^2_{1,\gamma}\leq C\,\left(||v_{|x_d=0}||^2_{0,\gamma}+||\mathcal{B}^\gamma(v,\psi)||^2_{0,\gamma}\right)\,. \tag{10}$$

We now turn to the study of the problem

$$\begin{cases}\dfrac{d\widehat{v}}{dx_d}=\mathcal{A}(\eta,\gamma)\,\widehat{v}+A_d^{-1}\,\widehat{F} & x_d>0\,,\\ \Pi(\eta,\gamma)M\,\widehat{v}(0)=\Pi(\eta,\gamma)\,\widehat{G}\,,\end{cases} \tag{11}$$

and make the following assumption:

Assumption 4 *If $\gamma>0$, the subspace*

$$\{Z\in\mathcal{E}^-(\eta,\gamma)\ \ s.t.\ \ \Pi(\eta,\gamma)M\,Z=0\}$$

is trivial for any $\eta\in\mathbb{R}^d$.

If $(\underline{\eta},0)\in\Sigma_+$ is a point where the subspace

$$\{Z\in\mathcal{E}^-(\underline{\eta},0)\ \ s.t.\ \ \Pi(\underline{\eta},0)M\,Z=0\}$$

is nontrivial, then there exists a neighbourhood $\mathcal{V}$ of $(\underline{\eta}, 0)$ in Σ_+ and a constant $c > 0$ such that

$$\forall\,(\eta,\gamma) \in \mathcal{V}\,, \quad \forall\, Z \in \mathcal{E}^-(\eta,\gamma)\,, \quad |\Pi(\eta,\gamma) M\, Z| \geq c\,\gamma\,|Z|\,. \tag{12}$$

Furthermore, there exists a C^∞ mapping $Q : \mathcal{V} \to Gl_{2N}(\mathbb{C})$ such that for all $z \in \mathcal{V}$, the matrix $Q(z)^{-1}\,\mathcal{A}(z)\,Q(z)$ is diagonal.

In other words, the shock wave u is not violently unstable, but there may exist some points of weak instability: those points correspond to regions of smoothness of eigenvalues of $\mathcal{A}$ and a formal way to explain (12) is to say that the neutrally unstable modes are simple roots of the Lopatinskii determinant. We refer to [4] for the verification of assumption 4 on a concrete example.

Under assumptions 1-4, we have the following result:

Theorem 1. *There exists a constant $C > 0$ such that for all $v \in H^2(\{x_d > 0\})$, for all $\psi \in H^2(\mathbb{R}^d)$ and for all $\gamma \geq 1$, the following estimate holds:*

$$\gamma |\!|\!| v |\!|\!|_{0,\gamma}^2 + \| v_{|x_d=0} \|_{0,\gamma}^2 + \|\psi\|_{1,\gamma}^2 \leq C \left(\frac{1}{\gamma^3} |\!|\!| \mathcal{L}^\gamma v |\!|\!|_{1,\gamma}^2 + \frac{1}{\gamma^2} \| \mathcal{B}^\gamma(v,\psi) \|_{1,\gamma}^2 \right) .$$

Compared to the maximal L^2 estimate derived under the uniform stability condition, we observe a loss of one derivative in the interior domain and a loss of one derivative on the boundary. Though some slight improvements can be achieved, it seems almost impossible to derive an energy estimate with no loss of derivative in the interior domain.

3 A sketch of proof

We now briefly indicate the way to deal with the new feature of our work, namely the occurence of neutrally unstable modes. If $(\underline{\eta}, 0)$ is such a neutrally unstable mode, then in a suitable neighbourhood $\mathcal{V}$ of $(\underline{\eta}, 0)$, we have

$$\mathcal{A}(\eta,\gamma) \sim \ \text{diag}\ (\omega_1,\dots,\omega_{2N}) \quad \text{with} \ \begin{cases} \text{Re}\ \omega_j \leq -c\gamma & j = 1,\dots,N-1, \\ \text{Re}\ \omega_j \geq c\gamma & j = N,\dots,2N, \end{cases}$$

and $\mathcal{E}^-(\eta,\gamma)$ is the vector space spanned by the $N-1$ first basis vectors.

In the uniformly stable case, a Kreiss' symmetrizer for (11) is an hermitian matrix valued mapping R such that

$$\text{Re}\ (R(\eta,\gamma)\,\mathcal{A}(\eta,\gamma)) \geq c\,\gamma\, I\,, \tag{13a}$$

$$R(\eta,\gamma) + C\,\beta(\eta,\gamma)^*\,\beta(\eta,\gamma) \geq c\, I\,, \tag{13b}$$

where $\beta(\eta,\gamma) := \Pi(\eta,\gamma) M$ is the boundary symbol. We refer to [3, 6] for the existence of such symmetrizers in the uniformly stable case. In our case, we define

$$R(\eta,\gamma) := \begin{pmatrix} -\gamma^2 I_{N-1} & \mathbf{0} \\ \mathbf{0} & \alpha I_{N+1} \end{pmatrix},$$

where α is a real number greater than 1 to be fixed. Using (12) and the existence of a smooth basis of $\mathcal{E}^-(\eta,\gamma)$, we can conclude that an appropriate choice of α yields an estimate of the type

$$\forall\,(\eta,\gamma) \in \mathcal{V}\,, \quad R(\eta,\gamma) + C\,\beta(\eta,\gamma)^*\,\beta(\eta,\gamma) \geq c\gamma^2 I_{2N}\,, \tag{14}$$

which is weaker than (13b). Our choice of R gives

$$\mathrm{Re}\ (R(\eta,\gamma)\,\mathcal{A}(\eta,\gamma)) \geq c\gamma \begin{pmatrix} \gamma I_{N-1} & \mathbf{0} \\ \mathbf{0} & \alpha' I_{N+1} \end{pmatrix}^2 \tag{15}$$

for some positive constant α'. We observe that this lower bound is weaker than the one in (13a). If we now fix a smooth function χ on Σ_+ with support in $\mathcal{V}$ and then extend χR as an homogeneous mapping of degree 2, we are able to derive an energy estimate for $\chi\widehat{v}$. First write the equality

$$\frac{d\chi\widehat{v}}{dx_d} = \mathcal{A}\,\chi\,\widehat{v} + A_d^{-1}\,\chi\,\widehat{F}\,,$$

then take the scalar product of this ODE with $\chi R\widehat{v}$ and integrate with respect to $\eta \in \mathbb{R}^d$ and $x_d > 0$. After a few calculations using (14) and (15), one gets the estimate

$$\gamma^3\,\|\!|\chi\,\widehat{v}|\!\|_{0,\gamma}^2 + \gamma^2\,\|\chi\,\widehat{v}_{|x_d=0}\|_{0,\gamma}^2 \leq C\,\left(\frac{1}{\gamma}\,\|\!|\chi\,\widehat{F}|\!\|_{1,\gamma}^2 + \|\chi\,\widehat{G}\|_{1,\gamma}^2\right).$$

The remaining part of the work is to glue this energy estimate with the maximal L^2 estimates obtained in the neighbourhoods of the "stable" modes by Kreiss' analysis. These esimates read

$$\begin{aligned}\gamma\,\|\!|\theta\,\widehat{v}|\!\|_{0,\gamma}^2 + \|\theta\,\widehat{v}_{|x_d=0}\|_{0,\gamma}^2 &\leq C\,\left(\frac{1}{\gamma}\,\|\!|\theta\,\widehat{F}|\!\|_{0,\gamma}^2 + \|\theta\,\widehat{G}\|_{0,\gamma}^2\right) \\ &\leq C\,\left(\frac{1}{\gamma^3}\,\|\!|\theta\,\widehat{F}|\!\|_{1,\gamma}^2 + \frac{1}{\gamma^2}\,\|\theta\,\widehat{G}\|_{1,\gamma}^2\right),\end{aligned}$$

where θ is a truncation function. Because Σ_+ is a C^∞ compact manifold, one can choose a finite covering and truncation functions subordinated to this covering. Summing the previously obtained inequalities, one gets a global energy estimate of the type

$$\gamma\|\!|v|\!\|_{0,\gamma}^2 + \|v_{|x_d=0}\|_{0,\gamma}^2 \leq C\,\left(\frac{1}{\gamma^3}\|\!|\mathcal{L}^\gamma v|\!\|_{1,\gamma}^2 + \frac{1}{\gamma^2}\|\mathcal{B}^\gamma(v,\psi)\|_{1,\gamma}^2\right).$$

One then uses (10) to conclude the proof.

References

1. S. Benzoni-Gavage. Stability of multi-dimensional phase transitions in a van der Waals fluid. *Nonlinear Anal.*, 31(1-2):243–263, 1998.
2. J. M. Bony. Calcul symbolique et propagation des singularités pour les équations aux dérivées partielles non linéaires. *Ann. Sci. École Norm. Sup. (4)*, 14(2):209–246, 1981.
3. J. Chazarain, A. Piriou. *Introduction to the theory of linear partial differential equations.* North-Holland Publishing Co., Amsterdam, 1982. Translated from the French.
4. J. F. Coulombel. Weak stability of non uniformly stable multidimensional shocks. *To appear in SIAM J. Math. Anal.*
5. R. Hersh. Mixed problems in several variables. *J. Math. Mech.*, 12:317–334, 1963.
6. H. O. Kreiss. Initial boundary value problems for hyperbolic systems. *Comm. Pure Appl. Math.*, 23:277–298, 1970.
7. A. Majda. The existence of multidimensional shock fronts. *Mem. Amer. Math. Soc.*, 43(281):v+93, 1983.
8. A. Majda. The stability of multidimensional shock fronts. *Mem. Amer. Math. Soc.*, 41(275):iv+95, 1983.
9. A. Majda. *Compressible fluid flow and systems of conservation laws in several space variables.* Springer-Verlag, New York, 1984.
10. G. Métivier. The block structure condition for symmetric hyperbolic systems. *Bull. London Math. Soc.*, 32(6):689–702, 2000.
11. G. Métivier. Stability of multidimensional shocks. In *Advances in the theory of shock waves*, pages 25–103. Birkhäuser, 2001.
12. Y. Meyer. Remarques sur un théorème de j.-m. bony. *Supplemento ai Rendiconti del Circolo Matematico di Palermo, Serie II*, 1:1–20, 1981.
13. A. Mokrane. Problèmes mixtes hyperboliques non-linéaires. *Ph.D. Thesis, Université de Rennes I*, 1987.
14. D. Serre. *Systems of conservation laws. 2.* Cambridge University Press, Cambridge, 2000.

Propagation and Interaction of Delta-Shock Waves of a Hyperbolic System of Conservation Laws

V. G. Danilov[1] and V. M. Shelkovich[⋆2]

[1] Moscow Technical University of Communication and Informatics, Russia
danilov@miem.edu.ru

[2] St.-Petersburg State Architecture and Civil Engineering University, Russia
shelkv@svm.abu.spb.ru

Summary. We propose a new regularization method – the *weak asymptotics method* – for investigating propagation and interaction of delta-shock waves of a hyperbolic system of conservation laws.

1 Introduction

1. In this paper, within the framework of the weak asymptotics method [1]–[5], we present formulas describing propagation and interaction of delta-shock waves for the system of equations

$$L_1[u] = u_t + \big(f(u)\big)_x = 0, \qquad L_2[u,v] = v_t + \big(g(u)v\big)_x = 0, \tag{1}$$

where $f(u), g(u) \in C^\infty$. We shall assume that the following conditions are satisfied (see [6]–[8])

$$f''(u) > 0, \quad g'(u) > 0, \quad f'(u) \le g(u). \tag{2}$$

Delta-shock wave type solution of system (1) corresponds to the initial data of the form

$$\begin{aligned} u^0(x) &= u_0^0(x) + u_1^0(x)H(-x), \\ v^0(x) &= v_0^0(x) + v_1^0(x)H(-x) + e^0\delta(x), \end{aligned} \tag{3}$$

where $u_0^0(x)$, $u_1^0(x) > 0$, $v_0^0(x)$, $v_1^0(x)$ are smooth functions, $H(x)$ is the Heaviside function, and δ is the Dirac delta function.

The interaction of delta-shock waves corresponds to the initial data of the form

$$\begin{aligned} u^0(x) &= u_0 + \sum_{k=1}^{2} u_k H(-x + x_k^0), \\ v^0(x) &= v_0 + \sum_{k=1}^{2} \big[v_k H(-x + x_k^0) + e_k^0 \delta(-x + x_k^0)\big]. \end{aligned} \tag{4}$$

⋆ Research supported by DFG Project 436 RUS 113/593/3-1 and by Grant 02-01-00483 of Russian Foundation for Basic Research.

In this paper we assume that u_0, v_0, e_k^0, $u_k > 0$, v_k, x_0^k, $k = 1, 2$, are constants, and $x_1^0 < x_2^0$.

For the first equation in (1), propagation and interaction of jump discontinuity type solutions (shock waves) was analytically studied in detail in [4].

2. The generalized solution of system (1), which preserves the structure of the initial condition (3),

$$\begin{aligned} u(x,t) &= u_0(x,t) + u_1(x,t)H(-x+\phi(t)), \\ v(x,t) &= v_0(x,t) + v_1(x,t)H(-x+\phi(t)) + e(t)\delta(-x+\phi(t)), \end{aligned} \tag{5}$$

is called a *delta-shock wave solution*, where the functions $u_i(x,t)$, $v_i(x,t)$, and $\phi(t)$ are to be found.

According to the weak asymptotics method, solution (5) is obtained as the weak limit of some of its regularizations, namely, of the *weak asymptotic solution* $u_\varepsilon(x,t)$, $v_\varepsilon(x,t)$ of system (1). Prior to defining such a solution, we recall the definition of the notion of smallness in the weak sense.

A family of functions $f_\varepsilon(x,t)$ is said to be an infinitesimal of order $O_{\mathcal{D}'}(\varepsilon^\alpha)$ if

$$\langle f_\varepsilon, \varphi \rangle = O(\varepsilon^\alpha),$$

for any test function $\varphi(x)$ and, moreover, the function $\langle f_\varepsilon, \varphi \rangle$ is continuous in t for any $\varphi(x)$.

Definition 1. A family of functions $u_\varepsilon(x,t)$, $v_\varepsilon(x,t)$ smooth for $\varepsilon > 0$ is called a weak asymptotic solution of the Cauchy problem (1), (3) if

$$\begin{aligned} L_1[u_\varepsilon] &= O_{\mathcal{D}'}(\varepsilon), & L_2[u_\varepsilon, v_\varepsilon] &= O_{\mathcal{D}'}(\varepsilon), \\ u_\varepsilon\big|_{t=0} &= u^0 + O_{\mathcal{D}'}(\varepsilon), & v_\varepsilon\big|_{t=0} &= v^0 + O_{\mathcal{D}'}(\varepsilon). \end{aligned}$$

The weak asymptotic solution u_ε, v_ε to the Cauchy problem (1), (3) is constructed in the form

$$\begin{aligned} u(x,t,\varepsilon) &= u_0(x,t) + u_1(x,t)H_u(-x+\phi(t),\varepsilon), \\ v(x,t,\varepsilon) &= v_0(x,t) + v_1(x,t)H_v\big(-x+\phi(t),\varepsilon\big) \\ &\quad + e(t)\delta_v\big(-x+\phi(t),\varepsilon\big) + R(t)\frac{1}{\varepsilon}\Omega''\big((-x+\phi(t))/\varepsilon\big). \end{aligned} \tag{6}$$

Here $H_j(z,\varepsilon) = \omega_{0j}\left(\frac{z}{\varepsilon}\right) = \int_{-\infty}^{z/\varepsilon} \omega_j(\eta)\,d\eta$ are regularizations of the Heaviside function $H(z)$, $j = u, v$, and $\delta_v(z,\varepsilon) = \frac{1}{\varepsilon}\omega_\delta(z/\varepsilon)$ is a regularization of the δ-function. The mollifiers $\omega_k(\eta)$, $k = u, v, \delta$, have the following properties: $\omega_k \geq 0$, $\omega_k \in \mathbb{S}(\mathbb{R})$, $\int \omega(\eta)\,d\eta = 1$. The function $\Omega(z)$ has the same properties as $\omega_k(z)$ except, generally speaking, the property of being nonnegative. It is easy to see that $\varepsilon^{-3}\Omega''(z/\varepsilon)$ is a regularization of $\delta''(z)$.

So we consider regularizations more general as compared with the vanishing viscosity method, but the outline of the definition of delta-shock wave solutions is preserved.

The generalized solution of system (1), which preserves the structure of the initial conditions (4),

$$
\begin{aligned}
u(x,t) &= u_0 + \sum\nolimits_{k=1}^{2} u_k H(-x+\phi_k(t)),\\
v(x,t) &= v_0 + \sum\nolimits_{k=1}^{2} \Big[v_k H(-x+\phi_k(t)) + e_k(t)\delta(-x+\phi_k(t))\Big],
\end{aligned}
\tag{7}
$$

describes interaction of delta-shock waves. Similarly to the case of a single delta-shock wave, solution (7) is obtained as the weak limit of the weak asymptotic solution of the Cauchy problem (1), (4), which is determined similarly to the solution of problem (1), (3).

In the case of interaction of two delta-shock waves, the regularization of solution (7) is constructed as follows:

$$
\begin{aligned}
u(x,t,\varepsilon) &= u_0(x,t) + \sum\nolimits_{k=1}^{2} u_k(x,t) H_{uk}\big(-x+\phi_k(t,\varepsilon),\varepsilon\big),\\
v(x,t,\varepsilon) &= v_0(x,t) + \sum\nolimits_{k=1}^{2} \Big[v_k(x,t) H_{vk}\big(-x+\phi_k(t,\varepsilon),\varepsilon\big)\\
&\quad + e_k(t,\varepsilon)\delta_{vk}\big(-x+\phi_k(t,\varepsilon),\varepsilon\big) + R_k(t,\varepsilon)\frac{1}{\varepsilon}\Omega_k''\Big(\frac{-x+\phi(t)}{\varepsilon}\Big)\Big],
\end{aligned}
\tag{8}
$$

where $H_{jk}(z,\varepsilon) = \omega_{0jk}(z/\varepsilon) = \int_{-\infty}^{z/\varepsilon}\omega_{jk}(\eta)\,d\eta$ are regularizations of the Heaviside function $H(z)$, $\delta_{vk}(z,\varepsilon) = \frac{1}{\varepsilon}\omega_{\delta k}(z/\varepsilon)$ are regularizations of the δ-function, and $\frac{1}{\varepsilon^3}\Omega_k''(z/\varepsilon)$ are regularizations of $\delta''(z)$, $j = u, v$, $k = 1,2$. Here the mollifiers have the same properties as in (6). We assume that $R_k(t,\varepsilon)$ are functions such that for all $t \geq 0$

$$
\begin{aligned}
R_k(t,\varepsilon)\frac{1}{\varepsilon}\Omega_k''\Big(\frac{-x+\phi(t)}{\varepsilon}\Big) &= O_{\mathcal{D}'}(\varepsilon),\\
\frac{d}{dt}\Big(R_k(t,\varepsilon)\frac{1}{\varepsilon}\Omega_k''\Big(\frac{-x+\phi(t)}{\varepsilon}\Big)\Big) &= O_{\mathcal{D}'}(\varepsilon), \quad \varepsilon \to +0.
\end{aligned}
\tag{9}
$$

The method for constructing weak asymptotic solutions consists in substituting formulas (6) or (8) into system of equations (1) and calculating the weak asymptotics of the result of substitution (see Secs. 2–4). In the case of a single equation (the first equation in system (1)), this was done in [3]–[5].

From the viewpoint of this technique, the second equation does not differ from the first equation. By Oleinik's conditions of well-posedness, the solutions of the first equation in the first and second cases are unique modulo $O_{\mathcal{D}}(\varepsilon)$. Hence the singular support of the weak limit of the weak asymptotic solution of the second equation should move along the trajectory determined by the first equation. So, to provide this matching of singularities, we introduce terms of the form $R\varepsilon^{-1}\Omega''\big((-x+\phi)/\varepsilon\big)$ in the formulas for the weak asymptotic solution. These terms do not make a contribution to the limit solution, but provide the desired matching of singularities.

The explicit formulas for the two-delta-shock-wave solution obtained in this paper allows us to describe some bifurcations arising after the interaction.

For example, it is easy to see that the interaction of two shock waves ($e_1(t) = e_2(t) = 0$ for $t < t^*$, where t^* is the instant of interaction) can create a single delta-shock wave ($e(t) \neq 0$) for $t > t^*$.

2 Propagation of a single delta-shock wave

To exclude possible disagreement of singularities in the first and second equations of system (1), we introduce the notion of admissible ansatz.

Definition 2. The weak asymptotic solution (6) is said to be *admissible* for the Cauchy problem (1), (3) if

$$R(t)c(t) = e(t)\left(\frac{f(u_0(x,t)+u_1(x,t)) - f(u_0(x,t))}{u_1(x,t)}\Big|_{x=\phi(t)} - a(t)\right), \qquad (10)$$

where $a(t) = \int g\big(u_0(\phi(t),t) + u_1(\phi(t),t)\omega_{0u}(\eta)\big)\omega_\delta(\eta)\,d\eta$ and $c(t) = \int g\big(u_0(\phi(t),t) + u_1(\phi(t),t)\omega_{0u}(\eta)\big)\Omega''(\eta)\,d\eta \neq 0$.

The following assertion holds.

Theorem 1. *Suppose that conditions* (2) *are satisfied. We assume that there exists* $T > 0$ *such that*

$$\begin{aligned} -\inf_{\xi>0}[u_0'(\xi)f''(u_0(\xi))] &< T^{-1}, \\ -\inf_{\xi<0}\Big[u_0(\xi)+u_1(\xi)\Big]' f''(u_0(\xi)+u_1(\xi))] &< T^{-1}. \end{aligned} \qquad (11)$$

Then there exists an admissible weak asymptotic solution of the Cauchy problem (1), (3) *and the functions* $u_i(x,t)$, $v_i(x,t)$, $i = 1,2$, $\phi(t)$, $e(t)$ *are determined by the following problems*

$$\begin{aligned} &L_1[u_0(x,t)] = 0, \quad x > \phi(t), \\ &L_1[u_0(x,t)+u_1(x,t)] = 0, \quad x < \phi(t), \\ &L_2[u_0(x,t), v_0(x,t)] = 0, \quad x > \phi(t), \\ &L_2[u_0(x,t)+u_1(x,t), v_0(x,t)+v_1(x,t)] = 0, \quad x < \phi(t), \\ &\dot\phi(t) = \frac{[f(u(x,t))]}{[u(x,t)]}\Big|_{x=\phi(t)}, \\ &\dot e(t) = \left([v(x,t)g(u(x,t))] - [v(x,t)]\frac{[f(u(x,t))]}{[u(x,t)]}\right)\Big|_{x=\phi(t)}, \end{aligned} \qquad (12)$$

and the function $R(t)$ *is determined by relation* (10).

To prove the theorem, we need to analyze the solvability of system (12) obtained by substituting into (1) the solution of the form (6) and calculating the weak asymptotics.

The conditions imposed on the functions $f(u)$ and $g(u)$ and the initial data allow us to prove the existence and uniqueness of the solution of system (12). The function $R(t)$ is determined by using solutions of system (12) according to formula (10).

3 Interaction of delta-shock waves

Similarly to the case of a single delta-shock wave, we define an admissible smooth ansatz for the two-delta-shock-wave solution.

Definition 3. The weak asymptotic solution (8) is said to be *admissible* for the Cauchy problem (1), (4) for $t \in [0, t^*)$ if

$$\begin{aligned} R_{10}(t)c_1 &= e_{10}(t)\Big(\frac{f(u_0+u_1+u_2)-f(u_0+u_2)}{u_1} - a_1\Big), \\ R_{20}(t)c_2 &= e_{20}(t)\Big(\frac{f(u_0+u_2)-f(u_0)}{u_2} - a_2\Big). \end{aligned} \tag{13}$$

Here $a_1 = \int g(u_0 + u_1\omega_{0u1}(\eta) + u_2)\omega_{\delta 1}(\eta)\,d\eta$, $c_1 = \int g(u_0 + u_1\omega_{0u1}(\eta) + u_2)\Omega_1''(\eta)\,d\eta \neq 0$, $a_2 = \int g(u_0 + u_2\omega_{0u2}(\eta))\omega_{\delta 2}(\eta)\,d\eta$, $c_2 = \int g(u_0 + u_2\omega_{0u2}(\eta))\Omega_2''(\eta)\,d\eta \neq 0$, and $R_{k0}(t) = \lim_{\varepsilon\to+0} R_k(t,\varepsilon)$, $e_{k0}(t) = \lim_{\varepsilon\to+0} e_k(t,\varepsilon)$, t^* is the instant of interaction of delta-shock waves, $k = 1, 2$. In addition,

$$\begin{aligned} C_{R_1}(\rho) &= \int g\big(u_0 + u_1\omega_{0u1}(\eta) + u_2\omega_{0u2}(\eta+\rho)\big)\Omega_1''(\eta)\,d\eta \neq 0, \\ C_{R_2}(-\rho) &= \int g\big(u_0 + u_1\omega_{0u1}(\eta-\rho) + u_2\omega_{0u2}(\eta)\big)\Omega_2''(\eta)\,d\eta \neq 0. \end{aligned} \tag{14}$$

The following assertion holds.

Theorem 2. *For all $t \in [0, +\infty)$ there exists an admissible weak asymptotic solution* (8) *of the Cauchy problem* (1), (4), *where $\phi_k(t,\varepsilon) = \phi_{k0}(t) + \psi_0(t)\phi_{k1}(\tau)\big|_{\tau=\psi_0(t)/\varepsilon}$, $e_k(t,\varepsilon) = e_{k0}(t) + \psi_0(t)e_{k1}(\tau)\big|_{\tau=\psi_0(t)/\varepsilon}$, $R_k(t,\varepsilon) = R_{k0}(t) + R_{k1}(\tau,t)\big|_{\tau=\psi_0(t)/\varepsilon}$, and $\psi_0(t) = \phi_{20}(t) - \phi_{10}(t)$. Here the functions*

$$\begin{aligned} \phi_{k0}(t) &= \phi_{k0}(0) + \frac{[f(u)]_k}{[u]_k}t, \qquad e_{k0}(t) = e_k^0 + \Big([g(u)v]_k - [v]_k\frac{[f(u)]_k}{[u]_k}\Big)t, \\ R_{k0}(t)c_k &= e_{k0}(t)\Big(\frac{[f(u)]_k}{[u]_k} - a_k\Big), \end{aligned} \tag{15}$$

are defined by Theorem 1 *and the admissibility conditions, $\phi_{k0}(0) = x_k^0$, $e_{k0}(0) = e_k^0$, $[r(u)s(v)]_1 = r(u_0+u_1+u_2)s(v_0+v_1+v_2) - r(u_0+u_2)s(v_0+v_2)$ and $[r(u)s(v)]_2 = r(u_0+u_2)s(v_0+v_2) - r(u_0)s(v_0)$ are jumps of the function*

$r(u)s(v)$ across the discontinuity lines $x=\phi_{10}(t)$ and $x=\phi_{20}(t)$, respectively, $k=1,2$. The corrections ϕ_{k1}, e_{k1}, R_{k1} have the form

$$\phi_{k1}(\tau)=\frac{(-1)^{k-1}u_{3-k}}{\big(f(u_0+u_1+u_2)-f(u_0+u_2)\big)u_2-\big(f(u_0+u_2)-f(u_0)\big)u_1} \times\frac{1}{\tau}\int_0^{\tau}B_2(-\rho(\tau'))\,d\tau' \tag{16}$$

$$e_{k1}(\tau)=\frac{(-1)^{k-1}u_1u_2}{\big(f(u_0+u_1+u_2)-f(u_0+u_2)\big)u_2-\big(f(u_0+u_2)-f(u_0)\big)u_1} \times\frac{1}{\tau}\int_0^{\tau}\widetilde{B}_2(-\rho(\tau'))\,d\tau'-v_k\phi_{k1}(\tau), \tag{17}$$

$$R_{k1}(\tau,t)=\frac{\widehat{e}_k(\tau,t)}{C_{Rk}\big((-1)^{k-1}\rho\big)}\bigg(\frac{f(u_0+u_k)-f(u_0)+B_k\big((-1)^{k-1}\rho\big)}{u_1} -A_k\big((-1)^{k-1}\rho\big)\bigg)-R_{k0}(t), \tag{18}$$

where $B_2(-\rho)$ is defined by (30), $\widetilde{B}_k\big((-1)^{k-1}\rho\big)$, $A_k\big((-1)^{k-1}\rho\big)$, and $C_{Rk}\big((-1)^{k-1}\rho\big)$ are defined by (32), $k=1,2$. Here $\rho=\rho(\tau)$ is the solution of the problem

$$\frac{d\rho}{d\tau}=F(\rho)=1+\frac{(u_1+u_2)B_2(-\rho)}{\big(f(u_0+u_2)-f(u_0)\big)u_1-\big(f(u_0+u_1+u_2)-f(u_0+u_2)\big)u_2}, \qquad \frac{\rho}{\tau}\Big|_{\tau\to+\infty}=1, \tag{19}$$

Proof. **1.** Substituting the admissible smooth ansatz (8) into the system (1) and using (29), (31), we find necessary and sufficient conditions for $L_1[u(x,t,\varepsilon)]=O_{\mathcal{D}'}(\varepsilon)$ and $L_2[u(x,t,\varepsilon),v(x,t,\varepsilon)]=O_{\mathcal{D}'}(\varepsilon)$:

$$\begin{aligned} u_k\dot{\phi}_k(t,\varepsilon)&=\big(f(u_0+u_k)-f(u_0)\big)+B_k\big((-1)^{k-1}\rho\big), \qquad (20)\\ v_k\dot{\phi}_k(t,\varepsilon)+\dot{e}_k(t,\varepsilon)&=g(u_0+u_k)(v_0+v_k)-g(u_0)v_0+\widetilde{B}_k\big((-1)^{k-1}\rho\big),\\ e_k(t,\varepsilon)\dot{\phi}_k(t,\varepsilon)&=e_k(t,\varepsilon)A_k\big((-1)^{k-1}\rho\big)+R_k(t,\varepsilon)C_{Rk}\big((-1)^{k-1}\rho\big), \end{aligned}$$

where $k=1,2$, $\rho=\psi(x,t)/\varepsilon$, $\psi=\big(\phi_2(t,\varepsilon)-\phi_1(t,\varepsilon)\big)$.

2. If $t\in[0,t^*)$, where $t=t^*>0$ is the instant of interaction, we have two propagating delta-shock waves which do not interact. By Theorem 1, their phases $\phi_{k0}(t)$ and the amplitudes of delta functions $e_{k0}(t)$ satisfy the system

$$\phi_{k0}(t)=\phi_{k0}(0)+\frac{[f(u)]_k}{[u]_k}t, \quad e_{k0}(t)=e_k^0+\bigg([g(u)v]_k-[v]_k\frac{[f(u)]_k}{[u]_k}\bigg)t, \tag{21}$$

and for $R_{k0}(t)$ the relations (13) hold, $k=1,2$. The instant of interaction $t=t^*>0$ is a root of the equation $\psi_0(t^*)=0$, where $\psi_0(t)=\phi_{20}(t)-\phi_{10}(t)$ is the distance between the fronts of *non-interacting* delta-shock waves.

We will seek functions $\phi_k(t,\varepsilon)$, $e_k(t,\varepsilon)$, $R_k(t,\varepsilon)$ as functions of the fast variable $\tau = \psi_0(t)/\varepsilon \in \mathbb{R}$ and the slow variable $t \geq 0$:

$$\begin{aligned} \phi_k(t,\varepsilon) &\stackrel{\text{def}}{=} \widehat{\phi}_k(\tau,t) = \phi_{k0}(t) + \psi_0(t)\phi_{k1}(\tau), \\ e_k(t,\varepsilon) &\stackrel{\text{def}}{=} \widehat{e}_k(\tau,t) = e_{k0}(t) + \psi_0(t)e_{k1}(\tau), \\ R_k(t,\varepsilon) &\stackrel{\text{def}}{=} \widehat{R}_k(\tau,t) = R_{k0}(t) + R_{k1}(\tau,t), \end{aligned} \tag{22}$$

where $\phi_{k0}(t)$, $e_{k0}(t)$, $R_{k0}(t)$ are *extended* by (21) for all $t \in [t^*, +\infty)$.

Before the interaction as $t < t^*$, we have $\tau > 0$; *after the interaction* as $t > t^*$, we have $\tau < 0$. We set the following boundary conditions:

$$\begin{aligned} &\phi_{k1}(\tau)\Big|_{\tau\to+\infty} = 0, \qquad e_{k1}(\tau)\Big|_{\tau\to+\infty} = 0, \qquad R_{k1}(\tau,t)\Big|_{\tau\to+\infty} = 0, \\ &\frac{d\phi_{k1}(\tau)}{d\tau}\Big|_{\tau\to-\infty} = o(\tau^{-1}), \qquad \frac{de_{k1}(\tau)}{d\tau}\Big|_{\tau\to-\infty} = o(\tau^{-1}), \end{aligned} \tag{23}$$

Finding the limit values of the phase and amplitude perturbations *after the interaction* $\phi_{k1}(\tau)\Big|_{\tau\to-\infty} = \phi_{k1,-}$, $e_{k1}(\tau)\Big|_{\tau\to-\infty} = e_{k1,-}$, we find the limit values

$$\begin{aligned} \widehat{\phi}_{k,-}(t) &= \widehat{\phi}_k(\tau,t)\Big|_{\tau\to-\infty} = \phi_{k0}(t) + \psi_0(t)\phi_{k1,-}, \\ \widehat{e}_{k,-}(t) &= \widehat{e}_k(\tau,t)\Big|_{\tau\to-\infty} = e_{k0}(t) + \psi_0(t)e_{k1,-}, \end{aligned} \tag{24}$$

and thus *determine "the result" of interaction* of delta-shock waves as $t > t^*$.

If $\psi_1(\tau) = \phi_{21}(\tau) - \phi_{11}(\tau)$, then $\psi(t,\varepsilon) = \psi_0(t)\big(1+\psi_1(\tau)\big)$ and $\rho(\tau) = \psi(t,\varepsilon)/\varepsilon = \tau\big(1+\psi_1(\tau)\big)$.

3. Substituting (22) into (20), for all $t \geq 0$ and $\tau \in \mathbb{R}$ we obtain

$$\dot{\phi}_{k0}(t) + \dot{\psi}_0(t)\frac{d}{d\tau}\Big(\tau\phi_{k1}(\tau)\Big) = \frac{f(u_0+u_k) - f(u_0) + B_k((-1)^{k-1}\rho)}{u_k}, \tag{25}$$

$$\begin{aligned} \dot{e}_{k0}(t) + \dot{\psi}_0(t)\frac{d}{d\tau}\Big(\tau e_{k1}(\tau)\Big) &= (v_0+v_k)g(u_0+u_k) - v_0 g(u_0) \\ &+ \widetilde{B}_k((-1)^{k-1}\rho) - v_k\Big(\frac{f(u_0+u_k) - f(u_0) + B_k((-1)^{k-1}\rho)}{u_k}\Big), \end{aligned} \tag{26}$$

$$\begin{aligned} &\Big(R_{k0}(t) + R_{k1}(\tau,t)\Big)C_{Rk}((-1)^{k-1}\rho) \\ &= \widehat{e}_k(\tau,t)\Big(\frac{f(u_0+u_k) - f(u_0) + B_k((-1)^{k-1}\rho)}{u_k} - A_k((-1)^{k-1}\rho)\Big), \end{aligned} \tag{27}$$

with the boundary conditions (23), $k = 1, 2$.

From (25) we obtain (19). Since $f''(u) > 0$ and $u_1, u_2 > 0$, calculating the limits of (30) (see Lemma 1 below), we find $F(\rho)\big|_{\rho\to+\infty} = 1$ and $F(\rho)\big|_{\rho\to-\infty} < 0$. Hence the equation $F(\rho) = 0$ has at least one root ρ_0.

Thus, applying Proposition 4.1 from [4] to the differential equation (19), we have $\rho(\tau) = \tau\big(1+\psi_1(\tau)\big) \to \rho_0$, $\tau \to -\infty$. That is, if $\tau \in [-\infty, +\infty]$ we have $\rho \in [\rho_0, +\infty]$. Integrating (19), we obtain $\rho = \rho(\tau)$. From (21), (25)–(27), we obtain (16)–(18).

Using the *weak asymptotic* solution constructed in Theorem 2, we obtain a *generalized solution.*

Theorem 3. *The weak limit of the weak asymptotic solution given by Theorem 2 has the form (4), where*

$$\begin{aligned}
\phi_k(t) &= x_k^0 + \frac{[f(u)]_k}{[u]_k} t \\
&\quad + (-1)^k \frac{(f(u_0+u_1+u_2) - f(u_0+u_2))u_2 - (f(u_0+u_2) - f(u_0))u_1}{(u_1+u_2)u_k} \\
&\quad \cdot (t-t^*)H(t-t^*), \qquad (28) \\
e_k(t) &= e_k^0 + \Big([vg(u)]_k - [v]_k \frac{[f(u)]_k}{[u]_k}\Big)t + (-1)^k \Big(\widetilde{B}_2(-\rho_0) \\
&\quad - v_k \frac{(f(u_0+u_1+u_2) - f(u_0+u_2))u_2 - (f(u_0+u_2) - f(u_0))u_1}{(u_1+u_2)u_k}\Big) \\
&\quad \cdot (t-t^*)H(t-t^*).
\end{aligned}$$

Thus, for $t \in (0, t^*)$ we have two delta-shock waves merging after the interaction (at the instant $t = t^*$) into one new delta-shock wave. Here the functions $\phi_k(t)$ and $e_k(t)$ are continuous with respect to t, but their derivatives have jumps at the point $t = t^*$.

Note that systems (25)–(27) with the boundary conditions (23), up to $O_{\mathcal{D}'}(\varepsilon)$, describe the process of merging two delta-shock waves into one.

4 Some lemmas on asymptotic expansions

Lemma 1 ([4], Lemma 1.1). *Let $f(u)$ be a smooth function, and let $u_0(x,t)$, $u_k(x,t)$ be bounded functions. Then*

$$\begin{aligned}
&f\big(u_0(x,t) + u_1(x,t)H_{u1}(-x,\varepsilon) + u_2(x,t)H_{u2}(-x+a,\varepsilon)\big) \\
&\quad = f(u_0(x,t)) + \Big[f(u_0(x,t)+u_1(x,t)) - f(u_0(x,t))\Big]H(-x) \\
&\quad + \Big[f(u_0(x,t)+u_2(x,t)) - f(u_0(x,t))\Big]H(-x+a) + B_1\Big(x,t,\frac{a}{\varepsilon}\Big)H(-x) \\
&\quad + B_2\Big(x,t,-\frac{a}{\varepsilon}\Big)H(-x+a) + O_{\mathcal{D}'}(\varepsilon), \quad \varepsilon \to +0, \qquad (29)
\end{aligned}$$

where the estimate $O_{\mathcal{D}'}(\varepsilon)$ is uniform in a, the "interaction switch functions" are

$$B_1(x,t,\rho) = \int \Big[f'\big(u_0(x,t) + u_1(x,t)\omega_{0u1}(-\eta) + u_2(x,t)\omega_{0u2}(-\eta+\rho)\big) - f'\big(u_0(x,t) + u_1(x,t)\omega_{0u1}(-\eta)\big)\Big] u_1(x,t)\omega_{u1}(-\eta)\,d\eta, \tag{30}$$

$$B_2(x,t,-\rho) = \int \Big[f'\big(u_0(x,t) + u_1(x,t)\omega_{0u1}(-\eta-\rho) + u_2(x,t)\omega_{0u2}(-\eta)\big) - f'\big(u_0(x,t) + u_2(x,t)\omega_{0u2}(-\eta)\big)\Big] u_2(x,t)\omega_{u2}(-\eta)\,d\eta.$$

Similarly to Lemma 1, one can prove the following lemmas.

Lemma 2. *Let $g(u)$ be a smooth function, and let $u_k(x,t)$, $v_k(x,t)$, $k=0,1$, be bounded functions. Then*

$$\begin{aligned} & g\big(u_0(x,t) + u_1(x,t)H_u(-x,\varepsilon)\big)\Big(v_0(x,t) + v_1(x,t)H_v(-x,\varepsilon) \\ &\quad + e(t)\delta_v(-x,\varepsilon) + R(t)\frac{1}{\varepsilon}\Omega''\Big(\frac{-x}{\varepsilon}\Big)\Big) = g\big(u_0(x,t)\big)v_0(x,t) \\ &\quad + \Big[g\big(u_0(x,t) + u_1(x,t)\big)\big(v_0(x,t) + v_1(x,t)\big) - g\big(u_0(x,t)\big)v_0(x,t)\Big]H(-x) \\ &\quad + \Big[e(t)a(t) + R(t)c(t)\Big]\delta(-x) + O_{\mathcal{D}'}(\varepsilon), \quad \varepsilon \to +0, \end{aligned}$$

where $a(t)$, $c(t)$ are defined in (10) *with $\phi(t)=0$.*

Lemma 3. *Let $f(u)$, $g(u)$ be a smooth functions. Then*

$$\begin{aligned} & g\big(u_0 + u_1H_{u1}(-x,\varepsilon) + u_2H_{u2}(-x+a,\varepsilon)\big)\Big(v_0 + v_1H_{v1}(-x,\varepsilon) \\ &\quad + v_2H_{v2}(-x+a,\varepsilon) + e_1(t)\delta_{v1}(-x,\varepsilon) + e_2(t)\delta_{v2}(-x+a,\varepsilon) \\ &\quad + R_1(t,\varepsilon)\frac{1}{\varepsilon}\Omega_1''\Big(\frac{-x}{\varepsilon}\Big) + R_2(t,\varepsilon)\frac{1}{\varepsilon}\Omega_2''\Big(\frac{-x+a}{\varepsilon}\Big)\Big) \\ &= g(u_0)v_0 + \Big[g(u_0+u_1)(v_0+v_1) - g(u_0)v_0\Big]H(-x) + \widetilde{B}_1\Big(\frac{a}{\varepsilon}\Big)H(-x) \\ &\quad + \Big[g(u_0+u_2)(v_0+v_2) - g(u_0)v_0\Big]H(-x+a) + \widetilde{B}_2\Big(-\frac{a}{\varepsilon}\Big)H(-x+a) \\ &\quad + \Big[e_1(t)A_1\Big(\frac{a}{\varepsilon}\Big) + R_1(t,\varepsilon)C_{R1}\Big(\frac{a}{\varepsilon}\Big)\Big]\delta(-x) \qquad (31) \\ &\quad + \Big[e_2(t)A_2\Big(-\frac{a}{\varepsilon}\Big) + R_2(t,\varepsilon)C_{R2}\Big(-\frac{a}{\varepsilon}\Big)\Big]\delta(-x+a) + O_{\mathcal{D}'}(\varepsilon), \quad \varepsilon \to +0, \end{aligned}$$

where the estimate $O_{\mathcal{D}'}(\varepsilon)$ is uniform in a, the "interaction switch functions" are

$$\begin{aligned} \widetilde{B}_1(\rho) = \int \Big\{\Big[g'\big(u_0 + u_1\omega_{0u1}(\eta) + u_2\omega_{0u2}(\eta+\rho)\big)\big(v_0 + v_1\omega_{0v1}(\eta) \\ + v_2\omega_{0v2}(\eta+\rho)\big) - g'\big(u_0 + u_1\omega_{0u1}(\eta)\big)\big(v_0 + v_1\omega_{0v1}(\eta)\big)\Big]u_1\omega_{u1}(\eta) \end{aligned}$$

$$+ \Big[g\big(u_0 + u_1\omega_{0u1}(\eta) + u_2\omega_{0u2}(\eta+\rho)\big) - g\big(u_0 + u_1\omega_{0u1}(\eta)\big)\Big] v_1\omega_{v1}(\eta)\Big\}\, d\eta,$$

$$\widetilde{B}_2(-\rho) = \int \Big\{\Big[g'\big(u_0 + u_1\omega_{0u1}(\eta-\rho) + u_2\omega_{0u2}(\eta)\big)\big(v_0 + v_1\omega_{0v1}(\eta-\rho)$$

$$+ v_2\omega_{0v2}(\eta)\big) - g'\big(u_0 + u_2\omega_{0u2}(\eta)\big)\big(v_0 + v_2\omega_{0v2}(\eta)\big)\Big] u_2\omega_{u2}(\eta)$$

$$+ \Big[g\big(u_0 + u_1\omega_{0u1}(\eta-\rho) + u_2\omega_{0u2}(\eta)\big) - g\big(u_0 + u_2\omega_{0u2}(\eta)\big)\Big] v_2\omega_{v2}(\eta)\Big\}\, d\eta,$$

$$A_1(\rho) = \int g\big(u_0 + u_1\omega_{0u1}(\eta) + u_2\omega_{0u2}(\eta+\rho)\big)\omega_{\delta 1}(\eta)\, d\eta, \tag{32}$$

$$A_2(-\rho) = \int g\big(u_0 + u_1\omega_{0u1}(\eta-\rho) + u_2\omega_{0u2}(\eta)\big)\omega_{\delta 2}(\eta)\, d\eta,$$

$$C_{R1}(\rho) = \int g\big(u_0 + u_1\omega_{0u1}(\eta) + u_2\omega_{0u2}(\eta+\rho)\big)\Omega_1''(\eta)\, d\eta,$$

$$C_{R2}(-\rho) = \int g\big(u_0 + u_1\omega_{0u1}(\eta-\rho) + u_2\omega_{0u2}(\eta)\big)\Omega_2''(\eta)\, d\eta.$$

References

1. Shelkovich, V.M. (1995): An associative-commutative algebra of distributions that includes multiplicators, generalized solutions of nonlinear equations. Mathematical Notices, **57**, no. 5, 765–783.
2. Danilov, V.G. (2002): Generalized solutions describing singularity interaction. Int. J. of Math. and Math. Sci., **29**, no. 8, 481–494.
3. Danilov, V.G., Shelkovich, V.M., (2001): Propagation and interaction of nonlinear waves to quasilinear equations. In: Hyperbolic problems: Theory, Numerics, Applications (Eighth International Conference in Magdeburg, February/March 2000, v.I). International Series of Numerical Mathematics, v. 140, Birkhäuser Verlag, Basel/Switzerland, pp. 267–276.
4. Danilov, V.G., Shelkovich, V.M., (2001): Propagation and interaction of shock waves of quasilinear equation. Nonlinear Studies, **8**, no. 1, 135–169.
5. Danilov, V.G., Shelkovich, V.M., (2002): Propagation and interaction of delta-shock waves. In: The Ninth International Conference on Hyperbolic Problems. Theory, Numerics, and Applications. Abstracts. California Institute of Technology, California USA, March 25–29, 2002, pp. 106–110.
6. Le Floch, Ph., (1990): An existence and uniqueness result for two nonstrictly hyperbolic systems. In: Nonlinear Evolution Equations That Change Type, Springer-Verlag, pp. 126–138.
7. Dechun Tan, Tong Zhang, Yuxi Zheng, (1994): Delta shock waves as limits of vanishing viscosity for hyperbolic systems of conservation laws. Journal of Differential Equations, **112**, 1–32.
8. Ercole, G., (2000): Delta-shock waves as self-similar viscosity limits. Quart. Appl. Math, **LVIII**, no. 1, 177–199.

Multidimensional Compressible Flows with Symmetry

David Hoff[1] and Helge Kristian Jenssen[2]

[1] Department of Mathematics, Indiana University, Rawles Hall, Bloomington, IN 47405, U.S.A. *hoff@indiana.edu*. Research supported in part by the NSF under Grant No. DMS-99866658.

[2] Department of Mathematics, Indiana University, Rawles Hall, Bloomington, IN 47405, U.S.A. *kjenssen@indiana.edu*. Research supported in part by the NSF under Grant No. DMS-99866658.

We prove the global existence of weak solutions of the Navier-Stokes equations for compressible, heat-conducting flow,

$$\begin{cases} \rho_t + \operatorname{div}(\rho\mathbf{U}) = 0 \\ (\rho\mathbf{U}^j)_t + \operatorname{div}(\rho\mathbf{U}^j\mathbf{U}) + P(\rho,\theta)_{x_j} = \mu\Delta\mathbf{U}^j + (\lambda+\mu)\operatorname{div}\mathbf{U}_{x_j} + \rho\mathbf{F}^j \\ (\rho E)_t + \operatorname{div}(\rho E\mathbf{U} + P(\rho,\theta)\mathbf{U}) = \Delta(\kappa\theta + \frac{1}{2}\mu|\mathbf{U}|^2) \\ \qquad\qquad +\mu\operatorname{div}((\nabla\mathbf{U})\mathbf{U}) + \lambda\operatorname{div}((\operatorname{div}\mathbf{U})\mathbf{U}) + \rho\mathbf{U}\cdot\mathbf{F}, \end{cases} \tag{1}$$

with initial data $(\rho_0, \mathbf{U}_0, \theta_0)$ and external force $\mathbf{F}$ which are large, discontinuous, and spherically or cylindrically symmetric. Here $x \in \mathbb{R}^3$ is the spatial coordinate, $t > 0$ is time, and $\rho, \mathbf{U}, \theta, E = \frac{1}{2}|\mathbf{U}|^2 + \theta$, and $P = K\rho\theta$ are respectively the density, velocity, temperature, energy density, and pressure of an ideal fluid (with unit specific heat), μ and λ are viscosity constants assumed to satisfy $\mu > 0$ and $3\lambda + 2\mu > 0$, and κ is a positive heat conduction coefficient.

For the symmetric cases under consideration here, (1) reduces to

$$\begin{cases} \rho_t + (\rho u)_\xi = 0 \\ (\rho u)_t + (\rho u^2)_\xi - \dfrac{\rho v^2}{r} + P_r = (\lambda+2\mu)u_{\xi r} + \rho f^1 \\ (\rho v)_t + (\rho uv)_\xi + \dfrac{\rho uv}{r} = \mu v_{\xi r} + \rho f^2 \\ (\rho w)_t + (\rho uw)_\xi = \mu w_{r\xi} + \rho f^3 \\ (\rho\theta)_t + (\rho u\theta)_\xi + Pu_\xi = \kappa\theta_{r\xi} + Q, \end{cases} \tag{2}$$

where $\dfrac{\partial}{\partial\xi} = \dfrac{\partial}{\partial r} + \dfrac{m}{r}$, $m = 1$ or 2, and Q is a positive definite, homogeneous quadratic in u_ξ, $\dfrac{u}{r}$, $v_r - \dfrac{mv}{r}$, and w_r. Specifically, in the spherically symmetric case, $m = 2, r(x) = |x|$,

$$\mathbf{U}(x,t) = u(r,t)\frac{x}{r}, \ \ \mathbf{F}(x,t) = f^1(r,t)\frac{x}{r},$$

and $v = w \equiv 0$. In the cylindrically symmetric case, $m = 1$, $r(x) = \sqrt{x_1^2 + x_2^2}$, and

$$\mathbf{U}(x,t) = u(r,t)\frac{(x_1, x_2, 0)}{r} + v(r,t)\frac{(x_2, -x_1, 0)}{r} + w(r,t)(0,0,1),$$

$\mathbf{F}$ being constructed from (f^1, f^2, f^3) in a similar way. In both cases we write $\rho(x,t) = \rho(r,t)$ and $\theta(x,t) = \theta(r,t)$.

We pose the system (1)-(2) in the region $\Omega = \{x \in \mathbb{R}^3 : r(x) < b\}$ with boundary conditions

$$\mathbf{U} = 0, \ \theta_r = 0 \text{ on } \partial\Omega, \tag{3}$$

and we obtain the solution as the limit as $a \to 0$ of corresponding solutions in the annular regions $\{x : a < r(x) < b\}$. This limit is particularly interesting near $r = 0$, which is the single point at which the problem is truly multidimensional. As we shall see, the analysis allows for the emergence of a vacuum state ($\rho = 0$) at the origin, and for the possibility that energy is absorbed into the point (or set) $r = 0$, and is lost in the limit as $a \to 0$.

We now give a more precise formulation of our results. We assume that the external force $\mathbf{F}$ satisfies

$$\mathbf{F} \in L^1([0,T]; L^\infty(\Omega)) \cap L^\infty([0,T]; L^2(\Omega)) \tag{4}$$

for each $T > 0$, and that the initial data $(\rho_0, \mathbf{U}_0, \theta_0)$ satisfies

$$\begin{cases} C_0^{-1} \le \rho_0 \le C_0, \ C_0^{-1} \le \theta_0 \text{ a.e.,} \\ \displaystyle\int_\Omega \rho_0 S(\rho_0, \mathbf{U}_0, \theta_0)dx \le C_0 \end{cases} \tag{5}$$

for a constant C_0. Here S is the entropy density

$$S(\rho, \mathbf{U}, \theta) = K\psi(\rho^{-1}) + \psi(\theta) + \tfrac{1}{2}|\mathbf{U}|^2,$$

where $\psi(s) = s - \log s - 1$. Observe that there are no smallness or regularity conditions imposed on $\mathbf{F}$ or on $(\rho_0, \mathbf{U}_0, \theta_0)$. We prove:

Theorem *Assume that the system parameters P, λ, μ, and κ, the external force $\mathbf{F}$, and the initial data $(\rho_0, \mathbf{U}_0, \theta_0)$ are as described above. Then there exist $\underline{r}, \rho, \mathbf{U}$, and θ such that:*

1. $\underline{r} : [0,\infty) \to [0,\infty)$ is a semicontinuous curve, and the set

$$\mathcal{F} = \{(x,t) : t > 0 \text{ and } \underline{r}(t) < r(x) < b\};$$

is open;

2. $\rho \in C([0,\infty); W^{-1,1}(\Omega)) \cap L^\infty_{loc}(\mathcal{F})$, and $\rho(\cdot,t) = 0$ a.e. in $\{x : r(x) \le \underline{r}(t)\}$ when $\underline{r}(t) > 0$;

3. $\mathbf{U}$ and θ are Hölder continuous on compact subsets of $\mathcal{F}$, and for test functions $\varphi \in W_0^{2,\infty}(\Omega)$ with supp $\varphi \times [t_1, t_2] \subseteq \mathcal{F}$, the mappings

$$t \mapsto \int_\Omega \rho \mathbf{U} \varphi \, dx \text{ and } t \mapsto \int_\Omega \rho \theta \varphi \, dx$$

are continuous on $[t_1, t_2]$;

4. if $\rho\mathbf{U}$ is taken to be zero outside $\mathcal{F}$, then the weak form

$$\int_\Omega \rho \varphi \, dx \, \Big|_{t_1}^{t_2} = \int_{t_1}^{t_2} \int_\Omega (\rho \varphi_t + \rho \mathbf{U} \cdot \nabla \varphi) \, dx dt$$

of the mass equation in (1) holds for test functions $\varphi \in C^1(\bar{\Omega} \times [t_1, t_2])$;

5. the weak forms of the momentum and energy equations,

$$\int_\Omega \rho \mathbf{U}^j \varphi \, dx \, \Big|_{t_1}^{t_2} - \int_{t_1}^{t_2} \int_\Omega (\rho \mathbf{U}^j \varphi_t + \rho \mathbf{U}^j \mathbf{U} \cdot \nabla \varphi + P(\rho, \theta) \phi_{x_j}) dx dt$$
$$= \int_{t_1}^{t_2} \int_\Omega (\mu \mathbf{U}^j \triangle \varphi + (\lambda + \mu) \mathbf{U} \cdot \nabla \phi_{x_j} + \rho \mathbf{F}^j \varphi) dx dt,$$

$$\int_\Omega \rho E \varphi \, dx \, \Big|_{t_1}^{t_2} - \int_{t_1}^{t_2} \int_\Omega (\rho E \varphi_t + \rho E \mathbf{U} \cdot \nabla \varphi + P(\rho, \theta) \mathbf{U} \cdot \nabla \varphi) dx dt$$
$$= \int_{t_1}^{t_2} \int_\Omega \left[(\kappa \theta + \tfrac{1}{2} \mu |\mathbf{U}|^2) \triangle \varphi + \rho \mathbf{U} \cdot \mathbf{F} \varphi \right] dx dt$$
$$- \int_{t_1}^{t_2} \int_\Omega \left[\mu \nabla \varphi \cdot ((\nabla \mathbf{U}) \mathbf{U}) + \lambda (\text{div } \mathbf{U}) \nabla \varphi \cdot \mathbf{U} \right] dx dt$$

hold for test functions $\varphi \in C^2(\bar{\Omega} \times [t_1, t_2])$ for which supp $\varphi \cap (\Omega \times [t_1, t_2]) \subset \mathcal{F}$;

6. in the spherically symmetric case, and in the cylindrically symmetric case with $v, w \equiv 0$, $\mathbf{U}$ may be defined as a distribution in all of $\Omega \times (0, \infty)$, the second equation in (1) holds in the sense of distributions on $\Omega \times (0, \infty)$, and $\rho \mathbf{U} \in C([0, \infty); (W_0^{2,\infty}(\Omega)^*)$;

7. the total energy associated to the solution $(\rho, \mathbf{U}, \theta)$, minus the work done by the external force, is nonincreasing in time:

$$\int_\Omega \rho_0 E_0 dx \geq \int_\Omega \rho(x,T) E(x,T) dx - \int_0^T \int_\Omega \rho \mathbf{U} \cdot \mathbf{F} dx dt$$
$$= \int_\Omega \rho_0 E_0 dx - \lim_{R \to 0} \lim_{a \to 0} \int_{\{a \leq r(x) \leq a + R\}} \rho^a(x,T) E^a(x,T) dx,$$

the equality on the right holding in the sense of distributions on $(0, \infty)$.

Complete details of the proof will be presented in [5].

We give here a brief description of the analysis, and we make a number of comments concerning the interpretation of the above results. As indicated earlier, the solution $(\rho, \mathbf{U}, \theta)$ is obtained as the limit as $a \to 0$ of solutions $(\rho^a, \mathbf{U}^a, \theta^a)$ in the annular region $\{x : a < r(x) < b\}$. These approximate solutions have been studied extensively by Frid and Shelukhin [2], who derive a number of a priori bounds, including bounds for the mass, energy, and entropy,

$$\int \rho^a dx, \int \rho^a E^a dx, \int \rho^a S(\rho^a, \mathbf{U}^a, \theta^a) dx \leq C, \tag{6}$$

where C is independent of a, as well as a number of pointwise and higher order regularity bounds which do depend on a. Among the latter are pointwise bounds $C_a^{-1} \leq \rho^a \leq C_a$, established by a rather clever and elegant version of an argument given in Kazhikov-Shelukhin [6]. In the present work we modify this argument in a nontrivial way to establish pointwise bound for ρ^a, *independent of* a, but only away from the origin of Lagrangean space. Specifically, we define the particle position $r_h^a(t)$ by

$$h = \int_{\{a \leq r(x) \leq r_h^a(t)\}} \rho^a(x,t) dx$$

for given $h \geq 0$, and we show that, for $h > 0$ and $T > 0$, there is a constant $C = C(h,T)$ such that

$$C(h,T)^{-1} \leq \rho^a(x,t) \leq C(h,T) \tag{7}$$

for $t \leq T$ and $r(x) \geq r_h^a(t)$. The bounds (7) then serve as the starting point for the derivation of higher-order regularity estimates for $\mathbf{U}^a$ and θ^a (away from $t = 0$) in which the bounding constants are independent of a, as in (7), but which hold in regions $\{x : r(x) \geq r_h^a(t)\}$ which do depend on a. On the other hand, $\dfrac{\partial r_h^a}{\partial t} = u^a(r_h^a, t)$, so that uniform estimates for $\mathbf{U}^a$ imply the convergence as $a \to 0$ (modulo a subsequence) of $r_h^a(\cdot)$ to Hölder continuous curves $r_h(\cdot)$, for $h > 0$. Since $r_h(t)$ is increasing in h, the limit

$$\underline{r}(t) = \lim_{h \to 0} r_h(t)$$

exists, and we may define the "fluid region"

$$\mathcal{F} = \{(x,t) : \underline{r}(x) < r(x) < b \text{ and } t > 0\}.$$

As asserted in (a) of the Theorem, $\underline{r}$ is semicontinuous, so that $\mathcal{F}$ is an open set. The sequences $\{\mathbf{U}^a\}$ and $\{\theta^a\}$ are then uniformly Hölder continuous on compact subsets of $\mathcal{F}$, and so have subsequences converging strongly, locally in $\mathcal{F}$. The pointwise bounds (7) yield weak limits $\rho^a(\cdot,t) \to \rho(\cdot,t)$ on $\{x : \underline{r}(t) + \delta \le r(x) \le b\}$ for $\delta > 0$, and it can be shown as well that $\rho^a(\cdot,t) \to 0$ in $L^1(\{x : r(x) \le \underline{r}(t)\})$.

We do not have sufficient information to infer that $\underline{r}(t) \equiv 0$, nor do we know whether solutions exist for which $\underline{r} \not\equiv 0$. The analysis simply shows that $\underline{r}(t)$ may be positive, and that, if it is, a vacuum state of radius $\underline{r}(t)$ centered at the origin has emerged.

The weak forms described in (4) and (5) of the Theorem follow in a straightforward way from the corresponding weak forms for $(\rho^a, \mathbf{U}^a, \theta^a)$. In the case $v, w_0 \equiv 0$, a little more can be said about the limit of $\mathbf{U}^a$ outside $\mathcal{F}$: we can invert the operator $L = \dfrac{\partial^2}{\partial r \partial \xi}$ on the interval $[a,b]$, subject to zero Dirichlet boundary conditions, to express u^a in terms of quantities known to be converging in $\mathcal{D}'((0,b) \times (0,\infty))$. The Green's function for L and its dependence on a are known explicitly, and can be used to obtain $\mathbf{U}$ in all of $\Omega \times (0,\infty)$ as the $\mathcal{D}'$-limit of $\mathbf{U}^a$. This gives the somewhat stronger result stated in (6). The argument fails for v and w, however, partly because of the different behavior of the Green's function for the operator $\dfrac{\partial^2}{\partial \xi \partial r}$ appearing in the w-equation. It fails as well for θ because there is no way to interpret the mechanical work term Q, which is quadratic in $\nabla \mathbf{U}$, in the complement of $\mathcal{F}$, where $\mathbf{U}$ is known only to be a distribution.

Viewing these results more generally, we may regard the restrictions in (5) that the test function be supported in $\mathcal{F}$ as reasonable, since there is no fluid outside $\mathcal{F}$, and the model is not really valid there. On the other hand, this failure to establish the weak forms (of the energy equation and of the momentum equation when $w \not\equiv 0$) in all of $\Omega \times (0,\infty)$ suggests that we may anticipate significant difficulty in extending these results to more general, nonsymmetrical flows, in which vacuum states may presumably arise spontaneously at unpredictable points of space-time. Additionally, the failure of the analysis to detect whether or not energy is lost (see (7)) calls into question the adequacy of the mass, energy, and entropy bounds (6), which are the only noncontingent a priori bounds known to hold in the multidimensional case, to yield a satisfactory understanding of the solution operator for these equations.

The theorem of the present paper extends the results of Hoff [3] for the barotropic case (in which $P = P(\rho)$ and the energy equation is omitted) in that more general initial data and completely general velocity directions may occur, and most important, that heat conduction and temperature-dependent equations of state are allowed. In the simpler barotropic case, the curve $\underline{r}$ is known to be Hölder continuous, and when vacuum states do occur, they occupy open sets in space-time, in which the velocity satisfies a simple homogeneous elliptic equation. As indicated in the discussion above, the situation is considerably more complicated in the nonbarotropic case considered here.

Other global existence results for the system (1) are given in Hoff [4] for two and three dimensional flows with *nonsymmetric* but *small* initial data, and by Lions (see Lions [7], Feireisl [1], and the references in these papers) for the corresponding barotropic system with $P(\rho) = K\rho^\gamma$ for certain γ, and for *large* data. As far as we are aware, there are no large-data existence results, other than those of the present paper, for the full nonbarotropic system (1) in a truly multidimensional setting.

References

1. Edward Feireisl, The dynamical system approach to the Navier-Stokes equations of compressible fluids. Advances in Math. Fluid Mech. (Paseky, 1999), 35-66, Springer, Berlin, 2000.
2. H. Frid, V. Shelukhin, Vanishing shear viscosity in the equations of compressible fluids for the flows with the cylinder symmetry, SIAM J. Math. Anal. **31** (2000), 1144–1156.
3. D. Hoff, Spherically symmetric solutions of the Navier-Stokes equations for compressible, isothermal flow with large, discontinuous initial data, Indiana Univ. Math. J. **41** (1992), 1225–1302.
4. D. Hoff, Discontinuous solutions of the Navier-Stokes equations for multidimensional flows of heat-conducting fluids, Arch. Rational Mech. Anal. **139** (1997), 303–354.
5. David Hoff and Helge Kristian Jenssen, Multidimensional heat-conducting compressible flows with large data and symmetry, in preparation.
6. A. V. Kazhikhov, V. V. Shelukhin, Unique global solution with respect to time of initial-boundary value problems for one-dimensional equations of a viscous gas, J. Appl. Math. Mech. **41** (1977), 273–282.; translated from Prikl. Mat. Meh. **41** (1977), 282–291 (Russian)
7. P. L. Lions, Mathematical topics in Fluid Mechanics, vol 2., Oxford, 1998.

Efficient Higher–Order Finite Volume Schemes for (Real Gas) Magnetohydrodynamics

Andreas Dedner, Christian Rohde, and Matthias Wesenberg

Institut für Angewandte Mathematik, Universität Freiburg, Freiburg, Germany
dedner|chris|wesenber@mathematik.uni-freiburg.de

Summary. The computational costs for solving the real gas Euler or MHD equations can be strongly reduced if we use an adaptive table instead of the full equation of state. We demonstrate this behaviour for an example from solar physics. Moreover, we introduce a new limiter suitable for second–order finite volume schemes which are based on linear reconstructions on unstructured triangular grids. This new limiter cures several problems of the approaches commonly used. Finally, we show that local grid adaption always seems to pay off in 1d and 2d, whereas a high–resolution first–order scheme can be more efficient (in terms of computational time versus error) than the second–order schemes.

1 Introduction

In a joint project with the group of M. Schüssler at the Max–Planck–Institut für Aeronomie (Katlenburg–Lindau, Germany) we develop a 3d finite volume code for simulations in both the convection zone and the photosphere of the sun. The physics in the convection zone can be modeled by the MHD equations for a perfect gas (cf. (1)–(4)) which fulfils $p = \frac{R}{\mu}\rho T$ for the temperature T, the general gas constant R, and the *constant* mean molecular weight μ. However, in the photosphere μ is a *function* of ρ and T due to the partial ionization of the plasma. This leads to a general equation of state (EOS) which is computationally expensive since its evaluation requires iterative procedures. In addition, we also have to consider the effects of radiation within the energy balance.

Our code incorporates locally adaptive tetrahedral grids with dynamic load balancing thus ensuring an efficient parallelization; results of a first realistic simulation are presented in [3]. During the different stages of implementation we often faced the situation that either suitable methods for our setting were not available or the existing methods could not be used. The latter happened if methods which had been developed and tested for simple model problems crashed if they were applied to realistic data or proved extremely inefficient. Thus we came up with a variety of new and improved methods which indeed work for our applications and which we tested thoroughly with respect to their efficiency. In [8] we introduced the MHD–HLLEM approximate Riemann solver; a new class of methods solving the radiation transport equation was suggested in [5]. Transparent boundary conditions for stratified atmospheres were derived in [4], and a new approach to divergence cleaning was presented in [2]. Moreover, two general approaches for extending perfect gas MHD solvers to the real gas case are considered in [6].

In this paper we present several results which we obtained with our MHD examples in mind, but which nevertheless cover a significantly broader range of applications. After the presentation of the system of equations in Sect. 2, we discuss three approaches which play a major role in enhancing the efficiency of our scheme. In Sect. 3 we show that for an expensive EOS the use of tabularized values in place of direct evaluations can reduce the computational costs of (magneto)hydrodynamic simulations considerably. However, due to the enormous memory requirements of a fixed table we suggest using *adaptively refined* tables. While in one space dimension it is straightforward to implement a scheme which is based on limited linear reconstructions and which shows second–order convergence in numerical tests, this is a delicate task in two space dimensions. We identified two main obstacles on unstructured (triangular) grids in 2D: inappropriate limiters and a tremendous influence of the grid structure. Both issues are addressed in Sect. 4. We suggest a new limiter which cures the shortcomings of the limiters usually used and compare the efficiency of the first and second order schemes in 2d. In Sect. 5 we investigate if local grid adaption always pays off in the sense that we obtain the same results in less time and if — in terms of computational efficiency — local grid adaption or a second–order extension is more important. Finally, we summarize our results in Sect. 6.

2 Governing Equations

The equations of magnetohydrodynamics (MHD) are a model for a compressible, inviscous, and electrically conducting gas in the presence of a magnetic field. They are derived from the Euler equations of gas dynamics and the Maxwell equations. Written in divergence form they read

$$\begin{aligned}
\partial_t \rho + \nabla \cdot (\rho \mathbf{u}) &= 0 \quad \textit{(conservation of mass)}, && (1a)\\
\partial_t (\rho \mathbf{u}) + \nabla \cdot (\rho \mathbf{u}\mathbf{u}^T + \mathcal{P}) &= 0 \quad \textit{(conservation of momentum)}, && (1b)\\
\partial_t \mathbf{B} + \nabla \cdot (\mathbf{u}\mathbf{B}^T - \mathbf{B}\mathbf{u}^T) &= 0 \quad \textit{(induction equations)}, && (1c)\\
\partial_t (\rho e) + \nabla \cdot (\rho e \mathbf{u} + \mathcal{P}\mathbf{u}) &= 0 \quad \textit{(conservation of energy)}, && (1d)\\
\nabla \cdot \mathbf{B} &= 0 \quad \textit{(divergence constraint)}. && (1e)
\end{aligned}$$

For the density ρ, the velocity $\mathbf{u} = (u_x, u_y, u_z)^T$, and the magnetic field $\mathbf{B} = (B_x, B_y, B_z)^T$ the total energy e is the sum of the internal, kinetic, and magnetic energies:

$$e = \varepsilon + \frac{1}{2}|\mathbf{u}|^2 + \frac{1}{8\pi\rho}|\mathbf{B}|^2. \tag{2}$$

If $\mathcal{I}$ denotes the unit tensor, the pressure tensor $\mathcal{P}$ is defined as

$$\mathcal{P} := \left(p + \frac{1}{8\pi}|\mathbf{B}|^2\right)\mathcal{I} - \frac{1}{4\pi}\mathbf{B}\mathbf{B}^T. \tag{3}$$

The gas pressure $p(\tau, \varepsilon)$ is given by a suitable *equation of state* (EOS), where $\tau := 1/\rho$ is the specific volume. The simplest example is the EOS of a perfect gas

$$p(\tau, \varepsilon) = (\gamma - 1)\frac{\varepsilon}{\tau}. \tag{4}$$

For $\gamma = 1.4$ this EOS is a model for air if the pressure and the temperature are moderate. For low temperatures and high pressures the EOS also has to respect the influence of intermolecular forces. One possible model in this situation is the van der Waals–EOS

$$p(\tau, \varepsilon) = \frac{R\left(\varepsilon - \varepsilon_0 + \frac{a}{\tau}\right)}{c_V(\tau - b)} - \frac{a}{\tau^2} \tag{5}$$

for suitably chosen constants a, b, R, c_V, and ε_0. We use (5) as a simple example for a general EOS. The system (1) has to be closed by suitable inital and boundary conditions.

3 Adaptive Tables for an Equation of State

The evaluation of the EOS which we have to use in the solar photosphere involves costly iterative procedures. This even holds for the most simple model of a pure, ionizing hydrogen gas. Hence it is not surprising that in one of our simulations we were able to save 90% of computational time by using interpolated values from a cartesian table instead of direct evaluations of the EOS. However, in [6] we have shown that in general the table and the compuational grid have to be refined *simultaneously* to obtain grid convergence (see also Fig. 2 left) and that too coarse tables may lead to strongly distorted solutions. On the other hand, a cartesian table with sufficient resolution is not always feasible: For a relative error of approximately 1.5% in the case of partial ionization we need a table with 25 million entries which requires several Gbytes of memory, see Fig. 1 left. With the adaptive strategy described below we maintain the same level of accuracy using 70% less memory (cf. Fig. 1 right). Thus we see that an *adaptive* table is a powerful tool for speeding up (magneto)hydrodynamic simulations with an expensive EOS.

The structure of the EOS can be significantly different if p is considered as a function of τ and ε or of ρ and ε. For example, the EOS of a perfect gas (4) is a linear function of ρ and ε, but not of τ. Therefore, the size of the adaptive table may vary significantly whether it is stored with respect to τ or ρ. In principle, it suffices to create a single table for p, since for all related quantities like the temperature, the speed of sound, or even the internal energy $\varepsilon(\tau, p)$ we can use approximated values. However, if possible we prefer using a second table for $\varepsilon(\tau, p)$ and storing also the temperature and

Algorithm 1 Adaptive table for EOS.

Require: $I > 1,\ J > 1,\ K > 2,\ L > 1$
create an *empty* $I \times J$–grid $\left(\tau_i, \varepsilon_j\right)_{1\le i\le I,\, 1\le j\le J}$ on level 0.
for all $p(\tau,\varepsilon)$ required during the simulation **do**
  compute (i_0, j_0) with $(\tau,\varepsilon) \in [\tau_{i_0}, \tau_{i_0+1}] \times [\varepsilon_{j_0}, \varepsilon_{j_l+1}]$
  $l \leftarrow 1$
  while the value of $p(\tau,\varepsilon)$ is unknown **do**
    if the subgrid at $(\tau_{i_0,\dots,i_{l-1}}, \varepsilon_{j_0,\dots,j_{l-1}})$ on level l exists **then**
      compute (i_l, j_l) with $(\tau,\varepsilon) \in [\tau_{i_0,\dots,i_l}, \tau_{i_0,\dots,i_l+1}] \times [\varepsilon_{j_0,\dots,j_l}, \varepsilon_{j_0,\dots,j_l+1}]$
      if the subgrid at $(\tau_{i_0,\dots,i_{l-1}}, \varepsilon_{j_0,\dots,j_{l-1}})$ on level l contains values **then**
        determine $p(\tau,\varepsilon)$ from $\left\{p(\tau_{i_0,\dots,i_l+r}, \varepsilon_{j_0,\dots,j_l+s})\middle| r, s \in \{0,1\}\right\}$.
      else
        $l \leftarrow l+1$
    else
      create *empty* $K \times K$–subgrid at $(\tau_{i_0,\dots,i_{l-1}}, \varepsilon_{j_0,\dots,j_{l-1}})$ on level l
      if $l = L$ or the error tolerance is reached on level $l-1$ **then**
        compute and store $\left\{p(\tau_{i_0,\dots,i_{l-1},r}, \varepsilon_{j_0,\dots,j_{l-1},s})\middle| r, s \in \{1,\dots K\}\right\}$

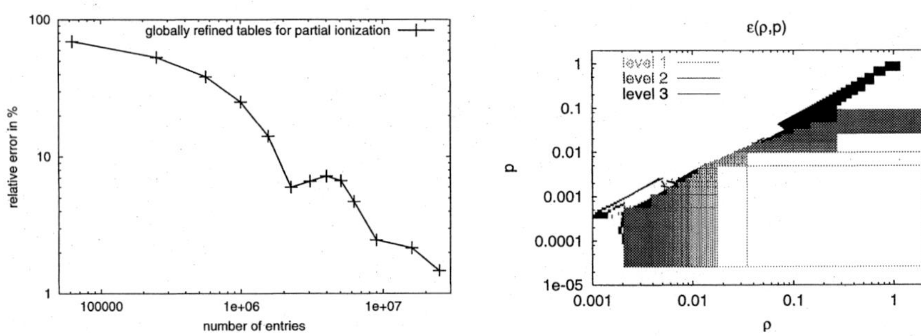

Fig. 1. Results for a pure, ionizing hydrogen gas. Left: Relative error versus required size of a cartesian EOS table. Right: Level distribution for a table of $\varepsilon(\rho, p)$.

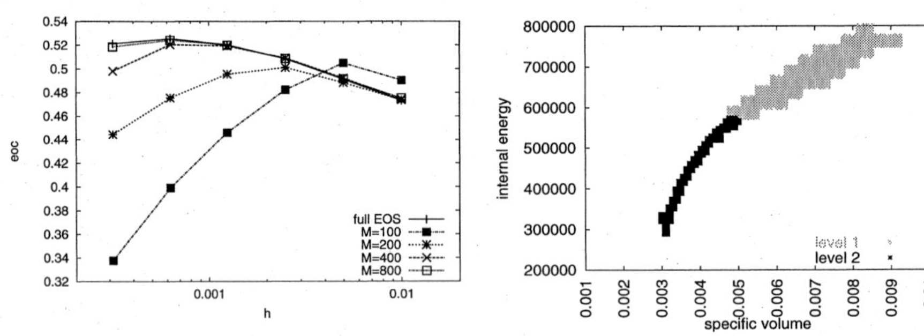

Fig. 2. Left: Results of the first order RGDW scheme (cf. [6]) for the van der Waals shock tube problem "Euler 1" from [6]; grid spacing h versus experimental order of convergence (eoc) for full EOS (solid line) and cartesian tables with resolution $M \times M$ (dotted/dashed lines). Right: Levels of refinement in an adaptive table for p as function of τ and ε for the same problem.

the speed of sound in both tables. The reduction of table size by using the adaptive algorithm is due to the fact that typically values are only required in the vicinity of the tables' diagonals, see Fig. 1 right and Fig. 2 right.

We implemented our adaptively refined tables for the EOS in the following manner: For example, if we want to tabularize $p(\tau, \varepsilon)$ in the prescribed domain $[\tau_{\min}, \tau_{\max}] \times [\varepsilon_{\min}, \varepsilon_{\max}]$, we start with an *empty* $I \times J$–grid on level $l = 0$. This means that we do not compute any values of $p(\tau, \varepsilon)$, but just choose uniformly distributed gridpoints τ_i and ε_j for $1 \leq i \leq I$ and $1 \leq j \leq J$ with $\tau_{i+1} - \tau_i = (\tau_{\max} - \tau_{\min})/(I-1)$ and $\varepsilon_{j+1} - \varepsilon_j = (\varepsilon_{\max} - \varepsilon_{\min})/(J-1)$. During a simulation we apply Algorithm 1 and successively build a hierarchical table of cartesian subtables. These tables contain only values in the vicinity of points (τ, ε) which actually occur in the computation. Since all (sub)tables are cartesian it is straightforward to compute the position of any value (τ, ε) within the tables (lines 3 and 7 of Algorithm 1). In general, only very few levels are required. For example, the finest level of an adaptive table with $I = 10$, $J = 6$, $K = 6$ and $L = 4$ equals an cartesian table with $10 \cdot 6 \cdot (5^4)^2 = 23437500$ entries. Note that we never compute values on level $l = 0$ since we do not assume that system (1) is hyperbolic in the whole domain $[\tau_{\min}, \tau_{\max}] \times [\varepsilon_{\min}, \varepsilon_{\max}]$. The table is refined until the maximal level L or the prescribed error tolerance tol is reached, see line 14 of Algorithm 1. To check the latter case we compute

$$\begin{aligned} \tau_{i_0,\ldots,i_{l-1}+\frac{1}{2}} &:= \frac{1}{2}\left(\tau_{i_0,\ldots,i_{l-1}} + \tau_{i_0,\ldots,i_{l-1}+1}\right), \\ \varepsilon_{j_0,\ldots,j_{l-1}+\frac{1}{2}} &:= \frac{1}{2}\left(\varepsilon_{j_0,\ldots,j_{l-1}} + \varepsilon_{j_0,\ldots,j_{l-1}+1}\right), \end{aligned}$$

and the interpolated value

$$\begin{aligned} p^{j_0,\ldots,j_{l-1}+\frac{1}{2}}_{i_0,\ldots,i_{l-1}+\frac{1}{2}} :=\frac{1}{4}\Big(& p\left(\tau_{i_0,\ldots,i_{l-1}}, \varepsilon_{j_0,\ldots,j_{l-1}}\right) + p\left(\tau_{i_0,\ldots,i_{l-1}+1}, \varepsilon_{j_0,\ldots,j_{l-1}}\right) \\ & + p\left(\tau_{i_0,\ldots,i_{l-1}}, \varepsilon_{j_0,\ldots,j_{l-1}+1}\right) + p\left(\tau_{i_0,\ldots,i_{l-1}+1}, \varepsilon_{j_0,\ldots,j_{l-1}+1}\right)\Big). \end{aligned}$$

We assume that the subtable on level l will be sufficiently accurate if

$$\frac{\left|p^{j_0,\ldots,j_{l-1}+\frac{1}{2}}_{i_0,\ldots,i_{l-1}+\frac{1}{2}} - p\left(\tau_{i_0,\ldots,i_{l-1}+\frac{1}{2}}, \varepsilon_{j_0,\ldots,j_{l-1}+\frac{1}{2}}\right)\right|}{p\left(\tau_{i_0,\ldots,i_{l-1}+\frac{1}{2}}, \varepsilon_{j_0,\ldots,j_{l-1}+\frac{1}{2}}\right)} \leq \mathrm{tol}\,.$$

4 Second Order by Limited Linear Reconstruction

In [8] we compared, among other solvers, the well–known HLLE scheme, the DW–solver based on [1], and our new MHD–HLLEM Riemann solver in the case of perfect gases. We found that for both smooth and non–smooth problems in 1d and 2d MHD–HLLEM is the most efficient solver in terms of computational time versus error. Furthermore, by means of componentwise linear reconstruction using the minmod limiter and a Runge–Kutta time integration it was straightforward to construct a 1d scheme which shows second–order convergence for smooth problems. In our tests the second–order scheme was *always* more efficient that the corresponding first–order scheme. After two remarks concerning the choice of variables we discuss the question whether the one–dimensional results mentioned above also hold in 2d.

Apart from the use of EOS tables as described in the previous section we encountered two further situations in which the results may differ significantly depending on whether we work in conservative or in primitive variables. On the one hand, if we follow the "direct Riemann solver approach" from [6] in the real gas case, the use of primitive variables within the solvers whereever possible saves (potentially) costly conversions between conservative and primitive variables. Thus the RGDW–solver introduced in [6] may be more efficient for some problems than our real gas MHD–HLLEM solver (RGHLLEM) since RGDW works well in primitive variables, whereas for the sake of robustness we implemented RGHLLEM completely in conservative variables. On the other hand, a reconstruction of the constant volume data in conservative variables can cause spurious oscillations, see Fig. 3.

The central part of extending a first order finite volume scheme to second order is the construction of linear functions for each element and each scalar quantity, which are then used for the evaluation of the one–dimensional fluxes at the interfaces. We follow the approach in [7], since it relies on a minimal domain of dependence, which is important for an efficient parallelization. If the constant volume value U_0 in T_0 is neither a local minimum nor a local maximum in comparison with the values U_1, U_2, and U_3 in T_1, T_2, and T_3, respectively (see Fig. 4 left), we proceed as follows: For $l \in \{1,2,3\}$ the values (U_0, U_i, U_j) and the corresponding barycenter triple (w_0, w_i, w_j) for $1 \le i < j \le 3$ with $\{i,j,l\} = \{1,2,3\}$ uniquely define a linear function

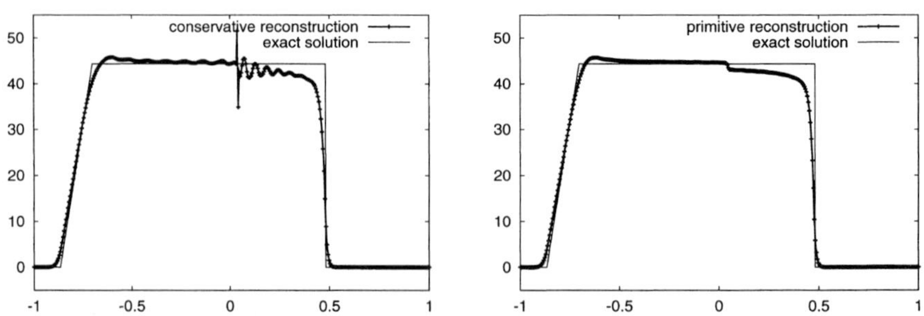

Fig. 3. Results of the second order RGDW scheme for the problem studied in Fig. 2; spatial position x versus u_x. Across the contact discontinuity u_x is constant (solid lines). The reconstruction in conservative variables introduces spurious oscillations (left), which do not occur if we reconstruct in primitive variables.

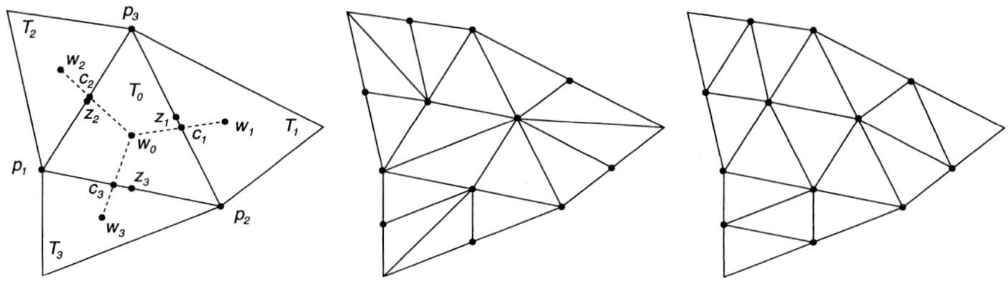

Fig. 4. Notation used for the DEO–reconstructions (left); grid after two bisections (middle) and after the quartering of each triangle (right).

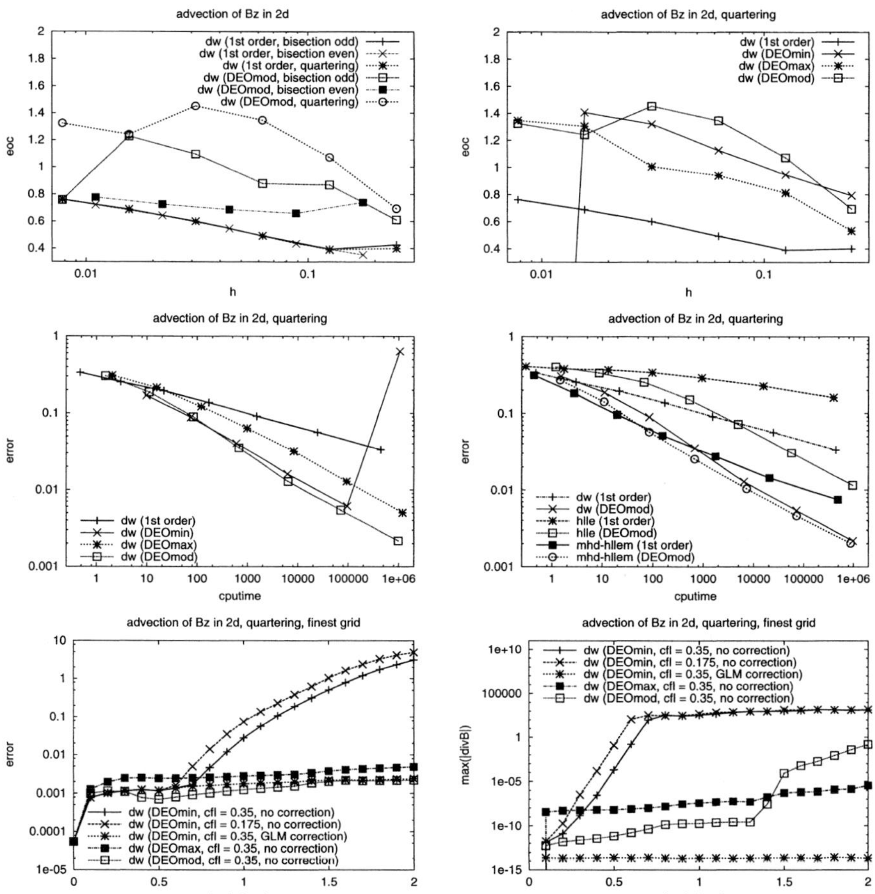

Fig. 5. Results for the 2d–advection of B_z. Top: The experimental order of convergence depends on the type of grid refinement and the limiter chosen. Middle: DEOmod is the most efficient limiter, MHD–HLLEM with DEOmod the most efficient scheme for this problem. Bottom: The breakdown of convergence of DW with DEOmin (see top right and middle left) is caused by a blowup of divergence errors.

D_l. After a suitable limitation (l) we select (s) a linear function L_0 on T_0:

DEOmin (l) $\tilde{D}_l :\equiv D_l$ for $1 \le l \le 3$

(s) $L_0 :\equiv \tilde{D}_{l_0}$ with $|\nabla \tilde{D}_{l_0}| = \min_{1 \le l \le 3}\{|\nabla \tilde{D}_l|\}$

DEOmax (l) $$\tilde{D}_l :\equiv \begin{cases} U_0 \text{ if } \exists j \in \{1,2,3\} \text{ with } D_l(c_j) \notin [U_0, U_j] \cup [U_j, U_0] \\ D_l \text{ otherwise} \end{cases}$$

(s) $L_0 :\equiv \tilde{D}_{l_0}$ with $|\nabla \tilde{D}_{l_0}| = \max_{1 \le l \le 3}\{|\nabla \tilde{D}_l|\}$

DEOmod (l) $\tilde{D}_l(\mathbf{x}) := U_0 + m_l \nabla D_l \cdot (\mathbf{x} - \mathbf{w}_0)$ for

$$g_l := \nabla D_l \cdot (\mathbf{w}_l - \mathbf{w}_0), \quad d_l := U_l - U_0, \qquad m_l := \begin{cases} 0 & \text{if } d_l g_l \le 0 \\ d_l / g_l & \text{if } d_l g_l > 0,\ |g_l| > |d_l| \\ 1 & \text{otherwise} \end{cases}$$

(s) $L_0 :\equiv \tilde{D}_{l_0}$ with $|\nabla \tilde{D}_{l_0}| = \max_{1 \le l \le 3}\{|\nabla \tilde{D}_l|\}$

DEOmin is taken directly from [7]. DEOmax is a modification of the second approach suggested in [7] since we check the values of D_l for oscillations in the intersection points of the barycenter connections with the corresponding sides c_j instead of the side midpoints z_j. In general these points do not coincide, cf. Fig. 4. Thus in the original version linear functions are not exactly recovered.

The choice of the linear function with the gentlest slope in the DEOmin–approach can be more restrictive than necessary, while at the same time the missing limitation can introduce oscillations. On the other hand, the DEOmax–limitation still admits oscillations since the values of the *reconstructions* in the neighbouring triangles are not taken into account. Moreover, DEOmax may switch to the constant data even if it would be sufficient just to *reduce* the slope. Our new DEOmod–approach cures all of these problems and is motivated as follows: For $1 \le j \le 3$ we limit the linear functions in the points c_j; the maximal value admitted is given by the linear interpolation of U_0 and U_j evaluated at c_j. If necessary, we reduce the slope exactly by the amount required to fulfil this condition. Finally we choose the limited function with the steepest gradient. For this approach we can prove that linear functions are exactly recovered and that it guarantees oscillation–free reconstructions in the sense $|U_0 - L_0(c_j)| + |L_j(c_j) - U_j| \le |U_0 - U_j|$ for $1 \le j \le 3$.

In Fig. 5 we summarize our results obtained for the advection of a smooth impulse in B_z (see [8] for more details). If we quarter all triangles within each refinement step as in Fig. 4 right, the experimental order of convergence (eoc) using the DW scheme with DEOmod is clearly larger than 1, which does not hold uniformly if the grid is refined by an odd or even number of bisections. Moreover, if compared with DEOmin, DEOmax, and the first order scheme, DEOmod has both the best eoc and computational efficiency. Altogether we find that the MHD–HLLEM scheme with DEOmod is the most efficient solver for this problem. The breakdown of convergence which occurs for the DW–solver with DEOmin is quite independent of the CFL–number. We observe that the errors grow *after* a blow–up of divergence errors. This does not happen if we use the GLM divergence cleaning introduced in [2]. We

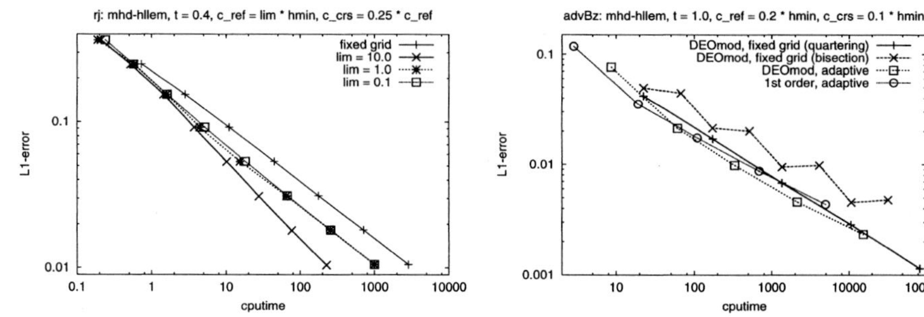

Fig. 6. Left: For the 2nd–order scheme in 1d and the Ryu–Jones Riemann problem from [8] local adaption saves up to 90% of CPU time. Right: For the 2d–advection of B_z the adaptive first *and* second order MHD–HLLEM schemes are highly efficient.

conjecture that the missing limitation within DEOmin leads to oscillations which cause a stronger amplification of roundoff errors in the (constant!) magnetic field components B_x and B_y than for DEOmax and DEOmod.

5 Local Grid Adaption

For the purpose of this paper we use simple grid indicators, which have proven to be quite efficient and which make it easy to consider an arbitrary set of quantities. Denote by I_j for $j \in \{1, \ldots, N\}$ the cells of a one–dimensional grid with midpoints x_j. Then we compute

$$\operatorname{ind}(I_j) := \max_{v \in \mathbb{V}} \left\{ \left| \frac{v_j - v_{j-1}}{x_j - x_{j-1}} - \frac{v_{j+1} - v_{j-1}}{x_{j+1} - x_{j-1}} \right|, \left| \frac{v_{j+1} - v_j}{x_{j+1} - x_j} - \frac{v_{j+1} - v_{j-1}}{x_{j+1} - x_{j-1}} \right| \right\}$$

for a given set $\mathbb{V} \subset \{\rho, \rho u_x, \rho u_y, \rho u_z, B_x, B_y, B_z, \rho e, u_x, u_y, u_z, e, p, s\}$. We split the cell I_j if $\operatorname{relind}(I_j) := \frac{\operatorname{ind}(I_j)}{\frac{1}{N}\sum_{k=1}^{N} \operatorname{ind}(I_k)} > c_{\text{ref}}$ and its diameter h_j fulfils $h_j \geq 2h_{\min}$ for prescribed constants $h_{\min}$ and c_{ref}. Conversely, we merge two cells I_j and I_{j+1} with a common parent cell if $\max\{\operatorname{relind}(I_j), \operatorname{relind}(I_{j+1})\} < c_{\text{crs}}$. In 2d we compare the discrete 1d-derivatives $(U_l - U_0)/|\mathbf{w}_l - \mathbf{w}_0|$ for $1 \leq l \leq 3$ with the directional derivatives of the linear function connecting the barycenters $\mathbf{w}_1$, $\mathbf{w}_2$, and $\mathbf{w}_3$. In 1d we obtained the best results if we refined not only the cells I_j according to $\operatorname{relind}(I_j)$, but also the four "neighbours" I_{j-2}, I_{j-1}, I_{j+1}, and I_{j+2}; in 2d the "refinement zone" enforced by grid conformity was sufficient. In our examples we used $\mathbb{V} := \{\rho, B_x, B_y, B_z\}$.

For the two examples shown in Fig. 6 we are able to obtain solutions of identical quality as on a uniform grid with $h = h_{min}$ in significantly less time. According to the 1d–results from [8] the 2nd–order schemes are always more efficient than their 1st–order counterparts. Thus Fig. 6 left indicates that in 1d the *combination* of 2nd order and local adaption yields the most efficient schemes. On the other hand, Fig. 6 right shows that in 2d the 1st–order adaptive scheme can be a highly efficient choice. Taking into account that local grid adaption necessarily disturbes the grid structure which may have a serious impact on the performance on the 2nd–order schemes (see Sect. 4), we conclude that in multiple space dimensions the use of a high–resolution 1st–order scheme like MHD–HLLEM in combination with local grid adaption is a very reasonable choice.

6 Conclusions

For real gas simulations the use of an adaptive table for the EOS can lead to a significant reduction of computational time. Our new limiter for linear reconstruction on unstructured triangular grids in 2d performs much better than the versions commonly used. However, since the performance of 2nd–order schemes in 2d is strongly dependent on the grid structure, whereas local grid adaption always seems to pay off, we favour the use of a high–resolution 1st–order scheme like our MHD–HLLEM solver in combination with local grid adaption for simulations in 2d and 3d.

Acknowledgement. The authors were partially supported by the DFG priority research program ANumE.

References

1. W. Dai and P.R. Woodward, *A simple Riemann solver and high–order Godunov schemes for hyperbolic systems of conservation laws*, J. Comput. Phys. **121** (1995), 51–65.
2. A. Dedner, F. Kemm, D. Kröner, C.-D. Munz, T. Schnitzer, and M. Wesenberg, *Hyperbolic divergence cleaning for the MHD equations*, J. Comput. Phys. **175** (2002), 645–673, doi:10.1006/jcph.2001.6961.
3. A. Dedner, D. Kröner, C. Rohde, and M. Wesenberg, *Efficient divergence cleaning in three–dimensional MHD simulations*, submitted to "High Performance Computing in Science and Engineering 2002", 2002.
4. A. Dedner, D. Kröner, I.L. Sofronov, and M. Wesenberg, *Transparent boundary conditions for MHD simulations in stratified atmospheres*, J. Comput. Phys. **171** (2001), 448–478.
5. A. Dedner and P. Vollmöller, *An adaptive higher order method for solving the radiation transport equation on unstructured grids*, J. Comput. Phys. **178** (2002), 263–289, doi:10.1006/jcph.2002.7001.
6. A. Dedner and M. Wesenberg, *Numerical methods for the real gas MHD equations*, Hyperbolic Problems: Theory, Numerics, Applications (H. Freistühler and G. Warnecke, eds.), ISNM, vol. 140, Birkhäuser, 2001, pp. 287–296.
7. L.J. Durlofsky, B. Engquist, and S. Osher, *Triangle based adaptive stencils for the solution of hyperbolic conservation laws*, J. Comput. Phys. **98** (1992), 64–73.
8. M. Wesenberg, *Efficient MHD Riemann solvers for simulations on unstructured triangular grids*, J. Numer. Math. **10** (2002), 37–71.

A New Approach to Divergence Cleaning in Magnetohydrodynamic Simulations

Andreas Dedner, Christian Rohde, and Matthias Wesenberg

Institut für Angewandte Mathematik, Universität Freiburg, Freiburg, Germany
dedner|chris|wesenber@mathematik.uni-freiburg.de

Summary. In this paper we present a number of approaches for reducing the divergence errors in magnetohydrodynamic simulations. The methods are derived from a general framework, which for example also includes the Hodge projection scheme. The corrections can be easily added to an existing scheme as is demonstrated for a finite–volume scheme. Numerical results in 2d and 3d confirm the advantages of our approach.

1 Introduction

A mathematical model for the interaction of a compressible, inviscous, and electrically conducting fluid and a magnetic field is given by the equations of magnetohydrodynamics (MHD). It is derived from the Euler equations of gas dynamics describing the evolution of the fluid's density ($\rho > 0$), momentum ($\rho\mathbf{u} \in \mathbb{R}^3$), and total energy density ($\rho e > 0$). This system is coupled with the Maxwell equations leading to an additional evolution equation for the magnetic field ($\mathbf{B} \in \mathbb{R}^3$). The full 8×8 system can be written as a hyperbolic system in divergence form

$$\begin{aligned}
\partial_t \rho + \nabla \cdot (\rho \mathbf{u}) &= 0 \quad \textit{(conservation of mass)}, && \text{(1a)}\\
\partial_t (\rho \mathbf{u}) + \nabla \cdot (\rho \mathbf{u}\mathbf{u}^T + \mathcal{P}) &= 0 \quad \textit{(conservation of momentum)}, && \text{(1b)}\\
\partial_t \mathbf{B} + \nabla \cdot (\mathbf{u}\mathbf{B}^T - \mathbf{B}\mathbf{u}^T) &= 0 \quad \textit{(induction equations)}, && \text{(1c)}\\
\partial_t (\rho e) + \nabla \cdot (\rho e \mathbf{u} + \mathcal{P}\mathbf{u}) &= 0 \quad \textit{(conservation of energy)}, && \text{(1d)}\\
\nabla \cdot \mathbf{B} &= 0 \quad \textit{(divergence constraint)}. && \text{(1e)}
\end{aligned}$$

The pressure tensor $\mathcal{P}$ is defined as

$$\mathcal{P} := \left(p + \frac{1}{8\pi}|\mathbf{B}|^2\right)\mathcal{I} - \frac{1}{4\pi}\mathbf{B}\mathbf{B}^T \tag{1f}$$

using the unit tensor $\mathcal{I}$. The gas pressure p is given by a suitable equation of state. The vector of conservative quantities is denoted by $\mathbf{U} = (\rho, \rho\mathbf{u}, \mathbf{B}, \rho e)^T$. At time $t = 0$ some initial data $\mathbf{U}_0$ has to be prescribed. The induction equation (1c) can be rewritten as $\partial_t \mathbf{B} + \nabla \times (\mathbf{B} \times \mathbf{u}) = 0$ and hence $\partial_t(\nabla \cdot \mathbf{B}) = 0$. Therefore on the analytical level (1e) is a condition only on the initial data $\mathbf{U}_0$. In a compact form we can write (1a)–(1d) as

$$\partial_t \mathbf{U} + \nabla \cdot \mathbf{F}(\mathbf{U}) = 0 \, .$$

Classical finite–volume schemes are frequently used for approximating the MHD equations. In their standard form these schemes can only take into account the evolutionary part of the system and totally ignore the divergence constraint (1e) since it is only an initial condition. Many numerical simulations have shown that this leads to unphysical results and unstable schemes. Numerous approaches have been suggested to reduce the error in the divergence of **B**. Some of these approaches are based on new schemes specially designed to reduce divergence errors; others can be seen as post processing for existing scheme (cf. [7] and the references therein). In this work we study a new technique to reduce divergence errors which can be easily integrated in existing numerical schemes. Our method is based on a modification of the MHD system coupling the divergence constraint to the hyperbolic system, which allows for many different correction mechanisms. It was first presented in [1] where four modifications were suggested and results for two of these were shown. Our numerical experiments have led us to favour the approach which combines transport and damping of divergence errors. In this work we compare this method with the remaining two approaches which include the well–known Hodge projection scheme.

In Sect. 2 we present the modified system. The implementation of the different correction mechanisms for the example of a standard finite–volume scheme is studied in Sects. 3 and 4. In addition to [1] some new aspects concerning the choice of parameters are discussed. In the last section we compare different modifications covering simple test–cases in 2d and a 3d application from solar physics.

2 GLM–MHD System

We couple the divergence constraint (1e) to the induction equation (1c) by introducing a new scalar function ψ and replacing (1e) and (1c) by

$$\partial_t \mathbf{B} + \nabla \times (\mathbf{B} \times \mathbf{u}) + \nabla\psi = 0 \quad \textit{(modified induction equations)}, \tag{2a}$$

$$\mathcal{D}\psi + \nabla \cdot \mathbf{B} = 0 \quad \textit{(modified divergence constraint)}. \tag{2b}$$

The linear differential operator $\mathcal{D}$ has a strong influence on the evolution of the divergence of **B**. This can be seen for smooth solutions by applying first the divergence and then the differential operator $\mathcal{D}$ to (2a) and then replacing ψ using (2b). This leads to an equation for $\nabla \cdot \mathbf{B}$:

$$\partial_t \mathcal{D}(\nabla \cdot \mathbf{B}) - \triangle(\nabla \cdot \mathbf{B}) = 0. \tag{3}$$

Note that for the original MHD system we have $\partial_t(\nabla \cdot \mathbf{B}) = 0$. In [1] we suggest four choices for $\mathcal{D}$ so that (3) is either the Laplace equation, the heat equation, the wave equation, or the telegraph equation. With the *elliptic* approach ($\mathcal{D} = 0$) we rediscover the well–known Hodge projection scheme. In the *parabolic* approach ($\mathcal{D} = \frac{1}{c_p^2}$) the auxiliary function ψ can be eliminated from the system leading to a second–order term in the induction equation. In the *hyperbolic* approach ($\mathcal{D} = \frac{1}{c_h^2}\partial_t$) the GLM–MHD system is purely hyperbolic thus preserving an important feature of the original MHD system. Finally, the *mixed* approach ($\mathcal{D} = \frac{1}{c_h^2}\partial_t + \frac{1}{c_p^2}$) results in a combination of transport and damping while still retaining the hyperbolicity of the system. Note that in all four cases the quantities in $\mathbf{U}$ are conserved.

3 Implementation

For the solution of the MHD system we use a standard explicit finite–volume scheme on unstructured triangular grids in 2d and tetrahedral grids in 3d. To achieve higher order we use a two step Runge–Kutta method for the time discretization and linear reconstruction for the spatial discretization [3, 4, 5]. In this section we describe the modifications of the first–order scheme required for implementing the different correction mechanisms. We focus on the hyperbolic, the mixed, and the parabolic correction. The elliptic correction is implemented as a post processing step where the magnetic field is projected onto its divergence free part by means of a standard solver for the Laplace equation.

We first fix some notation: The elements of the grid are polyhedral and are denoted by T_i where i is in some index set $\mathcal{I}$. $\boldsymbol{\omega}_i$ denotes the center of gravity of the element T_i. With $\mathcal{N}(i) \subset \mathcal{I}$ we denote the set of indices of those elements T_j which have a common interface S_{ij} with T_i. The unit outer normal at this interface with respect to T_i is denoted with $\mathbf{n}_{ij}$. With $h_{ij} := \max\{h : \mathbf{x} - h\mathbf{n}_{ij} \in T_i, \mathbf{x} \in S_{ij}\}$ we define the height of T_i over the face S_{ij}. Furthermore $\mathbf{p}_k$ for $k \in \mathcal{P}$ are the vertices of the grid where $\mathcal{P}$ is some index set. The set of all vertices which lie in T_i is $\mathcal{P}(i)$ and the set of all vertices on S_{ij} is $\mathcal{P}(i,j)$.

With $\mathbf{U}_i^n \approx \frac{1}{|T_i|}\int_{T_i} \mathbf{U}(\mathbf{x}, t^n)\, d\mathbf{x}$ our finite–volume scheme on an element T_i for $i \in \mathcal{I}$ can be written as

$$\mathbf{U}_i^{n+1} = \mathbf{U}_i^n - \frac{\Delta t^n}{|T_i|} \sum_{j\in\mathcal{N}(i)} \mathbf{g}_{ij}(\mathbf{U}_i^n, \mathbf{U}_j^n) \ , \tag{4}$$

with $\Delta t^n = t^{n+1} - t^n$. For $i \in \mathcal{I}$ and $j \in \mathcal{N}(i)$ the numerical flux function $\mathbf{g}_{ij} = (\mathbf{g}_{ij}^{\rho}, \mathbf{g}_{ij}^{\rho\mathbf{u}}, \mathbf{g}_{ij}^{\mathbf{B}}, \mathbf{g}_{ij}^{\rho e})^T$ is an approximation of $\int_{t^n}^{t^{n+1}} \int_{S_{ij}} \mathbf{F}(\mathbf{U}) \cdot \mathbf{n}_{ij}$. We define $\mathbf{U}_i^0$ as an approximation of $\frac{1}{|T_i|}\int_{T_i} \mathbf{U}_0(\mathbf{x})\, d\mathbf{x}$ using some quadrature rule. The timestep Δt^n is defined by a standard CFL condition

$$\Delta t^n = c_{cfl} \min_{i\in\mathcal{I}, j\in\mathcal{N}(j)} \frac{h_{ij}}{\lambda_{ij}} \ . \tag{5}$$

Here c_{cfl} is some constant in $(0, 1)$ and λ_{ij} is the fastest wave speed in the direction $\mathbf{n}_{ij}$ at the interface S_{ij} with data $\mathbf{U}_i$ (e.g. [1, 8]).

The hyperbolic, mixed, and parabolic corrections only require some simple modifications of the numerical flux function $\mathbf{g}_{ij}^{\mathbf{B}}$ for the induction equation.

3.1 Hyperbolic Correction

The modified induction and divergence equations are

$$\partial_t \mathbf{B} + \nabla \cdot (\mathbf{u}\mathbf{B}^T - \mathbf{B}\mathbf{u}^T) + \nabla\psi = 0 \ , \tag{6a}$$

$$\partial_t \psi + c_h^2 \nabla \cdot \mathbf{B} = 0 \ . \tag{6b}$$

with some $c_h > 0$. We need an additional scalar ψ_i^n on each element T_i. As initial condition we set $\psi_i^n = 0$ for all $i \in \mathcal{I}$. The finite–volume scheme for (6) can be written as

$$\mathbf{B}_i^{n+1} = \mathbf{B}_i^n - \frac{\Delta t^n}{|T_i|} \sum_{j \in \mathcal{N}(i)} \left(\mathbf{g}_{ij}^{\mathbf{B}}(\mathbf{B}_i^n, \mathbf{B}_j^n) + |S_{ij}| \psi_{ij}^n \mathbf{n}_{ij} \right) \ , \tag{7a}$$

$$\psi_i^{n+1} = \psi_i^n - \frac{\Delta t^n}{|T_i|} \sum_{j \in \mathcal{N}(i)} |S_{ij}| B_{ij}^n \ . \tag{7b}$$

Here ψ_{ij}^n and B_{ij}^n are some suitable values on the interface S_{ij} for ψ and $\mathbf{B} \cdot \mathbf{n}_{ij}$, respectively. To define these values we apply to (6) a rotation matrix which transforms the normal $\mathbf{n}_{ij}$ onto the unit vector in x–direction. This leads to a 2×2 system for $\mathbf{B} \cdot \mathbf{n}_{ij}$ and ψ which does not depend on the other MHD quantities

$$\partial_t (\mathbf{B} \cdot \mathbf{n}_{ij}) + \partial_x \psi = 0 \ ,$$

$$\partial_t \psi + c_h^2 \partial_x (\mathbf{B} \cdot \mathbf{n}_{ij}) = 0 \ .$$

For this linear system we solve the Riemann problem at $x = 0$ with left–hand state $(\mathbf{B}_i^n \cdot \mathbf{n}_{ij}, \psi_i^n)$ and right–hand state $(\mathbf{B}_j^n \cdot \mathbf{n}_{ij}, \psi_j^n)$. At $x = 0$ the solution is constant for all $t = 0$. We use this intermediate state in (7):

$$B_{ij}^n = \frac{1}{2} \left(\mathbf{B}_j^n - \mathbf{B}_i^n \right) \cdot \mathbf{n}_{ij} - \frac{1}{2c_h} \left(\psi_j^n - \psi_i^n \right) \ , \tag{9a}$$

$$\psi_{ij}^n = \frac{1}{2} \left(\psi_j^n - \psi_i^n \right) - \frac{c_h}{2} \left(\mathbf{B}_j^n - \mathbf{B}_i^n \right) \cdot \mathbf{n}_{ij} \ . \tag{9b}$$

Remark: Our numerical flux function is also based on the idea of rotated 1d Riemann solvers as used here for system (6). In 1d the divergence constraint and the evolution equation for the first component of the magnetic field only allow for a constant B_x. Thus a consistent application of the 1d solver would require $\mathbf{B} \cdot \mathbf{n}_{ij}$ to be constant over the face S_{ij} which is in general not the case. Thus some arbitrary choice for this value has to be used. In the case of the hyperbolic correction B_x can vary in space and time. System (6) offers a consistent choice for $\mathbf{B} \cdot \mathbf{n}_{ij}$, which we use in the computation of the flux for the other MHD quantities.

3.2 Mixed Correction

The only difference between the hyperbolic and the mixed correction is the additional source term in the evolution equation for ψ. We use a splitting method where we first solve the homogeneous hyperbolic GLM system and then solve the ODE

$$\partial_t \psi = -\frac{c_h^2}{c_p^2}\psi \ ,$$

with initial data for ψ given by the first step of our algorithm. Thus for the update of ψ we replace (7b) by

$$\psi_i^{n+1} = \exp\Big(-\frac{c_h^2}{c_p^2}\Delta t^n\Big)\Big(\psi_i^n - \frac{\Delta t^n}{|T_i|}\sum_{j\in\mathcal{N}(i)} |S_{ij}|B_{ij}^n\Big) \ . \tag{10}$$

3.3 Parabolic Correction

With $\mathcal{D}(\psi) = \frac{1}{c_p^2}\psi$ equation (2b) is reduced to the algebraic relation $\psi = -c_p^2\nabla\cdot\mathbf{B}$. Therefore the term $\nabla\psi$ in (2a) can be replaced by a second–order term

$$\partial_t\mathbf{B} + \nabla\cdot(\mathbf{u}\mathbf{B}^T - \mathbf{B}\mathbf{u}^T) = c_p^2\nabla(\nabla\cdot\mathbf{B}) \ . \tag{11}$$

In the finite–volume formulation we use Gauss' theorem to arrive at the following update for the magnetic field

$$\mathbf{B}_i^{n+1} = \mathbf{B}_i^n - \frac{\Delta t^n}{|T_i|}\sum_{j\in\mathcal{N}(i)}\big(\mathbf{g}_{ij}^{\mathbf{B}}(\mathbf{B}_i^n,\mathbf{B}_j^n) - c_p^2|S_{ij}|(\nabla\cdot\mathbf{B})_{ij}^n\mathbf{n}_{ij}\big) \ . \tag{12}$$

Here $(\nabla\cdot\mathbf{B})_{ij}^n$ is an approximation of $\nabla\cdot\mathbf{B}$ on the interface S_{ij}. We have tested a number of different techniques of reconstructing the magnetic field. A good choice which can be easily implemented on 2d/3d unstructured grids is the following: For each $k\in\mathcal{P}$ define the average values

$$\mathbf{B}_{\mathbf{p}_k}^n := \frac{\sum_{i\in\mathcal{I},k\in\mathcal{P}(i)}|T_i|\mathbf{B}_i^n}{\sum_{i\in\mathcal{I},k\in\mathcal{P}(i)}|T_i|} \ . \tag{13}$$

Then for each $j\in\mathcal{N}(i)$ define a polynomial interpolation $\mathbf{B}_{ij}^n$ of the values $\big\{\mathbf{B}_i,(\mathbf{B}_{\mathbf{p}_k}^n)_{k\in\mathcal{P}(i,j)}\big\}$ at the points $\big\{\boldsymbol{\omega}_i,(\mathbf{p}_k)_{k\in\mathcal{P}(i,j)}\big\}$. Using these functions we define

$$(\nabla\cdot\mathbf{B})_{ij}^n = \frac{1}{2}\nabla\cdot(\mathbf{B}_{ij}^n + \mathbf{B}_{ji}^n) \ . \tag{14}$$

4 Choice of Parameters

The evolution of divergence errors is influenced by the choice of the value for c_h (influencing the transport) and the value for c_p (influencing the damping). In each timestep of our method we define new constants $c_{h,n}$ and $c_{p,n}$ in such a way that our method introduces a maximum of transport and damping under the constraint that the scheme remains stable for the timestep Δt^n as defined in (5), which we use in the base scheme.

4.1 Hyperbolic Correction

In addition to the usual seven MHD wave speeds the hyperbolic correction introduces two new waves with speeds $\pm c_h$. To ensure that our CFL condition (5) remains valid we define as in [1]

$$c_{h,n} := c_{cfl} \min_{i\in\mathcal{I},j\in\mathcal{N}(i)} \frac{h_{ij}}{\Delta t^n} \ . \tag{15}$$

4.2 Mixed Correction

In the following we want to take a closer look at the choice of c_p^2 by studying the telegraph equation. Due to (3) $v := \nabla\cdot\mathbf{B}$ satisfies

$$\partial_{tt}v(\mathbf{x},t) + \frac{c_h^2}{c_p^2}\partial_t v(\mathbf{x},t) - c_h^2\Delta v(\mathbf{x},t) = 0. \tag{16}$$

For simplicity we use homogeneous Dirichlet boundary conditions. We study the time asymptotic stability since the aim of our method is to reduce divergence errors in time; a divergence–free magnetic field corresponds to $v\equiv 0$ so that we study under which conditions $v(\cdot,t)\to 0$ $(t\to\infty)$ for $v(\cdot,0)=v_0(\cdot)$ with $v_0\not\equiv 0$.

If v_0 is smooth we can expand it in the eigenfunctions of the Laplace operator, i.e., we assume that $v_0(\mathbf{x}) = \sum_m \alpha_m w_m(\mathbf{x})$ with $-\Delta w_m = \lambda_m w_m$. Direct calculations show that the solution to (16) has the form $v(\mathbf{x},t) = \sum_m v_m(t)\alpha_m w_m(\mathbf{x})$ where the functions $v_m(t)$ are given by

$$v_m(t) = \begin{cases} \exp\Big(-\frac{c_h^2}{2c_p^2}t\Big)\cos(q_m t)\ , & c_h^2\lambda_m \geq \frac{c_h^4}{4c_p^4}\ , \\ \exp\Big(-\frac{c_h^2}{2c_p^2}t\Big)\cosh(q_m t)\ , & c_h^2\lambda_m < \frac{c_h^4}{4c_p^4}\ , \end{cases}$$

with $q_m := \sqrt{\Big|c_h^2\lambda_m - \frac{c_h^4}{4c_p^4}\Big|}$. In the first case v_m decays to zero with the rate $\exp\Big(-\frac{c_h^2}{2c_p^2}t\Big)$ which is independent of λ_m. For the modes corresponding to small frequencies $\lambda_m < \frac{c_h^2}{4c_p^4}$ we find

$$v_m(t) = \frac{1}{2}\exp\Big(-\big(\frac{c_h^2}{2c_p^2} - q_m\big)t\Big) + \frac{1}{2}\exp\Big(-\big(\frac{c_h^2}{2c_p^2} + q_m\big)t\Big) .$$

Since $\frac{c_h^2}{2c_p^2} > q_m$ and $q_m > 0$ the function v_m also decays to zero for $t \to \infty$ but the rate decreases for decreasing λ_m. Therefore we choose c_p so that $\lambda_m \geq \frac{c_h^2}{4c_p^4}$ for all n. This leads to a lower bound on c_p^2. Since our aim is to introduce as much damping as possible we choose c_p^2 as small as possible

$$c_{p,n}^2 := \frac{c_{h,n}}{c_{rel}} \quad \text{with} \quad c_{rel} := 2\min_m \sqrt{\lambda_m} ,$$

with $c_{h,n}$ defined as in the hyperbolic approach by (15). For a rectangular domain in 3d with sidelengths $2L_x$, $2L_y$, and $2L_z$ we compute

$$c_{rel} = \pi\sqrt{\frac{1}{L_x^2} + \frac{1}{L_y^2} + \frac{1}{L_z^2}} . \tag{17}$$

Numerical experiments published in [1] show that fixing the quotient $\frac{c_p^2}{c_h}$ leads to a scheme for which the effects of the correction are independent of the grid resolution and the order of the scheme. The value in (17) corresponds quite well with the value we found by numerical experiments in [1].

4.3 Parabolic Correction

The choice of c_p^2 in the parabolic approach is restricted by the additional stability restriction introduced by the second–order term. For a stable scheme c_p^2 has to be chosen proportional to the spatial discretization. We have tested a number of different possibilities and have found that a good choice is

$$c_{p,n}^2 = 0.4 \min_{i\in\mathcal{I}, j\in\mathcal{N}(i)} h_{ij} c_{h,n} ,$$

where $c_{h,n}$ is again defined by (15). Note the similar definition of c_p in the parabolic and mixed correction. In the case of the mixed correction it seems reasonable to choose c_p^2 as small as possible so that the damping term $\exp\Big(-\frac{c_h^2}{c_p^2}\Delta t\Big)$ in (10) is as large as possible. The situation is different in the case of the parabolic correction where c_p^2 and not c_p^{-2} enters the finite–volume scheme in (12); thus smaller c_p^2 reduce the amount of damping introduced by our scheme.

5 Numerical Results

We compare the different approaches on 2d locally adapted, triangular grids using the second–order finite volume scheme by means of the 2d Riemann problem studied in [1]. For the numerical flux we use the MHD–HLLEM scheme presented in [8]. Details on the adaptation strategy and linear reconstruction are given in [5]. Since the results published in [1] show that the mixed approach leads to slightly better results than the hyperbolic approach with hardly any additional computational cost we only compare the mixed, the parabolic, and the elliptic approach using the base scheme and the *source term approach* (cf. [6, 7]) as reference methods as in [1].

In Fig. 1 the time evolution of $\nabla \cdot \mathbf{B}$ is plotted for the five methods. The results show that the mixed approach leads to the best results with a significant reduction in the divergence errors in both the maximum– and the L^1–norm. All GLM corrections lead to an improvement compared to both reference methods. Especially divergence errors are more or less constant with respect to time which is not true for the base scheme. A more quantitative result is shown in Fig. 2 where the solution is shown as 2d isoline plots and along 1d rays. The results from Fig. 1 are confirmed. Finally Fig. 3 shows the solution in the plane at $x = 0$ for the 3d fluxtube problem studied in [2]. Shortly after the simulation time shown the base scheme broke down due to negative pressure.

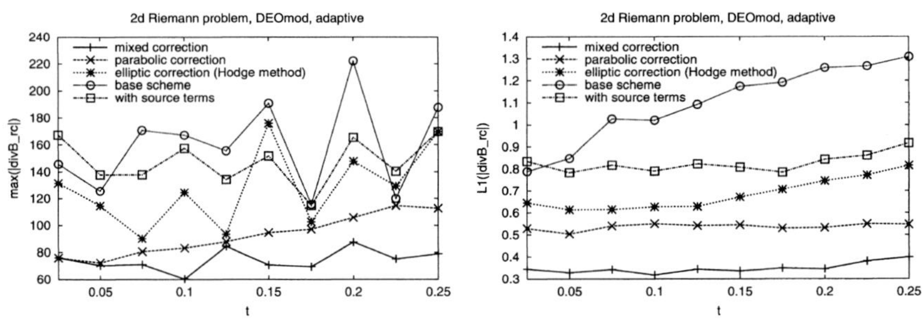

Fig. 1. Time evolution of divergence of **B** for 2d Riemann problem. Left: L^∞–norm in space. Right: L^1–norm in space.

6 Conclusions

We have derived a number of correction approaches to reduce divergence errors in MHD simulations, which are easily added to existing schemes. Due to our choice of the parameters the timestep of the base scheme can be used also for the modified schemes so that the number of timesteps required to reach the final time does not change. In the hyperbolic and the mixed approach this is achieved by choosing c_h as a function of the grid size and the timestep. In the mixed correction c_p^2 can be chosen arbitrarily since the splitting approach leads to no additional stability restriction. In the parabolic

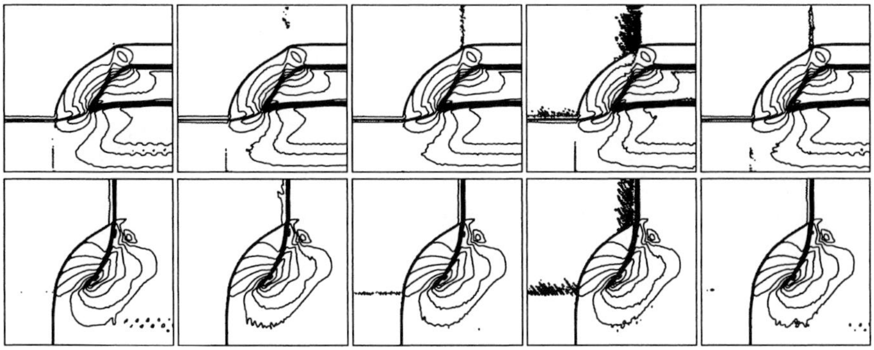

(a) Isolines of B_x and B_y

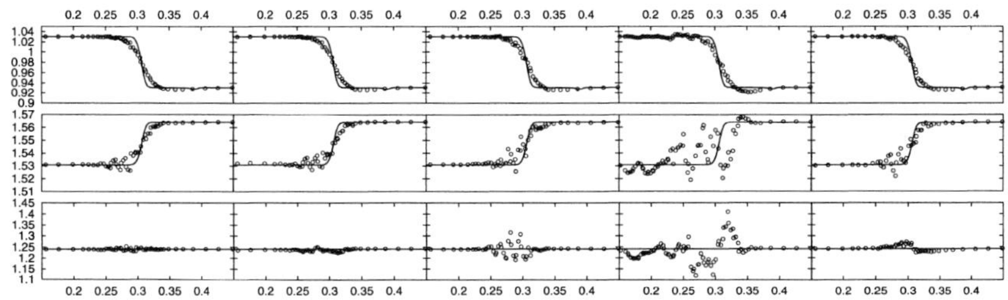

(b) ρ, u_x, and B_x along $x = 0.915$ with 1d reference solution

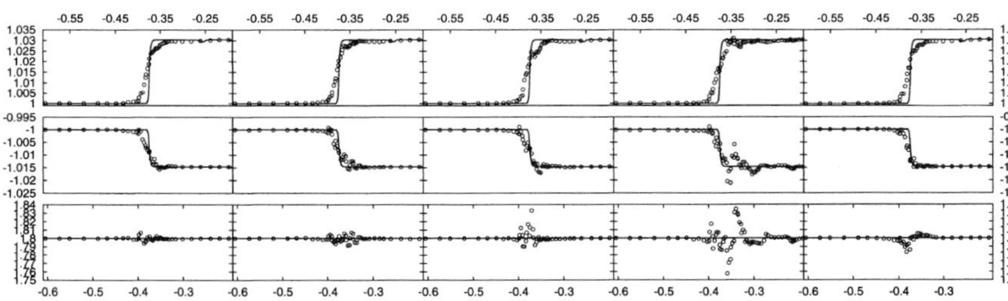

(c) ρ, u_y, and B_y along $y = -0.915$ with 1d reference solution

Fig. 2. Solution for 2d Riemann problem. Left to right: mixed, parabolic, and elliptic approach, base scheme, and with source terms.

approach the amount of damping is restricted by the grid size. Since $B_{ij}^n = B_{ji}^n$ and $\psi_{ij}^n = \psi_{ji}^n$ (cf. (9)) the modified scheme is conservative if the base scheme is conservative. This is also true for the parabolic approach since $(\nabla \cdot \mathbf{B})_{ij}^n = (\nabla \cdot \mathbf{B})_{ji}^n$ in (14).

The numerical results presented here and in [1] show that for all choices of $\mathcal{D}$ suggested the errors in the divergence of the magnetic field are significantly reduced. Most notably this increases the robustness of the scheme. Since the additional computational cost for the mixed approach is negligible and the method leads to the best results we favour this approach.

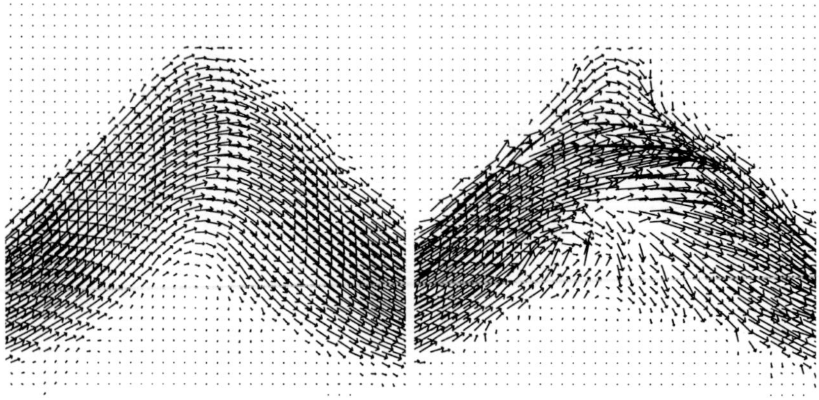

Fig. 3. Magnetic field for 3d fluxtube simulation at time $t = 1.5$ and $x = 0$. Left: mixed correction. Right: base scheme.

Acknowledgments

The authors were partially supported by the DFG priority research program ANumE.

References

1. A. Dedner, F. Kemm, D. Kröner, C.-D. Munz, T. Schnitzer, and M. Wesenberg, *Hyperbolic divergence cleaning for the MHD equations*, J. Comput. Phys. **175** (2002), no. 2, 645–673, doi:10.1006/jcph.2001.6961.
2. A. Dedner, D. Kröner, C. Rohde, and M. Wesenberg, *Efficient divergence cleaning in three–dimensional MHD simulations*, submitted to "High Performance Computing in Science and Engineering 2002", 2002.
3. A. Dedner, C. Rohde, B. Schupp, and M. Wesenberg, *A parallel, load balanced MHD code on locally adapted, unstructured grids in 3d*, in preperation, 2002.
4. A. Dedner, C. Rohde, and M. Wesenberg, *A MHD–simulation in solar physics*, Finite Volumes for Complex Applications II: Problems and Perspectives (Paris) (R. Vilsmeier, F. Benkhaldoun, and D. Hänel, eds.), Hermès Science Publications, 1999, pp. 491–498.
5. ———, *Efficient higher–order finite volume schemes for (real gas) magnetohydrodynamics*, submitted to the Ninth International Conference on Conservation Laws in Pasadena, March 2002, 2002.
6. K.G. Powell, *An approximate Riemann solver for magnetohydrodynamics (that works in more than one dimension)*, ICASE–Report 94–24 (NASA CR–194902), NASA Langley Research Center, Hampton, VA 23681–0001, 8. April 1994.
7. Gábor Tóth, *The* $\nabla \cdot B = 0$ *constraint in shock–capturing magnetohydrodynamics codes*, J. Comput. Phys. **161** (2000), 605–652, doi:10.1006/jcph.2000.6519.
8. M. Wesenberg, *Efficient MHD Riemann solvers for simulations on unstructured triangular grids*, J. Numer. Math. **10** (2002), 37–71.

Global Solutions for a Dusty Media Model

G. G. Doronin[1], N. A. Larkin[2], and A. J. Souza[3]

[1] Departamento de Matemática e Estatística, Universidade Federal de Campina Grande, Caixa Postal 10044, CEP 58109-970, Campina Grande, PB, Brazil. *gleb@dme.ufpb.br*

[2] Departamento de Matemática, Fundação Universidade Estadual de Maringá, Agência UEM, CEP 87020-900, Maringá, PR, Brazil. *nlarkine@uem.br*

[3] Departamento de Matemática e Estatística, Universidade Federal de Campina Grande, Caixa Postal 10044, CEP 58109-970, Campina Grande, PB, Brazil. *cido@dme.ufpb.br*

Summary. The initial-boundary value problem for the two-phase model of "fluid - solid particles" media is considered. Existence and uniqueness of global strong solutions for small initial data are proved.

1 Introduction

The interest in dusty media (i.e., fluids or gases with suspended small solid particles) dates from the 1940s. It was observed that adding dust to a gas flowing turbulently through a pipe results in a reduction in the pressure gradient required to maintain the flow at its original rate [1, 2]. Later, there was a number of works related to this phenomenon [3]-[8]. Nowadays, multiphase models of "fluid - solid particles" type include not only dusty gases, but also phenomena such as enhanced oil recovery, ecological catastrophes, new materials and others [8].

Mathematically, these models use either the Navier-Stokes equations for a fluid and hyperbolic equations for particles [3, 4], [9]-[12] or a nonviscous gas dynamics approach for both phases [5]-[8].

A type of equations simulating the carrier (gas or liquid) phase plays a crucial part in these studies. It was shown in [11] that if the equations, modelling the principal phase, do not have viscosity terms, then the Cauchy problem for this model does not possess even local in time solutions in classes of functions with a finite number of derivatives. Only analytic solutions do exist due to the Cauchy-Kowalevsky theorem. This phenomenon has the same nature as the Cauchy problem for the Laplace equation and makes it doubtful stability and convergence of numerical algorithms.

If the Navier-Stokes equations are used for the liquid phase, then it is possible to prove local-in-time solvability of the Cauchy or mixed problems [10]-[12]. However, results on the existence of global solutions are lacking. Such results seem indispensable not only mathematically (to justify the analysis of growth properties of solutions [4]), but also practically (in view of numerical simulation of smooth flows in an engine nozzle [7, 8]). Probably, a first step towards the global solvability was given in [9]. In that work the local concentration of particles was substituted by the averaged constant concentration that made it possible to prove the existence of global weak solutions to the mixed problem for the model system which did not include equations for the dust concentration.

What concerns the existence of global strong solutions, it is well-known that nonlinear hyperbolic equations, generally speaking, are solvable only locally in time. On the other hand, the presence of a lincar damping makes it possible to prove the existence of global strong solutions if the initial data are sufficiently small (see, for instance, [13]). The model to be studied here has an implicit damping (embedded into the phase interaction term), which allows us to prove the global solvability of the mixed problem.

Our approach in this paper is to consider a parabolic equation for the carrier phase and hyperbolic equations for particles. The flow is assumed to be one-dimensional, since almost all the basic mathematical difficulties that arise in the two-phase dusty gas study are presented already for the one-dimensional case; and it may be expected that solutions of multi-dimensional equations have essentially the same physical features as solutions of one-dimensional models [8].

We study initial-boundary value problem in a cylinder with boundary conditions for the velocity of the dust particles on both boundaries, left and right. The existence and uniqueness of global strong solutions for small initial data are proved. The regularization method, the method of Galerkin and compactness arguments are used.

2 The problem and main results

We consider a one-dimensional model based on the physical assumptions of [3, 4] which are summarized in the following. The dust particles are of uniform size, shape and mass. The carrier phase is assumed to be a viscous incompressible fluid. The dust concentration is small and sedimentation of the particles, buoyancy forces and any body forces acting on the fluid are neglected.

In order to obtain global solutions we replace in the Saffman model the local concentration of particles in the phase interaction term by its average concentration.

Under these assumptions, the model is written as follows:

$$u_t + uu_x - \mu u_{xx} + \alpha u = K(v-u)\int_0^1 m(x,t)\,dx, \tag{1}$$

$$v_t + vv_x = K(u-v), \tag{2}$$

$$m_t + (mv)_x = 0, \tag{3}$$

$$u(x,0) = u_0(x), \quad v(x,0) = v_0(x), \quad m(x,0) = m_0(x) \geq 0, \tag{4}$$

$$u(0,t) = u(1,t) = 0, \tag{5}$$

$$v(0,t) = v(1,t) = 0. \tag{6}$$

Here $(x,t) \in Q = \{(0,1) \times (0,T),\ T > 0\}$; $u(x,t)$ and $v(x,t)$ are velocities of the liquid medium and solid particles respectively; $m(x,t)$ is the local concentration of particles; μ and α are constant coefficients of viscosity and friction and the constant K is the coefficient of the phase interaction. The term $\int_0^1 m\,dx$ means the average concentration of solid particles.

Remark 1. It can be seen from (6) that $x = 0$ and $x = 1$ are characteristics of hyperbolic equations (2) and (3). This justifies the formulation of the mixed problem (1)-(6).

To formulate main results, we define real numbers m^0, E_0 and λ as follows:

$$m^0 = \int_0^1 m_0(x)\,dx,$$

$$E_0 = \|u_0\|^2 + m^0\|v_0\|^2,$$

$$\lambda = \Big[\|v_0'\|^2 + KE_0/2\mu\Big]^{1/2} + \Big[\|v_0''\|^2 + \tfrac{1}{2\mu^2}\Big(\mu K\|u_0'\|^2 + m^0K^2E_0\exp\{E_0/2\mu^2\}\Big)\Big]^{1/2}.$$

Hereafter $\|\cdot\|$ is the norm in $L^2(0,1):\ \|\cdot\|^2 = \int_0^1 |\cdot|^2\,dx$.

The main result of this paper is the following.

Theorem 1. *Let $\mu > 0$, $\alpha \geq 0$, $K > 0$; $u_0 \in H_0^1(0,1)$, $v_0 \in H^2(0,1) \cap H_0^1(0,1)$ and $m_0 \in H^1(0,1)$. If*

$$\lambda < K/7, \tag{7}$$

then for all $T > 0$ problem (1)-(6) has a unique strong solution:

$$u \in L^\infty(0,T;H_0^1(0,1)) \cap L^2(0,T;H^2(0,1) \cap H_0^1(0,1)),$$
$$u_t \in L^\infty(0,T;L^2(0,1)),$$
$$v \in L^\infty(0,T;H^2(0,1) \cap H_0^1(0,1)) \cap L^2(0,T;H^2(0,1) \cap H_0^1(0,1)),$$
$$v_t \in L^\infty(0,T;H_0^1(0,1)),$$
$$m \in L^\infty_{loc}(0,T;H^1(0,1)), \quad m \geq 0,$$
$$m_t \in L^\infty_{loc}(0,T;L^2(0,1)).$$

Proof. We prove this theorem in the following way. First, we observe that system (1)-(3) is splittable. Indeed, if functions $u(x,t)$ and $v(x,t)$ are determined, then the dust concentration $m(x,t)$ can be found by the formula:

$$m(x,t) = m_0\big(y(0;x,t)\big)\exp\left\{-\int_0^t \frac{\partial v}{\partial y}\big(y(\tau;x,t),\tau\big)\,d\tau\right\} \geq 0, \tag{8}$$

where $y(\tau;x,t)$ is a solution to the Cauchy problem

$$\frac{dy}{d\tau} = v(y,\tau); \quad y(t;x,t) = x.$$

Moreover, integrating (3) over Q, we obtain the conservation law

$$\int_0^1 m(x,t)\,dx = \int_0^1 m_0(x)\,dx = m^0,$$

hence, equation (1) can be written as

$$u_t + uu_x - \mu u_{xx} + \alpha u = m^0 K(v-u). \tag{9}$$

Thus, to prove the existence part of the theorem, it is sufficient to obtain solutions to (9) and (2) in Q which will be made by means of regularization of equation (2) by the term εv_{xx} with $\varepsilon > 0$. Global solutions to the regularized problem are constructed by the method of Faedo-Galerkin. Then we show that small (in the sense of (7)) initial data make it possible to obtain global a priori estimates independent on $\varepsilon > 0$. This allows us to pass to the limit as $\varepsilon \to 0$ and to obtain the required solution. Uniqueness is proved by the usual way.

3 Regularized problem

In order to prove global solvability of (1)-(6), we consider the following regularized problem:

$$L^u_\varepsilon u^\varepsilon \equiv u^\varepsilon_t + u^\varepsilon u^\varepsilon_x - \mu u^\varepsilon_{xx} + \alpha u^\varepsilon = m^0 K(v^\varepsilon - u^\varepsilon), \tag{10}$$

$$L^v_\varepsilon v^\varepsilon \equiv v^\varepsilon_t + v^\varepsilon v^\varepsilon_x - \varepsilon v^\varepsilon_{xx} = K(u^\varepsilon - v^\varepsilon), \tag{11}$$

$$u^\varepsilon(0,t) = u^\varepsilon(1,t) = 0, \tag{12}$$

$$v^\varepsilon(0,t) = v^\varepsilon(1,t) = 0, \tag{13}$$

$$u^\varepsilon(x,0) = u_0(x), \quad v^\varepsilon(x,0) = v_0(x), \tag{14}$$

where $\varepsilon > 0$ is an arbitrary real number.

Let $\{w_j(x)\}$ be L^2-orthonormal eigenfunctions of the Sturm-Liouville eigenvalue problem:

$$w_{jxx} + \xi_j w_j = 0, \quad x \in (0,1), \tag{15}$$
$$w_j(0) = w_j(1) = 0. \tag{16}$$

We seek approximate solutions to (10)-(14) in the form:

$$u^{\varepsilon N}(x,t) = \sum_{j=1}^{N} g_j^{\varepsilon N}(t) w_j(x), \tag{17}$$

$$v^{\varepsilon N}(x,t) = \sum_{j=1}^{N} h_j^{\varepsilon N}(t) w_j(x), \tag{18}$$

where $g_j^{\varepsilon N}(t)$ and $h_j^{\varepsilon N}(t)$ are solutions to the following Cauchy problem:

$$\big(L_\varepsilon^u u^{\varepsilon N}, w_j\big)(t) = m^0 K\big((v^{\varepsilon N} - u^{\varepsilon N}), w_j\big)(t), \tag{19}$$
$$\big(L_\varepsilon^v v^{\varepsilon N}, w_j\big)(t) = K\big((u^{\varepsilon N} - v^{\varepsilon N}), w_j\big)(t), \tag{20}$$
$$g_j^{\varepsilon N}(0) = (u_0, w_j), \tag{21}$$
$$h_j^{\varepsilon N}(0) = (v_0, w_j), \tag{22}$$
$$j = 1, ..., N.$$

Here and in the sequel $(u,v)(t) = \int_0^1 u(x,t)v(x,t)\,dx$. The norm in $H^k(0,1)$ is denoted by $||\cdot||_{H^k(0,1)}$ and

$$||u||_{L^\infty(0,T;H^k(0,1))} = \operatorname{ess}\sup_{t\in(0,T)} ||u||_{H^k(0,1)}.$$

4 A priori estimates

The problem (19)-(22) has solutions on some interval $(0, T_N)$, hence all the approximations (17), (18) are defined on $(0, T_N)$ for all $N \geq 1$. Now we need a priori estimates uniform in N and ε in order to pass to the limit as $N \to \infty$ and $\varepsilon \to 0$. To simplify the exposition, we omit the indices "N" and "ε" whenever it is unambiguous.

Substituting eigenfunctions w_j in (19), (20) by $u^{\varepsilon N}$ and $m^0 v^{\varepsilon N}$ correspondingly, we obtain

$$(L_\varepsilon^u u, u)(t) \equiv \frac{1}{2}\frac{d}{dt}||u||^2(t) + \alpha||u||^2(t) + \mu||u_x||^2(t)$$
$$= m^0 K\big((v-u), u\big)(t),$$
$$(L_\varepsilon^v v, m^0 v)(t) \equiv \frac{m^0}{2}\frac{d}{dt}||v||^2(t) + \varepsilon m^0 ||v_x||^2(t)$$
$$= m^0 K\big((u-v), v\big)(t).$$

Summing up the results, integrating over t and denoting

$$E_1^{\varepsilon N}(t) = ||u^{\varepsilon N}||^2(t) + m^0||v^{\varepsilon N}||^2(t),$$

we conclude that for all $t > 0,\ N \geq 1$ and $\varepsilon > 0$ the following inequality holds:

$$\begin{aligned} E_1^{\varepsilon N}(t) + 2\int_0^t \Big[\alpha||u^{\varepsilon N}||^2 + \mu||u_x^{\varepsilon N}||^2 + \varepsilon m^0||v_x^{\varepsilon N}||^2 \\ + m^0K||u^{\varepsilon N} - v^{\varepsilon N}||^2\Big](\tau)\,d\tau \leq E_1^{\varepsilon N}(0). \end{aligned} \tag{23}$$

Substituting in (19), and (20) w_j by $-u_{xx}$ and $-m^0v_{xx}$ respectively, summing up the results , we obtain

$$\begin{aligned} &\tfrac{1}{2}\tfrac{d}{dt}\Big[||u_x||^2(t) + m^0||v_x||^2(t)\Big] \\ &+ \Big(\mu||u_{xx}||^2 + m^0\varepsilon||v_{xx}||^2 + \alpha||u_x||^2\Big)(t) \\ &= \textstyle\int_0^1 \Big(m^0K(u-v)(u_{xx} - v_{xx}) + uu_xu_{xx} + m^0vv_xv_{xx}\Big)\,dx. \end{aligned} \tag{24}$$

It follows from the Steklov and Young inequalities that

$$\begin{aligned} \Big|\textstyle\int_0^1 uu_xu_{xx}\,dx\Big| &\leq \textstyle\int_0^1 \sup_{[0,1]}|u|\cdot|u_xu_{xx}|\,dx \\ &\leq ||u_x||^2||u_{xx}|| \leq \eta||u_{xx}||^2 + C_\eta||u_x||^4, \end{aligned} \tag{25}$$

where η is an arbitrary positive number and $C_\eta = 1/4\eta$.

Setting in (25) $\eta = \mu/4$, and integrating (24) over $t \in (0,T)$, we come to the inequality:

$$\begin{aligned} E_2^{\varepsilon N}(t) + \textstyle\int_0^t \Big(\mu||u_{xx}||^2 + m^0\varepsilon||v_{xx}||^2 + 2\alpha||u_x||^2\Big)(\tau)\,d\tau \\ \leq C_1 + C_2\textstyle\int_0^t \left(E_2^{\varepsilon N}(\tau)\right)^2 d\tau. \end{aligned} \tag{26}$$

Here we put

$$E_2^{\varepsilon N}(t) = ||u_x^{\varepsilon N}||^2(t) + m^0||v_x^{\varepsilon N}||^2(t),$$

$$C_1 = E_2^{\varepsilon N}(0) + \frac{\varepsilon\mu E_1^{\varepsilon N}(0)}{2K(\mu + m^0\varepsilon)},$$

$$C_2 = \max\left\{\frac{1}{\mu}, \frac{1}{m^0\varepsilon}\right\}$$

and estimate the term $\big(m^0K(u-v), (u_{xx} - v_{xx})\big)(t)$ by means of Young's inequality. Notice that the constants C_1 and C_2 in (26) depend only on m^0, μ, ε, K and do not depend on $T > 0$ and $N \geq 1$.

Since $E_2(t) \in L^1(0,T)$ and

$$\int_0^t E_2(\tau)\,d\tau \le \frac{E_1(0)}{2\min\{\mu,\varepsilon\}},$$

inequality (26) and Gronwall's lemma imply

$$E_2(t) \le C_1 \exp\{C_2 E_1(0)/2\min\{\mu,\varepsilon\}\}.$$

Therefore, for the function

$$E_3(t) = E_2(t) + \int_0^t \Big[\mu||u_{xx}||^2 + m^0\varepsilon||v_{xx}||^2 + 2\alpha||u_x||^2\Big](\tau)\,d\tau$$

we obtain from (26)

$$E_3(t) \le C_1 + C_2 \int_0^t E_2(\tau)E_3(\tau)\,d\tau$$

which gives

$$E_3^{\varepsilon N}(t) \le C_1 \exp\{C_2 E_1^{\varepsilon N}(0)/2\min\{\mu,\varepsilon\}\}. \tag{27}$$

Our aim now is to obtain uniform in ε a priori estimates on the whole interval $[0,T]$ and to pass to the limit as $N \to \infty$ and $\varepsilon \to 0$.

Due to Parseval's equality,

$$||\partial_x^i u^{\varepsilon N}||(0) \le ||\partial_x^i u_0(x)||, \quad i = 0, 1.$$

Substituting in (19) w_j by $-u_{xx}$ and acting in the same manner as by proving (25) and (26), we obtain

$$\begin{gathered}||u_x||^2 + \mu \textstyle\int_0^t ||u_{xx}||^2\,d\tau \le \\ ||u_0'||^2 + \frac{m^0 K E_0}{2\mu} + \frac{1}{\mu}\int_0^t ||u_x||^2\Big(||u_x||^2 + \mu\int_0^\tau ||u_{xx}||^2\,d\xi\Big)\,d\tau.\end{gathered}$$

This gives

$$||u_x^{\varepsilon N}||^2 + \mu \int_0^t ||u_{xx}^{\varepsilon N}||^2\,d\tau \le \left(||u_0'||^2 + \frac{m^0 K E_0}{2\mu}\right) \exp\left\{\frac{E_0}{2\mu^2}\right\}. \tag{28}$$

Substituting in (20) w_j by $v^{\varepsilon N}$ and integrating by parts, we get

$$\begin{aligned}&||v||^2(t) + 2\varepsilon \textstyle\int_0^t ||v_x||^2(\tau)\,d\tau + K\int_0^t ||v||^2(\tau)\,d\tau \\ &= ||v_0||^2 + K\textstyle\int_0^t ||u||^2(\tau)\,d\tau.\end{aligned} \tag{29}$$

Using repeatedly (15), (16), Young's inequality and neglecting nonnegative terms with $\varepsilon > 0$, we obtain from (20):

$$\begin{gathered}||v_x||^2(t) + \textstyle\int_0^t \Big(K - \sup_{x\in(0,1)}|v_x|\Big)||v_x||^2(\tau)\,d\tau \le \\ ||v_0'||^2 + K\textstyle\int_0^t ||u_x||^2(\tau)\,d\tau\end{gathered} \tag{30}$$

and

$$\|v_{xx}\|^2(t) + \int_0^t \Big(K - 7\sup_{x\in(0,1)} |v_x|\Big) \|v_{xx}\|^2(\tau)\, d\tau \leq \\ \|v_0''\|^2 + K \int_0^t \|u_{xx}\|^2(\tau)\, d\tau. \tag{31}$$

Proposition 1. *If $\lambda < K/7$, then*

$$\sup_{x\in(0,1)} |v_x^{\varepsilon N}(x,t)| < K/7 \quad \textit{for all } t > 0. \tag{32}$$

Proof. Indeed, when $t = 0$, $\sup |v_x^{\varepsilon N}(x,0)| \leq \|v_0'\| + \|v_0''\| \leq \lambda < K/7$. Suppose that $\sup |v_x^{\varepsilon N}(x,t_*)| = K/7$ for some $t = t_* > 0$. Then (23), (28)-(31) give

$$\begin{aligned} K/7 &= \sup_{x\in(0,1)} |v_x^{\varepsilon N}(x,t_*)| \leq \|v_x^{\varepsilon N}\|(t_*) + \|v_{xx}^{\varepsilon N}\|(t_*) \\ &\leq \Big[\|v_0'\|^2 + K \int_0^{t_*} \|u_x^{\varepsilon N}\|^2(\tau)\, d\tau\Big]^{1/2} + \Big[\|v_0''\|^2 + K \int_0^{t_*} \|u_{xx}^{\varepsilon N}\|^2(\tau)\, d\tau\Big]^{1/2} \\ &\leq \Big[\|v_x'\|^2 + KE_0/2\mu\Big]^{1/2} + \Big[\|v_{xx}''\|^2 + \Big(K\|u_0'\|^2 + \tfrac{m^0 K^2 E_0}{2\mu}\Big) \tfrac{\exp\{E_0/2\mu^2\}}{\mu}\Big]^{1/2} \\ &= \lambda < K/7. \end{aligned}$$

This contradiction proves the Proposition 1.

Thus, setting $\lambda < K/7$ and bearing in mind (28)-(31), we conclude that all the necessary norms of solutions to (19)-(22) are bounded by some constant independent on $T > 0$, $N \geq 1$ and $\varepsilon > 0$. This allows us to pass to the limit by diagonal process as $N \to \infty$, $\varepsilon \to 0$ and, taking into account (32) and (8), to obtain the required solution to problem (1)-(6).

5 Uniqueness

Let $(u^1, v^1, m^1)(x,t)$ and $(u^2, v^2, m^2)(x,t)$ be two solutions to (1)-(6). Then the functions $U = u^1 - u^2$ and $V = v^1 - v^2$ satisfy in Q the following problem:

$$U_t + u^1 U_x + u_x^2 U - \mu U_{xx} + \alpha U = m^0 K(V - U), \tag{33}$$

$$V_t + v^1 V_x + v_x^2 V = K(U - V), \tag{34}$$

$$U(x,0) = V(x,0) = 0, \tag{35}$$

$$U(0,t) = U(1,t) = 0, \tag{36}$$

$$V(0,t) = V(1,t) = 0. \tag{37}$$

Multiplying equations (33)-(34) by U and $m^0 V$ correspondingly and summing up the results, we obtain

$$\frac{1}{2}\frac{d}{dt}\Big(\|U\|^2+m^0\|V\|^2\Big)(t)+\int_0^1\Big\{\mu(U_x)^2+\alpha U^2+m^0K(U-V)^2\Big\}\,dx$$
$$=\int_0^1\Big\{\tfrac{1}{2}u_x^1U^2-u_x^2U^2+\tfrac{m^0}{2}v_x^1V^2-m^0v_x^2V^2\Big\}\,dx. \tag{38}$$

Since

$$\left|\int_0^1 u_x^iU^2\,dx\right|\le\|u_x^i\|^2\left(\xi\|U_x\|^2+C_\xi\|U\|^2\right),\quad i=1,2;\quad\forall\xi>0,$$

integrating (38) over $t\in(0,T)$ with $\xi\|u_x^i\|_{L^\infty(0,T;L^2(0,1))}=\mu/2$, we get

$$\Big(\|U\|^2+m^0\|V\|^2\Big)(t)+\int_0^t\int_0^1\Big\{\mu(U_x)^2+2\alpha U^2+2m^0K(U-V)^2\Big\}\,dx\,d\tau$$
$$\le 2\int_0^t C\Big\{\|U\|^2+m^0\|V\|^2\Big\}(\tau)\,d\tau,$$

where $C>0$ depends on μ, m^0 and on

$$\max\{\|u^i\|_{L^\infty(0,T;H^1(0,1))},\ \|v^i\|_{L^\infty(0,T;H^2(0,1))}\}$$

with $i=1,2$.

By Gronwall's lemma, we conclude that $U=V=0$ for a.e. $t\in(0,T)$. Consequently, by (8), $m(x,t)$ is defined uniquely. The proof of Theorem 1 is completed.

References

1. Sproull, W.T. (1961): Viscosity of dusty gases. Nature **190**, 976.
2. Kazakevich, F.P., Krapivin, A.M. (1958): Investigations of heat transfer and aerodynamical resistance in tube assemblies when the flow of gas is dustladen (in Russian). Izv. Vissh. Uchebn. Zavedenii. Energetica, **1**, 101.
3. Saffman, P.G. (1962): On the stability of laminar flow of a dusty gas. J. Fluid Mech., **13**, **1**, 120–128.
4. Crooke, P.S. (1972): On growth properties of solutions of the Saffman dusty gas model. Z. Angew. Math. Phys., **1**, **23**, 182–200.
5. Stewart, H.B., Wendroff, B. (1984): Two-phase flow: models and methods. J. Comput. Phys., **56**, 363–409.
6. Rakhmatulin, Kh.A. (1956): The basic gas dynamics of interpenetrating motions of compressible media (in Russian). Prikl. Mat. Mekh., **20**, 185–195.
7. Kliegel, J.R., Nickerson, G. (1962): Flow of a mixture of gas and solid particles in an axisymmetric nozzle. In: Detonation and two-phase flow, New York.
8. Nigmatulin, R.I. (1987): Dynamics of multiphase media (in Russian). Nauka, Moscow.
9. Larkin, N.A. (1978): A model system of equations of the mechanics of heterogeneous media (in Russian). Chisl. Metody Mekh. Sploshn. Sredy, **9**, **7**, 60–66.
10. Doronin, G.G. (1995): Correctness of the Saffman model of a dust-laden gas with allowance for compressibility. Differential Equations, **31**, **6**, 952–959.

11. Doronin, G.G., Larkin, N.A. (1996): Cauchy Problems for Equations of Two-Phase Flow. In: Blokhin, A.M. (ed) Differential Equation Theory. Nova Science Publishers Inc, New York, 32–40.
12. Doronin, G.G. (1999): On solvability of the Cauchy problem for the dusty gas equations. Comp. Appl. Math. **18**, **2**, 163–172.
13. Dafermos, C.M. (1995): A system of conservation laws with frictional damping. Zeit. Ang. Math. Phys., **46**, 294–307.

A Constrained Transport Upwind Scheme for Divergence-free Advection

Michael Fey[1] and Manuel Torrilhon[2]

[1] Seminar for Applied Mathematics, ETH Zurich, 8092 Zurich, Switzerland
fey@math.ethz.ch

[2] Seminar for Applied Mathematics, ETH Zurich, 8092 Zurich, Switzerland
manuel@math.ethz.ch

1 Introduction

Many transport equations in physics and engineering come along with intrinsic constraints. A standard example for an evolution equation with intrinsic constraint is given by

$$\partial_t u + \operatorname{curl} F(u) = 0, \qquad \operatorname{div} u = const. \tag{1}$$

where u is a space vector. The divergence constraint is intrinsic since it follows from the evolution equation and must not be viewed as an additional equation. There are several areas where equations like (1) arise: Maxwells equations, the transport of the magnetic field in magnetohydrodynamics (MHD), or the vorticity transport in case of incompressible flows. One can easily think of other constrained transport equations. In [8] a system that conserves vorticity is investigated.

Intrinsic constraints are expected to hold also in a numerical solution at least in a discrete manner. Nevertheless commonly available numerical methods do not care about intrinsic constraints of the equations. This paper will supply a general framework to derive explicit numerical methods which

- control the constraint locally in every single timestep,
- use only one finite volume grid for the equation and the constraint,
- take upwind directions into account,
- perform a truly multidimensional update

and apply the framwork to the case of divergence-free advection which is related to (1). The literature provides many works which deal with divergence constraints (e. g. [1]-[5] and [9]). In contrast to a global projection scheme like in [2], in a local approach usually the divergence is conserved by using different (staggered) grids for the equation and the constraint (introduced by [5]) or by using a FEM-type reconstruction [4].

In the following the known staggered schemes as well as a new advanced non-staggered upwind method are derived together in a general finite volume framework.

2 General framework

To set up a general framework for constrained transport schemes we consider $u \in \mathbb{R}^D$ and a generic transport operator $\mathcal{L}$ with a generic constraint $\mathcal{C}$. Hence, we have the equations

$$\partial_t u + \mathcal{L}(u) = 0, \qquad \mathcal{C}(u) = const, \tag{2}$$

where the constraint is intrinsic, i. e. the relation

$$\mathcal{C}(\mathcal{L}(u)) \equiv 0 \tag{3}$$

holds. In the following we assume that the constraint is linear. The computational domain is covered by a grid $\mathcal{T} = \{K_i\}_{i=1,2,...}$ with cells K_i whose size is bounded by h. A time discretization by Δt leads to a cell-wise constant gridfunction $\tilde{u}^m : \mathcal{T} \to \mathbb{R}^D$ which approximates u after m timesteps by cell mean values.

The central quantity of this paper is the so called "fluxdistribution".

Definition 1. *We define a "fluxdistribution"*

$$\Phi_K(\tilde{u}) : \mathcal{T} \to \mathbb{R}^D \tag{4}$$

with $\operatorname{supp} \Phi_K(\tilde{u}) = K \cup \mathcal{N}(K)$*, such that* $\Phi_K(\tilde{u})|_{\hat{K}}$ *gives the change of* u *at cell* $\hat{K}$ *caused by cell* K*.*

A fluxdistribution is assigned to each cell of the grid and depends on the solution in a local manner. The definition is more general than that of usual intercell fluxes, since it admits fluxes to any neighbouring cell, especially across corners. This follows the ideas of the method of transport [6]. Conservation of u is expressed by the statement that the integral over $\Phi_K(u)$ vanishes.

A certain fluxdistribution defines an evolution scheme by simply collecting contributions of all fluxdistributions. Written in complete gridfunctions we have

$$\tilde{u}^{m+1} = \tilde{u}^m + \sum\nolimits_{\hat{K}} \Phi_{\hat{K}}(\tilde{u}^m). \tag{5}$$

The restriction to a certain cell yields a local formulation, viz.

$$\tilde{u}^{m+1}(K) = \tilde{u}^m(K) + \sum_{\hat{K} \in K \cup \mathcal{N}(K)} \Phi_{\hat{K}}(\tilde{u}^m)|_K \tag{6}$$

To approximate the transport equation given in (2) consistency in the form

$$\mathcal{L}(u)|_K = - \lim_{\Delta t, h \to 0} \frac{1}{\Delta t} \sum_{\hat{K} \in K \cup \mathcal{N}(K)} \Phi_{\hat{K}}(u)|_K \tag{7}$$

is required.

Since the constraint is linear we expect a discretization which may be written as matrix operation

$$\mathcal{C}(u)|_K = \tilde{\mathcal{C}}_K \cdot \tilde{u} + O(h^n) \tag{8}$$

on the gridfunction $\tilde{u}$. If preservation of the constraint should be achieved for the scheme (5) the following quiete obvious statement leads the way.

Lemma 1. *¿From the definition of the evolution scheme we have*

$$\tilde{\mathcal{C}}_K \cdot \Phi_{\hat{K}}(\tilde{u}) = 0 \quad \forall K, \hat{K}, \tilde{u} \quad \Rightarrow \quad \tilde{\mathcal{C}}_K \cdot \tilde{u}^{m+1} = \tilde{\mathcal{C}}_K \cdot \tilde{u}^m \quad \forall K \tag{9}$$

as condition for the preservation of the discrete operator.

Since the condition is hard to control for any gridfunction $\tilde{u}$, we assume a decomposition

$$\Phi_K(\tilde{u}) = w_K(\tilde{u})\, \hat{\Phi}_K \tag{10}$$

into a weight $w_K(\tilde{u}) \in \mathbb{R}$ and a skeleton or shape function $\hat{\Phi}_K$. As indicated only the weight depends on the field $\tilde{u}$. Due to the linearity of the constraint the weight drops out of the condition in (9) and we obtain

$$\tilde{\mathcal{C}}_K \cdot \hat{\Phi}_{\hat{K}} = 0 \quad \forall K, \hat{K} \tag{11}$$

as purely geometric condition. To some extend this is the discrete analog to (3) which states that the constraint is intrinsic.

If a generic cell $\hat{K}$ is fixed, (11) gives a linear system of equations for the unknown components of the fluxdistribution. The crucial task of designing constrained transport schemes is to find non-trivial solutions of (11). The existence of solutions and their structure depends strongly on the used discretization of the constraint operator. In the next section we will discuss some examples for divergence-free transport.

In general there may exist several linearly independent solutions $\hat{\Phi}_K^{(g)}$ to (11) which all are preservative fluxdistributions. The final fluxdistribution has to be assembled from these solutions via

$$\Phi_K(\tilde{u}) = \sum_g w_K^{(g)}(\tilde{u})\, \hat{\Phi}_K^{(g)} \tag{12}$$

with appropriate coefficients $w_K^{(g)}$. This expression enters the scheme (6) and the remaining coeficients $w_K^{(g)}$ have to follow from consistency (7) and further considerations. Note, that the choice of $w_K^{(g)}$ does not affect the preservation of the constraint which is obtained by the skeletons of the fluxdistributions.

3 Divergence-free transport

As example for a transport equation with constraint we consider the equation

$$\partial_t u + \mathrm{curl}\,(u \times v) = 0, \qquad \mathrm{div}\, u = const \tag{13}$$

where v is a given velocity field. The constraint is obviously intrinsic due to $\mathrm{div}\,\mathrm{curl}\,\psi \equiv 0$. The transport equation in (13) reduces to an ordinary advection equation only if $\mathrm{div}\, u = 0$ and $v = const$. In all other cases the vector field u is mainly transported with the velocity field v, but such that its divergence is preserved. The equation occurs in the system of magnetohydrodynamics where u represents the magnetic field. The equation is also recovered in the vorticity method for calculating incompressible flow. There, u is given by the vorticity of the flow.

In this paper we restrict ourself to $\mathbb{R}^2$ where the vector u and the velocity v is given by $u = \left(u^{(x)}, u^{(y)}, 0\right)$ and $v = \left(v^{(x)}, v^{(y)}, 0\right)$. The transport equation in (13) reduces to

$$\begin{aligned} \partial_t u^{(x)} + \partial_y F\,(u,v) = 0 \\ \partial_t u^{(y)} - \partial_x F\,(u,v) = 0 \end{aligned} \tag{14}$$

with the single fluxfunction

$$F\,(u,v) = v^{(y)} u^{(x)} - v^{(x)} u^{(y)}. \tag{15}$$

We proceed with applying the general framework of the preceeding section to the system (14). The investigation will be presented in the case of an uniform rectangular grid with cell size $h \times h$. The generalization to anisotropic rectangular grids with $\Delta x \times \Delta y$ cells is straight forward. Note, that the general framework is valid for any kind of grid. The extension to more general grids is subject to future work.

3.1 Discrete divergence operators in 2d

Since the discrete version of the constraint operator influences the structure of the fluxdistribution, we present a certain class of discrete divergence operators. Each operator is obtained from a discretization of the first derivative in a finite difference approach. We require a symmetric 3×3 stencil and a approximation of second order. The following lemma is obtained by Taylor expansion.

Lemma 2. *Any second order approximation of the first derivative of a gridfunction φ in the center cell K of a 3×3 cartesian stencil has the form*

$$\left.\frac{\partial \varphi}{\partial x}\right|_K = \tilde{\mathcal{D}}_K\,(\alpha,\beta,\gamma) \cdot \varphi + O\left(h^2\right) \tag{16}$$

with arbitrary values for α, β, γ and

$$\tilde{\mathcal{D}}_K\,(\alpha,\beta,\gamma) = \frac{1}{2h}\begin{array}{|c|c|c|}\hline 0 & 0 & 0 \\ \hline -1 & 0 & 1 \\ \hline 0 & 0 & 0 \\ \hline\end{array} + \frac{\alpha}{h}\begin{array}{|c|c|c|}\hline -1 & 0 & 1 \\ \hline 2 & 0 & -2 \\ \hline -1 & 0 & 1 \\ \hline\end{array} + \frac{\beta}{h}\begin{array}{|c|c|c|}\hline 1 & -2 & 1 \\ \hline 0 & 0 & 0 \\ \hline -1 & 2 & -1 \\ \hline\end{array} + \frac{\gamma}{h}\begin{array}{|c|c|c|}\hline 1 & -2 & 1 \\ \hline -2 & 4 & -2 \\ \hline 1 & -2 & 1 \\ \hline\end{array} \tag{17}$$

as weights in the grid.

Note, that simple central differences, represented in the first box of $\tilde{\mathcal{D}}_K$ are not anymore the only second order symmetric approximation to $\partial_x \varphi$ in the above setting. The additional stencils with coefficients α, β and γ represent higher order cross derivatives which do not change second order approximation. A divergence operator $\tilde{\mathcal{C}}_K$ is now formed by

$$\operatorname{div} u|_K = \tilde{\mathcal{D}}_K (\alpha,\beta,\gamma) \cdot \tilde{u}^{(x)} - \tilde{\mathcal{D}}_K^T (\alpha,\beta,\gamma) \cdot \tilde{u}^{(y)} + O\left(h^2\right) \tag{18}$$

$$\equiv \tilde{\mathcal{C}}_K (\alpha,\beta,\gamma) \cdot \tilde{u} + O\left(h^2\right) \tag{19}$$

which leads to a three parameter family of operators. In the evaluation of the operator cell mean values are taken for point values which only introduces an error of second order. It is important to mention, that if any operator taken from (19) vanishes, all other operators give a result of $O\left(h^2\right)$. Hence, if a numerical scheme respects one operator exactly, all others will give only a discretization error.

3.2 Fluxdistributions

In order to derive divergence-preserving fluxdistributions we have to look for non-trivial solutions of (11) for a specific operator chosen from (19). The fluxdistribution skeleton $\hat{\Phi}_{\hat{K}}$ gives a vector in each cell, thus consists of 2×9 entries. If we fix $\hat{K}$ in (11) and evaluate the divergence operator around the fluxdistribution in cells K of a 5×5 area with center $\hat{K}$, we obtain a system of equations which describes the skeleton entries. The divergence of more remote cells is not influenced by the fluxdistribution at cell $\hat{K}$ and need not to be considered. The resulting system, of course, depends on the chosen operator.

Classical operator The classical discrete divergence operator $\tilde{\mathcal{C}}_K^{(0)}$ is obtained from (19) by setting $\alpha = \beta = \gamma = 0$, which leads to

$$\tilde{\mathcal{C}}_K^{(0)} \cdot \tilde{u} = \tilde{\mathcal{C}}_K (0,0,0) \cdot \tilde{u} = \frac{\tilde{u}_{i+1,j} - \tilde{u}_{i-1,j}}{2h} + \frac{\tilde{u}_{i,j+1} - \tilde{u}_{i,j-1}}{2h}. \tag{20}$$

This operator represents the best second order approximation to $\operatorname{div} u$ in the sense that the constant hidden in $O\left(h^2\right)$, i. e. the residual of the second order Taylor expansion is minimal. The linear system (11) for this operator has rang 17, thus admits one linearly independent solution denoted by $\hat{\Phi}_K^{(0)}$. The non-vanishing entries are given by

$$\left.\hat{\Phi}_K^{(0)}\right|_{i+1,j} = (0,1), \quad \left.\hat{\Phi}_K^{(0)}\right|_{i,j+1} = (-1,0), \tag{21}$$

$$\left.\hat{\Phi}_K^{(0)}\right|_{i,j-1} = (1,0), \quad \left.\hat{\Phi}_K^{(0)}\right|_{i-1,j} = (0,-1), \tag{22}$$

up to a constant factor. This fluxdistribution is sketched in the lower left corner of Fig. 1. The picture has to be interpreted as follows: A flux originating

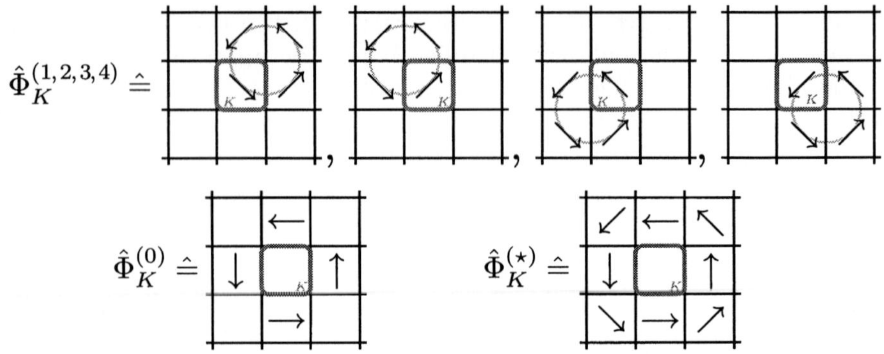

Fig. 1. Several shape functions for fluxdistributions for divergence-free transport. All these skeletons share the same physical interpretation: They have to approximate closed curves in order to avoid introducing sources into the vector field u and thus they preserve the divergence.

in K may change the value of $\tilde{u}$ in the right neighbouring cell only in the y-direction. Furthermore, if this neighbour is changed in that way, all other neighbouring cells have to be changed correspondingly as depicted in the figure. Of course, the value $\tilde{u}^{(x)}$ in the right neighbour of K does not remain constant, since it may be changed by fluxdistributions originating from other cells.

Consistency gives the expression

$$\varPhi_K^{(0)}(\tilde{u}) = \frac{\Delta t}{2} F(\tilde{u}, v)|_K \, \hat{\varPhi}_K^{(0)} \tag{23}$$

for the final fluxdistribution which turns (6) into a second order scheme for solving the system (14) and, simultaneously preserving the classical divergence operator. The resulting scheme introduces central differences for the derivatives in (14) and was proposed ad hoc by Toth [10] in a MHD setting. Since $\hat{\varPhi}_K^{(0)}$ is the only fluxdistribution respecting condition (11) with the classical operator, this scheme is the only second order scheme which preserves the divergence via the classical operator.

Extended operator In order to design a upwind scheme we look for a divergence operator which introduces more non-trivial solutions into the system (11). Empirical evaluations by computer algebra software suggest that the choice $\alpha = \frac{1}{8}$, $\gamma = 0$ admits four solutions for any value of β. The best approximation is then given by $\beta = 0$ and we obtain the extended operator $\tilde{\mathcal{C}}_K^{(\star)}$. It is defined by

$$\begin{aligned} \tilde{\mathcal{C}}_K^{(\star)} \cdot \tilde{u} &= \tilde{\mathcal{C}}_K\left(\tfrac{1}{8}, 0, 0\right) \cdot \tilde{u} \\ &= \frac{\{\tilde{u}_{i+1,j}\}_y - \{\tilde{u}_{i-1,j}\}_y}{2h} + \frac{\{\tilde{u}_{i,j+1}\}_x - \{\tilde{u}_{i,j-1}\}_x}{2h} \end{aligned} \tag{24}$$

where curled brackets stand for

$$\begin{aligned} \{\tilde{u}_{i,j}\}_y &= \tfrac{1}{4}\left(\tilde{u}_{i,j+1} + 2\tilde{u}_{i,j} + \tilde{u}_{i,j-1}\right) \\ \{\tilde{u}_{i,j}\}_x &= \tfrac{1}{4}\left(\tilde{u}_{i+1,j} + 2\tilde{u}_{i,j} + \tilde{u}_{i-1,j}\right) \end{aligned} \tag{25}$$

i. e. averaging in x and y-direction. The four admissible flux skeletons are displayed in the upper row of Fig. 1. In numbers the non-vanishing entries of the first one are given by

$$\hat{\Phi}_K^{(1)}\Big|_{i+1,j+1} = (-1,1), \ \hat{\Phi}_K^{(1)}\Big|_{i,j+1} = (-1,-1), \tag{26}$$

$$\hat{\Phi}_K^{(1)}\Big|_{i,j} = (1,-1), \ \hat{\Phi}_K^{(1)}\Big|_{i+1,j} = (1,1) \tag{27}$$

and the remaining three fluxdistributions are build analogously.

A first choice for a skeleton could be the symmetric distribution $\hat{\Phi}_K^{(\star)}$ which is given by

$$\hat{\Phi}_K^{(\star)} = \hat{\Phi}_K^{(1)} + \hat{\Phi}_K^{(2)} + \hat{\Phi}_K^{(3)} + \hat{\Phi}_K^{(4)} \tag{28}$$

and shown in the lower right corner of Fig. 1. The complete fluxdistribution

$$\Phi_K^{(\star)}(\tilde{u}) = \frac{\Delta t}{8} \left. F(\tilde{u}, v)\right|_K \hat{\Phi}_K^{(\star)} \tag{29}$$

follows by second order consistency. The resulting scheme is equivalent to the staggered grid schemes proposed for the magnetic field in MHD in [5], if the staggered variables are eliminated by interpolation and extrapolation as shown in [10]. Furthermore, we could show that the extended operator $\tilde{\mathcal{C}}_K^{(\star)}$ is the only second order operator which is preserved on the primary grid in a staggered grid scheme.

Finally, we introduce a scheme in which the different skeletons $\hat{\Phi}_K^{(i)}$ may be weighted differently according to the situation in the particular cell. First order consistency leads to

$$\Phi_K(\tilde{u}) = \sum_{g=1}^{4} \omega_K^{(g)} \frac{\Delta t}{2} \left. F(\tilde{u}, v)\right|_K \hat{\Phi}_K^{(g)} \tag{30}$$

where the weights $\omega_K^{(g)}$ are still free to choose provided they sum up to unity. These weights are used to incorporate upwind directions into the scheme. Each skeleton is associated with a direction vector, the first one with $n_1 = (1,1)$, etc. We propose to activate a certain fluxdistribution only if the wind, that is the velocity vector v, has a component in the associated direction. If the angle between direction and wind is denoted by $\varphi_g = \sphericalangle(n_g, v)$ a possible weight function is given by

$$\omega_K^{(g)}(v) = \frac{\cos^\kappa\left(\min\left(|\varphi_g|, \pi/2\right)\right)}{\sum_{\hat{g}=1}^{4} \cos^\kappa\left(\min\left(|\varphi_{\hat{g}}|, \pi/2\right)\right)} \tag{31}$$

with $\kappa \geq 1$.

In a second order scheme linear reconstruction in order to obtain derivatives of F becomes neccessary. Additionally, derivatives of the weight function are needed. Since the weight function with $\kappa = 1$ is not differentiable, $\kappa > 1$ is required for second order. Numerical evaluations, however, suggest higher values (≈ 6) for κ in order to recover full second order.

4 Numerical evaluation

As presented above the general framework recovers two schemes which has been proposed formerly in a rather ad hoc way (see e. g. [10]). Both schemes (23) and (29) rely on symmetric fluxdistributions and result in central differences for the fluxes in the system (14). In our numerical evaluations of the advection system (14) the schemes turned out to be unstable due to these central differences and the lack of upwinding. In a MHD context, however, they proved to be stable, see [1], [3], [5], [10]. The reason for this is most likely the use of predictor values as described in the cited works.

In the following sections we present numerical results for the upwind scheme with fluxdistribution (30) for the system (14). Due to space restriction only one test problem is presented.

4.1 Box test case

If initial conditions for the system (14) are given in the form

$$u_0\,(x,y) = (0, F'\,(x)\,F'\,(y)) \tag{32}$$

with derivatives of a function F and the velocity field points constantly in x-direction, i. e. $v\,(x,y) = (1,0)$, the exact solution is given by

$$u\,(x,y,t) = (F''\,(y)\,(F\,(x) - F\,(x-t))\,, F'\,(x-t)\,F'\,(y))\,. \tag{33}$$

As test example we choose

$$F\,(x) = \begin{cases} \frac{16}{5}s^5 - \frac{8}{3}s^3 + s & -\frac{1}{2} < s < \frac{1}{2} \\ \pm\frac{4}{15} & \text{else} \end{cases} \tag{34}$$

so the initial field u_0 is non-vanishing only inside the box $(-1/2, 1/2)^2$. Furthermore the system is rotated such that advection takes place in diagonal grid direction. The initial conditions are displayed in the upper right corner of Fig. 2. The center of the box is moved to $(-0.1, -0.1)$. The contour color represents values of $||u||$, which ranges form zero to 1.36. The lines in the plots represent field lines of u. Below the initial conditions numerical results for $t = 0.5$ are shown for two unifom meshes with different sizes. In the exact solutions the fieldlines are bended outside the box due to the advection. In

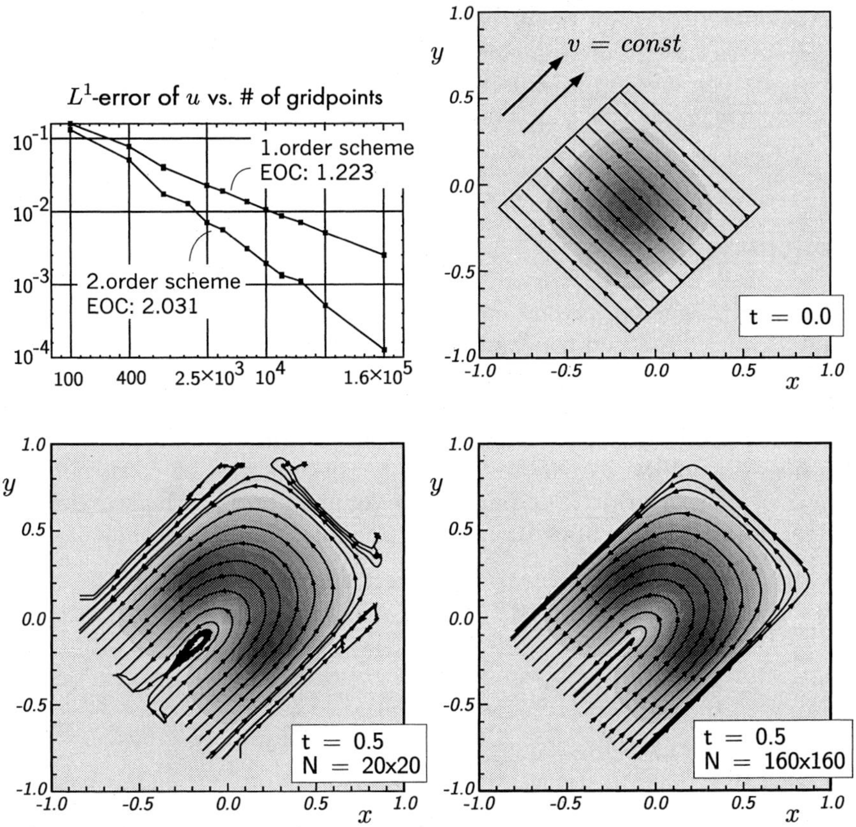

Fig. 2. Box test case for the advection system (14) calculated on a coarse (lower left) and a fine (lower right) grid. The initial conditions are shown in the upper right corner. Errors and empirical orders of convergence may be read off the upper left corner.

this particular examples the divergence of u does not vanish initially, thus the fieldlines fill up the way and their starting and ending points stay inside the initial box. On the coarse grid the solution is spoiled at the sides by artificial fieldlines. These fieldlines correspond to values of u in the magnitude of the truncation error which are introduced at the boundary of the box by the finite stencil. On the fine mesh the solution is well resolved. The upper left corner of Fig. 2 shows L^1-errors for this example in the cases of first and second order schemes. The solution was obtained with a weight function exponent $\kappa = 6$ in (31). Second order is well obtained, while the first order scheme exhibits a slight super-convergence. Small fluctuations in the error lines might be due to the lack of first derivatives of the solution at the boundary of the box.

In the numerical result the divergence, calculated with the extended operator (24), gives exactly the same result for any timestep, due to the design of the method. The method freezes the discrete divergence of the initial conditons. Of course, compared to the exact value of $\operatorname{div} u$ there is a discretization error, which decreases with second order if the mesh is refined.

5 Conclusions

The new non-staggered upwind constrained transport scheme is an easy to use and advanced method to simulate transport equations with constraints. In [7] this method was used to solve the magnetic evolution part of MHD. The characteristic directions of the Alven waves were used as velocity field in (13) for a good approximation of the rotational discontinuities. The general framework allows to derive schemes for other types of constraints as well as for different grids. The pure finite volume approach makes also the implementation in three dimensions and in an adaptivity environment easy.

References

1. Balsara, D. S. and Spicer, D. S., *A staggered mesh algorithm using high order Godunov fluxes to ensure solenoidal magnetic fields in magnetohydrodynamic simulations*, J. Comp. Phys. **149**, (1999) p.270
2. Brackhill, J. U. and Barnes, D. C., *The effect of nonzero $\nabla \cdot B$ on the numerical solution of the magnetohydrodynamic equations*, J. Comp. Phys. **35**, (1980) p.426
3. Dai, W. and Woodward, P. R., *On the divergence-free condition and conservation laws in numerical simulations for supersonic magnetohydrodynamic flows*, Astrophys. J. **494**, (1998) p.317
4. DeSterck, H., *Multi-Dimensional Upwind Constrained Transport on Unstructured Grids for Shallow Water Magnetohydrodynamics*, AIAA Paper 2001-2623, (2001)
5. Evans, C. R. and Hawley, J. F., *Simulation of Magnetohydrodynamic Flows: A Constrained Transport Method*, Astrophys. J. **332**, (1988) p.659
6. Fey, M., Jeltsch, R., and Morel, A.-T., *Multidimensional schemes for nonlinear systems of hyperbolic conservation laws*, technical report nr. 1995-11, Seminar for Applied Mathematics, (1995)
7. Limacher, R., *Simulation der MHD-Gleichungen mit der Transportmethode*, diploma thesis, ETH Zurich (2000)
8. Morton, K. W. and Roe, P. L., *Vorticity-Preserving Lax–Wendroff-Type Schemes for the System Wave Equation*, SIAM J. Sci. Comp. **23**/1, (2001) p.170
9. Powell, K. G., *An approximate Riemann solver for magnetohydrodynamics (that works in more than one dimension)*, ICASE Report No. 94-24, (1994)
10. Toth, G., *The $\nabla \cdot B$ Constraint in Shock-Capturing Magnetohydrodynamics Codes*, J. Comp. Phys. **161**, (2000) p.605

Green's Function Pointwise Estimates for the Modified Lax-Friedrichs Scheme

Pauline Godillon-Lafitte[1]

Unité de Mathématiques Pures et Appliquées (CNRS UMR # 5669),
ENS Lyon, 46 Allée d'Italie, 69364 Lyon Cédex 07
`http://www.umpa.ens-lyon.fr/ pgodillo`
Pauline.Godillon@umpa.ens-lyon.fr

Summary. Our aim is to find estimates of the Green's function of stationary discrete shock profiles by using techniques developed by Zumbrun and Howard [17] in the continuous viscous setting.

1 Introduction

Let $d \geq 1$ and consider a one-dimensional system of d conservation laws

$$u_t + f(u)_x = 0,\ t \geq 0,\ \ x \in \mathbf{R}, \tag{1}$$

where u maps $\mathbf{R} \times \mathbf{R}^+$ onto an open subset $\mathcal{U}$ of $\mathbf{R}^d$, $f : \mathcal{U} \subset \mathbf{R}^d \longrightarrow \mathbf{R}^d$ is smooth and the initial datum is given by

$$u(x,0) = \mathsf{u}_0,\ x \in \mathbf{R}. \tag{2}$$

We want to approximate (1) with the modified Lax-Friedrichs scheme:

$$u^n_{j+1} = (\mathcal{N}(u^n))_j = u_j - \frac{k}{h}(\mathcal{F}(u_j, u_{j+1}) - \mathcal{F}(u_{j-1}, u_j)),\ j \in \mathbf{Z},$$

where k is the time step, h the space step and the numerical flux $\mathcal{F}$ is

$$\mathcal{F}(u,v) = \frac{f(u) + f(v)}{2} + \mathbf{D}(u - v),$$

with $\mathbf{D}$ a scalar constant.
The numerical problem associated with (1)-(2) is

$$\begin{aligned} u^{n+1} &= \mathcal{N}(u^n),\ n \geq 1, \\ u^0_j &= \frac{1}{h}\int_{hj}^{h(j+1)} \mathsf{u}_0(x)\mathrm{d}x,\ j \in \mathbf{Z}. \end{aligned}$$

Let (u^-, u^+) be a stationary shock of (1), of *arbitrary strength*, that satisfies the following standard assumptions:

H1 the *Rankine-Hugoniot* condition: $f(u^+) = f(u^-)$;

H2 the *strict hyperbolicity* of (1) at $u^{\pm}$: $\mathrm{d}f(u^{\pm})$ is diagonalizable in $\mathbf{R}$ with simple eigenvalues, which, from now on, we denote by $a_1^{\pm} < \ldots < a_d^{\pm}$ with some associated eigenvectors $r_1^{\pm}, \ldots, r_d^{\pm}$;

H3 the shock is *non characteristic*: $a_k^{\pm} \neq 0$, $\forall k \in \{1, \ldots, d\}$;

H4 the *CFL* condition: $\mathsf{r}(\mathrm{d}f(u^{\pm})) < 2\mathbf{D} < \dfrac{h}{k}$, where r denotes the spectral radius;

H5 the shock is a *Lax-type p-shock*: there are $d+1$ ingoing characteristics and $d-1$ outgoing ones:

$$\begin{aligned} a_p^+ &< 0 < a_p^-, \\ a_{p-1}^- &< 0 < a_{p+1}^+. \end{aligned}$$

We are interested in the discrete shock profiles $\bar{u}$ that join asymptotically the end states $u^{\pm}$: in the particular case of stationary shocks, $\bar{u}$ is a fixed point of $\mathcal{N}$, that is it satisfies a second-order recurrence, whereas, for general shock speeds, it satisfies a functional differential equation. Note that, in the general case, the existence of such dicrete profiles remains open [15], although there are results for rational discrete speeds [10, 12, 14]. From now on, we make the following assumption:

H6 there exists a discrete profile $\bar{u}$ such that

$$\begin{aligned} \mathcal{N}(\bar{u}) &= \bar{u}, \\ \lim_{j \to \pm\infty} \bar{u}_j &= u^{\pm}. \end{aligned}$$

The critical issue is to determine the stability of such profiles. The first step is to study the linear stability: we linearize $\mathcal{N}$ about $\bar{u}$ and we denote the linearized operator by L. Given $\tilde{v}$ and $\mathbf{v}$, we then study the linearized numerical problem

$$\begin{aligned} v^{n+1} - v^n - Lv^n &= \tilde{v}^n, \ \forall n \geq 0, \\ \lim_{j \to \pm\infty} v_j^n &= 0, \quad \forall n \geq 0, \\ v^0 &= \mathbf{v}. \end{aligned} \tag{3}$$

A useful tool is the *discrete Green's function*, that is the sequence of matrices of $\mathcal{M}_d(\mathbf{R})$ $(G(j,n;l))_{j \in \mathbf{Z}, n \in \mathbf{N}; l \in \mathbf{Z}}$ that solves, given $l \in \mathbf{Z}$,

$$\begin{aligned} G(\cdot, n+1; l) - G(\cdot, n; l) - LG(\cdot, n; l) &= 0, \qquad \forall n \geq 0, \\ \lim_{j \to \pm\infty} G(j, n; l) &= 0, \qquad \forall n \geq 0, \\ G(j, 0; l) &= \delta_{jl} I_d, \ \forall j \in \mathbf{Z}. \end{aligned} \tag{4}$$

Indeed, the solutions of (3) can be expressed formally in terms of G, $\tilde{v}$ and $\mathbf{v}$:

$$v_j^n = \sum_{l=-\infty}^{+\infty} G(j,n;l)\mathbf{v}_l + \sum_{m=0}^{n-1} \sum_{l=-\infty}^{+\infty} G(j,n-m;l)\tilde{v}_l^m, \ j \in \mathbf{Z}, n \geq 1.$$

Thus, our aim is to find estimates of G that are analogous to Zumbrun and Howard's for the (continuous) viscous profiles [17] in order to be able to describe the solutions of (3). Since the presence of spectrum of L in $\{|\mathrm{Re}(\mu)+1| > 1\}$ means spectral instability [11], we assume that

H7 the operator L has no spectrum in $\{|\mathrm{Re}(\mu) + 1| \geq 1\} =: \Omega$ except at $\mu = 0$.

In fact, the study of the essential spectrum of L shows that the spectrum of L in Ω consists only of eigenvalues except at $\mu = 0$. Indeed, the essential spectrum touches the imaginary axis at $\mu = 0$ and $\mu = 0$ is also an eigenvalue of L. Therefore, the case of $\mu = 0$ must be dealt with very carefully [15, 3, 16]. This is the reason why we use the *discrete Evans function*, that we define by

$$\mathcal{E}(\mu,j) = \begin{vmatrix} w_1(\mu,j) & \dots & w_d(\mu,j) & w_{d+1}(\mu,j) & \dots & w_{2d}(\mu,j) \\ \Delta w_1(\mu,j) & \dots & \Delta w_d(\mu,j) & \Delta w_{d+1}(\mu,j) & \dots & \Delta w_{2d}(\mu,j) \end{vmatrix}, \quad (5)$$

where $\Delta w(\cdot,j) = w(\cdot,j) - w(\cdot,j-1)$ for $j \in \mathbf{Z}$. The Evans function was introduced by Evans [4], and its theory has developed a lot in the recent years both in the continuous setting [1, 5, 2, 8] and for numerical schemes [3, 16, 7]. The function $\mathcal{E}$ has the following properties, given $j \in \mathbf{Z}$:

- the zeros of $\mathcal{E}(\cdot,j)$ correspond to the eigenvalues of L,
- $\mathcal{E}(\cdot,j)$ can be chosen to be analytic in $\{|\mathrm{Re}(\mu) + 1| > 1\}$,
- $\mathcal{E}(\cdot,j)$ can be extended analytically to a neighborhood of $\mu = 0$ [5, 16, 7].

Using the properties of $\mathcal{E}$, we can give a new formulation of (H7):

H7' given $j \in \mathbf{Z}$, $\mathcal{E}(\cdot,j)$ does not vanish in Ω, except at $\mu = 0$.

Since the shock is of Lax-type and consequently $\mathcal{E}(0,\cdot) = 0$, we need an additional hypothesis:

H8 the degeneracy of the shock is minimum, that is

$$\frac{\partial \mathcal{E}}{\partial \mu}(0,j) \neq 0, \ \forall j \in \mathbf{Z}.$$

2 Estimates of G

We obtained the following theorem [6]

Theorem 1 *Assuming (H1-8), the Green's function of the linearized problem (3) behaves as follows, for* $\mathbf{e} \in \mathbf{R}^d$ *and* $l \in \mathbf{Z}$*:*

if $l \geq 0$,

$$G(j,n;l)\cdot \mathbf{e} = \chi_{|j|<|a_1^+|n}\mathcal{R}(l,j)\cdot \mathbf{e} + \sum_{q=p+1}^{d} \frac{1}{\sqrt{n}} O\left(\exp\left(-\frac{\left(hj - ka_q^+ n\right)^2}{Mh^2 n}\right)\right) r_q^+$$
$$+ \sum_{q=1}^{p+1} \frac{1}{n} O\left(\exp\left(-\frac{\left(hj - ka_q^- n\right)^2}{Mh^2 n}\right)\right) r_q^-$$
$$+ O\left(e^{-\gamma n} \exp\left(-\frac{(l-j)^2}{Mn}\right)\right), \quad j \in \mathbf{Z},$$

if $l < 0$,

$$G(j,n;l)\cdot \mathbf{e} = \chi_{|j|<a_d^- n}\mathcal{R}(l,j)\cdot \mathbf{e} + \sum_{q=1}^{p-1} \frac{1}{\sqrt{n}} O\left(\exp\left(-\frac{\left(hj - ka_q^- n\right)^2}{Mh^2 n}\right)\right) r_q^-$$
$$+ \sum_{q=p+1}^{d} \frac{1}{\sqrt{n}} O\left(\exp\left(-\frac{\left(hj - ka_q^+ n\right)^2}{Mh^2 n}\right)\right) r_q^+$$
$$+ O\left(e^{-\gamma n} \exp\left(-\frac{(l-j)^2}{Mn}\right)\right), \quad j \in \mathbf{Z},$$

where

- χ_E *is the characteristic function of the set* E,
- *the notation* $\Delta_j G(\cdot,\cdot,j)$ *refers to* $G(\cdot,\cdot,j) - G(\cdot,\cdot,j-1)$,
- *the residual term* $\mathcal{R}(l,j)$ *is a projection on the eigenspace of* L *associated with the eigenvalue* 0 *(see Equation (9)),*
- *all the constants are locally bounded in* l *and uniformly bounded in* n *and* j, *and the constants* M *and* γ *are positive.*

This description of the Green's function as a sum of three terms, that is an excited term appearing when the shock is reached and strongly linked to the kernel of L, a sum of Gaussian-type terms that are damped by viscosity and a fast time and space decaying term, is analogous to the result obtained in the continuous setting by Zumbrun and Howard [17], although they do not assume that y (the continuous version of our variable l) is bounded. Let us consider the case of an initial data, which is a Dirac mass in the direction $\mathbf{e}$, on the right-hand side of the shock. Theorem 1 states that this Dirac mass splits into p waves that propagate towards the shock along the entering characteristics, that is, the ones that correspond to negative eigenvalues. Then, each time the shock is reached by one of these waves, a stationary wave appears together with $d-1$ outgoing waves that are carried, on the right-hand side, by the characteristics corresponding to positive eigenvalues, and on the left-hand side, by the characteristics corresponding to negative eigenvalues.

In case the initial data is on the left-hand side of the shock, the interpretation is symmetric: there are $d-p+1$ waves that enter the shock and $d-1$ outgoing waves each time the shock is reached.
The bounds of the Green's function have been very useful in the search for nonlinear stability in the case of viscous boundary layers [9]. The very interesting work by Mascia and Zumbrun [13] on the stability of the relaxation shocks should allow the use of Theorem 1 to take a step further towards nonlinear stability.

3 Numerical simulations

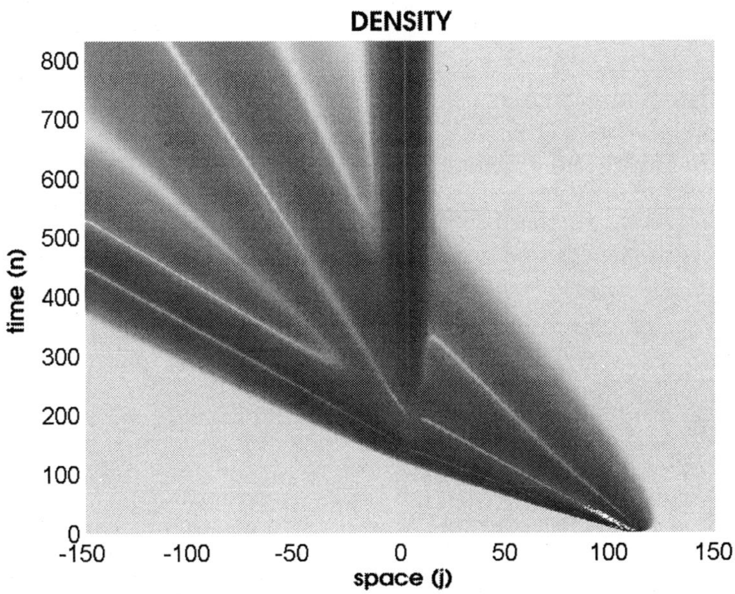

Fig. 1. $G(n, 112, j)$ with the initial data on the third vector of the canonical basis

Let us consider a perfect diatomic gas, that is, the adiabatic constant is equal to 1.4:

$$\begin{aligned}&\rho_t + (\rho v)_x = 0,\\&(\rho v)_t + (\rho(v^2 + (\gamma - 1)e))_x = 0,\\&(\rho(\tfrac{v^2}{2} + e))_t + (\rho v(\tfrac{v^2}{2} + \gamma e))_x = 0,\end{aligned}$$

where ρ represents the density, v the velocity and e the internal energy. The change to conservative variables is detailed in [3, 6]. For a Lax 3-shock, that is the eigenvalues of $\mathrm{d}f(u^{\pm})$ satisfying

$$a_1^+ < a_2^+ < a_3^+ < 0 \,,$$
$$a_1^- < a_2^- < 0 < a_3^- \,,$$

we obtained Figure 1. We took here the initial condition at $l > 0$: Figure 1 illustrates the fact that, at first, three waves of respective speeds a_1^+, a_2^+ and a_3^+ that correspond to the entering characteristics, go towards the shock and, each time the shock is reached, two waves of respective speeds a_1^- and a_2^- that correspond to outgoing characteristics, propagate away from the shock. There is also a stationary wave appearing in a neighborhood of the shock. As was predicted by Theorem 1, there is no outgoing wave on the right-hand side of the shock, since $\mathrm{d}f(u^+)$ has no positive eigenvalue. Unfortunately, we can see clearly only two sets of outgoing waves, because the third one is too damped by viscosity. More figures and animations are available at `http://www.umpa.ens-lyon.fr/~pgodillo/`.

4 Sketch of the proof

4.1 Estimates of the Laplace transform of G

A detailed version of this proof can be found in [6]. We use the Laplace transform with respect to n

$$w = (w^n)_{n \in \mathbf{N}} \mapsto \left(\lambda \in \mathcal{D} \mapsto \hat{w}(\lambda) := \sum_{n \in \mathbf{N}} e^{-\lambda n} w^n \right),$$

where

$$\mathcal{D} \subset \mathcal{S} := \{\lambda \in \mathbf{C} / -\pi \leq \mathrm{Im}(\lambda) \leq \pi\}.$$

For $\lambda \in \mathcal{D}$ and $l \in \mathbf{Z}$ being given, the Laplace transform of $G(j,\cdot;l)$, $G_\lambda(l;j)$, solves

$$\begin{aligned} (L - e^\lambda + 1) G_\lambda(j;l) &= -\delta_{jl}\, e^\lambda\, I_d,\ j \in \mathbf{Z}, \\ \lim_{j \to \pm\infty} G_\lambda(j;l) &= 0. \end{aligned} \tag{6}$$

The variable λ is linked to μ through the formula $\mu = e^\lambda - 1$. We rewrite $(L - e^\lambda + 1)w = 0$ as a dynamical system in $\mathbf{C}^{2d}$:

$$W_j = \mathbf{A}_j(\lambda) W_{j-1},\quad W_j = \begin{pmatrix} w_j \\ w_{j+1} - w_j \end{pmatrix},\quad j \in \mathbf{Z}. \tag{7}$$

To better understand the propagation away from the initial data, we show that, given $l \in \mathbf{Z}$ and $\mathbf{e} \in \mathbf{R}^d$, G_λ can be written as:

$$G_\lambda(j;l)\cdot \mathbf{e} = \sum_{q=1}^{d} \left[\nu_q(\lambda,l)\cdot \mathbf{e}\right] W_q(\lambda,j), \qquad \forall j \geq l,$$

$$= \sum_{q=1}^{d} \left[\nu_{d+q}(\lambda,l)\cdot \mathbf{e}\right] W_{d+q}(\lambda,j),\ \forall j \leq l,$$

where

- $\nu_q(\lambda,l)$ is a linear form on $\mathbf{C}^d$ for $1 \leq q \leq d$,
- $(W_q)_{q\in\{1,\dots,d\}}$ (resp. $(W_{d+q})_{q\in\{1,\dots,d\}}$) is a basis of the stable (resp. unstable) subspace of (7) as $j \to +\infty$ (resp. as $j \to -\infty$) such that

$$W(\lambda,j) = \rho^j(\lambda)V(\lambda,j),$$

  where $\rho(\lambda)$ is an eigenvalue of $\lim_{j\to+\infty}\mathbf{A}_j(\lambda)$ (resp. of $\lim_{j\to-\infty}\mathbf{A}_j(\lambda)$) and $V(\lambda,j)$ tends to an associated eigenvector $\mathbf{V}(\lambda)$ as $j \to +\infty$ (resp. as $j \to -\infty$).

At $j = l$, the compatibility equations given by (6) and (7) are

$$\begin{aligned}\sum_{q=1}^{d} \left[\nu_q(\lambda,l)\cdot \mathbf{e}\right] w_q(\lambda,l) &= \sum_{q=1}^{d} \left[\nu_{d+q}(\lambda,l)\cdot \mathbf{e}\right] w_q(\lambda,l),\\ \sum_{q=1}^{d} \left[\nu_q(\lambda,l)\cdot \mathbf{e}\right] \Delta w_q(\lambda,l) &= \sum_{q=1}^{d} \left[\nu_{d+q}(\lambda,l)\cdot \mathbf{e}\right] \Delta w_{d+q}(\lambda,l) - C_l e^\lambda \mathbf{e},\end{aligned} \tag{8}$$

with $C_l := 2(2\mathbf{D}I_d + \mathrm{d}f(\bar{u}_l))^{-1}$.
Note that the determinant of this system is the Evans function $\mathcal{E}(\lambda,l)$ (5). Its properties have the following consequences:

- Hypothesis (H7) implies that, thanks to the analytic extension, there exists $\eta > 0$ such that

$$\mathcal{E}(\lambda,l) \neq 0 \text{ for } \mathrm{Re}(\lambda) \geq -\eta \text{ et } \lambda \neq 0,$$

- Hypothesis (H8) implies that, for small λ,

$$|\mathcal{E}(\lambda,l)| \geq C(l)|\lambda|.$$

Therefore, the study of (8) shows that only $\nu_p(\cdot,l)$ et $\nu_{d+p}(\cdot,l)$ are meromorphic in $\{\mathrm{Re}(\lambda) > -\eta\}$, their pole being at $\lambda = 0$, and, for $q \neq p, d+p$, $\nu_q(\cdot,l)$ is holomorphic in $\{\mathrm{Re}(\lambda) > -\eta\}$. Thus, the function $\lambda \mapsto G_\lambda(l,j)$ is meromorphic in $\{\mathrm{Re}(\lambda) > -\eta\}$ with a pole at $\lambda = 0$ and its residue reads

$$\mathrm{Res}_{\lambda=0} G_\lambda(j;l)\cdot \mathbf{e} = \left[\phi_l \cdot \mathbf{e}\right]\Psi(j). \tag{9}$$

where ϕ_l is a linear form on $\mathbf{R}^d$ and Ψ is such that $\ker(L) = \mathbf{R}\Psi$.

4.2 Inverse Laplace transform

We use the formula

$$G(j,n;l) = \frac{1}{2i\pi}\int_{\Gamma} e^{\lambda n} G_{\lambda}(j;l)\mathrm{d}\lambda, \tag{10}$$

where $\Gamma = \{\mathrm{Re}(\lambda) = s; |\mathrm{Im}(\lambda)| \leq \pi\}$, $s > 0$.
But, thanks to the Cauchy formula, we are able to change Γ to better suit the behavior of G_{λ}.
We assume that $l \in \{-\ell, \dots, \ell\}$, with $\ell \in \mathbf{N}$ and we consider $|j-l|/n$:

- for $|j-l| \geq n+1$, $G(j,n;l) = 0$, since the modified Lax-Friedrichs scheme has a finite propagation speed;
- for intermediate $|j-l|/n$: $\|G(j,n;l)\| \leq C\exp\left(-\tau\frac{(l-j)^2}{n}\right)$,
- for small $|j-l|/n$:
  - if $|j-l| \leq \ell$, we obtain $G(j,n;l) = \chi_{n\geq N}\mathrm{Res}_{\lambda=0}G_{\lambda} + O(e^{-\eta n})$,
  - if $|j-l| > \ell$,
    - on the entering characteristics of speed a and direction r, we have $G(j,n;l)\cdot r = \chi_{n\geq|j/a|}\mathrm{Res}_{\lambda=0}G_{\lambda}\cdot r + O(e^{-\eta n})$,
    - on the outgoing characteristics of speed a and direction r, we obtain $G(j,n;l).r = O\left(\exp\left(-\frac{(j-an)^2}{Mn}\right)\right)$.

The paths we used in the formula (10) are precisely described in [6].

References

1. J. Alexander, R. Gardner, and C. Jones. A topological invariant arising in the stability analysis of travelling waves. *J. Reine Angew. Math.*, 410:167–212, 1990.
2. S. Benzoni-Gavage, D. Serre, and K. Zumbrun. Alternate Evans functions and viscous shock waves. *SIAM J. Math. Anal.*, 32(5):929–962 (electronic), 2001.
3. M. Bultelle, M. Grassin, and D. Serre. Unstable Godunov discrete profiles for steady shock waves. *SIAM J. Numer. Anal.*, 35(6):2272–2297 (electronic), 1998.
4. J. W. Evans. Nerve axon equations. I. Linear approximations. *Indiana Univ. Math. J.*, 21:877–885, 1971/72.
5. R. A. Gardner and K. Zumbrun. The gap lemma and geometric criteria for instability of viscous shock profiles. *Comm. Pure Appl. Math.*, 51(7):797–855, 1998.
6. P. Godillon. Green's function pointwise estimates for the modified Lax-Friedrichs scheme. *To appear in M2AN*, 2001.
7. P. Godillon. Necessary condition of spectral stability for a stationary Lax-Wendroff shock profile. *Submitted*, 2001.
8. P. Godillon. Linear stability of shock profiles for systems of conservation laws with semi-linear relaxation. *Phys. D*, 148(3-4):289–316, 2001.

9. E. Grenier and F. Rousset. Stability of one-dimensional boundary layers by using Green's functions. *Comm. Pure Appl. Math.*, 54(11):1343–1385, 2001.
10. G. Jennings. Discrete shocks. *Comm. Pure Appl. Math.*, 27:25–37, 1974.
11. C. K. R. T. Jones. Stability of the travelling wave solution of the FitzHugh-Nagumo system. *Trans. Amer. Math. Soc.*, 286(2):431–469, 1984.
12. A. Majda and J. Ralston. Discrete shock profiles for systems of conservation laws. *Comm. Pure Appl. Math.*, 32(4):445–482, 1979.
13. C. Mascia and K. Zumbrun. Pointwise green's function bounds and stability of relaxation shocks. *Preprint*, 2001.
14. D. Michelson. Discrete shocks for difference approximations to systems of conservation laws. *Adv. in Appl. Math.*, 5(4):433–469, 1984.
15. D. Serre. Remarks about the discrete profiles of shock waves. *Mat. Contemp.*, 11:153–170, 1996. Fourth Workshop on Partial Differential Equations, Part II (Rio de Janeiro, 1995).
16. D. Serre. Discrete shock profiles and their stability. In *Hyperbolic problems: theory, numerics, applications, Vol. II (Zürich, 1998)*, pages 843–853. Birkhäuser, Basel, 1999.
17. K. Zumbrun and P. Howard. Pointwise semigroup methods and stability of viscous shock waves. *Indiana Univ. Math. J.*, 47(3):741–871, 1998.

A Class of Global Non-smooth Axisymmetric Solutions to the Euler Equations of an Isentropic Perfect Gas in 2 Space Dimensions

Paul Godin[1]

Université Libre de Bruxelles, Département de Mathématiques, Campus Plaine CP 214, Boulevard du Triomphe, B - 1050 Bruxelles, Belgium.
pgodin@ulb.ac.be

1 Introduction

In papers by Serre [6], Grassin-Serre [3], and Grassin [2], global existence results were obtained for solutions to the Euler equations of a perfect gas, under some smoothness and growth assumptions on the initial data, provided the initial velocity is dispersive and the initial sound speed is small (in some norm). In this note we shall describe some results contained in [1]. We consider axisymmetric flows (for the isentropic Euler equations of a perfect gas) in 2 space dimensions, with non smooth initial data. More precisely, we consider the case that initial gradients may jump across a given circle (centered at 0); the tangential component of the initial velocity may also jump there. In each component of the complement of this circle, we assume that the initial data are restrictions of global rotation (round 0) invariant data which satisfy the assumptions of Grassin-Serre and Grassin. We describe results of [1] about global existence of a solution with the given initial data . The proof relies on some previous work of Li-Yu [4] and Li [5]. Our results are stated in Section 2. Some ideas of the proof are given in Section 3.

2 Statement of the results

We consider the compressible isentropic Euler equations in 2 space dimensions

$$\partial_t \rho + \sum_{k=1}^{2} \partial_k(\rho u_k) = 0, \tag{1}$$

$$\partial_t(\rho u_j) + \sum_{k=1}^{2} \partial_k(\rho u_j u_k) + \partial_j p = 0, \quad j = 1, 2, \tag{2}$$

where ρ is the density , $u =^t (u_1, u_2)$ the velocity and p the pressure. We assume that p is a function of ρ alone of the form $p(\rho) = q\rho^\gamma$ where $q > 0$ and $1 < \gamma \leq 2$. It is convenient to introduce $\pi = \dfrac{2}{\gamma - 1}c$, where $c = \sqrt{p'(\rho)}$

(cf. [3], [2]). Assuming u, ρ to be smooth and ρ to be > 0, (1), (2) can be rewritten as

$$\partial_t \pi + \sum_{k=1}^{2} u_k \partial_k \pi + C_1 \pi \sum_{j=1}^{2} \partial_j u_j = 0, \tag{3}$$

$$\partial_t u_j + \sum_{k=1}^{2} u_k \partial_k u_j + C_1 \pi \partial_j \pi = 0, \quad j = 1, 2, \tag{4}$$

where $C_1 = \frac{\gamma - 1}{2}$. (3), (4) is a symmetric hyperbolic system (well defined even when π vanishes), and one can consider initial conditions

$$u(x, 0) = u_0(x), \quad \pi(x, 0) = \pi_0(x). \tag{5}$$

In [3], [2], the Cauchy problem (3), (4), (5) was studied (actually for any space dimension) and global existence results (in positive time) were obtained under the following assumptions:

$$Du_0 \in L^\infty(\mathbb{R}^2), \qquad D^2 u_0 \in H^m(\mathbb{R}^2) \text{ for some } m > 1; \tag{6}$$

$$\begin{aligned} &\text{One can find } \delta > 0 \text{ such that, for all } x \in \mathbb{R}^2, \\ &\text{dist } (\text{sp } Du_0(x), \mathbb{R}_-) \geq \delta, \end{aligned} \tag{7}$$

$$\pi_0 \in H^{m+1}(\mathbb{R}^2) \text{ and } ||\pi_0||_{m+1} \text{ is small enough} \tag{8}$$

In the present work we want to examine situations where the intial data are not everywhere smooth. We shall restrict ourselves to axisymmetric flows, that is to solutions $(\pi, {}^t u)$ (or $(\rho, {}^t u)$) which have the following property :

$$\begin{aligned} &\text{for every rotation } S \text{ of } \mathbb{R}^2 \text{ with center } 0, \pi(Sx, t) = \pi(x, t) \text{ and} \\ &u(Sx, t) = (Su)(x, t). \end{aligned} \tag{9}$$

So the straight line $x = 0$ is a symmetry axis for the solution (in (x, t) space). If $x = (x_1, x_2)$, put $x^\perp = (-x_2, x_1)$, $r = |x|$. (9) means that for some functions $\Pi(r, t)$, $A(r, t)$, $B(r, t)$ we have $\pi(x, t) = \Pi(r, t)$ and $u(x, t) = A(r, t)\frac{x}{r} + B(r, t)\frac{x^\perp}{r}$. Then (3) becomes

$$\Pi_t + A\Pi_r + C_1 \Pi (A_r + \frac{A}{r}) = 0, \tag{10}$$

whereas (4) becomes

$$A_t + AA_r + C_1 \Pi \Pi_r - \frac{B^2}{r} = 0, \tag{11}$$

$$B_t + AB_r + \frac{AB}{r} = 0. \tag{12}$$

We shall consider initial conditions

$$\Pi(r,0)=\Pi_0(r),\ A(r,0)=A_0(r),\ B(r,0)=B_0(r) \tag{13}$$

such that (6), (7), (8) are satisfied except when $r=1$. (Π_0, A_0, B_0) will be a small perturbation of $(0, \overline{A}_0, \overline{B}_0)$, where $u_0(x)=\overline{A}_0(r)\frac{x}{r}+\overline{B}_0(r)\frac{x^\perp}{r}$ is a given function satisfying (6), (7). More precisely, let $u_0^{(j)}:\mathbb{R}^2\to\mathbb{R}^2$, $j=1,2$ be rotation invariant and satisfy (6) and be such that

$$\sum_{k\leq 1}|D^k(u_0^{(j)}-u_0)|+||D^2(u_0^{(j)}-u_0)||_m\leq\epsilon, \tag{14}$$

with $\epsilon>0$ small. Write $u_0^{(j)}(x)=\overline{A}_0^{(j)}(r)\frac{x}{r}+\overline{B}_0^{(j)}(r)\frac{x^\perp}{r}$. We assume that

$$\overline{A}_0^{(1)}(1)=\overline{A}_0^{(2)}(1). \tag{15}$$

Let $\Pi_0^{(j)}:[0,+\infty[\to]0,+\infty[$, $j=1,2$, be (nowhere vanishing) functions such that $x\mapsto\Pi_0^{(j)}(|x|)$ satisfy (8), and assume that

$$\Pi_0^{(1)}(1)=\Pi_0^{(2)}(1). \tag{16}$$

Finally we assume that

$$\begin{aligned}(\Pi_0,A_0,B_0)&=(\Pi_0^{(1)},A_0^{(1)},B_0^{(1)})\text{ if } r<1,\\ (\Pi_0,A_0,B_0)&=(\Pi_0^{(2)},A_0^{(2)},B_0^{(2)})\text{ if } r>1.\end{aligned} \tag{17}$$

Put $W={}^t(\Pi,A,B)$, $P(W)=\begin{pmatrix}A & C_1\Pi & 0\\ C_1\Pi & A & 0\\ 0 & 0 & A\end{pmatrix}$, $Q(W)=\begin{pmatrix}-C_1\Pi A\\ B^2\\ -AB\end{pmatrix}$.

(10), (11), (12) can be written as

$$W_t+P(W)W_r=\frac{1}{r}Q(W). \tag{18}$$

The eigenvalues of $P(W)$ are of course $\lambda_1(W)=A-c$, $\lambda_2(W)=A$, $\lambda_3(W)=A+c$ where $c=C_1\Pi$. The following local result (near $r=1$, $t=0$) can be proved with the methods of [4].

Theorem 1. *Under the preceding assumptions, one can find $T>0$, $r_2>1>r_1>0$, $X_j:[0,T)\to(r_1,r_2)$, $1\leq j\leq 3$ and $\Pi, A, B:(r_1,r_2)\times[0,T)\to\mathbb{R}$ with Π everywhere >0, such that*

(1) each X_j is smooth, $\frac{dX_j}{dt}(t)=\lambda_j(W(X_j(t),t))$, $X_j(0)=1$, that is $t\mapsto(X_j(t),t)$ parametrizes the j-th characteristic Γ_j through $(1,0)$ (for $P(W)$ in (18));

(2) (Π, A, B) *is smooth in the complement of and up to* $\Gamma_1 \cup \Gamma_2 \cup \Gamma_3$ *(actually* Π *and* A *are continuous in* $(r_1, r_2) \times [0, T)$*), and if* $C_2 = C_1^{\frac{1}{C_1}} (q\gamma)^{-\frac{1}{2C_1}}$*,* ${}^t(C_2 \Pi^{\frac{1}{C_1}}(|x|)), A\frac{x}{r} + B\frac{x^{\perp}}{r})$ *is a weak solution to (1), (2) if* $r_1 < |x| < r_2$*,* $t < T$*, with* $(\Pi, A, B)(r, 0) = (\Pi_0, A_0, B_0)(r)$ *if* $r_1 < r < r_2$ *and* $r \neq 1$*.*

Actually, in general, $\nabla \Pi$ and ∇A will jump across $\Gamma_1 \cup \Gamma_3$, B, ∇B and $\nabla \Pi$ will jump across Γ_2 (so the solution will present a contact discontinuity along Γ_2 and gradient discontinuities along Γ_1 and Γ_3).

Now we have the following global result, in which we put $\pi_0^{(j)}(x) = \Pi_0^{(j)}(|x|)$.

Theorem 2. *Assume furthermore that, for some* $C_1, C_2, C_3, C_4 > 0$, $|(\overline{B}_0^{(1)})^2(1) - (\overline{B}_0^{(2)})^2(1)| \leq C_1 \epsilon^2$, $||\pi_0^{(j)}||_{m+1} \leq C_2 \epsilon$ *for* $j = 1, 2$*, and* $\Pi_0(r) \geq C_3 \epsilon$ *for* $|r - 1| \leq C_4$*, with a small* ϵ *for which (14) holds. Then Theorem 2.1 holds with* $T = +\infty$*, the relation* $r_1 < r < r_2$ *being replaced everywhere by* $x \in \mathbb{R}^2$*.*

3 Some ideas of the proof of Theorem 2

If $r < X_1(t)$ or $r > X_3(t)$, we have a pure Cauchy problem and the results of Grassin-Serre and Grassin imply global existence and estimates.

Let $(0, \overline{A}, \overline{B})$ be the solution of (10), (11), (12) with initial data $(0, \overline{A}_0, \overline{B}_0)$. $\overline{A}, \overline{B}$ are global and one can check the following, where $\langle t \rangle = 1 + t$:

$$\left| \frac{\overline{A}}{r} - \frac{1}{\langle t \rangle} \right| + \left| \overline{A}_r - \frac{1}{\langle t \rangle} \right| \leq \frac{C}{\langle t \rangle^2},$$

$$\frac{|\overline{B}|}{r} + |\overline{B}_r| \leq \frac{C}{\langle t \rangle^2},$$

$$\overline{A} > 0 \text{ if } t \text{ is large.}$$

X_1, X_3 are global and

$$0 < C_1 \leq \frac{X_1(t)}{\langle t \rangle} < \frac{X_3(t)}{\langle t \rangle} \leq C_2.$$

From the results of Grassin-Serre and Grassin one can obtain that

$$|u - \bar{u}| + \Pi \leq \frac{C\varepsilon}{\langle t \rangle^{\gamma - 1}}$$

(and $\Pi > 0$) if $r < X_1(t)$ or $r > X_3(t)$, where $\bar{u} = \overline{A}\frac{x}{r} + \overline{B}\frac{x^\perp}{r}$.

It remains to study Π, A, B when $X_1(t) \leq r \leq X_3(t)$, and X_2. For t small enough, Π, A, B, X_2 exist by Theorem 2.1. We are going to prove global existence by a continuation method.

Put $R_1 = A - \Pi$, $R_2 = B$, $R_3 = A + \Pi$, $R = ^t (R_1, R_2, R_3)$. Then (18) can be written

$$R_t + \Lambda(R)R_r = \frac{1}{r}g(R),$$

where

$$\Lambda(R) = \begin{pmatrix} \alpha R_1 + \beta R_2 & 0 & 0 \\ 0 & \frac{R_1 + R_2}{2} & 0 \\ 0 & 0 & \beta R_1 + \alpha R_2 \end{pmatrix}, \qquad \alpha = \frac{\gamma+1}{4},\ \beta = \frac{3-\gamma}{4},$$

$$g(R) = \begin{pmatrix} \frac{\gamma-1}{8}(R_3^2 - R_1^2) + R_2^2 \\ -\frac{R_1 + R_3}{2}R_2 \\ \frac{\gamma-1}{8}(R_1^2 - R_3^2) + R_2^2 \end{pmatrix}.$$

The conditions to be satisfied on Γ_j are the following :

across Γ_1 : R is continuous and $[\partial_x R_j]_{\Gamma_1} = 0$ if $j \geq 2$;

across Γ_2 : $[R_1]_{\Gamma_2} = [R_3]_{\Gamma_2} = 0$;

across Γ_3 : R is continuous and $[\partial_x R_j]_{\Gamma_3} = 0$ if $j \leq 2$.

To prove global existence, we are going to show that if R is a solution for $0 \leq t < T$, then $|R| + |\nabla R|$ remains bounded for $t < T$ when $X_1(t) < r < X_3(t)$ and $r \neq X_2(t)$ (and Π remains > 0). Put $\overline{R}_1 = \overline{R}_3 = \overline{A}$, $\overline{R}_2 = \overline{B}$, $z_j = R_j - \overline{R}_j$ if $1 \leq j \leq 3$. If $z = ^t (z_1, z_2, z_3)$, we obtain

$$\frac{\partial z}{\partial t} + (\overline{A} + \Lambda(z))\frac{\partial z}{\partial r} + Mz = \frac{1}{r}g(z), \tag{19}$$

where

$$M = \begin{pmatrix} \alpha\overline{A}_r + \frac{\gamma-1}{4}\frac{\overline{A}}{r} & -\frac{2\overline{B}}{r} & \beta\overline{A}_r - \frac{\gamma-1}{4}\frac{\overline{A}}{r} \\ \frac{1}{2}(\overline{B}_r + \frac{\overline{B}}{r}) & \frac{\overline{A}}{r} & \frac{1}{2}(\overline{B}_r + \frac{\overline{B}}{r}) \\ \beta\overline{A}_r - \frac{\gamma-1}{4}\frac{\overline{A}}{r} & -\frac{2\overline{B}}{r} & \alpha\overline{A}_r + \frac{\gamma-1}{4}\frac{\overline{A}}{r} \end{pmatrix}$$

$$= \frac{1}{\langle t\rangle}\begin{pmatrix} \frac{\gamma}{2} & 0 & \frac{2-\gamma}{2} \\ 0 & 1 & 0 \\ \frac{2-\gamma}{2} & 0 & \frac{\gamma}{2} \end{pmatrix} + \mathcal{O}\left(\frac{1}{\langle t\rangle^2}\right) \quad \text{as } t \to +\infty.$$

Let $\mathcal{D}$ be the smallest dependence domain which contains $\Gamma_1 \cup \Gamma_3$. One can prove that $|r-1| \leq C\varepsilon$ if $(r,0) \in \overline{\mathcal{D}}$. The system in (19) has some dissipation properties, thanks to which one can show the following :

One can find $C > 0$ such that for any $T > 0$, the following holds : if, when $t < T$, z is a solution to (19) in $\mathcal{D}$, with initial data corresponding to (13), and under the assumptions of Theorem 2, the following estimates hold in $\mathcal{D} \setminus (\Gamma_1 \cup \Gamma_2 \cup \Gamma_2)$:

$$\begin{cases} |z_j| + \langle t\rangle|\partial_r z_j| \leq \dfrac{C\varepsilon}{\langle t\rangle^{\gamma-1}} \text{ if } j = 1,3, \\ |z_2| + \langle t\rangle|\partial_r z_2| \leq \dfrac{C\varepsilon}{\langle t\rangle}. \end{cases} \tag{20}$$

From (20) and (19), it follows at once that $\partial_t z_j$ satisfy estimates which are analogous with the ones satisfied by $\partial_r z_j$ in (20). From all this, it follows that X_2' never blows up and that ∇z_j $(j = 1,2,3)$ do not blow up in $\mathcal{D} \setminus (\Gamma_1 \cup \Gamma_2 \cup \Gamma_3)$. Hence the solution z exists (with the same structure) for all positive times.

Actually, if $(x,t) \notin \Gamma_2$, it can be checked that z_j does not jump along the j-th backward characteristic through (x,t). To prove the estimates on z in (20), one writes (19) as a system of integral equations along the backward characteristics. One can also use a system of integral equations to obtain the estimates for first order derivatives of (20), but the jumps of these derivatives must be taken into account.

References

1. P. GODIN, Article in preparation.

2. M. GRASSIN, Global smooth solutions to Euler equations for a perfect gas, Indiana University Math. J. 47, 4 (1998), 1397-1432.
3. M. GRASSIN and D. SERRE, Existence de solutions globales et régulières aux équations d'Euler pour un gaz parfait isentropique, C. R. Acad. Sci. Paris, 325, 1, (1997), 721-726.
4. T.T. LI and W.C. YU, Boundary value problems for quasilinear hyperbolic systems, Duke University Mathematics Series V, 1985.
5. T.T. LI, Global classical solutions for quasilinear hyperbolic systems, Wiley, Masson, 1993.
6. D. SERRE, Solutions classiques globales des équations d'Euler pour une fluide parfait incompressible, Ann. Inst. Fourier, Grenoble, 47, 1 (1997), 139-153.

A Numerical Method for Controllability Problems for the Wave Equation

Max D. Gunzburger[1], L. Steven Hou[1], and Lili Ju[1]

Department of Mathematics, Iowa State University, Ames IA 5011-2064, USA
{*gunzburg,hou,jllwyb*}*@iastate.edu*

1 Introduction

Let T denote a given positive number and let $u_0(\mathbf{x})$ and $u_1(\mathbf{x})$ denote given functions defined on $(0,1)$. Let $\Sigma = \{0,1\} \times (0,T)$, $Q = (0,1) \times (0,T)$ and $(u_0, u_1) \in L^2(0,1) \times H^{-1}(0,1)$. The *exact Dirichlet boundary controllability* problem for the wave equation is: *find a control function $g(\mathbf{x}, t)$ defined on Σ such that u satisfies*

$$\begin{cases} u_{tt} - u_{xx} = 0 \quad \text{in } Q \\ u|_{t=0} = u_0 \quad \text{and} \quad u_t|_{t=0} = u_1 \quad \text{in } (0,1) \\ u|_{t=T} = 0 \quad \text{and} \quad u_t|_{t=T} = 0 \quad \textit{in } (0,1) \\ u = g \quad \text{on } \Sigma\,. \end{cases} \tag{1}$$

It is well known that state and control functions u and g such that (1) is satisfied exist provided T is sufficiently large; see, e.g., [7, 8, 11, 10]. In general, solutions of the Controllability problem (1), when they exist, are not unique. However, it can be shown [7, 8] that a solution of the controllability problem (1) having minimum $L^2(\Sigma)$-norm is unique.

More generally, one can consider the *exact boundary controllability problem* (1) with the boundary condition $u = g$ on Σ is replaced by $B(u) = g$ on Σ, where B denotes a boundary operator. For example, for $B(u) = (\partial_n u)|_\Sigma = g$, one has the *Neumann* exact boundary controllability problem for the wave equation. One can also consider more general criteria for extracting a particular solution of the controllability problem from the set of all solutions. For example, one can replace the minimum $L^2(\Sigma)$-norm requirement for the control function g by requiring that the solution minimize a norm-like functional $\mathcal{H}(u)$ over an appropriate function space.

A numerical method was developed in [2, 7] to determine approximations of the minimum $L^2(\Sigma)$-norm solution of (1). However, the computational experiments in [2] showed that the approximate solutions obtained in this manner do not converge. Subsequently, a Tychonoff regularization method was added to the solution procedure after which convergent approximations were obtained, at least for smooth controls. Numerical approximation of (1) were also studied in [5, 12, 9]. In these papers, the convergence of the boundary controls for finite difference discretized equations to the control of the continuous 1-D wave equation as the mesh size tends to zero has been proved if the step size in time equals to that in space. Furthermore, if the step size in time is less than that in space, regularization techniques were used to obtain convergent approximations.

Another computational approach was proposed in [4] which obtained an approximation to the minimum $L^2(Q)$-norm (not $L^2(\Sigma)$-norm) solution of (1). Here, we develop that method so that controls satisfying other criteria, including minimizing the $L^2(\Sigma)$-norm can be obtained.

2 Coupled space-time finite difference discretization

Let $\Omega = (0,1)$. Then, the minimum value of T for which exact controllability holds is 1 so that, throughout, we assume that $T \geq 1$. We employ a finite difference discretization for the wave equation. Let $x_i = i/J$, $i,j = 0,1,\ldots,J$ and $t_k = kT/K$, $k = 0,1,\ldots,K$, where J and K are chosen positive integers, denote a partition of $\Omega \times (0,T)$, and let $u_{i,k}$ denote the approximation to $u(x_i,t_k)$. The grid sizes in space and time are then given by $h = 1/J$ and $\delta = T/K$, repectively. The wave equation (1) is discretized using standard three-point difference quotients so that, with $\lambda = \delta/h$, the $u_{i,k}$'s are required to satisfy: for $i = 1,2,\ldots,J-1$ and $k = 1,2,\ldots,K-1$,

$$u_{i,k+1} - 2(1-\lambda^2)u_{i,k} - \lambda^2(u_{i+1,k} + u_{i-1,k}) + u_{i,k-1} = 0\,. \qquad (2)$$

The standard CFL condition that guarantees the stability of the scheme (2) is given by $\lambda \leq 1$ which implies that $TJ/K \leq 1$. Then, since $T \geq 1$, we have $K \geq J$. The intial and terminal conditions are discretized by: for $i = 0,1,\ldots,J$,

$$u_{i,0} = u_0(x_i), \quad u_{i,1} = u_{i,0} + \delta u_1(x_i), \quad u_{i,K} = 0, \quad u_{i,K-1} = 0\,. \qquad (3)$$

The unknows associated with the domains of dependence of the initial and terminal data are uniquely determined, after computing (3), by an explicit marching procedure using some of the equations in (2):

$$\begin{array}{l} \text{for}\quad k = 1,\ldots,M \\ \quad k_1 = k,\ \ k_2 = K-k \\ \quad \text{for}\quad i = k,\ldots,J-k \\ \qquad u_{i,k_1+1} = 2(1-\lambda^2)u_{i,k_1} + \lambda^2(u_{i+1,k_1} + u_{i-1,k_1}) - u_{i,k_1-1} \\ \qquad u_{i,k_2-1} = 2(1-\lambda^2)u_{i,k_2} + \lambda^2(u_{i+1,k_2} + u_{i-1,k_2}) - u_{i,k_2+1} \\ \quad \text{end} \\ \text{end} \end{array} \qquad (4)$$

(Without loss of generality, we have assumed that $J = 2M+1$ for some integer $M > 0$; the case of J being an even integer can be treated in a similar manner.) The linear system formed by (2) and (3) has $p = (J -$

$1)(K-1)+4(J+1)$ equations and $q=(J+1)(K+1)$ unknowns. We symbollically express it in the form $A\mathbf{U}=\mathbf{F}$, where the matrix $A\in\mathbb{R}^{p\times q}$ has bandwidth proportional to $J+1$, $\mathbf{U}\in\mathbb{R}^q$ is a vector of the unknowns $u_{i,k}$, and $\mathbf{F}\in\mathbb{R}^p$ is a known vector determined from the intial data. It is also worth noting that the number of non-zero elements of each row or each column of A is less than or equal to 5. Furthermore, we have that $q-p=(J+1)(K-3)-(J-1)(K-1)=2(K-J)-4$ so that, if we choose $K>J+2$, then $q-p=2(K-J)-4>4-4>0$, i.e., the linear system $A\mathbf{U}=\mathbf{F}$ has more unknowns than equations. Let $\mathbf{V}$ be the vector of the unknowns determined by (3) and (4); it is easily determined that $\mathbf{V}\in\mathbb{R}^r$, where $r=4(J+1)+(J^2-1)/2$ is also the number of equations in (3) and (4) together.

Let $\widetilde{p}=p-r=(J-1)(K-1)-\frac{1}{2}(J^2-1)$ and $\widetilde{q}=q-r=(J+1)(K-3)-\frac{1}{2}(J^2-1)$. Clearly, $q-p=\widetilde{q}-\widetilde{p}$. Let $\widetilde{\mathbf{U}}\in\mathbb{R}^{\widetilde{q}}$ be the vector of $u_{i,k}$'s not belonging to $\mathbf{V}$. Then, the linear system $A\mathbf{U}=\mathbf{F}$ can be reduced to $\widetilde{A}\widetilde{\mathbf{U}}=\widetilde{\mathbf{F}}$, where $\widetilde{A}\in\mathbb{R}^{\widetilde{p}\times\widetilde{q}}$ and $\widetilde{\mathbf{F}}\in\mathbb{R}^{\widetilde{q}}$ are obtained by deleting the equations used in (3) and (4) to pre-determine $\mathbf{V}$ and then, in the remaining equations, moving the terms involving the components of $\mathbf{V}$ to the right-hand side. Note that we can re-express the linear system $\widetilde{A}\widetilde{\mathbf{U}}=\widetilde{\mathbf{F}}$ by the following recursive computations, i.e., by a marching procedure from the center of Q to the lateral surface Σ:

$$
\begin{aligned}
&\text{for}\quad i=M,\dots,1\\
&\qquad i_1=i,\, i_2=J-i\\
&\qquad \text{for}\quad k=i+1,\dots,K-i-1\\
&\qquad\qquad u_{i_1-1,k}=\tfrac{1}{\lambda^2}(u_{i_1,k+1}+u_{i_1,k-1})-2(\tfrac{1}{\lambda^2}-1)u_{i_1,k}-u_{i_1+1,k}\\
&\qquad\qquad u_{i_2+1,k}=\tfrac{1}{\lambda^2}(u_{i_2,k+1}+u_{i_2,k-1})-2(\tfrac{1}{\lambda^2}-1)u_{i_2,k}-u_{i_2-1,k}\\
&\qquad \text{end}\\
&\text{end}
\end{aligned}
\tag{5}
$$

It can be shown [6] that the matrix $\widetilde{A}$ (and therefore A as well) is of full row rank, i.e., the equations included in the linear system $\widetilde{A}\widetilde{\mathbf{U}}=\widetilde{\mathbf{F}}$ are linearly independent. Let $\mathbf{W}=\{u_{i,k}\,|\,i=,M,M+1,\ k=M+2,\dots,K-M-2\}$. Clearly, $\mathbf{W}\subset\mathbb{R}^s$, where $s=2(K-@M-3)$; also, the solution of the system $\widetilde{A}\widetilde{\mathbf{U}}=\widetilde{\mathbf{F}}$ can be uniquely from $\mathbf{W}$ by the marching procedure (5).

We now turn to the selection of a particular solution of the linear system $A\mathbf{U}=\mathbf{F}$, i.e., of the discretized wave equation and initial and terminal conditions, by requiring that solution to also solve a discrete optimization problem. We first focus on finding approximations of the minimum $L^2(\Sigma)$-norm solution of the controllability problem. Specifically, we use the trapezoidal rule to obtain the approximation

$$\begin{aligned}\mathcal{J}(u) \equiv \|g\|^2_{L^2(\Sigma)} &= \int_\Sigma u^2\, d\Gamma dt = \int_0^T \left[u^2(0,t) + u^2(1,t)\right] dt \\ &\approx \delta \sum_{k=0}^{K} \alpha_k \left({u_{0,k}}^2 + {u_{J,k}}^2\right) \equiv \mathcal{J}_h(\mathbf{U}),\end{aligned} \tag{6}$$

where $\alpha_0 = \alpha_K = 1/2$ and $\alpha_1 = \cdots = \alpha_{K-1} = 1$. Furthermore, we effect the partition $\widetilde{\mathbf{U}} = (\widetilde{\mathbf{U}}_1^T, \widetilde{\mathbf{U}}_2^T)^T$, where $\widetilde{\mathbf{U}}_1$ denotes the subvector of $\widetilde{\mathbf{U}}$ belonging to the boundary $\overline{\Sigma}$ and $\widetilde{\mathbf{U}}_2$ denotes the subvector of $\widetilde{\mathbf{U}}$ belonging to the interior Q. It is seen that $\widetilde{\mathbf{U}}_1 \in \mathbb{R}^{\widetilde{q}_1}$, where $\widetilde{q}_1 = 2(K-3)$, and $\widetilde{\mathbf{U}}_2 \in \mathbb{R}^{\widetilde{q}_2}$, where $\widetilde{q}_2 = \widetilde{q} - 2(K-3)$. We are then led to the following *discrete* quadratic optimization problem:

$$\min_{\widetilde{\mathbf{U}} \in \mathbb{R}^{\widetilde{q}}} \mathcal{J}_h(\widetilde{\mathbf{U}}) = \delta(\widetilde{\mathbf{U}}_1^T \widetilde{\mathbf{U}}_1 + b) \quad \text{subject to} \quad \widetilde{A}_1 \widetilde{\mathbf{U}}_1 + \widetilde{A}_2 \widetilde{\mathbf{U}}_2 = \widetilde{\mathbf{F}}, \tag{7}$$

where $\widetilde{A}_1 \in \mathbb{R}^{\widetilde{p}\times\widetilde{q}_1}$ and $\widetilde{A}_2 \in \mathbb{R}^{\widetilde{p}\times\widetilde{q}_2}$ are obained by appropriately rearranging and partitioning $\widetilde{A}$ corresponding to $\widetilde{\mathbf{U}}_1$ and $\widetilde{\mathbf{U}}_2$, and b is a constant depending on the components of $\mathbf{U}$ belonging to $\overline{\Sigma}$ that are pre-determined by (3). It should be pointed out that the problem (7) couples space and time together. Regarding the solution of the optimization system (7), we have the following result.

If u denotes the minimum $L^2(\Sigma)$-norm solution of (1) and u^h the corresponding discrete solution of (6) with $\delta = h$, i.e., $\lambda = 1$, then it is shown in [9] $g^h \to g$ strongly in $L^2(\Sigma)$, where $g = u|_\Sigma$ and $g^h = u^h|_\Sigma$.

We also consider finding the soultion of the exact controllability problem (1) that also has minimum $H^1(\Sigma)$-norm which results in more regular controls g. In this case, we use

$$\begin{aligned}\mathcal{K}(u) \equiv \|g\|^2_{H^1(\Sigma)} &= \int_0^T (u^2|_{(0,t)} + u_t^2|_{(0,t)} + u^2|_{(1,t)} + u_t^2|_{(1,t)})dt \approx \mathcal{K}_h(\mathbf{U}) \\ \approx \delta \sum_{k=0}^{K} \alpha_k \left(u_{0,k}^2 + u_{J,k}^2\right) &+ \frac{1}{\delta} \sum_{k=1}^{K} \Big((u_{0,k} - u_{0,k-1})^2 + (u_{J,k} - u_{J,k-1})^2 \Big)\end{aligned}$$

and are led to another discrete quadratic optimization problem:

$$\min_{\widetilde{\mathbf{U}} \in \mathbb{R}^{\widetilde{q}}} \mathcal{K}_h(\widetilde{\mathbf{U}}) = \delta(\widetilde{\mathbf{U}}_1^T R \widetilde{\mathbf{U}}_1 + b) \quad \text{subject to} \quad \widetilde{A}_1 \widetilde{\mathbf{U}}_1 + \widetilde{A}_2 \widetilde{\mathbf{U}}_2 = \widetilde{\mathbf{F}}, \tag{8}$$

where $R \in \mathbb{R}^{\widetilde{q}_1\times\widetilde{q}_1}$ is a tridiagonal, positive definite matrix. This discrete problem also has a unique solution; in fact, one may view (7) as the special case of (8) for which R is the identity matrix. If $\lambda = 1$, then again the convergence of the discrete control holds.

A Lagrange multiplier method. If $\widetilde{\mathbf{U}}$ is the solution of the optimization problems (7) or (8), then the Lagrange multiplier rule states that $\widetilde{\mathbf{U}}$ is a stationary point of the Lagrangian functional $\mathcal{L}(\widetilde{\mathbf{U}}) = \delta(\widetilde{\mathbf{U}}_1^T R \widetilde{\mathbf{U}}_1 + b) - \mathbf{S}^T(\widetilde{A}_1 \widetilde{\mathbf{U}}_1 +$

$\widetilde{A}_2\widetilde{\mathbf{U}}_2 - \widetilde{\mathbf{F}})$, where $\mathbf{S} \in \mathbb{R}^{\widetilde{p}}$ is the Lagrange multiplier introduced to enforce the constraint equations $\widetilde{A}_1\widetilde{\mathbf{U}}_1 + \widetilde{A}_2\widetilde{\mathbf{U}}_2 = \widetilde{\mathbf{F}}$ and $R = I$ in case of (7). Then, it is easily shown that $\widetilde{\mathbf{U}} = (\widetilde{\mathbf{U}}_1, \widetilde{\mathbf{U}}_2)$ and $\widetilde{\mathbf{S}} = \mathbf{S}/\delta$ are the solution of

$$\begin{pmatrix} 0 & \widetilde{A}_2^T \\ \widetilde{A}_2 & -\widetilde{A}_1^T R^{-1} \widetilde{A}_1 \end{pmatrix} \begin{pmatrix} \widetilde{\mathbf{U}}_2 \\ \widetilde{\mathbf{S}} \end{pmatrix} = \begin{pmatrix} 0 \\ \widetilde{\mathbf{F}} \end{pmatrix} \quad \text{and} \quad \widetilde{\mathbf{U}}_1 = -R^{-1}\widetilde{A}_1^T \widetilde{\mathbf{S}}. \tag{9}$$

Note that $\widetilde{\mathbf{U}}_1$ along with the boundary $u_{i,k}$'s found from (3) and (4) can be used to define the approximation to the exact Dirichlet boundary control $u|_\Sigma = g$. Clearly, the coefficient matrix of the linear system in the first equation in (9) is symmetric but it is not positive definite. Note that $\widetilde{A}_1$ and $\widetilde{A}_2$ are sparse matrices having, like A, at most 5 non-zero elements in each of their rows or columns. Thus, it is efficient to use iterative methods, e.g., BiCG, SYMMQL, or CGNR, to solve the first equation in (9). In higher dimensions, the application of a preconditioner may be useful since the dimension of the the first equation in (9) may be large. In a practical implementation, the matrices $\widetilde{A}_1$ and $\widetilde{A}_2$ need not be stored. It is also worth noting that, at each step of an iteration, we may only need to update $\widetilde{\mathbf{S}}$ and that part of $\mathbf{W}$ included in $\widetilde{\mathbf{U}}_2$; then the remainder of $\widetilde{\mathbf{U}}_2$ can be directly computed by the marching procedure (5).

Reduction to an unconstrained optimization problem. Another method for solving the optimization problems (7) or (8) uses the marching procedure (5) to determine $\widetilde{\mathbf{U}}_1$ from $\mathbf{W}$, provided that $\mathbf{V}$ is pre-calculated using (3) and (4). Thus, $\widetilde{\mathbf{U}}_1$ is a linear function of $\mathbf{W}$, i.e., there exists a matrix $D \in \mathbb{R}^{\widetilde{q}_1 \times s}$ such that $\widetilde{\mathbf{U}}_1 = D\mathbf{W} + \mathbf{C}$, where $\mathbf{C} \in \mathbb{R}^{\widetilde{q}_1}$ is determined from $\mathbf{V}$. Then, the optimization problem (8) (or (7) if we set $R = I$) is equivalent to $\min_{\mathbf{W}\in\mathbb{R}^r} \mathcal{K}_h(\widetilde{\mathbf{U}}_1)$ subject to $\widetilde{\mathbf{U}}_1 = D\mathbf{W} + \mathbf{C}$. Substituting $\widetilde{\mathbf{U}}_1 = D\mathbf{W} + \mathbf{C}$ into $\mathcal{K}_h(\widetilde{\mathbf{U}}_1)$, we obtain $\mathcal{K}_h(\widetilde{\mathbf{U}}_1) = \widetilde{\mathcal{K}}_h(\mathbf{W}) = \mathbf{W}^T(D^T R D)\mathbf{W} + 2\mathbf{C}^T R D\mathbf{W} + \mathbf{C}^T R\mathbf{C}$ so that the constrained optimization problem for $\mathcal{K}_h(\widetilde{\mathbf{U}}_1)$ is equivalent to the *one-dimensional unconstrained* optimization problem $\min_{\mathbf{W}\in\mathbb{R}^r} \widetilde{\mathcal{K}}_h(\mathbf{W})$. Then, $\mathbf{W}$ is a stationary point of $\widetilde{\mathcal{K}}_h(\cdot)$, we obtain $D^T R D\mathbf{W} = -2D^T R\mathbf{C}$. Since $D^T R D$ is a positive definite matrix, we can use, e.g., the conjugate gradient method, to solve this linear system.

The matrix $D \in \mathbb{R}^{\widetilde{q}_1 \times s}$ is a full matrix. In higher dimensions, $\widetilde{q}_1$ is a large number even for small J and K so that directly storing D requires large amounts of memory. However, recall that D embodies the marching procedure (5); that procedure is done level by level from the center of the space-time rectangle Q to its lateral sides Σ. In fact, the marching procedure can be expressed as

$$\begin{cases} \mathbf{U}_{M-1} = D_{3,M-1}\mathbf{W} + \mathbf{C}_{M-1} \\ \mathbf{U}_{M-2} = D_{2,M-2}\mathbf{U}_{M-1} + D_{3,M-2}\mathbf{W} + \mathbf{C}_{M-2} \\ \mathbf{U}_m = D_{1,m}\mathbf{U}_{m+1} + D_{2,m}\mathbf{U}_{m+2} + \mathbf{C}_m \quad \text{for} \quad m = M-3, \ldots, 0 \end{cases}$$

where $\mathbf{U}_m = \{u_{i,k} \in \widetilde{\mathbf{U}} \,|\, i = m \text{ or } i = J - m\}$ for $m = 0, 1, \dots, M-1$ and the $\{D_{i,m}\}$ are matrices. Clearly, $\widetilde{\mathbf{U}}_1 = \mathbf{U}_0$ and each $D_{i,m}$ is a bounded matrix having at most 5 non-zero elements in each of its rows or columns and D can be represented by the sequence of matrices $\{D_{i,m}\}$ corresponding to each level of the marching procedure (5). Consequently, we need only save $\{D_{i,m}\}$ to completely determine the action of D and D^T on a vector.

Note that this implementation cannot be directly applied to implicit (in time) approximations of the wave equation or to finite element approximations involving (unlumped) mass matrices. However, its extension to higher dimensions is straightforward.

3 Computational experiments

An example with known smooth exact solution, A smooth exact solution to the minimum $L^2(\Sigma)$-norm controllability problem can be constructed by using Fourier series in a similar way to that used in [2]. Suppose that $Q = (0,1) \times (0, 7/4)$ and $\Sigma = \{0,1\} \times (0, 7/4)$. Let $\psi_0(x,t) = -\sqrt{2}\pi \cos \pi(t - \frac{1}{4}) \cos 2\pi x$ and

$$\psi_1(x,t) = \left[2\sqrt{2}\pi (T-t) \sin(t - \frac{1}{4}) - \frac{10}{3}\sqrt{2} \sin \pi(t-T) \right] \sin \pi x$$
$$+ \sum_{p \geq 3 \text{ and } p \text{ odd}} \frac{4\sqrt{2}}{p^2-1} \left[\frac{3p}{p^2-4} \cos \pi(t - \frac{1}{4}) + \sin p\pi(t-T) \right] \sin p\pi x \,.$$

Then, set the initial conditions $u_0(x) = \psi_0(x,0) + \psi_1(x,0)$ and $u_1(x) = \partial\psi_0/\partial t(x,0) + \partial\psi_1/\partial t(x,0)$. It is worth noting that u_0 is a Lipschitz continuous function but does not belong to $C^1(\overline{\Omega})$ and u_1 is a bounded function but does not belong to $C^0(\overline{\Omega})$. For this initial data, it can be shown that $u(x,t) = \psi_0(x,t) + \psi_1(x,t)$ is the exact solution having minimum $L^2(\Sigma)$-norm of the controllability problem (1). The corresponding exact Dirichlet control g is then given by restricting $u(x,t)$ to the lateral sides Σ, i.e., $g(t) = u(0,t) = u(1,t) = -\sqrt{2}\pi \cos \pi(t - \frac{1}{4})$. For future reference, note that $\|g\|_{L^2(\Sigma)} \approx 6.1340$ and $\|u\|_{L^2(Q)} \approx 5.8656$.

We apply our numerical method to this example. Computational experiments were carried out for $h = 1/16$, $1/32$, $1/64$, and $1/128$ with $\lambda = 1$ and $\lambda = 7/8$ repectively, so that the stability condition $K \geq J$ is satisfied. The results of our computational experiments are summarized in Table 1, where u^h and g^h are the computed approximations of the exact solutions u and g, respectively. From this table, it seems that g^h converges to g in the $L^2(\Sigma)$-norm at a rate of roughly 1 as does u^h in the $L^2(Q)$-norm. In order to visualize the convergence of our method as h becomes smaller, in Figure 1 we provide plots of the exact solution g and the corresponding computed discrete solutions g^h for 1/64 and 1/128. It seems that our method produces (pointwise) convergent approximations for both $\lambda = 1$ and $\lambda = 7/8$ *without the need for regularization.* This should be contrasted with other methods, e.g., see [2], for which for $\lambda < 1$, regularization was needed in order to obtain convergence. It may at first seem that our results for $\lambda < 1$ are contradictory to the result of [9] which states that there exists a sequence of approximate controls that do not converge. However, that result says nothing about the *particular* sequence of minimum discrete $L^2(\Sigma)$-norm controls that is determined by our method.

Table 1. Results of computational experiments for the minimum $L^2(\Sigma)$-norm case for the example having smooth solutions for the control.

h		1/16	1/32	1/64	1/128
$\|g^h\|_{L^2(\Sigma)}$	$\lambda = 1$	5.9572	6.0313	6.0804	6.1008
	$\lambda = 7/8$	5.9803	6.0507	6.0890	6.1104
$\|g^h - g\|_{L^2(\Sigma)}/\|g\|_{L^2(\Sigma)}$	$\lambda = 1$	10.858%	4.625%	2.071%	1.051%
	$\lambda = 7/8$	9.662%	5.003%	2.137%	0.991%
$\|u^h\|_{L^2(Q)}$	$\lambda = 1$	5.6954	5.7873	5.8266	5.8442
	$\lambda = 7/8$	5.7249	5.7899	5.8327	5.8470
$\|u^h - u\|_{L^2(Q)}/\|u\|_{L^2(Q)}$	$\lambda = 1$	9.283%	4.153%	1.989%	1.012%
	$\lambda = 7/8$	7.636%	3.907%	1.864%	0.891%

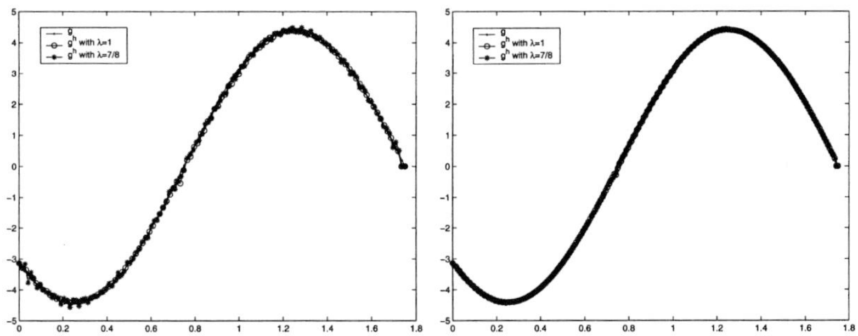

Fig. 1. The exact minimum $L^2(\Sigma)$-norm Dirichlet control g and the approximate controls g^h vs. time along $x = 0$ (and $x = 1$) for the example having smooth solutions for the control. Left: $h = 1/64$; right: $h = 1/128$.

Generic example with minimum $L^2(\Sigma)$-norm and minimum $H^1(\Sigma)$-norm boundary controls. In the example discussed above, the minimum $L^2(\Sigma)$-norm control is very smooth. Using our methods, we obtained good approximations for this example without the need for regularization. However, this is not the generic case. In general, even for smooth initial data, the minimum $L^2(\Sigma)$-norm Dirichlet control for the controllability problem (1) will not be smooth. In this section, we illustrate this point and also examine the performance of our method for the generic case.

We again choose $Q = (0,1) \times (0,7/4)$ and consider the following $C^\infty(\overline{\Omega})$ initial data: $u_0(x) = \sin(\pi x)$ and $u_1(x) = \pi \sin(\pi x)$. Note that the initial conditions vanish at the boundary and, that due to symmetry, we have that $u(0,t) = u(1,t)$, i.e., the control at the two sides of Q are the same.

We look at the case $\lambda = 1$. In Figure 2, we show the results for the control for several grid sizes ranging from $h = 1/32$ to $h = 1/256$. The (pointwise) convergence of the approximations is evident. Note that for these initial conditions, the minimum $L^2(\Sigma)$-norm control is seemingly piecewise smooth, i.e., it contains jump discontinuities. The pointwise convergence of the approximate control for the case of $\lambda = 1$ is probably a one-dimensional artifact; it is likely due to the fact that both the space and time variables in the wave equation in one dimension can act as time-like variables. are seemingly first order. Since the exact solutions corresponding to these initial conditions are not known, we will take the seemingly converged solutions obtained with $h = 1/256$ and $\lambda = 1$ as a baseline to which we compare our other computaional results. We will denote this baseline approximate control by g^*.

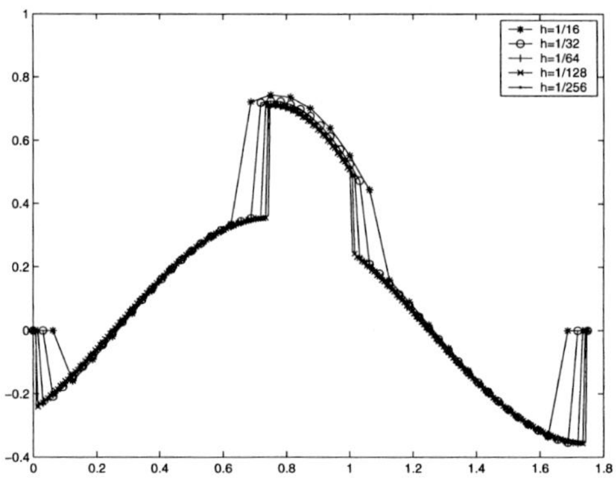

Fig. 2. The approximate minimum $L^2(\Sigma)$-norm Dirichlet control g^h vs. time for $\lambda = 1$ forthe example with piecewise smooth solutions at $x = 0$ (and $x = 1$).

Computational experiments were also carried out for $\lambda = 7/8$ and $\lambda = 7/12$ for several values of the grid size ranging from $h = 1/32$ to $h = 1/128$. The results are summarized in Table 2. In Figure 3, we provide plots of the computed discrete solution g^h for the two different values of λ and for different values of the grid size. From Figure 3, we see that the approximate minimum $L^2(\Sigma)$-norm Dirichlet controls obtained with values of $\lambda < 1$ are highly oscillatory. In fact, the frequencies of oscillation increase with decreasing grid size. However, is seems that the amplituded of the oscillationd do not increase as the grid size decreases. Furthermore, from the results in Table 2, is seems that for $\lambda < 1$, the approximate controls g^h, although oscillatory in a nature and not convergent in a pointwise sense, converge in an $L^2(\Sigma)$ sense. Also, the solution u^h is seemingly convergent in an $L^2(Q)$ sense. The results of Table 2 and Figure 3 indicate that for the generic case of non-smooth minimum $L^2(\Sigma)$ controls and for general $\lambda < 1$, our method produces convergent (in $L^2(Q)$ and $L^2(\Sigma)$) approximations *without the need of regulatization* but the approximations are not in general convergent in a pointwise sense. Of course, approximations that do not converge in a point-wise sense may be of no practical use, even if they converge in a root mean square sense.

Table 2. Results of computational experiments for the minimum $L^2(\Sigma)$-norm case for the example with piecewise smooth solutions and for $\lambda < 1$.

	λ	$h = 1/16$	$h = 1/32$	$h = 1/64$	$h = 1/128$
$\Vert g^h \Vert_{L^2(\Sigma)}$	7/8	0.6715	0.6354	0.6147	0.6036
	7/12	0.6550	0.6212	0.6085	0.5977
$\Vert u^h \Vert_{L^2(Q)}$	7/8	0.7546	0.7327	0.7204	0.7140
	7/12	0.7431	0.7234	0.7164	0.7124

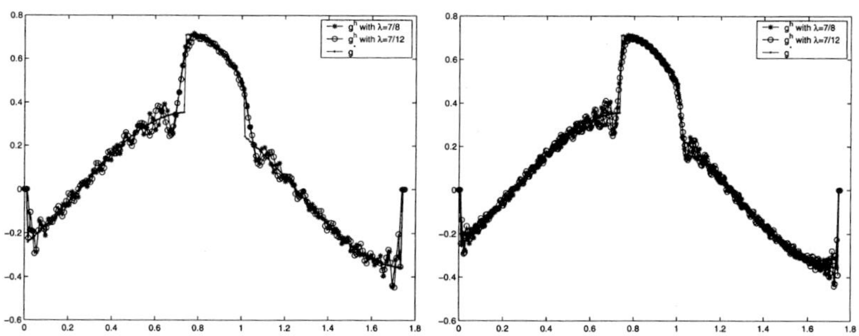

Fig. 3. The approximate minimum $L^2(\Sigma)$-norm Dirichlet controls g^h vs. time at $x = 0$ (and $x = 1$) for the example with piecewise smooth solutions. Left: $h = 1/64$; right: $h = 1/128$.

So far, the computational experiments indicate that, generically, minimum $L^2(\Sigma)$-norm boundary controls are only piecewise smooth and that approximations obtained by our method are in general not pointwise convergent so that they are not be realizable in practice. In order to obtain smoother controls, i.e., at least continuous in time, we consider the particular solution of the exact controllability problem (1) having minimum $H^1(\Sigma)$-norm. We use the same initial data as for Figure 3.

Figure 4 provides results of computational experiments for minimum $H^1(\Sigma)$-norm controls. It indicates that, in general, minimum $H^1(\Sigma)$-norm solutions of the controllability problem are smooth and that the approximate controls obtained by our method converge in a pointwise sense for both $\lambda = 1$ and $\lambda < 1$ without the need for regularization. Our results also indicate that our method produces convergent (in $L^2(Q)$ and $L^2(\Sigma)$) approximations.

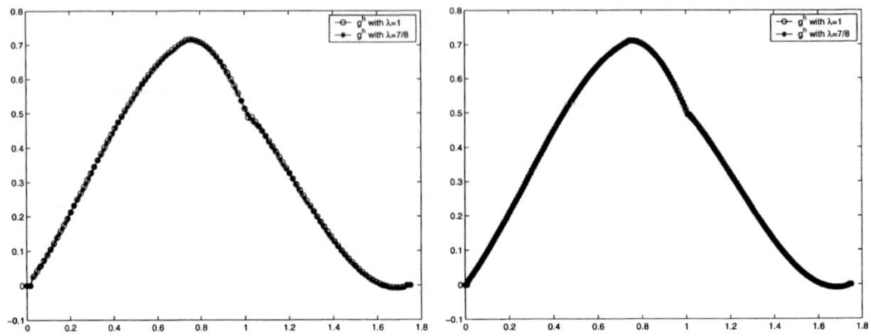

Fig. 4. The approximate minimum $H^1(\Sigma)$-norm Dirichlet controls g^h vs. time at $x = 0$ (and $x = 1$). Left: $h = 1/64$; right: $h = 1/128$.

References

1. R. Glowinski, R., Li, C.-H. (1990): On the numerical implementation of the Hilbert Uniqueness Method for the exact boundary controllability of the wave equation. C. R. Adad. Sc., Paris, **T. 311**, Serie I, 135–142.
2. Glowinski, R., Li, C.-H., Lions, J.-L. (1990): A numerical approach to the exact boundary controllability of the wave equation (I). Dirichlet controls: description of the numerical methods. Japan J. Appl. Math., **7**, 1–76.
3. Gunzburger, M., Hou, L., Ju, L.-L. (2002): An optimization-bsed numerical method for exact boundary Controllability problems for the wave equation. Control, Optimization and Calculus of Variations, submitted.
4. Gunzburger, M., Nicolaides, R. (1989): An algorithm for the boundary control of the wave equation. Appl. Math. Lett., **2**, 225–228.
5. Infante, J., Zuazua, E. (1999): Boundary observability for the space semi-discretizations of the 1-D wave equation. Math. Mod. Num. Anal., **33**, 407–438.
6. Ju, L.-L., Gunzburger, M., Hou, L. (2002): Approximation of exact boundary Controllability problems for the 1-D wave equation using optimization-based methods. preprint.
7. Lions, J.-L. (1988): Exact controllability and stabilization and perturbations for distributed systems. SIAM Review, **30**, 1–68.
8. Lions, J.-L. (1988): Controlabilité Exacte, Perturbation et Stabilisation des Systèmes Destribués, Vol. 1 and 2. Masson, Paris.
9. Negreanu, M., Zuazua, E. (2002): Uniform boundary controllability of a discrete 1-D wave equation. preprint.

10. Russell, D. (1973): A unified boundary controllability theory. Studies in Appl. Math., **52**, 189–211.
11. Russell, D. (1978): Controllability and stabilizability theory for linear partial differential equations: recent progress and open questions. SIAM Review, **20**, 639–739.
12. Zuazua, E. (1999): Boundary observability for the finite difference space semi-discretizations of the 2-D wave equation in the square. J. Math. Pures Appl., **78**, 523–563.

Numerical Study of a Viscous Consolidation Model

Katarina Gustavsson[1] and Björn Sjögreen[2]

[1] Royal Institute of Technology, Department of Numerical Analysis and Computing Science, S-100 44 Stockholm, Sweden *katarina@nada.kth.se*
[2] Royal Institute of Technology, Department of Numerical Analysis and Computing Science, S-100 44 Stockholm, Sweden *bjorns@nada.kth.se*

1 Introduction

In this paper we present a mathematical model and a numerical study of a one dimensional consolidation process of a flocculated suspension. The suspension is treated as a mixture of fluid and small, solid particles forming larger aggregates called flocs. The mathematical model is based on a macroscopic description of the process, so both phases are treated as separate interpenetrating continua. Each phase is kept track of by a scalar volume fraction field, ϕ, which indicates the proportion of the total volume occupied by particles. Equations for conservation of mass and momentum are formulated for each phase. Furthermore the suspension is characterized by constitutive relations correlating experimental data to three *material functions*: permeability, viscosity and a particle interaction pressure, yield pressure. These functions will be described in more details in section 6.

In the one dimensional case, we let the suspension be confined to a closed box of height H. Under the influence of gravitational forces acting in one direction, the suspension is irreversibly compresses and the solid phase is separated from the liquid phase. This is referred to as a consolidation process. The gravitational forces can either be natural gravity or centrifugal forces. In this case the process is driven by centrifugal forces, see figure 1. For further details about the one dimensional consolidation model see [1].

The purpose here is to investigate certain mathematical properties of the model and to develop a second order accurate numerical method for this problem. We treat only the one dimensional problem here, but the idea is to develop a method which can easily be extended to the two dimensional problem treated in [1].

Actually, we investigate and compare two models, the *viscous* model and the *inviscid* model. Both models are described in section 2. The main focus will be at the viscous model since it is directly derived from the two dimensional model mentioned above. For cases where both models are applicable, we can show that they give similar results. However we can also show that from a numerical point of view, the viscous model is better since it gives a less restrictive condition on the time step in the numerical computations.

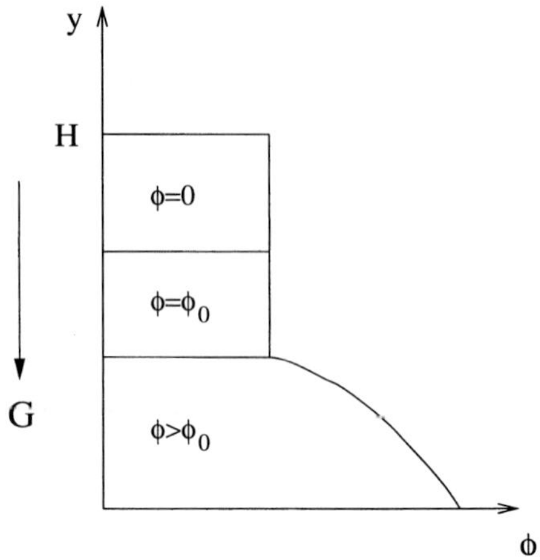

Fig. 1. One dimensional consolidation model. Distribution of ϕ at some time t. ϕ_0 is the initial concentration. The gravitational force is pointing in the negative y-direction.

2 Differential equation

The mathematical model is derived from the multi-dimensional model described in [1] and the equations obtained from conservation of mass and momentum reduces to the following system in one dimension,

$$\phi_t + (v\phi)_y = 0 \tag{1}$$

$$\frac{4}{3}(\eta(\phi)v_y)_y - \frac{v}{D(\phi)} = p_s(\phi)_y + c\phi \tag{2}$$

which we refer to as the *viscous* model. Here, $\phi = \phi(t,y)$ is the volume fraction of particles, and lies between zero and one and $v = v(y;\phi(y,t))$ is the velocity of the particle phase. The problem is defined on the domain $0 < y < H$, $t > 0$, with boundary conditions $v(0) = v(H) = 0$, and given initial data $\phi(0,y) = f(y)$. $c = \Delta\rho G$ is a given constant depending on the centrifugal force G and the density difference between the phases $\Delta\rho$. The material is characterized by the viscosity function, $\eta(\phi)$, the permeability, $D(\phi)$ and the yield pressure, $p_s(\phi)$. They are derived from material properties, and are all non-negative.

The viscous term in equation (2) is of order $D(\phi)\eta(\phi)/H^2$ and is usually very small compared to the other terms. This is because the suspension is highly concentrated and the permeability function $D(\phi)$ is very small. Therefore, it is more standard in consolidation modeling to solve the corresponding inviscid problem, obtained by setting $\eta(\phi) = 0$, see e.g. [2], [3] and [4] among others. With $\eta = 0$ we can eliminate v from (1), and obtain

$$\phi_t - (c\phi^2 D(\phi))_y = (D(\phi)p_s(\phi)_y)_y \tag{3}$$

which we recognize as a hyperbolic conservation law with a viscous term.

In this article we focus on the viscous model. One reason for treating the viscous problem is that it has features in common with the 2D consolidation model treated in [1]. Hence, the numerical methods developed here can easily be extended to the 2D consolidation model. Another reason is that the numerical approximation of the inviscid problem has a considerably more restrictive CFL-condition, than the viscous problem, thus making the viscous problem more suitable for a numerical solution. For cases where both the viscous and inviscid models are applicable we show that they give similar results, see section 7.

3 Well-posedness of the linearized problem

To learn more about the behavior of the solution to the viscous model we linearize the equations (1) and (2) around a solution $\phi_0(t, y)$ and $v_0(y; \phi_0)$. If we let the coefficient be frozen at a constant state ϕ_0, v_0, the equations are simplified to

$$\begin{aligned} \phi_t + v_0\phi_y + \phi_0 v_y &= 0 \\ \frac{4}{3}\eta(\phi_0)v_{yy} - \frac{v}{D(\phi_0)} &= p_s'(\phi_0)\phi_y + (F - \frac{D'(\phi_0)v_0}{D(\phi_0)^2})\phi \end{aligned} \tag{4}$$

We have used the notation ϕ, v for the small perturbation and the fact that $v_0 = -D(\phi_0)c\phi_0$.

For this linear constant coefficient problem we can use the Fourier transformation in the spatial variable, if we assume a Cauchy problem. Let $\hat{v}(\xi)$ and $\hat{\phi}(\xi, t)$ denote the Fourier transformations of v and ϕ respectively. We eliminate $\hat{v}(\xi)$ from the transformed system, to obtain the equation $\hat{\phi}_t = \hat{P}(\xi)\hat{\phi}$ where the Fourier symbol is given by

$$\hat{P}(\xi) = i\xi(v_0 + \frac{D'(\phi_0)v_0\phi_0/D(\phi_0)^2 - c\phi_0}{4\eta(\phi_0)\xi^2/3 + 1/D(\phi_0)}) - \xi^2\frac{p_s'(\phi_0)\phi_0}{4\eta(\phi_0)\xi^2/3 + 1/D(\phi_0)} \tag{5}$$

From this we conclude that the linearized equation for ϕ has two features. First a wave propagation seen as the first term in (5), the propagation speed varies with ξ, between v_0 and $v_0(2 + D'(\phi_0)\phi_0/D(\phi_0))$. Secondly, a term of the form $\xi^2/(\xi^2 + 1)$, which is a damping term if $p_s'(\phi_0) > 0$, $\eta(\phi_0) > 0$, $D(\phi_0) > 0$, which is the case for the physical problem.

The linearized problem is well-posed in the sense that

$$|e^{\hat{P}(\xi)t}| \leq 1,$$

but it is not parabolic. The definition of parabolic (see [5]), says that

$$\Re(\hat{P}(\xi)) \leq -\delta\xi^2$$

should hold for all ξ, where δ is a positive constant. This can obviously not be true for (5), when ξ is large.

In conclusion, ϕ obeys an equation which has the character of wave propagation and damping. The damping is similar to a parabolic equation (i.e., like the equation $\phi_t + a\phi_y = \epsilon\phi_{yy}$) for low frequencies, but behaves like a sink i.e., like the equation $\phi_t + a\phi_y = -\alpha\phi$ for high frequencies. Note that in the limit case of low frequency, the viscosity coefficient is $\epsilon = p_s'(\phi_0)\phi_0 D(\phi_0)$, and in the limit case of high frequency, the sink strength is $\alpha = 3p_s'(\phi_0)\phi_0/(4\eta(\phi_0))$, so that the viscosity coefficient η has little influence on the smoothing of ϕ, and for efficient damping of high modes it should even be small. The pressure derivative p_s', should be large for efficient damping and smoothing in the high and in the low limits respectively.

4 Difference Approximation

To treat the model numerically, we introduce grid points, $y_j = (j-1)\Delta y$, $j = 1, 2, \dots, N$, on the interval $[0, H]$ where $\Delta y = H/(N-1)$. Let v_j denote the numerical approximation of $v(y_j)$. For the volume fraction, we use a staggered approximation, and define $\phi_{j+1/2}$, $j = 1, 2, \dots, N-1$, the approximation of ϕ at the points $y_{j+1/2} = (y_j + y_{j+1})/2$. We will use the difference operators $D_+u_j = (u_{j+1} - u_j)/\Delta y$ and $D_-u_j = (u_j - u_{j-1})/\Delta y$.

We approximate (1) and (2) by the semi-discrete difference approximation

$$\frac{d\phi_{j+1/2}}{dt} = -D_+h_j \tag{6}$$

$$\frac{4}{3}D_+(\eta(\phi_{j-1/2})D_-v_j) - \frac{v_j}{D(\bar{\phi}_j)} = D_+p_s(\phi_{j-1/2}) + c\bar{\phi}_j \tag{7}$$

where $\bar{\phi}_j = (\phi_{j+1/2} + \phi_{j-1/2})/2$. The time integration can be done by a Runge-Kutta method, or by a linear multi step method. In the case of a second order central scheme, the numerical flux function, h_j, is

$$h_j = v_j\bar{\phi}_j. \tag{8}$$

This approximation is centered, and can perform badly in the neighborhood of steep gradients in ϕ.

We therefore next consider non-oscillatory TVD-type of schemes. One example is the scheme with constant artificial viscosity, defined by the numerical flux function

$$h_j = h(\phi_{j+1/2}, \phi_{j-1/2}) = v_j(\phi_{j+1/2} + \phi_{j-1/2})/2 - \frac{\varepsilon}{2\lambda}(\phi_{j+1/2} - \phi_{j-1/2}) \tag{9}$$

where ε is a constant. The time is divided into time steps $\Delta t = t_{n+1} - t_n$ and λ is the ratio $\Delta t/\Delta y$. We will use the notation $\phi^n_{j+1/2}$ for the approximation of $\phi(t_n, y_{j+1/2})$ and v^n_j for the approximation of $v(y_j, \phi(t_n, y))$. Taking $\varepsilon = 1$

in (9) gives the Lax-Friedrichs scheme. We will here consider the modified Lax-Friedrichs scheme, obtained when $\varepsilon < 1$. If v_j had been a constant, this scheme can be shown to give non-oscillatory solution for $\lambda v < \varepsilon < 1$. However, since v_j depends on all ϕ values at the same time level, the concept of upwind approximation is not meaningful, and since ϕ can grow in time, we can not define TVD properties either.

The scheme given by (6) and (9) is only first order accurate. In order to define a non-oscillatory second order accurate scheme, we introduce the piecewise linear interpolation

$$\phi(y) = \phi_{j+1/2} + s_{j+1/2}(y - y_{j+1/2})/\Delta y$$

in the cell $[y_j, y_{j+1}]$. The slope is defined through a non-oscillatory interpolation function

$$s_{j+1/2} = s(\Delta_+\phi_{j+1/2}, \Delta_-\phi_{j-1/2})$$

where we used

$$s(x, y) = (x(y^2 + \epsilon) + y(x^2 + \epsilon))/(x^2 + y^2 + 2\epsilon).$$

Here ϵ is a small constant to prevent division by zero when $x = y = 0$. The slopes are used to modify the numerical flux function into

$$h_j = h(\phi_{j+1/2} - s_{j+1/2}/2, \phi_{j-1/2} + s_{j-1/2}/2) \tag{10}$$

The modified flux function gives a second order accurate approximation away from smooth extrema. The complete method which we will use consists of (6) with the second order flux function (10) with h as defined in (9) together with a second order Runge-Kutta method in time,

$$\begin{aligned} \phi^{(1)}_{j+1/2} &= \phi^n_{j+1/2} - \lambda\Delta_+ h^n_j \\ \phi^{(2)}_{j+1/2} &= \phi^{(1)}_{j+1/2} - \lambda\Delta_+ h^{(1)}_j \\ \phi^{n+1}_{j+1/2} &= (\phi^n_{j+1/2} + \phi^{(2)}_{j+1/2})/2 \end{aligned} \tag{11}$$

here the superscript on the numerical flux function indicate at which Runge-Kutta stage it is evaluated. On each stage, the velocity v_j is found from the corresponding $\phi_{j+1/2}$, by solving the velocity equation (7). The v_j are found as the solution of a tridiagonal linear system of equations, which is inexpensive in one space dimension, but which requires additional effort in two and three space dimensions.

5 Stability Analysis

We get a necessary condition for the stability of the central difference approximations, by performing linear constant coefficient von Neumann analysis. This is done by applying the approximation (6) and (8) to the linearized problem. Under periodicity assumptions we transform to Fourier modes in space, in the standard way. Let $\hat{v}_\omega$ and $\hat{\phi}_\omega$ denote the Fourier coefficients. After elimination of $\hat{v}_\omega$, we obtain the equation $\frac{d\hat{\phi}_\omega}{dt} = \hat{Q}(\omega, \Delta y)\hat{\phi}_\omega$ with Fourier symbol

$$\hat{Q} = -i\frac{\sin \omega\Delta y}{\Delta y}(v_0 - \frac{c\phi_0 - D'(\phi_0)\phi_0 v_0/D(\phi_0)^2}{16\eta(\phi_0)\sin^2(\omega\Delta y/2)/(3\Delta y^2) + 1/D(\phi_0)}) - \frac{4\sin^2(\omega\Delta y/2)p_s'(\phi_0)\phi_0/\Delta y^2}{16\eta_0\sin^2(\omega\Delta y/2)/(3\Delta y^2) + 1/D(\phi_0)}$$

If we use a time integration method which encloses a square $-a < \Re(z) < 0, -b < \Im(z) < b$ in its region of absolute stability, the CFL-conditions

$$\lambda|v_0|(2 + |D'(\phi_0)\phi_0/D(\phi_0)|) < b \qquad 3\Delta t p_s'(\phi_0)\phi_0/(4\eta(\phi_0)) < a \tag{12}$$

are sufficient to guarantee linear stability. The following facts are observed

- It is possible to assure stability, even if we let $p_s'(\phi)$ blow up like $1/\Delta t$ at some point.
- In the inviscid model with $\eta = 0$, the more restrictive condition $4p_s'(\phi_0)\phi_0 D(\phi_0)(\Delta t/\Delta y^2) < a$ is obtained. This shows that the viscous model has better numerical stability properties.
- Here we can also see the advantage with a staggered grid. The approximation analogous to (6) and (8), but on a non-staggered grid, will lead to a Fourier symbol having the factor $\sin^2(\omega\Delta y)$ instead of $\sin^2(\omega\Delta y/2)$ in the numerator of the real part of $\hat{Q}$, showing the problem with odd-even modes for this problem, in analogy with the incompressible Navier-Stokes equations.

6 Material functions

The functions $D(\phi), \eta(\phi)$, and $p_s(\phi)$ are fitted to experimental data on industrial waste water and are modeled as follows

$$D(\phi) = (1-\phi)k_D\phi^{n_D} \tag{13}$$

$$\eta(\phi) = k_\eta\phi^{n_\eta} \tag{14}$$

$$p_s(\phi) = \begin{cases} k_p\phi^{n_p} & \phi > \phi^* \\ 0 & \phi \le \phi^* \end{cases} \tag{15}$$

The constants have been determined by experimental measurements to the following values,

$$\begin{aligned} k_D &= 4.6 \times 10^{-18} m^2\,Pa^{-1}\,s^{-1} & n_D &= -7.41 & \\ k_\eta &= 1.2691 \times 10^6\,Pas & n_\eta &= 3.4 & \\ k_p &= 5 \times 10^6\,Pa & n_p &= 4.846 & \phi^* = 0.1 \end{aligned}$$

For more details about the measurements and the data fitting see [6].

These functions might have to be modified somewhat to make them more suitable for mathematical and numerical treatment. In light of the previous section we should require that the derivative $p_s'(\phi)$ is bounded. The jump in the function $p_s(\phi)$ should be smoothed, so that the derivative $p_s'(\phi)$ is bounded. For example, taking

$$p_{s,K} = \frac{1}{2}(\tanh(K(\phi - \phi^*)) + 1)k_p\phi^{n_p} \tag{16}$$

gives the bound

$$p_{s,K}'(\phi) \leq (n_p + K\phi/2)k_p\phi^{n_p - 1}.$$

Furthermore, with this modification, in the inviscid model given by (3), ϕ will satisfy a convection-diffusion equation, with viscosity that never becomes zero.

The function $D(\phi)$ is zero for $\phi = 1$, but in practice it never happens that ϕ comes even close to 1. On the other hand, D becomes infinite for $\phi = 0$. This singularity can be removed by replacing the power ϕ^{n_D} (n_D is negative) by

$$\begin{array}{ll} \phi^{n_D} & \text{if } \phi \geq \phi_{lim} \\ (\phi_{lim})^{n_D}(1 + n_D(\phi - \phi_{lim})/\phi_{lim}) & \text{if } \phi < \phi_{lim} \end{array} \tag{17}$$

instead. Here ϕ_{lim} is a small constant. The viscosity coefficient can also be modified to avoid that it becomes zero when ϕ goes to zero. The previously suggested form, [1],

$$\eta_m(\phi) = \max(k_\eta\phi^{3.4}, \eta_{min})$$

has turned out to produce oscillations in the numerical solution. This is probably due to non-differentiability of η at the corner. In the second order accurate difference method used, the truncation error contains third derivatives of η. It is therefore more reasonable to aim for a modification which is at least in C^3. We have here used the C^∞ modification

$$\eta_m(\phi) = k_\eta(\phi^{3.4} + \eta_{min}) \tag{18}$$

where η_{min} is a very small constant.

7 Numerical results and discussion

In this section we will discuss some numerical experiments performed with the method described in section 4. First we present a grid convergence study and secondly a comparison between results obtained by the viscous and the inviscid model.

In the numerical experiments we used the initial data

$$\phi(0, y) = \phi_0(y) = (\phi_g - \phi_{min})\frac{1}{2}(1 - \tanh(K(y - y_0))) + \phi_{min} \tag{19}$$

where $\phi_g = 0.1$, $\phi_{min} = 0.001$ and K is a regularizing parameter. The constant c takes the value $1.0 \times 10^6 kgm^{-2}s^{-2}$ in the computations.

A grid convergence study of the scheme given by (6), (7), (9), (10) and (11) applied to the viscous model is presented in table 1. The table shows second order accuracy both for the velocity field, $\mathbf{v}$, as well as for the volume fraction, ϕ. Here we use $K = 100$ in the initial data function to obtain a solution sufficiently smooth to measure the order of accuracy. The convergence exponent, p was estimated by the formula

$$p = \log\left(\frac{\|\boldsymbol{f}_{\Delta y} - \boldsymbol{f}_{\Delta y/3}\|}{\|\boldsymbol{f}_{\Delta y/3} - \boldsymbol{f}_{\Delta y/9}\|}\right) / \log 3 \tag{20}$$

where $\boldsymbol{f}$ is a vector valued function and $\|\cdot\|$ is either the discrete $L1$-norm, $L2$-norm or the max-norm. The time step is chosen such that $\lambda = \Delta t/\Delta y$ is constant, where $\Delta t = 0.1$ on the coarsest mesh and $\Delta y = 0.1/36$.

No of grid points	*Order of convergence, p*					
N	$\|\mathbf{v}\|_1$	$\|\mathbf{v}\|_2$	$\|\mathbf{v}\|_\infty$	$\|\phi\|_1$	$\|\phi\|_2$	$\|\phi\|_\infty$
109	1.81	1.78	1.76	1.76	1.64	1.45
325	1.98	1.97	1.97	1.97	1.95	1.92
973	2.00	2.00	2.00	2.00	2.00	1.99

Table 1. Measured order of accuracy, p for the velocity field v and the volume fraction ϕ. $\varepsilon = 0.005$. Computations until $t = 4.0s$.

A sequence of solutions to the viscous model, (1) and (2), at different times is shown in figure 2.

We also performed numerical computations with the inviscid model and the same numerical scheme as in the viscous case.

In figure 2 we compare the solutions obtained by the two different models. The dashed line is the solution obtained using the viscous model and the solid line is the solution obtained by the inviscid model. In figure 2a we used the empirical Brinkmann model for the permeability

$$D(\phi) = K_D(1-\phi)\frac{(2-3\phi)^2}{\phi(3\phi + 4 + \sqrt{8\phi - 3\phi^2})} \tag{21}$$

where $K_D = 5 \times 10^{-9}$. In figure 2a, we can see that there is a difference close to the bottom where the viscosity is large and the permeability small. In Figure 2b we use the permeability model fitted to experimental data, (13). In this case the term $D(\phi)\eta(\phi)/H^2$ becomes so small that the difference between solutions obtained by the classical and the viscous model is very small. In Figure 3 we show a comparison of the term $D(\phi)\eta(\phi)/H^2$ for the two permeability functions. There is a large difference and that yields a large difference in time scales for the two cases.

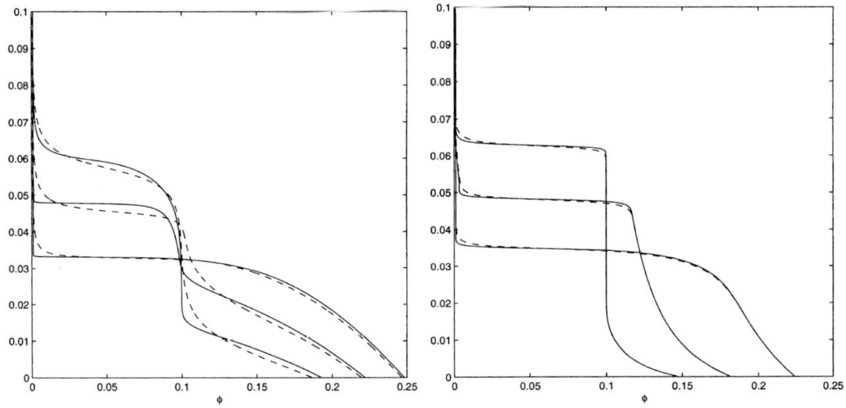

(a) ϕ at $4, 10, 20$ s, using permeability function given by equation (21)

(b) ϕ at $200, 2000, 20000$ s, using permeability function given by equation (13)

Fig. 2. Comparison between the viscous (dashed line) and the inviscid (solid line) model for two different permeability functions. Volume fraction, ϕ, as a function of y at different times. Note the different time scales given by the different permeability functions.

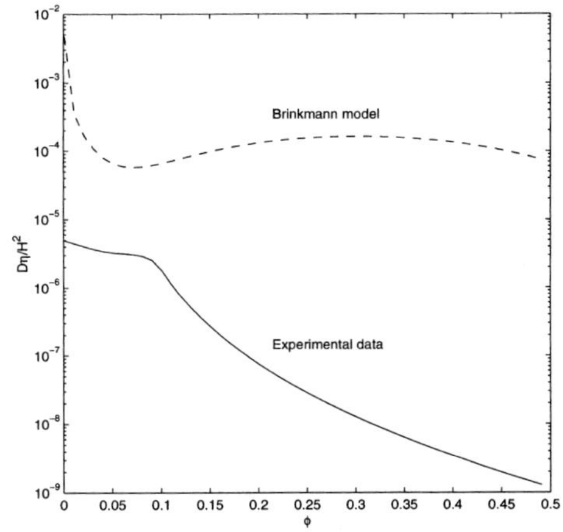

Fig. 3. Time scale of the problem is given by the factor $D(\phi)\eta(\phi)/H^2$. Two different permeability models.

References

1. Gustavsson, K. (1999): Simulation of consolidation processes by Eulerian two-fluid models. Licentiate's thesis. Royal Institute of Technology, Stockholm
2. Auzerais, F.M., Jackson R.J., Russel W.B. (1988): The resolution of shocks and the effect of compressible sediments in transient settling. J. of Fluid Mech., **195**, 437–462
3. Bürger, R., Wendland, W.L., Concha, F. (2000): Model equations for gravitational sedimentation-consolidation processes. ZAMM, Z. Angew. Math. Mech., **80,2**, 79–92
4. Dorobantu, M. (1997): One-dimensional consolidation models. Licentiate's thesis. Royal Institute of Technology, Stockholm
5. Kreiss, H-O., Lorentz, J. (1989): Initial-boundary value problems and the Navier-Stokes equations. Academic Press
6. Gustavsson, K., Oppelstrup, J. (2001): Numerical 2D models of consolidation of dense flocculated suspensions. J. of Engineering Math., **41**, 189–201

Goal-Oriented *A Posteriori* Error Estimation for Multiple Target Functionals

Ralf Hartmann[1,⋆] and Paul Houston[2,⋆⋆]

[1] Institute of Applied Mathematics, University of Heidelberg, Im Neuenheimer Feld 293, D-69120 Heidelberg, Germany *Ralf.Hartmann@iwr.uni-heidelberg.de*
[2] Department of Mathematics & Computer Science, University of Leicester, Leicester LE1 7RH, UK *Paul.Houston@mcs.le.ac.uk*

1 Introduction

In many applications the quantities of interest are a series of target functionals of the solution to the governing system of partial differential equations rather than the solution itself. For example, in the field of aerodynamics, examples include the drag and lift coefficients of an airfoil immersed into a fluid, the pressure difference between the leading and trailing edges of the airfoil and point evaluations of the density or pressure on the profile of the airfoil. While traditionally these quantities are measured in wind tunnel experiments, nowadays these experiments are increasingly replaced by numerical simulations aiming to predict these quantities to a high level of accuracy.

In the series of articles [2, 4, 5, 6, 7], we have developed the theory of goal–oriented *a posteriori* error estimation for discontinuous Galerkin methods applied to inviscid compressible fluid flows; see [10], for related work. On the basis of Type I *a posteriori* bounds, in [2, 4, 5, 6, 7], we considered the design of adaptive finite element algorithms that are capable of generating optimal meshes specifically tailored to the efficient computation of a *single* target functional of practical interest. The purpose of the current article is to extend this earlier work to the case when several target functionals of the solution need to be simultaneously approximated to a given level of accuracy.

2 Duality based *a posteriori* error estimation

Given that X and Y are two Hilbert spaces, $B(\cdot,\cdot)$ is a bounded bilinear functional defined on $X \times Y$ and $\ell(\cdot)$ is a bounded linear functional defined on Y, we suppose that u is the unique solution to the variational problem:

$$\text{find } u \in X \text{ such that } \quad B(u,v) = \ell(v) \quad \forall v \in Y\,. \tag{1}$$

Problem (1) can be thought of as the weak formulation of a linear partial differential equation on X whose unique solution is $u \in X$. In practice (1)

⋆ Supported by the DFG Priority Research Program and the SFB 359 at the IWR, University of Heidelberg.

⋆⋆ Paul Houston acknowledges the financial support of the EPSRC (GR/N24230).

cannot be solved in closed form but needs to be approximated numerically. In order to construct a Galerkin approximation to this problem, we consider a sequence of finite–dimensional spaces $\{X_h\}$ and $\{Y_h\}$, parameterised by $h > 0$; for simplicity we suppose that $X_h \subset X$ and $Y_h \subset Y$ for each h. For this paper, X_h and Y_h can be thought of as finite element spaces consisting of piecewise polynomial functions on a partition, of granularity h, of the computational domain. The Galerkin approximation u_h of u then satisfies:

$$\text{find } u_h \in X_h \text{ such that } \ B(u_h, v_h) = \ell(v_h) \quad \forall v_h \in Y_h\,. \tag{2}$$

Assuming there exists a unique solution u_h to (2), we note that since the discretisation method (2) is fully consistent with the original (primal) problem, the error $u - u_h$ satisfies the following Galerkin orthogonality property

$$B(u - u_h, v_h) = 0 \quad \forall v_h \in Y_h. \tag{3}$$

It is frequently the case in applications that, instead of accurately computing u_h everywhere within the computational domain, the focus of interest is a derived quantity $J(u_h)$ which should be computed to high accuracy, where $J(\cdot)$ is a given bounded linear functional defined on X. Indeed, it may happen that instead of a single linear functional $J(\cdot)$, several linear functionals $J_i(\cdot)$, $i = 1, \dots, N$, $N \geq 1$, are given, and one needs to compute each $J_i(u_h)$ to within a given user–defined tolerance $\mathtt{TOL_i}$, $i = 1, \dots, N$, respectively. More precisely, we consider the following *measurement problem*: find $J_i(u_h) \in \mathbb{R}$, $i = 1, \dots, N$, such that

$$|J_i(u) - J_i(u_h)| \leq \mathtt{TOL_i}\,, \quad \text{for } i = 1, \dots, N. \tag{4}$$

Our approach to the solution of this problem rests on a duality argument. For simplicity, we first outline the general technique for the case of a single target functional, i.e. $N = 1$, and we write $J(\cdot) \equiv J_1(\cdot)$. The starting point for the *a posteriori* error analysis is to introduce the following *dual* problem:

$$\text{find } z \in Y \text{ such that } \ B(w, z) = J(w) \quad \forall w \in X\,. \tag{5}$$

By using the linearity of $J(\cdot)$, together with (5), (3) and (1), we get

$$\begin{aligned} J(u) - J(u_h) &= J(u - u_h) = B(u - u_h, z) = B(u - u_h, z - z_h) \\ &= \ell(z - z_h) - B(u_h, z - z_h) \quad \forall z_h \in Y_h\,. \end{aligned} \tag{6}$$

As $v \mapsto \ell(v) - B(u_h, v)$ is a bounded linear functional on Y for each $u_h \in X_h$, there exists $R_p(u_h) \in Y'$ (where Y' denotes the dual space of Y) such that

$$\ell(z - z_h) - B(u_h, z - z_h) = \langle R_p(u_h), z - z_h \rangle\,; \tag{7}$$

here $\langle \cdot, \cdot \rangle$ denotes the duality pairing between Y' and Y. The element $R_p(u_h)$ will be referred to as the *primal residual*. Combining (6) and (7) gives

$$J(u) - J(u_h) = \langle R_p(u_h), z - z_h \rangle \qquad \forall z_h \in Y_h , \tag{8}$$

which is referred to as the *primal error representation formula.* On the basis of (8), Type I and Type II *a posteriori* error bounds may be derived, cf. [4].

Let us now consider the extension of the above analysis to the case when $N > 1$. The *direct method* for deriving an error representation formula analogous to (8) for each $J_i(\cdot)$ is to introduce the following N dual problems:

$$\text{find } z_i \in Y \text{ such that } \ B(w, z_i) = J_i(w) \quad \forall w \in X , \text{ for } i = 1, \ldots, N. \tag{9}$$

Thereby, repeating the above argument for each functional, we deduce that

$$J_i(u) - J_i(u_h) = \langle R_p(u_h), z_i - z_h \rangle \quad \forall z_h \in Y_h , \text{ for } i = 1, \ldots, N. \tag{10}$$

In practice, the dual solutions z_i, $i = 1, \ldots, N$, are unknown analytically and must be approximated numerically. Upon discretisation, this amounts to solving N systems of linear equations with the same matrix but N different right–hand side vectors. In view of the error representation formula

$$J_i(u - u_h) = B(u - u_h, z_i) = \langle R_p(u_h), z_i \rangle = \langle R_p(u_h), z_i - z_h \rangle , \tag{11}$$

an alternative approach consists of considering the following *error equation*:

$$\text{find } e \in X \text{ such that } \ B(e, w) = \langle R_p(u_h), w \rangle \quad \forall w \in Y , \tag{12}$$

whose solution is simply the discretisation error $e = u - u_h$. We remark that in the context of duality, (12) may be thought of as the *dual–dual* problem and (11) the *dual/dual–dual* equivalence relating (9) to (12), cf. [8]. Thereby, in practice, instead of solving N dual problems, cf. (9), for $z_i \in Y$ with data $J_i(\cdot)$ and then evaluating $\langle R_p(u_h), z_i - z_h \rangle$ to determine the size of the error in the target functional $J_i(\cdot)$, $i = 1, \ldots, N$, one can simply solve the error equation (12) for the error $e \in X$ and evaluate $J_i(e)$, for $i = 1, \ldots, N$. When $N \gg 1$ this approach is clearly much more computationally efficient than the direct method. However, a disadvantage of this second approach is that while solving the error equation (12) for e gives information concerning the size of the error in the computed target functionals $J_i(\cdot)$, $i = 1, \ldots, N$, it does not provide the necessary local information on each element in the computational mesh to guide adaptive mesh refinement when the desired level of accuracy has not been achieved on the current mesh. On the other hand, computing the solution z_i, $i = 1, \ldots, N$, to the N dual problems (9), the error representation formulae (10) provides not only information concerning the size of the error in the computed target functionals, but also upon localisation of the right–hand side of (10), local error indicators which can be employed for adaptive mesh design, cf. [4, 5], for example. In the following section we propose a strategy based on solving only two auxiliary problems (the *error equation* (12) and a *dual* problem subject to appropriate data which stems from a specific combined target functional, cf. (14) below) which provide all the necessary information needed to both estimate the size of the error in the computed target functionals, as well as provide local error indicators that can be used to design adaptively refined computational meshes.

3 *A posteriori* estimation for multiple target functionals

In this section we develop the *a posteriori* error analysis in the case when we wish to control the discretisation error in a number of target functionals $J_i(\cdot)$, $i = 1,\ldots,N$, of practical interest. To this end, we weaken the requirement (4), and simply insist that the sum of the relative errors in each of the target functionals $J_i(\cdot)$, $i = 1,\ldots,N$, is less than TOL. In practice, since $J_i(u)$, $i = 1,\ldots,N$, is unknown, we approximate the sum of the relative errors by

$$\sum_{i=1}^{N} |J_i(u) - J_i(u_h)|/|J_i(u_h)|\,,$$

assuming that $J_i(u_h) \neq 0$, for $i = 1,\ldots,N$.

Let us begin by assuming that the sign of the error in each target functional $J_i(\cdot)$, $i = 1,\ldots,N$, is known. For example, in some applications it may be known from either theoretical considerations or numerical experimentation that under mesh refinement the computed quantity of interest $J_i(u_h)$ is always either smaller or greater than the exact value $J_i(u)$, for $i = 1,\ldots,N$. Employing this *a priori* knowledge concerning the convergence of the target functionals, we introduce a *combined* target functional

$$J_c(v) = \sum_{i=1}^{N} \omega_i J_i(v)\,, \tag{13}$$

where $\omega_i = s_i/|J_i(u_h)|$ and s_i denotes the expected signs of the errors $J_i(u) - J_i(u_h)$, $i = 1,\ldots,N$, respectively. Thereby, we may now proceed as in Section 2 to derive an error representation formula for the error in the combined target functional $J_c(\cdot)$. To this end, we introduce the following dual problem:

$$\text{find } z_c \in Y \text{ such that } \; B(w, z_c) = J_c(w) \quad \forall w \in X\,. \tag{14}$$

Thus, as before, we now deduce the following error representation formula

$$J_c(u) - J_c(u_h) = \langle R_p(u_h), z_c - z_h\rangle \qquad \forall z_h \in Y_h\,. \tag{15}$$

Equation (15) now provides both global information concerning the size of the error in the combined target functional $J_c(\cdot)$, as well as local information necessary for adaptive mesh refinement.

In general, the signs s_i, $i = 1,\ldots,N$, will not be known *a priori*. Thereby, we must first solve the error equation (12) for e and evaluate $s_i = \mathrm{sgn}(J_i(e))$, $i = 1,\ldots,N$. Then, the dual problem (14) may be solved computationally using the predicted values of s_i, $i = 1,\ldots,N$, and the resulting error representation formula (15) computed. This approach leads to the solution of only two auxiliary problems, in comparison to the N required for the direct method. In the following section we consider the generalisation of this theory to the case when $J(\cdot)$ is nonlinear and $B(\cdot,\cdot)$ is a semilinear form; as a particular example, we consider the discontinuous Galerkin approximation to a first–order system of nonlinear hyperbolic conservation laws.

4 Application to nonlinear hyperbolic conservation laws

Given a bounded open polyhedral domain Ω in $\mathbb{R}^n$, $n \geq 1$, with boundary $\partial\Omega$, we consider the following problem: find $\mathbf{u} : \Omega \to \mathbb{R}^m$, $m \geq 1$, such that

$$\operatorname{div} \mathcal{F}(\mathbf{u}) = 0 \quad \text{in } \Omega \,, \tag{16}$$

subject to appropriate boundary/initial conditions, where $\mathcal{F} : \mathbb{R}^m \to \mathbb{R}^{m\times n}$ is continuously differentiable. To formulate the discontinuous Galerkin finite element method (DGFEM, for short) for (16), we first introduce some notation. Let $\mathcal{T}_h = \{\kappa\}$ be an admissible subdivision of Ω into open element quadrilateral domains κ; here h is a piecewise constant mesh function with $h(\mathbf{x}) = \operatorname{diam}(\kappa)$ when $\mathbf{x}$ is in element κ. For $p \in \mathbb{N}_0$, we define

$$S_{h,p} = \{\mathbf{v} \in [L_2(\Omega)]^m : \mathbf{v}|_\kappa \in [\mathcal{Q}_p(\kappa)]^m \quad \forall \kappa \in \mathcal{T}_h\} \,,$$

where $\mathcal{Q}_p(\kappa)$ denotes the set of tensor product polynomials of degree at most p in each coordinate direction over κ. Given that $\mathbf{v} \in \left[H^1(\kappa)\right]^m$ for each $\kappa \in \mathcal{T}_h$, we denote by $\mathbf{v}^+$ (resp., $\mathbf{v}^-$) the interior (resp., exterior) trace of $\mathbf{v}$ on $\partial\kappa$. The DGFEM for (16) is defined as follows: find $\mathbf{u}_h \in S_{h,p}$ such that

$$\begin{aligned}\mathcal{N}(\mathbf{u}_h, \mathbf{v}_h) \equiv \sum_{\kappa\in\mathcal{T}_h} \Big\{ -\int_\kappa \mathcal{F}(\mathbf{u}_h) \cdot \nabla \mathbf{v}_h \, \mathrm{d}\mathbf{x} + \int_{\partial\kappa} \mathcal{H}(\mathbf{u}_h^+, \mathbf{u}_h^-, \mathbf{n}_\kappa)\, \mathbf{v}_h^+ \, \mathrm{d}s \\ + \int_\kappa \varepsilon \nabla \mathbf{u}_h \cdot \nabla \mathbf{v}_h \, \mathrm{d}\mathbf{x} \Big\} = 0\end{aligned} \tag{17}$$

for all $\mathbf{v}_h \in S_{h,p}$, cf. [4, 9], for example. Here, $\mathcal{H}(\cdot,\cdot,\cdot)$ denotes a *numerical flux* function, assumed to be Lipschitz continuous, consistent and conservative. Further, ε is an artificial viscosity defined by $\varepsilon = C_\varepsilon \, h^{2-\beta} \, |\operatorname{div}\mathcal{F}(\mathbf{u}_h)|$, where $C_\varepsilon \geq 0$ and $0 < \beta < 1/2$ are constants. For $\kappa \in \mathcal{T}_h$ whose boundary intersects $\partial\Omega$, we replace $\mathbf{u}_h^-$ by appropriate boundary/initial conditions on $\partial\kappa \cap \partial\Omega$.

In order to apply the theory of duality–based *a posteriori* error estimation, we first need to introduce some notation. We write $\mathcal{M}(\mathbf{u}, \mathbf{u}_h; \cdot, \cdot)$ to denote the mean–value linearisation of the semi–linear form $\mathcal{N}(\cdot,\cdot)$ given by

$$\mathcal{N}(\mathbf{u}, \mathbf{v}) = \mathcal{N}(\mathbf{u}_h, \mathbf{v}) + \mathcal{M}(\mathbf{u}, \mathbf{u}_h; \mathbf{u} - \mathbf{u}_h, \mathbf{v}) \tag{18}$$

for all $\mathbf{v}$ in some suitably chosen space V such that $S_{h,p} \subset V$. With this notation, we note that the Galerkin orthogonality property of the DGFEM (17) may be expressed as follows: $\mathcal{M}(\mathbf{u}, \mathbf{u}_h; \mathbf{u} - \mathbf{u}_h, \mathbf{v}_h) = 0$ for all $\mathbf{v}_h$ in $S_{h,p}$.

Here, we shall be concerned with controlling the error measured in terms of a set of *nonlinear* target functionals $J_i(\cdot)$, $i = 1, \dots, N$. Assuming that $J_i(\cdot)$, $i = 1, \dots, N$, are differentiable, we write, for each i, $1 \leq i \leq N$, $\bar{J}_i(\cdot;\cdot)$ to denote the mean value linearisation of $J_i(\cdot)$ defined by

$$J_i(\mathbf{u}) = J_i(\mathbf{u}_h) + \bar{J}_i(\mathbf{u}, \mathbf{u}_h; \mathbf{u} - \mathbf{u}_h) \,. \tag{19}$$

With this notation, we introduce the dual problem: find $\mathbf{z}_c$ such that

$$\mathcal{M}(\mathbf{u}, \mathbf{u}_h; \mathbf{w}, \mathbf{z}_c) = \bar{J}_c(\mathbf{u}, \mathbf{u}_h; \mathbf{w}) \quad \forall \mathbf{w} \in V\,, \tag{20}$$

where $\bar{J}_c(\mathbf{u}, \mathbf{u}_h; \mathbf{w}) = \sum_{i=1}^{N} \omega_i \bar{J}_i(\mathbf{u}, \mathbf{u}_h; \mathbf{w})$, $\omega_i = s_i/|J_i(\mathbf{u}_h)|$ and $s_i = \mathrm{sgn}(\bar{J}_i(\mathbf{u}, \mathbf{u}_h; \mathbf{u}-\mathbf{u}_h))$, for $i = 1, \ldots, N$. Thereby, assuming the well–posedness of this problem, cf. [4], employing the Galerkin orthogonality property, we deduce the following error representation formula:

$$\begin{aligned} J_c(\mathbf{u}) - J_c(\mathbf{u}_h) &= \sum_{i=1}^{N} \omega_i (J_i(\mathbf{u}) - J_i(\mathbf{u}_h)) = \sum_{i=1}^{N} \omega_i \bar{J}_i(\mathbf{u}, \mathbf{u}_h; \mathbf{u} - \mathbf{u}_h) \\ &= \mathcal{M}(\mathbf{u}, \mathbf{u}_h; \mathbf{u} - \mathbf{u}_h, \mathbf{z}_c) = \mathcal{M}(\mathbf{u}, \mathbf{u}_h; \mathbf{u} - \mathbf{u}_h, \mathbf{z}_c - \mathbf{z}_h) \\ &= -\mathcal{N}(\mathbf{u}_h, \mathbf{z}_c - \mathbf{z}_h) \equiv \mathcal{E}_\Omega(\mathbf{u}, \mathbf{u}_h, h, \mathbf{z}_c - \mathbf{z}_h) = \sum_{\kappa \in \mathcal{T}_h} \eta_\kappa^{(\mathrm{I})} \end{aligned} \tag{21}$$

for all $\mathbf{z}_h$ in the finite element space $S_{h,p}$, where

$$\eta_\kappa^{(\mathrm{I})} = \int_\kappa R_h\,(\mathbf{z}_c - \mathbf{z}_h)\,\mathrm{d}\mathbf{x} + \int_{\partial\kappa} r_h\,(\mathbf{z}_c - \mathbf{z}_h)^+\,\mathrm{d}s - \int_\kappa \varepsilon \nabla \mathbf{u}_h \cdot \nabla(\mathbf{z}_c - \mathbf{z}_h)\,\mathrm{d}\mathbf{x}\,,$$

and $R_h|_\kappa = -\mathrm{div}\mathcal{F}(\mathbf{u}_h)$ and $r_h|_\kappa = \mathcal{F}(\mathbf{u}_h)\cdot \mathbf{n}_\kappa - \mathcal{H}(\mathbf{u}_h^+, \mathbf{u}_h^-, \mathbf{n}_\kappa)$ denote the internal and boundary finite element residuals, respectively, defined on each $\kappa \in \mathcal{T}_h$. Thereby, we deduce the following Type I *a posteriori* error bound.

Theorem 1. *Let* $\mathbf{u}$ *and* $\mathbf{u}_h$ *denote the solutions of (16) and (17), respectively, and suppose that the dual problem (20) is well–posed. Then,*

$$|J_c(\mathbf{u}) - J_c(\mathbf{u}_h)| \le \mathcal{E}_{|\Omega|}(\mathbf{u}, \mathbf{u}_h, h, \mathbf{z}_c - \mathbf{z}_h) \equiv \sum_{\kappa \in \mathcal{T}_h} |\eta_\kappa^{(\mathrm{I})}| \quad \forall \mathbf{z}_h \in S_{h,p}\,. \tag{22}$$

We note that the signs s_i, $i = 1, \ldots, N$, appearing in the definition of $J_c(\cdot)$ and $\bar{J}_c(\mathbf{u}, \mathbf{u}_h; \cdot)$ may be computed by solving the *error equation*:

$$\text{find } \mathbf{e} \text{ such that } \mathcal{M}(\mathbf{u}, \mathbf{u}_h; \mathbf{e}, \mathbf{w}) = -\mathcal{N}(\mathbf{u}_h, \mathbf{w}) \quad \forall \mathbf{w}\,, \tag{23}$$

where $\mathbf{e} = \mathbf{u} - \mathbf{u}_h$, and evaluating the signs of

$$J_i(\mathbf{u}) - J_i(\mathbf{u}_h) = \bar{J}_i(\mathbf{u}, \mathbf{u}_h; \mathbf{e}), \quad i = 1, \ldots, N. \tag{24}$$

As discussed in [4, 5], the local weighting terms involving the difference between the dual solution $\mathbf{z}_c$ and its projection/interpolant $\mathbf{z}_h$ onto $S_{h,p}$ appearing in the Type I bound (22) provide invaluable information concerning the global transport of the error and the interaction of different components of the error. Thereby, we refrain from eliminating the weighting terms involving the (unknown) dual solution $\mathbf{z}_c$ and approximate $\mathbf{z}_c$ numerically. To this end, we perform the linearisations leading to $\mathcal{M}(\mathbf{u}, \mathbf{u}_h; \cdot, \cdot)$ and $\bar{J}_c(\mathbf{u}, \mathbf{u}_h; \cdot)$ about $\mathbf{u}_h$ and approximate the dual solution $\mathbf{z}_c$ in (20) by a discontinuous Galerkin approximation $\hat{\mathbf{z}}_c$ computed on the same mesh used for $\mathbf{u}_h$, but with a higher degree polynomial, i.e. $\hat{\mathbf{z}}_c \in S_{h,\hat{p}_1}$ with $\hat{p}_1 > p$. Similarly, an approximation $\hat{\mathbf{e}} \in S_{h,\hat{p}_2}, \hat{p}_2 > p$, to $\mathbf{e}$ is computed in an analogous way by solving a linearised error equation. In the following section we set $p = 1$ and $\hat{p}_1 = \hat{p}_2 = 2$.

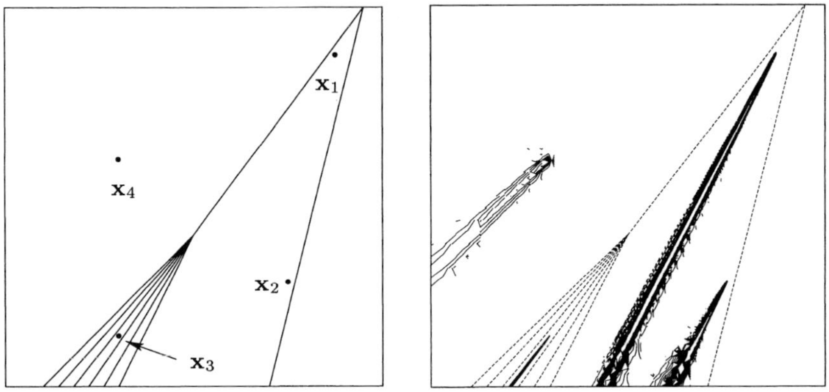

(a) (b)

Fig. 1. Burgers' equation: (a) Isolines of the exact (primal) solution u; (b) Isolines of the dual solution z_c

5 Numerical experiment

In this section we present a numerical example, based on [1], to highlight the key aspects of the proposed approach for the *a posteriori* error estimation of multiple target functionals and the subsequent design of adaptively refined computational meshes specifically tailored to the efficient computation of the selected functionals of interest. In particular, we highlight the following issues:

- We shall show that $\bar{J}_i(\mathbf{u}, \mathbf{u}_h; \hat{\mathbf{e}})$, $i = 1, \ldots, N$, evaluated by approximating (23) provides asymptotically sharp approximations to the true errors $J_i(\mathbf{u}) - J_i(\mathbf{u}_h)$, and thereby lead to the correct prediction of the signs s_i.
- The approximate error representation formula computed by replacing $\mathbf{z}_c$ in (21) by $\hat{\mathbf{z}}_c$, provides an asymptotic estimate of the exact error in $J_c(\cdot)$.
- An adaptive refinement algorithm designed using the *Type I* indicator $|\hat{\eta}_\kappa^{(\mathrm{I})}|$, cf. (22) with $\mathbf{z}_c$ replaced by $\hat{\mathbf{z}}_c$, results in meshes that are efficient for computing all the target functionals under consideration. Moreover, we demonstrate the superiority of this approach over a refinement algorithm based on the *Type II* indicator $\eta_\kappa^{(\mathrm{II})} = \|hR_h\|_{L_2(\kappa)} + \|h^{1/2}r_h\|_{L_2(\partial\kappa)}$, cf. [4].

5.1 Burgers' equation

Here, we consider the initial value problem for the inviscid (scalar) Burgers' equation $\partial_t u + \partial_x(u^2/2) = 0$ for $\mathbf{u} \equiv u$, subject to the initial condition

$$u(x,0) \equiv u_0(x) = \begin{cases} 1, & \text{for } x \le 0.1, \\ -2.5x + 1.25, & \text{for } 0.1 < x \le 0.3, \\ 0.5, & \text{for } 0.3 < x \le 0.7, \\ 0, & \text{for } x > 0.7; \end{cases}$$

Table 1. History of the true errors $J_i(e)$, $i = 1, \ldots, 4$, and their estimates $J_i(\hat{e})$ on locally refined meshes of 256 elements (Mesh 1) to 15202 elements (Mesh 9)

Mesh	$J_1(e)$	$J_1(\hat{e})$	$J_2(e)$	$J_2(\hat{e})$	$J_3(e)$	$J_3(\hat{e})$	$J_4(e)$	$J_4(\hat{e})$
1	-1.45e-01	-4.06e-01	7.32e-02	2.65e-02	-2.03e-02	-2.21e-02	-7.26e-06	-7.32e-06
2	-9.96e-03	-6.89e-02	4.83e-03	9.90e-03	-2.11e-03	-2.63e-03	-4.49e-06	-4.51e-06
3	4.96e-02	4.21e-02	-1.67e-02	-1.02e-02	1.50e-04	1.30e-04	-5.13e-07	-5.11e-07
4	-6.51e-03	-8.46e-03	4.53e-04	5.29e-04	4.75e-04	4.90e-04	-5.14e-07	-5.11e-07
5	-2.95e-04	-3.11e-04	-2.65e-05	-2.66e-05	8.45e-05	8.45e-05	-5.67e-07	-5.65e-07
6	-5.83e-06	-6.01e-06	-1.00e-05	-1.01e-05	6.61e-06	6.62e-06	-1.58e-07	-1.58e-07
7	1.63e-06	1.63e-06	4.60e-07	4.60e-07	-5.90e-07	-5.89e-07	-1.58e-07	-1.58e-07
8	4.43e-08	4.41e-08	1.61e-07	1.61e-07	5.33e-07	5.34e-07	9.99e-10	9.98e-10
9	-1.61e-08	-1.62e-08	-3.77e-08	-3.77e-08	1.26e-07	1.26e-07	9.99e-10	9.98e-10

Table 2. Adaptive algorithm for the combined target functional $J_c(\cdot)$

Mesh	# Ele	$J_c(e)$	$\hat{\mathcal{E}}_\Omega = \sum_\kappa \hat{\eta}_\kappa^{(\mathrm{I})}$	$\hat{\mathcal{E}}_\Omega / J_c(e)$	$\hat{\mathcal{E}}_{\lvert\Omega\rvert} = \sum_\kappa \lvert\hat{\eta}_\kappa^{(\mathrm{I})}\rvert$	$\hat{\mathcal{E}}_{\lvert\Omega\rvert} / J_c(e)$
1	256	4.642e-01	7.204e-01	1.55	9.140e-01	1.97
2	415	3.239e-02	1.586e-01	4.90	2.925e-01	9.03
3	712	1.327e-01	1.135e-01	0.86	1.967e-01	1.48
4	1192	1.456e-02	1.842e-02	1.27	5.975e-02	4.10
5	1981	7.563e-04	7.882e-04	1.04	5.951e-03	7.87
6	3316	4.066e-05	4.113e-05	1.01	5.953e-04	14.64
7	5608	5.121e-06	5.119e-06	1.00	1.646e-04	32.14
8	9175	1.124e-06	1.124e-06	1.00	6.345e-05	56.47
9	15202	2.771e-07	2.774e-07	1.00	2.573e-05	92.87

cf. [2, 3]. Fig. 1(a) shows the structure of the solution u: as time increases the slope of the linear declined part of the solution increases until a shock develops that moves to the right. This shock eventually merges with a second right moving shock that originates from the discontinuity in the initial data.

In this example, we shall be interested in the solution evaluated at the series of points $\mathbf{x}_1 = (0.875, 0.875)$, $\mathbf{x}_2 = (0.75, 0.275)$, $\mathbf{x}_3 = (0.3, 2/15)$ and $\mathbf{x}_4 = (0.3, 0.6)$; cf. Fig. 1(a). Defining the (linear) target functionals by $J_i(u) = u(\mathbf{x}_i)$, $i = 1, \ldots, 4$, we note that their true values are given by $J_1(u) = J_2(u) = 0.5$, $J_3(u) = 0.75$ and $J_4(u) = 1$. The dual solution z_c, cf. (20), is depicted in Fig. 1(b); here, we see that z_c consists of a series of 'spikes' or delta functions originating from each of the points of interest which are transported backwards along the single characteristic passing through each $\mathbf{x}_i$, $i = 1, \ldots, 4$, to the inflow boundary.

In Tables 1 and 2, we demonstrate the performance of the proposed adaptive algorithm with $C_\varepsilon = 1/4$ and $\beta = 1/10$, cf. [4]. From Table 1 we see that the values of $\bar{J}_i(u, u_h; \hat{e}) \equiv J_i(\hat{e})$ (since each $J_i(\cdot)$ is linear), $i = 1, \ldots, 4$, evaluated by approximating the error equation (23) provide asymptotically sharp

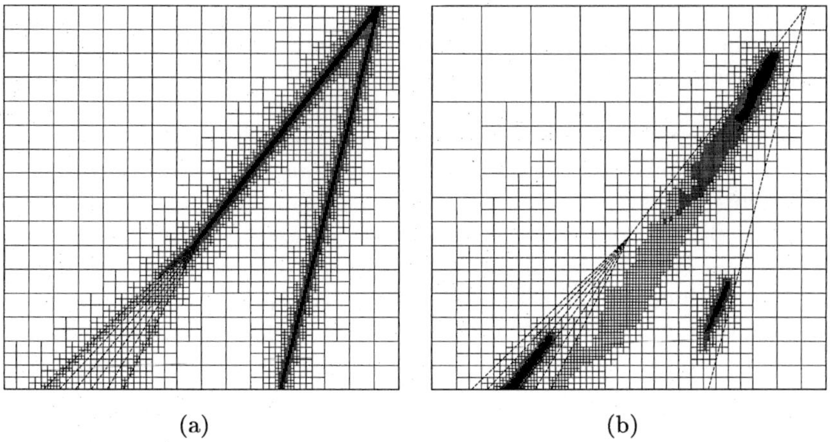

Fig. 2. (a) Mesh constructed using the Type II error indicator $\eta_\kappa^{(\mathrm{II})}$ with 17242 elements ($|J_c(u) - J_c(u_h)| = 2.937 \times 10^{-4}$); (b) Mesh constructed using the Type I indicator $|\hat{\eta}_\kappa^{(\mathrm{I})}|$ with 15202 elements ($|J_c(u) - J_c(u_h)| = 2.771 \times 10^{-7}$)

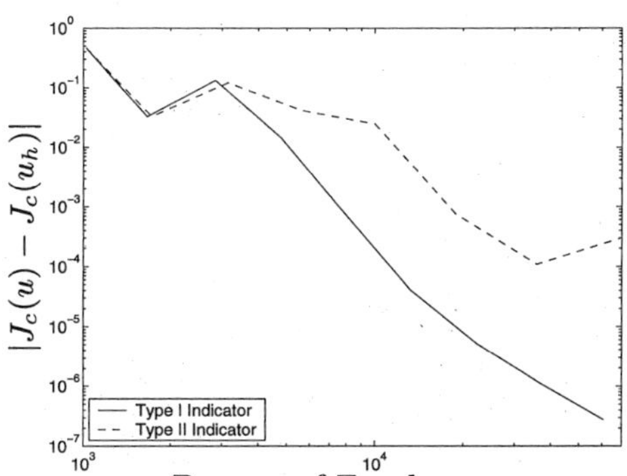

Fig. 3. Convergence of $|J_c(u) - J_c(u_h)|$ using the Type I and Type II indicators

approximations to the exact errors $J_i(e)$, and consequently the correct signs s_i, $i = 1, \ldots, 4$, needed for the computation of the combined target functional $J_c(\cdot)$. Furthermore, in Table 2 we see that the quality of the computed error representation formula $\hat{\mathcal{E}}_\Omega$ is extremely good, in the sense that the effectivity index $\hat{\mathcal{E}}_\Omega/(J_c(u) - J_c(u_h))$ is very close to one, even on relatively coarse meshes. On the other hand, we observe that the Type I *a posteriori* error bound is not sharp, in the sense that the second effectivity index $\hat{\mathcal{E}}_{|\Omega|}/|J_c(u) - J_c(u_h)|$ grows as the mesh is refined.

In Figs. 2(a) and 2(b) we show the meshes produced using the Type II and Type I indicators, respectively. From Fig. 2(a), we see that the mesh produced using $\eta_\kappa^{(\mathrm{II})}$ is concentrated in the region of the computational domain where the shocks are located, as well as in the vicinity of the top and bottom parts of the linear declined part of the solution. In contrast, from Fig. 2(b) we see that there is virtually no refinement in the regions of the computational domain where the shocks are located when $|\hat{\eta}_\kappa^{(\mathrm{I})}|$ is employed. Indeed, most of the refinement takes place upstream of each of the points of interest; this is due to the structure of the dual solution, cf. Fig. 1(b) above. We note that there is much less refinement upstream of the point $\mathbf{x}_4$ than for the other points, since $\mathbf{x}_4$ is located within a region of the computational domain where the primal solution is very smooth; hence, the primal residuals and the resulting local Type I error indicators are very small in this portion of the computational domain. For the same reason only a portion of the region upstream of point $\mathbf{x}_2$ has been refined by the adaptive algorithm.

Finally, in Fig. 3 we compare the true error in the computed functional $J_c(\cdot)$ using the two mesh refinement strategies. Here, we clearly observe the superiority of the Type I *a posteriori* error indicator; on the final mesh the true error in the combined functional is over three orders of magnitude smaller than $|J_c(e)|$ computed on the sequence of meshes produced using $\eta_\kappa^{(\mathrm{II})}$.

References

1. Bangerth, W., Hartmann, R., and Kanschat, G.: `deal.II` *Diff. Equ. Analysis Library, Tech. Rep.* IWR, University of Heidelberg. `http://www.dealii.org`.
2. Hartmann, R. (2001): Adaptive FE Methods for Conservation Equations. In: Freistühler, H., and Warnecke, G. (eds) Hyperbolic Problems: theory, numerics, applications: eighth international conference in Magdeburg, 2000. International series of numerical mathematics **141**:495–503. Birkhäuser, Basel.
3. Hartmann, R. (2002): Adaptive Finite Element Methods for the Compressible Euler Equations, PhD thesis, University of Heidelberg.
4. Hartmann, R., and Houston, P. (2002): Adaptive discontinuous Galerkin finite element methods for nonlinear hyperbolic conservation laws. SIAM J. Sci. Comp. (to appear).
5. Hartmann, R., and Houston, P. (2001): Adaptive discontinuous Galerkin finite element methods for the compressible Euler equations. Preprint 2001-42, (SFB 359), IWR Heidelberg, (submitted for publication).
6. Hartmann, R., and Houston, P. (2002): Goal–oriented a posteriori error estimation for compressible fluid flows. In *Proceedings of ENUMATH 2001*, submitted.
7. Houston, P., Hartmann, R., and Süli, E. (2001): Adaptive discontinuous Galerkin finite element methods for compressible fluid flows. In: Baines, M. (ed) Numerical methods for Fluid Dynamics VII, ICFD, pages 347–353.
8. Houston, P., Senior, B., and Süli, E. (2001): *hp*–Discontinuous Galerkin finite element methods for hyperbolic problems: error analysis and adaptivity. Int. J. Numer. Meth. Fluids (to appear).
9. Jaffre, J., Johnson, C., and Szepessy, A. (1995): Convergence of the discontinuous Galerkin finite element method for hyperbolic conservation laws. Math. Models Methods Appl. Sci., **5**(3):367–386.
10. Larson, M. and Barth, T.J. (2000): In: Cockburn, B., Karniadakis, G.E., Shu C.-W. (eds) Discontinuous Galerkin Finite Element Methods. Lecture Notes in Computational Science and Engineering, **11**:363–368. Springer–Verlag.

The Riemann Problem for a Phase Transition Problem in Thermoelastic Materials

Harumi Hattori

Department of Mathematics, West Virginia University, Morgantown, WV 26506, US *hhattori@wvu.edu*

1 Introduction

This paper is a simplified version of [7]. We consider the structure of the solutions to the Riemann problem for a system of conservation laws related to a phase transition problem. The system is nonisentropic and given by

$$\begin{aligned} v_t - u_x &= 0, \\ u_t - f_x &= 0, \\ \mathcal{E}_t - (uf)_x &= 0, \end{aligned} \tag{1}$$

where v, u, $\mathcal{E}$, and f are strain, velocity, total energy, and stress, respectively. The total energy is given by $\mathcal{E} = e + \frac{1}{2}u^2$, where e is the internal energy. We take strain and entropy s as state variables. Therefore, the stress and internal energy are expressed as $f = f(v,s)$ and $e = e(v,s)$, respectively. We assume that e is a smooth function of v and s, $e_s > 0$, and $e_{vs} \neq 0$. Specifically we assume that $e_{vs} < 0$. The thermodynamic relation is given by

$$de = f dv + \theta ds, \tag{2}$$

where θ is temperature. This implies that $\frac{\partial e}{\partial v} = f$, $\frac{\partial e}{\partial s} = \theta$, and $f_s < 0$. We also assume that there exit constants c for which the level curves $f(v,s) = c$ are not monotone in vs-plane as depicted in Figure 1. The curve $f_v = 0$ is also sketched. For $s < s_c$ the horizontal line $s = \text{const}$ intersects with $f_v = 0$ twice. If the value of v is in the region bounded by $s = s_c$ and $f_v = 0$ on the left and the right of C, v is in the α-phase and β-phase, respectively. In the region $s < s_c$, we assume that $f_{vv} < 0$ in the α-phase and $f_{vv} > 0$ in the β-phase. Normally we use v and θ as the state variables because often the temperature is the control variable for the phase transition problems. Nevertheless, there are a few reasons for choosing the entropy as the state variable. First, this makes the entropy condition easier to apply. Second, the relation between the thermodynamic relation and the Rankine-Hugoniot condition become clear; see Lemma 1 and Remark 1.

The Riemann problem is a special initial value problem in which the initial data are given by

$$U(x,0) = (v,u,s)(x,0) = \begin{cases} U_l = (v_l, u_l, s_l) & x < 0, \\ U_r = (v_r, u_r, s_r) & x > 0, \end{cases} \tag{3}$$

where (v_l, u_l, s_l) and (v_r, u_r, s_r) are two different constant states. As shown above we may use the vector notation and set $U = (v, u, s)$. We assume that the solution is self-similar and consists of constant states separated by the backward wave, the phase boundaries, the contact discontinuity, and the forward wave. The phase boundaries are jump discontinuities across which the phase changes. We discuss the case where the speeds of phase boundaries are much smaller than those of the forward and backward waves.

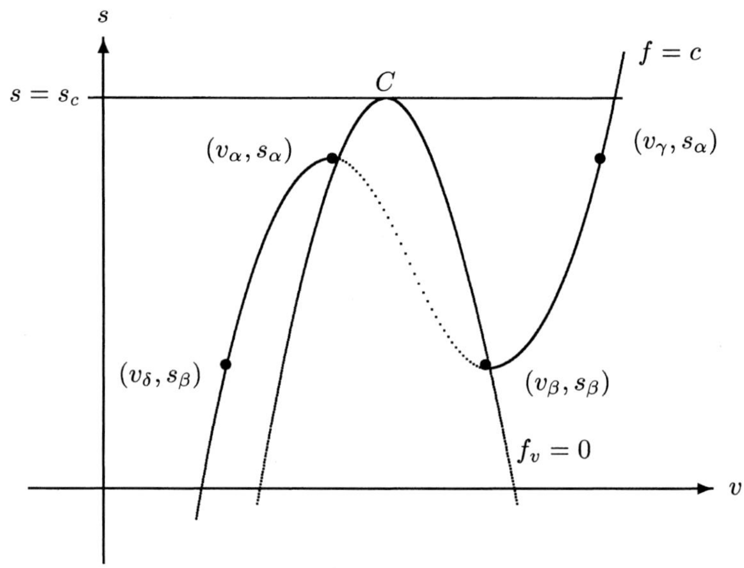

Figure 1

The goal of this paper is to examine the role of admissibility criteria in the context of the Riemann problem when the latent heat is taken into account. Namely, $e_{vs} \neq 0$. The solutions to the Riemann problem for (1) are not unique and the admissibility criteria are used to select a physically relevant solution. One criterion we use is the entropy condition. We use the "negative physical entropy" as the entropy. This criterion imposes that the entropy decreases across jump discontinuities. The rate of decay of the entropy is given by

$$E(v_-, s_-, v_+, s_+) = \sigma(v_-, v_+)(s_+ - s_-),$$

where $\sigma(v_-, v_+) = \pm\sqrt{\frac{f_+ - f_-}{v_+ - v_-}}$ is the speed of the jump discontinuity and the subscripts $-$ and $+$ denote the states to the left and right of discontinuity, respectively. The entropy condition requires that $E(v_-, s_-, v_+, s_+) \leq 0$ holds across each discontinuity. In Section 3, we examine the possible number of phase boundaries consistent with the entropy condition in the solution of the Riemann problem.

It turns out that the entropy condition itself is not enough to single out an admissible solution. Therefore, we need to impose further conditions. One such criterion is the kinetic relation proposed by Abeyaratne and Knowles [1, 2]. This condition postulates that there exists a nondecreasing function $\phi(g)$ of the driving traction g satisfying $\phi(0) = 0$ such that the speed of discontinuity is given by

$$\sigma = \phi(g).$$

They propose that

$$g = -(s_+ - s_-) \quad \text{or} \quad g = -(s_+ - s_-)(\theta_+ + \theta_-)$$

is a possible choice as the driving traction. In order that this relation is consistent with the entropy condition, we require that $\phi' > 0$ so that $\sigma g > 0$ holds. In this paper we choose $g = -(s_+ - s_-)$ for the driving traction. Another criterion is the entropy rate admissibility criterion which is proposed by Dafermos [4, 5]. This criterion roughly says that the rate of entropy decay is the fastest for the admissible solution. The entropy rate admissibility criterion postulates that the solution is admissible if it solves (1) and minimizes

$$\sum_{\text{jump discontinuities}} \sigma(s_+ - s_-). \tag{4}$$

The criterion of this type is applied to crack dynamics [10], where the underlying physics is similar to the phase transition problems. Therefore, the effort should be made to extend it to the conservation laws of mixed type. In Section 4 we study the consequence of the kinetic relation and entropy rate admissibility criteria. They are separately applied in addition to the entropy condition We observe that if v_l and v_r are specified in the same phase, these conditions generally select a unique solution. On the other hand, if v_l and v_r are specified in the different phases, there are initial data for which the kinetic relation may choose three different solutions. This is a sharp contrast to the isothermal case, where the kinetic relation singles out the admissible solutions in both cases [1].

Concerning the thermoelastic case, Hattori [6] considered the Riemann problem of the above system using the entropy rate admissibility criterion. He considered only the single phase boundary problem. Abeyaratne and Knowles [2] discussed the trilinear case where the Helmholtz free energy $\psi(v, \theta)$ satisfies $\psi_{v\theta} = 0$. Note that $\psi_{v\theta}(v, \theta) = e_{vs}(v, s)s_\theta$. Therefore, the condition $\psi_{v\theta} = 0$ is equivalent to $e_{vs} = 0$ provided that $s_\theta \neq 0$. In this case they show that the entropy condition eliminates the possibility of three phase boundary solutions. It is interesting to note that $e_{vs} \neq 0$ causes an essential difference. Asakura [3] studied the stability of Maxwell states. Slemrod [9] discussed the effects of viscosity and capillarity. Ngan and Truskinovsky [8] studied the effect of latent heat on traveling wave solutions in the presence of the viscosity and capillarity terms.

This paper consists of four sections. In Section 2, we summarize the properties of the phase boundary curves. In Section 3, we examine when and how many phase boundaries are consistent with the entropy condition in the solution of the Riemann problems. We discuss two cases where the v_l and v_r are specified in the same phase and in the different phases. Finally in Section 4, we study the consequence of the kinetic relation and entropy rate admissibility criterion.

2 Phase boundary curves

A phase boundary is a line of discontinuity in the xt-plane across which the phase changes. It satisfies the Rankine-Hugoniot condition

$$\sigma(v_+ - v_-) = -(u_+ - u_-), \tag{5}$$

$$\sigma(u_+ - u_-) = -(f_+ - f_-), \tag{6}$$

$$\sigma(\mathcal{E}_+ - \mathcal{E}_-) = -(f_+ u_+ - f_- u_-), \tag{7}$$

where σ is the speed of discontinuity and U_+ and U_- are the states connected by the phase boundary. The phase boundary curve $P(U_o)$ is the set of U connected to U_o by a phase boundary. One important difference between a phase boundary curve and a shock wave curve is that v_o and v are in the different phases. Therefore, unlike shock wave curves, U_o is not on the phase boundary curve $P(U_o)$. Let U_o^* be the state at which the phase boundary curve for U_o emanates. Denote by (v_α, s_α) and (v_β, s_β) the states in the vs-plane at which $f_v = 0$ on a level curve $f = f_o$; see Figure 1.1. In the following we assume that U_o and U_o^* are in the α-phase and β-phase, respectively.

Lemma 1. *If $e_\beta - e_o \leq f_o(v_\beta - v_o)$, then in the vs-plane the phase boundary curve for a given point (v_o, s_o) emanates from the point (v_o^*, s_o^*) with v_o^* in the other phase, satisfying the following two conditions*

$$f(v_o^*, s_o^*) = f_o, \quad e_o^* - e_o = f_o(v_o^* - v_o). \tag{8}$$

Also, $\int_C e_s ds = 0$ holds, where the integral is the path integral along $f = f_o$ from (v_o, s_o) to (v_o^*, s_o^*). If $e_\beta - e_o > f_o(v_\beta - v_o)$, the phase boundary curve emanates from

$$f_v(v_o^*, s_o^*) = 0, \quad e_o^* - e_o = \frac{1}{2}(f_o^* + f_o)(v - v_o), \tag{9}$$

provided that (v, s) satisfying (9) exists.

Proof. From the Rankine-Hugoniot condition, we obtain

$$e - e_o - \frac{1}{2}(f + f_o)(v - v_o) = 0 \tag{10}$$

holds for the phase boundary with nonzero speed. We require that this holds even when $\sigma = 0$, so that (10) is continuous at $\sigma = 0$. If $\sigma = 0$, $f = f_o$ and (10) imply

$$e_o^* - e_o = f_o(v_o^* - v_o). \tag{11}$$

On the other hand, integrating (2) along the level curve $f = f_o$ from (v_o, s_o) to (v, s), we see

$$\begin{aligned} e - e_o &= \int_C f dv + e_s ds \\ &= f_o(v - v_o) + \int_C e_s ds, \end{aligned}$$

where $\int_C$ denotes the above path integral. This shows that if $\int_C e_s ds \leq 0$ at $(v, s) = (v_\beta, s_\beta)$, $e_\beta - e_o \leq f_o(v_\beta - v_o)$ is satisfied, and (11) holds when $\int_C e_s ds = 0$. Since $e_s > 0$, the value of (v_o^*, s_o^*) satisfying (11) is uniquely determined. □

Remark 1. From (8) we have

$$e_o - f_o v_o = e_o^* - f_o^* v_o^*.$$

This shows that the enthalpy $e - fv$ is equal if the two sides of the stationary phase boundary is connected by the phase boundary curve only. Since in general a part of stationary phase boundary is a contact discontinuity, the above relation does not hold for an arbitrary stationary phase boundary.

Lemma 2. *If the level curve $f = f_o$ is not monotone, there exists a unique state (v_o^m, s_o^m) on $f = f_o$ with v_o^m in the α-phase such that the phase boundary emanates from (v_1^m, s_o^m) on $f = f_o$ with v_1^m in the β-phase. Since $e_{vs} < 0$, this also implies that if $s_o > s_o^m$, $s_o < s_o^*$ and if $s_o < s_o^m$, $s_o > s_o^*$.*

Remark 2. This is a consequence of $\int_C e_s ds = 0$ in Lemma 1. This provides an important clue for the theorems in the next section.

3 Riemann problem with entropy condition

First, we state the lemma limiting the number of phase boundaries allowed in the solution of the Riemann problem.

Lemma 3. *Suppose two phase boundaries move in the same direction. Then, the entropy condition is not satisfied across one of the phase boundaries.*

First we consider the case where v_l and v_r are specified in the same phase. Let U_1, U_2, U_3, and U_4 be the middle constant states from left to right with the backward phase boundary, the contact discontinuity, and the forward phase boundary in between. We have a two-parameter family of solutions and the speeds of phase boundaries σ_{p_1} and σ_{p_2} ($\sigma_{p_1} \leq 0 \leq \sigma_{p_2}$) can be chosen as the parameters. We study the case where the hyperbolic Riemann problem with the same initial data is constructed. This solution is the same as the double phase boundary solution with $\sigma_{p_1} = \sigma_{p_2} = 0$.

Theorem 1. *There are four possible solution configurations near $\sigma_{p_1} = \sigma_{p_2} = 0$ depending on the values of s_1 and s_4 at $\sigma_{p_1} = \sigma_{p_2} = 0$.*
(1) If $s_1 \leq s_1^m$ and $s_4 \leq s_4^m$ at $\sigma_{p_1} = \sigma_{p_2} = 0$, the hyperbolic Riemann solution is the only admissible solution.
(2) If $s_1 > s_1^m$ and $s_4 \leq s_4^m$ at $\sigma_{p_1} = \sigma_{p_2} = 0$, the speed of the backward phase boundary can be negative and the forward phase boundary has the zero speed.
(3) If $s_1 \leq s_1^m$ and $s_4 > s_4^m$ at $\sigma_{p_1} = \sigma_{p_2} = 0$, the backward phase boundary has the zero speed and the speed of the backward phase boundary can be negative.
(4) If $s_1 > s_1^m$ and $s_4 > s_4^m$ at $\sigma_{p_1} = \sigma_{p_2} = 0$, the backward phase boundary with the negative speed and the forward wave with the positive speed are possible.

Proof. We choose the speed of phase boundaries σ_{p_1} and σ_{p_2} as the parameters and see how U_i, ($i = 1, \ldots, 4$), change as we change σ_{p_1} and σ_{p_2} from zero. We derive the derivatives of U_i, ($i = 1, \ldots, 4$) with respect to σ_{p_1} and σ_{p_2}. We treat the case where both the backward and forward waves are rarefaction waves. The modifications necessary for the other cases are explained at the end of the proof. Differentiating the backward wave with respect to σ_{p_1} we have

$$\frac{\partial v_1}{\partial \sigma_{p_1}} = \frac{1}{\lambda_1}\frac{\partial u_1}{\partial \sigma_{p_1}}, \quad \frac{\partial s_1}{\partial \sigma_{p_1}} = 0. \tag{12}$$

For the backward phase boundary, from the Rankine-Hugoniot relation

$$f_2 - f_1 = \sigma_{p_1}^2 (v_2 - v_1), \tag{13}$$

$$\sigma_{p_1}(v_2 - v_1) = -(u_2 - u_1), \tag{14}$$

$$\frac{1}{2}(f_2 + f_1)(v_2 - v_1) = e_2 - e_1, \tag{15}$$

we obtain

$$\begin{bmatrix} \frac{\partial v_2}{\partial \sigma_{p_1}} \\ \frac{\partial s_2}{\partial \sigma_{p_1}} \end{bmatrix} = \frac{\begin{bmatrix} f_{2s} & -(f_{2s} - \frac{e_{2s}}{(v_2-v_1)}) \\ -(f_{2v} - \lambda_1 \sigma_{p_1}) & f_{2v} - \sigma_{p_1}^2 \end{bmatrix} \begin{bmatrix} \sigma_{p_1}(v_2 - v_1) \\ (\lambda_1+\sigma_{p_1})(v_2-v_1)+(\lambda_1-\sigma_{p_1})\frac{\partial u_2}{\partial \sigma_{p_1}} \end{bmatrix}}{(f_{2v} - \lambda_1 \sigma_{p_1})\frac{e_{2s}}{(v_2-v_1)} + \sigma_{p_1}(\lambda_1 - \sigma_{p_1}) f_{2s}}.$$

For the contact discontinuity we have

$$f_{3v}\frac{\partial v_3}{\partial \sigma_{p_1}} + f_{3s}\frac{\partial s_3}{\partial \sigma_{p_1}} = f_{2v}\frac{\partial v_2}{\partial \sigma_{p_1}} + f_{2s}\frac{\partial s_2}{\partial \sigma_{p_1}}, \tag{16}$$

$$\frac{\partial u_3}{\partial \sigma_{p_1}} = \frac{\partial u_2}{\partial \sigma_{p_1}}. \tag{17}$$

From the forward wave, we obtain

$$\frac{\partial v_4}{\partial \sigma_{p_1}} = -\frac{1}{\lambda_4}\frac{\partial u_4}{\partial \sigma_{p_1}}, \quad \frac{\partial s_4}{\partial \sigma_{p_1}} = 0. \tag{18}$$

For the forward phase boundary, from the Rankine-Hugoniot relation

$$f_4 - f_3 = \sigma_{p_2}^2(v_4 - v_3), \tag{19}$$

$$\sigma_{p_2}(v_4 - v_3) = -(u_4 - u_3), \tag{20}$$

$$\frac{1}{2}(f_4 + f_3)(v_4 - v_3) = e_4 - e_3, \tag{21}$$

we see

$$\begin{bmatrix} \frac{\partial v_3}{\partial \sigma_{p_1}} \\ \frac{\partial s_3}{\partial \sigma_{p_1}} \end{bmatrix} = \frac{\begin{bmatrix} f_{3s} & -(f_{3s} - \frac{e_{3s}}{(v_3-v_4)}) \\ -(f_{3v} + \lambda_4 \sigma_{p_2}) & f_{3v} - \sigma_{p_2}^2 \end{bmatrix} \begin{bmatrix} 0 \\ -(\lambda_4 + \sigma_{p_2})\frac{\partial u_3}{\partial \sigma_{p_1}} \end{bmatrix}}{(f_{3v} + \lambda_4 \sigma_{p_2})\frac{e_{3s}}{(v_3-v_4)} - \sigma_{p_2}(\lambda_4 + \sigma_{p_2}) f_{3s}}. \tag{22}$$

From this we have

$$\frac{\partial s_2}{\partial \sigma_{p_1}} = \frac{\lambda_1 \lambda_4 (v_2 - v_1)^2}{e_{2s}(\lambda_4 + \lambda_1)} + O(\sigma_{p_1}, \sigma_{p_2}),$$

$$\frac{\partial s_3}{\partial \sigma_{p_1}} = \frac{\lambda_1 \lambda_4 (v_2 - v_1)(v_3 - v_4)}{e_{3s}(\lambda_4 + \lambda_1)} + O(\sigma_{p_1}, \sigma_{p_2}).$$

This shows that s_2 and s_3 are increasing functions of σ_{p_1} near $\sigma_{p_1} = \sigma_{p_2} = 0$.

Similarly for the derivatives with respect to σ_{p_2}, we obtain

$$\frac{\partial s_2}{\partial \sigma_{p_2}} = -\frac{\lambda_1 \lambda_4 (v_2 - v_1)(v_3 - v_4)}{e_{2s}(\lambda_4 + \lambda_1)} + O(\sigma_{p_1}, \sigma_{p_2}),$$

$$\frac{\partial s_3}{\partial \sigma_{p_2}} = -\frac{\lambda_1 \lambda_4 (v_3 - v_4)^2}{e_{3s}(\lambda_4 + \lambda_1)} + O(\sigma_{p_1}, \sigma_{p_2}).$$

This result shows that s_2 and s_3 are decreasing functions of σ_{p_2} near $\sigma_{p_1} = \sigma_{p_2} = 0$.

The above results imply that the directional derivatives of s_2 and s_3 are negative as we decrease σ_{p_1} and increase σ_{p_2} from zero. This shows the following. If $s_1 \leq s_1^m$ and $s_4 \leq s_4^m$ at $\sigma_{p_1} = \sigma_{p_2} = 0$, then near $\sigma_{p_1} = \sigma_{p_2} = 0$ the hyperbolic Riemann solution is the only admissible solution. If $s_1 \leq s_1^m$ and $s_4 > s_4^m$ at $\sigma_{p_1} = \sigma_{p_2} = 0$, then near $\sigma_{p_1} = \sigma_{p_2} = 0$ the backward phase boundary has the speed zero. If $s_1 > s_1^m$ and $s_4 \leq s_4^m$ at $\sigma_{p_1} = \sigma_{p_2} = 0$, then near $\sigma_{p_1} = \sigma_{p_2} = 0$ the forward phase boundary has the speed zero. If $s_1 > s_1^m$ and $s_4 > s_4^m$ at $\sigma_{p_1} = \sigma_{p_2} = 0$, then the backward phase boundary with negative speed and the forward phase boundary with positive speed are both possible.

The modification necessary for the cases where the backward and/or forward waves are shocks are as follows. If the backward wave is a shock, we need to add $O(|v_1 - v_l|)$ terms to each derivative. Therefore, if the shocks are weak, the result is unaffected. □

Next, we study the case where v_l and v_r are specified in the different phases. We consider the case where the solution with the stationary phase boundary constructed. Let U_1 and U_4 be the middle constant states on the right of the backward and on the left of the forward waves, respectively. If there is only one phase boundary, we have one parameter family of solutions and we take its speed σ_p as a parameter.

Theorem 2. *If v_l and v_r are specified in the different phases, there are four different solution configurations near $\sigma_p = 0$ depending on the values of s_1 and s_3 at $\sigma_p = 0$.*
(1) If $s_1 \leq s_1^m$ and $s_4 \geq s_4^m$ at $\sigma_p = 0$, then the solution with the stationary phase boundary is the only one solution satisfying the entropy condition.
(2) If $s_1 > s_1^m$ and $s_4 \geq s_4^m$ at $\sigma_p = 0$, there is a one-parameter family of solutions with the backward phase boundary.
(3) If $s_1 \leq s_1^m$ and $s_4 < s_4^m$ at $\sigma_p = 0$, there is a one-parameter family of solutions with the forward phase boundary.
(4) If $s_1 > s_1^m$ and $s_4 < s_4^m$ at $\sigma_p = 0$, there is a one-parameter family of solutions with the backward phase boundary or the forward phase boundary. Also, it is possible to have the solution with three phase boundaries where the left phase boundary moves backward, the middle one is stationary, and the right phase boundary moves forward. In this case we have a two-parameter family of solutions.

Remark 3. Since the solution is continuous with respect to the initial data, the result (1) implies that if there exists neighborhood of U_l and U_r, where the stationary phase boundary is the only one satisfying the entropy condition.

4 The other admissibility criteria

The entropy condition itself is not enough to select a unique solution. In this section we examine the consequence of two admissibility criteria, the kinetic relation and the entropy rate admissibility criterion. We start from the kinetic relation.

The kinetic relation select a unique solution if v_l and v_r are specified in the same phase. On the other hand the kinetic relation may not select a unique solution if v_l and v_r are specified in the different phases. If $s_1 > s_1^m$ and $s_4 < s_4^m$, there are four solution configurations, the solution with a backward phase boundary, with a forward phase boundary, and with three phase boundaries.

Theorem 3. *If v_l and v_r are specified in the different phases and both $s_1 > s_1^m$ and $s_4 < s_4^m$ hold at $\sigma_p = 0$ (or $\sigma_{p_1} = \sigma_{p_2} = 0$), then there exists a positive ε such that, for every ϕ with $0 \leq \phi' \leq \varepsilon$, all three types solutions are admissible according to the kinetic relation of the form*

$$\sigma = \phi(s_- - s_+).$$

Concerning the entropy rate admissibility criterion, an interesting question in the Riemann problem is whether the criterion should be applied to each phase boundary separately or to all jump discontinuities. The reason for this distinction is that they produce different solutions. The former means that the entropy rate admissibility criterion is applied to each phase boundary so that each phase boundary is stable in the sense that if the constant states adjacent to the phase boundary are taken as the initial data for the Riemann problem, the solution consists of these initial data and the phase boundary with the same speed. In other words, no new waves are generated for this Riemann problem. This is justified if there is only one solution configuration. On the other hand, if there are more than one solution configuration, one way is to apply the criterion to all jump discontinuities for each configuration and choose the configuration minimizing the entropy rate. Then, it is applied to each phase boundary separately in the chosen configuration.

References

1. R. Abeyaratne and J. K. Knowles, Kinetic relations and the propagation of phase boundaries in solids, Arch. Rat. Mech. Anal. 114 (1991), 119-154.
2. R. Abeyaratne and J. K. Knowles, Dynamics of propagating phase boundaries: adiabatic theory for thermoelastic solids, Physica D 79 (1994) 269-288.
3. F. Asakura, Stability of Maxwell states in thermo-elasticity, preprint.
4. C.M. Dafermos, The entropy rate admissibility criterion for solutions of hyperbolic conservation laws, Jour. Diff. Eq. 14 (1973), 202-212.
5. C.M. Dafermos, The entropy rate admissibility criterion in thermoelasticity, Atti Accad. Naz. Lincei Rend. cl. Sci. Fis. Mat. Natur. 8 (1974), 113-119.

6. H. Hattori, The Riemann problem for a van der Waals fluid with entropy rate admissibility criterion: Non-isothermal case, J. Diff. Eqns. 65 (1986), 158-174
7. H. Hattori, The Riemann problem for thermoelastic materials with phase change, submitted for publication.
8. S.C. Ngan and L. Truskinovsky, Thermal trapping and kinetics of martensitic phase boundaries, Mech. Phys. Solids 47 (1999), 141-171.
9. M. Slemrod, Dynamic phase transitions in a van del Waals fluid, J. Diff. Eqns. 52 (1984), 1-23.
10. L.I. Slepyan, Principle of maximum energy dissipation rate in crack dynamics, J. Mech. Phys. Solids 41 (1993), pp.1019-1033.

The Global Existence and Large Time Behavior of Solutions to the Multidimensional Euler-Poisson Equations

Ling Hsiao*, Qiangchang Ju and Shu Wang**

Academy of Mathematics and Systems Sciences, CAS, Beijing 100080, P.R. China, hsiaol@math03.math.ac.cn, juqc@math08.math.ac.cn, shu.wang@univie.ac.at

Summary. The global existence of smooth solutions to the Cauchy problem for the multi-dimensional Euler-Poisson equations is established, when the initial data are perturbations of a stationary solutions. Moreover, it is proved that the solution decays asymptotically in time to the stationary solution exponentially fast.

1 Introduction

Our purpose is to investigate the large time behavior of solutions to the multi-dimensional Euler-Poisson equations

$$\begin{cases} n_t + \nabla \cdot (n\mathbf{u}) = 0, \\ (n\mathbf{u})_t + \nabla \cdot (n\mathbf{u} \bigotimes \mathbf{u}) + \nabla p(n) = n\nabla \Phi - \frac{n\mathbf{u}}{\tau}, \\ \Delta \Phi = n - b(x), \end{cases} \tag{1.1}$$

for $(x,t) \in R^N \times [0, +\infty), N = 2, 3$, with the initial data

$$n(x, 0) = n_0(x), \quad \mathbf{u}(x, 0) = \mathbf{u}_0(x). \tag{1.2}$$

The system (1.1), from the application point, can be used to describe the hydrodynamic model for semiconductors in the isentropic case, where $n > 0, \mathbf{u}, \Phi$ denote the electron density, the particle velocity vector and the electrostatic potential, respectively. $\tau > 0$ is the velocity relaxation time; the function $b(x)$ is the doping profile which describes the density of fixed, positively charged background ions, and the pressure-density function $p = p(n)$ has the property that $n^2p'(n)$ is strictly increasing function from $[0, +\infty)$ onto $[0, +\infty)$. A usual hypothesis is

$$p(n) = a^2 n^\gamma, \quad n > 0, \ a \neq 0, \ \gamma \geq 1. \tag{1.3}$$

* Supported by the Special Funds of State Major Basic Research Projects (Grant No. 1999075107) and the Innovation Funds of AMSS, CAS of China

** Supported by the National Youth Natural Science Fundation (Grant No. 10001034) and Postdoctoral Science Fundation of China

This model equation has been extensively studied for the Cauchy problem or the initial boundary value problem in one space dimension. In the stationary case, Degond -Markowich [1] discussed the existence and uniqueness of the steady state solution in the subsonic case. In the dynamic case, Zhang [2] and Marcati-Natalini [3] investigated the global existence of weak solutions of the initial-boundary value problem and the Cauchy problem respectively by using the tools of compensated compactness. Furthermore, Marcati-Natalini [3] and Hsiao-Zhang [4] studied the relaxation limit of the hydrodynamic model to the drift-diffusion equations for the Cauchy problem and initial-boundary value problem, respectively. As far as smooth solutions are concerned , Luo, Natalini and Xin [5], Hsiao and Yang [6] and Li, Markowich, and Mei [7] investigated the asymptotic behavior of solutions to the Cauchy problem and initial-boundary value problem of (1.1), respectively.

However, there are few results on the multi-dimensional Euler-Poisson (1.1) which is more significant and more interesting. The steady state in the subsonic case has been studied in [10]. To provide some insights into the multi-dimensional evolutionary problem, the global existence and asymptotic behavior of the spherically symmetrical solution of (1.1) were established in [11] with $\gamma = 1$ and $\Omega = \{x \in R^d : 0 < R_1 \leq |x| \leq R_2 < \infty\}$ and in [12] with $\gamma \geq 1$ and $\Omega = \{X \in R^d : |x| \geq R_1 > 0\}$ respectively, where the pressure-density function $p(n)$ is taken as (1.3). Recently, the solution without any symmetry is studied in [13], where the global existence and asymptotic behavior of smooth solutions to the Cauchy problem (1.1) are proved under the assumption (1.3) on $p(n)$ and the assumption $b(x) = constant$. Without the restriction of $b(x) = constant$, the corresponding result is established only for the case of $N = 2$ by Hsiao, Wang, and Zhao in [14]. However, the techniques used in [14] can't work for $N = 3$.

In the present paper, we will introduce the new progress made in [15] which establishes the global existence and asymptotic behavior of smooth solution to the Cauchy problem for both cases of $N = 2$ and $N = 3$, , with $b(x)$ satisfying the following condition

$$\lim_{|x|\to+\infty} b(x) = \bar{b} > 0, \tag{1.4}$$

$$b(x) > 0, \quad b(x) \in C^4(R^N), \text{ and } \nabla b(x) \in H^3(R^N). \tag{1.5}$$

For simplicity, we assume that $\tau > 0$ is a constant. Furthermore, we take $\tau = 1$ for convenience.

The stationary solution $(\mathcal{N}, \mathcal{U}, \mathcal{E})$ of (1.1)-(1.3) with $\mathcal{U} \equiv 0$ will be used. That is, we look for the solutions of the system

$$\begin{cases} \mathcal{N}\mathcal{E} = \nabla p(\mathcal{N}), \\ \operatorname{div}\mathcal{E} = \mathcal{N} - b(x), \end{cases} \tag{1.6}$$

under the condition

$$\mathcal{N} > 0, \ \ \mathcal{N} - b(x) \in H^4(R^N), \tag{1.7}$$

for which we have the following existence and uniqueness result.

Theorem 1. *If $b(x)$ satisfies the conditions (1.4)-(1.5), (1.6)-(1.7) has a unique classical solution $(\mathcal{N}, \mathcal{E})$.*

Remark. If $||\nabla b||_{H^2}$ is small enough, then it is shown in [15]

$$||\nabla \mathcal{N}||_{H^3} \leq O(1)||\nabla b||_{H^3}. \tag{1.8}$$

It can proved that this stationary solution $(\mathcal{N}, \mathcal{U}, \mathcal{E})$ with $\mathcal{U} \equiv 0$ is asymptotically stable. This is

Theorem 2. *Assume $b(x)$ satisfies the condition (1.4) -(1.5). Assume also that $n(\cdot, 0) - \mathcal{N} \in H^3(R^N) \cap L^1(R^N), \mathbf{u}(\cdot, 0) \in H^3(R^N)$. Then there exists a positive number δ, depending only on $b(x)$, such that if*

$$||n(\cdot,0) - \mathcal{N}, \mathbf{u}(\cdot,0), \nabla\Phi(\cdot,0) - \mathcal{E}||_{H^3} + ||(n_t, \mathbf{u}_t, \nabla\Phi_t)(\cdot,0)||_{H^2} + ||\nabla b||_{H^3} \leq \delta,$$

then the Cauchy problem (1.1)-(1.3) admits a unique global smooth solution $(n(x,t), \mathbf{u}(x,t), \ \Phi(x,t))$ for all $t \geq 0$. Moreover

$$\begin{aligned} &||n(\cdot,t) - \mathcal{N}, \mathbf{u}(\cdot,t), \nabla\Phi(\cdot,t) - \mathcal{E}||_{H^3} + ||(n_t, \mathbf{u}_t, \nabla\Phi_t)(\cdot,t)||_{H^2} \\ &\leq C_1\Big[||n(\cdot,0) - \mathcal{N}, \mathbf{u}(\cdot,0), \nabla\Phi(\cdot,0) - \mathcal{E}||_{H^3} + ||(n_t, \mathbf{u}_t, \nabla\Phi_t)(\cdot,0)||_{H^2}\Big] \exp(-\alpha_1 t) \end{aligned}$$

for some positive constants α_1 and C_1.

Remark. From the assumptions on initial data and the system (3.3), we can prove $\nabla\Phi(\cdot, 0) - \mathcal{E} \in H^3(R^N)$ by using Fourier Analysis.

2 Proof of Main Results

To prove Theorem 1.1, the crucial step is to establish the following a priori estimates.

Lemma 1. *[15] Suppose $b(x)$ satisfies the conditions (1.4) -(1.5). Let $(\mathcal{N}, \mathcal{E})$ be the solutions of (1.6)-(1.7) given by Theorem 1.1. Then*

$$\inf_{x \in R^N} b(x) \leq \mathcal{N}(x) \leq \sup_{x \in R^N} b(x), \tag{2.1}$$

and

$$||\nabla \mathcal{N}||_{H^3} \leq M, \tag{2.2}$$

where M depends on $||Db(x)||_{H^3}$.

With the help of Lemma 2.1, the standard iteration technique and the fixed point principle give the existence result for (1.6)-(1.7). The uniqueness of solution can be proved easily.

Next, we prove Theorem 1.2 by using the energy method. Let $(\mathcal{N}, \mathcal{E})$ be the solution to (1.6)-(1.7) given by Theorem 1.1. We write a solution to (1.1)-(1.2) as

$$n(x,t) = \mathcal{N} + v(x.t), \tag{2.3}$$

$$\nabla \Phi = \mathcal{E} + \mathbf{e}. \tag{2.4}$$

Then the function $(v, \mathbf{u}, \mathbf{e})$ satisfies the following system

$$\begin{cases} v_t + \operatorname{div}((\mathcal{N} + v)\mathbf{u}) = 0, \\ \mathbf{u}_t + (\mathbf{u} \cdot \nabla)\mathbf{u} + \nabla(h(\mathcal{N} + v) - h(\mathcal{N})) = \mathbf{e} - \mathbf{u}, \\ \operatorname{div}\mathbf{e} = v(x,t), \end{cases} \tag{2.5}$$

with the initial value

$$\begin{aligned} &v(x,0) = n_0(x) - \mathcal{N}, \qquad \mathbf{u}(x,0) = \mathbf{u}_0(x), \\ &\Big(\mathbf{e}(x,0) = \int_{R^N} \nabla_x G(x-y)(n_0(y) - \mathcal{N}(y))dy\Big), \end{aligned} \tag{2.6}$$

where $G(x,y)$ is a solution to $\Delta_x G(x,y) = \delta(x-y)$ with $x, y \in R^N$.

To prove Theorem 1.2, we need the existence of the local solution and the a priori estimates. The former is given by Lemma 2.2, and the latter is proved in Lemma 2.3.

Lemma 2. *Assume that $b(x)$ satisfies the condition (1.4)-(1.5) and $n(\cdot,0) - \mathcal{N} \in H^3(R^N) \cap L^1(R^N), \mathbf{u}(\cdot,0) \in H^3(R^N)$. Then there exists a unique smooth solution $(n(x,t), \mathbf{u}(x,t), \nabla\Phi(x,t))$ of (1.1)-(1.3) satisfying*

$$n(x,t), \mathbf{u}(x,t), \nabla\Phi(x,t) \in C^1(R^N \times [0, T_{max}))$$

and

$$n(x,t) - \mathcal{N}, \mathbf{u}(x,t), \nabla\Phi(x,t) - \mathcal{E} \in L^\infty(0,T; H^3(R^N))$$

defined on a maximal interval of existence $[0, T_{max})$. Moreover, if $T_{max} < \infty$, then

$$\begin{aligned} &\|(n(\cdot,t) - \mathcal{N}, \mathbf{u}(\cdot,t), \nabla\Phi(\cdot,t) - \mathcal{E})\|^2_{H^3(R^N)} + \|(n_t, \mathbf{u}_t, e_t)(\cdot,t)\|^2_{H^2(R^N)} \\ &+ \textstyle\int_0^t (\|(n(\cdot,t) - \mathcal{N}, \mathbf{u}(\cdot,t), e(\cdot,t))\|^2_{H^3(R^N)} + \|(n_t, \mathbf{u}_t, e_t)(\cdot,t)\|^2_{H^2(R^N)})dt \to \infty, \end{aligned}$$

as $t \to T_{max}$.

Proof. This local existence theorem can be established by a standard contraction mapping argument. For completeness, we give a sketch of the proof.

First, we deduce from $(2.5)_3$ and $(2.5)_1$ that

$$\nabla\Phi = \nabla\Phi(x,0) - \nabla\Delta^{-1}\mathrm{div}\int_0^t((\mathcal{N}+v)\mathbf{u})ds.$$

Thus, (2.5) can be rewritten in the following form with the nonlocal term.

$$\begin{cases} v_t + \mathrm{div}((\mathcal{N}+v)\mathbf{u}) = 0, \\ \mathbf{u}_t + \mathbf{u}\cdot\nabla\mathbf{u} + \nabla(h(\mathcal{N}+v) - h(\mathcal{N})) = \nabla\Phi(x,0) \\ \qquad -\nabla\Delta^{-1}\mathrm{div}\int_0^t((\mathcal{N}+v)\mathbf{u})ds - \mathcal{E} - \mathbf{u}, \\ v(x,0) = n_0(x) - \mathcal{N} \in H^3(R^N), \mathbf{u}(x,0) = \mathbf{u}_0(x) \in H^3(R^N). \end{cases}$$

Notice that the non-local term $\nabla\Delta^{-1}\mathrm{div}\int_0^t((\mathcal{N}+v)\mathbf{u})$ is a sum of the products of Riesz transform of $\int_0^t(\mathcal{N}+v)\mathbf{u}ds$. By the L^p boundedness of the Riesz transformation, see [16], we have

$$\|\nabla\Delta^{-1}\mathrm{div}\int_0^t((\mathcal{N}+v)\mathbf{u})ds\|_{H^3} \le C\|\int_0^t(\mathcal{N}+v)\mathbf{u}ds\|_{H^3}.$$

Then the standard contraction mapping gives the desired results, for details, see [8].

Lemma 3. *There exist positive constants δ_1, α and C_1 such that, for any $T > 0$, if*

$$\sup_{0\le t\le T}(\|(v,\mathbf{u},\mathbf{e})(\cdot,t)\|_{H^3} + \|(v_t,\mathbf{u}_t,\mathbf{e}_t)(\cdot,t)\|_{H^2}) + \|\nabla b\|_{H^3} \le \delta_1, \tag{2.7}$$

then

$$\begin{aligned} &\|(v,\mathbf{u},\mathbf{e})(\cdot,t)\|^2_{H^3} + \|(v_t,\mathbf{u}_t,\mathbf{e}_t)(\cdot,t)\|^2_{H^2} \\ &\le C_1(\|(v,\mathbf{u},\mathbf{e})(\cdot,0)\|^2_{H^3} + \|(v_t,\mathbf{u}_t,\mathbf{e}_t)(\cdot,0)\|^2_{H^2})\exp(-\alpha t), \end{aligned} \tag{2.8}$$

for any $t \in (0,T]$.

Proof. First, from (1.8), the assumption (2.7) and the Sobolev's inequality, we have

$$\begin{aligned} &\sup_{x\in R^N}|(v,\partial_x v, v_t, \mathbf{u}, \partial_x u, \mathbf{u}_t, \mathbf{e}, \partial_x\mathbf{e}, \mathbf{e}_t, \partial_x\mathcal{N}, \partial_x^2\mathcal{N})| \\ &\le C(\|(v,\mathbf{u},\mathbf{e})(\cdot,t)\|_{H^3} + \|(v_t,\mathbf{u}_t,\mathbf{e}_t)(\cdot,t)\|_{H^2}) + \|\nabla b\|_{H^3}) \le C\delta_1. \end{aligned} \tag{2.9}$$

(2.8) can be obtained by three steps as followed. The detailed discussion can be found in [15].

Step 1. Under the assumption (2.7), we can get the following estimates on v and $\mathbf{e}$,

$$||\partial_x^i v_t|| \le C\sum_{k=0}^{i+1}(||\partial_x^k \mathbf{u}|| + ||\partial_x^k v||), \quad i = 0, 1, 2. \tag{2.10}$$

$$||\partial_x^j \mathbf{e}||^2 \le C||\partial_x^{j-1} v||^2, \quad j = 1, 2, 3. \tag{2.11}$$

$$||\mathbf{e}_t|| \le C(||\mathbf{u}|| + ||v||), \tag{2.12}$$

$$||\partial_x^m \mathbf{e}_t|| \le C\sum_{k=0}^{m}(||\partial_x^k \mathbf{u}|| + ||\partial_x^k v||), \quad m = 1, 2. \tag{2.13}$$

Step 2. Under the assumption (2.7), we have

$$\begin{aligned}
&||(v, \mathbf{u}, \mathbf{e})(\cdot, t)||_{H^1}^2 + ||(v_t, \mathbf{u}_t, \mathbf{e}_t)(\cdot, t)||_{L^2}^2 \\
&\le C_1(||(v, \mathbf{u}, \mathbf{e})(\cdot, 0)||_{H^1}^2 + ||(v_t, \mathbf{u}_t, \mathbf{e}_t)(\cdot, 0)||_{L^2}^2)\exp(-\alpha_1 t),
\end{aligned}$$

for any $t \in (0, T]$.

We give a sketch to establish the above estimate next. We multiply $(2.5)_2$ by $\mathcal{N}\mathbf{u}$, and integrate it over R^N. We have, after integration by parts, that

$$\begin{aligned}
&\frac{d}{dt}\int_{R^N}\frac{\mathcal{N}}{2}|\mathbf{u}|^2 + \int_{R^N}\mathcal{N}|\mathbf{u}|^2 - \int_{R^N}(h(\mathcal{N}+v) - h(\mathcal{N}))\mathrm{div}(\mathcal{N}\mathbf{u}) \\
&+ \int_{R^N}\mathcal{N}((\mathbf{u}\cdot\nabla)\mathbf{u})\mathbf{u} - \int_{R^N}\mathcal{N}\mathbf{e}\mathbf{u} = 0.
\end{aligned} \tag{2.14}$$

First, we assume that δ_1 is chosen so small that

$$\frac{\mathcal{N}}{2} \le \mathcal{N} + v \le 2\mathcal{N}. \tag{2.15}$$

Then, it holds from (1.3) that

$$\begin{aligned}
&0 < D_1 \le h'(\mathcal{N}+v) \le D_2 < \infty, \\
&|h^{(k)}(\mathcal{N}+v)| \le D_3 < \infty
\end{aligned} \tag{2.16}$$

for some positive constants D_1, D_2, D_3, where k is an arbitrary positive integer.

Thus, in view of $(2.5)_1$ we get

$$\begin{aligned}
&-\int_{R^N}(h(\mathcal{N}+v) - h(\mathcal{N}))\mathrm{div}(\mathcal{N}\mathbf{u}) \\
&= \frac{d}{dt}\int_{R^N}\int_0^v(h(\mathcal{N}+s) - h(\mathcal{N}))dsdx + \int_{R^N}h'(\mathcal{N}+v\theta)v\mathrm{div}(\mathbf{u}v) \\
&\ge \frac{d}{dt}\int_{R^N}\int_0^v(h(\mathcal{N}+s) - h(\mathcal{N}+v) - h(\mathcal{N})(\mathcal{N}))dsdx \\
&\quad -C\delta_1\int_{R^N}(|u|^2 + |\nabla v|^2 + |v|^2),
\end{aligned} \tag{2.17}$$

for some positive constants $\theta : 0 < \theta < 1$, with the help of the smallness of $|v|$ and $|\nabla u|$.

By (2.4), $(2.5)_1$, and (2.9) we have

$$\begin{aligned} &-\int_{R^N} \mathcal{N}\mathbf{e}\mathbf{u} \\ &= \int_{R^N} \mathbf{e}\mathbf{e}_t + \int_{R^N} v\mathbf{e}\mathbf{u} + \int_{R^N} \mathcal{E}(\mathbf{e}_t + (\mathcal{N}+v)\mathbf{u}) \\ &\geq \frac{1}{2}\frac{d}{dt}\int_{R^N} |\mathbf{e}|^2 - C\delta_1 \int_{R^N} (|\mathbf{u}|^2 + |v|^2), \end{aligned} \tag{2.18}$$

where we have used the fact that

$$\int_{R^N} \partial_x^i \mathcal{E} \partial_x^i (\mathbf{e}_t + (\mathcal{N}+v)\mathbf{u}) = 0. \quad i = 0,1,2,3. \tag{2.19}$$

In fact, from $(2.5)_1$ and $(2.5)_3$, we can get

$$\partial_x^i \mathrm{div}(\mathbf{e}_t + (\mathcal{N}+v)\mathbf{u}) = 0.$$

Multiplying the above equality by $\partial_x^i h(\mathcal{N})$ and using the relation $\partial_x^i \nabla h(\mathcal{N}) = \partial_x^i (\frac{\nabla p(\mathcal{N})}{\mathcal{N}}) = \partial_x^i \mathcal{E}$, we can immediately get (2.19) by integration by parts.

Due to (2.9), it is easy to get

$$\int_{R^N} \mathcal{N}(\mathbf{u}\cdot\nabla\mathbf{u})\mathbf{u}dx \geq -C\delta_1 \int_{R^N} (|\mathbf{u}|^2 + |\nabla\mathbf{u}|^2). \tag{2.20}$$

Therefore, (2.14) together with (2.17), (2.18) and (2.20) implies

$$\begin{aligned} &\frac{d}{dt}\int_{R^N} \left[\frac{\mathcal{N}}{2}|\mathbf{u}|^2 + \int_0^v (h(\mathcal{N}+s) - h(\mathcal{N}))ds + \frac{1}{2}|\mathbf{e}|^2\right]dx + C\int_{R^N} |\mathbf{u}|^2 \\ &\leq C\delta_1 \int_{R^N} (|\nabla v|^2 + |v|^2 + |\mathbf{u}|^2). \end{aligned} \tag{2.21}$$

Differentiating $(2.5)_2$ with respect to t and then multiplying the resulting equation by $\mathcal{N}\partial_t \mathbf{u}$, we have

$$\begin{aligned} &\frac{1}{2}\frac{d}{dt}\int_{R^N} \mathcal{N}|\mathbf{u}_t|^2 + \int_{R^N} \mathcal{N}|\mathbf{u}_t|^2 + \int_{R^N} \partial_t \nabla(h(\mathcal{N}+v) - h(\mathcal{N}))\partial_t(\mathcal{N}\mathbf{u}) \\ &\quad - \int_{R^N} \mathcal{N}\mathbf{u}_t\mathbf{e}_t + \int_{R^N} \mathcal{N}\partial_t(\mathbf{u}\cdot\nabla\mathbf{u})\mathbf{u}_t = 0. \end{aligned} \tag{2.22}$$

First, integration by parts gives

$$\begin{aligned} &\int_{R^N} \partial_t \nabla(h(\mathcal{N}+v) - h(\mathcal{N}))\partial_t(\mathcal{N}\mathbf{u}) \\ &= -\int_{R^N} \mathcal{N}\partial_t(h(\mathcal{N}+v) - h(\mathcal{N}))\partial_t \mathrm{div}\mathbf{u} - \int_{R^N} \nabla\mathcal{N}\partial_t(h(\mathcal{N}+v) - h(\mathcal{N}))\partial_t\mathbf{u} \\ &\geq \frac{d}{dt}\int_{R^N} \frac{\mathcal{N}h'(\mathcal{N}+v)}{2(\mathcal{N}+v)} v_t^2 dx - C\delta_1 \int_{R^N} (|v_t|^2 + |u|^2) \end{aligned} \tag{2.23}$$

with the help of (2.9). Second, due to (2.4), $(2.5)_1$ and (2.9), we have

$$\begin{aligned} &-\int_{R^N} \mathcal{N}\mathbf{u}_t\mathbf{e}_t dx \\ &= \int_{R^N} \mathbf{e}_t\mathbf{e}_{tt} + \int_{R^N} \mathbf{e}_t(v\mathbf{u})_t \\ &\geq \frac{1}{2}\frac{d}{dt}\int_{R^N} |\mathbf{e}_t|^2 - C\delta_1 \int_{R^N} (|\mathbf{e}_t|^2 + |v|^2 + |\mathbf{u}_t|^2), \end{aligned} \tag{2.24}$$

and

$$\int_{R^N} \mathcal{N}\partial_t(\mathbf{u}\cdot\nabla\mathbf{u})\partial_t\mathbf{u} \geq -C\delta_1 \int_{R^N} (|\partial_t\mathbf{u}|^2 + |\nabla\mathbf{u}|^2) \tag{2.25}$$

.So, (2.22) together with (2.23)-(2.25), and (2.21) implies

$$\begin{aligned} &\frac{d}{dt}\int_{R^N} \Big[\frac{\mathcal{N}}{2}|\mathbf{u}|^2 + \frac{\mathcal{N}}{2}|\mathbf{u}_t|^2 + \int_0^v (h(\mathcal{N}+s) - h(\mathcal{N}))ds + \frac{\mathcal{N}h'(\mathcal{N}+v)}{2(\mathcal{N}+v)}v_t^2 \\ &\quad + \tfrac{1}{2}|\mathbf{e}|^2 + \tfrac{1}{2}|\mathbf{e}_t|^2\Big] dx \\ &\quad + C\int_{R^N} (|\mathbf{u}|^2 + |\mathbf{u}_t|^2)dx \leq C\delta_1 ||(v, \nabla v, \nabla\mathbf{u}, v_t, \mathbf{e}_t)||^2. \end{aligned} \tag{2.26}$$

Take div on the both sides of $(2.5)_2$ and multiply the resulting equation by div$\mathbf{u}$, then integrate it over R^N, using integration by parts, to get

$$\begin{aligned} &\frac{1}{2}\frac{d}{dt}||\mathrm{div}\mathbf{u}||^2 + ||\mathrm{div}\mathbf{u}||^2 \\ &= -\int_{R^N} \mathrm{div}\nabla(h(\mathcal{N}+v) - h(\mathcal{N}))\mathrm{div}\mathbf{u}dx \\ &\quad + \int_{R^N} \mathrm{div}\mathbf{e}\mathrm{div}\mathbf{u}dx - \int_{R^N} \mathrm{div}(\mathbf{u}\cdot\nabla\mathbf{u})\mathrm{div}\mathbf{u}dx. \end{aligned} \tag{2.27}$$

Now we estimate each term in the right side of (2.27). With the help of $(2.5)_1$ and (2.9), we have

$$\begin{aligned} &-\int_{R^N} \mathrm{div}\nabla(h(\mathcal{N}+v) - h(\mathcal{N}))\mathrm{div}\mathbf{u}dx \\ &\leq -\frac{d}{dt}\int_{R^N} \frac{h'(\mathcal{N}+v)}{2(\mathcal{N}+v)}|\nabla v|^2 - \int_{R^N} \frac{h'(\mathcal{N}+v)}{\mathcal{N}+v}\mathbf{u}\cdot\nabla(\frac{|\nabla v|^2}{2}) \\ &\quad - \int_{R^N} \frac{h'(\mathcal{N}+v) - h'(\mathcal{N})}{\mathcal{N}+v}\nabla\mathcal{N}\cdot\nabla v_t \\ &\quad - \int_{R^N} \Big[\frac{h'(\mathcal{N}+v) - h'(\mathcal{N})}{\mathcal{N}+v}\Big]\nabla\mathcal{N}\cdot\nabla(\mathbf{u}\cdot\nabla v) \\ &\quad + C\delta_1 \int_{R^N} (|\nabla v|^2 + |\mathbf{u}|^2 + |\nabla\mathbf{u}|^2 + |v|^2). \end{aligned} \tag{2.28}$$

The third and the forth integral in the right hand of (2.28) can be estimated as follows. Using integration by parts, it reads that

$$-\int_{R^N}[\frac{h'(\mathcal{N}+v)-h'(\mathcal{N})}{\mathcal{N}+v}]\nabla\mathcal{N}\nabla v_t \leq C\delta_1\int_{R^N}(v^2+v_t^2+|\nabla v|^2), \quad (2.29)$$

and

$$-\int_{R^N}\frac{h'(\mathcal{N}+v)-h'(\mathcal{N})}{\mathcal{N}+v}\nabla\mathcal{N}\nabla(\mathbf{u}\cdot\nabla v) \leq C\delta_1\int_{R^N}(|v|^2+|v_t|^2+|\nabla v|^2). \quad (2.30)$$

In addition, in view of $(2.5)_1$ and (2.9), we have

$$\int_{R^N}\text{dive}\text{div}\mathbf{u} \leq -\frac{d}{dt}\int_{R^N}\frac{1}{2\mathcal{N}}v^2 + C\delta_1\int_{R^N}(|v|^2+|\mathbf{u}|^2), \quad (2.31)$$

and

$$\int_{R^N}\text{div}(\mathbf{u}\cdot\nabla\mathbf{u})\text{div}\mathbf{u}dx \leq C\delta_1\int_{R^N}|\nabla u|^2 dx. \quad (2.32)$$

Thus, (2.27) together with (2.28)-(2.32) implies

$$\frac{1}{2}\frac{d}{dt}||(\text{div }\mathbf{u}, \left[\tfrac{h'(\mathcal{N}+v)}{\mathcal{N}+v}\right]^{1/2}\nabla v, (\tfrac{1}{\mathcal{N}})^{1/2}v)||^2 + ||\text{div}\mathbf{u}||^2 \quad (2.33)$$

$$\leq C\delta_1||(v,\nabla v, v_t, \mathbf{u}, \nabla\mathbf{u})||^2. \quad (2.34)$$

Similarly, we take *curl* of $(2.5)_2$ and multiply the resulting equation by *curl*$\mathbf{u}$ to get, by the same calculations as used in [13] , that

$$\frac{1}{2}\frac{d}{dt}||curl\mathbf{u}||^2 + ||curl\mathbf{u}||^2 \leq C\delta_1||\nabla\mathbf{u}||^2. \quad (2.35)$$

Combining (2.34) and (2.35), we have, with the help of the smallness of δ_1, that

$$\frac{d}{dt}\int_{R^N}\left[|\nabla\mathbf{u}|^2+\frac{h'(\mathcal{N}+v)}{\mathcal{N}+v})|\nabla v|^2, \frac{1}{\mathcal{N}}|v|^2\right]+C||\nabla\mathbf{u}||^2 \leq C\delta_1||(v,\nabla v, v_t,\mathbf{u})||^2. \quad (2.36)$$

On the other hand, Multiplying $(2.5)_2$ by ∇v, and integrating it over R^N, yield

$$\begin{aligned}&\int_{R^N}[(h'(\mathcal{N}+v)-h'(\mathcal{N}))\nabla\mathcal{N}+h'(\mathcal{N}+v)\nabla v]\cdot\nabla v dx\\ &=\int_{R^N}\mathbf{e}\nabla v-\int_{R^N}(\mathbf{u}_t+\mathbf{u})\nabla v dx-\int_{R^N}(\mathbf{u}\cdot\nabla\mathbf{u})\nabla v dx.\end{aligned} \quad (2.37)$$

Integration by parts leads to

$$\int_{R^N}\mathbf{e}\cdot\nabla v = -\int_{R^N}v\text{div}\mathbf{e} = -||v||^2,$$

$$-\int_{R^N}(\mathbf{u}\cdot\nabla\mathbf{u})\nabla v dx \leq C\delta_1||(\nabla v,\nabla\mathbf{u})||^2,$$

and

$$-\int_{R^N}(\mathbf{u}_t+\mathbf{u})\nabla v dx \leq \epsilon||\nabla v||^2 + C||(\mathbf{u},\mathbf{u}_t)||^2,$$

here, ϵ is an arbitrary positive constants. So, from the above inequalities, we conclude that

$$||(v,\nabla v)||^2 \leq C||(\mathbf{u},\mathbf{u}_t,\nabla\mathbf{u})||^2, \tag{2.38}$$

with the help of the smallness of ϵ and δ_1. We can also obtain easily from $(2.5)_2$ that

$$||\mathbf{e}||^2 \leq C||(\mathbf{u},\mathbf{u}_t,\nabla v,\nabla\mathbf{u})||^2. \tag{2.39}$$

On the other hand, using $(2.5)_1$, and $(2.5)_3$ we have

$$\begin{aligned}
&-\int_{R^N}\partial_x\mathbf{e}\partial_x(\mathcal{N}\mathbf{u})dx\\
&=\int_{R^N}\partial_x\mathbf{e}\partial_x\mathbf{e}_t+\int_{R^N}\partial_x\mathbf{e}\partial_x(v\mathbf{u})+\int_{R^N}\partial_x\mathcal{E}\partial_x(\mathbf{e}_t+(\mathcal{N}+v)\mathbf{u})\\
&=\frac{1}{2}\frac{d}{dt}\int_{R^N}|\partial_x\mathbf{e}|^2+\int_{R^N}\partial_x\mathbf{e}\partial_x(v\mathbf{u}),
\end{aligned} \tag{2.40}$$

where we have used (2.19) with $i=1$. (2.40) implies

$$\frac{d}{dt}\int_{R^N}\frac{1}{2}|\partial_x\mathbf{e}|^2 \leq ||(\partial_x\mathbf{e},\partial_x\mathbf{u},\partial_x v)||^2. \tag{2.41}$$

Thus, (2.26), together with (2.36), (2.38), (2.39), (2.40) and (2.11) with $j=1$, implies

$$\begin{aligned}
&\frac{d}{dt}\int_{R^N}\Big[|\mathbf{u}|^2+|\mathbf{u}_t|^2+\frac{2}{\mathcal{N}}\int_0^v(h(\mathcal{N}+s)-h(\mathcal{N}))ds+\frac{h'(\mathcal{N}+v)}{\mathcal{N}+v}v_t^2\\
&+\tfrac{1}{\mathcal{N}}(|\mathbf{e}|^2+|\mathbf{e}_t|^2)+\lambda|\nabla\mathbf{e}|^2+\tfrac{2}{\mathcal{N}}|\nabla\mathbf{u}|^2+\tfrac{2h'(\mathcal{N}+v)}{\mathcal{N}(\mathcal{N}+v)}|\nabla v|^2+\tfrac{2}{\mathcal{N}^2}v^2\Big]dx+\\
&C||(v,\nabla v,v_t,\mathbf{u},\nabla\mathbf{u},\mathbf{u}_t,\mathbf{e},\mathbf{e}_t)||^2\leq 0
\end{aligned} \tag{2.42}$$

for some positive constant $\lambda>0$. So we get the desired estimates.

Step 3. By the similar but more complicated calculation, we can get the second and third order energy estimates, combined to (2.42) which proves the Lemma 3.2.

References

1. Degond, P. , Markowich, P. A.(1990): On a one-dimensional steady-state hydrodynamic model for semiconductors. Appl. Math. Letters, **31**, 25-29
2. Zhang, B.(1993): Convergence of the Godunov scheme for a simplified one-dimensional hydrodynamic model for semiconductor devices. Comm. Math. Phys. **157**, 1-22.

3. Marcati, P. A., Natalini R.(1995): Weak solutions to hydrodynamic model for semiconductors and relaxation to the drift-diffusion equation. Arch. Rational Mech. Anal., **129**, 129-145
4. Hsiao, L., Zhang, K. J.(2000): The relaxation of the hydrodynamic model for semiconductors to the drift-diffusion equations. J. Diff. Eqns., **165** 315-354
5. Luo, T., Natalini, R., Xin, Z. P.(1998): Large-time behavior of the solutions to a hydrodynamic model for semiconductors. SIAM J. Appl. Math., **59**, 810-830
6. Hsiao, L., Yang, T.(2001): Asymptotics of initial boundary value problems for hydrodynamic and drift diffusion models for semiconductors. J. Diff. Eqns., **170**, 472-493
7. Li, H. L., Markowich, P. A., Mei M. Asymptotic behavior of solutions to a hydrodynamic model for semiconductors. Preprint
8. Kato, T.(1975): The Cauchy problem for quasilinear symmetric hyperbolic systems. Arch. Rational Mech. Anal., **58** 181-205
9. Majda, A. (1984): Compressible Fluid Flow and Systems of Conservation laws in Several Space Variables. Springer-Verlag, Berlin/New York
10. Degond, P., Markowich, P. A.(1993): A steady-state potential flow model for semiconductors. Ann. Mat. Pura Appl., **IV**: 87-98
11. Hsiao, L., Wang, S. The asymptotic behavior of global solutions to the hydrodynamic model with spherical symmetry. Accepted by Nonlinear Analysis.
12. Hsiao, L., Wang S. The asymptotic behavior of global solutions to the hydrodynamic model in the exterior domain. Accepted by Acta Mathematica Scientia.
13. Hsiao, L., Markowich, P., Wang, S. The asymptotic behavior of global smooth solutions to the multidimensional hydrodynamic model for semiconductors. Preprint.
14. Hsiao, L., Wang, S., Zhao, H. Asymptotic of Global Solutions to the Multidimensional Hydrodynamic Model for Semiconductors. Accepted by Math. Meth. Appl. Sci.
15. Hsiao, L., Ju, Q., Wang, S. The Asymptotic Behavior of Global Smooth Solutions to the Multidimensional Hydrodynamic Model for Semiconductors. Preprint.
16. Stein, E. M.(1970): Singular Integrals and Differentiability Properties of Functions. Princeton University Press,Princeton, New Jersey
17. Hsiao, L., Luo, T.(1996): Nonlinear diffusive phenomena of solutions for the system of compressible adiabatic flow through porous media. J. Diff. Eqns., **125**, 329-365

Relaxation Schemes for Conservation Laws with Discontinuous Coefficients

K. H. Karlsen[1], C. Klingenberg[2], and N. H. Risebro[3]

[1] Department of Mathematics, University of Bergen, Johs. Brunsgt. 12 N–5008 Bergen, Norway, *kennethk@mi.uib.no*
[2] University of Würzburg, Department of Applied Mathematics and Statistics, Am Hubland, D–97074 Würzburg, Germany, *Christian.Klingenberg@IWR.Uni-Heidelberg.De*
[3] Department of Mathematics, University of Oslo, P.O. Box 1053, Blindern, N–0316 Oslo, Norway, *nilshr@math.uio.no*

Summary. We study relaxation schemes for conservation laws with discontinuous coefficients, and compare these numerically with an Engquist–Osher type scheme and with a front tracking method. If the coefficients are of bounded variation, the first order relaxation scheme, the Engquist–Osher scheme, and the front tracking method converge to a weak solution.

1 Introduction

Over the years, several numerical methods have been constructed and analyzed for conservation laws with a discontinuous coefficient $k(x)$:

$$u_t + f(k(x), u)_x = 0, \tag{1}$$

where $u(x,t)$ is the scalar unknown function that is sought; and the flux function $f(k,u)$ and the coefficient $k(x)$ are given functions. In particular, the Glimm scheme [20] and the front tracking method [12] (and the references therein), as well as a class of monotone difference schemes [22] has been shown to produce a strongly compact sequence of approximate solutions, whose limit points are weak solutions of (1). Recently, also a relaxation scheme was shown to have this property [8]. Strongly degenerate convection-diffusion equations have been treated in [9]. These schemes are (formally) only first order in the discretization parameters, and no second order scheme has been shown to produce convergent sequences. The relaxation scheme is particularly simple to generalize in order to produce a (formally) second order scheme. In this note we shall compare the results produced by the above schemes on two model problems.

Nonlinear PDEs of the form (1) occur in several applications. We mention here briefly flow in porous media [5], sedimentation processes [1, 3, 4], and traffic flow on a highway [24, 6]. They also arise in radar shape-from-shading problems [18]. In view of their applications, there is great demand for accurate, efficient, and, at the same time, “easy to implement” numerical methods for conservation laws with discontinuous coefficients.

Independently of the smoothness of $k(x)$, solutions to (1) are in general not smooth and weak solutions must be sought. A *weak solution* is here defined as a function $u \in L^\infty$ which satisfies (1) in the sense of distributions, i.e., in $\mathcal{D}'$. If $k(x)$ is "smooth", a weak solution u of (1) satisfies the *entropy condition* if for all convex entropy/entropy flux pairs (η, q)

$$\eta(u)_t + q(k(x), u)_x + k'(x)\big(\eta'(u) f_k(k(x), u) - q_k(k(x), u)\big) \leq 0 \quad \text{in } \mathcal{D}'. \tag{2}$$

If f and k are sufficiently smooth functions and $u_0 \in L^\infty$, Kružkov's theory [14] tells us that there exists a unique weak solution to the initial value problem for (1) which satisfies the entropy condition (2).

In the case of a discontinuous coefficient $k(x)$, the very notion of entropy solution (2) breaks down (but see [13, 11, 22, 10, 19]). In this case, (1) is often been written as a 2×2 system of equations:

$$k_t = 0, \qquad u_t + f(k, u)_x = 0. \tag{3}$$

If $u \mapsto f_u(k, u)$ changes sign, then this system is non-strictly hyperbolic, a situation described as resonance. An important consequence of resonance is that no a priori bound on the spatial total variation of the conserved quantity u is available [20, 23].

Since there is generally no spatial total variation bound for the conserved variable u itself, the "singular mapping" approach has been used up to now as the analytical tool for proving convergence of numerical methods. Temple [20] was the first to use this approach when he established convergence of the Glimm scheme for a 2×2 resonant system of conservation laws modeling the displacement of oil in a reservoir by water and polymer, which is equivalent to a conservation law with a discontinuous coefficient (see, e.g., [12]). Convergence has been established for the 2×2 Godunov method by Lin, Temple, and Wang [15, 16], while the 2×2 front tracking method has been analyzed by Gimse and Risebro [5] and Klingenberg and Risebro [13, 12]. In Bürger, Karlsen, and Klingenberg [2], the 2×2 front tracking method is applied to a model of continuous sedimentation in ideal clarifier-thickener units.

The 2×2 Glimm, Godunov, and front tracking methods are very accurate since they rely on an exact 2×2 Riemann solver. However, the price to pay for using a 2×2 Riemann solver is that the numerical methods become complicated to implement. This is even more pertinent what regards front tracking, as the whole front tracking machinery also has to be implemented. As simpler alternatives to these methods, Towers [22, 21] devised appropriate scalar versions of the Godunov and Engquist–Osher methods. He also established convergence of these methods by the singular mapping approach. Extensions to strongly degenerate convection-diffusion equations can be found in [9].

The purpose of the present paper is to compare an alternative "simple" class of numerical methods for (1) with some of the established methods. Following Jin and Xin [7], the starting point herein is to approximate (1) by a 2×2 semilinear hyperbolic system with a stiff relaxation term containing the discontinuous flux function $f(k(x), u)$:

$$u_t^\tau + v_x^\tau = 0, \qquad v_t^\tau + a^2 u_x^\tau = \frac{1}{\tau}\left(f(k(x), u^\tau) - v^\tau\right), \tag{4}$$

where $\tau > 0$ is the relaxation parameter and a satisfies the subcharacteristic condition. Note that the variable v in (4) can be eliminated. The result is a conservation law (with a discontinuous coefficient) that has been regularized by a wave operator:

$$u_t^\tau + f(k(x), u^\tau)_x = -\tau\left(u_{tt}^\tau - a^2 u_{xx}^\tau\right).$$

Hence we expect (4) to be a first order approximation to (1) as $\tau \downarrow 0$. An excellent overview of the "relaxation approach" to hyperbolic problems can be found in Natalini [17]. To build a numerical scheme, we now discretize (4) by an upwind scheme. The resulting scheme, which is called the relaxation scheme, has the advantage of not relying on a Riemann solver. This a consequence of the special semilinear structure of (4). In characteristic variables, (4) reduces to a diagonal system which is trivial to discretize with a upwind scheme without resorting to a Riemann solver. The stiff relaxation term is discretized implicitly. In [8], we have analyzed this relaxation scheme using the compensated compactness method. A similar relaxation scheme is obtained by starting with the system (3), replacing this by

$$\begin{cases} u_t^\tau + v_x^\tau = 0, \\ v_t^\tau + a^2 u_x^\tau = \dfrac{1}{\tau}\left(f\left(k^\tau, u^\tau\right) - v^\tau\right), \end{cases} \qquad \begin{cases} k_t^\tau + l_x^\tau = 0, \\ l_t^\tau + a_k^2 k_x^\tau = -\dfrac{l^\tau}{\tau}, \end{cases} \tag{5}$$

where a_k is a constant (also satisfying the subcharacteristic condition). The resulting relaxation scheme was not studied in [8].

In Section 2 we describe the numerical methods in more detail, while numerical examples are presented and discussed in Section 3. Finally, we make a few concluding remarks in Section 4.

2 The numerical schemes

In this section we present the various schemes under investigation, and we start with the relaxation schemes.

The relaxation schemes. These schemes are based on replacing (1) with the system (4), where $\tau > 0$ is the relaxation parameter and a satisfies the subcharacteristic condition $0 < \max_{k,u} |f_u(k, u)| < a$.

For (4), we specify the following initial data:

$$u^\tau(x, 0) = u_0(x), \qquad v^\tau(x, 0) = f(k(x), u_0(x)).$$

In characteristic variables $w = u + \frac{v}{a}$, $z = u - \frac{v}{a}$ the system (4) simplifies to a diagonal system:

$$\begin{aligned} w_t^\tau + a w_x^\tau &= \frac{1}{a\tau}\left(f\left(k(x), \tfrac{1}{2}(w^\tau + z^\tau)\right) - \tfrac{a}{2}(w^\tau - z^\tau)\right), \\ z_t^\tau - a z_x^\tau &= \frac{-1}{a\tau}\left(f\left(k(x), \tfrac{1}{2}(w^\tau + z^\tau)\right) - \tfrac{a}{2}(w^\tau - z^\tau)\right). \end{aligned} \tag{6}$$

Let $h > 0$ and $\Delta t > 0$ be the spatial and temporal discretization parameters, respectively. The spatial domain $\mathbf{R}$ is discretized into grid cells $I_j = \left(x_{j-1/2}, x_{j+1/2}\right)$, $j \in \mathbf{Z}$, where $x_k = kh$ for $k = 0, \pm\frac{1}{2}, \pm 1, \pm\frac{3}{2}, \dots$. Similarly, we discretize the time interval $(0, T)$ via $t_n = n\Delta t$ for $n = 0, 1, \dots, N$, where the integer N is chosen such that $N\Delta t = T$, resulting in the time strips (t_n, t_{n+1}), $n = 0, 1, \dots, N-1$. We let u_j^n v_j^n, w_j^n, and z_j^n denote the approximate solutions at $x = x_j$ and $t = t_n$, and consider a semi-implicit upwind scheme discretization of (6):

$$\begin{cases} \dfrac{1}{\Delta t}\left(w_j^{n+1} - w_j^n\right) + \dfrac{a}{h} D_- w_j^n = \dfrac{1}{a\tau} g\left(k_j, u_j^{n+1}, v_j^{n+1}\right), \\ \dfrac{1}{\Delta t}\left(z_j^{n+1} - z_j^n\right) - \dfrac{a}{h} D_+ z_j^n = -\dfrac{1}{a\tau} g\left(k_j, u_j^{n+1}, v_j^{n+1}\right), \end{cases} \tag{7}$$

where $D_\pm b_j = \mp(b_j - b_{j\pm 1})$ for a quantity b_j, and $g(k, u, v) = f(k, u) - v$. In the original variables, this finite difference scheme reads

$$\begin{aligned} &\frac{1}{\Delta t}\left(u_j^{n+1} - u_j^n\right) + \frac{1}{2h} D v_j^n - \frac{a}{2h} D_+ D_- u_j^n = 0, \\ &\frac{1}{\Delta t}\left(v_j^{n+1} - v_j^n\right) + \frac{a^2}{2h} D u_j^n - \frac{a}{2h} D_+ D_- v_j^n = \frac{1}{\tau} g\left(k_j, u_j^{n+1}, v_j^{n+1}\right), \end{aligned} \tag{8}$$

where the central difference $D = D_- + D_+$. Note that although this is an implicit method, we do not have to solve a system of equations in order to update u^n and v^n, since we can use the first equation to update u^n, and then the source term is linear in v^{n+1}. We start the iterations (8) by defining

$$u_j^0 = \frac{1}{h}\int_{I_j} u_0(x)\,dx, \quad v_j^0 = \frac{1}{h}\int_{I_j} f(k(x), u_0(x))\,dx, \quad k_j = \frac{1}{h}\int_{I_j} k(x)\,dx.$$

We can also device a similar first order relaxation scheme by starting with the system (3) and replace this by (5), where a_k is a positive constant. This leads to the numerical scheme

$$\begin{aligned} &\frac{1}{\Delta t}\left(u_j^{n+1} - u_j^n\right) + \frac{1}{2h} D v_j^n - \frac{a}{2h} D_- D_+ u_j^n = 0, \\ &\frac{1}{\Delta t}\left(k_j^{n+1} - k_j^n\right) + \frac{1}{2h} D l_j^n - \frac{a_k}{2h} D_- D_+ k_j^n = 0, \\ &\frac{1}{\Delta t}\left(v_j^{n+1} - v_j^n\right) + \frac{a^2}{2h} D u_j^n - \frac{a}{2h} D_- D_+ v_j^n = \frac{1}{\tau} g\left(k_j, u_j^{n+1}, v_j^{n+1}\right), \\ &\frac{1}{\Delta t}\left(l_j^{n+1} - l_j^n\right) + \frac{a_k^2}{2h} D k_j^n - \frac{a_k}{2h} D_- D_+ l_j^n = -\frac{l_j^{n+1}}{\tau}. \end{aligned} \tag{9}$$

To obtain a formally second order scheme we start by using van Leer's MUSCL scheme. This scheme uses a piecewise linear interpolation of the characteristic variables defined as

$$\begin{aligned}(v+au)_{j+1/2} &= (v+au)_j + \frac{1}{2}h_j\sigma_j^+,\\ (v-au)_{j+1/2} &= (v-au)_j - \frac{1}{2}h_{j+1}\sigma_{j+1}^-,\end{aligned}$$

where $\sigma_j^\pm$ is the (limited) slope of $u \pm av$ in the jth cell;

$$\begin{aligned}\sigma_j^\pm &= \frac{1}{h}\left(v_{j+1} \pm au_{j+1} - v_j \mp au_j\right)\phi\left(\theta_j^\pm\right),\\ \theta_j^\pm &= \frac{v_j \pm au_j - v_{j-1} \mp au_{j-1}}{v_{j+1} \pm au_{j+1} - v_j \mp au_j},\end{aligned}$$

where ϕ is a slope limiter. We use the van Leer limiter given by $\phi(\theta) = \frac{\theta+|\theta|}{1+|\theta|}$. This results in the time stepping formula,

$$\begin{aligned}u_j^{n+1} &= u_j^n - \frac{\lambda}{2}Dv_j^n + \frac{a\lambda}{2}D_-D_+u_j^n + \frac{\lambda}{4}\left(D\sigma_j^- + D_+\sigma_j^+\right)\\ &=: u_j^n + \Delta t F\left(u^n, v^n\right),\\ v_j^{n+1} &= v_j^n - \frac{a^2\lambda}{2}Du_j^n + \frac{a\lambda}{2}D_-D_+v_j^n + \frac{\lambda}{4}\left(D\sigma_j^- - D_+\sigma_j^+\right)\\ &\quad + \frac{\Delta t}{\tau}g\left(k_j, u_j^{n+1}, v_j^{n+1}\right) =: v_j^n + \Delta t G(u^n, v^n, v^{n+1}),\end{aligned} \tag{10}$$

where $\lambda = \Delta t/h$. The scheme defined in (10) is formally second order in space and first order in time. In order to attain second order in time as well, we use a Heun type scheme defined as

$$\begin{aligned}\tilde{u} &= u^n + \Delta t F(u^n, v^n), \qquad \tilde{v} = v^n + \Delta t G(\tilde{u}, v^n, \tilde{v}),\\ u^{n+1} &= u^n + \frac{\Delta t}{2}\left(F(u^n, v^n) + F(\tilde{u}, \tilde{v})\right),\\ v^{n+1} &= v^n + \frac{\Delta t}{2}\left(G(\tilde{u}, v^n, \tilde{v}) + G(u^{n+1}, \tilde{v}, v^{n+1})\right).\end{aligned} \tag{11}$$

The second order version of (9) is defined in an analogous manner.

The numerical schemes devised here by discretizing the 4×4 system (5) (and not the 2×2 system (4)) were not studied in [8].

The monotone difference schemes. For simplicity, we shall restrict this presentation to the prototype monotone first order scheme for (1), namely the Engquist–Osher scheme, as defined in [22, 21]. Let $f^{\mathrm{EO}}(k,u,v)$ denote the Engquist–Osher flux defined by

$$f^{\mathrm{EO}}(k,u,v) = \frac{1}{2}\left(f(k,u) + f(k,v) - \int_u^v \left|f_u(k,z)\right|\,dz\right).$$

Then "trick" employed in [22, 21] is to discretize k staggered with respect to u, i.e., set

$$k_{j+1/2} = \frac{1}{h} \int_{I_{j+1/2}} k(x)\, dx. \tag{12}$$

Now the scheme reads

$$u_j^{n+1} = u_j^n - \lambda \left(f^{\mathrm{EO}} \left(k_{j+1/2}, u_j^n, u_{j+1}^n \right) - f^{\mathrm{EO}} \left(k_{j-1/2}, u_{j-1}^n, u_j^n \right) \right), \tag{13}$$

where $\lambda = \Delta t/h \leq \max |f_u(k,u)|$. The advantage of this scheme is its upwind character and the ease of implementation.

The front tracking method. Front tracking for (1) is strictly speaking not a difference scheme. Instead of solving (1) we solve

$$u_t^\delta + f^\delta(k^\delta, u^\delta)_x = 0, \qquad u^\delta(x,0) = u_0^\delta(x). \tag{14}$$

Here f^δ is carefully constructed so that we can construct the exact weak solution. The approximate flux function f^δ is chosen so that $u \mapsto f^\delta(k,u)$ is piecewise linear and continuous for each fixed k, and $f^\delta \to f$ as $\delta \to 0$. The functions $k^\delta = k^\delta(x)$ and $u_0^\delta(x)$ are step functions that are chosen so that $u_0^\delta(x)$ takes its values among the breakpoints (the discontinuity points of $f_u^\delta(k^\delta, \cdot)$) of f^δ, and $k^\delta \to k$, $u_0^\delta \to u_0$ in L^1 as $\delta \to 0$. Now the weak solution of (14) will be a piecewise constant function, constant on a finite set of polygons in the (x,t) plane. The discontinuities of u^δ are referred to as *fronts*, and u^δ is constructed by tracking these fronts, and solving the Riemann problems defined by their interactions. The details of this algorithm are too complicated to present here, in order to obtain the complete picture see [5, 13, 12]. In these papers it was also demonstrated that the sequence (u^δ) is strongly compact in L^1, and that any limit is a weak solution to (1) for several types of flux functions.

The main advantage of the front tracking algorithm is its accuracy in representing discontinuities in the solution, its ability to capture the large time behavior accurately, and its execution speed. The main disadvantage is its complexity. It requires an exact Riemann solver for (14), and it is probably safe to say that unless an extremely large number of calculations are to be made, the extra time spent at programming, compared with e.g., (13), far outweighs the time gained during calculations.

In order to compare the front tracking method with the difference methods, we have set $\delta = h$, and projected the final result of front tracking onto the grid used by the difference methods.

Remark 1. The first order relaxation scheme (4) [8], the Engquist–Osher scheme (13) [22, 21], and the front tracking method [5, 13, 12] have all been shown to produce limits that are weak solutions to (1). This has not been demonstrated for the second order schemes.

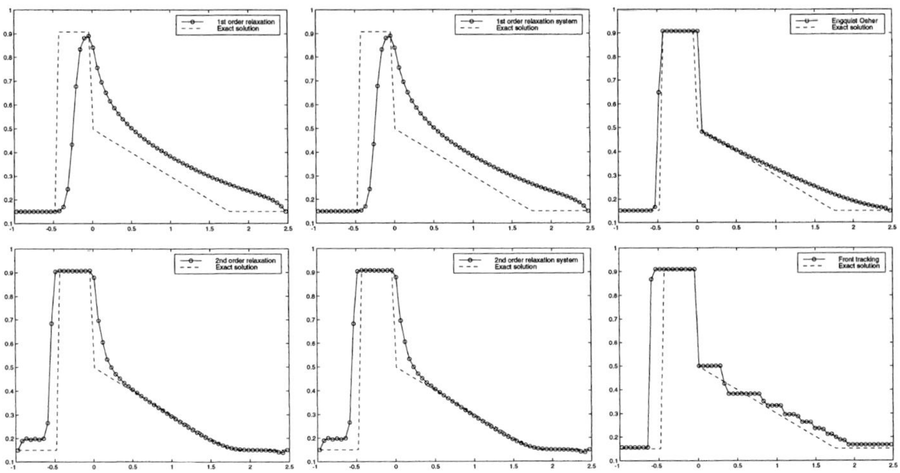

Fig. 1. The approximate solutions for the initial data (15).

3 Numerical results

We present two numerical examples, in which all the above methods are tested. In both examples we use the flux function $f(k,u) = ku(1-u)$. The first example is a Riemann problem with initial data given by

$$u(x,0) = 0.15, \qquad k(x) = 1 + 2\chi_{x<0}(x) \tag{15}$$

The exact solution of this Riemann problem consists of a shock moving to the left separating u values $(0.15, 0.933)$, then a stationary discontinuity at $x = 0$ separating u values $(0.933, 0.5)$, and finally a rarefaction wave with positive speed from $u = 0.5$ to $u = 0.15$. For all the difference methods we used $\Delta t = 0.95 \max|k|\, h$, and for the relaxation methods we used $\tau = 10^{-11}$, $a = \max|k|$ and $a_k = 0.00125$.

In Figure 1 we show the solutions at $t = 3.0$ computed by all the methods using $h = 3.5/63$ in the interval $x \in [-1, 2.5]$. From this figure it seems that there is very little to be gained by relaxing the coefficient, and the error table, Table 1 confirms this. The other notable feature of Figure 1 is the "step function" character of the front tracking approximation. This is due to the construction of f^δ, and in general the solution of the Riemann problem for (14) is $\mathcal{O}\left(\sqrt{\delta}\right)$ away from the solution of the Riemann problem for (1) near $x = 0$.

We have also compared the relative L^1 errors defined by $100\times\frac{\sum_j |u(x_j)-u_j^M|}{\sum_j |u(x_j)|}$, where u denotes the exact solution. This is shown in Table 1. The second example has periodic initial data given in the interval $[-1/2, 1/2]$, with

$$u(x,0) = 0.5, \qquad k(x) = \cos^2(\pi x) + 1 + 8\,|x|\,\chi_{[-1/4,1/4]}(x), \tag{16}$$

N	R	R2	RS	RS2	EO	FT
16	42.27	41.42	42.27	41.43	9.96	13.62
32	40.11	21.20	40.11	21.20	13.19	13.20
64	31.88	8.01	31.88	7.99	5.02	9.52
128	24.57	6.49	24.57	6.49	2.57	2.13
256	14.56	2.13	14.56	2.12	1.70	1.32
512	8.77	0.91	8.77	0.92	0.85	1.38
1024	5.52	0.79	5.52	0.79	0.41	0.40

Table 1. 100 times relative L^1 errors for the various methods, N: number of grid points, R: 1st order relaxation, R2: second order relaxation, RS: 1st order system relaxation, RS2: 2nd order system relaxation, EO: Engquist–Osher, FT: front tracking.

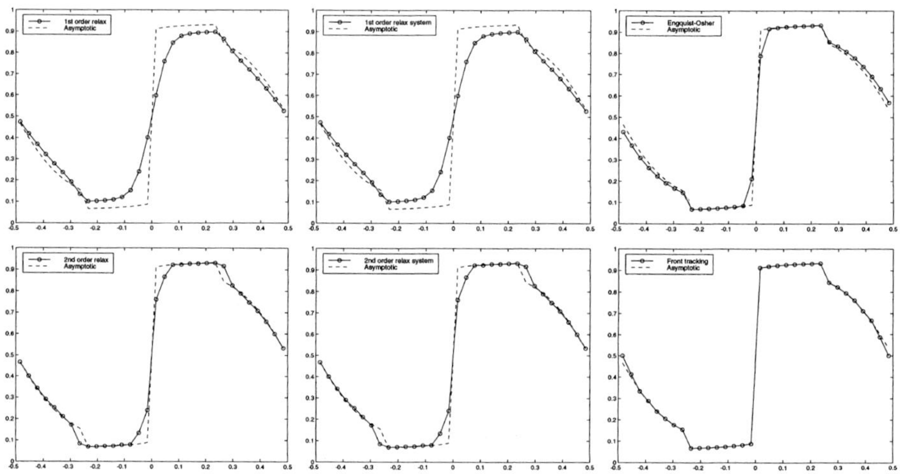

Fig. 2. The approximate solutions for the initial data (16).

In this case one can calculate the asymptotic solution as $t \to \infty$ analytically, see [13], and we have calculated the approximate solution at $t = 5.0$. At this time the approximate solution is almost stationary, and therefore we have used the asymptotic solution as a reference solution to compute L^1 errors. In Figure 2 we show the relaxation approximation using $h = 1/31$ at $t = 5.0$, as well as the asymptotic solution to the conservation law. In Table 2 we show the relative errors produced by the various methods for this problem. Again we see that there is not much, if anything, to be gained by using the system relaxation. Also for this problem front tracking performs better than all the other methods. Furthermore, all the methods have "numerical" convergence rates of one.

N	R	R2	RS	RS2	EO	FT
16	20.55	8.04	20.55	8.04	7.44	2.03
32	11.72	4.00	11.72	4.00	3.79	0.78
64	6.83	1.74	6.83	1.74	1.91	0.29
128	3.87	0.87	3.87	0.87	0.96	0.19
256	2.09	0.43	2.09	0.43	0.48	0.12
512	1.09	0.21	1.09	0.21	0.24	0.06
1024	0.56	0.10	0.56	0.10	0.14	0.03

Table 2. 100 times relative L^1 errors for the various methods, with initial data given in (16).

4 Conclusion

Numerical examples given here (and elsewhere) demonstrate that the front tracking method as well as the monotone difference and relaxation schemes can all be used as efficient and accurate simulation tools for conservation laws with discontinuous coefficients. The front tracking method, the monotone (Engquist-Osher and Godunov) difference schemes, and the first order version of the relaxation schemes are all supported by theoretical convergence analysis. The front tracking method is very accurate since it does not contain any numerical dissipation due to the use of an exact 2×2 Riemann solver. In particular, discontinuous waves are accurately resolved. Moreover, the front tracking method is unconditionally stable in the sense that there is no time step associated with the method. The monotone difference and relaxation schemes, on the other hand, contain numerical dissipation. Consequently, discontinuous waves may be (slightly) smeared out. Also there is a time step restriction (the CFL condition) since the method considered here is explicit. However, compared with the front tracking method, the monotone difference and relaxation schemes are very easy to implement. Let us also add that the monotone (Engquist-Osher) scheme seems to be less dissipative than the relaxation schemes.

References

1. R. Bürger, K. H. Karlsen, , N. H. Risebro, and J. D. Towers. Numerical methods for the simulation of continuous sedimentation in ideal clarifier-thickener units. Preprint, Department of Mathematics, University of Bergen, 2001.
2. R. Bürger, K. H. Karlsen, C. Klingenberg, and N. H. Risebro. A front tracking approach to a model of continuous sedimentation in ideal clarifier-thickener units. Nonlinear Analysis: Real World Applications. To appear.
3. S. Diehl. On scalar conservation laws with point source and discontinuous flux function. *SIAM J. Math. Anal.*, 26(6):1425–1451, 1995.
4. S. Diehl. A conservation law with point source and discontinuous flux function modelling continuous sedimentation. *SIAM J. Appl. Math.*, 56(2):388–419, 1996.

5. T. Gimse and N. H. Risebro. Solution of the Cauchy problem for a conservation law with a discontinuous flux function. *SIAM J. Math. Anal.*, 23(3):635–648, 1992.
6. H. Holden and N. H. Risebro. A mathematical model of traffic flow on a network of unidirectional roads. *SIAM J. Math. Anal.*, 26(4):999–1017, 1995.
7. S. Jin and Z. P. Xin. The relaxation schemes for systems of conservation laws in arbitrary space dimensions. *Comm. Pure Appl. Math.*, 48(3):235–276, 1995.
8. K. H. Karlsen, C. Klingenberg and N. H. Risebro A relaxation scheme for conservation laws with a discontinuous coefficient Preprint, 2002.
9. K. H. Karlsen, N. H. Risebro, and J. D. Towers. On an upwind difference scheme for degenerate parabolic convection-diffusion equations with a discontinuous coefficient. IMA J. Numer. Anal. To appear.
10. K. K. Karlsen, N. H. Risebro, and J. D. Towers. L^1 stability for piecewise smooth solutions of degenerate parabolic convection-diffusion equations with discontinuous coefficients. Work in preparation (July 29, 2003).
11. R. A. Klausen and N. H. Risebro. Stability of conservation laws with discontinuous coefficients. *J. Differential Equations*, 157(1):41–60, 1999.
12. C. Klingenberg and N. H. Risebro. Stability of a resonant system of conservation laws modeling polymer flow with gravitation. *J. Differential Equations*, 170(2):344–380, 2001.
13. C. Klingenberg and N. H. Risebro. Convex conservation laws with discontinuous coefficients. Existence, uniqueness and asymptotic behavior. *Comm. Partial Differential Equations*, 20(11-12):1959–1990, 1995.
14. S. N. Kružkov. First order quasi-linear equations in several independent variables. *Math. USSR Sbornik*, 10(2):217–243, 1970.
15. L. W. Lin, B. J. Temple, and J. H. Wang. A comparison of convergence rates for Godunov's method and Glimm's method in resonant nonlinear systems of conservation laws. *SIAM J. Numer. Anal.*, 32(3):824–840, 1995.
16. L. W. Lin, B. J. Temple, and J. H. Wang. Suppression of oscillations in Godunov's method for a resonant non-strictly hyperbolic system. *SIAM J. Numer. Anal.*, 32(3):841–864, 1995.
17. R. Natalini. Recent results on hyperbolic relaxation problems. In *Analysis of systems of conservation laws (Aachen, 1997)*, pages 128–198. Chapman & Hall/CRC, Boca Raton, FL, 1999.
18. D. N. Ostrov. Viscosity solutions and convergence of monotone schemes for synthetic aperture radar shape-from-shading equations with discontinuous intensities. *SIAM J. Appl. Math.*, 59(6):2060–2085 (electronic), 1999.
19. N. Seguin and J. Vovelle. Analysis and approximation of a scalar conservation law with a flux function with discontinuous coefficients. Preprint 2002.
20. B. Temple. Global solution of the Cauchy problem for a class of 2×2 nonstrictly hyperbolic conservation laws. *Adv. in Appl. Math.*, 3(3):335–375, 1982.
21. J. D. Towers. A difference scheme for conservation laws with a discontinuous flux - the nonconvex case. *SIAM J. Numer. Anal.* To appear.
22. J. D. Towers. Convergence of a difference scheme for conservation laws with a discontinuous flux. *SIAM J. Numer. Anal.*, 38(2):681–698 (electronic), 2000.
23. A. Tveito and R. Winther. The solution of nonstrictly hyperbolic conservation laws may be hard to compute. *SIAM J. Sci. Comput.*, 16(2):320–329, 1995.
24. G. B. Whitham. *Linear and nonlinear waves.* Wiley-Interscience [John Wiley & Sons], New York, 1974. Pure and Applied Mathematics.

Relaxation Models and Finite Element Schemes for the Shallow Water Equations

Theodoros Katsaounis[1] and Charalambos Makridakis[2]

[1] Département de Mathématiques et Applications, Ecole Normale Supérieure, 45, rue d'Ulm, 75230 Paris, France, *Theodoros.Katsaounis@ens.fr*
[2] Department of Applied Mathematics, University of Crete, and Institute of Applied and Computational Mathematics, FORTH, Heraklion, Greece. *makr@math.uoc.gr*

1 Introduction

We consider the one-dimensional system of shallow water equations (or Saint-Venant system) with a source term

$$h_t + (hu)_x = 0, \tag{1a}$$

$$(hu)_t + (hu^2 + \frac{g}{2}h^2)_x = -ghZ', \tag{1b}$$

which describes the flow at time $t \geq 0$ at point $x \in \mathbb{R}$, where $h(x,t) \geq 0$ is the height of water, $u(x,t)$ is the velocity, $Z(x)$ is the bottom height and g the gravity constant. In the sequel will denote $Q = hu$ the discharge. System (1) belongs in the more general class of hyperbolic systems with source terms

$$u_t + f(u)_x = q(u), \tag{2}$$

where u is a vector valued function and f, q are the given flux and source functions. In this paper we propose relaxation models and corresponding time discrete and finite element schemes for approximating (1). Our schemes can be formulated for the more general system (2) and special attention is given in the steady state approximations and their relation to the exact steady states especially for (1).

The system (1) was originally derived by A. de Saint–Venant [22], while recently Gerbeau and Perthame [10] derived a more complete system starting from the Navier–Stokes equations. The above system is a hyperbolic system with a source term, due to the topography of the bottom, which includes the unknown quantity h. In recent years many methods were proposed for the numerical approximation of solutions of hyperbolic conservations laws with source terms. The main difficulty being here, obtaining a scheme that is able to resolve shocks accurately and at the same time to have a reasonable steady state behavior. This is not a trivial matter since straight forward adaptations of upwind schemes have bad steady state performance. Hence particular attention has to be given in the numerical treatment of the source terms. In this direction finite volumes are widely used, and in particular the well balanced schemes introduced by Greenberg and LeRoux, [11], Gosse and LeRoux [14] and further developed by Gosse [12],[13]. Another approach is based on wave propagation algorithms, [20], [5] and Godunov type schemes [9],[20]. Recently Jin [18] introduced a new method based on averaging the values at the interfaces. An alternative class of schemes were considered in [3]. These schemes are based on the kinetic interpretation of the system. An entropy satisfying kinetic scheme which also accurately computes the steady states is presented in [21]. Finite difference discretizations of (1) based on relaxation models are consider in [2], [4].

In the sequel we consider relaxation models and finite element relaxation schemes for (2), (1). We discuss first certain relaxation models and their time discrete counterparts. These models include the standard relaxation approximation for problems of balance laws considered e.g. in [6] as well as new relaxation approximations that we believe that are better suited to designing schemes for problems with non zero source terms. In addition we propose new relaxation approximations for the shallow water equations that admit as solutions exactly the steady states of the original system (1). We then apply the ideas of [1] discretizing by finite elements these models. Standard finite element spaces are used for the spatial discretization while implicit-explicit Runge–Kutta methods provide the time stepping mechanisms. A special variant of the resulting schemes which satisfies a discrete version of the steady state equation for (1) is also proposed.

The resulting schemes

- are based on a regularization mechanism with finite speed of propagation (no extra diffusion is added in the discretization of the relaxation model)
- do not need the solution of approximate local Riemann problems
- are formulated as low order, or high order schemes, simply by selecting the discretization space appropriately
- can be extended in multi-dimensions in a straightforward manner by using the finite element framework.

Since this method has no other regularizing mechanism besides that of the relaxation it should be combined with adaptive mesh refinement (close to the shocks) as in [1].

This paper is organized as follows: The needed preliminary material, the relaxation models and the properties of their time discrete counterparts are introduced in Section 2. In Section 3 we present the finite element schemes. Finally in Section 4 a series of experiments displaying the features of the method are presented.

2 Relaxation Models

For simplicity we consider first the scalar 1-D conservation law. The extension to the case of systems, if is done according to [19] is straightforward. Special relaxation models for the shallow water system (1) will be discussed in the sequel.

In the case of a conservation law with a source term

$$\begin{aligned} u_t + f(u)_x &= q(u), \quad x \in \mathbb{R},\ t > 0, \\ u(x,0) &= u_0(x), \quad x \in \mathbb{R}, \end{aligned} \tag{3}$$

a relaxation system that corresponds to the one of [19] for $q = 0$, considered e.g. in [6, 18], takes the form

$$\begin{aligned} u_t + v_x &= q(u), \\ v_t + c^2 u_x &= -\frac{1}{\epsilon}(v - f(u)), \end{aligned} \tag{4}$$

yielding the following regularization of (3),

$$u_t + f(u)_x = q(u) + \epsilon q(u)_t - \epsilon(u_{tt} - c^2 u_{xx}). \tag{5}$$

Hence (4) introduces a wave type regularization of (3) and the additional term $\epsilon q(u)_t$. We do not expect that (4) will preserve the general steady states of (3) unless they (or q) have a special structure that, allows that as we will see in the sequel for the shallow water system. Nevertheless the error terms in the steady state regime are $O(\epsilon)$, and since computations with relaxation schemes are done with very small ϵ (or even with $\epsilon = 0$ in the relaxed case) (4) will yield a reasonable model for computations in this regime. Another important feature of the relaxation model (4) in the homogeneous case is the implicit-explicit time discretization: Consider time discrete approximations $U^n \cong u(t^n), V^n \cong V(t^n)$ where $t^n = n\delta$, δ being the temporal stepsize. It is known that the stiff source term in (4) should be descritized implicitly,

$$\begin{aligned} \frac{U^{n+1} - U^n}{\delta} + V_x^n &= 0, \\ \frac{V^{n+1} - V^n}{\delta} + c^2 U_x^n &= -\frac{1}{\epsilon}(V^{n+1} - f(U^{n+1})). \end{aligned} \tag{6}$$

Obviously the system (6) can be decoupled, and since U^{n+1} is known from the first equation, the solution of a nonlinear system is not needed. However for (4) this is not the case. Indeed, discretizing the source terms implicitly we get

$$\begin{aligned} \frac{U^{n+1} - U^n}{\delta} + V_x^n &= q(U^{n+1}), \\ \frac{V^{n+1} - V^n}{\delta} + c^2 U_x^n &= -\frac{1}{\epsilon}(V^{n+1} - f(U^{n+1})), \end{aligned} \tag{7}$$

which is fully coupled.

As an alternative approach we consider the following relaxation system

$$\begin{aligned} u_t + v_x &= 0, \\ v_t + c^2 u_x &= -\frac{1}{\epsilon}(v - f(u)) - \frac{1}{\epsilon}R(u), \end{aligned} \tag{8}$$

where $R(u)$ is an antiderivative of $q(u)$,

$$R(u(x,t)) = \int_a^x q(u(s,t))ds\,.$$

In this case (8) provides exactly a wave-type regularization of (3), indeed we have

$$u_t + f(u)_x = q(u) - \epsilon(u_{tt} - c^2 u_{xx}). \tag{9}$$

An implicit-explicit time discretization is now possible when we treat the source terms implicitly:

$$\begin{aligned} \frac{U^{n+1} - U^n}{\delta} + V_x^n &= 0, \\ \frac{V^{n+1} - V^n}{\delta} + c^2 U_x^n &= -\frac{1}{\epsilon}(V^{n+1} - f(U^{n+1})) - \frac{1}{\epsilon}R(U^{n+1}). \end{aligned} \tag{10}$$

This idea is applied successfully in the sequel in the case of shallow water system, yielding efficient schemes with nice steady-state behavior.

2.1 Relaxation models for the shallow water system

Next we discuss the properties of certain relaxation models for for the Saint-Venant equations corresponding to (4) and (8) respectively. In addition we introduce a relaxation system that preserves the steady states of (1). The steady state solution of (1) satisfies

$$(hu)_x = 0 \quad (hu^2 + \frac{g}{2}h^2)_x = -ghZ', \tag{11}$$

i.e.,

$$\begin{aligned} Q &= hu = \text{constant}, \\ \frac{u^2}{2} &+ g(h + Z) = \text{constant}. \end{aligned} \tag{12}$$

A particular case that provides a benchmark for many approximating schemes is the flow at rest, i.e., when $u = 0$, and $h + Z = \text{constant}$.

The relaxation models for (1) corresponding to (4) and (8) respectively are

$$h_t + v_x = 0, \tag{13a}$$

$$Q_t + w_x = -ghZ', \tag{13b}$$

$$v_t + c_1^2 h_x = -\frac{1}{\epsilon}(v - Q), \tag{13c}$$

$$w_t + c_2^2 Q_x = -\frac{1}{\epsilon}(w - (\frac{Q^2}{h} + \frac{g}{2}h^2)), \tag{13d}$$

and

$$h_t + v_x = 0, \tag{14a}$$
$$Q_t + w_x = 0, \tag{14b}$$
$$v_t + c_1^2 h_x = -\frac{1}{\epsilon}(v - Q), \tag{14c}$$
$$w_t + c_2^2 Q_x = -\frac{1}{\epsilon}(w - (\frac{Q^2}{h} + \frac{g}{2}h^2)) + \frac{1}{\epsilon}R(Z;h), \tag{14d}$$

where

$$R(Z;h)(x,t) = \int_a^x gh(y,t)Z'(y)dy\,.$$

The constants c_1, c_2 should be chosen appropriately, so that certain subcharacteristic conditions hold true. Formally in the relaxation limit $\epsilon \to 0$ we recover (1), for both relaxation systems (13) and (14). Notice that in the case where $Z \equiv 0$ then the two systems are identical. The above systems are combined in the sequel with implicit-explicit finite element discretizations yielding a reliable class of schemes for the approximation of (1). Note that due to the special form of the source term in the shallow water system, cf. (13a), even the first system, (13), admits decoupled implicit-explicit discretizations. Although the resulting schemes perform in a nice way in all the test problems, the exact steady states of (1) are not obtained as solutions of (13) and (14) with the exception of the flow at rest with $Z'' = 0$. Overall the qualitative behavior of (14) is better, cf. Section 4.

3 Finite Element Schemes

We consider now the finite element discretizations of (13), and (14). We assume for simplicity that the data our problem and its solution in $[0, T]$ has compact support in a set $\Omega_1 \subset \mathbb{R}$. Let Ω be an appropriate set such that $\Omega_1 \subset \Omega$ where our numerical solution will be sought. We consider a classical one dimensional C^0 finite element space defined on an element decomposition $\mathcal{T}_\kappa$ of Ω:

$$\mathcal{S}_\kappa = \mathcal{S}_\kappa^1 = \{\phi \in C(\mathbb{R}) : \phi|_T \in \mathbb{P}_1(T), T \in \mathcal{T}_\kappa, \phi|_{\Omega^c} \equiv 0\}\,.$$

The schemes are extended in a straightforward manner using finite element spaces of arbitrary degree p:

$$\mathcal{S}_\kappa = \mathcal{S}_\kappa^p = \{\phi \in C(\mathbb{R}) : \phi|_T \in \mathbb{P}_p(T), T \in \mathcal{T}_\kappa, \phi|_{\Omega^c} \equiv 0\}\,.$$

We denote by T the elements, intervals in the one dimensional case, of the partition $\mathcal{T}_\kappa$, and κ_T the length of the interval T. Then the finite element discretization of (14) is : we seek functions $h_\kappa, Q_\kappa, v_\kappa, w_\kappa \in \mathcal{S}_\kappa$ so that,

$$(\partial_t h_\kappa, \phi) + (\partial_x v_\kappa, \phi) = 0, \quad \forall \phi \in \mathcal{S}_\kappa, \tag{15a}$$
$$(\partial_t Q_\kappa, \psi) + (\partial_x w_\kappa, \psi) = 0, \quad \forall \psi \in \mathcal{S}_\kappa, \tag{15b}$$
$$(\partial_t v_\kappa, \phi) + c_1^2(\partial_x h_\kappa, \phi) = -\frac{1}{\epsilon}(v_\kappa - Q_\kappa, \phi), \quad \forall \phi \in \mathcal{S}_\kappa, \tag{15c}$$
$$\begin{aligned}(\partial_t w_\kappa, \psi) + c_2^2(\partial_x Q_\kappa, \psi) = &-\frac{1}{\epsilon}(w_\kappa - F(h_\kappa, Q_\kappa), \psi) \\ &+ \frac{1}{\epsilon}(R(Z; h_\kappa), \psi), \quad \forall \psi \in \mathcal{S}_\kappa,\end{aligned} \tag{15d}$$

where

$$F(h, Q) = \frac{Q^2}{h} + \frac{g}{2}h^2.$$

These schemes are high order and are combined with corresponding implicit-explicit Runge-Kutta time discretizations as in [1]. The behavior of the resulting schemes in several test problems is presented in Section 4.

The qualitative features of our finite element schemes as well as their relation to the classical finite difference schemes of [19] were discussed in detail in [1]. We briefly recall here that the effective equation corresponding to these schemes has a dispersion term coming from the finite element discretization. If this term is dominant, the approximate solution will have oscillatory character close to the shocks. Hence in order to exclude this possibility, as it was observed in [1], a condition of the form

$$\kappa < \gamma\epsilon$$

should be respected in a neighborhood of the shock, where γ is a constant. Hence, if the the local mesh size κ close to the shock is small enough the schemes are free of oscillations, cf. [1] for details.

In order to locate the shock and to use finer mesh there we use the *estimator function* g introduced in [1]. This is a piecewise linear function with point values

$$\kappa_j |u(x_j, t^m) - u(x_{j-1}, t^m)|$$

where $\kappa_j = x_j - x_{j-1}$ and has the property to *locate accurately the (approximate) shocks.* Its construction was motivated by the a posteriori estimates of [15, 16]. This function is used in the mesh refinement shock computations in the next section. We note though that the proposed schemes are not necessarily linked to a particular method of the mesh selection. This can be done in many alternative ways. The only requirement of our method is to have fine enough mesh size close to the shocks. (Note that in practice γ ranges from $10^2 - 10^3$).

4 Numerical experiments

We present the results of a series of numerical experiments illustrating the various features of the schemes. We have tested the fully discrete analogs, of the finite element method for (13), (14) in the piecewise linear case. We still refer these methods as (13), (14). Overall the qualitative behavior of (14) is better than of (13). So we display results of (13) only in the first example—the flow at rest. A general observation is that, although our schemes do not satisfy exactly the steady state equation (12) but rather approximations of it, the results in the standard steady state benchmarks presented in the sequel are very promising. Issues related to fine $h-p$ adaptive computations as well as computations in more than one dimension will be the subject of a forthcoming work.

In all the experiments reported here the time step δ is computed according to the *CFL*-condition

$$\max\{c_1, c_2\}\frac{\delta}{\kappa} \le \frac{1}{2}.$$

The adaptive strategy followed in these experiments is very similar to the one reported in [1] : while the number of nodes is kept fixed throughout the computation, their locations are changing accordingly.

4.1 Flow at rest

We consider the system (1) with initial conditions

$$\begin{aligned} u(x,0) &= 0 \quad \forall x \in \mathbb{R}, \\ h(x,0) + Z(x) &= H \quad \forall x \in \mathbb{R}, \end{aligned}$$

then clearly

$$\begin{aligned} u(x,t) &= 0 \quad \forall x \in \mathbb{R}, t \ge 0, \\ h(x,t) + Z(x) &= H \quad \forall x \in \mathbb{R}, t \ge 0, \end{aligned}$$

is a solution to (1).

We test our scheme to this steady state flow where the bathymetry is non-trivial and is given by

$$Z(x) = \begin{cases} 0.2 - 0.05(x-10)^2, & 8 \le x \le 12, \\ 0, \quad \text{otherwise}, \end{cases} \tag{16}$$

in channel of length $L = 20m$ and $H = 2m$, $\epsilon = 5.E-4$, $c_1 = c_2 = 5$.

Figures 1, and 2 show a magnified view of the results obtained by these schemes. The variance of the values of the water level $H = 2m$, are of $O(\epsilon)$. The differences are become more clear in the comparison of discharge in figure 2. There clearly (14) is better than (13), $(Q = 0)$.

4.2 Transcritical flow with shock

In this test case the upstream boundary condition for the discharge is $Q = 0.18m^2/s$ and the downstream boundary condition for the water level is $h =$

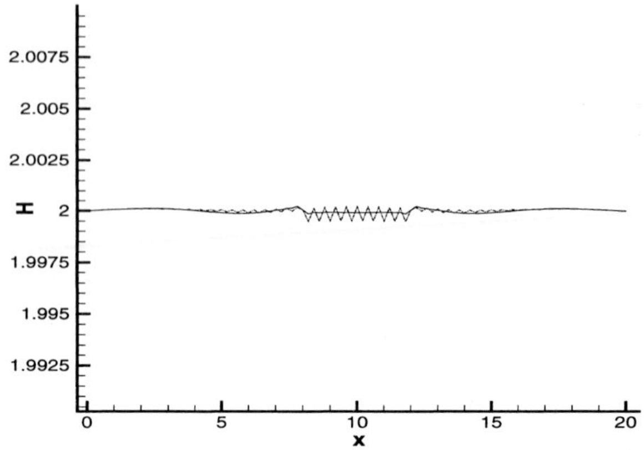

Fig. 1. *Magnified view.* Water level: (13)(thicker solid line), (14)(fine solid line)

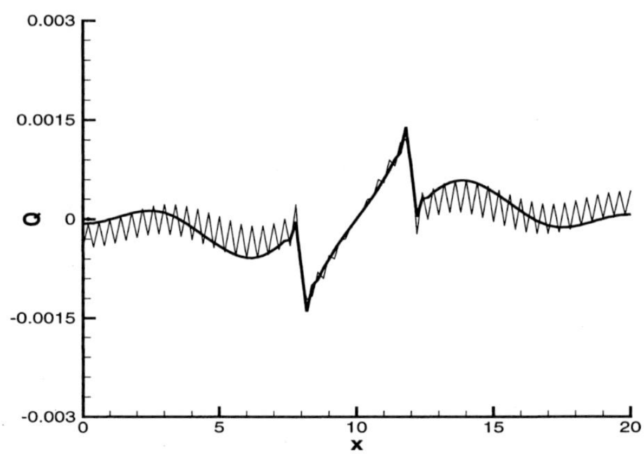

Fig. 2. *Magnified view.* Discharge: (13)(fine solid line),(14)(thicker solid line)

$0.33m$. Local refinement was applied to obtain the results. The final water level of (13), (14) is displayed in Figure 3. Figure 4 displays the discharge given by the scheme (14).

Acknowledgment. We would like to thank Prof. B. Perthame for several helpful discussions.

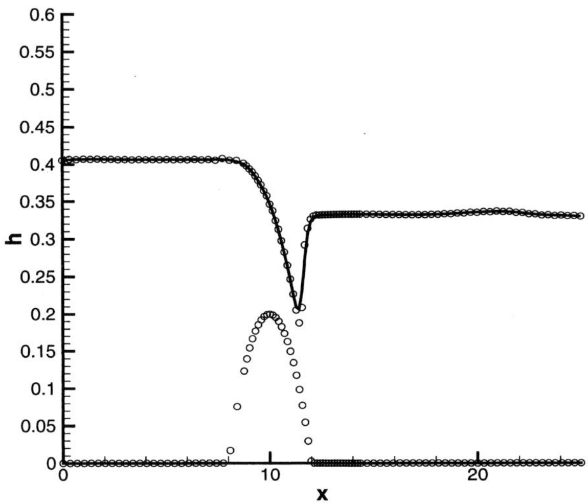

Fig. 3. Transcritical flow with shock over a hump : Water level. (14)(circles), (13)(solid line).

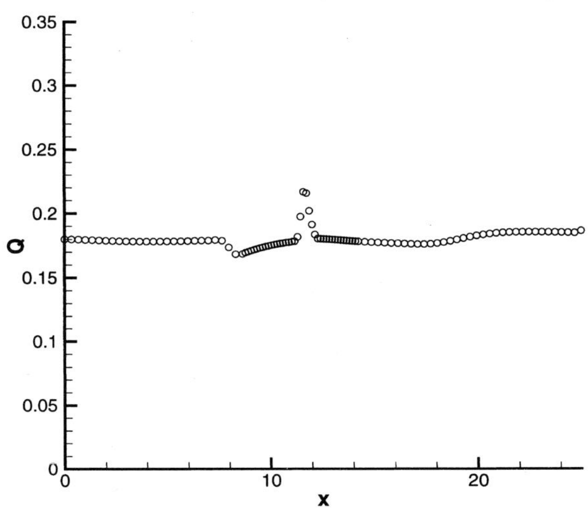

Fig. 4. Transcritical flow with shock over a hump : Discharge,(14).

References

1. Arvanitis, Ch., Katsaounis, T., Makridakis, Ch. (2001): Adaptive finite element relaxation schemes for hyperbolic conservations laws, Math. Model. Numer. Anal., **35**, 17–33
2. Aral, M., Zhang, Y., Jin, S., (1998): Application of relaxation scheme to wave propagation simulation in open channel networks, J. Hydraulic Engineering, **124**, 1125-1133
3. Audusse, E., Bristeau, M.O., Perthame, B. (2000): Kinetic schemes for Saint–Venant equations with source terms on unstructured grids, INRIA Report RR–3989
4. Delis, A., Katsaounis, Th., (2002): Relaxation schemes for the shallow water equations, Report DMA-02-02, ENS, Paris, France
5. Bale, D.S., LeVeque, R.J., Mitran, S., Rossmanith, J.A. (2001): A wave propagation method for conservation laws and balance laws with spatially varying flux functions, Univ. of Washington, Seattle, Applied Math. Tech. Rep.
6. Chalabi, A. (1999): Convergence of relaxation schemes for hyperbolic conservation laws with stiff source terms, Math. Comp., **68**, pp.955–970.
7. Coquel, F., Perthame, B., (1998): Relaxation of energy and approximate Riemann solvers for general pressure laws in fluid dynamics, SIAM J. Numer. Anal., **35**, 2223–2249
8. Dafermos, C. (1999): Hyperbolic conservation laws in continuum physics, Springer-Verlag
9. Gallouët, T., Herard, J.H., Seguin, N. (2000): Some approximate Godunov schemes to compute shallow water equations with topography, AIAA-2001
10. Gerbeau, J.F., Perthame, B. (2001): Derivation of viscous Saint–Venant system for laminar shallow water, Discrete Contin. Dynam. Systems, Series B, **1**, 89–102
11. Greenberg, J.M., LeRoux, A.Y. (1996): A well balanced scheme for the numerical processing of source terms in hyperbolic equations, SIAM J. Num. Anal., **33**, 1–16
12. Gosse, L. (2000): A well–balanced flux vector splitting scheme designed for hyperbolic systems of conservation laws with source terms, Comput. Math. Appl., **39**, 135–159
13. Gosse, L. (2001): A well balanced scheme using non-conservative products designed for hyperbolic systems of conservation laws with source terms Math. Mod. Meth. Appl. Sci., **11**, 339–365
14. Gosse, L., LeRoux, A.Y. (1996): A well–balanced scheme designed for inhomogeneous scalar conservation laws C.R. Acad. Sci. Paris Sér. I Math., **323**, 543-546
15. Gosse, L., Makridakis, Ch. (1998): A-posteriori error estimates for numerical approximations to scalar conservations laws: schemes satisfying strong and weak entropy inequalities, IACM-FORTH, Technical Report 98-4
16. Gosse, L., Makridakis, Ch. (2000): Two a-posteriori error estimates for one dimensional scalar conservation laws, SIAM J. Numer. Anal., **38**, 964-988
17. Gosse, L., Tzavaras, A.E. (2001): Convergence of relaxation schemes to the equations of elastodynamics, Math. Comp. **70**, 555–577
18. Jin, S. (2001): A steady-state capturing method for hyperbolic systems with geometrical source terms, Math. Model. Numer. Anal., **35**, 631-645

19. Jin, S., Xin, Z. (1995): The relaxing schemes of conservations laws in arbitrary space dimensions, Comm. Pure Appl. Math, **48**, 235–277
20. LeVeque, R.J. (1998): Balancing source terms and flux gradients in high resolution Godunov methods: the quasi–steady wave propagation algorithm, J. Comput. Phys., **146**, 346-365
21. Perthame, B., Simeoni, C. (2001): A kinetic scheme for the Saint–Venant system with a source term, Calcolo, **38**, 201–231
22. de Saint–Venant, A.J.C. (1871): Théorie du mouvement non–permanent des eaux, avec application aux crues des rivière et à l'introduction des mareé dans leur lit, C.R.Acad. Sci. Paris, **73**, 147-154
23. Shu, C.W. (1988): Total-variation-diminishing time discretizations, SIAM J. Sc. Comp., **9**, 1073–1084
24. Stoker, J.J. (1986): Water waves, Interscience Publishers Inc., New York

Second Order Approximation of the Viscous Saint-Venant System and Comparison with Experiments

Theodoros Katsaounis and Chiara Simeoni

Département de Mathématiques et Applications - École Normale Supérieure
45, rue d'Ulm - 75230 Paris Cedex 05 - France
katsaoun@dma.ens.fr, simeoni@dma.ens.fr

Summary. We present numerical simulations of the Saint-Venant system for shallow waters, including small friction and viscosity, motivated by the interest in recovering the results of experimental studies on the free-surface flows over an obstacle. We use the kinetic scheme "with reflexions" formulated in [17], appropriately extended to obtain second order accuracy according to the theory developed in [11].

1 Introduction

The Saint-Venant equations for shallow waters were originally written by A. de Saint-Venant in 1871 from heuristic considerations about the mechanisms governing physical phenomena such as the flows in rivers or coastal areas (see [20]).

In a one-dimensional framework, these equations constitute simple mathematical model for the flow in ideal rectangular rivers, described at time $t \geq 0$ and at point $x \in \mathbb{R}$ through the height of water $h(t,x) \geq 0$ and its velocity $u(t,x) \in \mathbb{R}$ by means of the hyperbolic system

$$\frac{\partial h}{\partial t} + \frac{\partial}{\partial x}(hu) = 0, \tag{1}$$

$$\frac{\partial}{\partial t}(hu) + \frac{\partial}{\partial x}\left(hu^2 + \frac{g}{2}h^2\right) + ghZ' = 0, \tag{2}$$

where g denotes the gravity intensity and $Z(x)$ is the bottom topography; therefore $h+Z$ is the level of the water surface and, in what follows, we also denote the discharge by $q = hu$.
Besides, other terms can be added to the right-hand side of equation (2) to take into account further natural features of the physical context, for instance friction on the bottom and viscosity inside the fluid.

The question to introduce more complete systems becomes crucial when one deals with the experimental verification of situations typically occurring in hydraulics, to provide a classification of different flow regimes in presence of an obstacle, for which the description based on the model (1)-(2) is unsatisfactory. Indeed, for dam breaks or hydraulic jumps, it does not allow to recover mathematically the right position with respect to the topography (see [12], for example) and the interaction of the source terms corresponding to bottom slope and friction in the shallow water equations is dominant for characterizing the steady states.

Despite its simple configuration, the shallow water flow in channels with nontrivial topography present a wide variety of regimes, producing some peculiar behaviours of the free-surface (wave trains, hydraulic jumps, turbulent profiles), which have not yet been fully examined. In [1], the results of classical analytical theories are reviewed, even though these models are validated only in the weakly nonlinear and weakly dispersive limits. Based on two-dimensional numerical simulations of the nonlinear uniform potential flow around moving obstacles, an accurate description of the different types of breaking waves is proposed in [14]. However, due to the inherent limitations of the theoretical formulations, their predictions have to be interpreted in terms of experimental verifications.

We refer to [26] for a survey of the experimental study of free-surface flows over an obstacle: the behaviour of an incident subcritical channel flow is investigated for various blocking factors (namely, obstacle shape and stationary water depth); the results are analyzed in comparison with the classification schemes proposed in the previous works (see the references in that paper).

We present in this paper some numerical simulations of the Saint-Venant system, according to the experimental configuration set in [26]. To take into account dissipative effects in the physical phenomenon, we consider a modified equation for the momentum including small friction and viscosity, then the system under analysis reads

$$\frac{\partial h}{\partial t} + \frac{\partial}{\partial x}(hu) = 0, \tag{3}$$

$$\frac{\partial}{\partial t}(hu) + \frac{\partial}{\partial x}\left(hu^2 + \frac{g}{2}h^2\right) + ghZ' = -\frac{g}{K^2}\frac{u|u|}{h^{1/3}} + \mu\frac{\partial}{\partial x}\left(h\frac{\partial u}{\partial x}\right), \tag{4}$$

where K is the Strickler's coefficient of the Manning's equation and μ denotes the kinematic viscosity of the fluid. The particular form of the source terms is suggested by empirical laws, which were originally obtained for steady state flows (refer to [25]). Similar models have been derived from the Navier-Stokes system for incompressible flows with a free moving boundary (see [5] and its references) and more complex laws for the friction term can also be formulated to modelize analogous problems in the case of granular media (snow avalanches, for example, as referred in [15]).

The numerical approximation of system (3)-(4) is carried out by means of the kinetic scheme proposed and extensively studied in [17]. In fact, for all experimental tests performed in the works quoted above, the flow was observed to be stationary and a zone of supercritical flow downstream the obstacle always occurs, though the cases examined concern the subcritical regime. So, we use a method which preserves the free-surface profile of steady states with nontrivial bottom topography and which is able to deal with transcritical regimes.

The discretization of the additional source terms in (4) is rather standard, based on a semi-implicit approach for the friction term and direct integrations by the finite volume method for the diffusive term. Moreover, when friction and viscosity are neglected, we recover the system (1)-(2). Appropriate extensions of the primitive algorithm are thus considered, in order to improve the numerical accuracy.

The paper is organized as follows. In Section 2, we recall some specific notations of the Upwind Interface Source method for hyperbolic conservation laws with geometrical source term, illustrated in [18], by extending its general formalism to the system (1)-(2). We also introduce the approximation of the dissipative terms, for treating the system (3)-(4). Connected with the numerical approach developed in [17], the question to derive a second order scheme is addressed in Section 3. We describe the experimental configuration underling our analysis in Section 4 and we present the results of numerical simulations made according to the experimental data provided in [26]. Some remarks are discussed to justify theoretical tests in comparison with the experiments.

2 Formalism of the numerical method

The Saint-Venant system for shallow waters (1)-(2) belongs to the class of the hyperbolic systems of balance laws, with a geometrical source term, and can be written in the equivalent form

$$\frac{\partial \mathcal{U}}{\partial t} + \frac{\partial}{\partial x} A(\mathcal{U}) = B(x, \mathcal{U}), \tag{5}$$

where $\mathcal{U} = (h, hu)$ represents the vector of conservative variables, the flux function is given by $A(\mathcal{U}) = \left(hu, hu^2 + \frac{g}{2}h^2\right)$ and $B(x,\mathcal{U}) = (0, -ghZ')$ indicates the external term.

The equation (5) reproduces the general formalism introduced in [18] for the particular case of scalar conservation laws, so the numerical theory stated in that context formally extends to system (1)-(2), to characterize approximations with suitable theoretical properties.

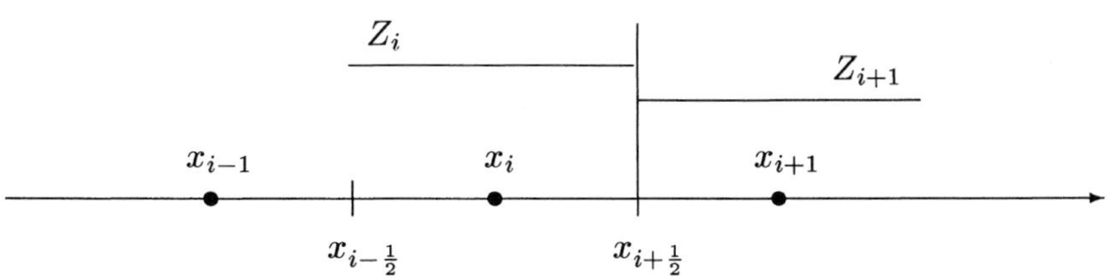

We set up a mesh on $\mathbb{R}$, whose central vertices are x_i, $i \in \mathbb{Z}$, made of cells $C_i = [x_{i-\frac{1}{2}}, x_{i+\frac{1}{2}})$ with nonuniform length Δx_i and the points $x_{i+\frac{1}{2}}$ indicate the cell interfaces. We also consider a time discretization t^n, $n \in \mathbb{N}$, with variable time-step Δt. Then we construct a piecewise constant representation of the function $Z(x)$ on the mesh, with coefficients $Z_i = \frac{1}{\Delta x_i} \int_{C_i} Z(x)\,dx$ for example.

A classical approach to nonlinear hyperbolic problems consists in using finite volume methods, which are designed for computations with arbitrary meshes (refer to [4], for instance). Taking into account the source term directly in the definition of the numerical fluxes, the fully explicit finite volume scheme for equation (5) is written in the compact form

$$\mathcal{U}_i^{n+1} - \mathcal{U}_i^n + \frac{\Delta t}{\Delta x_i}\left(\mathcal{A}_{i+\frac{1}{2}}^{n,-} - \mathcal{A}_{i-\frac{1}{2}}^{n,+}\right) = 0, \tag{6}$$

where $\mathcal{A}_{i+\frac{1}{2}}^{n,\pm} = \mathcal{A}^{\pm}(\mathcal{U}_i^n, \mathcal{U}_{i+1}^n, \Delta Z_{i+\frac{1}{2}})$, with $\Delta Z_{i+\frac{1}{2}} = Z_{i+1} - Z_i$, are defined by means of appropriate numerical functions $\mathcal{A}^{\pm} = (\mathcal{A}_h^{\pm}, \mathcal{A}_q^{\pm})$ for each component of (5) and the following consistency properties are required,

$$\mathcal{A}_h^+(\mathcal{U}, \mathcal{V}, \Delta Z) = \mathcal{A}_h^-(\mathcal{U}, \mathcal{V}, \Delta Z), \tag{7a}$$

$$\mathcal{A}_q^+(\mathcal{U}, \mathcal{V}, \Delta Z) - \mathcal{A}_q^-(\mathcal{U}, \mathcal{V}, \Delta Z) = -gh\Delta Z + \mathcal{O}\left(|\Delta Z| + |\mathcal{U} - \mathcal{V}|\right), \tag{7b}$$

$$\mathcal{A}^+(\mathcal{U}, \mathcal{U}, 0) = \mathcal{A}^-(\mathcal{U}, \mathcal{U}, 0) = A(\mathcal{U}). \tag{7c}$$

We note that condition (7a) ensures that the numerical fluxes are conservative for the equation (1). Moreover, we have $\mathcal{A}_{i+\frac{1}{2}}^{n,\pm} = \mathcal{A}_{i+\frac{1}{2}}^n + (\mathcal{A}_{i+\frac{1}{2}}^{n,\pm} - \mathcal{A}_{i+\frac{1}{2}}^n)$ and the quantity in parentheses holds for discrete contributions of the source term at the cell interfaces, according to the Upwind Interface Source method; therefore, the relation (7b) guarantees consistency with the continuous model, as readily obtained by standard asymptotic expansions. We deduce from (7c) that the numerical scheme (6) satisfies the classical definitions for homogeneous problems.

The kinetic scheme for the Saint-Venant system proposed in [17] is compatible with the above formalism and endowed with further stability properties associated to the physical model (it preserves the steady state of still water, satisfies a discrete entropy inequality and makes non-negative water height).

In order to perform numerical simulations with experimental data, we consider the modified shallow water equations (3)-(4), for which that scheme applies to terms corresponding to the hyperbolic system (1)-(2).
The discretization of the friction term is implicit (see [16], for example) and splitted into two steps, only concerning the equation (4), which include the approximation of the viscous term,

$$q_i^{n+\frac{1}{2}} - q_i^n + \frac{\Delta t}{2\Delta x_i}\left(\mathcal{A}_{q,\,i+\frac{1}{2}}^{n,-} - \mathcal{A}_{q,\,i-\frac{1}{2}}^{n,+}\right) = \mu \frac{\Delta t}{2\Delta x_i}\left(V_{i+\frac{1}{2}}^n - V_{i-\frac{1}{2}}^n\right), \tag{8}$$

$$q_i^{n+1} - q_i^{n+\frac{1}{2}} = -\frac{\Delta t}{2}\frac{g}{K^2}\frac{q_i^{n+1}|q_i^{n+\frac{1}{2}}|}{\left(h_i^{n+1}\right)^{\frac{7}{3}}}, \tag{9}$$

where the numerical formulas used for calculating viscosity,

$$V^n_{i+\frac{1}{2}} = \frac{h_i + h_{i+1}}{2} \frac{u_{i+1} - u_i}{\Delta x_{i+\frac{1}{2}}}, \tag{10}$$

with $\Delta x_{i+\frac{1}{2}} = \frac{\Delta x_i}{2} + \frac{\Delta x_{i+1}}{2}$, are derived by means of simple finite volume integrations on the mesh cells and appropriate approximations of the resulting interfacial values (we note that the discretization (10) can be reinterpreted according to the classical first order finite element method). Some different methods for processing friction terms are proposed in [3], [6] and [7].

3 Second order schemes

To obtain second order extensions of finite volume schemes in form (6), a rather geometrical approach is based on *slope limiter* techniques.
We construct a piecewise linear approximation of the function $Z(x)$ on the mesh, whose coefficients are $Z_i + (x - x_i)Z_i'$, $x \in \mathcal{C}_i$, and we denote by Z_i' the numerical derivatives computed by applying an appropriate *slope limiter* (we refer to [8] and [13] for a survey of these discrete operators).

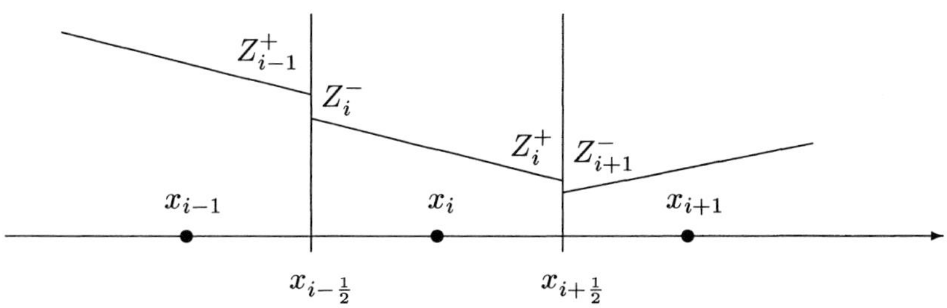

According to the arguments in [11], a second order scheme for the Interfacial Source method formally reads

$$\mathcal{U}_i^{n+1} - \mathcal{U}_i^n + \frac{\Delta t}{\Delta x_i}\left(A^{n,-}_{i+\frac{1}{2}} - A^{n,+}_{i-\frac{1}{2}}\right) + B_i^n = 0, \tag{11}$$

where the numerical fluxes $A^{n,\pm}_{i+\frac{1}{2}} = \mathcal{A}^{\pm}(\mathcal{U}_i^{n,+}, \mathcal{U}_{i+1}^{n,-}, \Delta Z_{i+\frac{1}{2}})$ use the interfacial values of piecewise linear reconstructions of the numerical functions,

$$\mathcal{U}_i^{n,\pm} = \mathcal{U}_i^n \pm \frac{\Delta x_i}{2}\mathcal{U}_i^{n\,\prime}, \qquad Z_i^{\pm} = Z_i \pm \frac{\Delta x_i}{2} Z_i', \tag{12}$$

and therefore $\Delta Z_{i+\frac{1}{2}} = Z^-_{i+1} - Z^+_i$ in this case.
The additional discrete source term is defined by $B_i^n = (0, g Z_i' h_i^n)$. Although other methods exhibit noticeable improvements, it was shown in [11] that the centered term B_i^n is necessary to achieve second order accuracy for the

Upwind Interface Source method, if the *slope limiter* used to construct the values (12) is correctly defined (see [21], [22] and [23]).

For the sake of simplicity, we consider in (11) only the first order discretization in time. It is standard to obtain higher order accuracy by applying Runge-Kutta methods for instance (see [9], [10] and its references).
The second order scheme (11) is validated by the numerical results obtained for the steady states of the Saint-Venant system (1)-(2).

4 Experimental configuration and numerical results

The situation studied in [26] is the one-dimensional flow of an incompressible fluid over an obstacle on the bottom of a smooth rectangular channel (see Figure 1).
The parameters of the problem are the stationary water depth H and the mean velocity U, the kinematic viscosity of the fluid μ, the characteristic length of the obstacle L and its height Z_0. By combining these values, we define some specific dimensionless numbers: the Froude number $F_0 = \frac{U}{\sqrt{gH}}$, which relates the depth-averaged flow velocity to the characteristic wave propagation speed, in the long waves approximation; the blocking factor $\alpha = \frac{Z_0}{H}$, controlling the flow linearity in the absence of other perturbations (when α tends to zero, the flow becomes linear); the obstacle ratio $\beta = \frac{Z_0}{L}$, which can be interpreted as a control on the flow hydrostaticity (by analogy with the long waves approximation); the Reynolds number $Re = \frac{HU}{\mu}$, for the simple case of an ideal rectangular channel.
We note that the definitions above correspond to simplifications adapted to the theoretical system described by (3)-(4), when the fluid density and the channel width are formally reduced.

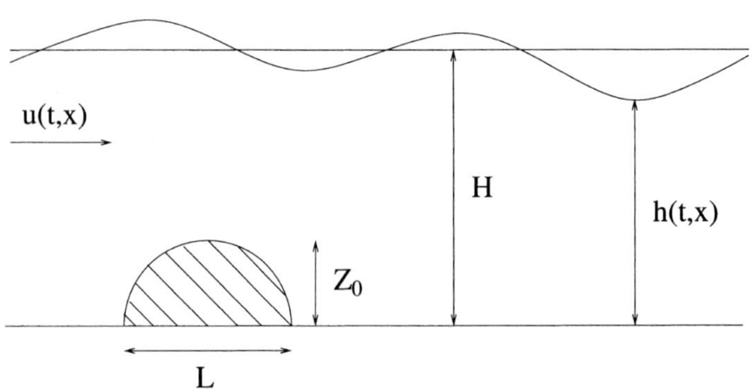

Fig. 1. Experimental configuration

The experiments have been carried out in a channel of length $12\,m$, inclined at slope 0.002 and entirely made of glass. Two obstacle shapes are considered, a smooth Gaussian bump given by $Z(x) = Z_0 \exp\left(-\frac{x^2}{2L^2}\right)$, with ratio $\beta = 0.23$, and a semi-circular bump with ratio $\beta = 0.58$ (this last shape is commonly used for experimental analyses). Two sizes $Z_0 = 1.7\,cm$ and $Z_0 = 4.1\,cm$ are fixed for each shape, which allow to access a wider range of α values ($0.147 \leq \alpha \leq 0.7$). The tests are performed in subcritical regime ($F_0 < 1$) and the obstacles are placed in the channel so that fully developed turbulent flow conditions are attained before the obstacle.

Table 1.

F_0	Z_0	α	β	Re
0.62	$1.7cm$	0.179	0.58	$3.08 * 10^4$
0.66	$1.7cm$	0.26	0.58	$1.51 * 10^4$
0.64	$4.1cm$	0.68	0.58	$1.32 * 10^4$

We present some numerical results corresponding to the test cases illustrated in Table 1, for which experimental data are available to make direct comparisons. The pairs of figures reproduce the level of the water surface and the local Froude number in the vicinity of the obstacle, obtained at time $T = 150\,s$ (when the flow has become completely stationary), normalized with respect to the obstacle size. According to the experiments, the flow is classified by means of the free-surface profile and three different regimes are observed as function of the blocking factor. The boundaries of each regime, in terms of α values, are essentially independent of the obstacle shape.
Several tests have been performed also varying the Strickler's coefficient K introduced in (4), to evaluate the physical adequacy of the friction law (refer to [19] for a more precise discussion). The nonuniform mesh used for the numerical simulations is refined around the obstacle or in regions of stiff topographical variations, with a minimal mesh size of $5 * 10^{-4}\,m$.
The subcritical flow downstream of the obstacle displays many of the features of a supercritical flow behind a sluice gate with ensuing hydraulic jumps, implying a certain independence of the upstream conditions. The treatment of the boundary conditions, which turns out to be crucial for the numerical accuracy, is provided by the method developed in [2].

Regime I ($\alpha \leq 0.25$)

The graphics of the local Froude number show a transition from supercritical to subcritical flow downstream of the obstacle, through a hydraulic jump (Figure 3); the flow remains subcritical downstream the jump. The lower boundary of this regime is expected to be the value of α below which a classical subcritical regime should occur.

The results of the experimental test reveal the presence of a wave train on the free-surface profile (Figure 2). Moreover, some photographs taken

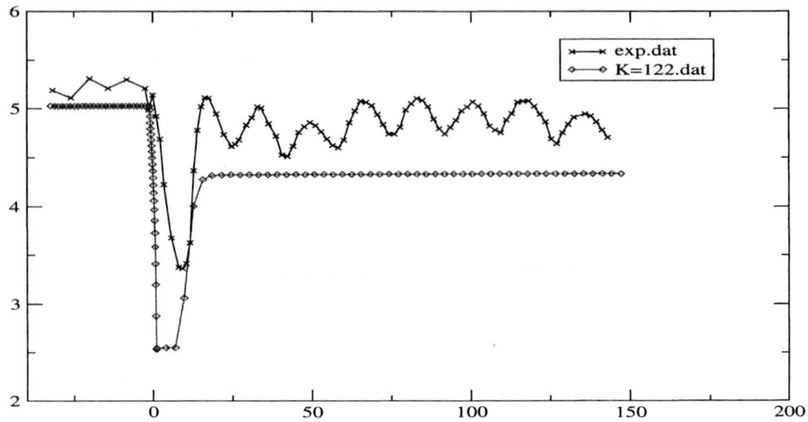

Fig. 2. Water surface ($\alpha = 0.179$)

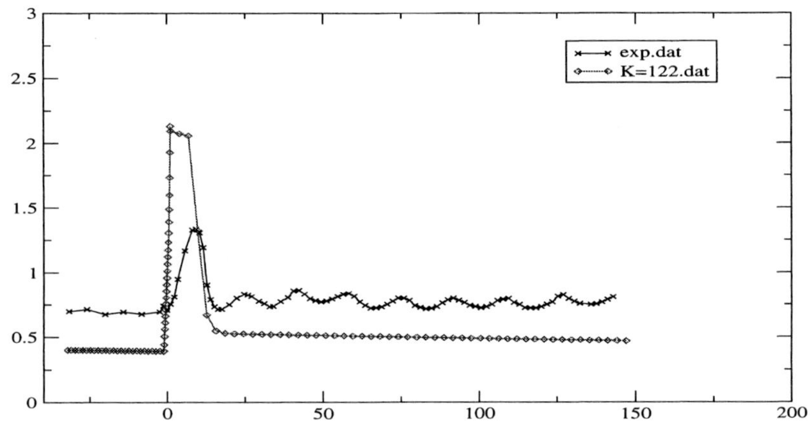

Fig. 3. Local Foude number

during the experiments show bubbles arising on the crest of the hydraulic jump, as consequence of the turbulent motion. These phenomena are not taken into account in the mathematical model (3)-(4) and cannot be obtained numerically (we refer to [24] for further analysis).

Regime II $(0.26 \leq \alpha < 0.68)$

The graph of the local Froude number shows that a region of supercritical flow appears downstream the obstacle, followed by a transition from supercritical to subcritical flow through a hydraulic jump (Figure 5); however, in this regime, the local Froude number undergoes further transitions and its values remain close to $F=1$ downstream the jump.

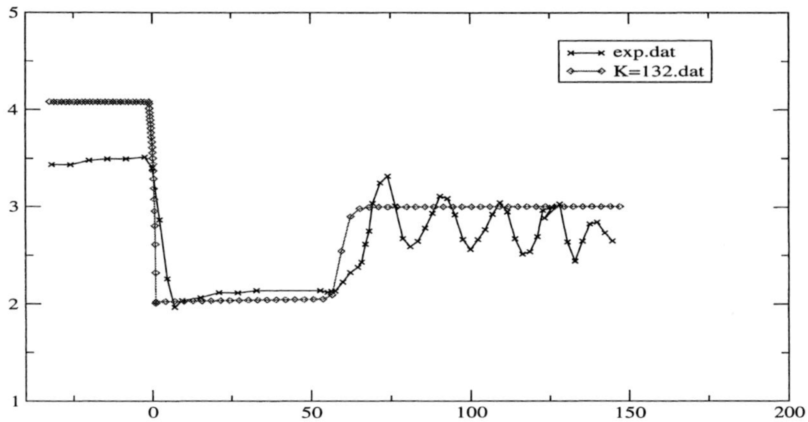

Fig. 4. Water surface ($\alpha = 0.26$)

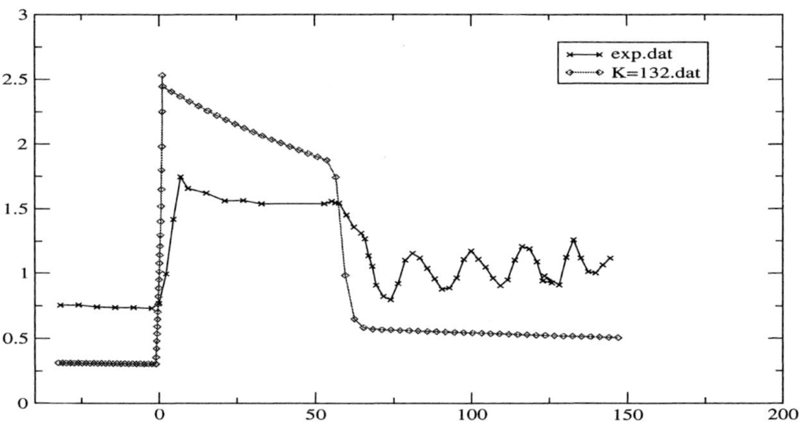

Fig. 5. Local Foude number

In the experimental tests, the local Froude number is oscillating around the critical value and the wave train on the free-surface profile becomes a series of triangular hydraulic jumps. This flow regime is essentially three-dimensional, the friction on the channel walls plays a significant role for the energy dissipation across the jumps (inducing water deceleration) and confinement effects seem to control the behaviour of the flow. Theoretical values obtained for the water depth (Figure 4) are higher than the experimental ones, suggesting that the numerical results are necessarily inaccurate.
Nevertheless, as two-dimensional side wall effects have not been considered in the model (3)-(4), it seems impossible to reproduce all the phenomena with one-dimensional simulations.

Regime III ($\alpha \geq 0.68$)

The experiments performed until $\alpha = 0.7$ show that the flow remains supercritical downstream the obstacle (Figure 7); in other words, no hydraulic jumps arise in this regime. Small perturbations appear on the free-surface profile, in form of a wave train located downstream the obstacle (Figure 6).

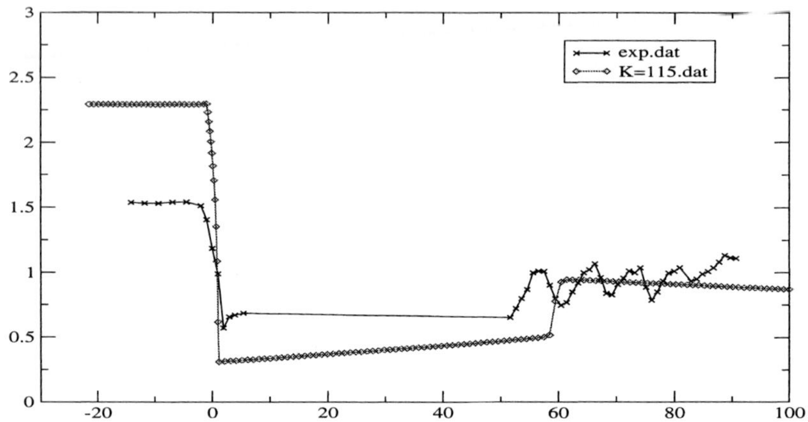

Fig. 6. Water surface ($\alpha = 0.68$)

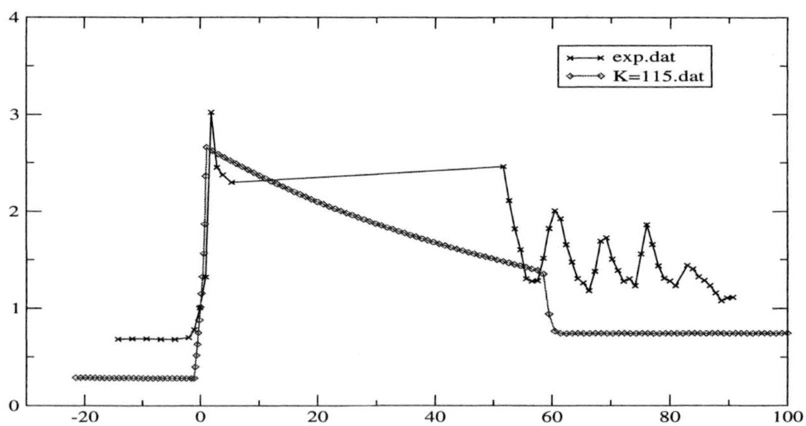

Fig. 7. Local Foude number

We remark a rather good agreement with the experimental results concerning the position of hydraulic jumps and the length of the supercritical region predicted by the shallow water equations (3)-(4), when friction and viscosity are included.

Acknowledgments

This work was partially supported by the ACI "Modélisation de processus hydrauliques à surface libre en présence de singularités" (Ministre de la Recherche - France).

We would like to thank François Bouchut (DMA-ENS) and Marie-Odile Bristeau (INRIA) for several helpful discussions.

References

1. P.G. Baines, *Topographic effects in stratified flows*, Cambridge Monographs on Mechanics, Cambridge University Press, Cambridge, 1995
2. M.O. Bristeau, B. Coussin, Boundary conditions for the shallow water equations solved by kinetic schemes, *INRIA Report*, **4282** (2001)
3. A. Chinnayya, A.Y. Le Roux, A new general Riemann solver for the shallow water equations with friction and topography, preprint
4. R. Eymard, T. Gallouët, R. Herbin, *Finite Volume Methods*, Handbook of numerical analysis, vol. VIII, P.G.Ciarlet and J.L.Lions editors, Amsterdam, North-Holland, 2000
5. J.F. Gerbeau, B. Perthame, Derivation of viscous Saint-Venant system for laminar shallow water; numerical validation, *Discrete Contin. Dyn. Syst. Ser.B*, **1** (2001), no. 1, 89-102
6. P. Glaister, The efficient prediction of shallow water flows. I. Theory, *Comput. Math. Appl.*, **32** (1996), no. 12, 129-143
7. P. Glaister, The efficient prediction of shallow water flows. II. Application, *Comput. Math. Appl.*, **33** (1997), no. 9, 115-141
8. E. Godlewski, P.A. Raviart, *Hyperbolic systems of conservation laws*, Mathmatiques & Applications, no. 3/4, Ellipses, Paris, 1991
9. S. Gottlieb, C.W. Shu, Total variation diminishing Runge-Kutta schemes, *Math. Comp.*, **67** (1998), no. 221, 73-85
10. S. Gottlieb, C.W. Shu, E. Tadmor, Strong stability-preserving high-order time discretization methods, *SIAM Rev.*, **43** (2001), no. 1, 89-112
11. T. Katsaounis, C. Simeoni, First and second order error estimates for the Upwind Interface Source method, preprint
12. G.A. Lawrence, Steady flow over an obstacle, *J. Hydraul. Engng*, ASCE, **113** (1987), no. 8, 981-991
13. R.J. LeVeque, *Numerical methods for conservation laws*, Lectures in Mathematics ETH Zürich, Birkhäuser Verlag, Basel, 1990
14. K. Lowery, S. Liapis, Free-surface flow over a semi-circular obstruction, *Internat. J. Numer. Methods Fluids*, **30** (1999), 43-63
15. A. Mangeney-Castelnau, J.P. Vilotte, M.O. Bristeau, B. Perthame, C. Simeoni, S. Yernini, Numerical modelling of avalanches based on Saint-Venant equations using kinetic schemes, submitted to *Journal of Geophysical Research*
16. A. Paquier, *Modélisation et simulation de la propagation de l'onde de rupture de barrage*, PhD Thesis, Université J. Monnet, St-Étienne (France), 1995
17. B. Perthame, C. Simeoni, A kinetic scheme for the Saint-Venant system with a source term, *Calcolo*, **38** (2001), no. 4, 201-231

18. B. Perthame, C. Simeoni, Convergence of the Upwind Interface Source method for hyperbolic conservation laws, preprint
19. Pukhnacheva, T.P., The problem of determining the roughness coefficient for a flow in an open channel, *Prikl. Mekh. Tekhn. Fiz.*, **38** (1997), no. 3, 93-98; translation in *J. Appl. Mech. Tech. Phys.*, **38** (1997), no. 3, 412-416
20. A. J.C. de Saint-Venant, Théorie du mouvement non-permanent des eaux, avec application aux crues des rivières at à l'introduction des marées dans leur lit, *C.R. Acad. Sci. Paris*, **73** (1871), 147-154
21. C.W. Shu, S. Osher, Efficient implementation of essentially nonoscillatory shock-capturing schemes, *J. Comput. Phys.*, **77** (1988), no. 2, 439-471
22. C.W. Shu, S. Osher, Efficient implementation of essentially nonoscillatory shock-capturing schemes II, *J. Comput. Phys.*, **83** (1989), no. 1, 32-78
23. C.W. Shu, High order ENO and WENO schemes for computational fluid dynamics. High-order methods for computational physics, pp. 439-582, *Lect. Notes Comput. Sci. Eng.*, **9**, Springer, Berlin, 1999
24. S. Soares-Frazão, Y. Zech, Undular bores and secondary waves. Experiments and hybrid finite-volume modelling, *Journal of Hydraulic Research*, 2002
25. P.L. Viollet, J.P. Chabard, P. Esposito, D. Laurance, *Mécanique des fluides appliquée*, Presses des Ponts et Chaussées, Paris, 1998
26. F. Vigie, O. Eiff, D. Astruc, Experimental study of the flow regimes of subcritical channel flow over an obstacle, submitted to *Riverflow 2002*

Dynamics of Multilayer Shear Flows

Igor L. Kliakhandler

Department of Mathematical Sciences Michigan Technological University
1400 Townsend Drive Houghton, MI 49931-1295 *igor@mtu.edu*

Summary. The multi-layer shear flows are one of the most graphic and experimentally accessible examples of conservation laws system. The flows are interesting due to their conceptual simplicity, rich dynamical phenomenology, and technological relevance. The simplest system is multilayer thin film flowing down an inclined plane - the process used in manufacturing of color photographic films and papers. Conservation of mass in each layer leads to a system of coupled PDEs. Its hyperbolic part corresponds to a kinematic part of the motion. Investigation of its structure reveals a very nontrivial types of behaviour that might appear in such system. Namely, in addition to hyperbolic waves, the system may also exhibit elliptic behaviour. The influence of dissipative effects results in a simple and remarkable manifestation of the result obtained by Majda and Pego (1985): that the dissipative terms may in fact destabilize the hyperbolic system, and that the solutions of viscosity-smoothed system do not converge to the solutions of the corresponding hyperbolic system.

1 Introduction

The multilayer shear flows comprise such flows as Poiseuille, Couette, film flows, etc. For simplicity of the presentation, only multilayer thin films will be discussed in the present paper; the discussed phenomena are however typical for all multilayers shear flows.

The precision coating of colour photographic films and papers is especially important application of multilayer liquid films. The cross-sectional picture of Fuji color film is presented on figure 1. Modern photographic films often involve more than 10 different layers; during the process of their deposition, the multilayer "sandwich" flow is used. In chemical technology the interfacial waves enhance the transport of mass, heat and momentum across the film, whereas formation of the waves in coating layers is highly undesirable.

Another application of multilayer films is explanation of generation of transverse ridges occurring on the surfaces of glaciers, figure 2. The appearance of regular periodic "ice waves" on the extent of many kilometers is clearly seen. Notice the moraines just below the bedrock peak. The waves are generated due to complex process of ice stratification, and interaction of different ice "layers".

To model these systems and wavy phenomena, we consider a two-dimensional multiple-layer stratified film of incompressible Newtonian liquids, involving n layers with viscosities μ_1, μ_2, etc. and densities ρ_1, ρ_2, etc. flowing down an inclined plane under the gravity, figure 1. The associated

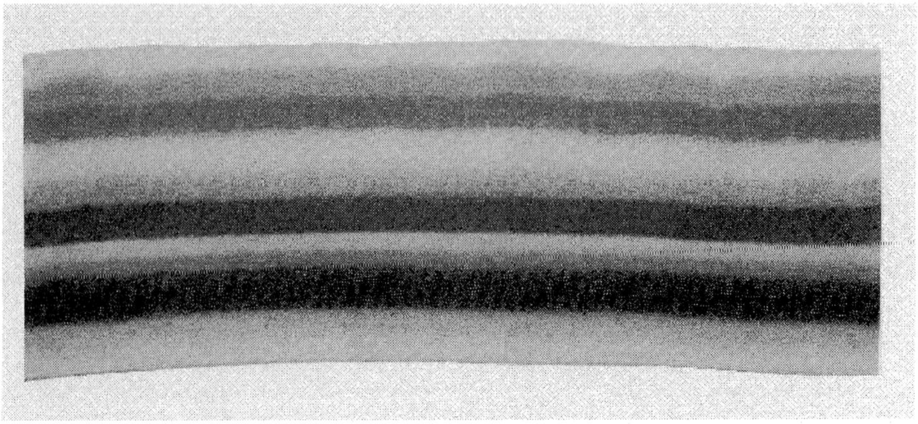

Fig. 1. Cross-sectional view of color negative film. Many layers of liquid are coated. COPYRIGHT: FUJI PHOTO FILM CO., LTD. Courtesy of C. Kobayashi.

Fig. 2. Waves on the surface of glaciers in Orgives, Juno Icefield, Alaska. The image by Mark Dubois.

steady-state planar interfaces are marked as H_0^1, H_0^2, etc. Here x, y are the streamwise and spanwise coordinates, respectively. The surrounding atmosphere is assumed to be a weightless, inviscid quiescent gas. Motion of the liquids in such "sandwich" is described by Navier-Stokes equations. The perti-

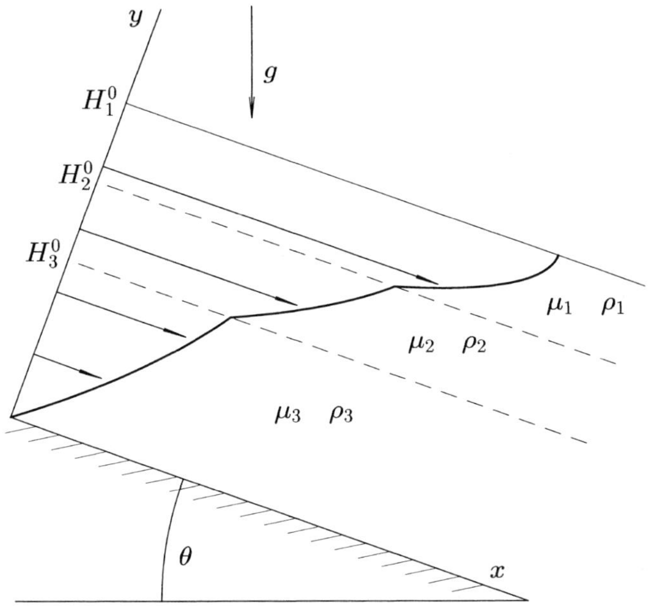

Fig. 3. Geometry of multilayer downflowing film

nent boundary conditions imply (i) no-slip on the solid bottom, (*ii*) continuity of velocities and tangent stresses on the inner interfaces, (*iii*) jump im normal stresses due to surface tension, and (*iv*) important kinematic conditions on the interfaces. Note that most of the boundary conditions are prescribed on free unknown surfaces separating the layers.

The original Navier-Stokes problem with boundary conditions on unknown interfaces is too hard for direct numerical simulations or analytic treatment. As a result, asymptotic methods have been developed for this type of problems. The interfacial variations are assumed to be small, and waves are considered to be slowly varying both in space and time. The result is so-called evolution equations for the deviation of interfaces from their steady-state locations. The kinematic boundary conditions, expressing conservation of mass in each layer, are the key conditions for the derivation.

The details of the derivation are exposed in Kliakhandler and Sivashinsky (1997), and Kliakhandler (1999). Here we will discuss the interesting and nontrivial features shown by the equations.

2 Simplest kinematic model

First, let us consider the simplest case of quasi-unidirectional and quasi-steady flow. We are concerned therefore with the most basic kinematic effects. In this section, we omit surface tension and density stratification which provide smoothing similar to that of viscosity.

The kinematic effects of interfacial dynamics are described by the coupled set of equations,

$$\frac{\partial H_i}{\partial t} + \frac{\partial Q_i}{\partial x} = 0 . \tag{1}$$

Here $H_i(x,t)$ is the ith interface, and Q_i is the mass flux in ith layer. The fluxes Q_i depend on layer thicknesses H_i and on viscosities μ_i.

In the case of a homogeneous film, equations (1) are reduced to one equation, where $Q = CH^3$ with some constant C. This is the typical hyperbolic equation. The conventional wisdom is to assume the same hyperbolic type of behaviour for the *coupled* system (1). However, the careful investigation of system (1) reveals its unexpected and remarkable behaviour.

To make things even simpler, consider linearized version of (1):

$$\frac{\partial h_i}{\partial t} + \alpha_{ij}\frac{\partial h_j}{\partial x} = 0 . \tag{2}$$

Here $h_i(x,t) = H_i(x,t) - H_i^0$ is deviation of ith interface $H_i(x,t)$ from its steady location H_i^0, $\alpha_{ij} = \partial Q_i/\partial H_j$ is matrix associated with kinetic (transport, advective, drift) effects. For repeated subscripts the summation convention is adopted. Linear system (2) describes dynamics of small perturbations of multilayer film.

The unexpected outcome is that matrix α_{ij} may have *complex* eigenvalues for a quite broad range of layer viscosities and thicknesses. Let us represent h_i in the form

$$h_i = a_i e^{ikx+\omega t} \tag{3}$$

with k being wavenumber of perturbations, and ω being frequency of perturbations. For real eigenvalues of matrix α, ω is pure imaginary, and corresponds to traveling waves. But for parameters of the flow with complex eigenvalues of matrix α, there appear ω's with complex conjugated pairs. Therefore, some ω's have positive real part in this case, and there is an *instability* of underlying parallel multilayer unidirectional flow. In this case, $Max\,(Re[\omega]) \sim k$.

This type of instability is termed as α-effect. Alpha-effect appears in a broad range of hydrodynamical situations (see Frisch, She and Sulem, 1987). Being identified in the simplest situation, the α-effect is expected to arise in more sophisticated models of stratified shear flows.

In the case of α-effect the system (1) is expected to be ill-posed, being of elliptic-hyperbolic type. Otherwise, being hyperbolic, the kinetic model (1) provides shock waves. In both cases the result obtained can be thereby taken as an indication that the model should be improved by incorporation of the dissipative mechanisms yielding damping of short-scale disturbances. Those mechanisms include surface tension, and density stratification, omitted in this Section.

3 Surface-tension effects

If the surface tension between interfaces on figure 3 is taken into account, equations (1) should be modified. Impact of surface tension leads to an addition of new terms (Kliakhandler, 1999):

$$\frac{\partial H_i}{\partial t} + \frac{\partial Q_i}{\partial x} + \frac{\partial}{\partial x}\left[G_{ij}\frac{\partial^3 H_j}{\partial x}\right] = 0 \,. \tag{4}$$

Here G_{ij} is the impact function of surface tension at jth interface on dynamics at ith interface. All G_{ij} depend on layer thicknesses H_i and on viscosities μ_i.

As in previous case, it is convenient to consider a simplified version of (4). Expanding all Q_i in Taylor series up to quadratic terms, and evaluating G_{ij} on the undisturbed interfaces H_i^0, we will obtain:

$$\frac{\partial h_i}{\partial t} + \alpha_{ij}\frac{\partial h_j}{\partial x} + \beta_{ijk}\frac{\partial h_j h_k}{\partial x} + \chi_{ij}\frac{\partial^4 h_j}{\partial x^4} = 0 \,. \tag{5}$$

Equations (5) is the main result of the present paper. In a sense, equations (5) are a counterpart to the famous Kuramoto-Sivashinsky (KS) equation:

$$h_t + hh_x + h_{xx} + h_{xxxx} = 0. \tag{6}$$

The KS equation appears also in many other physical contexts such as drift waves in plasma (Cohen *et al.*, 1976), spatial patterns in chemical oscillations (Kuramoto & Tsuzuki, 1976), and unstable flame fronts (Sivashinsky, 1977). The KS equation has smooth solutions and due to its ability to display turbulent-like behaviour has enjoyed much attention (Cross & Hohenberg, 1993).

In addition to typical patterns displayed by KS equation, system (5) shows very interesting and non-trivial features of hyperbolic waves propagation and stability, which are discussed further. Our study will be based on simplest and transparent linear stability features, which are the key to the study of dynamic behaviour and stability. In addition to equations (5), we will investigate their linearized version:

$$\frac{\partial h_i}{\partial t} + \alpha_{ij}\frac{\partial h_j}{\partial x} + \chi_{ij}\frac{\partial^4 h_j}{\partial x^4} = 0 \,. \tag{7}$$

4 Surface-tension and alpha effect

Let us consider system (5) for the alpha-type of instability. In this case, short waves are damped by surface tension. This means that eigenvalues of matrix χ_{ij} are real and positive. Substitution (3) results in a simple matrix eigenvalue problem; typical solution is shown on figure 4. Note that region with positive $Re[\omega]$ corresponds to instability, whereas negative $Re[\omega]$ corresponds

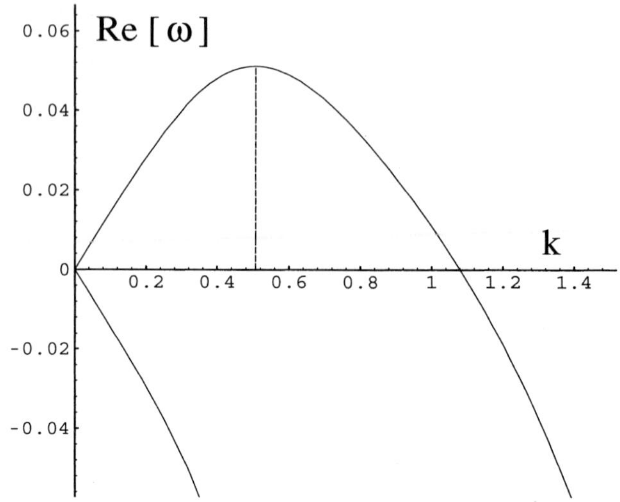

Fig. 4. Typical results of linear stability problem for the alpha-type of instability.

to stability. As may be seen, for large k the system is stable: all wavenumbers are damped, $Re[\omega] < 0$. The same may be seen directly from (7): surface tension suppresses fast oscillations.

The most direct way to access interfacial evolution driven by alpha-type of instability is to simulate equations (5) numerically. The instabilities induce patterns; their typical length-scale is of great interest. Usually, the most unstable wavenumber k_c with maximal $Re[\omega]$ (about 0.5 on figure 4) controls the wavescale of the patterns, with typical length λ being

$$\lambda = \frac{2\pi}{k_c}. \tag{8}$$

Conventional pseudo-spectral method was used for numerical simulations of equations (5). Typical results of the simulations are shown of figure 5. The well settled flow assumes the form of an irregular highly intermittent wavy structure whose typical scale markedly exceeds that suggested by the linear stability analysis. This extension of the nonlinear structures compared with linear most unstable wavelength is well-known in other contexts as "inverse cascade", and is subject of intensive investigation (Frisch, She, Sulem, 1987). It is remarkable that simple model equations (5) with complex eigenvalues for matrix α_{ij} captures such a non-trivial dynamic effect.

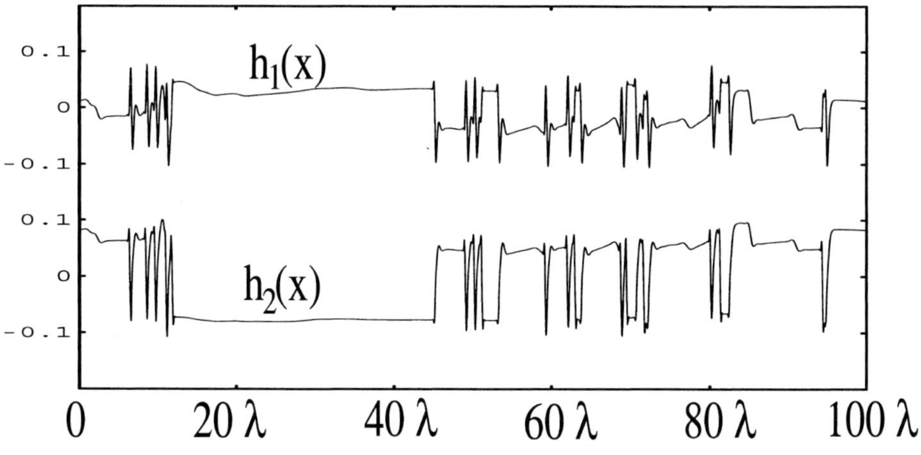

Fig. 5. Typical results of numerical simulation for the alpha-type of instability.

5 Surface-tension-induced instability and Majda-Pego (1985) result

In the absence of α-effect the eigenvalues ω of the truncated equations (1) are imaginary. This, however, does not imply that the same is true for the complete system (5) incorporating the surface tension. In the approximation of small k, the asymptotic solution of matrix eigenvalue problem for linearized system (7) may be obtained:

$$\omega_j \simeq -ikM_j + k^4 L_j, \quad j = 1, 2 \tag{9}$$

where

$$L_j = \frac{\alpha_{12}\chi_{21} - \alpha_{22}\chi_{11} + \alpha_{21}\chi_{12} - \alpha_{11}\chi_{22} + M_j(\chi_{11} + \chi_{22})}{\alpha_{11} + \alpha_{22} - 2M_j} \tag{10}$$

and M_j are the eigenvalues of the matrix α_{mn}. Matrices α_{ij} and χ_{ij} both depend of parameters of the flow. It turns out that in a certain (μ_i, H_i^0) parameters range L_j may become positive implying instability caused by such damping effect as surface tension ! Figure 6 exhibit such typical behaviour. The mathematical feasibility of such an outcome has already been ascertained by Majda and Pego (1985) for second-order parabolic systems with viscous-like damping. Equations (5) may serve as a simple illustration of their theory.

To clarify the emergence of surface-tension-driven instability, consider the following example (Kliakhandler, 1999):

$$\frac{\partial}{\partial t}\begin{pmatrix} h_1 \\ h_2 \end{pmatrix} + \begin{pmatrix} a & b \\ 0 & 0 \end{pmatrix} \frac{\partial}{\partial x}\begin{pmatrix} h_1 \\ h_2 \end{pmatrix} + \begin{pmatrix} \varepsilon & 0 \\ c & \varepsilon \end{pmatrix} \frac{\partial^4}{\partial x^4}\begin{pmatrix} h_1 \\ h_2 \end{pmatrix} = 0. \tag{11}$$

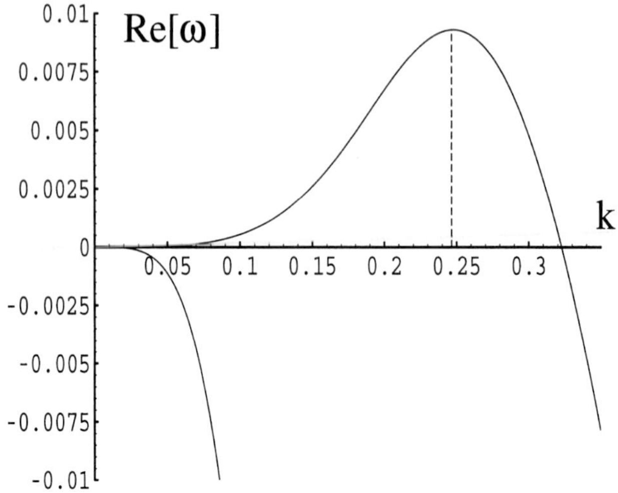

Fig. 6. Typical results of linear stability problem for the surface-tension-induced type of instability.

This deliberately simple form of matrices α_{ij} and χ_{ij} is minimally sufficient for the generation of surface-tension-driven instability with damping of short modes. Simple analysis shows that two modes exist in (11). The first mode is mostly travelling and may be associated with first interface, whereas the second mode is mostly standing and may be associated with second interface. The crucial point here is that $L = |bc/a\varepsilon| > 1$. The numerator of L measures the coupling between interfaces, since b and c are on the skew diagonals in matrices α_{ij} and χ_{ij}, respectively. The denominator of L measures the self-interaction of the interfaces. Hence, if $L > 1$, or the coupling is sufficiently strong, the system becomes unstable.

In the absence of surface tension the hyperbolic part of (5) will generate shock solutions. The surface tension may be regarded as a mechanism which smooths the shocks. However, in the presence of the surface-tension-driven instability found, the solutions of (5) with vanishing surface tension do not converge to the solutions of the pertinent hyperbolic system.

Numerical simulation of (5) for the surface-tension-induced instability is shown on figure 7. Here the typical length-scale of the waves is close to λ obtained from linear analysis.

6 Conclusion

The system of equations describing the motion of interfaces in multilayer downflowing films is derived. The equations are a counterpart to the famous

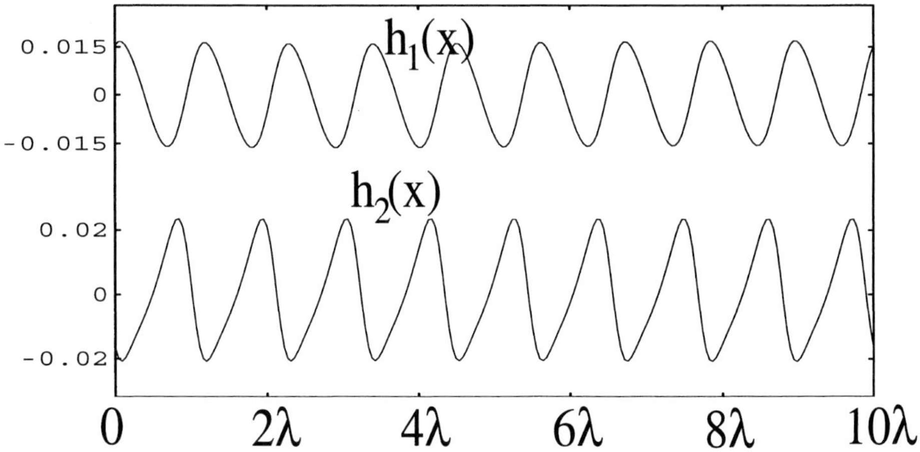

Fig. 7. Typical results of nonlinear dynamics for the surface-tension-induced type of instability.

Kuramoto-Sivashinsky equation. Their analysis shows that for physically interesting range of parameters, the equations display two interesting phenomena: alpha-effect, and surface-tension-driven instability. Alpha-effect appears due to emergence of complex eigenvalues in linear transport matrix. Surface-tension-driven instability is a manifestation of Majda-Pego (1985) result that damping effects may in fact destabilize the underlying hyperbolic system.

References

1. Cohen, B.I., Krommes, J.A., Tang, W.M., Rosenbluth, M.N. (1976): Non-linear saturation of the dissipative trapped-ion mode by mode coupling. Nuclear Fusion, **16**, 971–992.
2. Cross, M.C., Hohenberg, P.C. (1993): Pattern formation outside of equilibrium. Reviews of Modern Physics **65**, 851–1112.
3. Frisch, U., She, Z.S., Sulem, P.L. (1987): Large-scale flow driven by the anisotropic kinetic alpha effect. Physica D, **28**, 382–392.
4. Kliakhandler, I.L., Sivashinsky, G.I. (1997): Viscous damping and instabilities in stratified liquid film flowing down a slightly inclined plane. Physics of Fluids, **9**, 23–30.
5. Kliakhandler, I.L. (1999): Long interfacial waves in multilayer thin films and coupled Kuramoto-Sivashinsky equations. Journal of Fluid Mechanics, **391**, 45–65.
6. Kuramoto, Y., Tsuzuki T. (1976): Persistent propagation of concentration waves in dissipative media far from thermal equilibrium. Prog. Theor. Phys. **55**, 356–369.
7. Majda, A., Pego, R. L. (1985): Stable viscosity matrices for systems of conservation laws. J. Diffl Equat. **56**, 229-262.

8. Sivashinsky, G.I. (1977): Nonlinear analysis of hydrodynamic instability of laminar flames. Part 1. Derivation of basic equations. Acta Astronautica, **4**,1177–1206.

Effects of Viscosity on a Shock Wave Solution of the Euler Equations

Marco Kupiainen[1] and Gunilla Kreiss[2]

[1] Department of Numerical Analysis and Computer Science, Royal Institute of Technology, Sweden *marcok@nada.kth.se*
[2] Department of Numerical Analysis and Computer Science, Royal Institute of Technology, Sweden *gunillak@nada.kth.se*

1 Introduction

Hyperbolic conservation laws are used to model many interesting phenomena such as gasdynamics, traffic flow, magneto-hydrodynamics etcetera. In general discontinuities will develop in the solution even when the initial data is smooth. In many cases dissipative effects have been neglected in the modelling. However, in most numerical computations of discontinous solutions numerical dissipation is present. One hopes that the solution will mainly be effected in the vicinity of the discontinuitites.

There has been considerable interest in analysing how the solution and its dynamical properties are effected by the dissipation. In Goodman & Xin [3], Efraimsson & Kreiss [4] and Rousset [5] the effect of dissipation on travelling wave solution is considered. Conditions for stability of strong viscous shock waves have been derived in Kreiss & Kreiss [2] and Howard & Zumbrun [6].

In this paper we consider the Euler equations with an equation of state admitting multiple solutions to Riemann problems. In Wendroff [1] several types of instabilities for this problem are investigated numerically using the Lax-Friedrichs scheme on a staggered grid [8]. Here we re-examine one case, a Riemann problem where the end states can be connected with a 3-shock or with two shockwaves and an intermediate state. With discontinous initial data the Lax-Friedrichs scheme produces the solution with the intermediate state. Further computations show that with sufficiently smooth initial data the Lax-Friedrichs scheme produces the discrete 3-shock wave solution. Similar results are obtained with other schemes. owever, we note that the smoothness requirement (in order to obtain a 3-shock solution) of the initial profile is dependent on the numerical method used.

To understand this behaviour we investigate the stability of a corresponding viscous problem. We find that the travlling wave solution of the viscous problem corresponfing to the 3-shock satisfies the stability condition of Kreiss & Kreiss [2].

2 Governing equations

The Euler gas dynamic equations expressed in Lagrangian coordinates are:

$$v_t - u_x = 0, \tag{1}$$

$$u_t + p_x = 0, \tag{2}$$

$$(e + 1/2u^2)_t + (pu)_x = 0. \tag{3}$$

The independent variable x denotes mass and the dependent variables are the specific volume $v(x,t)$, the particle velocity $u(x,t)$ and the specific entropy $s(x,t)$. The internal energy e is related to v and s by the *equation of state* (EOS):

$$e \equiv e(v,s),$$

and the pressure p is defined by

$$p \equiv -\frac{\partial e}{\partial v} = -e_v.$$

2.1 The Equation of state

As in [1] we consider an equation of state, constructed as in Smith [7], to allow multiple solutions. It has the form:

$$e(v,s) = \gamma(v)\theta(s,4)\epsilon(v,s),$$

where

$$\gamma(v) = \begin{cases} 1, & \text{for } v \geq 1 - 0.5\delta, \\ 1 + \left(\frac{1-0.5\delta - v}{v}\right)^4, & \text{otherwise,} \end{cases}$$

$$\theta(s,k) = \begin{cases} 1, & \text{for } s \geq -0.5\delta \quad \text{or } k = 0, \\ \exp\left(10\delta^{-1}(s+0.5\delta)\right) \sum_{i=0}^{k} \frac{(-10\delta^{-1}(s+0.5\delta))^i}{i!}, & \text{otherwise,} \end{cases}$$

$$\epsilon(v,s) = \exp(1-v) + \alpha(1 - \exp(-\psi(s)))\exp(\alpha^{-1}(1 - v + \psi(s))),$$

$$\psi(s) = \begin{cases} s, & \text{for } s \geq -0.5\delta, \\ \delta(1 - 0.1\sum_{k=0}^{4}\theta(s, 4-k)), & \text{otherwise.} \end{cases}$$

The parameters α and δ are

$$\alpha = 0.1,$$

$$\delta = 10^{-6}.$$

3 Problem presentation

3.1 The Hugoniot locus

In order to calculate shock-curves one must use formulation (1)-(3) of the Euler equations since by definition shocks are not smooth. We get a relation from the Rankine-Hugoniot jump condition and the definition of e,

$$e(v(s),s) - \underbrace{e(v_R,s_R)}_{e_R} + \frac{1}{2}(p(v(s)),s) + \underbrace{p(v_R,s_R)}_{p_R}(v(s)-v_R) = 0. \quad (4)$$

The shock curves in the up-plane are given by

$$u(s) = u_R \pm \sqrt{(p(v(s),s) - p_R)(v_R - v(s))}.$$

The sign of the above expression is given by the fact that the subsonic-supersonic condition

$$-p_v(v,s) > \sigma^2 > -p_v(v_R,s_R)$$

holds everywhere, where σ is the shockspeed. This gives that the 3-shock curve is

$$u(s) = u_R + \sqrt{(p(v(s),s) - p_R)(v_R - v(s))}, \qquad s > s_R. \quad (5)$$

In figure 1 we have illustrated the velocity-pressure and the volume-pressure locus for a 3-shock of possible left-states, with $s > s_R$, for a given right-state, given by (6).

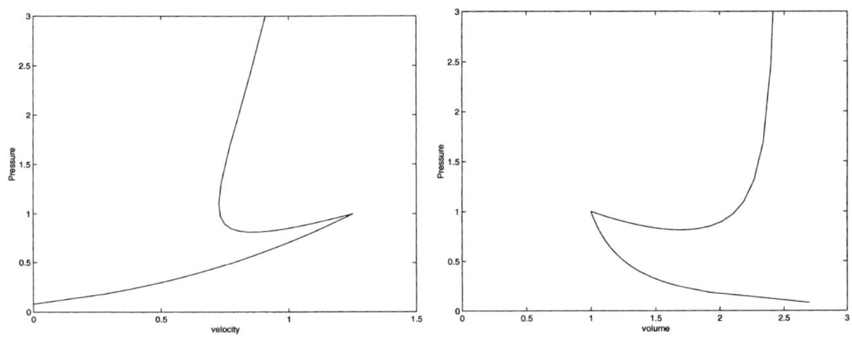

(a) velocity-pressure (b) volume-pressure

Fig. 1. The locus for a 3-shock.

3.2 The Riemann Problem

In our calculations we have used, as in Wendroff [1],

$$\mathbf{w}_R = \begin{pmatrix} v_R \\ u_R \\ s_R \end{pmatrix} = \begin{pmatrix} 2.7028 \\ 0 \\ -10^{-6} \end{pmatrix}. \quad (6)$$

At the left-state we use $s_L = -7.78 \cdot 10^{-7}$ which from (1)-(3), (4) and (5) uniquely determines the left-state in our Riemann-problem.

3.3 Numerical calculations using the Lax-Friedrichs scheme

By the construction of our Riemann-problem we expect to see a shock moving with speed $\sigma = \frac{p_L - p_R}{u_L - u_R}$, but as we can see in figure 2 the Lax-Friedrichs scheme with a staggered grid [8] produces a solution where the shock splits up into to two moving fronts. This is in agreement with computations in [1]. We have used 200 points in the space discretization. The time-step Δt is chosen $\Delta t = 0.5 \max(u \pm \sqrt{-p_v})$. Note that the discontinuities are smeared out. This is to be expected, since the scheme is very viscous.

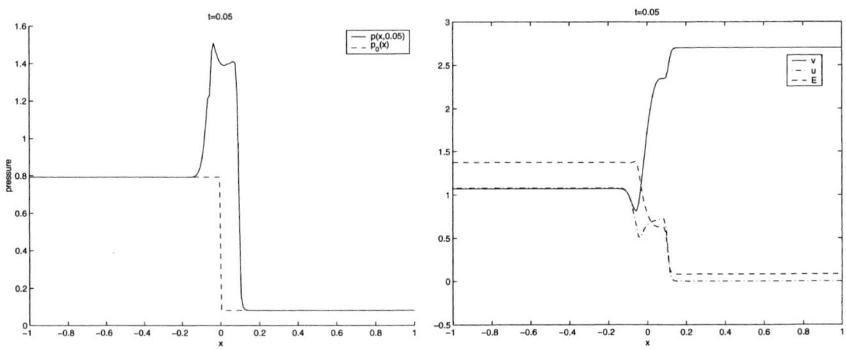

(a) Shock splitting (b) Solution profiles

Fig. 2. Numerical solutions at time=0.05. $p_0(x)$ is the pressure at $t = 0$.

4 Stability Investigation

We will now investigate the stability of a viscous travelling wave solution corresponding to the 3-shock. Consider the Euler equations with viscosity

$$w_{\hat{t}} + f(w)_{\hat{x}} = \nu w_{\hat{x}\hat{x}} \tag{7}$$

with discontinous initial data as above. The viscosity is motivated by the numerical viscosity introduced by the Lax-Friedrichs scheme. Note that νI is a simplification of the viscosity matrix in the modified equation for the Lax-Friedrichs scheme. We scale the problem and change variables $(x := \frac{\hat{x}}{\nu}, t := \frac{\hat{t}}{\nu})$ to get

$$w_t + f(w)_x = w_{xx},$$

with boundary conditions $\lim_{x\to-\infty} w = w_L$, $\lim_{x\to\infty} w = w_R$. Let $W(\xi)$, where $\xi = x - \sigma t$, be a smooth travelling wave solution connecting the right-

and left-states (w_L and w_R) corresponding to the 3-shock presented above, and let $w_0(\xi) = W(\xi) + \varepsilon h_0(\xi)$. We also introduce the corresponding eigenvalue problem

$$\lambda\phi + [(f'(W) - \sigma I)\phi]_\xi = \phi_{\xi\xi}, \qquad \phi \in L_2 \tag{8}$$

It is easily verified that zero is an eigenvalue of (8). Clearly W cannot be a stable solution if there are eigenvalues with positive real part. It is proven in Kreiss & Kreiss [2] that a travelling wave solution of Lax-type[1] is nonlinearily stable[2] under zero mass perturbations[3] for sufficiently small ε, if there are no eigenvalues in $\{z \in C : Re(z) \geq 0\} \setminus \{0\}$. We note that the requirement that $\text{span}(v_1^L, v_2^L, w_R - w_L) = \Re^3$, where v_i^L is the eigenvector corresponding to $\lambda_i^L \geq \lambda_{i-1}^L$, is fulfilled. We have investigated numerically the eigenvalue condition. The smooth travelling wave solution is needed for this investigation.

Integrating once we get the travelling wave solution by solving the corresponding ODE

$$W'(\xi) = f(W(\xi)) - \sigma W(\xi) + C \tag{9}$$

with the fourth order Runge-Kutta method with the left-state as initial data. The constants are determined using the fact that the left- and right-states are fixed points. In order to center the shock we add points of constant state U_L to the solution.

Before we discretize (8) we need to truncate the domain from $\Re$ to a bounded domain $[0, L]$ which is equivalent to viewing the problem on $[-L/2, L/2]$, due to the translational invariace of a travelling shockwave solution. The truncation of the domain must come with boundary conditions. We are interested to find eigenvalues with $Re(\lambda) > 0$ and if there are any such eigenvalues the corresponding eigenfunction decays exponentially with $|\xi|$. This motivates our homogeneous boundary conditions at both ends of our computational domain.

$$\phi_0 = 0, \qquad \phi_{N+1} = 0.$$

We discretize (8) as

$$\frac{\phi_{n+1} - 2\phi_n + \phi_{n-1}}{\Delta\xi^2} - \frac{A_{n+1}\phi_{n+1} - A_{n-1}\phi_{n-1}}{2\Delta\xi} = \lambda\phi_n, \qquad n = 1, \ldots, N. \tag{10}$$

---

[1] A travelling wave solution is of Lax-type if it connects the right- and left-states such that, for a system with n characteristics, there are $n+1$ characteristics entering a shock. Further we require that $\text{span}(w_R - w_L, v_1, \ldots, v_{n-1}) = \Re^n$, where $\{v_i\}_{i=1}^{n-1}$ are the eigenvectors corresponding to characteristics pointing away from the shock.

[2] Problem (7) is nonlinearily stable if $w(x,t)$ tends to $W(x - \sigma t)$ as $t \to \infty$ for sufficiently small ε.

[3] If $\int_{-\infty}^{\infty} h_0'(x)dx = 0$ for a perturbation $h_0'(x)$ then it is called a zero mass perturbation.

Here $W_i = (v_i, u_i, E_i)^T$ is the discrete travelling wave solution to (9) which connects W_R with W_L and $A_i = f'(W_i) - \sigma I$.

The calculation of eigenvalues using a shift and invert strategy with the restarted Arnoldi method produces no positive eigenvalues as seen in figure 3. The zero eigenvalue corresponds to a translation of the shockprofile. We

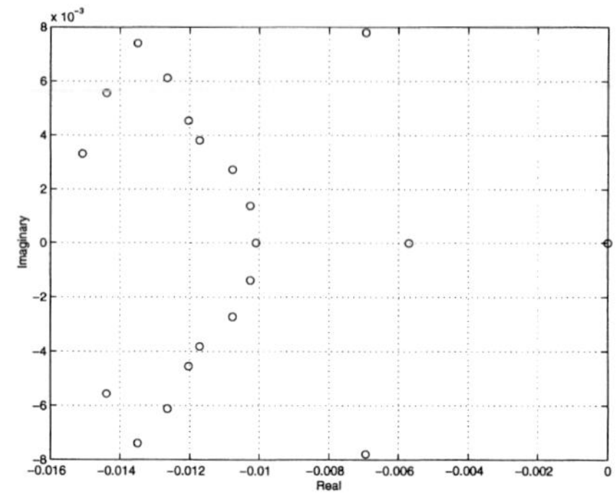

Fig. 3. The 20 eigenvalues with largest real part.

have investigated numerically how the truncation of $\Re$ to $[-L/2, L/2]$ affects the eigenvalues and found using different $L = 1, \ldots, 30000$ that there are no eigenvalues with positive real part. Using a fixed L, we investigated the asymptotic behaviour of the eigenvalues when $\Delta\xi \to 0$ and found no positive eigenvalues. We conclude that we have resolved the eigenvalue problem and that we have no positive eigenvalues and hence implying stability of the problem.

4.1 The viscous problem

In the stability investigation we simplified the viscosity of the Lax-Friedrichs scheme to be a scaled unit-matrix. We will make a "parameter-study" of

$$w_{\hat{t}} + f(w)_{\hat{x}} = c\nu w_{\hat{x}\hat{x}}, \tag{11}$$

where c is the parameter to be varied. The space-discretization is

$$\frac{\partial}{\partial t} w_i + D_0 f(w_i) = c\nu D_+ D_- w_i, \tag{12}$$

and march this in time with the second order Rung-Kutta method. We have used zero:th order extrapolation as a numerical boundary condition at the left and right boundary. As initial data we used the Riemann-data. We have not seen any sign of these solutions converging to the 3-shock wave solution. Initially the solution produces a large pressure spike as seen in figure 4 and then evolves to be similar to the solution in figure 2.

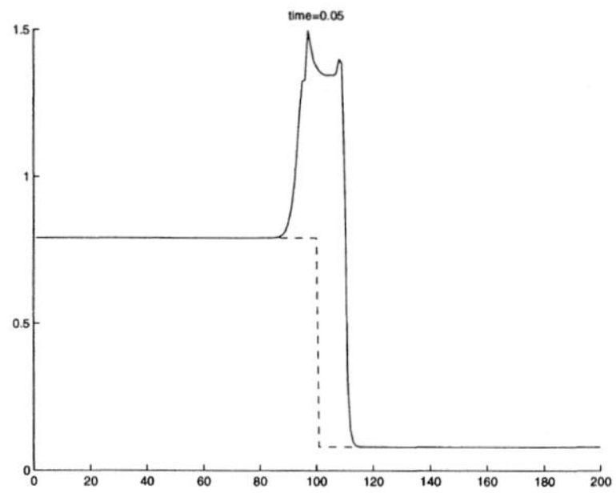

(a) $c = 1$

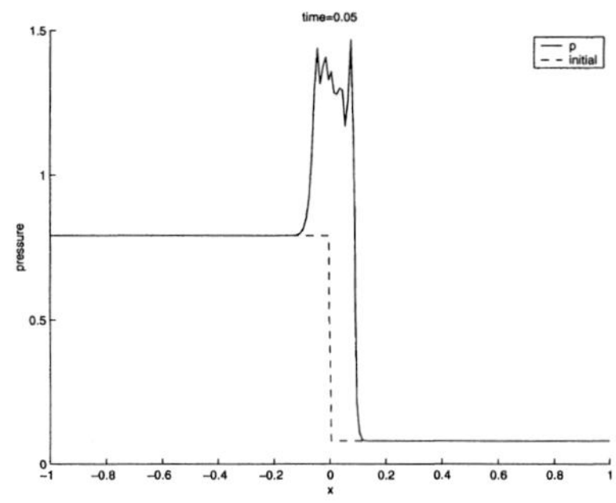

(b) $c = 1/2$

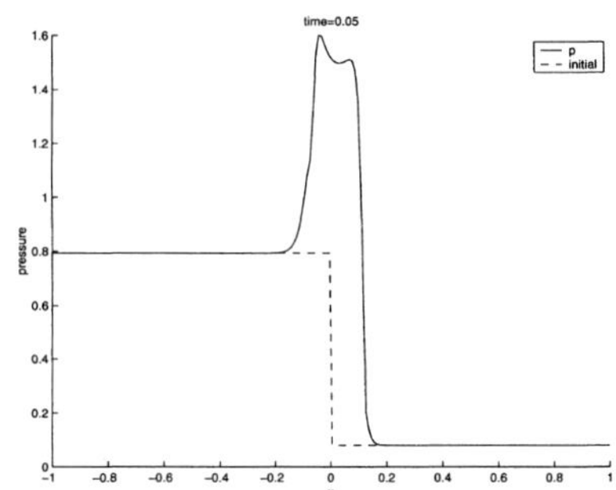

(c) $c = 2$

Fig. 4. Solution with discontinous initial data at time=0.05.

5 Influence of the Initial State

From the results above we believe that the failure of the numerical solution to develop a 3-shock is due to the initial data. The stability of the viscous profile only guarantees that it will develop if the initial data is sufficiently close to it. We have performed calculations to verify this by:

- using the travelling wave solution to $W' = \frac{1}{\nu}[f(W) - \sigma W] + C$ as initial data and
- constructing an initial profile from the known left and right-states with a smooth transition between the states.

Travelling wave solution as initial data: We use (12) with the travelling wave solution as initial data with $c = 0.25,\ 1,\ 4$. We see in figure 5 that for each $c > 0$ there is a travelling wave solution corresponding to a 3-shock. The shockprofiles steepness increases with decreasing c. For $c < 1$ the viscous shock profile is steeper than initial data, and the time-dependent solution converges to the viscous shock profile when $t \to \infty$. For $c >> 1$ the initial data is steeper than the shock profile and in such cases the solution does not converge to the viscous profile of the 3-shock. We also use the Lax-Friedrichs scheme and as we can see in figure 5, a discrete travelling wave solution corresponding to the 3-shock has developed.

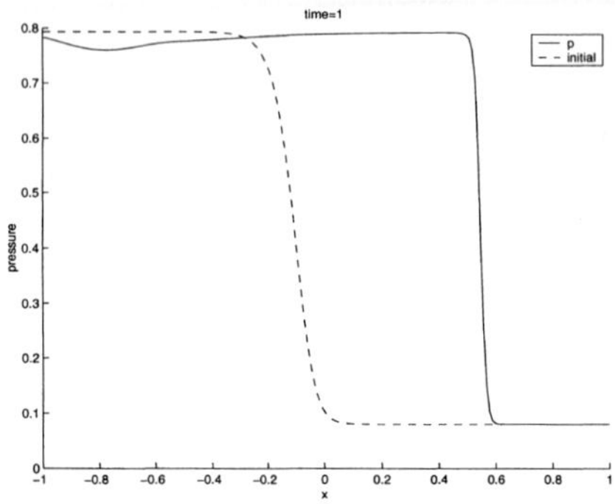

(a) $c = 1/4$

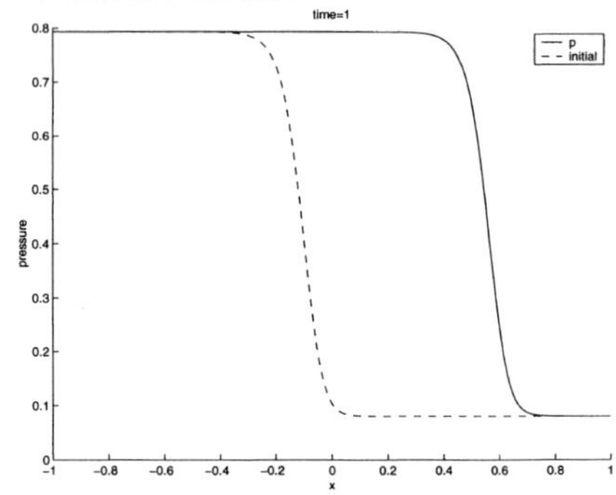

(b) $c = 1$

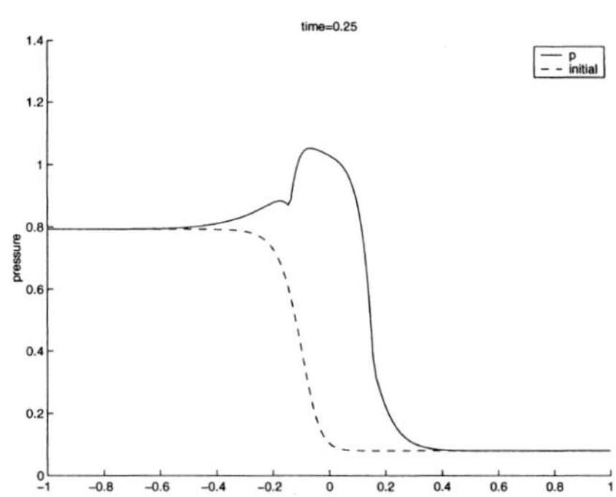

(c) $c = 4$

Fig. 5. Solution with travelling wave solution as initial data.

Constructed initial profile: We construct the profile from the known left and right-states of the entropy ($s_L = -7.78 \cdot 10^{-7}, s_R = -10^{-6}$) by using the following formula to get a smooth transition between the values.

$$s(n,x) = \begin{cases} s_L & \text{if } x < -0.5n\Delta x \\ e^{-\left(\frac{x+0.5n\Delta x}{x-0.5n\Delta x}\right)^2}(s_L - s_R) + s_R & \text{if } x \in [-0.5n\Delta x, 0.5n\Delta x] \\ s_R & \text{if } x > 0.5n\Delta x \end{cases} \tag{13}$$

By using this transition we can decide approximately how many points n we want in the transition between s_L and s_R. After constructing a smooth s we can construct the corresponding $W_0 = (u_0, v_0, E_0)^T$. The solution converges to the viscous 3-shock profile if we choose n such that the steepness of the profile is small enough. If we on the other hand choose n such that the initial profile is too steep i.e. the problem approaches the Riemann problem, we get other waves as seen in figure 6.

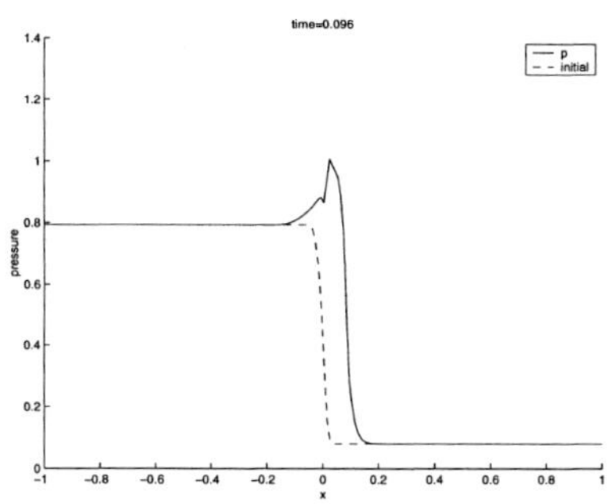

(a) $n = 10$

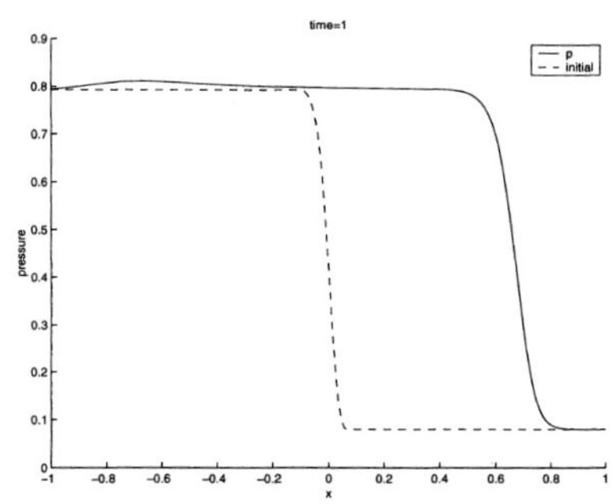

(b) $n = 20$

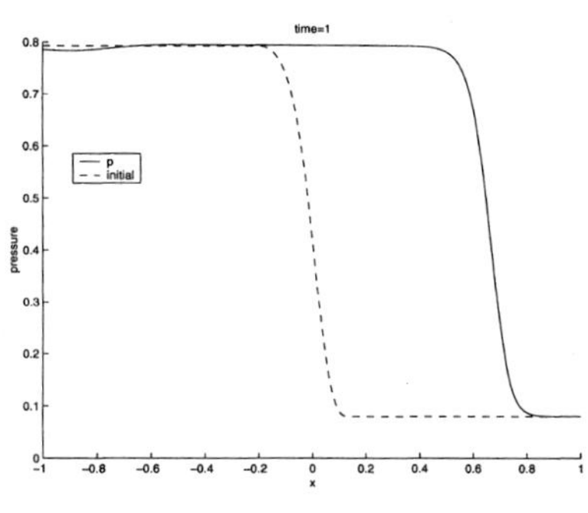

(c) $n = 40$

Fig. 6. Solution of (11) with constructed initial data.

6 Conclusions

We have considered a Rieman problem allowing multiple solutions, a solution with 3-shock and a solution with two shock waves. Numerical computations never yield the 3-shock solution. Our first hypothesis was that the 3-shock was unstable. Since numerical viscosity is present in the computtions we analyzed a corresponding viscous problem. We found that sufficient conditions for non linear stability are satisfied.

In another sct of experiments we used smoother initial data. Then a discrete 3-shock does develop. The smoothness requirement is related to the steepness of the discrete shock profile and thus it depends on the numerical method. This is disturbing.

Since reality is viscous it may not be realistic to use discontinous initial data. However, if the physiscal viscosity is much smaller than the numerical viscosity using a physical shock as initial data may yield the wrong answer.

References

1. Burton Wendroff (1990): A study of Non-uniqueness and Instability for Convex Materials. Third International Conference on Hyperbolic Problems, Theory, Numerical Methods and Applications, 957–973, Studentlitteratur, Lund
2. Kreiss, G., Kreiss, H-O. (1998): Stability of systems of viscous conservation laws. Comm. Pure Appl. Math., **51**, 1397–1424
3. Goodman J., Xin Z. (1992): Viscous Limits for Piecewise Smooth to Systems of Conservation Laws. Arch. Rational Mech. Anal., 121:235-265
4. Efraimsson G., Kreiss G. (2001): Approximate Solutions to Slightly Viscous Conservation Laws. Technical Report TRITA-NA 0140, NADA, KTH
5. Rousset F. (2001): Viscous Limits for Strong Shocks of System of Conservation Laws, Preprint
6. Zumbrun K., Howard P. (1998): Pointwise semigroup methods and stability of viscous shock waves. Indiana Univ. Math. J., 47(3):741-871
7. Smith, Randolph G. (1979): The Riemann Problem in Gas Dynamics. Trans. Amer. Math Soc., **249**, 1–50
8. Hirsch C. (1992): Numerical Computation of Internal and External Flows, Volume 2. Wiley, New York

An Accurate Deterministic Projection Method for Hyperbolic Systems with Stiff Source Term

Alexander Kurganov[1]

Department of Mathematics, Tulane University, 6823 Saint Charles Avenue, New Orleans, LA 70118, USA *kurganov@math.tulane.edu*

1 Introduction

We study numerical methods for one-dimensional hyperbolic systems of balance laws ($x \in \mathbb{R}; u \in \mathbb{R}^N, N \geq 1$) with very stiff source terms:

$$u_t + f(u)_x = \frac{1}{\varepsilon} S(u), \qquad 0 < \varepsilon << 1. \tag{1}$$

In particular, we are interested in an inviscid, compressible, reacting flow, described by the reactive Euler equations:

$$\begin{pmatrix} \rho \\ \rho u \\ E \\ \rho z \end{pmatrix}_t + \begin{pmatrix} \rho u \\ \rho u^2 + p \\ u(E+p) \\ \rho u z \end{pmatrix}_x = \frac{1}{\varepsilon} \begin{pmatrix} 0 \\ 0 \\ 0 \\ -\rho z K(\tau) \end{pmatrix}. \tag{2}$$

Here, the dependent variables ρ, u, E, and z are the density, velocity, total energy, and the fraction of unburnt gas, respectively. The pressure is given by the following equation of state (EOS),

$$p = (\gamma - 1) \cdot \left[E - \frac{\rho u^2}{2} - q_0 \rho z \right], \tag{3}$$

where the parameter q_0 corresponds to chemical heat release, and $\gamma = \text{Const}$. On the right-hand side of (2), $\tau := p/\rho$ is the temperature and ε is the reaction time. Finally, the Arrhenius kinetics term, [9], is

$$K(\tau) = e^{-\tau_c/\tau}, \tag{4}$$

where τ_c is the ignition temperature.

Typically, chemical reaction time scales are much faster than the fluid dynamical ones. Therefore, in order to fully (numerically) resolve detonation waves, one has to take both temporal (Δt) and spatial (Δx) grid scales to be proportional to ε. This may be extremely computationally expensive, or, in the case of higher spatial dimensions, practically impossible. This is the reason why we are interested in *underresolved* numerical methods, where $\Delta t, \Delta x >> \varepsilon$.

In such a case, the chemical reaction may be considered infinitely fast, and thus the Arrhenius kinetics term, (4), may be replaced with (even stiffer) Heaviside kinetics term, [19],

$$K(\tau) = H(\tau - \tau_c) := \begin{cases} 1, & \text{if } \tau \geq \tau_c, \\ 0, & \text{otherwise.} \end{cases} \tag{5}$$

Designing an accurate underresolved numerical method for a very stiff system (2)–(3),(5) or, in general, for the system (1) with a very small ε, is a rather challenging problem. Since the system is stiff, it is natural that one may wish to use an operator splitting (fractional step) method. Then, the step of solving the ODE,

$$u_t = \frac{1}{\varepsilon} S(u),$$

reduces to the projection of the computed solution onto an equilibrium state: $u \mapsto \mathcal{P}u$, where $S(\mathcal{P}u) \equiv 0$, while the corresponding homogeneous system of hyperbolic conservation laws,

$$u_t + f(u)_x = 0, \tag{6}$$

can be solved by any (stable) shock-capturing method.

However, if the so-called *deterministic projection operator* is used, this approach may lead to a spurious weak detonation wave that travels with an *unphysical* propagation speed of one grid cell per time step. This occurs since shock-capturing methods smear shock profiles, and as soon as the unphysical value of the temperature in this numerical shock layer is above the ignition temperature, a certain part of the gas gets burnt prematurely. This peculiar numerical phenomenon was first observed by Colella, Majda and Roytburd [8] in 1986, and since then it has attracted lots of attention (see, e.g., [5, 7, 12, 16, 17]). In order to fix this numerical problem, the ignition temperature was artificially increased in [6], or replaced with uniformly distributed random variable (random projection method by Bao and Jin, [2, 3]). Other, more complicated, but rather successful approaches have been proposed in [10, 13, 18].

In this paper, we introduce an **accurate deterministic projection (ADP)** method for balance laws with stiff source terms, which may be considered as a simple and robust alternative to the aforementioned approaches. The key idea is to evolve u according to the homogeneous system (6) (this will guarantee the correct propagation speed!), while using the projected values of $\widetilde{u} := \mathcal{P}u$ only whenever they are required (for example, for computing p in the EOS, (3), when the system (2)–(3),(5) is considered).

The paper is organized as follows. In §2, we introduce the ADP approach for a model scalar problem. In §3, we apply the ADP method to the reactive Euler equations.

2 Scalar Problem

We consider the following scalar hyperbolic balance law with a stiff source term, studied in [16, 2, 4, 11],

$$u_t + f(u)_x = \frac{1}{\varepsilon}(u - \alpha)(1 - u^2), \qquad -1 < \alpha < 1, \tag{7}$$

subject to a "prepared", piecewise constant initial data,

$$u(x, 0) =: u_0(x) = \begin{cases} 1, & \text{if } x \leq x_0, \\ -1, & \text{if } x > x_0. \end{cases} \tag{8}$$

Here, $0 < \varepsilon \leq 1$ is a small parameter, f is a convex flux function, and x_0 is a given point. This is a "toy model", where one can easily see the major difficulty we encounter while dealing with hyperbolic problems with stiff source terms.

The source in (7) admits three equilibria: two of them, $u = \pm 1$, are stable, and the third one, $u = \alpha$, is unstable. The exact solution is the shock discontinuity,

$$u(x, t) = \begin{cases} 1, & \text{if } x \leq x_0 + st, \\ -1, & \text{if } x > x_0 + st, \end{cases} \tag{9}$$

where the shock speed, determined by the Rankine-Hugoniot condition, is

$$s = \frac{f(1) - f(-1)}{2}. \tag{10}$$

Note that the speed is independent of α, and that the solution of the initial value problem (IVP) (7)–(8) is identical to the solution of the corresponding homogeneous equation,

$$u_t + f(u)_x = 0, \tag{11}$$

with the same initial data (8).

Let us now consider the operator splitting method for the IVP (7)–(8). We denote by $\mathcal{S}_C(\cdot)$ the solution operator, associated with the homogeneous convection equation (11), and by the $\mathcal{S}_R(\cdot)$ the solution operator of the stiff (reaction) ODE,

$$u_t = \frac{1}{\varepsilon}(u - \alpha)(1 - u^2). \tag{12}$$

Assume that the solution of (7)–(8) at time level $t = t^n$, $u(x, t^n)$, is given. According to the operator splitting technique, the solution at time $t^{n+1} = t^n + \Delta t$, $u(x, t^{n+1})$, is then approximated by

$$u^{n+1}(x) = \mathcal{S}_R(\Delta t)\mathcal{S}_C(\Delta t)u(x, t^n).$$

In fact, if one uses the exact solution operators, $\mathcal{S}_R$ and $\mathcal{S}_C$, this approximation will be exact, that is $u^{n+1}(x) \equiv u(x, t^{n+1})$.

However, in practice one has to use approximate solution operators instead of the exact ones. Let $\widetilde{\mathcal{S}}_C(\Delta t)$ corresponds to one step of a shock-capturing method. In the studied case of the underresolved numerical method ($\varepsilon << \Delta t$), the second approximate solution operator, $\widetilde{\mathcal{S}}_R(\Delta t)$, will be reduced to the following projection operator:

$$\widetilde{\mathcal{S}}_R(\Delta t)w \equiv \mathcal{P}_\alpha w := \begin{cases} 1, & \text{if } w > \alpha, \\ -1, & \text{if } w \le \alpha, \end{cases} \qquad \forall w \in \mathbb{R}. \tag{13}$$

When $\widetilde{\mathcal{S}}_C(\Delta t)$ is applied to the solution $\left\{u_j^n \approx u(x_j,t^n)\right\}_{j=1}^J$, computed at time t^n, the resulting shock profile, $u^* = \widetilde{\mathcal{S}}_C(\Delta t)u^n$, will be smeared. If the projection operator is applied to u^*, it will result in the following step function (assuming, for simplicity, that the approximate solution operator $\widetilde{\mathcal{S}}_C$ is monotone):

$$u_j^{n+1} = \begin{cases} 1, & \text{if } u_j^* > \alpha, \\ -1, & \text{if } u_j^* \le \alpha, \end{cases} \qquad \forall j.$$

Let us compare $\left\{u_j^n \approx u(x_j,t^{n+1})\right\}_{j=1}^J$ with $\left\{u_j^n \approx u(x_j,t^n)\right\}_{j=1}^J$. Depending on α and on the values of u_j^* in the numerical shock layer, the shock location may be shifted by several grid cells to the right (or to the left), or it may remain at the same location as at time t^n. Next time step of such a *"standard" deterministic projection method* will result in **exactly** the same move of the shock, and so on. As a result, we obtain a shock that propagates with an artificial speed of several cells per time step, which is, in general, not equal to the physically correct speed given by (10).

We propose a very simple ADP method that allows to capture the location of the discontinuity accurately (within the accuracy of a shock-capturing scheme used). Our approach can be schematically presented in the following operator form:

$$u^N = \mathcal{P}_\alpha \widetilde{\mathcal{S}}_C(T)u^0, \tag{14}$$

where $T = N\Delta t$ is a final time, and $\widetilde{\mathcal{S}}_C(T) = \widetilde{\mathcal{S}}_C(\Delta t) \circ \ldots \circ \widetilde{\mathcal{S}}_C(\Delta t) \circ \widetilde{\mathcal{S}}_C(\Delta t)$.

If $\widetilde{\mathcal{S}}_C(\cdot)$ corresponds to a convergent method for the homogeneous IVP (11),(8), then the intermediate solution, $u^* := \widetilde{\mathcal{S}}_C(T)u^0$, will be a (smeared) approximation to the exact solution, (9), where the width of the numerical shock layer is typically of size $\mathcal{O}(\Delta x)$. The projection, $\mathcal{P}_\alpha u^*$, will then result in a step function with a discontinuity, located within $\mathcal{O}(\Delta x)$ from the exact location ($x = x_0 + sT$).

In Figures 1a,b, we present both accurate solutions, computed by the proposed ADP method, (14), and inaccurate solutions, computed by the "standard" deterministic projection method,

$$u^{n+1} = \mathcal{P}_\alpha \widetilde{\mathcal{S}}_C(\Delta t)u^n,$$

with the same approximate solution operator $\widetilde{\mathcal{S}}_C$. In these examples, the flux is $f(u) = u + u^2/2$, and α is taken to be 0.5 and -0.75, respectively. The initial discontinuity is placed at $x_0 = 0.3$; $\Delta x = 0.02$, $\Delta t = 0.01$, the final time is $T = 0.6$ (60 time steps), the correct shock speed is 1/2.

These numerical examples clearly demonstrate the accuracy and robustness of the ADP method for the considered scalar problem.

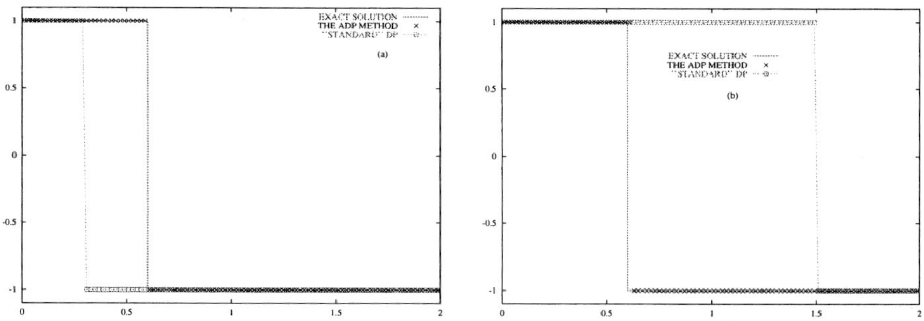

Fig. 1. The solutions of the IVP (7)–(8) with (a) $\alpha = 0.5$, (b) $\alpha = -0.75$

3 Detonation Waves

In this section, we develop the ADP method for the reactive Euler equations. First, note that system (2) can be viewed as the homogeneous system,

$$\begin{pmatrix} \rho \\ \rho u \\ E \end{pmatrix}_t + \begin{pmatrix} \rho u \\ \rho u^2 + p \\ u(E+p) \end{pmatrix}_x = 0, \tag{15}$$

coupled (through the EOS (3)) with the inhomogeneous equation, describing the propagation of the interface between the burnt and unburnt species,

$$(\rho z)_t + (\rho u z)_x = -\frac{1}{\varepsilon}\rho z K(\tau). \tag{16}$$

Thus, when solving system (15)–(16), the operator splitting (and hence, a projection step) should be applied only to equation (16). It is known, [8], that the "standard" projection may lead to spurious, unphysical shock waves traveling with an artificial speed (similarly to the scalar case in §2). Utilizing the ADP method allows one to avoid such an undesirable situation.

3.1 The ADP Method for Stiff Detonation Waves

Unfortunately, the method, described in §2, cannot be applied to the reactive Euler equations straightforwardly, since at every time step we need to know the projected values of z (which have to be equal to either 0 or 1) for computing p in the EOS (3)). In this case, the accurate projection procedure can be implemented as follows.

Assume that we have computed a solution at time level $t = t^n$. After the convection step, we obtain $\{z_j^*\}_{j=1}^J$, which are then projected by the deterministic projection operator:

$$\tilde{z}_j^{n+1} = \mathcal{P}_{\tau_c}(z_j^*) := \begin{cases} 1, & \text{if } \tau_j^n \geq \tau_c, \\ 0, & \text{if } \tau_j^n < \tau_c. \end{cases} \tag{17}$$

These values (not the smeared values of z_j^*, computed by the shock-capturing method!) are to be used in the EOS, and thus, for the evolution of $\rho, \rho u$, and E by solving (15). At the same time, in order to avoid appearance of nonphysical waves, the non-projected values,

$$z_j^{n+1} := z_j^*,$$

should be used for the evolution of ρz via the corresponding homogeneous equation,

$$(\rho z)_t + (\rho u z)_x = 0. \tag{18}$$

Remark 1. Note that at every time level we keep two sets of values of z: $\{z_j^n\}_{j=1}^J$ and $\{\tilde{z}_j^n\}_{j=1}^J$.

Remark 2. The ADP method for the reactive Euler equations resembles the level set method, used in multifluid computations (see, e.g., the review paper [1] and the references therein): we use equation (18) to track the interface between the burnt and unburnt species. However, unlike the multifluid situation, here both z and u are discontinuous at the interface.

Remark 3. In fact, due to the specific structure of the stiff system (15)–(16), the ADP method for the reactive Euler equations can be simplified even further. Note that the values of $z_j^{n+1} \equiv z_j^*$ are used neither in the homogeneous system (15), nor in the EOS (3) that couples system (15) with equation (18). Indeed, one may completely avoid solving equation (18), since the projection, carried out in (17), also does not employ smeared (and thus, unphysical) values of z_j^*. The latter ensures that the computed location of the interface will always be $\mathcal{O}(\Delta x)$ away from the exact one, provided a numerical method, used for solving the homogeneous system (15), converges to its correct entropy solution.

Remark 4. The proposed method can be implemented with one's favorite hyperbolic solver. In the presented numerical examples (both in §2 and §3.2), we have used the second-order central-upwind scheme from [15].

3.2 Numerical Examples

In this section, we demonstrate the performance of the proposed ADP method by applying it in a rather complicated situation, where a detonation wave collides with a shock, a contact discontinuity, and a rarefaction wave. This example is taken from [14, 3].

The parameters are:

$$\gamma = 1.2, \quad q_0 = 50, \quad \tau_c = 3, \quad \frac{1}{\varepsilon} = 230.75.$$

The initial data are:

$$(\rho, u, p, z)(x, 0) = \begin{cases} (\rho_l, u_l, p_l, 0), & \text{if } x \le 10, \\ (\rho_m, u_m, p_m, 1), & \text{if } 10 < x \le 40, \\ (\rho_r, u_r, p_r, 1), & \text{if } x > 40, \end{cases}$$

where $\rho_l = 3.64282, u_l = 6.2489, p_l = 54.8244; \rho_m = 1, u_m = 0, p_m = 1$; and $\rho_r = 4, u_r = 0, p_r = 10$. These data correspond to a right moving detonation, a left moving shock, a stationary contact, and a right moving rarefaction. A series of collisions occur after the detonation catches up with the other waves.

We compare the solutions, computed by the ADP method and by the "standard" deterministic projection method. In both cases, we take $\Delta x = 0.125$ and $\Delta t = 0.005$. The reference solution is computed using the fully resolved calculation with $\Delta x = 0.005$ and $\Delta t = 0.00025$.

In Figure 2, we show the computed solutions (density, pressure, temperature, and fraction of unburnt gas) at time $T = 2$ (before collisions). At this time, both methods provide rather accurate approximations.

The results at a later time $T = 4$ (after the collision with the shock, but before the collision with the rarefaction wave) are shown in Figure 3. At this time, only the density is sufficiently accurately captured by the "standard" deterministic projection method, while an unphysical shock wave has already appeared in the other components. At the same time, the location and amplitude of discontinuities, obtained by the ADP method, are correct.

Finally, in Figure 4, we present the results obtained at time $T = 8$ (after all collisions). The computations with the "standard" method are now completely wrong, while the resolution, achieved by the ADP method, is as high as at the smaller times.

Remark 5. We have also tested the ADP method on a variety of examples taken from [2, 3]. The physically correct solutions have been accurately captured in all the performed numerical experiments.

Acknowledgement. The author thanks the participants of the applied and computational mathematics seminar at Tulane University for a number of stimulating discussions. This work was supported in part by the NSF Grant DMS-0073631.

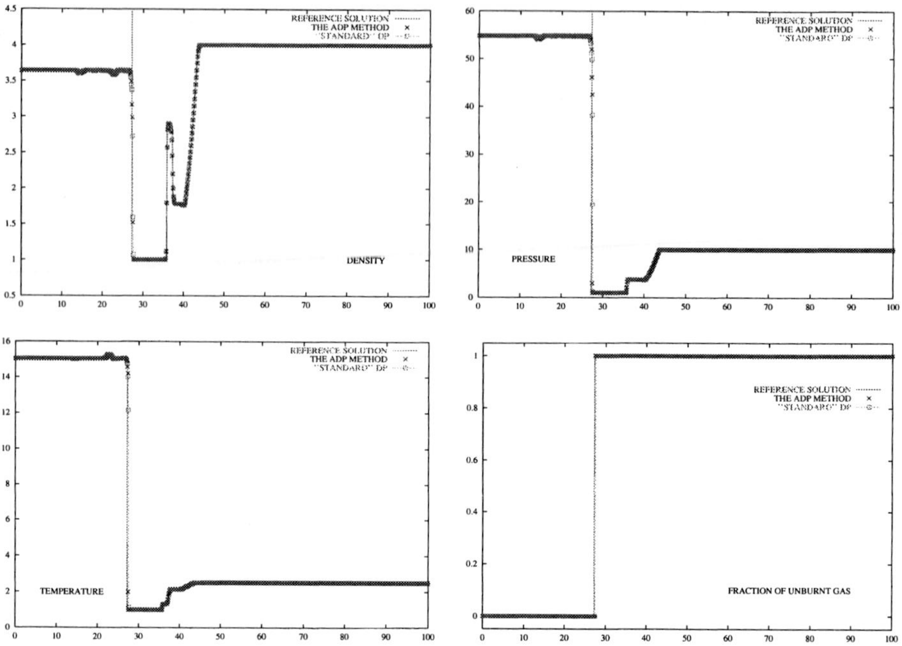

Fig. 2. $T = 2$

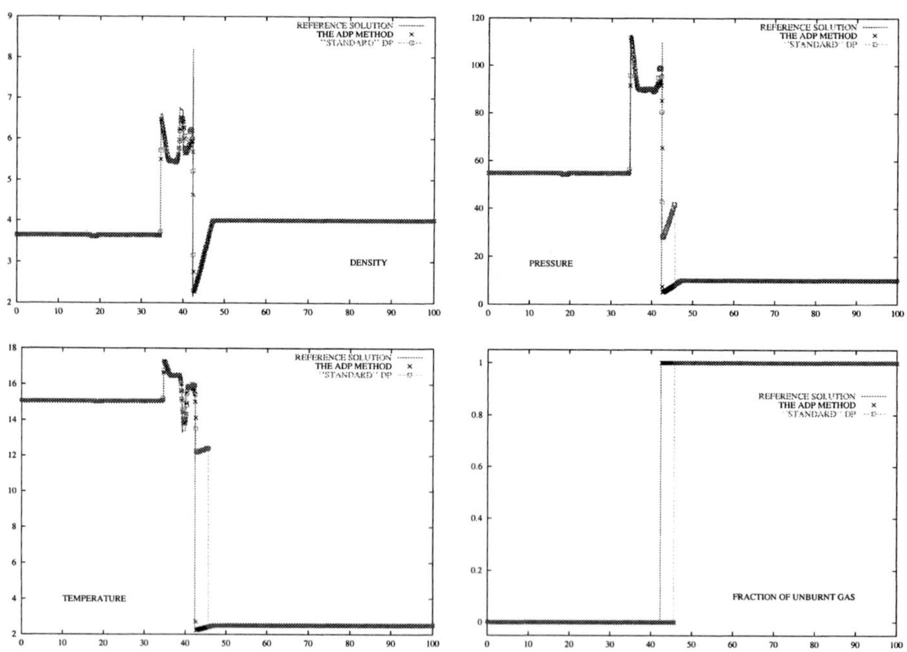

Fig. 3. $T = 4$

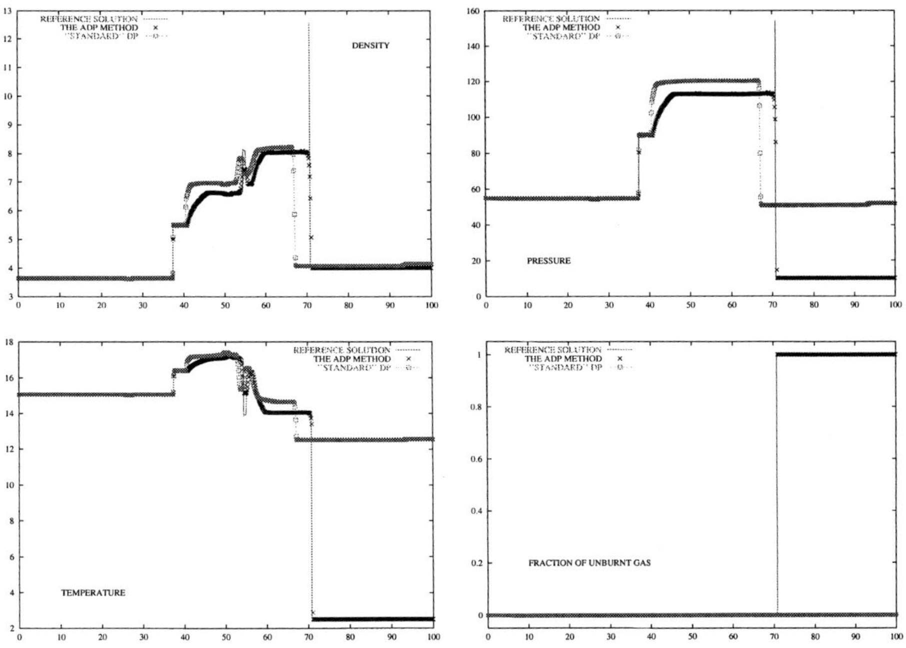

Fig. 4. $T = 8$

References

1. Abgrall, R., Karni, S. (2001): Computations of compressible multifluids. J. Comput. Phys., **169**, 594–623
2. Bao, W., Jin, S. (2000): The random projection method for hyperbolic conservation laws with stiff reaction terms. J. Comput. Phys., **163**, 216–248
3. Bao, W., Jin, S. (2001): The random projection method for stiff detonation capturing. SIAM J. Sci. Comput., **23**, 1000-1026
4. Bao, W., Jin, S. (2002): Error estimates on the random projection methods for hyperbolic conservation laws with stiff reaction terms. Appl. Num. Math., to appear.
5. Ben-Artzi, M. (1989): The generalized Riemann problem for reactive flows. J. Comput. Phys., **81**, 70–101
6. Berkenbosch, A.C., Kaasschieter, E.F., Klein R. (1998): Detonation capturing for stiff combustion chemistry. Combust. Theory Model., **2**, 313–348
7. Bourlioux, A., Majda, A., Roytburd, V. (1991): Theoretical and numerical structure for unstable one-dimensional detonations. SIAM J. Appl. Math., **51**, 303–343
8. Colella, P., Majda, A., Roytburd, V. (1986): Theoretical and numerical structure for reacting shock waves. SIAM J. Sci. Statist. Comput., **7**, 1059–1080
9. Courant, R., Friedrichs, K.O. (1971): Supersonic Flow and Shock Waves. Interscience, New York
10. Engquist, B., Sjogreen, B. (1991): Robust difference approximations of stiff inviscid detonation waves. CAM Report 91-03, UCLA

11. Fan, H., Jin, S., Teng, Z.H. (2000): Zero reaction limit for hyperbolic conservation laws with source terms. J. Differential Equations, **168**, 270–294
12. Griffiths, D.F., Stuart A.M., Yee, H.C. (1992): Numerical wave propagation in an advection equation with a nonlinear source term. SIAM J. Numer. Anal., **29**, 1244–1260
13. Helzel, C., LeVeque, R.J., Warnecke, G. (2000): A modified fractional step method for the accurate approximation of detonation waves. SIAM J. Sci. Comput., **22**, 1489–1510
14. Hwang, P., Fedkiw, R.P., Merriman, B., Karagozian, A.K., Osher, S.J. (2000): Numerical resolution of pulsating detonation waves. Combust. Theory Model., **4**, 217–240
15. Kurganov, A., Noelle, S., Petrova, G. (2000): Semi-discrete central-upwind schemes for hyperbolic conservation laws and Hamilton-Jacobi equations. SIAM J. Sci. Comput., **22**, 1461–1488
16. LeVeque, R.J., Yee, H.C. (1990): A study of numerical methods for hyperbolic conservation laws with stiff source terms. J. Comput. Phys., **86**, 187–210
17. Pember, R.B. (1993): Numerical methods for hyperbolic conservation laws with stiff relaxation I. Spurious solutions. SIAM J. Appl. Math., **53**, 1293–1330
18. Sussman, M.A. (1993): Source term evaluation for combustion modeling. AIAA paper 93-0239
19. Williams, F.A. (1965): Combustion Theory. Addison-Wesley, Reading, MA

Adaptive Second Order Central Schemes on Unstructured Staggered Grids

Marc Küther[1] and Mario Ohlberger[2]

[1] Institut für Angewandte Mathematik, Universität Freiburg, Germany
marc@mathematik.uni-freiburg.de
[2] Institut für Angewandte Mathematik, Universität Freiburg, Germany
mario@mathematik.uni-freiburg.de

Summary. This communication is devoted to the derivation and numerical implementation of higher order central schemes on adaptive unstructured grids for approximating nonlinear hyperbolic conservation laws in several spatial dimensions.

1 Introduction

We consider the following Cauchy problem for the unknown function $u : \mathbb{R}^d \times [0, \infty) \to \mathbb{R}$, $d \geq 1$:

$$\partial_t u(x,t) + \operatorname{div} \mathrm{f}(\mathrm{u}(\mathrm{x},\mathrm{t})) = 0 \quad (x,t) \in \mathbb{R}^d \times (0,\infty), \tag{1}$$

$$u(.,0) = u_0 \qquad \text{in } \mathbb{R}^d, \tag{2}$$

where $f \in C^1(\mathbb{R}, \mathbb{R}^d)$ denotes some nonlinear flux function and $u_0 \in L^\infty(\mathbb{R}^d) \cap BV(\mathbb{R}^d)$ some initial data. Here $BV(\mathbb{R}^d)$ is the space of functions with bounded variation (cf. [4]).

In this contribution we focus on the derivation of second order central schemes on a general class of unstructured grids in arbitrary spatial dimensions. The only assumption we use on two subsequent grids is an *overlap assumption*, see below. The second order accuracy is achieved by using a reconstruction and limitation procedure. Furthermore we propose an adaption strategy for the first and second order methods where we use the theoretical a posteriori result of the first order scheme to derive appropriate refinement indicators. We finally implemented the adaptive second order scheme on a particular choice of meshes and demonstrate the performance of the method by numerical experiments.

First and second order central schemes on staggered grids for conservation laws were introduced by Nessyahu and Tadmor in 1990 [8] in one spatial dimension and generalized to particular unstructured grids in two space dimensions by Arminjon and Viallon in 1999 [1]. In [7] first order central schemes were generalized to arbitrary unstructured staggered grids. In addition a priori and a posteriori error estimates were obtained, by interpreting the central scheme on staggered grids as a upwind finite volume scheme in conservation form on the intersection grid accompanied by suitable prolongation and restriction steps. Hence the a priori and a posteriori theory developed in [2] and [6] could be applied.

2 Central schemes on arbitrary unstructured grids

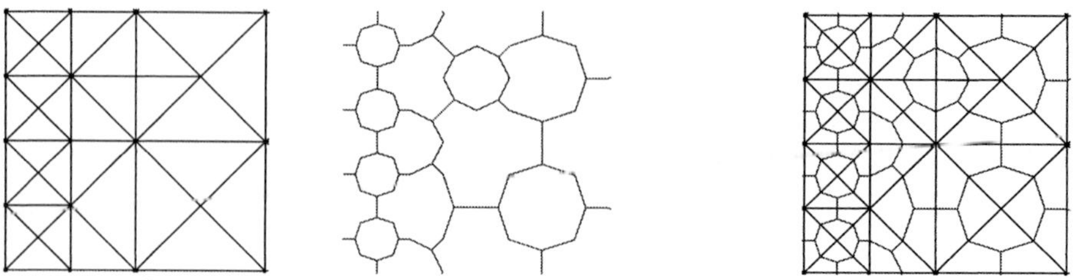

Fig. 1. A possible choice of grids at even and odd timesteps and their intersection.

For the discretization of (1),(2) let a sequence of unstructured grids $(\mathcal{T}_h^n)_{n\in\mathbb{N}}$ be given. Here n corresponds to the time level. We assume that each grid $\mathcal{T}_h^n$ is is a non-overlapping partition of $\mathbb{R}^d$ consisting of polyhedrons T_i^n, $i \in I^n \subset \mathbb{N}$, of finite diameter. The sequence of meshes has to satisfy the following regularity *uniformly*:
There exists $\alpha > 0$ such that for all $n \in \mathbb{N}$ and all $i \in I^n$

$$\alpha\,\mathrm{diam}(\mathrm{T}_\mathrm{i}^\mathrm{n})^d \leq |T_i^n|, \quad |\partial T_i^n| \leq \frac{1}{\alpha}\mathrm{diam}(\mathrm{T}_\mathrm{i}^\mathrm{n})^{d-1}.$$

One possible choice of such grids is illustrated in Fig. 1. In order to go from one grid to another we have to link two consecutive grids $\mathcal{T}_h^n$ and $\mathcal{T}_h^{n+1}$. Thus, for $i \in I^{n+1}$ we define:

$$K^{n,n+1}(i) := \{j \in I^n |\ T_j^n \cap T_i^{n+1} \neq \emptyset\},$$
$$K_\partial^{n,n+1}(i) := \{j \in I^n |\ S_{ij}^{n,n+1} := T_j^n \cap \partial T_i^{n+1} \neq \emptyset\}.$$

The scaled outer normal $\nu_{ij}^{n,n+1}$ to $S_{ij}^{n,n+1}$ is piecewise defined (as the sum of the scaled outer normals on each part of $S_{ij}^{n,n+1}$ contained in a $(d-1)$ dimensional hyperplane) and has the length of $|S_{ij}^{n,n+1}|$.

Two consecutive grids have to satisfy the following overlap assumption which excludes the case that two consecutive grids are equal.

There exists a constant $C_{\mathrm{ov}} > 0$ such that for all $n \in \mathbb{N}$, $i \in I^{n+1}$, $j \in K^{n,n+1}(i)$

$$C_{\mathrm{ov}} \leq \frac{|T_j^n \cap T_i^{n+1}|}{|T_i^{n+1}|} \leq 1, \tag{3}$$

$$\partial T_i^{n+1} \cap \partial T_j^n \quad \text{has dimension at most } d-2. \tag{4}$$

As a theoretical tool we shall need an *intersection grid* $\mathcal{T}_h^{n,n+1}$ given as $(T_i^{n,n+1})_{i\in I^{n,n+1}}$ obtained by intersecting each $T_i^n \in \mathcal{T}_h^n$ with each $T_i^{n+1} \in$

$\mathcal{T}_h^{n+1}$. Note that the intersection of two polyhedrons is again a polyhedron. For $i \in I^{n,n+1}$ we denote the set of indices corresponding to neighboring polyhedrons $T_l^{n,n+1}$ by $\mathcal{N}^{n,n+1}(i)$ and the common interface by $S_{il}^{n,n+1}$.

The time axis $[0,\infty)$ is partitioned into intervals $[t^n, t^{n+1}[$, $n \in \mathbb{N}$, of length $k^n := t^{n+1} - t^n$.

2.1 A review of first order central schemes on arbitrary grids

The idea to derive a first order central scheme on the given sequence of unstructured grids for approximating the solution of (1), (2) is the following, see [7]:
Let piecewise constant values $(u_i^n)_{i\in I^n}$, $n \in \mathbb{N}$, be given.

1. Prolongate the values trivially to the intersection grid $\mathcal{T}_h^{n,n+1}$.
2. Perform a upwind finite volume step, e.g. with the Godunov scheme, on the intersection grid.
3. Perform the L^2 projection of these values onto $\mathcal{T}_h^{n+1}$ and get piecewise constant values $(u_i^{n+1})_{i\in I^{n+1}}$.

When considering the overall algorithm the numerical fluxes introduced in the second step reduce to evaluations of the continuous flux f in (1) because of the consistency and conservation of the numerical flux and the overlap assumption on two consecutive grids. The resulting first order Lax-Friedrichs scheme is given explicitly as follows. For given values u_i^n, $i \in I^n$ define values u_i^{n+1}, $i \in I^{n+1}$ by

$$u_i^{n+1} = \sum_{j\in K^{n,n+1}(i)} \frac{|T_j^n \cap T_i^{n+1}|}{|T_i^{n+1}|} u_j^n - \frac{k^n}{|T_i^{n+1}|} \sum_{j\in K_\partial^{n,n+1}(i)} f(u_j^n)\nu_{ij}^{n,n+1}.$$

The advantage of this point of view is that one can analyze central schemes on staggered grids within the same framework as finite volume schemes on a fixed grid with appropriate pre– and post– processing. Particularly one can prove the following a posteriori error estimate (see [7]).

Theorem 1 (A posteriori error estimate).
Assume that the assumptions on the data and the meshes stated above are satisfied and that the CFL-condition

$$k^n V_{[U_m,U_M]} \le \frac{1}{2}(1-\xi)\alpha^2 h_i^{n,n+1} \quad \xi \in]0,1[, \tag{5}$$

is met, where $V_{[U_m,U_M]}$ is a constant such that $|f'(s)| \le V_{[U_m,U_M]}$, for all $s \in [U_m, U_M]$. Let $K \subset\subset \mathbb{R}^d \times \mathbb{R}^+$, $\omega = V_{[U_m,U_M]}$ and choose $T, R > 0$ and $x_0 \in \mathbb{R}^d$ such that $T \in]0, R/\omega[$ and $K \subset \bigcup_{0\le t\le T} B_{R-\omega t}(x_0) \times \{t\}$. Then we have

$$||u - u_h||_{L^1(K)} \le T(||u_0 - u_h(.,0)||_{L^1(\{|x-x_0|<R+1\})} + aQ + b\sqrt{Q}), \tag{6}$$

where a and b are computable constants (see [7] for details) and

$$Q := \frac{1}{2}\sum_{n=0}^{N_0}\sum_{i\in I_D^{n+1}} h_i^{n+1}|T_i^{n+1}| \sum_{j,l\in K^{n,n+1}(i)} \frac{|T_j^n\cap T_i^{n+1}|}{|T_i^{n+1}|}\frac{|T_l^n\cap T_i^{n+1}|}{|T_i^{n+1}|}|u_j^n - u_l^n|$$
$$+\sum_{n=0}^{N_0-1} k^n \sum_{i\in I_D^{n+1}} |T_i^{n+1}| \left|u_i^{n+1} - \sum_{j\in K^{n,n+1}(i)} \frac{|T_j^n\cap T_i^{n+1}|}{|T_i^{n+1}|}u_j^n\right|$$
$$+6V_{[U_m,U_M]}\sum_{n=0}^{N_0} k^n \sum_{i\in I_D^{n+1}} (h_i^{n+1}+k^n) \sum_{(j,l)\in\mathcal{E}^{n,n+1}(i)} |S_{jl}^{n,n+1}(i)|\,|u_j^n - u_l^n|.$$

2.2 Derivation of the higher order central scheme

In what follows, we will use the same point of view as above to derive second order central schemes on general unstructured staggered grids.

First, we introduce linear reconstruction operators $L^n : PC(\mathcal{T}_h^n) \to PL(\mathcal{T}_h^n)$, where the spaces $PC(\mathcal{T}_h^n)$ and $PL(\mathcal{T}_h^n)$ are defined as

$$PC(\mathcal{T}_h^n) := \{\varphi \in BV(\mathbb{R}^d)|\ \varphi|_{T_i} = c_i \in \mathbb{R}, T_i \in \mathcal{T}_h^n\},$$
$$PL(\mathcal{T}_h^n) := \{\varphi \in BV(\mathbb{R}^d)|\ \varphi|_{T_i} \in \mathbb{P}_1, T_i \in \mathcal{T}_h^n\}.$$

Here $\mathbb{P}_1$ denotes the space of linear functions. In addition we assume that L^n fulfills the following properties for all $v_h \in PC(\mathcal{T}_h^n)$, $n \in \mathbb{N}$:

Conservation:

$$\int_{T_i^n} v_h = \int_{T_i^n} L^n v_h, \quad \text{for all } T_i^n \in \mathcal{T}_h^n.$$

Non oscillatory:

$$|L^n v_h|_{BV} \le |v_h|_{BV}, \qquad ||L^n v_h||_{L^\infty} \le ||v_h||_{L^\infty}.$$

Such reconstruction operators are for example the reconstruction of Durlofsky, Engquist, Osher [3] or the reconstruction of Wierse [10] together with some suitable limiters.

Let us derive a continuous version of the second order central scheme which uses generalized Riemann problems, see [4].

Let piecewise constant function u_h^n on $\mathcal{T}_h^n$, $n \in \mathbb{N}$, be given.

1. Form the functions $L^n u_h^n$ which are piecewise linear on $\mathcal{T}_h^n$.
2. Prolongate the functions trivially to the intersection grid $\mathcal{T}_h^{n,n+1}$.
3. Let v denote the solution of the generalized Riemann problem

$$\partial_t v + \mathrm{div}\, \mathrm{f(v)} = 0 \quad \text{in } \mathbb{R}^d \times (\mathrm{t^n}, \mathrm{t^{n+1}}), \quad \mathrm{v}(.,\mathrm{t^n}) = \mathrm{u_h^n} \quad \text{in } \mathbb{R}^d.$$

Integration of this PDE gives for $i \in I^{n,n+1}$

$$v_i^{n+1} := \frac{1}{|T_i^{n,n+1}|} \int_{T_i^{n,n+1}} v(\cdot, t^{n+1}) = \frac{1}{|T_i^{n,n+1}|} \int_{T_i^{n,n+1}} L^n u_h^n - \frac{k^n}{|T_i^{n,n+1}|} \sum_{l \in \mathcal{N}^{n,n+1}(i)} \int_{t^n}^{t^{n+1}} \int_{S_{il}^{n,n+1}} f(v(\sigma,t)) n_{il} \, d\sigma dt,$$

where n_{il} denotes the unit outer normal to $S_{il}^{n,n+1}$.

4. Perform the L^2 projection of these values onto $\mathcal{T}_h^{n+1}$ and get piecewise constant values $(u_i^{n+1})_{i \in I^{n+1}}$.

By the same reasoning as in the first order case we get for the overall scheme the following formula:

$$u_i^{n+1} = \frac{1}{|T_i^{n+1}|} \sum_{j \in K^{n,n+1}(i)} \int_{T_j^n \cap T_i^{n+1}} L^n u_h^n - \frac{1}{|T_i^{n+1}|} \sum_{j \in K_\partial^{n,n+1}(i)} \int_{t^n}^{t^{n+1}} \int_{S_{ij}^{n,n+1}} f(v(\sigma,t)) n_{ij} \, d\sigma dt.$$

In order to get a feasible numerical method we have to approximate the flux integrals by appropriate quadrature rules. Similar to Nessyahu and Tadmor [8] we assume the time step to be small enough such that the solution of the generalized Riemann problem is smooth across the edges $S_{ij}^{n,n+1}$. Hence, the solution satisfies the PDE in a classical pointwise sense on those interfaces. Therefore, we can use the midpoint rule to approximate the time integral

$$\int_{t^n}^{t^{n+1}} \int_{S_{ij}^{n,n+1}} f(v(\sigma,t)) n_{ij} \, d\sigma dt = k^n \int_{S_{ij}^{n,n+1}} f(v(\sigma, t^n + 0.5k^n)) n_{ij} \, d\sigma + O((k^n)^3).$$

In order to express $v(\sigma, t^n + 0.5k^n)$ in terms of values at time level t^n we use Taylor's expansion and get for $x \in S_{ij}^{n,n+1}$

$$\begin{aligned} v(x, t^n + 0.5k^n) &= v(x,t^n) + \frac{1}{2} k^n \partial_t v(x,t^n) + O((k^n)^2) \\ &= v(x,t^n) - \frac{1}{2} k^n \mathrm{div} f(v(x,t^n)) + O((k^n)^2) \\ &= v(x,t^n) - \frac{1}{2} k^n \sum_{i=1}^{d} f_i'(L^n u_h^n(x)) \partial_{x_i} L^n u_h^n(x) + O((k^n)^2). \end{aligned}$$

Using again an appropriate quadrature rule (e.g. trapezoidal rule) to approximate the spatial integral we end up with the following predictor-corrector form of our second order central scheme which is the generalization of the Nessyahu-Tadmor scheme to unstructured staggered grids in any spatial dimension.

Definition (Second order staggered Lax-Friedrichs scheme) Define an approximation u_h to the solution of (1), (2) by the following scheme.
For $i \in I^0$ set

$$u_i^0 := \frac{1}{|T_i^0|} \int_{T_i^0} u_0(x) \mathrm{d}x.$$

For given values u_i^n, $i \in I^n$ define values u_i^{n+1}, $i \in I^{n+1}$ by

$$\begin{aligned}
u_{ijk}^{n+\frac{1}{2}} &= u_j^n - \frac{1}{2} k^n \sum_{i=1}^{d} f_i'(L^n u_h^n(z_{ijk})) \partial_{x_i} L^n u_h^n(z_{ijk}), \\
u_i^{n+1} &= \frac{1}{|T_i^{n+1}|} \sum_{j \in K^{n,n+1}(i)} \int_{T_j^n \cap T_i^{n+1}} L_j^n u \\
&\quad - \frac{k^n}{|T_i^{n+1}|} \sum_{j \in K_\partial^{n,n+1}(i)} |S_{ij}^{n,n+1}| \sum_{k=1}^{L(i,j)} \omega_{ijk} f(u_{ijk}^{n+\frac{1}{2}}) \nu_{ij}^{n,n+1},
\end{aligned}$$

where for all $i \in I^{n+1}$ and $j \in K_\partial^{n,n+1}(i)$ $L(i,j)$ denotes the number of integration points z_{ijk} on $S_{ij}^{n,n+1}$ with positive weights ω_{ijk} such that $\sum_{k=1}^{L(i,j)} \omega_{ijk} = 1$. Finally, also in the second order case we define the discrete solution as

$$u_h(x,t) = u_i^n \quad \text{for } t \in [t^n, t^{n+1}[\text{ and } x \in T_i^n.$$

3 Implementation and numerical experiments

We finally implemented the second order scheme in two spatial dimensions based on a primal triangular mesh for even time steps and on a dual - Donald type - mesh for odd time steps (see Fig. 1). The grid adaption is performed solely at the even timesteps on the primal triangular grid where we use the theoretical a posteriori result of the first order method for our adaption strategy (see also [6]). The corresponding adaptive dual mesh is then implicitly given as the dual mesh of the adaptive triangulation. This particular choice of meshes allows us to use the well known MUSCL–type reconstruction of Durlofsky, Engquist and Osher [3] together with some modified superbee limiter on the primal triangulation and a reconstruction of Sonar [9] with the Barth-Jesperson limiter on the dual mesh in order to obtain the second order method.

The following numerical experiments show that the second order central scheme produces at least as good results as an second order upwind method does (see also [5]). In addition, a comparison with calculations on a uniform grid demonstrate the efficiency of the adaptive scheme.

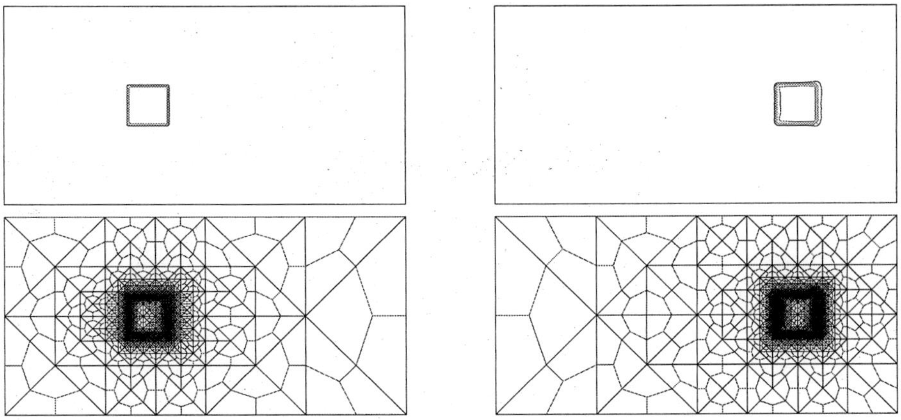

Fig. 2. Linear advection of the characteristic function of a square ($t = 0.0$ on the left hand side and $t = 0.4$ on the right hand side). The upper pictures show the isolines of the solution, while the staggered adaptive computational grids are shown at the bottom.

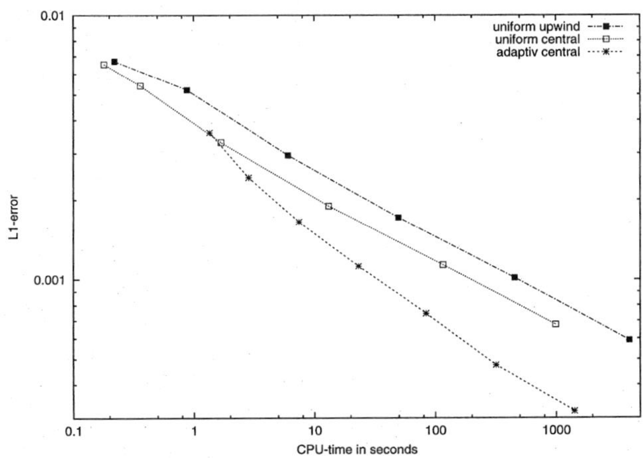

Fig. 3. L^1 error versus CPU time for the linear advection of a characteristic function. Comparison between uniform second order upwind method, uniform second order central scheme and adaptive second order central scheme.

3.1 Linear advection problems

Let us consider the linear conservation law (1) where $f(u) = (u, 0)^\top$ together with either discontinuous or smooth initial data. For the discontinuous case we choose the characteristic function of a square given by

$$u_0(x) = \begin{cases} 1, \text{ if } & 0.3 \le x_1 \le 0.4, 0.2 \le x_2 \le 0.3, \\ 0, \text{ else} & \end{cases},$$

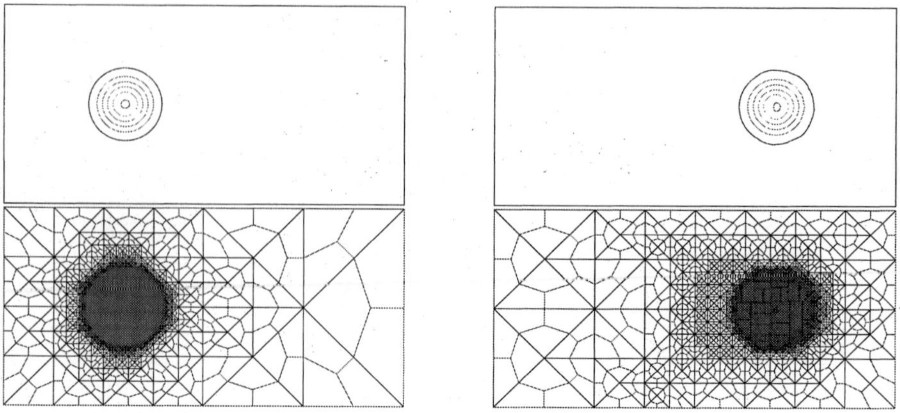

Fig. 4. Linear advection of a smooth (C^1) function ($t = 0.0$ on the left hand side and $t = 0.4$ on the right hand side). The upper pictures show the isolines of the solution, while the staggered adaptive computational grids are shown at the bottom.

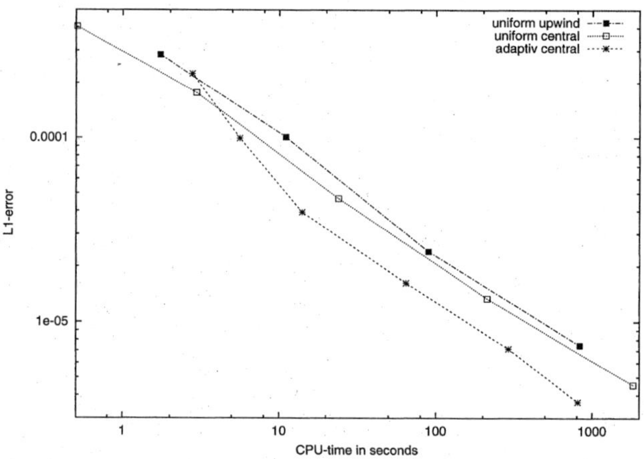

Fig. 5. L^1 error versus CPU time for the linear advection of a smooth function. Comparison between uniform second order upwind method, uniform second order central scheme and adaptive second order central scheme.

while in the smooth case we choose a global C^1–function given by

$$u_0(x) = \begin{cases} 0.1(2r^3 - 3r^2 + 1), \text{ if } & r := 10\sqrt{(x_1 - 0.3)^2 + (x_2 - 0.25)^2} \leq 1 \\ 0, & \text{else} \end{cases} .$$

In both cases the exact solution is $u(x, t) = u_0(x_1 - t, x_2)$.

Figures 2 and 4 show the numerical solutions of the adaptive higher order central scheme in both cases, when the initial condition is discontinuous or smooth. In the corresponding diagrams (Fig. 3 and Fig. 5) we compare the

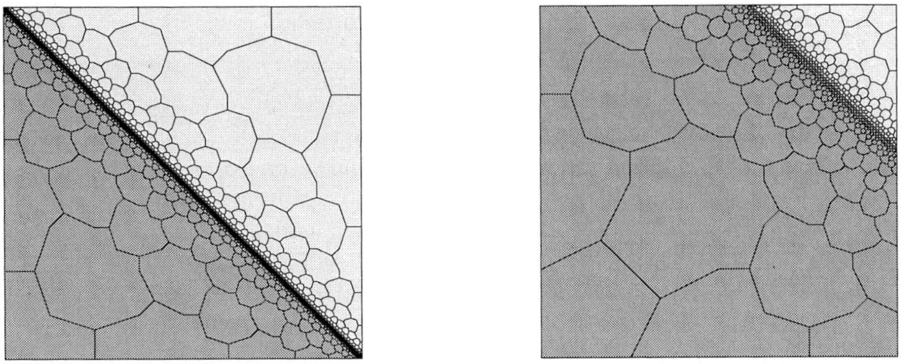

Fig. 6. Burgers type convection of a shock ($t = 0.0$ on the left hand side and $t = 0.1$ in the middle). The pictures show a color shading of the solution (dark grey = 2.0, light grey = 1.0) together with the adaptive dual computational grid.

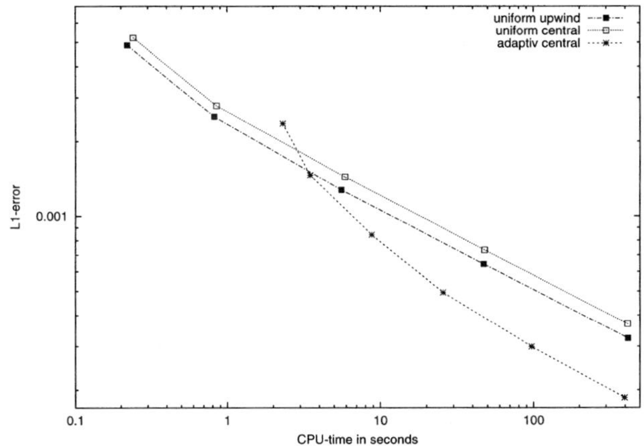

Fig. 7. L^1 error versus CPU time for the Burgers type convection of a step function. Comparison between uniform second order upwind method, uniform second order central scheme and adaptive second order central scheme.

uniform and adaptive higher order central scheme with an higher order upwind scheme. The results demonstrate, that the central scheme on uniform meshes behaves at least as good as an second order upwind method does. In addition, one can gain a lot in efficiency by using adaptive methods.

3.2 A Burgers type problem

Let us consider the nonlinear conservation law (1) together with the initial condition (2) in $\mathbb{R}^2$, where

$$f(u) = \begin{pmatrix} u^2 \\ u^2 \end{pmatrix} \qquad \text{and} \qquad u_0(x) = \begin{cases} 2, \text{ if } \frac{x_1+x_2}{2} - 0.5 \leq 0 \\ 1, \text{ if } \frac{x_1+x_2}{2} - 0.5 > 0 \end{cases}.$$

Then the exact solution of this Burgers type problem is

$$u(x,t) = \begin{cases} 2, \text{ if } \frac{x_1+x_2}{2} - 0.5 \leq 3t \\ 1, \text{ if } \frac{x_1+x_2}{2} - 0.5 > 3t \end{cases}.$$

In Figure 6 the adaptive refined dual grid of the second order central scheme together with a color shading of the discrete solution is shown at $t = 0$ and $t = 0.1$. In Figure 7 we compare the uniform and adaptive second order central scheme with an second order upwind scheme. Also in this case the efficiency of the adaptive central scheme is shown.

References

1. P. Arminjon and M.-C. Viallon (1999): Convergence of a finite volume extension of the Nessyahu-Tadmor scheme on unstructured grids for a two-dimensional linear hyperbolic equation. *SIAM J. Numer. Anal.*, **36**, 738–771
2. C. Chainais-Hillairet (1999): Finite volume schemes for a nonlinear hyperbolic equation. Convergence towards the entropy solution and error estimates. *M2AN Math. Model. Numer. Anal.*, **33**, 129–156
3. L.J. Durlofsky, B. Engquist, and S. Osher (1992): Triangle based adaptive stencils for the solution of hyperbolic conservation laws. *Journal of Computational Physics*, **98**, 64–73
4. E. Godlewski and P.-A. Raviart (1991): *Hyperbolic systems of conservation laws*, volume 3/4 of *Mathematiques & Applications*. Ellipses, Paris
5. R. Klöfkorn, D. Kröner, and M. Ohlberger (2002): Local adaptive methods for convection dominated problems. *To appear in Internat. J. Numer. Methods Fluids*
6. D. Kröner and M. Ohlberger (2000): A-posteriori error estimates for upwind finite volume schemes for nonlinear conservation laws in multi dimensions. *Math. Comput.*, **69**, 25–39
7. M. Küther (2001): Error estimates for the staggered Lax-Friedrichs scheme on unstructured grids. *SIAM J. Numer. Anal.*, **39**, 1269–1301
8. H. Nessyahu and E. Tadmor (1990):. Non-oscillatory central differencing for hyperbolic conservations laws. *J. Comput. Phys.*, **87**, 408–463
9. Th. Sonar (1993): On the design of an upwind scheme for compressible flow on general triangulations. *Numer. Algorithms*, **4**, 135–149
10. M. Wierse (1997):. A new theoretically motivated higher order upwind scheme on unstructured grids of simplices. *Adv. Comput. Math.*, **7**, 303–335

Physical Symmetries and Hyperbolic GLM Divergence Correction Scheme for Maxwell and MHD Equations

Y.J. Lee[1], F. Kemm[1], C.-D. Munz[1] and R. Schneider[2]

[1] Institut für Aerodynamik und Gasdynamik, Universität Stuttgart, Pfaffenwaldring 21, D-70550 Stuttgart, Germany
lee@iag.uni-stuttgart.de, kemm@iag.uni-stuttgart.de, munz@iag.uni-stuttgart.de

[2] Forschungszentrum Karlsruhe, Institut für Hochleistungsimpuls- und Mikrowellentechnik,Postfach 3640, D-76021 Karlsruhe, Germany
rudolf.schneider@ihm.fzk.de

Summary. The charge conservation laws in general are not strictly obeyed in computational electromagnetics and MHD, due to the presence of various types of numerical errors. The violation of the charge conservation laws in field theoretical viewpoint is a consequence of the *broken* gauge symmetry in the computational space. In this paper, we present a new field theoretical method for the treatment of the often violated charge conservation laws in computational electrodynamics and MHD, which is consistent with the symmetries of Maxwell theory, namely the Lorentz-, gauge- and duality symmetries. This method under particular constraints reduces to the existing GLM scheme which has so far lacked a concrete theoretical framework, and thereby provides a new insight into the GLM scheme. The central idea of our divergence correction scheme is the implementation of the physically consistent counter term Ansätze to Maxwell and MHD equations, for the restoration of the gauge symmetry. One of the main advantages of our method is that the divergence conditions for the charge conservations are implemented into the Maxwell and MHD equations in a *hyperbolic* form, rather than the genuine elliptic form, and it can be easily implemented into the existing codes for Maxwell and MHD solvers via operator splitting Ansatz.

1 Introduction

The physics of electromagnetism is described by Maxwell equations,

$$-\partial_t \mathbf{E} + \nabla \times \mathbf{B} = \mathbf{j}_e, \tag{1}$$

$$\partial_t \mathbf{B} + \nabla \times \mathbf{E} = 0, \tag{2}$$

$$\nabla \cdot \mathbf{E} = \rho_e, \tag{3}$$

$$\nabla \cdot \mathbf{B} = 0, \tag{4}$$

for the electric field $\mathbf{E}$ and the magnetic induction $\mathbf{B}$, with external electric charge density ρ_e and current density $\mathbf{j}_e$. Here, the unit light velocity $c = 1$ and Heaviside's units (with the Coulomb force given by $qq'/4\pi r^2$) are taken.

The important consequences of Maxwell equations (1)-(4) to which we attend are charge conservation laws. The conservation of the electric charge is described by

$$\partial_t \rho_e + \nabla \cdot \mathbf{j}_e = 0, \tag{5}$$

whereas the conservation of the magnetic charge is described by the elliptic equation (4), due to the permanent absence of magnetic monopoles. These charge conservation constraints must be satisfied by the solutions of hyperbolic subsystem Eqs. (1) and (2) of Maxwell equations, provided that the initial field configurations satisfy the divergence conditions (3) and (4). However, due to the presence of various types of numerical insufficiencies, the numerical solution of the hyperbolic part of the Maxwell system alone often do not obey the charge conservation constraints (4) and (5) and, consequently, inconsistencies and non-physical field configurations may be generated.

In this paper, we approach the problem of the numerical violation of charge conservation laws in computational electrodynamics and MHD from a field theoretical point of view, and present a numerical scheme which is consistent with the symmetries of Maxwell theory, namely the Lorentz-, gauge- and duality symmetries. It has been shown that this new scheme under particular constraints reduces to the existing GLM method [1, 2] which has so far lacked a concrete theoretical framework. In this respect, this paper also presents a new insight into the GLM scheme.

2 Symmetries in Maxwell theory

Symmetry plays an important role in understanding characteristic nature of given physical theory and conservation laws. In this regard, we shall review the these fundamental symmetries of Maxwell theory in this section.

The basic notations we shall use in this section can be summarized as follows. The space-time coordinates in a certain inertial frame are denoted by x^μ, where the Greek indices (μ, ν) always run over the coordinate labels 0, 1, 2, 3. The metric tensor for flat Minkowski space-time is specified by $g_{\mu\nu} = \mathsf{Diag}[-1, +1, +1, +1]$. Furthermore, we adopt Einstein's summation convention: the repeated indices in the same term, once upstairs and once downstairs, are summed over (for details, see Refs. [4, 5]).

2.1 Lorentz symmetry

The Maxwell theory obeys the physical law of Einstein's special theory of relativity, which states that the physics is invariant under a special class of space-time coordinate transformations, called Lorentz transformations. This symmetry property of Maxwell theory becomes transparent when the basic equations (1)-(4) are written in term of compact notation for vector and tensor fields which exhibits relativistic covariance. The Lorentz covariant compact form of Maxwell equations is given by

$$\partial_\mu F_e^{\mu\nu} = j_e^\nu , \quad \partial_\mu F_g^{\mu\nu} = j_g^\nu , \tag{6}$$

for the external electric and magnetic four-current densities defined by

$$j_e^\mu \equiv (\rho_e, \mathbf{j}_e), \quad j_g^\mu \equiv (\rho_g, \mathbf{j}_g), \tag{7}$$

where $F_e^{\mu\nu}$ is the Lorentz covariant antisymmetric rank-2 field strength tensor given in terms of electric and magnetic fields,

$$F_e^{\mu\nu} = \begin{pmatrix} 0 & -E^1 & -E^2 & -E^3 \\ E^1 & 0 & -B^3 & B^2 \\ E^2 & B^3 & 0 & -B^1 \\ E^3 & -B^2 & B^1 & 0 \end{pmatrix}, \tag{8}$$

and $F_g^{\mu\nu}$ the corresponding dual field strength tensor defined as

$$F_g^{\mu\nu} \equiv \frac{1}{2}\epsilon^{\mu\nu}{}_{\rho\sigma} F_e^{\rho\sigma} = \begin{pmatrix} 0 & -B^1 & -B^2 & -B^3 \\ B^1 & 0 & E^3 & -E^2 \\ B^2 & -E^3 & 0 & E^1 \\ B^3 & E^2 & -E^1 & 0 \end{pmatrix}, \tag{9}$$

where $\epsilon^{\mu\nu\rho\sigma}$ is the Levi-Civita antisymmetric symbol with $\epsilon^{0123} = 1$.

Note that in presenting Eq. (6), we have not assumed vanishing magnetic charge and currents for reasons which will become clear later. In the presence of magnetic currents, the Lorentz invariant Lagrangian density for Eq. (6) has been given by Zwanziger [6]

$$\mathcal{L}_{Zw} = \mathcal{L}_{free} + \mathcal{L}_{int.} \tag{10}$$

$$\begin{aligned} \mathcal{L}_{free} &= \frac{1}{2}\hat{n}_\mu \hat{n}^\kappa \Big[f^{\mu\nu}(A)\tilde{f}_{\kappa\nu}(B) - f^{\mu\nu}(B)\tilde{f}_{\kappa\nu}(A) \Big] \\ &+ \frac{1}{2}\hat{n}_\mu \hat{n}^\kappa \Big[f^{\mu\nu}(A) f_{\kappa\nu}(A) + f^{\mu\nu}(B) f_{\kappa\nu}(B) \Big], \end{aligned} \tag{11}$$

$$\mathcal{L}_{int.} = j_e^\mu A_\mu + j_g^\mu B_\mu , \tag{12}$$

with the definitions

$$f_{\mu\nu}(A) \equiv \partial_\mu A_\nu - \partial_\nu A_\mu , \quad f_{\mu\nu}(B) \equiv \partial_\mu B_\nu - \partial_\nu B_\mu , \tag{13}$$

$$\tilde{f}_{\mu\nu}(A) \equiv \frac{1}{2}\epsilon_{\mu\nu}{}^{\rho\sigma} f_{\rho\sigma}(A), \quad \tilde{f}_{\mu\nu}(B) \equiv \frac{1}{2}\epsilon_{\mu\nu}{}^{\rho\sigma} f_{\rho\sigma}(B), \tag{14}$$

where $\hat{n}^\mu$ is a constant four-vector with $\hat{n}_\mu \hat{n}^\mu = 1$, and the two Lorentz covariant vector fields A^μ and B^μ describe the interaction with electric and magnetic charges, respectively. [Note that A^μ and B^μ are not the conventional vector potentials encountered in classical electrodynamics.] The corresponding expressions for the field-strength tensor $F_e^{\mu\nu}(A, B)$ and its dual counterpart $F_g^{\mu\nu}(A, B)$ are given by

$$F_e^{\mu\nu}(A,B) = \hat{n}^\mu \hat{n}_\kappa f^{\kappa\nu}(A) - \hat{n}^\nu \hat{n}_\kappa f^{\kappa\mu}(A) - \frac{1}{2}\epsilon^{\mu\nu}{}_{\rho\sigma}\left[\hat{n}^\rho \hat{n}_\kappa f^{\kappa\sigma}(B) - \hat{n}^\sigma \hat{n}_\kappa f^{\kappa\rho}(B) - f^{\rho\sigma}(B)\right] \quad (15)$$

$$F_g^{\mu\nu}(A,B) = \hat{n}^\mu \hat{n}_\kappa f^{\kappa\nu}(B) - \hat{n}^\nu \hat{n}_\kappa f^{\kappa\mu}(B) + \frac{1}{2}\epsilon^{\mu\nu}{}_{\rho\sigma}\left[\hat{n}^\rho \hat{n}_\kappa f^{\kappa\sigma}(A) - \hat{n}^\sigma \hat{n}_\kappa f^{\kappa\rho}(A) - f^{\rho\sigma}(A)\right], \quad (16)$$

in terms of gauge fields A^μ and B^μ. The two seemingly different vector fields A^μ and B^μ collectively describe a single photon interacting both with electric and magnetic charges, under the imposition of proper constraints (for details, see Ref. [6]). Given the full description of the field strengths $F_e^{\mu\nu}(A,B)$ and $F_g^{\mu\nu}(A,B)$, the variation of the action functional for the Lorentz invariant Lagrange density $\mathcal{L}_{Zw}$ yields the electromagnetic field equations (6).

2.2 Gauge symmetry and charge conservation

In the absence of the external source terms ($j_e^\mu = 0$ and $j_g^\mu = 0$), the action for the Lagrangian density $\mathcal{L}_{free}$ given by Eq. (11) is invariant under local gauge transformations defined by

$$A_\mu \to A_\mu - \partial_\mu \chi_A, \quad B_\mu \to B_\mu - \partial_\mu \chi_B, \quad (17)$$

where χ_A and χ_B are arbitrary differentiable scalar functions. This gauge symmetry is also obeyed by all the experimentally observed electromagnetic phenomena, also for the non-vanishing external sources ($j_e^\mu \neq 0$ and $j_g^\mu \neq 0$). [Indeed, in the context of the second quantized theory for the electromagnetism, the gauge symmetry is indeed a necessary consequence of the Lorentz symmetry; see Ref. [5].] It is an easy exercise to check that the invariance of the action for $\mathcal{L}_{Zw}$ under gauge transformations (17) necessarily implies the conservation of the electric and magnetic currents, which is expressed by

$$\partial_\mu j_e^\mu = 0, \quad \partial_\mu j_g^\mu = 0. \quad (18)$$

[Also refer Eq. (5).] That is, the charge conservation laws are the consequence of gauge symmetry of Maxwell theory.

2.3 Duality symmetry

It is easy to verify that the extended Maxwell equations (6) for non-vanishing electric and magnetic currents are covariant under global phase rotations represented by

$$(F^{\mu\nu} + i\tilde{F}^{\mu\nu}) \to e^{i\alpha}(F^{\mu\nu} + i\tilde{F}^{\mu\nu}), \quad (j_e^\mu + i\, j_g^\mu) \to e^{i\alpha}(j_e^\mu + i\, j_g^\mu), \quad (19)$$

where α is a real constant. This illustrates that, in the presence of non-vanishing magnetic current density, there is a duality symmetry between the electricity and magnetism. The reason why we need this seemingly non-physical duality symmetry in computational electromagnetism will become clear later, in the context.

3 Field theoretical method for divergence correction

It is by now clear that the numerical violation of the charge conservation laws is a consequence of the broken gauge symmetry in the discretized spacetime. In this section, we shall derive a Lorentz covariant divergence correction strategy based on field theoretical principles. This section forms the core of this paper.

3.1 Symmetry-based corrections for non-conserving charges

We consider the extended Maxwell equations (6) with duality symmetry, and let $\bar{F}_e^{\mu\nu}$ and $\bar{F}_g^{\mu\nu}$ be the associated discretized solutions. These numerical solutions $\bar{F}_e^{\mu\nu}$ and $\bar{F}_g^{\mu\nu}$ indeed solve Eqs. (6) only *approximately*, due to the presence of various types of numerical errors. If one expresses the electric and magnetic numerical errors respectively with D_e^μ and D_g^μ, the discretized solutions $\bar{F}_e^{\mu\nu}$ and $\bar{F}_g^{\mu\nu}$ actually satisfy the equations given by

$$\bar{\partial}_\mu \bar{F}_e^{\mu\nu} + D_e^\nu[\hat{L}; \bar{F}^{\mu\nu}, x^\mu] = \tilde{j}_e^\nu, \quad \bar{\partial}_\mu \bar{F}_g^{\mu\nu} + D_g^\nu[\hat{L}; \bar{F}^{\mu\nu}, x^\mu] = \tilde{j}_g^\nu, \tag{20}$$

where $\bar{\partial}_\mu$ represents discretized partial derivative, and $\tilde{j}_e^\mu$ and $\tilde{j}_g^\mu$ the desired conserved electric and magnetic current densities. The error terms D_e^μ and D_g^μ, in general, depend on the lattice structure parameterized here by $\hat{L}$, the field strengths $\bar{F}_e^{\mu\nu}$ and $\bar{F}_g^{\mu\nu}$ and the spacetime coordinate x^μ of the relevant grid point.

Due to the presence of these error terms, the numerical electric and magnetic current densities $\bar{j}_e^\mu$ and $\bar{j}_g^\mu$ which can be expressed by

$$\bar{j}_e^\mu \equiv \tilde{j}_e^\mu - D_e^\mu, \quad \bar{j}_g^\mu \equiv \tilde{j}_g^\mu - D_g^\mu, \tag{21}$$

do not form conserved currents

$$\bar{\partial}_\mu \bar{j}_e^\mu = -\bar{\partial}_\mu D_e^\mu \neq 0, \quad \bar{\partial}_\mu \bar{j}_g^\mu = -\bar{\partial}_\mu D_g^\mu \neq 0. \tag{22}$$

We call this violation of the charge conservation laws the *electric* and *magnetic lattice anomalies*, respectively. These anomalies stem from the artificial generation of the external source terms D_e^μ and D_g^μ to the electric and magnetic current densities, due to numerical errors.

The origin of the lattice anomalies can be explained as follows. Recall that the charge conservation laws (18) can be deduced from Maxwell equations (6), using the *commutativity* of differential operators and the *antisymmetricity* of the field strength tensors $F_e^{\mu\nu}$ and $F_g^{\mu\nu}$. From this, we deduce that the observed numerical violations of the charge conservation laws are mainly due to the fact that the discretized differential operators are not strictly commutative

$$[\bar{\partial}_\mu, \bar{\partial}_\nu] \equiv \bar{\partial}_\mu \bar{\partial}_\nu - \bar{\partial}_\nu \bar{\partial}_\mu \neq 0 \tag{23}$$

for $\mu \neq \nu$, while the field strengths $F_e^{\mu\nu}$ and $F_g^{\mu\nu}$ by definition remain strictly antisymmetric. For non-commuting discretized differential operators, equations (22) for the lattice anomalies take the following forms:

$$\bar{\partial}_\mu D_e^\mu = -\frac{1}{2}\left[\bar{\partial}_\mu, \bar{\partial}_\nu\right] \bar{F}_e^{\mu\nu} \neq 0, \quad \bar{\partial}_\mu D_g^\mu = -\frac{1}{2}\left[\bar{\partial}_\mu, \bar{\partial}_\nu\right] \bar{F}_g^{\mu\nu} \neq 0. \tag{24}$$

Apparently, the presence of the non-vanishing collective error terms D_e^μ and D_g^μ break gauge symmetry, which is crucial for the charge conservation laws. The general strategy to cure these anomalies would be to introduce counter terms $\mathcal{L}_{ec}$ and $\mathcal{L}_{gc}$ to the Lagrangian density $\mathcal{L}_{Zw}$ given by Eqs. (10)-(12), such that the counter terms cancel the anomalous contributions from D_e^μ and D_g^μ, in order to restore gauge symmetry. The Lagrangian density which includes these counter terms reads

$$\mathcal{L} = \mathcal{L}_{Zw} + \mathcal{L}_{ec}(A^\mu, D_e^\mu) + \mathcal{L}_{gc}(B^\mu, D_g^\mu). \tag{25}$$

The next logical step is to determine $\mathcal{L}_{ec}$ and $\mathcal{L}_{gc}$ in terms of known field configurations. However, the analytic determination of these counter Lagrangian density terms is a very difficult task. Instead, we shall look for an indirect method for the determination of these counter terms.

Note that the counter terms, upon variation of the corresponding action, should provide the external current densities which cancel the anomalous contributions $-D_e^\mu$ to $\bar{j}_e^\mu$ and $-D_g^\mu$ to $\bar{j}_g^\mu$; refer Eq. (21). In this regard, the counter Lagrangian densities can be given by

$$\mathcal{L}_{ec} = A_\mu D_e^\mu, \quad \mathcal{L}_{gc} = B_\mu D_g^\mu, \tag{26}$$

treating D_e^μ and D_g^μ as if these are independent degrees of freedom. Explicitly, the introduction of these counter terms modifies the numerical non-conserving electric and magnetic current densities $\bar{j}_e^\mu$ and $\bar{j}_g^\mu$ to the conserved ones respectively represented by $\hat{j}_e^\mu$ and $\hat{j}_g^\mu$,

$$\bar{j}_e^\mu \to \hat{j}_e^\mu \equiv \bar{j}_e^\mu + D_e^\mu, \quad \bar{j}_g^\mu \to \hat{j}_g^\mu \equiv \bar{j}_g^\mu + D_g^\mu. \tag{27}$$

Now, the corrected electric and magnetic current densities $\hat{j}_e^\mu$ and $\hat{j}_g^\mu$ are in principle anomaly free

$$\bar{\partial}_\mu \hat{j}_e^\mu = \bar{\partial}_\mu(\bar{j}_e^\mu + D_e^\mu) = 0, \quad \bar{\partial}_\mu \hat{j}_g^\mu = \bar{\partial}_\mu(\bar{j}_g^\mu + D_g^\mu) = 0, \tag{28}$$

which means the restoration of once broken gauge symmetry and thus the charge conservation laws.

Still, D_e^μ and D_g^μ have not been determined yet. The idea for the determination of these counter terms is to solve the conservation equations (28) with respect to D_e^μ and D_g^μ. However, in order to solve Eq. (28) for D_e^μ and D_g^μ, we do not have sufficient information. That is, we have *eight* unknowns (D_e^μ and D_g^μ) for *two* equations. We overcome this problem by making Ansätze for D_e^μ and D_g^μ respectively in terms of only one unknown. In addition, we require that the Ansätze for D_e^μ and D_g^μ be Lorentz covariant four-vectors, in order to keep Lorentz symmetry. The simplest Ansätze which satisfy these two conditions can be given by

$$D_e^\mu(\Phi) = \lambda_e\,\bar{\partial}^\mu \Phi, \quad D_g^\mu(\Psi) = \lambda_g\,\bar{\partial}^\mu \Psi \tag{29}$$

where λ_e and λ_g are real constants and Φ and Ψ real scalar fields. Note that, in the context of variational principle, λ_e and λ_g are Lagrange multipliers; refer Eqs. (26) and (29). We can now determine the scalar fields Φ and Ψ from the conservation equations (28) which now become

$$\bar{\partial}_\mu(\bar{j}_e^\mu + \lambda_e\,\bar{\partial}^\mu \Phi) = 0, \quad \bar{\partial}_\mu(\bar{j}_g^\mu + \lambda_g\,\bar{\partial}^\mu \Psi) = 0. \tag{30}$$

Once Φ and Ψ are computed from Eq. (30) with the knowledge of $\bar{j}_e^\mu$ and $\bar{j}_g^\mu$, one can, in principle, algebraically solve the modified Maxwell equations,

$$\bar{\partial}_\mu \hat{F}_e^{\mu\nu} - \lambda_e \partial^\nu \Phi = \bar{j}_e^\nu, \quad \bar{\partial}_\mu \hat{F}_g^{\mu\nu} - \lambda_g \partial^\nu \Psi = \bar{j}_g^\nu, \tag{31}$$

for the corrected electric and magnetic field strengths $\hat{F}_e^{\mu\nu}$ and $\hat{F}_g^{\mu\nu}$, for each time step. This in principle completes the derivation of field theoretical divergence correction scheme.

The counter terms indeed can be more conveniently implemented by directly solving modified Maxwell equations,

$$\partial_\mu F_e^{\mu\nu} - \lambda_e \partial^\nu \Phi = j_e^\nu, \quad \partial_\mu F_g^{\mu\nu} - \lambda_g \partial^\nu \Psi = j_g^\nu, \tag{32}$$

which is the analytic counterpart to Eq. (31). The direct application of the counter terms to the Maxwell equations reduces considerable amount of computational effort, and such simplification can be justified for reasons stated below.

Recall that the conventional Maxwell equations (1)-(4) are comprised of *eight* equations for *six* unknowns, and the numerical violation of the charge conservation laws is a result of solving only *six* hyperbolic equations (1) and (2), while leaving the two divergence constraints (3) and (4) not directly imposed. On the other hand, note that the modified Maxwell equations are comprised of *eight* equations for *eight* unknowns, thanks to the appearance of additional two scalar fields Φ and Ψ, and the corresponding numerical solver therefore should attain a good control over the two divergence constraints at each time step. Furthermore, the divergence constraints in this case naturally lead to inhomogeneous Klein-Gordon wave equations (30) which serve as modified charge conservation laws. This guarantees the fact that the Lagrange multiplier terms $D_e^\mu = \lambda_e \partial^\mu \Phi$ and $D_g^\mu = \lambda_g \partial^\mu \Psi$ indeed provide the counter terms to the non-vanishing numerical anomalies $\partial_\mu j_e^\mu \neq 0$ and $\partial_\mu j_g^\mu \neq 0$, locally in time. As a consequence, the numerical solutions of Eq. (31) preserve the symmetries of Maxwell theory, and therefore the charge conservation laws.

3.2 A field theoretical interpretation of Φ and Ψ fields

We attend to the fact that Eq. (30) is mathematically the inhomogeneous Klein-Gordon wave equations with the source terms provided by $\partial_\mu j_e^\mu$ and $\partial_\mu j_g^\mu$, which can be obtained by varying the action for the Lagrangian densities given by

$$\mathcal{L}_\Phi = \frac{\lambda_e}{2}\partial_\mu \Phi \partial^\mu \Phi + \Phi \partial_\mu j_e^\mu, \quad \mathcal{L}_\Psi = \frac{\lambda_g}{2}\partial_\mu \Psi \partial^\mu \Psi + \Psi \partial_\mu j_g^\mu. \tag{33}$$

These Lagrangian densities describe the conventional massless scalar field theories, where the scalar fields Φ and Ψ respectively couples to the non-vanishing divergence errors, $\partial_\mu j_e^\mu \neq 0$ and $\partial_\mu j_g^\mu \neq 0$; see Ref. [5]. This fact physically implies that the Φ and Ψ wavelets (generated by the error source terms $\partial_\mu j_e^\mu$ and $\partial_\mu j_g^\mu$) transport the divergence errors away from the computational domain, via the wave equations (30). In this context, in order to properly implement our divergence correction scheme to computational electromagnetism for initial divergence free configurations, it is logical to set the transmissive boundary conditions for Φ and Ψ at the computational boundaries, with the initial conditions given by $\Psi|_{t=0} = \Phi|_{t=0} = 0$.

3.3 Broken Lorentz symmetry and extended GLM scheme

In the derivation of the counter term Ansätze to the electric and magnetic lattice anomalies, we have emphasized on the preservation of the Lorentz symmetry of Maxwell theory. However, the Lorentz symmetry is only an approximate symmetry in discretized space-time, which should be recovered in the continuum limit; refer for instance Ref. [7]. Therefore, one may break (at least slightly) the Lorentz covariance of the counter term Ansätze to electric and magnetic anomalies in the discretized space-time, for further extension of the presented symmetry-based divergence correction scheme.

An intentional breaking of Lorentz symmetry is achieved by taking following Ansätze for the anomaly canceling counter currents

$$D_e^\mu(\Phi) = C_e^{\mu\nu}\partial^\nu \Phi, \quad D_g^\mu(\Psi) = C_g^{\mu\nu}\partial^\mu \Psi, \tag{34}$$

where $C_e^{\mu\nu} = C_{e\,\mu\nu}$ and $C_g^{\mu\nu} = C_{g\,\mu\nu}$ are Lorentz non-covariant real matrices which are to be chosen for the optimization of the numerical performance of the divergence clearance. Upon implementing the modified counter terms (34), the divergence correction scheme discussed so far becomes equivalent to numerically solving the following set of modified Maxwell equations:

$$\partial_\mu F^{\mu\nu} - C_e^{\mu\nu}\partial^\nu \Phi = j_e^\nu, \quad \partial_\mu \tilde{F}^{\mu\nu} - C_g^{\mu\nu}\partial^\nu \Psi = 0. \tag{35}$$

If the matrices $C_e^{\mu\nu}$ and $C_g^{\mu\nu}$ are set to be diagonal such that $C_e^{\mu\nu} = c_e^\mu \delta^{\mu\nu}$ and $C_g^{\mu\nu} = c_g^\mu \delta^{\mu\nu}$, then Eq. (35) becomes equivalent to

$$-\partial_t \mathbf{E} + \nabla \times \mathbf{B} - \mathrm{diag}[c_e^1, c_e^2, c_e^3] \cdot \nabla \Phi = \mathbf{j}_e, \tag{36}$$
$$\partial_t \mathbf{B} + \nabla \times \mathbf{E} + \mathrm{diag}[c_g^1, c_g^2, c_g^3] \cdot \nabla \Psi = 0, \tag{37}$$
$$c_e^0\, \partial_t \Phi + \nabla \cdot \mathbf{E} = \rho_e, \tag{38}$$
$$c_g^0\, \partial_t \Psi + \nabla \cdot \mathbf{B} = 0, \tag{39}$$

in conventional terms, for vanishing external magnetic charge and currents. These modified Maxwell equations (36)-(39) keep the hyperbolicity of genuine Maxwell equations (1)-(4), and reduces to GLM-Maxwell equations of Munz et. al. [1, 2] upon choosing particular matrix elements $c_e^k = c_g^k = 1$ for $k =$ 1, 2, 3.

4 Application to Shallow Water MHD

The symmetry-based divergence cleaning method can also be applied to MHD equations. A toy (but realistic) MHD model upon which our extended GLM concept can be implemented is "shallow water" MHD (SMHD) [8]. The corresponding SMHD equations with the implementation of a proper counter term to the magnetic anomaly are given by

$$\frac{\partial}{\partial t}\begin{bmatrix} h \\ h\mathbf{v} \\ h\mathbf{B} \\ \Psi \end{bmatrix} + \nabla \cdot \begin{bmatrix} h\mathbf{v} \\ h\mathbf{v}\mathbf{v} - h\mathbf{B}\mathbf{B} + \mathsf{I}(gh^2/2) \\ h\mathbf{v}\mathbf{B} - h\mathbf{B}\mathbf{v} + \mathrm{diag}[\cos^2\theta, \sin^2\theta]\,\Psi \\ \chi^2 h\mathbf{B} \end{bmatrix} = 0, \tag{40}$$

where h represents the height of the shallow water system, $\mathbf{v}$ the two dimensional velocity and the $\mathbf{B}$ the two dimensional magnetic field which must satisfy the divergence constraint $\nabla\cdot(h\mathbf{B}) = 0$. The degree of broken Lorentz symmetry has been parameterized by θ via $C_g = \mathrm{diag}[\,\chi^2, \cos^2\theta, \sin^2\theta\,]$. We illustrate the new effect of this broken Lorentz symmetry in Fig. 1 which shows the propagation of the initial divergence error $\nabla \cdot (h\mathbf{B})|_{t=0} = 1/[1 + (10\rho)^2]$ with $\rho \equiv \sqrt{x^2 + y^2}$, for $\chi = 2$ and for various values of θ. The method used is the HLL-based finite volume solver presented in Ref. [9]. The illustrated θ-dependent error distributions at $t = 0.5$ indicates that the direction of divergence error transport can, in principle, be controlled very precisely by choosing a proper value for θ. In some practical applications — for instance, if the divergence perturbation occurs near a corner of the computational domain — such a speed-up of the information transport into a certain direction seems to be very favorable; for further details, see Ref. [9].

References

1. Munz, C.-D., Schneider, R., Sonnendrücker, E., Voss, U. (1999): Maxwells equations when the charge conservations is not satisfied. C. R. Acad. Sci. Paris, p. 328, 431-436.

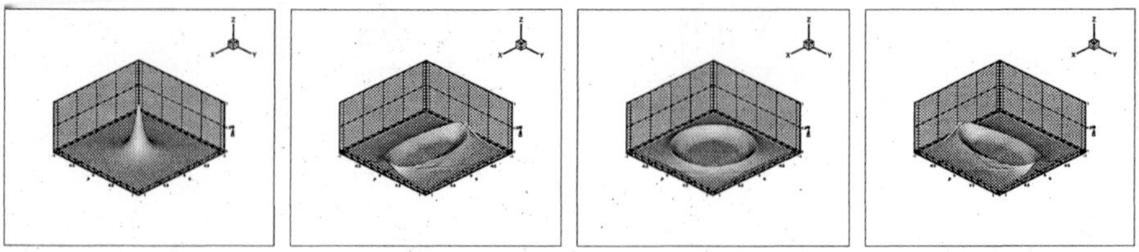

Fig. 1. Different GLM correction results for $\nabla \cdot (\sigma \mathcal{B})$. The leftmost figure is the initial error distribution at $t = 0$. The rest shows the divergence error distributions at $t = 0.5$ for (from left to right) $\theta = \pi/8$, $2\pi/8$ and $3\pi/8$, respectively.

2. Munz, C.-D., Omnes, P., Schneider, R., Sonnendrücker, E., Voss, U. (2000): Divergence correction techniques for Maxwell solvers based on a hyperbolic model. J. Comput. Phys. **161** 484-511
3. Dedner, A., Kemm, F., Kröner, D., Munz, C.-D., Schnitzer, T., Wesenberg, M. (2001): Hyperbolic Divergence Cleaning for the MHD Equations. J. Comput. Phys., 645-673.
4. Weinberg, S. (1971): Gravity and Cosmology. John Wiley & Sons, New York
5. Weinberg, S. (1995): The Quantum Theory of Fields. Cambridge University Press, Cambridge
6. Zwanziger, D. (1971): Local-Lagrangian Quantum Field Theory of Electric and Magnetic Charges. Phys. Rev. D., **3**, 880-891
7. Rothe, H.J. (1992): Lattice Gauge Theories. World Scientific, Singapore
8. Gilman, P.A. (2000): Magnetohydrodynamic "Shallow Water" Equations for the Solar Tacholine. The Astrophys. J., **544**, L79-L82
9. Lee, Y.J., Munz, C.-D., Schneider, R. (2002): Field Theoretical Modeling of Divergence Cleaning for Computational Electromagnetics and Magnetohydrodynamics. In preparation.

Mathematical Modelling of Traffic Flows

Tong Li

Department of Mathematics, University of Iowa, Iowa City, IA 52242,
tli@math.uiowa.edu

Summary. We are concerned with mathematical modeling of of traffic flow. In particular, well-posedness theory of various nonequilibrium continuum models of traffic flow are established.

One such a model is an inhomogeneous traffic flow model which allows road situation changes. The road situation changes arise naturally from obstacles, curvature, number of lanes, speed limit and others. Our proposed inhomogeneous model is a system of nonlinear hyperbolic equations in which the flux and the source terms depend on the space variable. We show the global existence of solutions and their convergence to a solution of the equilibrium equation as the relaxation time goes to zero.

Furthermore, a traffic flow model with a nonconcave fundamental diagram is studied. The model is a system of nonconcave hyperbolic conservation laws with relaxation. A nonconcave fundamental diagram is observed in real traffic flow.

Other more realistic traffic flow models are discussed.

1 Introduction

Traffic flow exhibits extremely complex behavior. Experimental observations of real traffic have revealed physical phenomena such as shock waves, stop-and-go waves, hysteresis, phase transitions and so on, see [5], [6], [8], [21], [22]. There are various important approaches toward the modeling of traffic phenomena: microscopic models which explain traffic phenomena on the basis of the behavior of single vehicles [2], macroscopic models which describe traffic phenomena through parameters which characterize the collective traffic properties [1] [15] [17] [19] [22] [23] [24] and the Boltzmann-like models [5] [7] [18].

The first continuum theory of traffic flow is the LWR theory developed independently by Lighthill and Whitham [15] and Richards [19]. The LWR theory assumes that there exists an *equilibrium* speed-concentration relationship $v = v_e(\rho)$. The LWR model is a scalar nonlinear conservation law

$$\rho_t + (\rho v_e(\rho))_x = 0. \tag{1}$$

It is assumed that $v_e'(\rho) < 0$. Let $v_e(0) = v_f$ and $v_e(\rho_j) = 0$, then v_f is the free flow speed and ρ_j is the jam concentration.

A lot of experimental data have been collected and plotted in the (ρ, q) plane, where

$$q(\rho) = \rho v_e(\rho) \tag{2}$$

is called a fundamental diagram in traffic flow, see Fig.1 and Fig.2.

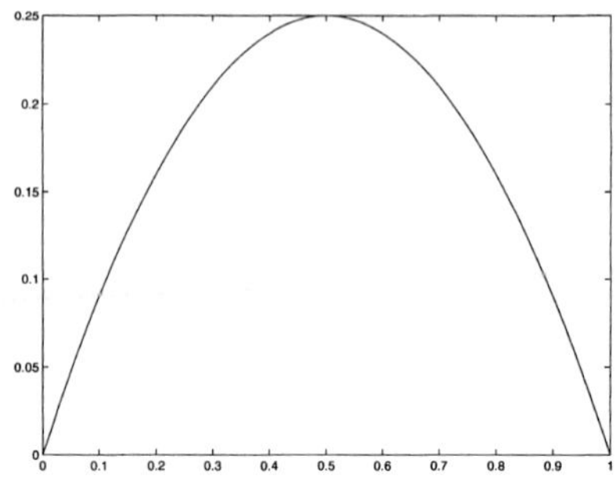

Fig. 1. A Fundamental Diagram (Concave)

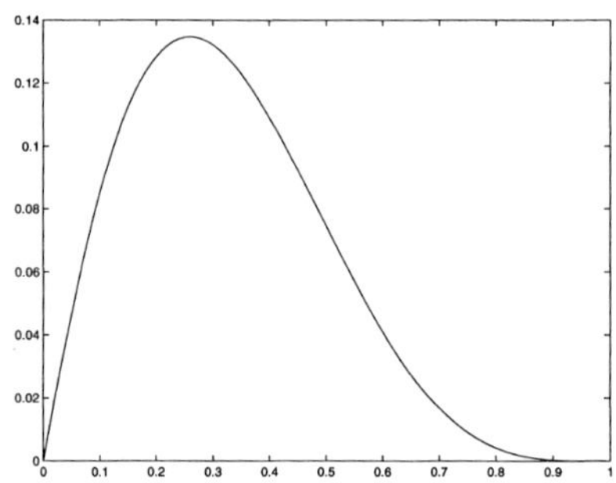

Fig. 2. A Fundamental Diagram (Nonconcave)

The initial data is

$$\rho(x, 0) = \rho_0(x). \tag{3}$$

The LWR model can explain the formation of shock waves which corresponds to congestion formation in traffic flow but fails in describing more complicated traffic flow patterns.

Higher order models, or nonequilibrium models, consist of the continuity equation and an equation to describe acceleration behavior, were developed in the literature [17] [22] [23] [24]. The acceleration consists of a relaxation to the static equilibrium speed-concentration relation and an anticipation which expresses the effect of drivers reacting to conditions downstream. The relaxation time summarizes the characteristic times for acceleration, brake and reacting.

Zhang [23] derived such a higher order continuum model of traffic flow. The model is a system of hyperbolic equations with relaxation which is marginally stable and is non-diffusive. Using a generalized Glimm scheme, we established [9] the global existence of BV solution and the zero relaxation limit. A comparison between our results and data measured on I-880 freeway in California, see Skabardonis etc. (1996), [20], showed qualitative agreement between the analytic results and the data.

Since L^1 topology is natural from point view of the conservation laws, we established [10] the L^1 well-posedness theory. In particular, we obtained the continuous dependence of the solution on its initial data in L^1 topology. We also showed that the equilibrium shock waves are nonlinearly stable in L^1 norm.

The previous nonequilibrium models were derived assuming homogeneous road conditions. In reality, the roadway is rarely homogeneous. The inhomogeneities of roadway include obstacles, curvature, number of lanes, speed limit and so on. These motivate our study [11] of inhomogeneous models of traffic flows which take road situation changes into account. Our proposed model is a system of conservation laws with relaxation and sources, the flux and the source terms depend on the space variable. The L^1 well-posedness theory for the model was established in [11]. Moreover, the L^1-convergence to the unique zero relaxation limit was proved.

The nonconcave fundamental diagram is suggested from the experiment data, see [5] [6] and Fig. 2. It is interesting to note that in some traffic flow models a nonconcave fundamental diagram is used in obtaining complicated traffic flow patterns including clusters [6]. We established [12] global solutions of a system of nonconcave hyperbolic conservation laws with relaxation arising from traffic flow.

The plan of the paper is the following. Section 2 contains derivation of a nonequilibrium model and results on the BV solutions. In Section 3, we present the L^1 well-posedness theory. Section 4 is devoted to the inhomogeneous model. We study a traffic flow model with a nonconcave fundamental diagram in Section 5. Section 6 is a discussion.

2 Global BV Solutions

In this section, we study [9] the nonequilibrium model of Zhang [23]

$$\rho_t + (\rho v)_x = 0 \tag{4}$$

$$v_t + (\frac{1}{2}v^2 + g(\rho))_x = \frac{v_e(\rho) - v}{\tau} \tag{5}$$

with initial data

$$(\rho(x,0), v(x,0)) = (\rho_0(x), v_0(x)). \tag{6}$$

The model was derived based on the assumption that the time needed for a *following* vehicle to assume a certain speed is determined by *leading* vehicles. The dynamics of the velocity is due to anticipation g and relaxation $\frac{v_e(\rho)-v}{\tau}$. $g'(\rho) = \rho(v_e'(\rho))^2$.

Derivation of the model For $\tau > 0$ and $\Delta x > 0$,

$$\frac{dx}{dt}(t+\tau) = v_e(\rho(x+\Delta x, t)).$$

To leading order

$$v + \tau\frac{dv}{dt} = v_e(\rho(x,t)) + \Delta x \rho_x v_e'(\rho(x,t)).$$

That is

$$\frac{dv}{dt} = \frac{v_e(\rho(x,t)) - v}{\tau} + \frac{\Delta x}{\tau}\rho_x v_e'(\rho(x,t)). \tag{7}$$

Letting

$$\frac{\Delta x}{\tau} = -(\lambda_*(\rho) - v_e(\rho)) = -\rho v_e'(\rho),$$

the relative wave propagating speed to the car speed at the equilibrium, we arrive at (5).

The system (4) (5) is strictly hyperbolic provided $\rho > 0$. The characteristic fields genuinely nonlinear provided that the fundamental diagram is concave. A marginal stability condition

$$\lambda_1 = \lambda_* < \lambda_2 \tag{8}$$

is satisfied, where λ_* is the equilibrium characteristic speed

$$\lambda_*(\rho) = q'(\rho) = v_e(\rho) + \rho v_e'(\rho). \tag{9}$$

There is no diffusion correction in the higher order Chapman-Enskog expansion.

Using Glimm scheme construction, we obtained the following results in [9].

Theorem 1. *Let the initial data be of bounded total variation and of small distance to the equilibrium curve. Under the hypotheses that $v_e(\rho)$ is linear, the initial value problem (4) (5) (6) has a solution $u(x,t) = (\rho, v)(x,t)$ for $t \geq 0$ satisfying*

$$TVu(\cdot, t) = O(1)[TVu(\cdot, 0)] \tag{10}$$

for some $O(1)$ depending only on the system (4) (5), $TVu(\cdot, 0)$ and the initial distance to the equilibrium curve.

Moreover, there is a sequence of solutions obtained for the relaxed system converge to a solution of the equilibrium equation (1) with data (3) as the relaxation parameter τ goes to zero.

The assumption that the equilibrium velocity is linear comes from an actual measurement [4]. Under this assumption, the fundamental diagram $q = q(\rho)$ defined in (2) is a quadratic function. This is a typical example of a concave fundamental diagram, see Fig. 1. In this case, the characteristic fields of the nonlinear conservation laws (4) (5) are genuinely nonlinear. The shock wave curve coincides with the rarefaction wave curve in each characteristic field. It was shown that the distance to the equilibrium curve decays exponentially with increasing t. Therefore the accumulative effects of the source term is under control. We proved the boundedness of the total variation of solutions using Glimm scheme construction.

Now we show that the solutions obtained for the relaxed system converge to a solution of the equilibrium equation (1) as the relaxation parameter τ goes to zero.

Let us denote the solutions to (4) (5) (6) by (ρ^τ, v^τ). ¿From the Glimm scheme construction, we obtain that the total variations of (ρ^τ, v^τ) are independent of the relaxation parameter τ. Hence there is a subsequence of (ρ^τ, v^τ), still denoted as (ρ^τ, v^τ), that converges in $L^1_{loc}(R^1 \times (0, +\infty))$ as $\tau \to 0$. Let the limit be (ρ, v).

$$\int_0^{+\infty} \int_{R^1} (\rho^\tau \phi_t + \rho^\tau v^\tau \phi_x) \, dxdt = 0, \tag{11}$$

$$\int_0^{+\infty} \int_{R^1} \left(v^\tau \phi_t + \left(\frac{1}{2}(v^\tau)^2 + g(\rho^\tau) \right) \phi_x + \frac{v_e(\rho^\tau) - v^\tau}{\tau} \phi \right) dxdt = 0 \tag{12}$$

for any $\phi \in C_0^1(R^1 \times (0, +\infty))$.

Multiplying (12) by τ and letting $\tau \to 0$ in $\tau \cdot (12)$, we arrive at that

$$\int_0^{+\infty} \int_{R^1} (v_e(\rho) - v) \phi dxdt = 0$$

for any $\phi \in C_0^1(R^1 \times (0, +\infty))$.

Hence

$$v = v_e(\rho), \quad a.e..$$

Letting $\tau \to 0$ in (11) and using the uniform boundedness of (ρ^τ, v^τ) again, we conclude that the limit ρ satisfies the equilibrium equation (1) with initial data $\rho_0(x)$.

3 L^1 Well-posedness

We establish the L^1 well-posedness theory for (4) (5) in this section.

The continuous dependence of the solution on its initial data in L^1 topology was obtained in [10]. The uniqueness of solutions is a corollary of the continuous dependence of the solution on its initial data. We constructed a functional for two solutions such that it is equivalent to the L^1 distance between the two solutions and it is time-decreasing. The construction makes use of the L^1 contraction semigroup property for the scalar conservation laws and the exponential decay property of the source term. We showed an L^1-contractive property of the entropy solution operator in the Riemann invariant coordinate.

Theorem 2. *If $(\rho_1(x,t), v_1(x,t))$ and $(\rho_2(x,t), v_2(x,t))$ are entropic solutions of (4) (5) for all x and $t > 0$, with initial data $(\rho_{10}(x), v_{10}(x))$, $(\rho_{20}(x), v_{20}(x))$ such that $\rho_{10} - \rho_{20}, v_{10} - v_{20} \in L^1$, then*

$$\begin{aligned} &||\rho_1(\cdot,t) - \rho_2(\cdot,t)||_{L^1} + ||v_1(\cdot,t) - v_2(\cdot,t)||_{L^1} \\ &\le C(||\rho_{10}(\cdot) - \rho_{20}(\cdot)||_{L^1} + ||v_{10}(\cdot) - v_{20}(\cdot)||_{L^1}) \end{aligned} \tag{13}$$

where C is a constant independent of t and the initial data.

We then studied the large-time behavior of the entropy solutions. The equilibrium shock waves (ρ_{sh}, v_{sh}) were shown to be nonlinearly stable in L^1 norm in [10]. The main tools used in the proof are the L^1 contractivity, a comparison principle and the exponentially decay property of the source term.

Theorem 3. *Let the initial data (6) be certain bounded perturbation of an equilibrium shock wave satisfying*

$$(\rho_0 - \rho_{sh}, v_0 - v_{sh}) \in (L^\infty)^2 \cap (L^1)^2$$

and let the initial distance to the equilibrium curve be small. Then the global bounded entropy solution of (4) (5) (6) exists and tends to a shifted equilibrium shock wave in L^1 norm as $t \to \infty$,

$$\lim_{t\to+\infty} (||\rho(\cdot,t) - \rho_{sh}(\cdot + k)||_{L^1} + ||v(\cdot,t) - v_{sh}(\cdot + k)||_{L^1}) = 0 \tag{14}$$

where the shift k is given by

$$k = \frac{1}{\rho_{sh}(+\infty) - \rho_{sh}(-\infty)} \int_R (\rho(x,0) - \rho_{sh}(x))dx. \tag{15}$$

4 An Inhomogeneous Model

Now we model [12] traffic flows with road situation changes such as obstacles. Waves for such a system propagate in a more complicated way than those do for models with homogeneous road conditions.

We propose the following inhomogeneous model of traffic flows

$$\rho_t + (\rho v)_x = 0 \tag{16}$$

$$v_t + vv_x + \rho v_{*\rho}^2 \rho_x = \frac{v_*(\rho, x) - v}{\tau} + \rho v_{*x} v_{*\rho} - v_* v_{*x} \tag{17}$$

where $v = v_*(\rho, x)$ is the equilibrium velocity which depends on both density ρ and location x. Assumed that $v_*(\rho, x) = -a\rho + b(x)$ where $a > 0$ and $b(x)$ accounts for velocity changes due to road situation changes.

When the state is in equilibrium, i.e., when $v = v_*(\rho, x)$, our model (16) (17) is reduced to the inhomogeneous equilibrium LWR model

$$\partial_t \rho + \partial_x(\rho v_*(\rho, x)) = 0. \tag{18}$$

The fundamental diagram $q(\rho, x) = \rho v_*(\rho, x)$ is a function of both ρ and x.

The inhomogeneous model (16) (17) is derived based on the assumption that the time needed for a *following* vehicle to assume a certain speed is determined by *leading* vehicles while taking the road changes into account. For $\tau > 0$ and $\Delta x > 0$,

$$\frac{dx}{dt}(t + \tau) = v_*(\rho(x + \Delta x, t), x + \Delta x + \tau v_*)$$

where τv_* accounts for the road changes on interval $(x, x + \Delta x)$.

To leading order

$$v + \tau \frac{dv}{dt} = v_*(\rho, x) + \Delta x \rho_x v_{*\rho}(\rho, x) + \tau v_* v_{*x}.$$

That is

$$\frac{dv}{dt} = \frac{v_*(\rho, x) - v}{\tau} + \frac{\Delta x}{\tau} \rho_x v_{*\rho}(\rho, x) + v_* v_{*x}. \tag{19}$$

Letting

$$\frac{\Delta x}{\tau} = -(\lambda_*(\rho, x) - v_*(\rho, x)) = -\rho v_{*\rho}(\rho, x),$$

the relative wave propagating speed to the car speed at the equilibrium, we obtain (17).

For the Cauchy problem of (16) (17), we established its global existence and the unique zero relaxation limit of solutions.

Theorem 4. *Let $(\rho_0, v_0) \in (L^\infty)^2$, $v_0 - v_*(\rho_0, x) \in L^1$ and the initial data be of small distance to the equilibrium curve. The unique entropy solution (ρ^τ, v^τ) of the Cauchy problem (16) (17) (6) exists globally. Moreover, (ρ^τ, v^τ) converges in $(L^1)^2$ to $(\rho, v_*(\rho, x))$ as $\tau \to 0$ for any $t > 0$. ρ is the unique entropy solution of the equilibrium equation (18) with initial data ρ_0.*

We carried out [11] the asymptotic states of a general solution whose initial data tend to constant states as $|x| \to +\infty$. A time-asymptotic state is a combination of nonlinear waves that tends to exact solution of (16) (17) and becomes noninteractive as $t \to +\infty$. Such wave patterns represent the large-time states for general solutions.

5 A Nonconcave Model

Nonconcave fundamental diagrams arise from the traffic data fitting, see [5] [6] and see Fig. 2.

The traffic flow model we studied [12] is a system of nonconcave hyperbolic conservation laws with relaxation.

$$\rho_t + (\rho v)_x = 0 \tag{20}$$

$$v_t + v v_x + \rho v_*'(\rho) v_x = \frac{v_*(\rho) - v}{\tau}. \tag{21}$$

$v = v_*(\rho)$ is the equilibrium velocity.

In addition being marginally stable and non-diffusive, the model (20) (21) is anisotropic, i.e., characteristics travel no faster than traffic. The fundamental diagram $q(\rho) = \rho v_*(\rho)$ is allowed to be nonconcave, see Fig. 2. Thus one of the characteristic fields of the system is neither linearly degenerate nor genuinely nonlinear.

Zhang [24], Aw and Rascle [1], Klar and Wegener [7], J. Greenberg [3] studied the concave case of the model.

The equation that describes speed dynamics is derived from a car-following model [24]

$$\tau x_n''(t) = x_{n-1}'(t) - x_n'(t)$$

where $\tau > 0$ is the driver response time and $\Delta x = x_{n-1}(t) - x_n(t) > 0$.

$$\tau \frac{dv}{dt}(x, t) = v(x + \Delta x, t) - v(x, t).$$

To leading order, we have

$$\tau(v_t + v v_x) = v_x(x, t) \Delta x.$$

To obtain closed equations for ρ and v, we have to specify the coefficient $\frac{\Delta x}{\tau}$. To find the closure relation for the balance law, we let the disturbance propagation speed

$$\frac{\Delta x}{\tau} = -(\lambda_*(\rho) - v_*(\rho)) = -\rho v_*'(\rho) > 0,$$

be the relative wave propagating speed to the car speed at the equilibrium. We obtain (21) by adding the relaxation term.

We established the global existence of the solutions to (20) (21) (6) in [12].

Theorem 5. *Under appropriate conditions including* $\rho_0(x) \geq \epsilon > 0$, ρ_0, v_0 *have bounded total variations, there exists a subsequence of the finite difference approximation solutions that converges in* $(L^1_{loc})^2$ *to an entropy solution of the Cauchy problem (20) (21) (6) with bounded total variations, as the mesh sizes* Δt, $\Delta M \to 0$.

6 Discussion

One of the most important characters of traffic flow is described by the fundamental diagram (2). The introduction of the desired flow-concentration relationship is mathematical idealization of the goals that the driver population strives to realize. This idealization may be perturbed as a consequence of various inhomogeneities: inhomogeneous roadways conditions, weather conditions, interactions with other drivers, traffic congestion managements and so on. The inhomogeneities may depend on space, time or on both. We studied a traffic flow model with a space-dependent fundamental diagram in Section 4. [13] is devoted to modeling the traffic flow with a time-dependent fundamental diagram.

The multiple-branched speed-concentration relation is to capture the phase transitions depicted by the observed speed-concentration plots. We plan to model more traffic phenomena such as hysteresis and phase transition in the future.

Acknowledgments. This paper was finished while the author was visiting the Institute for Advanced Study and was supported by the Career Development Awards of University of Iowa.

References

1. Aw, A. and Rascle, M. (2000), Resurrection of "second order" models of traffic flow, *SIAM J. Appl. Math.* **60**, 916–938.
2. Gazis, D. C., Herman, R., and Rothery, R. W. (1961), Nonlinear Follow-the-leader Models of Traffic Flow. *Operations Research* **9**, 545–567.
3. Greenberg, J. M. (2001), Extension and amplifications of a traffic model of Aw and Rascle, *SIAM J. Appl. Math.* **62**, 729–745.

4. Greenshields, B. (1933), A study of traffic capacity, *Highway Research Board, Proceedings,* **14**, 468–477.
5. Helbing, D., Hennecke, A., Shvetsov, V. and Treiber, M. (2001), MASTER: macroscopic traffic simulation based on a gas-kinetic, non-local traffic model, *Transportation Research Part B: Methodological*, **35**, 183–211.
6. Kerner, B. S. and Konhaüser, P. (1994), Structure and parameters of clusters in traffic flow. *Physical Review B*, Vol. **50**, 54–83.
7. Klar, A. and Wegener, R. (2000), Kinetic derivation of macroscopic anticipation models for vehicular traffic, *SIAM J. Appl. Math.* **60**, 1749–1766.
8. R. D. Kühne, Macroscopic freeway model for dense traffic-stop-start waves and incident detection, *Ninth International Symposium on Transportation and Traffic Theory*, VNU Science Press, 1984, pp. 21-42.
9. Li, T. (2000), Global Solutions And Zero Relaxation Limit For A Traffic Flow Model, *SIAM J. Appl. Math.* **61**, 1042–1061.
10. Li, T. (2001), L^1 stability of conservation laws for a traffic flow model, *Electron. J. Diff. Eqns.* **2001**, 1–18.
11. Li, T. (2002), Well-posedness Theory of An Inhomogeneous Traffic Flow Model, *Discrete and Continuous Dynamical Systems*, Series B, **2**, 401–414.
12. Li, T. (2001), Global Solutions of Nonconcave Hyperbolic Conservation Laws with Relaxation Arising from Traffic Flow, submitted.
13. Li, T. (2002), Modeling Traffic Flow with a Time-Dependent Fundamental Diagram, submitted.
14. Li, T. and Zhang, H. M. (2001), The Mathematical Theory of an Enhanced Nonequilibrium Traffic Flow Model, *Network and Spatial Economics, A Journal of Infrastructure Modeling and Computation*, Special Double Issue on Traffic Flow Theory, **1**, 167–179.
15. Lighthill, M. J. and Whitham, G. B. (1955), On kinematic waves: II. A theory of traffic flow on long crowded roads, *Proc. Roy. Soc., London*, Ser. **A229**, 317–345.
16. Oleinik, O. A. (1963), Discontinuous solutions of non-linear differential equations, *Uspehi. Mat. Nauk.* **12**(1957), 3-73. English transl., *Amer. Math. Soc. Transl.*, **26**, 95–172.
17. Payne, H. J. (1971), Models of freeway traffic and control, *Simulation Councils Proc. Ser. : Mathematical Models of Public Systems*, Editor G.A. Bekey, **1**, 51–60.
18. Prigogine, I. and Herman, R. (1971), *Kinetic Theory of Vehicular Traffic*, American Elsevier Publishing Company Inc., New York.
19. Richards, P. I. (1956), Shock waves on highway, *Operations Research*, **4**, 42–51.
20. Skabardonis, A., Petty, K., Noeimi, H. Rydzewski, D. and Varaiya, P. (1996). I-880 field experiment: data-base development and incident delay estimation procedures. Transportation Research Record 1554, 204–212.
21. Treiterer, J. and Myers, J. A. (1974), The hysteresis phenomenon in traffic flow, in *Transportation and Traffic Theory*, Proceedings of the Sixth International Symposium on Transportation and Traffic Theory, Edited by D. J. Buckley.
22. Whitham, G. B. (1974), *Linear and Nonlinear Waves*, Wiley, New York.
23. Zhang, H. M. (1998), A theory of nonequilibrium traffic flow, *Transportation Research, B.*, **32(7)**, 485–498.
24. Zhang, H. M. (2002), A non-equilibrium traffic model devoid of gas-like behavior, *Transportation Research, B.*, **36**, 275–290.

New High-Resolution Central-Upwind Schemes for Nonlinear Hyperbolic Conservation Laws

Chi-Tien Lin*

Department of Applied Mathematics, Providence University, Shalu 43301, Taiwan. *ctlin@pu.edu.tw*

Summary. The main feature of central schemes is that they are Riemann-solver free. The original high-order central scheme by Nessyahu and Tadmor [NeTa] assumes a global speed of wave propagation which introduces a large amount of numerical viscosity and lowers the resolution of discontinuities. Later, Kurganov and Tadmor [KurTa] proposed a new version of central schemes. In their construction, the whole space is divided into smooth and singular rectangular controlled volumes according to local speed of wave propagation. The updated values at each gridpoints are obtained from more precise computation on each region resulting in sharper resolution and less numerical viscosity. In this paper, we shall construct new fully-discrete central-upwind scheme by replacing the rectangular singular regions with smaller triangular regions. Numerical simulation will be present to show lower numerical viscosity and hence sharper results, as expected.

1 Introduction

Many of the modern high-resolution approximations for nonlinear conservation laws

$$\begin{cases} u_t + f(u)_x = 0 \\ u(x,t) = u_0(x) \end{cases} \tag{1}$$

employ the Godunov approach (also called the finite volume method) which consists of three consecutive steps: *reconstruction*, *evolution* and *projection*. To this end, we assume a uniform grid, $\Delta x = x_{j+1} - x_j$ for simplicity and utilize the sliding average of the solution $u(x,t)$,

$$\bar{u}(x,t) \equiv \frac{1}{|I_x|}\int_{I_x} u(s,t)ds, \qquad I_x \equiv \{s : |s-x| \leq \frac{\Delta x}{2}\}$$

so that the integration of the conservation laws (1) over the rectangle $I_x \times [t, t+\Delta t]$ yields an equivalent reformulation

* This research was supported by NSC grants 88-2119-M-126-001 and 89-2115-M-126-007. Part of the work is done while the author visited the National Center of Theoretical Sciences (CTS), 18 June - 14 Sep, 2001

$$\bar{u}(x,t+\Delta t) = \bar{u}(x,t) - \frac{1}{\Delta x}\left[\int_t^{t+\Delta t} f(u(x+\frac{\Delta x}{2},\tau))d\tau - \int_t^{t+\Delta t} f(u(x-\frac{\Delta x}{2},\tau))d\tau\right] \quad (2)$$

Assume that at time t^n, we have computed the cell average over $I_j \equiv \{x : x_{j-\frac{1}{2}} \le x \le x_{j+\frac{1}{2}}\}$, $\bar{w}_j^n$, where $x_{j+\frac{1}{2}} = (j+\frac{1}{2})\Delta x$. To compute the cell average over $I_{j+\frac{1}{2}} \equiv \{x : x_j \le x \le x_{j+1}\}$ at the next time step t^{n+1}, we first reconstruct an piecewise polynomial interpolant based on the known cell-average $\bar{w}_j^n$ at time t^n, which may result discontinuities at the points $\{x_{j+\frac{1}{2}}\}$, center of $I_{j+\frac{1}{2}}$. This interpolant is then evolved exactly in time and projected onto the cell $I_{j+\frac{1}{2}}$ at the next time level t^{n+1} according to (2), resulting with

$$\begin{aligned}\bar{w}_{j+\frac{1}{2}}^{n+1} &\equiv \bar{w}(x_{j+\frac{1}{2}},t^{n+1}) \\ &= \frac{1}{\Delta x}[\int_{x_j}^{x_{j+1}} w(x,t^n)dx - \int_{t^n}^{t^{n+1}} f(w(x_{j+1},\tau)d\tau - \int_{t^n}^{t^{n+1}} f(w(x_j,\tau)d\tau]\end{aligned} \quad (3)$$

Here $\bar{w}(x_{j+\frac{1}{2}},t^n)$ is the average of the interpolant $w(x,t^n)$ over the cell $I_{j+\frac{1}{2}}$. In this context, we may distinguish between the two main classes of Godunov methods, according the location of possible discontinuities: upwind and central schemes.

Godunov's original scheme is the forerunner of all upwind schemes. Its higher-order and multidimensional generalizations were constructed, analyzed, and implemented with great success during 1980s and 1990s; consult [LeVe] and the reference therein. Upwind schemes evaluate their cell averages over the same spatial cells at all time steps. This in turn requires characteristic information along the discontinuous interfaces of these spatial cells. It is the need to trace the characteristic fans– using approximate Riemann solvers, field decomposition, etc –that greatly complicates the upwind algorithms.

The Lax-Friedrichs (LxF) scheme is the other canonical first-order scheme, which is the forerunner of all central schemes. Like the Godunov scheme, it is based on piecewise constant approximate solution; its Riemann-solver-free recipe, however, is considerably simpler. Unfortunately, the excessive numerical viscosity in the LxF scheme yields a relatively poor resolution. A second-order sequel to the LxF scheme was introduced by Nessyahu and Tadmor, the Nessyahu-Tadmor (NT) scheme [NeTa] by replacing the piecewise constant approximation with van Leer's MUSCL-type piecewise linear interpolation. This is then followed by an exact evolution–LxF solver– which avoids using time-consuming approximate Riemann solver. Thus, the NT scheme retains the advantage of a simple, Riemann-solver-free recipe, and at the same time it enjoys high resolution comparable to the upwind results.

Although numerical viscosity of the second-order NT scheme is considerably lower than in the first-order LxF scheme, unfortunately, this does not circumvent the difficulties with small time steps which arise, e.g., with convection-diffusion equations. To overcome this difficulty, a new class of central schemes, *central-upwind* (KT for short) schemes, was first introduced by Kurganov and Tadmor [KurTa]. The main idea in their construction is to use more precise information about the local speed of wave propagation, in order to average the nonsmooth parts of the computed solution over a smaller controlled rectangle of variable size of order $O(\Delta t)$. See §2.3 or [KurTa] for details. In this paper, we propose another construction. That is, to average the nonsmooth parts of the computed solution over a smaller triangular controlled volume, resulting a high resolution method with lower numerical viscosity. (See §3.)

2 LxF, NT, and KT schemes

In this section, we review briefly on the framework of central schemes for hyperbolic conservation laws (1), we shall address on the construction of the family of LxF, NT and KT schemes.

2.1 LxF schemes

The LxF approximation to (1), can be derived by constructing a piecewise-constant interpolant through the given cell-averages,

$$w(x,t^n) = \sum_j \bar{w}_j^n \chi_j(x), \tag{4}$$

whose possible discontinuities are secured at the center of the cell $\{x_j \leq \xi < x_{j+1}\}$. Here, $\chi_j(x)$ is the characteristic function over $I_j = \{x_{j-1/2} \leq \xi < x_{j+1/2}\}$. For the CFL restriction, $\Delta t \max_u |f'(u)| \leq \frac{\Delta x}{2}$, the exact solution of (1) remains constant at the integer grid points, x_{j-1}, x_j. Consequently, the required integrals of the fluxes in (3) are evaluated in terms of their constant initial data, without using any (approximate) Riemann solvers. The *staggered* LxF scheme then amounts to

$$\bar{w}_{j+\frac{1}{2}}^{n+1} = \frac{1}{2}(\bar{w}_j^n + \bar{w}_{j+1}^n) - \lambda\left[f(\bar{w}_{j+1}^n) - f(\bar{w}_j^n)\right]. \tag{5}$$

That is, we have computed an approximate solution at the next time level $t = t^{n+1}$, a solution which is realized by its cell average over $I_{j+\frac{1}{2}}$ rather than the original cell I_j. Staggered schemes may increase the difficulties in the treatment of numerical boundary conditions. However, this staggered scheme can be changed into non-staggered one by re-averaging over a piecewise constant reconstruction of the underlying staggered values $\bar{w}^{n+1}_{j+\frac{1}{2}}$ at the t^{n+1} time level, which leads to the modified LxF scheme [JLLOT],

$$\bar{w}^{n+1}_j = \frac{1}{4}(\bar{w}^n_{j-1} + 2\bar{w}^n_j + \bar{w}^n_{j+1}) - \frac{\lambda}{2}\left[f(\bar{w}^n_{j+1}) - f(\bar{w}^n_{j-1})\right] \tag{6}$$

2.2 NT schemes

To construct the second-order NT schemes, we first reconstruct a piecewise-linear interpolant from the known cell-average $\bar{w}^n_j$ at time t^n,

$$w(x, t^n) = \sum_j [\bar{w}^n_j + (w_x)_j(\frac{x - x_j}{\Delta x})]\chi_j(x). \tag{7}$$

Here $(w_x)_j$ denotes the discrete slope. A possible computation of these slopes, which results in an overall non-oscillatory scheme, is given by the minmod limiter. (Consult [JLLOT] for others.) Note that possible discontinuities are secured at the center of the cell $I_{j+\frac{1}{2}} \equiv \{x_j < x < x_{j+1}\}$, i.e., $x_{j+\frac{1}{2}}$.

Like the construction of LxF schemes, this interpolant is then evolved in time. Under CFL restriction, the exact solution of (1) remains smooth at the integer grid points, x_{j-1}, x_j and the required integrals of the fluxes in (3) can be approximated by the midpoint rule, resulting with the staggered NT scheme,

$$\bar{w}^{n+1}_{j+\frac{1}{2}} = \frac{1}{2}(\bar{w}^n_j + \bar{w}^n_{j+1}) + \frac{1}{8}((w_x)_j - (w_x)_{j+1}) - \lambda[f(w^{n+\frac{1}{2}}_{j+1}) - f(w^{n+\frac{1}{2}}_j)], \tag{8}$$

where the midpoint values, $w^{n+\frac{1}{2}}_j$, are predicted by Taylor expansion,

$$w^{n+\frac{1}{2}}_j = \bar{w}^n_j - \frac{\lambda}{2}(f_x)_j. \tag{9}$$

The discrete derivatives of the flux, $(f_x)_j$, can be computed, by the minmod limiter to each of the components of f. This component-wise approach is one of the main advantage offered by the central NT schemes over the corresponding characteristic decompositions required by upwind schemes - consult the discussion in [NeTa].

To obtain a non-staggered NT scheme, we can re-average over a piecewise linear reconstruction of the underlying staggered values, $\bar{w}^{n+1}_{j+\frac{1}{2}}$, which leads to, [JLLOT],

$$\begin{aligned} \bar{w}^{n+1}_j = &\frac{1}{4}(\bar{w}^n_{j-1} + 2\bar{w}^n_j + \bar{w}^n_{j+1}) - \frac{1}{16}((w_x)_{j+1} - (w_x)_{j-1}) \\ &-\frac{\lambda}{2}[f(w^{n+\frac{1}{2}}_{j+1}) - f(w^{n+\frac{1}{2}}_{j-1})] - \frac{1}{8}((w_x)_{j+\frac{1}{2}} - (w_x)_{j-\frac{1}{2}}). \end{aligned} \tag{10}$$

Here, $(w_x)_j$ and $(w_x)_{j+\frac{1}{2}}$ are, respectively, the discrete derivatives at time level t^n and t^{n+1}, and the midpoint value $w_j^{n+\frac{1}{2}}$ is predict at the time level $t^{n+\frac{1}{2}}$ according to (9).

2.3 KT schemes

The second-order NT scheme and its extensions owe their superior resolution to the lower amount of numerical dissipation, considerably lower than in the first-order LxF scheme. For problems with small time steps which arise , e.g., with convection-diffusion equations, the influence of the numerical dissipation accumulated over the many steps of the NT scheme which smears solutions. To circumvent the difficulties, a new class of central scheme was introduced by Kurganov and Tadmor, KT for short [KurTa]. The main idea in their construction is *localization.* Namely, use more precise information about local speed of wave propagation, in order to average the nonsmooth parts of the computed solution over smaller cells of variable size of order $O(\Delta t)$.

To construct the second-order KT schemes, we again reconstruct the piecewise-linear interpolant (7), $w(x, t^n)$, at time t^n based on the computed cell average $\bar{w}_j^n$. Before this interpolant is evolved in time, we estimate for the local speed of propagation at the center $x_{j+\frac{1}{2}}$ and denote the upper bound by $a_{j+\frac{1}{2}}^n$, i.e., the maximum speed of propagation. Moreover, denote

$$x_{j+\frac{1}{2},l}^n = x_{j+\frac{1}{2}} - \phi_{j+\frac{1}{2}}^n \Delta t; \quad x_{j+\frac{1}{2},r}^n = x_{j+\frac{1}{2}} + \phi_{j+\frac{1}{2}}^n \Delta t; \tag{11}$$

with a free parameter $\phi_{j+\frac{1}{2}}^n$ satisfying the CFL condition

$$a_{j+\frac{1}{2}}^n \leq \phi_{j+\frac{1}{2}}^n \leq \frac{1}{2\lambda}. \tag{12}$$

Due to finite speed of propagation, the points $x_{j+\frac{1}{2},l}^n, x_{j+\frac{1}{2},r}^n$ separate between smooth and possible singular regions $[x_{j-\frac{1}{2},r}^n, x_{j+\frac{1}{2},l}^n] \times [t^n, t^{n+1}]$ and $[x_{j+\frac{1}{2},l}^n, x_{j+\frac{1}{2},r}^n] \times [t^n, t^{n+1}]$. Under to the CFL restriction, (12), the required integrals of the fluxes in (3) over each anrrower rectangular controlled volumes can be approximated by the midpoint rule and the midpoint values, $w_{j+\frac{1}{2},l}^{n+\frac{1}{2}}$ and $w_{j+\frac{1}{2},r}^{n+\frac{1}{2}}$, can be obtained from the corresponding Taylor expansions with respect to $x_{j+\frac{1}{2},l}^n$ and $x_{j+\frac{1}{2},r}^n$, resulting with

$$\begin{aligned} w_{j+\frac{1}{2}}^{n+1} = & \frac{\bar{w}_j^n + \bar{w}_{j+1}^n}{2} + \frac{\Delta x - \phi_{j+\frac{1}{2}}^n \Delta t}{4} ((w_x)_j - (w_x)_{j+1}) \\ & - \frac{1}{2\phi_{j+\frac{1}{2}}^n} \left[f(w_{j+\frac{1}{2},r}^{n+\frac{1}{2}}) - f(w_{j+\frac{1}{2},l}^{n+\frac{1}{2}}) \right]; \end{aligned}$$

$$w_j^{n+1} = \bar{w}_j^n + \frac{\Delta t}{2}\left(\phi_{j-\frac{1}{2}}^n - \phi_{j+\frac{1}{2}}^n\right)(w_x)_j$$
$$-\frac{\lambda}{1-\lambda(\phi_{j-\frac{1}{2}}^n - \phi_{j+\frac{1}{2}}^n)}\left[f(w_{j+\frac{1}{2},l}^{n+\frac{1}{2}}) - f(w_{j-\frac{1}{2},r}^{n+\frac{1}{2}})\right];$$

$$w_{j+\frac{1}{2},l}^{n+\frac{1}{2}} := w_{j+\frac{1}{2},l}^n - \frac{\Delta t}{2} f(w_{j+\frac{1}{2},l}^n)_x,\ w_{j+\frac{1}{2},l}^n := \bar{w}_j^n + (\frac{\Delta x}{2} - \phi_{j+\frac{1}{2}}^n \Delta t)(w_x)_j;$$
$$w_{j+\frac{1}{2},r}^{n+\frac{1}{2}} := w_{j+\frac{1}{2},r}^n - \frac{\Delta t}{2} f(w_{j+\frac{1}{2},r}^n)_x,\ w_{j+\frac{1}{2},r}^n := \bar{w}_{j+1}^n - (\frac{\Delta x}{2} - \phi_{j+\frac{1}{2}}^n \Delta t)(u_x)_{j+1}.$$

To obtain the cell averages over the original grid of the uniform, non-staggered cells $[x_{j-\frac{1}{2}}, x_{j+\frac{1}{2}}]$, we consider the piecewise-linear construction at the time level t^{n+1}

$$w(x,t^{n+1}) :=$$
$$\sum_j \left\{ \left[w_{j+\frac{1}{2}}^{n+1} + (w_x)_{j+\frac{1}{2}}^{n+1}(x - x_{j+\frac{1}{2}})\right] \chi_{[x_{j+\frac{1}{2},l}^n, x_{j+\frac{1}{2},r}^n]} + w_j^{n+1} \chi_{[x_{j-\frac{1}{2},r}^n, x_{j+\frac{1}{2},l}^n]} \right\} \tag{13}$$

Here $\chi_{[x_{j+\frac{1}{2},l}^n, x_{j+\frac{1}{2},r}^n]}$ is the characteristic function on $[x_{j+\frac{1}{2},l}^n, x_{j+\frac{1}{2},r}^n]$ and the exact spatial derivatives, $w_x(x_{j+\frac{1}{2}}, t^{n+1})$, can be approximated by

$$(w_x)_{j+\frac{1}{2}}^{n+1} = \frac{2}{\Delta x} MM \left(\frac{w_{j+1}^{n+1} - w_{j+\frac{1}{2}}^{n+1}}{1+\lambda(\phi_{j+\frac{1}{2}}^n - \phi_{j+\frac{3}{2}}^n)}, \frac{w_{j+\frac{1}{2}}^{n+1} - w_j^{n+1}}{1+\lambda(\phi_{j+\frac{1}{2}}^n - \phi_{j-\frac{1}{2}}^n)} \right). \tag{14}$$

Finally, the desired cell averages, $\bar{w}_j^{n+1}$, are obtained by averaging the reconstructed approximate solution in (13) over the interval $[x_{j-\frac{1}{2}}, x_{j+\frac{1}{2}}]$, which leads to the final form of the second-order KT schemes

$$\bar{w}_j^{n+1} = \lambda\phi_{j-\frac{1}{2}}^n w_{j-\frac{1}{2}}^{n+1} + \lambda\phi_{j+\frac{1}{2}}^n w_{j+\frac{1}{2}}^{n+1} + \left[1 - \lambda(\phi_{j+\frac{1}{2}}^n - \phi_{j-\frac{1}{2}}^n)\right] w_j^{n+1}$$
$$+\frac{\Delta x}{2}\left[(\lambda\phi_{j-\frac{1}{2}}^n)^2 (w_x)_{j-\frac{1}{2}}^{n+1} - (\lambda\phi_{j+\frac{1}{2}}^n)^2 (w_x)_{j+\frac{1}{2}}^{n+1}\right].$$

3 New Schemes: Triangle version

We now demonstrate the construction of our new central-upwind scheme for nonlinear hyperbolic conservation laws (1). The main idea in the construction of our new central scheme is that instead of using a rectangular controlled volume determined by the local speed of wave propagation as in the construction of the KT scheme, we take a smaller triangular controlled volume which leads to less numerical viscosity when compared to the KT scheme. For simplicity, we demonstrate a second-order construction for one-dimensional case only. The construction may be extended to high-order or higher-dimensional cases. [KL]

As usual, assume that we have already computed the piecewise-linear solution, $w(x,t^n)$ and the local speed of wave propagation, $a^n_{j+\frac{1}{2}}$, at the time level t^n. We also choose two free parameters $\phi_{j+\frac{1}{2}}$ and $\epsilon > 0$ satisfying the CFL condition

$$a^n_{j+\frac{1}{2}} \leq \phi^n_{j+\frac{1}{2}} \leq \frac{1}{2\lambda} - \frac{\epsilon}{\Delta t}. \tag{15}$$

We then evolve this piecewise-linear interpolant in time and project at the next time level t^{n+1} onto two different kinds of regions: smooth and possible nonsmooth. In the possible non-smooth region, the interpolant is evolved in time starting from the interval $[x_{j+\frac{1}{2}} - \epsilon, x_{j+\frac{1}{2}} + \epsilon]$ and is projected onto the interval $[x^{n+1}_{j+\frac{1}{2},l} - \epsilon, x^{n+1}_{j+\frac{1}{2},r} + \epsilon]$ at the time level t^{n+1}. Here,

$$x^{n+1}_{j+\frac{1}{2},l} = x_{j+\frac{1}{2}} - \phi^n_{j+\frac{1}{2}}\Delta t; \quad x^{n+1}_{j+\frac{1}{2},r} = x_{j+\frac{1}{2}} + \phi^n_{j+\frac{1}{2}}\Delta t; \tag{16}$$

Exact computation of spatial integrals yield

$$\begin{aligned}
w^{n+1}_{j+\frac{1}{2}} &:= \frac{1}{\Delta x_{j+\frac{1}{2}} + 2\epsilon} \int_{x^{n+1}_{j+\frac{1}{2},l}-\epsilon}^{x^{n+1}_{j+\frac{1}{2},r}+\epsilon} w(x,t^{n+1})dx \\
&= \frac{1}{\Delta x_{j+\frac{1}{2}} + 2\epsilon} \Bigg[\int_{x_{j+\frac{1}{2}}-\epsilon}^{x_{j+\frac{1}{2}}+\epsilon} w(x,t^n)dx + \int_0^{\Delta t} w(y_{j+\frac{1}{2},r}(t),t)\phi^n_{j+\frac{1}{2}}dt \\
&\quad + \int_0^{\Delta t} w(y_{j+\frac{1}{2},l}(t),t)\phi^n_{j+\frac{1}{2}}dt \\
&\quad + \int_0^{\Delta t} f(w(y_{j+\frac{1}{2},l}(t),t))dt - \int_0^{\Delta t} f(w(y_{j+\frac{1}{2},r}(t),t))dt \Bigg]
\end{aligned} \tag{17}$$

Here $\Delta x_{j+\frac{1}{2}} := x^n_{j+\frac{1}{2},r} - x^n_{j+\frac{1}{2},l} = 2\phi^n_{j+\frac{1}{2}}\Delta t$. $y_{j+\frac{1}{2},r}(t) = x_{j+\frac{1}{2}} + \phi^n_{j+\frac{1}{2}}t + \epsilon$ and $y_{j+\frac{1}{2},l}(t) = x_{j+\frac{1}{2}} - \phi^n_{j+\frac{1}{2}}t - \epsilon$

Similarly, for the smooth parts, the interpolant is evolved in time starting from the interval $[x_{j-\frac{1}{2}} + \epsilon, x_{j+\frac{1}{2}} - \epsilon]$ and is projected onto the interval $[x^{n+1}_{j-\frac{1}{2},r} + \epsilon, x^{n+1}_{j+\frac{1}{2},l} - \epsilon]$ at time t^{n+1}, resulting in

$$\begin{aligned}
w^{n+1}_{j} &:= \frac{1}{\Delta x_j - 2\epsilon} \int_{x^{n+1}_{j-\frac{1}{2},r}+\epsilon}^{x^{n+1}_{j+\frac{1}{2},l}-\epsilon} w(x,t^{n+1})dx \\
&= \frac{1}{\Delta x_j - 2\epsilon} \Bigg[\int_{x_{j-\frac{1}{2}}+\epsilon}^{x_{j+\frac{1}{2}}-\epsilon} w(x,t^n)dx - \int_0^{\Delta t} w(y_{j+\frac{1}{2},l}(t),t)\phi^n_{j+\frac{1}{2}}dt
\end{aligned}$$

$$
+\int_0^{\Delta t} w(y_{j-\frac{1}{2},r}(t),t)\phi^n_{j-\frac{1}{2}}\,dt
- \int_0^{\Delta t} f(w(y_{j+\frac{1}{2},l}(t),t))dt - \int_0^{\Delta t} f(w(y_{j-\frac{1}{2},r}(t),t))dt\Bigg] \tag{18}
$$

where $\Delta x_j := x_{j+\frac{1}{2},l} - x_{j-\frac{1}{2},r} = \Delta x - (\phi^n_{j-\frac{1}{2}} + \phi^n_{j+\frac{1}{2}})\Delta t$.

Under the CFL restriction, (15), the required integrals in (17, 18) can be approximated by the midpoint rule and the midpoint values, $w^{n+\frac{1}{2}}_{j+\frac{1}{2},l}$ and $w^{n+\frac{1}{2}}_{j+\frac{1}{2},r}$, can be obtained from the corresponding Taylor expansions with respect to $x^n_{j+\frac{1}{2},l}$ and $x^n_{j+\frac{1}{2},r}$, resulting with (take $\epsilon \to 0$)

$$
\begin{aligned}
&w^{n+1}_{j+\frac{1}{2}} \\
&= \frac{\Delta t}{2\phi^n_{j+\frac{1}{2}}}\left[\left(w^{n+\frac{1}{2}}_{j+\frac{1}{2},r} + w^{n+\frac{1}{2}}_{j+\frac{1}{2},l}\right)\phi^n_{j+\frac{1}{2}} - f(w^{n+\frac{1}{2}}_{j+\frac{1}{2},r}) + f(w^{n+\frac{1}{2}}_{j+\frac{1}{2},l})\right];
\end{aligned}
$$

$$
w^{n+1}_j = \frac{\Delta t}{\Delta x_j}\left[(\Delta x)w^n_j - w^{n+\frac{1}{2}}_{j-\frac{1}{2},r}\phi^n_{j-\frac{1}{2}} - w^{n+\frac{1}{2}}_{j+\frac{1}{2},l}\phi^n_{j+\frac{1}{2}} - f(w^{n+\frac{1}{2}}_{j+\frac{1}{2},l}) + f(w^{n+\frac{1}{2}}_{j-\frac{1}{2},r})\right]; \tag{19}
$$

$$
\begin{aligned}
w^{n+\frac{1}{2}}_{j+\frac{1}{2},l} &:= w^n_{j+\frac{1}{2},l} - \frac{\Delta t}{2} f(w^n_{j+\frac{1}{2},l})_x,\ w^n_{j+\frac{1}{2},l} := \bar{w}^n_j + (\frac{\Delta x}{2} - \phi^n_{j+\frac{1}{2}}\Delta t)(w_x)_j; \\
w^{n+\frac{1}{2}}_{j+\frac{1}{2},r} &:= w^n_{j+\frac{1}{2},r} - \frac{\Delta t}{2} f(w^n_{j+\frac{1}{2},r})_x,\ w^n_{j+\frac{1}{2},r} := \bar{w}^n_{j+1} - (\frac{\Delta x}{2} - \phi^n_{j+\frac{1}{2}}\Delta t)(w_x)_{j+1}.
\end{aligned}
$$

At this stage, we realize the solution at the time level t^{n+1} in terms of the approximate cell averages $w^{n+1}_{j+\frac{1}{2}}$ and w^{n+1}_j. To obtain the cell averages over the original grid of the uniform, non-staggered cells $[x_{j-\frac{1}{2}}, x_{j+\frac{1}{2}}]$, we consider the re-projection mechanism at time level t^{n+1} as proceeded in the construction of the KT schemes, (13, 14), and results in the final form of our second-order schemes

$$
\begin{aligned}
&\bar{w}^{n+1}_j \\
&= \frac{\phi^n_{j-\frac{1}{2}}\Delta t}{\Delta x} w^{n+1}_{j-\frac{1}{2}} + \frac{\phi^n_{j+\frac{1}{2}}\Delta t}{\Delta x} w^{n+1}_{j+\frac{1}{2}} + \left[1 - \lambda(\phi^n_{j+\frac{1}{2}} - \phi^n_{j-\frac{1}{2}})\right] w^{n+1}_j \\
&\quad + \frac{1}{2\Delta x}\left[(\phi^n_{j-\frac{1}{2}}\Delta t)^2 (w_x)^{n+1}_{j-\frac{1}{2}} - (\phi^n_{j+\frac{1}{2}}\Delta t)^2 (w_x)^{n+1}_{j+\frac{1}{2}}\right].
\end{aligned}
$$

We end this section with three remarks

- Both KT and our schemes reduce to the NT scheme if we take $\phi^n_{j+\frac{1}{2}} = \frac{1}{2\lambda}$, hence to LxF schemes if we take furthermore, the numerical derivatives $(w_x)^{n+1}_{j+\frac{1}{2}} = 0 = (w_x)^n_j$.

- Both KT and our scheme enjoy the same semi-discrete formula when we take $\Delta t \to 0$. See [KL] for details.
- Numerical viscosity of our scheme is less than that of the KT scheme as it is designed to be so. Since the difference is quite small to see from most numerical simulation. However, this fact is confirmed numerically by Example 3 in the next section.

4 Numerical experiments

In this section, we present and compare numerical results via KT and our new schemes. In Example 1, we show the order test for our scheme. In Example 2, we show a standard test on the shock tube problem. Most numerical simulation by KT and our schemes are hard to distinguish each other. However, we present in Example 3 a sharper resolution of our scheme against the KT scheme.

Example 1. Burgers' equation.
Here, we solve the inviscid Burgers' equation

$$u_t + \left(\frac{u^2}{2}\right)_x = 0, \qquad -1 \le x \le 1, \tag{20}$$

subject to the 2-periodic initial data $u_0(x) = 0.25+0.5\sin(\pi x)$. Table 1 shows the error and second-order order of our schemes at time $T = 1.5$, before the shock develops.

Table 1. Error and order in L_1 norm

	CFL = .45		CFL = .25	
No	L1 norm	L1 order	L1 norm	L1 order
40	0.002070	1.857520	0.002834	1.890214
80	0.000548	1.916432	0.000735	1.947920
160	0.000145	1.922300	0.000194	1.918088
320	0.000039	1.900322	0.000051	1.919836
640	0.000010	1.926276	0.000013	1.937500

Example 2. 1D Shock tube problem.
In this example, we apply our schemes to the 1D shock tube problem governed by the Euler system, where

$$\begin{aligned} u &= (\rho, \rho u, E)^T, \\ f(u) &= (\rho u, P + \rho u^2, u(E+P))^T, \end{aligned} \tag{21}$$

augmented with $P = (\gamma - 1)(E - \frac{1}{2}\rho u^2)$. Here ρ, u, P and E are respectively, the density, velocity, pressure and total specific energy. The system is complete with Sod's initial data

$$(\rho_L, u_L, P_L) = (1, 0, 1), x < 0 \qquad (\rho_R, u_R, P_R) = (0.125, 0, 0.1), x > 0.$$

In Figure 1, we show the density, velocity and pressure plots obtained by our schemes.

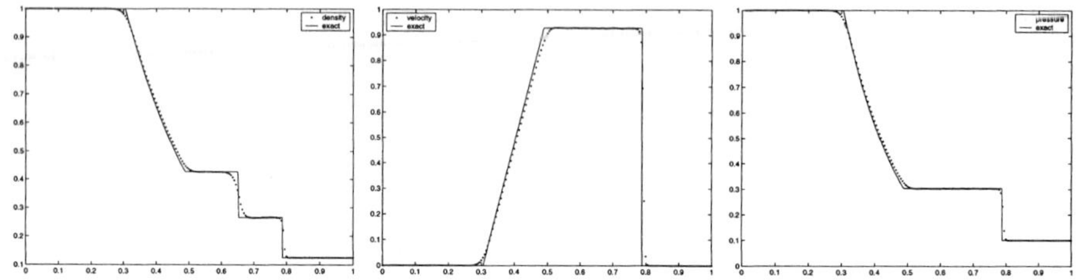

Fig. 1. Shock tube problem: Sod's initial data

Example 3. Steady contact discontinuity.
We solve the 1D Euler system (21) on the interval $[-0.2, 0.2]$, subject to the initial data

$$\mathbf{u}(x,0) = \begin{cases} \mathbf{u}_L = (1, 0, 2.5)^T, & x < 0, \\ \mathbf{u}_R = (0.5, 0, 2.5)^T, & x > 0, \end{cases} \tag{22}$$

which corresponds to a single steady contact discontinuity. We use 100 grid points and compute the solution at a final time $T = 5$. Figure 2 shows the density, computed using the KT and our new schemes. The excessive numerical viscosity, present in all these schemes, does not allow a sharp resolution of the discontinuity. However, our scheme gives a better resolution of the steady contact wave.

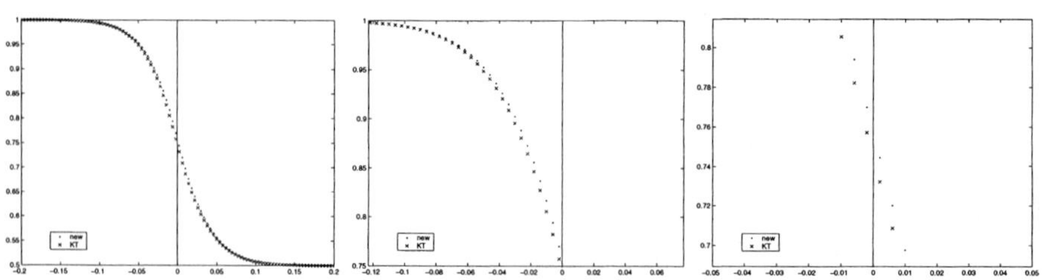

Fig. 2. Steady contact discontinuity

References

[JLLOT] Jiang, G.-S., Levy, D., Lin,C.-T., Osher, O., Tadmor, E., (1998): High-Resolution Non-Oscillatory Central Schemes with Non-Staggered Grids for Hyperbolic Conservation Laws. SIAM J. Numer. Anal.,

[KL] Kurganov, A., Lin, C.-T., (2003): On the Reduction of Numerical Dissipation in Central-Upwind Schemes. In preparation.

[KurTa] Kurganov, A., Tadmor, E., (2000): New High-Resolution Central Schemes for Nonlinear Conservation Laws and Convection-Diffusion Equations. JCP, **160**, 241–282.

[NeTa] Nessyahu, H., Tadmor, E., (1990): Non-oscillatory Central Differencing for Hyperbolic Conservation Laws. JCP, **87**, 408–463.

[LeVe] LeVeque, R. (2002): Finite Volume methods for Hyperbolic Problems. Cambridge University Press.

Afternotes on PHM: Harmonic ENO Methods

Antonio Marquina[1] and Susana Serna[2]

[1] Departamento de Matemática Aplicada, Universitat de València, Avda. Dr. Moliner,50 46100 Burjassot, Spain *Antonio.Marquina@uv.es*
[2] Departamento de Matemática Aplicada, Universitat de València, Avda. Dr. Moliner,50 46100 Burjassot, Spain *Susana.Serna@uv.es*

Summary. PHM methods have been used successfully as reconstruction procedures to design high-order Riemann solvers for nonlinear scalar and systems of conservation laws, (see [8], [1], [4]). We introduce a new class of polynomial reconstruction procedures based on the harmonic mean of the absolute values of finite differences used as difference-limiter, following the original idea used before to design the piecewise hyperbolic method, introduced in [8]. We call those methods 'harmonic ENO methods', (HENO). Furthermore, we give analytical and numerical evidence of the good behavior of these methods used as reconstruction procedures for the numerical approximation by means of shock-capturing methods for scalar and systems of conservation laws in 1D. We discuss, in particular, the behavior of a fourth order harmonic ENO method,(HENO4 in short), compared with PHM, ENO3 and WENO5 methods, (see [2], [10], [3]).

1 Introduction

In this paper, we shall consider numerical approximations to nonlinear conservation laws of the form:

$$\frac{\partial \mathbf{u}}{\partial t} + \sum_{i=1}^{d} \frac{\partial \mathbf{f}_i(\mathbf{u})}{\partial x_i} = 0 \tag{1}$$

where $\mathbf{u}$ is a m-dimensional vector of unknowns and $\mathbf{f}_i(\mathbf{u})$ are d vector-valued functions called *fluxes*, We assume strong hyperbolicity, i.e. the Jacobian matrices of the former system

$$\mathbf{A}_i = \frac{\partial \mathbf{f}_i(\mathbf{u})}{\partial \mathbf{u}} \tag{2}$$

locally diagonalize with real eigenvalues and a complete system of eigenvectors. The one-dimensional case of the system will be in the form

$$\mathbf{u}_t + (\mathbf{f}(\mathbf{u}))_x = 0 \tag{3}$$

with the initial value condition

$$\mathbf{u}(x,0) = \mathbf{u}_0(x) \tag{4}$$

Weak solutions of nonlinear conservation laws are piecewise smooth with jump discontinuities. High-order accurate numerical approximations to these functions are such that they achieve high accuracy on smooth regions and sharpen profiles of discontinuities, without spurious oscillations. Essentially non oscillatory, (ENO), polynomial reconstruction procedures were designed to accomplish this purpose, ([2]). ENO methods are high-order accurate on smooth regions and appear to be very robust on shocks. However, several drawbacks became relevant after some experimentation with ENO methods took place, from which we mention:

1. Loss of accuracy on smooth regions with specific input data, ([7]),
2. Smearing of certain discontinuities ([3]),
3. Smoothing up of corners, (discontinuities of the first derivative), ([8]),
4. Too wide stencil to get high order accuracy, ([3]).

In order to overcome those difficulties several remedies were proposed. Shu proposed to design a more centered ENO selection to reduce the loss of accuracy. PHM methods were introduced to improve the resolution of corners. WENO methods were designed to get optimal accuracy for a specific stencil, degenerating to a classical ENO method at discontinuities.

In this paper we introduce an extended class of ENO methods, based on the harmonic mean, used to design the PHM methods, (see [8], [1], [4]). We focus our study on ENO parabolas and we apply the harmonic limiters on neighboring second order differences. In particular, we discuss the fourth order accurate Harmonic ENO4, obtained from a parabola chosen from three different parabolas, with the same stencil as the ENO3. In section 3 we test our HENO4 method on several 1D model problems, for scalar and systems of conservation laws, and we compare them with ENO3, PHM and WENO5 schemes to analyze their ability in resolving shocks and complex flow problems. Further study and 2D experiments are presented in [9].

2 Piecewise Polynomial Reconstructions: Harmonic ENO methods

We use the example of a scalar conservative law to settle our notation and computational framework:

$$v_t + f(v)_x = 0, \tag{5}$$

$$v(x,t) = v_0(x), \tag{6}$$

where v_0 is a periodic or compactly supported piecewise smooth function.

We consider the following computational grid: $x_j = jh$, (h is the spatial step), $t_n = n\Delta t$, (Δt is the time step), $I_j = [x_{j-\frac{1}{2}}, x_{j+\frac{1}{2}}]$ is the spatial cell, where $x_{j+\frac{1}{2}} = x_j + \frac{h}{2}$ is the cell interface, $t_n = n\Delta t$, is the time discretization and $C_j^n = [x_{j-\frac{1}{2}}, x_{j+\frac{1}{2}}] \times [t_n, t_{n+1}]$ is the computational cell. Let v_j^n be an aproximation of the mean value in I_j, $\frac{1}{h}\int_{x_{j-\frac{1}{2}}}^{x_{j+\frac{1}{2}}} u(x,t_n)dx$, of the exact solution $u(x,t_n)$ of the initial value problem (5) and (6), obtained from a finite difference scheme in conservation form:

$$v_j^{n+1} = v_j^n - \frac{\Delta t}{h}(\tilde{f}_{j+\frac{1}{2}} - \tilde{f}_{j-\frac{1}{2}}), \tag{7}$$

where the numerical flux, $\tilde{f}$, is a function of $k+l$ variables

$$\tilde{f}_{j+\frac{1}{2}} = \tilde{f}(v_{j-k+1}, \cdots v_{j+l}), \tag{8}$$

which is consistent with the flux of the equation (5),

$$\tilde{f}(v, \cdots, v) = f(v) \tag{9}$$

For our purposes, a reconstruction procedure is an algorithm to obtain point values at the cell interfaces from cell averages and its differences corresponding to a set of discrete variables or fluxes, up to a degree of accuracy.

We want to reconstruct a function $u(x)$ from its mean values given at cells:

$$v_j = \frac{1}{h}\int_{x_{j-\frac{1}{2}}}^{x_{j+\frac{1}{2}}} u(x)dx \tag{10}$$

such that u is a piecewise smooth function associated to the spatial grid defined above, i.e., u restricted to each cell I_j is smooth, and, therefore, possible jump discontinuities are located at cell interfaces.

Our grid data are:

– (i) for every j the mean value of $u(x)$ in C_j, v_j is given to satisfy (10).
– (ii) for every j, $d_{j+\frac{1}{2}}$ is given by the undivided first order difference:

$$d_{j+\frac{1}{2}} = v_{j+1} - v_j \tag{11}$$

All the reconstruction methods analyzed in this paper have the same stencil as the classical ENO3 method and based on parabolas of the form:

$$p_j(x) = a_j + (x - x_j)\left[b_j + \frac{c_j}{2}(x - x_j)\right] \tag{12}$$

defined on C_j where a_j, b_j and c_j are determined from the grid data.

We use the following notations:

$$d_j = \frac{d_{j+\frac{1}{2}} + d_{j-\frac{1}{2}}}{2} \tag{13}$$

$$D_j = d_{j+\frac{1}{2}} + d_{j-\frac{1}{2}} \tag{14}$$

The classical ENO3 method is based on a selection procedure that chooses one parabola from three candidates:

$$p_{j-1}(x) = v_j - \frac{D_{j-1}}{24} + \frac{x - x_j}{h}\left[d_{j-\frac{1}{2}} + \frac{D_{j-1}}{2} + \frac{D_{j-1}}{2}\left(\frac{x - x_j}{h}\right)\right] \quad (15)$$

$$p_j(x) = v_j - \frac{D_j}{24} + \frac{x - x_j}{h}\left[d_j + \frac{D_j}{2}\left(\frac{x - x_j}{h}\right)\right] \quad (16)$$

$$p_{j+1}(x) = v_j - \frac{D_{j+1}}{24} + \frac{x - x_j}{h}\left[d_{j+\frac{1}{2}} - \frac{D_{j+1}}{2} + \frac{D_{j+1}}{2}\left(\frac{x - x_j}{h}\right)\right] \quad (17)$$

which correspond to the left, central and righthandside choice, respectively. The ENO3 selection procedure to get the ENO parabola for the computational cell $C_j = [x_{j-\frac{1}{2}}, x_{j+\frac{1}{2}}]$ reads as follows:

if $|d_{j-\frac{1}{2}}| \leq |d_{j+\frac{1}{2}}|$ **then**
    if $|D_{j-1}| \leq |D_j|$ **then**
        Left choice
    else
        Central choice
    end
else
    if $|D_j| \leq |D_{j+1}|$ **then**
        Central choice
    else
        Right choice
    end
end

The truncation error expressions at the right interface of the above parabolas are:

$$p_{j-1}(x_{j+\frac{1}{2}}) - u = -12\left(\frac{h}{2}\right)^3 \frac{u'''}{6} + O(h^4) \quad (18)$$

$$p_j(x_{j+\frac{1}{2}}) - u = 4\left(\frac{h}{2}\right)^3 \frac{u'''}{6} + O(h^4) \quad (19)$$

$$p_{j+1}(x_{j+\frac{1}{2}}) - u = -4\left(\frac{h}{2}\right)^3 \frac{u'''}{6} + O(h^4) \quad (20)$$

To explore new ways to design ENO methods such that we get better behavior near discontinuities, we make to play in our study two more parabolas, using an intermediate value between two neighboring second order differences. If we choose the harmonic mean, i.e., $\mathrm{mean}(x, y) := \frac{2xy}{x+y}$, then, we can define the **harmonic limiters**, (used before for the PHM methods):

$$\mathrm{harmod}(x, y) = \frac{(\mathrm{sign}(x) + \mathrm{sign}(y))}{2} \frac{2|x||y|}{|x| + |y|} \quad (21)$$

$$\mathrm{hareno}(x, y) = \mathrm{minsign}(x, y) \frac{2|x||y|}{|x| + |y|}, \quad (22)$$

where minsign(x, y) = sign(x) if $|x| <= |y|$, and minsign(x, y) = sign(y) otherwise.

Then, we have the following parabolas:

$$p^H_{j-\frac{1}{2}}(x) = v_j - \frac{H_{j-\frac{1}{2}}}{24} + \frac{x - x_j}{h}\left[d_{j-\frac{1}{2}} + \frac{H_{j-\frac{1}{2}}}{2} + \frac{H_{j-\frac{1}{2}}}{2}\left(\frac{x - x_j}{h}\right)\right]$$

$$p^H_{j+\frac{1}{2}}(x) = v_j - \frac{H_{j+\frac{1}{2}}}{24} + \frac{x - x_j}{h}\left[d_{j+\frac{1}{2}} - \frac{H_{j+\frac{1}{2}}}{2} + \frac{H_{j+\frac{1}{2}}}{2}\left(\frac{x - x_j}{h}\right)\right]$$

where $H_{j-\frac{1}{2}} :=$ mean(D_{j-1}, D_j) and $H_{j+\frac{1}{2}} :=$ mean(D_j, D_{j+1}), where "mean" is either the harmod or the hareno limiter.

Proposition 1. *We have the following truncation error expressions at the right interface, $x_{j+\frac{1}{2}}$, for the above parabolas:*

$$p^H_{j-\frac{1}{2}}(x_{j+\frac{1}{2}}) - u = -4\left(\frac{h}{2}\right)^3 \frac{u'''}{6} + O(h^4) \tag{23}$$

$$p^H_{j+\frac{1}{2}}(x_{j+\frac{1}{2}}) - u = O(h^4) \tag{24}$$

Theorem 1. *The harmod limiter is total variation diminishing, (TVD).*

Proof. The harmod limiter satisfies the inequality (16.26), from LeVeque [5], (p. 180), (see [9] for details).

Next, we construct the following new three parabolas:

Proposition 2. *The following three parabolas are fourth order accurate at $x_{j+\frac{1}{2}}$:*

– *The right choice*

$$p^H_{j+\frac{1}{2}} \tag{25}$$

– *the central choice*

$$0.5 \cdot p^H_{j-\frac{1}{2}}(x_{j+\frac{1}{2}}) + 0.5 \cdot p_j(x_{j+\frac{1}{2}}) \tag{26}$$

– *the left choice*

$$-0.5 \cdot p_{j-1}(x_{j+\frac{1}{2}}) + 1.5 \cdot p^H_{j-\frac{1}{2}}(x_{j+\frac{1}{2}}) \tag{27}$$

Proof.

We can eliminate the $O(h^3)$ terms by computing a convex combination of p_{j-1} and $p_{j-\frac{1}{2}}$. Indeed, from

$$\alpha p_{j-1} + \beta p_{j-\frac{1}{2}} \tag{28}$$

we get the system

$$\begin{aligned} \alpha + \beta &= 1 \\ -12\alpha - 4\beta &= 0 \end{aligned} \tag{29}$$

where the solution is

$$\begin{aligned} \alpha &= -0.5 \\ \beta &= 1.5 \end{aligned}$$

and from

$$\gamma p_{j-\frac{1}{2}} + \delta p_j \tag{30}$$

we obtain

$$\begin{aligned} \gamma + \delta &= 1 \\ -4\gamma + 4\delta &= 0 \end{aligned} \tag{31}$$

whence

$$\begin{aligned} \gamma &= 0.5 \\ \delta &= 0.5 \end{aligned}$$

¿From (24) we observe that $p_{j+\frac{1}{2}}$ has fourth order. The fourth order combinations for the aproximation at the left interface are symetric.

We design our fourth order Harmonic ENO method (HENO4) by using the left, central or right choice mencioned above, depending on the selection procedure of the classical ENO3 method described above.

The following features were found for our HENO4 method:

1. The HENO4 method is formally fourth order accurate with the same stencil as the classical ENO3 method.
2. The resolution near discontinuities is better than ENO methods.
3. We observe an improvement of resolution of corners, (discontinuities in the first derivative).
4. We observe an improvement in the resolution of local extrema.

3 Numerical Experiments

Example 1: Linear advection

We solve the linear equation $u_t + u_x = 0$, $-1 \leq x \leq 1$ with $u(x,0) = u_0(x)$ periodic

$$u_0(x) = \begin{cases} 1 & -0.7 \leq x \leq -0.2 \\ \sin\left(\pi \frac{x-0.2}{0.6}\right) & 0.2 \leq x \leq 0.8 \\ 0 & \text{otherwise} \end{cases}$$

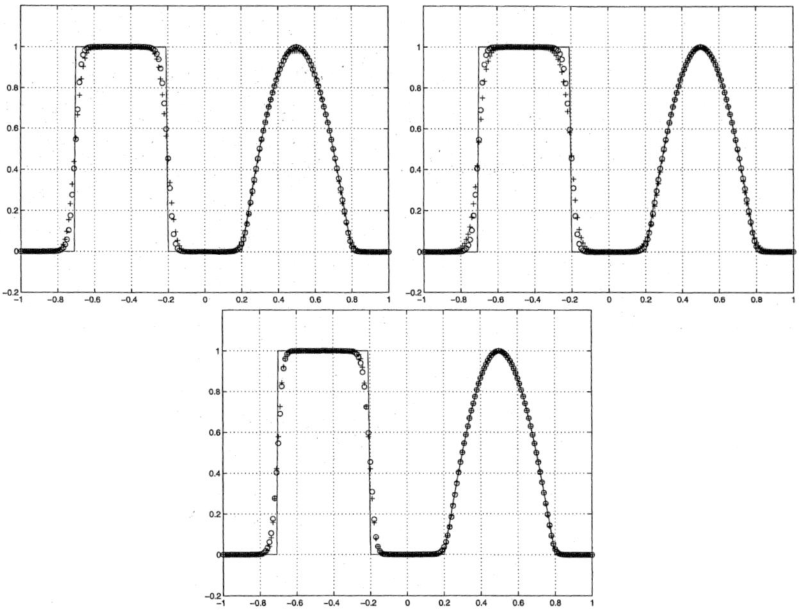

Fig. 1. Example 1. Top left: HENO4 "o" versus PHM "+", top right: HENO4 "o" versus ENO3 "+", bottom: HENO4 "o" versus WENO5 "+"

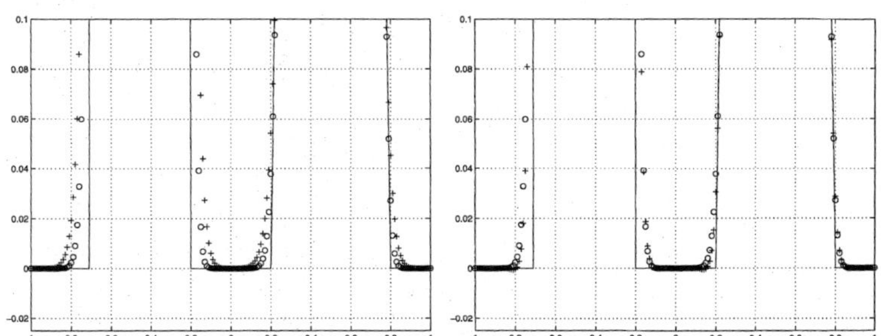

Fig. 2. Zoomed regions: left, HENO4 "o" versus ENO3 "+", right, HENO4 "o" versus WENO5 "+"

We compute the approximate solution for different methods presented here using a grid of 200 points, with a CFL number ($\Delta t/h$) of 0.5 and total time of two periods. We use the Godunov upwind method for all our calculations with the linear advection and Runge-Kutta 3 integration in time.

In Figure 1 we display the numerical approximation of Example 1 using the PHM, ENO3, WENO5 and HENO4 methods. At the top-left, we compare HENO4 and PHM methods and we can observe the better resolution of the local extrema of our HENO4 method.

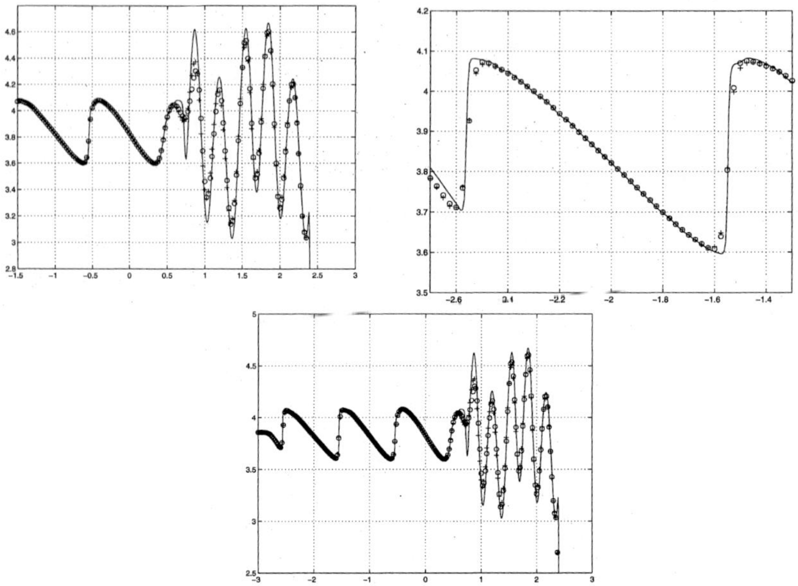

Fig. 3. Zoomed regions of Shock entropy test. HENO4 "o" versus WENO5 "+"

We compare HENO4 and ENO3 in Figure 1 top-right and Figure 2 left (zoomed region), and we observe the better behavior at corners of HENO4. In Figure 1 bottom and the zoomed region in Figure 2 right, we observe the similar behavior of HENO4 and WENO5 method.

The experiments performed with Euler equations were done using Marquina's flux formula (see [1]) applying the reconstruction procedure on the characteristic fluxes and evolving in time by means of the third order accurate Shu-Osher Runge-Kutta method.

Example 2: 1D Shock Entropy Wave Interaction

We consider the 1D Euler equations for gas dynamics for the following initial condition:

$$u_0(x) = \begin{cases} (\rho_L, v_L, P_L) & -5 \leq x < -4 \\ (\rho_R, v_R, P_R) & -4 \leq x \leq 5 \end{cases}$$

$$(\rho_L, v_L, P_L) = (3.85714, 2.62936, 10.33333)$$
$$(\rho_R, v_R, P_R) = (1 + 0.2\sin(5x), 0, 1)$$

We compute the numerical approximation for 400 points at time=1.8 and CFL=0.1 for WENO5 and HENO4. We observe in Figure 3 that the behavior near shocks is better for our HENO4 method and we may observe a little better resolution of the local extrema by zooming that region.

References

1. R. Donat and A. Marquina, *Capturing Shock Reflections: An improved Flux Formula* J. Comput. Phys., v. 125, (1996) pp. 42-58.
2. A. Harten, B. Engquist, S. Osher and S. Chakravarthy, *Uniformly High Order Accurate Essentially Non-oscillatory Schemes III,* J. Comput. Phys., v. 71, No. 2, (1987), pp. 231-303.
3. G.S. Jiang and C. W. Shu, *Efficient Implementation of weighted ENO schemes,*, J. Comput. Phys., 126, (1996), p. 202.
4. S. Li and L. Petzold, *Moving Mesh Methods with Upwinding Schemes for Time-Dependent PDEs,* J. Comput. Phys., v. 131, (1997), pp. 368-377.
5. R.J. LeVeque *Numerical methods for Conservation Laws,* Birkhauser Verlag, Zuerich, (1990).
6. X-D. Liu and S. Osher and T. Chan *Weighted essentially non-oscillatory schemes,* J. Comput. Phys., v. 115, (1994), pp. 200-212.
7. A. Rogertson and E. Meiburg, *A Numerical Study of the Convergence of ENO schemes,* J. Sci. Comp., v. 5, (1990). pp. 127-150.
8. A. Marquina, *Local Piecewise Hyperbolic Reconstructions for Nonlinear Scalar Conservation Laws,* SIAM J. Sci. Comp., v. 15, (1994) pp. 892-915.
9. S. Serna and A. Marquina, *Power ENO methods,* preprint.
10. C. W. Shu and S. J. Osher, *Efficient Implementation of Essentially Non-Oscillatory Shock Capturing Schemes II,* J. Comput. Phys., v. 83, (1989) pp. 32-78.

One-Dimensional Stability of Viscous Shock and Relaxation Profiles

Corrado Mascia[1] and Kevin Zumbrun[2]

[1] Università "La Sapienza", Roma, Italy *mascia@mat.uniroma1.it*
[2] Indiana University, Bloomington, USA *kzumbrun@indiana.edu*

Under the weak assumption of spectral stability, or stable point spectrum of the linearized operator about the wave, we establish sharp pointwise Green's function bounds and consequent linear and nonlinear stability for shock profiles of relaxation and real viscosity systems satisfying the dissipativity condition of Zeng/Kawashima. These include in particular compressible Navier–Stokes and MHD equations, and essentially all standard relaxation models: in particular, the discrete kinetic models of Broadwell, Jin-Xin, Natalini, Bouchut, Platkowski–Illner, and the moment closure models of Grad, Levermore, Müller–Ruggeri. A consequence is stability of small-amplitude profiles of Broadwell and Jin-Xin models and of general real viscosity systems, for each of which spectral stability has been verified in other works. These are the first complete stability results for profiles of a real viscosity system, and the first for relaxation models with nonscalar equilibrium equations [1]. Our results apply also in principle to large-amplitude shocks, an important direction for future investigation.

As pointed out by Zeng [21], the two problems (relaxation/real viscosity) are essentially dual in the linear constant–coefficient case. A similar duality, at the structural level, may be seen in the variable-coefficient case [23]. In the following we deal with the relaxation case.

Consider, then, a general reaction-convection system of form

$$U_t + F(U)_x = Q(U) \qquad U \in \mathbf{R}^N$$

with $\sigma(dF(U))$ real, semisimple with constant multiplicity; and assume that it has the *relaxation structure:*

$$\begin{cases} u_t + f(u,v)_x = 0, \\ v_t + g(u,v)_x = \tau^{-1} q(u,v), \end{cases} \tag{1}$$

where u, $f \in R^n$, v, g, $q \in \mathbf{R}^r$, and $\tau > 0$ is a (usually small) parameter determining relaxation time. We assume that the *equilibrium manifold* is smooth and of the form

$$\mathcal{E} = \{q(u,v) = 0\} = \{v = v^*(u)\} \tag{2}$$

[1] However, we note a previous HYP conference proceedings (Beijing) in which Szepessy and Xin sketch a quite different argument for stability of small-amplitude Broadwell shocks.

and that the *stability condition* $\operatorname{Re}\sigma\big(q_v(u,v^*(u))\big) < 0$ is satisfied. This condition means that, looking for space independent solutions, i.e. solving the system of o.d.e. $U_t = Q(U)$, all of the solutions converge to $\mathcal{E}$ as $t \to +\infty$. Such models occur, for example, non-thermal equilibrium gas dynamics, traffic dynamics, and multiphase flow.

System (1) can be viewed "to zeroth order" as a regularization of the associated *equilibrium*, or "relaxed" system of conservation laws

$$u_t + f^*(u)_x = 0, \qquad \text{where} \quad f^*(u) := f(u, v^*(u)), \tag{3}$$

which system may or may not be hyperbolic. To "first order," it may be approximated, at least formally, by an associated parabolic system of conservation laws

$$u_t + f^*(u)_x = \tau(B^*(u)u_x)_x, \tag{4}$$

where $B^*(u) = -f_v q_v^{-1}(g_u - g_v q_v^{-1} q_u + q_v^{-1} q_u(f_u - f_v q_v^{-1} q_u)$, with $g^* := g(u, v^*(u))$, is determined by Chapman–Enskog expansion as described in, e.g., [18, 8]. Our interest is in behavior of solutions for a *fixed* relaxation time $\tau = 1$, as time t goes to infinity, a roughly analogous situation as pointed out by T.-P. Liu [8].

A particularly interesting phenomenon, suggested by (3)–(4), is the existence of smooth *traveling front solutions*

$$(u,v)(x,t) = \Phi(x - st) = (\bar{u}, \bar{v})(x - st), \qquad \lim_{z\to\pm\infty} \Phi(z) = (u_\pm, v_\pm), \tag{5}$$

of (1), where by necessity $v_\pm = v^*(u_\pm)$ and $u_\pm$ corresponds to a shock solution of (3), i.e. $u_\pm \in \mathbf{R}^n$, $s \in \mathbf{R}$, are such that

$$u(x,t) = \begin{cases} u_- & x < st, \\ u_+ & x > st, \end{cases}$$

is a (laxian) *shock wave* for the reduced system $u_t + f^*(u)_x = 0$. See [8] for a treatment of existence in the general case $n = r = 1$, and [19, 1] for generalizations in case n or $r > 1$. Such traveling waves are known as *relaxation shocks* or *relaxation profiles.*

The question of their stability was investigated in [8] in the general case $n = r = 1$, for which the equilibrium system is scalar, under the assumption of "weak," or small-amplitude shocks. This analysis has been generalized to strong shocks in the special case of the 2×2 Jin–Xin model [11]. For the 3×3 Broadwell model, a standard discrete kinetic model for which $n = 2$, $r = 1$, Szepessy and Xin [15, 17] have announced the result of linear and nonlinear stability of weak shocks, extending partial results of [5]: to our knowledge, the only such result for the system case $n > 1$ occurring in most physical applications. However, up to now there was no complete analysis of stability for any other case, in particular for the case $n > 1$, $r >> 1$ arising through approximation of Boltzmann or Vlasov–Poisson equations by moment closure or discretization. For strong shocks, there was no complete analysis for *any* system such that n or $r > 1$.

Assume, without loss of generality, Φ stationary, i.e. $s = 0$. Due to translation invariance,

$$\Phi_\delta(x) := \Phi(x + \delta)$$

is a relaxation shock itself with same asymptotic states. Then the question is the following: given W_0 small, does there exist a global solution to $U_t + F(U)_x = Q(U)$, with initial datum $\Phi(x) + W_0(x)$? If this is the case,

$$\textit{orbital stability}: \qquad \lim_{t\to\infty} \text{dist}\big(U\,,\,\{\Phi(x+\delta)\,:\,\delta \in \mathbf{R}\}\big) = 0 \quad ?$$

Consider the linearized operator

$$W_t = \mathcal{L}W := -(dF(\Phi)\,W)_x + dQ(\Phi)\,W.$$

Again by translation invariance, $\mathcal{L}\Phi_x = 0$, so that $0 \in \sigma(\mathcal{L})$. Hence, at this level, we look for *linearized orbital stability*:

$$\lim_{t\to+\infty} \text{dist}(W, \text{span}\,\{\Phi_x\}) = 0.$$

A necessary condition for linearized orbital stability is *strong spectral stability*:

$$\sigma(\mathcal{L}) \cap \{\text{Re}\,\lambda \geq 0\} = \{0\}.$$

Our main result is to determine additional assumptions guaranteeing, together with strong spectral stability assumption, that the relaxation shock is asymptotically stable. The classical strategy for such a kind of results is in the spirit of the finite-dimensional o.d.e. property: if $\sigma(\mathcal{L}) \cap \{\text{Re}\,\lambda \geq -\alpha\} = \{0\}$, then $\text{dist}(W, \mathcal{V}) \leq Ce^{-\alpha t}$, where $\mathcal{V}$ is the eigenspace of the eigenvalue 0. So the main point is to take advantage from the gap between the eigenvalue 0 and the remaining part of the spectrum to prove exponentially fast convergence to the manifold of the wave. In our case, as in the problem of stability of viscous shock waves (see [22] and referencede therein), 0 is not an isolated eigenvalue, hence the above strategy is not applicable. Moreover, algebraic fast decay, rather than exponential, is expected.

Indeed, decompose the spectrum of $\mathcal{L}$ as $\sigma = \sigma_p \cup \sigma_{\text{ess}}$ where point spectrum σ_p is composed by isolated eigenvalues with finite multiplicity and essential spectrum $\sigma_{\text{ess}} = \sigma \setminus \sigma_p$. Then, the essential spectrum of $\mathcal{L}$ is determined by the bounded solutions of the asymptotics operators $\mathcal{L}_\pm W := -dF(\Phi_\pm)\,W_x + dQ(\Phi_\pm)\,W$. A straightforward calculation shows that $0 \in \sigma_{\text{ess}}$

In the following, let us assume that, for some $\theta > 0$,

$$\text{Re}\,\sigma\,(i\xi\,dF(\Phi_\pm) + dQ(\Phi_\pm)) \leq -\frac{\theta|\xi|^2}{1+|\xi|^2},$$

for any $\xi \in \mathbf{R}$. Then, if $X + iY \in \sigma_{\text{ess}}(\mathcal{L})$, it holds $X \leq -\frac{\theta Y^2}{1+Y^2}$, that gives us a complete control of the location of the essential spectrum (note that the operator $\mathcal{L}$ is not sectorial, hence it does not generate an analytic semigroup as in the viscous case, but a C^0 semigroup).

Example. Consider the 2×2 Jin-Xin system, under the subcharacteristic condition $\alpha = a/|h'| > 1$,

$$\begin{cases} u_t + v_x = 0 \\ v_t + a^2 u_x = h(u) - v. \end{cases} \qquad (a > 0)$$

$\lambda \in \sigma\left(i\xi\, dF(\Phi_\pm) + dQ(\Phi_\pm)\right)$ if $\lambda^2 + \lambda + a^2\xi^2 - ih'(u_\pm)\xi = 0$. Hence, setting $\lambda = X + iY$, define

$$X := F_\pm(Y) = \frac{1}{2}\left(-1 \pm \sqrt{\frac{1+4Y^2}{1+4\alpha^2Y^2}}\right)$$

Then, the essential spectrum $\sigma_{\text{ess}}(\mathcal{L})$ is contained in the region with right boundary given by the rightmost of the graphs $X = F_+(Y)$.

Our philosophy is to study stability under assumptions on point spectrum. To this aim, we introduce the Evans function $D(\lambda)$, which has the property that, on the resolvent set of $\mathcal{L}$, its zeroes agree (in both location and multiplicity) with the eigenvalues of $\mathcal{L}$. Then we assume:

[D] D have precisely one zero on $\text{Re}\,\lambda \geq 0$ (necessarily located at $\lambda = 0$).

Before stating the results, let us recall the assumptions:

[H1] $\sigma\left(dF(U)\right)$ real, $\neq s$, semisimple with constant multiplicity;
[H2] $\sigma\left(df^*(u_\pm)\right)$ real, distinct, $\neq s$;
[H3] $\exists\, \theta > 0$ such that, for any $\xi \in \mathbf{R}$,

$$\text{Re}\ \sigma\left(i\xi dF(\Phi_\pm) + dQ(\Phi_\pm)\right) \leq \frac{-\theta|\xi|^2}{1+|\xi|^2}$$

[D] D have precisely one zero on $\text{Re}\,\lambda \geq 0$.

Theorem 1. *(Linearized stability) Assume [H1–3]. Let W be the solution of*

$$W_t = \mathcal{L}W := -(dF(\Phi)\,W)_x + dQ(\Phi)\,W,$$

with initial condition $W(x,0) = W_0(x)$. Then condition [D] is equivalent to linear orbital stability $L^1 \cap L^p \to L^p$. More precisely, condition [D] implies

$$dist_{L^p}\left(W, span\,\{\Phi'\}\right) \leq \frac{C}{(1+t)^{\alpha_p}} |W_0|_{L^1 \cap L^p},$$

where $\alpha_p = \frac{1}{2}(1 - \frac{1}{p})$.

Note that the value of the rate of decay α_p is the same of 1-D heat kernel, consistent with the presence of diffusion waves, whose asymptotical presence is consistent with Chapman-Enskog expansion.

Our main result concerns the nonlinear case and is as follows. A similar result is established for viscous profiles of systems with real viscosity.

Theorem 2. *(Nonlinear stability) Assume [H1-3] and [D]. Let U be the solution of*

$$U_t + F(U)_x = Q(U), \qquad U(x,0) = \Phi(x) + W_0(x).$$

If, in addition, the system (1) is simultaneously symmetrizable, Φ is small (i.e. $|u_+ - u_-|$ small) and $||W_0||_{L^1\cap H^2} \leq \zeta_0$ sufficiently small, then for any $p \geq 2$ and $\alpha_p = \frac{1}{2}(1 - \frac{1}{p})$

$$|U - \Phi(\cdot - \delta(t))|_{L^p} \leq C\zeta_0(1+t)^{-\alpha_p}, \tag{6}$$

with $\delta(t)$ satisfying

$$|\dot{\delta}(t)| \leq C\zeta_0(1+t)^{-\frac{1}{2}} \quad e \quad |\delta(t)| \leq C\zeta_0.$$

In particular, Φ is orbitally stable $L^1 \cap H^2 \to L^p$.

For discrete kinetic models, not necessarily simultaneously symmetrizable, and data small in $W^{3,1} \cap W^{3,\infty}$, we obtain the rates (6) for all $1 \leq p \leq \infty$, for shocks of arbitrary strength.

This result is obtained using a modified version of the spectral arguments introduced in [22, 24] to treat the strictly parabolic case. Namely, we first obtain detailed pointwise bounds on the Green's function of the linearized perturbation equation by a detailed asymptotic analysis of the resolvent equation combined with stationary phase type estimates in the spectral resolution (Laplace inversion) formula describing the Green's function. However, substantial new complications arise from the need to estimate (at the linearized level), and control (at the nonlinear level) the contribution of the singular component of the Green's function, corresponding to hyperbolic nature of the high frequency ($\sim$ short time) behavior of equation (1).

Theorem 2 in particular shows that *strong spectral stability implies nonlinear stability*, for weak shocks of general simultaneously symmetrizable models, or for strong shocks of discrete kinetic models. As noted above, strong spectral stability follows for weak relaxation profiles of the Jin–Xin and Broadwell models from the results of [5] and [6, 2], respectively, thus yielding complete nonlinear stability results for these two cases. A very interesting open problem is spectral stability of weak shocks of general dissipative relaxation systems, as identified by Zeng/Kawashima [4, 21], or, likewise, spectral stability of strong shocks of special discrete kinetic models.

References

1. Freistühler, H., Zeng Y. (1998): Shock profiles for systems of balance laws with relaxation, preprint.
2. Humpherys, J. (2000): Stability of Jin–Xin relaxation shocks, preprint.
3. Jin, S., Xin Z. (1995): The relaxation schemes for systems of conservation laws in arbitrary space dimensions, Comm. Pure Appl. Math. **48**, no. 3, 235–276.
4. Kawashima S. (1983): Systems of a hyperbolic–parabolic composite type, with applications to the equations of magnetohydrodynamics, thesis, Kyoto University.
5. Kawashima, S., Matsumura A. (1985): *Asymptotic stability of traveling wave solutions of systems for one-dimensional gas motion,* Comm. Math. Phys. **101**, no. 1, 97–127.
6. Liu, H. (2000): Asymptotic stability of relaxation shock profiles for hyperbolic conservation laws, preprint.
7. Liu, T.-P. (1997): Pointwise convergence to shock waves for viscous conservation laws, Comm. Pure Appl. Math. **50**, no. 11, 1113–1182.
8. Liu, T.-P. (1987): Hyperbolic conservation laws with relaxation, Comm. Math. Phys. **108**, no. 1, 153–175.
9. Liu, T.-P., Zeng, Y. (1997): Large time behavior of solutions for general quasilinear hyperbolic–parabolic systems of conservation laws, AMS memoirs **599**.
10. Lighthill, M.J., Whitham, G.B. (1955): On kinematic waves. II. A theory of traffic flow on long crowded roads, Proc. Roy. Soc. London. Ser. A. **229**, 317–345.
11. Mascia, C., Natalini, R. (1996): L^1 nonlinear stability of traveling waves for a hyperbolic system with relaxation, J. Differential Equations **132**, no. 2, 275–292.
12. Matsumura, A., Nishihara, K. (1985): On the stability of travelling wave solutions of a one-dimensional model system for compressible viscous gas, Japan J. Appl. Math. **2**, no. 1, 17–25.
13. Müller, I., Ruggeri, T. (1998) Rational extended thermodynamics, Second edition, with supplementary chapters by H. Struchtrup and Wolf Weiss, Springer Tracts in Natural Philosophy, **37**, Springer-Verlag, New York.
14. Platkowski, T., Illner, R. (1988): Discrete velocity models of the Boltzmann equation: a survey on the mathematical aspects of the theory, SIAM Review **30**, no. 2, 213–255.
15. Szepessy, A. (1997): On the stability of Broadwell shocks, Nonlinear evolutionary partial differential equations (Beijing, 1993), 403–412, AMS/IP Stud. Adv. Math., **3**, Amer. Math. Soc., Providence, RI.
16. Szepessy, A., Xin, Z. (1993): Nonlinear stability of viscous shock waves, Arch. Rat. Mech. Anal. **122**, 53–103.
17. Szepessy, A., Xin. Z.: Stability of Broadwell shocks, unpublished manuscript.
18. Whitham, G.B. (1999): Linear and nonlinear waves, Reprint of the 1974 original, Pure and Applied Mathematics, A Wiley-Interscience Publication, John Wiley & Sons, Inc., New York.
19. Yong W.-A., Zumbrun, K. (2000): Existence of relaxation shock profiles for hyperbolic conservation laws, SIAM J. Appl. Math. **60**, no. 5, 1565–1575.
20. Zeng, Y. (1994): L^1 asymptotic behavior of compressible, isentropic, viscous 1-d flow, Comm. Pure Appl. Math. **47**, 1053–1092.

21. Zeng, Y. (1999): Gas dynamics in thermal nonequilibrium and general hyperbolic systems with relaxation, Arch. Ration. Mech. Anal. **150**, no. 3, 225–279.
22. Zumbrun, K., Howard, P. (1998): Pointwise semigroup methods and stability of viscous shock waves, Indiana Mathematics Journal **47**, no. 4, 741–871.
23. Zumbrun, K. (2001): Multidimensional stability of planar viscous shock waves. In: Advances in the theory of shock waves, Progr. Nonlinear Differential Equations Appl., **47**, Birkhäuser Boston, Boston, MA, 2001.
24. Zumbrun, K. (2000): Refined Wave–tracking and Nonlinear Stability of Viscous Lax Shocks, Methods Appl. Anal. **7**, no. 4, 747–768.

Simulation of Cavitation in Thermodynamic Equilibrium

A. Meister[1], A. Berg[2] and U. Iben[2]

[1] Institute of Mathematics, Medical University of Lübeck, Wallstraße 40, D-23560 Lübeck, Germany, *meister@math.mu-luebeck.de*

[2] Robert Bosch GmbH, Dept. FV/FLM, P.Box 106050, D-70059 Stuttgart, Germany, *Anselm.Berg@de.bosch.com, Uwe.Iben@de.bosch.com*

Summary. A method for the numerical simulation of cavitation phenomena is described. Thereby, a homogeneous cavitation model based on a thermodynamic equilibrium model of liquid and steam is considered. Besides a brief introduction into the mathematical model described in [1] we focus on the discretization technique. The ability of the scheme is proven by numerical experiments.

1 Introduction

Cavitation denotes local evaporation and condensation processes. Two causes for evaporation are external heating which increases the evaporation pressure and a decreasing pressure generated by a fast closing valve or a local acceleration of the flow field. The necessity to simulate cavitating flows arises from the demand to prevent damages and to understand the dynamic properties of the flow field in the early development phase of hydraulic tools.

Many papers deal with this topic, see for example [8], [11]. In general, a combination of a continuous and a discrete model is used. The flow is described by the Euler or the Navier-Stokes equations, representing the continuous model, whereas the cavitation is modeled by discrete bubble models where oscillations of single bubbles or bubble clouds are considered, cf. [3], [11], [13]. These models include many unknown physical values and properties, such as the bubble diameter, the bubble number density, and the extreme complicated bubble interactions. Due to this reason, these models are difficult to handle for real industrial applications. Moreover, the experimental and numerical study of single phenomena using these models are extremely time consuming. Recently in [12] single bubble phenomena were investigated. For bubble clouds and interaction of bubbles, experimental studies were published in [6].

A first approach to model flow with cavity formation could be to start with the conservation equations of the homogeneous mixture as done in [8] or [4]. In this case the evaporation and condensation process is described by the equation of state, i.e. $\rho = \rho(p, T = \text{fixed})$. This approach can lead to stability problems due to the large range of the mixture density and propagation speed. In [4] a combination of an explicit and locally implicit scheme is successfully used to avoid these stability problems. As a next step one can split the continuity equation into one for the gaseous phase and one for the liquid phase. Thereby, both phases are coupled by a source term determining the phase transition. This strategy has recently been realized by [16] for isotropic water flow which means that the energy equation is neglected analogously to the ansatz by [8] and [4]. In order to avoid this assumption an extension of the model described in [16] is derived in [1]. Thereby, an additional energy equation is introduced for the homogeneous mixture in connection with a thermodynamic equilibrium model for the phase transition.

In the present paper, a numerical method based on the model derived in [1] is presented. We start with a brief introduction into the governing equations. Thereafter, a finite difference method using an operator splitting technique is described. The ability of the method to simulate cavity flow is shown by two examples relevant in technical applications.

Throughout the paper we shall make use of the following notation: The index G denotes the gaseous and steam phase, the index L denotes the liquid phase. Physical values denoting a mixture are written without an index.

2 Governing Equations

As the velocities of the fluid v_{L} and gaseous/steam v_{G} in a pipe are approximately the same, a homogeneous mixture is considered, i.e. $v_{\mathrm{L}} = v_{\mathrm{G}} = v$. This assumption is sensible for acoustic cavitation and low speeds of the mixture. The cavitation provoked by pressure waves appears as small bubbles or steam clouds. In order to apply a numerical scheme we assume the steam to be equally distributed within the computational cells. It is worth mentioning that the wave motion in a pipe can not lead to a complete evaporation. In order to reach this physical state an additional heat flux to the mixture is necessary. For this reason, the mass fraction as well as the volume fraction of the steam and unsolved air (cf. Section 3) are always strictly smaller than one. The mass fraction μ and the volume fraction ε are defined by

$$\mu = \frac{m_{\mathrm{G}}}{m} \quad \text{and} \quad \varepsilon = \frac{V_{\mathrm{G}}}{V} = \mu \frac{\rho}{\rho_{\mathrm{G}}}, \tag{1}$$

where m and V denote the mass and the volume of a computational cell, respectively. Therefore the mixture density ρ is given by

$$\rho = \varepsilon \rho_{\mathrm{G}} + (1-\varepsilon)\rho_{\mathrm{L}} \quad \overset{(1)}{\Leftrightarrow} \quad \frac{1}{\rho} = \frac{\mu}{\rho_{\mathrm{G}}} + \frac{1-\mu}{\rho_{\mathrm{L}}}. \tag{2}$$

By following the described approach of splitting the continuity equation and coupling the resulting equations by a source term, but leaving the momentum and energy equation of the mixture unchanged we get the two phase Euler equations in one space dimension. For a more detailed derivation of the system confer [10], [16]:

$$\frac{\partial}{\partial t}\mathbf{u} + \frac{\partial}{\partial z}\mathbf{f}(\mathbf{u}) = \mathbf{g}(\mathbf{u}), \tag{3}$$

where $\mathbf{u} = \left(\varepsilon\rho_{\mathrm{G}}, (1-\varepsilon)\rho_{\mathrm{L}}, \rho v, \rho(e+\frac{v^2}{2})\right)^T$ denotes the vector of unknowns, $\mathbf{f}(\mathbf{u}) = \left(\varepsilon\rho_{\mathrm{G}}v, (1-\varepsilon)\rho_{\mathrm{L}}v, \rho v^2 + p, \rho v(e+\frac{v^2}{2}+\frac{p}{\rho})\right)^T$ represents the convective flux function and $\mathbf{q}(\mathbf{u}) = \left(\rho\frac{d\mu}{dt}, -\rho\frac{d\mu}{dt}, 0, 0\right)^T$ describes the source term due to evaporation and condensation. The inner energy e and the enthalpy h are connected by the usual definition $h = e + \frac{p}{\rho}$. A comprehensive derivation of the source term based on the relation (1) between the voulme fraction and the mass fraction can be found in [1].

The material derivative within this expression emphasizes that evaporation ($\rho\frac{d\mu}{dt} < 0$) and condensation ($\rho\frac{d\mu}{dt} > 0$) is strongly coupled to the fluid elements and take place on fluid pathes. The four eigenvalues of the Jacobian of the flux vector $\mathbf{f}$ are:

$$\lambda_1 = v - c, \quad \lambda_2 = \lambda_3 = v, \quad \text{and} \quad \lambda_4 = v + c, \tag{4}$$

where c is the speed of sound of the homogeneous mixture. The eigenvalues λ_1 and λ_4 represent the characteristics of the system (3), while the eigenvalues $\lambda_2 = \lambda_3$ represent the path. Following the usual definition of the speed of sound as the derivative of the pressure by the density along an isentrope one can employ (3) to compute the dependence of the speed of sound in the two phase region on the speeds of sound of the liquid and gaseous phase; in particular one will find the expression (see [15])

$$\frac{1}{\rho c^2} = \frac{\varepsilon}{\rho_{\mathrm{G}} c_{\mathrm{G}}^2} + \frac{(1-\varepsilon)}{\rho_{\mathrm{L}} c_{\mathrm{L}}^2}. \tag{5}$$

3 Thermodynamic model for the evaporation

For a comprehensive description of the fluid flow the governing equations (3) need to be closed. Therefore one of the unknowns must be modeled and expressed in other physical quantities. The requirement is usually fullfilled by introducing an equation of state representing the properties of the liquid or gas considered. In our particular situation of a water-gas mixture we have to define two equations of state as well as a model for the phase transition.

Following [1] we use a modified ideal gas law for the steam phase, i.e. $p = \rho_{\mathrm{G}} R_{H_2O} T$. For the liquid phase, the algorithms $\rho_{\mathrm{L}} = \rho_{\mathrm{L}}(p,T)$ and $e_{\mathrm{L}} = e_{\mathrm{L}}(p,T)$, derived in [14], describing the properties of water are applied. These functions can also be used to compute the mass fraction of steam for a given pressure $p_{tr} < p \leq p_{\mathrm{evap}}(T)$ and temperature T with p_{tr} denoting the pressure at the triple point of the fluid, cf. [2]. For $p_{tr} < p \leq p_{evap}(T)$ the density of the fluid is fixed by a constant due to the density function of Wagner et.al. representing the mixture density ρ for this range.

As mentioned we restrict our attention to a thermodynamic equilibrium model for the evaporation. This model does not contain any additional parameter which would be difficult to obtain such as the diameter of steam bubbles or the number of nucleis. In the presented context the fluid reaches the vapor pressure $p_{\text{evap}}(T)$ due to a pressure loss or a rarefaction wave and starts to evaporate, cf. [5] or [10]. To calculate the mass fraction μ_e of evaporated liquid we employ

$$\mu_e(p) = \frac{h(p_{evap}(T)) - h_{\mathrm{G}}(p)}{h_{\mathrm{L}}(p) - h_{\mathrm{G}}(p)} \quad \text{for } p_{tr} < p \leq p_{evap}(T). \tag{6}$$

This formular can be derived by assuming that evaporation takes place on pathes of constant enthalpy (see [1]), which represents a suitable approximation in the field of tiny mass fraction as considered in our study. The presented model (6) of thermodynamic equilibrium can be used for studies of cavitation and basic phenomena of two phase flow which will be described in Section 5. Note, that the steam pressure p_{evap} depends on the temperature T as shown in Figure 1 for the medium of water.

Every fluid contains unsolved and solved air. Speaking about cavitation we need to distinguish between gas-cavitation and "real" cavitation. The so-called gas cavitation constitutes a diffusion process. Simplifying the model we assume a constant mass fraction of unsolved air μ_{air} to be homogeneously distributed in the whole computational domain. Therefore the total mass fraction of the gaseous phase, i.e. the steam and unsolved air, is given by $\mu = \mu_e + \mu_{\text{air}}$ with $\mu_{\text{air}} = 10^{-8}$.

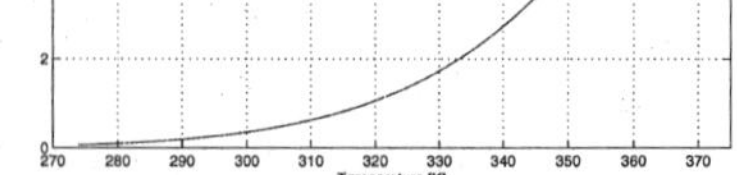

Fig. 1. Steam pressure for water accoring to [14]

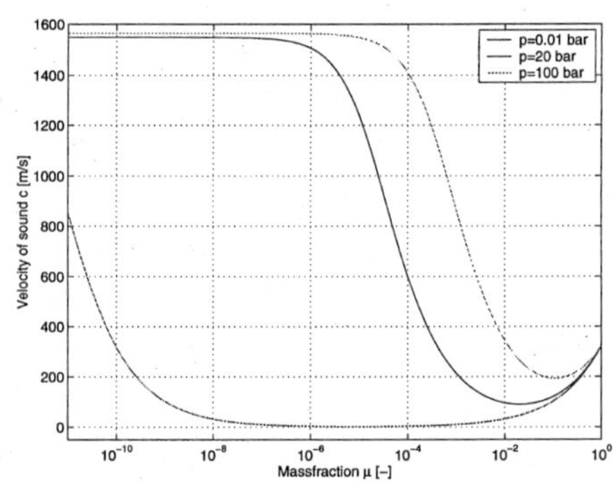

Fig. 2. Dependence of the speed of sound c on the massfraction μ for constant pressure p and temperature T.

Recalling equations (5) and (1) we need to point out the strong dependence of the speed of sound on the value of the mass fraction of steam and unsolved air in the two phase region (cf. Figure 2) which leads to the fact, that the wave motion appears to be a discontinuous and non-linear process. In particular for small pressure values the decrease becomes apparent, cf. [15].

4 Numerical Method

With respect to the numerical approximation the spatial part of the computational domain $[0, 0.1] \times [0, t_{\text{end}}]$ is divided equidistantly, i.e. $z_i = i\Delta z = i\frac{0.1}{N}$. Regarding the time step technique, we write $t^{n+1} = t^n + \Delta t^n$, $t^0 = 0$ and apply the formula

$$\Delta t^n = \lambda_{CFL} \frac{\Delta z}{\max_{k,i}(|\ \lambda^n_{k,i}\ |)}, \qquad k \in \{1, 2, 3, 4\},\ i \in \{1, 2, \ldots, N\}, \tag{7}$$

where the variable λ_{CFL} denotes the well-known CFL-number, chosen suitably with $0 < \lambda_{CFL} < 1$. A CFL-number close to one again requires an accurate estimation of the propagation speed.

To solve (3) numerically, the first order HLL-scheme in connection with operator splitting is used. We introduce the notation $\mathbf{u}^n_i = \left(u^n_{1,i}, u^n_{2,i}, u^n_{3,i}, u^n_{4,i}\right)^T$ with $u^n_{k,i} = u_k(z_i, t^n)$, $k = 1, 2, 3, 4$. Within the first step of the operator splitting technique we neglect the source term. Introducing a simple forward Euler time discretization, we obtain

$$\mathbf{u}^*_i = \mathbf{u}^n_i + \frac{\Delta t^n}{\Delta z} \left(\mathbf{F}^{\text{HLL}}(\mathbf{u}^n_{i+1}, \mathbf{u}^n_i) - \mathbf{F}^{\text{HLL}}(\mathbf{u}^n_i, \mathbf{u}^n_{i-1})\right). \tag{8}$$

Thereby, the numerical flux function of the HLL-scheme writes

$$\mathbf{F}^{\text{HLL}}(\mathbf{u}_{i+1}, \mathbf{u}_i) = \frac{a_{\text{R},i+1}\mathbf{f}(\mathbf{u}_i) - a_{\text{L},i}\mathbf{f}(\mathbf{u}_{i+1})}{a_{\text{R},i+1} - a_{\text{L},i}} + \frac{a_{\text{L},i}a_{\text{R},i+1}(\mathbf{u}_{i+1} - \mathbf{u}_i)}{a_{\text{R},i+1} - a_{\text{L},i}}. \tag{9}$$

In this context, $a^n_{\text{R},i}$ and $a^n_{\text{L},i}$ denote the largest and the smallest propagation speed in the computational cell i respectively.

Obviously, we get the numerical flux function of the Lax-Friedrichs method if the grid speed $\frac{\Delta z}{\Delta t^n}$ is used instead of $a^n_{\text{R},i+1}$ and $a^n_{\text{L},i}$. Exactly this numerical propagation speed is employed in the case that condensation takes place which corresponds to a decreasing mass fraction. The reason for this strategy is associated with the behaviour of the speed of sound in dependence on the mass fraction. In the case of a tiny mass fraction we recognize that a decreasing mass fraction may cause a steep increasing of the speed of sound (cf. Figure 2). Thus, the disspation in the HLL-scheme may be too small and consequently oscillations within the numerical approximation may occur as reported by [4] and [16]. We easily get rid of this drawback by using the Lax-Friedrichs scheme locally at each grid point with a negative source Γ. This technique is often used for processes with highly variable propagation speeds, cf. [9], [7].

Whereas equation (8) yields a first order approximation of the momentum and total energy per unit volume at the time level t^{n+1} we obtain usually unphysical quantities with respect to the density parts. Hence, we introduce the auxiliary variable $\mathbf{u}^*$ into the right hand side of the correction step

$$\mathbf{u}_i^{n+1} = \mathbf{u}_i^* + \Delta t^n \mathbf{Q}(\mathbf{u}_{i+1}^*, \mathbf{u}_i^*, \mathbf{u}_{i-1}^*). \tag{10}$$

In advance to the derivation of the numerical approximation of the phase transition it is sensible to rewrite the source term in a more convenient manner with respect to the numerical method. Using the expression

$$h = \mu h_{\mathrm{G}} + (1-\mu)h_{\mathrm{L}} \tag{11}$$

in combination with a vanishing material derivative of the enthalpy we obtain

$$0 = \frac{dh_{\mathrm{L}}}{dt} + \mu\left(\frac{dh_{\mathrm{G}}}{dt} + \frac{dh_{\mathrm{L}}}{dt}\right) + (h_{\mathrm{G}} - h_{\mathrm{L}})\frac{d\mu}{dt}. \tag{12}$$

Due to the fact that the mass fraction is always below 10^{-2} we nelect in (12) the term multiplied by the mass fraction. Consequently, the source term can be modeled as

$$\rho\frac{d\mu}{dt} = -\frac{\rho}{h_{\mathrm{G}} - h_{\mathrm{L}}}\frac{dh_{\mathrm{L}}}{dt} = -\frac{\rho}{h_{\mathrm{G}} - h_{\mathrm{L}}}\frac{dh_{\mathrm{L}}}{dp}\frac{dp}{dt}. \tag{13}$$

The source term exclusively effects the flow field if the mass fraction of evaporated liquid μ_e changes. A necessary condition for this is $p_i^* < p_{\mathrm{evap}}(T_i^*)$. Thus we define the numerical source term due to the phase transition in the form

$$\mathbf{Q}(\mathbf{u}_{i+1}, \mathbf{u}_i, \mathbf{u}_{i-1}) = \begin{cases} 0, & \text{if} \quad p_i \geq p_{\mathrm{evap}}(T_i) \\ \Gamma, & \text{otherwise,} \end{cases}$$

$$\text{with} \quad \Gamma = \rho_i \cdot \frac{1}{h_{\mathrm{L}}(p_i) - h_{\mathrm{G}}(p_i)} \cdot \frac{h_{\mathrm{G}}(p_i) - h_{\mathrm{G}}(p_i^n)}{p_i - p_i^n} \cdot \left(\frac{p_i - p_i^n}{\Delta t^n} + v_i \frac{p_{i+1} - p_{i-1}}{2\Delta z}\right).$$

Following the discussion of the incorporated equation of state, we note that the pressure p is a function of the density and the temperature which is given implicitly only. This reveals a main difference to the modeling of problems in gas dynamics. Nevertheless, it is obvious that both the evaluation of the numerical source term $\mathbf{Q}$ and the visualization of the time dependent evolution of different physical quantities, like pressure p, mass fraction μ, temperature T, requires the calculation of these values from the conserved variables computed within the numerical scheme. While the computation of the density ρ, velocity v and inner energy e is straightforward it is much more complicated to ascertain the pressure p and temperature T. Using the functional relationships $(1-\mu)\rho = (1-\varepsilon)\rho_{\mathrm{L}}$, $p = \rho_{\mathrm{G}} R_{H_2O} T$, $\varepsilon\rho_{\mathrm{G}} = \mu\rho$, $h_{\mathrm{G}} = c_V T$ and equation (11) one can derive the system of non-linear equations

$$\mathbf{r}(p,T) = \begin{pmatrix} r_1(p,T) \\ r_2(p,T) \end{pmatrix} = \begin{pmatrix} (1-\mu)\rho - \left(1 - \frac{\mu\rho R_{H_2O}T}{p}\right)\rho_{\mathrm{L}} \\ \left(e + \frac{p}{\rho}\right) - \mu c_V T - (1-\mu)h_{\mathrm{L}} \end{pmatrix} = \begin{pmatrix} 0 \\ 0 \end{pmatrix} \tag{14}$$

which determines the pressure and temperature by using $e = \frac{u_4}{u_1+u_2} - \frac{1}{2}\left(\frac{u_3}{u_1+u_2}\right)^2$ and $\rho = u_1 + u_2$. This system has to be solved at each grid point in advance to the evaluation of the source term as well as after each time step to achieve the physical quantities with respect to their visualization. Often, a Newton-type scheme is employed to solve such non-linear systems of equations. Unfortunately, a variety of numerical experiments shows that a Newton-type method always starts to diverge during the first two or three steps. This behaviour is probably justified by the course of the function r_1, where the gradient with respect to the pressure and temperature is almost everywhere vanishing or at least very small. Moreover, the isoline associated with a vanishing funtion r_2 is for small μ almost parallel to the pressure axis showing that the solution of the equation $r_2(p,T) = 0$ is mainly a one-dimensional problem of the temperature. Based on this knowledge we decided to solve the system (14) by means of two one-dimensional bisection approaches. The whole numerical method can now be written schematically in the form

- Given the initial condition $\mathbf{u}^0 = (\mathbf{u}_0^0, \ldots, \mathbf{u}_N^0)^T$.
- Set $t^0 = 0$ and $n = 0$.
- While $t^n < t_{\text{end}}$
  - Compute $t^{n+1} = t^n + \Delta t^n$ by means of (7).
  - Compute $\mathbf{u}^*$ using the predictor step (8).
  - Solve the non-linear system (14) to obtain p^* and T^*.
  - Apply the correction step (10) to compute $\mathbf{u}^{n+1}$
  - Solve the non-linear system (14) to obtain p^{n+1} and T^{n+1}.

This diagram proves that the fast solution of the non-linear system is important with respect to the performance of the whole numerical method. Although, this procedure converges reliable it is highly time consuming compared to other parts of the algorithm. Therefore, the improvement of this solution technique is one of the main tasks of the future.

5 Test cases

The influence of the fast closing and opening valves on the fluid is an important process to be studied in the context of cavitation in pipe flows in injection systems and hydraulic pipelines.

For the numerical simulations we investigate the inviscid flow in a pipe. Thereby we consider both, a constant cross sectional area $A = 10^{-6}\,\mathrm{m}^2$ and a constant length $l = 0.1\,\mathrm{m}$ of the pipe. These values are typical for a pipe of an injection system. In all calculations we assume the pipe to be completely filled with a gas-liquid mixture, which is defined by the properties of the water-gas mixture. The CFL number is fixed to 0.9 and $N = 100$ for both simulations. The required accuracy with respect to the solution of the non-linear system was always set to be $|r_i(p,T)| \leq 10^{-11}$.

The first example deals with open ends of the pipe at the beginning of the simulation, while the right side is closed after 10^{-5} s. Obviously, closing the outflow part will cause an icrease of the pressure such that no evaporation will take place. A sufficient accurate simulation of this so-called Joukowski-shock is important since the physical quantities influence the flow field during the opening phase significantly. A further advantage of this configuration is the existence of an analytic value for the pressure rise. Concerning the initial conditions $v = 1\text{m/s}$, $T = 313\text{K}$ and $p = 10^5\text{Pa}$ we obtain $\Delta_{\text{ana}}p = \rho v c = 1.519 \cdot 10^6\text{Pa}$, which corresponds well with the numerical result $\Delta_{\text{num}}p = 1.511 \cdot 10^6\text{Pa}$.

The second test case differs from the first one only by the fact that the inflow valve closes instead of the outflow one. Consequently, due to the inertia of the fluid a rarefaction wave appears, the pressure may drop under the steam pressure and the fluid starts to evaporate. The aim of this configuration is to investigate the wave motion and the physical relevant parameters for different initial temperatures which lead to different values of the evaporation pressure. Therefore, the temperature is varied between 293K and 353K such that the evaporation pressure grows from 2339Pa up to 47415Pa. Thus, the time of evaporation depends significantly on the initial temperature. Thereby, we would like to answer the questions:

- Is there a simple conjunction between the volume fraction ε and the mass fraction μ?
- How does the pressure and velocity depend on the initial temperature?
- Is the area of cavitation strongly influenced by the initial temperature?

At first, we consider an initial temperature of 293K. The velocity distribution show the expansion fan, see Figure 3. The pressure drops under the steam pressure (Figure 4) and consequently, the fluid starts to evaporate in the vicinity of the valve. Thereby, the maximum value of mass and the volume fraction are approximately $3 \cdot 10^{-7}$ and 10^{-2}, respectively, see Figures 5 and 6. Increasing the initial temperature leads to similar results. The area of cavitation is mainly constant and the mass fraction increases while the volume fraction decreases. This phenomena can be observed for all initial distributions considered. Thus, we can state that the study does not show a simple conjunction between the behaviour of the volume and the mass fraction. This effect has a great meaning for experimental investigations since the volume fraction is the physical quantity visible in experiments and one can conclude on the behaviour of the mass fraction only by the knowledge of the temperature and the pressure.

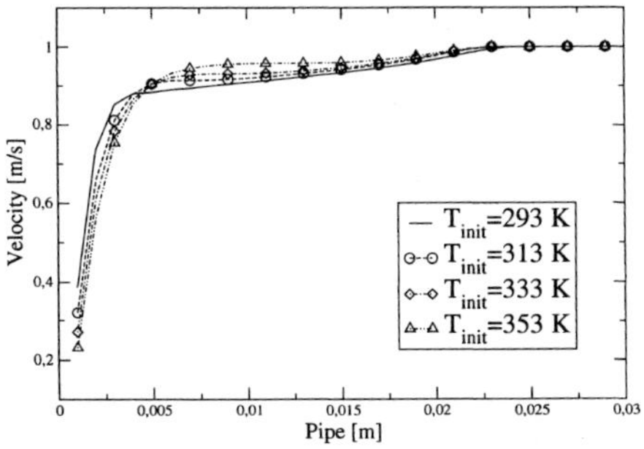

Fig. 3. Velocity distribution for varying initial temperatures

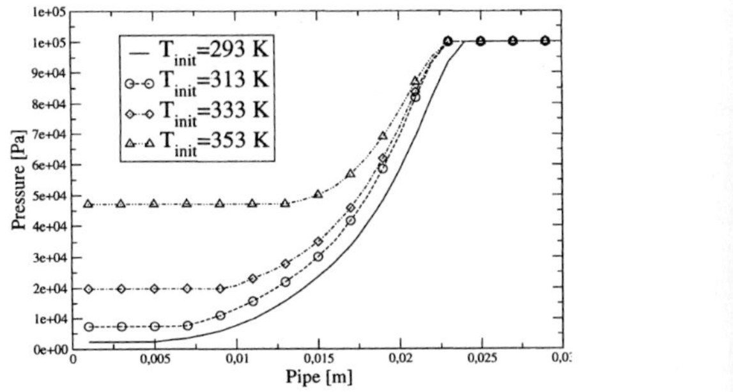

Fig. 4. Pressure distribution for varying initial temperatures

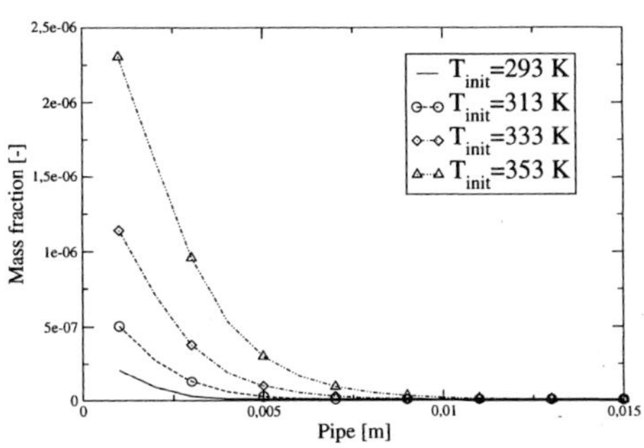

Fig. 5. Distribution of the mass fraction for varying initial temperatures

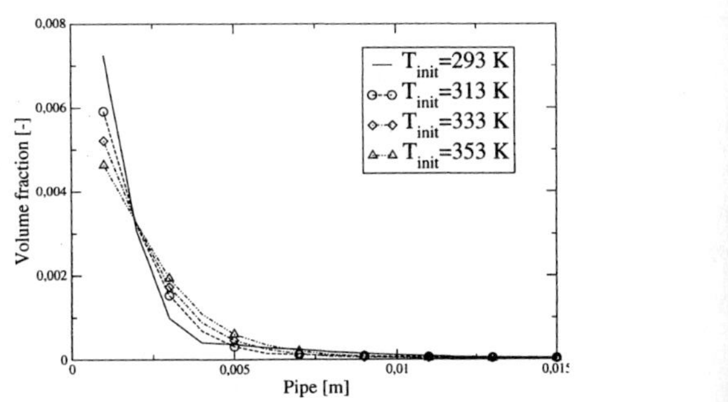

Fig. 6. Distribution of the volume fraction for varying initial temperatures

Conclusion

In this paper, a homogeneous cavitation model on the basis of mechanical and thermodynamic equilibrium derived in [1] is briefly introduced. Thereby a complete system of conservation equations consisting of one mass balance equation for each phase coupled by a source term, the momentum equation and the energy equation for the mixture were used. To close this hyperbolic system of equations, the thermodynamic and caloric equation of state for water and steam were employed. The system of equations is solved by a finite difference method using a modified HLL-scheme in combination with an operator splitting technique which incorporate the source term due to evaporation and condensation in a proper manner.

Finally two test cases marking different steps towards the simulation of pipe flow with fast opening and closing valves have been presented. Both show a good match with the expected results.

References

1. A. Berg, U. Iben, A. Meister, and J. Schmidt. Simulation of Cavitation in Hydraulic Pipelines based on the Thermodynamic and Caloric Properties of Liquid and Steam. *Preprint, Hamburger Beiträge zur Angewandte Mathematik, Reihe F, (16)*, 2002.
2. H.D. Baehr and K. Stephan. *Wärme- und Stoffübergang.* Springer-Verlag, 1994.
3. T. Baur. *Auswirkungen der Diffusion gelöster Gase auf das Wachstum von Kavitationsblasen*, 1997. Diplomarbeit am Institut für Wasserbau und Wasserwirtschaft.
4. M. Beck, U. Iben, N. Mittwollen, H.-K. Iben, and C.-D. Munz. On solution of conservation equations in cavitated hydraulic pipelines. In *3rd International Symposium on Computational Technologies for Fluid/Thermal/Chemical Systems with Industrial Applications.* 2001 ASME Pressure Vessels and Piping Conference July 22-26, Hyatt Regency, Atlanta, Georgia, USA, 2001.
5. A. Berg. *Ein Beitrag zur numerischen Simulation von Kavitation in ein und zwei Raumdimensionen*, 2001. Diplomathesis, Universität Hamburg.
6. C. Brückner. Structure and dynamics of the wake bubbles and its relevance for bubble interaction. *Physics of fluid*, 11(7), 1999.
7. B. Einfeldt, C.-D. Munz, P.L. Roe, and B. Sjögreen. On Godunov-type methods near low densities. *J. Comput. Phys.*, 92:273–295, 1991.
8. T. Hapke and H.-K. Iben. Modellierung und Berechnung der Kavitation in instationären Leitungsströmungen. *MTZ*, 59, 1998.
9. A. Harten, P. Lax, and B. van Leer. On upstream differencing and Godunov-type schemes for hyperbolic conservation laws. *SIAM Rev.*, 25:35–62, 1983.
10. U. Iben. *Entwicklung und Untersuchung von Kavitationsmodellen im Zusammenhang mit transienten Leitungsströmungen*, 2002. in preparation.
11. H. Nguyen-Schaefer and H. Sprafke. Numerical study on interaction effects of bubbles induced by air-release and cavitation in hydraulic systems. In *Tenth Bath International Fluid Power Workshop.* Research Studies Press, John Wiley and Sons, 1998.
12. C.-D. Ohl and E. Ory. Aspherical bubble collaps - comparison with simulations. In W. Lauterborn and T. Kurz, editors, *Nonlinear Acoustics at the Turn of the Millennium*, pages 393–396. American Institute of Physics, 2000.
13. G. Regner. *Blasendynamisches Kavitationsmodell für eine eindimensionale Strömungssimulation.* PhD thesis, Technische Universität Graz, 1999.
14. W. Wagner, S. Span, and T. Bonsen. *Wasser und Wasserdampf.* Springer Verlag, 1999.
15. W. Wallis. *One dimensional two-phase flow.* McGraw Hill, 1975.
16. U. Iben, F. Wrona, C.-D. Munz and M. Beck. *Cavitation in hydraulic tools based on thermodynamic properties of liquid and gas.* accepted in Multiphase Flows, ASME, 2002.

Semi-discrete Finite Element Method for a Class of Visco-elastic Problems with Long Memory under Condition of Friction

Pradeepa Nair and Amiya K. Pani

Department of Mathematics, Indian Institute of Technology Bombay, Powai-Mumbai, 400 076. India.
pradeepa@math.iitb.ac.in
akp@math.iitb.ac.in

1 Introduction

Viscoelasticity means that the material under consideration exhibits both creep and relaxation phenomena, i.e., for a fixed stress, the strain in a sample grows in time, whereas for a fixed strain, the stress decreases monotonously. Visco-elastic materials have properties between those of purely elastic materials and viscous fluids. A family of such materials are obtained by allowing the dependence of stress on the current values of both the strain and the velocity gradient. Such materials have memory in the sense that the stress depends not only on the present values of the strain and /or velocity gradient, but on the entire temporal history of the motion. Typically the memory fades with time, i.e., disturbances which occurred in the distant past have less influence on the present stress than those which occurred in the more recent past. In addition, the visco-elastic materials are called linear, if the constituent law establishes a linear relation between the stress tensor σ and the strain tensor ϵ.

When the state of stress at an instant t depends on the deformation at that instant of time t and also on all times preceding it, then we say that the material has a long memory. On the other hand, when the state of stress at an instant t depends only on the deformation at the instant t and the immediate previous moments, then we say that the material has a short memory.

In this short article, a finite element method is proposed to analyze the dynamic behaviour of the visco-elastic bodies when the frictions are imposed on the boundaries. This numerical scheme depends on the variational formulation techniques formulated by Duvaut and Lions [2] in the early seventies. They formulated the problem in a variational setting and proved existence and uniqueness results for this class of problems by using some *a priori* bounds and compactness arguments.

For purely elastic bodies, Oden *et al.* proposed finite element Galerkin methods and established *a priori* error estimates for the dynamics [4] and static case [1]. The book by Oden and Kikuchi [3] gives a complete description and analysis on the contact problems with friction. Talaslidis and Panagiotopoulos [7] derived variational inequalities for the dynamics of solids with unilateral contact with friction. Earlier, Nair and Pani [5] used the convex regularization techniques to study the finite element analysis of visco-elastic problems with a short memory under conditions of friction.

The outline of the article is as follows. In the beginning of Section 2, we formulate the problem as a variational inequality and state only the existence and uniqueness result. In section 3, we demonstrate how the frictional effects show up in the formulation as a non differentiable functional and then we regularize the problem by considering an approximation of the friction functional by a convex differentiable functional. Moreover, we derive an estimate between the solutions of the original problem and the regularized problem. By discretizing in the spatial direction with help of a finite element Galerkin procedure, we obtain the semi discrete problem and then discuss the related error estimates in Section 4.

Through out the article, C denotes various positive constants which may possibly depend on the given data, but is independent of the penalty parameter ε, the spatial discretization parameter h and which may have different values at different places of its occurrence.

2 Problem Formulation

In this section, based on the techniques used by Duvaut and Lions [2], we formulate the problem for the dynamic case of materials with long memory under friction.

Let $\Omega \subset \mathbb{R}^3$ denote an open bounded Lipschitz domain which represents the interior of a linear visco-elastic body and let $x = (x_1, x_2, x_3)$ be the cartesian coordinates of a point x in Ω. Further, let $dx = dx_1\, dx_2\, dx_3$ denote the volume measure in $\mathbb{R}^3$. Let the term $u(x,t)$ denote the displacement defined on $\Omega \times [0,T]$.

The linear momentum equation in terms of stress and strain is given as :

$$\frac{\partial}{\partial x_j} \sigma_{ij}(u) + f_i = u_{itt} \quad \text{in} \quad \Omega \times (0,T), \tag{1}$$

with the following boundary conditions and initial conditions

Boundary Conditions :

$$\begin{aligned} u_i &= U_i \quad \text{on} \quad \Gamma_U \times (0,T) \\ \sigma_N &= F_N \quad \text{on} \quad \Gamma_{\mathcal{F}} \times (0,T) \end{aligned}$$

$$|\sigma_T| < g = \nu\,|F_N| \Rightarrow u_{tT} = 0$$
$$|\sigma_T| = g \Rightarrow \exists\lambda > 0 \text{ such that } u_{tT} = -\,\lambda\,\sigma_T$$

Initial Conditions :

$$u(x,0) = u_0(x), \qquad u_t(x,0) = u_1(x), \qquad \forall x \in \Omega.$$

Here, σ_{ij} denotes the stress tensor and f_i represents the ith components of the body force, which is assumed to be given functions defined in $\Omega \times [0,T]$. σ_N denotes the normal stress, F_N, the normal force, σ_T, the tangential stress and ν denotes the coefficient of friction. Γ_U is a part of the boundary, where displacements/velocities are prescribed and $\Gamma_{\mathcal{F}}$ is the remaining part of the boundary, where we impose the conditions of friction. The function U_i is the ith component of the prescribed displacement which is assumed to be given functions of $x \in \Omega$ and $t \in [0,T]$.

The constituent law for a visco-elastic body with a long memory is given by

$$\sigma_{ij}(t) = a_{ijkl}\,\epsilon_{kl}(u(t)) + \int_0^t b_{ijkl}(t-s)\,\epsilon_{kl}(u(s))\,ds, \tag{2}$$

where, a_{ijkl} denotes the coefficients of instantaneous elasticity and b_{ijkl} represents the effects of the memory of the material.

The coefficients a_{ijkl} are bounded in x and satisfy

$$\begin{aligned} &a_{ijkl}\,\epsilon_{ij}\,\epsilon_{kl} \geq \alpha\,\epsilon_{ij}\epsilon_{ij}, \qquad \text{for some} \quad \alpha > 0, \\ &a_{ijkl} = a_{jikl} = a_{klij}. \end{aligned} \tag{3}$$

Also, the coefficients $b_{ijkl}(t)$ are bounded in both x and t and satisfy

$$\begin{aligned} &b_{ijkl} = b_{jikl}, \\ &b_{ijkl},\ \frac{\partial}{\partial t} b_{ijkl},\ \frac{\partial^2}{\partial t^2} b_{ijkl} \in L^\infty(\Omega \times (0,T)). \end{aligned} \tag{4}$$

Without loss of generality, we assume that $U_i = 0$ on $\Gamma_U \quad \forall t \in (0,T)$. Let $V = \{v = (v_1, v_2, v_3) \in (H^1(\Omega))^3 \;:\; v = 0 \;\text{ on }\; \Gamma_U\}$. Let $H = (L^2(\Omega))^3$. For the variational formulation of the problem, we introduce the following notations:

$$\left.\begin{aligned} a(u,v) &= \int_\Omega a_{ijkl}\,\epsilon_{kl}(u)\,\epsilon_{ij}(u)\,dx, \\ b(u,v) &= \int_\Omega b_{ijkl}\,\epsilon_{kl}(u)\,\epsilon_{ij}(u)\,dx, \end{aligned}\right\} \tag{5}$$

Using (3), it is easy to check that $a(\cdot,\cdot)$ satisfy the following properties :

When $m(\Gamma_U) = 0$, there is a positive constant λ such that

$$a(u,u) + \lambda ||u||^2 \geq \alpha_0 \, ||u||_1^2, \tag{6}$$

In case $m(\Gamma_U) > 0$, using Korn's inequality, it is easy to obtain

$$a(u,u) \geq \alpha_0 \, ||u||_1^2, \tag{7}$$

Let

$$j(v) = \int_{\Gamma_{\mathcal{F}}} g|v_T| \, d\Gamma \tag{8}$$

denote the virtual work done by the frictional forces on the velocity v for $g = \nu|F_N|$.

Then, the problem (1)–(2) with initial and boundary conditions can be written as:

Find $u(t) \in V, \;\; t \in [0,T]$ such that for $v \in V$

$$\begin{aligned}(u_{tt}(t), v - u_t(t)) + a(u(t), v - u_t(t)) + \int_0^t b(t-s; u(s), v - u_t(t)) \, ds \\ \geq j(u_t(t)) - j(v) + (f(t), v - u_t(t)) + \langle F_N, v_N \rangle \,, \end{aligned} \tag{9}$$

with the initial conditions

$$u(0) = u_0 \quad \text{and} \quad u_t(0) = u_1,$$

where $\langle F_N, v_N \rangle = \displaystyle\int_{\Gamma_{\mathcal{F}}} F_N(t) \, v_N \, d\Gamma.$

Below, we give a set of sufficient conditions for the existence and uniqueness of weak solution to the variational inequality (9). For a proof, see ([6])

Theorem 1. *Under the following assumptions:*

- $f, f_t, f_{tt} \in L^2(0,T;H), \;\; F_N, F_{tN}, F_{ttN} \in L^2(\Gamma_{\mathcal{F}} \times (0,T)),$
- g *is independent of* t,
- $u_0 \in (H^2(\Omega))^3$ *and* $u_1 \in V, \;\; u_{1T} = 0$ *on* $\Gamma_{\mathcal{F}}$ *and*
- $(L(0), v) - a(u_0, v) - b(u_1, v) = (u_2, v) \quad u_2 \in H,$

there exists a unique solution of (9) *satisfying* $u, \;\; u_t \in L^\infty(0,T;V)$ *and* $u_{tt} \in L^\infty(0,T;H)$.

3 Convex Regularization

In order to apply Galerkin method, it is convenient to work on a variational equality formulation rather than the inequality formulation (9). To this end, we first approximate the friction functional j, which is not differentiable by a convex differentiable functional.

For a given $\varepsilon > 0$, set

$$j_\varepsilon(v) = \int_{\Gamma_{\mathcal{F}}} g\, \phi_\varepsilon(|v_T|)\, d\Gamma, \tag{10}$$

where $\phi_\varepsilon : V \mapsto I\!R$ is defined by

$$\phi_\varepsilon(|v_T|) = \begin{cases} |v_T| - \dfrac{\varepsilon}{2}, & |v_T| > \varepsilon, \\ \dfrac{|v_T|^2}{2\varepsilon}, & |v_T| \leq \varepsilon. \end{cases} \tag{11}$$

The graph of ϕ_ε is given in Fig. 1.

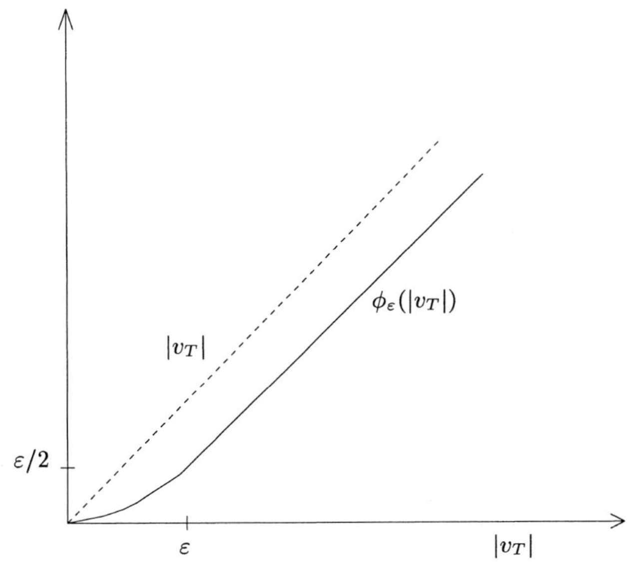

Fig. 1. Graph of the function ϕ_ε.

From the graph, it is clear that ϕ_ε is a convex differentiable function and its derivative φ_ε is given by

$$\varphi_\varepsilon(|v_T|) = \begin{cases} 1, & |v_T| > \varepsilon, \\ \dfrac{|v_T|}{\varepsilon}, & |v_T| \leq \varepsilon. \end{cases} \tag{12}$$

For the graph of φ_ε, see Fig. 2.

Note that j_ε is a convex Gâteaux differentiable functional and its Gâteaux derivative at $u(t)$ in the direction of v is given by

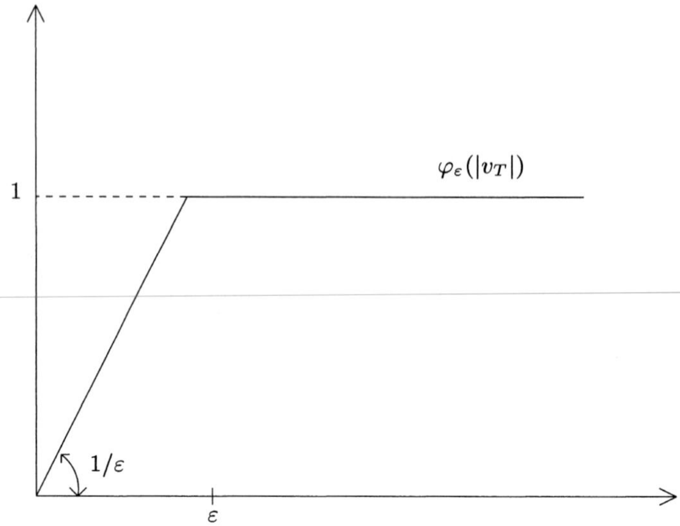

Fig. 2. Graph of the function φ_ε.

$$\langle Dj_\varepsilon(u), v\rangle = \int_{\Gamma_{\mathcal{F}}} g\,\varphi_\varepsilon(|u_T|)\,\frac{u_T\cdot v_T}{|u_T|}\,d\Gamma, \qquad u, v \in V, \tag{13}$$

where φ_ε is as in (12).

Now, the variational inequality (9) can be written in the form of the following variational formulation. Find $u_\varepsilon(t) \in V$, for each $t \in [0,T]$ such that

$$\begin{aligned}(u_{\varepsilon tt}(t), v) + a(u_\varepsilon(t), v) + \int_0^t b(t-s; u_\varepsilon(s), v)\,ds + \langle Dj_\varepsilon(u_{\varepsilon t}(t)), v\rangle \\ = (f(t), v), + \langle F_N, v_N\rangle, \qquad \forall v \in V, \end{aligned}\tag{14}$$

with the initial conditions

$$u(0) = u_0 \quad \text{and} \quad u_t(0) = u_1.$$

Note that, for a given $\varepsilon > 0$, there exists a unique solution u_ε to (14).

The existence uniqueness result that is Theorem 1, presented in [6] is based on the convex regularization and compactness argument.
Before, we prove a theorem which indicates the convergence of the solution u_ε of the regularized problem (14) to the solution of (9) as ε tends to zero, we list some properties of j_ε. For a proof, we refer to [6].

Lemma 1. *Let* $j, j_\varepsilon : V \mapsto I\!R$ *be given by* (8) *and* (10), *respectively, and let* $g \in L^\infty(\Gamma_{\mathcal{F}})$ *and* $g \geq 0$ *a.e. Then, for all* $v, w \in V$, *the following estimates hold:*

$$|\varphi_\varepsilon(|v_T|) - |v_T|| \le \varepsilon,$$
$$|j_\varepsilon(v) - j(v)| \le m(\Gamma_{\mathcal{F}})\; ||g||_{L^\infty(\Gamma_{\mathcal{F}})}\; \varepsilon,$$
$$|\varphi_\varepsilon(|v_T|) - \varphi_\varepsilon(|w_T|)| \le |v_T - w_T|,$$
$$|j_\varepsilon(v) - j_\varepsilon(w)| \le ||g||_{H^{1/2}(\Gamma_{\mathcal{F}})}\; ||v_T - w_T||_{H^{-1/2}(\Gamma_{\mathcal{F}})}.$$

Below, we now prove the main theorem of this section.

Theorem 2. *Let u be the solution of problem (9) and let u_ε be the solution of (14) for a fixed $\varepsilon > 0$. Then there exists a positive constant C, independent of ε, such that*

$$||u_{\varepsilon t}(t) - u_t(t)|| + ||u_\varepsilon(t) - u(t)||_1 \le C\,\varepsilon^{1/2}.$$

Proof. To prove this, first set $v = u_{\varepsilon t}(t) - u_t(t)$ in (14) and $v = u_{\varepsilon t}(t)$ in (9). Now, subtracting the these equations and using the convexity of j_ε, it follows that

$$\frac{1}{2}\frac{d}{dt}\left[||u_{\varepsilon t}(t) - u_t(t)||^2 + a(u_\varepsilon(t) - u(t), u_\varepsilon(t) - u(t))\right]$$
$$\le -\int_0^t b(t-s; u_\varepsilon(s) - u(s), u_{\varepsilon t}(t) - u_t(t))\, ds + j_\varepsilon(u_t(t)) - j_\varepsilon(u_{\varepsilon t}(t))$$
$$+ j(u_{\varepsilon t}(t)) - j(u_t(t)).$$

An integration from 0 to t yields

$$\begin{aligned} ||u_{\varepsilon t}(t) - u_t(t)||^2 &+ a(u_\varepsilon(t) - u(t), u_\varepsilon(t) - u(t)) \\ &\le -2\int_0^t\int_0^t b(s-\tau; u_\varepsilon(\tau) - u(\tau), u_{\varepsilon t}(s) - u_t(s))\, d\tau\, ds \\ &+ 2\int_0^t [j_\varepsilon(u_t(s)) - j_\varepsilon(u_{\varepsilon t}(s)) + j(u_{\varepsilon t}(s)) - j(u_t(s))]\, ds. \end{aligned}$$

A use of Lemma 1 with the estimate for $b(\cdot,\cdot)$ yields

$$\begin{aligned} ||u_{\varepsilon t}(t) - u_t(t)||^2 + a(u_\varepsilon(t) - u(t), u_\varepsilon(t) - u(t)) &\le C\int_0^t ||u_\varepsilon(s) - u(s)||_1^2\, ds \\ + C\,||u_\varepsilon(t) - u(t)||_1 \int_0^t ||u_\varepsilon(s) - u(s)||_1\, ds &+ 4\,\varepsilon T\,||g||_{L^\infty(\Gamma_{\mathcal{F}})}\, m(\Gamma_{\mathcal{F}}). \end{aligned}$$

Using Young's inequality and the coercivity of $a(\cdot,\cdot)$ yields

$$\begin{aligned} ||u_{\varepsilon t}(t) - u_t(t)||^2 + \frac{\alpha}{2}||u_\varepsilon(t) - u(t)||_1^2 &\le C\int_0^t ||u_\varepsilon(s) - u(s)||_1^2 ds \\ &+ C(T)\,\varepsilon\,||g||_{L^\infty(\Gamma_{\mathcal{F}})}\, m(\Gamma_{\mathcal{F}}). \end{aligned}$$

Finally, a use of Gronwall's lemma yields the desired error estimate and this completes the proof.

4 Semi-discrete Galerkin Method

We now consider the finite element Galerkin method for approximating the solution of (14). We use the standard procedure of partitioning the domain into finite elements over which we introduce piecewise polynomial approximations of the displacement field u_ε at each time t. Hence, we obtain a family $\{V_h\}$ of finite dimensional subspaces of the space V of admissible displacements, with the parameter h being the mesh size. We, further, assume that the space V_h satisfy the following approximation property

$$\inf_{\chi \in V_h} \{\| v - \chi \| + h \| \nabla(v - \chi) \|\} \leq Ch^2 \| v \|_2. \tag{15}$$

Further, we assume the spaces V_h are such that the traces of functions on the boundary $\Gamma_{\mathcal{F}}$ satisfy

$$\inf_{\chi \in V_h} ||\gamma(u) - \gamma(\chi)||_{-1/2,\Gamma_{\mathcal{F}}} \leq C\, h^2\, ||\gamma(u)||_{3/2,\Gamma_{\mathcal{F}}} \leq C\, h^2\, ||u||_{2,\Omega}, \tag{16}$$

where $|| \cdot ||_{-1/2,\Gamma_{\mathcal{F}}}$ denotes the norm on the fractional order Sobolev space $\left(H^{-1/2}(\Gamma_{\mathcal{F}})\right)^3$ and γ is the trace operator.

We formulate the semi-discrete finite element approximation of the regularized problem (9) as :
Find $u_\varepsilon^h(t) \in V_h$ such that

$$(u_{\varepsilon tt}^h(t), \chi) + a(u_\varepsilon^h(t), \chi) + \int_0^t b(t-s; u_\varepsilon^h(s), \chi)\, ds + \langle Dj_\varepsilon(u_{\varepsilon t}^h(t)), \chi \rangle$$
$$= (f(t), \chi) + \langle F_N, \chi \rangle, \qquad \forall \chi \in V_h, \tag{17}$$

with the initial conditions

$$u_\varepsilon^h(0) = I_h u_0 \quad \text{and} \quad u_{\varepsilon t}^h(0) = I_h u_1,$$

where I_h is the standard interpolation operator.

Since, V_h is of finite dimension, (17) leads to a system of nonlinear ordinary differential equations. Using Picard's theorem, there exists a solution u_ε^h in $[0, t_n]$ for some $t_n > 0$. Therefore, the solution u_ε^h exists for $t \in [0, T]$ using continuation argument.

As in the case of $u_\varepsilon(t)$, we can prove similar *a priori* bounds for $u_\varepsilon^h(t)$. Below, we derive an *a priori* error estimate for the semi-discrete finite element approximation u_ε^h of u_ε.

Theorem 3. *Under the following assumptions that* $u_\varepsilon \in L^2(0,T; \left(H^2(\Omega)\right)^3)$ *and* $u_{\varepsilon t} \in L^2(0,T; \left(H^2(\Omega)\right)^3)$, *there exists a constant* $C > 0$ *independent of* ε *and* h *such that*

$$||u_{\varepsilon t}(t) - u_{\varepsilon t}^h(t)|| + ||u_\varepsilon(t) - u_\varepsilon^h(t)||_1 \leq C\, h. \tag{18}$$

Proof. Subtracting (17) from (9), it follows that

$$(u_{\varepsilon tt}(t) - u^h_{\varepsilon tt}(t), \chi) + a(u_\varepsilon(t) - u^h_\varepsilon(t), \chi) + \int_0^t b(t-s; (u_\varepsilon(s) - u^h_\varepsilon(s), \chi)\, ds \\ + \langle Dj_\varepsilon(u_{\varepsilon t}(t)) - Dj_\varepsilon(u_{\varepsilon t}h(t)), \chi\rangle = 0.$$

Using interpolation operator, we first split $e(t)$ as

$$\begin{aligned} e(t) &= u_\varepsilon(t) - u^h_\varepsilon(t) = (u_\varepsilon(t) - I_h(u_\varepsilon(t))) - (u^h_\varepsilon(t) - I_h(u_\varepsilon(t))) \\ &= \eta(t) - \theta(t). \end{aligned}$$

Since the estimates of η are known, it is enough to estimate θ. Now the equation in θ becomes

$$(\theta_{tt}(t), \chi) + a(\theta(t), \chi) + \int_0^t b(t-s; \theta(s), \chi)\, ds = (\eta_{tt}(t), \chi) + a(\eta(t), \chi) \\ + \int_0^t b(t-s, \eta(s); \chi)\, ds + \left\langle Dj_\varepsilon(u_{\varepsilon t}(t)) - Dj_\varepsilon(u^h_{\varepsilon t}(t)), \chi \right\rangle. \quad (19)$$

Setting $\chi = \theta_t(t) \in V_h$ in (19), we obtain

$$\begin{aligned} &\frac{1}{2}\frac{d}{dt}\left[||\theta_t(t)||^2 + a(\theta(t), \theta(t)) \right] + \int_0^t b(t-s, \theta(s); \theta_t(t))\, ds \\ &= (\eta_{tt}(t), \theta_t(t)) + a(\eta(t), \theta_t(t)) + \int_0^t b(t-s; \eta(s), \theta_t(t))\, ds \\ &+ \left\langle Dj_\varepsilon(u_{\varepsilon t}(t)) - Dj_\varepsilon(u^h_{\varepsilon t}(t)), \theta_t(t) \right\rangle \\ &= (\eta_{tt}(t), \theta_t(t)) + \frac{d}{dt} a(\eta(t), \theta(t)) - a(\eta_t(t), \theta(t)) + \int_0^t b(t-s; \eta(s), \theta_t(t))\, ds \\ &+ \left\langle Dj_\varepsilon(u_{\varepsilon t}(t)) - Dj_\varepsilon(u^h_{\varepsilon t}(t)), \theta_t(t) \right\rangle. \end{aligned}$$

An integration from 0 to t yields

$$\begin{aligned} &||\theta_t(t)||^2 + a(\theta(t), \theta(t)) \le 2\int_0^t (\eta_{tt}(s), \theta_t(s))\, ds + 2a(\eta(t), \theta(t)) \\ &+2\int_0^t a(\eta_t(s), \theta_t(s)) ds + 2\int_0^t \int_0^s b(s-\tau; \eta(\tau) - \theta(\tau), \theta_t(s))\, d\tau\, ds \\ &+2\int_0^t \left\langle Dj_\varepsilon(u_{\varepsilon t}(s)) - Dj_\varepsilon(u^h_{\varepsilon t}(s)), \theta_t(s) \right\rangle ds. \end{aligned} \quad (20)$$

Using estimate of $b(\cdot, \cdot)$, we obtain

$$\begin{aligned} &2\int_0^t \int_0^s b(s-\tau; \theta(\tau), \theta_t(s))\, d\tau\, ds + 2\int_0^t \int_0^s b(s-\tau; \eta(\tau), \theta_t(s))\, d\tau\, ds \\ &\le \frac{\alpha}{2} ||\theta(t))||_1^2 + C(\alpha) \int_0^t ||\theta(s)||_1^2\, ds + C(\alpha) \int_0^t ||\eta(s)||_1^2\, ds. \end{aligned} \quad (21)$$

Using (21) in (20) and by Young's inequality, it follows that

$$\begin{aligned}||\theta_t(t)||^2 + a(\theta(t),\theta(t)) &\le C(\alpha)\left[\int_0^t \left(||\eta_{tt}(s)||^2 + ||\eta_t(s)||_1^2\right) ds + ||\eta(s)||_1^2\right] \\ &+ C(\alpha)\int_0^t \left(||\theta_t(s)||^2 + ||\theta(s)||_1^2\right) ds \\ &+ C\,h^2 \int_0^t ||u_{\varepsilon t}(s)||_2\, ds + \frac{\alpha}{2}\,||\theta(t)||_1^2.\end{aligned}$$

Now by using the coercivity property (3) of $a(\cdot,\cdot)$ for $m(\Gamma_U) > 0$ as well as $m(\Gamma_U) = 0$, the approximation property, and a use of Gronwall's lemma completes the remaining part of the proof.

Finally, coupling the mesh parameter h and the regularization parameter ε as $\varepsilon = O(h^2)$, we obtain the main result of this section.

$$||u_t(t) - u_{\varepsilon t}^h(t)|| + ||u(t) - u_\varepsilon^h(t)||_1 \le C\,h.$$

References

1. Campos, L. T., Oden, J. T. and Kikuchi, N., A Numerical analysis of a class of contact problems with friction in elastostatics, Comput. Methods Appl. Mech. Engrg., 34 (1982), pp. 821–845.
2. Duvaut, G. and Lions, J. L., Inequalities in Mechanics and Physics, Springer-Verlag, Berlin (1972).
3. Kikuchi, N. and Oden, J. T., Contact Problems in Elasticity: A Study of Variational Inequalities and Finite Element Methods, SIAM, Philadelphia (1988).
4. Martins, J. A. C. and Oden, J. T., A Numerical analysis of a class of problems in elastodynamics with friction, Comput. Methods Appl. Mech. Engrg., 40 (1983), pp. 327–360.
5. Nair, P. and Pani, A. K., Finite element analysis of a class of visco-elastic materials with short memory under conditions of friction, Research Report, Industrial Mathematics Group, Indian Institute of Technology, Bombay (2002).
6. Nair, P., Finite element approximations to evolutionary variational inequalities, Ph. D Thesis, Indian Institute of Technology, Bombay (2002).
7. Talaslidis, D. and Panagiotopoulos, P. D., A linear finite element approach to the solution of the variational inequalities arising in contact problems of structural dynamics, Internat. J. Numer. Meth. Engrg., 8 (1982) pp. 1505–1520.

Global Solutions for a Free Boundary Problem of Hyperbolic Type in the Case of Non-negative Peeling Speed

Kazuaki Nakane[1] and Tomoko Shinohara[2]

[1] Osaka Institute of Technology, Faculty of Engineering, *nakane@ge.oit.ac.jp*
[2] Faculty of Integrated Human Studies, Kyoto University, *sinohara@math.h.kyoto-u.ac.jp*

1 Introduction

Let us consider the following one-dimensional free boundary problem

$$(P)\quad \begin{cases} u_{xx} - u_{tt} = 0 & \text{in} \quad (0,\infty) \times \{t > 0\} \cap \{u > 0\}, \\ u_x^2 - u_t^2 = Q^2 & \text{on} \quad (0,\infty) \times \{t > 0\} \cap \partial\{u > 0\}, \end{cases}$$

with the initial conditions

$$(I)\quad \begin{cases} u(x,0) = e(x) & \text{on} \quad (-l_0, 0), \\ u_t(x,0) = g(x) & \text{on} \quad (-l_0, 0), \end{cases}$$

and the boundary condition

$$(B)\quad u(-l_0, t) = f(t) \qquad \text{on} \qquad [0,\infty),$$

where $e(x)$, $g(x)$ and $f(t)$ are given functions, Q and l_0 are positive constants.

This problem arises from a variational problem which is related to the following physical model: " A thin film is pasted on the plate. By lifting up the edge of the film in the vertical direction, the film is peeled from the plate. We would like to know the behavior of the peeling front." The shape of the film is described by the graph of a function $u : \Omega \to \mathbf{R}$. It can be assumed that the effect of the peeling front on the Lagrangian is constant Q^2. Then, our problem is to find a stationary point of the functional

$$(J)\quad J(u) = \int_0^{T^*} \int_\Omega \Big(\tau|\nabla u|^2 - \rho(D_t u)^2 + Q^2\Big)\chi_{u>0}\,dxdt \qquad u \in \mathrm{K}$$

where Ω is a domain in $\mathbf{R}^n$, T^* is a positive constant, $\chi_{u>0}$ is a characteristic function of the set $\{(x,t) \in \Omega \times (0,T^*); u(x,t) > 0\}$ and K is a suitable function space. Here the constants τ and ρ are the tension, and the line density, respectively. To approach this problem, we assume that a stationary point is sufficiently smooth. Then we can derive (1) and (2) as the Euler-Lagrange equations from the functional (J) (cf. [1], [2] and [5]),

$$\tau \Delta u - \rho u_{tt} = 0 \qquad \text{in} \quad \Omega \times \{t > 0\} \cap \{u > 0\}, \tag{1}$$

$$\frac{\tau}{2}|\nabla u|^2 - \frac{\rho}{2}u_t^2 = \frac{Q^2}{2} \qquad \text{on} \quad \Omega \times \{t > 0\} \cap \partial\{u > 0\}. \tag{2}$$

In this note, as a first step, the one-dimensional problem will be analyzed. By using change of the variable t, $\tau = 1$ and $\rho = 1$ can be assumed. Therefore, we will consider the problem (P).

The initial condition (I) implies that the thin film has been already peeled from the plate on the interval $(-l_0, 0)$ and the boundary condition (B) corresponds to the situation in which the edge of the film is lifted up by $f(t)$. Kikuchi and Omata [5] showed the existence of time-local solutions under several conditions which were imposed on the functions $e(x)$, $g(x)$ and $f(t)$. On the global existence of solutions, however, they have not stated. In [3], numerical experiments were carried out. A sufficient condition for the global existence was given by Nakane [4] by using an iteration method. This method, however, can not be applied to the case the peeling speed is zero.

In section 2, the solutions can be constructed by pasting the local solutions, inductively. In this note, it is shown that if the peeling speed f' is non-negative then the pasting interval does not decrease (cf. **Main Theorem**). It implies that time global solutions for problem (P), (I) and (B) are constructed. It is remarked here that this condition is an improvement on the previous one in [4].

For any initial conditions, a non-negative peeling speed is a sufficient conditions for global existence of solutions. On the other hand, by admitting $f' < 0$, there is an example that the solution can not be constructed globally, nevertheless f is non-negative. The concrete example will be given in section 3.

2 Construction of solutions

To solve our problem, two variables ξ and η are introduced by

$$t = (\xi + \eta)/2, \quad x = (\xi - \eta)/2.$$

Since the initial values e and g are given on the line $\{(x,t); t = 0\}$, they are given on the line $\{(\xi,\eta); \xi + \eta = 0\}$. We regard them as the functions of ξ and rewrite them e and g again. Similarly, the boundary value f is given on the line $\{(\xi,\eta); \xi - \eta + 2l_0 = 0\}$ and is a function of η. Therefore (P), (I) and (B) are transformed into

$$(P') \qquad \begin{cases} u_{\xi\eta} = 0 & \text{in} \quad \{u > 0\}, \\ -4u_\xi u_\eta = Q^2 & \text{on} \quad \partial\{u > 0\}, \end{cases}$$

$$(I') \qquad \begin{cases} u(\xi, -\xi) = e(\xi) & \text{on} \quad (-l_0, 0), \\ u_\eta(\xi, -\xi) + u_\xi(\xi, -\xi) = g(\xi) & \text{on} \quad (-l_0, 0), \end{cases}$$

$$(B') \qquad u(\eta - 2l_0, \eta) = f(\eta - l_0) \qquad \text{on} \quad [l_0, \infty).$$

For these equations, we consider the following problem:

Problem 2.1 *Let T be a positive constant. Find a pair of functions $u \in C^0(\{(\xi,\eta); \xi \geq \eta - 2l_0, \xi \geq -\eta\})$ and $l \in C^0([0,T)) \cap C^1((0,T))$ which satisfies (P'), (I') and (B') for $\eta < T$ and*
(i) $l(0) = 0$,
(ii) $u \in C^2(\{(\xi,\eta); \eta - 2l_0 < \xi < l(\eta), \xi > -\eta\}) \cap C^1(\{(\xi,\eta); \eta - 2l_0 < \xi \leq l(\eta), \xi \geq -\eta\})$,
(iii) $u > 0$ *in* $\{(\xi,\eta); \eta - 2l_0 \leq \xi < l(\eta), \xi > -\eta\}$,
(iv) $u(\xi,\eta) = 0$ *in* $\{(\xi,\eta); \xi \geq l(\eta)\} \cup \{(\xi,\eta); \xi \geq -\eta, \eta < 0\}$.

Assumption 2.1 *The functions $f(\eta - l_0) \in C^2([l_0,\infty))$, $e(\xi) \in C^2([-l_0,0])$ and $g(\xi) \in C^1([-l_0,0])$ satisfy*

$$(\text{A.0}) \qquad \begin{cases} e(\xi) > 0 & \text{on} \quad [-l_0, 0), \\ g(\xi) > 0 & \text{on} \quad (-l_0, 0), \end{cases}$$

$$(\text{A.1}) \qquad \begin{cases} f(0) = e(-l_0), \\ f'(0) = g(-l_0), \\ f''(0) = e''(-l_0), \end{cases}$$

$$(\text{A.2}) \qquad \begin{cases} e(0) = 0, \\ e'(0)^2 - g(0)^2 = Q^2, \\ (e''(0) + g'(0))(e'(0) - g(0))^4 = Q^4(e''(0) - g'(0)), \end{cases}$$

$$(\text{A.3}) \qquad \begin{cases} e'(\xi) < g(\xi) & \text{on} \quad [-l_0, 0], \\ f'(\xi + l_0) - (e'(\xi) + g(\xi))/2 > 0 & \text{on} \quad [-l_0, 0], \end{cases}$$

$$(\text{A.4}) \qquad f'(\eta - l_0) \geq 0 \quad \text{on} \quad [l_0, \infty).$$

The following Lemma is originally given in [5]. To confirm the global existence of the solution, we will modify its proof.

Lemma 2.1 *Let c be a positive number and $I = [0,c)$ an interval on the η-axis. Let $\gamma(\eta) \in C^2(I)$ be a function which satisfies*
(i)$\gamma(0) = 0$,

(ii)$\gamma'(\eta) > 0$ on I.
Then the following functions are well-defined

$$\begin{aligned} \psi(\eta) &= \gamma(\eta) && \text{on} \quad I, \\ l(\eta) &= \frac{4}{Q^2}\int_0^\eta \psi'(s)^2 ds && \text{on} \quad I, \\ \phi(\xi) &= -\psi(l^{-1}(\xi)) && \text{on} \quad [0, l(c)), \end{aligned}$$

and a pair of functions $u(\xi,\eta) = \phi(\xi) + \psi(\eta) \in C^2(\{(\xi,\eta); \eta \in I, 0 < \xi < l(\eta)\}) \cap C^1(\{(\xi,\eta); \eta \in I, 0 \le \xi \le l(\eta)\})$ *and* $l(\eta) \in C^2(I)$ *satisfies*

$$(PL) \begin{cases} u_{\xi\eta} = 0 & \text{in} \quad \{(\xi,\eta); 0 < \xi < l(\eta), \eta \in I\}, \\ -4u_\xi u_\eta = Q^2 & \text{on} \quad \{(l(\eta),\eta); \eta \in I\}, \\ u(l(\eta),\eta) = 0 & \text{on} \quad I, \\ u(0,\eta) = \gamma(\eta) & \text{on} \quad I, \\ l(0) = 0. \end{cases}$$

Moreover, a pair of functions (u, l) which satisfies (PL) are uniquely determined.

Proof. Because of (ii), $l^{-1}(\xi)$ exists in $[0, l(c))$. Then the function ϕ is well-defined. It is not difficult to make sure that u and l satisfy (PL).

To show the uniqueness of the solution, we suppose that there exists another pair of functions $(\tilde{u}(\xi,\eta), \tilde{l}(\eta))$ satisfying (PL) which has the form $\tilde{u}(\xi,\eta) = \tilde{\phi}(\xi) + \tilde{\psi}(\eta)$. It follows that

$$\tilde{u}(\tilde{l}(\eta),\eta) = \tilde{\phi}(\tilde{l}(\eta)) + \tilde{\psi}(\eta) = 0. \tag{3}$$

By differentiating both sides of (3) and from the fact $-4\tilde{u}_\xi\tilde{u}_\eta = Q^2$, we have

$$\tilde{l}'(\eta) = \frac{4}{Q^2}\tilde{\psi}'(\eta)^2.$$

Since $l(0) = 0$, it holds that

$$\tilde{l}(\eta) = \frac{4}{Q^2}\int_0^\eta \tilde{\psi}'(s)^2 ds.$$

Because of $\tilde{u}(0,\eta) = \gamma(\eta)$, we obtain $\tilde{\psi}(\eta) = \gamma(\eta) + \tilde{\phi}(0)$ and

$$\tilde{\psi}'(\eta) = \gamma'(\eta).$$

It implies that $\tilde{l}$ is equal to l. Moreover, since $l^{-1}(\xi)$ exists, we have from (3)

$$\tilde{\phi}(\xi) = -\tilde{\psi}(l^{-1}(\xi)) = -\gamma(l^{-1}(\xi)) + \tilde{\phi}(0).$$

Therefore, $\tilde{u}(\xi,\eta) = \tilde{\phi}(\xi) + \tilde{\psi}(\eta)$ coincides with $u(\xi,\eta) = \phi(\xi) + \psi(\eta)$, it is a contradiction proving our assertion. □

Main Theorem *For any $T > 0$, there exists a unique solution to Problem 2.1.*

Proof. Firstly, let ϕ_0 and ψ_1 be functions such that

$$\begin{cases} \phi_0(\xi) = \frac{1}{2}(e(\xi) + \int_0^{\xi} g(s)ds) & \text{on} \quad [-\lambda_1, 0), \\ \psi_1(\eta) = \frac{1}{2}(e(-\eta) + \int_{-\eta}^{0} g(s)ds) & \text{on} \quad [0, \lambda_1), \end{cases}$$

where

$$\lambda_1 = l_0.$$

Evidently, $u(\xi,\eta) = \phi_0(\xi) + \psi_1(\eta)$ is a unique solution to the initial value problem in $\{(\xi,\eta); -l_0 < -\eta < \xi < 0\}$. Let us define the free boundary l_1 by

$$l_1(\eta) = \frac{4}{Q^2} \int_0^{\eta} \psi_1'(s)^2 ds.$$

By (A.3), $\psi_1'(\eta)$ is positive on $[0, \lambda_1)$. Therefore there exists $l_1^{-1}(\xi)$ on $[0, l_1(\lambda_1))$ and we can define ϕ_1 by

$$\phi_1(\xi) = -\psi_1(l_1^{-1}(\xi)) \qquad \text{on} \quad [0, l_1(\lambda_1)).$$

By using these functions, we define the functions u and l by

$$u(\xi,\eta) = \begin{cases} \phi_0(\xi) + \psi_1(\eta) & \text{in} \quad D_{0,1}, \\ \phi_1(\xi) + \psi_1(\eta) & \text{in} \quad D_{1,1}, \\ 0 & \text{in} \quad D_{e,1}, \end{cases}$$

$$l(\eta) = l_1(\eta) \qquad \text{on} \quad [0, \lambda_1),$$

where

$$\begin{aligned} D_{0,1} &= \{(\xi,\eta); \xi \geq -\eta, \xi < 0, 0 \leq \eta < \lambda_1\}, \\ D_{1,1} &= \{(\xi,\eta); 0 \leq \xi < l_1(\eta), 0 \leq \eta < \lambda_1\}, \\ D_{e,1} &= \{(\xi,\eta); \xi \geq l_1(\eta), 0 \leq \eta < \lambda_1, \}. \end{aligned}$$

(A.2) guarantees the regularity at $\xi = 0$ of ϕ_0 and ϕ_1. By Lemma 2.1, $(u(\xi,\eta), l(\eta))$ is a unique solution to Problem 2.1 for $T := \lambda_1$.

Nextly, we define ψ_2 by

$$\psi_2(\eta) = \begin{cases} f(\eta - l_0) - \phi_0(\eta - 2l_0) & \text{on} \quad [\lambda_1, 2\lambda_1), \\ f(\eta - l_0) - \phi_1(\eta - 2l_0) & \text{on} \quad [2\lambda_1, \lambda_2), \end{cases}$$

where

$$\lambda_2 = l_1(\lambda_1) + 2l_0.$$

Since ϕ_0 and ϕ_1 are connected smoothly, $\psi_2(\eta)$ is of C^2-class. Because the part of the boundary known so far ends in $(\xi,\eta) = (l_1(\lambda_1), l_1)$, l_2 is defined by

$$l_2(\eta) = \frac{4}{Q^2}\int_{\lambda_1}^{\eta} \psi_2'(s)^2 ds + l_1(\lambda_1) \qquad \text{on} \quad [\lambda_1, \lambda_2).$$

Obviously, l_2 is of C^2-class. Because it yields that

$$\psi_2'(\eta) = \begin{cases} f'(\eta - l_0) - \dfrac{1}{2}(e'(\eta - 2l_0) + g(\eta - 2l_0)) & \text{on} \quad [\lambda_1, 2\lambda_1), \\ f'(\eta - l_0) + \dfrac{Q^2}{4}\dfrac{1}{\psi_1'(l_1^{-1}(\eta - 2l_0))} & \text{on} \quad [2\lambda_1, \lambda_2), \end{cases}$$

$\psi_2'(\eta) > 0$ holds on $[\lambda_1, \lambda_2)$ by (A.3) and (A.4). Then there exists $l_2^{-1}(\xi)$ on $[l_2(\lambda_1), l_2(\lambda_2))$. Hence we can define

$$\phi_2(\xi) = -\psi_2(l_2^{-1}(\xi)) \quad \text{on} \quad [l_2(\lambda_1), l_2(\lambda_2)).$$

The functions $(u(\xi,\eta), l(\eta))$ are extended as the following

$$u(\xi,\eta) = \begin{cases} \phi_0(\xi) + \psi_2(\eta) & \text{in} \quad D_{0,2}, \\ \phi_1(\xi) + \psi_2(\eta) & \text{in} \quad D_{1,2}, \\ \phi_2(\xi) + \psi_2(\eta) & \text{in} \quad D_{2,2}, \\ 0 & \text{in} \quad D_{e,2}, \end{cases}$$

$$l(\eta) = l_2(\eta) \qquad \text{on} \quad [\lambda_1, \lambda_2),$$

where

$$\begin{aligned} D_{0,2} &= \{(\xi,\eta); \xi \geq \eta - 2l_0, \xi < 0, \quad \lambda_1 \leq \eta < 2\lambda_1\}, \\ D_{1,2} &= \{(\xi,\eta); \xi \geq \eta - 2l_0, 0 \leq \xi < l_1(\lambda_1), \quad \lambda_1 \leq \eta < \lambda_2\}, \\ D_{2,2} &= \{(\xi,\eta); l_1(\lambda_1) \leq \xi < l_2(\eta), \quad \lambda_1 \leq \eta < \lambda_2\}, \\ D_{e,2} &= \{(\xi,\eta); \xi \geq l_2(\eta), \lambda_1 \leq \eta < \lambda_2\}. \end{aligned}$$

It follows from (A.1) that ψ_1 and ψ_2 are connected smoothly at $\eta = \lambda_1$, concurrently ϕ_1 and ϕ_2 are smooth at $\xi = \lambda_2 - 2l_0$. By applying Lemma 2.1 again, we can see the pair of functions $(u(\xi,\eta), l(\eta))$ is a unique solution to Problem 2.1 for $T := \lambda_2$.

Inductively, for $j \geq 3$, we define the functions ψ_j, l_j and ϕ_j by

$$\begin{aligned} \psi_j(\eta) &= f(\eta - l_0) - \phi_{j-1}(\eta - 2l_0) && \text{on} \quad [\lambda_{j-1}, \lambda_j), \\ l_j(\eta) &= \frac{4}{Q^2}\int_{\lambda_{j-1}}^{\eta} \psi_j'(s)^2 ds + l_{j-1}(\lambda_{j-1}) && \text{on} \quad [\lambda_{j-1}, \lambda_j), \\ \phi_j(\xi) &= -\psi_j(l_j^{-1}(\xi)) && \text{on} \quad [l_j(\lambda_{j-1}), l_j(\lambda_j)), \end{aligned}$$

where

$$\lambda_j = l_{j-1}(\lambda_{j-1}) + 2l_0.$$

It can be seen that $\psi_3(\eta)$ and $l_3(\eta)$ are well-defined and of C^2-class. Because it holds that

$$\psi_3'(\eta) = f'(\eta - l_0) - \phi_2'(\eta - 2l_0) = f'(\eta - l_0) + \frac{Q^2}{4}\frac{1}{\psi_2'(l_2^{-1}(\eta - 2l_0))},$$

we obtain $\psi_3'(\eta) > 0$ on $[\lambda_2, \lambda_3)$ by (A.4) and the fact $\psi_2'(\eta) > 0$ on $[\lambda_1, \lambda_2)$. So the function l_3^{-1} exists on $[\ l_3(\lambda_2), l_3(\lambda_3))$, ϕ_3 is well-defined and of C^2-class. Suppose that it has been already shown that $\psi_j(\eta)$, $l_j(\eta)$ and $\phi_j(\xi)$ are well-defined and of C^2-class, and $\psi_j'(\eta) > 0$ on $[\lambda_{j-1}, \lambda_j)$. It can be obtained that ψ_{j+1} and l_{j+1} are well-defined and of C^2-class, immediately. We have

$$\psi_{j+1}'(\eta) = f'(\eta - l_0) - \phi_j'(\eta - 2l_0) = f'(\eta - l_0) + \frac{Q^2}{4}\frac{1}{\psi_j'(l_j^{-1}(\eta - 2l_0))}, \tag{4}$$

it follows $\psi_{j+1}'(\eta) > 0$ on $[\lambda_j, \lambda_{j+1})$ by (A.4) and induction assumption. Therefore, there exists l_{j+1}^{-1} on $[l_{j+1}(\lambda_j), l_{j+1}(\lambda_{j+1}))$, it implies that ϕ_{j+1} is well-defined and of C^2-class.

The regularity of ψ_2 and ψ_3 at $\eta = \lambda_2$ can be shown, since ϕ_1 and ϕ_2 are connected smoothly. Accordingly, we obtain the regularity ϕ_2 and ϕ_3 at $\xi = \lambda_3 - 2l_0$. It is supposed that ϕ_{j-1} and ϕ_j are connected smoothly at $\xi = \lambda_j - 2l_0$. From the definition, ψ_j and ψ_{j+1} are connected smoothly at $\eta = \lambda_j$. Therefore ϕ_j and ϕ_{j+1} are connected smoothly at $\xi = \lambda_{j+1} - 2l_0$.

By combining these functions for $j \geq 3$, the pair of functions $(u(\xi,\eta), l(\eta))$ can be extended as the following

$$u(\xi,\eta) = \begin{cases} \phi_{j-1}(\xi) + \psi_j(\eta) & \text{in} \quad D_{j-1,j}, \\ \phi_j(\xi) + \psi_j(\eta) & \text{in} \quad D_{j,j}, \\ 0 & \text{in} \quad D_{e,j}, \end{cases}$$

$$l(\eta) = l_j(\eta) \qquad \text{on} \quad [\lambda_{j-1}, \lambda_j),$$

where

$$\begin{aligned} D_{j-1,j} &= \{(\xi,\eta); \xi \geq \eta - 2l_0, \xi < l_{j-1}(\lambda_{j-1}), \lambda_{j-1} \leq \eta < \lambda_j\}, \\ D_{j,j} &= \{(\xi,\eta); l_{j-1}(\lambda_{j-1}) \leq \xi < l_j(\eta), \lambda_{j-1} \leq \eta < \lambda_j\}, \\ D_{e,j} &= \{(\xi,\eta); \xi \geq l_j(\eta), \lambda_{j-1} \leq \eta < \lambda_j\}. \end{aligned}$$

By the same argument as for $j = 2$, $(u(\xi,\eta), l(\eta))$ is a unique solution to Problem 2.1 for $T := \lambda_j$.

Finally, we shall show the sequence $\{\lambda_j\}_{j=1}^{\infty}$ goes to infinity. We remark that $l_{j-1}(\lambda_{j-1}) = l_j(\lambda_{j-1})$, it holds that $\lambda_{j+1} - \lambda_j = l_j(\lambda_j) - l_j(\lambda_{j-1})$. Then the following inequation holds,

$$\lambda_{j+1} - \lambda_j = \frac{4}{Q^2}\int_{\lambda_{j-1}}^{\lambda_j} \psi_j'(s)^2 ds$$

$$= \frac{4}{Q^2}\int_{\lambda_{j-1}}^{\lambda_j}\left(f'(s-l_0)+\frac{Q^2}{4}\frac{1}{\psi'_{j-1}(l_{j-1}^{-1}(s-2l_0))}\right)^2 ds$$

$$\geq \frac{4}{Q^2}\int_{\lambda_{j-1}}^{\lambda_j}\left(\frac{Q^2}{4}\frac{1}{\psi'_{j-1}(l_{j-1}^{-1}(s-2l_0))}\right)^2 ds. \tag{5}$$

By using the change of variable

$$y = l_{j-1}^{-1}(s-2l_0),$$

the right hand side of (5) is reduced

$$\frac{4}{Q^2}\int_{\lambda_{j-1}}^{\lambda_j}\left(\frac{Q^2}{4}\frac{1}{\psi'_{j-1}(l_{j-1}^{-1}(s-2l_0))}\right)^2 ds = \int_{\lambda_{j-2}}^{\lambda_{j-1}} dy = \lambda_{j-1}-\lambda_{j-2}.$$

Then we obtain

$$\lambda_{j+1}-\lambda_j \geq \lambda_{j-1}-\lambda_{j-2},$$

it implies that $\{\lambda_j\}$ diverges. □

3 Necessity of the condition (A.4)

It is enough to suppose that

$$\text{(A.4)} \quad f'(\eta - l_0) \geq 0 \quad \text{for} \quad \eta \in [l_0, \infty),$$

to construct global solutions for any initial conditions. This condition guarantees that each piece of local solution can be constructed and pasting intervals do not converge. On the other hand, by admitting $f' < 0$, we can construct a solution whose sequence $\{\lambda_j\}$ exists and converges to a bounded constant with a pair of particular initial and boundary conditions. Here, a concrete example will be given.

Example 3.1 We set $Q = 3$, $l_0 = \lambda_1 = 1$. As an initial condition, the functions g, e are given by

$$\begin{cases} e(\xi) = -5\xi & \text{on} \quad [-l_0, 0], \\ g(\xi) = 4 & \text{on} \quad [-l_0, 0]. \end{cases}$$

Initially, $f'(\eta - l_0)$ is defined on $[\lambda_1, \lambda_3)$:

$$f'(\eta - l_0) = \begin{cases} 4 & \text{on} \quad [\lambda_1, 2\lambda_1) = [1, 2), \\ p(\eta) & \text{on} \quad [2\lambda_1, \lambda_2) = [2, 11), \end{cases}$$

Here, p is a decreasing C^1 function which satisfies

$$(C)\begin{cases} p(2\lambda_1) = 4, & p(\lambda_2) = 0, \\ p'(2\lambda_1) = 0, & p'(\lambda_2) = 0. \end{cases}$$

By integrating f' with $f(\eta - l_0)|_{\eta=l_0} = 5$, the boundary condition f is given. From a direct calculation, we have the following:

$$\phi_0(\xi) = -\xi/2 \quad \text{on} \quad [-l_0, 0),$$

$$\begin{cases} \psi_1(\eta) = 9\eta/2, & l_1(\eta) = 9\eta \quad \text{on} \quad [0, l_0), \\ \phi_1(\xi) = -\xi/2 & \quad \text{on} \quad [0, l_1(\lambda_1)), \end{cases}$$

$$\begin{cases} \psi_2'(\eta) = \begin{cases} 9/2 & \text{on} \quad [\lambda_1, 2\lambda_1), \\ p(\eta) + 1/2 & \text{on} \quad [2\lambda_1, \lambda_2), \end{cases} \\ l_2(\eta) = \dfrac{4}{9}\displaystyle\int_{\lambda_1}^{\eta} \psi_2'(s)^2 ds + l_1(\lambda_1). \end{cases}$$

It is not so difficult to make sure that the functions $e, g \in C^2([-l_0, 0])$ and $f \in C^2([\lambda_1, \lambda_2))$ satisfy the conditions $(A.0) - (A.3)$.

Nextly, $f'(\eta - l_0)$ on $[\lambda_2, \lambda_3)$ is defined by

$$f'(\eta - l_0) = \frac{1}{2} - \frac{9}{4\psi_2'(l_2^{-1}(\eta - 2l_0))}.$$

From the condition (C), one can see that f is well-defined and of $C^2([\lambda_1, \lambda_3))$. Then we have

$$\begin{aligned} \psi_3'(\eta) &= f'(\eta - l_0) + \frac{9}{4}\frac{1}{\psi_2'(l_2^{-1}(\eta - 2l_0))} \\ &= \frac{1}{2} - \frac{9}{4}\frac{1}{\psi_2'(l_2^{-1}(\eta - 2l_0))} + \frac{9}{4}\frac{1}{\psi_2'(l_2^{-1}(\eta - 2l_0))} = \frac{1}{2} \qquad \text{on} \quad [\lambda_2, \lambda_3). \end{aligned}$$

Moreover, we define f' on $[\lambda_3, \infty)$ by

$$\begin{aligned} f'(\eta - l_0) &= \frac{1}{2} - \frac{9}{4}\frac{1}{\psi_3'(l_3^{-1}(\eta - 2l_0))} \\ &= -4. \end{aligned}$$

Because ψ_2 and ψ_3, concurrently l_2 and l_3, are connected smoothly, f is of C^2 on $[\lambda_1, \infty)$. Therefore, for $n \geq 3$, it holds that

$$\begin{cases} \psi_n'(\eta) = 1/2 & \text{on} \quad [\lambda_{n-1}, \lambda_n), \\ l_n(\eta) \ = (\eta - \lambda_{n-1})/9 + l_{n-1}(\lambda_{n-1}) & \text{on} \quad [\lambda_{n-1}, \lambda_n). \end{cases}$$

Then we obtain

$$\lambda_{n+1} - \lambda_n = \frac{4}{9}\int_{\lambda_{n-1}}^{\lambda_n} \psi_n'(s)^2 ds = \frac{1}{9}(\lambda_n - \lambda_{n-1}) \quad \text{for} \quad n \geq 3.$$

Hence, the following relation is yielded

$$\begin{aligned}\lim_{n\to\infty}\lambda_n &= \lim_{n\to\infty}\{\lambda_2 + (\lambda_3 - \lambda_2) + \cdots + (\lambda_n - \lambda_{n-1})\}\\ &= \lim_{n\to\infty}\{\lambda_2 + (\lambda_3 - \lambda_2)(1 + 1/9 + 1/9^2 + \cdots 1/9^{n-3})\}\\ &= \lambda_2 + \frac{9}{8}(\lambda_3 - \lambda_2).\end{aligned}$$

It means that the sequence $\{\lambda_n\}$ converges to a bounded constant λ_∞.

Additionally, by a direct calculation, it follows that

$$\begin{aligned}f(\lambda_\infty - l_0) &= f(\eta - l_0)|_{\eta=\lambda_1} + \int_{\lambda_1}^{\lambda_\infty} f'(s - l_0)ds\\ &= f(0) + \int_{\lambda_1}^{2\lambda_1} 4ds + \int_{2\lambda_1}^{\lambda_2} p(s)ds\\ &+\int_{\lambda_2}^{\lambda_3}\left(\frac{1}{2} - \frac{9}{4\psi_2'(l_2^{-1}(s - 2l_0))}\right) ds + \int_{\lambda_3}^{\lambda_\infty}(-4)ds.\end{aligned}$$

By using the change of variable $t = l_2^{-1}(s - 2l_0)$, we have

$$\int_{\lambda_2}^{\lambda_3}\left(\frac{9}{4\psi_2'(l_2^{-1}(s - 2l_0))}\right) ds = \int_{\lambda_1}^{2\lambda_1}\frac{9}{2}dt + \int_{2\lambda_1}^{\lambda_2}\left(p(t) + \frac{1}{2}\right) dt.$$

We remark that $\lambda_1 = 1$, $\lambda_2 = 11$, $f(0) = 5$ and $\lambda_\infty - \lambda_3 = (\lambda_3 - \lambda_2)/8$, then it holds that

$$\begin{aligned}f(\lambda_\infty - l_0) &= 5 + 4\lambda_1 + \int_{2\lambda_1}^{\lambda_2} p(s)ds\\ &+\frac{1}{2}(\lambda_3 - \lambda_2) - \frac{9}{2} - \int_{2\lambda_1}^{\lambda_2} p(t)dt - \frac{1}{2}(\lambda_2 - 2\lambda_1) - \frac{1}{2}(\lambda_3 - \lambda_2) = 0.\end{aligned}$$

It implies that the free boundary and the peeling edge cross on the prescribed boundary $\{(\xi,\eta); \xi - \eta + 2l_0 = 0\}$.

References

1. H. W. Alt and L. A. Caffarelli (1981): Existence and regularity for a minimum problem with free boundary, J. Reine Angew. Math., **325**, 105–144
2. H. W. Alt, L. A. Caffarelli and A. Friedman (1984): A free boundary problem for quasi-linear elliptic equations, Ann. Scuola Norm. Sup. Pisa Cl. Sci., **11**, 1–44
3. H. Imai, K. Kikuchi, K. Nakane, S. Omata and T. Tachikawa (2001): A Numerical Approach to the Asymptotic Behavior of Solutions of a One-Dimensional Free Boundary Problem of Hyperbolic Type, J.J.I.A.M., **18-1** , 43–58
4. K. Nakane (2001): Global existence of solutions to a one-dimensional free boundary problem of hyperbolic type, Proceedings of Japan Academy, **Vol. 77, Ser. A, No. 7** , 103–107
5. K. Kikuchi and S. Omata (1999): A Free boundary problem for a one dimensional hyperbolic equation, Adv. Math. Sci. Appl. **9**(2) , 775–786

Stability Analysis for Periodic Solution Waves in Viscous Conservation Laws

Myunghyun Oh

Ohio State University *myoh@math.ohio-state.edu*

1 Introduction

Periodic traveling waves are a frequently observed class of solutions of diverse types of nonlinear evolutionary PDE in one space variable. They also frequently arise as solutions of higher order dispersive equations. Our main interest is in periodic solutions arising in conservation laws of the form

$$u_t + f(u)_x = (B(u)u_x)_x, \tag{1}$$

where $u, f \in \Re^n$. For certain conservation laws involving phase transition, for example, van der Waals fluids, or oil recovery models, there can appear pattern formation/complex asymptotic behavior. Related to these phenomena is the appearance of new, traveling periodic solutions along with the more familiar traveling front (viscous shock) solutions. Particular examples arise from homoclinic cycles, 2-cycles, and 3-cycles of shock waves in viscous profiles [AMPZ]. There is an infinite family of periodic orbits in the interior of each cycle (Fig. 1).

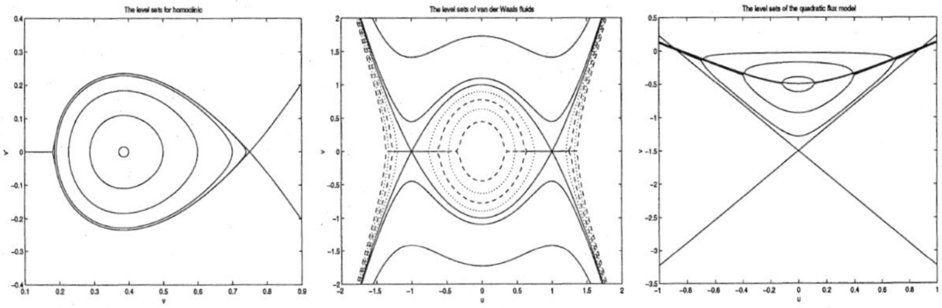

(a) (b) (c)

Fig. 1. Periodic orbits inside cycles

Certain interesting oscillatory behavior detected by Frid-Liu [FL] and Čanić-Peters [CP] seems possibly related to the periodic waves. On the other hand, an alternative explanation was offered by [AMPZ] involving metastable collections of slowly interacting shocks. Either scenario is quite different from the usual 'hyperbolic' behavior; to determine which is the governing mechanism, a starting point is stability of periodic patterns. A positive stability result would indicate a periodic mechanism, while a negative stability result would eliminate that possibility, giving evidence for the nonstandard picture of [AMPZ]. Note that in the standard strict hyperbolic setting, neither type of solution can even exist. Thus, hyperbolic intuition does not shed light on their stability or instability, which must instead be determined through a direct investigation.

The analysis of spectra of periodic waves is less completely developed than is the case for asymptotically constant waves. Several analyses have been carried out in a general setting where the periodic waves have a distinguished asymptotic limit such as a bounding heteroclinic cycle or (small amplitude) limiting constant solution (see [Ec], [G.1], [M.1], and [M.2]); however, there is no machinery for general intermediate amplitude waves. In [G.2], Gardner proved a stability result for the large wavelength periodic waves but he employed the stability of the limiting homoclinic wave. Gardner's analytic framework is general enough to include waves occurring in a variety of different equations but it is difficult to compute the stability practically. Our goal is to establish a general theory and method of analysis for determining the stability of these periodic waves, analogous to that recently developed in the viscous shock (front) case: specially, the Evans function and stability index calculations developed in [GZ], and the pointwise bounds of Green's function developed in [ZH]. Also we derive a numerical test of periodic wave stability. The second topic has bearing also on general reaction–diffusion equations, sharpening recent results obtained by Schneider via weighted norm methods. Together, the first and third decide the question of stability of periodic waves, at least for the special model systems considered by Frid–Liu and Čanić–Peters in the numerical experiments. This has the implication that the oscillatory patterns they observed are formed by a mechanism substantially different from that of the reaction diffusion/pattern formation setting.

2 Preliminaries

A key construction in the stability analysis of the wave is the Evans function, $D(\lambda)$, which is an analytic function of the eigenvalue parameter λ, whose roots coincide, counting multiplicities, with the eigenvalues of the associated linearized operator about the wave $\bar{u}(\cdot)$,

$$Lv := (B(x)v_x)_x - A(x)v_x, \tag{2}$$

where $A(x) = f'(\bar{u}) - B'(\bar{u})\bar{u}'$ and $B(x)$ are periodic matrices in the system case. The Evans function for periodic waves was developed in [G.1]. In the present setting,

$$Lw = \lambda w, \tag{3}$$

$$W' = \varLambda(\lambda, x)W, \tag{4}$$

where

$$W = (w, w'), \quad ' = \frac{\partial}{\partial x},$$

and

$$\varLambda = \begin{pmatrix} 0 & I \\ B^{-1}(\lambda I + A') & B^{-1}(A - B') \end{pmatrix}_{2n\times 2n}.$$

The Evans function is then defined as

$$D(\lambda, \gamma) = \det\,(M(\lambda) - \gamma I). \tag{5}$$

Here, $\gamma = e^{ikT}$, $k \in \Re$, and M is the monodromy matrix $M(\lambda) = \varPhi(T, \lambda)$ where T is a period of the wave and $\varPhi(\xi, \lambda)$ is the fundamental matrix for the linearized equation (4) about the periodic wave. $D(\lambda, \gamma) = 0$ precisely when γ is an eigenvalue of M, in which case one readily sees that $\lambda(k)$ trace out the spectrum of the linearized operator L about the traveling wave. Each zero corresponds to a solution $v(x,t) = e^{ikx+\lambda(k)t}z(x)$ of the linearized evolution equation $v_t = Lv$, where z is periodic on the same period T as the underlying traveling wave. Fixing $k = 0$, for example, we obtain the discrete (point) spectrum of the operator L considered as acting on the periodic functions on $[0, T]$. Stability of this ($k = 0$)-spectrum corresponds to stability with respect to periodic perturbations; more generally, the k-spectrum for different k reflect stability with respect to perturbations that are periodic on periods different from T. Note that in the periodic case D depends on a second parameter $\gamma = e^{ikT}$, whereas it depends only on λ for the traveling front case.

3 Evans function calculations

As in the front case, there is always a ($k = 0$)-eigenvalue at $\lambda = 0$, with eigenfunction equal to the derivative $\bar{u}_x$ of the underlying traveling wave $\bar{u}$; this corresponds with translational invariance of the dynamics. Thus, D vanishes at $(\lambda, k) = (0, 0)$. What is done in the front case is to evaluate the first nonvanishing derivative of D with respect to λ, and compare the sign to that approached by D as λ approached real plus infinity, thus obtaining a ***stability index*** that can detect existence of a root on the positive real axis: by the intermediated value theorem, opposite sign implies the existence of such a root. In Gardner–Zumbrun, it was shown by this technique that homoclinic profiles occurring in these model systems are unstable. By a general theorem of Gardner, relating spectra of periodic solutions to that of a bounding homoclinic cycle, one can then conclude that periodic orbits near such a boundary are also unstable. On the other hand, the numerically observed patterns seem to lie, rather, near bounding poly- (two- and three-) cycles, so this result does not directly apply.

The first, "front-type" Evans function calculation is to fix $k = 0$ and to compute directly an analogous stability index for the periodic waves, bypassing the argument of Gardner. The computation is rather more complicated than in the front case. It turns out to be possible for the same class of models considered in Gardner–Zumbrun to evaluate the stability index in the large-amplitude limit as periodic waves approach either 1-, 2-, or 3-cycles. This computation recovers the known result for homoclinics, but, interestingly, gives an indefinite, *positive sign* in the large-amplitude limit for the two- and three-cycle cases, corresponding to an even, and possibly vanishing, number of roots on the positive real axis. Likewise, even in the homoclinic case, the calculation gives an indefinite result also in the *small-amplitude* limit; and, our numerics confirm that in all the described cases, the number of unstable roots is in fact *zero*, i.e. the waves are *stable* with respect to periodic perturbations. But, the small-amplitude limit in particular is known to be unstable, by the instability of the nonlinear center to which they converge, considered as a constant solution (direct computation by Fourier Transform). So, the general approach of Gardner–Zumbrun is clearly unsuitable for these cases.

The resolution is simple for, the instability in the constant-coefficient limit is obtained from unstable spectra connected with *nonzero* k. Thus, we should Taylor expand D in both λ and k and try to determine the local dispersion relations $\lambda = \lambda_j(k)$ associated with the roots at k, $\lambda = 0$ (of multiplicity $n+1$, as it turns out, where $u \in \Re^n$). The appearance of the k variable makes our calculations look like those for a 2-dimensional traveling front problem but the result is 1-dimensional. When $n = 2$,

$$D(\lambda, \gamma) \approx \bar{\delta}\, \lambda \;\det\,(\,-\lambda \frac{\partial \bar{u}_m}{\partial u_-} A(u_-)^{-1} - ikT),$$

where $\bar{\delta} = det\,([\tilde{z}_1], [w_2])$, $\tilde{z}_1$ and w_2 are particular solutions of variational equations, and $\bar{u}_m$ is the average of the wave $\bar{u}$. Here and below, $[f]$ denotes $f(T) - f(0)$ for arbitrary function f. Unfortunately, the resulting formulae are quite complicated in general, and seem to yield little information. However, they are evaluable precisely for the class of model systems under consideration. And, in the large-amplitude limit, there turns out to be a beautiful dichotomy: either the stability index is negative, or else $d\lambda/dk$ has nonzero real part at k, $\lambda = 0$. From either possibility, there follows instability, though of different types.

There are various interesting transitions of the number of unstable eigenvalues of the periodic waves with respect to different amplitude inside a homoclinic orbit (Fig. 2): There are typically three roots at the origin, $\lambda = 0$, and a fourth stable root near the origin. As the amplitude goes to zero, the fourth root goes to the origin. As the amplitude becomes large, the fourth root of D moves from the left to the right half-plane of $Re(\lambda)$, becoming unstable. This work is described in [OZ.1].

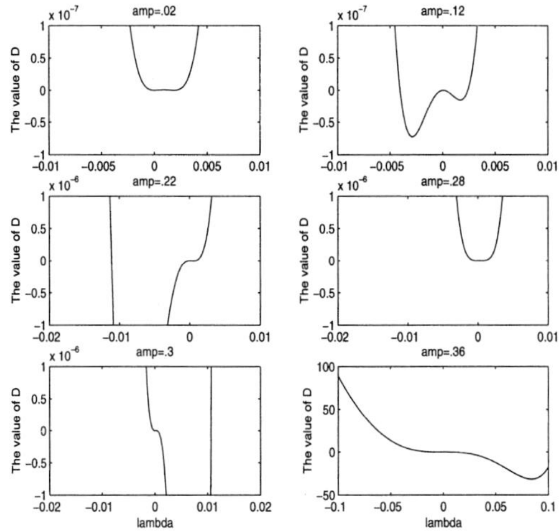

Fig. 2. Transition of the number of unstable eigenvalues

4 Pointwise bounds

We study the detailed behavior of the linearization of (1) about a stationary solution, assuming spectral stability. This is complementary to the section 3, in that our results are designed to show linearized and nonlinear stability rather than instability. We point out that it is not clear that all periodic solutions should be unstable in the conservation law setting. In [ZH], pointwise bounds are established on the time propagating Green's function, $G(x, t; y)$, for (2) in traveling front viscous shock case. The analysis in the front case hinges on employing the resolvent kernel, $G_\lambda(x, y)$, associated with the operator, $L - \lambda I$, and integrating over appropriate moving contours,

$$G(x, t; y) = \frac{1}{2\pi i} \int_\Gamma e^{\lambda t} G_\lambda(x, y) d\lambda. \tag{6}$$

This yields estimates on G similar to those appearing in Fourier transform analysis of the constant coefficient case, $Lv := -av_x + v_{xx}$ in [LZ]. A key step was to extend the Resolvent operator $(L - \lambda I)^{-1}$ *outside of the Resolvent set*, i.e. across the essential spectrum boundary; in doing so, one encounters poles, which yield residues corresponding to eigenfunctions/time-asymptotic states. Noth that in the periodic case, there are no eigenvalues, and no poles of the Resolvent.

Observing that "a" Resolvent is defined on each component of the complement of curves representing the essential spectrum, one can move the contour in the inverse Laplace Transform formula again to the stable complex half plane $Re\lambda < 0$ to obtain a negligible error term, plus the integral over all spectral curves of the jump in the Resolvent. But, the integral over each such curve, near $\lambda = 0$, turns out to be just an oscillatory "stationary phase" integral quite similar to that arising in the Fourier Transform analysis of the constant coefficient case. These can then be well estimated by deforming the spectral curves and continuing the integrand in an obvious (and analytic) way; at this point the analysis becomes a simplified version of that of [ZH]. Surprisingly, though the spectrum/Evans function theory is much more complicated in the periodic case, the analogy to the constant-coefficient case is much stronger in the derivation of pointwise bounds.

We illustrate this for the scalar constant coefficient case. By the Fourier transform, the spectrum lies on a dispersion curve

$$\Gamma' : \lambda = -iak - k^2. \tag{7}$$

Likewise, the characteristic equation determining asymptotic decaying rates μ of solutions w of the eigenvalue equation (3) is $\mu^2 - a\mu - \lambda = 0$, where

$$\mu = \frac{a \pm \sqrt{a^2 + 4\lambda}}{2}.$$

If $a > 0$, then $Re(\mu_1) < 0 < Re(\mu_2)$ to the right (i.e. one decaying solution at $-\infty$ and one decaying solution at ∞), and $Re(\mu_1), Re(\mu_2) > 0$ to the left (i.e. two decaying solutions at $-\infty$ and two growing solutions at ∞). Thus,

$$G_\lambda = \begin{cases} \rho_1^+(x)\tilde{\rho}_1^+(y) & \text{if } x > y \\ -\rho_2^-(x)\tilde{\rho}_2^-(y) & \text{if } x < y \end{cases}$$

to the right,

$$G_\lambda = \begin{cases} 0 & \text{if } x > y \\ -[\rho_1^-(x)\tilde{\rho}_1^-(y) + \rho_2^-(x)\tilde{\rho}_2^-(y)] & \text{if } x < y \end{cases}$$

to the left, and

$$[G_\lambda] = \begin{cases} \rho_1^+(x)\tilde{\rho}_1^+(y) & \text{if } x > y \\ \rho_1^-(x)\tilde{\rho}_1^-(y) & \text{if } x < y \end{cases}$$

where $\rho_j^\pm$ are decaying solutions of L, $\tilde{\rho}_j^\pm$ are those dual modes for $j = 1, 2$. Moving Γ to $-\infty$ (in the real direction) in (6), for $x > y$, we obtain

$$\begin{aligned}\int_{\Gamma} e^{\lambda t} G_\lambda(x,y) d\lambda &= \int_{\Gamma'} e^{\lambda t}[G_\lambda(x,y)] d\lambda \\ &= \int_{\Gamma'} e^{\lambda t} \rho_1^+(x) \tilde{\rho}_1^+(y) d\lambda \\ &= \int_{\Gamma'} e^{\lambda t} e^{\mu_1 x} e^{-\mu_1 y} \rho_1^+(0) \tilde{\rho}_1^+(0) d\lambda \\ &= \int_{\Gamma'} e^{\lambda t} e^{\mu_1 (x-y)} \rho_1^+(0) \tilde{\rho}_1^+(0) d\lambda \\ &= \int_{-\infty}^{\infty} e^{-iak-k^2} e^{ik(x-y)} dk, \end{aligned} \tag{8}$$

where Γ' is the dispersion curve (7). Note that $\rho_1^+(0)\tilde{\rho}_1^+(0) = \frac{1}{\Delta\mu} = \frac{dk}{d\lambda}$. Thus, (8) gives back the Fourier transform about which we already know the estimates for G. In the variable coefficient case, the result is, rather

$$G \approx \sum_{j=1}^{N} \frac{c_j}{\sqrt{4\pi b_j t}} e^{-\frac{|x-y-a_j t|^2}{4 b_j t}} q_j(0,x) \tilde{q}_j^*(0,y)$$

where $\lambda_j(k) = -ia_j k - b_j k^2 + \cdots$ denote the spectral curves passing through $\lambda = 0$, and all other spectrum is assumed strictly stable with q_j and $\tilde{q}_j$ periodic right and left eigenfunctions of L; and c_j are unspecified constants (see [OZ.2] for more details).

In the reaction–diffusion setting, our linear estimates refine Schneider's, which were obtained by quite different weighted norm/Bloch decomposition methods, as hinted above. On the other hand, Schneider's results are more general, applying also to the multi-dimensional case; likewise, Schneider carries out a full nonlinear stability analysis, which we do not. We hope to address these issues in future work.

5 Numerical computations

The analytic instability results obtained for large- and small-amplitude periodic waves by Evans function calculations are already of interest, since numerically observed oscillatory patterns in fact lie near the large-amplitude boundary (either a two- or a three-cycle of heteroclinic orbits, depending on the context). However, they leave open the possibility of stable intermediate amplitude waves. Here, we investigate numerically the stability over the full range of possible periodic orbits generalizing work done by Brin in the traveling front case. This involved: (i) Numerically determining the range of existence of periodic orbits, and solving for profiles on a representative selection of parameters; (ii) Numerical evaluation of the Evans function for a fixed profile and fixed frequencies k, λ; (iii) Determination of spectral curves $\lambda(k)$, i.e. zero-level sets of $D(\lambda, k)$; (iv) Winding number calculations for fixed k on judiciously chosen contours; and (v) Careful error estimates for all of the above.

The issues in the periodic case are rather different than those in the front case. In the periodic case, it is tricky to find the zero set of the two-parameter periodic-type Evans function than of one-parameter front-type Evans function. Adding to the difficulty is the fact that $D(\lambda, k)$ vanishes to $(n+1)$st order in λ at $(0,0)$, where n is the dimension of variable u, so that there are $n+1$ spectral curves bifurcating through the origin.

The resulting pictures are extremely illuminating, showing in a way that calculations can not the detailed evolution of the spectrum as the amplitude of the underlying periodic orbit is varied from zero ($\sim$ constant-coefficient case) to the limiting amplitude as orbits approach their bounding polycycle. They both verify and greatly amplify the observations obtained earlier by Evans function calculations. In particular, one immediately sees that instability is rather dramatic, whereas the analytical results via Taylor expansion about the origin imply only that there exist infinitesimally unstable spectra; moreover, this instability is not confined to small or large-amplitude limits, but persists across all amplitudes.

This suggests the strategy implemented in (iv) for numerical verification of instability; namely, to fix k at the value k_0 corresponding to the maximally unstable spectrum $\lambda_0 = \lambda(k_0)$ obtained by graphical examination, then perform a winding number calculation on a well-chosen contour about λ_0. This contour should lie well away from the root at λ_0, but must also be bounded away from the *stable* complex half-plane $Re\lambda \leq 0$; finally, a practical upper bound on $|\lambda|$ is enforced by the need to obtain good error estimates, which degrade rapidly for high frequencies.

Fig. 3. The number of unstable eigenvalues

Since the computation of D involves the numerical solutions of an ODE, we must determine how accurate the computed values are. There is no analytic solution of the ODE for (1). We have to rely entirely on numerical estimates to gauge the accuracy of the Evans function method [Br]. In the tables below, third order polynomial extrapolation was used to compute the values for $\tau = 0$ where τ is a local truncation error depending on the step size h, i.e. $\tau := Ch^4$ for some positive constant C (recall that the error to lowest order comes from solution of the profile equation by the fourth order Runge–Kutta scheme). The bounds for the winding number calculations were found conservatively (i.e. by first estimate) to be of order 10^{-2}, and likely much better, comfortably below the bound 1 needed to apply Rouchés Theorem.

Table 1. Error analysis for van der Waals fluids

τ	$D(.52 - .1i)$	extrapolated relative error
10^{-5}	$-4.636345 - 12.13182i$	.0082597282
10^{-6}	$-4.636959 - 12.13066i$	.0082772553
10^{-7}	$-4.637134 - 12.13046i$	.0006327496
10^{-8}	$-4.722007 - 12.19752i$	.0000185195
.	.	
.	.	
.	.	
0	$-4.721931 - 12.19775i$	

References

[AMPZ] Azevedo, A., Marchesin, D., Plohr, B., and Zumbrun, K. (1999): Capillary instability in multiphase flow models. preprint

[Br] Brin, L. Q. (1998): Numerical testing of the stability of viscous shock waves. Ph.D. dissertation, Indiana University

[CP] Čanić, S. and Peters, G. (2000): On the oscillatory solutions in hyperbolic conservation laws. Nonlinear Anal. Real World Appl. 1, no. 2, 287–314

[Ec] Eckhaus, W. (1965): Studies in nonlinear stability theory. Springer Tracts in Natural Philosophy, Vol. 6 Springer-Verlag New York, New York, Inc. viii+117 pp

[FL] Frid, H. and Liu, I.-S. (1998): Oscillation waves in Riemann problems for phase transitions. Quart. Appl. Math. 56, no. 1, 115–135

[G.1] Gardner, R. A. (1993): On the structure of the spectra of periodic traveling waves. J. Math. Pures Appl. 72, 415-439

[G.2] Gardner, R. A. (1997): Spectral analysis of long wavelength periodic waves and applications. J. reine angew. Math. 491, 149-181

[GZ] Gardner, R. A. and Zumbrun, K. (1998): The Gap lemma and geometric criteria for instability of viscous shock profiles. CPAM, Vol.LI, 0789-0847

[LZ] Liu, T. P. and Zumbrun, K. (1995): Nonlinear stability of an undercompressive shock for complex Burgers equation. Comm. Math. Phys. 168, no. 1, 163–186

[M.1] Maginu, K. (1980): Existence and stability of periodic traveling wave solutions to Nagumo's nerve equation. J. MATH. Biol. 10, 133-153

[M.2] Maginu, K. (1981): Stability of periodic traveling wave solutions with large spatial periods in reaction-diffusion systems. J. Diff. Equation, 39, 73-99

[OZ.1] Oh, M. and Zumbrun, K. (2001): Stability of periodic solutions of viscous conservation laws - Analysis of the Evans function. preprint

[OZ.2] Oh, M. and Zumbrun, K. (2001): Stability of periodic solutions of viscous conservation laws - Pointwise bounds on the Green's function. preprint

[ZH] Zumbrun, K. and Howard, P. (1998): Pointwise semigroup methods and stability of viscous shock waves. Indiana Univ. Math. J. 47, No. 3, 741-871

Scalar Conservation Laws with Boundary Effect

Tao Pan[1] and Hongxia Liu[2]

[1] Dept. of Mathematics and Information Sciences, Guangxi University, Nanning, 530004, P.R.China, and Lab. of Bioinformation and Food Engineering, Faculty of Bioresources, Mie University, Tsu, Mie, 514-8507, Japan, pan@bife.bio.mie-u.ac.jp

[2] Dept. of Mathematics, Jinan University, Guangzhou, 510632, P.R.China, hongxia-liu@163.net

Summary. This paper reviews the recent work on the global weak entropy solutions of scalar conservation laws with boundary effect and the global error estimate for viscosity methods to initial-boundary value problems of scalar conservation laws. In section 2, we state how to obtain the existence of weak entropy solution for general nonconvex scalar conservation laws by Dafermos's polygonal approximations method which can be used to simulate geometric structure of the weak entropy solution. We also state the large time behaviors of weak entropy solution by discussing the interaction of elementary waves for scalar convex conservation laws on a bounded domain. In section 3, we are concerned with the global error estimate for viscous approximations to inviscid solution to initial-boundary value problems of convex scalar conservation laws by the matching traveling wave solutions method.

1. Introduction Consider the scalar conservation laws on $(0,\infty)\times(0,\infty)$

$$u_t + f(u)_x = 0 \tag{1.1}$$

with the initial condition

$$u(x,0) = u_0(x), \quad x > 0, \tag{1.2}$$

and the boundary condition

$$u(0,t) = u_b(t), \quad t > 0. \tag{1.3}$$

The viscosity method approximating the initial-boundary value problem (1.1)-(1.3) is to solve the viscous scalar conservation laws on $[0,\infty)\times[0,\infty)$

$$(v_\varepsilon)_t + f(v_\varepsilon)_x = \varepsilon(v_\varepsilon)_{xx} \tag{1.4}$$

with the initial data

$$v_\varepsilon(x,0) = v_0(x), \quad x \geq 0, \tag{1.5}$$

and the boundary data

$$v_\varepsilon(0,t) = v_b(t), \quad t \geq 0, \tag{1.6}$$

where $\varepsilon > 0$ is a small viscosity parameter, and the weak entropy solution of (1.1)-(1.3) $u(x,t)$ can be constructed as the limit of solutions to the initial-boundary value problems (1.4)-(1.6) for the parabolic equation.

The initial-boundary value problem of scalar conservation laws plays important roles in the mathematical modeling and simulation of the practical problem of the one-dimensional sedimentation processes and traffic flow on highways [2,3,4,32,44]. For the initial-boundary value problems (1.1)-(1.3) on a bounded domain, the vanishing viscosity method proving the existence for the weak entropy solution in the BV-setting was established by Bardos-Leroux- Nedelec [1]. They also showed the uniqueness of the weak entropy solution by Kruzkov's method. The main difficulty for scalar conservation laws with boundary is to have a good formation of the boundary condition. Namely, for a fixed initial value as (1.2), we really cannot impose such a condition at the boundary as (1.3), and the boundary condition is necessarily linked to the entropy condition. In other words, the weak entropy solution of (1.1)-(1.3) $u(x,t)$ does not admit a trace at the boundary, namely, $u(0,t)$ does not always equal $u_b(t)$. Specically, Bardos-Leroux-Nedelec [1] proved that (1.3) can be replaced by the weaker condition

$$(\mathrm{sgn}(u(0+,t)-k)-\mathrm{sgn}(u_b(t)-k))(f(u(0+,t))-f(k)) \leq 0 \quad \text{for all } k \in (-\infty,\infty) \tag{1.7}$$

where $\mathrm{sgn}(x) = 1$ if $x > 0$, $\mathrm{sgn}(x) = 0$ if $x = 0$, $\mathrm{sgn}(x) = -1$ if $x < 0$. The approach dealing with the boundary in [1] has been followed by many authors. The condition (1.7) was used by LeFloch [14] in order to extend Lax's explicit formula [13] to the initial-boundary value problems on $[0,+\infty) \times [0,+\infty)$. By using the vanishing viscosity method and Hopf-Cole transformation, Joseph [9,10] extended Lax's formula for the inviscid Burgers equation. Anther derivation is given by Joseph-Veerappa Gowda [11]; see also LeFloch-Nedelec [15] and Terracina [43]. Also refer to the work [34] by Szepessy for a very general result of the existence and uniqueness.

In section 2, we state how to obtain the existence of weak entropy solution of the initial-boundary value problem for general nonconvex scalar conservation laws on $[0,+\infty) \times [0,+\infty)$ by Dafermos's polygonal approximations method [5], which was developed by Pan-Lin [29] where they showed that the approximations on $[0,+\infty) \times [0,+\infty)$ can be constructed in a finite number of step, which was left by Dafermos in [5] (for the initial value problem). So the approximations can be used to take numerical computation, and simulate the geometric structure of the weak entropy solution.

The paper [4] constructed the global weak entropy solution to the initial-boundary problem on a bounded interval for some special initial-boundary data corresponding the practical problem of continuous sedimentation of an ideal suspension, but they have not obtained the general result for all cases. We guess that the large time behaviors of weak entropy solution with two-side boundary effect will depend only on the boundary data, and be independent of the initial data, which is similar to that described by Glimm-Lax [6] and Liu [19], namely, the asymptotic stability of the global weak entropy solution is related to the solution of Riemann problem. In order to clarify this phenomena, Liu-Pan [16] investigated the structure and large time behaviors of weak entropy solution to convex scalar conservation laws on a bounded domain $[0,1] \times [0,+\infty)$ with three pieces of constant data by studying the interaction of elementary waves and two-side boundary ($x = 0, 1$). It is of interest to note that there is a new phenomona on the geometric structure of weak entropy solution, that is, if the central rarefaction wave collides with the boundary $x = 0$ or $x = 1$, then the boundary will reflects a shock wave. In this note, we will state the results of [16].

About the global error estimate for viscous solution to inviscid solution of the initial value problem of scalar conservation laws, Kuznetsov [12] was the first to establish the half-order rate of L^1-convergence for viscosity approximation and monotone scheme for BV entropy solution,. Tang-Teng [38] and Sabac [33] proved that this half-order rate of convergence is optimal in BV solution class. However, for convex conservation laws with piecewise smooth solution, the L^1-convergence rate can be improved to first-order, see e.g., Teng-Zhang [42] for monotone scheme, Tang-Teng [39] for viscosity approximation and Teng [41] for relaxation approximation, where the matching asymptotic method, developed by Goodman-Xin [8], was used to get the first-order error estimates. This method relies on L^1-stability properties of the approximation equations and nonlinear large time asymptotic stability of viscous shock profiles. For the initial value problems of systems of viscous conservation laws, the stability theory of the rarefaction waves and viscous shock waves has been extensively studies by many authors in the past decade, see e.g., [7,25,20,35] (For the initial-boundary problems, see also [23,22,21,28,24,30,26,31] for recent progress). We also point out that there have some first-order *pointwise* convergence results for viscous approximations to convex scalar conservation laws, see e.g., Tadmor-Tang [36, 37] where they used the energy method with some bootstrap extrapolation technique to obtain some first-order *pointwise* convergence results for viscous approximations to convex scalar conservation laws with piecewise smooth solutions. Recently, for nonconvex scalar conservation laws whose weak entropy solution is in the piecewise smooth solution class consisting of finitely many discontinuities, Tang-Teng-Xin [40] proved that the optimal rate of L^1-convergence of viscous approximations is a fractional number α satisfying $\frac{1}{2} < \alpha \leq 1$.

According to (1.7), the weak entropy solution $u(x,t)$ for (1.1)-(1.3) does not admit a trace at the boundary, namely, $u(0,t)$ does not always equal $u_b(t)$. Whereas, as viscosity approximation of the weak entropy solution for (1.1)-(1.3), the solution of initial-boundary value problems (1.4)-(1.6) for parabolic equation does admit a fixed trace at the boundary. Therefore, it is very interesting to consider the error estimates for the viscosity approximation for the initial-boundary value problems of scalar conservation laws. Liu-Pan [17] showed that the rate of L^1-convergence of viscous approximations to the scalar conservation laws on $[0,+\infty]\times[0,+\infty)$ with three pieces of constant data by the matching method [8,39]. They proved that the rate of L^1-convergence of viscous approximations is the half-order if the inviscid solution includes the interaction that the central rarefaction wave collides with the boundary $x=0$ and the boundary reflects a shock wave. Otherwise, the L^1-convergence rate is similar to that of the initial value problem in [39]. Very recently, Liu-Pan [18] have extended the result of [17] to more general initial-boundary value problems of the convex conservation laws whose weak entropy solution is in the piecewise smooth solution class consisting of finitely many discontinuities. In section 3, we will state the work of [18].

2. Polygonal Approximations and Large Time Behavior of Solution In this section, we first state how to obtain the existence of weak entropy solution of the initial-boundary value problem for general nonconvex scalar conservation laws on $[0,+\infty)\times[0,+\infty)$ by Dafermos's polygonal approximations method, and also state how to prove that the approximations in $[0,+\infty)\times[0,+\infty)$ can be constructed in a finite number of step, which was developed by Pan-Lin [29].

Definition 2.1 *A bounded and local bounded variation function $u(x,t)$ on $[0,\infty)\times[0,\infty])$ is called a weak entropy solution of the initial-boundary problems (1.1)-(1.3), if for each $k\in(-\infty,\infty)$, and for any nonnegative test function $\phi\in C_0^\infty([0,\infty)\times[0,\infty))$, it satisfies the following inequality*

$$\begin{aligned}&\int_0^\infty\int_0^\infty\{|u-k|\phi_t+\operatorname{sgn}(u-k)(f(u)-f(k))\phi_x\}dxdt\\&+\int_0^\infty|u_0(x)-k|\phi(x,0)dx+\int_0^\infty\operatorname{sgn}(u_b(t)-k)(f(u(0,t))-f(k))\phi(0,t)dt\\&\geq 0.\end{aligned}\tag{2.1}$$

By Definition 2.1, the weak entropy solution satisfies the following boundary entropy condition.

Lemma 2.2 *If $u(x,t)$ is a weak entropy solution of (1.1)-(1.3), then,*

$$\begin{aligned}&u(0+,t)=u_b(t)\quad\text{or}\\&\tfrac{f(u(0+,t))-f(k)}{u(0+,t)-k}\leq 0,\quad k\in I(u(0+,t),u_b(t)),\quad k\neq u(0+,t),\quad\text{a.e. }\ t\geq 0,\end{aligned}\tag{2.2}$$

where $I(u(0+,t),u_b(t))=[\min\{u(0+,t),u_b(t)\},\max\{u(0+,t),u_b(t)\}]$.

The following lemma can be proved by Definition 2.1 and Lemma 2.2, which will be used to construct the piecewise smooth solution to (1.1)-(1.3).

Lemma 2.3 *A piecewise smooth function $u(x,t)$ with piecewise smooth discontinuity curves is a weak entropy solution to (1.1)-(1.3) in the sense of (2.1), if and only if the following conditions are satisfied:*

(1) $u(x,t)$ satisfies equation (1.1) *on its smooth domains.*

(2) If $x = x(t)$ is a weak discontinuity of $u(x,t)$, then $\frac{dx(t)}{dt} = f'(u(x(t),t))$. If $x = x(t)$ is a strong discontinuity of $u(x,t)$, then

$$\frac{dx(t)}{dt} = \frac{f(u^+) - f(u^-)}{u^+ - u^-}, \qquad \text{(Rankine – Hugoniot condition)}$$

and for any u between u^- and u^+

$$\frac{f(u^+) - f(u)}{u^+ - u} \geq \frac{f(u^+) - f(u^-)}{u^+ - u^-}, \qquad \text{(Oleinik's entropy condition)}$$

where $u^- = u(x(t) - 0, t), u^+ = u(x(t) + 0, t)$.

(3) The boundary entropy condition (2.2) holds.

(4) $u(x,0) = u_0(x)$ a.e. $x \geq 0$.

The main result of [29] is the following theorem.

Theorem 2.4 *Assume that $u_0(x), u_b(t)$ are bounded variation function on $[0, +\infty)$, and*

$$m \leq u_0(x), u_b(t) \leq M, \qquad \text{for } x, t \in [0, +\infty] \tag{2.3}$$

where m, M are constants. And let $f(u)$ is locally Lipschitz continuous function on $(-\infty, +\infty)$, and

$$|f(u) - f(u')| \leq K|u - u'|, \qquad \text{for } u, u' \in [m, M] \tag{2.4}$$

where K is a positive constant. Then there exists a weak entropy solution $u(x,t)$ to (1.1)-(1.3), which has the following properties: For every fixed $t \in [0, +\infty]$, $u(\cdot, t)$ is a bounded variation function on $[0, +\infty]$ and bounded from above by M, from below by m.

For the case that $f(u)$ is a piecewise linear function, $u_0(x)$ is a step function and $u_b(t) \equiv u_b$ is a constant, following Dafermos [5], we can construct a piecewise constant solution $u(x,t)$ on $(-\infty, \infty) \times [T_0, T_1)$ for the initial value problem

$$\begin{cases} u_t + f(u)_x = 0, \\ u(x,0) = \begin{cases} u_b & x < 0, \\ u_0(x) & x > 0, \end{cases} \end{cases}$$

where, t-axis is regarded as a line discontinuity, $T_0 = 0, T_1$ is the earlist intersection time of any two or more neighboring line discontinuities in the time interval (T_0, ∞). Let

$$u(x, T_1) =: \begin{cases} u_b & x < 0 \\ u(x, T_1-) & x > 0, \end{cases}$$

then repeat the process using $u(x, T_1)$ as new initial value, we can extend the solution into $(-\infty, \infty) \times [T_0, T_2)$, where T_2 is the earlist intersection time of any two or more neighboring line discontinuities in the time interval (T_1, ∞), and so on, we can construct a piecewise constant solution $u(x,t)$ on $(-\infty, \infty) \times [0, \infty)$ layer by layer. We call these horizotal lines $T = T_k (k = 0, 1, 2, \cdots)$ the initial lines, here, T_k is the earlist intersection time of any two or more neighboring line discontinuities in the time interval (T_{k-1}, ∞).

Definition 2.5 *(1) Any two lines discontinuities are called* same direction discontinuous lines *if the difference of their lefe state and right state has the same sign. Otherwise, they are called* contrary direction discontinuous lines.

(2) The cross point of same direction discontinuous lines (two lines or more lines) is called a 1-cross point. *The cross point of lines discontinuities including the contrary direction discontinuous lines is called a* 2-cross point.

(3) If a line discontinuity extends upwards and hits the boundary $x = 0$, the cross point is called an incident point, *and the line discontiuity is called an* incident line.

(4) If the incident lines incide at same point on the boundary $x = 0$, and on the left of these incident lines, all lines discotinuities which lie in the second quadrant are the same direction discontinus lines, then the incident lines are called 1-incident lines. *The others are called* 2-incident lines. *The corresponding incident point is called a* 1-incident *point or* 2-incident point, *respectively.*

Definition 2.6 *The initial line which has only 1-cross point or 1-incident point is called a* 1-initial line. *The others are called* 2-initial lines.

Definition 2.7 *(1) The number of all discontinuous points of $u(\cdot,t)$ in $(-\infty, +\infty)$ is denoted by $D(t), t \geq 0$, and the number of all discontinuous points of $u(\cdot,t)$ in $(0, +\infty)$ is denoted by $D_1(t), t \geq 0$;*

(2) The number of vertices of the polygonal line $v = f(u)$ between $u(x_k - 0, t)$ and $u(x_k + 0, t)$ is denoted by $N(x_k, t)$, where x_k is the discontinuous point of $u(\cdot,t)$ in $(-\infty, +\infty)$.

(3) Denote $A(t) = D(t) + \sum_{x_k} N(x_k, t), t \geq 0$.

We have

Lemma 2.8 *(1) $A(t_k - 0) \geq A(T_k + 0), A(t) = A(T_k)$ for $t \in [T_k, T_{k+1})$;*
(2) If T_k is a 1-initial line, then $D_1(T_k - 0) > D_1(T_k)$;
(3) If T_k is a 2-initial line, then $A(T_k - 0) > A(T_k)$.

Lemma 2.9 *the number of initial lines $t = T_k$ is finite.*

In the case that $u_0(x), u_b(t)$ are all step functions, following the above method, we can construct the solution $u(x,t)$ on $(-\infty, \infty) \times [0, \infty)$, and in the whole process we only take a finite number of initial lines. We restrict the constructed solution $u(x,t)$ on $[0, \infty) \times [0, \infty)$, still denoted by $u(x,t)$, then we can easily verify that $u(x,t)$ is the weak entropy solution to (1.1)-(1.3), and have

Lemma 2.10 *Let $f(u)$ be a piecewise linear function with vertices at the point $(u^{(1)}, f(u^{(1)})), \cdots, (u^{(r)}, f(u^{(r)}))$ satisfying (2.4). Assume that $u_0(x)$ is a step function with values $u_0^{(1)}, \cdots, u_0^{(p)} \in [m, M]$, and $u_b(t)$ is a step function with values $u_b^{(1)}, \cdots, u_b^{(q)} \in [m, M]$. Then the initial-boundary value problem (1.1)-(1.3) has a piecewise constant weak entropy solution $u(x,t)$, which satisfies the following properties: for any fixed $t \in [0, +\infty), u(\cdot, t)$ is a step function with values in $J = \{u^{(1)}, \cdots, u^{(r)}, u_0^{(1)}, \cdots, u_0^{(p)}, u_b^{(1)}, \cdots, u_b^{(q)}\} \subset [m, M]$, moreover*

$$\operatorname*{Var}_{[0,+\infty)} u(\cdot,t) \leq \operatorname*{Var}_{(-\infty,+\infty)} U_0(x) \tag{2.5}$$

$$\int_0^{+\infty} |u(x,t) - u(x,t'))| dx \leq K|t - t'| \operatorname*{Var}_{(-\infty,+\infty)} U_0(x), \quad \text{for } t, t' \in [0, +\infty) \tag{2.6}$$

where

$$U_0(x) = \begin{cases} u_b(-x) & x < 0 \\ u_0(x) & x \geq 0. \end{cases}$$

In general case, we approximate $f(u)$ by a sequence of piecewise linear functions and $u_0(x), u_b(t)$ by a sequence of step functions, respectively. Then by an analogous compactness argument as in [27], we can establish the existence.

Now we state the structure and large time behaviors of weak entropy solution to scalar conservation laws on a bounded spacial domain $[0, 1]$, with the initial data $u(x, 0) = u_m$ and the boundary data $u(0,t) = u_-, u(1,t) = u_+$, where $u_\pm, u_m$ are constants, which are not equivalent, and $f \in C^2$ satisfies $f'' > 0,\ f(0) = f'(0) = 0$.

In this case, the boundary entropy condition is changed to the following

$$\begin{aligned} &u(0,t) = u_0^b(t) \ \text{ or} \\ &\frac{f(u(0+,t)) - f(k)}{u(0+,t) - k} \leq 0, \ k \in I(u(0+,t), u_0^b(t)), \ k \neq u(0+,t), \ \text{a.e. } t \geq 0, \end{aligned} \tag{2.7}$$

$$\begin{aligned} &u(1,t) = u_1^b(t) \ \text{ or} \\ &\frac{f(u(1-,t)) - f(k)}{u(1-,t) - k} \geq 0, \ k \in I(u(1-,t), u_1^b(t)), \ k \neq u(1-,t), \ \text{a.e. } t \geq 0. \end{aligned} \tag{2.8}$$

Denote by $u_{\max}$ the maximum point of $f(u)$ on $\{u_-, u_+\}$, and by $\overline{u}_{\min}$ or $\overline{u}_{\max}$ the minimum or maximum point of $|f'(u)|$ on $\{u_-, u_+\}$, respectively, then we can obtain the following large time behaviors of weak entropy solution to this problem.

Theorem 2.11 *(1) For the case of $u_- \leq u_+$,*
if $f'(u_-) \cdot f'(u_+) > 0$, then $\lim_{t\to\infty} u(x,t) = \overline{u}_{\min}$ $(x \in [0,1])$;
if $f'(u_-) \cdot f'(u_+) \leq 0$, then $\lim_{t\to\infty} \|u(\cdot,t)\|_{L^1[0,1]} = 0$.

(2) For the case of $u_- > u_+$,
if $f'(u_-) \cdot f'(u_+) \geq 0$, *then* $\lim\limits_{t \to \infty} u(x,t) = \overline{u}_{\max}$ $(x \in [0,1])$;
if $f'(u_-) \cdot f'(u_+) < 0$, $f(u_-) \neq f(u_+)$, *then* $\lim\limits_{t \to \infty} u(x,t) = u_{\max}$ $(x \in [0,1])$;
if $f'(u_-) \cdot f'(u_+) < 0$, $f(u_-) = f(u_+)$, *then*

$$\lim_{t \to \infty} u(x,t) = u_+ \ (x \in [0,1]) \ \text{ for } u_m \leq u_+;$$

$$\lim_{t \to \infty} u(x,t) = u_- \ (x \in [0,1]) \ \text{ for } u_m \geq u_-;$$

$$\lim_{t \to \infty} u(x,t) = \begin{cases} u_-, & x \in [0,a_1) \\ u_+, & x \in (a_1,1] \end{cases} \text{ for } u_- < u_m < u_+$$

where $a_1 = \frac{f(u_-)-f(u_m)}{u_- - u_m}\left(\frac{f(u_-)-f(u_m)}{u_- - u_m} - \frac{f(u_m)-f(u_+)}{u_m - u_+}\right)^{-1}$.

By studying the interaction of elementary waves and two-side boundary $(x = 0,1)$, we gave a construction proof of this Theorem in [16]. Here we only point out that there is a new phenomona on the geometric structure of weak entropy solution, that is, if the central rarefaction wave collides with the boundary $x = 0$ or $x = 1$, then the boundary will reflects a shock wave, for example, in the case of $u_+ < 0 \leq u_- < u_m, f(u_m) > f(u_+) > f(u_-)$ or $u_-, u_+ < 0 < u_m, f(u_+) < f(u_m)$ (see figure 2.1-2.2).

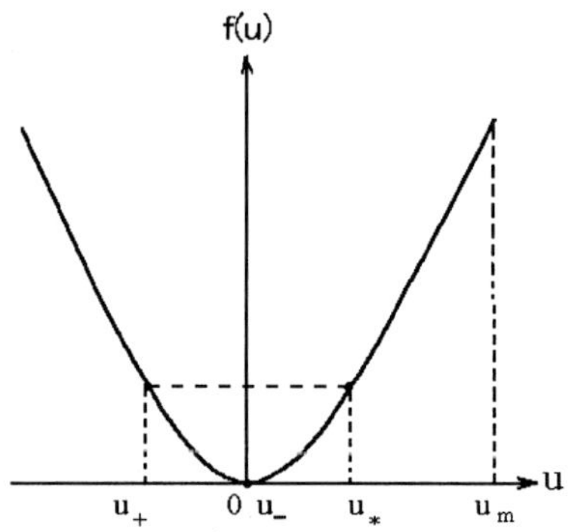

Figure 2.1

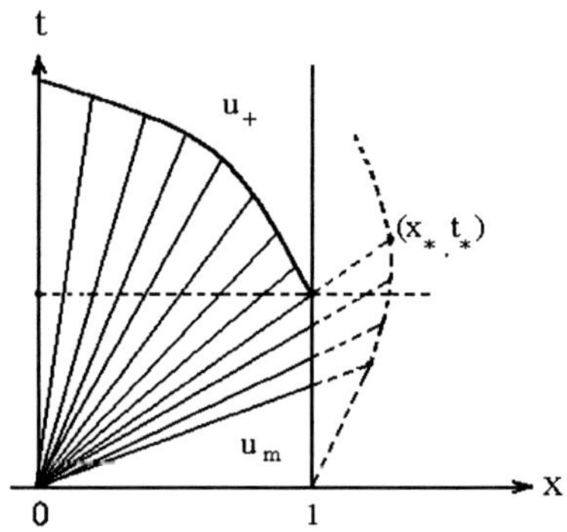

Figure 2.2

We guess that the conclusion in Theorem 2.11 can be extended to more general initial-boundary problems, such as the problems with piecewise smooth solutions, even those with general BV initial-boundary data. These will be expected in our problems, which will be investigated in a forthcoming paper.

3. Error Estimate for Viscosity Methods In this section, we will give the error estimates for the viscosity approximation for the initial-boundary value problem (1.1)-(1.3) whose weak entropy solution is in the piecewise smooth solution class consisting of finitely many discontinuities.

Throughout this section, the norm $||\cdot||$ denotes the standard L^1-norm, $||\cdot||_{L^1[0,\infty)}$; $C(t)$ denotes a positive constant independent of ε .

Before stating our main result, we make the following assumptions on the flux f and the initial-boundary data.

(A_1) $f \in C^2$ satisfies

$$f'' \geq \alpha > 0, \qquad f(0) = f'(0) = 0.$$

(A_2) $u_0(x)(x>0)$ is bounded and piecewise C^2-smooth function with a finite number of discontinuious points γ_i, $u_0(\gamma_i \pm 0)$ and $\dot{u}_0(\gamma_i \pm 0)$ exist and are finite.

(A_3) $\ddot{u}_0 \in L^1(R)$ and $\lim_{x\to\infty} \dot{a}(u_0(x)) = 0$.

(A_4) $a(u_0(x))$ has a finite number of inflection points and $\dot{a}(u_0)$ is sufficiently smooth near its negative minimum points.

(A_5) $u_b(t)(t>0)$ is a constant function.

Where $a(u) = f'(u)$, $\dot{a}(u_0(x)) := \frac{d}{dx}(a(u_0(x)))$, $\dot{u}_0(x) := \frac{du_0(x)}{dx}$, $\ddot{u}_0(x) := \frac{d^2u_0(x)}{dx^2}$.

Theorem 3.1 *Suppose that $v_0(x) - u_0(x) \to 0$ (as $x \to \infty$) and $v_0(\cdot) - u_0(\cdot) \in L^1$. Under the assumptions $(A_1) - (A_5)$, if v_ε is the smooth solution of (1.4)-(1.6) and u is the weak entropy solution of (1.1)-(1.3), then the following error estimate holds for any $T \geq 0$:*

$$\sup_{0\leq t\leq T} ||v_\varepsilon(\cdot,t) - u(\cdot,t)|| \leq ||v_0(\cdot) - u_0(\cdot)|| + C(T)(\varepsilon^{1/2} + \varepsilon|\ln\varepsilon| + \varepsilon).$$

In order to complete the proof of this Theorem, we must clarify the geometric structure of the weak entropy solution of the problem (1.1)-(1.3). Being motivated by the work of Pan-Lin [29], under the assumptions of this Theorem, we constructed the weak entropy solution of (1.1)-(1.3) in the piecewise smooth solution class consisting of finitely many discontinuities. The geometric structure of weak entropy solution are described in details in [17,18]. After clarifying the solution structures, we can use the matching method to prove this Theroem. As stated in the introduction, this method relies on the behavior of the traveling solutions of the approximation equation and the L^1-stability properties of the nonhomogeneous viscous equations with initial and boundary data . We first introduce the traveling wave solution lemma and the L^1-stability lemma, which corrresponds to that of the initail value problem in [39].

Lemma 3.2 *Let $v^{(i)}(x,t)$ $(i=1,2)$ be continuous and piecewise smooth solutions of the following equations*

$$(v^{(i)})_t + f(v^{(i)})_x - \varepsilon(v^{(i)})_{xx} = g_i(x,t), \quad x>0,\ t>d\geq 0,\ i=1,2.$$

The above equations hold for all values of $x>0$ except on some curves $X_m(t)$, $1 \leq m \leq M$, where $v_x^{(i)}$ may not exist. If $\omega := v^{(1)} - v^{(2)} \to 0$ as $x \to \infty$, then

$$\begin{aligned}||\omega(\cdot,t)|| \leq ||\omega(\cdot,d)|| &+ \int_d^t ||g_1(\cdot,\tau)-g_2(\cdot,\tau)||d\tau \\ &+\int_d^t \mathrm{sgn}\omega(0,\tau)(f(v^{(1)}(0,\tau))-f(v^{(2)}(0,\tau)))d\tau \\ &+\varepsilon\int_d^t |\omega_x(0+,\tau)|d\tau \\ &+\varepsilon\sum_{m=1}^{M}\int_d^t |\omega_x(X_m(\tau)+0,\tau)-\omega_x(X_m(\tau)-0,\tau)|d\tau.\end{aligned}$$

Lemma 3.3 *Let $u^-(t) > u^+(t)$ be two given functions, $X(t)$ be defined by the equation*

$$\frac{dx(t)}{dt} = \frac{f(u^-)-f(u^+)}{u^- - u^+}.$$

Then there is a unique traveling wave $V(x-X(t);u^-,u^+)$ of (1.4) taking on $V(-\infty;u^-,u^+)= u^-, V(0;u^-,u^+)= \frac{1}{2}(u^-+u^+), V(\infty;u^-,u^+) = u^+$, which has the following properties:

(1) *$V(\xi\ ;u^-,u^+)$ is a decreasing function with respect to ξ.*
(2) $|V(\xi\ ;u^-,u^+) - H(\xi;u^-,u^+)| \leq (u^- - u^+)\exp\{-\alpha(u^- - u^+)|\xi|/2\varepsilon\}$.
(3) $||H(\cdot\ ;u^-,u^+) - V(\cdot\ ;u^-,u^+)||_{L^1(-\infty,\infty)} \leq c\varepsilon$.
(4) $||V_{u^-}(\cdot\ ;u^-,u^+)\dot{u}^- + V_{u^+}(\cdot\ ;u^-,u^+)\dot{u}^+ - H(\cdot\ ;\dot{u}^-,\dot{u}^+)||$
$\leq c\varepsilon(|u_x(X(t)+0,t)| + |u_x(X(t)-0,t)|)$.

where $\dot{u}^{\pm} = d(u^{\pm})/dt$, H is the so-called Heaviside function defined by

$$H(x;u^-,u^+) = \begin{cases} u^-, & x<0 \\ u^+, & x>0, \end{cases}$$

and α is defined by (A_1).

A complete proof of Theorem 3.1 can be found in [18]. Here we only give the main clue of proof. if the weak entropy solution $u(x,t)$ of (1.1)-(1.3) contains no shock discontinuity for $t \in (\tau_1,\tau_2)$ $(0<\tau_1<\tau_2<T<\infty)$, then in view of continuity of u in the time interval (τ_1,τ_2), we can directly apply Lemma 3.1 to $v_\varepsilon(x,t)$ and $u(x,t)$; Otherwise, we construct an auxiliary continuous approximation $\overline{v}_\varepsilon$ by replacing jumps of all shocks in the weak entropy solution u at each fixed time t in (τ_1,τ_2) with their corresponding traveling wave solutions of the approximation equation. Using the traveling wave solution Lemma 3.2 estimates the L^1-error of $\overline{v}_\varepsilon - u$. As for the L^1-error bound of $v_\varepsilon - \overline{v}_\varepsilon$, we apply the L^1-stability Lemma 3.1 to $\overline{v}_\varepsilon(x,t)$ and $v_\varepsilon(x,t)$.

Remark 3.3 *If the inviscid solution includes no interaction that the rarefaction wave collides with the boundary $x=0$ and the boundary reflects a shock wave, then the error of the viscosity solution to the inviscid solution is bounded by $O(\varepsilon|\ln\varepsilon|+\varepsilon)$ in L^1-norm (see also [18]).*

Remark 3.4 *The conclusion in Theorem 3.1 can be extended to more general initial-boundary problems, in which $u_b(t)(t>0)$ is a piecewise constant function with a finite number of discontinuous points.*

We also point out that the error estimate for the viscous approximations obtained here is *global*. By the method used in [36,37] and the obtained global L_1-error bound, we derive some new results on the *pointwise* error estimate for the viscous approximations near the boundary. For abbreviation, we omit the detailed statement.

References

1. C.Bardos, A.Y.Leroux and J.C.Nedelec, *First order quasilinear equations with boundary conditions*, Commun. in Partial Diff. Eqs., **4**(1979), 1017-1034.
2. M.C.Bustos, F.Concha and W.L.Wendland, *Global weak solution to the problem of continuous sedimentation of an ideal suspension*, Math. Methods in the Appl. Sci., **13**(1990), 1-22.
3. M.C.Bustos, F.Paiva and W.L.Wendland, *Control of continuous sedimentation of an ideal suspension as an initial and boundary value problem*, Math. Methods in the Appl. Sci., **12**(1990), 533-548.
4. M.C.Bustos, F.Paiva and W.L.Wendland, *Entropy boundary condition in the theory of sedimentation of ideal suspension*, Math. Methods in the Appl. Sci., **19**(1996), 679-697.
5. C.M. Dafermos, *Polygonal Approximations of solutions of the initial value problem for a conservation law*, J. Math. Anal. Appl., **38**(1972), 33-41.
6. J.Glimm and P.D.Lax, *Decay of solution of systems of nonlinear hyperbolic conservation laws*, Amer. Math. Soc. Memoir, **101**(1970), A. M. S., Providence.
7. J.Goodman, *Nonlinear asymptotic stability of viscous shock profiles for conservation laws*, Arch. Rat. Mech. Anal., **95**(1986), 325-344.
8. J.Goodman and Z.Xin, *Viscous limits for piecewise smooth solutions to systems of conservation laws*, Arch. Rat. Mech. Anal., **121**(1992), 235-265.
9. K.T.Joseph, *Burgers equation in the quarter plane: a formula for the weak limit*, Comm. Pure Apll. Math. **41**(1988), 133-149.
10. K.T.Joseph, *Boundary layer in the approxomate solution*, Trans. Amer. Math. Sco. **314**(1989), 709-726.
11. K.T.Joseph and G.D.Veerappa Gowda, *Explicit formula for the solution of convex conservation laws with boundary condition*, Duke Math. J., **62**(1991), 401-416.
12. N.N.Kuznetsov, *Accuracy of some approximate methods for computing the weak solutions of a first-order quasi-linear equation*, USSR Comput. Math. Math. Phys., **16**(1976), 105-119.
13. P.D. Lax, *Hyperbolic systems of conservation laws II*, Comm. Pure Appl. Math., **10**(1957), 537-566.
14. P.LeFloch, *Explicit formula for scalar nonlinear conservation laws with boundary conditions*, Math. Methods in the Appl. Sci., **10**(1988), 265-287.
15. P.LeFloch and J.C.Nedelec, *Explicit formula for weighted scalar nonlinear conservation laws*, Trans. Amer. Math. Sco., **314**(1989), 709-726.
16. H.X.Liu and T.Pan, *Interaction of Elementary waves for scalar conservation laws on a bounded domain*, Math. Methods in the Appl. Sci., to appear
17. H.X.Liu and T.Pan, *L^1-convergence rate of viscosity methods for scalar conservation laws with the interaction of elementary waves and the boundary*, Q. Appl. Math., to appear.

18. H.X.LIU AND T.PAN, *L^1-error bound for viscous approximation for scalar conservation laws with boundary effect*, to appear.
19. T.-P.LIU, *Linear and nonlinear large time behavior of general systems of hyperbolic conservation laws*, Comm. Pure. Appl. Math., **30**(1977), 767-796.
20. T.P.LIU, *Nonlinear stability of shock waves for viscous conservation laws*, Mem. Amer. Math. Soc., **328**(1985), AMS, Providence, RI.
21. T.P.LIU, A.MATSUMURA AND K.NISHIHARA, *Behavior of solutions for the Burgers equations with boundary corresponding to rarefaction waves*, SIAM J. Math. Anal., **29**(1998), 293-308.
22. T.P.LIU AND K.NISHIHARA, *Asymptotic behavior for svalar viscous conservation laws with boundary effect*, J. Diff. Equs., **133**(1997), 296-320.
23. T.P.LIU AND S.H.YU, *Propagation of a stationary shock layer in the presence of a boundary*, Arch. Rat. Mech. Anal., **139**(1997), 57-82.
24. A.MATSUMURA AND M.MEI, *Convergence to traveling fronts of solutions of the p-system with viscosity in the presence of a boundary*, Arch. Rat. Mech. Anal., **146**(1999), 1-22.
25. A.MATSUMURA AND K.NISHIHARA, *Asymptotic stability of traveling waves of a one-dimensional model system for compressible viscous gas*, Japan J. Appl. Math., **2**(1985), 17-25.
26. A.MATSUMURA AND K.NISHIHARA, *Large-time behaviors of solutions to an inflow problem in the half space for a one-dimensional system of compressible viscous gas*, Comm. Math. Phys. **222**(2001), 449-474.
27. O.A.OLEINIK, *Discontinuous solutions of nonlinear linear equation*, Usp. Mat.Nauk., **12**(1957), 3-73 (in Russian).
28. T.PAN, AND H.X.LIU, *Asymptotic Behaviors of the Solution to A Initial-Boundary Value Problem for Scalar Viscous Conservation Laws*, Appl. Math. Letter, **15**(2002), 727-734.
29. T.PAN AND L.W.LIN, *The global solution of the scalar nonconvex conservation law with boundary condition I; II*, J. Partial Diff. Eqs., **8**(1995), 371-383; **11**(1998), 1-8.
30. T.PAN, H.X.LIU AND K.NISHIHARA, *Asymptotic stability of the rarefaction wave of a one-dimensional model system for compressible viscous gas with boundary*, Japan J. Indust. Appl. Math., **16**(1999), 431-441.
31. T.PAN, H.X.LIU AND K.NISHIHARA, *Asymptotic Behavior of a One-dimensional Compressible Viscous Gas with Free Boundary*, SIAM J. Math. Anal., to appear.
32. P.I.RICHARDS, *Shock waves on highways*, Oper. Res., **4**(1956), 42-51.
33. F.SABAC, *The optimal convergence rate of monotone finite differential methods for hyperbolic conservation laws*, SIAM J. Numer. Anal., **34**(1997), 2306-2318.
34. A. SZEPESSY, *Measure-value solution to scalar conservation laws with boundary condition*, Arch. Rat. Mech. Anal., **139**(1989), 181-193.
35. A.SZEPESSY AND Z.P.XIN, *Nonlinear stability of viscous shock waves*, Arch. Rat. Mech. Anal., **122**(1993), 53-103.
36. E.TADMOR AND T.TANG, *Pointwise error estimates for scalar conservation laws with piecewise smooth solutions*, SIAM J. Numer. Anal. **36**(1999), 1739-1758.
37. E.TADMOR AND T.TANG, *Pointwise error estimates for relaxation approximations to conservation laws*, SIAM J. Math. Anal. **32**(2001), 870-886.
38. T.TANG AND Z.H.TENG, *The sharpness of Kuznetsov's $O(\sqrt{\Delta x})$ L^1-error estimate for monotone difference schemes*, Math. Comp., **64**(1995), 581-589.

39. T.Tang and Z.H.Teng, *Viscosity methods for piecewise smooth solutions to scalar conservation laws*, Math. Comp., **66**(1997), 495-526.
40. T.Tang, Z.H.Teng and Z.P.Xin, *Fractional rate of convergence for viscous approximation to noncovex conservation laws*, SIAM J. Math. Anal., to appear.
41. Z.H.Teng, *First-order L^1-convergence for relaxation approximations to conservation laws*, Comm. Pure Appl. Math., **51**(1998), 857-895.
42. Z.H.Teng and P.W.Zhang, *Optimal L^1-rate of convergence for viscosity method and monotone scheme to piecewise constant solutions with shocks*, SIAM J. Numer. Anal., **34**(1997), 959-978.
43. A.Terracina, *Comparison properties for scalar conservation law with boundary condition*, Nonlinear Anal., **28**(1997), 633-653.
44. G.B.Whitham, *Linear and nonlinear waves*, Wiley-Interscience, New York, 1974.

To the Theory of Generalized Entropy Solutions of the Cauchy Problem for a First Order Quasilinear Equation in the Class of Locally Integrable Functions

E.Yu. Panov

Novgorod State University, Velikiy Novgorod, Russia, *pey@novsu.ac.ru*

Summary. We construct the theory of locally integrable generalized entropy solutions (g.e.s.) of the Cauchy problem for a first order nonhomogeneous quasilinear equation in the case when the flux is only continuous and satisfies the linear growth condition. We prove existence of maximal and minimal g.e.s., deduce sufficient conditions for uniqueness of g.e.s. Variants of comparison principle are proved, estimates of L^p-norms of g.e.s. with respect to spatial variables are found. Uniqueness of g.e.s. is established in the case when input functions are periodic in space variables. It is also shown that the linear growth restriction for the flux vector seems to be necessary for well-posedness of the Cauchy problem in classes of unbounded g.e.s.

1 Introduction

Consider the Cauchy problem for a first order quasilinear equation

$$u_t + \operatorname{div}_x \varphi(u) = g(t,x,u) \; , \tag{1}$$

$\varphi = (\varphi_1, \ldots, \varphi_n)$, $u = u(t,x)$, $(t,x) \in \Pi_T = (0,T) \times \mathbb{R}^n$, $T > 0$, with the initial condition

$$u(0,x) = u_0(x) \in L^1_{loc}(\mathbb{R}^n) \; . \tag{2}$$

We assume that the flux vector $\varphi(u)$ is only continuous and satisfies the following growth restriction

$$|\varphi(u)| \le C_0(1+|u|) \; , \; C_0 = \text{const} \; , \tag{3}$$

here and below $|\varphi|$ is the Euclid norm of a vector $\varphi \in \mathbb{R}^n$.

Let us denote $\bar{\Pi}_T = [0,T) \times \mathbb{R}^n$ and suppose that the source function $g = g(t,x,u) \in L^1_{loc}(\bar{\Pi}_T)$ for any fixed $u \in \mathbb{R}$, is continuous with respect to the variable u for all $(t,x) \in \Pi_T$ and for some constant $L \ge 0$

$$g(t,x,u) - g(t,x,v) \le L(u-v) \quad \forall u,v \in \mathbb{R} \; , \; u > v \; . \tag{4}$$

Definition 1 (S.N. Kruzhkov, [9]). *A locally integrable function $u = u(t,x) \in L^1_{loc}(\Pi_T)$ is called a generalized entropy solution (briefly -* g.e.s. *) of the Cauchy problem (1), (2) if $g = g(t,x,u(t,x)) \in L^1_{loc}(\Pi_T)$ and* *a)* $\forall k \in \mathbb{R}$

$$|u-k|_t + \mathrm{div}_x[\mathrm{sign}\,(u-k)(\varphi(u)-\varphi(k))] \le \mathrm{sign}\,(u-k)g$$

in the sense of distributions on Π_T *(in* $\mathcal{D}'(\Pi_T)$ *);*
b) $\operatorname*{ess\,lim}_{t\to 0+} u(t,\cdot) = u_0$ *in* $L^1_{loc}(\mathbb{R}^n)$*.*

One could introduce a g.e.s. $u \in L^p_{loc}$ for $p > 1$ replacing (3) to the requirement $|\varphi(u)| \le C_0(1+|u|^p)$ (without any restrictions for $p=\infty$). But, in this case, the Cauchy problem seems to be ill-posed (see [7, 8]), even if $n=1$, $g\equiv 0$ none of the positive results (established below for $p=1$) remains valid. Consider, for instance, the equation $u_t + (|u|^{p-1}u)_x = 0$, $p>1$ and let $r_0 = (p-1)^{-1}$. Then for any $r > r_0$ a nontrivial locally bounded g.e.s. $u = u(t,x)$ could be constructed for this equation with zero initial data, which satisfies the additional growth restriction

$$|u| \le \mathrm{const}(1+|x|^r)\,. \tag{5}$$

Thus, even the zero g.e.s. seems not to be unique in the case under consideration. It is interesting that existence and uniqueness of g.e.s. could be established in the class of functions satisfying the restriction (5) with $r = r_0$ (thus this exponent r_0 is "critical").

Denote by $C_0^\infty(\bar{\Pi}_T)$ the subspace of functions in $C^\infty(\bar{\Pi}_T)$ with a compact support (which therefore may intersect with the initial hyperspace $\{t=0\}$). We give the following usefull reformulation of Definition 1.

Proposition 1. *A function* $u = u(t,x) \in L^1_{loc}(\bar{\Pi}_T)$ *is a* g.e.s. *of the problem (1), (2) if and only if for any* $k \in \mathbb{R}$ *and any test function* $f \in C_0^\infty(\bar{\Pi}_T)$*,* $f \ge 0$

$$\int_{\Pi_T} \{|u-k|f_t + \mathrm{sign}\,(u-k)[(\varphi(u)-\varphi(k),\nabla_x f) + gf]\}dtdx + \int_{\mathbb{R}^n} |u_0(x)-k|f(0,x)dx \ge 0\,.$$

Corollary 1. *If a function* $u = u(t,x) \in L^1_{loc}(\bar{\Pi}_T)$ *is a* g.e.s. *of the problem (1), (2) then* $g = g(t,x,u(t,x)) \in L^1_{loc}(\bar{\Pi}_T)$ *and for any test function* $f = f(t,x) \in C_0^\infty(\bar{\Pi}_T)$

$$\int_{\Pi_T} [uf_t + (\varphi(u),\nabla_x f) + gf]dtdx + \int_{\mathbb{R}^n} u_0(x)f(0,x)dx = 0\,. \tag{6}$$

In particular the identity (6) with $f = f(t,x) \in C_0^\infty(\Pi_T)$ *implies that*

$$u_t + \mathrm{div}_x\varphi(u) = g \quad in\ \mathcal{D}'(\Pi_T)$$

and $u(t,x)$ *satisfies the equation (1) in the sense of distributions (* i.e. *it is really a generalized solution).*

We need notions a generalized entropy sub- and super- solution (see [2]-[6],[12]). Let $f^+ = \max(f,0)$, $f^- = \max(-f,0)$. Denote by $\chi_{u>v}$ the indicator function of the set of independent variables where the relation $u > v$ holds for two given functions u and v. In the same way we define functions $\chi_{u<v}$, $\chi_{u=v}$ and so on. Let $\operatorname{sign}^+(r) = \chi_{r>0}$ be the Heaviside function and $\operatorname{sign}^-(r) = -\operatorname{sign}^+(-r)$.

Definition 2. *A function $u(t,x) \in L^1_{loc}(\Pi_T)$ is called a generalized entropy sub-solution (briefly -* g.e.sub-s. *) of the problem (1), (2) iff*

1) $\forall k \in \mathbb{R}$ the function $\operatorname{sign}^+(u(t,x)-k)g(t,x,u(t,x)) \in L^1_{loc}(\Pi_T)$ and

$$((u-k)^+)_t + \operatorname{div}_x[\operatorname{sign}^+(u-k)(\varphi(u)-\varphi(k))] \le \operatorname{sign}^+(u-k)g(t,x,u) \quad \textit{in } \mathcal{D}'(\Pi_T);$$

2) $\operatorname*{ess\,lim}_{t\to 0+}(u(t,x) - u_0(x))^+ = 0$ *in* $L^1_{loc}(\mathbb{R}^n)$.

Definition 3. *A function $u(t,x) \in L^1_{loc}(\Pi_T)$ is called a generalized entropy super-solution (briefly -* g.e.super-s. *) of the problem (1), (2) iff*

1) $\forall k \in \mathbb{R}$ $\quad\operatorname{sign}^-(u(t,x)-k)g(t,x,u(t,x)) \in L^1_{loc}(\Pi_T)$ and

$$((u-k)^-)_t + \operatorname{div}_x[\operatorname{sign}^-(u-k)(\varphi(u)-\varphi(k))] \le \operatorname{sign}^-(u-k)g(t,x,u) \quad \textit{in } \mathcal{D}'(\Pi_T);$$

2) $\operatorname*{ess\,lim}_{t\to 0+}(u(t,x) - u_0(x))^- = 0$ *in* $L^1_{loc}(\mathbb{R}^n)$.

Remark 1. a) A function $u = u(t,x) \in L^\infty(\Pi_T)$ is a g.e.super-s. to the problem (1), (2) if and only if the function $-u$ is a g.e.sub-s. to the Cauchy problem $u_t - \operatorname{div}_x \varphi(-u) = -g(t,x,-u)$, $u(0,x) = -u_0(x)$;

b) A function $u(t,x) \in L^\infty(\Pi_T)$ is a g.e.s. of the problem (1), (2) if and only if it is a g.e.sub-s. and a g.e.super-s. of this problem simultaneously;

c) Definition 2 (and 3) can be reformulate in the form of a single integral inequality in the way as in Proposition 1.

2 Existence of g.e.s.

Consider firstly the case when $g = g(t,x) \in L^1_{loc}(\bar{\Pi}_T)$.

Theorem 1. *There exists a map $\mathcal{F} : L^1_{loc}(\mathbb{R}^n) \times L^1_{loc}(\bar{\Pi}_T) \mapsto L^1_{loc}(\bar{\Pi}_T)$, which associates to the input data (u_0, g) a* g.e.s. *$u = \mathcal{F}(u_0, g)$ of the problem (1), (2) so that if $u = \mathcal{F}(u_0,g)$, $v = \mathcal{F}(v_0,h)$ and $(u_0 - v_0)^+ \in L^1(\mathbb{R}^n)$, $(g-h)^+ \in L^1(\Pi_T)$ then for* a.e. *$\tau \in (0,T)$*

$$\int_{\mathbb{R}^n} (u(\tau,x) - v(\tau,x))^+ dx$$

$$\le \int_{\mathbb{R}^n} (u_0(x) - v_0(x))^+ dx + \int_{\Pi_\tau} \chi_{u\ge v}(g(t,x) - h(t,x))^+ dt dx \,.$$

Corollary 2. *1) if* $u_0 \le v_0$ a.e. *on* $\mathbb{R}^n$, $g \le h$ a.e. *on* Π_T *and* $u = \mathcal{F}(u_0, g)$, $v = \mathcal{F}(v_0, h)$ *then* $u \le v$ a.e. *on* Π_T *(comparison principle);*

2) if $u_0 - v_0 \in L^1(\mathbb{R}^n)$, $g - h \in L^1(\Pi_T)$, *and* $u = \mathcal{F}(u_0, g)$, $v = \mathcal{F}(v_0, h)$ *then for* a.e. $\tau \in (0,T)$ $u(\tau,\cdot) - v(\tau,\cdot) \in L^1(\mathbb{R}^n)$ *and*

$$\|u(\tau,\cdot) - v(\tau,\cdot)\|_1 \le \|u_0 - v_0\|_1 + \|g - h\|_{L^1(\Pi_\tau)} \,.$$

In the general case denote by $S \subset L^1_{loc}(\bar{\Pi}_T \times \mathbb{R})$ a set of functions $g = g(t,x,u)$ that satisfy the condition (4) (with a constant $L = L(g)$ depending on g).

Theorem 2. *There exists a map* $\mathcal{F} : L^1_{loc}(\mathbb{R}^n) \times S \mapsto L^1_{loc}(\bar{\Pi}_T)$, *which associates to the input data* (u_0, g) *a* g.e.s. $u = \mathcal{F}(u_0, g)$ *of the problem (1), (2) so that the following comparison principle holds*

if $u = \mathcal{F}(u_0, g)$, $v = \mathcal{F}(v_0, h)$, *and* $u_0 \le v_0$ a.e. *on* $\mathbb{R}^n$, $g \le h$ a.e. *on* $\Pi_T \times \mathbb{R}$ *then* $u \le v$ a.e. *on* Π_T.

3 Existence of the maximal and the minimal g.e.s.

In the case under consideration when the property of finite speed of propagation for initial perturbation can be violated uniqueness of g.e.s. to the problem (1), (2) is no longer valid even in the class of bounded g.e.s. (see examples in [12, 13]). In this connection the result about existence of the maximal and the minimal solution plays important role. Thus, based on this result in the last section we establish uniqueness of g.e.s. for periodic input data. In the class of bounded g.e.s. existence of the maximal and the minimal g.e.s. was proved before under various additional restrictions in the papers [5, 1, 16]. We use the approach developed in [15] similar to well-known Perron's method in the proof of solvability of Dirichlet problem for the Laplace equation. Firstly, we prove that the supremum of any family of g.e.sub-s. (the infimum of any family of g.e.super-s.) is a g.e.sub-s. (respectively - a g.e.super-s.) of our problem as well. Therefore there exist the maximal g.e.sub-s. and respectively the minimal g.e.super-s. of the problem (1), (2) and it only remains to prove that they are indeed g.e.s. of our problem. Thus we have the following

Theorem 3. *There exist the maximal* g.e.sub-s. $v = J^+(u_0, g)$ *and the minimal* g.e.super-s. $w = J^-(u_0, g)$ *of the problem (1), (2) and they are* g.e.s. *as well. In particular,* $v = J^+(u_0, g)$ *and* $w = J^-(u_0, g)$ *are respectively the maximal and the minimal* g.e.s. *of the problem (1), (2).*

Corollary 3. *Let* $u_0, v_0 \in L^1_{loc}(\mathbb{R}^n)$, $g, h \in S$, $u_0 \le v_0$ a.e. *on* $\mathbb{R}^n$, $g \le h$ a.e. *on* $\Pi_T \times \mathbb{R}$, *and* $u^\pm = J^\pm(u_0, g)$, $v^\pm = J^\pm(v_0, h)$. *Then* $u^+ \le v^+$, $u^- \le v^-$ a.e. *on* Π_T.

4 Comparison principle and uniqueness of g.e.s.

As we said before, a g.e.s. of the problem (1), (2) isn't generally unique. First sufficient conditions for uniqueness of g.e.s. in the class of bounded integrable functions were found in [10, 11] and extended to the class of arbitrary bounded g.e.s. in [12, 13]. These conditions are formulated in terms of continuity modulus of flux functions $\varphi_i(u)$, $i = 1, \dots, n$. After some corrections the conditions in [12, 13] seem to be suitable for the proof of uniqueness of g.e.s. (more generally - the comparison principle) also in the general case of the problem (1), (2). We make use of the following key result from [5] (see also [9, 4, 6, 2]).

Lemma 1. *Let* u *be a* g.e.sub-s., v *be a* g.e.super-s. *of the problem (1), (2) with input data* (u_0, g), (v_0, h) *respectively,* $g = g(t,x)$, $h = h(t,x)$, $g, h \in L^1_{loc}(\Pi_T)$. *Then*

$$\frac{\partial}{\partial t}(u-v)^+ + \operatorname{div}_x[\operatorname{sign}^+(u-v)(\varphi(u)-\varphi(v))]$$
$$\leq (g-h)\chi_{u>v} + (g-h)^+\chi_{u=v} \leq (g-h)^+\chi_{u\geq v} \quad \text{in } \mathcal{D}'(\Pi_T) .$$

Theorem 4. *Suppose that there exist nondecreasing subadditive functions* $\omega_i(r)$ *on* $\mathbb{R}_+ = (0, +\infty)$, $\omega_i(r) > 0$ *for* $r > 0$ *such that*

$$|\varphi_i(u) - \varphi_i(v)| \leq \omega_i(|u-v|) \quad \forall u, v \in \mathbb{R} \ , \ i = 1, \dots, n \ , \tag{7}$$

$$\liminf_{r\to 0+} r^{1-n} \prod_{i=1}^{n} \omega_i(r) < \infty \ . \tag{8}$$

Let $u = u(t,x)$ *be a* g.e.sub-s. *and* $v = v(t,x)$ *be a* g.e.super-s. *of the problem (1), (2) with input data* (u_0, g), (v_0, h) *respectively such that* $g(t,x,u), h(t,x,v) \in L^1_{loc}(\Pi_T)$, $(u_0 - v_0)^+ \in L^1(\mathbb{R}^n)$, *and*

$$q = q(t,x,u,v) = \chi_{u\geq v}\min\{(g(t,x,v) - h(t,x,v))^+ \ ,$$
$$(g(t,x,u) - h(t,x,u))^+\} \in L^1(\Pi_T) \ .$$

Then for a.e. $\tau \in (0, T)$

$$\int_{\mathbb{R}^n} (u(\tau,x) - v(\tau,x))^+ dx$$
$$\leq e^{L\tau}\left(\int_{\mathbb{R}^n} (u_0(x) - v_0(x))^+ dx + \int_{\Pi_\tau} q(t,x,u,v)dtdx\right) ,$$

where $L = \max(L(g), L(h))$.

Remark that the condition (7) means uniform continuity of the flux vector $\varphi(u)$. In particular the restriction (3) is fulfilled automatically under the condition (7).

Corollary 4 (comparison principle). *Let* $u_0, v_0 \in L^1_{loc}(\mathbb{R}^n)$, $g, h \in S$, $u_0(x) \le v_0(x)$ a.e. *on* $\mathbb{R}^n$, $g(t,x,u) \le h(t,x,u)$ a.e. *on* $\Pi_T \times \mathbb{R}$. *Suppose that the conditions (7), (8) are fulfilled and let* $u(t,x)$ *be a* g.e.sub-s., $v(t,x)$ *be a* g.e.super-s. *of the problem (1), (2) with input data* (u_0, g), (v_0, h) *respectively. Then* $u \le v$ a.e. *on* Π_T.

Evidently Corollary 4 implies *uniqueness* of a g.e.s. of the problem (1), (2).

It turns out that in the case when one of the functions u or v is bounded and doesn't depend on spatial variables Theorem 4 remains valid without any restriction on the flux vector $\varphi(u)$. That can be proved in similar way as the proof of Theorem 2.2 from [5]. On the base of this result we get following estimates of L^p-norms of g.e.s. (cf. [14, 5]).

Theorem 5. *Suppose that for* $1 \le p \le \infty$ $u_0(x) \in L^p(\mathbb{R}^n)$, $g_0(t,x) = g(t,x,0) \in L^1((0,T), L^p(\mathbb{R}^n))$; $u = u(t,x)$ *is a* g.e.s. *of the problem (1), (2). Then* $u(\tau,\cdot) \in L^p(\mathbb{R}^n)$ *for* a.e. $\tau \in (0,T)$ *and*

$$\|u(\tau,\cdot)\|_p \le \mathrm{e}^{L\tau} \left(\|u_0\|_p + \int_0^\tau \|g_0(t,\cdot)\|_p dt \right), \quad L = L(g).$$

In particular for $u_0 \in L^\infty(\mathbb{R}^n)$, $g(t,x,0) \in L^1((0,T), L^\infty(\mathbb{R}^n))$ the g.e.s. $u(t,x) \in L^\infty(\Pi_T)$ and a.e. on Π_T

$$|u(t,x)| \le M = \|u\|_\infty + \int_0^T \|g(t,\cdot,0)\|_\infty dt.$$

If v is a g.e.sub-s. of the problem (1), (2) then $v \le u^+ \le M$ a.e. on Π_T, where $u^+ = J^+(u_0, g)$ is the maximal g.e.s. Similarly if v is a g.e.super-s. of the problem (1), (2) then $v \ge u^- \ge -M$ a.e. on Π_T, where $u^- = J^-(u_0, g)$ is the minimal g.e.s. Therefore, Theorem 4 can be revised in the case under consideration: it is enough that the requirement (7) holds only for $u, v \in [-M, M]$. Remark that existence of the functions $\omega_i(r)$ in (7) such that $\omega_i(r) \to 0$ as $r \to 0$ follows from uniform continuity of the flux vector on $[-M, M]$. In particular for $n = 1$, the condition (8) is always satisfied and we derive uniqueness of g.e.s. to the problem (1), (2), without any restrictions. But, already for $n = 2$ uniqueness of g.e.s. can be violated, see examples in [12, 13].

5 The case of periodic input data

Let $\xi = \{\xi_1, \ldots, \xi_n\}$ be a basis on $\mathbb{R}^n$. Functions $u(x) \in L^1_{loc}(\mathbb{R}^n)$, $u(t,x) \in L^1_{loc}(\Pi_T)$ are called ξ-periodic if $u(x + \xi_i) = u(x)$ a.e. on $\mathbb{R}^n$ and respectively $u(t, x + \xi_i) = u(t,x)$ a.e. on Π_T for all $i = 1, \ldots, n$. In the same way we define ξ-periodicity of functions $g(t,x,u) \in S$.

Suppose that input data $u_0(x)$, $g(t,x,u)$ of the problem (1), (2) are ξ-periodic. It turns out that in this case *a g.e.s. of the problem* (1), (2) *is unique.* The proof is based on the simple observation that the maximal and the minimal g.e.s. are ξ-periodic functions that easy implies their coincidence. More generally we derive the following comparison principle

Theorem 6. *Let* $v(t,x)$ *be a* g.e.sub-s., *and* $w(t,x)$ *be a* g.e.super-s. *of the problem (1), (2) with input data* (v_0,g) *and* (w_0,h) *respectively. Assume that* $v_0(x) \le w_0(x)$ a.e. *on* $\mathbb{R}^n$, $g(t,x,u) \le h(t,x,u)$ a.e. *on* $\Pi_T \times \mathbb{R}$, *and one of the data* (v_0,g) *or* (w_0,h) *are* ξ*-periodic. Then* $v(t,x) \le w(t,x)$ a.e. *on* Π_T.

Acknowledgement. The research was supported by Russian Foundation for Basic Research (grants RFBR no. 00-01-00921 and no. 02-01-00483).

References

1. Andreianov, B.P., Benilan, Ph., Kruzhkov, S.N. (2000): L^1-theory of scalar conservation law with continuous flux function. J. of Functional Analysis, **171**, 15–33.
2. Barthélemy, L. (1988): Probléme d'obstacle pour une équation quasilinéar du premier order. Sci. Toulouse, **9**(2), 137–159.
3. Barthélemy, L., Bénilan, Ph. (1992): Subsolution for abstract evolution equations. Potential Analysis, **1**, 93–113.
4. Benilan, Ph. (1972): Equation d'evolution dans un space de Banach quelconque et applications. These de Doctorat d'Etat, Centre d'Orsey. Universite de Paris-Sud.
5. Benilan, Ph., Kruzhkov, S.N. (1996): Conservation laws with continuous flux functions. Nonlinear Differ. Equations and Appl., **3**, 395–419.
6. Crandall, M.G. (1972): The semigroup approach to first order quasilinear equations in several space variables. Israel J. Math., **12**, 108–122.
7. Goritsky, A.Yu., Panov, E.Yu. (1999): Example of nonuniqueness of entropy solutions in the class of locally bounded functions. Russian Journal of Mathematical Physics, **6**(4), 492–494.
8. Goritsky, A.Yu., Panov, E.Yu. (2002): Locally bounded generalized entropy solutions to the Cauchy problem for a first-order quasilinear equation. Proceedings of the Steklov Institute of Mathematics, **236**, 110–123.
9. Kruzhkov, S.N. (1970): First order quasilinear equations in several independent variables. Math. USSR Sb., **10**(2), 217–243.
10. Kruzhkov, S.N., Hildebrand, F. (1974): The Cauchy problem for quasilinear first order equations in the case the domain of dependence on initial data is infinite. Moscow Univ. Math. Bull., **29**(1), 75–81.
11. Kruzhkov, S.N., Andreyanov, P.A. (1975): On the nonlocal theory of the Cauchy problem for quasi-linear equations of first order in the class of locally summable functions. Soviet Math. Dokl., **16**, 16–20.
12. Kruzhkov, S.N., Panov, E.Yu. (1991): First order quasilinear conservation laws with an infinite domain of dependence on the initial data. Soviet Math. Dokl., **42**(2), 316–321.

13. Kruzhkov, S.N., Panov, E.Yu. (1995): Osgood's type conditions for uniqueness of entropy solutions to Cauchy problem for quasilinear conservation laws of the first order. Annali Univ. Ferrara-Sez., **XL**, 31–53.
14. Panov, E.Yu. (1996): On measure valued solutions of the Cauchy problem for a first order quasilinear equation. Izvestiya: Mathematics, **60**(2), 335–377.
15. Panov, E.Yu. (2001): A remark on the theory of generalized entropy sub- and super-solutions of the Cauchy problem for a first-order quasilinear equation. Differential Equations, **37**(2), 272–280.
16. Panov, E.Yu. (2002): On maximal and minimal generalized entropy solutions to Cauchy problem for a first-order quasilinear equation. Sbornik: Mathematics, **193**(5), 727–743.

Pressure Linearization Method for the Computation of Real Fluids

Marica Pelanti

Department of Applied Mathematics, University of Washington, Box 352420, Seattle, WA 98195-2420, USA. *pelanti@amath.washington.edu*

Summary. We present a method for the numerical solution of the Euler equations for fluids governed by a general equation of state, in the framework of finite volume schemes. In the evolution step, the idea is to replace the original pressure law with a locally-parameterized linear one. Then, the original EOS is used in the projection step to update the cell values of the pressure. A validation of the proposed scheme is provided by its interpretation as a relaxation method for the Euler equations. In the present paper we specialize our approach to a Roe-type method, and some two-dimensional numerical results are presented. Finally, we propose a modification of our scheme that allows us to ensure the invariance of the pressure and the velocity across contact discontinuities, for those cases in which difficulties arise in updating the pressure through the original nonlinear equation of state, as a consequence of the cell-based discretization of finite volume methods.

1 Introduction

We consider the numerical solution of the Euler equations for fluids governed by a general equation of state (EOS) by finite volume schemes. In the one-dimensional case, the Euler system takes the conservative form

$$q_t + f(q)_x = 0\,, \tag{1a}$$

with the vector q of the conserved variables and the corresponding flux function $f(q)$ given by

$$q = \begin{pmatrix} \rho \\ m \\ E^{\mathrm{t}} \end{pmatrix}, \qquad f(q) = \begin{pmatrix} m \\ \dfrac{m^2}{\rho} + p \\ \dfrac{m}{\rho}(E^{\mathrm{t}} + p) \end{pmatrix}, \qquad E^{\mathrm{t}} = E + \frac{m^2}{2\rho}\,. \tag{1b}$$

Here ρ represents the density, $m = \rho u$ the momentum, with u flow velocity along the x-axis, E^{t} the total energy per unit volume, E the internal energy per unit volume, and p the pressure. The definition of the problem is completed by specifying a relation between the pressure and the conserved variables through an equation of state. Under the assumption of local thermodynamic equilibium, the pressure can be expressed as a function of any two independent thermodynamic variables, and in the following we will assume a (convex) EOS in the form $p = p(E, \rho)$. Hence, the closure of the system (1) is obtained through the relation $p = p(E^{\mathrm{t}} - m^2/(2\rho), \rho)$.

Introducing the thermodynamic derivatives associated to $p(E, \rho)$,

$$\kappa(E,\rho) \equiv \frac{\partial p(E,\rho)}{\partial E} \quad \text{and} \quad \chi(E,\rho) \equiv \frac{\partial p(E,\rho)}{\partial \rho}, \tag{2}$$

the speed of sound can be expressed through these quantities as $c = \sqrt{\kappa h + \chi}$, where $h = (E + p)/\rho$ is the specific enthalpy.

We propose here a method with the purpose, analogous to the idea in [3], of introducing a general and simple procedure for solving (1) for a real fluid by a finite volume method. In the evolution step, the idea is to replace the original pressure law with a locally-parameterized one, linear in the thermodynamic variables E, ρ. Then, in the projection step, we use the original equation of state $p(E, \rho)$ to define the updated cell-averaged values of the pressure.

The method is described in Section 2. In Section 3 we then explain how our scheme can be interpreted as a relaxation method for the solution of the Euler system (1), thus providing a validation of our approach. Note that, from this point of view, we find close similarity between our method and the energy relaxation idea of Coquel and Perthame [3]. In Section 4 we apply our scheme to a Roe-type method and report some numerical results in two dimensions. Finally, in Section 5, we briefly describe a modification of our method that allows us to guarantee constant velocity and pressure across constant discontinuities, in those cases in which finite volumes methods fail to preserve the invariance of u, p, for reasons related to the cell-based description of the solution.

The subject of current investigation is the extension of this modified version of our method to *real* multicomponent flow computations, whose study has been the starting point for the idea of linearizing the pressure law.

2 Pressure Linearization Method

2.1 Linear pressure law

Our approach for the solution of the Euler equations with a general pressure law $p = p(E, \rho)$ is based on the replacement in the evolution step of the original equation of state with one that is linear in the selected thermodynamic variables E, ρ:

$$p^{\mathrm{L}} = \bar{\kappa} E + \bar{\chi} \rho + \beta, \tag{3}$$

with $\bar{\kappa}$, $\bar{\chi}$, β locally defined for each cell of the computational domain. Assuming $\bar{\kappa}$, $\bar{\chi}$, β frozen at some values, the speed of sound corresponding to the pressure law (3) is given by

$$c^{\mathrm{L}} = \sqrt{\bar{\kappa} h^{\mathrm{L}} + \bar{\chi}}\,, \tag{4}$$

with $h^{\mathrm{L}} = (E + p^{\mathrm{L}})/\rho$. Let us consider a discretization with volumes indexed in space by $i \in \mathbb{Z}$, and in time by $n \in \mathbb{N}$. The following *consistency condition* for the pressure must hold in each cell:

$$p(E_i^n, \rho_i^n) = \bar{\kappa}_i^n E_i^n + \bar{\chi}_i^n \rho_i^n + \beta_i^n\,. \tag{5}$$

This equality ensures consistency also for the enthalpy: $h_i^n = (h^{\mathrm{L}})_i^n$. In order to guarantee the correct approximation of the propagation speeds of the problem, we require that the speed of sound associated in each cell to p^{L}, $(c^{\mathrm{L}})_i^n = \sqrt{\bar{\kappa}_i^n h_i^n + \bar{\chi}_i^n}$, is the same as we would obtain from the original EOS. To this aim, we define $\bar{\kappa}$ and $\bar{\chi}$ as the cell values of the thermodynamic derivatives (2):

$$\bar{\kappa}_i^n = \kappa(E_i^n, \rho_i^n) \qquad \text{and} \qquad \bar{\chi}_i^n = \chi(E_i^n, \rho_i^n)\,, \quad \forall i, n\,. \tag{6}$$

The consistency condition (5) then determines uniquely

$$\beta_i^n = p(E_i^n, \rho_i^n) - \bar{\kappa}_i^n E_i^n - \bar{\chi}_i^n \rho_i^n\,. \tag{7}$$

Geometrically, in the coordinate space (E, ρ, p), we are locally approximating the graph of the function $p(E, \rho)$ at the point $(E_i^n, \rho_i^n, p(E_i^n, \rho_i^n))$ with the tangent plane at this point.

2.2 The scheme

The problem we face in the evolution step consists in solving the Euler equations with a pressure law linear in E, ρ, with parameters $\bar{\kappa}$, $\bar{\chi}$, β varying in the computational domain. This is analogous to a multifluid problem, where there are fluid components governed by different equations of state. By virtue of this similarity, we will solve our problem by following some approaches [1, 12] proposed in the study of multicomponent flows.

First, let us rewrite (3) in the form

$$p^{\mathrm{L}} = \frac{1}{\psi}(E + \eta\rho + \mu) = p^{\mathrm{L}}(E, \psi, \eta\rho, \mu)\,, \tag{8}$$

where

$$\psi = \frac{1}{\bar{\kappa}}\,, \qquad \eta = \frac{\bar{\chi}}{\bar{\kappa}}\,, \quad \text{and} \quad \mu = \frac{\beta}{\bar{\kappa}}\,. \tag{9}$$

The system we solve to approximate the original one consists of the Euler equations with the pressure law $p^{\mathrm{L}}(E, \psi, \eta\rho, \mu)$, together with three equations for ψ, $\eta\rho$ and μ, that here are considered as additional variables. We write these supplementary equations governing ψ, $\eta\rho$ and μ by extending the idea of the method proposed by Shyue [12] for multicomponent flow computations. The new system takes the form:

$$q_t + f^{\mathrm{L}}(q, \psi, \eta\rho, \mu)_x = 0\,, \tag{10a}$$
$$(\eta\rho)_t + (\eta\rho u)_x = 0\,, \tag{10b}$$
$$\psi_t + u\psi_x = 0\,, \tag{10c}$$
$$\mu_t + u\mu_x = 0\,, \tag{10d}$$

where

$$f^{\mathrm{L}}(q, \psi, \eta\rho, \mu) = \begin{pmatrix} m \\ \dfrac{m^2}{\rho} + p^{\mathrm{L}} \\ \dfrac{m}{\rho}(E^{\mathrm{t}} + p^{\mathrm{L}}) \end{pmatrix}, \tag{10e}$$

with p^{L} as in (8). In the solution of the system above, we associate to the pressure law (8) the speed of sound (4), that in terms of ψ, η is given by $c^{\mathrm{L}} = \sqrt{h^{\mathrm{L}}/\psi + \eta/\psi}$. This definition assumes frozen conditions for the quantities η, ψ, μ. Finally, the proposed algorithm consists of the following steps:

i. Given q^n, we compute ψ^n, $(\eta\rho)^n$ and μ^n by using the local values of the thermodynamic derivatives and the consistency condition, through (6), (7) and (9). With these initial values at time t^n we solve the new system (10) by a given finite volume method to obtain the updated vector q^{n+1}.
ii. We update the pressure through the original EOS: $p^{n+1} \equiv p(E^{n+1}, \rho^{n+1})$.

Note that, since in (*ii*) we just need the updated values of the vector $q = (\rho, m, E^{\mathrm{t}})^{\mathrm{T}}$, and not those of the other variables, we don't have to track ψ, $\eta\rho$ and μ, that are instead reset each time step as explained in (*i*). Thus, in the implementation of our method, we can work with vectors with the same dimension as q. The effect of the equations (10b), (10c), (10d) in the algorithm is the definition of a modified eigenstructure with respect to the original Euler system, which plays a role in the updating of the vector q.

3 Relaxation method

We now explain how our scheme can be interpreted as a relaxation method. Let us consider the following system with a relaxation source term:

$$q_t + f^{\mathrm{L}}_{\vartheta}(q, \vartheta)_x = 0\,, \tag{11a}$$
$$\vartheta_t + u\,\vartheta_x = -\frac{1}{\tau}(\vartheta - \vartheta_{\mathrm{eq}}(E, \rho))\,, \tag{11b}$$

where $\tau > 0$ is the *relaxation time*, $\vartheta \equiv (\bar{\kappa}, \bar{\chi}, \beta)^{\mathrm{T}}$, and $f^{\mathrm{L}}_{\vartheta}(q, \vartheta) = f^{\mathrm{L}}$ as in (10e), with p^{L} expressed in terms of ϑ in the form (3), i.e. $p^{\mathrm{L}} = \bar{\kappa}E + \bar{\chi}\rho + \beta$. Moreover, $\vartheta_{\mathrm{eq}} \equiv (\bar{\kappa}_{\mathrm{eq}}, \bar{\chi}_{\mathrm{eq}}, \beta_{\mathrm{eq}})^{\mathrm{T}}$, where we define $\bar{\kappa}_{\mathrm{eq}}(E, \rho) \equiv \kappa(E, \rho)$, $\bar{\chi}_{\mathrm{eq}}(E, \rho) \equiv \chi(E, \rho)$, and finally

$$\beta_{\mathrm{eq}}(E, \rho) \equiv p(E, \rho) - \bar{\kappa}_{\mathrm{eq}}(E, \rho)E - \bar{\chi}_{\mathrm{eq}}(E, \rho)\rho\,. \tag{12}$$

Note that the homogeneous system corresponding to (11) is equivalent to (10), since the advection equations for $\bar{\kappa}$, $\bar{\chi}$, β given by the homogeneous part of (11b) are equivalent to (10b), (10c), (10d).

Formally, when $\tau \to 0^+$ (*equilibrium limit*), $\vartheta = \vartheta_{\text{eq}}$, so that we recover the Euler system (1) with the original EOS since $p^{\text{L}}|_{\vartheta=\vartheta_{\text{eq}}} = p(E,\rho)$ by virtue of the above relation (12).

One way to solve (11) is to employ a fractional step method (as suggested for instance in [8]). First, we apply some finite volume scheme to the corresponding homogeneous system for advancing the solution over a time step Δt. Then we consider the equations obtained from (11) setting the spatial gradients to zero. The exact solution of these equations over time Δt gives $q^{(t+\Delta t)} = q^{(t)}$ and $\vartheta^{(t+\Delta t)} = \vartheta_{\text{eq}}^{(t+\Delta t)} + e^{-\Delta t/\tau}(\vartheta^{(t)} - \vartheta_{\text{eq}}^{(t+\Delta t)})$. Since we are only concerned with the equilibrium limit $\tau \to 0^+$, we simply set, instead of the latter equation, $\vartheta^{(t+\Delta)} = \vartheta_{\text{eq}}^{(t+\Delta t)}$. Employing this equilibrium condition allows us to avoid the need to update ϑ in the first step.

In summary, our relaxation scheme consists of the following steps:

i'. Update q^n to q^{n+1} by solving the homogeneous relaxation system (which is equivalent to (10)). We observe that this step amounts to considering the *frozen conditions limit* $\tau \to \infty$. Therefore it is appropriate here to employ the *frozen speed of sound* c^{L} given by (4), whose definition considers the nonequilibrium variable ϑ held constant.

ii'. Set $\vartheta^{n+1} = \vartheta_{\text{eq}}(E^{n+1}, \rho^{n+1})$ (*equilibrium limit* $\tau \to 0^+$).

Note that step (*ii'*) implies the computation of the updated value of the pressure p^{n+1} by means of the original EOS since $p^{n+1} = p^{\text{L}}|_{\vartheta^{n+1}=\vartheta_{\text{eq}}^{n+1}} = p(E^{n+1}, \rho^{n+1})$ through (12). It is then clear that the relaxation scheme (*i'*)-(*ii'*) we have described here gives the same algorithm we presented previously in (*i*)-(*ii*).

3.1 First order asymptotic approximation

By using an asymptotic expansion of the Chapman–Enskog type for $\bar{\kappa}$, $\bar{\chi}$ and β, the first order approximation of the relaxation system (11) is found to be

$$q_t + f(q)_x = -\begin{pmatrix} 0 \\ \tau\{\rho[(c^{\text{L}})^2 - c^2]u_x\}_x \\ \tau\{\rho[(c^{\text{L}})^2 - c^2]u\, u_x\}_x \end{pmatrix}, \tag{13}$$

with c^{L} as in (4). In order to have a nonnegative diffusion matrix, we deduce from (13) that a stability criterion is

$$c^{\text{L}} \geq c\,. \tag{14}$$

The above relation represents a *subcharacteristic condition* for the problem, expressing the requirement that the speed of sound c^{L} of the homogeneous relaxation system is at least as large as the speed of sound c of the original Euler equations. The parameters of the linear pressure p^{L} have been defined in (6) so that in each cell (14) is satisfied in the equality limit. This choice minimizes the numerical diffusion, since, the greater c^{L} is with respect to c, the higher the numerical viscosity. Note that in our scheme we can also employ global discrete values for $\bar{\kappa}$ and $\bar{\chi}$ satisfying (14) to avoid the computation of the derivatives of the pressure, thus following the idea of [3]. However, in this case we would need to estimate *a priori* the solution, and also we would have the drawback of a higher numerical diffusion.

4 Application to the Roe Method

We now describe how our approach applies to the Roe method [11]. Let us express the system (10) in the fully quasi-linear form $z_t + A^{\text{L}}(z) z_x = 0$, with $z = (\rho, m, E^{\text{t}}, \eta\rho, \psi, \mu)^{\text{T}}$ and the matrix $A^{\text{L}}(z)$ given by:

$$A^{\text{L}}(z) = \begin{pmatrix} 0 & 1 & 0 & 0 & 0 & 0 \\ \frac{1}{\psi}\frac{u^2}{2} - u^2 & \left(2 - \frac{1}{\psi}\right) u & \frac{1}{\psi} & \frac{1}{\psi} & -\frac{\zeta}{\psi^2} & \frac{1}{\psi} \\ u\left(\frac{1}{\psi}\frac{u^2}{2} - h^{\text{t}}\right) & h^{\text{t}} - \frac{1}{\psi}u^2 & \left(1 + \frac{1}{\psi}\right) u & \frac{1}{\psi}u & -\frac{\zeta}{\psi^2}u & \frac{1}{\psi}u \\ -\eta u & \eta & 0 & u & 0 & 0 \\ 0 & 0 & 0 & 0 & u & 0 \\ 0 & 0 & 0 & 0 & 0 & u \end{pmatrix}, \tag{15}$$

where $h^{\text{t}} = h + u^2/2$ is the specific total enthalpy and $\zeta \equiv E + \eta\rho + \mu$.

Following the approach of Roe [11], we solve Riemann problems at cell interfaces between neighbouring cell averages with a locally linearized version of the system above, with a constant matrix $\hat{A}^{\text{L}}(z_\ell, z_r)$ depending on the values of the left and right states z_ℓ and z_r. After Roe, we define the matrix $\hat{A}^{\text{L}}$ as the original matrix A^{L} evaluated at an average state $(\hat{u}, \widehat{h^{\text{t}}}, \hat{\eta}, \hat{\psi}, \hat{\zeta})$, involving only those variables explicitly appearing in A^{L}: $\hat{A}^{\text{L}} \equiv A^{\text{L}}(\hat{u}, \widehat{h^{\text{t}}}, \hat{\eta}, \hat{\psi}, \hat{\zeta})$.

Let us now introduce the operator $\Delta(\cdot) \equiv (\cdot)_r - (\cdot)_\ell$, and the Roe-average operator:

$$\tilde{(\cdot)} \equiv \frac{\sqrt{\rho_\ell}\,(\cdot)_\ell + \sqrt{\rho_r}\,(\cdot)_r}{\sqrt{\rho_\ell} + \sqrt{\rho_r}}. \tag{16}$$

In order to guarantee conservation, we need to satisfy the condition $(\hat{A}^{\text{L}} \Delta z)_i = (\Delta g^{\text{L}})_i$, for $i = 1, 2, 3, 4$, where $g^{\text{L}} \equiv (f^{\text{L}}, \eta\rho)^{\text{T}}$. We can ensure this property by defining the average variables as follows:

$$\hat{u} = \tilde{u}\,, \quad \widehat{h^{\text{t}}} = \widetilde{h^{\text{t}}}\,, \quad \hat{\eta} = \tilde{\eta}\,, \quad \hat{\psi} = \frac{\psi_\ell\,\psi_r}{\tilde{\psi}}\,, \quad \hat{\zeta} = \frac{\psi_\ell\,\psi_r}{\tilde{\psi}^2}\tilde{\zeta}\,. \tag{17}$$

The average speed of sound associated to $(\hat{u}, \widehat{h^{\text{t}}}, \hat{\eta}, \hat{\psi}, \hat{\zeta})$ is then given by:

$$\hat{c} = \sqrt{\frac{1}{\hat{\psi}}\left(\hat{h}^{\mathrm{t}} - \frac{\hat{u}^2}{2}\right) + \frac{\hat{\eta}}{\hat{\psi}}}\,. \tag{18}$$

The matrix of the linearized right eigenvectors takes the form

$$\hat{R} = \begin{pmatrix} 1 & 1 & 0 & 0 & 0 & 1 \\ \hat{u} - \hat{c} & \hat{u} & 0 & 0 & 0 & \hat{u} + \hat{c} \\ \hat{h}^{\mathrm{t}} - \hat{u}\hat{c} & \hat{h}^{\mathrm{t}} - \hat{\psi}\hat{c}^2 & -1 & \frac{\hat{\zeta}}{\hat{\psi}} & -1 & \hat{h}^{\mathrm{t}} + \hat{u}\hat{c} \\ \hat{\eta} & \hat{\eta} & 1 & 0 & 0 & \hat{\eta} \\ 0 & 0 & 0 & 1 & 0 & 0 \\ 0 & 0 & 0 & 0 & 1 & 0 \end{pmatrix}, \tag{19}$$

and the corresponding linearized eigenvalues (wave speed) are

$$\hat{\lambda}_1 = \hat{u} - \hat{c}\,, \qquad \hat{\lambda}_2 = \hat{\lambda}_3 = \hat{\lambda}_4 = \hat{\lambda}_5 = \hat{u}\,, \qquad \hat{\lambda}_6 = \hat{u} + \hat{c}\,. \tag{20}$$

Moreover, the wave strengths $\alpha = \hat{R}^{-1}\Delta z$, that is the coefficients of the projection of Δz onto $\hat{R}$, can be expressed through (4), (18), as:

$$\alpha_1 = \frac{\Delta p - \hat{\rho}\,\hat{c}\,\Delta u}{2\,\hat{c}^2}\,, \quad \alpha_2 = \Delta\rho - \frac{\Delta p}{\hat{c}^2}\,, \quad \alpha_6 = \frac{\Delta p + \hat{\rho}\,\hat{c}\,\Delta u}{2\,\hat{c}^2}\,, \tag{21}$$

$$\alpha_3 = \hat{\rho}\,\Delta\eta\,, \quad \alpha_4 = \Delta\psi\,, \quad \alpha_5 = \Delta\mu\,, \tag{22}$$

where $\hat{\rho} \equiv \sqrt{\rho_\ell}\,\sqrt{\rho_r}$.

We remark that the updating of q employs only the upper half of the matrix $\hat{R}$. Moreover, the waves with speed $\hat{u}$ that are used to update q can be combined together. See [10] for details.

4.1 Numerical results

Several numerical experiments have been performed in one dimension employing the ideal polytropic gas law, the Tamman EOS, the two-molecular vibrating gas EOS, and the Osborne model. Here we present the results of the computation (using CLAWPACK [7]) for a test in two dimensions involving the reflection of a planar shock wave by an oblique surface. This problem is well known in the literature, and we refer the reader especially to [2] for any detail, and for instance to [6] for some experimental results.

Following [9], we compute the solution for this test with both the ideal polytropic gas EOS, $p = (\gamma - 1)E$, $\gamma = 1.4$ (air), and the two-molecular vibrating gas EOS, given by $p(E,\rho) = R\rho T(E)$, with the temperature defined by the implicit equation $E = \rho c_v^{t,r} T + \rho R \Theta_v / (e^{\Theta_v/T} - 1)$. We set $R = 287.086$, $c_v^{t,r} = 5/2$, and $\Theta_v = 10^3$. The (second order) results are displayed in Figure 1. From them we can see that our scheme is able to capture the distinguishing features related to the different form of the equation of state. In particular, we observe that in the computation with the two-molecular vibrating gas the main shock is less curved at the compression corner, and that the Mach stem near the wall is pushed forward by the jet.

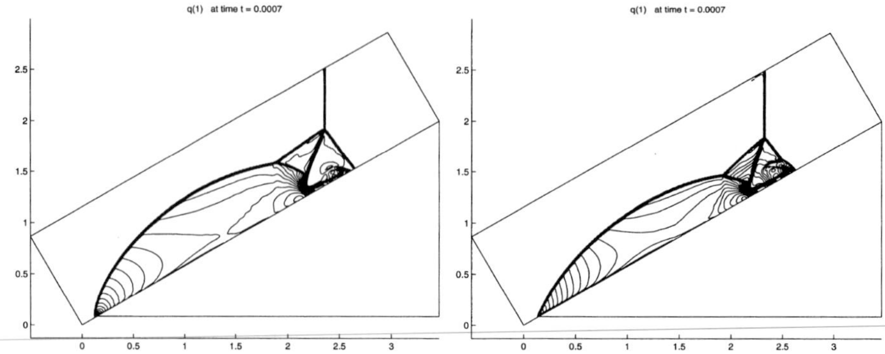

Fig. 1. Double-Mach Reflection. Left: Ideal polytropic gas. Right: Two-molecular vibrating gas. Density contours at $t = 0.0007$. $\Delta x = 1/480$, $\Delta y = 1/120$, CFL $= 0.9$.

5 Modified PLM

In the solution of the Euler equations, a finite volume method that updates the pressure by evaluating the EOS $p(E, \rho)$ in the cell values of E, ρ obtained from the conservative variables may fail to preserve the invariance of the pressure and the velocity across contact discontinuities, if p is nonlinear in E, ρ. These difficulties are related to the cell-based description of the solution. Obviously, also the proposed Pressure Linearization Method (PLM), consisting in the algorithm (*i*)-(*ii*), will suffer from the same problem. We propose here a modification (in one dimension) of our method that allows to maintain u and p constant across a contact discontinuity.

Our approach consists in updating the pressure in a different way around contact discontinuities. The crucial point is that the cells where the discontinuity is spread by the numerical method cannot be regarded as systems in local thermodynamic equilibrium, hence the equation of state $p(E, \rho)$ loses here its physical meaning. We adopt the framework of Section 2, that is we use the conservative Euler equations (1), and we relate the values of p, E, ρ through the linear equation (3) (or (8)). The formulation is augmented by a conservative advection equation for a tracer variable Y that locates the contact discontinuity, and it is initialized to zero and one on opposite sides of the interface. Far from the contact discontinuity ($Y = 0$ or $Y = 1$, up to a tolerance) we still update the pressure through $p(E, \rho)$, otherwise ($0 < Y < 1$) we define the new cell value of the pressure by one of the two following methods:

1. Use the linear EOS (8): $p^{n+1} \equiv (E^{n+1} + (\eta\rho)^{n+1} + \mu^{n+1})/\psi^{n+1}$. Note that in this case we need the updated values of $\eta\rho$, ψ, μ. Referring to the Roe-type method of Section 4, we employ the full structure of $\hat{R}$.
2. Use the updating formula based on the evolution equation for the pressure $p_t + u\, p_x + \rho(c^{\mathrm{L}})^2 u_x = 0$. This equation can be derived by differentiating the *linear* EOS, and replaces the equation for μ in the formulation (10). In

the specialization of this approach to the Roe method, the updating of q is performed by using the same eigenstructure reported in Section 4, while p is updated by employing the decomposition $\Delta p = \alpha_1 (c^{\mathrm{L}})^2 + \alpha_6 (c^{\mathrm{L}})^2$.

In this modified version of our scheme, the consistency condition (5) that defines locally β, given the local values of $\bar{\kappa}$ and $\bar{\chi}$, is written in the more general form

$$p_i^{n+1} = \bar{\kappa}_i^n E_i^n + \bar{\chi}_i^n \rho_i^n + \beta_i^n \,, \tag{23}$$

which must hold whether p_i^{n+1} is defined as $p(E_i^n, \rho_i^n)$ or according to the updating procedures (1) or (2).

Switching to one of the updating methods described above allows to maintain the equilibrium of the pressure and the velocity across an interface. See [10] for more detail.

5.1 Numerical results

We compute the solution for a Riemann shock tube problem considered in [4]. The fluid is modeled by the van der Waals EOS, $p(E, \rho) = (R/c_v)(E + a\rho^2)/(1 - \rho b) - a\rho^2$, with $R = 461.5$, $c_v = 1401.88$, $a = 1684.54$, and $b = 0.001692$. The initial data are $\rho_\ell = 333.33, u_\ell = 0, p_\ell = 3.7311358 \times 10^7$ for the left state, and $\rho_r = 111.11$, $u_r = 0$, $p_r = 2.1770768 \times 10^7$ for the right state. The exact solution of this problem consists in a rarefaction wave moving to the left, a contact discontinuity and a shock wave moving to the right. In the upper part of Figure 2 we present the results (first order accurate) computed by the method of Roe [11], extended to real gases by following the approach of Glaister [5]. The same results are obtained with the PLM defined by the algorithm (i)-(ii). We observe the unphysical behaviour of the computed pressure and velocity around the contact discontinuity. The lower part of Figure 2 shows the results computed by the modified PLM with the updating approach (2) (approach (1) gives the same plots). We observe that the scheme is able to preserve the invariance of u, p across the contact discontinuity.

Acknowledgments. This work was supported in part by DOE grant DE-FG03-96ER25292 and NSF grant DMS-0106511.

References

1. Abgrall, R., Saurel, R. (1999): A Simple Method for Compressible Multifluid Flows. SIAM J. Sci. Comput., **21**, 1115–1145
2. Colella, P., Woodward, P. (1984): The piecewise-parabolic method (PPM) for gasdynamical simulations. J. Comput. Phys., **54**, 174–201
3. Coquel, F., Perthame, B. (1998): Relaxation of Energy and Approximate Riemann Solvers for General Pressure Laws in Fluid Dynamics. SIAM J. Numer. Anal., **35**, 2223–2249

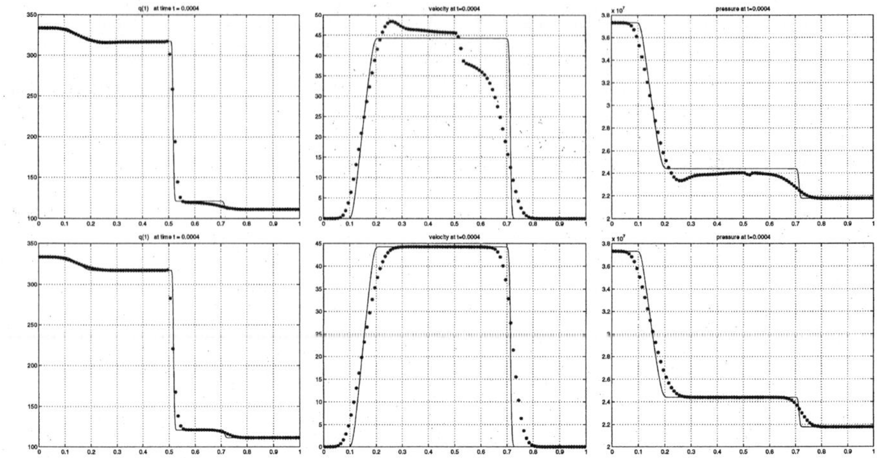

Fig. 2. Density, velocity, pressure at time $t = 0.0004$. $\Delta x = 1/100$, CFL $= 0.8$. Above: Results obtained by a Roe-type method with standard updating of the pressure. Below: Results obtained by the modified PLM.

4. Gallouët, T., Haerard, J.-M., Seguin, N.: Some recent Finite Volume schemes to compute Euler equations using real gas EOS, preprint, Marseille
5. Glaister, P. (1988): An Approximate Linearised Riemann Solver for the Euler Equations for Real Gases. J. Comput. Phys., **74**, 382–408
6. Glaz, H.M., Colella, P., Glass, I.I., Deschambault, R.L. (1985): A Numerical Study of Oblique Shock-Wave Reflections with Experimental Comparisons. Proceedings of the Royal Society of London. Series A, Mathematical and Physical Sciences, **398**, 117–140
7. LeVeque, R.J.: Conservation Law Package (CLAWPACK) software, avalaible at `http://www.amath.washington.edu/~claw`
8. LeVeque, R.J., Pelanti, M. (2001): A Class of Approximate Riemann Solvers and Their Relation to Relaxation Schemes. J. Comput. Phys., **172**, 572–591
9. Montarnal, P., Shu, C.W. (1999): Real Gas Computation Using an Energy Relaxation Method and High-Order WENO Schemes. J. Comput. Phys., **148**, 59–80
10. Pelanti, M.: Pressure Linearization Method for the Computation of Real Fluid Flows, in preparation.
11. Roe, P.L. (1981): Approximate Riemann solvers, parameter vectors and difference schemes. J. Comput. Phys., **43**, 357–372
12. Shyue, K.M. (1998): An Efficient Shock-capturing Algorithm for Compressible Multicomponent Problems. J. Comput. Phys., **142**, 208–242

Modeling and Computing Geophysical Mass Flows

E. Bruce Pitman, Abani Patra, Andy Bauer, Michael Sheridan and Marcus Bursik

University at Buffalo, Buffalo, NY 14260

1 Introduction

Avalanches and debris flows pose a significant risk to those living on the flanks of hills and volcanoes. In 1985, debris flows originating from the summit of Ruiz volcano in Colombia overran entire villages, killing more than 20,000. In 1998, a mudflow at the Casita Volcano in Nicargua caused thousands of deaths. Pyroclastic and debris flows produced during the 1991 eruption of Mt. Pinatubo in the Phillipines destroyed cities and villages some 15 km. from the summit, and forced the evacuation of hundreds of thousands. Residents in the state of Washington remember the 1980 Mt. St. Helens eruptions, and look warily at Mt. Ranier.

Volcanic activity can cause a variety of gravity-driven mass flows. Hot pyroclastic flows can develop from the rock and ash fountaining from the vent and rush downhill at speeds as high as 250 meters/second. These hot flows can also melt snow on the mountain, creating a muddy mix of ash, water and rock. Seismic activity at a volcano can trigger the failure of an entire flank, generating a giant debris avalanche.

At present, we cannot predict when volcanic activity might lead to a large-scale eruption. Nor can we predict when heavy rains might trigger a mudslide. However we can address the question of what area is most at risk from a mass flow initiated from a particular location. This is the issue we address here. We describe efforts in developing models of geophysical mass flows. We present a numerical algorithm for solving a particular system of model equations, and we describe our implementation of a parallel, adaptive grid Godunov solver to simulate flows. Unique in this project is the incorporation of elevation data into simulations, allowing us to (try to) reproduce historical flow events.

2 Thin layer modeling

In this section we summarize one approach to modeling geophysical mass flows. These efforts began with a paper by Savage and Hutter [18]. Starting from equations of mass and momentum balance on an inclined plane, using a Coulomb constitutive relation and making scaling arguments, they developed a one-dimensional model system formally similar to the shallow water equations. Savage and Hutter wrote a code to numerically solve the resulting equations, but reported some difficulties with their Lagrangian-based algorithm. They also performed a series table-top experiments, to test their predictions. Savage and Hutter later extended their work to two dimensions, found a similarity solution to the governing equations, and examined flows over more general basal surfaces.

The effort was taken up by Iverson [9, 10, 4], who tried to extend the modeling to include the effects of pore fluid, and employed modern shock-capturing numerical techniques to solve the model system.

Gray [6, 7] also re-derived the governing equations for general basal surfaces, and carefully examined the special flow of avalanches in a rotating drum.

Pitman [15] wrote a Godunov solver for one-dimensional flows that incorporates general topographical features into a simulation.

We now present a heuristic derivation of the model system, an approach that highlights the fundamental ideas required to develop a hydraulic model from a physical system governed by Coulomb frictional plasticity. We refer the reader to the original work by Savage and Hutter, to gain an appreciation of their scaling arguments [18, 8].

It is of important to recognize that the fundamental evolution equations for an incompressible Coulomb material are linearly ill-posed [19]. That is, the basic balance laws, when linearized around a constant state, exhibit runaway growth. This in spite of the widespread use - and presumed validity - of the steady-state Coulomb model system in engineering practice. Some degree of well-posedness can be recovered by allowing slight compressibility [14, 20], but the fact remains that we do not understand the basic physics of time dependent granular flow.

2.1 A heuristic derivation

In this subsection, we limit the discussion to flow in two dimensions. Consider a thin layer of granular material flowing over a smooth basal surface described, in a coordinate system aligned with the bed slope, by the graph $z = b(x)$. That is, the position $(x, b(x))$ makes an angle θ with the horizontal; see Figure 1.

The general, free surface flow of this material down an incline is far too difficult a problem to study. Instead, we will derive a "depth-averaged" model of the flowing layer, yielding a system of equations in one space dimension (plus time).

A slice of this layer between the points $x - \frac{\Delta x}{2}$ and $x + \frac{\Delta x}{2}$ is subject to forces in the x and z directions due to gravity, friction, and material deformation. We will assume a Mohr-Coulomb consitiutive relation. The Mohr-Coulomb theory generalizes the notion of simple sliding friction to a continuum. We will assume that all the material is flowing, and begin with the constitutive assumptions of the theory:

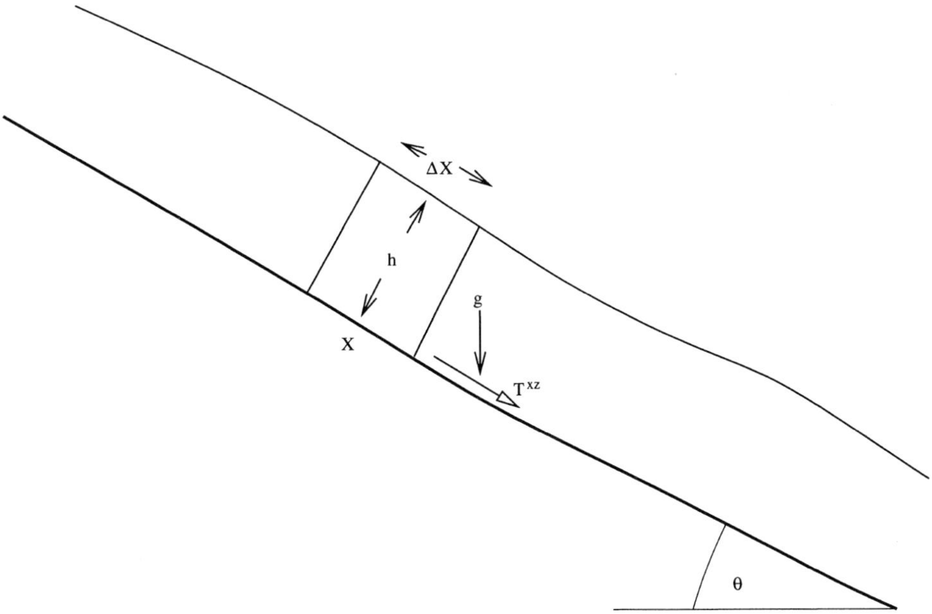

Fig. 1. A schematic diagram of the average forces acting on a mass of granular material flowing down a hillside.

- Material deforms when stresses reach yield, described by the condition $||dev(T)|| = \kappa\, tr(T)$, where T is the 2×2 stress tensor, dev is the deviator of the tensor $dev(T) = T - \frac{1}{2}tr(T)I$, and $tr(T)$ is the trace of the stress $tr(T) = \sum_{i=1}^{2} T^{x_i x_i}$.
- As material deforms, the stress and strain-rate tensors are aligned, $dev(T) = \lambda\, dev(V)$, where the strain-rate $V = -\frac{1}{2}(\partial_{x_i} u_{x_j} + \partial_{x_j} u_{x_i})$, where $\mathbf{u} = (u_1, u_2)$ is the velocity.
- Material is rigid if stresses are below yield.

Let us develop balance laws for $\bar{T}$, the stresses averaged over the layer depth h. Now if motion normal to the basal surface is negligible, the only forces in the z-direction are lithostatic

$$\partial_z \bar{T}^{zz} = -\rho g \cos(\theta)$$

Because the stress vanishes on the top surface, we find $\bar{T}^{zz} = \rho g \cos(\theta)(h - z)$.

In the x-direction, the total time rate of change of the momentum of the slice is given by $\frac{d}{dt}(\rho h u \Delta x)$, and is balanced by the derivative of the down-slope stress $\partial_x \bar{T}^{xx}$, the basal friction $\bar{T}^{xz}|_b$, the lateral component of gravity $\rho g h \sin(\theta)$, and acceleration due to curvature of the basal surface, which we model as $\frac{db}{dx}$.

In its full detail, Coulomb plasticity is too complex a theory to apply here. Instead we make a series of simplifying assumptions that allow us to derive a tractable set of equations. First, at the basal surface we assume perfect alignment of the normal stress and the shear stress, so $\bar{T}^{xz}|_b = \mu \bar{T}^{zz}|_b$. We further assume that the stresses remained aligned throughout the layer thickness; any misalignment is likely to be small if the layer is not very thick. Finally, we assume that the xx-stress and the zz-stress are proportional throughout the layer, $\bar{T}^{xx} = K\bar{T}^{zz}$, where K is the passive earth coefficient, a classical factor that is widely used [21] and whose origins extend back to Rankine [16]. Bringing all of these terms together, and using the hydrostatic relation for the normal stress (and thus for all the stresses), the balance law reads

$$\partial_t \rho h u + \partial_x \rho h u^2 + \frac{1}{2}\partial_x \beta \rho g h^2 \tag{1}$$
$$= \rho g h \sin(\theta) - sgn(u)\cos(\theta)\tan(\delta)\rho g h - \rho g h \cos(\theta)\tan(\delta) K \frac{db(x)}{dx}$$

Here, the coefficient $\beta = K\cos(\theta)$. The general approach suggested by this derivation, and the specific alignment and friction assumptions, are reminiscent of the theory of bin loads due to Janssen [11]; also see [13].

To better appreciate the governing system, the equations should be scaled in both dependent and independent variables. Clearly, overall stresses can be scaled by the lithostatic force $\rho g H \cos(\theta)$, where H is a characteristic thickness of the flowing layer. This scaling removes the density as a parameter, and clears some of the trigonometric functions. But more important is a scaling of the independent variables. We will make a long wave assumption, scaling x by L, a characteristic downslope length, z by H, and assume $H/L = \epsilon << 1$. We also scale t by $\sqrt{L/g}$. With these scalings, the equations as presented above are modified by the introduction of important coefficients into the system; we recap the governing equations below (generalized there to the case of two dimensional flows).

Similar to the shallow water equations in structure, this system of equations is strictly hyperbolic and geniunely nonlinear, with characteristic speeds $\lambda_\pm = u \pm \sqrt{\beta h}$.

3 Computing flow in two dimensions

The heuristic approach sketched above, or a more formal derivation of model equations based on scaling arguments, can be extended to flow over a two dimensional surface [8]. As recognized by Iverson and Denlinger [10], for the model system to apply to flow over terrain, there can be no preferential direction, as is present in earlier derivations. The governing equations must possess rotational invariance. Rather than re-derive the governing equations, we simply present the governing equations for a layer of thickness h and the two-dimensional velocity $\mathbf{u}$

$$\partial_t h + \nabla h\mathbf{u} = 0 \tag{2}$$
$$\partial_t h\mathbf{u} + \nabla(h\mathbf{u}\mathbf{u} + \frac{\beta}{2}\nabla h^2 \mathbf{I}) = h\,\frac{\mathbf{g}}{|\mathbf{g}|} - h\, sgn(\mathbf{u})\mathbf{I}\,\tan(\delta)\,[1 + \nabla b]$$

Here, $\beta = \epsilon K$ is small, $\mathbf{I}$ is the identity matrix, and the unit vector $\frac{g}{|g|}$ is the local (scaled) gravitational acceleration. This system is again hyperbolic, with two nonlinear wave families similar to the one-dimensional case, and a linear family whose wavespeed is the convection speed.

3.1 A finite volume Godunov solver

To solve this hyperbolic system of equations, we use a parallel, adaptive mesh, Godunov solver. The basic ingredient in the method is an approximate Riemann solver, due originally to Davis [3]. To summarize the method briefly, dependent variables are considered as cell-averages. Their values are advanced by a predictor-corrector method; source terms are included in these updates, and no splitting – neither for the source nor the multiple dimensions – is necessary. Slope-limiting is used to prevent unphysical oscillations.

The principal ingredient of the method is a centered, approximate Riemann solver, akin to one introduced by Rusanov. Given left and right states U^l, U^r and a physical flux f, the numerical flux F is given by

$$F(U^l, U^r) \;=\; \frac{1}{2}(f(U^r) + f(U^l)) - \frac{\alpha}{2}(U^r - U^l)\ .$$

Here α is an upper bound on the magnitude of the characteristic speeds of all waves, when evaluated at both the l and the r states. The Davis method does not provide the high resolution of some other solvers. However its ease of use recommends it for our model equations.

3.2 Adaptive Gridding

Parallel computing and adaptive gridding provide users with a high degree of resolution of solutions to hyperbolic PDEs [1]. For our thin layer equations, we have implemented a simple adaptive grid structure We refine cells based on indicators derived from the solution at the previous timestep. Likewse, when our indicators suggest that coarser discretization will not adversely affect solution quality, we unrefine clusters of cells. When coarse and fine grids meet at a cell edge, we impose a conservation requirement, namely that the total numerical flux across the unrefined grid cell edges to equal (in magnitude) the total numerical flux across all refined grid cell edges.

In the calculations reported here, we use a simple scaled L^2 norm of the flux around the boundary of each cell as the refinement indicator. That is, the refinement indicator for the kth cell is

$$\eta_k = \frac{1}{d_k}\oint_{\partial\Omega_k} |F|^2 dx$$

where d_k is the diameter of cell k, and $\partial\Omega_k$ is the boundary of the cell. The cells to be refined are those for which η_k is the largest $p\%$ of all cells, where, of course, p is adjusted by the user.

We also monitor the change of the pile height on every time step: $h(t_n) - h(t_{n-1})$; when this change is below a user defined threshold, we unrefine the grid locally. Both the refinement and unrefinement inidicators are heuristic, and based on experience and physical intuition.

We must also define the frequency of grid adaption. In the calculations reported here, we examine the indicators every two timesteps to decide whether or not to refine or unrefine. Again, this condition is heuristic and not scientific. We are investigating more sophisticated strategies, based on estimates of local error.

3.3 Parallelization

Parallelization of adaptively refined grids has been addressed by several researchers over the last several years; see for example Flaherty et al.[5], Edwards and Stewart [2] and Laszloffy, Patra and Long [12]. To take full advantage of multiple processing, the critical issues to balance are

- the number and location of cells created and deleted during grid refinement
- processor load
- efficiency of storage and data access.

For our model system, we extend the approach described in our earlier work on adaptive steady-state finite element codes [12]. Two modifications to this earlier work have been necessary to accomodate the time-dependent hyperbolic nature of our equations:

1. the addition of modules to enable unrefinement of cells
2. the addition of data storage for a layer of "ghost cells" on each subdomain.

These "ghost cells" are a layer of cell data for all cells adjacent to the line (in two-dimensions) that partitions the grid among some pair of processors. The data of a cell adjecent to the partition and belonging to one processor is replicated on the other processor. This replication enables us to peform the computations necessary for the Godunov scheme for every cell, without explicit communication. However, the use of ghost cells requires us to synchronize the necessary data among the processors at the end of each time step. During the unrefinement process, refined cells are removed and the parent cells reactivated.

3.4 Data Management and Load Balancing

Although we used only regular Cartesian grids in the simulations reported here, unstructured grids often arise in general adaptive refinitement applications. Thus we have designed to allow for general grids, and our data structure are not simple arrays. All data access operations involve some kind of search. For efficient computation we need to make this search procedure fast and efficient.

Our data storage model is a distributed hash table Data is indexed by an ordering along a space filling curve (SFC), and is then stored in the table [12]. Each cell is mapped to a unique key using its location along the SFC passing through its centroid. This key also provides a unique identifier that is easily generated (typically it is generated directly from geometric coordinates by using bit manipulations). To obtain a decomposition of the problem, we introduce a partitioning of this key space. This induces a distribution of the data sets, i.e., a decomposition of the problem for distributed memory computing. When the grid is changed by refinement or unrefinement, a redistribution of cells among the processors is performed by adjusting this key space partition. This redistribution helps maintain load balance.

Using the infrastructure just described, the parallel solution of our model partial differential equations is easily implemented. After a decomposition of the domain based on the key space partitioning, we introduce one layer of ghost cells on each processor. Each cell also maintains a list of the keys of its neighbors. These features, together with the limited range of influence of any individual cell, a characteristic of hyperbolic problems, make access to on-processor and off-processor data for the Godunov solver relatively simple.

4 Results

4.1 Inclined Plane

This is a simple test of the numerical methodology descrbed here. Laboratory experiments using this type of geometry have been described by Iverson and Denlinger [4] among others. We show in Fig 2 flow of an initially parabolic cap down a plane inclined at a 45° angle. At 1000 m., the plane becomes horizontal. Savage and Hutter [18] show that the paraboloid is a similarity solution of the governing system; our simulations give a qualitative indication of this feature. Notice how, after the hump reaches the horizontal section, flow quickly ceases. For this run, a free-flowing material was used, with an internal friction angle of 30°, and a bed friction of 20°.

4.2 Realistic terrain

Our efforts in this project are motivated by the desire to simulate real debris and avalanche flows. To this end, we have selected a small set of volcano sites at which we have detailed information about historical mass flow events, information either from real-time reports of the event or from geological field work and reconstruction. Our simulations, then, will attempt to reproduce these flows with a minimum of parameter adjustment. To this end, the model

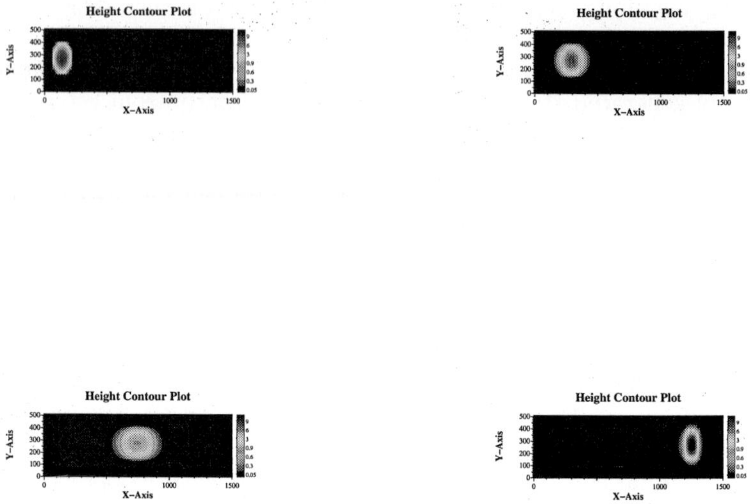

Fig. 2. Granular avalanche down an inclined plane. The plane makes an angle of 45° until 1000 m., at which point the surface rapidly flattens to become horizontal. The internal friction angle of the flowing mass is 30°, and the bed friction is 20°.

introduced in Section 2 is well-suited – there are only two parameters to choose, the internal and basal friction angles. Here we report on a simulation at Volcan de Colima in Mexico.

The computation begins by loading topographical data, the elevation of Colima and its surroundings. For the simulation reported here, this elevation data is gridded by cells $100\,m \times 100\,m$. The vertical accuracy $\pm 30\,m$.

In 1991, a block and-ash flow of about $\sim 10^6\,m^3$ initiated from the SW face of a lava dome located at the summit. We do not know the exact geometry of the failure surface, but for our simulation we assume a paraboloid-shaped mass of $10^6\,m^3$ located nearly at the top of the cone. This shape exaggerates the initial inclination of the flowing material, and makes the very early timecourse of our simulation highly questionable. The contour maps of Fig. 3 show a history of the flow simulation as it proceeds down the mountainside. The simulated flow flow does not channelize quite as did the '91 flow. But interestingly it does split along a ridge between barancas, similar to the

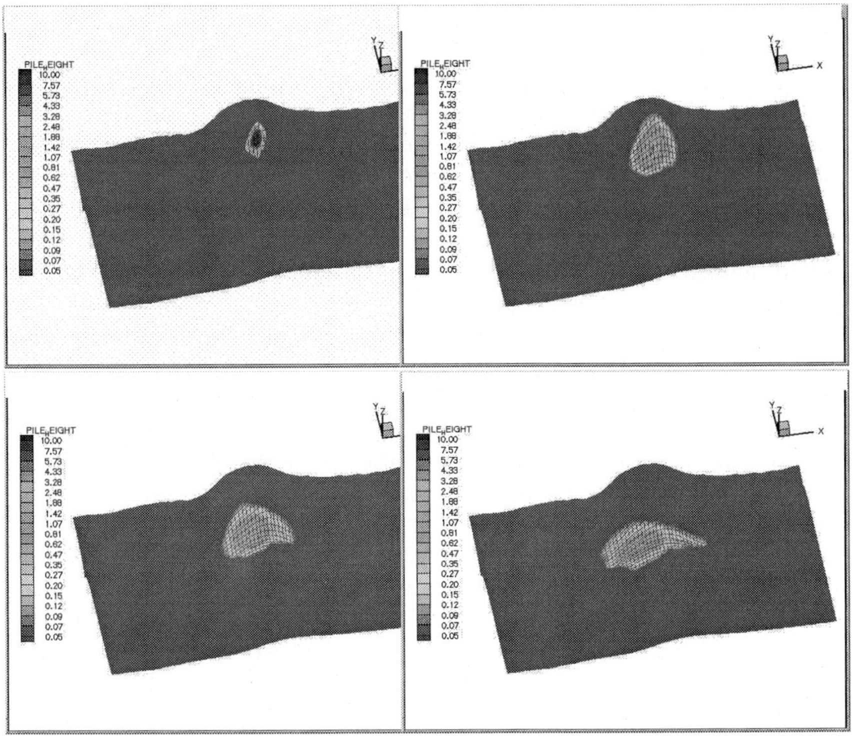

Fig. 3. The spreading of avalanche modeled using the thin layer system of equations and terrain data from Volcan de Colima in Mexico.

behavior of the real flow. Although there are differences between our simulated flow and the 1991 event, nonetheless the general characteristics of the simulation qualitatively match reports of this avalanche [17].

5 Discussion

We now have in place a computational environment for simulating the system of equations modeling single-phase geophysical mass flows, a compute environment that integrates accurate topographical data into a parallel, adaptive mesh Godunov solver. Simulation results are qualitatively correct, but detailed experimental work, especially large-scale fieldwork against which to compare simulations, is still required. Although the Coulomb theory at the heart of the physical modeling presented here requires only a small number of physical parameters (internal and basal friction angles), these parameters are only loosely defined, and measurements of these quantities is crude.

On the one hand, our model system is flexible enough to allow several candidate constitutive relations (which may not suffer the ill-posedness of the

Coulomb theory) to be used. On the other hand, any 'thin layer' model of flow, by necessity, hides important information through the averaging process. An open problem is to recover some of that hidden information - for example, to understand the internal dynamics of recirculation - without recourse to a direct simulation of the complete Coulomb model equations. Another important issue in this discussion is obvious: Does the specific constitutive theory employed in such a thin layer model qualitatively change simulation results?

Particle methods for the solution of the original mass and momentum balance laws (with appropriate constitutive relations) may provide an alternative approach to simulation, and would allow investigation of internal structures that are important in these flows.

More accurate topographical data - potentially available down to $5\,m$-scale - will provide better fidelity of the computations. In particular, we are investigating the marriage of computational mesh refinement grids to a hierarchy of refined topographical mappings, a union that may alleviate artifacting due to numerical errors.

Finally, the Coulomb model used here accounts only for the deformation of solids material. Missing is the effect of interstitial fluid - either air, in dilute flows, or water, in muddy flows. Iverson [9] has taken an important first step in including fluid into a thin layer model, but more work is required before we can be confident of any approach.

6 Acknowledgements

An earlier version of this report was submitted to *Physics of Fluids* consequent to a Symposium on Multi-Component and Multiphase Fluid Dynamics, held in conjuction with the 14th U.S. National Congress of Applied Mechanics. Computations results were performed at the Center for Computational Research at the University at Buffalo. This project is supported in part by grants NSF 0121254 and NSF 0087665 from the National Science Foundation.

References

1. Berger, M. and Colella, P. (1989) "Local Adaptive Mesh Refinement for Shock Hydrodynamics" *JCP* **82** 64-84
2. Carter, H.E. and Stewart, J.F., (2001) "The SIERRA Computational mechanmics framework", SAND2001-2720P, Sandia National Laboratories, Albuquerque, NM
3. Davis, S.F. (1988) "Simplified second-order Godunov-type methods", *SIAM J. Sci. Stat. Comput.* **9** 445-473
4. Denlinger, R.P and Iverson, R.M (2001) "Flow of variably fluidized granular material across three-dimensional terrain 2. Numerical predictions and experimental tests", *J. Geophys. Res.* **106** 553-566

5. Flaherty, J.E., Loy, R.M., Shephard,M.S., Simone, M.L., Szymanksi, B.K., Teresco, J.D., and Ziantz, L.H., (1999) "Distribuited Octree Data Structures and Local Refinement Method For the Parallel Solution of Three-Dimensional Conservation Laws", in *Grid Generation and Adaptive Algorithms*, IMA vol. 113, Springer-Verlag
6. Gray, J.M.N.T. (2001) "Granular flow in partially filled slowly rotating drums" *J. F. M.* **441** 1-29
7. Gray, J.M.N.T., Wieland, M. and Hutter, K. (1999) "Gravity-driven free surface flow of granular avalanches over comlex basal terrain" *Proc. Roy. Soc. London Serial A* **455** 1841-1874
8. Hutter, K. Siegel, M., Savage, S.B., and Nohguchi, Y. (1993) "Two dimensional spreading of a granular avalanche down an inclined plane; Part 1: Theory" *Acta Mech.* **100** 37-68
9. Iverson, R.M (1997) "The physics of debris flows", *Rev. Geophys.* **35** 245-296
10. Iverson, R.M and Denlinger, R.P, (2001) "Flow of variably fluidized granular material across three-dimensional terrain 1. Coulomb mixture theory", *J. Geophys. Res.* **106** 537-552
11. Janssen, H.A (1895) *Z. Ver. Dtsch. Ing.* **39** 1045.
12. Laszloffy, A., Long, J. and Patra, A. (2000) "Simple data management schemes and scheduling schemes For managing the irregularities in parallel adaptive *hp* finite element simulations" *Parallel Computing* **26** 1765-1788.
13. Nedderman, R. (1992) *Statics and Kinematics of Granular Materials* Cambridge University Press
14. Pitman, E.B. and Schaeffer, D.G. (1987) "Stability of Time Dependent Compressible Granular Flow in Two Dimensions", *Comm. Puer Appl. Math.* **40** 421-447
15. Pitman, E. B. and Tulyakov, S. (1998) "Granular debris flows in 1 dimension" UB Math Department report
16. Rankine, W.J.M. (1857) "On the stability of loose earth", *Philos. Trans. Roy. Soc. London* **147** 9-27
17. Rodriguez-Elizarraras, S., Siebe, C., Komorowski, J.-C., Espindola, J.M., and Saucedo, R. (1991) "Field observations of pristine block-and-ash flow deposits emplaced April 16-17, 1991 at Volcan de Colima, Mexico" *J. Volcanology and Geothermal Research* **48** 399-412.
18. Savage, S.B. and Hutter, K. (1989) "The motion of a finite mass of granular material down a rough incline", *J.F.M.* **199** 177-215
19. Schaeffer, D.G. (1987) "Instability in the evolution equations describing granular flow", *J. Diff. Eqn.* **66** 19-50.
20. Schaeffer, D.G. and Pitman, E.B. (1988) "Illposedness in Three Dimensional Plastic Flow" *Comm. Pure Appl. Math.* **41** 879-890
21. Terzaghi, K. (1936) "The shearing resistance of saturated soils and the angle between planes of shear", *Proc. 1st Int. Conf. Soil Mech.* 54-56

Adaptive Application of Characteristic Projection for Central Schemes

Gabriella Puppo[1]

Dip. di Matematica, Politecnico di Torino, Corso Duca degli Abruzzi 24, 10129 Torino, Italy *puppo@calvino.polito.it*

Summary. In this work, the application of central schemes to systems of conservation laws is studied. A comparison between the componentwise implementation of these schemes and the algorithm exploiting characteristic projection is carried out, studying the behaviour of the weights appearing in the reconstruction. Next, a more efficient adaptive strategy that employs characteristic projections only close to discontinuities is proposed.

1 Introduction

Central schemes based on staggered grids have proved to be effective and efficient tools for the integration of systems of conservation laws. Their main advantage is the simplicity of the resulting codes and their high speed. As a consequence, they have reached a wide range of applications, see for instance the review in [5].

The price paid for these advantages is weaker monotonicity preserving properties, with respect to the best upwind schemes, at least in some test cases for high order central schemes: compare for instance the tests in [1] and in [2]. An effective cure was recently proposed in [4]. It consists in applying central schemes projecting the equations along characteristic directions. This device however considerably slows down the code, and it also mars the simplicity of central schemes: while before it was only necessary to know the flux function of the system of equations, now it is necessary to know and compute the eigenstructure of the system.

In this work, we first investigate the behaviour of the reconstruction appearing in [4], and next we propose an adaptive strategy which permits to recover the speed of traditional central schemes, while enjoying the lack of oscillations obtained with the scheme in [4].

We apply these ideas to the recently proposed Central Runge Kutta scheme of order 4, [3]. The same techniques can be extended to other central schemes.

2 Description of the scheme

We consider the system of equations:

$$u_t + f_x(u) = 0, \tag{1}$$

with $u \in R^m$, $f : R^m \to R^m$ continuously differentiable. We suppose that the Jacobian of f, $A(u) = f'(u)$ has real eigenvalues and a complete set of eigenvectors.

Let Δt and h be the mesh widths in time and space, with $\lambda = \Delta t/h$. Let $I_j = [x_j - \frac{h}{2}, x_j + \frac{h}{2}]$ be the interval of width h centered around the grid point x_j. Let $\bar{u}_j^n$ be the cell average of the function $u(x,t)$ on the interval I_j at time $t^n = t_0 + n\Delta t$.

We briefly describe the 4-th order Central Runge Kutta scheme on which we will perform all numerical tests. For more details, see [3].

Given the cell averages $\{\bar{u}_j^n\}$ of the solution at time t^n, we reconstruct the function $u(x, t^n)$ with a suitable fourth order non oscillatory reconstruction operator R:

$$u^n(x) = R(x; \{\bar{u}^n\}) = \sum_j P_j^d(x)\chi_{I_j}(x), \tag{2}$$

where $P_j^d(x)$ are polynomials of degree $d = 2$ (for CRK4). The reconstruction is piecewise smooth, with jumps at the end points of the interval I_j.

Following the philosophy of Central Schemes, [5], we integrate the conservation law (1) on the *staggered* interval $I_{j+1/2} = [x_j, x_{j+1}]$. We obtain the exact equation:

$$\left.\frac{d\bar{u}}{dt}\right|_{j+1/2} = -\frac{1}{h}\left[f(u(x_{j+1},t)) - f(u(x_j,t))\right]. \tag{3}$$

As in all CRK schemes [3], this equation is first discretized in time with a Runge-Kutta scheme. The updated solution is given by:

$$\bar{u}_{j+1/2}^{n+1} = \bar{u}_{j+1/2}^{n} - \lambda \sum_{i=1}^{\nu} b_i K_{j+1/2}^{(i)}, \tag{4}$$

where ν is the number of steps of the RK scheme, b and a (appearing below) are the vector and the matrix defining the RK scheme, and:

$$K_{j+1/2}^{(i)} = f(u_{j+1}^{(i-1)}) - f(u_j^{(i-1)}) \qquad \text{with} \quad u_j^{(0)} = u^n(x_j). \tag{5}$$

The term $\bar{u}_{j+1/2}^n$ in (4) is computed directly using the reconstruction (2). To evaluate the intermediate states, $u_j^{(i)}$, we exploit the fact that the reconstruction $u^n(x)$ is smooth except for jumps at the cell walls, $x_{j\pm1/2}$. Thus, $u(x_j,t)$ remains smooth for $t \in [t^n, t^n + \Delta t]$, if Δt is small enough. As in all central schemes based on staggered grids, we can therefore evaluate the intermediate states $u_j^{(i)}$ integrating the conservation law (1) in its differential form, namely: $u_t = -f_x(u)$. Thus:

$$u_j^{(i)} = u_j^{(0)} + \Delta t \sum_{l=1}^{i} a_{i,l}\hat{K}_j^{(l)}, \qquad \hat{K}_j^{(l)} = -\left.\frac{\partial f(u^{l-1})}{\partial x}\right|_j \tag{6}$$

for $i = 1, \cdots, \nu - 1$. Here, the standard fourth order Runge-Kutta scheme is used; with our notation: $b = (1/6, 1/3, 1/3, 1/6)$, while a is a diagonal matrix, with $a(1,1) = a(2,2) = 1/2$ and $a(3,3) = 1$.

We now consider space discretization, which plays a particularly important role in the extension to systems of equations.

At the beginning of each time step, we need a reconstruction algorithm to compute $\bar{u}^n_{j+1/2}$ and $u^{(0)}_j$. Next, we need a reconstruction scheme which computes $f_x(u^{(i)})$, starting from the point values $f(u^{(i)})$, for $i = 0, \cdots, \nu - 1$. We use the CWENO reconstruction [2], outlined below.

The reconstruction from cell averages at the beginning of each time step is a non-linear convex combination of three parabolas. On the interval I_j:

$$u^n(x)|_{I_j} = R_j(x) = \omega_j^{-1} P_{j-1}(x) + \omega_j^0 P_j(x) + \omega_j^{+1} P_{j+1}(x), \tag{7}$$

where the ω_j^k are the non linear weights, which satisfy $\sum_k \omega_j^k = 1$. The coefficients of the parabola P_{j+k} are computed interpolating the data $\bar{u}^n_{j+k+l}$, $l = -1, 0, 1$ in the sense of cell averages.

The weights ω_j^k are determined in order to maximize accuracy in smooth regions and to prevent the onset of spurious oscillations. As in [1], we define the weights with the following formulae:

$$\omega_j^k = \frac{\alpha_j^k}{\sum_{l=-1}^{1} \alpha_j^l} \qquad \alpha_j^k = \frac{C^k}{\left(\epsilon + \mathrm{IS}_j^k\right)^2}. \tag{8}$$

In the present case, $C^{-1} = C^1 = 3/16$, while $C^0 = 5/8$. The parameter ϵ prevents a vanishing denominator. For the CRK4 scheme, we always have $\epsilon = 10^{-6}$.

Finally, $\mathrm{IS}_j^k, k = -1, 0, 1$ are called *smoothness indicators.* They measure the regularity of the polynomial P_{j+k} on the interval I_j. Following [1], we take:

$$\mathrm{IS}_j^k = \sum_{l=1}^{2} \int_{x_{j-1/2}}^{x_{j+1/2}} h^{2l-1} \left(\frac{d^l P_{k+j}}{dx^l} \right)^2 dx, \qquad k = -1, 0, 1. \tag{9}$$

Thus $\omega_j^k = C^k + O(h^2)$ on smooth regions, while $\omega_j^k = O(h^4)$ if the data in the stencil of the polynomial P_{j+k} contain a jump discontinuity.

Finally, for flux derivatives, the polynomials P_{j+k} are determined interpolating the data $f(u(x_{j+k+l}))$ in the sense of point values. The reconstruction in this case must yield accurate estimates of the derivatives of f. The structure of the non linear weights does not change, but we must use a different set of constants, namely: $C^{-1} = C^1 = 1/6$, while $C^0 = 1/3$, see [2].

3 Systems of equations

We consider two extensions of central schemes to systems of equations. We test the resulting algorithms on two classical problems from 1D gas dynamics, namely: Lax' Riemann Problem and Woodward and Colella blast wave problem, which describes the interaction of two Riemann problems, characterized by strong shocks. These test problems have been widely used in the literature. The initial and boundary conditions and the details of the implementation for central WENO schemes can be found in [2].

We now describe the two algorithms we are compairing. The simplest approach to the solution of systems of equations is to apply the scheme component by component, following the scalar recipe. It was found in [2] that the results improve if all components of the solution share the same smoothness indicator. At the beginning of each time step a Global Smoothness indicator is computed, namely:

$$IS_k^j = \frac{1}{m}\sum_{r=1}^{m}\frac{1}{||\bar{u}_r||_2}\left(\sum_{l=1}^{2}\int_{I_j} h^{2l-1}\left(\frac{d^l P_{j+k,r}}{dx^l}\right)^2 dx\right), \tag{10}$$

for $k=-1,0,1$. Here r denotes the r-th component of the solution and of the vector-valued interpolating polynomial. Compairing with (9), we see that the Global Smoothness Indicator is just a weighted average of the Smoothness Indicators given by each component.

These quantities are then used to compute the non linear weights throughout the time step. This will be called *Componentwise Implementation (CW)* of Central schemes.

In the *CW* approach, very little information on the physics of the system of equations is required. It is enough to provide the flux function f and an estimate of the characteristic speeds of propagation, to satisfy the stability condition, which, for the CRK4 scheme, is $\lambda \leq 12/25 * 1/\max|\mu(u)|$, where μ are the eigenvalues of the Jacobian of f, see [3].

The solution computed in this fashion is fast and simple to program. However small wiggles may appear. These spurious oscillations decrease in amplitude as the grid is refined. See for instance the left part of Fig. 1.

A second approach was suggested in [4]. It consists in applying to the central framework a technique that is widely employed in schemes based on flux splitting. Specifically, the computation of the weights in the WENO reconstruction is performed component by component but using characteristic directions, instead of conserved variables.

Let $Q(u)$ be the non-singular matrix whose columns are the right eigenvectors of $A(u)$. At each grid point j, compute $Q(\bar{u}_j^n)$. Let $\{\bar{u}_{j+l}^n\}$ be the data in the stencil of the jth cell. For CRK4, the stencil of the reconstruction consists of 5 points; thus $l=-2,\cdots,2$. Map each cell average belonging to the stencil of the jth cell along the characteristic directions, i.e. compute the vectors:

$$\bar{v}_{j+l} = Q^{-1}(\bar{u}_j^n)\bar{u}_{j+l}^n \qquad l=-2,\cdots,2.$$

Construct the conservative interpolant Π_v^j for the data $\{\bar{v}_{j+l}\}$, using the CWENO recipe described above *componentwise*. Then define the interpolant for the conserved variables u through the equation:

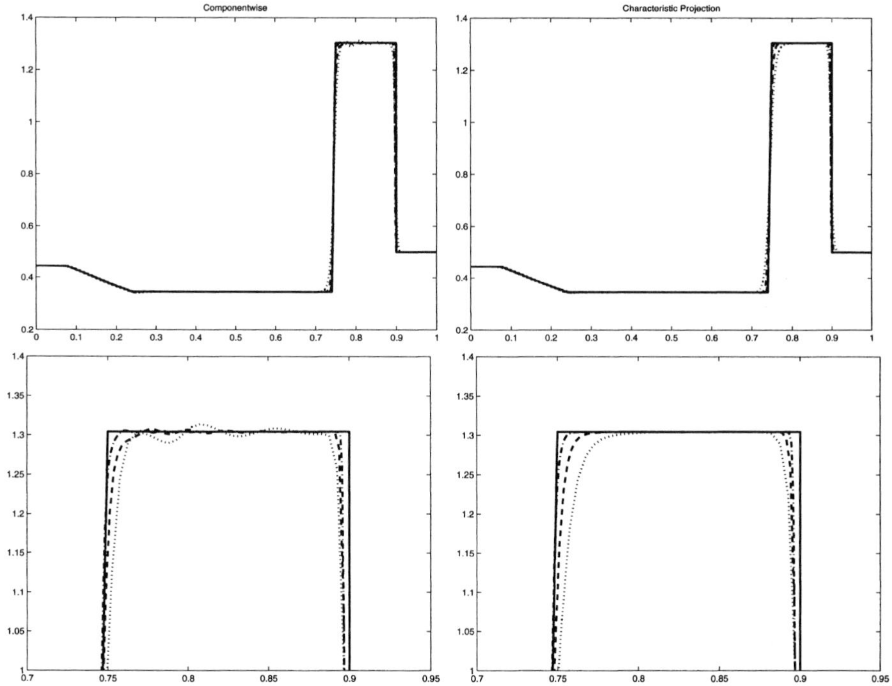

Fig. 1. Lax' Riemann problem. Componentwise (left) and Characteristic Projection (right) implementation of CRK4 scheme, for various grid sizes. **Bottom:** Detail of the density peak. $N = 200$ (dots), $N = 400$ (dash), $N = 800$ (dash-dot).

$$\Pi_u^j(x - x_j) = Q(\bar{u}_j^n)\Pi_v^j(x - x_j).$$

Finally, use this function to evaluate the two staggered half-cell averages:

$$\bar{u}_{j,-}^n = \frac{1}{h}\int_{x_j-\frac{h}{2}}^{x_j} \Pi_u^j(x - x_j)\,dx \qquad \bar{u}_{j,+}^n = \frac{1}{h}\int_{x_j}^{x_j+\frac{h}{2}} \Pi_u^j(x - x_j)\,dx,$$

with $\bar{u}_{j+1/2}^n = \bar{u}_{j,+}^n + \bar{u}_{j+1,-}^n$ and the point value $u(x_j, t^n) = \Pi_u^j(0)$. Notice that because the mapping between conservative and characteristic variables is linear, then a conservative reconstruction in v will provide a conservative reconstruction in u.

According to the results in [4] it is necessary to perform this computation only at the beginning of each time step, to evaluate $\bar{u}_{j+1/2}^n$. Our results confirm this finding. When we use characteristic projection, the derivatives of the fluxes are still computed with the non linear weights obtained with the Global Smoothness indicator. This algorithm will be called *Characteristic Projection (CP)* scheme.

The results obtained with this recipe can be seen on the right of Fig. 1. The spurious oscillations in the density peak of the solution of Lax' Riemann Problem have disappeared.

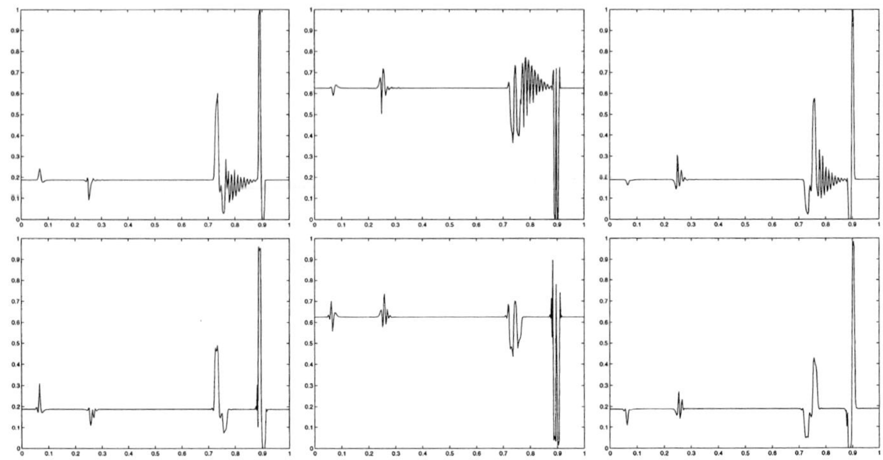

Fig. 2. Global Weights for $N = 400$. From left to right: Left, Central and Right weight. Solution computed with *CW* (top) and *CP* (bottom) algorithms.

The improvement is due to a better separation of the effects of the contact and the shock wave on the weights. Fig. 2 shows a comparison of the non linear global weights which result from the solution to Lax' problem computed with the *CW* (top) and the *CP* (bottom) algorithms. The weights shown in the figure are computed using (10), and they are needed to evaluate the flux derivatives, in both strategies. If we compare with the corresponding solution given in Fig. 1, we see that the fluctuations in all weights clearly signal the location of the shock, the contact, and the head and tail of the rarefaction wave. However in the *CW* implementation, between the shock and the contact the weights have a very irregular behaviour. This effect is absent in the results obtained with the *CP* implementation. This is probably the origin of the wiggles appearing on the left of Fig. (1).

Further insight in the behaviour of the weights can be obtained plotting the weights computed with the *CP* algorithm along each characteristic field, see Fig. 3. Here correctly the contact wave appears only in the second characteristic field, belonging to the eigenvalue $\mu = v$ of the Jacobian $A(u)$, where v is the gas velocity. The head and tail of the rarefaction appear only in very small wiggles in the weights computed for the first characteristic field, belonging to the eigenvalue $\mu = v - c$ (here c is the sound speed). Thus we see that, even if the *CP* algorithm corresponds to a simple local linearization of the system of equations, still it is quite effective in separating all waves thus preventing the oscillatory behaviour of the *CW* scheme.

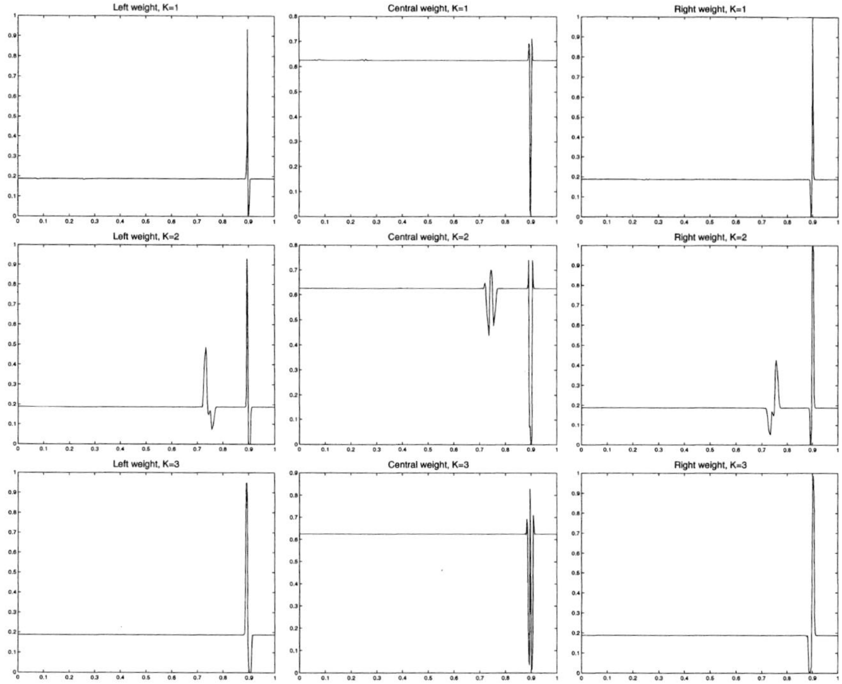

Fig. 3. Weights computed with characteristic projection for $N = 400$. From left to right: Left, Central and Right weight. From top to bottom: first, second and third characteristic field.

4 Adaptive characteristic projection

Table 1. CPU time

N	*CW*	*CP*	Adaptive
200	5.69	15.74	6.50
400	22.45	62.74	25.44
800	90.34	250.34	101.90
1600	359.24	1002.66	409.32

We have seen that the *CP* implementation is quite effective in the prevention of spurious oscillations. However this procedure is quite costly, as is apparent from reading the first two columns of Table 1. Here the CPU time required for the solution of Lax' Riemann Problem is reported for the *CW* and the *CP* algorithms for various grid sizes. These data were obtained running the code on a Pentium II 266 Mz processor, with the F77 Linux compiler. It is clear that *CP* is roughly 3 times as expensive as *CW*. However Fig. 1 and Fig. 2 show that the *CW* algorithm needs to be improved only close to discontinuities.

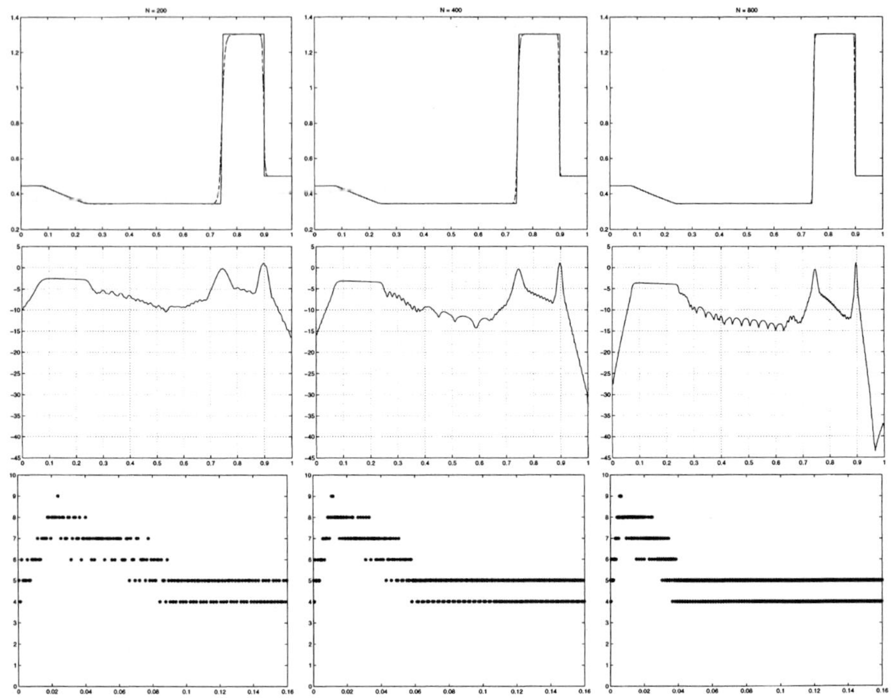

Fig. 4. Top: solution with selective application of characteristic projection, for $N = 200$, $N = 400$ and $N = 800$. **Middle:** Logarithmic plot of $\mathrm{IS}_j^{\mathrm{TOT}}$ vs x at final time. **Bottom:** Number of cells where characteristic projection is used vs time.

Thus it appears that the *CP* scheme can be made much more efficient, if projection along characteristic directions is applied selectively, only in those cells close to discontinuities, and computing the solution in the remaining cells with the *CW* recipe.

The adaptive scheme we propose computes the Global Smoothness indicator in each cell, at the beginning of each time step. Then we obtain a measure of the smoothness of the data in the whole stencil, by evaluating:

$$\mathrm{IS}_j^{\mathrm{TOT}} = \sum_{l=-1}^{1} \mathrm{IS}_j^l,$$

where IS_j^l is given in (10). As for the Smoothness Indicator, IS_j^l, this quantitity is $O(h^2)$ in smooth regions, while it is $O(1)$ if a jump discontinuity is present somewhere in the whole stencil on which the reconstruction is based. Thus, we apply the *CP* scheme only in those cells for which:

$$\mathrm{IS}_j^{\mathrm{TOT}} \geq 1, \tag{11}$$

while we use the *CW* scheme in the remaining cells. The resulting scheme will be called Adaptive scheme.

We show the results obtained with this strategy in Fig. 4 for Lax' Riemann Problem. In the first row, we show a comparison of the solution computed with the *CP* (dashed line) and the Adaptive (dotted line) schemes. The solid line is the exact solution. As is apparent from the figure, for all grids tested, the two numerical solutions almost coincide.

The second row in Fig. 4 shows plots of the logarithm of $\mathrm{IS}_j^{\mathrm{TOT}}$ for various grid sizes at final time. Compairing with the exact solution in the row above, it is apparent that the peaks of $\mathrm{IS}_j^{\mathrm{TOT}}$ coincide with the location of the discontinuities.

Finally, the third row of Fig. 4 shows the number of cells $N_{CP}(t)$ on which projection along characteristic directions was actually used, at time t, as a function of time for various grid sizes. It is interesting to note that $N_{CP}(t)$ does not depend on the number of grid points. The quantity $N_{CP}(t)$ is larger for small values of t, when the waves issueing form the initial discontinuity have not yet separated, and it decreases as the waves separate. It is clear from these results that a considerable saving in computing time can be obtained through an adaptive application of projection along characteristic directions.

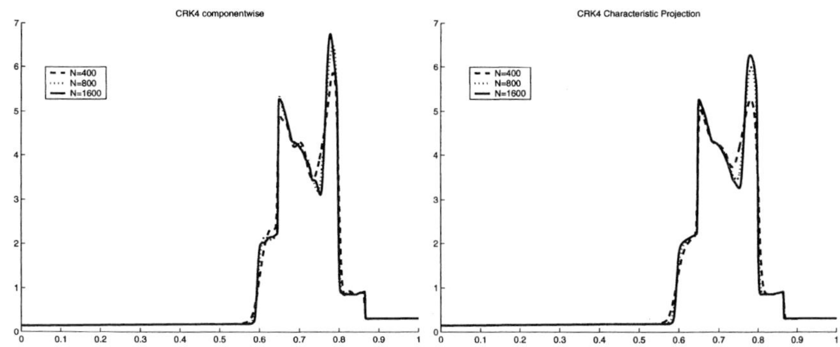

Fig. 5. Solution to Woodward and Colella blast problem, at $T = 0.038$. **Left:** *CW* scheme. **Right:** Adaptive *CP* scheme.

The last column of Table 1 shows the CPU time required by the Adaptive strategy just described. The computational cost of the Adaptive scheme is comparable to the *CW* case, but now the control of spurious oscillations is analogous to what is achieved by the *CP* scheme.

To check the robustness of the adaptive strategy, we compute the solution of Woodward and Colella blast wave problem. The density obtained at $T =$

0.038 is shown in Fig. 5, for the *CW* (left) and the Adaptive (right) strategies. Again, the *CW* scheme results in spurious oscillations, which disappear when the Adaptive characteristic projection is used. The numerical results obtained by the Adaptive and the *CP* scheme are indistinguishable.

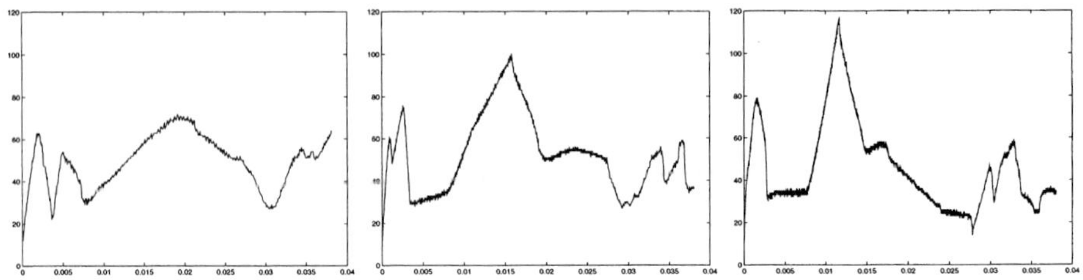

Fig. 6. Woodward and Colella blast problem. Number of cells labelled for projection along characteristic directions as a function of time. From left to right: $N = 400$, $N = 800$, $N = 1600$.

Finally, Fig. 6 shows the values of $N_{CP}(t)$ in the case of Woodward and Colella problem as a function of time, for various grid sizes. These plots are quite irregular, due to the complex behaviour in time of the solution. However we note that even in this case $N_{CP}(t)$ depends only weekly on the number of grid points. This is even more apparent if we consider the average of $N_{CP}(t)$, $< N_{CP} >$, over time. We find: $< N_{CP} >= 48.6$, $< N_{CP} >= 51.9$, $< N_{CP} >= 46.5$ for $N = 400$, $N = 800$ and $N = 1600$ respectively.

We conclude that the adaptive strategy based on applying characteristic projection only in those cells for which: $\mathrm{IS}_j^{\mathrm{TOT}} \geq 1$ is effective in preventing the onset of spurious oscillations, without increasing significantly the computational cost of the componentwise algorithm. Moreover the strategy just described is robust, since it remains effective even on a problem involving a complex solution, with several wave interactions.

References

1. Jiang G.-S., Shu C.-W.: (1996): Efficient Implementation of Weighted ENO Schemes. J. Comput. Phys., **126**, pp.202–228.
2. Levy D., Puppo G., Russo G.: (1999) Central WENO Schemes for Hyperbolic Systems of Conservation Laws. Math. Model. and Numer. Anal., **33**, no. 3, pp.547–571.
3. Pareschi L., Puppo G., Russo G., (2002): Central Runge-Kutta Schemes for Conservation Laws, in preparation.
4. Qiu J, and Shu C.W.: (2002) On the construction, comparison, and local characteristic decomposition for high order central WENO schemes, submitted to J. Comput. Phys.

5. Tadmor E.: (1998): Approximate Solutions of Nonlinear Conservation Laws. In: Quarteroni A. (ed) Advanced Numerical Approximation of Nonlinear Hyperbolic Equations, Lecture Notes in Mathematics, Springer, Berlin.

Comparison of Several Difference Schemes for the Euler Equations in 1D and 2D

Richard Liska[1] and Burton Wendroff[2]

[1] Czech Technical University in Prague, Faculty of Nuclear Sciences and Physical Engineering, Břehová 7, 115 19 Prague 1, Czech Republic
liska@siduri.fjfi.cvut.cz

[2] Los Alamos National Laboratory, Los Alamos, New Mexico 87544, USA
bbw@lanl.gov

Summary. Representative ten finite difference schemes are applied on a suite of one-dimensional and two-dimensional test problems for the Euler equations. Sample results are presented in this paper while the full results are available on the web at `http://www-troja.fjfi.cvut.cz/~liska/CompareEuler/compare`.

1 Introduction

Hyperbolic conservation laws, and the Euler equations of compressible fluid dynamics in particular, have been the subject of intensive research for at least the past five decades, and with good reason. The applications are many - aircraft design, stellar formation, weather prediction to name only a few. There are some theoretical results however even if the theory were perfect the applications would not be possible without methods for obtaining approximate solutions. The unfortunate situation here is that rigorous error estimates for supposed approximate solutions are almost entirely nonexistent. So, it is universally recognized that tests of methods on difficult problems are essential. Our concern here is to compare the behavior of some Eulerian methods to each other on problems that seem to us to be sufficiently difficult and representative to enable the reader to draw some conclusions about the applicability of these methods. The now classic work of this nature is the paper by Gary Sod [1] or Woodward Colella review [2]. Several difficult tests are discussed in [3].

2 Finite Difference Schemes

We have chosen ten methods that we feel are representative of the different basic finite difference approaches to solving hyperbolic conservation laws:

A composite scheme (CFLF) called LWLFn consists of a cycle of $n-1$ time steps of some version of the Lax-Wendroff (LW) scheme followed by one step with Lax-Friedrichs (LF). The LF step acts as a consistent (with the differential equations) filter to reduce the oscillations of LW. Here we use the 2D version called CFLF [4].

A CFLF hybrid scheme (CFLFh) is also as the previous one a combination of second order LW and first order LF type schemes. The numerical flux is given by a weighted average of LF diffusive and LW oscillatory fluxes. The weight is chosen in such a way that the scheme is second order on smooth solutions, but becomes sufficiently dissipative in shocks. We use the Harten weight [5] (for overview of weights see [6]) with the CF and LF numerical fluxes [4].

Centered scheme with limiter (JT) uses discontinuous limited piecewise linear reconstruction from cell averages to get fluxes at cell edges. It uses neither dimensional splitting nor eigenvector decomposition nor any overt Riemann solver. We use the 2D version of this non-oscillatory central scheme by Jiang and Tadmor [7] which is the 2D successor of the Nessyahu-Tadmor 1D NT scheme [8] (see also [9]).

Positive scheme (LL) by Liu and Lax [10, 11] is an explicit two-level method giving the new value of the solution vector as a linear combination of values at the previous time with coefficients that are positive symmetric matrices summing to the identity. It uses eigenvector decomposition and limiting.

Clawpack wave propagation scheme (CLAW) is a sophisticated flux splitting scheme developed by LeVeque [12], based on advection and wave propagation ideas[13]. From the many options permitted by Clawpack [14] we use the non-split version with monotonized centered limiter using the Roe Riemann solver with 4 waves (separate shear and entropy waves).

Weighted average flux scheme (WAFT) developed by Toro [15] defines the flux at the cell boundary as a spatial weighted average over the states of an approximate Riemann solver. A limiter is employed in the computation of the weights. We use the 2D non-split version from the Numerica library [16] with HLLC approximate Riemann solver using the Rankine-Hugoniot condition for evaluating the middle fluxes.

WAF scheme with HLLE speeds (WAFC) is a WAF split scheme using the HLLC approximate Riemann solver with Einfeldt [17] speeds and fluxes evaluated at the middle states.

Weighted essentially non-oscillatory scheme (WENO5) is a weighted combination of essentially non-oscillatory schemes on several stencils. Weights used with upwind biased spatial differencing produce high order accuracy for smooth flows but become low order and dissipative for shocks. We use the fifth order version from [18] applying the WENO procedure after eigenvector decomposition with Runge-Kutta time integration.

Component wise WENO scheme (CWENO3) is the third order variant of WENO applied directly to the conserved variables without eigenvector decomposition. Its simple description has been presented in [4].

Piecewise parabolic scheme (PPM) is a third order Godunov method using piecewise parabolic limited reconstruction to obtain states to use in the Riemann problems defining the fluxes [19]. Dimensional splitting is used. Here we use PPMLib [20] version [1].

These finite difference methods are compared on a set of 1D and 2D problems.

Typical features included in many schemes are dimensional splitting, approximate Riemann solvers, eigenvector decomposition and limiters. Which of these ingredients are employed by a particular scheme is summarized in the table 1.

Table 1. Basic ingredients of difference schemes

	dim. splitting	Riemann solver	eigenvector dec.	limiter
CFLF				
CFLFh				
WAFT		•		•
WAFC	•	•		•
CLAW		•	•	•
PPM	•	•		•
WENO5			•	
CWENO3				
LL			•	•
JT				•

3 Test Problems in 1D

Most of the 1D problems, except the last one, are Riemann problems (five of them, tests *1,2,4,5,6*, are from [15]) for which we have exact solutions and we can compare errors of the different numerical schemes. Table 2 presents L^1 relative errors for all schemes and all 1D Riemann problems.

1D problems include:

Test 1 is Toro's [15] version of Sod's Riemann problem [1] with a sonic point in the rarefaction. Results of all schemes for this problem are presented in Fig. 1. All ten methods resolve the shock very well without oscillation with the exception of CFLF, although CWENO3 and LL are more dissipative than the others. PPM is outstanding on the contact, with WAFC and WAFT not far behind. Many of the schemes tested in [15] develop the so-called sonic glitch in the rarefaction, but this is not present in any of our ten schemes, but WAFC, PPM, CLAW, and JT have a dip at the base of the rarefaction wave.

[1] Recently we have found a very nice PPM Lagragian remap code VH-1 [21] including force sources and all standard geometries in 1D, 2D and 3D. Take a look, it is public domain with all source files.

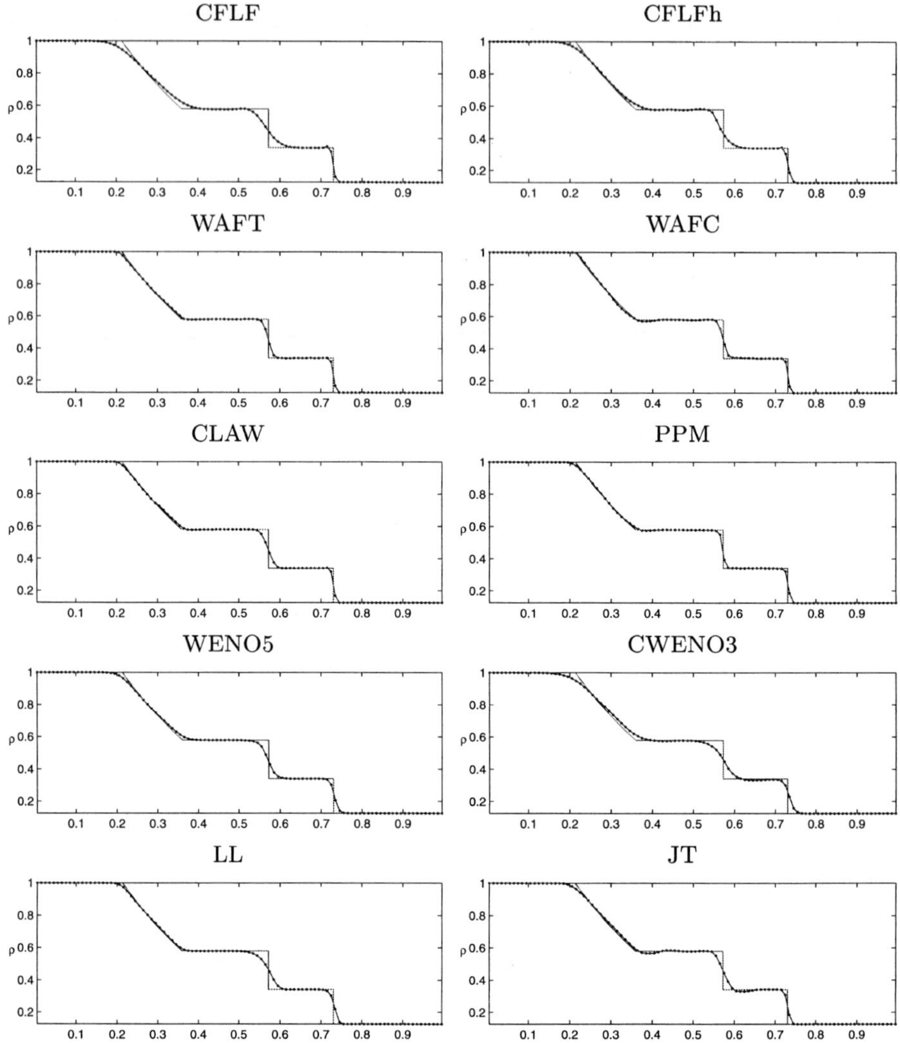

Fig. 1. 1D results (density) for the test 1 problem by all ten schemes with exact solution by solid line.

Test 1-tvj is the same problem as Test 1 but with a large jump in transverse velocity (all other 1D tests have zero transverse velocity everywhere). In this problem we use 2D capability of all codes by using non-zero transversal velocity on 1D mesh. The transversal velocity jump should have no influence on the solution which should be identical to the solution of test 1 with transversal velocity jump advected with the contact wave. The results of this problem by WAFT, CLAW, WENO5 and PPM schemes are presented in Fig.

2. Results of other schemes are similar with density local minimum around the contact. The jump produces an error in the kinetic energy and destroys the solution for all schemes. However, if we run WAFC for this problem with removed transverse kinetic energy then we get no effect from the non-constant transverse velocity, obtaining the same result as for test 1. This is an option only available to dimensionally split schemes and it works well in 2D only for jumps aligned with the grid. When we tried WAFC with the transverse kinetic energy removed for the 2D Noh problem (described later), it gives the wrong shock speed. For that problem, Paul Woodward [22] pointed out to us that conversion of kinetic energy to heat is the critical mechanism, therefore in that case ignoring some kinetic energy is not a good idea. But in [23] the transverse kinetic energy removal is part of a scheme for two-fluid problems and seems to be very effective.

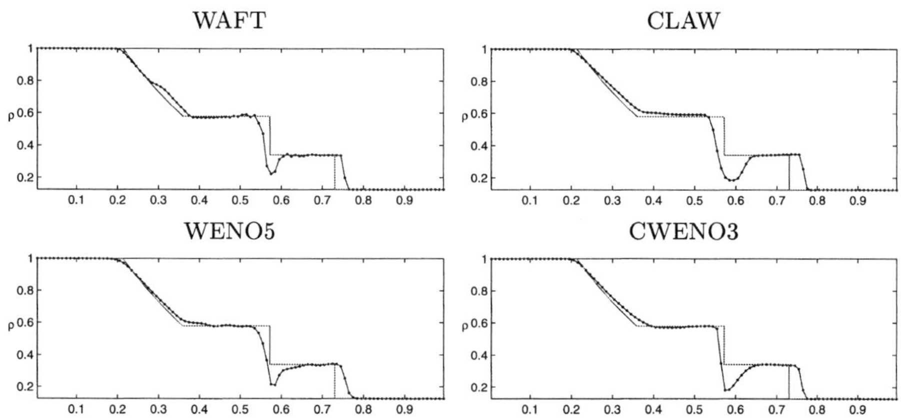

Fig. 2. 1D results (density) for the test 1-tvj problem (with transversal velocity jump) by WAFT, CLAW, WENO5 and PPM schemes with exact solution by solid line.

Test 2 develops a near vacuum in the central state with both density and pressure very close to zero, but the internal energy is far from zero.

1D-Noh is the classical 1D Noh problem [24]. The solution of this problem consists of two infinite strength shocks moving out from the center, leaving a constant density and pressure state behind. The results of this problem by WAFT, CLAW, WENO5 and PPM schemes are presented in Fig. 3. Other tests have nice results similar to WENO5 result. CFLF has a slight overshoot at the shocks, but only a very small dip in density at the center, as does JT. PPM has a poorly resolved shock and a significant dip at the center, and is not symmetric about the center. The WAFT run was with the superbee limiter and is also not symmetric, while a run using minmod limiter was symmetric (superbee produces better results than minmod for other tests).

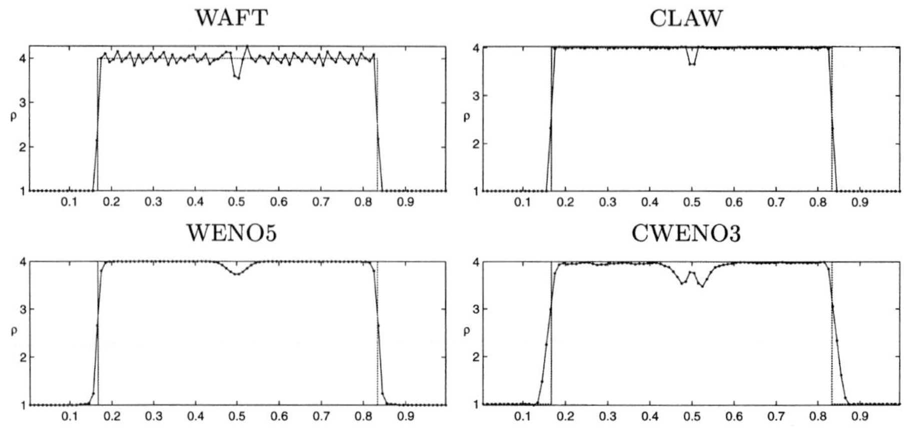

Fig. 3. 1D results (density) for the Noh problem by WAFT, CLAW, WENO5 and PPM schemes with exact solution by solid line.

Test 3a is a variant of Toro's Test 3 [15] with a stationary contact.

Test 4 has two strong shocks with a strong contact wave.

Test 5 has only one wave, a stationary contact and distinguishes which schemes resolve it exactly. Stationary contact is resolved exactly by schemes WAFT, WAFC, CLAW, PPM and WENO5.

Test 6 has also only one wave, a slowly moving contact.

Peak produces a very narrow strong peak in density between the shock and contact and rarefaction with small change in density and large change in velocity.

Blast is the classical Woodward-Collela blast wave problem [2] including interaction of two Riemann problems with reflecting boundary conditions.

4 Test Problems in 2D

The accuracy of the numerical schemes is numerically determined for the smooth exact 2D solution ($\rho(x,y,t) = 1 + 0.2\sin(\pi(x+y-t(u+v)))$, u, v, p constants) with periodic boundary conditions. The most accurate is 5-th order WENO5 followed by 3-rd order PPM. Most schemes are second order and CFLF is just first order.

The first set of 2D test problems consists of six cases from the collection of 2D Riemann problems proposed by [25] and used by others [11, 26], namely, cases *3,4,6,12,15* and *17* from [11] (which are configurations *3, 4, B, F, G, K* from [25]). These 2D Riemann problems are solved on a square divided into four quadrants. The initial conditions, constant states in each quadrant, were chosen in such a way that each of four 1D interfaces between the quadrants

Table 2. L^1 errors in % of 1D Riemann problems. For test 2 errors are in internal energy, for peak in velocity and for all the other tests in density. Tests 3a and 4 use 200 cells, test peak uses 800 cells and all other tests use 100 cells. Fail means that the scheme has failed to solve given test.

Test	1	1-tvj	2	Noh	3a	4	5	6	peak
CFLF	2.0	7.3	16.4	1.4	12.0	3.8	1.5	0.7	3.0
CFLFh	1.5	8.1	10.2	1.9	10.3	2.7	0.7	0.8	1.9
WAFT	0.7	3.3	21.9	2.8	2.6	1.4	0	0.3	1.0
WAFC	0.8	7.8	fail	1.4	2.8	1.4	0	0.3	1.1
CLAW	0.8	5.5	fail	1.3	3.1	1.7	0	0.4	fail
PPM	0.5	3.9	6.3	4.6	9.4	1.1	0	0.1	1.3
WENO5	1.3	3.5	23.7	2.0	9.2	2.2	0	0.4	2.4
CWENO3	2.2	6.8	28.2	2.5	14.7	3.9	1.1	1.1	2.4
LL	1.3	10.0	31.3	1.5	5.2	2.4	0.5	0.7	0.8
JT	1.3	3.9	6.4	1.7	8.1	2.3	0.6	0.6	1.1

defines 1D Riemann problem whose solution includes only one wave, a shock, contact or rarefaction.

The second set of 2D problems consists of four tests:

2D Noh is a classical test [24] with exact solution being an infinite strength circularly symmetric shock reflecting from the origin. Behind the shock (i.e. inside the circle) the density is 16, the velocity is 0 and the pressure is 16/3. The shock speed is 1/3 and ahead of the shock, that is for $\sqrt{x^2+y^2} > t/3$, the density is $(1 + t/\sqrt{x^2+y^2})$ while velocity and pressure remain the same as initially, i.e. velocity is directed towards the origin and the pressure is zero. For an analysis of this problem see [27].

Few results for this problem are shown on Fig. 4 where we see that LL scheme was completely wrong. CFLF scheme was best. Schemes WAFT, CLAW, WENO5 failed to run. Scheme PPM required CFL number 0.2 on 400×400 grid however run with 0.8 on 100×100 grid. JT run on 100×100 but not on 400×400. CFLFh gave quite good result, compare scatter plots of CFLF and CFLFh on Fig. 4. WAFC was very noisy CWENO3 gave a nice result. All results are available on the web [29].

Rayleigh-Taylor instability is a physical phenomenon appearing when a layer of heavier fluid is placed on top of a layer of lighter fluid. This is the only problem having a source in the Euler equations. A gravitational source term appears in the momentum equation. The typical mushroom structure results.

Implosion is defined on a square box with reflecting boundary conditions. Initially the gas inside a smaller square, placed in the center of the box and rotated by $\pi/4$, has smaller density and pressure than the rest of the box. The initial data are the Sod problem data [1]. At later stages of this problem a rich structure of waves reflected from the boundaries develops and

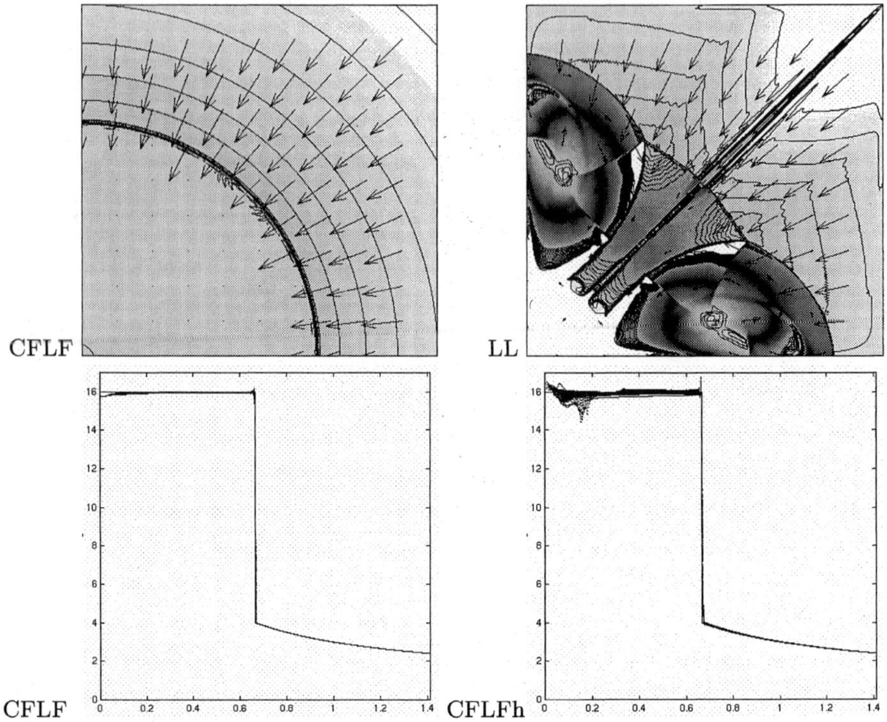

Fig. 4. Results for the Noh problem by the CFLF, LL and CFLFh schemes. Density color map is overlayed by density contours and velocity arrows on the top. Scatter plot of density versus radius together with the exact solution is on the bottom. The computations were done on the square $(x, y) \in (0, 1) \times (0, 1)$ with 400×400 grid to time $T = 2$.

the split dimension PPM and WAFC schemes, which both use alternating Strang splitting [28] (which is not symmetric), are losing the symmetry as well as WAFT using superbee limiter (with minmod limiter WAFT is keeping symmetry).

Explosion problem [15] starts with higher density and pressure circular area inside a lower density and pressure gas. An unstable contact develops at later times.

For all 2D problems, except the 2D Noh problem, we no longer have exact solutions available, so a definitive objective evaluation of the validity of the solutions obtained is not possible.

5 Final remarks

We have applied a representative set of ten finite difference schemes approximating Euler equations to a suite of problems in one and two dimensions.

It is clear that some methods appear to work better than others on a specific problem, but no one scheme has shown itself to be superior on all of them, which should come as no surprise. Typically schemes have more troubles with resolution of contact waves than that of shock waves. In most cases PPM and/or WAFs schemes work best on contacts.

For more details see [29] where all results are presented.

Acknowledgement. Many people have helped along the way in this project. Particular thanks are owing to Paul Woodward who presented a careful appraisal of some of the problems and of the behavior of PPM. Eleuterio Toro generously shared with us the WAF code from the Numerica library. We also would like to acknowledge Tariq Aslam, Guan-Shan Jiang, Milan Kuchařík, Alexander Kurganov, Randall LeVeque, Xu-Dong Liu, Len Margolin, James Quirk, Misha Shashkov, Chi-Wang Shu, and Eitan Tadmor.

This research was supported by the U. S. Department of Energy under contract W-7405-ENG-36. R. Liska was supported in part by the Czech Grant Agency grant 201/00/0586 and would like to thank the Institute for Geophysics and Planetary Physics (IGPP) for hosting his visits at Los Alamos National Laboratory.

References

1. G. A. Sod. A survey of several finite difference schemes for hyperbolic conservation laws. *J. Comp. Phys.*, 27:1–31, 1978.
2. P. Woodward and P. Colella. The numerical simulation of two-dimensional fluid flow with strong shocks. *J. Comp. Phys.*, 54:115–173, 1984.
3. J.J. Quirk. A contribution to the great riemann solver debate. *Int. J. For Num. Meth. in Fluids*, 18:555–574, 1994.
4. R. Liska and B. Wendroff. Composite schemes for conservation laws. *SIAM J. Numer. Anal.*, 35(6):2250–2271, 1998.
5. A. Harten. The artificial compression method for computation of shocks and contact discontinuities: Iii self adjusting hybrid schemes. *Mathematics of Computation*, 32:363–389, 1978.
6. C.B. Laney. *Computational Gasdynamics.* Cambridge University Press, Cambridge, UK, 1998.
7. G.S. Jiang and E. Tadmor. Non-oscillatory central schemes for multidimensional hyperbolic conservation laws. *SIAM Journal on Scientific Computing*, 19:1892–1917, 1998.
8. H. Nessyahu and E. Tadmor. Non-oscillatory central differencing for hyperbolic conservation laws. *J. Comp. Phys.*, 87(2):408–463, 1990.
9. P. Arminjon, D. Stanescu, and M.C. Viallon. A two-dimensional finite volume extension of the lax-friedrichs and nessyahy-tadmor schemes for compressible flows. In M. Hafez and K. Oshima, editors, *Proc. of the 6-th Internat, Symp. on Computational Fluid Dynamics*, volume 4, pages 7–14, Lake Tahoe, NV, 1995.
10. X.D. Liu and P.D. Lax. Positive schemes for solving multi-dimensional hyperbolic systems of conservation laws. *CFD Journal*, 5(1):1–24, 1996. `http://www.math.ucsb.edu/~xliu/publication/paper/positive1.pdf`.

11. P. D. Lax and X.-D. Liu. Solution of two dimensional Riemann problem of gas dynamics by positive schemes. *SIAM J. on Scientific Comp.*, 19(2):319–340, 1998.
12. R.J. LeVeque. Wave propagation algorithms for multi-dimensional hyperbolic systems. *J. Comp. Phys.*, 131:327–353, 1997.
13. R.J. LeVeque. High-resolution conservative algorithms for advection in incompressible flow. *SIAM J. Numer. Anal.*, 33:627–665, 1996.
14. R.J. LeVeque. Clawpack version 4.0 user's guide. Technical report, University of Washington, Seatle, 1999. `http://www.amath.washington.edu/~claw/`.
15. E.F. Toro. *Riemann Solvers and Numerical Methods for Fluid Dynamics.* Springer-Verlag, Berlin, Heidelberg, 1997.
16. E.F. Toro. Numerica a library of source codes for teaching, research and applcations:, 2000. `http://www.numeritek.com/numerica_software.html`.
17. B. Einfeldt. On godunov-type methods for gas dynamics. *SIAM J. Numer Anal.*, 25:294–318, 1988.
18. Guan-Shan Jiang and Chi-Wang Shu. Efficient implementation of weighted eno schemes. *J. Comp. Phys.*, 126:202–228, 1996.
19. P. Colella and P. Woodward. The piecewise parabolic method (ppm) for gas-dynamical simulations. *J. Comp. Phys.*, 54:174–201, 1984.
20. Ppmlib library. University of Minnesota. `http://www.lcse.umn.edu/PPMlib/`.
21. John Blondin et.al. VH-1 The Virginia Numerical Bull Session ideal hydrodynamics PPMLR. `http://wonka.physics.ncsu.edu/pub/VH-1/index.html`.
22. P. Woodward. personal communication. 2001.
23. R. Saurel and R. Abgrall. A simple method for compressible multifluid flows. *SIAM J. Sci. Comput.*, 21:1115–1145, 1999.
24. W. F. Noh. Errors for calculations of strong shocks using an artificial viscosity and artificial heat flux. *J. Comp. Phys.*, 72:78–120, 1987.
25. C. W. Schulz-Rinne, J. P. Collins, and H. M. Glaz. Numerical solution of the Riemann problem for two-dimensional gas dynamics. *SIAM J. Sci. Comput.*, 14:1394–1414, 1993.
26. A. Kurganov and E. Tadmor. Solution of two-dimensional riemann problems for gas dynamics without riemann problem solvers. CAM 00-34, UCLA, Los Angeles, 2000.
27. William J. Rider. Revisiting wall heating. *J. Comp. Phys.*, 162:395 – 410, 2000.
28. G. Strang. On the construction and comparison of difference schemes. *SIAM J. Numer. Anal.*, 5(3):506–517, 1968.
29. R. Liska and B. Wendroff. Comparision of several difference schemes on 1d and 2d test problems for the euler equations. Technical Report LA-UR-01-6225, Los Alamos National Laboratory, Los Alamos, 2001. Comparison project web page with animations is at `http://www-troja.fjfi.cvut.cz/~liska/CompareEuler/compare`.

Adaptive Time Integration for Hyperbolic Conservation Equations

William J. Rider, Len G. Margolin, and James R. Kamm

Los Alamos National Laboratory, Los Alamos, NM 87545, USA
wjr@lanl.gov,len@lanl.gov,kammj@lanl.gov

Summary. We introduce a fundamentally new time integration method for hyperbolic conservation laws based on self-adaptivity of the temporal method itself. The adaptivity is based upon the smoothness of the solution measured locally in time. Our approach can be contrasted with the usual global selection of a time integration methods and error-based time step selection methodology. A challenge to this approach is maintaining the adaptivity and the conservation form. These methods are challenged with several standard problems as well as high-resolution experimental data of shock-driven mixing (Richtmyer-Meshkov).

1 Introduction

There have been two major approaches to improving time accuracy in numerical methods for hyperbolic conservation laws: the Lax-Wendroff [3] approach, and the method-of-lines [8]. Both of these are "linear" schemes in the sense that the same time differencing method is applied everywhere in the computational domain, usually for each time-step[1]. We distinguish these from "nonlinear" schemes, implying that the differencing depends on the local flow variables. In this sense, any nonlinear character in standard approaches to time differencing arises from the spatial differencing.

Nonlinear differencing in space has become a standard methodology during the last twenty years, and is generally considered to be essential in effectively solving a broad range of applications. In this research, we explore the utility of nonlinear differencing in time, distinct from that used in space. We describe a methodology that allows the time and space differencing to adapt independently, and locally rather than globally. Finally, we examine the performance of this approach and show a situation where this method appears to yield greatly improved results.

We note that robust software for ordinary differential equations often adjusts the order of the integration scheme based on the local characteristics of the solution [1]. However, in that case the selected order of integration is typically applied to the entire system of ODEs. Extending this approach to a method-of-lines for hyperbolic conservation laws leads to a differencing that is

[1] Runge-Kutta methods are nonlinear in terms of Δt, that is their effective errors are nonlinear in terms of the step size, but linear in the sense that the method is not adapted to the nature of the local solution.

global in space and time with temporal adaptivity, i.e., the same differencing is applied globally in space during a time-step, but varies from one time-step to the next. For example, the order is typically lowered when the estimates of the ordered expansion coefficients do not form a convergent sequence. On the other hand, if the coefficients indicate that the solution is smooth, the order might be increased. Such an approach would forfeit the advantages of modern shock capturing methods, where the order of integration is also spatially dependent upon the local (spatial) character of the solution. In particular, a shock wave is a local phenomenon whose presence should not necessitate the degradation of accuracy in smoother areas of the flow.

Lax-Wendroff techniques implicitly invoke the assumption of space-time similarity. It is assumed that the spatial truncation error will dominate the higher order temporal errors, guaranteeing a dissipative approximation and hence computational stability. While this situation may hold in many circumstances where the phenomena are (approximately) self-similar, it is not general. In particular, near a critical point, where the characteristic speed is nearly zero, the dissipation may vanish, thereby allowing higher-order terms to dominate. In our time adaptive methods, we define new "limiters" that depend on the time derivatives in a manner similar to the manner in which spatial limiters depend on spatial derivatives. In particular, we employ a nonlinear combination of time-differencing methods in the same manner that "upwind" and "downwind" differences are used to achieve a desired property such as monotonicity or positivity preservation[2].

Our recent experience has shown that some difficulties may arise from other issues besides the breakdown of space-time similarity. Numerical algorithms can be viewed as dynamical systems [9]. One must be careful, therefore when implementing a time-differencing scheme, not to introduce spurious fixed points, which would be problematic for a wide variety of applications, including both steady and dynamic problems. Error control, upon which we base our adaptive time algorithms, has been shown to be an effective manner of combating spurious fixed points in solutions.

In the following section, we will describe a basic design of adaptive time integration methods starting with the blending of a linear multistep method with a forward-in-time integrator. Next, we show some results obtained with this method in Section 3. Finally, we will extend this approach introducing a temporal integrator that blends Runge-Kutta methods with weighted ENO ideas. This is described in Section 4.

[2] One must take care when reducing the order of the temporal differencing; when coupled with high-order spatial differencing, the potential for oscillatory behavior arises. Where this is an issue, we drop the spatial accuracy as necessary to maintain stability.

2 Initial Design

In designing our new methodology, we require that each cell edge allow an independent temporal integrator so that the resulting methods remain in conservation form. Thus, in one dimension we choose

$$u_j^{n+1} = u_j^n - \lambda \left[f\left(u\right)_{j+\frac{1}{2}} - f\left(u\right)_{j-\frac{1}{2}} \right], \tag{1}$$

where $f\left(u\right)_{j+\frac{1}{2}}$ has an order of time accuracy that is independent of any other edges. Here, u is the discretized dependent variable and f is the associated flux. Subscripts indicate the cell index and superscripts indicate discretized time level. In general, our scheme will depend on the temporal behavior in the zones that determine the flux; e.g., in the simplest case, the limiter will be constructed from u_j and u_{j+1} and its temporal derivatives. Although the spatial limiters will be evaluated at time n, as in the Lax-Wendroff or Runge-Kutta type approaches, we will now need to save previous fluxes or time-derivatives in an Adams-Bashforth fashion.

We have constructed our method in a simple form where temporal differences are tested, through a limiter, in the same manner as spatial differences are used for a Lax-Wendroff-style scheme. The temporal limiter is defined as

$$\widetilde{u_t} = \text{minmod}\left[\left(u_t\right)^n, 2\left(u_t\right)^{n-1} \right], \text{ or } = \text{minmod}\left[2\left(u_t\right)^n, \left(u_t\right)^{n-1} \right], \tag{2}$$

where in the most straightforward implementation, $\left(u_t\right)^{n-1} = \left(u^n - u^{n-1}\right)/\Delta t$ and $\left(u_t\right)^n = -\left[\left(f\left(u\right)\right)_x\right]^n$, as in the Lax-Wendroff approach to time differencing. This is a straightforward form, resulting from using the backwards and forward time derivatives in place of spatial differences in a standard minmod limiter. We note that the extra time differences could be used to raise the temporal accuracy to third order

$$\widetilde{u_t} = \text{minmod}\left[\frac{5}{3}\left(u_t\right)^n - \frac{2}{3}\left(u_t\right)^{n-1}, 2\left(u_t\right)^n, 2\left(u_t\right)^{n-1} \right]. \tag{3}$$

There are two distinct outputs resulting from this approach: the actual update of the solution from the previous time step, and the identification of the presence of a critical point (i.e., a sign change in the time derivative). We will examine two cases, using the model equation,

$$\frac{\partial u}{\partial t} + u\frac{\partial u}{\partial x} = g\left(x,t\right), \tag{4}$$

i.e., inviscid Burgers' equation with a source term. The first case is a quasi-steady condition, where $\frac{\partial u}{\partial t} \approx 0$ so that $u\frac{\partial u}{\partial x} \approx g\left(x,t\right)$. This balance must be maintained for the computations to be well-behaved [5] and achieve the proper steady state. The second, perhaps more interesting case is where $u\frac{\partial u}{\partial x} \approx 0$, implying the balance $\frac{\partial u}{\partial t} \approx g\left(x,t\right)$. This indicates the presence of a stagnation point. Here, the balance between the source term and time derivative must be preserved by the numerical method. The proper evolution of the flow through dynamic critical points is crucial to the overall accuracy of the solution.

3 Tests of The Initial Method

Our initial example is a scalar wave equation where one expects self-similarity to hold. We require from the outset that our methods should perform well in this important cases. The results are shown in Figure 1. In this case the flow is completely self-similar and the new method does not adversely effect the results. In other words, the time-limiter does not turn on except for where the spatial limiting has been activated. We also will examine this method's performance on Burgers' equation.

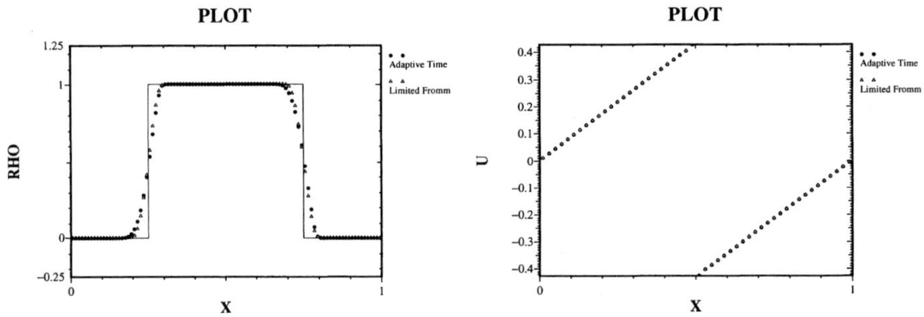

Fig. 1. These two figures basically demonstrate that when self-similarity should hold that the time-limiter does not undermine the method's performance.

For the Euler equations we will start in one dimension. Considering Sod's shock tube, we again see that our new method does not adversely influence the results. This is displayed for the density and velocity fields in Figure 2. Positive aspects of the new method are slightly reduced post shock oscillations and better resolution of the head of the rarefaction. In general, our new approach does not lower the resolution of solutions where self-similarity holds.

Although we expect that multidimensional examples of Euler or Navier-Stokes flows may exhibit more complex behavior than the examples above, we believe it is a necessary prerequisite for our schemes to perform well in these more idealized situations. We conclude by presenting a case where our approach provides a significant improvement for a nontrivial problem. The experiment [7] is a complex example of the evolution of a multidimensional fluid instability. Our attempts to analyze this experiment has provided both

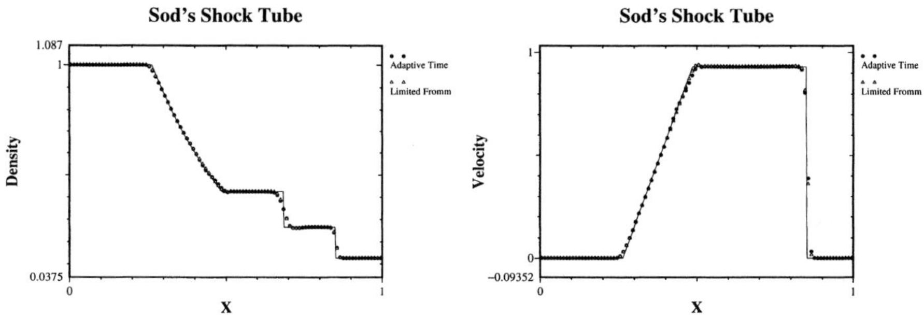

Fig. 2. Results for Sod's shock tube are shown for a limited Fromm's scheme and a spatially limited Fromm's scheme with a time-limiter. The exact solution is given by the solid line.

the initial impetus for our research as well as a framework to judge its utility. We have found that standard high resolution methods such as MUSCL, WENO, DG, TVD, etc., all produce solutions that are not consistent with the *statistical* behavior of the experimental data (as quantified by the fractal dimension, the wavelet spectra and the correlation-based measures of the density field). Of greater concern (and interest), we have found that a first-order Godunov method produces results more consistent with the observed statistical behavior. Further analysis of the simulations has localized the discrepancy to the behavior of the acoustic waves. We have found–and will document–that our adaptive time approach, as discussed above, leads to greatly improved results, in terms of their qualitative and quantitative agreement with the experimental data.

As shown in Figures 3 and 4, the method produces results that are dramatically different than the standard method. This is confirmed by the statistical analysis of these results. The adaptive time method arguably produces the best match to the experimental data both quantitatively and qualitatively as determined by statistical analysis.

4 Extending the Initial Method: Weighted Adaptive Runge-Kutta (WARK)

Buoyed by our initial success, we can extend our method using ideas borrowed from weighted ENO (WENO) methods [4, 2]. Rather than choose weights for a set of spatial stencils, we will choose weights for a series of temporal stencils. The design of the method will follow the general details of a WENO method. The key to this method is the definition of the weights used in the scheme. If the time behavior of the flow is smooth the method will reproduce the high-order result. We name our methods weighted adaptive Runge-Kutta methods (WARK).

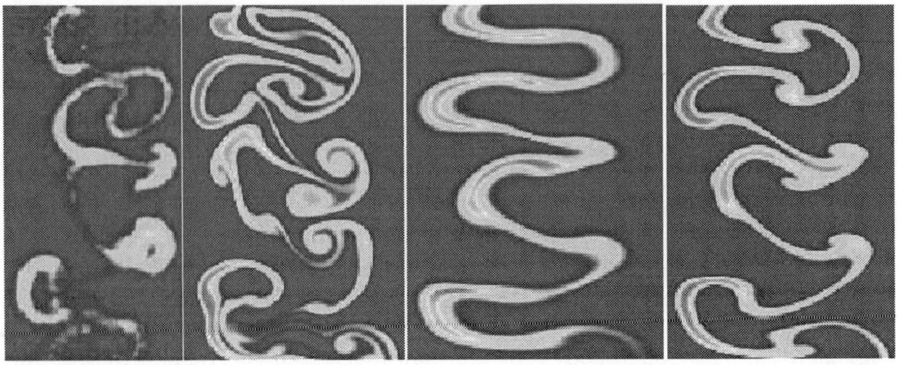

Fig. 3. A montage of images from the gas curtain configuration showing the experimental data at 400 μs, and three simulations initialized from the experimental measurements. From left to right are the experimental images, a standard high-resolution Godunov method, a first-order Godunov method and a high-resolution Godunov method with time limiting. The visual differences in the simulations are striking.

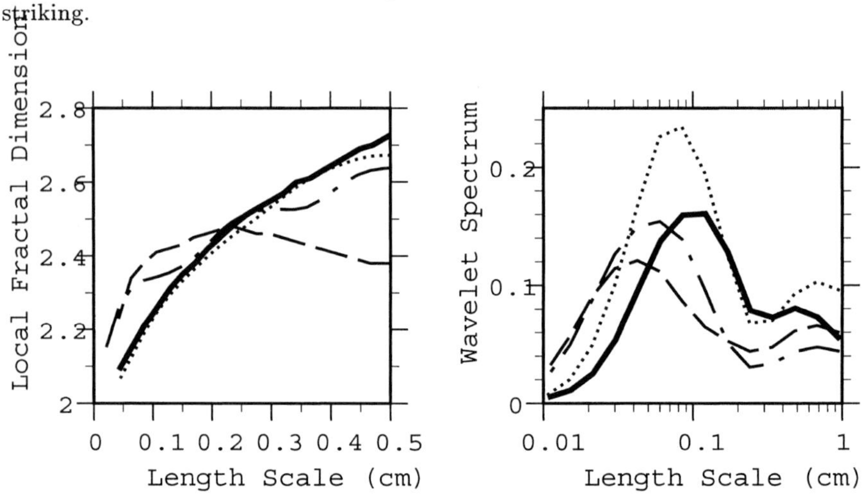

Fig. 4. The statistical analysis of the images shown in Figure 3. The analysis uses fractals and wavelets to measure the properties of the images. The time-limited method is arguably the best match with the experimental statistics. The solid line shows the experimental data, the long dashed line is the standard high-resolution method, the dotted line is the first-order Godunov, and the dot-dashed line is the adaptive time integration.

The steps of this method are identical to other Runge-Kutta methods until the final step of the method where the values are updated to the new time step. For example consider a second-order method (Heun's method) for advancing $u_t = \Psi(u)$:

1. Starting with initial data, $u^0 = u^n$,
2. advance using forward Euler, $u^1 = u^0 + \Delta t \Psi(u^0)$,
3. also advance the solution for a half time-step (for improved Euler), $u^2 = u^0 + \frac{1}{2}\Delta t \Psi(u^0)$.
4. before advancing to the new time value of u determine the nonlinear weights, $u^{n+1} = u^n + \Delta t \left\{\frac{1}{2}\omega^0 \left[\Psi(u^0) + \Psi(u^1)\right] + \omega^1 \Psi(u^2)\right\}$.

Note that we have chosen a weighted average of two separate means of advancing the solution at second-order accuracy. The key to this method is defining the weights, ω^k, where $\omega^0 + \omega^1 = 1$. As we will see the general form for the method is

$$u^{n+1} = u^n + \sum_k \omega^k S\left[\Psi(u^k)\right], \tag{5}$$

where $S(\Psi)$ is a stencil that advances the data with a certain accuracy and $\sum_k \omega_k = 1$ is a constraint. The key is the definition of the nonlinear weights and the self-adaptivity coming from the smoothness indicators, IS^k. In this case both weights have the same linear value of $\bar{\omega} = \left(\frac{1}{2}, \frac{1}{2}\right)$, and we use the time derivative associated with each as IS, thus

$$\hat{\omega}^k = \frac{1}{\left(IS^k + \epsilon\right)^2},$$

where $IS^k = \Psi\left(u^k\right)^2$ and $\epsilon = \times 10^{-6}$. Finally, the weights are normalized using $\omega^k = \hat{\omega}^k / \sum_m \hat{\omega}^m$.

The only remaining issue to rectify is how to define these weights on the fluxes rather than the function evaluations. This is required because we must maintain a conservative discretization in space. We will explore the option where one is to average the two time derivatives in the zones adjacent to a flux,

$$IS^k_{j+\frac{1}{2}} = S\left[\Psi\left(u^k_j\right)\right]^2 + S\left[\Psi\left(u^k_{j+1}\right)\right]^2.$$

A variety of other methods can easily be converted into this general format and made adaptive. For example the popular third-order TVD Runge-Kutta integrator and the classical fourth-order Runge-Kutta methods have been experimented with. For the third-order TVD the only major change necessary is the final step which is modified to read,

$$u^{n+1} = u^n + \Delta t \sum_0^1 \omega^k \Psi\left(u^k\right),$$

with linear weights, $\bar{\omega} = \left(\frac{1}{3}, \frac{2}{3}\right)$. We start with the usual sequence of operations,

$$u^1 = u^0 + \Delta t\Psi\left(u^0\right); u^2 = \frac{3}{4}u^0 + \frac{1}{4}u^1 + \Delta t\Psi\left(u^1\right).$$

The linear stencils are two second-order methods

$$S\left[\Psi\left(u\right)\right]^1 = \frac{1}{2}\left[\Psi\left(u^0\right) + \Psi\left(u^1\right)\right] ,$$

and

$$S\left[\Psi\left(u\right)\right]^2 = \Psi\left(u^2\right) .$$

Using modified equation analysis, we can see the differences in the properties of the two second-order methods [6]. Applying this approach to examine the time-differencing errors for $\Psi\left(u\right) = -f\left(u\right)_x$ we find the following, for $S\left[\Psi\left(u\right)\right]^1$ the error is

$$\left(\Delta t f'\right)^2 \left(\frac{1}{4}f''\left(u_x\right)^2 + \frac{1}{6}f' u_{xx}\right)_x ,$$

where $f' = \partial f\left(u\right)/\partial u$, and $f'' = \partial^2 f\left(u\right)/\partial u^2$ and for $S\left[\Psi\left(u\right)\right]^2$,

$$\left(\Delta t f'\right)^2 \left(\frac{1}{8}f''\left(u_x\right)^2 + \frac{1}{12}f' u_{xx}\right)_x .$$

Both the dissipative and dispersive errors are opposite in sign (required to cancel to achieve third-order), the schemes can successfully adapt to the flow. In particular, the terms proportional to $f''\left(u_x\right)^2$ are (anti)dissipative depending upon the sign of u_x. With this behavior the scheme can adapt locally so that appropriate dissipation can be applied via the time integration scheme.

A classical fourth-order Runge-Kutta can be made adaptive as well. Runge-Kutta methods have been experimented with. For the third-order TVD the only major change necessary is the final step which is modified to read,

$$u^{n+1} = u^n + \Delta t \sum_0^2 \omega^k \Psi\left(u^k\right) ,$$

with linear weights, $\bar{\omega} = \left(\frac{1}{3}, \frac{1}{3}, \frac{1}{3}\right)$. We start with the usual sequence of operations,

$$u^1 = u^0 + \frac{1}{2}\Delta t\Psi\left(u^0\right); u^2 = u^0 + \frac{1}{2}\Delta t\Psi\left(u^1\right); u^3 = u^0 + \Delta t\Psi\left(u^2\right).$$

The linear stencils are two second-order methods

$$S\left[\Psi\left(u\right)\right]^0 = \frac{1}{2}\left[\Psi\left(u^0\right) + \Psi\left(u^3\right)\right] ,$$

$$S\left[\Psi\left(u\right)\right]^{1} = \Psi\left(u^{1}\right) ,$$

and

$$S\left[\Psi\left(u\right)\right]^{2} = \Psi\left(u^{2}\right) .$$

Now we will test these methods on a simple, self-similar case. As shown in Figure 5 the method produce small changes over the standard methods in this case.

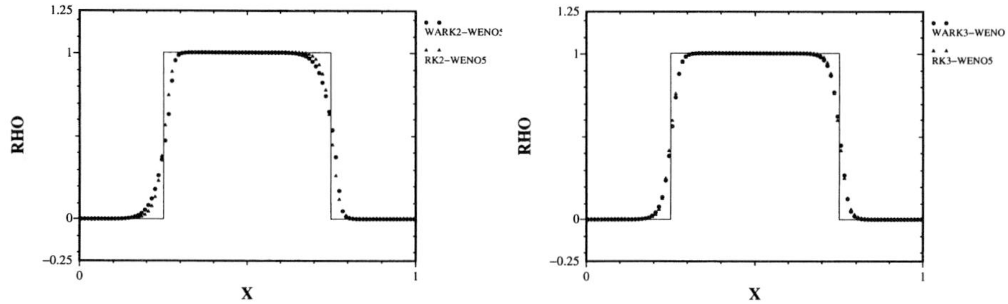

Fig. 5. The basic performance of the WARK method for a scalar wave equation and 5th order WENO differencing in space as compared with the "linear" time differencing that forms the basis of the scheme used.

We find that for a self similar flow, the solutions are not perturbed by the adaptivity in time.

5 Summary

In summary, we have only begun to explore the possible formulations of time adaptivity. We believe that our preliminary results illustrate the power and flexibility of the approach, as well as its effectiveness. In particular, the ability to deal with problems that do not exhibit hyperbolic self-similarity should provide a new approach to attacking this difficult and important class of problems.

Acknowledgments

Also available as Los Alamos National Laboratory Report, LA–UR–02–4089. This work performed under the auspices of the U.S. Department of Energy by Los Alamos National Laboratory under Contract W-7405-ENG-36. Direct support for this research was provided by the Laboratory Directed Research and Development (LDRD) program.

References

1. U. M. Ascher and L. R. Petzold, *Computer Methods for Ordinary Differential Equations and Differential-Algebraic Equations*, SIAM, Philadelphia, PA, 1996.
2. G.-S. Jiang and C.-W. Shu, Efficient of Weighted ENO Methods, *J. Comp. Phys.*, **126**, 202-228, 1996.
3. P. D. Lax and B. Wendroff, Systems of Conservation Laws, *Communications in Pure and Applied Mathematics*, **13**:217–237, 1960.
4. X.-D. Liu, S. Osher and T. Chan, Weighted Essentially Non-Oscillatory Schemes, *J.Comp. Phys.*, **115**, 200–212, 1994.
5. L. G. Margolin and D. A. Jones, An Approximate Inertial Manifold for Computing Burgers' Equation, *Physica D*, **60**:175–184, 1992
6. W. J. Rider, L. G. Margolin and J. R. Kamm, Techniques for Analysis of Nonlinear Approximations to Nonlinear Equations, submitted for review, Los Alamos National Laboratory Reportm LA–UR–02–3215, 2002.
7. P. M. Rightley, P. Vorobieff, R. Martin & R. F. Benjamin, Experimental-observations of the mixing transition in a shock-accelerated gas curtain, *Phys. Fluids*, **11**:186–200, 1999.
8. C.-W. Shu,Total-Variation-Diminishing Time Discretizations, *SIAM J. Sci. Comp.*, **9**:1073–1084, 1988.
9. H. C. Yee, N. D. Sandham and M. J. Djomehri, On Spurious Behavior in CFD Simulations, *Int. J. Num. Meth. Fluids*, **30**: 675-711, 1999.

A Constrained Transport Method for the Shallow Water MHD Equations

James A. Rossmanith

Department of Applied Mathematics, University of Washington, Box 352420, Seattle, WA 98195, USA. *jrossman@amath.washington.edu*

Summary. We present in this paper a finite volume method for solving the shallow water magnetohydrodynamic equations. This method uses a constrained transport approach in order to guarantee that the magnetic field satisfies a discrete divergence-free constraint. We briefly review three versions of the constrained transport method that appear in the literature and discuss some of their shortcomings. In order to overcome these difficulties, a new constrained transport method based on advecting the magnetic potential is introduced. Using a common base scheme, these four methods are then applied to a two-dimensional test problem.

1 Introduction

The ideal magnetohydrodynamic (MHD) equations are important in modeling phenomena in a wide range of applications including space weather, solar physics, laboratory plasmas, and general relativistic fluid flows. Apart from the specific challenges that each one of these applications provides, there are two main difficulties that arise when integrating the MHD equations:

1. The magnetic field should be kept divergence-free from time step to time step in accordance with Maxwell's equations.
2. The numerical update should produce a non-oscillatory solution near discontinuities.

It is well known in the literature that failure to accomplish either one of these can produce unphysical solutions (see [6]). Developing numerical methods that address these issues is made even more difficult by the fact that the MHD system has a complicated eigenstructure. For this reason, it is desirable to have a model system that retains the divergence-free and non-oscillatory challenges from the MHD equations, but has a simpler eigenstructure.

Gilman [3] introduced a simplified version of the MHD equations called the *shallow water magnetohydrodynamic* (SMHD) equations. Not only do these equations retain the need for a numerical method that satisfies the two conditions above, but the SMHD equations are important as a mathematical model in their own right. They have been used by Gilman [3] as a model of the solar tachocline, a thin layer of thickness of about 2%-5% of the solar radius that separates the convective zone from the radiative zone. It is believed that magnetic fields originating in this thin layer are responsible for solar flares and sunspots; and furthermore, it is believed that mathematical models such as the SMHD equations can be used to gain insight in the dynamics of the tachocline.

The SMHD equations model the dynamics of a constant density, shallow, and electrically conducting fluid that is hydrostatically balanced in the vertical direction. This system can be written as

$$\partial_t \begin{bmatrix} h \\ hu_1 \\ hu_2 \\ hB_1 \\ hB_2 \end{bmatrix} + \partial_x \begin{bmatrix} hu_1 \\ hu_1^2 - hB_1^2 + \frac{1}{2}gh^2 \\ hu_1u_2 - hB_1B_2 \\ 0 \\ hu_1B_2 - hu_2B_1 \end{bmatrix} + \partial_y \begin{bmatrix} hu_2 \\ hu_1u_2 - hB_1B_2 \\ hu_2^2 - hB_2^2 + \frac{1}{2}gh^2 \\ hu_2B_1 - hu_1B_2 \\ 0 \end{bmatrix} = 0 \qquad (1)$$

$$\nabla \cdot (h\mathbf{B}) = \partial_x (hB_1) + \partial_y (hB_2) = 0 \,, \qquad (2)$$

where g is the gravitational constant. The unknowns in the SMHD system are the height h, the momentum $h\mathbf{u} = (hu_1, hu_2)$, and the magnetic field $h\mathbf{B} = (hB_1, hB_2)$.

We present in this paper a constrained transport method for solving the SMHD equations. The general constrained transport approach on Cartesian grids as well as three versions of this approach that appear in the literature are described in the next section. In Section 4 we develop a new constrained transport method. All these methods are applied to a specific test problem in Section 5. The method described in this paper was developed in the author's thesis [6], where an extension to the sphere is also presented. More details will appear in other papers that are currently in preparation.

2 The constrained transport approach

In the case of continuous variables, the divergence-free constraint (2) is automatically satisfied for all time if the initial magnetic field is also divergence-free. If the SMHD equations are discretized using a standard cell-centered finite volume discretization, however, the absolute value of the divergence can become large, especially near shocks and other discontinuities (see [6]). The idea behind the constrained transport approach is to stagger the magnetic field in such a way that a discrete divergence of the magnetic field is identically zero. Unlike many of the methods that have been developed for solving the MHD equations, this method does not require the introduction of modified or additional equations. Several variants of the constrained transport method have been considered in the literature. In the current paper, we will only consider versions of the method that are directly compatible with the LeVeque wave propagation method on Cartesian grids [4]. For a review of currently available constrained transport methods as well as a more detailed list of references see [6].

2.1 Solving the induction equation

In (1), the equation governing the time evolution of the magnetic field is the *induction equation*:

$$\partial_t \begin{bmatrix} hB_1 \\ hB_2 \end{bmatrix} + \partial_x \begin{bmatrix} 0 \\ hu_1B_2 - hu_2B_1 \end{bmatrix} + \partial_y \begin{bmatrix} hu_2B_1 - hu_1B_2 \\ 0 \end{bmatrix} . \qquad (3)$$

Because the magnetic field is divergence-free, we can define a *magnetic potential* $\mathbb{A}$ such that

$$h\mathbf{B} = \nabla \times \mathbb{A} = \begin{bmatrix} \partial_y \mathbb{A} \\ -\partial_x \mathbb{A} \end{bmatrix} . \tag{4}$$

Using the magnetic potential, the induction equation can be written as

$$\partial_t \mathbb{A} + \Omega = 0 \tag{5}$$
$$hB_1 = \partial_y \mathbb{A} \tag{6}$$
$$hB_2 = -\partial_x \mathbb{A} , \tag{7}$$

where

$$\Omega = -u_1 hB_2 + u_2 hB_1 \tag{8}$$

is the vertical component of the electrical field given by Ohm's law (see [6]).

Let us momentarily assume that the electric field Ω is known and our goal is to numerically solve for the magnetic field in such a way that some discrete version of (2) is exactly satisfied at each time step in the update. The constrained transport approach to this problem is to stagger the numerical grid so that the 1-component of the magnetic field lives on the left and right grid interfaces (hb_1), the 2-component on the top and bottom grid interfaces (hb_2), and the magnetic potential on the grid corners ($\mathbb{A}$). This staggering is depicted in Figure 1. The so-called *constrained transport* (CT) update for the magnetic field then takes the form

$$\mathbb{A}^{n+1}_{i-1/2,j-1/2} = \mathbb{A}^{n}_{i-1/2,j-1/2} - \Delta t \, \Omega^{n+1/2}_{i-1/2,j-1/2} \tag{9}$$
$$[hb_1]^{n+1}_{i-1/2,j} = \frac{1}{\Delta y}\left(\mathbb{A}^{n+1}_{i-1/2,j+1/2} - \mathbb{A}^{n+1}_{i-1/2,j-1/2}\right) \tag{10}$$
$$[hb_2]^{n+1}_{i,j-1/2} = \frac{1}{\Delta x}\left(\mathbb{A}^{n+1}_{i-1/2,j-1/2} - \mathbb{A}^{n+1}_{i+1/2,j-1/2}\right) . \tag{11}$$

It is easy to verify that this update will numerically conserve the following discrete divergence at each time step:

$$[\nabla \cdot (h\mathbf{b})]^{n+1}_{ij} \equiv \frac{[hb_1]^{n+1}_{i+1/2,j} - [hb_1]^{n+1}_{i-1/2,j}}{\Delta x} + \frac{[hb_2]^{n+1}_{i,j+1/2} - [hb_2]^{n+1}_{i,j-1/2}}{\Delta y} . \tag{12}$$

Hence if $[\nabla \cdot (h\mathbf{b})]^0 \equiv 0$ initially then $[\nabla \cdot (h\mathbf{b})]^n \equiv 0$ for all time.

2.2 Solving the SMHD equations

The following question now presents itself: How can the above update for the induction equation be coupled to the update for the full SMHD equations? For the grid cell centered at (x_i, y_j), we define the following quantities:

$$\begin{aligned} \text{at Cell Centers:} &\quad h_{ij},\ [u_1]_{ij},\ [u_2]_{ij},\ [hB_1]_{ij},\ [hB_2]_{ij} \\ \text{at Cell Corners:} &\quad \mathbb{A}_{i-1/2,j-1/2} \\ \text{at Left/Right Edges:} &\quad [hb_1]_{i-1/2,j} \\ \text{at Top/Bottom Edges:} &\quad [hb_2]_{i,j-1/2} . \end{aligned}$$

Note that we have both cell centered ($h\mathbf{B}$) and a staggered ($h\mathbf{b}$) magnetic field values. With these definitions, the following algorithm can be used to obtain a update for the SMHD equations that maintains a divergence-free magnetic field:

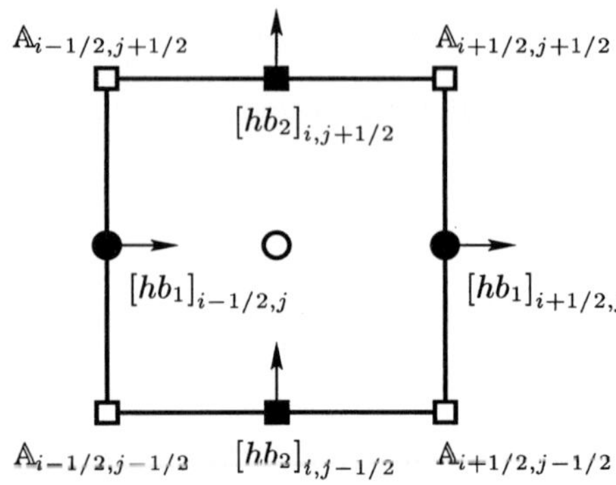

Fig. 1. The constrained transport grid staggering.

Step 0. Primary quantities: $\boxed{h^n,\ u_1^n,\ u_2^n,\ hb_1^n,\ hb_2^n,\ \mathbb{A}^n}$

Step 1. Average $h\mathbf{b}^n$ to get cell centered values: $\boxed{h^n,\ u_1^n,\ u_2^n,\ hB_1^n,\ hB_2^n}$

$$\Longrightarrow [hB_1]_{ij}^n = \tfrac{1}{2}\Big([hb_1]_{i-1/2,j}^n + [hb_1]_{i+1/2,j}^n\Big)$$

$$\Longrightarrow [hB_2]_{ij}^n = \tfrac{1}{2}\Big([hb_2]_{i,j-1/2}^n + [hb_2]_{i,j+1/2}^n\Big)$$

Step 2. Solve the SMHD equations: $\boxed{h^{n+1},\ u_1^{n+1},\ u_2^{n+1},\ hB_1^\star,\ hB_2^\star}$

Step 3. Average $\mathbf{u}^{n+1}$ to get staggered values: $\boxed{U_1^\star,\ U_2^\star,\ \mathbb{A}^n}$

$$\Longrightarrow [U_1]_{i,j-1/2}^\star = \tfrac{1}{2}\Big([u_1]_{i,j-1}^{n+1} + [u_1]_{ij}^{n+1}\Big)$$

$$\Longrightarrow [U_2]_{i-1/2,j}^\star = \tfrac{1}{2}\Big([u_2]_{i-1,j}^{n+1} + [u_2]_{ij}^{n+1}\Big)$$

Step 4. Solve induction equation (9): $\boxed{\mathbb{A}^{n+1}}$

Step 5. Compute $\nabla \times \mathbb{A}$ via (10)-(11): $\boxed{hb_1^{n+1},\ hb_2^{n+1}}$

In words, the algorithm says that we begin with *primary* quantities h^n, $\mathbf{u}^n$, $h\mathbf{b}^n$, and $\mathbb{A}^n$. The magnetic field is averaged to obtain a cell centered value. Using standard high-resolution finite volume methods the SMHD equations are solved. The updated velocities at cell centers are then averaged to obtain staggered velocities. The induction equation is solved at the cell corners in order to update the magnetic potential. Finally, these magnetic potential values are differenced and we obtain updated magnetic field values.

In the above description of constrained transport, each step is clearly defined with the exception of **Step 4**. In **Step 4** there are many options for the integration of the induction equation. We consider in detail the approaches of [1, 2, 7] in the next section.

3 Methods for integrating the induction equation

The most difficult question in the constrained transport method is the question of how to integrate the induction equation (9). In particular, the question becomes how to construct the electric field Ω from known information at time $t = t_n$ and $t = t_\star$. In this section we consider the approaches of Dai and Woodward [2], Balsara and Spicer [1], and Ryu, Miniati, Jones, and Frank [7].

3.1 The DW scheme

The first CT method we consider is taken from Dai and Woodward [2]. Their updating procedure for the magnetic potential first involves spatially averaging cell centered electric field values to obtain cell corner values. The spatial averaging is done for both the $t = t_n$ values and the $t = t_{n+1}$ values obtained after taking a full step with a finite volume method. These two values are then averaged in time to obtain an approximation to the electric field at $t = t_{n+1/2}$. Mathematically this can be written as

$$\begin{aligned} \Omega_c^{n+1/2} \equiv \frac{1}{2}\Bigg[& \frac{1}{4}\left(\Omega_{ij}^n + \Omega_{i-1,j}^n + \Omega_{i,j-1}^n + \Omega_{i-1,j-1}^n\right) \\ & + \frac{1}{4}\left(\Omega_{ij}^\star + \Omega_{i-1,j}^\star + \Omega_{i,j-1}^\star + \Omega_{i-1,j-1}^\star\right)\Bigg], \end{aligned} \tag{13}$$

where $\Omega_c \equiv \Omega_{i-1/2,j-1/2}$. The value for $\Omega^n = -u_1^n hB_2^n + u_2^n hB_1^n$ is obtained from the previous time step $t = t_n$ and $\Omega^\star = -u_1^{n+1} hB_2^\star + u_2^{n+1} hB_1^\star$ is obtained from the results of the cell centered finite volume method. The value $\Omega^{n+1/2}$ is then used in the induction equation update (9)-(11) to obtain $\mathbb{A}^{n+1}$, hb_1^{n+1}, and hb_2^{n+1}.

The accuracy of this scheme suffers from the fact that the update for the magnetic potential is not TVD. This update requires averaging in both space and time; and therefore, can introduce strong spurious oscillations in the vicinity of discontinuities (see [6]).

3.2 The BS scheme

The second CT method we consider is taken from Balsara and Spicer [1]. Their update for the magnetic potential is based on the realization that Ω is proportional to both the 5th component of the SMHD flux function in the x-direction and the 4th component of the SMHD flux function in the y-direction:

$$\Omega = -f_5 = g_4 = -u_1 hB_2 + u_2 hB_1 \,. \tag{14}$$

Furthermore, high-resolution methods produce at each time step numerical approximations to the flux function of the form:

$$[F_5]_{i-1/2,j}^{n+1/2} = -\frac{1}{2}\left(\Omega_{ij}^n + \Omega_{i-1,j}^n\right) + \Phi_{i-1/2,j}^x \tag{15}$$

$$[G_4]_{i,j-1/2}^{n+1/2} = +\frac{1}{2}\left(\Omega_{ij}^n + \Omega_{i,j-1}^n\right) + \Phi_{i,j-1/2}^y \,, \tag{16}$$

where Φ^x and Φ^y include *upwind* and other *high-resolution* corrections in the x- and y- directions, respectively. To obtain a value of Ω at cell corners, the following spatial interpolation is applied:

$$
\begin{aligned}
\Omega_c^{n+1/2} &\equiv \frac{1}{4}\Big([G_4]_{i,j-1/2} + [G_4]_{i-1,j-1/2} - [F_5]_{i-1/2,j} - [F_5]_{i-1/2,j-1}\Big) \\
&= \frac{1}{4}\left(\Omega_{ij}^n + \Omega_{i-1,j}^n + \Omega_{i,j-1}^n + \Omega_{i-1,j-1}^n\right) \\
&+ \frac{1}{2}\left[\frac{1}{2}\left(\Phi_{i,j-1/2}^y + \Phi_{i-1,j-1/2}^y\right) - \frac{1}{2}\left(\Phi_{i-1/2,j}^x + \Phi_{i-1/2,j-1}^x\right)\right] \\
&\equiv \bar{\Omega}_c^n + \frac{1}{2}\left(\bar{\Phi}_{i-1/2,j-1/2}^y - \bar{\Phi}_{i-1/2,j-1/2}^x\right).
\end{aligned}
\tag{17}
$$

The value $\Omega^{n+1/2}$ is then used in the induction equation update (9)-(11) to obtain $\mathbb{A}^{n+1}$, hb_1^{n+1}, and hb_2^{n+1}.

3.3 The RMJF scheme

The third CT method we consider is taken from Ryu, Miniati, Jones, and Frank [7]. This method approximates the value for Ω via the following formula:

$$
\Omega_c^{n+1/2} \equiv \frac{1}{2}\Big(\big[\check{G}_4\big]_{i,j-1/2} + \big[\check{G}_4\big]_{i-1,j-1/2} - \big[\check{F}_5\big]_{i-1/2,j} - \big[\check{F}_5\big]_{i-1/2,j-1}\Big), \tag{18}
$$

where

$$
\big[\check{F}_5\big]_{i-1/2,j} \equiv \frac{1}{2}\Big([u_1 h B_2]_{ij}^n + [u_1 h B_2]_{i-1,j}^n\Big) + \Phi_{i-1/2,j}^x \tag{19}
$$

$$
\big[\check{G}_4\big]_{i,j-1/2} \equiv \frac{1}{2}\Big([u_2 h B_1]_{ij}^n + [u_2 h B_1]_{i,j-1}^n\Big) + \Phi_{i,j-1/2}^y. \tag{20}
$$

The motivation for this update is that when either $\partial_x q \equiv 0$ or $\partial_y q \equiv 0$, the RMJF scheme exactly reduces to a 1D high-resolution method (see [6]). Plugging definitions (19) and (20) back into (18) and making use of Ohm's law (8) yields an equivalent formulation:

$$
\begin{aligned}
\Omega_c^{n+1/2} &\equiv \frac{1}{4}\left(\Omega_{ij}^n + \Omega_{i-1,j}^n + \Omega_{i,j-1}^n + \Omega_{i-1,j-1}^n\right) \\
&+ \frac{1}{2}\left(\Phi_{i,j-1/2}^y + \Phi_{i-1,j-1/2}^y\right) - \frac{1}{2}\left(\Phi_{i-1/2,j}^x + \Phi_{i-1/2,j-1}^x\right) \\
&\equiv \bar{\Omega}_c^n + \bar{\Phi}_{i-1/2,j-1/2}^y - \bar{\Phi}_{i-1/2,j-1/2}^x.
\end{aligned}
\tag{21}
$$

This method is identical to the BS scheme with the exception that the RMJF scheme adds approximate values of both Φ^x and Φ^y into the update, while the BS scheme averages these two values. In this formulation it is clear that when either Φ^x or Φ^y are identically zero, the RMJF scheme will reduce to a 1D high-resolution method.

Recall that Φ^x and Φ^y contain, among other things, the first order upwind corrections. This implies that the BS and RMJF schemes differ in their update for the magnetic potential by $\mathcal{O}(1)$! Therefore, at best only one of these methods can be a consistent numerical update for the magnetic potential. In fact, it is shown in [6] that both the BS and the RMJF schemes as formulated here are inconsistent numerical methods. The reason these inconsistencies are not so apparent when solving the ideal MHD equations are discussed in Section 6.

4 MPACT: a new constrained transport method

The three methods described above for computing the magnetic potential produce either an update with spurious oscillations in the case of the DW scheme or an inconsistent update in the case of the BS and RMJF schemes. In this section we introduce a novel approach for computing the electric field that will be shown to produce a consistent and significantly less oscillatory numerical update.

4.1 The magnetic potential equation

From the induction equation we have shown that the magnetic potential obeys an equation of the form

$$\partial_t \mathbb{A} + \Omega = \partial_t \mathbb{A} - u_1 h B_2 + u_2 h B_1 = 0 \,. \tag{22}$$

If in this equation we replace the magnetic field by the appropriate derivatives of the magnetic potential ($h\mathbf{B} = \nabla \times \mathbb{A}$), then we obtain

$$\partial_t \mathbb{A} + u_1 \partial_x \mathbb{A} + u_2 \partial_y \mathbb{A} = 0 \,. \tag{23}$$

This is a linear advection equation and is strictly hyperbolic for all $(u_1, u_2)^t \in \mathbb{R} \times \mathbb{R}$. In this section we introduce a new constrained transport method based on coupling the numerical update for the magnetohydrodynamic equations with an update for the magnetic potential equation (23). Because this method is based on the advection of the magnetic potential, we refer to this scheme as the MPACT (Magnetic Potential Advection Constrained Transport) scheme.

4.2 An operator splitting approach

By writing the full magnetic potential equation for the magnetic potential we are now able to update the magnetic potential given a velocity field. The simplest way to solve both the SMHD equations and the magnetic potential equation is by operator splitting techniques. The two operators are

$$\begin{aligned} \mathcal{L}_1 &\equiv \text{ the SMHD equations} \\ \mathcal{L}_2 &\equiv \text{ the magnetic potential equation (23).} \end{aligned}$$

Step 2 and **Step 4** in the constrained transport method as outlined in Section 2.2 can now be written as

$$\begin{aligned} &\textbf{Step 2:} \text{ (full step with } \mathcal{L}_1) \qquad & q^\star &= \mathcal{L}_1^{\Delta t} q^n \\ &\textbf{Step 4:} \text{ (full step with } \mathcal{L}_2) \qquad & q^{n+1} &= \mathcal{L}_2^{\Delta t} q^\star \,. \end{aligned}$$

The SMHD equations can again be solved via a standard cell centered high-resolution method. We address the question of how to update the magnetic potential below.

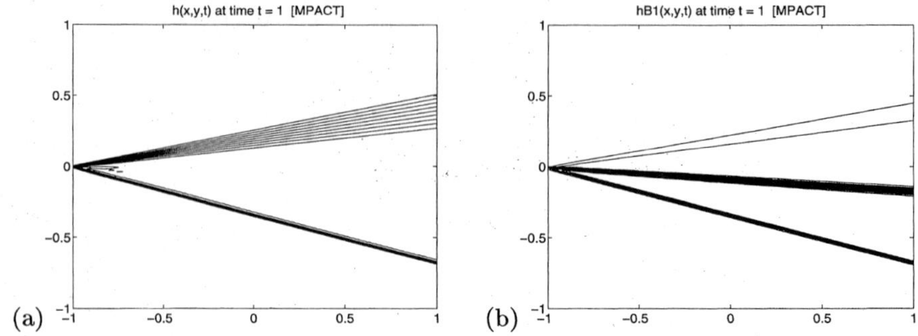

Fig. 2. Solution obtained from the MPACT scheme for the multidimensional shock problem. Shown are (a) the height at time $t = 1$ and (b) hB_1 at time $t = 1$.

4.3 Keeping the magnetic field non-oscillatory

Although the magnetic potential equation (23) is a simple advection equation, care must be taken when solving this equation. We would like to apply a method such as the LeVeque wave propagation method to the advection equation, in which case wave limiters would be applied to the solution of $\mathbb{A}$ [4]. At the end of the day, however, we are interested in ensuring a non-oscillatory magnetic field. This is not guaranteed by applying limiters to only $\mathbb{A}$. One option would be to apply limiters when differencing the magnetic potential:

$$h\mathbf{B} = \nabla \times \mathbb{A}\,. \tag{24}$$

This is in general a bad idea, since we require *central* finite difference operators when approximating $h\mathbf{B} = \nabla \times \mathbb{A}$ in order for the exact divergence-free condition to hold (see Section 2.1).

Instead, we propose in this paper an alternative approach based on composite schemes [5]. The idea is to add a judicious amount of diffusion into the solution of the magnetic potential equation so that the magnetic field remains non-oscillatory, but the shock capturing ability of overall method is not destroyed. Therefore, we replace **Step 4** above by a partial diffusive Godunov step and a partial high-resolution wave propagation step:

Step 4a: (partial step with Godunov) $\quad q^{n+\alpha} = \mathcal{L}_2^{\alpha\Delta t} q^{\star}$

Step 4b: (partial step with Wave Propagation) $\quad q^{n+1} = \mathcal{L}_2^{(1-\alpha)\Delta t} q^{n+\alpha}$,

where $0 \leq \alpha \leq 1$. We have found in numerical experiments that a value of $\alpha = 0.3$ gives a good balance between shock resolution and oscillation reduction.

5 A test problem

We consider a test problem from De Sterck [8]. The computational domain is the square box $(x, y) \in [-1, 1] \times [-1, 1]$. The initial conditions are discontinuous across the line $y = 0$:

(a) (b)
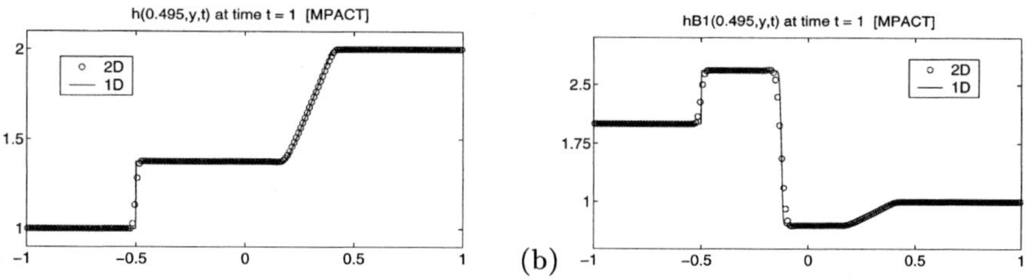

Fig. 3. An x = constant slice of the 2D solution shown in Figure 2 compared against a 1D solution of the steady-state equations.

$$[h, u_1, u_2, B_1, B_2] = \begin{cases} [1, 4.5, 0, 2, 0] & \text{if} \quad y < 0 \\ [2, 5.5, 0, 0.5, 0] & \text{if} \quad y > 0\,. \end{cases} \tag{25}$$

The boundary conditions on the top, bottom, and right sides are set to outflow, while at the left boundary we impose an inflow condition that is also given by (25). The computation is run on a 100×100 grid until time $t = 1$, at which point the system has reached its steady-state.

The LeVeque wave propagation method is used to solve the SMHD equations in **Step 2** of the constrained transport update. We test on this problem the wave propagation method without constrained transport (the base scheme) as well as the DW, BS, RMJF, and MPACT schemes. This test problem was chosen because all of the above methods except for the MPACT scheme crash at various points in the calculation because they produce negative heights (see [6]). The base scheme crashes because $\nabla \cdot h\mathbf{B}$ becomes very large near the singularity at $(x, y) = (-1, 0)$. The DW scheme crashes because large unphysical oscillations are generated across the multidimensional shock. The BS and RMJF schemes crash because they are inconsistent methods on this multidimensional test problem. The results from the MPACT scheme are shown in Figure 2 and compare well to the results found in [8].

In order to better quantify the performance of the 2D MPACT scheme, we solve the steady state SMHD equations by treating it as a 1D "time" dependent equation where x is the time-like coordinate and y is the space-like coordinate (see [6] for details). The inflow conditions at $x = -1$ act as "initial conditions" for this system. For simplicity we can solve this equation using a highly resolved (20,000 grid points) local Lax-Friedrichs method. The 1D solution is compared to an x = constant slice of the 2D MPACT calculation in Figure 3.

6 Why do the DW, BS, and RMJF schemes work so poorly on SMHD?

Because the divergence-free constraint is ignored in the base scheme, it is not surprising that this approach produces negative heights. However, it is somewhat surprising that the RMJF, BS, and DW schemes do not produce better results. Although flaws in these methods were discussed in Section 3, all three constrained transport schemes have been used quite successfully on the ideal MHD equations [1, 2, 7]. Why then do these methods behave so poorly on the SMHD equations?

The answer to this question lies in the Courant number. Constrained transport is a process which occurs on the time scale of the fluid velocity ($\mathbf{u}$):

$$\partial_t \mathbb{A} + u_1 \partial_x \mathbb{A} + u_2 \partial_y \mathbb{A} = 0\,. \tag{26}$$

In MHD, the fast magnetosonic waves propagate on a time scale much faster than the fluid velocity, and thus, these waves dictate the time step restrictions necessary for a stable numerical update. This implies that a significant amount of numerical diffusion is added to the much slower waves in the MHD system. In SMHD, the fastest waves are the magnetogravity waves which propagate much slower than the fast magnetosonic waves. This implies that less numerical diffusion is being added to the constrained transport part of the update. Therefore, any problems that occur due to constrained transport are significantly enhanced by the SMHD equations (see [6] for an example). Because of this and because the SMHD equations have a less complicated eigenstructure than the MHD equations, the SMHD system can and should be used as a standard test problem for validating algorithms designed for the full MHD equations.

Acknowledgments. This work was supported in part by NSF grants DMS-9803442 and DMS-0106511, DOE grant DE-FG03-00ER2592, and a VIGRE Fellowship supported by NSF grant DMS-9810726.

References

1. Balsara, D.S., Spicer, D. (1999): A staggered mesh algorithm using high order Godunov fluxes to ensure solenoidal magnetic fields in magnetohydrodynamic simulations. J. Comp. Phys., **149**, 270–292
2. Dai, W., Woodward, P.R. (1998): A simple finite difference scheme for multidimensional magnetohydrodynamic equations. J. Comp. Phys., **142**, 331–369
3. Gilman, P.A. (2000): Magnetohydrodynamic "shallow water" equations for the solar tachocline. Astrophys. J., **544**, L79–L82
4. LeVeque, R.J. (1997): Wave propagation algorithms for multi-dimensional hyperbolic systems. J. Comp. Phys., **131**, 327–335
5. Liska, R., Wendroff, B. (1998): Composite schemes for conservation laws. SIAM J. Numer. Anal., **35**, 2250–2271
6. Rossmanith, J.A. (2002): A wave propagation method with constrained transport for ideal and shallow water magnetohydrodynamics. PhD Thesis, University of Washington, Seattle, Washington
7. Ryu, D.S., Miniati, F., Jones, T.W., Frank, A. (1998): A divergence-free upwind code for multidimensional magnetohydrodynamic flows. Astrophys. J., **509**, 244–255
8. De Sterck, H. (2001): Multi-dimensional upwind constrained transport on unstructured grids for "shallow water" magnetohydrodynamics. AIAA 2001-2623.

Solutions with Linear Profile of Velocity to the Euler Equations in Several Dimensions

Olga S. Rozanova

Department of Mathematics and Mechanics, Moscow State University, Vorobiovy Gory, 119899 Moscow, Russia
rozanova@mech.math.msu.su, rozanova@bsu.edu.ru

We study the solutions to the Euler equations with the right-hand sides describing diverse external forces:

$$\partial_t \rho + \operatorname{div}(\rho \mathbf{V}) = 0, \tag{1}$$

$$\partial_t(\rho \mathbf{V}) + (\rho \mathbf{V}, \nabla)\,\mathbf{V} + \nabla p = \rho \mathbf{f}(\mathbf{x}, t, \mathbf{V}, \rho, S), \tag{2}$$

$$\partial_t S + (\mathbf{V}, \nabla S) = 0 \tag{3}$$

with the state equation $p = e^S \rho^\gamma$, $\gamma = const > 1$, the function $\mathbf{f}$ is supposed to be smooth with respect to all arguments. Here $\rho(t,\mathbf{x})$, $\mathbf{V}(t,\mathbf{x})$, $S(t,\mathbf{x})$ are components of the solution, given in $\mathbb{R}_+ \times \mathbb{R}^n$, $n \geq 1$ (density, velocity and entropy, correspondingly), $p(t,\mathbf{x})$ is the pressure.

Put the Cauchy problem for (1–3):

$$\rho(0,\mathbf{x}) = \rho_0(\mathbf{x}) \geq 0,\ \mathbf{V}(0,\mathbf{x}) = \mathbf{V}_0(\mathbf{x}),\ S(0,\mathbf{x}) = S_0(\mathbf{x}). \tag{4}$$

Denote $p_0(\mathbf{x}) = e^{S_0(\mathbf{x})} \rho_0^\gamma(\mathbf{x})$, and remark that the first of the Cauchy conditions can be replaced to $p(0,\mathbf{x}) = p_0(\mathbf{x}) \geq 0$.

We consider the classical solutions to (1–3) with the density sufficiently quickly decreasing as $|\mathbf{x}| \to \infty$ to guarantee a convergency of the integral $\int\limits_{\mathbb{R}^n} \rho |\mathbf{x}|^2 d\mathbf{x}$ (so called solutions with a finite moment).

As well known, solutions to the Cauchy problem for system (1–3) may lose the initial smoothness for a finite time. Sometime there is a possibility to estimate the time of a singularity formation from above (see, f.e., [1] and references therein). Moreover, in the case $\mathbf{f} = 0$ the singularity appears in any solution with compactly supported initial data (f.e., [2]).

At the same time, there exist some nontrivial classes of globally smooth solutions. Our special interest will be *the solutions with linear profile of velocity* $\mathbf{V} = A(t)\mathbf{r} + \mathbf{b}(t)$, where $A(t)$ and $\mathbf{b}(t)$ are a matrix $(n \times n)$ and a n - vector, dependent on time, $\mathbf{r}$ is a radius-vector of the point.

Definition 1. *We shall call the global-in-time classical solution* $(\bar\rho(t,\mathbf{x})$, $\bar{\mathbf{V}}(t,\mathbf{x})$, $\bar S(t,\mathbf{x}))$ *to the system (1–3) the interior solution, if* $(\bar p^{\frac{\gamma-1}{2\gamma}}, \bar{\mathbf{V}}, \bar S) \in C^1(\mathbb{R}_+ \times \mathbb{R}^n)$, *where* $\bar p = e^{\bar S} \bar\rho^\gamma$, *and any solution* $(\rho(t,\mathbf{x}), \mathbf{V}(t,\mathbf{x}), S(t,\mathbf{x}))$ *to the Cauchy problem (1–4) with the sufficiently small norm*

$$||(p_0^{(\gamma-1)/2\gamma}(\mathbf{x}) - \bar{p}^{(\gamma-1)/2\gamma}(0,\mathbf{x}),\ \mathbf{V}_0(\mathbf{x}) - \bar{\mathbf{V}}(0,\mathbf{x}),\ S_0(\mathbf{x}) - \bar{S}(0,\mathbf{x})||_{H^m(\mathbb{R}^n)},$$

$m > n/2+1$, is also smooth, such that $(p^{\frac{\gamma-1}{2\gamma}},\ \mathbf{V},\ S) \in C^1(\mathbb{R}_+ \times \mathbb{R}^n)$.

Note that the trivial solution is not interior at least for $\mathbf{f} = \mathbf{0}$. But a set of the interior solutions is not empty. In [4] (a generalization of [3]) for $\mathbf{f} = \mathbf{0}$ it was shown that the solution $(0,\ \bar{\mathbf{V}}(t,\mathbf{x}),\ const)$ to (1–4) is interior, provided $\bar{\mathbf{V}}(t,\mathbf{x})$ is a globally smooth solution to the equation $\partial_t \mathbf{V} + (\mathbf{V},\nabla)\,\mathbf{V} = 0$ such that the spectrum of its Jacobian is separated initially from the real negative semi-axis, $D\bar{\mathbf{V}}(0,\mathbf{x}) \in L^\infty(\mathbb{R}^n)$, $D^2\bar{\mathbf{V}}(0,\mathbf{x}) \in H^{m-1}(\mathbb{R}^n)$ (we denote by D^k the vector of all spatial derivatives of the order k).

The result is clear from a physical point of view: the velocity field with a positive divergency spreads the initially concentrated small mass, that prevents the singularity formation.

The question arises: can one construct the interior solution with the density not close to zero? It occurs that the answer is positive.

Below we *suppose* firstly the existence of global-in-time smooth solution with linear profile of velocity having certain properties and show that this solution is interior. Then we construct the interior solutions for $n = 2$ and $n = 3$ in some physically important cases.

1 What global solution with linear profile of velocity is interior?

For the solutions to (1–3) with a finite moment we have $\inf \rho = \inf p = 0$, thus we need to use the symmetrization proposed in [5]. After involving the new variable $\Pi = \kappa p^{\frac{\gamma-1}{2\gamma}}$, $\kappa = \dfrac{2\sqrt{\gamma}}{\gamma-1}$, we obtain the symmetric form of the system (1–3) with the solution $\mathbf{U} = (\Pi,\ \mathbf{V},\ S)$:

$$\exp(\frac{S}{\gamma})(\partial_t + (\mathbf{V},\nabla))\Pi + \frac{\gamma-1}{2}\exp(\frac{S}{\gamma})\Pi\, div\,\mathbf{V} = 0, \tag{1.1}$$

$$(\partial_t + (\mathbf{V},\nabla))\mathbf{V} + \frac{\gamma-1}{2}\exp(\frac{S}{\gamma})\Pi\nabla\Pi = \mathbf{f}_1(t,\mathbf{x},\Pi,\mathbf{V},S), \tag{1.2}$$

$$(\partial_t + (\mathbf{V},\nabla))S = 0, \tag{1.3}$$

where $\mathbf{f}_1 = \mathbf{f}(t,\mathbf{x},e^{-\frac{S}{\gamma}}\,(\frac{\Pi}{\kappa})^{\frac{2}{\gamma-1}},\mathbf{V},S)$. Let $\Pi_0 := \kappa p_0^{\frac{\gamma-1}{2\gamma}}$, $\mathbf{U}_0 := (\Pi_0,\mathbf{V}_0,S_0)$. In this way, the Cauchy problem (1–4) is transformed to the problem (1.1–1.3), with the initial data $\mathbf{U}(0,\mathbf{x}) = \mathbf{U}_0(\mathbf{x})$. Further, suppose that the system (1.1–1.3) has a classical solution $\bar{\mathbf{U}} := (\bar{\Pi},\ \bar{\mathbf{V}},\ \bar{S}))$, where $\bar{\mathbf{V}} := A(t)\mathbf{r} + \mathbf{b}(t)$, $\bar{\Pi} := \kappa\bar{p}^{\frac{\gamma-1}{2\gamma}}$. Denote also $\mathbf{U} - \bar{\mathbf{U}} := \mathbf{u} := (\pi,\ \mathbf{v},\ s)$.

Before formulating Theorem 1 we perform certain transformations. Firstly we go on to the system with the solution $\mathbf{u}$. Then, following [3], we carry out the nondegenerate change of variables such that the infinite semi-axis of time turns to semi-interval.

Assume that there exists a nondegenerate square $(n \times n)$ matrix $A_1(t)$ such that $A_1(t)(A_1^{-1}(t))' = A(t)$. Choose a nonnegative decreasing function $\lambda(t)$ such that the integral $\int_0^\infty \lambda(\tau)d\tau$ converges to the finite value σ_∞, and set $\sigma(t) = \int_0^t \lambda(\tau)d\tau$. Let $(\sigma, \mathbf{y}) := (\sigma(t), A_1(t)\mathbf{x})$ be new variables. Then $\nabla_{\mathbf{x}} = A_1^* \nabla_{\mathbf{y}}$, $\mathrm{div}_{\mathbf{x}} \mathbf{V} = \mathrm{div}_{\mathbf{y}} A_1 \mathbf{V}$, $\partial_\sigma = \lambda^{-1}(t)(\partial_t + A\mathbf{r}\nabla_{\mathbf{x}})$. In that way, the semi-infinite axis of time goes to the semi-interval $[0, \sigma_\infty)$. Note that there exists the inverse function $t = t(\sigma)$. Further, involve the variables $\mathbf{W} = \lambda^{-1}(t)A_1(t)\mathbf{v}$, $P = \lambda^{-q}(t)\pi$, the constant q will be defined below. Let $\mathcal{U}(\sigma, \mathbf{y}) := (P, \mathbf{W}, s)$, $\bar{P} := \lambda^{-q}\bar{\Pi}$, $\bar{\mathbf{W}} := \lambda^{-1}A_1\bar{\mathbf{V}}$, $\bar{\mathcal{U}} := (\bar{P}, \bar{W}, \bar{S})$.

So we get a system

$$(\partial_\sigma + (\mathbf{W}, \nabla_{\mathbf{y}}))P + \frac{\gamma - 1}{2}(P + \bar{P})\mathrm{div}_{\mathbf{y}}\, \mathbf{W} =$$

$$-(\mathbf{W}, \nabla_{\mathbf{y}})\bar{P} - PQ_1 - \lambda^{-1}(A_1\mathbf{b}, \nabla_{\mathbf{y}} P), \tag{1.4}$$

$$(\partial_\sigma + (\mathbf{W}, \nabla_{\mathbf{y}}))\mathbf{W} + \frac{\gamma - 1}{2}\Psi(S, \sigma)(P + \bar{P})\nabla_{\mathbf{y}} P =$$

$$-\frac{\gamma - 1}{2}\Psi(S, \sigma)(P + \bar{P})\nabla_{\mathbf{y}}\bar{P} + \frac{\gamma - 1}{2}\Psi(\bar{S}, \sigma)\bar{P}\nabla_{\mathbf{y}}\bar{P} -$$

$$-\mathbf{W}Q_2 + G - \lambda^{-1}(A_1\mathbf{b}, \nabla_{\mathbf{y}})\mathbf{W}, \tag{1.5}$$

$$(\partial_\sigma + (\mathbf{W}, \nabla_{\mathbf{y}}))s = -(\mathbf{W}, \nabla_{\mathbf{y}})\bar{S} - \lambda^{-1}(A_1\mathbf{b}, \nabla_{\mathbf{y}} s). \tag{1.6}$$

Here we take into account that $t = t(\sigma)$, $\mathbf{x} = A_1^{-1}(t(\sigma))\mathbf{y}$, $\lambda = \lambda(t(\sigma)), A = A(t(\sigma)), A_1 = A_1(t(\sigma)), \mathbf{b} = \mathbf{b}(t(\sigma))$ and denote

$$Q_1 = Q_1(t(\sigma)) := \lambda(t)^{-1}(\frac{\gamma - 1}{2}\mathrm{tr}A(t) + q(\ln \lambda(t))'),$$

$$Q_2 = Q_2(t(\sigma)) := \lambda(t)^{-1}((\ln \lambda(t))'E + A(t) + A_1(t)A(t)A_1^{-1}(t)),$$

$$\Psi(S, \sigma) := \exp(\frac{S}{\gamma})RA_2,$$

$$A_2 = A_2(t(\sigma)) := A_1(t)A_1^*(t)(\mathrm{det}A_1(t))^{-2/n}$$

$$R = R(t(\sigma)) := \lambda^{2q-2}(t)(\det A_1(t))^{2/n},$$

$$G = G(\sigma, \mathbf{y}, \mathcal{U}, \bar{\mathcal{U}}) := \lambda^{-2}(t)A_1(t)(\mathbf{f}_1(t, \mathbf{x}, \lambda^q(t)(\bar{P} + P),$$

$$\lambda(t)A_1^{-1}(t)(\bar{\mathbf{W}} + \mathbf{W}), (\bar{S} + s)) - \mathbf{f}_1(t, \mathbf{x}, \lambda^q(t)\bar{P}, \lambda(t)A_1^{-1}(t)\bar{\mathbf{W}}, \bar{S})).$$

Note that A_2 is a family of bounded invertible matrices.

The initial data $\mathbf{U}_0$ and the vector-function $\bar{\mathbf{U}}$ induce the Cauchy data $\mathcal{U}_0(\mathbf{y})$ for the system (1.4 – 1.6) as follows: $\mathcal{U}_0(\mathbf{y}) = \lambda^{-q}(\Pi_0(A_1^{-1}\mathbf{y}) - \bar{\Pi}(0, A_1^{-1}\mathbf{y}), \lambda^{-1}A_1(\mathbf{V}_0(A_1^{-1}\mathbf{y}) - \bar{\mathbf{V}}(0, A_1^{-1}\mathbf{y})), S_0(A_1^{-1}\mathbf{y}) - \bar{S}(0, A_1^{-1}\mathbf{y}))|_{\sigma=0}$.

After multiplying (1.5) by $\Psi^{-1}(S,\sigma)$ the system becomes symmetric hyperbolic. We can apply the theorem on a local existence of the unique solution to the Cauchy problem for symmetric hyperbolic systems [7] to the problem (1.4 –1.6), $\mathcal{U}(0,\mathbf{y}) = \mathcal{U}_0(\mathbf{y})$, provided $\Psi^{-1}(S,\sigma)$ is uniformly positive. Sometimes $\Psi^{-1}(S,\sigma)$ is really uniformly positive (see [3]), and in that cases we can conclude that if $\mathcal{U}_0 \in H^m(\mathbb{R}^n), m > 1+n/2$, then $\mathcal{U} \in \cap_{j=0}^m C^j([0,\sigma_*); H^{m-j}(\mathbb{R}^n)), \sigma_* > 0$. Since $\sigma_* \to \infty$, as $\|\mathcal{U}_0\|_{H^m} \to 0$, then choosing the initial data $\mathcal{U}_0$ small in the Sobolev H^m- norm, we can extend the time of existence of the smooth solution $\mathcal{U}$ to the Cauchy problem up to $\sigma_* \ge \sigma_\infty$. In this case we conclude that $\mathcal{U} \in \cap_{j=0}^m C^j([0,\sigma_\infty); H^{m-j}(\mathbb{R}^n))$ and $\mathbf{u} \in \cap_{j=0}^m C^j([0,\infty); H^{m-j}(\mathbb{R}^n))$.

However, generally speaking, $\Psi^{-1}(S,\sigma) \to 0$ as $t \to \infty$, therefore we can extend the time of existence of the smooth solution $\mathcal{U}$ to the Cauchy problem up to $[0,\sigma_*)$, for any $\sigma_* < \sigma_\infty$, choosing the initial data small in the Sobolev H^m- norm, but we cannot guarantee the existence of the solution for $[0,\sigma_\infty)$. Thus, we need to act differently.

If t_* is the supremum of the time of existence of the smooth solution $\mathbf{U}$ for the Cauchy problem (1.1–1.3), $\mathbf{U}(0,\mathbf{x}) = \mathbf{U}_0(\mathbf{x})$ (the solution $\mathbf{u}$ preserves the smoothness during the same time), then $\sigma_* = \sigma(t_*)$. If $\sigma_* < \sigma_\infty (t_* < \infty)$, then $\limsup_{\sigma\to\sigma_*} (\|\mathcal{U}\|_{L_\infty}(\sigma) + \|\nabla_{\mathbf{y}}\mathcal{U}\|_{L_\infty}(\sigma)) = +\infty$ $(\limsup_{t\to t_*} (\|\mathbf{u}\|_{L_\infty}(t) + \|\nabla_{\mathbf{x}}\mathbf{u}\|_{L_\infty}(t)) = \infty)$. Therefore, if we prove that $\|\mathcal{U}\|_{H^m}(\sigma) < +\infty, \sigma \in [0,\sigma_\infty)$, then we establish that $\sigma_* = \sigma_\infty (t_* = \infty)$.

Let

$$D^J = \left(\frac{\partial}{\partial y_1}\right)^{j_1} \cdots \left(\frac{\partial}{\partial y_n}\right)^{j_n}, \sum_i j_i = |J|.$$

Further, $\mathbf{f_V}$ stands for the matrix $\|\frac{\partial f_i}{\partial V_j}\|$, $\nabla_\Pi \mathbf{f}_1$ and $\nabla_S \mathbf{f}_1$ for the derivatives of $\mathbf{f}_1$ with respect to Π and S, correspondingly.

Theorem 1. *Let the function $\mathbf{f}_1(t,\mathbf{x},\Pi,\mathbf{V},S)$ have the derivatives with respect to all arguments up to the order $m+1, m > n/2+1$. Suppose that the system (1–3) has a global in time classical solution $(\bar\rho, \bar{\mathbf{V}}, \bar S)$ with linear profile of velocity $\bar{\mathbf{V}} = A(t)\mathbf{r} + \mathbf{b}(t)$ such that*

a) $\bar p^{\frac{\gamma-1}{2\gamma}}(t,\mathbf{x}) \in \cap_{j=0}^{m+1} C^j([0,\infty); H^{m-j}(\mathbb{R}^n))$;
$D\,\bar S(t,\mathbf{x}) \in \cap_{j=0}^m C^j([0,\infty); H^{m-j}(\mathbb{R}^n))$;

b)there exists a matrix $A_1(t)$ such that $A(t) = A_1(t)(A_1^{-1}(t))'$, $\det A_1(t) > 0$, for $t \ge t_0 \ge 0$.

Let us assume that there exist a smooth real-valued decreasing nonnegative function $\lambda(t)$, constant q and a skew-symmetric matrix $U_\phi(t)$ with real coefficients, having the following properties:

$$\int_{t_0}^{+\infty} \lambda(\tau)d\tau < \infty, \qquad \int_{t_0}^{+\infty} \lambda^q(\tau)(\det A_1(\tau))^{1/n}\, d\tau < \infty, \qquad (1.7),(1.8)$$

$$R(t)' \geq 0 \quad and \quad Q_1(t) \geq 0 \quad for \quad t \geq t_0 \geq 0,$$

the following functions, vector-functions and matrices are bounded in $[t_0, \infty) \times \mathbb{R}^n$ *(under bounded* $(\Pi, \mathbf{V}, S)$*):*

$$\lambda^{-q}(t) \exp(-\frac{\gamma - 1}{2} \mathrm{tr} A(t)), \quad \exp(-\int_{t_0}^{t} A(\tau) d\tau), \quad \lambda^{-q}(t) \mathbf{b}(t), \tag{1.10}$$

$$\lambda^{-1} D^J (\nabla_\Pi \mathbf{f}_1), \quad \lambda^{-(q+1)} D^J (\nabla_S \mathbf{f}_1), \tag{1.11}$$

$$Q_3 := (Q_2(t) - \lambda^{-1}(t) D^J (A_1(t) \nabla_{\mathbf{V}} \mathbf{f}_1) A_1^{-1}(t) - U_\phi(t)), \; |J| = 0, 1, \ldots, m.$$

Then the solution $(\bar{\rho}, \bar{\mathbf{V}}, \bar{S})$ *is interior. Moreover,*

$$(p^{(\gamma-1)/2\gamma} - \bar{p}^{(\gamma-1)/2\gamma}, \, \mathbf{V} - \bar{\mathbf{V}}, \, S - \bar{S}) \in \cap_{j=0}^{m} C^j([0, \infty); H^{m-j}(\mathbb{R}^n)).$$

Remark 1. If Q_3 depends only on time, then we can require only the nonnegativity of this expression for $t \geq t_0$.

Sketch of the proof. Let $\sigma \in [0, \sigma_*)$. Suppose for the sake of simplicity that $t_0 = 0$, otherwise we can consider the Cauchy problem at $\sigma_0 = \sigma(t_0) < \sigma_*$ with the initial data induced by the Cauchy data $\mathcal{U}_0$. Note that choosing $\mathcal{U}_0$ sufficiently small in the H^m - norm we can prolong the smooth solution $\mathcal{U}$ up to σ_0. Denote $(X, \mathbf{Y}, Z) : R^n \to R \times R^n \times R$ a vector-function from $L_2(\mathbb{R}^{n+2})$. Introduce a norm

$$[X, \mathbf{Y}, Z]^2(\sigma) := \int_{\mathbb{R}^n} ((X^2 + Z^2) + \mathbf{Y}^* \Psi^{-1}(S, \sigma) \mathbf{Y}) d\mathbf{y}.$$

It is equivalent to the usual $L^2(\mathbb{R}^{n+2})$ - norm, but not uniformly in time. For any $p \in N$ we define

$$E_p(\sigma) = \frac{1}{2} \sum_{|J|=p} [D^J P, \, D^J \mathbf{W}, \, D^J s]^2(\sigma), \qquad F_m(\sigma) = \sum_{p=0}^{m} E_p(\sigma).$$

Let us compute

$$\frac{dE_p}{d\sigma} = \sum_{|J|=p} [\int_{\mathbb{R}^n} (D^J P D^J \partial_\sigma P d\mathbf{y} + \int_{\mathbb{R}^n} D^J s D^J \partial_\sigma s d\mathbf{y} +$$

$$+ \int_{\mathbb{R}^n} D^J \mathbf{W} \Psi^{-1}(S, \sigma) D^J \partial_\sigma \mathbf{W} d\mathbf{y} +$$

$$+ \int_{\mathbb{R}^n} (D^J \mathbf{W})^* (\frac{\partial \Psi^{-1}(S, \sigma)}{\partial \sigma} - \frac{1}{\gamma} \Psi^{-1}(S, \sigma) \partial_\sigma S) D^J \mathbf{W} d\mathbf{y}] =$$

$$= I_1 + I_2 + I_3 + I_4,$$

the integrals I_k, $k = 1,2,3,4$, are numbered in the consecutive order.

We estimate every of the integrals in the standard way using the Hölder and the Galiardo-Nirinberg inequalities (see [3],[4]) under assumption (1.9 – 1.11) of Theorem 1. We denote by c_i certain positive constants, which do not depend on σ. Omitting the details we get $I_1 \le c_1(E_p+R^{1/2}F_m^{1/2}E_p+R^{1/2}F_m)$, $I_2 \le c_2(R^{1/2}F_m^{1/2}E_p + R^{1/2}F_m)$, $I_3 \le c_3(F_m + R^{1/2}F_m^{1/2}E_p + R^{1/2}F_m + R^{1/2}E_p^{1/2}\sum_{i=1}^4 F_m^{1+\gamma_i})$, $\gamma_i > 0$, $I_4 \le c_5(1+R^{1/2}F_m + R^{1/2}F_m^{1/2})E_p$. Using all the estimates we get $F_m' \le c_6(F_m + R^{1/2}(F_m + \sum_{i=1}^6 F_m^{3/2+\tilde\gamma_i}))$, $\tilde\gamma_i \ge 0$.

Set $\Lambda_m(\sigma) = e^{-c_6\sigma}F_m$. Then

$$\Lambda_m' \le c_7(\Lambda_m + \sum_{i=1}^{6}(\Lambda_m)^{3/2+\tilde\gamma_i})R^{1/2}, \tilde\gamma_i \ge 0, \tag{1.12}$$

where the constant c_7 depends on c_6, σ_∞. Let $\Theta(\tilde g) := \int\limits_\delta^{\tilde g} \frac{dg}{g+\sum_{i=1}^6 g^{3/2+\tilde\gamma_i}}$, $\delta > 0$. The integral diverges in the zero, such that $\Theta(0) = -\infty$. Integrating inequality (1.12) over σ we obtain

$$\Theta(\Lambda_m(\sigma)) \le \Theta(\Lambda_m(0)) + c_8 \int\limits_0^\sigma R^{1/2}(\tilde\sigma)d\tilde\sigma,$$

moreover, as follows from (1.8), the integral in the right-hand side of the last inequality converges as $\sigma \to \sigma_\infty$ to the constant C, depending only on the initial data. Choosing $\Lambda_m(0)$, (that is the H_m- norm of the initial data) sufficiently small, one can make the value $\Theta(\Lambda_m(0)) + C$ later then $\Theta(+\infty)$, it signifies that $\Lambda_m(\sigma)$ and $F_m(\sigma)$ are bounded from above for all $\sigma \in [0, \sigma_\infty)$ and $\sigma_* = \sigma_\infty$. So, $\mathbf{u} \in \cap_{j=0}^m C^j([0,\infty); H^{m-j}(\mathbb{R}^n))$ and Theorem 1 is proved.

Remark 2. The solutions with linear profile of velocity, satisfying the conditions of Theorem 1, exist. In the case investigated in [3], $\mathbf{f} = 0$, $\gamma \le 1 + \frac{2}{n}$, $A(t) = (E + tA(0))^{-1}A(0)$, $\mathrm{Sp}A(0) \notin \mathbb{R}_-$, $\mathbf{b}(t) = 0$, $A_1(t) = A(t)$, $\bar\rho = 0$, $\bar S = const$. Here $\lambda = (1+t)^{-2}$, $q = \frac{n(\gamma-1)}{4}$.

The following Corollary helps us to prove that the solutions to the Euler system constructed in Section 2 are interior.

Corollary 1. *Let $\mathbf{f} = L(t)\mathbf{V}$, where $L(t) = -\mu E + U_1(t)$, μ is a nonnegative constant, $U_1(t)$ is a one-parameter family of skew-symmetric matrices with the smooth coefficients. Assume that the system (1–3) has a global in time classical solution $(\bar\rho, A(t)\mathbf{r}, \bar S)$, satisfying the condition a) of Theorem 1 and $A(t) = \alpha(t)E + \beta(t)U_2$, where $\alpha(t)$ and $\beta(t)$ are the real-valued functions, U_2 is a skew-symmetric matrix with constant coefficients. Moreover, suppose that $\alpha(t) \sim \frac{\Delta}{t}$, as $t \to \infty$, $\Delta = const$, such that $\Delta > 0$ for $\mu > 0$, and $\Delta > \frac{1}{2}$ for $\mu = 0$. Then this solution is interior.*

Proof. Here $\mathbf{b}(t) = 0$. The matrix $A_1(t) = \exp(-\int_{t_0}^t A(\tau)d\tau)$, satisfying the condition b) of Theorem 1, exists, $\det A_1(t) = \exp(-n\int_{t_0}^t \alpha(\tau)d\tau)$, as the eigenvalues of a skew-symmetric real-valued matrix have the real part equal to zero. The first and second conditions in (1.11) hold, since $\mathbf{f}$ is independent on ρ and S, Q_3 depends only on t(see Remark 1). We choose as $U_\phi(t)$ the skew-symmetric matrix $A_1(t)(\beta(t)U_2 - U_1(t))A_1^{-1}(t)$ to obtain $Q_3 = \lambda^{-1}(t)((\ln\lambda(t))' + 2\alpha(t) + \mu)E$. If $\lambda(t) = \exp(-\varepsilon\int_{t_0}^t \alpha(\tau)d\tau), \varepsilon = const > 0, \varepsilon\Delta > 1$, then both integrals in (1.7), (1.8) converge. To guarantee the nonnegativity of $R'(t), Q_1(t), Q_3(t)$ we must satisfy the following inequalities for ε and q: $q < 1$, $\varepsilon \geq \frac{1}{1-q}, \varepsilon \leq \frac{(\gamma-1)n}{2q}$, $\varepsilon \leq 2$ (the last inequality arises only in the case $\mu = 0$), the conditions (1.8) hold for $\varepsilon \leq \frac{(\gamma-1)n}{2q}$. For $\mu > 0$, it implies $\frac{1}{1-q} \leq \varepsilon \leq \frac{(\gamma-1)n}{2q}$, $\frac{1}{\Delta} < a \leq \frac{(\gamma-1)n}{2q}$, $q < \bar{q} := \min\{\frac{\Delta n(\gamma-1)}{2}, \frac{n(\gamma-1)}{2+n(\gamma-1)}\}$, and we set $q \in (0,\bar{q})$. If $\mu = 0$, then additionally we have inequalities $\frac{1}{\Delta} < \varepsilon \leq 2$, $\frac{1}{1-q} \leq \varepsilon \leq 2$. Since $\Delta < \frac{1}{2}$, then we can choose ε to satisfy the first inequality, for the second we have $q \leq \frac{1}{2}$. In this case we set $q \in (0, \min\{\bar{q}, \frac{1}{2}\})$. Thus all conditions of Theorem 1 are satisfied and the Corollary is proved.

2 Construction of some interior solutions

2.1 $n = 2$

Now we construct some interior solutions for the important case $\mathbf{f} = L\mathbf{V}$, with matrix $L = \begin{pmatrix} -\mu & -l \\ l & -\mu \end{pmatrix}$, $\mu = const \geq 0$, $l = const$. This right-hand side describes in the simplest way the Coriolis force and the Rayleigh friction in the meteorological model neglecting the vertical processes.

We consider only the smooth solutions to (1–3) with the density (and pressure) vanishing at $|\mathbf{x}| \to \infty$ rather quickly to guarantee the convergency of all integrals involved (whereas the velocity components may even grow). For the solution the total mass $\int\limits_{\mathbb{R}^2} \rho d\mathbf{x}$ is conserved, the total energy $\mathcal{E}(t) = \int\limits_{\mathbb{R}^2} \left(\frac{\rho|\mathbf{V}|^2)}{2} + \frac{p}{\gamma-1}\right) d\mathbf{x} = \mathcal{E}_k(t) + \mathcal{E}_p(t)$ is nonincreasing function. Let us involve integral functionals $G(t) = \frac{1}{2}\int\limits_{\mathbb{R}^2} \rho|\mathbf{r}|^2\, d\mathbf{x}$, $F_i(t) = \int\limits_{\mathbb{R}^2} (\mathbf{V}, \mathbf{X}_i)\rho\, d\mathbf{x}$, $i = 1, 2$, where $\mathbf{X}_1 = \mathbf{r} = (x, y)$, $\mathbf{X}_2 = \mathbf{r}_\perp = (y, -x)$. Note that $G(t) > 0$.

For the *classical* solutions to $(1-3)$, the following relations hold [6]:

$$G'(t) = F_1(t), \qquad F_1'(t) = 2(\gamma-1)\mathcal{E}_p(t) + 2\mathcal{E}_k(t) - lF_2(t) - \mu F_1(t),$$

$$F_2'(t) = lF_1(t) - \mu F_2(t), \qquad \mathcal{E}'(t) = -2\mu\mathcal{E}_k(t). \tag{2.1}$$

Generally speaking, the functions $\mathcal{E}_p(t)$ and $\mathcal{E}_k(t)$ cannot be expressed through $G(t), F_1(t), F_2(t)$. But if we choose the velocity field with the linear profile

$\mathbf{V} = A(t)\mathbf{r}$, we obtain the *closed* system of ODE for finding the coefficients of the matrix $A(t)$. In general case it is rather complicated (see [8]) and can be solved only numerically.

The simplest result can be obtained if

$$\mathbf{V} = \alpha(t)\mathbf{r} + \beta(t)\mathbf{r}_\perp = (\alpha(t)E + \beta(t)\begin{pmatrix} 0 & 1 \\ -1 & 0 \end{pmatrix})\mathbf{r}.$$

It is easy to see that in this case $F_1(t) = 2\alpha(t)G(t)$, $F_2(t) = 2\beta(t)G(t)$, $\mathcal{E}_k(t) = (\alpha^2(t) + \beta^2(t))G(t)$, $\mathcal{E}_p(t) = KG^{1-\gamma}(t), K = \mathcal{E}_p(0)G^{\gamma-1}(0)$. In that way, all function involved in the system (2.1) are expressed through $G(t), \alpha(t), \beta(t)$. For the convenience denote $G_1(t) = 1/G(t)$ and obtain the system

$$\begin{aligned} G_1'(t) &= -2\alpha(t)G_1(t), \\ \beta'(t) &= \alpha(t)(l - 2\beta(t)) - \mu\beta(t), \\ \alpha'(t) &= -\alpha^2(t) + \beta^2(t) - l\beta(t) - \mu\alpha(t) + (\gamma - 1)KG_1^\gamma(t). \end{aligned} \tag{2.2}$$

For $\gamma > 1$, the functions $\alpha(t)$ and $\beta(t)$ are bounded, it follows from the expression for the total energy, consequently, the density and pressure are bounded for all solutions of the class we want to construct.

If $\mu = 0$, the system (2.2) can be integrated. In the cases $\mu > 0$ and $\mu = l = 0$, there is a unique stable equilibrium in the origin. We have the following asymptotics of the solution components as $t \to \infty$:

$$\text{if } \mu = l = 0, \text{ then } \alpha(t) \sim \frac{1}{t}, \ G_1(t) \sim \alpha^2(t), \ \beta(t) = const \cdot G_1(t);$$

$$\text{if } \mu > 0, \ l = 0, \text{ then } \quad \alpha(t) \sim \frac{1}{2\gamma t}, \ \beta(t) \sim \frac{const}{t^{1/\gamma}e^{\mu t}}, \ G_1(t) \sim \frac{const}{t^{1/\gamma}};$$

$$\text{if } \mu > 0, l \neq 0, \quad \text{then } \quad \alpha(t) \sim \frac{1}{2\gamma t}, \ \beta(t) \sim \frac{\mu}{2l\gamma t}, \ G_1(t) \sim \frac{const}{t^{1/\gamma}}.$$

Thus, as follows from Corollary 1, in the cases $\mu = l = 0$ and $\mu > 0$ we have constructed the velocity field for the interior solution. For $\mu = l = 0$, we have $\alpha(t) \sim t^{-1}, t \to \infty$, $\Delta = 1 > \frac{1}{2}$. For $\mu > 0$, we get $\alpha(t) \sim \frac{1}{2\gamma}t^{-1}$, $t \to \infty, \Delta = \frac{1}{2\gamma} > 0$.

Knowing $\alpha(t)$ and $\beta(t)$, in the standard way we can find components of density and entropy from the equations (1), (3), linear with respect to ρ and S, respectively. From (2) and (2.2) we obtain that for the smooth solutions the condition $\nabla p = -(\gamma - 1)G_1^{1-\gamma}(0)E_p(0)G_1^\gamma(t)\rho\mathbf{r}$ holds. It follows that the functions ρ_0 and $p_0 = e^{S_0}\rho_0^\gamma$ must be axially symmetric and compatible, i.e.

$$\nabla p_0 = -(\gamma - 1)G_1(0)E_p(0)\rho_0\mathbf{r}. \tag{2.3}$$

Note also that for constructing the interior solution we must choose the initial data ρ_0 and S_0 to satisfy the condition a) of Theorem 1. So, $\rho(t, |r|) = \exp(-2\int_0^t \alpha(\tau)d\tau)\rho_0(|r|\exp(-\int_0^t \alpha(\tau)d\tau))$, $S(t, |r|) = S_0(|r|\exp(-\int_0^t \alpha(\tau)d\tau))$.

Remark 3. One can show that for the velocity field with the linear profile of general form $A(t)\mathbf{r}$ in the cases $\mu = l = 0$ there is the asymptotics $A(t) \sim \frac{1}{t}E$, $t \to \infty$ (see [8] for the details), that is accordingly to Corollary 1 the corresponding solution with linear profile of velocity is interior. The analogues result takes place in the case $\mu > 0$, $l = 0$, as well.

2.2 $n = 3$

Now we consider the right-hand side of (2), corresponding to the Coriolis force and the Rayleigh friction in the real physical space. Namely,

$$\mathbf{f} = -\mu\mathbf{V} + \delta[\mathbf{V} \times \omega] = (-\mu E + \delta \begin{pmatrix} 0 & \omega_3 & -\omega_2 \\ -\omega_3 & 0 & \omega_1 \\ \omega_2 & -\omega_1 & 0 \end{pmatrix})\mathbf{V}, \qquad (2.4)$$

where $\delta = 0$ or 1, ω is a constant vector $(\omega_1, \omega_2, \omega_3)$, $\mu \geq 0$ is a constant, $[\mathbf{a} \times \mathbf{b}]$ is the cross product of the vectors $\mathbf{a}$ and $\mathbf{b}$.

Besides of $G(t) = \frac{1}{2}\int\limits_{\mathbb{R}^3} \rho|\mathbf{r}|^2\, d\mathbf{x}$ and $F_1(t) = \int\limits_{\mathbb{R}^3} (\mathbf{V}, \mathbf{r})\rho\, d\mathbf{x}$, the generalization of the functionals, involved in 2.1, to the 3D case, we consider $\tilde{F}_2(t) = \int\limits_{\mathbb{R}^3} (\mathbf{V}, [\omega \times \mathbf{r}])\rho\, d\mathbf{x}$, $F_3(t) = \int\limits_{\mathbb{R}^3} ([\mathbf{V} \times \omega], [\omega \times \mathbf{r}])\rho\, d\mathbf{x}$, $H(t) = \frac{1}{2}\int\limits_{\mathbb{R}^3} ([\omega \times \mathbf{r}])^2 \rho\, d\mathbf{x}$. For the smooth solutions to the system (1–3) with the right-hand side (2.4) we have

$$G'(t) = F_1(t),\ F_1'(t) = 2\mathcal{E}_k(t) + 3(\gamma - 1)\mathcal{E}_p(t) + \delta\tilde{F}_2(t) - \mu F_1(t),$$

$$\tilde{F}_2'(t) = \delta F_3(t) - \mu\tilde{F}_2(t),\ H'(t) = 0.$$

We seek the solution with a special linear velocity profile

$$\mathbf{V} = \alpha(t)\mathbf{r} + \beta(t)[\mathbf{r} \times \omega] = (\alpha(t)E + \beta(t) \begin{pmatrix} 0 & \omega_3 & -\omega_2 \\ -\omega_3 & 0 & \omega_1 \\ \omega_2 & -\omega_1 & 0 \end{pmatrix})\mathbf{r}.$$

In this case $F_1(t) = 2\alpha(t)G(t), \tilde{F}_2(t) = -2\beta(t)H(0), F_3(t) = -2\alpha H(0), \mathcal{E}_k(t) = \alpha^2(t)G(t) + \beta^2(t)H(0)$, $\mathcal{E}_p(t) = const \cdot G^{-\frac{3(\gamma-1)}{2}}(t)$. Changing $G^{-1}(t) = G_1(t)$, we obtain the closed system of ODE

$$\alpha'(t) = -\alpha^2(t) - \mu\alpha(t) + \beta^2(t)G_1(t)H(0) - \delta\beta(t)G_1(t)H(0) + \frac{3(\gamma-1)}{2}\tilde{K}G_1^{\frac{3\gamma-1}{2}}(t),$$

$$\beta'(t) = \delta\alpha(t) - \mu\beta(t), \quad G_1'(t) = -2\alpha(t)G_1(t),$$

with $\tilde{K} = \mathcal{E}_p(0)G^{\frac{3(\gamma-1)}{2}}(0)$. Since the total energy is nonincreasing, we can draw similarly to the 2D case that $|\alpha(t)| < \infty, |\beta(t)| < \infty$.

In the situation where $\mu = \delta = 0$, where $\mu > 0, \delta = 0$ and where $\mu > 0, \delta = 1$, there is a stable equilibrium in the origin $\alpha(t) = \beta(t) = G_1(t) = 0$.

If $\mu = 0, \delta = 0$, we have

$$\alpha(t) \sim t^{-1}, \beta(t) = \beta(0) = const, G_1(t) \sim const \cdot t^{-2}, t \to \infty.$$

If $\mu > 0, \delta = 0$, then

$$\alpha(t) \sim \frac{1}{3\gamma - 1} t^{-1}, \beta(t) = \beta(0) \exp(-\mu t), G_1(t) \sim const \cdot t^{-\frac{2}{3\gamma-1}}, t \to \infty.$$

In the case $\mu > 0, \delta = 1$,

$$\alpha(t) \sim \frac{1}{3\gamma - 1} t^{-1}, \beta(t) \sim \frac{1}{\mu(3\gamma - 1)} t^{-1}, G_1(t) \sim const \cdot t^{-\frac{2}{3\gamma-1}}, t \to \infty.$$

Thus, according to Corollary 1, we construct the velocity field for the interior solution. What about other component of solution, they can be found by the same formulas, as in the 2D case, and must satisfy the compatibility condition (2.3).

Acknowledgments. The work is partially supported by RFBR Grant 00-02-16337.

References

1. Rozanova, O.S. (1998): Generation of singularities of compactly supported solutions of the Euler equations on a rotated plane. Differ. Equat., **34**(8), 1118-1123
2. Xin, Z. (1998): Blowup of smooth solutions to the compressible Navier-Stokes equation with compact density. Comm.Pure Appl.Math., **51**, 229–240
3. Serre, D. (1997): Solutions classiques globales des équations d'Euler pour un fluide parfait compressible. Annales de l'Institut Fourier, **47**, 139–153
4. Grassin, M. (1998): Global smooth solutions to Euler equations for a perfect gas. Indiana University Math.J.,**47**(4), 1397–1432
5. Makino, T., Ukai, S., Kawashima, S.(1986): Sur la solutions à support compact de l'équations d'Euler compressible. Japan J. Appl. Math., **3**, 249–257
6. Rozanova, O.S. (1998): Energy balance in a model of the dynamics of a two-dimensional baroclinic atmosphere. Izvestiya, Atmospheric and Oceanic Physics, **34**(2), 186-196
7. Kato, T.(1975): The Cauchy problem for quasilinear symmetric hyperbolic systems. Arch.Ration.Math.Anal.,**58**, 181–205
8. Rozanova, O.S. (2002): On classes of globally smooth solutions to the Euler equations in several dimensions. LANL e-print math.AP/0203230.

An Adaptive Order Godunov Type Central Scheme

Eitan Tadmor[1] and Jared Tanner[2]

[1] Department of Mathematics, Institute for Physical Science & Technology and Center for Scientific Computation And Mathematical Modeling (CSCAMM), University of Maryland College Park, MD 20742-3289 *tadmor@cscamm.umd.edu*

[2] Department of Mathematics University of California Davis, CA 95616 *jtanner@math.ucdavis.edu*

Summary. Traditionally, high order Godunov-type central schemes employ local polynomial reconstructions. These reconstructions avoid transfer of information across discontinuities through nonlinear limiters which act as local edge detectors. Here we introduce an adaptive method which employ global edge detection to ensure that information is extracted in the direction of smoothness while maintaining computational stability. Additionally, the global edge detection substantially reduces the computational cost. The reconstruction incorporates the largest symmetric stencil possible without crossing discontinuities. Consequently, the spatial order of accuracy is proportional to the number of cells to the nearest discontinuity, reaching exponential order at the interior of regions of smoothness.

1 Godunov Type Schemes

We seek an approximate solution of the hyperbolic conservation law

$$u_t + f(u)_x = 0 \qquad x \in [0, 2\pi], \;\; t \geq 0. \tag{1}$$

Of particular interest are solutions with contact discontinuities, shocks, and rarefaction waves. To better handle these interesting phenomena (1) is transferred to the exact integral formulation by integrating over a control volume in space time, $[x - \Delta x/2, x + \Delta x/2] \times [t, t + \Delta t]$

$$\begin{aligned}\bar{u}(x, t + \Delta t) = \bar{u}(x, t) &- \frac{1}{\Delta x}\int_{\tau}^{\tau+\Delta t} f(u(x + \Delta x/2, \tau))d\tau \\ &+ \frac{1}{\Delta x}\int_{\tau}^{\tau+\Delta t} f(u(x - \Delta x/2, \tau))d\tau\end{aligned} \tag{2}$$

were $\bar{u}(x,t)$ is the spatial average of $u(x,t)$ over $[x - \Delta x/2, x + \Delta x/2]$

$$\bar{u}(x, t) := \frac{1}{\Delta x}\int_{x-\Delta x/2}^{x+\Delta x/2} u(y, t)dy. \tag{3}$$

Godunov type schemes are constructed under the integral framework, (2), which yields a relationship between the evolution of cell averages and the flux evaluations along the spatial cell boundaries.

Unlike finite difference methods which work strictly with point values Godunov methods begin with a collection of cell averages,

$$\{\bar{u}(x_j, n)\}_j \qquad x_j := (j - 1/2)\Delta x, \qquad j = 1, 2, \ldots 2\pi/\Delta x, \tag{4}$$

and form a global reconstruction, $w(x, t^n)$, to approximate $u(x, n)$,

$$w(x, t^n) := \sum_j R_j(x, t^n)\chi_{[x_j - \Delta x/2, x_j + \Delta x/2]}. \tag{5}$$

This global reconstruction is then evolved in time by relationship (2).

Two frameworks that have been used for this evolution step are the upwind and central schemes. The upwind schemes maintain the original cells for each time evolution, as a result the flux integrals in (2) are evaluated at a point of discontinuity in the global reconstruction. Consequently (approximate)Riemann solvers are needed to determine the flux integral. Although this is not a severe limitation in one dimension it complicates multidimensional calculations.

Central schemes use a staggered cell average. Given the cell averages $\{\bar{u}_j^n\}_j$ the global reconstruction is formed, (8), and from it the staggered cell averages are computed

$$\bar{w}_{j+1/2}^n := \int_{x_j}^{x_j + \Delta x/2} R_j(x, t^n)dx + \int_{x_{j+\Delta x/2}}^{x_{j+1}} R_{j+1}(x, t^n)dx. \tag{6}$$

The evolution of these staggered cell averages

$$\begin{aligned} \bar{w}_{j+1/2}^{n+1} = \bar{w}_{j+1/2}^n - \frac{1}{\Delta x}\int_\tau^{\tau+\Delta t} & f(w_{j+1}^n(x(\tau)))d\tau \\ + \frac{1}{\Delta x}\int_\tau^{\tau+\Delta t} & f(w_j^n(x(\tau)))d\tau \end{aligned} \tag{7}$$

involves flux integrals computed at the smooth midpoints of the reconstruction, $w(x, t^n)$. This allows for the flux integrals to be calculated from quadrature formulas and consequently yields a method free of Riemann solvers. We now focus our attention on central schemes which emphasize the global reconstruction.

Extensive work has been done in determining the desired properties of such reconstructions, see [Le79, HEOC85, NT90, Sh97, LT98] and the references therein. In the original work by Godunov [Go59] a piecewise constant reconstruction was used, $w(x, t) := \sum_j \bar{u}(x_j, t)\chi_{[x_j - \Delta x/2, x_j + \Delta x/2]}$, yielding a first order upwind method. Since Godunov's original paper both upwind and central schemes have been extended to second order [Le79, NT90] and higher. As the evolution equation is exact the overall accuracy of a scheme is dictated by the accuracy of the global reconstruction. Traditionally local polynomial reconstructions have been used,

$$w(x,t^n) := \sum_j P_j(x,t^n)\chi_j(x), \quad \chi_j := \chi_{[x_j-\Delta x/2, x_j+\Delta x/2]}, \tag{8}$$

where $P_j(x,t^n)$ is a fixed order polynomial. We now focus on the reconstruction of piecewise smooth solutions common in hyperbolic problems.

It is well known that approximations of piecewise smooth functions suffer from oscillation and poor accuracy when the approximation crosses a discontinuity. For Godunov schemes the introduction of these oscillations causes the method to be inaccurate and often times unstable. In order to avoid the introduction of oscillations methods were introduced with properties such as Total Variation Diminishing(TVD) and Number of Extrema Diminishing(NED). To satisfy these properties with more than first order accuracy the global reconstruction necessarily became nonlinear. A prototype example is provided by the minmod *limiter* for piecewise linear reconstruction

$$\begin{aligned} P_j^{mm}(x,t^n) &= \bar{u}_j^n + (x-x_j)w'(x_j,t^n) \\ w'(x_j,t^n) &:= \begin{cases} 0 & (\bar{u}_{j+1}^n - \bar{u}_j^n)(\bar{u}_j^n - \bar{u}_{j-1}^n) \le 0 \\ (\bar{u}_{j+1}^n - \bar{u}_j^n)/\Delta x & |\bar{u}_{j+1}^n - \bar{u}_j^n| \le |\bar{u}_j^n - \bar{u}_{j-1}^n| \\ (\bar{u}_j^n - \bar{u}_{j-1}^n)/\Delta x & |\bar{u}_j^n - \bar{u}_{j-1}^n| \le |\bar{u}_{j+1}^n - \bar{u}_j^n| \end{cases} \end{aligned} \tag{9}$$

This reconstruction is *limited* in the sense that which cell averages are used is determined by the local smoothness. To avoid oscillations the cell averages used are taken in the direction of greater smoothness.

Godunov's method [Go59] was extended by van Leer[Le79] who replaced the piecewise constant reconstruction by (9) yielding a second order upwind method. As with the piecewise constant reconstruction the minmod reconstruction satisfies the three properties

1. Conservation — the reconstruction retains the same average

$$\frac{1}{\Delta x}\int_{x_j-\Delta x/2}^{x_j+\Delta x/2} P_j^{mm}(x,t^n) \equiv \bar{u}_j^n \tag{10}$$

2. Non-Oscillatory — furnished for example, by the TVD property $\|\sum P_j(\cdot,t^n)\chi_j(x)\|_{BV} \le \|\sum \bar{u}_j\chi_j(\cdot)\|_{BV}$.
3. Accuracy — a finite polynomial accuracy

$$|P_j(x,t^n) - u(x,t^n)| \le O(h^r), \;\; r \text{ fixed } x \in \left[x_j - \frac{\Delta x}{2}, x_j + \frac{\Delta x}{2}\right]. \tag{11}$$

The second order accuracy of (9) can be increased by using higher order polynomial reconstructions where care is taken not to introduce oscillations. Direct extensions can be found in [Sh97, LT98]... As with (9), limiters are applied so that the stencil changes in the direction of smoothness and spurious oscillations are avoided (or at least, significantly reduced).

The nonlinear limiters used in these polynomial reconstructions can be seen as *local edge detectors*, as they encourage the reconstruction to use information which does not cross a discontinuity. In the context of the Weakly Non-Oscillatory (WENO) reconstructions, [LPR99], the limiters are referred to as "smoothness indicators". As each cell recomputes the smoothness indicators, the action of these local edge detectors is redundant, consult e.g., [LPR99, Pu02]. Indeed, though solutions of (1) are sought within the class of BV functions. In actual simulations, however, these solutions are realized as piecewise smooth solutions, where finitely many jumps separate between regions of smoothness, [Ta02]. It is therefore makes sense to adopt a global point of view: to use a global sweep to detect *all* edges and to reconstruct highly accurate polynomials in between those edges.

In the next section we use the adaptive mollifier of [TT02] yielding a symmetric reconstruction of order proportional to the number of cells to the nearest discontinuity. Unlike the traditional local finite-order polynomial reconstructions, the present stencils are not restricted to a fixed size; instead, the usage of the whole region of smoothness enables us to achieve higher – in fact exponential accuracy.

2 An adaptive order central scheme

Below we construct a central scheme with explicit global edge detection and an adaptive order symmetric reconstruction. The global edge detector is more efficient than the polynomial limiters as it locates all discontinuities at once. Once these locations are known we form a global reconstruction with accuracy proportional to the number of cells to the nearest discontinuity. We conclude by investigating this schemes treatment of contact discontinuities, shocks and rarefaction waves in scalar advection equations.

2.1 Global Edge Detection

Accurate reconstructions of piecewise smooth data require the discontinuity locations. In [GeTa99, GeTa00b] Gelb and Tadmor analyzed and constructed a family of *global* edge detectors which use a function's (pseudo)spectral information. For our Godunov type scheme we used the exponential concentration kernel introduced in [GeTa00b]

$$K_N^{\sigma}(t) := -\sum_{k=1}^{N} \sigma(\frac{k}{N}) \sin kt \qquad \sigma(\eta) := Const.\eta e^{1/\eta(\eta-1)}. \tag{12}$$

Under proper normalization convolving the concentration kernel with the spectral projection, $K_N^{\sigma} \star S_N f(x)$, converges to the jump function $[f](x) := f(x^+) - f(x^-)$. Automated use of edge detection for the reconstruction method requires a simple indication whether a particular cell contain a discontinuity. This can be achieved by a new procedure for separation of scales advocated in [GeTa02]. Here, one implement a minmod limiter, minmod$(K_N^\sigma \star S_N f(x_j), f(x_{j+1}) - f(x_{j-1})$ followed by a simple thresholding of order $\Delta x \max|f(x)|$.

2.2 Adaptive Order Reconstruction

Spectral projections of smooth functions posses exponential convergence; however the presence of a single jump discontinuity reduces the convergence to at most first order. In [GoTa85] Gottlieb and Tadmor developed a two parameter, (p, θ) mollifier, composed of a smooth localization function ρ_c and the Dirichlet kernel of order p

$$\psi_{p,\theta}(x) := \frac{1}{\theta}\rho_c(\frac{x}{\theta})D_p(\frac{x}{\theta}), \quad \rho_c := e^{\left(\frac{cx^2}{x^2-\pi^2}\right)}, \ c > 0. \tag{13}$$

This mollifier recovers the exponential convergence for (pseudo)spectral projections of *piecewise* smooth data. In [TT02] by further analyzing (13) we determined the optimal parameters to be *adaptive*, depending on the distance to the nearest discontinuity, namely

$$\begin{aligned} \theta = \theta(x) &:= \frac{d(x)}{\pi}, \quad d(x) = dist(x, sing\ suppf) \\ p = p(x) &\sim \kappa \cdot \theta(x) N, \quad 0 < \kappa = \beta \log \beta < 1. \end{aligned} \tag{14}$$

Of particular interest for our Godunov type scheme is the manipulation of the equidistant sampling of piecewise smooth. The corresponding result from [TT02] is the error bound for the adaptive mollifier where the two parameters are given by (14) with $\kappa = 1$,

$$\left|\frac{\pi}{N}\sum_{\nu=0}^{2N-1} \psi_{p,\theta}(x-y_\nu)f(y_\nu) - f(x)\right| \leq Const_c \cdot (d(x)N)^2 \left(\frac{\beta}{e}\right)^{2\sqrt{\beta\eta_c\theta(x)N}}. \tag{15}$$

When normalized to have unit mass the adaptive mollifier possesses first order accuracy in the $O(1/N)$ neighborhood of discontinuities and exponential accuracy away. It should also be noted that this mollifier can be used to compute derivatives from equidistant stencils. The derivative of the mollifier applied to the data converges to the derivative of the data with exponential accuracy similar to (15).

Using the adaptive mollifier as a tool for the reconstruction step in our Godunov type scheme results in an adaptive order proportional to the number of cells to the nearest discontinuity.

2.3 The numerical scheme

Godunov schemes that are second order and higher involve local edge detection through the use of nonlinear limiters and smoothness indicators. This implicit edge detection is used to form polynomial reconstructions of finite order. In section 2.1 we discussed the global edge detection which utilize the distance function, $d(x)$ dictated in the adaptive mollifier (13). Using these tools we form a semi-global central scheme with accuracy proportional to the number of cells to the nearest discontinuity.

The adaptive order reconstruction begins by forming the primitivc of $w(x,t^n)$ from the given cell averages, (4),

$$W(x_j,t^n) := \sum_{k=0}^{j} \bar{w}_j^n \equiv \int_0^{x_j+\Delta x/2} w(x,t^n)dx \tag{16}$$

This converts the cell averages into pointvalues of the primitive. The global reconstruction is then formed by adaptive differentiation of the primitive

$$w(x,t^n) := -\frac{\pi}{N}\sum_{j=0}^{2N-1} W(y_j,t^n)\frac{d}{dx}\psi_{Nd(x)-1}(x-y_j) \qquad d(x) \geq m\Delta x. \tag{17}$$

To limit oscillations in the immediate vicinity of the discontinuities, $d(x) < m\Delta x$, a fixed number, say m neighboring cells are treated with a local, user defined polynomial reconstructions such as those in [NT90, LT98].

The staggered cell averages are computed by "integrating" the reconstruction (17) which corresponds to a divided difference of the primitive midpoints

$$W(x_j-\Delta x/2,t^n) \sim \sum_{j=0}^{2N-1} W(y_j,t^n)\psi_{Nd(x)}\left(x_j-\frac{\Delta x}{2}-y_j\right) \qquad d(x) \geq m\Delta x. \tag{18}$$

For the fixed number of cells near discontinuities, $d(x) < m\Delta x$, the primitive midpoints can be calculated by adding the exact integration of the polynomial reconstruction to the nearest primitive value to the left. The staggered cell averages are then computed by

$$\bar{w}_{j+1/2}^n := \frac{1}{\Delta x}(W(x_j+\Delta x/2,t^n) - (W(x_j-\Delta x/2,t^n)). \tag{19}$$

We note that $w(\cdot,t^n)$ is not an exact interpolant of $W(\cdot,t^n)$. Consequently, the conservation property, (10), is satisfied modulo the exponentially small and hence negligible error (15).

3 Numerical Experiments

To examine the accuracy of our adaptive order method we compute approximate solutions to two prototypical problems, linear advection and Burgers' equation. In these problems we focus our attention on the local accuracy and the evolution of the largest errors. This is of paramount importance to determine if the low order regions pollute the overall accuracy.

At each step Godunov type schemes project the approximate solution into cell averages. The approximate cell averages can be decomposed into the true average and the local error, $\bar{w}_j^n = \bar{u}_j^n + e_j^n$. The evolution of the approximate solution (7) can be viewed as evolving both the exact solution and the local error. As the evaluation step propagates the exact solution along its characteristics part of the local error must similarly follow. The most that can be hoped for in our adaptive order method is that the error does not spread from low order regions near discontinuities to the smooth regions.

In the following examples we examine how our adaptive central scheme treats the three primary phenomena of scalar hyperbolic conservation laws; contact discontinuities, shocks, and rarefaction waves. In all cases $\Delta x := \pi/N$ and $\Delta t \sim \Delta x$. Near discontinuities, $d(x) \leq 2\Delta x$, the global reconstruction, (17), reduces to the piecewise linear reconstruction (9) and consequently to the NT scheme [NT90].

We begin our examination of discontinuities by computing an approximate solution to the linear advection equation, $u_t + u_x = 0$, with periodic boundary conditions. The initial condition $u(x,0) = f(x)$,

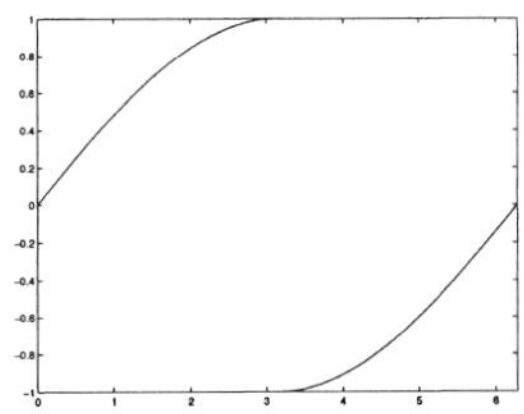

$$f(x) = \begin{cases} \sin(x/2) & x \in [0,\pi) \\ -\sin(x/2) & x \in [\pi, 2\pi) \end{cases} \tag{20}$$

has a discontinuity at $x = \pi$ which is advocated to $x = \pi + 2$. In figure 1(b) we see that the initial error committed at $x = \pi$ is advected with the discontinuity leaving high order accuracy in the smooth region, including the path of the discontinuity. Observe the local $l_1[\frac{7}{8}\pi, \frac{9}{8}\pi]$ error in table 1. Slight blurring about the contact discontinuity can be seen in figure 1(a). This can be reduced by increasing the local polynomial reconstruction from second order [NT90] to third [LT98].

To examine a moving shock we turn to the non-linear periodic Burgers' Equation,

$$u_t + \left(\frac{u^2}{2}\right)_x = 0, \tag{21}$$

with periodic initial conditions $u(x,0) = 1 + f(x)$, (20). As in the case of the contact discontinuity, the initial discontinuity at $x = \pi$ is advected to $x = \pi + 2$ without contaminating the regions it moves through, see figure 2(b) and table 2. In fact, the self focusing helps to narrow the low order region surrounding the discontinuity.

The final phenomena we examine is rarefaction waves. We again examine the periodic Burgers' equation but with initial conditions $u(x,0) = -f(x)$, (20), evolved to the final time $T = 1$. In this case the initial discontinuity at $x = \pi$, opens to a linear entropy solution in $[\pi - 1, \pi + 1]$, see figure 3(a). The characteristics used to fill this domain originate from the initially low order region, $x = \pi$, and consequently the fan's region also suffers from low order accuracy, figure 3(b). As for the linear advection equation the low order accuracy does not flow against the characteristics and hence leaves high resolution in the rest of the domain.

Table 1. Error measurements for figure 1(b), $\Delta x = \pi/N$, $\Delta t = \Delta x/2$. The flux evaluations were evaluated with Simpson's quadrature were the necessary midpoints were computed using the traditional RK4.

N	$l_1[0, 2\pi]$	Rate (2^{-x})	Ave $\log_{10}(error)$	$l_1[\frac{7}{8}\pi, \frac{9}{8}\pi]$	Rate (2^{-x})
40	0.351	0.97	-3.86	8.55×10^{-6}	6.86
80	0.179	0.64	-5.59	7.38×10^{-8}	5.63
160	0.115		-6.78	1.49×10^{-9}	

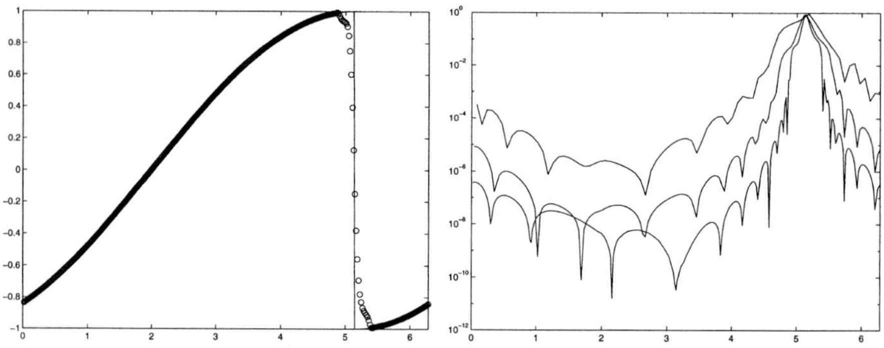

(a) (b)

Fig. 1. Approximate solution(circle) and exact solution(line) of the periodic linear advection equation at $T = 2$ with $N = 160$ (a) and the $\log_{10}$(error) for $N = 40, 80, 160$, (b).

Table 2. Error measurements for figure 2(b), $\Delta x = \pi/N$, $\Delta t = \Delta x/4$. Flux evaluations as per table 1.

N	$l_1[0, 2\pi]$	Rate (2^{-x})	Ave $\log_{10}(error)$	$l_1[\frac{7}{8}\pi, \frac{9}{8}\pi]$	Rate (2^{-x})
40	0.097	0.96	-4.39	2.09×10^{-6}	6.76
80	0.050	1.25	-6.14	1.93×10^{-8}	6.84
160	0.021		-7.97	1.69×10^{-10}	

To summarize, accuracy at a given point is determined by the accuracy used to compute points along its analytical domain of dependence. Very high order is achieved in the trails of contact discontinuities and shocks. the accuracy deteriorates, however, with low order boundary calculations and inside rarefaction fans and future work is required to study the latter.

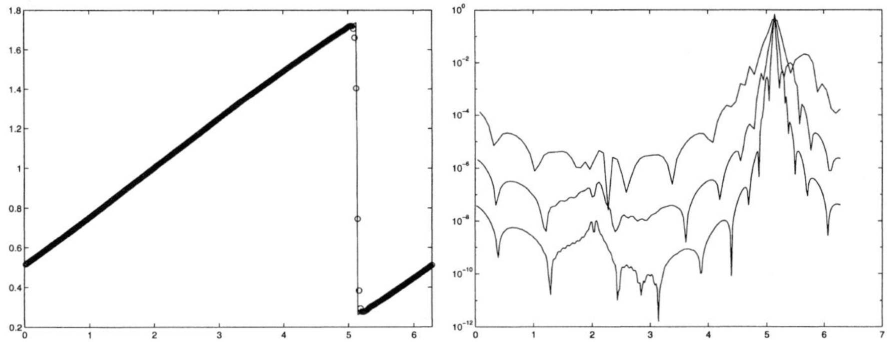

(a) (b)

Fig. 2. Approximate solution(circle) and exact solution(line) for Burgers' equation at $T = 2$, $N = 160$ (a) and the $\log_{10}$(error) for $N = 40, 80, 160$, (b).

Table 3. Error measurements for figure 3(b), $\Delta x = \pi/N$, $\Delta t = \Delta x/2$. Flux evaluations as per table 1.

N	$l_1[0, 2\pi]$	Rate (2^{-x})	Ave $\log_{10}(error)$	$l_1[\frac{7}{8}\pi, \frac{9}{8}\pi]$	Rate (2^{-x})
80	0.055	1.33	-4.13	9.83×10^{-3}	2.02
160	0.022	1.07	-5.37	2.43×10^{-3}	0.86
320	0.010		-6.55	1.34×10^{-3}	

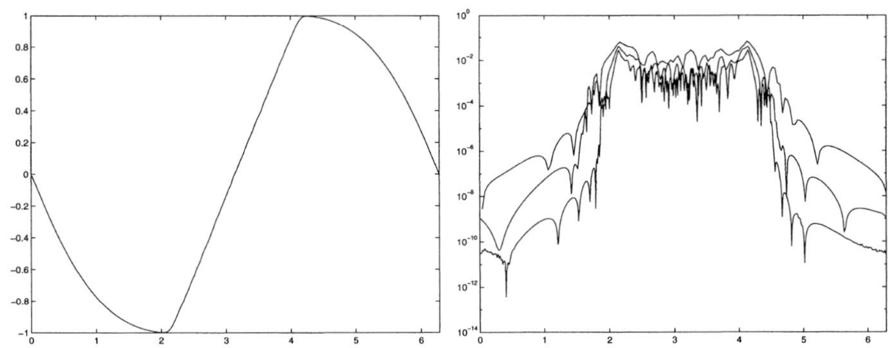

(a) (b)

Fig. 3. Approximate solution for Burgers' equation at $T = 2$ with $N = 320$ (a) and $\log_{10}$(error) for $N = 80, 160, 320$, (b).

References

[Ge00] A. Gelb, *A Hybrid Approach to Spectral Reconstruction of Piecewise Smooth Functions*, Journal of Scientific Computing, October 2000.

[GeTa99] A. Gelb and E. Tadmor, *Detection of Edges in Spectral Data*, Applied Computational Harmonic Analysis 7, (1999) 101-135.

[GeTa00b] A. Gelb and E. Tadmor, *Detection of Edges in Spectral Data II. Nonlinear Enhancement*,SIAM Journal of Numerical Analysis, 38 (2000), 1389-1408.

[GeTa02] A Gelb and E. Tadmor, *An Adaptive Edge Detection Method for Piecewise Smooth Functions*, in preparation.

[Go59] S.K. Godunov, *A finite difference method for the numerical computation of discontinuous solutions of the equations of fluid dynamics.*, Mat. Sb, 47 (1959), 271-290.

[GoTa85] D. Gottlieb and E. Tadmor, *Recovering pointwise values of discontinuous data within spectral accuracy*, in "Progress and Supercomputing in Computational Fluid Dynamics", Proceedings of 1984 U.S.-Israel Workshop, Progress in Scientific Computing, Vol. 6 (E. M. Murman and S. S. Abarbanel, eds.). Birkhauser, Boston, 1985, 357-375.

[HEOC85] A. Harten, B. Engquist, S. Osher and S.R. Chakravarthy, *Uniformly high order accurate essentially non-oscillatory schemes. III*, Jour. Comput. Phys. 71, 1982, 231-303.

[Le79] B. van Leer, *Towards the ultimate conservative difference scheme, V. A second order sequel to Godunov's method.* JCP 32 (1979), 101-136.

[LPR99] D. Levy, G. Puppo, and G. Russo, *Central WENO schemes for hyperbolic systems of conservation laws.* Math. Mod. and Numerical Analysis (1999), 547-571.

[LT98] X.D. Liu and E. Tadmor, *Third order nonoscillatory central scheme for hyperbolic conservation laws.* Numer. Math. 79 (1998), 397-425.

[NT90] H. Nessyahu and E. Tadmor, *Non-oscillatory central differencing for hyperbolic conservation laws.* JCP 87 (1990), 408-448.

[Pu02] G. Puppo, *Adaptive application of characteristic projection for central schemes*, preprint.

[Sh97] C.-W. Shu, *Essentially non oscillatory and weighted essentially non oscillatory schemes for hyperbolic conservation laws*, in "Advanced Numerical Approximation of Nonlinear Hyperbolic Equations" (A. Quarteroni, ed), Lecture Notes in Mathematics #1697, Cetraro, Italy, 1997.

[Ta94] E. Tadmor, *Spectral Methods for Hyperbolic Problems*, from "Lecture Notes Delivered at Ecole Des Ondes", January 24-28, 1994. Available at http://www.math.umd.edu/~tadmor/pub/spectral-approximations/Tadmor.INRIA-94.pdf

[Ta02] E. Tadmor, *High Resolution Methods for Time Dependent Problems with Piecewise Smooth Solutions*, "International Congress of Mathematicians" Proceedings of ICM02 Beijing 2002 (Li Tatsien, ed.), Vol.III: Invited Lectures, Higher Education Press, 2002, pp. 747-757.

[TT02] E. Tadmor and J. Tanner, *Adaptive Mollifiers - High Resolution Recovery of Piecewise Smooth Data from its Spectral Information*, J. Foundations of Comp. Math. 2 (2002), 155-189.

Numerical Study of Dynamic Phase Transitions in 2-D with a Relaxed Scheme

Shaoqiang Tang

Dept. of Mech. & Engng. Sci., Peking University, Beijing 100871, China
maotang@pku.edu.cn

1 Introduction

In two space dimensions, a van der Waals fluid is governed by

$$\begin{cases} \rho_t + (\rho u)_x + (\rho v)_y = 0, \\ (\rho u)_t + (\rho u^2 + p(\rho,T))_x + (\rho uv)_y = 0, \\ (\rho v)_t + (\rho uv)_x + (\rho v^2 + p(\rho,T))_y = 0, \end{cases} \tag{1}$$

with equation of state

$$p(\rho,T) = \frac{8\rho T}{3-\rho} - 3\rho^2. \tag{2}$$

Here ρ is the density, p the pressure, u and v the velocities in x and y directions respectively.

At a subcritical temperature $T < 1$, the pressure is monotone increasing for density ρ in α-phase $(0,\alpha)$ or β-phase $(\beta,3)$, making the system hyperbolic. If $T = 0.9$, as we shall use in this study, it is found that $\alpha = 0.654236$ and $\beta = 1.3916$. On the other hand, the pressure is monotone decreasing if the density lies in (α,β), where the system becomes elliptic. An elliptic evolutionary problem is ill-posed, due to instability of Hadamard's type [5]. Therefore, a physical solution should stay within the stable α-phase and β-phase.

A special category of physical solutions and elementary waves are *subsonic phase boundary* solutions. Such a phase boundary connects an α-phase state with a β-phase state, and propagates slower than the sound-speeds at both sides. They are essential in solving Riemann problems, yet do not satisfy traditional entropy conditions for hyperbolic equations.

In one space dimension, there have been continuous efforts in resolving this difficulty. One may either employ certain generalized entropy condition [8], or include more dissipation mechanisms besides viscosity, such as capillarity [10], or mass-viscosity [6]. It was proposed that dynamic phase transitions are not completely described by the equation of state, but also further conditions supplied by the materials, i.e. kinetic relations and nucleation criteria [1].

In a previous study, we have demonstrated that low-order dissipation is capable of balancing with the instability in van der Waals fluid, and constructed a category of discrete kinetic models. These models supply a big variety of kinetic relations, and nucleation criteria that may be relevant to different materials [9]. For a special case, namely the Suliciu model, we established nonlinear stability of stationary phase boundaries via Laplace transform method recently [11].

In two dimensional case, the understanding is much more limited. To our knowledge, there are only a few substantial results, including a linear stability theory [3], and a numerical study on Hsieh-Wang's mass-viscosity model [4]. In the meantime, most physically interesting phenomena appear in more than one space dimension. It is highly demanded to make theoretical and numerical explorations on multi-dimensional phase transition.

In the following, we shall first describe our relaxation model and scheme for solving (1), then present some one-dimensional and two-dimensional numerical results. Further remarks are also made.

2 Relaxation model

For each equation in the form of

$$f_t + A(f)_x + B(f)_y = 0, \tag{3}$$

a relaxation model is constructed as follows [7].

$$\left\{\begin{aligned} g_t^+ + \lambda g_x^+ &= \frac{M^+(f)-g^+}{\epsilon}, \\ g_t^- - \lambda g_x^- &= \frac{M^-(f)-g^-}{\epsilon}, \\ h_t^+ + \omega h_x^+ &= \frac{N^+(f)-h^+}{\epsilon}, \\ h_t^- - \omega h_x^- &= \frac{N^-(f)-h^-}{\epsilon}, \end{aligned}\right. \quad \text{with} \quad \left\{\begin{aligned} M^+(f) &= \frac{f}{4}+\frac{A(f)}{2\lambda}, \\ M^-(f) &= \frac{f}{4}-\frac{A(f)}{2\lambda}, \\ N^+(f) &= \frac{f}{4}+\frac{B(f)}{2\omega}, \\ N^-(f) &= \frac{f}{4}-\frac{B(f)}{2\omega}. \end{aligned}\right. \tag{4}$$

We recover f from $f = g^+ + g^- + h^+ + h^-$. Each equation in (4) is one-dimensional, and we apply a splitting technique [2]. To see this, we consider the first equation of (4). Knowing $g^+(t,x,y) = g_{i,j}^{+,n}$ at time $t = t^n$ and position $(x,y) = (x_i, y_j)$, we first solve an ordinary differential equation

$$g_t^+ = \frac{M^+(f)-g^+}{\epsilon}. \tag{5}$$

Because f remains unchanged in this step, namely,

$$\begin{aligned} \frac{df}{dt} &= \frac{dg^+}{dt} + \frac{dg^-}{dt} + \frac{dh^+}{dt} + \frac{dh^-}{dt} \\ &= \frac{1}{\epsilon}(M^+ + M^- + N^+ + N^- - g^+ - g^- - h^+ - h^-) \\ &= 0, \end{aligned} \tag{6}$$

the exact solution at time $t = t^n + \Delta t$ is

$$g_{i,j}^{+,*} = g_{i,j}^{+,n}(1 - e^{\epsilon \Delta t}) + M^+(f_{i,j}^n)e^{\epsilon \Delta t}. \tag{7}$$

As a matter of fact, numerical experiments show better results by a *relaxed scheme*, with less smearing effect in resolving a discontinuity. This is done by putting to the limit $\epsilon \to 0$. Then we have $g_{i,j}^{+,*} = M^+(f_{i,j}^n)$.

After resolving the source term, we solve an advection step, i.e.

$$g_t^+ + \lambda g_x^+ = 0, \qquad g_{i,j}^+(t^n) = g_{i,j}^{+,*}, \tag{8}$$

by a standard second order MUSCL scheme with minmod limiter. More precisely, denoting $\zeta = \dfrac{\lambda \Delta t}{\Delta x}$, we have,

$$\begin{aligned} g_{i,j}^{+,n+1} &= \zeta\, g_{i-1,j}^{+,*} + (1-\zeta)\, g_{i,j}^{+,*} \\ &+ \frac{\zeta(1-\zeta)}{2}\left[\mathrm{minmod}(g_{i,j}^{+,*} - g_{i-1,j}^{+,*}\,,\ g_{i-1,j}^{+,*} - g_{i-2,j}^{+,*})\right. \\ &\left.-\mathrm{minmod}(g_{i+1,j}^{+,*} - g_{i,j}^{+,*}\,,\ g_{i,j}^{+,*} - g_{i-1,j}^{+,*})\right]. \end{aligned} \tag{9}$$

Treating other equations in (4) similarly, we obtain a relaxed scheme

$$\begin{aligned} f_{i,j}^{n+1} &= f_{i,j}^n + \zeta(M_{i-1,j}^+ + M_{i+1,j}^- - f_{i,j}^n/2) \\ &+ \frac{\zeta(1-\zeta)}{2}\left[\mathrm{minmod}(M_{i,j}^+ - M_{i-1,j}^+\,,\ M_{i-1,j}^+ - M_{i-2,j}^+)\right. \\ &-\mathrm{minmod}(M_{i+1,j}^+ - M_{i,j}^+\,,\ M_{i,j}^+ - M_{i-1,j}^+) \\ &+\mathrm{minmod}(M_{i+1,j}^- - M_{i,j}^-\,,\ M_{i,j}^- - M_{i-1,j}^-) \\ &\left.-\mathrm{minmod}(M_{i+2,j}^- - M_{i+1,j}^-\,,\ M_{i+1,j}^- - M_{i,j}^-)\right] \\ &+\text{y-direction terms}, \end{aligned} \tag{10}$$

where $M_{i,j}^{\pm} = M^{\pm}(f_{i,j}^n)$.

Now we perform limiting process $\Delta t \to 0$, and a semi-continuous form of the relaxed scheme reads

$$\begin{aligned} \frac{df_{i,j}}{dt} &= \frac{A_{i-1,j} - A_{i+1,j}}{2\Delta x} + \frac{\lambda}{2\Delta x}\left[\frac{f_{i-1,j} + f_{i+1,j}}{2} - f_{i,j}\right. \\ &+\mathrm{minmod}(M_{i,j}^+ - M_{i-1,j}^+\,,\ M_{i-1,j}^+ - M_{i-2,j}^+) \\ &-\mathrm{minmod}(M_{i+1,j}^+ - M_{i,j}^+\,,\ M_{i,j}^+ - M_{i-1,j}^+) \\ &+\mathrm{minmod}(M_{i+1,j}^- - M_{i,j}^-\,,\ M_{i,j}^- - M_{i-1,j}^-) \\ &\left.-\mathrm{minmod}(M_{i+2,j}^- - M_{i+1,j}^-\,,\ M_{i+1,j}^- - M_{i,j}^-)\right] \\ &+\text{y-direction terms}. \end{aligned} \tag{11}$$

This set of ordinary differential equations are integrated by a second-order Runge-Kutta method.

3 Numerical results in 1-D

To solve a 1-D van der Waals fluid problem, we may either directly use the relaxed scheme described in the previous section, or construct a 1-D relaxed scheme in the same manner. For the later choice, we solve a scalar conservation law

$$f_t + A(f)_x = 0, \tag{12}$$

by a relaxation model

$$\begin{cases} p_t^+ + \xi p_x^+ = \dfrac{P^+(f) - p^+}{\epsilon}, \\ p_t^- - \xi p_x^- = \dfrac{P^-(f) - p^-}{\epsilon}, \end{cases} \quad \text{with} \quad \begin{cases} P^+(f) = \dfrac{f}{2} + \dfrac{A(f)}{2\xi}, \\ P^-(f) = \dfrac{f}{2} - \dfrac{A(f)}{2\xi}. \end{cases} \tag{13}$$

It is straightforward to show that a travelling wave of this system satisfies the same equation as a viscous approximation

$$f_t + A(f)_x = \nu f_{xx}. \tag{14}$$

As the travelling wave profile determines the candidates of discontinuities in the limit $\epsilon \to 0$, it is conceivable that our scheme may be comparable to the following model with both mass and momentum viscosities.

$$\begin{cases} \rho_t + (\rho u)_x = \nu \rho_{xx}, \\ (\rho u)_t + (\rho u^2 + p(\rho, T))_x = \nu (\rho u)_{xx}. \end{cases} \tag{15}$$

We mention that this *conservative mass-viscosity model* differs from Hsieh-Wang's mass viscosity model, which is not in conservation form for the momentum equation. Nevertheless, a stationary discontinuity in either model, or Slemrod's viscosity-capillarity model, satisfies the Maxwell construction of equal area law, and therefore connects $(\rho_1, u) = (0.42574158, 0)$ with $(\rho_2, u) = (1.6572702, 0)$.

Taking the arguments as before, we have a relaxed scheme in semi-continuous form as

$$\begin{aligned} \frac{df_i}{dt} = \frac{A_{i-1} - A_{i+1}}{2\Delta x} + \frac{\xi}{2\Delta x}\Big[& f_{i-1} + f_{i+1} - 2f_i \\ & + \text{minmod}(P_i^+ - P_{i-1}^+ \,,\; P_{i-1}^+ - P_{i-2}^+) \\ & - \text{minmod}(P_{i+1}^+ - P_i^+ \,,\; P_i^+ - P_{i-1}^+) \\ & + \text{minmod}(P_{i+1}^- - P_i^- \,,\; P_i^- - P_{i-1}^-) \\ & - \text{minmod}(P_{i+2}^- - P_{i+1}^- \,,\; P_{i+1}^- - P_i^-)\Big]. \end{aligned} \tag{16}$$

This is exactly the same as (11) if we take $\xi = \lambda/2$, and accordingly $P^\pm = 2M^\pm$.

We now illustrate a shock reflection problem. Initially, an incidental shock (in α-phase) at $x = 0$ with

$$(\rho, u) = \begin{cases} (0.5, 8.080194), & \text{for} \quad x < 0, \\ (0.01, 0), & \text{for} \quad x > 0, \end{cases} \tag{17}$$

propagates to the right at speed $s_1 = 8.2451$. Assume there is a solid boundary at $L = 5$, where $u(t, L) = 0$. Numerical simulation with $\Delta x = 0.0025, \Delta t = 0.000125, \lambda = 20$ of the reflection is depicted in Figure 1. It is observed that a new phase with $(\rho, u) = (2.690092, 0)$ is generated after the shock hits the solid end. The new phase boundary propagates to the left at speed $s_2 = -1.8847$.

4 Patterns in 2-D

Now we present our numerical results for 2-D dynamic phase transitions. Periodic boundary conditions are used. We take $\Delta x = \Delta y = 0.05$ for the space domain $[-5, 5] \times [-5, 5]$, and time step size $\Delta t = 0.001$.

First, we take initial data

$$(\rho, u, v) = (0.6 + 1.5\mathrm{e}^{-(x^2+y^2)}, 0, 0), \tag{18}$$

and compute with relaxation speeds $\lambda = \omega = 1$. Some snapshots are shown in Figure 2. Subplot (b) clearly shows the gas-liquid phase separation, as well as interaction between the wave front and periodic boundaries. From subplot (c), we observe a sharp transition between two states that are close to ρ_1 and ρ_2, respectively.

In the second example, we start with initial profile

$$(\rho, u, v) = (1, 0.01\mathrm{e}^{-(x^2+y^2)}x, 0.01\mathrm{e}^{-(x^2+y^2)}y). \tag{19}$$

As the pattern moves faster than the previous case, we take relaxation speeds $\lambda = \omega = 5$. The evolution is illustrated by some snapshots of the density distribution in Figure 3. In this figure, white regions correspond to β-phase with density close to ρ_2; black ones correspond to α-phase with density close to ρ_1; and the gray ones in two top subplots correspond to $\rho = 1$. Noticing that $\rho = 1$ lies in the unstable phase, the non-equilibrium velocity field immediately triggers a separation of phase near the origin. Due to an anti-symmetry, a double spiral forms, and further develops until it touches the boundaries. Afterwards a coarsening process starts, and the double spiral rotates and disappears.

5 Conclusion

In this paper, we develop a semi-continuous relaxed scheme and apply it to dynamic phase transitions in 1-D and 2-D. Preliminary numerical experiments demonstrate that the nonlinear interaction between dissipation and instability may generate patterns for different initial profiles. This nonlinear interaction is complicated, demanding an intensive theoretical study. Different schemes and parameters lead to different diffusion effects, and possibly different evolution results. This may reflect the high sensitivity in physical situations of dynamic phase transitions.

Acknowledgements

The author would like to thank Professor R. Natalini and Professor J. Knowles for their stimulating discussions. This research is partially supported by Chinese Special Funds for Major State Basic Research Project 'Nonlinear Science', and NSFC under grant number 10002002.

References

1. Abeyaratne, R., Knowles, J.K. (1991): Kinetic relations and the propagation of phase boundaries in solids. Arch. Rational Mech. Anal., **114**, 119–154.
2. Aregba-Driollet, D., Natalini, R. (2000): Discrete kinetic schemes for multidimensional conservation laws. SIAM J. Num. Anal., **37**, 1973–2004.
3. Benzoni, S. (1998): Stability of multi-dimensional phase transitions in a van der Waals fluid. Nonlinear Analysis T.M.A., **31**, 243–263.
4. He, C. (1998): The Study Phase Transition in van der Waals Fluid. MPhil Thesis, Hong Kong Univ of Sci & Tech, Hong Kong.
5. Hsieh, D.Y., (1994): Elemental mechanisms of hydrodynamic instabilities. Acta Mechanica Sinica, **10**, 193–202.
6. Hsieh, D.Y., Wang, X.P. (1997): Phase transitions in van der Waals fluid. SIAM J. Appl. Math., **57**, 871–892.
7. Jin, S., Xin, Z.P. (1995): The relaxation schemes for systems of conservation laws in arbitrary space dimensions. Comm. Pure Appl. Math., **48**, 235–278.
8. Keyfitz, B.L., Shearer, M. (1990): Nonlinear Evolution Equations That Change Type. Springer, New York.
9. Natalini, N., Tang, S. (2000): Discrete kineic models for dynamic phase transitions. Comm. Appl. Nonlinear Anal., **7**, 1–32.
10. Slemrod, M. (1983): Dynamical phase transitions in a van der Waals fluid. Arch. Rational Mech. Anal., **81**, 301–315.
11. Tang, S., Zhao, H. (2002): Nonlinear stability of Suliciu model for dynamic phase transitions. Preprint (May be found in the website http://www.math.ntnu.no/conservation/2002/014.html).

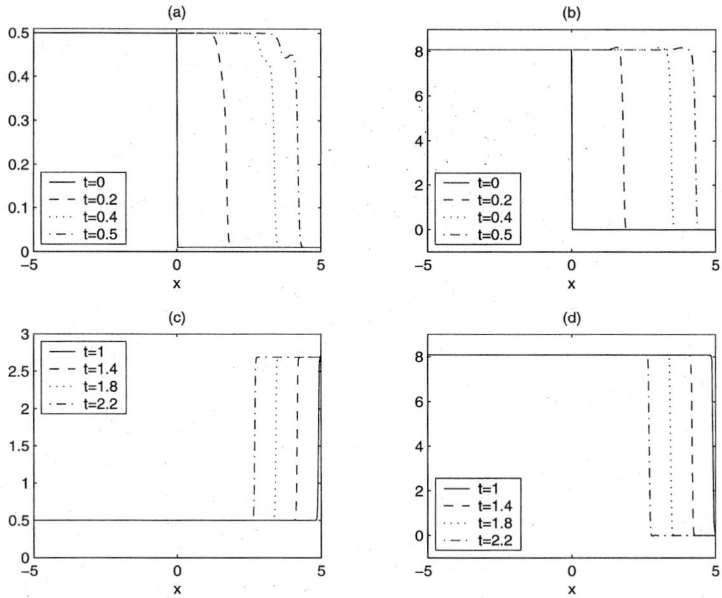

Fig. 1. Shock reflection in 1-D: (a) ρ before reflection; (b) u before reflection; (c) ρ after reflection; (d) u after reflection.

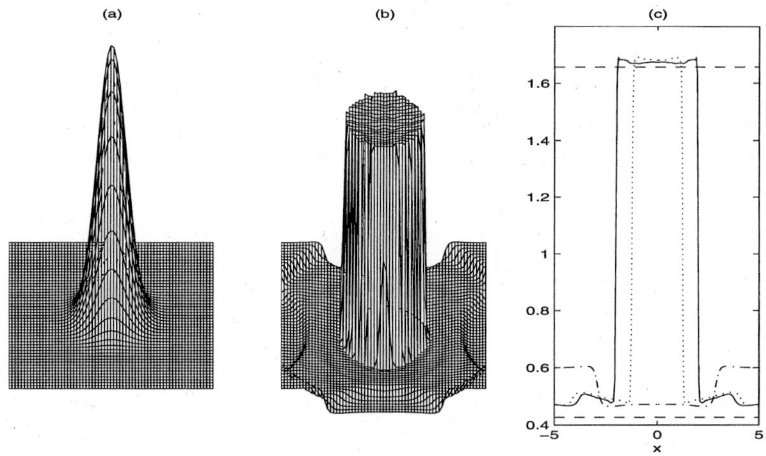

Fig. 2. Snapshots in Example 1 for 2-D dynamic phase transition: (a) top-front view of ρ at $t = 0$; (b) top-front view of ρ at $t = 20$; (c) $\rho(20, x, y)$ at $y = 0$ (solid), $y = 1.5$ (dotted) and $y = 5$ (dash-dotted), with level lines $\rho = \rho_1, \rho = \rho_2$ (dashed).

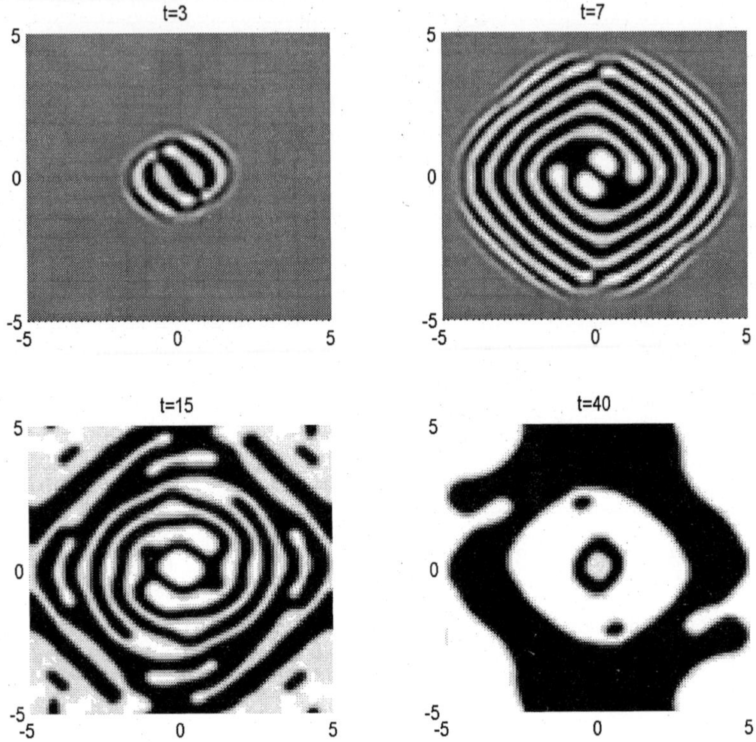

Fig. 3. Snapshots in Example 2 for 2-D dynamic phase transition: top view of ρ at various time.

Particle Method for Simulation of Free Surface Flows

S. Tiwari[1] and J. Kuhnert[2]

[1] Fraunhofer Institut Techno- und Wirtschaftmathematik
Gottlieb-Daimler-Strasse, Geb. 49
D-67663 Kaiserslautern, Germany
tiwari@itwm.fhg.de
[2] *kuhnert@itwm.fhg.de*

1 Introduction

In this paper we present a particle method for simulations of free surface flows. This is a fully Lagrangian and grid free method. A fluid domain is first replaced by a discrete number of points, which are referred to particles. Each particle carries all fluid informations, like density, velocity, temperature etc and moves with fluid velocity. Therefore, particles themselves can be considered as geometrical grids of the fluid domain. This method has some advantages over grid based techniques, for example, it can handle fluid domains which change naturally, whereas grid based techniques require additional computational effort.

Numerical simulations of free surface flows have many industrial applications like casting, tank filling and others. Many methods have been developed to simulate free surface flows[5, 6, 9]. A classical grid free Lagrangian method is the Smoothed Particle Hydrodynamics (SPH) method, which was originally introduced to solve problems in astrophysics [8, 3]. It has since been extended to simulate the compressible Euler equations in fluid dynamics and applied to a wide range of problems, see [11, 12]. The implementation of the boundary conditions is the main problem of the SPH method. Our approach differs from the method of SPH in approximations of derivatives and the treatment of boundary conditions. We approximate the derivatives of a function by the weighted least squares method. This is similar to the moving least squares method, see [1, 2, 7] for detail references. In [7] the boundary conditions are well treated by the moving least squares technique. Like in [7] we have replaced the boundary by particles and prescribe the boundary condition on the boundary particles.

These approaches are similar to the central difference rule in the finite difference discretization and have well known problems of stability. Some viscosity is required to stabilize the numerical scheme. In [15] a weighted least squares method is proposed to solve the full Navier-Stokes equations. The artificial viscosity is not required to stabilize the scheme but the natural Navier-Stokes viscous term is used. In this approach the solutions of the compressible Euler equations can be obtained from the Navier-Stokes equations by letting viscosity and heat conductivity tend to zero [15]. Furthermore, in [16] the method is extended to solve incompressible viscous Navier-Stokes equations and shown that the Navier-Stokes equations can be solved quite accurately.

In this paper and in [16] we have simulated the incompressible free surface flows as a limit of the compressible, viscous Navier-Stokes equations with the equation of state such that the flow is weakly incompressible. This type of equation of state was first used by Monaghan [11] to simulate incompressible free surface flows by SPH. Such an equation of state in the framework of SPH has been further used to simulate incompressible viscous flows in [12].

The paper is organized as follows. In section 2 we introduce the mathematical model, boundary conditions and the determination of the free surface particles. In section 3 the least squares method to approximate the spatial derivatives is described and the discrete system of equations are derived. Finally, in section 4 some numerical tests are presented.

2 Mathematical Model

2.1 Governing Equations

We consider the compressible Navier-Stokes equations in the Lagrangian form

$$\frac{D\rho}{Dt} = -\rho \nabla \cdot \mathbf{v} \tag{1}$$

$$\rho \frac{D\mathbf{v}}{Dt} = -\nabla p + \mu \nabla \cdot \sigma(\mathbf{v}) + \rho \mathbf{g}, \tag{2}$$

where ρ is the mass density, $\mathbf{v}$ is the velocity vector, $\mathbf{g}$ is the body force acceleration vector, μ is the dynamic viscosity, p the dynamic pressure, and σ is the strain tensor

$$\sigma_{ij} = \frac{\partial v_i}{\partial x_j} + \frac{\partial v_j}{\partial x_i} - \frac{2}{3} \delta_{ij} \nabla \cdot \mathbf{v}.$$

The equations (1 - 2) are closed by the state equation $p = p(\rho)$, which we will discuss in detail in the following subsection. Furthermore, the equations (1 - 2) must be supplemented with appropriate boundary and initial conditions.

2.2 Equation of State

The fluid is assumed to be weakly compressible. For weakly compressible flows, the following equation of state is used [12, **?**],

$$p - p_0 = c^2 (\rho - \rho_0), \tag{3}$$

where c is the speed of sound, p_0 is the reference pressure, is considered as to be zero here and ρ_0 is the reference density. In the incompressible limit a large value of c is chosen such that the Mach number is of order ≈ 0.1. This guarantees density variations being in the range of ≈ 1 percent [11]. A larger value of c requires a smaller time step Δt since the CFL-condition has to be satisfied. This is the main disadvantage of this approach.

2.3 Boundary Conditions

In our numerical simulations we have considered three types of boundary conditions: solid wall, inflow and free surface. At a solid wall, fluid may slip freely along the wall or it may stick to it. If the viscosity is too low, a free slip condition is used.

In an inflow boundary the velocities and density are prescribed.

In the absence of surface tension forces between two fluids, the surface stress boundary condition at the interface between two fluids is given by

$$[\mu\sigma \cdot \mathbf{n} - p\mathbf{n}] = 0, \tag{4}$$

where [.] denotes the jump on the free surface boundary between two fluids and $\mathbf{n}$ the unit normal vector on the interface.

Suppose that the viscosity of the fluid adjacent to the free surface is negligible and has the pressure p_0, then the normal and tangential component of (4) is given by

$$p - \mu n_i \sigma_{ij} n_j = p_0 \tag{5}$$

$$t_i \sigma_{ij} n_j = 0, \tag{6}$$

where $\mathbf{t}$ denotes the unit tangent vector on the interface.

Substituting the equation of state (3) into (5) we obtain

$$\rho = \rho_0 + \frac{1}{c^2} \mu n_i \sigma_{ij} n_j. \tag{7}$$

In the case of low Mach number limit, the sound speed c is large and for small $\mu\sigma$ the second term in (7) vanishes and we simply obtain $\rho = \rho_0$ on the free surface. This yields $p = p_0$, that is, the liquid pressure is equal to the gas pressure on the free surface.

The implementation of the boundary conditions (5) is straightforward and (6) is included in the approximation of the derivatives of velocity fields, which will be discussed in section 3.

2.4 Determination of the free surface particles

In this section we would like to give a brief description of the strategy how to indicate particles belonging to a free surface. We would like to remind the reader, that these particles are not known a priori, however it is important to have a very accurate selection of them, otherwise the whole numerical procedure and application of boundary conditions is likely to fail. For the determination of the free-surface-particles, we come up with a definition. We say that a particle with the position $\mathbf{x}_i$ belongs to some free surface, if we can place a sphere in the neighborhood of the particle such that

- $\mathbf{x}_i$ lies on the surface of the sphere (i.e. it is not the center)
- the radius of the sphere is $r_S = \alpha \cdot h$ where h is the smoothing length and α is a constant, preferably in the range $\alpha = [0.7...1.0]$
- no other particle lies inside of the sphere

This definition is rather theoretical, however it makes sense. If a particle is really at the free surface, then there will be indeed such a sphere, because one half-space is more or less empty for surface particles. An interior particle, however, should not find such a sphere, or, in other words, if it would find a sphere meeting the above conditions, then this would mean there is a big hole in the interior of the flow domain, and this is not acceptable from the point of view of computation accuracy. Consequently, this means that interior holes have to be stuffed with particles before their radius tends to reach the magnitude of r_S.

Obeying these rules, we have a unique description of particles at the free surface. More problematic is the implementation of the whole idea. To search for appropriate holes for one particle (for instance for the particle at position $\mathbf{x}_i$), it takes about $25 \cdot M^2$ floating point operations, where M is the number of relevant neighbor particles related to the position $\mathbf{x}_i$. However, M is usually in the range of $M = [20..50]$ for $2D$-applications and $M = [40..90]$ for $3D$-applications, depending on the particle configuration. Hence, the effort of searching surface particles is huge and can take 10 percent of the over-all-computation-time. The idea to reduce that effort is to

- consider only those particles as candidates for being at the free surface at time level t_n, if they are in the neighborhood of a free surface particle at time level t_{n-1} (this reduces the number of particles to be checked)
- do the search for the free surface particles not for each time step

Both methods mentioned above have shown excellent applicability.

3 Least squares discretization

3.1 Least squares approximation of the derivatives

Let $f(t, \mathbf{x})$ be a scalar function and $f_i(t)$ its values at $\mathbf{x}_i$ for $i = 1, 2, \ldots, N$ and time t. Consider the problem of approximating the spatial derivatives of a function $f(t, \mathbf{x})$ at particle position $\mathbf{x}_0$ in terms of the values of its neighboring points. In order to restrict the number of points we associate a weight function $w_i = w(\mathbf{x}_i - \mathbf{x}_0; h))$ with small compact support, where h determines the size of the support. In the classical smoothed particle hydrodynamics method, h is known as the smoothing length. The weight function can be quite arbitrary but in our computation, we consider a Gaussian weight function in the following form

$$w(\mathbf{x}_i - \mathbf{x}_0; h) = \begin{cases} exp(-\alpha \frac{\|\mathbf{x}_i - \mathbf{x}_0\|^2}{h^2}), & \text{if } \frac{\|\mathbf{x}_i - \mathbf{x}_0\|}{h} \leq 1 \\ 0, & \text{else,} \end{cases}$$

with α positive constant. Different weight functions have been tested and found that the Gaussian one with $\alpha = 6.25$ gives the best approximation of derivatives. The smoothing length defines a set of neighboring particles around $\mathbf{x}$. Let $P(\mathbf{x}_0, h) = \{\mathbf{x}_i : i = 1, 2, \ldots, n\}$ be the set of n neighboring points of $\mathbf{x}_0$ in a ball of radius h. For consistency reasons some obvious restrictions are required, for example, in $2D$ there should be at least 5 particles and they should neither be all on the same line nor on a circle.

Consider Taylor's expansion of $f(t, \mathbf{x_i})$ around $\mathbf{x}_0$

$$f(t, \mathbf{x}_i) = f(t, \mathbf{x}_0) + \nabla f \cdot (\mathbf{x}_i - \mathbf{x}_0) + \frac{1}{2}(\mathbf{x}_i - \mathbf{x}_0)^T D(\mathbf{x}_i - \mathbf{x}_0) + e_i \quad (8)$$

where e_i is the error in the Taylor's expansion at the point $\mathbf{x}_i$. We note that $f(t, \mathbf{x}_0) = f_0$ is the value of a function at the particle position $\mathbf{x}_0$ and is known. The coefficients $a = [\nabla f, D]^T$ are computed by minimizing

$$J = \sum_{i=1}^{n} w_i e_i^2 = (M\mathbf{a} - \mathbf{b})^T W (M\mathbf{a} - \mathbf{b}), \quad (9)$$

where M is the geometrical matrix and is given by

$$\begin{pmatrix} dx_1 & dy_1 & dz_1 & \frac{1}{2}dx_1^2 & dx_1 dy_1 & dx_1 dz_1 & \frac{1}{2}dy_1^2 & dy_1 dz_1 & \frac{1}{2}dz_1^2 \\ \vdots & \vdots & \vdots & \vdots & \vdots & \vdots & \vdots & \vdots & \vdots \\ dx_n & dy_n & dz_n & \frac{1}{2}dx_n^2 & dx_n dy_n & dx_n dz_n & \frac{1}{2}dy_n^2 & dy_n dz_n & \frac{1}{2}dz_n^2 \end{pmatrix},$$

$W = diag\,(w_1, \cdots, w_n)\,,\ \mathbf{a} = [f_x, f_y, f_z, f_{xx}, f_{xy}, f_{xz}, f_{yy}, f_{yz}, f_{zz}]^T$
and $\mathbf{b} = [f_1 - f_0, f_2 - f_0, \ldots, f_n - f_0]^T$.
In the matrix M we have used the symbols $dx_i = x_i - x_0$, $dy_i = y_i - y_0$ and $dz_i = z_i - z_0$, where $\mathbf{x}_i = (x_i, y_i, z_i)$ and $\mathbf{x}_0 = (x_0, y_0, z_0)$.

For $n > 9$, this system is over-determined for the nine unknowns f_k and f_{kl} for $k, l = 1, 2, 3$. The coefficients $\mathbf{a}$ is obtained by minimization of J and is given by $\frac{\partial J}{\partial \mathbf{a}} = 0$. This yields

$$\mathbf{a} = (M^T W M)^{-1} (M^T W)\mathbf{b}. \quad (10)$$

3.2 Approximation of derivatives of velocities on the free surface

On the free surface we have to apply the boundary condition (6) while computing the spatial derivatives. For the sake of simplicity we consider the two dimensional computation. The boundary condition (6) can be explicitly written as

$$\frac{2}{3}\frac{\partial u}{\partial x}n_x n_y + \frac{\partial u}{\partial y}(n_y^2 - n_x^2) + \frac{\partial v}{\partial x}(n_y^2 - n_x^2) + \frac{2}{3}\frac{\partial v}{\partial y}n_x n_y = 0, \tag{11}$$

where u, v denotes the x and y components of the velocity and n_x, n_y the x and y components the unit normal vector.

The inclusion of (11) in the approximation of derivatives of velocities on the free surface is not straightforward. The equation (11) contains the first derivatives of both components u and v. On the other hand, for example, the derivatives of u are obtained by the Taylor's series expansion, which contains only the derivatives of u but not of v. Therefore, we have to compute derivatives of u and v together. The method is as follows: Suppose that, we want to approximate the derivatives of u and v on the free surface particle (x_0, y_0) from its neighboring values. Let $(u_0, v_0) = (u, v)(x_0, y_0), (u, v)_i = (u, v)(x_i, y_i), dx_i = x_i - x_0, dy_i = y_i - y_0$ for $i = 1, \cdots n$. Consider the Taylor's series expansion of u and v around (x_0, y_0)

$$u_i = u_0 + \frac{\partial u}{\partial x}dx_i + \frac{\partial u}{\partial y}dy_i + \frac{1}{2}\frac{\partial^2 u}{\partial x^2}dx_i^2 + \frac{\partial^2 u}{\partial x \partial y}dx_i dy_i + \frac{1}{2}\frac{\partial^2 u}{\partial y^2}dy_i^2 + e_i \tag{12}$$

$$v_i = v_0 + \frac{\partial v}{\partial x}dx_i + \frac{\partial v}{\partial y}dy_i + \frac{1}{2}\frac{\partial^2 v}{\partial x^2}dx_i^2 + \frac{\partial^2 v}{\partial x \partial y}dx_i dy_i + \frac{1}{2}\frac{\partial^2 v}{\partial y^2}dy_i^2 + e_i. \tag{13}$$

Similarly, we can obtain the derivatives of u and v by minimizing the errors, where the geometrical matrix M takes the form

$$\begin{pmatrix} dx_1 & dy_1 & \frac{1}{2}dx_1^2 & dx_1dy_1 & \frac{1}{2}dy_1^2 & 0 & 0 & 0 & 0 & 0 \\ \vdots & \vdots & \vdots & \vdots & \vdots & \vdots & \vdots & \vdots & \vdots & \vdots \\ dx_n & dy_n & \frac{1}{2}dx_n^2 & dx_ndy_n & \frac{1}{2}dy_n^2 & 0 & 0 & 0 & 0 & 0 \\ 0 & 0 & 0 & 0 & 0 & dx_1 & dy_1 & \frac{1}{2}dx_1^2 & dx_1dy_1 & \frac{1}{2}dy_1^2 \\ \vdots & \vdots & \vdots & \vdots & \vdots & \vdots & \vdots & \vdots & \vdots & \vdots \\ 0 & 0 & 0 & 0 & 0 & dx_n & dy_n & \frac{1}{2}dx_n^2 & dx_ndy_n & \frac{1}{2}dy_n^2 \end{pmatrix}, \tag{14}$$

$\mathbf{a} = [u_x, u_y, u_xx, u_{xy}, u_{yy}, v_x, v_y, v_{xx}, v_{xy}, v_{yy}]^T$ and
$\mathbf{b} = [u_1 - u_0, u_2 - u_0, \ldots, u_n - u_0, v_1 - v_0, v_2 - v_0, \ldots, v_n - v_0]^T$.

We add the equation (11) to the above equations (12 - 13), where the matrix (14) is enhanced by the $(2n+1)$th row

$$\tfrac{2}{3}n_x n_y \;\; (n_y^2 - n_x^2) \;\; 0 \;\; 0 \;\; 0 \;\; (n_y^2 - n_x^2) \;\; \tfrac{2}{3}n_x n_y \;\; 0 \;\; 0 \;\; 0$$

and the vector $\mathbf{b} = [u_1 - u_0, u_2 - u_0, \ldots, u_n - u_0, v_1 - v_0, v_2 - v_0, \ldots, v_n - v_0, 0]^T$.

The above system has $2n+1$ equations with 10 unknowns. The minimization of the error gives the derivatives of both velocity components together. But one has to invert the 10×10 matrix instead of 5×5 matrix. We solve this larger system only for few free surface particles.

The minimization process is the same as above. We choose the weight to be equal 1 for the $(2n+1)$th equation.

3.3 Space and time discretization

By using the least square approximation of the spatial derivatives in the system (1-2) at the point $\mathbf{x}_i$ we get

$$\frac{D\rho_i(t)}{Dt} = -\rho_i(t)\, \Pi\nabla \cdot \mathbf{v}(t, \mathbf{x}_i) \tag{15}$$

$$\rho_i(t)\, \frac{D\mathbf{v}_i(t)}{Dt} = -\Pi\nabla p(t, \mathbf{x}_i) + \mu\, \Pi\nabla \cdot \sigma(\mathbf{v}(t, \mathbf{x}_i)) + \rho_i(t) g_i\,. \tag{16}$$

for all $i = 1, 2, \ldots, N$, where $\Pi\nabla p$ is denoted by the least squares approximation of ∇p.

In addition to (15-17) the equations that determine the particle positions should be included as

$$\frac{d\mathbf{x}_i}{dt} = \mathbf{v}_i, \text{ for } \qquad i = 1, \cdots, N. \tag{17}$$

This means that the particles move with their own fluid velocities.

Hence the system of ordinary differential equations (15-17) can be solved by a simple integration scheme. One can use the explicit Euler scheme, but it requires a very small time step. Thus, if possible, higher order methods should be used. Here we have used a two step time integration method [14].

4 Numerical Tests

4.1 The broken dam problem

This is a very popular and simple test case to validate many numerical schemes for the simulations of free surface flows. It consists of simple initial configurations and simple initial and boundary conditions. In [10] the experimental results are reported and several authors have reported their numerical results[6, 9, 11].

Consider a rectangular column of water with a width of $a = 0.06m$ and a height of $0.12m$. The left side and the bottom line in Fig. 1 consist of the solid wall. In the simulation, the upper and the right boundary of the water columns considered as the free surface boundary. Initially, 512 particles are distributed uniformly with initial spacing $\Delta x = 0.00375$. The gravity with

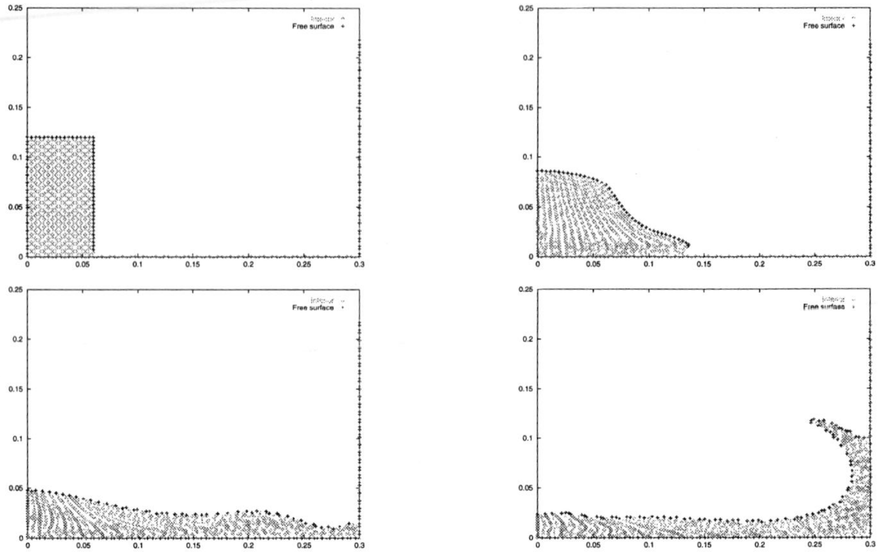

Fig. 1. The broken dam problem: particle positions at times $0s$, $0.1s$, $0.2s$ and $0.315s$

$g = 9.81m/s^2$act downwards. When the right wall (dam) is removed, the water column collapses under the influence of the gravity. The density and dynamic viscosity of the fluid are $\rho_0 = 1000kg/m^3$ and $\mu = 0.5kg/(ms)$, respectively. The equation of state is considered as $p = p_0 + 150(\rho - \rho_0)$ such that the density fluctuation is always less than 1 percent. In Fig. 1 we have plotted the free surface and interior particles. The free surface particles are the darker ones. The free slip boundary condition is used on the solid wall. Using the no slip boundary condition gives the same value of the surge front, but not the height of the water column.

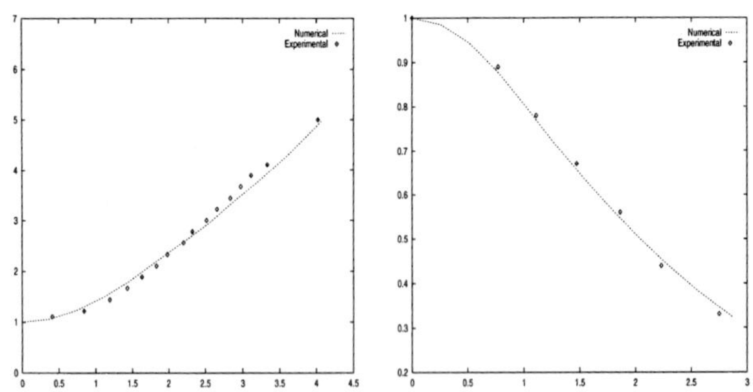

Fig. 2. Left: dimensionless front position x/a versus dimensionless time $t\sqrt{2g/a}$. Right: dimensionless column height $x/(2a)$ versus dimensionless time $t\sqrt{g/a}$

In Fig.2 the position of the leading fluid front versus time as well as the height of the water column versus time are compared with experimental results provided by [10]. The data are plotted before the particles hit the right wall. These figures show a good agreement between the numerical and experimental results.

4.2 The circular disc with core

We consider the filling of a circular disc with a circular core. Several authors have considered this test case since the experimental results are available [4, 9, 13]. The diameter of the disc and core are $140mm$ and $45mm$, respectively. The water entering through the gate of width $45mm$ with a velocity of $18m/s$. We consider the density and Raynold's number $\rho_0 = 1000kg/m^3$, $Re = 250$. We have used no slip boundary conditions on the core and disc and free slip on the tube. On the free surface, we apply similar boundary conditions as for the previous examples. In Fig. 3, we have ploted the experimental and numerical results at times $t = 8.82ms$, $11.67ms$ $16.17ms$. The figures show a close agreement between the experimental and numerical results. At time $t = 16.17ms$ the total number of particles are approximately 7500.

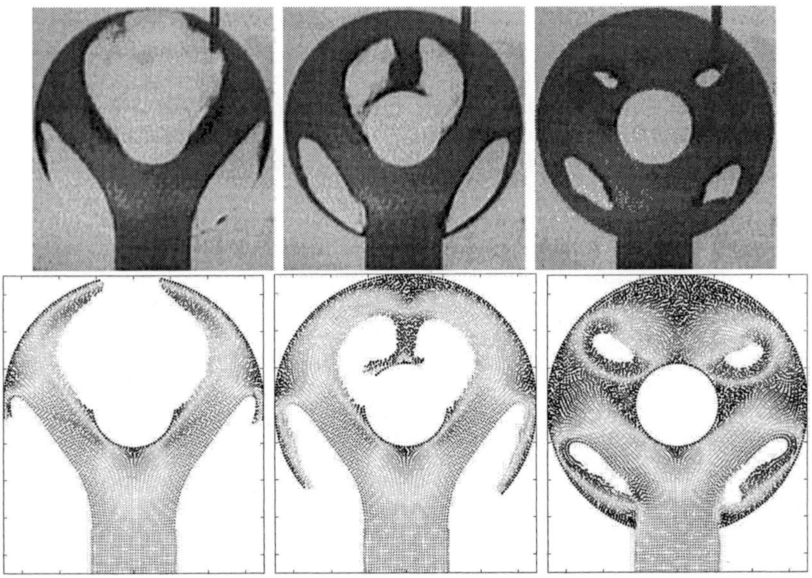

Fig. 3. Comparison of experimental and numerical solutions in a circular disc with core. First row: experimental, second row: Numerical results at times $8.82ms$,$11.76ms$, $16.17ms$ (left to right)

Acknowledgments This work is supported by DFG - Priority Research Program: Analysis and Numerics for Conservation Laws. The first author would like to thank Deutsche Forschungsgemeinschaft (DFG) for the financial support.

References

1. T. Belytschko, Y. Krongauz, M. Flemming, D. Organ, W. K. S. Liu, *Smoothing and accelerated computations in the element free Galerkin method*; J. Comp. Appl. Maths. **74**, 111-126, (1996)
2. G.A Dilts, *Moving least squares particle hydrodynamics I, consistency and stability*; Hydrodynamics methods group report, Los Alamos National Laboratory (1996)
3. R. A. Gingold and J. J. Monaghan, *Smoothed particle hydrodynamics: theory and application to non-spherical stars*; Mon. Not. Roy. Astron. Soc. **181**, 375-389, (1977)
4. J. Ha and P. W. Cleary *Comparison of SPH simulations of high pressure die casting with the experiments and VOF simulations of Schmid and Klein*, J. Cast Metals Res., **12**, 409 - 418 (2000).
5. F. H. Harlow and J. E. Welch, *Numerical study of large amplitude free surface motions*; Phys. Fluids **8**, 2182 (1965).
6. F. J. Kelecy and R. H. Pletcher, *The development of free surface capturing approach for multi dimensional free surface flows in closed containers*; J. Comput. Phys. **138**, 939 (1997)
7. J. Kuhnert, *General smoothed particle hydrodynamics*; Ph.D. thesis, Kaiserslautern University, Germany (1999)
8. L. B. Lucy, *A numerical approach to the testing of the fission hypothesis*; Astron. J., **82**, 1013, (1977)
9. V. Maronnier, M. Picasso and J. Rappaz, *Numerical simulation of free surface flows*; J. Comput. Phys. **155**, 439 (1999).
10. J. C. Martin and W. J. Moyce, *An experimental study of the collapse of liquid columns on a liquid horizontal plate*; Philos. Trans. Roy. Soc. London Ser. A 244, 312 (1952).
11. J.J.Monaghan, *Simulating free surface flows with SPH*; J. Comput. Phys., **110**, 399, (1994)
12. J.P.Morris, P.J.Fox and Y.Zhu, *Modeling Low Reynolds Number Incompressible Flows Using SPH*; J. Comput. Phys., **136**, 214-226, (1997)
13. M. Schmid and F. Klein, *Fluid flow in die cavities - experimental and numerical simulation*, NADCA 18. International Die Casting Congress and Exposition (Indianapolis, 1995) 93-99.
14. J. Stoer and R. Bulirsch, *Introduction to numerical analysis*; Spring-Verlag, New York (1980)
15. S. Tiwari, *A LSQ-SPH approach for solving compressible viscous flows*, International series of numerical mathematics Vol. 141, ed. by H. Freistuehler, G. Warnecke, Birkhaueser, 2000.
16. S. Tiwari and S. Manservisi *Modeling incompressible Navier-Stokes flows by LSQ-SPH*, preprint ITWM, Kaiserslautern, 2000.

Three Dimensional Numerical Modelling of Convective Instability by Supernova Explosion with Nested Grids Scheme on Multiprocessors Systems

Sergey D. Ustyugov and Alexander N. Andrianov

Keldysh Institute of Applied Mathematics, Russian Academy of Sciences, Miusskaya 4, 125047, Russia *ustyugs@spp.keldysh.ru* *and@a5.kiam.ru*

We carried out three dimensional hydrodynamical simulations to study development large scale convective instability into a protoneutron star by supernova expolsion II type. The initial state is rotating protoneutron star with excess entropy at the centre. On the last stage of collapse of the iron core of a star the process of nonequilibrium neutronization of matter provides high entropy and great mean energy of the neutrinos in the central part of the protoneutron star. The hydrodynamically unstable configuration of a protoneutron star is formed. The statement of the problem consists in studing the development of the nonuniformity scale of the entropy distribution. The central density and central temperature was taken to be $\rho = 10^{14}\mathrm{g/cm^3}$ and $T = 10^{11}K$. On the boundary of numerical region we keep initial equilibrium values of all variables. The weak solid-body rotation of protoneutron star was taken into account, with the ratio of the rotational kinetic energy T to the gravitational energy $|W|$ being $T/|W| = 0.01$. The coordinate system was chosen so that the plane of rotation of the star coinced with the xy plane. It was assumed that relativistic degenerate electrons and an ideal nonrelativistic Fermi gas of nucleons contribute to the equation of a state of substance inside the star. The excess entropy was produced close to the center of the star and had functional form $S = S_0 + (S' - S_0)exp\{-|\mathbf{r} - \mathbf{r}_0|/|b|\}$, $S' = 3S_0$ and $r_0 = 0$. The equations of hydrodynamics, in the adiabatic approximation $dS/dt = 0$, with gravitation was used to model the convective processes. In the adiabatic formulation, the entropy shows the evolution of the dredge-up of hot, lighter matter. The hydrodynamic equations was written in conservative form

$$\frac{\partial \mathbf{U}}{\partial t} + \frac{\partial \mathbf{F}}{\partial x} + \frac{\partial \mathbf{G}}{\partial y} + \frac{\partial \mathbf{H}}{\partial z} = \mathbf{S}$$

The numerical method which we used was an explicit Godunov-type conservative TVD difference scheme with second order by space and time

$$U^{n+1}_{i,j,k} = U^n_{i,j,k} - \Delta t L(U^n_{i,j,k}),$$

where $\Delta t = t^{n+1} - t^n$ and L is

$$L(U_{i,j,k}) = \frac{\tilde{F}_{i+1/2,j,k} - \tilde{F}_{i-1/2,j,k}}{\Delta x_i} + \frac{\tilde{G}_{i,j+1/2,k} - \tilde{G}_{i,j-1/2,k}}{\Delta y_j}$$
$$+ \frac{\tilde{H}_{i,j,k+1/2} - \tilde{H}_{i,j,k-1/2}}{\Delta z_k} + S_{i,j,k}$$

Fluxes along each direction, for example x, was defined by local-characteristic method [1] as follows

$$\tilde{F}_{i+1/2,j,k} = \frac{1}{2}\left[F_{i,j,k} + F_{i+1,j,k} + R_{i+1/2}W_{i+1/2}\right]$$

where $R_{i+1/2}$ is matrix right eigenvectors of Jacobian matrix $A(U) = \partial F/\partial U$.

The elements l of vectors $W_{i+1/2}$ is

$$(\phi^l_{i+1/2})^U = \frac{1}{2}\psi(a^l_{i+1/2})(g^l_{i+1} + g^l_i) - \psi(a^l_{i+1/2} + \gamma^l_{i+1/2})\alpha^l_{i+1/2}$$

where

$$\gamma^l_{i+1/2} = \frac{1}{2}\psi(a^l_{i+1/2})\begin{cases} (g^l_{i+1} - g^l_i)/\alpha^l_{i+1/2} & ,\alpha^l_{i+1/2} \neq 0 \\ 0 & ,\alpha^l_{i+1/2} = 0 \end{cases}$$

The amplitude of waves $\alpha^l_{i+1/2}$ is elements of vectors

$$\alpha_{i+1/2} = R^{-1}_{i+1/2}(U_{i+1,j,k} - U_{i,j,k}).$$

The function ψ is

$$\psi(z) = \begin{cases} |z| & ,|z| \geq \delta_1 \\ (z^2 + \delta_1^2)/2\delta_1 & ,|z| < \delta_1 \end{cases}$$

and define entropy correction by $|z|$, where δ_1-small positive parameter. The function-limiter g^l_i was defined as follows:

$$g^l_i = minmod(\alpha^l_{i-1/2}, \alpha^l_{i+1/2})$$
$$g^l_i = (\alpha^l_{i+1/2}\alpha^l_{i-1/2} + |\alpha^l_{i+1/2}\alpha^l_{i-1/2}|)/(\alpha^l_{i+1/2} + \alpha^l_{i-1/2})$$

The one step of time integration is defined by Runge-Kutta method [2] as

$$U^{(1)} = U^n + \Delta t L(U^n)$$

$$U^{(2)} = \frac{3}{4}U^n + \frac{1}{4}U^{(1)} + \frac{1}{4}\Delta t L(U^{(1)})$$

$$U^{n+1} = \frac{1}{3}U^n + \frac{2}{3}U^{(2)} + \frac{2}{3}\Delta t L(U^{(2)})$$

The time step is restricted to satisfy the Courant condition along each row in three directions, and is calculated with

$$\Delta t = CFL \times min\left(\frac{\Delta x}{|v_x| + c_s}, \frac{\Delta y}{|v_y| + c_s}, \frac{\Delta z}{|v_z| + c_s}\right)$$

and $CFL = 0.2$ typically. v_x,v_y,v_z are the velocity of matter in each direction and c_s is the speed of sound.

For higher resolution in the numerical modelling we have used three level of nested rectangular grids with 128 cells in each directions. Solution results for above described problem were obtained due to the NORMA system application [3] on the system with distributed memory multiprocesseres (two Alpha 21264/667 MHz in node, memory 1 Gb in node, SAN Myrinet to communication, 384 nodes). Norma program was compiled in Fortran with MPI library. The speedup of computation in 122.8 times on 128 processors were obtained during solution process.

The evolution of entropy distributions are shown in figure 1 and 2. Two cross sections of the star are chosen as images. In the first cross section (fig.1), the horizontal axis corresponds to the x axis, while the vertical axes corresponds to the z axis of our coordinate system. The second cross section(fig.2) correspond to a view from above onto the plane of rotation, corresponding to the plane of the equator. The initial configuration is chosen at time t=0.075 ms, and the final configuration is shown for time t=6.31 ms. The complete evolution lasted 20 ms.

The two bubbles initially appear (at about 3 ms), elongated in opposite directions along the axis of rotation. Four additional bubbles appear soon afterwards (at about 5 ms) and lie in the plane of rotation of the protoneutron star. The bubbles located along the axis of rotation broken off the hot core first and float to the surface. This occurs because the density varies more rapidly along the axis of rotation. The bubbles located in the plane of rotation broken off the hot nucleus core later and also float to the surface. Our calculations show that, following the first bubbles, additional bubbles with significantly less volume are formed, which also begin to float to the surface.

The first stage of the development of the peturbations is the growth of asymmetry along the axis of rotation of the star, which lasts for 3 ms. The asymmetry is still very weakly expressed in the plane of rotation, and the evolution proceeds peacefully. The asymmetry of the distribution of the substance is negligible during this stage, and bubbles of hot matter have not yet been formed. In the next millisecond bubbles moving along the axis of rotation have already been formed and bubbles in the plane of rotation begin to be distinguished at this stage. On the last stage to the moment time 6 ms take place final formation of bubbles in the plane of the equator and motion their to the surface neutrinosphere. The bubbles along the axis of rotation have already escaped from the calculated region.

Figure 3 shows the three-dimensional profile of the entropy distribution initially (t=0.075 ms) and when the bubbles have already formed (t=3.65 ms). Figure 4 shows velocity field distribution, convective cells are clearly visible that show both the rise of hot matter to the surface and the fall of cold matter to the center of the protoneutron star.

The calculations showed that at a distance corresponding to the maximum gravity, a large scale bubble with a radius $\sim 2.0 \cdot 10^6$cm, density $\sim 4.0 \cdot 10^{11}$ g/cm^3 and temperature $\sim 5.0 \cdot 10^9$K emerges. The mass of one dredged-up element is $0.02 M_\odot$. The calculations indicated that the neutrino-rich matter reached density of $\sim 10^{11}$g/cm^3 in $\sim 10^{-2}$s, implying that the neutrinos contained in the matter freely escape from it in the regime of volume radiation. The shock from the initial bounce when the collapse in the stellar core stops will then be supported by the neutrino emission, resulting in the ejection of an envelope. Les us now estimate the intensity of the netrino emission. At the initial time, the mass of matter with entropy excess ($S_{max} = 1.4$) is $0.07 M_\odot$. In 4 ms, $0.02 M_\odot$ of this matter enters the region of densities lower than $\sim 10^{11}$g/cm^3, and the matter becomes transparent for the neutrinos in it. The number of these neutrinos is comparable to the number of electrons with a mean energy 30Mev. The intensity of the neutrino emission is $\sim 2.5 \cdot 10^{52}$ erg/s. The fraction of the energy of this neutrino emission absorbed by the matter per gram in shock is $\sim 2.0 \cdot 10^{24}$ erg/g s. This value is comparable to the neutrino losses from the shock front, i.e., the mechanism of explosive convection could support the diverging shock and lead to the ejection of a supernova envelope[4]. This work was supported by the Russian Foundation for Basic Research (project code 01-01-00411).

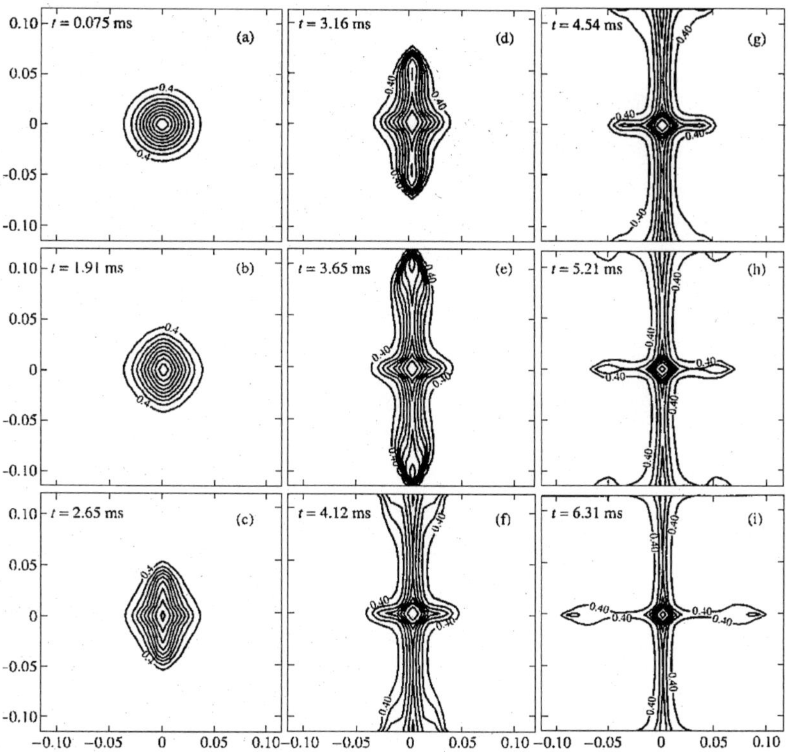

Fig.1 Levels of constant entropy in the Ozx plane. Both axes have scales in fractions of the characteristic length; 0.1 corresponds to 20 km. The entropy is expressed in dimensionless units, normalized to the Boltzmann constant and the nucleon density. The initial background entropy is 1.6327 k_B/nucl (corresponding to the normalized value 0.37).

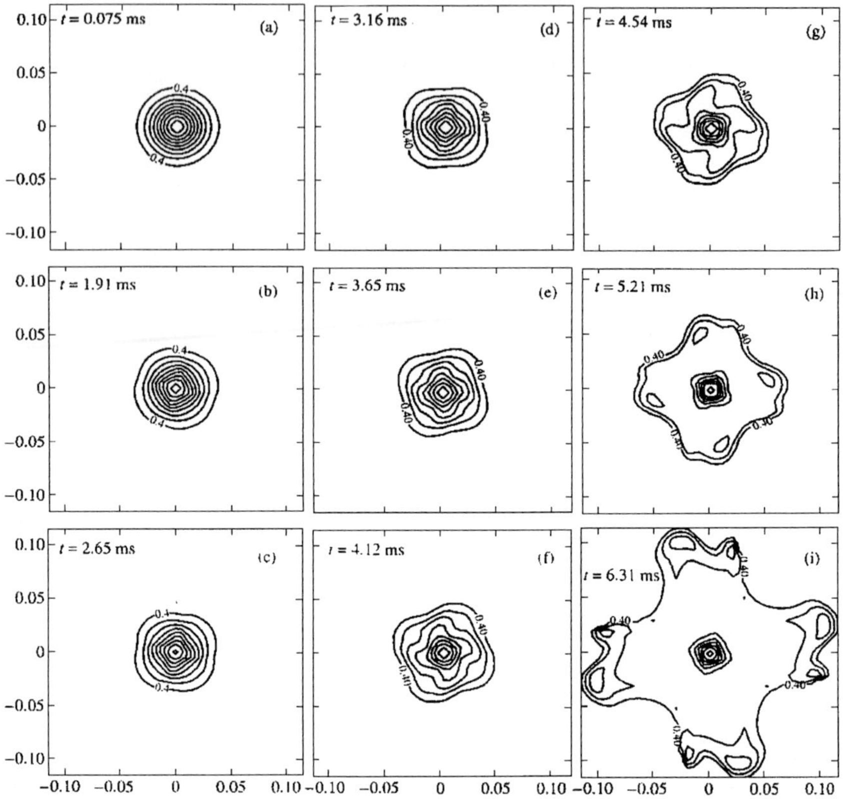

Fig.2 Same as for Fig. 1, but in the equatorial plane.

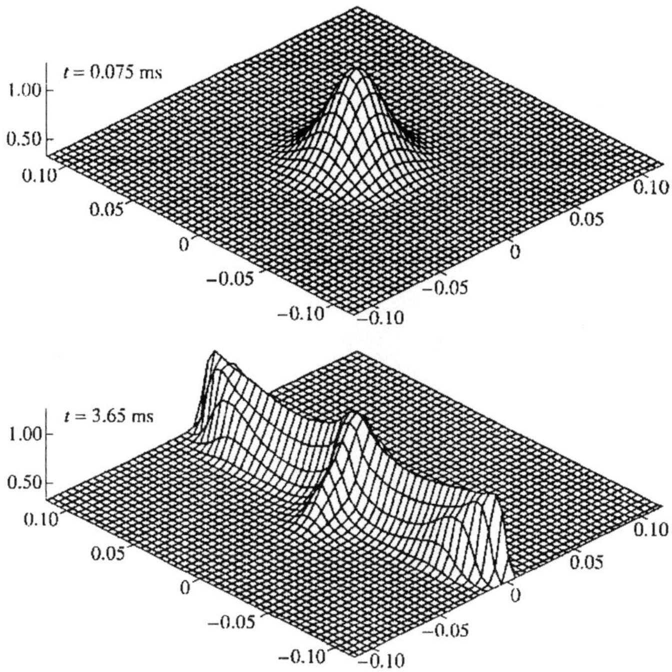

Fig.3 Three-dimensional entropy profile for two characteristic times. The axis scales are the same as in Fig.1.

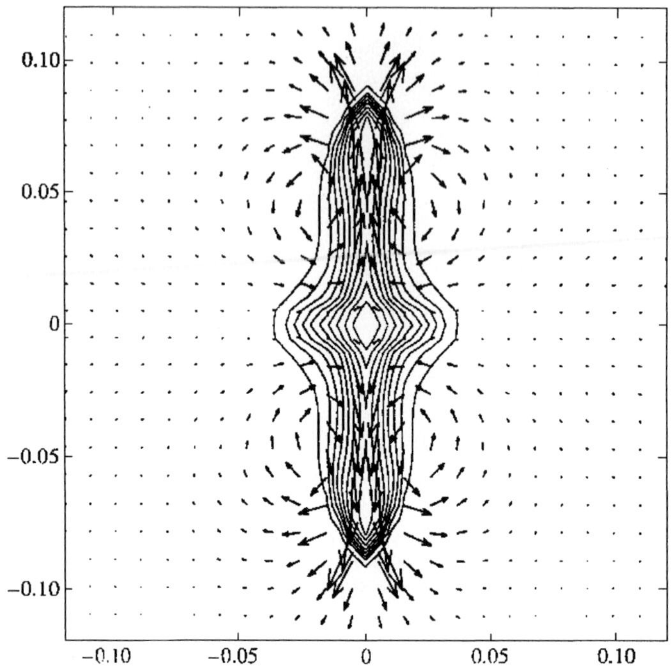

Fig.4 Distributions of the entropy and velocity field in the Ozx plane during the formation and initial motion of a bubble toward the neutrinosphere. The axis scales are the same as in Fig.1.

References

1. Yee H.C., Klopfer G.H., Montagne J.L., JCP, **88**, p.31, 1990.
2. Shu C.W., Osher S., JCP, **77**, p.439, 1988.
3. A.N.Andrianov, K.N.Efimkin, V.Y.Levashov, I.N.Shishkova., Lecture Notes in Computer Science, v.2073, July, 2001.
4. Ustyugov S.D., Chechetkin V.M., Astronomy Reports, **43**, p.718, 1999

A Local Level-Set Concept for Front Tracking on Arbitrary Grids

F. Völker, R.Vilsmeier, and D.Hänel

Institute of Combustion and Gasdynamics (IVG), University of Duisburg, 47048 Duisburg, Germany
E-mail:(voelker,hj000vi,hj454ha)@vug.uni-duisburg.de

Summary. This paper proposes a general multi-dimensional front tracking concept for arbitrary physical problems. The tracking method is based on the level-set approach with a restricted dynamic definition range in the vicinity of the fronts. Special attention is drawn to the problem of the classical level-set method, i.e. accuracy issues and topological restrictions. In this concern, a less sensitive time integration is introduced and the problem of interacting discontinuities is addressed. The concept is integrated in the basic modular Finite-Volume solution package MOUSE [1] for systems of conservation laws on arbitrary grids.

1 Introduction

In many flow problems discontinuous flow features, as material interfaces, shocks or detonations, occur. In most cases, such flow features are resolved numerically by capturing methods. These captured discontinuities become smeared out by artificial intermediate states and falsify the solution in problems of strong interaction. Besides adaptivity, reducing the spacial range of intermediate states, tracking methods for embedded discontinuities are of gaining interest.

Many pioneering developments have to be mentioned in this context, such as the work of Richtmyer and Morton [2] or Moretti [3]. One of the most promising approaches however, is the newer level-set method [4, 5, 6, 7], also used in the present paper. In this approach, the front position is defined as an iso-value of a smooth scalar function. Besides its strong assets, the classical level-set method shows some severe drawbacks. First of all, topological restrictions apply, as only closed, differentiable interfaces can be described. Also, most implementations claim accuracy problems concerning the preservation of the exact front position, when preparing the level-set function for further physical motion steps.

The present paper attempts to overcome the drawbacks and restrictions of the classical method, however introducing further algorithmic difficulties. Object oriented programming allows modular and flexible program developments, reusable for arbitrary physical and topological problems.

2 Basic Numerical Concept

A general, time dependent system of conservation laws in integral form reads:

$$\frac{\partial}{\partial t}\int\limits_V \mathbf{Q}\,dV + \oint\limits_A \mathbf{H}\cdot\mathbf{n}\,dA = \int\limits_V \mathbf{S}\,dV \qquad (1)$$

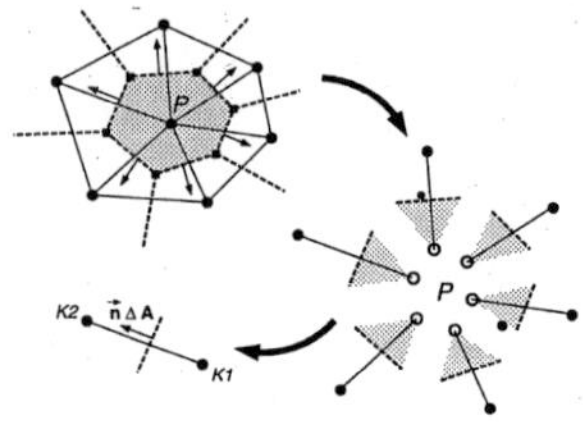

Figure 1: Control volume

where Q, H and S represent the variables, fluxes and source terms respectively. The system of equations (1) is solved on arbitrary grids employing a nodal finite volume method, figure 1. For a discrete control volume V_d the above system reads:

$$\frac{\Delta\mathbf{Q}}{\Delta t}|_{Vd} + Res_{\Delta,Vd} = 0 \quad \text{with:} \quad Res_{\Delta,Vd} = \frac{1}{V_{Vd}}\sum_{i=1}^{nv}\mathbf{H}_i\,\mathbf{n}_i\,\Delta A_i \qquad (2)$$

where Vd is a control volume with the size V_{Vd} and the sum is carried out over all bounding segments. For the developments in the present paper, a fully explicit time integration method is used throughout.

3 Treatment of Discontinuities

According to their physical behaviour, discontinuous fronts can be grouped in two distinct classes:

- **Actively** transported fronts, e.g. shocks, detonation or deflagration waves,...
- **Passively** transported fronts, e.g. material interface, shear layers,...

While passively transported (passive) discontinuities float along with a carrier speed, the propagation of actively transported (active) discontinuities is ruled by additional physical relations, e.g. the Rankine Hugoniot equations. For active discontinuities the propagation speed is thus only defined directly, while for passive discontinuities it is available in the whole field.

To calculate both, passive and active discontinuities, a local level-set formulation is chosen. As all methods should qualify for arbitrary physical problems, the more difficult class of active discontinuities must be considered throughout. The basic concept has previously been introduced in [8] and [9]. Therefore, only a brief repetition is outlined here, while new extensions are issued in more detail. Methods concerning the back influence of discontinuities onto the surrounding continuous field are completely omitted.

3.1 The Local Level-Set Function

A level-set is described by a scalar function $G(\mathbf{x}, t)$ and the front is represented by a constant iso-value, chosen $G = 0$ here. In an Eulerian frame, the transport of a level-set moving with a propagation speed $\mathbf{c}(\mathbf{x}, t)$ reads:

$$\frac{\partial G}{\partial t} + \mathbf{c} \cdot \nabla G = 0 \qquad (3)$$

Discrete values for the function G are stored at the nodes of the mesh, while the definition range is restricted to a given number of neighborship levels, figure 2. Points and edges are marked with P and K respectively and the number index refers to the neighborship level. Edges of type K_0 refer to those hitting the front and the exact front position is described by the $G = 0$ value on these edges upon linear reconstruction. The discrete definition range is dynamic and must be adjusted, if nodes are overrun by the front.

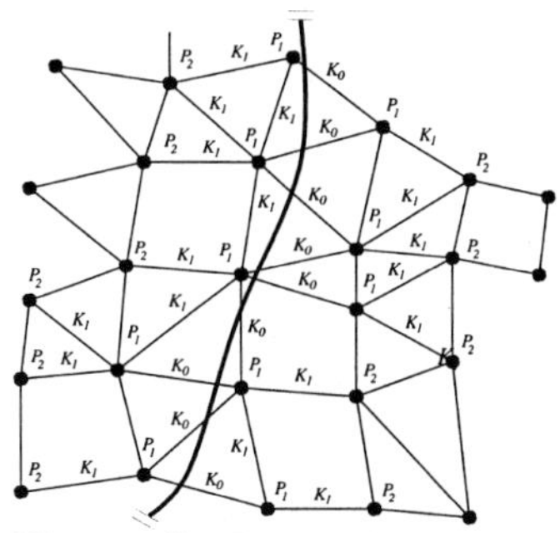

Figure 2: Discontinuity on a 2-D mesh and membership of nodes and edges to corresponding levels.

In the present work, multiple fronts play an essential role. Above basic narrow band description applies to each of the fronts independently and their discrete definition ranges may overlap in arbitrary ways without restriction.

3.2 Front Motion with Level-Set

The transport of the front in addition with a two step normalization procedure forms the front tracking algorithm for one time step. For reasons of accuracy the level-set function is normalized to approximately satisfy the slope condition $\mid \nabla G \mid = 1$, corresponding to a distance function to the front. The normalization procedure involves iterations by pointwise least square approaches and proceeds in two steps: In the first step, the direction of the gradient ∇G^k is computed. For a node i with a restricted number of neighbors $\tilde{j} \in E_{par}(i)$ the error approach reads:

$$\Phi_{par} = \sum_{\tilde{j} \in E_{par}(i)} \left(\frac{\nabla G_i^k \cdot \mathbf{m}_{i\tilde{j}} - (G_{\tilde{j}} - G_i)}{|\mathbf{m}_{i\tilde{j}}|^{q/2}} \right)^2 \rightarrow \text{minimum} \qquad (4)$$

with $\mathbf{m}_{i\tilde{j}} = \mathbf{X}_{\tilde{j}} - \mathbf{X}_i$ and a distance weighting exponent $q \geq 0$. In the second step

$$\Phi_{hyp} = \sum_{j \in E_{hyp}(i)} \left(\frac{\frac{\nabla G_i^k}{|\nabla G_i^k|} \cdot \mathbf{m}_{ij} - (G_j - G_i^{k+1})}{|\mathbf{m}_{ij}|^{q/2}} \right)^2 \rightarrow \text{minimum} \qquad (5)$$

the functional value of G^{k+1} is set to satisfy the slope condition. Due to dependencies with neighboring nodes, the method is used iteratively and k is an iteration counter. In above equations, the indices *par* and *hyp* refer to restricted ranges of influence. Further details can be found in [8].

Unfortunately, it is not possible to satisfy the slope condition without slightly altering the front position in a linear representation on the edges of the mesh. Additional corrections are thus performed to preserve the exact front position after normalization. Doing so, the exact slope condition is again voided, which is however more tolerable than a falsified front position itself.

The motion of the front is carried out by an explicit time discretization of equation 3. Assuming a constant slope, the propagation step reads:

$$G^{n+1} = G^n - \mathbf{c} \cdot \nabla G^n \Delta t \tag{6}$$

This equation is solved under the assumption of frozen velocity $\mathbf{c}$ during a time step. Due to the inaccuracies of the constant slope condition, an additional translation error is introduced. To reduce this error, above equation is extended by a Taylor expansion for the gradient:

$$G^{n+1} = G^n - \mathbf{c} \cdot \nabla G^n \Delta t + \frac{1}{2} \Delta t^2 \mathbf{c}^\mathbf{T} (\nabla \otimes (\nabla G)) \mathbf{c} \tag{7}$$

According to this modification, the constant slope condition $|\nabla G| = const$ of the level set function is less important, thus also lowering the need for large iteration numbers within the normalization procedure. Note however, that this is not a second order accurate time integration, as the speed $\mathbf{c}$ remains frozen. It only allows to preserve the accuracy on a sloppier field of G.

At this point, the discussion about the number of neighborship levels in the local representation needed, must be recalled. Assuming, at most one neighborship level can be overrun by the front in a time step, already two point levels are required. With above extended equation 7 however, a larger stencil to compute the second gradient $\nabla \otimes (\nabla G)$ is needed and thus an additional neighbor level, the third, is required. Finally, for practical reasons, a fourth level is introduced. This is done to avoid an influence of new nodes onto the computations in the case the dynamic definition range is adjusted.

Until now, no spacial discretization of the semi-discrete equations 6 and 7 is given yet. Several possibilities exist. Considering the finite volume structure of the basic concept, chapter 2, a corresponding spacial discretization for the level-set equations is an obvious trial. However, this variant turned out to be less useful for active discontinuities as a propagation speed is needed far apart the discontinuity.

A second variant is to compute the terms $\mathbf{c}\nabla G$ and $\mathbf{c}^\mathbf{T}(\nabla \otimes (\nabla G))\mathbf{c}$ directly on all edges, hitting a discontinuity in a differencing approach. However, since all values of G are stored at the nodes, a re-distribution of edgewise contributions onto these is required. This variant allows an upwind formulation for the first derivative term, but a distribution onto nodes of the second neighborship level is critical.

For the above reasons, a nodal approach is chosen here. ∇G and $\nabla \otimes (\nabla G)$ are computed with a second order least square approach on the nodes itself. As the propagation speed $\mathbf{c}$ is not directly available at the nodes but in general only at edges hitting the front, a corresponding extrapolation is performed.

3.3 Transport Velocity

For the above nodal discretization of equations 6 and 7 a propagation speed vector $\mathbf{c}$ is needed at the nodes. This speed is obtained in a two stage process. In the first stage, a speed vector $\mathbf{c}_{i,j}$ at all edges K_0 is computed. In the second stage this speed is redistributed onto the nodes of the definition range. According to physical jump conditions, a front propagation speed is directly accessible at an interface. As only the normal propagation of the interface is of interest, a direction perpendicular to the front is assigned to the edges in K_0. This perpendicular direction is given by the gradient ∇G and interpolated onto the actual position of the front:

$$\nabla G_{i,j} = \nabla G_j \frac{\mid \mathbf{m_{i,o}} \mid}{\mid \mathbf{m_{i,j}} \mid} + \nabla G_i \left(1 - \frac{\mid \mathbf{m_{i,o}} \mid}{\mid \mathbf{m_{i,j}} \mid}\right) \quad (8)$$

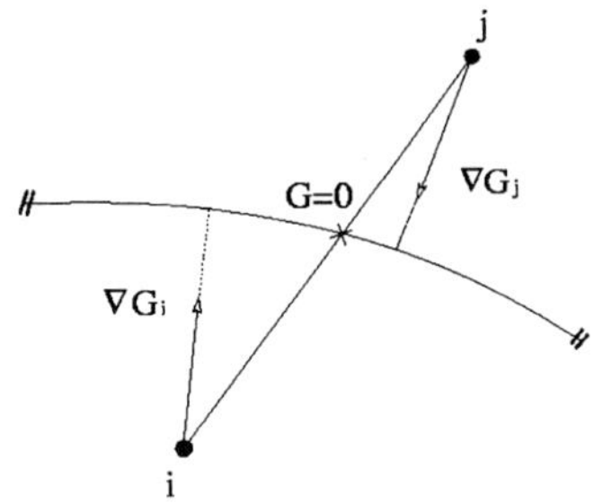

Figure 3: Discontinuity and edge with the vertices i and j and their G-gradients

with $\mathbf{m_{i,0}} = \mathbf{x_i} - \mathbf{x_0}$ and $\mathbf{m_{i,j}} = \mathbf{x_i} - \mathbf{x_j}$. The physical jump conditions are formulated in the direction of $\nabla G_{i,j}$, then yielding $\mathbf{c}_{i,j}$. In general, several edges hitting a front may be appended to a node in the first neighborship level P_1. In this case a weighted interpolation is performed. Figure 4 sketches such a configuration. Node i has three discontinuous edges (k_1, k_2 and k_3) appended, each of which with its own representative edge-wise propagation speed. The edges bounding the element containing the direction ∇G_i stored at node i are taken for interpolation.

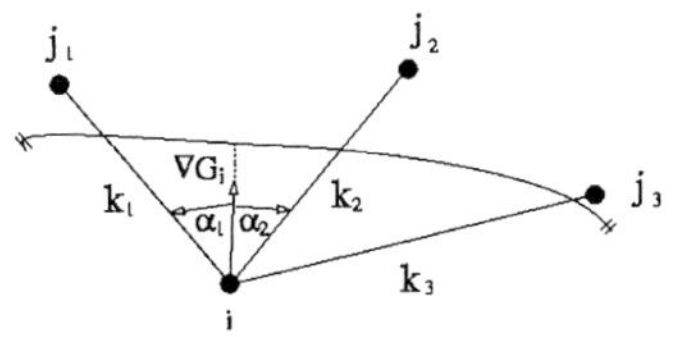

Figure 4: Discontinuity and discrete edges with the vertices i and j_n

If no interpolation is possible the propagation speed of the edge with the smallest angle between itself and ∇G_i is taken.

4 Topological aspects

A discrete iso-value of a smooth scalar function $G(\mathbf{x}, t)$ may only describe closed differentiable curves (surfaces). Due to the bounded domain, also curves (surfaces) hitting the boundaries are possible. In this classical formulation of the level-set method, the strict separation of regions as well as the intrinsic differentiability causes severe restrictions. Obviously, methods to overcome these restrictions are sought. The attempts presented here relate to work in progress. Developments are currently restricted to 2-D although a generalization seams possible.

4.1 Regions

As a matter of principle, the iso-value of a level-set $L(A, B)$ will separate a field in two regions A and B, while a point $P : (\mathbf{x}, t)$ can only be either in region A or in region B. Consider a point $P_0 : (\mathbf{x}_0, t_0)$ in the coordinate space located in region A: Starting at P_0 any trajectory in the coordinate space will yield to the region B, if the discontinuity is crossed an odd number of times. This holds true for multiply connected domains.

Above topological constraint is acceptable for many passive and active discontinuous problems. Examples in fluid mechanics are:

- Two non-miscible fluids: A point $P : (\mathbf{x}, t)$ is **either** in the region of a fluid A **or** a fluid B.
- Phase transitions: ... **either** evaporated **or** liquid (solid).
- Discontinuous combustion model: ... **either** non-burned **or** burned.

However, especially in compressible gas-dynamics, above restriction is not always permissible. As example, consider a shock wave to be tracked, that may end somewhere in the field and degrade to a compression wave.

4.2 Differentiability

Consider two points D_1 and D_2 located at an interface, and let δ be the distance between these. Interfaces, described with the classical level-set approach must be differentiable in the sense that:

$$\lim_{\delta \to 0} (\nabla G|_{D_1} - \nabla G|_{D_2}) \quad = \quad 0 \quad . \tag{9}$$

Again, for most physical cases in fluid dynamics intrinsic differentiability is an acceptable constraint. As example, surface tension effects acting in the interface between two non-miscible fluids or between phases can be considered. In such cases a non-differentiable interface is impossible. However, the physical curvatures might be essentially smaller than those resolvable on a given computational grid. In compressible gas dynamics, due to the possible interactions of discontinuities, problems get worse. As example, consider the transient detonation wave, as given in chapter 5, figure 7. In this example the front shock was tracked by a single level set. The front deformation is however ruled by transverse shocks in the appended reaction zone. These shocks are attached to the front shock, forming the well known detonation cells. Accordingly, the "corners" in the front shock are non-differentiable and the classical level set method smears these out to a moderate curvature radius, depending on the resolution of the grid.

4.3 Attempts for non-closed discontinuities

A possible way to describe an open discontinuity (an open level-set) is to limit its active range by intersection with other transverse level-sets. Following, the actual discontinuity will be called a master and its limiting level-sets slaves. The slaves separate the master in two distinct regions: valid as a discontinuity or not valid. In 2-D the intersection with a slave thus creates an ending point of the discontinuity.

Master and slaves are moved independently and a common speed vector $\mathbf{c}$ is formulated. The component of $\mathbf{c}$ normal to the discontinuity rules its displacement as in the classical method, while the movement of the slaves is responsible for alterations of its extension in time. For simplicity, level-set gradients of master and slaves are orthogonal at the intersection, which is ensured by a reconstruction step of all slaves after motion.

4.4 Bifurcating discontinuities and intersection points

A possible way to create interacting discontinuities is to decompose these in several open sections, whose ending points are attached to a common point of intersection. As several discontinuities are attached to a single intersection point and as this point may move according to the dynamics of the problem, it has to be ensured that the ending points of the appended discontinuities follow synchronously. In first tests, it was found to be useful to formulate a correction speed at all ending points to follow an intersection. With a given intersection point location $P_{Int} : (\mathbf{x}_{Int}, [t])$ and the ending point location $P_i : (\mathbf{x}_i, [t])$ of a discontinuity i, the coupling speed $\mathbf{c}_i$ is defined as:

$$\mathbf{c}_i = \frac{\mathbf{x}_{Int} - \mathbf{x}_i}{\Delta\tau} \tag{10}$$

where $\Delta\tau$ is an iteration time step, actually chosen as the true time step in explicit calculations. As alternate, additional post-iterations to satisfy the matching condition with better accuracy are possible, if needed. As the values for the level-set functions involved are stored at the fixed nodes of a given mesh, it is useful to relax the above coupling speed over a certain range of influence. For first tests a linear approach in a relaxation radius R_i was chosen. For a node P_i within a level-set i the correction speed reads:

$$\mathbf{c}_{P,i} = \begin{cases} \mathbf{c}_i \cdot (1 - \frac{|\mathbf{x}_P - \mathbf{x}_i|}{R_i}) & \text{for} \quad |\mathbf{x}_P - \mathbf{x}_i| < R_i \\ 0 & \text{for} \quad |\mathbf{x}_P - \mathbf{x}_i| \geq R_i \end{cases} \tag{11}$$

4.5 Extended scope of applications

Not all of the above mentioned is yet available as a reliable program code. In fact many questions remain, such as priority rules for intersection point motion or precision limits. In consequence however, the formulation allows the extension of the level-set method to a large number of further physical problems, for example:

- Three fluid problems: These might be three non-mixing phases or two phase problems (gas / liquid) including two non-mixing liquids.

- Shock detection [1]: It cannot be assumed a shock is found as a closed line at once.
- Interactions of a discontinuity with itself yielding sharp corners.

5 Computational Examples

Above explained methods are demonstrated by several test cases.

3D bow shock on a sphere: In this test case, the bow shock of the inviscid flow past a sphere is computed as a single closed surface. The unstructured grid consists of 216940 tetrahedral elements. The free stream Mach number is $Ma_\infty = 2$. Figure 5 shows the initialization of the shock front as a planar surface in front of the sphere, figure 6 shows a converged solution. Broken partitions of the discrete definition range and iso-colors of pressure on a cut plane are shown in both figures.

Figure 5: Initialization of a 3D bow shock with shock, subset grid, inner sphere and constant pressure values

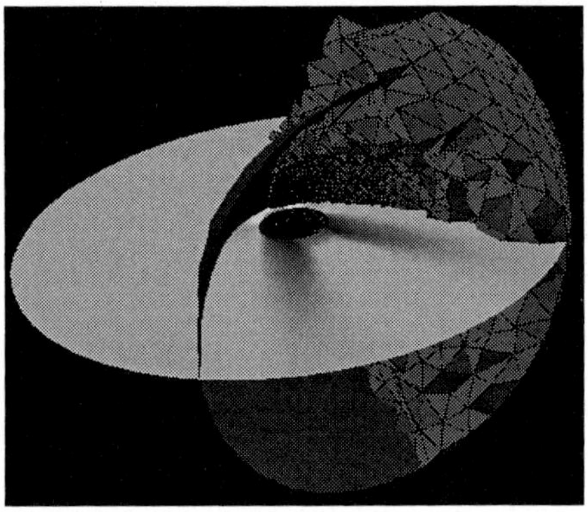

Figure 6: Final state of the 3D bow shock calculation with front, subset mesh and pressure values in the cut plane

Unstable inviscid 2D Chapman-Jouget detonation: The second example is a snapshot of an unstable inviscid 2D CJ-detonation on a triangulated grid, figure 7. Chemical source terms are calculated with a one step reaction model. The detonation is initialized as a steady CJ-detonation with an incoming flow of Ma=8.3 from the right side. The onset of detonation cells can be observed. In theory, the front shock has kinks at the position where transverse shocks in the reaction zone (captured here) are attached. The classical level-set theory fails to describe the effect with sharp corners.

[1] Besides topological requisites, shock detection itself is a challenging problem.

Inviscid, transonic flow past a NACA 0012 airfoil: The transonic, inviscid flow past a NACA 0012 airfoil is considered at $Ma_\infty = 0.75$ and $\alpha_{attack} = 2^o$. The level-set propagation was combined with an explicit multi-grid method for the flow field [10]. In this test case, a single, limited re-compression shock wave on top of the airfoil appears. It is simulated here by a single ending front, described by a master-slave combination as can be seen in figure 8. At present, no shock detection and transverse shock growth algorithms are available yet. Therefore, the master-slave combination has been initialized near the leading edge with an estimated altitude above the surface. While the shock moves towards the final steady state position, the limiting slave level-set has been frozen. Up to the ending point, inner boundary conditions on the shock apply, while degrading to the capturing method beyond.

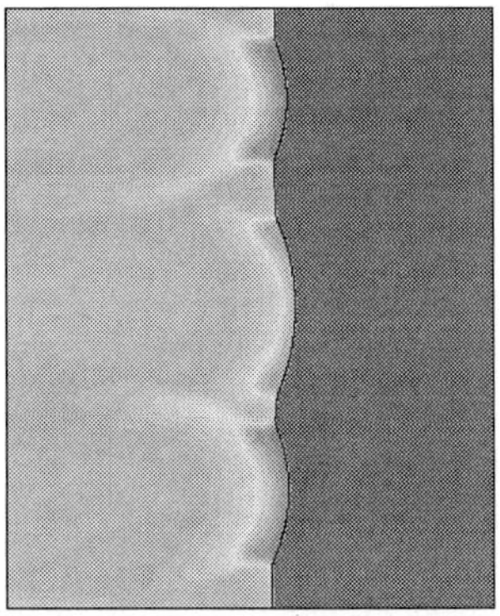

Figure 7: Snapshot of an inviscid unstable 2D CJ-detonation with pressure values calculated on an unstructured grid

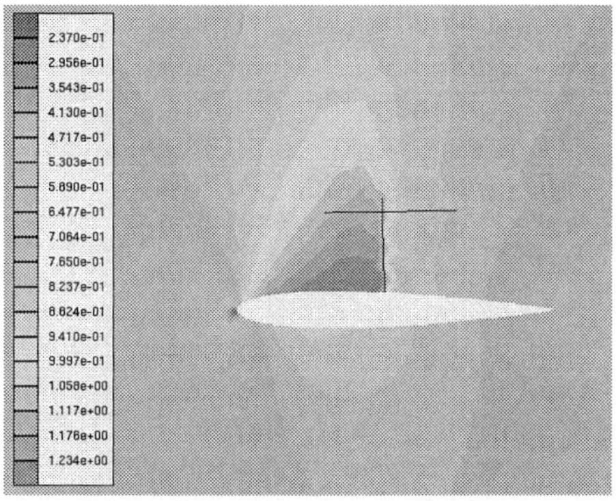

Figure 8: Master-slave combination for a restricted re-compression shock wave on a transonic airfoil: Colors of constant Mach number and front position.

Passive transport of a master-slave combination: The next two figures (figure 9 and figure 10) show an example for a master-slave combination on a triangulated grid in a phantasy domain. The flow from the left is inviscid at $Ma_\infty = 0.5$. Both level-sets are transported with the local fluid velocity. In this case the fluid velocity causes the slave to extend the master.

Level-sets attached to moving points: The last example is a test calculation of a master-slave combination guided by three moving points. The master level-sets have zero speed, while the three guiding nodes march with constant velocity to the right side. This is a purely academic case, testing the ability of restricted fronts to follow intersection points. The computation is performed on a structured grid. A sequence in time is given in figure 11.

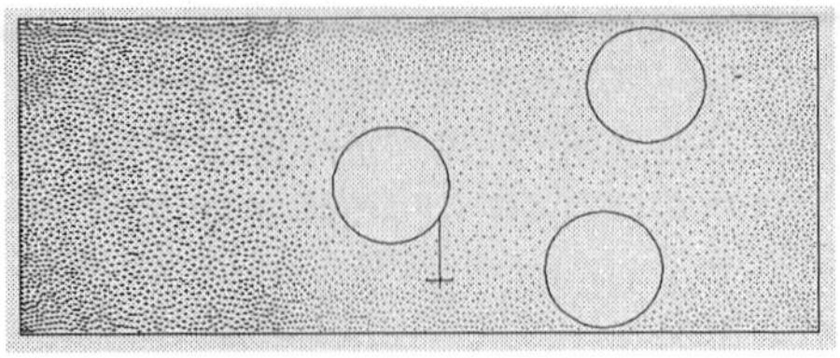

Figure 9: Initialization of a passively transported master-slave combination.

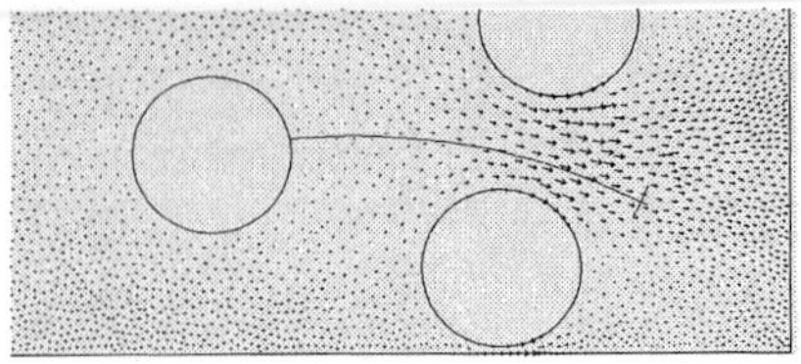

Figure 10: Passive master-slave combination after several time steps

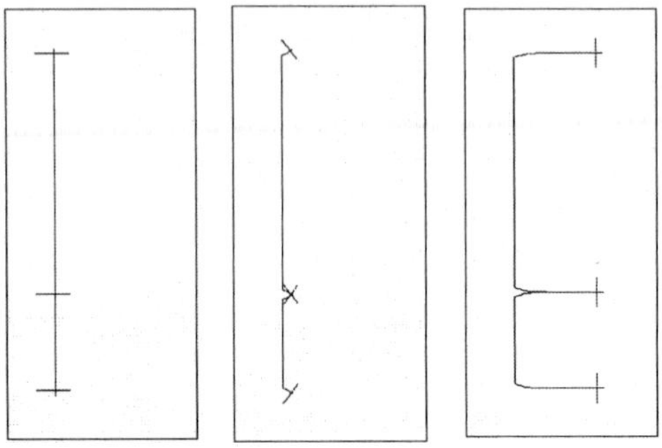

Figure 11: Two clipped fronts with ends following three moving points in the field. Sequence, showing all level-sets involved (two masters, four slaves)
left: initial condition
middle: after 10 time steps
right: after 100 time steps

References

1. O. Gloth and R. Vilsmeier and D.Hänel, Homepage of Project MOUSE, *Online: http://www.vug.uni-duisburg.de/MOUSE* (periodically updated)
2. R.Richtmyer, K.Morton (1967): Difference Methods for Initial value Problems. Interscience, New York
3. G.Moretti (1987): A Technique for Integrating Two-Dimensional Euler Equations. Computers and Fluids 15: 59-75
4. AJ. Sethian (1999): Level Set Methods and Fast Marching Methods: Evolving Interfaces in Computational Geometry, Fluid Mechanics. Computer Vision and Materials Science Cambridge: Cambridge University Press
5. A. Kerstein, W. Ashurst, F. Williams (1988): Field Equation for Interface Propagating in an Unsteady Homogeneous Flow Field. Phys. Rev. A, vol. 37, pp2728-2731
6. Mulder, Osher, Sethian (1992): Computing Interface Motion in Compressible Gas Dynamics. J. Comp. Phys. , vol. 100, pp 209-228
7. M. Sussman, P. Smereka, S. Osher (1994): A Level Set Approach for Computing Solutions to Incompressible Two-Phase Flow. J. of Comp. Physics, vol.114, pp 146-159
8. L.Tran, R.Vilsmeier, D.Hänel (1999): A local level set method for the treatment of discontinuities on unstructured grids. FVCA-II 99, Duisburg, Germany
9. R. Vilsmeier, D.Hänel, L. Tran (2001) A Front Tracking Method on Unstructured Grids. Accepted paper, Computer & Fluids, (2002).
10. L. Fournier and O. Gloth (1999): An Attempt to Develop a Multi Purpose FAS Multigrid Algorithm. FVCA-II 99, Duisburg, Germany

Large-Time Behavior of Solutions to the Multi-Dimensional Euler Equations with Damping

Dehua Wang

Department of Mathematics, University of Pittsburgh, Pittsburgh, PA 15260, USA. *dwang@math.pitt.edu*

1 Introduction

We consider the three-dimensional compressible Euler equations of isentropic gases with damping:

$$\begin{aligned} \rho_t + \nabla \cdot (\rho u) &= 0, \\ \rho(u_t + u \cdot \nabla u) + \nabla P &= -c\rho u, \end{aligned} \tag{1}$$

where $\rho(x,t) \in \mathbb{R}$, $u(x,t) \in \mathbb{R}^3$ represent the density, velocity of the gas, respectively; $x = (x_1, x_2, x_3) \in \mathbb{R}^3$ is the space variable, $t > 0$ is the time variable; the pressure is $P = P(\rho) = \rho^\gamma/\gamma$ with $\gamma \geq 1$ as the adiabatic exponent; $c > 0$ is a constant, and $\varepsilon = 1/c$ may be called the relaxation time for some physical flows. We study the Cauchy problem of (1) with the initial condition:

$$(\rho, u)|_{t=0} = (\rho_0(x), u_0(x)). \tag{2}$$

We are interested in the damping effect on the regularity and large-time behavior of the smooth solutions. It will be proved that the size of the smooth initial data in certain norms plays the key role. If the initial data are small in an appropriate norm, the damping can prevent the development of singularities in smooth solutions and the Cauchy problem has a unique global smooth solution. If the initial data are large in certain form, the damping is not strong enough to prevent the formation of singularities even though the initial data are smooth. This indicates that the smoothness of the solution will break down in a finite time.

The smoothing effect induced by damping was studied by Nishida [36] and many other authors for the one-dimensional case. We notice that the multi-dimensional case is much more complicated than the one-dimensional case. In this paper we discuss our techniques and ideas to prove the global existence and decay of the multi-dimensional smooth solutions when the initial data are small. A theorem on formation of singularities will also be presented. Our main results show that, when the initial data are small in H^3, the global smooth solution to the Cauchy problem (1) and (2) exists uniquely. A theorem will be given to show that the L^2 norm of the density generally does not decay exponentially. The analysis is based on the energy estimates directly on the system (1). A different approach is adopted in the joint work [39] with T. Sideris and B. Thomases. The paper is organized as follow. We first review some related studies and results for the one-dimensional problems in Section 2. Then we review related results about the multi-dimensional Euler equations in Section 3. Some results from the joint papers [8, 9, 43] with G.-Q. Chen are discussed in the reviews. Finally we show our results on the multi-dimensional Euler equations with damping in Section 4. The theorems and only outline of proofs are given.

2 One-dimensional problems

Consider the one-dimensional hyperbolic system of n conservation laws:

$$u_t + f(u)_x = 0, \quad u \in \mathbb{R}^n, \tag{3}$$

where $u \in \mathbb{R}^n$ is the unknown variables and $f(u)$ is the flux. The main feature of this system is the finite-time formation of shock waves no matter how smooth and small the initial data are, generally. This classical result has been proved by many authors using the method of characteristics, e.g., Lax [26], John [24], Liu [31], Klainerman-Majda [25], Dafermos [13], etc. This property shows that one has to seek weak solutions including shock waves to the general hyperbolic system (3). The BV solution of (3) with small total variation was constructed by Glimm [20] using the random choice method; by Dafermos [11] for scalar conservation laws, DiPerna [16] for 2×2 systems, by Bressan [3] and Risebro [37] for general $n \times n$ systems, using the wave-front tracking approximations; and recently, by Bianchini-Bressan [2] using the vanishing viscosity method. For weakly linearly degenerate hyperbolic systems, there exists global C^1 solution with small initial data ([27]).

Certain dissipations, such as damping, heat diffusion, and memory effect, can prevent the formation of shock waves (or singularities) for smooth small initial data, i.e., there exists global smooth solutions if the initial data are smooth and small (see Dafermos [12]). The one-dimensional Euler equations with damping has been studied by many authors; see Nishida [36] and Hsiao [23] as well as the references cited therein. The one-dimensional Euler equations with heat diffusion was studied by Slemrod [40] for global existence of small smooth solution and by Dafermos-Hsiao [15] and Hrusa-Messaoudi [22] for the development of singularities. For the memory effects, see Dafermos [14]. For a full system of Euler-Poisson equations which has two damping terms, one heat diffusion term, and some elliptic mode from the Poisson equation, Chen-Wang [43] proved that all these smoothing effects together are still not strong enough to prevent the formation of singularities in smooth solutions with large initial data. Some critical threshold phenomena for pressureless Euler-Poisson equations were obtained by Engelberg-Liu-Tadmor [18].

3 Multi-dimensional Euler equations

In (1), when $c = 0$ we obtain the isentropic three-dimensional Euler equations without damping. There have been lots of studies on the multi-dimensional Euler equations. For the non-isentropic three-dimensional Euler equations, Sideris in [38] proved the finite-time formation of singularities using an average quantity. See also Makino-Ukai-Kawashima [33] on a compressible fluid body surrounded by vacuum. It is believed that in fact only $\nabla\rho$ and ∇u blow up in most cases; see a proof in Alinhac [1] for the two-dimensional case with axisymmetric initial data. For the existence of global smooth solution under certain assumptions on the initial data, see [21].

There have been many studies on the weak entropy solutions to the Euler equations. For the one-dimensional case, the global weak entropy solution in L^∞ was constructed in DiPerna [17], Chen [5], and Lions-Perthame-Souganidis-Tadmor [29, 30]. For the multi-dimensional Euler flows with spherical symmetry and other geometric structures as well as transonic nozzle flows, the global weak entropy solution in L^∞ was constructed in Chen-Glimm [7] and Chen-Wang [8, 9]. See their references for other related studies.

For potential compressible fluid flows, recent mathematical efforts have been made to establish the existence and behavior of solutions. See Chen [10], Chen-Feldman [6], Gamba-Morawetz [19], Lien-Liu [28], Morawetz [34, 35], Zhang [44], and the references cited therein. See also the book [45]. A related model, called the unsteady transonic small disturbance equations, has been analyzed in Canic-Keyfitz-Lieberman [4]. Also see their references for related discussions.

4 Multi-dimensional Euler equations with damping

We now discuss the existence, decay, and formation of singularities of smooth solutions to the Cauchy problem (1) and (2). Denote the norm of $H^m(\mathbb{R}^3)$ by $\|\cdot\|_{H^m}$ and the norm of $L^2(\mathbb{R}^3)$ by $\|\cdot\|$. Use D_x^m for the vector of the m-th spatial derivative and C for a generic positive constant independent of the initial data. For any given constant $\bar\rho > 0$, set

$$E(t) = \|(\rho(\cdot,t) - \bar\rho, u(\cdot,t))\|^2_{H^3(\mathbb{R}^3)} + \|(\rho_t, u_t)(\cdot,t)\|^2_{H^2(\mathbb{R}^3)}. \qquad (4)$$

Then, using (1) to define $(\rho_t, u_t)(x, 0)$,

$$E(0) = \|(\rho_0 - \bar\rho, u_0)\|^2_{H^3} + \|(\nabla\cdot(\rho_0 u_0), u_0\cdot\nabla u_0 + \rho_0^{\gamma-2}\nabla\rho_0 + cu_0)\|^2_{H^2}.$$

We have the following global existence of smooth solution.

Theorem 1. *Suppose that* $(\rho_0 - \bar\rho, u_0) \in H^3(\mathbb{R}^3)$ *and* $E(0)$ *is sufficiently small. Then the Cauchy problem* (1)-(2) *has a unique global smooth solution* (ρ, u) *satisfying the following estimate:*

$$E(t) + C\int_0^t (\|u(\cdot,\tau)\|^2_{H^3} + \|(\nabla\rho, \rho_t, u_t)(\cdot,\tau)\|^2_{H^2})\, d\tau \le CE(0), \tag{5}$$

for some positive constants C independent of the initial data.

The outline of the proof of Theorem 1 is as follows. The main step is the global a priori estimates thanks to the local existence ([32]). Set $v = \rho - \bar{\rho}$. Take $\bar{\rho} = c = 1$ without loss of generality. Then (v, u) satisfies the following equations:

$$v_t + u \cdot \nabla v + \rho\, \mathrm{div} u = 0, \tag{6a}$$

$$u_t + u \cdot \nabla u + \nabla p(\rho) = -u, \tag{6b}$$

where $\rho = v + 1$ and $p(\rho)$ is defined by $p'(\rho) = \rho^{-1}P'(\rho) = \rho^{\gamma-2}$. From (4),

$$E(t) = \|(v,u)(\cdot,t)\|^2_{H^3(\mathbb{R}^3)} + \|(v_t,u_t)(\cdot,t)\|^2_{H^2(\mathbb{R}^3)}.$$

For any given $T > 0$ we will estimate $(v,u)(x,t)$ for $x \in \mathbb{R}^3$ and $t \in [0,T]$ under the a priori assumption: $\sup_{t\in[0,T]} E(t) \le \delta^2$ for some small constant $0 < \delta \ll 1$. Sobolev's inequality and (6) imply that

$$\sup_x |(v, \nabla v, v_t, u, \nabla u, u_t)|^2 \le CE(t) \le C\delta^2. \tag{7}$$

For sufficiently small δ, one has $r_1 \le \rho, p'(\rho) \le r_2$ for some positive constants r_1 and r_2. We first have the following lemmas with outline of proofs.

Lemma 1.

$$\frac{1}{2}\frac{d}{dt}\left(\|u\|^2 + \int_{\mathbb{R}^3} V_1\, dx\right) + C\|u\|^2 \le C\delta\|(\nabla v, \nabla u)\|^2, \tag{8}$$

where $V_1 = 2\int_0^v (p(\xi+1) - p(1))/(\xi+1)d\xi \ge 0$.

Proof. Rewrite the equation (6b) as $u_t + u\cdot\nabla u + \nabla(p(v+1) - p(1)) = -u$. Multiply it by u and integrate to obtain,

$$\frac{1}{2}\frac{d}{dt}\|u\|^2 + \|u\|^2 - \int_{\mathbb{R}^3} (p(v+1) - p(1))\mathrm{div} u\, dx + \int_{\mathbb{R}^3} (u\cdot\nabla u)u\, dx = 0. \tag{9}$$

From (6a), $\mathrm{div} u = -(v_t + u\cdot\nabla v)/\rho$. Then, using (7), one has

$$-\int_{\mathbb{R}^3} (p(v+1) - p(1))\mathrm{div} u\, dx \ge \frac{1}{2}\frac{d}{dt}\int_{\mathbb{R}^3} V_1\, dx - C\delta\int_{\mathbb{R}^3} (|u|^2 + |\nabla v|^2)\, dx,$$

and

$$\int_{\mathbb{R}^3} (u\cdot\nabla u)u\, dx \ge -C\delta\int_{\mathbb{R}^3} (|u|^2 + |\nabla u|^2)\, dx.$$

Thus (9) implies (8) for sufficiently small δ.

Lemma 2.

$$\frac{1}{2}\frac{d}{dt}\left(\|(u_t,\nabla u)\|^2+\|\rho^{\frac{\gamma-3}{2}}(v_t,\nabla v)\|^2\right)+C\|(u_t,\nabla u)\|^2\le C\delta\|(v_t,\nabla v)\|^2. \tag{10}$$

Proof. Differentiating (6b) with respect to t and multiplying it by u_t,

$$\frac{1}{2}\frac{d}{dt}\|u_t\|^2+\|u_t\|^2-\int_{\mathbb{R}^3}p(\rho)_t(\mathrm{div}u)_t\,dx+\int_{\mathbb{R}^3}(u\cdot\nabla u)_t u_t\,dx=0.$$

Using the equation (6a) and the inequality (7), one has

$$\frac{1}{2}\frac{d}{dt}\left(\|u_t\|^2+\int_{\mathbb{R}^3}\rho^{\gamma-3}v_t^2\,dx\right)+\|u_t\|^2\le C\delta\|(v_t,u_t,\nabla u)\|^2. \tag{11}$$

Differentiating (6b) with respect to x and integrating its product with ∇u, we have

$$\frac{1}{2}\frac{d}{dt}\|\nabla u\|^2+\|\nabla u\|^2+\int_{\mathbb{R}^3}\nabla(\nabla p(\rho))\nabla u\,dx+\int_{\mathbb{R}^3}\nabla(u\cdot\nabla u)\nabla u\,dx=0.$$

From (6a), (7), and integration by parts, we obtain the following:

$$\frac{1}{2}\frac{d}{dt}\left(\|\nabla u\|^2+\int_{\mathbb{R}^3}\rho^{\gamma-3}|\nabla v|^2\,dx\right)+\|\nabla u\|^2\le C\delta\|(\nabla v,\nabla u)\|^2. \tag{12}$$

Adding (11) and (12) leads to (10) for sufficiently small δ.

Lemma 3. *For some constant* $\lambda_1>0$,

$$\begin{aligned}&\frac{1}{2}\frac{d}{dt}\left(\|(u,u_t,\nabla u)\|^2+\int_{\mathbb{R}^3}(V_1+\rho^{\gamma-3}|(v_t,\nabla v)|^2)\,dx\right)\\&+C\|(u,u_t,\nabla u)\|^2+\lambda_1\|(v_t,\nabla v)\|^2\le 0.\end{aligned} \tag{13}$$

Proof. First, it follows directly from (6a) that

$$\|v_t\|^2\le C\|\nabla v,\nabla u\|^2. \tag{14}$$

Integrating the product of (6b) with ∇v and using (7) and the Young's inequality, we have, for any constant $\lambda>0$,

$$\int_{\mathbb{R}^3}p'(\rho)|\nabla v|^2\,dx\le\lambda\|\nabla v\|^2+C(\lambda)\|(u,u_t,\nabla u)\|^2.$$

Then, the positive lower bound of $p'(\rho)$ and choosing a sufficiently small λ yield,

$$\|\nabla v\|^2\le C\|(u,u_t,\nabla u)\|^2. \tag{15}$$

Multiplying (14) and (15) by λ_1, adding them together with (8) and (10), and using (14) and (15) for the terms on the right side, we obtain (13) for small δ and λ_1.

Lemma 4. *For some constant* $\lambda_2 > 0$,

$$\begin{aligned}&\frac{1}{2}\frac{d}{dt}\left(\|(\nabla u_t, D_x^2 u)\|^2 + \int_{\mathbb{R}^3} \left(\rho^{\gamma-3}|(\nabla v_t, D_x^2 v)|^2 + V_2\right) dx\right)\\ &+ C\|(\nabla u_t, D_x^2 u)\|^2 + \lambda_2\|(\nabla v_t, D_x^2 v)\|^2 \le C(\delta+\lambda_2)\|(u, u_t, \nabla u)\|^2,\end{aligned} \tag{16}$$

where $V_2 = 2(\gamma-2)\rho^{\gamma-4}\sum_{i,j} v_{x_i} v_{x_j} v_{x_i x_j}$.

Proof. Differentiating (6b) with respect to x twice and integrating its product with $D_x^2 u$ yield, after intergration by parts,

$$\frac{1}{2}\frac{d}{dt}\|D_x^2 u\|^2 + \|D_x^2 u\|^2 - \int_{\mathbb{R}^3} D_x^2 p(\rho) D_x^2(\mathrm{div} u)\, dx + \int_{\mathbb{R}^3} D_x^2(u\cdot\nabla u) D_x^2 u\, dx = 0.$$

From (6a) and (7), one has, for sufficiently small δ,

$$\begin{aligned}&\frac{1}{2}\frac{d}{dt}\int_{\mathbb{R}^3}\left(|D_x^2 u|^2 + \rho^{\gamma-3}|D_x^2 v|^2 + 2(\gamma-2)\rho^{\gamma-4}\sum_{i,j} v_{x_i} v_{x_j} v_{x_i x_j}\right) dx\\ &+ C\|D_x^2 u\|^2 \le C\delta\|(\nabla v, \nabla v_t, D_x^2 v, \nabla u)\|^2.\end{aligned} \tag{17}$$

It follows directly from differentiating (6a) with respect to x that

$$\|\nabla v_t\|^2 \le C\|(\nabla v, \nabla u)\|_{H^1}^2, \tag{18}$$

and differentiating (6b) with respect to x leads to

$$\|D_x^2 v\|^2 \le C\|(\nabla v, u, \nabla u, \nabla u_t, D_x^2 u)\|^2. \tag{19}$$

By differentiating (6b) with respect to x and t and integrating its product with ∇u_t, one obtains

$$\frac{1}{2}\frac{d}{dt}\|\nabla u_t\|^2 + \|\nabla u_t\|^2 - \int_{\mathbb{R}^3} \nabla(p'(\rho) v_t)\nabla(\mathrm{div} u_t)\, dx + \int_{\mathbb{R}^3} \nabla(u\cdot\nabla u)_t \nabla u_t\, dx = 0.$$

From (6a) and (7), one has, by some straightforward calculations, for sufficiently small δ,

$$\frac{1}{2}\frac{d}{dt}\int_{\mathbb{R}^3}\left(|\nabla u_t|^2 + \rho^{\gamma-3}|\nabla v_t|^2\right) dx + C\|\nabla u_t\|^2 \le C\delta\|(\nabla v, D_x^2 v, \nabla u, D_x^2 u)\|^2. \tag{20}$$

Multiplying (18) and (19) by λ_2 and adding them together with (17) and (20), and using (15), (18), and (19) again, we obtain (16).

Lemma 5. *For some small constants* $\lambda > 0$ *and* $\lambda_3 > 0$,

$$\begin{aligned}&\frac{1}{2}\frac{d}{dt}\left(\|(D^2 u_t, D_x^3 u)\|^2 + \int_{\mathbb{R}^3}\left(\rho^{\gamma-3}|(D_x^2 v_t, D_x^3 v)|^2 + V_3\right) dx\right)\\ &+ C\|(D_x^2 u_t, D_x^3 u)\|^2 + \lambda_3\|(D_x^2 v_t, D_x^3 v)\|^2\\ &\le C(\delta+\lambda+\lambda_3)\|(u, u_t, \nabla u, \nabla u_t, D_x^2 u)\|^2.\end{aligned} \tag{21}$$

where $V_3 = 2(\gamma-2)\rho^{\gamma-5}\sum_{i,j,k} v_{x_i x_j x_k}\left(3\rho v_{x_i} v_{x_j x_k} + (\gamma-3) v_{x_i} v_{x_j} v_{x_k}\right)$.

Proof. Take the third derivative of (6b) and integrating its product with $D_x^3 u$ to get,

$$\frac{1}{2}\frac{d}{dt}\|D_x^3 u\|^2 + \|D_x^3 u\|^2 - \int_{\mathbb{R}^3} D_x^3 p(\rho) D_x^3 \mathrm{div} u \, dx + \int_{\mathbb{R}^3} D_x^3(u\cdot\nabla u) D_x^3 u \, dx = 0.$$

Using (6a) and (7) and from Gagliado-Nirenberg's inequality and some tedious calculations, one has, for sufficiently small δ and λ,

$$\begin{aligned} &\frac{1}{2}\frac{d}{dt}\left(\|D_x^3 u\|^2 + \int_{\mathbb{R}^3} \rho^{\gamma-3}|D_x^3 v|^2 \, dx + \int_{\mathbb{R}^3} V_3 \, dx\right) dx + C\|D_x^3 u\|^2 \\ &\le C(\lambda+\delta)\left(\|(\nabla v, \nabla u)\|_{H^1}^2 + \|D_x^2 u_t\|^2\right). \end{aligned} \tag{22}$$

Differentiating (6b) twice with respect to x and once with respect to t and integrating its product with $D_x^2 u_t$ lead to

$$\begin{aligned} &\frac{1}{2}\frac{d}{dt}\left(\|D_x^2 u_t\|^2 + \int_{\mathbb{R}^3} \rho^{\gamma-3}|D_x^2 v_t|^2 \, dx\right) + C\|D_x^2 u_t\|^2 \\ &\le C\delta\left(\|(\nabla v, \nabla u)\|_{H^1}^2 + \|(\nabla u_t, D_x^3 u)\|^2\right). \end{aligned} \tag{23}$$

Then estimate (21) follows from (22) and (23), for sufficiently small constants δ and $\lambda_3 > 0$. The proof of Lemma 5 is complete.

To prove Theorem 1 we add (13), (16), and (21) to obtain, for sufficiently small $\delta, \lambda, \lambda_i, i = 1, 2, 3$,

$$\begin{aligned} &\frac{1}{2}\frac{d}{dt}\int_{\mathbb{R}^3} F \, dx + C\left(\|u\|_{H^3}^2 + \|u_t\|_{H^2}^2\right) \\ &+ \lambda_1\|(v_t, \nabla v)\|^2 + \lambda_2\|\nabla(v_t, \nabla v)\|^2 + \lambda_3\|D_x^2(v_t, \nabla v)\|^2 \le 0, \end{aligned} \tag{24}$$

where

$$\begin{aligned} F =& |(u, u_t, \nabla u, \nabla u_t, D_x^2 u, D_x^2 u_t, D_x^3 u)|^2 \\ &+ \rho^{\gamma-3}|(v_t, \nabla v, \nabla v_t, D_x^2 v, D_x^2 v_t, D_x^3 v)|^2 + V_1 + V_2 + V_3. \end{aligned}$$

From (7), $C_1 E(t) \le \int_{\mathbb{R}^3} F \, dx \le C_2 E(t)$ for some positive constants C_1 and C_2. Integrating (24) with respect to t yields

$$\int_{\mathbb{R}^3} F(x,t) \, dx + C\int_0^t \left(\|u\|_{H^3}^2 + \|(\nabla v, v_t, u_t)\|_{H^2}^2\right)(\tau) d\tau \le \int_{\mathbb{R}^3} F(x,0) \, dx.$$

Then (5) follows and the proof of Theorem 1 is complete.

We notice that $\|\rho - \bar{\rho}\|^2$ generally does not decay exponentially, as the following theorem shows.

Theorem 2. *Suppose that the initial value $(\rho_0 - \bar{\rho}, u_0) \in H^3$ and $E(0)$ is sufficiently small so that Theorem 1 holds. If supp $(\rho_0 - \bar{\rho}, u_0) \subset \{|x| \le R\}$ for some $R > 0$ and $\int_{\mathbb{R}^3} (\rho_0 - \bar{\rho}) \, dx \ne 0$, then*

$$\|\rho - \bar{\rho}\| \ge C(R + \bar{\sigma} t)^{-3/2}, \tag{25}$$

for some constant $C > 0$ and $\bar{\sigma} = \bar{\rho}^{\frac{\gamma-1}{2}}$.

Theorem 1 shows that the damping in (1) can prevent the formation of singularities in smooth solutions with small initial data. If the initial data are large, the solution of the Cauchy problem (1) and (2) will still develop singularities in a finite time even if the initial data are smooth. Define

$$F(t) = \int_{\mathbb{R}^3} x \cdot \rho u \, dx, \quad A(t) = (R + \bar{\sigma}t)^2 \int_{\mathbb{R}^3} (\rho_0 - \bar{\rho}) dx + \frac{4\pi\bar{\rho}}{3}(R + \bar{\sigma}t)^5,$$

Then we have the following theorem.

Theorem 3. *Suppose that the initial data* (ρ_0, u_0) *are in* $C^1(\mathbb{R}^3)$ *and satisfy* *supp* $(\rho_0 - \bar{\rho}, u_0) \subset \{|x| \leq R\}$ *and* $\int_{\mathbb{R}^3}(\rho_0 - \bar{\rho})dx \geq 0$, *and* $(\rho, u) \in C^1([0, L) \times \mathbb{R}^3)$ *is a solution of the Cauchy problem* (1) *and* (2) *for some* $L > 0$. *For any fixed* $T > 0$, *if* $F(0) > \exp(cT)A(T)/T$, *then the life span* L *of the* C^1 *solution* (ρ, u) *is less than* T.

The proofs of Theorems 2 and 3 were established in Sideris-Thomases-Wang [39]. The global L^∞ weak entropy solution with spherical symmetry of (1) can be constructed as in Chen-Glimm [7] and Chen-Wang [8, 9].

Acknowledgments

The results discussed here are from several joint papers with G.-Q. Chen and one joint paper with T. Sideris and B. Thomases. My research was supported in part by the National Science Foundation and the Office of Naval Research.

References

1. Alinhac, S. (1993): Temps de vie des solutions régulières des équations d'Euler compressibles axisymétriques en dimension deux. Invent. Math., **111**, 627-670
2. Bianchini, S., Bressan, A. (2001): Vanishing viscosity solutions of nonlinear hyperbolic systems. preprint
3. Bressan, A. (1992): Global solutions of systems of conservation laws by wave-front tracking. J. Math. Anal. Appl., **170**, 414-432
4. Canic, S., Keyfitz, B., Lieberman, G.M. (2000): A proof of existence of perturbed steady transonic shocks via a free boundary problem. Comm. Pure Appl. Math., **53**, 484-511
5. Chen, G.-Q. (1986): Convergence of the Lax-Friedrichs scheme for isentropic gas dynamics (III). Acta Math. Sci., **6**, 75-120
6. Chen, G.-Q., Feldman, M. (2001): Multidimensional transonic shocks and free boundary problems for nonlinear equations of mixed type. Preprint, Northwestern University
7. Chen, G.-Q., Glimm, J. (1996): Global solution to the compressible Euler equations with geometrical structure. Commun. Math. Phys., **179**, 153-193
8. Chen, G.-Q., Wang, D. (1996): Convergence of shock capturing schemes for the compressible Euler-Poisson equations. Commun. Math. Phys., **179**, 333-364

9. Chen, G.-Q., Wang, D. (1998): Shock capturing approximations to the compressible Euler equations with geometric structure and related equations. Z. Angew. Math. Phys., **49**, 341-362
10. Chen, S.-X. (1998): Asymptotic behavior of supersonic flow past a convex combined wedge. Chinese Annals Math., **19B**, 255-264
11. Dafermos, C.M. (1972): Polygonal approximations of solutions of the initial-value problem for a conservation law. J. Math. Anal. Appl., **38**, 33-41
12. Dafermos, C.M. (1981): Can dissipation prevent the breaking of waves ?. In: Transactions of the Twenty-Sixth Conference of Army Mathematicians, 187-198. ARO Rep. **81. 1**, U. S. Army Res. Office, Research Triangle Park, North Carolina, USA
13. Dafermos, C.M. (1985): Development of singularities in the motion of materials with fading memory. Arch. Rational Mech. Anal., **91**, 193-205
14. Dafermos, C.M. (1988): Solutions in L^∞ for a conservation law with memory. In: Analyse Mathématique et Applications, 117-128. Gauthier-Villars: Paris
15. Dafermos, C.M., Hsiao, L. (1986): Development of singularities in solutions of the equations of nonlinear thermoelasticity. Q. Appl. Math., **44**, 463-474
16. DiPerna, R. (1976): Global existence of solutions to nonlinear hyperbolic systems of conservation laws. J. Diff. Eqs., **20**, 187-212
17. DiPerna, R. (1983): Convergence of viscosity method for isentropic gas dynamics. Commun. Math. Phys., **91**, 1-30
18. Engelberg, S., Liu, H., Tadmor, E. (2001): Critical thresholds in Euler-Poisson equations. Indiana Univ. Math. J., **50**, 109-157
19. Gamba, I.M., Morawetz, C.S. (1996): A viscous approximation for a 2-D steady semiconductor or transonic gas dynamic flow: existence theorem for potential flow. Comm. Pure Appl. Math., **49**, 999-1049
20. Glimm, J. (1965): Solutions in the large for nonlinear hyperbolic system of equations. Comm. Pure Appl. Math., **18**, 95-105
21. Grassin, M., Serre, D. (1997): Existence de solutions globales et régulières aux équations d'Euler pour un gaz parfait isentropique. C. R. Acad. Sci. Paris Ser. I, **325**, 721-726
22. Hrusa, W.J., Messaoudi, S.A. (1990): On formation of singularities in one-dimensional nonlinear thermoelasticity. Arch. Rational Mech. Anal., **111**, 135-151
23. Hsiao, L. (1997): Quasilinear Hyperbolic Systems and Dissipative Mechanisms. World Scientific Publishing, Singapore
24. John, F. (1974): Formation of singularities in one-dimensional nonlinear wave propagation. Comm. Pure Appl. Math., **27**, 377-405
25. Klainerman, S., Majda, A. (1980): Formation of singularities for wave equations including the nonlinear vibrating string. Comm. Pure Appl. Math., **33**, 241-263
26. Lax, P.D. (1964): Development of singularities in solutions of nonlinear hyperbolic partial differential equations. J. Math. Phys., **5**, 611-613
27. Li, T.-T. (1994): Global Classical Solutions for Quasilinear Hyperbolic Systems. John Wiley & Sons, Ltd., Chichester
28. Lien, W.-C., Liu, T.-P. (1999): Nonlinear stability of a self-similar 3-dimensional gas flow. Commun. Math. Phys., **204**, 525-549
29. Lions, P.-L., Perthame, B., Souganidis, P. (1996): Existence and stability of entropy solutions for the hyperbolic systems of isentropic gas dynamics in Eulerian and Lagrangian coordinates. Comm. Pure Appl. Math., **49**, 599-638

30. Lions, P.-L., Perthame, B., Tadmor, E. (1994): Kinetic formulation of the isentropic gas dynamics and p-systems. Commun. Math. Phys., **163**, 169-172
31. Liu, T.-P. (1979): The development of singularities in the nonlinear waves for quasilinear hyperbolic partial differential equations. J. Diff. Eqs., **33**, 92-111
32. Majda, A. (1984): Compressible fluid flow and systems of conservation laws in several space variables. Applied Mathematical Sciences **53**. Springer-Verlag, New York
33. Makino, T., Ukai, S., Kawashima, S. (1986): Sur la solution à support compact de l'iquations d'Euler compressible. Japan J. Appl. Math., **3**, 249-257
34. Morawetz, C.S. (1985): On a weak solution for a transonic flow problem. Comm. Pure Appl. Math., **38**, 797-818
35. Morawetz, C.S. (1995): On steady transonic flow by compensated compactness. Meth. Appl. Anal., **2**, 257-268
36. Nishida, T. (1978): Nonlinear Hyperbolic Equations and Related Topics in Fluid Dynamics. Publications Mathematiques D'Orsay **78-02**. Université de Paris-Sud, Orsay
37. Risebro, N.H. (1993): A front-tracking alternative to the random choice method. Proc. Amer. Math. Soc., **117**, 1125-1139
38. Sideris, T.C. (1985): Formation of singularities in three-dimensional compressible fluids. Commun. Math. Phys., **101**, 475-485
39. Sideris, T.C., Thomases, B., Wang, D. (2003): Decay and singularity formation for solutions to the three-dimensional Euler equations with damping. To appear in Commun. Partial Differential Equations.
40. Slemrod, M. (1981): Global existence, uniqueness, and asymptotic stability of classical smooth solutions in one-dimensional nonlinear thermoelasticity. Arch. Rational Mech. Anal., **76**, 97-133
41. Tadmor, E. (1986): A minimum entropy principle in the gas dynamics equations. Appl. Numer. Math., **2**, 211-219
42. Wang, D. (2001): Global solutions and relaxation limits of Euler-Poisson equations. Z. angew. Math. Phys., 52, 620-630
43. Wang, D., Chen, G.-Q. (1998): Formation of singularities in compressible Euler-Poisson fluids with heat diffusion and damping relaxation. J. Diff. Eqs., **144**, 44-65
44. Zhang, Y.-Q. (1999): Global existence of steady supersonic potential flow past a curved wedge with a piecewise smooth boundary. SIAM J. Math. Anal., **31**, 166-183
45. Zheng, Y. (2001): Systems of Conservation Laws: Two-Dimensional Riemann Problems. Birkhäuser, Boston

Isentropic Gas Dynamics with Large Data

Robin Young

University of Massachusetts, Amherst
young@math.umass.edu

Introduction

We consider the equations of isentropic gas dynamics in a Lagrangian frame, also known as the p-system

$$\begin{pmatrix} v \\ u \end{pmatrix}_t + \begin{pmatrix} -u \\ p(v) \end{pmatrix}_x = 0, \tag{1}$$

where v, u and p are the specific volume, velocity and pressure, respectively. We assume that the pressure function satisfies the physical conditions

$$p'(v) < 0, \quad \lim_{v \to 0} p(v) = \infty \quad \text{and} \quad \lim_{v \to \infty} p(v) = 0, \tag{2}$$

which correspond to hyperbolicity, infinite compression, and the vacuum, respectively. We are interested in solutions including or near vacuum, which can be equally characterized as infinite volume or vanishing pressure, density or sound speed. It is well known [9] that a necessary condition for the formation of a vacuum, say in the solution to the Riemann problem, is

$$\int^{\infty} \sqrt{-p'(\nu)}\, d\nu < \infty, \tag{3}$$

and we assume this to be the case.

Much is known about solutions to (1); however, constructive theories rely on local asymptotic estimates for wave interactions, which fail near the vacuum [8]. Here we treat waves and their interactions exactly, so that wave strengths can be large. We solve the Riemann problem including the vacuum, in which the specific volume is a Radon measure whose singular part is a Dirac δ-function. At first glance, the vacuum seems less singular in Eulerian coordinates (with $\rho = 0$) than in Lagrangian (with $v = \delta$), but the solutions are equivalent [10], and we prefer the simpler wave structure of the Lagrangian point of view.

Our long-term goal is to understand solutions with strong waves, including a constructive proof of the existence and decay of such solutions. In this paper, we analyze the structure of solutions, assuming the existence and the usual regularity of solutions constructed by Glimm's scheme or front

[0] This work was partially supported by the National Science Foundation under grant number DMS-0104485.

tracking [4, 3, 1]. In particular, we are interested in solutions near the vacuum, for which there is no constructive proof as yet. The vacuum has been the subject of several studies, and in this paper we verify certain conjectures on the (lack of) formation and extinction of the vacuum [8, 6, 7].

We begin by describing the nonlinear waves and the solution of the Riemann problem. In our development it is critical that all waves can be described in terms of a single nonlinear function $g(v_1, v_2)$. This representation works for shocks, rarefactions and composite waves of arbitrary strength, and for nonconvex pressures. Wave interactions are then understood by studying the function g in detail.

We then give a well-known condition on Riemann data which leads to a vacuum, namely when the velocity difference is large enough. This naturally extends to a condition on general Cauchy data which is a perturbation of such Riemann data. Next, after a careful analysis of wave interactions, we show that this is the only way that a vacuum may develop, which was conjectured in [7]. The idea is fairly simple: a vacuum could arise at finite times from the interaction of two rarefactions, but as the interaction occurs and the solution gets closer to vacuum, so wavespeeds and the rate of further interaction slow down. Indeed, to reach the vacuum would require zero characteristic speed, but if a characteristic originally has nonzero speed, then (by uniqueness for ODE) it always has nonzero speed. A refinement of this argument provides time dependent bounds for the specific volume.

We also examine the structure of solutions with a vacuum: in [8] it is shown that an oncoming shock always meets the vacuum in finite time, and we reprove this in our context. We show that the shock disappears into the vacuum, but the velocity of the edge of the vacuum is reduced as a result. As a result of many such interactions, the vacuum may eventually be extinguished, and two shocks are emitted from the point of extinction. We describe the solution, thus confirming a conjecture in [8].

1 The Riemann problem

We begin by recalling the solution to the Riemann problem, which is well known [11, 9, 15, 2]. Writing the system (1) in quasilinear form, we get

$$\begin{pmatrix} v \\ u \end{pmatrix}_t + \begin{pmatrix} 0 & -1 \\ -c^2(v) & 0 \end{pmatrix} \begin{pmatrix} v \\ u \end{pmatrix}_x = 0, \tag{4}$$

where the sound speed c is defined by

$$c(v) = \sqrt{-p'(v)}, \tag{5}$$

which is defined and nonzero for all v. The (continuous) wavespeeds are the eigenvalues $\mp c(v)$ of the flux matrix, so we can clearly identify backward and forward waves. Simple waves are described by integrating the eigenvectors, and these integrals can be written

$$u - u_0 = \pm \int_{v_0}^{v} c(\nu)\, d\nu. \tag{6}$$

Equation (6) describes the change in state across a backward or forward simple wave, respectively. The Rankine-Hugoniot conditions have the solution

$$\sigma = \mp \sqrt{\frac{p(v_0) - p(v)}{v - v_0}} \quad \text{and} \quad u - u_0 = \pm \sigma\,(v - v_0), \tag{7}$$

(where σ is the shock speed), giving the backward and forward shock curves, respectively. Both curves (6) and (7) are globally defined, so we are not restricted to small waves.

We piece the simple wave and Hugoniot curves together to construct self-similar centered waves, with which we will solve the Riemann problem. By considering the Lax/Liu entropy conditions for both convex and non-convex pressure functions, we obtain the following concise description of the composite waves, a version of which appears in [11]. Given fixed specific volumes v_1 and v_2 with $v_1 \le v_2$ (resp. $v_1 \ge v_2$), define $\widehat{p}$ to be the lower (resp. upper) convex envelope of the pressure p between v_1 and v_2. Now define the local sound speed

$$\widehat{c}\,(v) = \sqrt{-\widehat{p}\,'(v)} \tag{8}$$

for v between v_1 and v_2, and define

$$g(v_1, v_2) = \int_{v_1}^{v_2} \widehat{c}\,(\nu)\, d\nu = \int_{v_1}^{v_2} \sqrt{-\widehat{p}\,'(\nu)}\, d\nu. \tag{9}$$

Note that g is not a simple integral as the convex function $\widehat{p}$ varies with v_1 and v_2. Using the function g, we can describe both forward and backward waves by the single equation

$$u_r - u_l = g(v_a, v_b), \tag{10}$$

where the subscripts refer to the right, left, ahead and behind states, respectively. The wave consists of jumps or characteristics with speeds satisfying

$$\frac{dx}{dt} = \mp \widehat{c}\,(v), \tag{11}$$

and along which we have

$$u - u_a = \pm g(v_a, v), \tag{12}$$

where v can be taken to vary continuously between v_a and v_b, and jumps or shocks occur when $\widehat{c}$ is constant on a subinterval. Note that when p is convex, (10) reduces to (6) or (7) for rarefactions or shocks, respectively.

We now combine our descriptions of the forward and backward waves to solve the Riemann problem: for details, see [12, 13]. Given left and right states $(v_l\; u_l)^\dagger$ and $(v_r\; u_r)^\dagger$, we denote the outgoing middle state by $(v_*\; u_*)^\dagger$, which connects backward and forward waves from the outer states. According to (10), we have

$$u_* - u_l = g(v_l, v_*) \quad \text{and} \quad u_r - u_* = g(v_r, v_*), \tag{13}$$

so we must solve the single nonlinear equation

$$u_r - u_l = g(v_l, v_*) + g(v_r, v_*) \equiv \phi(v_*) \tag{14}$$

for the middle volume v_*. The function g increases in the second variable, so there is a unique solution to (14) provided the velocity difference $u_r - u_l$ is in the range of ϕ. This range is an interval $(-\infty, K)$ where $K = \phi(\infty)$ is finite by (3), and if $u_r - u_l \geq K$, a vacuum forms. The vacuum can be described in Lagrangian coordinates as a δ-function in specific volume, where the weight of the δ-function gives the width of the vacuum in Eulerian (spatial) coordinates. For the Riemann problem, the vacuum is described by adding the singular measure

$$\nu_S = (u_+ - u_-)\, t\, \delta(x) \tag{15}$$

to the specific volume, where

$$u_- = u_l + g(v_l, \infty) \quad \text{and} \quad u_+ = u_r - g(v_r, \infty) \tag{16}$$

are the fluid velocities at the left and right edges of the vacuum, respectively. The solution thus consists of two infinite rarefactions, with a (stationary) Dirac measure across which the velocity increases from u_- to u_+. The following theorem is proved in detail in [12].

Theorem 1. *Given arbitrary states $(v_l\; u_l)^\dagger$ and $(v_r\; u_r)^\dagger$, there is a unique solution $(v(x,t), u(x,t))^\dagger$ to the Riemann problem, with*

$$u(x,t) \in L^\infty \cap BV \cap L^1_{loc} \quad \text{and} \quad v(x,t) \in \mathcal{M} \cap L^1_{loc}, \tag{17}$$

the singular part of v being given by (15), (16).

1.1 Interactions

Before continuing, we must understand wave interactions in detail. We consider Glimm interactions, which are not exact solutions but give the the asymptotic states resulting from the interaction, and are those used in constructing random choice or front tracking approximations [4, 1]. Thus we consider two incident waves separated by a common state, and we wish to resolve the Riemann problem using the exterior states, and give the outgoing wave strengths in terms of the incoming strengths.

By symmetry, there are only two cases to consider, namely a forward wave passing through a backward wave, and two forward waves colliding. As in our solution of the Riemann problem, we use (10) to describe the waves and eliminate the u's to derive a single nonlinear equation for the outgoing middle specific volume. For two forward incident waves, say spanning left, middle and right states, (10) gives

$$u_m - u_l = g(v_m, v_l) \quad \text{and} \quad u_r - u_m = g(v_r, v_m), \tag{18}$$

which together with (13) reduces to

$$g(v_m, v_l) + g(v_r, v_m) = g(v_l, v_*) + g(v_r, v_*), \tag{19}$$

in which we must solve for v_*. Similarly, for interacting forward and backward waves, say spanning west, south and east states, we have

$$u_s - u_w = g(v_s, v_w) \quad \text{and} \quad u_e - u_s = g(v_s, v_e), \tag{20}$$

before the interaction, and (setting $v_n = v_*$) after interaction we have

$$u_n - u_w = g(v_w, v_n) \quad \text{and} \quad u_e - u_n = g(v_e, v_n), \tag{21}$$

so we must solve

$$g(v_s, v_w) + g(v_s, v_e) = g(v_w, v_n) + g(v_e, v_n) \tag{22}$$

for the outgoing state v_n.

Clearly the key to understanding these interactions is a sharp analysis of the nonlinear function g. To this end, recalling (9), we define auxiliary functions $h(v)$ and $\theta(v_1, v_2)$ by

$$\begin{aligned} h(v) &= \int_1^v c(\nu)\, d\nu \quad \text{and} \\ \theta(v_1, v_2) &= \frac{1}{2}\,(g(v_1, v_2) - h(v_2) + h(v_1)), \end{aligned} \tag{23}$$

so that we have the identity

$$g(v_1, v_2) = h(v_2) - h(v_1) + 2\,\theta(v_1, v_2) \tag{24}$$

for all v_i. Here h is an exact integral, and θ is a measure of how far g is from being an integral; in other words, according to (6) and (7), θ is a measure of the total shock strength in a wave. Indeed, when p is convex, θ is zero across rarefactions. It is shown in [12, 13] that each of the functions $g(v_1, v_2)$, $\theta(v_1, v_2)$ and $h(v_2) - h(v_1)$ are increasing in the second and decreasing in the first variables, so that combinations of these may be used to measure wave strength. It turns out that the best measure of wave strength is

$$\gamma(v_1, v_2) \equiv h(v_2) - h(v_1) + \theta(v_1, v_2). \tag{25}$$

Indeed, after substituting (24) into (22) and simplifying, we get

$$\gamma(v_w, v_n) + \theta(v_e, v_n) = \gamma(v_s, v_e) + \theta(v_s, v_w), \tag{26}$$

and similarly

$$\gamma(v_e, v_n) + \theta(v_w, v_n) = \gamma(v_s, v_w) + \theta(v_s, v_e), \tag{27}$$

with similar relations holding for (19). In particular, in the convex case, $\theta(v_1, v_2) = 0$ across a rarefaction, and we conclude that when measured as in (25), passage through a rarefaction has no effect on wave strength. By similarly analyzing properties of g, h and θ, in [12] we deduce the following facts about interactions in the p-system, independent of the size of waves:

- A wave always has the same type (shock or rarefaction) after crossing a wave of the opposite family;
- The strength of a wave is unchanged if it passes through a continuous wave of the opposite family;
- The state behind a shock wave is always further from the vacuum state.

Here and in the sequel we use the terms shock and rarefaction to include composite waves in the non-convex case.

2 The vacuum

We now examine the conditions under which a vacuum may arise, and describe the effects of waves interacting with the vacuum. Some of our results were anticipated in [8] and conjectured in [7]. We have already seen that a vacuum may form in the Riemann problem if the velocity difference is larger than the capacity for the gas to expand. This is similarly true for general initial data which can be locally regarded as a perturbation of Riemann data, and we get a similar condition for vacuum formation. We refer to [12, 14] for details.

Theorem 2. *For any $x \in \mathbf{R}$ such that*

$$u_0(x+) - u_0(x-) \geq g(v_0(x-), \infty) + g(v_0(x+), \infty) \tag{28}$$

holds, a vacuum develops immediately at time $t = 0+$. The vacuum is given as a δ-function living at x analogous to (15).

If (28) holds, we say that the vacuum is embedded in the data. Note that this condition is violated if the initial velocity is continuous. We shall show that (28) is the only way to generate a vacuum: that is, if the initial data contains no embedded vacuum, then no vacuum forms in finite time. A similar result was conjectured in [7]. Note that we are here assuming the prior existence of a solution which can be approximated by front tracking or random choice: a harder problem is to prove a constructive existence result which relies on bounds for approximate solutions.

Our proof relies on a study of wave interactions: having noted that the state behind a shock wave is always further from the vacuum, our first observation is that if a vacuum were to develop at some positive time, it would have to come from the interaction of forward and backward rarefactions. Such an interaction can easily be found: according to (22), the incident rarefactions (joining west, south and east states) should be strong enough that

$$g(v_s, v_w) + g(v_s, v_e) \geq g(v_w, \infty) + g(v_e, \infty). \tag{29}$$

It is apparent that the vacuum can form, and indeed it does appear asymptotically. However, the vacuum does not develop in finite time. To see this, we examine the characteristics emerging from the interaction.

Since we are considering the interaction of forward and backward rarefactions, we know that the waves leaving the interaction region are also rarefactions. Consider any curve which crosses the outgoing forward rarefaction wave, say $(X(\eta), T(\eta))$. Across this curve, the states $(v(\eta)\ u(\eta))^\dagger$ satisfy (12). Also, the wave is simple, so we can trace the characteristics backwards to the (northeast) edge of the interaction region, say $(x(\eta), t(\eta))$. These characteristics satisfy $dx/dt = \widehat{c}$, so that

$$X(\eta) - x(\eta) = \widehat{c}(\eta)\ (T(\eta) - t(\eta)). \tag{30}$$

Also, since the curve $(x(\eta), t(\eta))$ is the edge of a forward simple wave, it must be a backward characteristic [5, 3], and so satisfies

$$\frac{dx}{d\eta} = -\,\widehat{c}(v(\eta))\ \frac{dt}{d\eta}. \tag{31}$$

Now regarding $(X(\eta), T(\eta))$ and $v(\eta)$ as data, we combine (30) and (31) to get the differential equation

$$\frac{d}{d\eta}(\widehat{c}\ t) + H'(\eta) = -\,\widehat{c}(v(\eta))\ \frac{dt}{d\eta}, \tag{32}$$

where $H(\eta) = X(\eta) - \widehat{c}(\eta)\ T(\eta)$ is monotone decreasing as $v(\eta)$ increases from v_e up to v_n (or ∞). Solving for $t(\eta)$, we get

$$\sqrt{c(\eta)}\, t(\eta) - \sqrt{c(\eta_0)}\, t(\eta_0) = \int_{\eta_0}^{\eta} \frac{-H'}{2\sqrt{c}}\, d\eta > 0. \tag{33}$$

We conclude that $\sqrt{c(\eta)}\,t(\eta)$ is bounded away from zero for all η, and in particular, if our state v_n approaches vacuum, then $c \to 0$ and $t(\eta) \to \infty$, so the time at which state v_n emerges from the interaction region becomes infinite.

Theorem 3. *A vacuum cannot form from wave interactions in finite time, so any vacuum in the solution must be embedded in the initial data. In particular, if the initial velocity is continuous, there is no vacuum in the solution.*

3 Extinction of the vacuum

We now discuss the evolution of the vacuum state. If a vacuum is embedded in the data, the singular measure (15) appears in the specific volume. The weight of the measure, $(u_+ - u_-)\,t$, gives the width of the vacuum in physical space (Eulerian coordinates), which initially increases at a constant rate. We have already seen that a rarefaction will not reach the vacuum in finite time, so will not affect the rate of expansion. We now consider the effect of a shock on the vacuum. This problem has been studied in [8] (in Eulerian coordinates).

Our approach is similar to that used for rarefactions: we begin with a backward shock approaching the vacuum from the right, thus passing through the infinite forward rarefaction which borders the vacuum. For brevity, we take the rarefaction to be centered at the origin. Suppose the shock follows the curve $(x(\eta), t(\eta))$, with states $(v_-\ u_-)^\dagger$ on the left (ahead) and $(v_+\ u_+)^\dagger$ on the right. Since the shock passes through a centered rarefaction, we have

$$x(\eta) = \widehat{c}\,(v_-(\eta))\,t(\eta), \tag{34}$$

while the shock path satisfies

$$\frac{dx}{d\eta} = -\sigma(\eta)\,\frac{dt}{d\eta}, \tag{35}$$

σ being given by (7). Differentiating (34) and using (35), we again get a differential equation

$$(\sigma + c)\,\frac{dt}{d\eta} + t(\eta) = 0, \tag{36}$$

whose solution

$$\log t(\eta) = \log t(\eta_0) - \int_{\eta_0}^{\eta} \frac{d\eta}{\sigma + c} \tag{37}$$

together with (34) gives the shock trajectory. Note that σ depends on $v_+(\eta)$, which is given in terms of $v_-(\eta)$ and the initial shock by (22). The shock meets the vacuum in finite time provided the integral converges (as $c \to 0$), and this follows from the properties of g, see [12, 14].

We have shown that any shock meeting the rarefaction bordering a vacuum will meet the vacuum in finite time, and we now examine what happens next. We have observed that the strength of the shock, when measured according to (25), doesn't change across a rarefaction, and so is nonzero when the shock meets the vacuum. Thus when the shock meets the vacuum, we expect some effect on the vacuum. What happens is the shock disappears into the vacuum, while the velocity of the right edge of the vacuum state is decreased by an appropriate amount. Indeed, ahead of the shock, the edge velocity is given by (13):

$$u_+^a = u_s - g(v_s, \infty), \tag{38}$$

while behind the shock, the vacuum is reached by an infinite rarefaction from state $(v_e \; u_e)^\dagger$, so that

$$u_+^b = u_e - g(v_e, \infty). \tag{39}$$

On the other hand, for the initial shock we have

$$u_e - u_s = g(v_e, v_s) < 0, \tag{40}$$

so that the difference in edge speed of the vacuum is

$$u_+^b - u_+^a = g(v_s, v_e) - g(v_e, \infty) + g(v_s, \infty) < 0, \tag{41}$$

and indeed in the convex case, using (24) and (25) we get exactly

$$u_+^b - u_+^a = 2\,(h(v_e) - h(v_s) + \theta(v_s, v_e)) = 2\,\gamma(v_s, v_e). \tag{42}$$

We now describe the global evolution of the vacuum: we have seen that it starts out expanding at a constant rate (in spatial coordinates), but each time a shock meets the vacuum, the edge velocity decreases, so the rate of expansion decreases. After enough shocks have met the vacuum, the edge velocity on the right may be smaller than that on the left, so the width of the vacuum decreases (this is a compressive vacuum in [8]). Eventually the edges of the vacuum will themselves meet, and the vacuum is extinguished. It was conjectured in [8] that two shocks form at the point of extinction, and we now describe why this is the case; see [12] for details.

We would like to solve the Riemann problem at the point of extinction of the vacuum, but cannot because the specific volume is infinite there, although the left and right velocities are finite. We instead solve the Riemann problem with these left and right velocities, but with finite specific volume, say $\bar{v}$. For $\bar{v}$ large enough, (14) yields a finite value of v_* and two shocks are emitted. Now taking the limit as $\bar{v} \to \infty$, we obtain both $v_* \to \infty$ and $\sigma \to 0$, so in the limit these shocks have zero strength. However, by considering the interaction of the (infinite) forward rarefaction on the right of the vacuum with the forward shock which is emitted, we see that a backward compressive wave is reflected. This in turn runs into the weak backward shock, thus strengthening it. Similar remarks hold for the backward emitted shock; indeed, for large $\bar{v}$, the interaction is symmetric. As the shocks are strengthened, their speeds increase and the two shocks are bent outwards. Thus the emitted shocks start with zero strength and increase in strength as they bend away from the point of extinction. The two shock trajectories together form a cusp at the point of extinction.

4 Time-dependent bounds

We have assumed the existence of a solution to (1) in order to understand the structure of solutions. Actually proving the existence of such solutions usually relies on total variation bounds. Clearly as we approach the vacuum, the total variation of the specific volume is unbounded. We have seen, however, that the vacuum forms only in infinite time, so we can look for bounds which are time dependent. Here we describe one such time dependent bound (still assuming prior existence).

Theorem 4. *Suppose that the initial data* $(v_0(x)\ u_0(x))^\dagger$ *have bounded variation, and suppose also that for every* x,

$$u_0(x+) - u_0(x-) < g(v_0(x-), \infty) + g(v_0(x+), \infty). \tag{43}$$

Then there is some $\delta > 0$ *such that*

$$c(v(x,t)) \geq \frac{\delta}{T^2} \quad \textit{for} \quad t \leq T. \tag{44}$$

Proof. Define the quantity

$$\begin{aligned} \Delta x = \inf \{ & x_2 - x_1 \mid x_1 \leq x_2, \\ & u_0(x_2) - g(v_0(x_2, \infty)) < u_0(x_1) + g(v_0(x_1), \infty)\}, \end{aligned} \tag{45}$$

which is positive by our assumption. We interpret Δx as the minimum distance between approaching rarefactions. Also, define

$$c_{min} = \min_x \ c(v_0(x)) \quad \text{and} \quad c_{max} = \max_x \ c(v_0(x)), \tag{46}$$

which are the minimum and maximum initial wavespeeds, respectively. Now the minimum time until any two rarefactions meet satisfies

$$c_{max}\, T_{min} \leq \frac{\Delta x}{2}, \quad \text{or} \quad T_{min} \leq \frac{1}{2} \frac{\Delta x}{c_{max}}. \tag{47}$$

As discussed above, the specific volume increases only when opposite rarefactions interact, so we again use our analysis of such interactions. According to (33), across any rarefaction leaving such an interaction, we have

$$\sqrt{\hat{c}}\, t \geq \sqrt{c_0}\, t_0, \tag{48}$$

and so for $t \leq T$, we have

$$\sqrt{\hat{c}} \geq \frac{\sqrt{c_0}\, t_0}{T} \geq \frac{\sqrt{c_{min}}\, T_{min}}{T}. \tag{49}$$

The conclusion of the theorem now follows.

References

1. A. Bressan. *Hyperbolic Systems of Conservation Laws: The one-dimensional Cauchy Problem.* Oxford University Press, 2000.
2. Constantine M. Dafermos. *Hyperbolic Conservation Laws in Continuum Physics.* Springer, 2000.
3. R. DiPerna. Singularities of solutions of nonlinear hyperbolic systems of conservation laws. *Arch. Rat. Mech. Anal.*, 60:75–100, 1975.
4. J. Glimm. Solutions in the large for nonlinear hyperbolic systems of equations. *Comm. Pure Appl. Math.*, 18:697–715, 1965.
5. P.D. Lax. Hyperbolic systems of conservation laws, II. *Comm. Pure Appl. Math.*, 10:537–566, 1957.
6. Long-Wei Lin. On the vacuum state for the equations of isentropic gas dynamics. *J. Math. Anal. Appl.*, 121:406–425, 1987.
7. Longwei Lin. Survey of vacuum theory and conjecture. *AMS/IP Stud. Adv. Math.*, 3:293–305, 1997.
8. T.-P. Liu and J.A. Smoller. On the vacuum state for the isentropic gas dynamics equations. *Adv. Appl. Math.*, 1:345–359, 1980.
9. J. Smoller. *Shock Waves and Reaction-Diffusion Equations.* Springer-Verlag, New York, 1982.
10. D. Wagner. Equivalence of the Euler and Lagrangian equations of gas dynamics for weak solutions. *Jour. Diff. Equations*, 68:118–136, 1987.
11. B. Wendroff. The Riemann problem for materials with nonconvex equation of state. *J. Math. Anal. Appl.*, 38:454–466, 640–658, 1972.
12. Robin Young. The *p*-system with large data. In preparation.
13. Robin Young. The *p*-system I: The Riemann problem. *Contemp. Math.*, 2001. to appear.
14. Robin Young. The *p*-system II: The vacuum. In *Nonlinear Evolution Equations, Warsaw*, 2001. to appear.
15. Tong Zhang and Ling Hsiao. *The Riemann Problem and Interaction of Waves in Gas Dynamics.* Longman, New York, 1989.

Third and Fourth Order Weighted ENO Schemes for Hamilton-Jacobi Equations on 2D Unstructured Meshes

Yong-Tao Zhang and Chi-Wang Shu

Division of Applied Mathematics, Brown University, Providence, RI 02912, *zyt@cfm.brown.edu*, *zyt@cfm.brown.edu*

Summary. We present our recent work on designing third and fourth order accurate weighted essentially non-oscillatory (WENO) schemes for solving the nonlinear Hamilton-Jacobi equations on two-dimensional unstructured meshes. The main ideas are nodal based approximations, the usage of monotone Hamiltonians as building blocks on unstructured meshes, nonlinear weights using smooth indicators of second and higher derivatives, and a strategy to choose diversified smaller stencils to make up the bigger stencil in the WENO procedure. Both third-order and fourth-order WENO schemes use combinations of second-order approximations with nonlinear weights. A brief introduction to the methods and a selected few numerical experiments are included here.

1 Introduction

In this presentation, we consider the numerical solution of two-dimensional Hamilton-Jacobi (H-J) equations

$$\phi_t + H(\phi_x, \phi_y) = 0, \qquad \phi(x, y, 0) = \phi_0(x, y). \tag{1}$$

Such equations appear often in many applications. Typical solutions of H-J equations are continuous with possibly discontinuous derivatives, so in certain sense H-J equations are easier to solve than their conservation laws counterparts.

For structured mesh, finite difference methods similar to those developed for conservation laws could be easily designed. However, for unstructured meshes there are conceptional difficulties in designing schemes such as finite volume schemes and finite element schemes, which are very useful for conservation laws. The problem arises because the H-J equations cannot be written in a "conservation form", suitable for integration by parts which is the backbone of finite volume and finite element methods. Thus algorithms, especially high order algorithms for H-J equations on unstructured meshes are relatively few.

In this presentation, we give a brief introduction and several additional numerical examples to our recent work [11] on discretizing directly H-J on a 2D unstructured mesh to construct high order WENO schemes. The adopted

WENO methodology can be traced back to the earlier work [5, 7], and the techniques most relevant to the current paper are the third and fourth order finite volume WENO schemes for 2D conservation laws in general triangulations [3] and the finite difference WENO schemes for Hamilton-Jacobi equations [4]. However, significant new components of the algorithms have been developed in [11] to deal with the additional difficulty caused by the "non-conservation" form of the H-J equations and unstructured meshes. These include nodal based approximations, the usage of monotone Hamiltonians [1] as building blocks on unstructured meshes, nonlinear weights using smooth indicators of second and higher derivatives, and a strategy to choose diversified smaller stencils to make up the bigger stencil in the WENO procedure. Both third-order and fourth-order WENO schemes using combinations of second-order approximations with nonlinear weights are described.

We outline our schemes in section 2. Section 3 contains several selected numerical examples verifying the stability, convergence and accuracy of the schemes. High-order accuracy in smooth regions, good resolution of derivative singularities, and convergence to viscosity solutions are observed. More detailed description of the schemes and more extensive numerical examples are presented in [11].

2 Outlines of WENO schemes for 2D unstructured mesh

In this section we outline third and fourth order WENO schemes on unstructured meshes in 2D for the H-J equations.

2.1 Building blocks

The equation (1) is solved in the domain Ω, which has a triangulation $\mathcal{T}_h$ consisting of triangles. The nodes will be named by their indices $0 \le i \le N$, with a total of $N+1$ nodes. For every node i, we define the $k_i + 1$ angular sectors $T_0, \cdots, T_{k_i}$ meeting at the point i; they are the inner angles at node i of the triangles having i as a vertex. The indexing of the angular sectors is ordered counterclockwise. $\mathbf{n}_{l+\frac{1}{2}}$ is the unit vector of the half-line $D_{l+\frac{1}{2}} = T_l \bigcap T_{l+1}$, and θ_l is the inner angle of sector T_l, $0 \le l \le k_i$; see Figure 1.

We will denote by ϕ_i the numerical approximation to the viscosity solution of (1) at node i. $(\nabla\phi)_0, \cdots, (\nabla\phi)_{k_i}$ will respectively represent the numerical approximation of $\nabla\phi$ at node i in each angular sector $T_0, \cdots, T_{k_i}$.

An important building block for the WENO algorithms is the Lax-Friedrichs type monotone Hamiltonian for arbitrary triangulations developed by Abgrall in [1], which is a generalization of the Lax-Friedrichs monotone Hamiltonian for Cartesian meshes in Osher and Shu [8]. This monotone Hamiltonian is given by

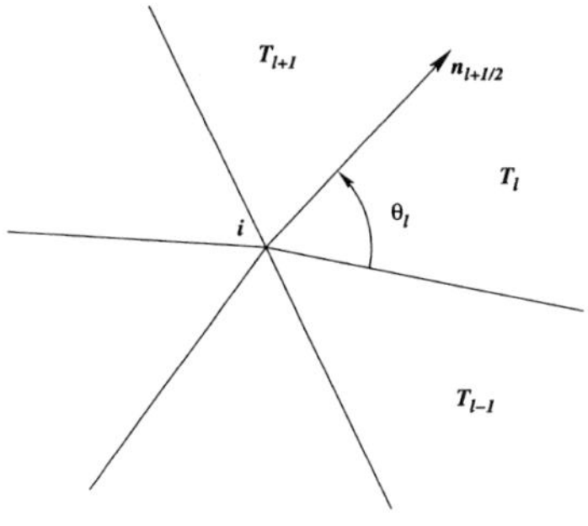

Fig. 1. node i and its angular sectors.

$$\hat{H}((\nabla\phi)_0,\cdots,(\nabla\phi)_{k_i}) = H\left(\frac{\sum_{l=0}^{k_i}\theta_l(\nabla\phi)_l}{2\pi}\right) - \frac{\alpha}{\pi}\sum_{l=0}^{k_i}\beta_{l+\frac{1}{2}}\left(\frac{(\nabla\phi)_l+(\nabla\phi)_{l+1}}{2}\right)\cdot \mathbf{n}_{l+\frac{1}{2}} \quad (2)$$

where

$$\beta_{l+\frac{1}{2}} = \tan\left(\frac{\theta_l}{2}\right)+\tan\left(\frac{\theta_{l+1}}{2}\right), \quad \alpha = \max\{\max_{\substack{A\le u\le B\\ C\le v\le D}}|H_1(u,v)|, \max_{\substack{A\le u\le B\\ C\le v\le D}}|H_2(u,v)|\}.$$

Here H_1 and H_2 are the partial derivatives of H with respect to ϕ_x and ϕ_y, respectively, or the Lipschitz constants of H with respect to ϕ_x and ϕ_y, if H is not differentiable. $[A,B]$ is the value range for $(\phi_x)_l$, and $[C,D]$ is the value range for $(\phi_y)_l$, over $0\le l\le k_i$ for the local Lax-Friedrichs Hamiltonian, and over $0\le l\le k_i$ and $0\le i\le N$ for the global Lax-Friedrichs Hamiltonian.

The $\hat{H}$ in (2) defines a monotone Hamiltonian. It is Lipschitz continuous in all arguments and is consistent with H, i.e., $\hat{H}(\nabla\phi,\cdots,\nabla\phi) = H(\nabla\phi)$. Hence if we have high-order approximations to $\nabla\phi$ at node i in every angular sector, the numerical Hamiltonian $\hat{H}$ will be a high-order approximation to H.

The semi-discrete scheme is thus given by:

$$\frac{d}{dt}\phi_i(t) + \hat{H}((\nabla\phi)_0,\cdots,(\nabla\phi)_{k_i}) = 0 \quad (3)$$

and it is discretized in time by the high order nonlinearly stable Runge-Kutta time discretization in [10]. If the spatial dimension reduces to one, the numerical Hamiltonian (2) will reduce to the local or global Lax-Friedrichs Hamiltonian [8]:

$$\hat{H}(u^-,u^+) = H\left(\frac{u^-+u^+}{2}\right) - \frac{1}{2}\alpha(u^+-u^-) \quad (4)$$

with $\alpha = \max_{A \le u \le B} |H'(u)|$. If the maximum for computing α is taken over the range covered by u^- and u^+, we get the local Lax-Friedrichs Hamiltonian; if it is also taken over all grid points, we get the global Lax-Friedrichs Hamiltonian.

2.2 Linear schemes

Now we need to construct a high-order approximation for $\nabla\phi$ in every angular sector of every node. Lagrange interpolations are used to obtain a cubic polynomial $p^3(x,y)$ for third-order scheme or fourth degree polynomial $p^4(x,y)$ for fourth-order scheme on every triangle Δ, then ∇p^3 will be a third-order and ∇p^4 will be a fourth-order approximation for $\nabla\phi$ on Δ. Interpolation nodes for p^3 or p^4 form our big stencils. They are taken from some or all of the nodes shown in Figure 2. See [11] for the detailed procedures to determine the big stencils.

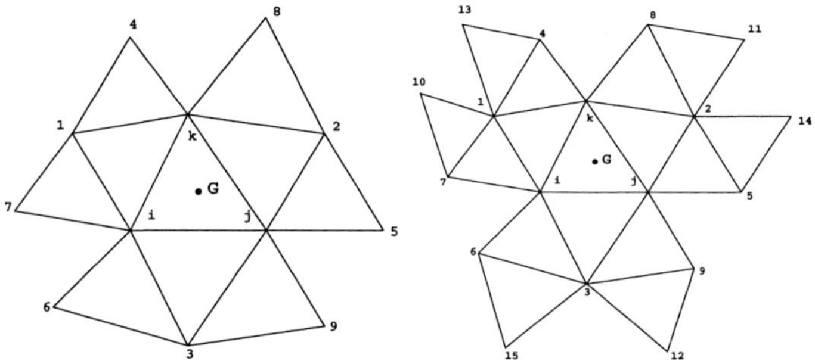

Fig. 2. The nodes used for the big stencils. Left: third-order scheme; right: fourth-order scheme.

In order to construct a third-order or fourth-order WENO scheme, an important step is to obtain a third-order or fourth-order approximation using a linear combination of lower order approximations. For each node (x_l, y_l) in Δ, we will use the linear combinations of second-order approximations to get the same third-order or fourth-order approximation to $\nabla\phi(x_l, y_l)$ as $\nabla p^3(x_l, y_l)$ or $\nabla p^4(x_l, y_l)$, i.e., we require

$$\frac{\partial}{\partial x} p^3(x_l, y_l) = \sum_{s=1}^{q} \gamma_{s,x} \frac{\partial}{\partial x} p_s(x_l, y_l), \qquad \frac{\partial}{\partial y} p^3(x_l, y_l) = \sum_{s=1}^{q} \gamma_{s,y} \frac{\partial}{\partial y} p_s(x_l, y_l) \tag{5}$$

for the third-order scheme or

$$\frac{\partial}{\partial x} p^4(x_l, y_l) = \sum_{s=1}^{q} \gamma_{s,x} \frac{\partial}{\partial x} p_{s,x}(x_l, y_l), \qquad \frac{\partial}{\partial y} p^4(x_l, y_l) = \sum_{s=1}^{q} \gamma_{s,y} \frac{\partial}{\partial y} p_{s,y}(x_l, y_l) \tag{6}$$

for the fourth-order scheme, where $p_s, p_{s,x}, p_{s,y}$ are quadratic interpolation polynomials, and $\gamma_{s,x}$ and $\gamma_{s,y}$ are the linear weights for the x-directional derivative and the y-directional derivative respectively, for $s = 1, \cdots, q$, $q = 5$ for third-order schemes and $q = 10$ for fourth-order schemes. The linear weights are constants depending only on the local geometry of the mesh. The equalities in (5) and (6) should hold for any choices of the function ϕ. The stencils which these quadratic interpolation polynomials use are called small stencils. The detailed procedures to choose diversified small stencils to make up the big stencil are described in [11]. If we compute the third-order or fourth-order approximation $(\nabla\phi)_0, \cdots, (\nabla\phi)_{k_l}$ for all the mesh nodes l, by the linear combinations of second-order approximations, we can put them into the scheme (3) to get a third-order or fourth-order linear scheme.

2.3 WENO schemes

Now we present the WENO schemes based on non-linear weights. The resulting schemes will be suitable to compute the H-J equations whose solutions are not smooth. We only present the case of WENO approximation for the x-directional derivative at vertex i of the target cell $\triangle_l$. Other cases are similar. In order to compute the non-linear weights, we need to compute the smoothness indicators first.

For a polynomial $p(x, y)$ defined on the target cell $\triangle_0$ with degree up to k, we take the smoothness indicator β as:

$$\beta = \sum_{2\leq|\alpha|\leq k} \int_{\triangle_0} |\triangle_0|^{|\alpha|-2} \left(D^\alpha p(x,y)\right)^2 dxdy \tag{7}$$

where α is a multi-index and D is the derivative operator.

Now we define the non-linear weights as:

$$\omega_j = \frac{\widetilde{\omega}_j}{\sum_m \widetilde{\omega}_m}, \qquad \widetilde{\omega}_j = \frac{\gamma_j}{(\varepsilon + \beta_j)^2} \tag{8}$$

where γ_j is the jth linear weight (e.g. the $\gamma_{s,x}$ in our linear schemes), β_j is the smoothness indicator for the jth interpolation polynomial $p_j(x, y)$ (e.g. the p_s in equation (5) for the third-order case and the $p_{s,x}$ in equation (6) for the fourth-order case) associated with the jth small stencil, and ε is a small positive number to avoid the denominator to become 0. We take $\varepsilon = 10^{-6}$ for all the computations in this presentation. The final WENO approximation for the x-directional derivative at vertex i of target cell $\triangle_l$ is given by

$$(\phi_x)_i = \sum_{j=1}^{q} \omega_j \frac{\partial}{\partial x} p_j(x_i, y_i) \tag{9}$$

where (x_i, y_i) are the coordinates of vertex i, $q = 5$ for the third-order schemes and $q = 10$ for the fourth-order schemes. If there are negative linear weights, we adopt the splitting technique of treating negative weights in

WENO schemes developed in [9]. We can then put the WENO approximations (9) into (3) to get our WENO schemes.

3 Numerical examples

In this section, we present a few selected numerical examples of applying the WENO schemes described in the previous section to a set of one and two dimensional problems. More examples can be found in [11]. The CFL number is taken as 0.5 in all the cases, except for the accuracy test where it is taken to be smaller if necessary to guarantee that spatial errors dominate. The local Lax-Friedrichs flux is used for all the test cases. For the temporal discretization, we use the third-order TVD Runge-Kutta scheme of Shu and Osher in [10].

We have not described the details of the one dimensional algorithm in the previous section. The algorithm in 1D is just the same 2D algorithm with only two "angular sectors". The WENO interpolation follows along the lines of [5] and [2]. In all the one-dimensional numerical examples, we use non-uniform meshes. These non-uniform meshes are obtained by randomly shifting the cell boundaries in a uniform mesh in the range $[-0.1h, 0.1h]$ where h is the uniform mesh size. When we perform the accuracy test, the refinement of the meshes is achieved by cutting each cell into two smaller equal sized ones.

Example 3.1. One-dimensional Burgers equation:

$$\begin{cases} \phi_t + \frac{(\phi_x+1)^2}{2} = 0, & -1 \le x < 1, \\ \phi(x,0) = -\cos(\pi x) \end{cases} \tag{10}$$

with periodic boundary conditions.

At $t = 0.5/\pi^2$, the solution is still smooth. We list the L^∞ errors and the numerical orders of accuracy in Table 1, using linear and WENO schemes. We can see that the correct orders of accuracy are obtained.

Example 3.2. One-dimensional equation with a non-convex flux:

$$\begin{cases} \phi_t - \cos(\phi_x + 1) = 0, & -1 \le x < 1, \\ \phi(x,0) = -\cos(\pi x) \end{cases} \tag{11}$$

with periodic boundary conditions.

At $t = 1.5/\pi^2$, the solution has developed a corner-like discontinuity in the derivative. The numerical result with 41 points is shown in Figure 3.

Example 3.3. Two dimensional Burgers equation:

$$\begin{cases} \phi_t + \dfrac{(\phi_x + \phi_y + 1)^2}{2} = 0, & -2 \le x < 2, -2 \le y < 2, \\ \phi(x,y,0) = -\cos(\frac{\pi(x+y)}{2}). \end{cases} \tag{12}$$

Table 1. Accuracy for 1D Burgers equation, Linear and WENO Schemes, $t = 0.5/\pi^2$.

Linear	3rd order		5th order		7th order	
N	L^∞ error	order	L^∞ error	order	L^∞ error	order
10	8.12E-03	—	3.89E-03	—	1.75E-03	—
20	1.46E-03	2.47	4.29E-04	3.18	1.14E-04	3.95
40	2.52E-04	2.54	2.33E-05	4.20	4.78E-06	4.57
80	3.83E-05	2.72	8.94E-07	4.70	8.54E-08	5.81
160	5.23E-06	2.87	2.94E-08	4.93	8.57E-10	6.64
320	6.82E-07	2.94	9.39E-10	4.97	7.33E-12	6.87

WENO	3rd order		5th order		7th order	
N	L^∞ error	order	L^∞ error	order	L^∞ error	order
10	5.29E-02	—	5.21E-03	—	2.92E-03	—
20	1.93E-02	1.45	3.78E-04	3.78	3.02E-04	3.27
40	4.82E-03	2.00	3.05E-05	3.63	5.58E-06	5.76
80	1.05E-03	2.20	1.26E-06	4.60	5.66E-08	6.62
160	2.08E-04	2.33	4.48E-08	4.82	5.29E-10	6.74
320	2.31E-05	3.17	1.47E-09	4.93	4.38E-12	6.92

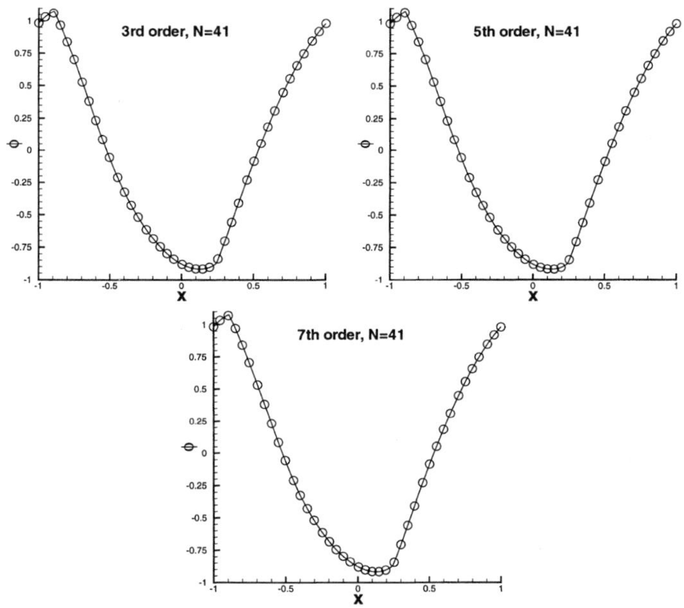

Fig. 3. One dimension, non-convex, $H(u) = -\cos(u+1), t = 1.5/\pi^2$. Solid line: the exact solution; circles: numerical solutions. Non-uniform mesh with 41 points. Top left: third order WENO scheme; top right: fifth order WENO scheme; bottom: seventh order WENO scheme.

with periodic boundary conditions.

At $t = 0.5/\pi^2$, the solution is still smooth. We first use uniform triangular meshes, shown in Figure 4, left, for the coarsest case $h = \frac{2}{5}$ where h is the length of the right angled side, to test the accuracy for the third and fourth order linear schemes and WENO schemes. We then use non-uniform meshes, shown in Figure 4, right, for the coarsest case $N = 105$ where N is the total number of nodes in the mesh. The refinement of the non-uniform meshes is done in a uniform way, namely by cutting each triangle into four smaller similar ones. The accuracy results are shown only for the non-uniform meshes, to save space, in Table 2. The expected orders of accuracy are obtained.

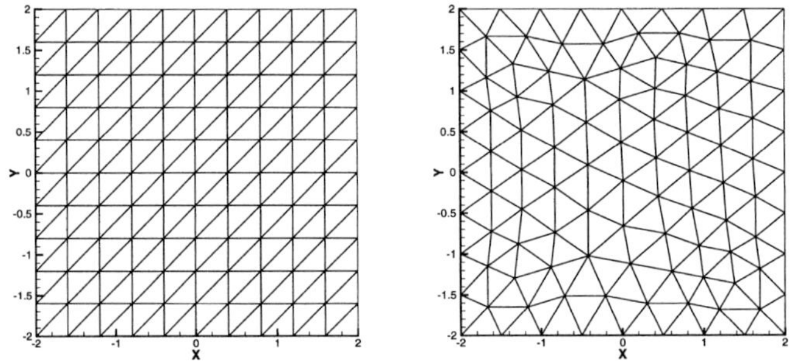

Fig. 4. Left: coarsest uniform mesh with $h = \frac{2}{5}$; right: coarsest non-uniform mesh with $N = 105$ nodes.

Table 2. Accuracy for 2D Burgers equation, Non-uniform Meshes, Linear and WENO Schemes, $t = 0.5/\pi^2$.

	3rd order				4th order			
	Linear		WENO		Linear		WENO	
N	L^∞ error	order	L^∞ error	order	L^∞ error	order	L^∞ error	order
105	2.73E-02	—	1.03E-01	—	2.72E-02	—	1.20E-01	—
385	5.29E-03	2.37	4.63E-02	1.15	5.73E-03	2.25	4.35E-02	1.46
1473	1.27E-03	2.06	1.35E-02	1.78	6.51E-04	3.14	2.05E-02	1.09
5761	1.92E-04	2.72	3.25E-03	2.06	4.21E-05	3.95	2.47E-03	3.05
22785	2.73E-05	2.81	4.44E-04	2.87	2.38E-06	4.14	1.76E-04	3.81

Example 3.4. Two-dimensional equation with a non-convex flux:

$$\begin{cases} \phi_t - \cos(\phi_x + \phi_y + 1) = 0, & -2 \le x < 2, -2 \le y < 2, \\ \phi(x,y,0) = -\cos(\frac{\pi(x+y)}{2}). \end{cases} \tag{13}$$

with periodic boundary conditions.

For this example we use a non-uniform mesh (Figure 4, right) with $N = 1473$ nodes. At $t = 1.5/\pi^2$, the solution develops a discontinuous derivative, and we show the results using the third and fourth order linear and WENO schemes in Figure 5. We can see that for this example which is not very demanding, the linear schemes can also obtain non-oscillatory results. In [11], we give more demanding cases where linear schemes fail to converge to the correct viscosity solution while WENO schemes converge very well.

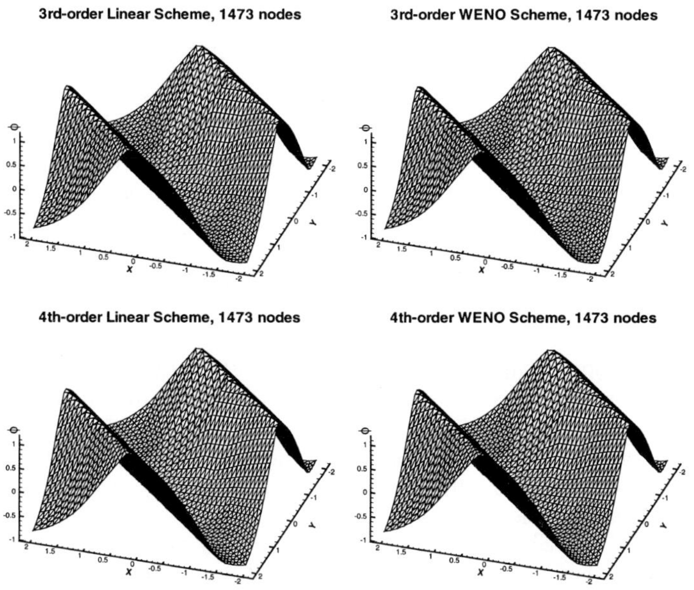

Fig. 5. Two dimensions, $H(u, v) = -\cos(u + v + 1)$, $t = 1.5/\pi^2$. Nonuniform mesh with $N = 1473$ nodes. Top left: third order linear scheme; top right: third order WENO scheme; bottom left: fourth order linear scheme; bottom right: fourth order WENO scheme.

Example 3.5. 2D eikonal equation with a non-convex Hamiltonian, which arises in geometric optics [6]:

$$\begin{cases} \phi_t + \sqrt{\phi_x^2 + \phi_y^2 + 1} = 0, & 0 \le x < 1, 0 \le y < 1, \\ \phi(x, y, 0) = 0.25(\cos(2\pi x) - 1)(\cos(2\pi y) - 1) - 1. \end{cases} \tag{14}$$

In Figure 6, we show the results at $t = 0.6$ with the third-order and fourth order WENO schemes on a uniform mesh (Figure 4, left) with $h = \frac{1}{50}$. The results with a non-uniform mesh are shown in [11].

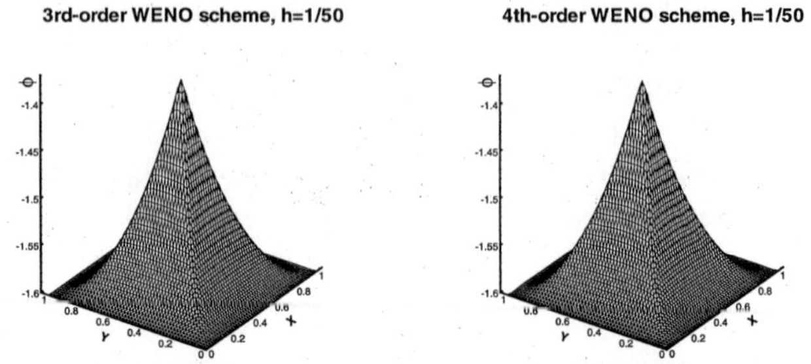

Fig. 6. The 2D eikonal equation, $H(u,v) = \sqrt{u^2 + v^2 + 1}$, $t = 0.6$, uniform mesh with $50 \times 50 \times 2$ elements.

Acknowledgments: This research is supported by ARO grant DAAD19-00-1-0405, NSF grants DMS-9804985 and ECS-9906606, NASA Langley grant NCC1-01035 and AFOSR grant F49620-99-1-0077.

References

1. R. Abgrall (1996): Numerical discretization of the first-order Hamilton-Jacobi equation on triangular meshes. Communications on Pure and Applied Mathematics, **49**, 1339–1373.
2. D. Balsara and C.-W. Shu (2000): Monotonicity preserving weighted essentially non-oscillatory schemes with increasingly high order of accuracy. Journal of Computational Physics, **160**, 405–452.
3. C. Hu and C.-W. Shu (1999): Weighted essentially non-oscillatory schemes on triangular meshes. Journal of Computational Physics, **150**, 97–127.
4. G. Jiang and D.-P. Peng (2000): Weighted ENO schemes for Hamilton-Jacobi equations. SIAM Journal on Scientific Computing, **21**, 2126–2143.
5. G. Jiang and C.-W. Shu (1996): Efficient implementation of weighted ENO schemes. Journal of Computational Physics, **126**, 202–228.
6. S. Jin and Z. Xin (1998): Numerical passage from systems of conservation laws to Hamilton-Jacobi equations and relaxation schemes. SIAM Journal on Numerical Analysis, **35**, 2385–2404.
7. X.-D. Liu, S. Osher and T. Chan (1994): Weighted essentially non-oscillatory schemes. Journal of Computational Physics, **115**, 200–212.
8. S. Osher and C.-W. Shu (1991): High-order essentially nonoscillatory schemes for Hamilton-Jacobi equations. SIAM Journal on Numerical Analysis, **28**, 907–922.
9. J. Shi, C. Hu and C.-W. Shu (2002): A technique of treating negative weights in WENO schemes. Journal of Computational Physics, **175**, 108–127.
10. C.-W. Shu and S. Osher (1988): Efficient implementation of essentially non-oscillatory shock capturing schemes. Journal of Computational Physics, **77**, 439–471.
11. Y.-T. Zhang and C.-W. Shu (2001): High order WENO schemes for Hamilton-Jacobi equations on triangular meshes. SIAM Journal on Scientific Computing, submitted.

High Resolution Conjugate Filters for Hyperbolic Conservation Laws

Y. C. Zhou[1] and G. W. Wei[2]

[1] Department of Mathematics, Michigan State University, East Lansing, MI 48824, *zhouyon1@math.msu.edu*

[2] Department of Mathematics, Michigan State University, East Lansing, MI 48824, *wei@math.msu.edu*

Summary. A new realization of the conjugate filter oscillation (CFOR) scheme based on the Hermite kernel is proposed for hyperbolic conservation laws. Such a realization is found to give a nearly optimal resolution and a better approximation to the low-pass filter than the previous kernels. Typical one- and two-dimensional numerical examples, with or without shocks, are employed to explore the utility, test the resolution and examine the stability of the present CFOR-Hermite scheme. Small ratio of point-per-wavelength (PPW) is achieved in advancing a wavepacket and in resolving the interaction of shock/entropy wave. The present results for the advection of an isotropic vortex compare very favorably to those in the literature.

1 Introduction

One of the most fascinating phenomena of hyperbolic systems is the presence of shock waves. Numerous numerical schemes have been developed for shock capturing since von Neumann and Richtmyer's genius conceiving of using artificial viscosity to eliminate spurious oscillations near a shock. Most of prevalent shock-capturing methods have inherited their idea of shock capturing by using additional viscosity. The incorporation of additional viscosity basically can be realized by one of two means, one is to explicitly add a diffusion term to the original or discretized governing equations, and the other is to discretized the equation with an upwind type of biased stencils which lead to a distinct dissipation term in the discretized equation [1]. Although the additional viscosity makes it possible to reduce the spurious oscillations near the discontinuity, it also degrades the formal numerical order and resolution of a scheme. To judiciously control the viscosity within an appropriate level is a nontrivial task, particularly in the situation where a shock is interacting with other complex flow phenomena such as acoustic waves, vortices or turbulence eddies. The difficulty in handling these situations has stimulated the continuous research interest in shock-capturing.

High-order shock-capturing schemes [2, 3] usually have higher resolutions with respect to the Fourier analysis than low-order ones. In the Fourier domain, high-order schemes give a wider region of correct wavenumber than low-order schemes. This observation initiates the work in optimizing the basic (compact) finite difference schemes with respect to an effective wavenumber range [4]. In the Fourier analysis, it is easy to see that beyond an effective wavenumber range, a numerical scheme will lose its accuracy. The Fourier spectrum of a shock is scattered over the whole Fourier domain. Therefore, the error in predicting the high frequency components will be accumulated and amplified during the time integration and lead to spurious oscillations near the shock. As aforementioned, the suppression of these oscillations is generally achieved by introducing some dissipative terms in the numerical schemes. The use of filters is an interesting approach [5, 6, 7, 8]. However, there is no general rule about the selection of optimal filters. In this work, we discuss an alternative filter approach, the conjugate filter oscillation reduction (CFOR) scheme[9] for hyperbolic conservation laws.

2 DSC algorithm and CFOR-Hermite scheme

The basic algorithm of CFOR scheme is a discrete singular convolution (DSC) [10], which is originated from the distribution theory. In distribution theory, a function or its nth order derivative can be given by the convolution of the function with the delta distribution or the nth order derivative of the delta distribution

$$f^{(n)} = f * \delta^{(n)} \qquad (n = 0, 1, 2, \cdots). \tag{1}$$

The discrete singular convolution is discrete realization of this convolution, with some appropriate approximation to the singular kernel $\delta(x)$. In other words, DSC algorithm seeks the following approximation

$$f^{(n)}(x) \approx \sum_k \delta_\sigma^{(n)}(x - x_k) f(x_k), \qquad (n = 0, 1, 2, \cdots) \tag{2}$$

where $\delta_\sigma^{(n)}(x - x_k)$ are sequences of approximations to $\delta^{(n)}(x - x_k)$, characterized by σ. Here x_k is an appropriate set of discrete points centered around the point x.

From the point view of signal processing, the expression $\delta_\sigma(x)$ can be identified as a low-pass filter and its high-order derivatives $\frac{d^n}{dx^n}\delta_\sigma(x), \quad (n = 1, 2)$ can be identified as high-pass filters. In our CFOR scheme [9], the high-pass filters are used to approximate the spatial derivatives required for solving differetial equations while the low-pass filter is utilized to clean up the high frequency oscillations produced by the spatial derivatives. The low-pass filtering is implemented via the prediction and restoration at time steps selected by a TVD switch [9]. Only when the absolute difference of the TVD between two consequent time steps is larger than a prescribed threshold, which depends on the problem, is the low-pass filtering activated. The standard forth-order Runge-Kutta method is adopted for the time integration.

The success of the CFOR scheme depends on the DSC kernel δ_σ. Many kernels have been proposed [10]. In this work, we explore the performance of the CFOR scheme with a different DSC kernel, the Hermite kernel

$$\delta_\sigma(x-x_k) = \frac{1}{\sigma}\exp\left(-\frac{(x-x_k)^2}{2\sigma^2}\right)\sum_{k=0}^{n/2}\left(-\frac{1}{4}\right)^k \frac{1}{\sqrt{2\pi}k!}H_{2k}\left(\frac{x-x_k}{\sqrt{2}\sigma}\right), \quad (3)$$

where $H_{2k}(x)$ is the usual Hermite polynomial. The frequency response of the high-pass and low-pass filters constructed by using the Hermite kernel is plotted in Fig. (1).

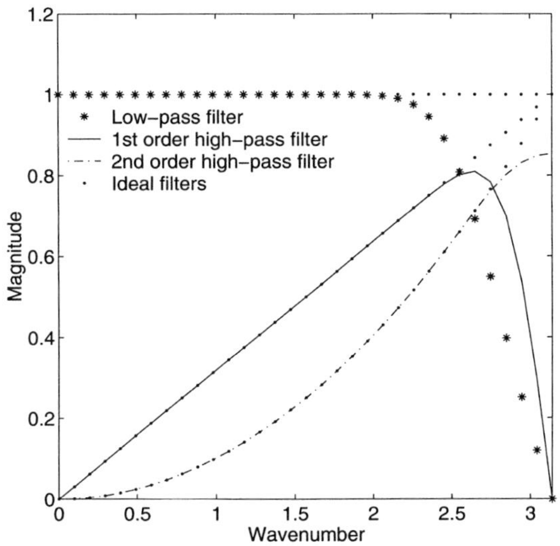

Fig. 1. Frequency responses of the conjugate Hermite filers (in the unit of $1/\Delta$). The maximum amplitude of the filters is normalized to the unit.

The CFOR-Hermite scheme has a few distinct features. First, the scheme is based on the DSC algorithm, a local method as shown in Eq. (2), which provides a controllable accuracy for the spatial discretization in a smooth region. The approximation errors of the DSC algorithm decay exponentially with respect the increase of grid points for certain bandlimited functions [11]. Therefore, it is a *local spectral method.* In fact, the accuracy of the DSC is very competitive with the Fourier pseudo-spectral method. Secondly, the low-pass filter here is a localized filter, both in physical domain and frequency domain. In physical domain, the effective support of the filter is controlled by r ($r = \frac{\sigma}{\Delta}$), which also controls the effective bandwidth of the filter in the frequency domain. Thus, it has time-frequency localization, like a wavelet scaling function. Thirdly, the conjugate high-pass and low-pass filters are constructed form a single generating function. Therefore, they have essentially the same support, regularity, smoothness, time-frequency localization and numerical resolution. It is in this sense that they are *conjugate filters.* Finally, the effective frequency bandwidths of both high-pass and low-pass filters are independently adjustable. This makes it possible to design a low-pass filter which can conjugate with a high-pass filter on a wavenumber region as wider as possible, while most effectively suppress spurious oscillations.

Numerical experiments are carried out in the rest of this paper with $n = 88$ and $r = 3.05$. Three examples are employed to access the accuracy, stability and resolution of the CFOR scheme with the CFOR-Hermite scheme. Our numerical tests include a 1D advection equation, a 1D compressible Euler system and a 2D compressible Euler system.

3 Numerical experiments

EXAMPLE 1 *Advection of a sine-Gaussian wavepacket*

In this problem, a narrow Gaussian envelope modulated by a sine wave moves inside the computation domain [-1,1], governed by the linear wave equation $u_t + u_x = 0$ with periodic boundary conditions. The sine-Gaussian wavepacket is defined as

$$u(x, t_0) = \sin[2\pi k(x - x_0)]e^{-\frac{(x-x_0)^2}{2\sigma_0^2}}, \tag{4}$$

where $x_0 = 0$ is the location of the wave centered at the initial time $t_0 = 0$ and $\sigma_0 = \sqrt{2}/10$. The errors of the numerical results with respect to the exact solution is listed in Table 1. The time increment is chosen such that the overall error is dominated by the CFOR-Hermite spatial differentiation. We compute the CFOR-Hermite results at a variety of k values. The ratio of

Time	k(PPW)					
	5(20)	10(10)	15(6.7)	20(5)	25(4)	30(3.3)
2	1.17E-14	4.77E-14	4.23E-14	5.86E-12	2.21E-8	4.70E-5
4	2.11E-14	8.86E-14	8.23E-14	1.17E-11	4.43E-8	9.41E-5
6	2.46E-14	1.36E-13	1.11E-13	1.76E-11	6.64E-8	1.41E-4
8	3.19E-14	1.79E-13	1.46E-13	2.35E-11	8.86E-8	1.88E-4
10	4.01E-14	2.27E-13	1.73E-13	2.93E-11	1.11E-7	2.36E-4

Table 1. L_1 error for advancing the sine-Gaussian wavepacket ($\Delta t = 5\times10^{-6}$, $\Delta x = 0.01$).

grid points per wavelength (PPW) is calculated for each case. For a successful scheme, a smaller PPW indicates a higher numerical resolution. Extremely high accuracy is obtained for $k \leq 20$. For larger k values, numerical errors are limited by the differentiation errors. However, even at 3.3 PPW, the present errors remain very small.

EXAMPLE 2 *Free advection of a 2D isentropic vortex*

This example describes the free evolution of a 2D vortex described by the compressible Euler equations with periodic boundary conditions. For the detail definition of the problem please refer to [1]. The numerical errors and the formal order of the CFOR-Hermite scheme are tabulated and compared those of Garnier *et al* [1]. We also list the errors at $t = 10, 20, 50$ and 100 for the mesh of 80×80. In such a long time evolution, the CFOR-Hermite scheme is used to stabilize the nonlinear growth of errors. Obviously, the errors reflect the extremely small disparity between computed vortex and the idea vortex, which is supposed to be invariant during the time evolution. It can be seen that the CFOR-Hermite scheme is highly accurate and stable.

N		CFOR	C4	ENO	MUSCL	WENO	ENO^A	WENO^A
40	error	2.37E-5	1.13E-3	1.28E-3	2.39E-3	9.39E-4	7.81E-4	6.11E-4
80	error	4.73E-9	5.78E-5	2.08E-4	5.99E-4	7.07E-5	6.68E-5	4.58E-4
	order	12.29	4.29	2.62	1.99	3.73	3.55	3.74
160	error	3.34E-10	3.79E-6	3.01E-5	1.26E-4	2.46E-6	7.84E-6	2.95E-6
	order	3.82	3.93	2.79	2.25	4.84	3.09	3.97
320	error	5.12E-11	2.41E-7	4.07E-6	2.26E-5	8.52E-8	6.82E-7	2.13E-7
	order	2.71	3.97	2.89	2.47	4.85	3.52	3.79

Table 2. L_2 error for the density at $t = 2$; The CFL number is 0.5; C4: fourth-order accurate, conservative centered scheme; ENO: third-order; MUSCL: third-order; WENO: fifth-order; XXX^A denotes that the C4 scheme is used as the basic scheme with the XXX being a characteristic-based filter, while Harten's artificial compression method is used as a sensor to indicate the local numerical dissipation, see Ref. [1].

Time	2	10	50	100
L_1	4.73E-9	1.23E-8	4.58E-8	1.05E-7
L_2	1.41E-8	3.64E-8	1.41E-7	3.17E-7

Table 3. Errors for the density at different times (CFL=0.5, $N = 80$).

EXAMPLE 3 *Shock/entropy wave interaction*

Finally, we use a benchmark example [2] to demonstrate the shock-capturing ability of the CFOR-Hermite scheme with natural high frequency oscillations. In this problem, a small amplitude entropy wave is compressed by a moving shock. After the shock, the amplitude is amplified and the frequency is increased. Linear analysis provides quantitative results which are used to calibrate our numerical results. The problem is defined in a computation domain [0,5] and the initial condition for a 1D compressible Euler system is

$$(\rho, u, p) = \begin{cases} (3.85714, & 2.629369, 10.33333); \; x \leq 0.5 \\ (e^{-\epsilon \sin(kx)}, & 0, \qquad 1.0); \; x > 0.5 \end{cases}. \tag{5}$$

The results are evaluated when the shock moves to $x = 4.5$ [2]. The entropy

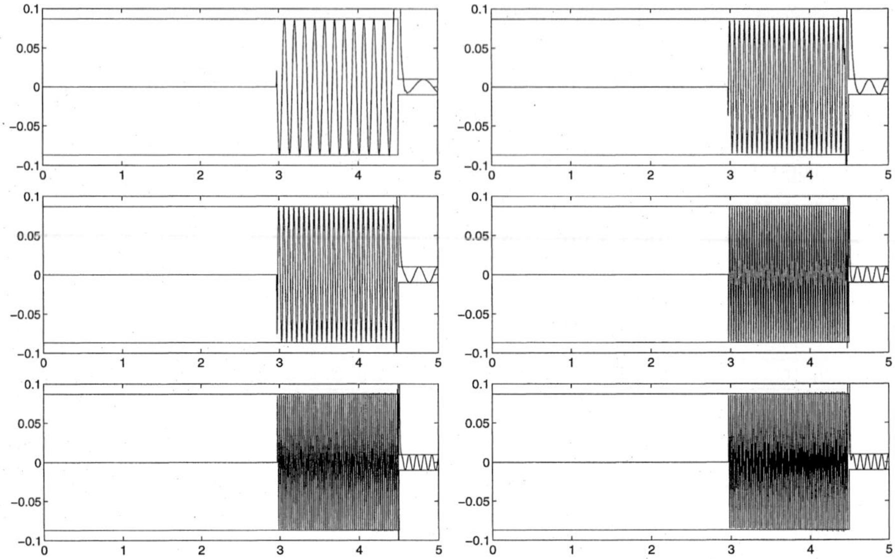

Fig. 2. Entropy waves for different pre-shock wavenumber. Upper left: $k = 13, N = 400$; Upper right: $k = 26, N = 400$; Middle left: $k = 26, N = 800$; Middle right: $k = 52, N = 1200$; Low left: $k = 65, N = 1200$; Low right: $k = 70, N = 1200$.

waves of a variety of wavenumbers (k=13, 26, 52, 65 and 70) are computed with the meshes of $N = 400, 800$ or 1200, respectively. The corresponding PPW values for the generated waves are 20 and 6, respectively. Two solid lines wrapping the high frequency wave are the bounds from the linear analysis. It is seen that the present results agree very well with the linear analysis. The high frequency waves at $k = 65$ are well resolved. It is very difficult for most existing shock-capturing schemes to resolve such high frequency waves with such a small PPW ratio due to their intrinsic narrow effective wavenumber range and excessive dissipation.

4 Conclusion remarks

The spectral property of the conjugate filter oscillation reduction (CFOR) scheme with a Hermite kernel is studied. The low-pass filters constructed by the Hermite kernel provide better approximations to the ideal low-pass filter, thus it is suitable for the suppression of the high frequency oscillations arising with the high-resolution differentiation scheme. Three representative numerical experiments are designed to test the performance of the proposed scheme. The high resolution of the present Hermite-CFOR scheme is highlighted by the interaction of the shock and high-frequency entropy waves, which is difficult for other existing shock-capturing schemes. As other state of art schemes, such as WENO and central schemes, do not depend on adjustable parameters, further investigation is required to make the present scheme parameter free.

References

1. E. Garnier, P. Sagaut and M. Deville, 'A class of explicit ENO filters with application to unsteady flows', *J. Comput. Phys.*, **170**, 184-204, 2001.
2. C.-W. Shu, 'Essentially non-oscillatory and weighted essentially non-oscillatory schemes for hyperbolic conservation laws', ICASE TR-97-65, 1997.
3. X.-D. Liu and E. Tadmor, 'Third order non-oscillatory central scheme for hyperbolic conservation laws', *Numer. Math.*, **79**, 397-425, 1998.
4. S. K. Lele, 'Compact Finite Difference Scheme with spectral-like Resolution'. *J. Comput. Phys.*, **103**, 16-42, 1992.
5. B. Engquist, P. Lötstedt and B. Sjögreen, 'Nonlinear filters for efficient shock computation', *Math. Comput.*, **186** 509-537, 1989.
6. D. Gottlieb and E. Tadmor, 'Recovering pointwise values of discontinuous data within spectral accuracy', in *Progress and Supercomputing in Computational Fluid Dynamics*, Earll M. Murman and Saul S. Abarbanel eds., (Birkh äuser, 357-375, 1985).
7. D. Gottlieb and J. S. Hesthaven, 'Spectral methods for hyperbolic problems', *J. Comput. Appl. Math.*, **128**, 83-131, 2001.
8. W. S. Don and D. Gottlieb, 'Spectral simulation of supersonic reactive flows', *SIAM J. Numer Anal.*, **35**, 2370-2384, 1998.
9. G. W. Wei and Yun Gu, 'Conjugated filter approach for solving Burgers' equation', *J. Comput. Appl. Math.*, **149**, 439-456, 2002.
10. G. W. Wei, 'Discrete singular Convolution for the solution for the Fokker-Planck equations', *J. Chem. Phys.*, **110**, 8930-8942, 1999.
11. G. Bao, G. W. Wei, and A. H. Zhou, 'Analysis of regularized Whittaker-Kotel'nikov-Shannon sampling expansion', preprint.

List of Participants

Authors are denoted by an asterisk*.

Abouziarov, Mike
Ancona, Fabio*
Andreae, Sigrid*
Andrianov, Alexander*
Andrianov, Nikolay*
Antonenko, Maxim
Appelö, Daniel*
Arminjon, Paul*
Artebrant, Robert*
Asakura, Fumioki*
Audusse, Emmanuel
Ayodeji, Matthew Bankole
Babin, Anatoli
Babovsky, Hans K.
Bagnerini, Patrizia*
Balbas, Jorge
Bartholomeeusen, Gert*
Ben-Artzi, Matania*
Bianchini, Stefano*
Bobba, Kumar
Bou-Rabee, Nawaf
Bressan, Alberto*
Breuß, Michael*
Bryson, Stephen*
Calhoun, Donna
Calhoun-Lopez, Marcus
Canic, Suncica
Cecil, Thomas
Chalons, Christophe*
Chan, Tony
Chen, Gui-Qiang*
Chen, C. L.
Chen, Shu-Xing*
Chertock, Alina
Coclite, Giuseppe Maria*
Corli, Andrea*
Coulombel, Jean-Francois*
da Mota, Jesus Carlos
Dafermos, Constantine
Danila, Robert Iulian
Dartey, Godfred
Davis, Stephen
de Souza, Aparecido
Dedner, Andreas*
Degond, Pierre*
Doronin, Gleb
Engquist, Bjorn
Fatai, Olabooye Kolawole
Fedkiw, Ron*
Fey, Michael*
Franklin, Joel
Freistuehler, Heinrich
Gandlin, Rima*
Giles, Mike*
Godillon-Lafitte, Pauline*
Godin, Paul*
Green, Mark
Gremaud, Pierre
Grote, Marcus*
Gunzbrger, Max*
Gustavsson, Katarina*
Ha, Seung Yeal
Hariharan, S. I.
Hartmann, Ralf*
Hattori, Harumi*
Heiermann, Joerg
Helzel, Christiane
Hill, David
Hirani, Anil
Hoff, David*
Hou, Thomas
Hsiao, (Xiao) Ling*
Huber, Nikolay
Humpherys, Jeffrey
Hwang, Seok
Hyman, James (Mac)
Ismagilov, Timur
Jabin, Pierre-Emmanuel
Jenssen, Kristian*
Jokhadze, Otari

Ju, Qiangchang
Kamm, James*
Kanevsky, Alex
Katsaounis, Theodoros*
Keller, Herb
Kim, Eun Heui
Kivotides, Demosthenes
Kliakhandler, Igor*
Klingenberg, Christian*
Kovenya, Viktor
Kreiss, Gunilla*
Kupiainen, Marco*
Kurganov, Alexander*
Laforest, Marc
Lee, Yong Joong*
LeVeque, Randall*
Levy, Doron*
Lew, Adrian
Li, Daqian (Tatsien)
Li, Xiantao
Li, Tong*
Li, Xiaolin*
Li, Shengtai
Lin, Chi-Tien*
Liska, Richard*
Liu, Hailiang*
Liu, Tai-Ping*
Lukacova, Maria
Lyng, Gregory
Malcolm, Abdul Razak
Mao, De-kang
Marchesin, Dan
Markovich, Peter*
Marquina, Antonio*
Marson, Andrea*
Mascia, Corrado*
Mauser, Norbert*
Meister, Andreas*
Mitran, Sorin
Mueller, Siegfried*
Nair, Pradeepa*
Nakaki, Tatsuyuki
Nakane, Kazuaki*
Natalini, Roberto
Nezzar, Suzanne
Nishibata, Shinya
Nussenzveig Lopes, Helena
O'Reilly, Gerard
Ofeliya, Adamyan
Ogundepo, Olufemi
Oh, Myunghyun*
Ohlberger, Mario*
Osher, Stanley
Othmer, Hans
Pan, Ronghua
Pan, Tao*
Panov, Evgeniy*
Papalexandris, Miltiadis
Papanicolaou, George
Pareschi, Lorenzo*
Pelanti, Marica*
Perthame, Benoit*
Petzold, Linda
Pierce, Niles
Pitman, E. Bruce*
Pogorelov, Nikolai
Pohlman, Matthew
Ponce, Gustavo
Popov, Bojan
Pullin, Dale
Puppo, Gabriella*
Rascle, Michel*
Rider, William*
Robinson, Allen*
Rossi, Gianni
Rossmanith, James*
Rousset, Frederic
Rozanova, Olga*
Rozovskii, Boris
Rubino, Bruno
Russell, Thomas
Rykov, Yuri
Ryzhik, Lenya
Schroll, Achim*
Serna, Susana*
Serre, Denis
Shearer, Michael
Shelkovich, Vladimir*

Shu, Chi-Wang*
Simeoni, Chiara*
Sjögreen, Björn*
Sohn, Sung-Ik
Song, Y. Tony
Strinopoulos, Theofilos
Su, Bo*
Tadmor, Eitan*
Tang, Shaoqiang*
Tang, Tao
Tanner, Jared*
Tiwari, Sudarshan*
Torrilhon, Manuel*
Towers, John*
Trivisa, Konstantina*
Tsai, Richard
Tzavaras, Athanasios*
Ukai, Seiji*
Vigbedorh, Victor Kojo
Voelker, Frank*
Wang, Shu*
Wang, Dehua*
Wang, Ke
Warnecke, Gerald*
Wei, Guo Wei*
Wesenberg, Matthias*
Westhead, Andrew
Wheatley, Vincent
Wu, Sijue
Xu, Kun
Yakubov, Yakov
Yamazaki, Mitsuru*
Yang, Danping
Yong, Wen-An
Yong, Darryl*
Young, Robin*
Yström, Jacob
Yu, Xinwei
Yuhui, Sun
Zhang, Yongtao*
Zhang, Zhengru
Zhao, Ning
Zhou, Yongcheng*
Zhou, Hao-Min

Druck: Strauss Offsetdruck, Mörlenbach
Verarbeitung: Schäffer, Grünstadt